ENCYCLOPEDIA OF
Physical Science
AND Technology

THIRD EDITION

FI–GI

Volume 6

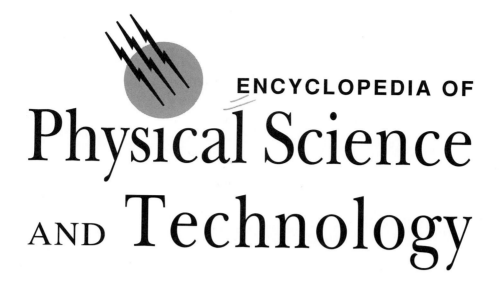

ENCYCLOPEDIA OF
Physical Science
AND Technology

THIRD EDITION

FI–GI
Volume 6

Editor-in-Chief

Robert A. Meyers, Ramtech, Inc.

ACADEMIC PRESS
A Harcourt Science and Technology Company

San Diego   San Francisco   Boston   New York   London   Sydney   Tokyo

Copyright © 2002 by ACADEMIC PRESS

All Rights Reserved.
No part of this publication may be reproduced or transmitted in any form or by any
means, electronic or mechanical, including photocopy, recording, or any information
storage and retrieval system, without permission in writing from the publisher.

Requests for permission to make copies of any part of the work should be mailed to:
Permissions Department, Harcourt Inc., 6277 Sea Harbor Drive,
Orlando, Florida 32887-6777

Academic Press
A Harcourt Science and Technology Company
525 B Street, Suite 1900, San Diego, California 92101-4495, USA
http://www.academicpress.com

Academic Press
Harcourt Place, 32 Jamestown Road, London NW1 7BY, UK
http://www.academicpress.com

Library of Congress Catalog Card Number: 2001090661

International Standard Book Number:

0-12-227410-5 (Set)	0-12-227420-2 (Volume 10)
0-12-227411-3 (Volume 1)	0-12-227421-0 (Volume 11)
0-12-227412-1 (Volume 2)	0-12-227422-9 (Volume 12)
0-12-227413-X (Volume 3)	0-12-227423-7 (Volume 13)
0-12-227414-8 (Volume 4)	0-12-227424-5 (Volume 14)
0-12-227415-6 (Volume 5)	0-12-227425-3 (Volume 15)
0-12-227416-4 (Volume 6)	0-12-227426-1 (Volume 16)
0-12-227417-2 (Volume 7)	0-12-227427-X (Volume 17)
0-12-227418-0 (Volume 8)	0-12-227429-6 (Index)
0-12-227419-9 (Volume 9)	

PRINTED IN THE UNITED STATES OF AMERICA
01 02 03 04 05 06 MM 9 8 7 6 5 4 3 2 1

Contents

Foreword

The editors of the *Encyclopedia of Physical Science and Technology* had a daunting task: to make an accurate statement of the status of knowledge across the entire field of physical science and related technologies.

No such effort can do more than describe a rapidly changing subject at a particular moment in time, but that does not make the effort any less worthwhile. Change is inherent in science; science, in fact, seeks change. Because of its association with change, science is overwhelmingly the driving force behind the development of the modern world.

The common point of view is that the findings of basic science move in a linear way through applied research and technology development to production. In this model, all the movement is from science to product. Technology depends on science and not the other way around. Science itself is autonomous, undisturbed by technology or any other social forces, and only through technology does science affect society.

This superficial view is seriously in error. A more accurate view is that many complex connections exist among science, engineering, technology, economics, the form of our government and the nature of our politics, and literature, and ethics.

Although advances in science clearly make possible advances in technology, very often the movement is in the other direction: Advances in technology make possible advances in science. The dependence of radio astronomy and high-energy physics on progress in detector technology is a good example. More subtly, technology may stimulate science by posing new questions and problems for study.

The influence of the steam engine on the development of thermodynamics is the classic example. A more recent one would be the stimulus that the problem of noise in communications channels gave to the study of information theory.

As technology has developed, it has increasingly become the object of study itself, so that now much of science is focused on what we have made ourselves, rather than only on the natural world. Thus, the very existence of the computer and computer programming made possible the development of computer science and artificial intelligence as scientific disciplines.

The whole process of innovation involves science, technology, invention, economics, and social structures in complex ways. It is not simply a matter of moving ideas out of basic research laboratories, through development, and onto factory floors. Innovation not only requires a large amount of technical invention, provided by scientists and engineers, but also a range of nontechnical or "social" invention provided by, among others, economists, psychologists, marketing people, and financial experts. Each adds value to the process, and each depends on the others for ideas.

Beyond the processes of innovation and economic growth, science has a range of direct effects on our society.

Science affects government and politics. The U.S. Constitution was a product of eighteenth century rationalism and owes much to concepts that derived from the science of that time. To a remarkable extent, the Founding Fathers were familiar with science: Franklin, Jefferson, Madison, and Adams understood science, believed passionately in empirical inquiry as the source of truth, and felt that government should draw on scientific concepts for inspiration. The concept of "checks and balances" was borrowed from Newtonian physics, and it was widely believed that, like the orderly physical universe that science was discovering, social relations were subject to a series of natural laws as well.

Science also pervades modern government and politics. A large part of the Federal government is concerned either with stimulating research or development, as are the National Aeronautics and Space Administration (NASA) and the National Science Foundation, or with seeking to regulate technology in some way. The reason that science and technology have spawned so much government activity is that they create new problems as they solve old ones. This is true in the simple sense of the "side effects" of new technologies that must be managed and thus give rise to such agencies as the Environmental Protection Agency (EPA). More importantly, however, the availability of new technologies makes possible choices that did not exist before, and many of these choices can only be made through the political system.

Biotechnology is a good example. The Federal government has supported the basic science underlying biotechnology for many years. That science is now making possible choices that were once unimagined, and in the process a large number of brand new political problems are being created. For example, what safeguards are necessary before genetically engineered organisms are tested in the field? Should the Food and Drug Administration restrict the development of a hormone that will stimulate cows to produce more milk if the effect will be to put a large number of dairy farmers out of business? How much risk should be taken to develop medicines that may cure diseases that are now untreatable?

These questions all have major technical content, but at bottom they involve values that can only be resolved through the political process.

Science affects ideas. Science is an important source of our most basic ideas about reality, about the way the world is put together and our place in it. Such "world views" are critically important, for we structure all our institutions to conform with them.

In the medieval world view, the heavens were unchanging, existing forever as they were on the day of creation. Then Tycho Brahe observed the "new star"—the nova of 1572—and the inescapable fact of its existence forced a reconstruction of reality. Kepler, Galileo, and Newton followed and destroyed the earth- and human-centered universe of medieval Christianity.

Darwin established the continuity of human and animal, thus undermining both our view of our innate superiority and a good bit of religious authority. The germ theory of disease-made possible by the technology of the microscope destroyed the notion that disease was sent by God as a just retribution for unrepentant sinners.

Science affects ethics. Because science has a large part in creating our reality, it also has a significant effect on ethics. Once the germ theory of disease was accepted, it could no longer be ethical, because it no longer made sense, to berate the sick for their sins. Gulliver's voyage to the land of the Houynhyms made the point:

By inventing a society in which individuals' illnesses were acts of free will while their crimes were a result of outside forces, he made it ethical to punish the sick but not the criminal.

Knowledge—most of it created by science—creates obligations to act that did not exist before. An engineer, for instance, who designs a piece of equipment in a way that is dangerous, when knowledge to safely design it exists, has violated both an ethical and a legal precept. It is no defense that the engineer did not personally possess the knowledge; the simple existence of the knowledge creates the ethical requirement.

In another sense, science has a positive effect on ethics by setting an example that may be followed outside science. Science must set truth as the cardinal value, for otherwise it cannot progress. Thus, while individual scientists may lapse, science as an institution must continually reaffirm the value of truth. To that extent science serves as a moral example for other areas of human endeavor.

Science affects art and literature. Art, poetry, literature, and religion stand on one side and science on the other side of C. P. Snow's famous gulf between the "two cultures." The gulf is largely artificial, however; the two sides have more in common than we often realize. Both science and the humanities depend on imagination and the use of metaphor. Despite widespread belief to the contrary, science does not proceed by a rational process of building theories from undisputed facts. Scientific and technological advances depend on imagination, on some intuitive, creative vision of how reality might be constructed. As Peter Medawar puts it: [Medawar, P. (1969). Encounter 32(1), 15-23]: All advance of scientific understanding, at every level, begins with a speculative adventure, an imaginative preconception of what might be true—a preconception that always, and necessarily, goes a little way (sometimes a long way) beyond anything that we have logical or factual authority to believe in.

The difference between literature and science is that in science imagination is controlled, restricted, and tested by reason. Within the strictures of the discipline, the artist or poet may give free rein to imagination. Although we may critically compare a novel to life, in general, literature or art may be judged without reference to empirical truth. Scientists, however, must subject their imaginative construction to empirical test. It is not established as truth until they have persuaded their peers that this testing process has been adequate, and the truth they create is always tentative and subject to renewed challenge.

The genius of science is that it takes imagination and reason, which the Romantics and the modern counterculture both hold to be antithetical, and combines them in a synergistic way. Both science and art, the opposing sides of the "two cultures," depend fundamentally on the creative use of imagination. Thus, it is not surprising that many mathematicians and physicists are also accomplished musicians, or that music majors have often been creative computer programmers.

Science, technology, and culture in the future. We can only speculate about how science and technology will affect society in the future. The technologies made possible by an understanding of mechanics, thermodynamics, energy, and electricity have given us the transportation revolutions of this century and made large amounts of energy available for accomplishing almost any sort of physical labor. These technologies are now mature and will continue to evolve only slowly. In their place, however, we have the information revolution and soon will have the biotechnology revolution. It is beyond us to say where these may lead, but the implications will probably be as dramatic as the changes of the past century.

Computers affected first the things we already do, by making easy what was once difficult. In science and engineering, computers are now well beyond that. We can now solve problems that were only recently impossible. Modeling, simulation, and computation are rapidly becoming a way to create knowledge that is as revolutionary as experimentation was 300 years ago.

Artificial intelligence is only just beginning; its goal is to duplicate the process of thinking well enough so that the distinction between humans and machines is diminished. If this can be accomplished, the consequences may be as profound as those of Darwin's theory of evolution, and the working out of the social implications could be as difficult.

With astonishing speed, modern biology is giving us the ability to genuinely understand, and then to change, biological organisms. The implications for medicine and agriculture will be great and the results should be overwhelmingly beneficial.

The implications for our view of ourselves will also be great, but no one can foresee them. Knowledge of how to change existing forms of life almost at will confers a fundamentally new power on human beings. We will have to stretch our wisdom to be able to deal intelligently with that power.

One thing we can say with confidence: Alone among all sectors of society and culture, science and technology progress in a systematic way. Other sectors change, but only science and technology progress in such a way that today's science and technology can be said to be unambiguously superior to that of an earlier age. Because science progresses in such a dramatic and clear way, it is the dominant force in modern society.

Erich Bloch
National Science Foundation
Washington, D.C.

Preface

We are most gratified to find that the first and second editions of the *Encyclopedia of Physical Science and Technology* (1987 and 1992) are now being used in some 3,000 libraries located in centers of learning and research and development organizations world-wide. These include universities, institutes, technology based industries, public libraries, and government agencies. Thus, we feel that our original goal of providing in-depth university and professional level coverage of every facet of physical sciences and technology was, indeed, worthwhile.

The editor-in-chief (EiC) and the Executive Board determined in 1998 that there was now a need for a Third Edition. It was apparent that there had been a blossoming of scientific and engineering progress in almost every field and although the World Wide Web is a mighty river of information and data, there was still a great need for our articles, which comprehensively explain, integrate, and provide scientific and mathematical background and perspective. It was also determined that it would be desirable to add a level of perspective to our Encyclopedia team, by bringing in a group of eminent Section Editors to evaluate the existing articles and select new ones reflecting fields that have recently come into prominence.

The Third Edition Executive Board members, Stephen Hawking (astronomy, astrophysics, and mathematics), Daniel Goldin (space sciences), Elias Corey (chemistry), Paul Crutzen (atmospheric science), Yuan Lee (chemistry), George Olah (chemistry), Melvin Schwartz (physics), Edward Teller (nuclear technology), Frederick Seitz (environment), Benoit Mandelbrot (mathematics), Allen Bard (chemistry) and Klaus von Klitzing (physics)

concurred with the idea of expanding our coverage into molecular biology, biochemistry, and biotechnology in recognition of the fact that these fields are based on physical sciences. Military technology such as weapons and defense systems was eliminated in concert with present trends moving toward emphasis on peaceful uses of science and technology. Aaron Klug (molecular biology and biotechnology) and Phillip Sharp (molecular and cell biology) then joined the board to oversee their fields as well as the overall Encyclopedia. The Advisory Board was completed with the addition of John Bollinger (engineering), Michael Buckland (library sciences), Jean Carpentier (aerospace sciences), Ludwig Faddeev (physics), Herbert Friedman (space sciences), R. A. Mashelkar (chemical engineering), Karl Pister (engineering) and Gordon Slemon (engineering).

A 40 page topical outline of physical sciences and technology was prepared by the EiC and then reviewed by the board and modified according to their comments. This formed the basis for assuring complete coverage of the physical sciences and for dividing the science and engineering disciplines into 50 sections for selection of section editors. Six of the advisory board members decided to serve also as section editors (Allen Bard for analytical chemistry, Elias Corey for organic chemistry, Paul Crutzen for atmospheric sciences, Yuan Lee for physical chemistry, Phillip Sharp for molecular biology, and Melvin Schwartz for physics). Thirty-two additional section editors were then nominated by the EiC and the board for the remaining sections. A listing of the section editors together with their section descriptions is presented on p. v.

The section editors then provided lists of nominated articles and authors, as well as peer reviewers, to the EiC based on the section scopes given in the topical outline. These lists were edited to eliminate overlap. The Board was asked to help adjudicate the lists as necessary. Then, a complete listing of topics and nominated authors was assembled. This effort resulted in the deletion of about 200 of the Second Edition articles, the addition of nearly 300 completely new articles, and updating or rewrite of approximately 480 retained article topics, for a total of over 780 articles, which comprise the Third Edition. Examples of the new articles, which cover science or technology areas arising to prominence after the second edition, are: molecular electronics; nanostructured materials; image-guided surgery; fiber–optic chemical sensors; metabolic engineering; self-organizing systems; tissue engineering; humanoid robots; gravitational wave physics; pharmacokinetics; thermoeconomics, and superstring theory.

Over 1000 authors prepared the manuscripts at an average length of 17-18 pages. The manuscripts were peer reviewed, indexed, and published. The result is the eighteen volume work, of over 14,000 pages, comprising the Third Edition.

The subject distribution is: 17% chemistry; 5% molecular biology and biotechnology; 11% physics; 10% earth sciences; 3% environment and atmospheric sciences; 12% computers and telecommunications; 8% electronics, optics, and lasers; 7% mathematics; 8% astronomy, astrophysics, and space technology; 6% energy and power; 6% materials; 7% engineering, aerospace, and transportation. The relative distribution between basic and applied subjects is: 60% basic sciences, 7% mathematics, and 33% engineering and technology. It should be pointed out that a subject such as energy and power with just a 5% share of the topic distribution is about 850 pages in total, which corresponds to a book-length treatment.

We are saddened by the passing of six of the Board members who participated in previous editions of this Encyclopedia. This edition is therefore dedicated to the memory of S. Chandrasekhar, Linus Pauling, Vladimir Prelog, Abdus Salam, Glenn Seaborg, and Gian-Carlo Rota with gratitude for their contributions to the scientific community and to this endeavor.

Finally, I wish to thank the following Academic Press personnel for their outstanding support of this project: Robert Matsumura, managing editor, Carolan Gladden and Amy Covington, author relations; Frank Cynar, sponsoring editor; Nick Panissidi, manuscript processing; Paul Gottehrer and Michael Early, production; and Chris Morris, Major Reference Works director.

Robert A. Meyers, Editor-in-Chief
Ramtech, Inc.
Tarzana, California, USA

FROM THE PREFACE TO THE FIRST EDITION

In the summer of 1983, a group of world-renowned scientists were queried regarding the need for an encyclopedia of the physical sciences, engineering, and mathematics written for use by the scientific and engineering community. The projected readership would be endowed with a basic scientific education but would require access to authoritative information not in the reader's specific discipline. The initial advisory group, consisting of Subrahmanyan Chandrasekhar, Linus Pauling, Vladimir Prelog, Abdus Salam, Glenn Seaborg, Kai Siegbahn, and Edward Teller, encouraged this notion and offered to serve as our senior executive advisory board.

A survey of the available literature showed that there were general encyclopedias, which covered either all facets of knowledge or all of science including the biological sciences, but there were no encyclopedias specifically in the physical sciences, written to the level of the scientific community and thus able to provide the detailed information and mathematical treatment needed by the intended readership. Existing compendia generally limited their mathematical treatment to algebraic relationships rather than the in-depth treatment that can often be provided only by calculus. In addition, they tended either to fragment a given scientific discipline into narrow specifics or to present such broadly drawn articles as to be of little use to practicing scientists.

In consultation with the senior executive advisory board, Academic Press decided to publish an encyclopedia that contained articles of sufficient length to adequately cover a scientific or engineering discipline and that provided accuracy and a special degree of accessibility for its intended audience.

This audience consists of undergraduates, graduate students, research personnel, and academic staff in colleges and universities, practicing scientists and engineers in industry and research institutes, and media, legal, and management personnel concerned with science and engineering employed by government and private institutions. Certain advanced high school students with at least a year of chemistry or physics and calculus may also benefit from the encyclopedia.

Robert A. Meyers
TRW, Inc.

Guide to the Encyclopedia

Readers of the *Encyclopedia of Physical Science and Technology (EPST)* will find within these pages a comprehensive study of the physical sciences, presented as a single unified work. The encyclopedia consists of eighteen volumes, including a separate Index volume, and includes 790 separate full-length articles by leading international authors. This is the third edition of the encyclopedia published over a span of 14 years, all under the editorship of Robert Meyers.

Each article in the encyclopedia provides a comprehensive overview of the selected topic to inform a broad spectrum of readers, from research professionals to students to the interested general public. In order that you, the reader, will derive the greatest possible benefit from the *EPST*, we have provided this Guide. It explains how the encyclopedia was developed, how it is organized, and how the information within it can be located.

LOCATING A TOPIC

The *Encyclopedia of Physical Science and Technology* is organized in a single alphabetical sequence by title. Articles whose titles begin with the letter A are in Volume 1, articles with titles from B through Ci are in Volume 2, and so on through the end of the alphabet in Volume 17.

A reader seeking information from the encyclopedia has three possible methods of locating a topic. For each of these, the proper point of entry to the encyclopedia is the Index volume. The first method is to consult the alphabetical Table of Contents to locate the topic as an article title; the Index volume has a complete A-Z listing of all article titles with the appropriate volume and page number.

Article titles generally begin with the key term describing the topic, and have inverted word order if necessary to begin the title with this term. For example, "Earth Sciences, History of" is the article title rather than "History of Earth Sciences." This is done so that the reader can more easily locate a desired topic by its key term, and also so that related articles can be grouped together. For example, 12 different articles dealing with lasers appear together in the La- section of the encyclopedia.

The second method of locating a topic is to consult the Contents by Subject Area section, which follows the Table of Contents. This list also presents all the articles in the encyclopedia, in this case according to subject area rather than A-Z by title. A reader seeking information on nuclear technology, for example, will find here a list of more than 20 articles in this subject area.

The third method is to consult the detailed Subject Index that is the essence of the Index volume. This is the best starting point for a reader who wishes to refer to a relatively specific topic, as opposed to a more general topic that will be the focus of an entire article. For example, the Subject Index indicates that the topic of "biogas" is discussed in the article Biomass Utilization.

CONSULTING AN ARTICLE

The First Edition of the *Encyclopedia of Physical Science and Technology* broke new ground in scholarly reference publishing through its use of a special format for articles.

The purpose of this innovative format was to make each article useful to various readers with different levels of knowledge about the subject. This approach has been widely accepted by readers, reviewers, and librarians, so much so that it has not only been retained for subsequent editions of *EPST* but has also been adopted in many other Academic Press encyclopedias, such as the *Encyclopedia of Human Biology*. This format is as follows:

- Title and Author
- Outline
- Glossary
- Defining Statement
- Main Body of the Article
- Cross References
- Bibliography

Although it is certainly possible for a reader to refer only to the main body of the article for information, each of the other specialized sections provides useful material, especially for a reader who is not entirely familiar with the topic at hand.

USING THE OUTLINE

Entries in the encyclopedia begin with a topical outline that indicates the general content of the article. This outline serves two functions. First, it provides a preview of the article, so that the reader can get a sense of what is contained there without having to leaf through all the pages. Second, it serves to highlight important subtopics that are discussed within the article. For example, the article "Asteroid Impacts and Extinctions" includes subtopics such as "Cratering," "Environmental Catastophes," and "Extinctions and Speciation."

The outline is intended as an overview and thus it lists only the major headings of the article. In addition, extensive second-level and third-level headings will be found within the article.

USING THE GLOSSARY

The Glossary section contains terms that are important to an understanding of the article and that may be unfamiliar to the reader. Each term is defined in the context of the article in which it is used. The encyclopedia includes approximately 5,000 glossary entries. For example, the article "Image-Guided Surgery" has the following glossary entry:

Focused ultrasound surgery (FUS) Surgery that involves the use of extremely high frequency sound targeted to highly specific sites of a few millimeters or less.

USING THE DEFINING STATEMENT

The text of most articles in the encyclopedia begins with a single introductory paragraph that defines the topic under discussion and summarizes the content of the article. For example, the article "Evaporites" begins with the following statement:

EVAPORITES are rocks composed of chemically precipitated minerals derived from naturally occurring brines concentrated to saturation either by evaporation or by freeze-drying. They form in areas where evaporation exceeds precipitation, especially in a semiarid subtropical belt and in a subpolar belt. Evaporite minerals can form crusts in soils and occur as bedded deposits in lakes or in marine embayments with restricted water circulation. Each of these environments contains a specific suite of minerals.

USING THE CROSS REFERENCES

Though each article in the *Encyclopedia of Physical Science and Technology* is complete and self-contained, the topic list has been constructed so that each entry is supported by one or more other entries that provide additional information. These related entries are identified by cross references appearing at the conclusion of the article text. They indicate articles that can be consulted for further information on the same issue, or for pertinent information on a related issue. The encyclopedia includes a total of about 4,500 cross references to other articles. For example, the article "Aircraft Aerodynamic Boundary Layers" contains the following list of references:

Aircraft Performance and Design ● Aircraft Speed and Altitude ● Airplanes, Light ● Computational Aerodynamics ● Flight (Aerodynamics) ● Flow Visualization ● Fluid Dynamics

USING THE BIBLIOGRAPHY

The Bibliography section appears as the last element in an article. Entries in this section include not only relevant print sources but also Websites as well.

The bibliography entries in this encyclopedia are for the benefit of the reader and are not intended to represent a complete list of all the materials consulted by the author in preparing the article. Rather, the sources listed are the author's recommendations of the most appropriate materials for further research on the given topic. For example, the article "Chaos" lists as references (among others) the works *Chaos in Atomic Physics*, *Chaos in Dynamical Systems*, and *Universality in Chaos*.

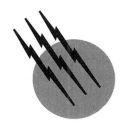

Flight (Aerodynamics)

John D. Anderson, Jr.
National Air and Space Museum, Smithsonian Institution

GLOSSARY

Drag Component of aerodynamic force in the freestream direction.

Euler equations Governing equations of fluid flow, not including viscous effects.

Hypersonic flow Flow at high Mach numbers, generally greater than 5.

Inviscid flow Flow wherein viscous effects are negligible.

Lift Component of aerodynamic force perpendicular to the freestream direction.

Navier–Stokes equations Governing equations of fluid flow including all the viscous effects.

Shock waves Very thin compression regions across which the flow properties change almost discontinuously.

Subsonic flow Flow where the Mach number is everywhere less than 1.

Supersonic flow Flow where the Mach number is everywhere greater than 1.

Transition Phenomena associated with the change from laminar to turbulent flow.

Transonic flow Flow involving mixed regions of subsonic and supersonic flow.

Viscous flow Flow where the transport phenomena of mass diffusion, viscosity, and thermal conductivity are important.

AERODYNAMICS is the study of the motion of a fluid, usually air, and its associated interactions with solid surfaces in the flow. These surfaces may be aerodynamic bodies such as airplanes and missiles or the inside walls of ducts such as inside rocket nozzles and wind tunnels. In this article, some of the basic nomenclature and principles from aerodynamics are presented and discussed. The content emphasizes aerodynamic theory, but also examines some experimental aspects. The purpose is to present an overall "bird's-eye" view of aerodynamics in a self-contained manner.

Encyclopedia of Physical Science and Technology, Third Edition, Volume 6

I. HISTORICAL PERSPECTIVE

After centuries of false starts and misguided ideas, the concept of the modern airplane was finally conceived by the Englishman Sir George Cayley (1773–1857) in 1799. Cayley proposed a *fixed* wing for generating lift, another *separate* mechanism for propulsion, and a combined horizontal and vertical tail for stability. This was in contrast to the prevailing concepts of machines or humans with flapping wings to emulate birds. Cayley's ideas were adopted and nurtured by scores of would-be aviators during the 19th century; however, all attempts to design a powered, manned, heavier-than-air aircraft were unsuccessful until December 17, 1903. On that date, Orville (1871–1948) and Wilbur (1867–1912) Wright achieved the first successful powered flight of a manned airplane on the sand dunes of Kill Devil Hills, 4 miles south of Kitty Hawk, NC. The first flight lasted for only 12 sec, and it covered a distance over the ground of 120 ft—far less than the length of a football field—but it initiated a period of intense aeronautical development carrying through the biplanes of World War I, the high-performance propeller-driven airplanes of World War II, the advent of jet propulsion in the 1940s, and the spectacular development of rockets and spacecraft that now marks the last quarter of the 20th century.

The parallel development of aerodynamics, the science on which successful flight was achieved, began in a serious fashion with the classical mechanics of Isaac Newton (1642–1727) as described in his famous "Principia" of 1687. However, Newton considered a fluid flow as a uniform, rectilinear stream of particles, which, upon striking a surface, would transfer their normal momentum to the surface but would preserve their tangential momentum. This led to Newton's famous sine-squared law, which stated that the aerodynamic force on an inclined surface varies as $\sin^2 \theta$, where θ is the angle between the incoming fluid stream and the surface. Newton's fluid model and the sine-squared law were not accurate for most fluid flow problems (although it is a reasonable approximation for some of the very high speed, hypersonic flows encountered in modern aerodynamics), and the science of aerodynamics had to wait for the pioneering theoretical insights of Daniel Bernoulli (1700–1783) and Leonard Euler (1707–1783). Beginning with Bernoulli and Euler, and expanded by the work of Louis M. Navier (1785–1836) in France and Sir George G. Stokes (1819–1903) in England, the fundamental mathematical equations of fluid flow on which aerodynamics is built were well understood by the end of the 19th century. The 20th century has witnessed a virtual explosion of aerodynamic theory and experiments, carrying from the low-speed flows associated with the Wright Brothers' airplane through the hypersonic flows associ-

ated with the modern space shuttle. This article will, in part, survey such aerodynamic principles and theory.

II. BASIC APPLICATIONS

The practical objectives of aerodynamics are (1) the prediction of forces and moments on, and heat transfer to, bodies moving through a fluid (usually air) and (2) the determination of flows moving internally through ducts, such as flow through wind tunnels and jet engines. Applications in item (1) come under the heading of *external aerodynamics* and in item (2) under the heading of *internal aerodynamics*.

The definitions of aerodynamic forces and moments are illustrated in Fig. 1. The way that nature imposes a force on a body moving through a stationary fluid (or alternatively a stationary body immersed in a moving fluid) is by means of the *pressure distribution* exerted by the fluid and acting in a direction locally perpendicular to the surface and the *shear stress distribution* created by the frictional nature of the fluid and acting in a direction locally tangential to the surface. The pressure and shear stress distributions, integrated over the entire surface of the body, yield a net resultant aerodynamic force R and moment M on the body, as shown in Fig. 1. The line connecting the leading and trailing edges of the body is the *chord line*, and the distance from the leading to the trailing edge measured

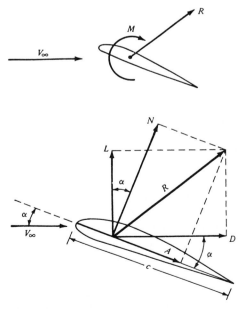

FIGURE 1 Aerodynamic forces and moments, illustrating the resolution of the resultant force into normal and axial forces and also into lift and drag. [From Anderson, J. D., Jr. (2001). "Fundamentals of Aerodynamics," 3rd ed., McGraw-Hill, New York.]

along the chord line is simply called the *chord c*. The direction and magnitude of the flow far ahead of the body is the relative wind, or simply the freestream velocity V_∞. The angle between c and V_∞ is the *angle of attack α*. The resultant aerodynamic force R is frequently resolved into two components, the *normal force N* and the *axial force A*, acting perpendicular and parallel to the chord, respectively. Even more frequently, R is resolved into the aerodynamic *lift L* and *drag D*, acting perpendicular and parallel to V_∞, respectively. The moment M can be taken about any point, but is commonly taken about the leading edge (M_{LE}) or the quarter chord (a point $0.25c$ from the leading edge) ($M_{C/4}$).

The science of aerodynamics is more fundamentally based on *dimensionless parameters* involving force and moment coefficients rather than the forces and moments themselves. If ρ_∞ represents the freestream fluid density, such dimensionless parameters are

$$\text{Lift coefficient} \equiv C_L \equiv L/q_\infty S$$

$$\text{Drag coefficient} \equiv C_D \equiv D/q_\infty S$$

$$\text{Moment coefficient} \equiv C_M \equiv M/q_\infty S c$$

where q_∞ is the freestream *dynamic pressure*, defined as $\frac{1}{2}\rho_\infty V_\infty^2$, and S is a suitable reference area of the body. These force and moment coefficients are in turn functions of (1) the shape of the body, (2) the angle of attack, and (3) various dimensionless similarity parameters. The most important similarity parameters are

$$\text{Mach number} = M_\infty/a_\infty$$

$$\text{Reynolds number} = \text{Re}_\infty = \rho_\infty V_\infty c/\mu_\infty$$

$$\text{Prandtl number} = \text{Pr}_\infty = \mu_\infty c_p/k_\infty$$

where a_∞ is the speed of sound in the free stream, μ_∞ is the freestream viscosity coefficient, c_p is the specific heat at constant pressure of the free stream, and k_∞ is the freestream thermal conductivity. For a body of given shape and angle of attack, C_L, C_D, and C_M are different functions of M_∞, Re_∞, and Pr_∞.

The importance of these dimensionless parameters is seen in the principle of *flow similarity*. By definition, two different flows are dynamically similar if (1) the streamline patterns are geometrically similar; (2) the distributions of V/V_∞, p/p_∞, T/T_∞, etc. throughout the flow field are the same when plotted against common nondimensional spatial coordinates (where V, p, and T are the local values of velocity, pressure, and temperature throughout the flow); and (3) C_L, C_D, and C_M are the same. Actually, item (3) is a consequence of item (2): if the nondimensional pressure and shear stress distributions over different bodies are the same, then the nondimensional force coefficients will be the same. The criteria that ensures that two different flows

are dynamically similar are (1) the bodies and any other solid boundaries are geometrically similar for both flows and (2) the similarity parameters (such as M_∞, Re_∞, and Pr_∞) are the same for both flows. The principle of flow similarity is used, for example, to compare results obtained in different wind tunnels at different speeds using different gases for different sized models: if the models are geometrically similar and at the same α, and if the similarity parameters such as M_∞, Re_∞, and Pr_∞ are the same for the two wind tunnels, then the measurements of C_L, C_D, and C_M from the two different wind tunnels will be the same. The same comparison can be made between wind-tunnel results for a model of a flight vehicle and the full-scale results for the actual flight vehicle moving through the atmosphere. If the free-flight values of M_∞, Re_∞, and Pr_∞ can be simultaneously simulated in the wind tunnel, then the measured values of C_L, C_D, and C_M in the wind tunnel will be the same as for full-scale flight in the atmosphere. This is the underlying justification for wind-tunnel testing in aerodynamics. Unfortunately, many wind tunnels are not able to simultaneously match the full-scale flight values of M_∞, Re_∞, and Pr_∞, due to the many conflicting design variables necessary to achieve such simultaneous simulation. Instead, one wind tunnel may be able to simulate just M_∞ for free flight; another wind tunnel may be able to simulate just Re_∞ for full-scale flight. This is one of the reasons for the vast number of different types of wind tunnels that exists throughout the world today.

With the advent of very high speed flight, beginning in the 1950s with intercontinental ballistic missiles and continuing with the manned space program including the space shuttle and other hypersonic vehicles, the aerodynamic heating of the vehicle surface became a dominant aspect. Due to the vast amount of energy dissipated by friction in such flows, and the high temperatures encountered behind strong shock waves on such vehicles, heat-transfer rates to the vehicle surface can be very high. Aerodynamic heating, like aerodynamic forces and moments, is also governed by dimensionless parameters. If Q is the rate of heat added to the vehicle, then a dimensionless heat-transfer coefficient, the Stanton number, can be defined as

$$\text{Stanton number} \equiv C_H = Q/\rho_\infty V_\infty C_p(T_{aw} - T_w),$$

where T_w is the wall temperature of the surface and T_{aw} is the adiabatic wall temperature—temperature the wall would have if the heating rate were zero. In turn, $C_H = f(M_\infty, \text{Re}_\infty, \text{Pr}_\infty, T_w/T_\infty)$, i.e., the heat-transfer coefficient is a function of the similarity parameters, Mach number, Reynolds number, Prandtl number, and the wall-to-freestream temperature ratio. Hence, the principle of dynamic similarity also holds for aerodynamic heating.

III. GENERAL GOVERNING EQUATIONS

The governing equations of aerodynamics are based on three basic physical principles:

1. Mass is conserved.
2. Newton's second law:
   force = (mass) × (acceleration).
3. Energy is conserved.

These physical principles are applied to various models of a fluid flow, for example, an infinitesimally small fluid element moving along a streamline. This leads to a series of partial differential equations relating the flowfield variables of density ρ; pressure p; three components of velocity in the x, y, and z directions in Cartesian space u, v, and w, respectively; and internal energy e (where e depends in part on temperature, e.g., for a calorically perfect gas $e = c_v T$ where c_v is the specific heat at constant volume). These equations are

Continuity

$$\frac{\partial \rho}{\partial t} + \nabla \cdot (\rho \mathbf{V}) = 0 \tag{1}$$

Momentum (x component)

$$\rho \frac{Du}{Dt} = -\frac{\partial p}{\partial x} + \frac{\partial \tau_{xx}}{\partial x} + \frac{\partial \tau_{yx}}{\partial y} + \frac{\partial \tau_{zx}}{\partial z} \tag{2}$$

Momentum (y component)

$$\rho \frac{Dv}{Dt} = -\frac{\partial p}{\partial y} + \frac{\partial \tau_{xy}}{\partial x} + \frac{\partial \tau_{yy}}{\partial y} + \frac{\partial \tau_{zy}}{\partial z} \tag{3}$$

Momentum (z component)

$$\rho \frac{Dw}{Dt} = -\frac{\partial p}{\partial z} + \frac{\partial \tau_{xz}}{\partial x} + \frac{\partial \tau_{yz}}{\partial y} + \frac{\partial \tau_{zz}}{\partial z} \tag{4}$$

Energy

$$\rho \frac{D(e + V^2/2)}{Dt} = p\dot{q} + \frac{\partial}{\partial x}\left(k\frac{\partial T}{\partial x}\right) + \frac{\partial}{\partial y}\left(k\frac{\partial T}{\partial y}\right)$$

$$+ \frac{\partial}{\partial z}\left(k\frac{\partial T}{\partial z}\right) - \nabla \cdot p\mathbf{V} + \frac{\partial(u\tau_{xx})}{\partial x}$$

$$+ \frac{\partial(u\tau_{yx})}{\partial y} + \frac{\partial(u\tau_{zx})}{\partial z} + \frac{\partial(v\tau_{xy})}{\partial x}$$

$$+ \frac{\partial(v\tau_{yy})}{\partial y} + \frac{\partial(v\tau_{zy})}{\partial z} + \frac{\partial(w\tau_{xz})}{\partial x}$$

$$+ \frac{\partial(w\tau_{yz})}{\partial y} + \frac{\partial(w\tau_{zz})}{\partial z} \tag{5}$$

where

$$\tau_{xy} = \tau_{yx} = \mu\left(\frac{\partial v}{\partial x} + \frac{\partial u}{\partial y}\right)$$

$$\tau_{yz} = \tau_{zy} = \mu\left(\frac{\partial w}{\partial y} + \frac{\partial v}{\partial z}\right)$$

$$\tau_{zx} = \tau_{xz} = \mu\left(\frac{\partial u}{\partial z} + \frac{\partial w}{\partial x}\right)$$

$$\tau_{xx} = \lambda(\nabla \cdot \mathbf{V}) + 2\mu\frac{\partial u}{\partial x}$$

$$\tau_{yy} = \lambda(\nabla \cdot \mathbf{V}) + 2\mu\frac{\partial v}{\partial y}$$

$$\tau_{zz} = \lambda(\nabla \cdot \mathbf{V}) + 2\mu\frac{\partial w}{\partial z}$$

$\dot{q}$ = rate of volumetric heat added per unit mass of gas (say by radiation)

k = thermal conductivity

$V = u\mathbf{i} + v\mathbf{j} + w\mathbf{k}$

$\dfrac{D}{Dt}$ = substantial derivative operator

$$= \frac{\partial}{\partial t} + u\frac{\partial}{\partial x} + v\frac{\partial}{\partial y} + w\frac{\partial}{\partial z}$$

These equations describe the general unsteady, three-dimensional flow of a compressible viscous fluid. They are the foundations of theoretical aerodynamics. Historically, Eqs. (2)–(4) are called the *Navier–Stokes equations*; however, in recent aerodynamic literature, especially dealing with computational solutions, the entire system of Eqs. (1)–(5) has been called the Navier–Stokes equations.

Equations (1)–(5) are nonlinear, coupled, partial differential equations with no known analytical solution. Therefore, the historical development of theoretical aerodynamics has utilized various simpler models of flows, which are governed approximately by special, simplified forms of Eqs. (1)–(5). For this reason and others, aerodynamics is conveniently subdivided into various types of flows, as described in the next section.

IV. HOW AERODYNAMICS IS SUBDIVIDED

An understanding of aerodynamics, like that of any other physical science, is obtained through a "building-block" approach—we dissect the discipline, form the parts into nice polished blocks of knowledge, and then later attempt to reassemble the blocks to form an understanding of the whole. An example of this process is the way that different types of aerodynamic flows are categorized and visualized. Although nature has no trouble setting up the most detailed and complex flow with a whole spectrum of interacting physical phenomena, we must attempt to understand such flows by modeling them with less detail and neglecting some of the (hopefully) less significant phenomena. As a result, a study of aerodynamics has evolved into a study of numerous and distinct types of flow. The purpose of this section is to itemize and contrast these types of flow and to briefly describe their most important physical phenomena.

A. Continuum versus Free-Molecule Flow

Consider the flow over a body, say, for example, a circular cylinder of diameter d. Also, consider the fluid to consist of individual molecules, which are moving about in random motion. The mean distance that a molecule travels between collisions with neighboring molecules is defined as the mean free path λ. If λ is orders of magnitude smaller than the scale of the body measured by d, then the flow appears to the body as a continuous substance. The molecules impact the body surface so frequently that the body cannot distinguish the individual molecular collisions, and the surface feels the fluid as a continuous medium. Such flow is called continuum flow. The other extreme is where λ is on the same order as the body scale; here, the gas molecules are spaced so far apart (relative to d) that collisions with the body surface occur only infrequently, and the body surface can feel distinctly each molecular impact. Such flow is called free molecular flow. For manned flight, vehicles such as the space shuttle encounter free molecular flow at the extreme outer edge of the atmosphere, where the air density is so low that λ becomes on the order of the shuttle size. There are intermediate cases, where flows can exhibit some characteristics of both continuum and free-molecule flows; such flows are generally labeled "low-density flows," in contrast to continuum flow. By far, the vast majority of practical aerodynamic applications involves continuum flows. Low-density and free-molecule flows are just a small part of the total spectrum of aerodynamics.

B. Inviscid versus Viscous Flow

A major facet of a gas or liquid is the ability of the molecules to move rather freely. When the molecules move, even in a very random fashion, they obviously transport their mass, momentum, and energy from one location to another in the fluid. This transport on a molecular scale gives rise to the phenomena of mass diffusion, viscosity (friction), and thermal conduction. All real flows exhibit the effects of these transport phenomena; such flows are called viscous flows. In contrast, a flow that is assumed to involve no friction, thermal conduction, or diffusion is called an inviscid flow. Inviscid flows do not truly exist

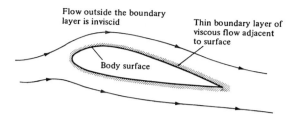

FIGURE 2 Division of an aerodynamic flow into a viscous boundary layer adjacent to the surface and an inviscid outer flow. [From Anderson, J. D., Jr. (2001). "Fundamentals of Aerodynamics," 3rd ed., McGraw-Hill, New York.]

in nature; however, there are many practical aerodynamic flows (more than you would think) where the influence of transport phenomena is small, and we can model the flow as being inviscid. For this reason, more than 70% of aerodynamic theory deals with inviscid flows.

Theoretically, inviscid flow is approached in the limit as the Reynolds number goes to infinity. However, for practical problems, many flows with high but finite Re can be assumed to be inviscid. For such flows, the influence of friction, thermal conduction, and diffusion is limited to a very thin region adjacent to the body surface (the boundary layer), and the remainder of the flow outside this thin region is essentially inviscid. This division of the flow into two regions is illustrated in Fig. 2. For flows over slender bodies, such as the airfoil sketched in Fig. 2, inviscid theory adequately predicts the pressure distribution and lift on the body and gives a valid representation of the streamlines and flow field away from the body. However, because friction (shear stress) is a major source of aerodynamic drag, inviscid theories by themselves cannot adequately predict total drag.

In contrast, there are some flows that are dominated by viscous effects. For example, if the airfoil in Fig. 2 is inclined to a high incidence angle to the flow (high angle of attack), then the boundary layer will tend to separate from the top surface, and a large wake is formed downstream. The separated flow is sketched at the top of Fig. 3; it is characteristic of the flow field over a "stalled" airfoil. Separated flow also dominates the aerodynamics of blunt bodies, such as the cylinder at the bottom of Fig. 3. Here, the flow expands around the front face of the cylinder, but separates from the surface on the rear face, forming a rather fat wake downstream. The types of flow illustrated in Fig. 3 are dominated by viscous effects: no inviscid theory can independently predict the aerodynamics of such flows.

C. Incompressible versus Compressible Flows

A flow in which the density ρ is constant is called incompressible. In contrast, a flow where the density is variable

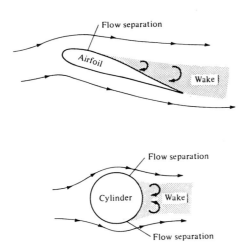

FIGURE 3 Illustration of flow separation. [From Anderson, J. D., Jr. (2001). "Fundamentals of Aerodynamics," 3rd ed., McGraw-Hill, New York.]

is called compressible. For our purposes here, we will simply note that all flows, to a greater or lesser extent, are compressible: truly incompressible flow, where the density is precisely constant, does not occur in nature. However, analogous to our discussion of inviscid flow, there are a number of aerodynamic problems that can be modeled as being incompressible without any detrimental loss of accuracy. For example, the flow of homogenous liquids is treated as incompressible, and hence, most problems involving hydrodynamics assume ρ is a constant. Also, the flow of gases at low Mach number is essentially incompressible; for $M < 0.3$, it it always safe to assume ρ is a constant. This was the flight regime of all airplanes from the Wright Brothers' first flight in 1903 to just prior to World War II. It is still the flight regime of most small, general aviation aircraft of today. Hence, there exists a large bulk of aerodynamic experimental and theoretical data for incompressible flows. On the other hand, high-speed flow (near Mach 1 and above) must be treated as compressible; for such flows, ρ can vary over wide latitudes.

D. Mach-Number Regimes

Of all the ways of subdividing and describing different aerodynamic flows, the distinction based on Mach number is probably the most prevalent. If M is the local Mach number at an arbitrary point in a flow field, then by definition the flow is locally

Subsonic if $M < 1$

Sonic if $M = 1$

Supersonic if $M > 1$

Looking at the whole flow field simultaneously, four different speed regimes can be identified using Mach number as the criterion.

1. Subsonic flow ($M < 1$ everywhere). A flow field is defined as subsonic if the Mach number is less than 1 at every point. Subsonic flows are characterized by smooth streamlines (no discontinuity in slope), as sketched in Fig. 4a. Moreover, since the flow velocity is everywhere less than the speed of sound, disturbances in the flow (say the sudden deflection of the trailing edge of the airfoil in Fig. 4a) propagate both upstream and downstream and are felt throughout the entire flow field. Note that a freestream Mach number M_∞ less than 1 does not guarantee a totally subsonic flow over the body. In expanding over an aerodynamic shape, the flow velocity increases above the freestream value, and if M_∞ is close enough to 1, the local Mach number may become supersonic in certain regions of the flow. This gives rise to a rule of thumb that $M_\infty < 0.8$ for subsonic flow over slender bodies. For blunt bodies, M_∞ must be even lower to insure totally subsonic flow. (Again, emphasis is made that the above is just a loose rule of thumb and should not be taken as a precise quantitative definition.) Also, note that incompressible

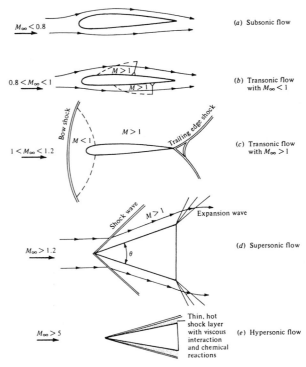

FIGURE 4 Characterization of flow fields based on the Mach number range. [From Anderson, J. D., Jr. (2001). "Fundamentals of Aerodynamics," 3rd ed., McGraw-Hill, New York.]

flow is a special limiting case of subsonic flow where $M \to 0$.

2. Transonic flow (mixed regions where $M < 1$ and $M > 1$). As stated above, if M_∞ is subsonic but near unity, the flow can become locally supersonic ($M > 1$). This is sketched in Fig. 4b, which shows pockets of supersonic flow over both the top and the bottom surfaces of the airfoil, terminated by weak shock waves behind which the flow becomes subsonic again. Moreover, if M_∞ is increased slightly above unity, a bow shock wave is formed in front of the body; behind this shock wave the flow is locally subsonic, as shown in Fig. 4c. This subsonic flow subsequently expands to a low supersonic value over the airfoil. Weak shock waves are usually generated at the trailing edge, sometimes in a "fishtail" pattern as shown in Fig. 4c. The flow fields shown in Figs. 4b and 4c are characterized by mixed subsonic–supersonic flows and are dominated by the physics of both types of flows. Hence, such flow fields are called transonic flows. Again, as a rule of thumb for slender bodies, transonic flows occur for freestream Mach numbers in the range $0.8 < M_\infty < 1.2$.

3. Supersonic flow ($M > 1$ everywhere). A flow field is defined as supersonic if the Mach number is greater than 1 at every point. Supersonic flows are frequently characterized by the presence of shock waves across which the flow properties and streamlines change discontinuously (in contrast to the smooth, continuous variations in subsonic flows). This is illustrated in Fig. 4d for supersonic flow over a sharp-nosed wedge; the flow remains supersonic behind the oblique shock wave from the tip. Also shown are distinct expansion waves, which are common in supersonic flow. (Again, the listing of $M_\infty > 1.2$ is strictly a rule of thumb. For example, in Fig. 4d, if θ is made large enough, the oblique shock wave will detach from the tip of the wedge and will form a strong, curved bow shock ahead of the wedge with a substantial region of subsonic flow behind the wave. Hence, the totally supersonic flow sketched in Fig. 4d is destroyed if θ is too large for a given M_∞. This shock detachment phenomenon can occur at any value of $M_\infty > 1$, but the value of θ at which it occurs increases as M_∞ increases. In turn, if θ is made infinitesimally small, the flow field in Fig. 4d holds for $M_\infty \geq 1.0$. However, the above discussion clearly shows that the listing of $M_\infty > 1.2$ in Fig. 4d is a very tenuous rule of thumb and should not be taken literally. In a supersonic flow, because the local flow velocity is greater than the speed of sound, disturbances created at some point in the flow cannot work their way upstream (in contrast to subsonic flow). This property is one of the most significant physical differences between subsonic and supersonic flows. It is the basic reason why shock waves occur in supersonic flows but do not occur in steady subsonic flow.

4. Hypersonic flow (very high supersonic speeds). Refer again to the wedge in Fig. 4d. Assume θ is a given, fixed value. As M_∞ increases above 1, the shock wave moves closer to the body surface. Also, the strength of the shock wave increases, leading to higher temperatures in the region between the shock and the body (the shock layer). If M_∞ is sufficiently large, the shock layer becomes very thin, and interactions between the shock wave and the viscous boundary layer on the surface occur. Also, the shock layer temperature becomes high enough that chemical reactions occur in the air. The O_2 and N_2 molecules are torn apart; that is, the gas molecules dissociate. When M_∞ becomes large enough such that viscous interaction and/or chemically reacting effects begin to dominate the flow (Fig. 4e), the flow field is called hypersonic. (Again, a somewhat arbitrary but frequently used rule of thumb for hypersonic flow is $M_\infty > 5$.) Hypersonic aerodynamics received a great deal of attention during the period 1955–1970 because atmospheric entry vehicles encounter the atmosphere at Mach numbers between 25 (ICBMs) and 36 (the Apollo lunar return vehicle.) From 1985 to the present, this attention has shifted to air-breathing hypersonic cruise vehicles and single-stage-to-orbit vehicles. Today, hypersonic aerodynamics is just part of the whole spectrum of realistic flight speeds.

In summary, we attempt to organize our study of aerodynamic flows according to one or more of the various categories discussed in this section. The block diagram in Fig. 5 is presented to help emphasize these categories and to show how they are related.

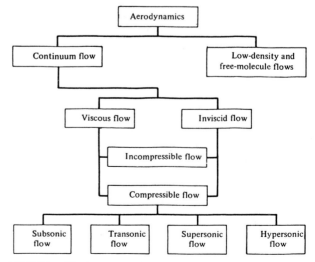

FIGURE 5 Block diagram for the various categories of aerodynamic flows. [From Anderson, J. D., Jr. (2001). "Fundamentals of Aerodynamics," 3rd ed., McGraw-Hill, New York.]

There are several additional types of flows frequently assumed in aerodynamics.

1. Irrotational flow. Flow where the individual elements of fluid moving along a streamline are in translational motion only and are not rotating. For irrotational flow, a scaler potential function $\phi = \phi(x, y, z)$ can always be defined in such a fashion that

$$\mathbf{V} = \nabla \phi.$$

This is an important aspect of theoretical aerodynamics; ϕ is called the *velocity potential.*

2. Adiabatic flow. Flow where no heat is added or taken away.

3. Isentropic flow. Flow undergoing a thermodynamic process that is both adiabatic and reversible. For such flow, the thermodynamic entropy is constant along a streamline.

V. INVISCID INCOMPRESSIBLE FLOW

Incompressible flow is constant-density flow. Inviscid flow is flow where the transport phenomena of mass diffusion, viscosity, and thermal conduction are negligibly small. Under these two assumptions, the governing equations of motion, Eqs. (1–5), become considerably simpler. Equation (1) reduces to

$$\nabla \cdot \mathbf{V} = 0, \tag{6}$$

which is the continuity equation for incompressible flow. Eqs. (2)–(4) combine-and reduce to

$$p_1 + \tfrac{1}{2}\rho V_1^2 = p_2 + \tfrac{1}{2}\rho V_2^2. \tag{7}$$

Equation (7) is the famous Bernoulli equation, and it relates the pressure and velocity at two different points along a streamline. Another way of writing the Bernoulli equation is

$$p + \tfrac{1}{2}\rho V^2 = \text{constant} \tag{8}$$

along a streamline. If all the streamlines of the flow originate from a uniform reservoir (such as the reservoir of a wind tunnel or the atmosphere far ahead of a moving flight vehicle), then the constant in Eq. (8) is the same for all streamlines, and Eq. (7) then relates the pressure and velocity at *any* two points in the flow, not necessarily on the same streamline. For inviscid incompressible flow, the energy equation [Eq. (5)] will also reduce to the Bernoulli equation; hence, the energy equation for such a flow is redundant and is not needed. Therefore, the Bernoulli equation can be viewed as both a statement of Newton's second law, $F = ma$, and a statement of the conservation of mechanical energy in an inviscid incompressible flow.

The Bernoulli equation is used in many applications of low-speed aerodynamics: the flow over airfoils, the flow through ducts, and the utilization of the Pitot tube for airspeed measurements, to name just a few. Let us examine more closely the last item mentioned above, namely, the Pitot tube.

In 1732, the Frenchman Henri Pitot was busy trying to measure the flow velocity of the Seine River in Paris. One of the instruments he used was his own invention—a strange-looking tube bent into an $\angle$ shape, as shown in Fig. 6. Pitot oriented one of the open ends of the tube so that it faced directly into the flow. In turn, he used the pressure inside this tube to measure the water flow velocity. This was the first time in history that a proper measurement of fluid velocity was made, and Pitot's invention has carried through to the present day as the Pitot tube—one of the most common and frequently used instruments in any modern aerodynamic laboratory. Moreover, a Pitot tube is the most common device for measuring flight velocities of airplanes. The purpose of this section is to describe the basic principle of the Pitot tube.

Consider a flow with pressure p_1 moving with velocity V_1, as sketched at the left of Fig. 6. Let us consider the significance of the pressure p_1 more closely. The pressure is associated with the time rate of change of momentum of the gas molecules impacting on or crossing a surface; that is, pressure is clearly related to the motion of the molecules. This motion is very random, with molecules moving in all direction with various velocities. Now imagine that you hop on a fluid element of the flow and ride with it at the velocity V_1. The gas molecules, because of their random motion, will still bump into you, and you will

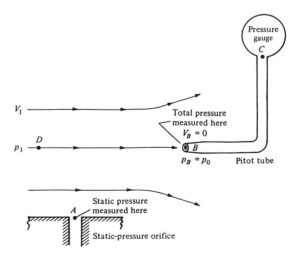

FIGURE 6 Illustration of Pitot and static pressure measurements. [From Anderson, J. D., Jr. (2001). "Fundamentals of Aerodynamics," 3rd ed., McGraw-Hill, New York.]

feel the pressure p_1 of the gas. We now give this pressure a specific name: the static pressure. Static pressure is a measure of the purely random motion of molecules in the gas; it is the pressure you feel when you ride along with the gas at the local flow velocity. All pressures used in this article so far have been static pressures; the pressure p appearing in all our previous equations has been the static pressure. In engineering, whenever a reference is made to "pressure" without further qualification, that pressure is always interpreted as the static pressure. Furthermore, consider a boundary of the flow, such as a wall, where a small hole is drilled perpendicular to the surface. The plane of the hole is parallel to the flow, as shown at point A in Fig. 6. Because the flow moves over the opening, the pressure felt at point A is due only to the random motion of the molecules; that is, at point A, the static pressure is measured. Such a small hole in the surface is called a static pressure orifice, or a static pressure tap.

In contrast, consider that a Pitot tube is now inserted into the flow, with an open end facing directly into the flow. That is, the plane of the opening of the tube is perpendicular to the flow, as shown at point B in Fig. 6. The other end of the Pitot tube is connected to a pressure gauge, such as point C in Fig. 6; that is, the Pitot tube is closed at point C. For the first few milliseconds after the Pitot tube is inserted into the flow, the gas will rush into the open end and will fill the tube. However, the tube is closed at point C; there is no place for the gas to go, and hence after a brief period of adjustment, the gas inside the tube will stagnate; that is, the gas velocity inside the tube will go to zero. Indeed, the gas will eventually pile up and stagnate everywhere inside the tube, including at the open mouth at point B. As a result, the streamline of the flow that impinges directly at the open face of the tube (streamline DB in Fig. 6) sees this face as an obstruction to the flow. The fluid elements along streamline DB slow down as they get closer to the Pitot tube and go to zero velocity right at point B. Any point in a flow where $V = 0$ is called a stagnation point of the flow; hence, point B at the open face of the Pitot tube is a stagnation point, where $V_B = 0$. In turn, from Bernoulli's equation we know the pressure increases as the velocity decreases. Hence, $p_B > p_1$. The pressure at a stagnation point is called the stagnation pressure, or total pressure, denoted by p_0. Hence, at point B, $p_B = p_0$.

From the above discussion, we see that two types of pressure can be defined for a given flow: static pressure, which is the pressure you feel by moving with the flow at its local velocity V_1, and total pressure, which is the pressure that the flow achieves when the velocity is reduced isentropically to zero.

How is the Pitot tube used to measure flow velocity? To answer this question, first note that the total pressure p_0 exerted by the flow at the tube inlet (point B) is impressed throughout the tube (there is no flow inside the tube; hence, the pressure everywhere inside the tube is p_0). Therefore, the pressure gage at point C reads p_0. This measurement, in conjunction with a measurement of the static pressure p_1 at point A, yields the difference between total and static pressure, $p_0 - p_1$, and it is this pressure difference that allows the calculation of V_1 via Bernoulli's equation. In particular, apply Bernoulli's equation between point A, where the pressure and velocity are p_1 and V_1, respectively, and point B, where the pressure and velocity are p_0 and $V = 0$, respectively:

$$p_A + \tfrac{1}{2}\rho V_A^2 = p_B + \tfrac{1}{2}\rho V_B^2$$

or

$$p_1 + \tfrac{1}{2}\rho V_1^2 = p_0 + 0.$$

Solving for V_1, we have

$$V_1 = \sqrt{\frac{2(p_0 - p_1)}{\rho}}.$$

The above equation allows the calculation of velocity simply from the measured difference between total and static pressure. The total pressure p_0 is obtained from the Pitot tube, and the static pressure p_1 is obtained from a suitably placed static pressure tap.

It is possible to combine the measurement of both total and static pressure in one instrument, a Pitot-static probe. A Pitot-static probe measures p_0 at the nose of the probe and p_1 at a suitably placed static pressure tap on the probe surface downstream of the nose.

In Eq. (8), the grouping $\tfrac{1}{2}\rho V^2$ is called the *dynamic pressure* by definition and is used in all flows, incompressible to hypersonic:

$$\text{Dynamic pressure} \equiv \tfrac{1}{2}\rho V^2.$$

However, for incompressible flow, the dynamic pressure has special meaning; it is precisely the difference between total and static pressure:

$$q = p_0 - p.$$

It is important to keep in mind that this relation comes from Bernoulli's equation and thus holds for incompressible flow only. For compressible flow, where Bernoulli's equation is not valid, the pressure difference $p_0 - p$ is not equal to q. The velocities of compressible flows, both subsonic and supersonic, can be measured by means of a Pitot tube, but the equations are different from the above. (Velocity measurements in subsonic and supersonic compressible flows are discussed in Section VI.)

In aerodynamics, it is useful to define a dimensionless pressure as

$$C_p \equiv \frac{p - p_\infty}{q_\infty}, \qquad (9)$$

where $q_\infty = \frac{1}{2}\rho_\infty V_\infty^2$ and where C_p is defined as the *pressure coefficient* used throughout aerodynamics, from incompressible to hypersonic flow. In the aerodynamic literature it is very common to find pressures given in terms of C_p rather than the pressure itself. For *incompressible flow*, C_p can be expressed in terms of velocity only. By applying the Bernoulli equation to Eq. (9), the result for incompressible flow is

$$C_p = 1 - (V/V_\infty)^2. \qquad (10)$$

The pressure coefficient is more than just a dimensionless pressure; it is a similarity parameter in the same sense as C_L and C_D, and for a given body shape, angle of attack, and geometrically similar location in the flow, $C_p = f(M_\infty, \mathrm{Re}_\infty, \mathrm{Pr}_\infty)$.

In general treatments of inviscid, incompressible, irrotational flows, the basic governing equations reduce to Laplace's equation in terms of the velocity potential. Combining Eq. (6) with the irrotational expression $\mathbf{V} = \nabla\phi$, we directly obtain

$$\nabla^2\phi = \frac{\partial^2\phi}{\partial x^2} + \frac{\partial^2\phi}{\partial y^2} + \frac{\partial^2\phi}{\partial z^2} = 0, \qquad (11)$$

which is *Laplace's equation*—one of the most familiar equations from mathematical physics. Hence, solutions of incompressible, inviscid, irrotational flows are called *potential flow solutions*.

In theoretical aerodynamics, the concept of the stream function is sometimes employed. By definition, the stream function $\bar{\psi}$ is a constant along a given streamline, and the change in $\bar{\psi}$, $\triangle\bar{\psi}$, between two streamlines is equal to the *mass* flow between these streamlines. For incompressible flow, a modified stream function ψ is defined such that $\triangle\bar{\psi}$ between two streamlines equals the *volume flow* between the streamlines. From the definition of the stream function, it can be shown that the flow velocity can be found by differentiating ψ, for example,

$$\nabla^2\psi = \frac{\partial^2\psi}{\partial x^2} + \frac{\partial^2\psi}{\partial y^2} + \frac{\partial^2\psi}{\partial z^2} = 0. \qquad (12)$$

Hence, for such flows, Laplace's equation governs both ϕ and ψ. Moreover, in a flow field, lines of constant ψ (i.e., streamlines) are everywhere orthogonal to lines of constant ϕ (i.e., equipotential lines).

Laplace's equation is linear, and hence, any number of particular solutions, say ϕ_1, ϕ_2, ϕ_3, and ϕ_4, can be added to obtain another solution, say ϕ_5:

$$\phi_5 = \phi_1 + \phi_2 + \phi_3 + \phi_4.$$

This allows the solution of a given potential flow to be synthesized by superimposing a number of other, more elementary, flows. This is the major underlying strategy of the solution of inviscid, incompressible, irrotational flows. In the synthesis of such flows, the following elementary flows are useful.

1. Uniform flow. Constant-property flow at velocity V_∞, with straight streamlines oriented in a single direction (say, the x direction),

$$\phi_1 = V_\infty x = V_\infty r \cos\theta$$
$$\psi_1 = V_\infty y = V_\infty r \sin\theta$$

where r and θ are polar coordinates.

2. Source flow (in two dimensions). Streamlines are straight lines emanating from a central point (say, the origin of the coordinate system) where the velocity along each streamline varies inversely with distance from the central point. In polar coordinates,

$$\phi_2 = \frac{\Lambda}{2\pi}\ln r \qquad \psi_2 = \frac{\Lambda}{2\pi}\theta,$$

where Λ is the source strength defined as the rate of volume flow from the source. A negative value of Λ denotes a *sink*, where the flow moves toward the central point.

3. Source flow (in three dimensions). A flow with straight streamlines emanating in three dimensions from an origin, where the velocity varies inversely as the square of the distance from the origin, and

$$\phi_3 = -\frac{\lambda}{4\pi r},$$

where λ is the source strength, that is, the rate of volume flow from the origin. For a sink, λ is negative.

4. Doublet flow (in two dimensions). The superposition of a source and sink of equal but opposite strength, Λ and $-\Lambda$, where the distance l between the two approaches zero at the same time that the product Λl remains constant. The strength of the doublet is $\kappa \equiv \Lambda l$. In polar coordinates,

$$\phi_4 = \left(\frac{\kappa}{2\pi}\right)\frac{\cos\theta}{r} \qquad \psi_4 = -\left(\frac{\kappa}{2\pi}\right)\frac{\sin\theta}{r}.$$

5. Doublet flow (in three dimensions). The superposition of a three-dimensional source and sink of equal but opposite strength, λ and $-\lambda$, where the distance l between the two approaches zero at the same time that the product $\mu = \lambda l$ remains constant. In spherical coordinates,

$$\phi_5 = -\left(\frac{\mu}{4\pi}\right)\frac{\cos\theta}{r^2}.$$

6. Vortex flow (in two dimensions). Streamlines are concentric circles about a given central point, where the velocity along any given circular streamline is constant

but varies from one streamline to another inversely with distance from the common center. In polar coordinates with an origin at the central point,

$$\phi_6 = -\frac{\Gamma}{2\pi}\theta \qquad \psi_6 = \frac{\Gamma}{2\pi}\ln r,$$

where Γ is the vortex strength, defined as the circulation about the vortex.

The term "circulation" introduced above is a general concept in aerodynamics and is particularly useful in the analysis of low-speed airfoils and wings. Circulation is defined as

$$\Gamma = -\oint \mathbf{V} \cdot \mathbf{ds}.$$

That is, circulation is the line integral of flow velocity integrated about the closed curve C drawn in the flow. For vortex flow, the vortex strength Γ is the circulation taken about any closed curve that encloses the central point.

The six elementary flows described above, by themselves, are not practical flow fields. However, they can be superimposed in various ways to synthesize practical flows in two and three dimensions, such as flows over cylinders, spheres, airfoils, wings, and whole airplanes. For example, three such flows are described below. Keep in mind that the expressions given for ϕ and ψ for the elementary flows are solutions of Laplace's equation for those flows, and therefore, such solutions can be added together (superimposed) to produce other solutions.

1. Nonlifting flow over a circular cylinder. This symmetrical flow is synthesized by the superposition of a uniform flow with a doublet, yielding the stream function for a cylinder of radius R in a free stream of velocity V_∞:

$$\psi = (V_\infty r \sin\theta)\left(1 - \frac{R^2}{r^2}\right).$$

The resulting pressure-coefficient variation over the surface of the cylinder is symmetrical and is given by

$$C_p = 1 - 4\sin^2\theta.$$

Because the pressure variation is symmetrical, there is no lift or drag theoretically predicted for the cylinder.

2. Lifting flow over a circular cylinder. In actual experience, a circular cylinder spinning about its axis and immersed in a flow with velocity V_∞ will experience a lift force. This is because the frictional effect between the fluid and the spinning cylinder tends to increase the flow velocity on one side of the cylinder and to decrease the flow velocity on the other side. From the Bernoulli equation, these unequal velocities lead to unequal pressures on both sides of the cylinder, causing a lift force to be produced.

This phenomenon is sometimes called the "Magnus effect." The curve of a spinning baseball and the hook or slice of a spinning golf ball are examples of the Magnus effect. In this case, for a cylinder of radius R in a flow with velocity V_∞, the lifting flow is synthesized by the superposition of a uniform flow, a doublet, and a vortex, yielding

$$\psi = (V_\infty r \sin\theta)\left(1 - \frac{R^2}{r^2}\right) + \frac{\Gamma}{2\pi}\ln\frac{r}{R}.$$

The resulting pressure coefficient distribution is not symmetrical and is given by

$$C_p = 1 - \left[4\sin^2\theta + \frac{2\Gamma\sin\theta}{\pi R V_\infty} + \left(\frac{\Gamma}{2\pi R V_\infty}\right)^2\right].$$

In turn, this yields a theoretical value for lift and drag per unit span of the cylinder as

$$L = \rho_\infty V_\infty \Gamma \qquad D = 0,$$

where Γ is the value of the circulation about the cylinder and is also equal to the strength of the elementary vortex used in the superposition process.

3. Nonlifting flow over a sphere. This flow is synthesized by superposition of a uniform flow and a three-dimensional doublet, yielding a result for the pressure-coefficient variation over the spherical surface as

$$C_p = 1 - \frac{9}{4}\sin^2\theta.$$

This is a symmetrical pressure distribution and leads to the theoretical results of zero lift and zero drag.

The theoretical result that drag is zero for all the above flows is called *d'Alembert's paradox* and is a consequence of neglecting friction in the theory. In reality, skin friction and flow separation from the surface—both viscous effects—create a finite drag on any real aerodynamic body.

However, the theoretical result for lift obtained for the lifting cylinder is quite real, namely,

$$L = \rho_\infty V_\infty \Gamma.$$

This result is a general result for the inviscid incompressible flow over a cylindrical body of any arbitrary shape and is called the *Kutta–Joukowski theorem*. It states that the lift per unit span along the body is directly proportional to the circulation about the body. This result is the major focus of the circulation theory of lift, first developed in the early 1900s, and is still in use today for the prediction of the lifting characteristics of low-speed bodies.

For the source, vortex, and doublet flows described above, the center of the flow can be considered a *point* source, *point* vortex, or *point* doublet, where the point is

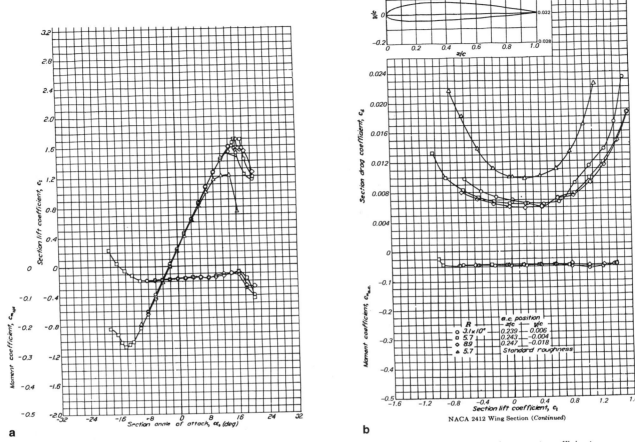

FIGURE 7 Lift and moment coefficients vs angle of attack for a conventional airfoil. (b) Drag and moment coefficients
vs lift coefficient for a conventional airfoil.

a singularity since the velocity goes to infinity at the center. This leads to a general theoretical approach wherein the flow field over a body of general shape is calculated by distributing such singularities over the surface and calculating the strength of these singularities such that, in combination with the uniform free stream, the body shape becomes a streamline of the flow. Moreover, for an airfoil with a sharp trailing edge, the flow must leave smoothly at the trailing edge; this is in comparison to other possible solutions where it can leave the surface at some other point on the airfoil. Indeed, where the flow leaves the surface is a function of the amount of circulation Γ present around the airfoil. Nature chooses just the right value of Γ so that the flow always leaves smoothly at the sharp trailing edge. This is called the *Kutta condition*. If the airfoil is reasonably thin and the angle of attack is small, an approximate theory based on the idea of distributed singularities mentioned above (thin airfoil theory) leads to the following result for the rate of change of lift coefficient of c_1 vs angle of attack: $dc_1/d\alpha = 2\pi$ (per radian).

That is, for small angles of attack, the lift coefficient varies linearly with angle of attack—a result that is verified by experiment.

To further illustrate the aerodynamic properties of airfoils at low speed, Figs. 7a and 7b show experimental data for a National Advisory Committee for Aeronautics (NACA) 2412 airfoil; here, lift coefficient c_1 and moment coefficient about the quarter chord $c_{m_{c/4}}$ are shown as a function of angle of attack α (Fig. 7a), and the drag coefficient c_d and moment coefficient about the aerodynamic center $c_{m_{a.c.}}$ are shown as functions of c_1 (Fig. 7b). The aerodynamic center is defined as that point on the airfoil about which moments do not vary with angle of attack, as is clearly seen in Fig. 7b.

For three-dimensional bodies. the concept of distributed singularities over the body surface is also employed. Historically, for a finite wing, these singularities took the form of a number of vortex filaments (discrete lines of vorticity) in the shape of a horseshoe, with one segment running along the span of the wing (the so-called "lifting line") and the other segments trailing downstream into the wake of the wing. This lifting-line theory, due to Ludwig Prandtl (1875–1953), resulted in the first reasonable calculation of lift and drag on an airplane wing. It also identified a component of drag called *induced drag*, which is due to an alteration of the pressure distribution over the wing caused by strong vortices trailing downstream from the wing tips. These wing-tip vortices induce a general downward component of velocity over the wing (called downwash), which in turn changes the pressure distribution in such a fashion as to increase the drag. This increase in drag is induced drag. Induced drag is directly proportional to the square of the lift coefficient; therefore, induced drag rapidly increases as the lift increases. For a complete airplane, the total drag coefficient C_D at low speeds can be expressed as

$$C_D = C_{D_0} + C_{D_i}, \tag{13}$$

where

$$C_{D_i} = C_L^2 / \pi \mathrm{AR}.$$

In Eq. (13), which is called the *drag polar*, C_{D_0} is the parasite drag coefficient at zero lift; parasite drag is produced by the net effect of skin friction over the body surface plus the extra pressure drag produced by regions of flow separation over the surface (sometimes called form drag). Also in Eq. (13), C_{D_i} is the drag coefficient due to lift and is due to the increment of parasite drag (above the zero lift value) that occurs when the airplane is at the higher angles of attack associated with the production of lift, plus the induced drag due to the tip vortices, hence down-wash. In the above, AR is the aspect ratio, $\mathrm{AR} = b^2/S$, where b is the wing span and S is the platform area of the wing. Also, e is an efficiency factor, where $e \leq 1$, and is, in part, associated with the mathematical shape of the spanwise distribution of lift over the wing.

In modern aerodynamics, the three-dimensional potential flow over low-speed bodies is calculated by means of singularities distributed over small, flat panels, and these panels in turn cover the surface of the body. The development of the "panel method," using either, or a combination of, source, vortex, or doublet panels, has been well developed since 1965.

VI. INVISCID COMPRESSIBLE FLOW

Compressible flow differs from incompressible flow in at least the following respects:

1. The density of the flow becomes a variable.
2. The flow speeds are high enough that the flow kinetic energy becomes important and therefore energy changes in the flow must be considered. This couples the science of thermodynamics into the aerodynamic considerations.
3. Shock waves can occur, which can completely dominate the flow.

The governing equations for inviscid, compressible flow are obtained from Eqs. (1)–(5) as

Continuity

$$\frac{\partial \rho}{\partial t} + \nabla \cdot (\rho \mathbf{V}) = 0 \tag{14}$$

Momentum (x component)

$$\rho \frac{Du}{Dt} = -\frac{\partial p}{\partial x} \tag{15}$$

Momentum (y component)

$$\rho \frac{Dv}{Dt} = -\frac{\partial p}{\partial y} \tag{16}$$

Momentum (z component)

$$\rho \frac{Dw}{Dt} = -\frac{\partial p}{\partial z}$$

Energy

$$\rho \frac{D(e + V^2/2)}{Dt} = p\dot{q} - \nabla \cdot (p\mathbf{V}) \tag{17}$$

Equations (14)–(17) are called the Euler equations. Note that for a compressible flow, Bernoulli's equation [Eq. (7) or (8)] does *not* hold.

The speed of sound is an important parameter for compressible flow. The speed of sound a is given by $a^2 = (\partial p/\partial \rho)_s$, where the subscript indicates constant entropy. For a calorically perfect gas (a gas with constant specific heats and that obeys the gas law $p = \rho RT$),

$$a = \sqrt{\gamma RT},$$

where $\gamma = c_p/c_v$ and R is the specific gas constant. In turn, the speed of sound is used to define the *Mach number* as

$$M = V/a,$$

where V is the flow velocity.

Two defined properties particularly important to the analysis of compressible flow are (1) total temperature T_0, defined as that temperature that would exist if the flow were brought to rest adiabatically, and (2) total pressure p_0, defined as that pressure that would exist if the flow were brought to rest isentropically. For a calorically perfect gas, the relation between total and static properties is a function of γ and M only, as

$$\frac{T_0}{T} = 1 + \frac{\gamma - 1}{2}M^2$$

and

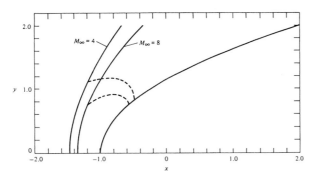

FIGURE 8 Supersonic flow over a blunt body; shock waves and sonic lines for two different freestream March numbers. [From Anderson, J. D., Jr. (2001). "Fundamentals of Aerodynamics," 3rd ed., McGraw-Hill, New York.]

$$\frac{p_0}{p} = \left(1 + \frac{\gamma - 1}{2} M^2\right)^{\gamma/(\gamma-1)}. \qquad (18)$$

Eq. (18) is particularly useful for the measurement of Mach number in subsonic compressible flow. A Pitot tube in the flow will sense p_0; if the static pressure p is known at the same point, then the Pitot measurement of P_0 will directly yield the Mach number via Eq. (18).

In any steady flow where $M > 1$, shock waves can occur. For example, consider the blunt-nosed, parabolically shaped body shown at the right in Fig. 8. Flow is moving from left to right.If the upstream flow is supersonic ($M_\infty > 1$), then a curved shock wave will exist slightly upstream of the blunt nose. Two such shocks are shown to the left of the body; the leftmost shock illustrates the case for $M_\infty = 4$, and the rightmost shock pertains to the case for $M_\infty = 8$. The shock waves slow down the flow. The lower part of the flow field behind the shock wave is locally subsonic, and the upper part, after sufficient expansion around the body, is locally supersonic, albeit at a Mach number lower than M_∞. The dashed lines shown in Fig. 8 divide the subsonic and supersonic regions behind the shock and are called the *sonic lines*, since the local Mach number $M = 1$ along these lines. Note from Fig. 8 that as the freestream Mach number increases, the shock wave moves closer to the body and the sonic line moves down closer to the centerline of the flow. An essential ingredient of the understanding of supersonic flow is the calculation of the shape and strength of shock waves, such as those illustrated in Fig. 8. Therefore, let us examine shock waves in more detail.

A shock wave is an extremely thin region, typically on the order of 10^{-5} cm, across which the flow properties can change drastically. The shock wave is usually at an oblique angle to the flow, such as sketched in Fig. 9a; however, there are many cases where we are interested in a shock wave normal to the flow, as sketched in Fig. 9b. For example, referring again to Fig. 8, the portion of the bow shock wave directly in front of the nose is normal, whereas the shock wave is oblique away from the nose. In both cases of a normal or an oblique shock, the shock wave is an almost explosive compression process, where the pressure increases almost discontinuously across the wave. As shown in Fig. 9, in region 1 ahead of the shock, the Mach number, flow velocity, pressure density, temperature, entropy, total pressure, and total enthalpy (defined as $e + p/\rho + V^2/2$) are denoted by M_1, V_1, p_1, ρ_1, T_1, s_1, $p_{0,1}$, and $h_{0,1}$, respectively. The analogous quantities in region 2 behind the shock are denoted by a subscript 2. The qualitative changes across the wave are noted in Fig. 9. The pressure, density, temperature, and entropy increase across the shock, whereas the total pressure, Mach number, and velocity decrease. Physically, the flow across a shock wave is adiabatic, which leads to a constant total enthalpy across the wave. For a perfect gas, this also means that the total temperature is constant across the shock, i.e., $T_{0,1} = T_{0,2}$. In both oblique shock and normal shock cases, the flow ahead of the shock must be supersonic, that is, $M_1 > 1$. Behind the oblique shock, the flow usually remains supersonic, that is, $M_2 > 1$, but at a reduced Mach number,

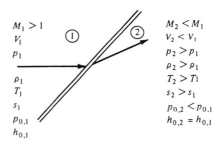

(*a*) Oblique shock wave

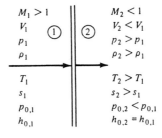

(*b*) Normal shock wave

FIGURE 9 Oblique and normal shock waves. [From Anderson, J. D., Jr. (2001). "Fundamentals of Aerodynamics," 3rd ed., McGraw-Hill, New York.]

$M_2 < M_1$. However, there are a few special cases where the oblique shock is strong enough to decelerate the downstream flow to a subsonic Mach number. For the normal shock, as sketched in Fig. 9b, the downstream flow is always subsonic, that is, $M_2 < 1$.

The quantitative changes across a normal shock wave can be obtained from Eqs. (14)–(17) specialized for steady one-dimensional flow and integrated to obtain the following basic normal shock equations:

Continuity

$$\rho_1 V_1 = \rho_2 V_2$$

Momentum

$$p_1 + \rho_1 V_1^2 = p_1 + \rho_2 V_2^2$$

Energy

$$e_1 + p_1/\rho_1 + V_1^2/2 = e_2 + p_2/\rho_2 + V_2^2/2$$

Along with the equations of state $p = \rho RT$ and $e = c_v T = RT/(\gamma - 1)$ for a calorically perfect gas, the basic normal shock equations can be algebraically manipulated to obtain the following changes of flow properties across the normal shock:

$$\frac{p_2}{p_1} = 1 + \frac{2\gamma}{\gamma + 1}\left(M_1^2 - 1\right) \qquad (19)$$

$$\frac{\rho_2}{\rho_1} = \frac{V_1}{V_2} = \frac{(\gamma + 1)M_1^2}{2 + (\gamma - 1)M_1^2} \qquad (20)$$

$$M_2^2 = \frac{1 + [(\gamma - 1)/2]M_1^2}{\gamma M_1^2 - (\gamma - 1)/2} \qquad (21)$$

Note that the changes across a normal shock wave depend only on the value of γ and the upstream Mach number M_1.

For an oblique shock wave, let β be the angle between the shock wave and the upstream flow direction: β is called the wave angle. Also, note in Fig. 9a that the flow, in crossing the oblique shock, is bent in the upward direction behind the shock. Let θ be the angle between the downstream and upstream flow directions: θ is called the deflection angle. Then, Eqs. (19)–(21) hold for an oblique shock if M_1, V_1, M_2, and V_2 are replaced by $M_1 \sin\beta$, $V_1 \sin\beta$, $M_2 \sin(\beta - \theta)$, and $V_2 \sin(\beta - \theta)$, respectively. Moreover, for an oblique shock, the wave angle, deflection angle, and upstream Mach number are related through the equation

$$\tan\theta = 2\cot\beta \frac{M_1^2 \sin^2\beta - 1}{M_2^2(\gamma + \cos 2\beta) + 2}. \qquad (22)$$

From Eqs. (19)–(21) written appropriately for an oblique shock, combined with Eq. (22), we see that if we know any two quantities about the shock, say β and M_1, all

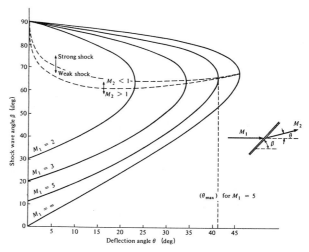

FIGURE 10 Oblique shock wave relations. [From Anderson, J. D., Jr. (1990). "Modern Compressible Flow," 2nd ed., McGraw-Hill, New York.]

other quantities such as θ, M_2, p_2/p_1, etc. are determined. A plot of Eq. (22) for $\gamma = 1.4$ (air at standard conditions) is given in Fig. 10. From this figure, we can deduce the following physical characteristics about oblique shock waves:

1. For any given M_1, there is a maximum deflection angle θ_{max}. If the physical geometry is such that $\theta > \theta_{max}$, then no solution exists for a straight oblique shock wave; in such a case, the flow field will adjust in such a fashion to curve and detach the shock wave.

2. For any given $\theta < \theta_{max}$, there are two values of β predicted by Fig. 10 for a given M_1. The larger value of β is called the strong shock solution, and the smaller β is the weak shock solution. In nature, the weak shock solution is favored and is the one that usually occurs.

3. The downstream Mach number M_2 is subsonic for the strong shock solutions, whereas M_2 is supersonic for the weak shock solutions, except for the narrow band between the two dashed lines in Fig. 10, where $M_2 < 1$ even for the weak oblique shock case.

A Pitot tube can be used to measure the Mach number in a supersonic flow. However, in contrast to our previous discussions of the use of a Pitot tube, for the supersonic flow case a normal shock wave will exist in front of the mouth of the tube. Hence, the Pitot tube will measure the total pressure behind the normal shock, *not* the total pressure of the flow itself as in the subsonic case. By appropriate manipulation of the basic normal shock equations, the following relation can be found for a Pitot tube in supersonic flow:

$$\frac{p_{0.2}}{p_1} = \left[\frac{(\gamma + 1)^2 M_1^2}{4\gamma M_1^2 - 2(\gamma - 1)}\right]^{\gamma/(\gamma-1)} \left(\frac{1 - \gamma + 2\gamma M_1^2}{\gamma + 1}\right), \tag{23}$$

where $p_{0.2}$ is the total pressure behind the normal shock (the pressure measured by the tube), p_1 is the static pressure upstream of the normal shock, and M_1 is the Mach number upstream of the shock. If a Pitot tube measurement is taken at a given point in a supersonic flow, and if the static pressure is known at that point, then the Pitot measurement will directly yield the local Mach number at that point via Eq. (23).

An oblique shock, in the limit of infinitely weak strength, becomes a *Mach wave,* where $\theta = 0$, and $\beta = \mu$, where μ is called the *Mach angle:*

$$\mu = \sin^{-1}(1/M).$$

For all oblique shocks of a finite strength, $\beta > \mu$.

Oblique shock waves are created when a supersonic flow is bent into itself, such as the flow over a concave corner as shown in Fig. 11a. In contrast, when a supersonic flow is bent away from itself, such as the flow over a convex corner as sketched in Fig. 11b, an *expansion wave* is created. Expansion waves are the opposite of shock waves. Expansions are composed of an infinite number of Mach waves and are a region of smooth continuous change through the expansion fan. Across the expansion wave, the Mach number increases, and the pressure, temperature, and density decrease. The flow through an expansion wave is isentropic (constant entropy), and hence, both the total pressure and the total temperature are constant through the wave. For a given Mach number M_1 ahead of the expansion wave and a given deflection angle θ, the Mach number M_2 behind the expansion can be found from

$$\theta = v(M_2) - v(M_1),$$

where v is the *Prandtl–Meyer function,* given by

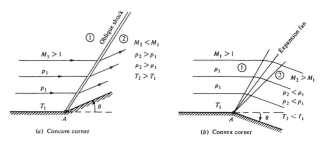

$$v(M) = \sqrt{\frac{\gamma + 1}{\gamma - 1}} \tan^{-1} \sqrt{\frac{\gamma - 1}{\gamma + 1}(M^2 - 1)} - \tan^{-1}\sqrt{M^2 - 1}$$

For a calorically perfect gas. The corresponding temperature and pressure changes can be obtained from

$$\frac{T_2}{T_1} = \frac{1 + [(\gamma - 1)/2]M_1^2}{1 + [(\gamma - 1)/2]M_2^2}$$

and

$$\frac{p_2}{p_1} = \left(\frac{T_2}{T_1}\right)^{\gamma/(\gamma-1)}.$$

The external supersonic flow fields over some aerodynamic bodies can sometimes be synthesized in an approximate fashion by a proper combination of local oblique shock and expansion wave solutions. Such approaches are called *shock-expansion solutions.*

Compressible flows through ducts (i.e., internal compressible flows) are of great importance in the design of high-speed wind tunnels, jet engines, and rocket engines, to name just a few applications. Consider a duct with a local cross-sectional area A. The area may change with length along the duct. By a proper combination of the continuity, momentum, and energy equations for a compressible flow, the following relation can be extracted, which governs the compressible flow in a variable area duct, where the flow properties are assumed to be uniform across any cross section but can change from one cross section to another (so-called quasi-one-dimensional flow):

$$\frac{dA}{A} = (M^2 - 1)\frac{dV}{V}. \tag{24}$$

In Eq. (24), dA is the local change in area, dV is the corresponding change in velocity, and M is the local Mach number. From Eq. (24), we see that

1. For *subsonic flow* $(0 \leq M < 1)$, V increases as A decreases, and V decreases as A increases. Therefore, to increase the velocity, a *convergent* duct must be used, whereas to decrease the velocity, a *divergent* duct must be employed.

2. For *supersonic flow* $(M > 1)$, V increases as A increases, and V decreases as A decreases. Hence, to increase the velocity, a *divergent* duct must be used, whereas to decrease the velocity, a *convergent* duct must be employed.

3. When $M = 1$, $dA = 0$. Hence, sonic flow will occur in that location inside a variable-area duct where the area variation is a local minimum. Such a location is called a sonic throat.

FIGURE 11 Oblique shock and expansion waves. [From Anderson, J. D., Jr. (2001). "Fundamentals of Aerodynamics," 3rd ed., McGraw-Hill, New York.]

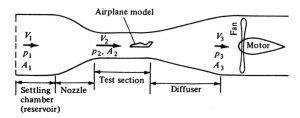

FIGURE 12 Sketch of a subsonic wind tunnel. [From Anderson, J. D., Jr. (2001). "Fundamentals of Aerodynamics," 3rd ed., McGraw-Hill, New York.]

Clearly, subsonic and supersonic flows through a changing-area duct behave differently. This is why subsonic wind tunnels are configured differently than supersonic wind tunnels. A sketch of a basic subsonic tunnel is shown in Fig. 12. Here, we see a convergent duct for speeding up the subsonic flow (the nozzle), a constant-area section where a test model is placed in the high-speed (but subsonic) flow, and then a divergent duct (the diffuser) for slowing the flow down before exhausting it to the surroundings. In contrast, a sketch of a basic supersonic tunnel is shown in Fig. 13. Here, the initially low-speed subsonic flow from a reservoir is speeded up in a convergent section, reaching Mach 1 at the throat, and then the now-supersonic flow is further speeded up in a divergent section. Hence, a nozzle designed to produce a supersonic flow is a *convergent–divergent* nozzle, sometimes called a *de Laval nozzle*, after Carl G. P. de Laval (1845–1912), who first employed such nozzles in a steam turbine. The high-speed supersonic flow then enters a constant-area test section where a test model is placed. Downstream of the model, the flow, along with the shock wave system produced by the model, enters a diffuser designed to slow the flow to a low subsonic speed before exhausting to the surroundings. To accomplish this, the initially supersonic flow must first be slowed in a convergent section to Mach 1 at a minimum area (the "second" throat) and then further slowed subsonically in a divergent section. Hence, the supersonic diffuser is also a convergent–divergent duct, just as in the case of the supersonic nozzle; however, their functions are completely opposite.

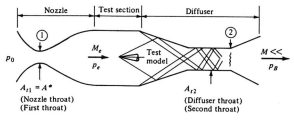

FIGURE 13 Sketch of a supersonic wind tunnel. [From Anderson, J. D., Jr. (2001). "Fundamentals of Aerodynamics," 3rd ed., McGraw-Hill, New York.]

The flow in a properly designed supersonic nozzle is isentropic. For this case, and for a calorically perfect gas, the Mach number that exists at a given cross section with area A is simply a function of the ratio of this area to the throat area, denoted by A^*. That is,

$$\left(\frac{A}{A^*}\right)^2 = \frac{1}{M^2}\left[\frac{2}{\gamma+1}\left(1 + \frac{\gamma-1}{2}M^2\right)\right]^{(\gamma+1)/(\gamma-1)}$$

This is a double-valued function; for a given A/A^*, two solutions exist for M—a subsonic value in the convergent portion of the nozzle and a supersonic value in the divergent portion.

Let us return to the consideration of external compressible flows. If the flow is subsonic or supersonic, irrotational, and involves the flow over slender bodies at small angle of attack, Eqs. (14)–(17) reduce to a single linearized equation:

$$\left(1 - M_\infty^2\right)\frac{\partial^2 \phi}{\partial x^2} + \frac{\partial^2 \phi}{\partial y^2} + \frac{\partial^2 \phi}{\partial z^2} = 0, \qquad (25)$$

where ϕ is a *perturbation* velocity potential defined such that $\partial\phi/\partial x = u$, $\partial\phi/\partial y = v$, and $\partial\phi/\partial z = w$, where u, v, and w are small perturbation velocities superimposed on the uniform free stream in such a manner that $u = V_\infty + u$, $v = v$, and $w = w$. Equation (25) does *not* hold for transonic or hypersonic Mach numbers.

A solution of Eq. (25) for a subsonic flow leads to a method for correcting low-speed, incompressible flow data to take into account compressibility effects—so-called *compressibility corrections*. For example, if C_{po} denotes the low-speed incompressible flow value of the pressure coefficient at a certain point on a two-dimensional airfoil, then the value of C_p at the same point for a high-speed subsonic freestream Mach number M_∞ is given approximately by $C_p = C_{po}/\sqrt{1 - M_\infty^2}$. This compressibility correction is called the *Prandtl–Glauert* rule and is generally valid for $M_\infty < 0.7$. In turn, the lift and moment coefficients for the airfoil are similarly related to their corresponding low-speed incompressible values c_{l_0} and c_{m_0} as

$$c_l = \frac{c_{l_0}}{\sqrt{1 - M_\infty^2}} \qquad c_m = \frac{c_{m_0}}{\sqrt{1 - M_\infty^2}}.$$

Equation (25) can also be solved for supersonic flows, leading to a simple expression for the pressure coefficient for two-dimensional flow:

$$C_p = \frac{2\theta}{\sqrt{M_\infty^2 - 1}}, \qquad (26)$$

where θ is the local angle between the tangent to the body and the freestream direction and M_∞ is the freestream Mach number. In turn, this result from linearized theory can be used to calculate the lift and wave drag coefficients for thin airfoils at a small angle of attack. For example, for a flat plate at angle of attack α, Eq. (26) leads to

$$c_l = \frac{4\alpha}{\sqrt{M_\infty^2 - 1}}, \qquad (27)$$

$$c_d = \frac{4\alpha^2}{\sqrt{M_\infty^2 - 1}}. \qquad (28)$$

The wave drag coefficient c_d is a ramification of the high pressure behind a shock wave and can be associated with the entropy increase across a shock wave. We can view wave drag as that component of drag that is associated with shock waves on the body; obviously, wave drag exists only at supersonic and hypersonic speeds. For thin airfoils at small α, Eq. (27) still holds; however, Eq. (28) is modified to take into account the finite thickness and curvature (camber) of the airfoil.

Modern exact calculations of inviscid compressible flows—subsonic, transonic, supersonic, and hypersonic—are made by means of sophisticated numerical solutions of Eqs. (14)–(17) on a high-speed digital computer.

As a final note in this section, a comment is made about hypersonic flow. Such flow, which by rule of thumb is denoted as the regime where $M_\infty > 5$, is characterized by several important physical phenomena already described in Section IV. Among these is the existence of thin shock layers around the body (the shock waves lie close to the body). For such a flow, the physical flow picture looks very much like the original model proposed by Newton in 1687 (as described in Section I). Hence, Newton's sine-squared law is a reasonable prediction for the pressure coefficient at hypersonic speeds, that is,

$$C_p = 2\sin^2\theta,$$

where θ is the angle between a tangent to the surface and the freestream direction. A *modified* Newtonian formula actually gives improved results; that is,

$$C_p = C_{p_{\max}}\sin^2\theta,$$

where $C_{p_{\max}}$ is the pressure coefficient at the stagnation point after the flow has passed through a normal shock wave at the freestream Mach number.

VII. VISCOUS FLOW

The aerodynamic results discussed in the previous sections dealt with inviscid flows. In contrast, in this section we examine the impact of transport phenomena that exert dissipative effects in a flow. In particular, we examine the influence of viscosity and thermal conduction; mass diffusion will not be considered here.

In the governing equations for fluid dynamics. Eqs. (1)–(5), the viscous effects are present through the shear stress, normal stress, and heat conduction terms. Shear stress and normal stress are directly proportional to the velocity gradients in the flow, for example, $\tau_{yx} = \mu(\partial u/\partial y)$; hence, viscosity effects are large in regions of large velocity gradients. Similarly, thermal conduction involves the transport of heat in the opposite direction of a temperature gradient, for example, $\dot{q}_y = -k(\partial T/\partial y)$, where $\dot{q}_y$ is the heat flux (energy per unit area per unit time) in the y direction, k is the thermal conductivity, and $\partial T/\partial y$ is the temperature gradient in the y direction. Therefore, thermal conduction effects are large in regions of large temperature gradients.

The major practical effects of viscous flow on aerodynamic problems are the following:

1. The action of viscosity generates a shear stress at a solid surface, which in turn creates a drag called skin friction drag.

2. Shear stress acting on a surface tends to slow the flow velocity near the surface. If the flow is experiencing an adverse pressure gradient (a region where the pressure increases in the flow direction), then the low-energy fluid elements near the surface cannot negotiate the adverse pressure gradient; as a result, the flow separates from the surface. Flow separation alters the pressure distribution over the surface in such a fashion to increase the drag; this is called "pressure drag due to flow separation," or "form drag." In addition, if the body is producing lift, then flow separation can greatly reduce the lift. This is the mechanism that limits the lift coefficient on an airfoil, wing, or lifting body to some maximum value. For example, returning to Fig. 7a, note that c_l increases with α until a maximum value is achieved. As α is further increased, massive flow separation occurs, which causes the lift to rapidly decrease. Under this condition, the airfoil is said to be "stalled."

3. The kinetic energy of the fluid elements near the surface is reduced due to the retarding effects of friction; in turn, this energy is converted to thermal energy in the flow near the surface, thus increasing the temperature of the flow. For high-speed flows, particularly at high supersonic and hypersonic speeds, this dissipative phenomenon can create very high temperatures near the surface. Through the mechanism of thermal conduction at the surface, large aerodynamic heating rates can result.

The magnitude of the skin friction and aerodynamic heating and the extent of flow separation are greatly influenced by the nature of the viscous flow; that is, whether the flow is laminar or turbulent. In a laminar flow, the path lines of various fluid elements are smooth and regular. In contrast, if the motion of a fluid element is very irregular and tortuous, the flow is called turbulent flow. The net effect of turbulence is to pump some of the higher energy flow that exists far away from the surface to a region closer the surface. This increases the velocity and temperature gradients at the surface, hence increasing skin friction and heat transfer. From this point of view, turbulent flow is a detriment. On the other hand, flow separation is delayed by turbulent flow, hence reducing the pressure drag due to flow separation. In this fashion, turbulent flow is advantageous. In most aerodynamic problems, the viscous flow first begins as laminar; then at some downstream location the flow experiences a transition to turbulent. There is a basic principle that nature always moves toward the state of maximum disorder; in terms of viscous flow, this means that nature is always driving toward turbulent flow. The location of the transition region is influenced by various phenomena. For example, if the Reynolds number Re_∞ is increased, transition moves forward. If the Mach number M_∞ is increased, transition moves rearward. If the surface is roughened, transition moves forward. If the surface is highly cooled, transition moves rearward. If the amount of natural turbulence inherently present in the free stream is increased, transition moves forward. These are just a few of the many phenomena that influence transition from laminar to turbulent flow.

Viscous flow problems are most exactly analyzed by solving the complete Navier–Stokes equations. Eqs. (1)–(5). However, these nonlinear equations are so complex that no general analytical solution exists, and exact solutions for practical aerodynamic applications are only presently being obtained through detailed finite-difference numerical solutions on a high-speed computer. Even in these cases, some Navier–Stokes solutions require many hours of computer time to go to completion. This is a current state-of-the-art problem in the analysis of viscous flow.

Faced with these difficulties in solving Eqs. (1)–(5), Ludwig Prandtl in 1904 introduced the *boundary layer concept*. Referring again to Fig. 1, in many problems most of the viscous effects are limited to a thin region adjacent to the solid surface. This thin viscous region is called the *boundary layer*. For such a region, the governing equations, Eqs. (1)–(5), can be reduced to the *boundary layer equations*, which in two-dimensional flow are

Continuity

$$\frac{\partial(\rho u)}{\partial x} + \frac{\partial(\rho v)}{\partial y} = 0$$

x Momentum

$$\rho u \frac{\partial u}{\partial x} + \rho v \frac{\partial u}{\partial y} = -\frac{dp_e}{dx} + \frac{\partial}{\partial y}\left(\mu \frac{\partial u}{\partial y}\right)$$

y Momentum

$$\frac{\partial p}{\partial y} = 0$$

Energy

$$\rho u \frac{\partial h}{\partial x} + \rho v \frac{\partial h}{\partial y} = \frac{\partial}{\partial y}\left(k \frac{\partial T}{\partial y}\right) + u \frac{dp_e}{dx} + \mu \left(\frac{\partial u}{\partial y}\right)^2$$

where x and y are coordinates tangential and normal to the surface respectively. Since $\partial p/\partial y = 0$ through the boundary layer, then the pressure distribution at the outer edge of the boundary layer, $p_e = f(x)$, is impressed directly through the boundary layer to the surface.

The boundary layer equations can be readily solved for flow over a flat plate. Classical solutions for incompressible laminar flow yield

$$c_f = \frac{0.664}{\sqrt{\mathrm{Re}_x}}$$

and

$$\delta = \frac{5.0x}{\sqrt{\mathrm{Re}_x}},$$

where c_f is the skin friction coefficient defined as $c_f \equiv \tau_w / \frac{1}{2}\rho_\infty V_\infty^2$, Re_x is the Reynolds number based on running length from the leading edge, $\mathrm{Re}_x \equiv \rho_\infty V_\infty x/\mu_\infty$, and δ is the thickness of the boundary layer. Note from these results that, for a laminar boundary layer, c_f and hence τ_w varies as $x^{-1/2}$, and δ grows as $\sqrt{x}$.

Results for a laminar compressible flow over a flat plate indicate that c_f and δ are functions not only of Re_x but of $M_\infty, \mathrm{Pr}_\infty$, and T_w/T_∞ as well, where Pr_∞ is the freestream Prandtl number, T_w is the surface temperature, and T_∞ is the freestream temperature. We can write, for compressible laminar flow,

$$c_f = \frac{0.664}{\sqrt{\mathrm{Re}_x}} f(M_\infty, \mathrm{Pr}_\infty, T_w/T_\infty)$$

and

$$\delta = \frac{5.0}{\sqrt{\mathrm{Re}_x}} g(M_\infty, \mathrm{Pr}_\infty, T_w/T_\infty).$$

Results for incompressible turbulent flow cannot be obtained exactly from the boundary layer equations. Assumptions must be made about a model for the turbulence, and empirical data is always needed to some stage in a turbulent-flow analysis. However, approximate results for a flat plate are

$$c_f = \frac{0.058}{\mathrm{Re}_x^{0.2}}$$

and

$$\delta = \frac{0.37x}{\mathrm{Re}^{0.2}}.$$

Here, c_f is seen to vary as $x^{-0.2}$, and the turbulent boundary layer thickness grows as $x^{0.8}$. As in the case of laminar flow, these incompressible flow results become modified for compressible flow, where c_f and δ are both functions of M_∞, Pr_∞, and T_w/T_∞, in addition to Re_x.

VIII. TWO MODERN AERODYNAMIC INVENTIONS

The drag coefficient for an airfoil as a function of Mach number at subsonic speeds is sketched qualitatively in Fig. 14. Over a large part of the Mach number range, c_d is relatively constant. This pertains to the completely subsonic flow over the airfoil, as sketched in Fig. 4a. As the freestream Mach number is increased, the flow Mach number on the top surface of the airfoil also increases. That freestream Mach number at which sonic flow is first achieved somewhere on the airfoil surface is called the *critical Mach number*, denoted by M_{cr} in Fig. 14. If M_∞ is further increased, an entire pocket of supersonic flow develops on the top (and even possibly the bottom) surface of the airfoil, as sketched in Fig. 4b. This corresponds to point b in Fig. 14. With the appearance of the shock waves at the end of these supersonic pockets, the drag coefficient begins to rapidly increase. The freestream Mach number at which this major drag increase begins is called the *drag-divergence Mach number*, also shown in Fig. 14. The major increase in drag near a freestream Mach number of 1 was once thought to be so severe that airplanes would never fly faster than sound. This myth of the "sound barrier" was supported by theoretical results such as Eq. (28), which shows c_d going to infinity as M_∞ goes to 1. How-

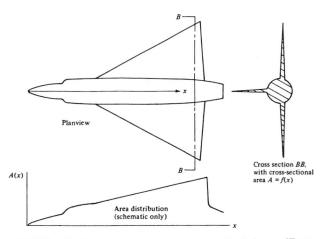

FIGURE 15 Example of a non-area-ruled airplane. [From Anderson, J. D., Jr. (2001). "Fundamentals of Aerodynamics," 3rd ed., McGraw-Hill, New York.]

ever, the linear theory that produces this result is not valid near Mach 1. In reality, the drag coefficient will peak at some finite value around Mach 1, as shown in Fig. 14; as long as the airplane has excess thrust from the engines to overcome this peak drag, the aircraft can easily fly into the supersonic regime.

Two major practical aerodynamic inventions since 1950 have been (1) the area rule and (2) the supercritical airfoil. The former is an effort to reduce the peak value of c_d shown in Fig. 14, and the latter attempts to push the drag-divergence Mach number to a higher value.

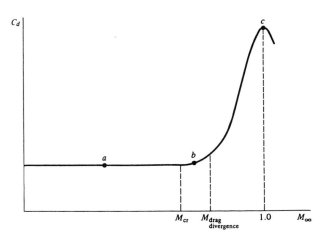

FIGURE 14 Schematic of the variation of drag coefficient for an airfoil as a function of freestream Mach number at subsonic and transonic speeds. [From Anderson, J. D., Jr. (2000). "Introduction to Flight," 4th ed., McGraw-Hill, New York.]

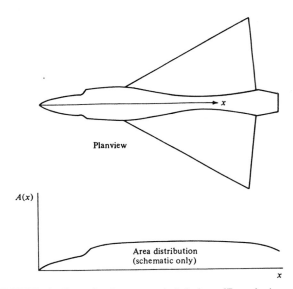

FIGURE 16 Example of an area-ruled airplane. [From Anderson, J. D., Jr. (2001). "Fundamentals of Aerodynamics," 3rd ed., McGraw-Hill, New York.]

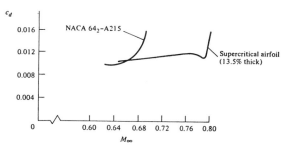

FIGURE 17 Effect of a supercritical airfoil on the drag-divergence Mach number. [From Andreson, J. D., Jr. (2001). "Fundamentals of Aerodynamics," 3rd ed., McGraw-Hill, New York.]

The area rule states that the distribution of cross-sectional area of the body should be smooth and gradual. This implies that the fuselage area must be decreased in the vicinity of the wing. Qualitative sketches of non-area-ruled and area-ruled aircraft are shown in Figs. 15 and 16, respectively. The peak transonic drag coefficient is less for the area-ruled airplane in Fig. 16 than for the non-area-ruled aircraft in Fig. 15. Thus, the area rule allows high-performance airplanes to routinely break the speed of sound.

The supercritical airfoil is designed to avoid the large pocket of supersonic flow over the airfoil until larger values of M_∞ are acheived. In other words, the supercritical airfoil increases the drag-divergence Mach number, as shown in Fig. 17.

The development of both the area rule in the 1950s and the supercritical airfoil in the 1960s has greatly increased the speed and efficiency of transonic and supersonic airplanes. It is noteworthy that both of these major developments stemmed from the same person. Richard T. Whitcomb (1921–   ), who was an aerodynamacist for the National Advisory Committee for Aeronautics and later for the National Aeronautics and Space Administration (NASA).

IX. EPILOGUE

This short article on aerodynamics has attempted to present a bird's-eye view of the subject. Many details and other important subjects are not discussed. Therefore, the reader is encouraged to examine the complete books on aerodynamics listed in the bibliography for a considerable amplification of material on the subject.

SEE ALSO THE FOLLOWING ARTICLES

AIRCRAFT AERODYNAMIC BOUNDARY LAYERS • AIRCRAFT PERFORMANCE AND DESIGN • AIRCRAFT SPEED AND ALTITUDE • COMPUTATIONAL AERODYNAMICS • FLOW VISUALIZATION • FLUID DYNAMICS • HEAT TRANSFER • JET AND GAS TURBINE ENGINES

BIBLIOGRAPHY

Anderson, J. D., Jr. (1990). "Modern Compressible Flow; with Historical Perspective," 2nd ed., McGraw-Hill, New York.

Anderson. J. D., Jr. (2000). "Introduction to Flight," 4th ed., McGraw-Hill, New York.

Anderson, J. D., Jr. (2001). "Fundamentals of Aerodynamics," 3rd ed., McGraw-Hill, New York.

Bertin, J. J., and Smith, M. L. (1979). "Aerodynamics for Engineers," Prentice Hall, Englewood Cliffs, NJ.

Katz, J., and Plotkin, A. (1991). "Low-Speed Aerodynamics: From Wing Theory to Panel Methods," McGraw-Hill, New York.

Kuethe, A. M., and Chow, C. Y. (1986). "Foundations of Aerodynamics," 4th ed., Wiley, New York.

Marthy, T., and Brebbia, C. (1990). "Computational Methods in Viscous Aerodynamics," Computational Mechanics, Southampton.

McCormick., B. W. (1979). "Aerodynamics, Aeronautics, and Flight Mechanics," Wiley, New York.

Shivell, R. S. (1983). "Fundamentals of Flight," Prentice-Hall, Englewood Cliffs, NJ.

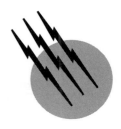

Flow Visualization

Wolfgang Merzkirch

Universität Essen

GLOSSARY

Image processing Digitization of a recorded flow picture and subsequent evaluation of the pattern by a computer to determine quantitative data.

Laser-induced fluorescence Visualization by tracer material that emits fluorescent radiation upon excitation by laser light.

Line-of-sight method Visualization by transmitting a light wave through the fluid flow.

Tomography Computer-aided reconstruction of the three-dimensional, refractive-index distribution in a flow to which a line-of-sight method has been applied in various viewing directions (projections).

Tracer particles Foreign particles with which a fluid flow is seeded for the purpose of flow visualization by light scattering.

Whole-field method Diagnostic method providing the information for a whole field of view (photograph) at a specific instant of time.

THE METHODS OF FLOW VISUALIZATION are diagnostic tools for surveying and measuring the flow of liquids and gases. They make visible the motion of a fluid that is normally invisible because of its transparency. By applying one of the methods of flow visualization, a flow picture can be directly observed or recorded with a camera. The information in the picture is available for a whole field, that is, the field of view, and for a specific instant of time (*whole-field method*). The information can be either qualitative, thus allowing for interpreting the mechanical and physical processes involved in the development of the flow, or quantitative, so that measurements of certain properties of the flow field (velocity, density, and so forth) can be performed. The techniques of flow visualization, which are used in science and industry, can be classified according to three basic principles: light scattering from tracer particles; optical methods relying on refractive index changes in the flowing fluid; and interaction of the fluid flow with a solid surface.

I. PRINCIPLES OF FLOW VISUALIZATION TECHNIQUES

The methods of visualizing a flowing fluid are based on the interaction of the flow with light. A light wave incident into the flow field (*illumination*) may interact with the fluid in two different ways: (1) light can be scattered from the fluid molecules or from the tracer particles with which the fluid is seeded; and (2) the properties of the light

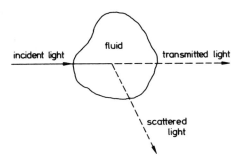

FIGURE 1 Light incident into a fluid flow is scattered from the fluid molecules or from tracer particles in various directions. The light wave transmitted through the transparent fluid is altered in comparison to the incident light, if the flow field exhibits changes of the fluid's refractive index.

wave can be changed, because of a certain optical behavior of the fluid, so that the light wave transmitted through the flow is different from the incident light (Fig. 1). The visualization methods based on these two interaction processes are totally different in nature and apply to different flow situations.

Since the light scattered from the fluid molecules (*Rayleigh scattering*) is extremely weak, the flow is seeded with small tracer particles (dust, smoke, dye, and so forth), and the more intense radiation scattered from these tracers is observed instead. It is thereby assumed that the motion of the tracer is identical with the motion of the fluid, an assumption that does not always hold, for example, in nonstationary flows. The scattered light carries information on the state of the flow at the position of the tracer particle, that is, the recorded information is local. For example, if the light in Fig. 1 is incident in form of a thin light sheet being normal to the plane of the figure, an observer could receive and record information on the state of the flow (e.g., the velocity distribution) in the respective illuminated plane.

The signal-to-noise ratio in this type of flow visualization can be improved if the tracer does not just rescatter the incident light but emits its own, characteristic radiation (*inelastic scattering*). This principle is realized by fluorescent tracers (e.g., iodine), which may emit bright fluorescing light once the fluorescence is induced by an incident radiation with the appropriate wavelength (*laser-induced fluorescence*).

An optical property of a fluid that may change the state of the transmitted light wave (Fig. 1) is its index of refraction. This index is related to the fluid density, so that flows with varying density, temperature, or concentration may affect the transmitted light. Two effects are used for such an optical flow visualization: the deflection of the light beams from their original direction (*refraction*) and the change of the phase distribution of the wave. The latter

is measured by means of optical interferometry; shadowgraph and schlieren methods are sensitive to the refractive light deflection, and they are used for qualitative optical flow visualization.

The information on the state of the flow or the density field is integrated along the path of the transmitted light, that is, the information is not local as in the case of the techniques using light scattering. For the purpose of a quantitative information on the three-dimensional flow field, it is necessary to desintegrate (invert) the data, which is recorded in two-dimensional (plane) form, for example, on a photograph. This requires the application of methods known as computer tomography.

The result of a flow visualization experiment is a flow picture that may be recorded with a camera (for example, a photographic camera, movie camera, or video camera). The information on the state of flow is available in the recording plane (x–y plane) for a specific instant of time, t_i; or for a number of discrete instants of time in the case of kinematic recording. This makes the information different from that obtained by probe measurements (e.g., hot wire anemometer, laser-Doppler velocimeter), where the data is measured only at one specific location (coordinate x, y, z), but as a continuous function of time t.

For the purpose of generating quantitative data on the flow, the recorded visual pattern must be identified, evaluated, and interpreted. In order to objectify the evaluation and make it independent of a viewer's interpretation, the pattern can be recognized by electro-optical devices and transformed into digital data, which are then processed by a computer (*image processing*).

II. FLOW VISUALIZATION BY TRACER MATERIAL

That a flow becomes visible from foreign particles that are floating on a free water surface or suspended in the fluid is a fact of daily experience. This crude approach has been refined for laboratory experiments. The methods of flow visualization by adding a tracer material to the flow are not real science but art that concerns the selection of the appropriate tracers, their concentration in the fluid, and the systems for illumination and recording.

The trace material, after being released, is swept along with the flow. If one does not resolve the motion of single particles, qualitative information on the flow structure (streamlines, vortices, separated flow regimes) becomes available from the observed pattern. The identification of the motion of individual tracers provides quantitative information on the flow velocity, provided that there is no velocity deficit between the tracer and the fluid. Only in the case of fluorescent (or phosphorescent) tracers is it

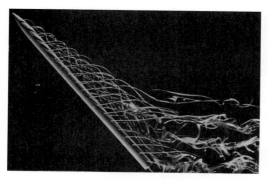

FIGURE 2 Visualization of the swirling flow behind an inclined cylinder by dye lines in water. Original dye lines are alternatingly red and blue. (Courtesy of Dr. H. Werlé, ONERA, Châtillon, France.)

possible to deduce data on quantities other than velocity (density, temperature).

Besides some general properties that any seed material for flow visualization should have (e.g., nontoxic, noncorrosive), there are mainly three conditions the tracers should meet; they are neutral buoyancy, high stability against mixing, and good visibility. The first requirement

is almost impossible to meet for air flows. Smoke or oil mist are the most common trace materials in air, with the particle size of these tracers being so small (<1 μm) that their settling velocity is minimized. A number of neutrally buoyant dyes are known for the visualization of water flows, the colors introducing an additional component of information (Fig. 2).

Special arrangements for illumination and recording as well as timing are necessary if the goal is to measure the velocity of individual tracer particles. A time exposure is a possible way for visualizing the instantaneous velocity distribution in a whole field (plane) of the flow. Each particle appears in the form of a streak whose length is a measure of the velocity vector in the plane (Fig. 3). An alternate way is to take a double exposure produced by two very short light pulses with a definite time interval between the two pulses. An optical or numerical Fourier analysis of the field of particle double images provides the distribution of the velocity vectors ("particle image velocimetry": PIV). The plane section in the flow is realized by expanding a thin laser light beam in one plane by means of a cylindrical lens, so that all tracer particles in this plane light sheet are illuminated. The velocity component normal to the plane

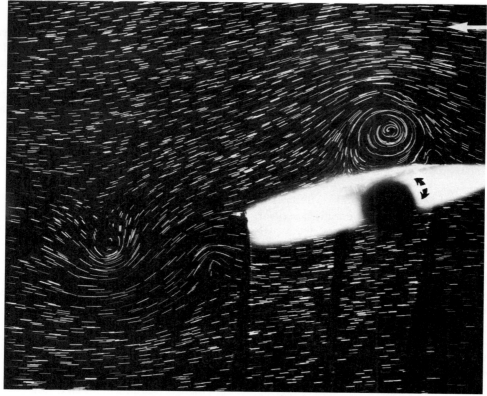

FIGURE 3 Time exposure of the water flow around a pitching airfoil shape. The water is seeded with tracer particles that appear as streaks of finite length. The pattern of the streaks is a measure of the distribution of the velocity vector in the illuminated plane. (Courtesy of Dr. M. Coutanceau, Université de Poitiers, France.)

FIGURE 4 Smoke lines around a car in a full-scale wind tunnel. (Courtesy of Volkswagenwerk AG, Wolfsburg, Federal Republic of Germany.)

is not recovered. Flow visualization by tracer materials is a standard technique in wind tunnels, water tunnels, other flow facilities, and field studies. A typical application is the study of the flow around car bodies with the aim of improving the aerodynamic characteristics of the shape (Fig. 4).

III. VISUALIZATION BY REFRACTIVE INDEX CHANGES

The refractive index of a (transparent) fluid is a function of the fluid density. The relationship is exactly described by the Clausius-Mosotti equation; for gases, this equation reduces to a simple, linear relationship between the refractive index, n, and the gas density, ρ, known as the Gladstone-Dale formula. Therefore, refractive index variations occur in a fluid flow in which the density changes, for example, because of compressibility (high-speed aerodynamics or gas dynamics), heat release (convective heat transfer, combustion), or differences in concentration (mixing of fluids with different indices of refraction).

A light wave transmitted through the flow with refractive index changes is affected in two different ways: it is deflected from its original direction of propagation and its optical phase is altered in comparison to the phase of the undisturbed wave. In a recording plane at a certain distance behind the flow field under study, three different quantities can be measured (Fig. 5). Each quantity defines a group of optical visualization methods that depend in a different way on the variation of the refractive index, n in the flow field (Table I). A particular method requires the use of an optical apparatus transforming the measurable quantity (light deflection, optical phase changes) into a visual pattern in the recording plane. The pattern is either qualitative (shadowgraph, schlieren) or quantitative (moiré, speckle photography, interferometry), thus allowing for

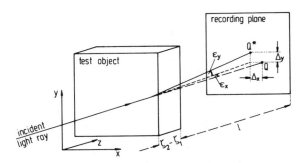

FIGURE 5 Interaction of a light ray with the refractive index field of a fluid flow (test object). The information on light deflection and optical phase changes is recorded in a plane at distance l behind the flow field. [From Merzkirch, W. (1987). "Flow Visualization," 2nd ed., Academic Press, San Diego.]

TABLE I Optical Methods for Line-of-Sight Flow Visualization

Optical method	Quantity measured (see Fig. 5)	Sensitive to changes of	Information on refractive index or density
Shadowgraph	$\overline{\Delta x + \Delta y}$	$\left(\dfrac{\partial^2 n}{\partial x^2} + \dfrac{\partial^2 n}{\partial y^2}\right)$	Qualitative
Schlieren	ε_x or ε_y	$\dfrac{\partial n}{\partial x}$ or $\dfrac{\partial n}{\partial y}$	Qualitative
Moiré deflectometry	ε_x or ε_y	$\dfrac{\partial n}{\partial x}$ or $\dfrac{\partial n}{\partial y}$	Quantitative
Speckle deflectometry	ε_x or ε_y	$\dfrac{\partial n}{\partial x}$ and $\dfrac{\partial n}{\partial y}$	Quantitative
Schlieren interferometry	Optical phase change in Q^*	$\dfrac{\partial n}{\partial x}$ or $\dfrac{\partial n}{\partial y}$	Quantitative
Reference beam interferometry	Optical phase change in Q^*	n	Quantitative

a deduction of data of the refractive index or density distribution in the flow.

A standard case of application of optical flow visualization is the air flow around a projectile or aerodynamic shape flying at high velocity (Fig. 6). This application to experimental ballistic studies can be traced back to the middle of the 19th century when the methods had been invented in their simplest form. A tremendous push forward in the development of these methods was the availability of laser light, from which particularly the interferometric methods benefitted (Fig. 7). Finally, the combination of holography with interferometry facilitated the mechanical design of practical interferometers.

A major problem with the optical visualization methods, which rely on the transmittance of light (*line-of-sight* methods), is the integration of the information on the refractive index (or density) along the path of the light, in terms of Fig. 5 along the z-direction. The data received in the recording plane are functions of only x and y, whereas in the general case, refractive index or fluid density in the flow are functions of all three space coordinates, x, y, and z.

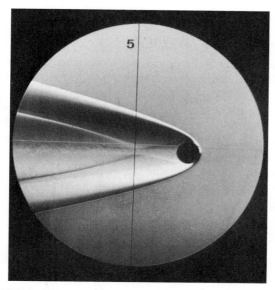

FIGURE 6 Schlieren photograph of a sphere flying at high supersonic velocity from left to right. [From Merzkirch, W. (1987). "Flow Visualization," 2nd ed., Academic Press, San Diego.]

FIGURE 7 Shearing (schlieren) interferogram of the hot gases rising from a candle flame. (Photographed by H. Vanheiden, Universität Essen, Federal Republic of Germany.)

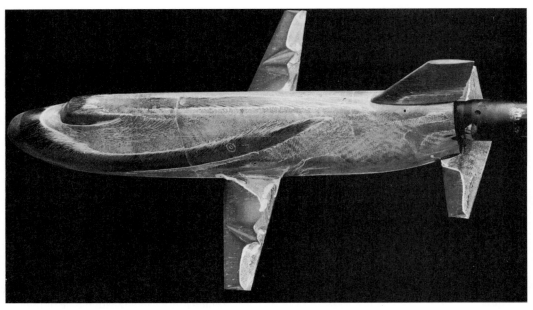

FIGURE 8 Oil film pattern of the flow around an orbiter model that was tested in a wind tunnel. (Courtesy of L. H. Seegmiller, NASA Ames Research Center, Moffet Field, California.)

This three-dimensional distribution of the fluid density can be resolved by recording with the optical setup several projections in different directions through the flow and processing the obtained data with the methods of computer tomography. If the flow has rotational symmetry, one projection only is sufficient, and the axisymmetrically distributed fluid density can be determined by applying to the data an inversion scheme (*Abel inversion*).

IV. SURFACE FLOW VISUALIZATION

The interaction of a fluid flow with the surface of a solid body is a subject of great interest. Many technical measurements are aimed to determine the shear forces, pressure forces, or heating loads applied by the flow to the body. A possible means of estimating the rates of momentum, mass, and heat transfer is to visualize the flow pattern very close to the body surface. For this purpose, the body surface can be coated with a thin layer of a substance that, upon the interaction with the fluid flow, develops a certain visible pattern. This pattern can be interpreted qualitatively, and in some cases, it is possible to measure certain properties of the flow close to the surface. Three different interaction processes can be used for generating different kinds of information.

1. Mechanical Interaction

In the most common technique, which applies to air flows around solid bodies, the solid surface is coated with a thin layer of oil mixed with a finely powdered pigment. Because of frictional forces, the air stream carries the oil with it, and the remaining streaky deposit of the pigment gives information on the direction of the flow close to the surface. The observed pattern may also indicate positions where the flow changes from laminar to turbulent and positions of flow separation and attachment (Fig. 8). Under certain circumstances the wall shear stress can be determined from a measurement of the instantaneous oil film thickness.

2. Chemical Interaction

The solid surface is coated with a substance that changes color upon the chemical reaction with a material with which the flowing fluid is seeded. The reaction, and thereby the color change, is the more intense, the higher the mass transfer from the fluid to the surface. Separated flow regimes with little mass transfer rates can therefore be well discriminated from regions of attached flow. Coating substances are known that change color upon changes of the surface pressure. This is an elegant way for determining pressure loads on the surface of aerodynamic bodies.

3. Thermal Interaction

Coating materials that change color as a function of the surface temperature (temperature sensitive paints, liquid crystals) are known. Observation of the respective color changes allows for determining the instantaneous positions of specific isothermals and deriving the heat transfer rates to surfaces, which are heated up or cooled down in a fluid flow. Equivalent visible information is available, without the need of surface coating, by applying an infrared camera.

SEE ALSO THE FOLLOWING ARTICLES

FLUID DYNAMICS ● HEAT FLOW ● IMAGING OPTICS ● LIQUIDS, STRUCTURE AND DYNAMICS ● IMAGING OPTICS

BIBLIOGRAPHY

Adrian, R. J. (1991). "Particle-imaging techniques for experimental fluid mechanics," *Annu. Rev. Fluid Mechan.* **23,** 261–304.

Emrich, R. J., ed. (1981). "Fluid Dynamics," Vol. 18 of *Methods of Experimental Physics*. Academic Press, New York.

Fomin, N. A. (1998). "Speckle Photography for Fluid Mechanics Measurements," Springer-Verlag, Berlin.

Goldstein, R. J., ed. (1983). "Fluid Mechanics Measurements," Hemisphere, Washington, DC.

Japan Society of Mechanical Engineers, ed. (1988). "Visualized Flow," Pergamon Press, Oxford.

McLachlan, B. G., Kavandi, J. L., Callis, J. B., Gouterman, M., Green, E., Khalil, G., and Burns, D. (1993). "Surface pressure field mapping using luminescent coatings," *Exp. Fluids* **14,** 33–41.

Merzkirch, W. (1987). "Flow Visualization," 2nd ed., Academic Press, San Diego.

Raffel, M., Willert, C., and Kompenhans, J. (1998). "Particle Image Velocimetry," Springer-Verlag, Berlin.

Van Dyke, M., ed. (1982). "An Album of Fluid Motion," Parabolic Press, Stanford, CA.

Yang, W. J., ed. (1989). "Handbook of Flow Visualization," Hemisphere, Washington, DC.

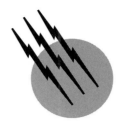

Fluid Dynamics

Elaine S. Oran
Jay P. Boris
Naval Research Laboratory

I. Fundamental Concepts of Fluid Dynamics
II. Further Complexities of Fluid Systems
III. Forefronts in Fluid Dynamics
IV. Approaches for Studying Fluid Dynamics

GLOSSARY

Compressibility The variation of fluid density with time in a frame of reference moving with the flow. Compressible flows can support sound waves and shock waves. In incompressible flows, the density is constant in an element of fluid moving with the flow.

Euler equations The set of fluid equations for conservation of mass, momentum, and energy without terms representing viscosity. The vector momentum equation with no viscous term is often called the Euler equation.

Computational fluid dynamics The discipline and techniques by which solutions of discretized versions of the fluid dynamic equations are simulated on digital computers.

Mach number The ratio of the fluid velocity to the speed of sound in a fluid. The Mach number depends on the frame of reference and thus is not a property of the fluid.

Navier-Stokes equations The set of fluid equations for conservation of mass, momentum, and energy including viscosity terms. The momentum equation with the viscous terms is often called the Navier-Stokes equation.

Non-Newtonian flow Flows such as occur in polymer solutions, foams, slurries, or blood, in which Newton's Law, expressing the proportionality of the shear stress to the shear rate, does not hold. Description of such flows requires understanding the interactions of complex materials with viscosities that are not constant and may have hysteresis effects.

Reynolds number A dimensionless ratio of inertial forces to viscous forces. This quantity is often used to characterize the stability of a flow. Laminar flows have a low Reynolds number, and turbulent flows have a high Reynolds number.

Sound wave, also acoustic wave Small, compressible perturbations in the fluid variables, particularly in the pressure, that travel at Mach one, the speed of sound of the fluid.

Shock wave A surface in a flow that propagates faster than the speed of sound of the material ahead of the surface, that is, at Mach number greater than one. A shock is a form of gas-dynamic discontinuity across which the pressure, temperature, density, fluid velocity, and sometimes the equation of state change discontinuously. The flow behind a shock wave is compressed.

Supersonic flow A flow in which structures or waves move at velocities greater than the speed of sound in the fluid. In this regime, shock waves are likely to occur. In a subsonic flow, the Mach number is less than one, and in a transonic flow, the Mach number is approximately one.

Turbulence Flows or regions in a flow with vorticity that appear highly disordered, and whose disorder extends over a range of physical scale lengths. Large identifiable vorticity structures in the flow are often called coherent structures.

Vorticity A vector quantity, defined as $\nabla \times \mathbf{v}$, that describes the rate of rotation of a parcel of fluid. The compressibility ($\nabla \cdot \mathbf{v}$) and the vorticity are both required to specify a general flow.

FLUID DYNAMICS is the study of the motion of material that can be represented as a fluid. A gas, such as the sun or the earth's atmosphere, or a liquid, such as the water in the oceans, can usually be represented as a fluid. Even the earth's continents behave as a fluid on long enough time and space scales. The criterion is that the material can be represented as a deforming continuous medium, even though it consists of individual atoms and molecules. Any sizable volume of a fluid contains so many molecules that a description can be based on statistical averaging over their individual small-scale motions. The continuum approximation is valid when the mean free path traveled by a particle between collisions is small compared to the macroscopic dimensions of the system. Then average macroscopic quantities such as density, velocity, and energy are well defined at each point. Using the short mean free path assumption, a set of partial differential fluid equations can be derived rigorously from considerations of the individual particle collisions which conserve mass, momentum, and energy.

Fluid dynamics underlies many areas of science, including astrophysics, aeronomy, meteorology, aeronautics, combustion, hydrodynamics, plasma physics, ocean sciences, and hydrology. The equations governing fluid motion express the conservation of mass, momentum, and energy in terms of corresponding fluxes that leave or enter a volume of space. The form of these equations is well known, but they are notoriously difficult to solve. Applications of new experimental and computational methods have led to a major increase in our ability to observe and predict complex fluid flows, and so better understand our environment.

I. FUNDAMENTAL CONCEPTS OF FLUID DYNAMICS

A. Basic Equations of Fluid Dynamics

Three partial differential equations describe the dynamics of a gas composed of only one chemical species with no external forces:

$$\frac{\partial \rho}{\partial t} = -\nabla \cdot (\rho \mathbf{v}), \tag{1}$$

$$\frac{\partial \rho \mathbf{v}}{\partial t} = -\nabla \cdot (\rho \mathbf{v} \mathbf{v}) - \nabla \cdot \mathbf{P}, \tag{2}$$

$$\frac{\partial E}{\partial t} = -\nabla \cdot (E \mathbf{v}) - \nabla \cdot (\mathbf{v} \cdot \mathbf{P}) - \nabla \cdot \mathbf{q}. \tag{3}$$

Here ρ is the mass density, $\mathbf{v}$ is the fluid velocity, and E and is the total energy density, defined as

$$E = \rho \epsilon + \frac{1}{2} \rho v^2, \tag{4}$$

where ϵ is the internal energy of the fluid, comprised of rotational, vibrational, and random kinetic (thermal) motions of the constituent atoms and molecules. The heat flux due to thermal conduction is

$$\mathbf{q} = -\lambda \nabla T, \tag{5}$$

where T is the temperature. The pressure tensor is

$$\mathbf{P} \equiv P(N, T)\mathbf{I} + \tau, \tag{6}$$

where $\mathbf{I}$ is a unit tensor. The viscous forces are usually derived from a viscous stress tensor

$$\tau \equiv -\mu[\nabla \mathbf{v} + (\nabla \mathbf{v})^T] + \left(\frac{2}{3}\mu - \kappa\right)(\nabla \cdot \mathbf{v})\mathbf{I}, \tag{7}$$

where the superscript T indicates the transpose matrix and μ and κ are the coefficient of shear viscosity and bulk viscosity, respectively. For an ideal gas

$$P = \rho R T = (\gamma - 1)\rho \epsilon, \tag{8}$$

where R is the gas constant, and γ is the ratio of specific heats of the gas, C_p/C_v.

Equations (1)–(8) are the basic equations discussed in most textbooks on fluid dynamics of gases. Equations (1)–(3) describe the behavior of the mass density, vector momentum density, and internal energy density as functions of time and position. Each equation consists of two parts: the convective transport terms and the diffusion and source terms. Convective transport terms in fluid dynamics have the generic form

$$\frac{\partial \varphi}{\partial t} + \nabla \cdot \mathbf{f} = 0 \tag{9}$$

which describes the conservation of the quantity φ, which may be ρ, $\rho \mathbf{v}$, or ϵ, in a volume element of space. The quantity $\mathbf{f}$ is the flux of φ. The source and diffusive terms are, for example, the heat flux term in Eq. (3) and the divergence of $\mathbf{P}$ in Eq. (2).

Equation (1) is the mass-density continuity equation and Eq. (2) is the *Navier-Stokes equation*. Sometimes the set of Eqs. (1)–(3) is referred to as the Navier-Stokes equations. The *Euler equation* is obtained from Eq. (2) by setting κ and μ equal to zero. Equations (1)–(3) are also called the Euler equations when all of the diffusive transport terms are set to zero.

The pressure tensor, Eq. (5), has two parts: $P(N, T)$ is associated with compression and τ is associated with viscous dissipation. In general, the bulk viscosity κ is negligible except in problems of shock-wave structure and absorption and attenuation of acoustic waves. Because κ is generally very small, it is often dropped from consideration. The shear viscosity μ, however, is often very important, as explained in the discussions below.

Equations (1)–(3) are written in terms of the "primitive" variables ρ, $\rho\mathbf{v}$, and E. Other formulations of fluid dynamics are often used. One important derived quantity in multidimensional flow is the *vorticity*, ω, defined as

$$\omega \equiv \nabla \times \mathbf{v}. \tag{10}$$

Vorticity is a vector quantity describing the rates of rotation of a fluid parcel. It satisfies an evolution equation that can be derived from Eqs. (1)–(2) using Eq. (10),

$$\frac{\partial \omega}{\partial t} + \omega \nabla \cdot \mathbf{v} = \omega \cdot \nabla \mathbf{v} + \frac{(\nabla\rho \times \nabla \cdot \mathbf{P})}{\rho^2}. \tag{11}$$

A net *circulation* is defined for an area as the line integral of the fluid velocity around the closed perimeter of that area

$$\Gamma = \oint \mathbf{v} \cdot d\mathbf{r} = \int_S \omega \cdot d\mathbf{S} \tag{12}$$

where $\mathbf{S}$ is a closed surface. The evolution of the net circulation is specified by

$$\frac{d\Gamma}{dt} = \int_S \left[\frac{(\nabla\rho \times \nabla \cdot \mathbf{P})}{\rho^2} \right] d\mathbf{S}, \tag{13}$$

an integration of the vorticity source term in Eq. (11) over the area $\mathbf{S}$.

When $\omega = 0$, there is no rotation in the fluid. In this case the flow can be described entirely by a *velocity potential*, ϕ, where

$$\mathbf{v} = \nabla\phi. \tag{14}$$

When the flow described by Eq. (14) is also *incompressible*, that is, when $\nabla \cdot \mathbf{v} = 0$,

$$\nabla \cdot \mathbf{v} = \nabla^2\phi = 0, \tag{15}$$

which is the Laplace equation for determining ϕ. This type of flow is called *potential flow*.

In the general case, the velocity of a fluid $\mathbf{v}(\mathbf{r}, t)$, can be expanded as the sum of three terms

$$\mathbf{v} = \nabla \times \psi + \nabla\phi + \nabla\phi_p. \tag{16}$$

The vector *stream function* ψ for the rotational component satisfies

$$\nabla \times (\nabla \times \psi) = \omega \tag{17}$$

and the compressional component is no longer zero

$$\nabla \cdot \nabla\phi = \nabla \cdot \mathbf{v} \neq 0. \tag{18}$$

The third term, $\nabla\phi_p$, is a particular potential solution, which can be added to the rotational and compressional contributions in Eq. (12) as both its curl and divergence are zero. This particular solution is usually chosen to adapt the composite solution for $\mathbf{v}$, given by Eq. (13), to prescribed boundary conditions.

B. Regimes of Flow and Dimensionless Numbers

Different regimes of fluid flow are often characterized by the ratio of the *sound speed*, defined as the velocity at which a small pressure disturbance moves in the fluid, to the flow speed. This ratio is the *Mach number* of the flow,

$$M \equiv \frac{|\mathbf{v}|}{c_s}. \tag{19}$$

The sound speed, c_s, is as

$$c_s = \sqrt{\left.\frac{\partial P}{\partial\rho}\right|_s}, \tag{20}$$

where the derivative is taken at constant entropy. For an ideal gas, this reduces to

$$c_s = \sqrt{\frac{\gamma P}{\rho}}. \tag{21}$$

Qualitatively different flow regimes are identified by the value of the flow Mach number, as shown in Table I.

General properties of fluid systems are also discussed in terms of other dimensionless ratios of physical variables. These ratios express properties of different systems that scale with each other. Dimensionless numbers are derived either from dimensional analysis when the governing equations are not known, or from inspection analysis when the equations are known. For example, the Reynolds number, derived from the momentum equation, is given by

TABLE I Mach Numbers Characterizing Flow Regimes

Flow regime	Mach range	Comment
Incompressible	$M < 0.4$	No dynamic compression; density and pressure are essentially constant moving with flow
Subsonic	$0.4 < M < 1$	Compression becomes important
Transonic	$M \sim 1$	Weak shock waves occur
Supersonic	$1 < M \leq 3$	Strong shock waves occur
Hypersonic	$M \geq 3$	Very strong shocks occur, most of the energy is kinetic energy

TABLE II Dimensionless Numbers

Name	Symbol	Definition	Significance
Atwood	A	$\Delta\rho/(\rho_1 + \rho_2)$	Density difference/density sum
Brinkman	Br	$\mu v^2/T\lambda$	Heat from viscous dissipation/heat from thermal conduction
Euler	Eu	$P/\rho v^2$	Pressure force/inertial force
Drag coefficient	C_D	$(\rho' - \rho)/lg\rho'v^2$	Drag force/inertial force
Froude	Fr	v^2/lg	Kinetic energy/gravitational energy
Grashof	Gr	$(g\Delta Tl^3)/T\mu^3$	Buoyancy forces/viscous forces
Knudsen	Kn	δ/l	Convection time/collision time
Mach	M	v/c_s	Magnitude of compressibility
Newton	Nt	$F/\rho l^2 v^2$	Imposed force/inertial force
Peclet	Pe	$vl\rho c_p/\lambda$	Convective heat transport/heat transport by conduction
Poisseuille	Po	$l\Delta p/\mu v$	Pressure force/viscous force
Prandtl	Pr	$c_p\mu/\lambda$	Momentum transport/thermal conduction
Reynolds	Re	$lv\rho/\mu$	Inertial forces/viscous forces
Richardson	Ri	$gl/(\Delta v)^2$	Buoyancy/vertical shear
Strouhal	Sr	vl/v	Vibrational rate/convective flow rate
Weber	W	$\rho lv^2/\sigma$	Inertial force/surface tension forces

$$\frac{\rho vl}{\mu} \equiv \mathrm{Re}, \tag{22}$$

which is the ratio of inertial to viscous forces. Several other dimensionless numbers commonly used in fluid dynamics are given in Table II.

The basic equations of fluid flow appear deceptively simple. The richness, broad applicability, and challenges of fluid dynamics stem from the strong nonlinearities and from additional physical complications added to the equations. Each type of complication is characteristic of a class of flows, and typically has an entire scientific community studying it. Complications arise, for example, through boundary conditions, external forces (such as, for example, gravity or electromagnetic forces), nonideal equations of state relating P, ρ, γ, ϵ, and T, disparate time and space scales in the problem, compressibility leading to shocks, chemical reactions and energy release, presence of multiple species or phases, or complications in transport processes.

C. Compressible Flow

When a flow is compressible, $\nabla \cdot \mathbf{v} \neq 0$, density can vary dynamically in the flow, and the thermodynamics and equations of the state of the system must be included in descriptions of the flow field. Practically, this means that fluctuations of physical variables such as pressure, density, and momentum can be substantial and shock waves can develop. Compressible flow is the general case with incompressible flow being a very idealized case. In the incompressible limit, acoustic waves are infinitely fast and shock waves do not exist. Nonetheless, incompressibility is a useful approximation mathematically because the fluid density can be treated as constant moving with the flow.

Compressibility effects become progressively more important for $M \geq 0.4$. As indicated in Table I, these effects are important from the subsonic through the hypersonic velocity ranges. Even at very low velocities, heat addition in a gas or energy release from chemical reactions can lead to large compressibility effects, although the pressure remains essentially constant (with no shock or acoustic waves of consequence).

An important consequence of compressible fluctuations in flow variables is that the level of sound or noise in the system can be high. Sound waves are small fluctuations in the system that propagate at $M = 1$. The equation describing the propagation of acoustic waves may be derived by expanding the flow variables as a constant or slowly varying part plus a small variation, $\varphi = \bar{\varphi} + \varphi'$, in the density and momentum equations, Eqs. (1) and (2). Linearizing and combining the result gives

$$\nabla^2\varphi' - \frac{1}{c_s}\frac{\partial^2\varphi'}{\partial t^2} = 0, \tag{23}$$

where φ' represents the linear perturbation in ϕ', v', or p'. This acoustic wave equation is strictly only valid for small disturbances in nonreacting media. In many flows, particularly highly compressible flows, strong effects arise from the nonlinear interactions among the waves and the convective flow in the system.

Shocks are surfaces in a flow that propagate at $M > 1$ into the undisturbed fluid. A shock is a form of gas-dynamic discontinuity across which fluid properties, from a macroscopic viewpoint, change discontinuously, and across which there is some sort of fluid flow. Across a shock, the pressure, temperature, density, fluid velocity, and perhaps the equation of state and thermodynamic properties of the system can change. This type of discontinuity is distinguished from contact discontinuities across which the pressure is continuous but other fluid variables, such as temperature and density, can change. Microscopically, the changes across a shock front are due to changes in average molecular kinetic energy from collisions, and the rate of these changes is determined by the finite energy transfer of a molecular collision. The directed kinetic energy of the gas motion relative to the shock degrades into random thermal energy, thus changing the internal energy of the system. This process is irreversible, which means that there is a net increase of entropy across a shock front. Changes in physical variables across a real shock front are fast but finite. In the continuum, fluid-dynamics limit, shocks are very thin, their width determined by the viscosity of the system. The flow conditions and the velocity of a shock in an idealized one-dimensional planar flow can be determined by solving a set of relations based on conservation laws called the Riemann equations.

The properties of compressible flows are extremely important in, for example, combustion, aeronautical engineering, and astrophysics. They are crucial to the development of high-speed aircraft, where the problems range from optimizing aerodynamic shapes, to determining and controlling noise levels, to optimizing the design of internal combustion and jet engines. In all of these flows, there are large density gradients, shocks, and fluctuations of physical variables. An important aspect of shock behavior in such an environment is the interaction of shocks with density gradients to produce vorticity, an important mechanism for the generation of turbulence in a flow.

Explosions and detonations involve strongly compressible flows that are important in considerations ranging from safety in mines and transport of fuels to designing munitions. Here a basic fluid dynamic problem is determining the multidimensional structure of multiple shock interactions for which there is rarely a clean analytic solution. Recent approaches to these problems have combined numerical solutions of the flow equations with experimental observations.

II. FURTHER COMPLEXITIES OF FLUID SYSTEMS

In the five areas discussed in this section, a major additional process must be added to Eqs. (1)–(8). Often these appear simply as extra terms in the equations, as in the case when chemical reactions are present, and sometimes these changes reflect fundamental deficiencies of the formulation, as in the case of non-Newtonian fluid dynamics.

A. Non-Newtonian Flow

An inherent assumption in the Navier-Stokes equations is the Newtonian constitutive equation, which is Eq. (7) with μ constant. For an incompressible flow ($\nabla \cdot \mathbf{v} = 0$) in two dimensions, this reduces to

$$\tau_{yx} = -\mu \frac{dv_x}{dy}. \tag{24}$$

The quantity τ_{yx} is the shear stress and the derivative on the right side is the shear rate or shear strain. This equation, which says that stress is proportional to the strain, is Newton's Law and describes a Newtonian fluid. This is valid for many usual fluids under normal conditions. However, for a large class of important flows and for normal fluids under extreme conditions when the shear stress is very high, this assumption does not hold, and new constitutive relations are needed.

The important idea here is that a basic assumption of fluid dynamics, the proportionality of the flux of a quantity to a gradient that defines its transport, has broken down. In Eq. (24), μ must be replaced by some function μ_a that may depend on various terms in Eq. (6) or may even retain a memory of previous states of the fluid. The origin of this behavior is in the microscopic behavior of the molecules of the fluid, and understanding why these phenomena occur can only be done through this microscopic level of understanding. The study of the deformation of flow of matter is called rheology, and the constitutive equations are also called the rheological equations of state.

Some examples of non-Newtonian fluids include polymer solutions, foams, slurries, suspensions, and blood. Non-Newtonian fluids have been classified by whether or not they show shear thinning or thickening, their time-dependence, their viscoelastic properties, and the extent to which they show normal stress (the size of the diagonal terms in the pressure tensor). The problem is that these behaviors are difficult to quantify and non-Newtonian materials often behave more than one way depending on the conditions.

An important problem in non-Newtonian flows is drag reduction. By adding minute concentrations of specific long-chained polymers to ordinary Newtonian fluids such as water, the pressure losses in pipes or around surfaces in a turbulent flow can be greatly reduced. The measured viscosity of the Newtonian fluid may then be slightly higher, but it is essentially unchanged. However, the material behaves as if the viscosity were much lower. This phenomenon is not completely understood, but its explanation is rooted in the behavior of the long polymer chains in the water, how they unravel and the long times it takes them to unravel and then relax to their original state. At a certain strain rate, the shape of the polymer molecule becomes anisotropic and partially aligns with the flow. This anisotropy produces stresses along the backbone of the aligned molecule. In shearing flows, certain components of these stresses act in a direction normal to the deformation gradient, so they are called normal stresses. The large number of atoms in a polymer molecule makes possible large deformation and increases the time required for the polymer to diffuse back to equilibrium. Thus, polymer solutions have relaxation times on the order of milliseconds to seconds, which are many orders of magnitude greater than the corresponding relaxation times of Newtonian fluids.

Non-Newtonian fluids often have very interesting and startling properties. For example, the presence of small amounts of a polymer in an otherwise Newtonian fluid, such as water, can allow a fluid to move upward, against the force of gravity, and actually climb over the lip and out of a beaker. Addition of polymers to fluids has been shown to delay the onset of turbulence. Nonetheless, most attempts to use the seemingly odd properties of non-Newtonian fluids in practical engineering problems have not been successful.

B. Chemically Reacting Flow

In chemically reacting flows, additional equations describing the production and loss of chemical species and heat must be added to Eqs. (1)–(8). In the simplest case, this means adding the set of equations

$$\frac{\partial n_i}{\partial t} = -\nabla \cdot (n_i \mathbf{v}) - \nabla \cdot (n_i \mathbf{v}_{di}) + Q_i - L_i n_i,$$
$$i = 1, \ldots, N_s, \quad (25)$$

which describes the change in number density of species $\{n_i\}$ through production $\{Q_i\}$ and loss $\{L_i\}$ terms. Adding chemical reactions also requires expanding the definition of the heat flux vector $\mathbf{q}$ in Eq. (3) as

$$\mathbf{q}(N, T) \equiv -\lambda_m \nabla T + \sum_i n_i h_i \mathbf{v}_{di}, \quad (26)$$

here λ_m is the thermal conductivity coefficient of the mixture, h_i is the enthalpy of species i, and $\mathbf{v}_{di}$ is the diffusion velocity of species i.

Fluid dynamics and chemistry interact in several fundamental ways that become crucial aspects of reactive flows. Fluid-dynamic convection moves a fluid from one location to another and allows it to penetrate other elements of fluid, or to change density in response to changing flow conditions along its path. Each of these three convection effects interact differently with ongoing chemical reactions in the system. For example, in potential flow, nearby fluid elements remain near each other, so the main effect of potential flow on chemical reactions is to move the reactions with the associated fluid from one place to another. However, when the vorticity is not zero, these rotational (or shearing) motions allow fluid from one place to penetrate and convectively mix with other portions of the fluid. Shearing, in turn, enhances species gradients and causes faster molecular mixing. When the flow is curl free but not divergence free ($\nabla \times \mathbf{v} = 0$ but $\nabla \cdot \mathbf{v} \neq 0$), there are velocity gradients in the direction of motion leading to compressions and rarefactions. Compression generally accelerates chemical reactions and expansion generally retards them. A runaway effect is possible, in which higher densities increase the rate of chemical reactions, which in turn increase pressure even faster. This effect can also lead to unstable oscillations.

Fluid motions arising directly from local expansions can be violent, but they decrease in strength with distance from the source and they stop when the energy release is complete. The indirect effects of expansion are longer lived and can be more important. Once expansion is complete, it leaves low density pockets of hot reacted gas in the midst of unreacted fuel or oxidizer. Gradients in temperature and density become frozen into the fluid and move with it. Embedded density gradients interact with dynamic pressure gradients to generate additional vorticity. Vorticity is generated slowly by this passive influence of reaction kinetics on the flow, but its integrated effect can be much larger than even the long-lived vortices generated directly during the expansion. Although expansion-generated vortices have flow velocities that are only a few percent of the expansion velocities that produced them, they persist for long times after the expansion has stopped. Unlike direct expansion, these rotational flows generated by chemical energy release cause additional mixing.

Flames and detonations are chemically reacting flows with local heat release. The fundamental difference between them is in the rate of energy release which is related to the rate of propagation of the reaction front. A flame front propagates subsonically and a detonation front propagates supersonically. A detonation is a supersonic wave structure consisting of a leading shock wave, followed by a reaction zone, followed by a region of reacted fluid. The shock wave heats and compresses the undisturbed reactive mixture as it passes through it. Then the raised temperature triggers chemical reactions and eventually energy release. The detonation is controlled by convective transport and chemical energy release. In the region of the shock front itself, the timescales are so short that diffusive transport effects, such as thermal conduction and molecular diffusion, are not important. Near walls, however, diffusive transport effects and boundary layers can be important. A flame is a subsonic compressible flow with the structure of an internal boundary layer. The flame propagation is controlled by a balance of diffusive and convective transport processes in which molecular diffusion and thermal conduction play crucial roles.

C. Multiphase Flow

Multiphase flow covers a wide range of problems, including suspended grain dust or coal dust, droplets and sprays, propellant burning, charring, soot, smoke formation, slurries, bubbles in liquids, rain, and sedimentation. Because each of these subjects has its own distinguishing characteristics, different scientific communities have formed that use their own specific formulations, approximations, and measurement methods. A major problem and point of continuing confusion in multiphase flows is the lack of a unique set of equations and supporting assumptions that fit every multiphase situation.

Multiphase flow is intrinsically a nonequilibrium process. Generally it is assumed that each phase is in local thermodynamic equilibrium, but that the different phases are not necessarily in equilibrium with each other. In the corresponding mathematical models, it is necessary to specify the rates of transfer of mass, momentum, and energy among the phases in order to close the set of equations. Multiphase flow equations may be derived from continuum mechanics constraints based on conservation of mass, momentum, and energy. However, these equations require additional phenomenological terms to described the additional phases and their interactions with each other. For example, we could write a set of density equations for each phase i as

$$\frac{\partial \alpha_i \rho_i}{\partial t} + \nabla \cdot \alpha_i \rho_i \mathbf{v}_i = \Gamma_i \tag{27}$$

where the $\{\Gamma_i\}$ are mass transfer terms representing the rate of production of phase i and α_i could, for example, be the volume fraction of phase i. Similar generalizations could be made for Eqs. (2) and (3) that would include addition coupling terms proportional to $\mathbf{v}_i - \mathbf{v}_j$ and $T_i - T_j$.

There are four types of differences between a single phase fluid and one with two or more coexisting phases. There are chemical changes that occur when chemical reactions change the relative numbers of molecules of each phase. These can also include special surface reactions at the phase interfaces. Then it is often necessary to treat the phases and species individually as well as consider their interactions. There are thermal differences, which occur when the different phases have different temperatures and because velocity equilibration can be much faster than temperature equilibration. There are dynamic differences, which occur when it is necessary to describe each phase by its own separate velocity field. Finally, there are spatial separations, which means that the granularity of the different phases is large enough that it must be taken into consideration.

Many important multiphase flow problems involve the flow of "obstacles," such as drops, bubbles, or solid particles, in a background gas or liquid flow. A number of levels of complexity have been postulated to treat these types of problems. In single-fluid models, the different phases are considered a single fluid whose properties are composites of those of the various phases present. All phases of the flow are assumed to be closely coupled and thus move with the same velocity. Flows containing very small droplets or particles can often be treated this way, but such a model improperly describes rain for which the relative motion between air and water droplets is important. In two-fluid models, it is assumed that the different phases move at very nearly the same mean velocity. However, the small velocity difference may be significant in differential compression or expansion or to describe settling of heavier particles over an extended period. In multifluid models, each volume of fluid is characterized by a distribution of particle or droplet sizes and characteristics. Solving these equations is difficult and expensive because separate momentum equations are needed for each phase or particle size.

Liquid droplets in a background gas describe a wide range of problems from rain to atomization of liquid fuel. Types and levels of models used to study these problems include locally homogeneous flow models, which are similar to the single-fluid models, separated flow models, which are multifluid models, and drop life history models, which describe the detailed behavior of individual droplets. In locally homogeneous flow models, the gas and liquid phases are assumed to be in dynamic and thermodynamic equilibrium, so that the phases have the same velocity and temperature. This limiting case only accurately represents a spray consisting of very small drops. In separated flow models, the effects of finite rates of transport between the phases are considered. This is really a very broad category of models, ranging from those in which particulates are treated as Lagrangian particles in a continuum background to the case of full multiphase, multifluid continuum models such as those based on the types of equations shown above. The third type of model, drop-life history models, focuses on individual droplets and attempts to calculate their dynamic behavior self-consistently with changes in their environments.

D. Turbulence

Turbulence persists as one of the major theoretical and computational problems of fluid dynamics. This is not only because of its theoretical complexity and richness, but also because of its prevalence and therefore importance in many flow systems. It is not even exactly clear how to define fluid turbulence. Flows with vorticity that appear highly disordered, and whose disorder ranges over many physical scale lengths, are called "turbulent." Large, persistent structures observed in such flows are called *coherent structures*. The disorder develops from the nonlinear evolution of fluid instabilities that occur due to velocity or density differences in local regions of the flow. However, as more and more complex "order" is discovered in these flows, what is called turbulent has changed somewhat. In general, the higher the Reynolds number, the more turbulent the flow. At low Reynolds number, the viscosity damps the instabilities and reduces the range of scale lengths over which turbulence can exist.

If it were possible to solve the full set of time-dependent, three-dimensional fluid dynamic equations for mass, momentum, and energy, for the range of time and space scales from the size of the system down to the Kolmogorov scale (the small scale at which turbulent structures are dissipated into heat), turbulence would appear naturally in the detailed solutions. However, because the range of important scales is continuous and extensive, direct computation is generally impractical. Many approaches to studying or trying to quantify turbulence have been based on various approximations to separate the scales. The problem is that there is no clearly defined way to separate scales, either theoretically or computationally (see Sections IV.A and B).

The standard statistical theories divide the solution into a mean, time-averaged flow and a component that fluctuates in time,

$$f(\mathbf{x}, t) = \bar{f}(\mathbf{x}) + f'(\mathbf{x}, t), \qquad (28)$$

where the bar indicates the Reynolds average of that variable and $\bar{f}' = 0$. When variables written as a mean and fluctuating part are substituted in the incompressible, single-component mass and momentum equations, the result is

$$\frac{\partial \bar{\rho}}{\partial t} + \nabla \cdot \bar{\rho} \, \bar{\mathbf{v}} = 0, \qquad (29)$$

and

$$\frac{\partial}{\partial t}(\rho \, \bar{\mathbf{v}}) + \nabla \cdot (\rho \, \overline{\mathbf{v}\mathbf{v}}) = -\nabla \bar{P} + \nabla \cdot (\bar{\tau} - \rho \, \overline{\mathbf{v}'\mathbf{v}'}). \quad (30)$$

These are the Reynolds-averaged Navier-Stokes equations, which have the same general form as Eqs. (1) and (2), but with an additional Reynolds stress term,

$$\rho \overline{\mathbf{v}'\mathbf{v}'} = \rho \overline{(\mathbf{v} - \bar{\mathbf{v}}) \, (\mathbf{v} - \bar{\mathbf{v}})}. \qquad (31)$$

A major difficulty in this approach is "closing" the set of equations, which is done by postulating models relating the extra terms, such as Eq. (31), to known quantities. Further, the basic assumption of Eq. (28) that a suitable average $\bar{f}(x)$ exists is invalidated in studies of time-dependent phenomena with excitation at all spatial and temporal scales. Finally, this approach tends to average over and thus obscure the coherent structures in the flow.

To circumvent these difficulties, there have been calculations of transitional and turbulent flows that attempt to resolve the time-dependent structure of the turbulent flows. This approach does not involve the time averages of macroscopic flow variables used in statistical turbulence models. The problem with this approach is that the fastest and largest computers are neither fast nor large enough to solve most turbulent flow problems from first principles, even though the fundamental set of equations is generally agreed to be adequate. In some idealized low Reynolds number flows, simulations have been able to resolve the full range of scales down to the Kolmogorov scale. In higher Reynolds numbers, only the large scales can be calculated in detail and the small scales are modeled phenomenologically with subgrid models. The computation fluid dynamics (CFD) of modeling turbulence is discussed further in Section IV.B. The basic idea of subgrid modeling is to represent phenomenologically the macroscopic effects of fluid flow occurring on space scales which are not resolved. This commonly means representing length scales which are smaller than a computational cell by a phenomenological model, though it can also mean representing what is happening in a second or third dimension not modeled in detail. The approach that calculates the large scales and models the small scales, sometimes called large-eddy simulation, shows promise for describing turbulent flows.

E. Nonequilibrium Gas Flows

The interactions of atoms, ions, and molecules in a gas or liquid may be described by a hierarchy of mathematical models, ranging from fundamental solutions of sets of elementary interactions of particles (such as molecular dynamics methods) to approximations of systems in which the individual particles are replaced by continuum fluid elements (such as the Navier-Stokes equations). In between these extremes, there are various levels of statistical and particle-based theories that account for the nonequilibrium or particulate nature of matter.

Gases may be characterized by the Knudsen number of the flow, defined as $Kn = \lambda/L$, the ratio of λ, the mean free path between molecular collisions, and L, the characteristic length of the system. When the value of Kn is typically less than 0.05 or even 0.1, that is, there are sufficient collisions between the molecules, we can assume that the gas behaves like a continuum fluid. In these cases, the Navier-Stokes equations are an excellent model of the physics, given the required input, initial, and boundary conditions. However, as Kn increases, averaging assumptions that produce the fluid approximation are not valid, and representing the gas as a continuum fluid becomes a poor approximation. Some account must then be taken of the nonequilibrium, or particle nature of the material.

There are now many applications of high Kn flows that are of practical scientific and engineering importance. Spacecraft reentry into planetary atmospheres, the function of thrusters used on spacecraft to adjust orbits, and the behavior of out-gased plumes are all space-related problems involving high Kn gases. A variety of vapor-phase processing methods are used to produce thin films, and plasma-etching techniques are used to produce semiconductor components. These material processing applications also involve high Kn gases. In both space dynamics and materials processing, the densities of the gases are quite low (a few torr or less). However, in the relatively new development of microsystems, such as microelectromechanical systems (MEMS), gases that flow in micronsized channels at relatively high densities (atmospheric pressure or higher) are still characterized by high values of Kn because of the very small size.

Statistically based Monte Carlo methods offer alternate approaches to the direct solution of such nonequilibrium flow problems. These approaches have been extremely successful for predicting the behavior of a wide range of nonequilibrium gas flows. Because of their computational simplicity, they are now being used extensively for atmospheric and space predictions, as well as for materials processing.

III. FOREFRONTS IN FLUID DYNAMICS

A. Vortex Persistence, Collapse, and Bursting

Large-scale macroscopic vortices are created in violent weather phenomena and in the wake of aircraft, where they are called *trailing vortices*. Their basic structure, persistence, and ultimate collapse are classical fluid-dynamic subjects that are not yet qualitatively understood, let alone quantified. Though long and thin, trailing vortices have three-dimensional character including an axial flow that

is fastest at the core. Angular momentum flux, rather than the circulation, is strictly conserved. The resulting small radial convection of vorticity is quite different from vorticity spreading due to "turbulent diffusion." Because the presence of these vortices can be dangerous to aircraft take off, flight, and landing, they are now the subject of intense current research.

The manner in which trailing vortices collapse and dissipate is still the subject of research and debate. One explanation, called "predictable decay," holds that the vortex dissipation is controlled by turbulence in the cores and in the background fluid. The interaction between background turbulence and the vortices means that vorticity is gradually and steadily being shed from the vortex into the surrounding flow. Another explanation, "stochastic collapse," postulates that a fluid dynamic instability is initiated by fluctuations, and this causes the macroscopic vortex to collapse faster than what would normally be predicted by shedding and diffusion. The latter explanation is gaining credibility, but has not yet been convincingly proven. The notion of relatively slow decay of trailing vortices, or alternately, the wide variability of the delay before the onset of collapse, forms the basis of current air traffic control rules for spacing aircraft takeoffs and landings. If the stochastic collapse viewpoint is correct, the perceived slow decay of trailing vortices is really a statistical averaging effect resulting from the wide variability of the time delay before onset of collapse.

"Bursting" seems to be a different phenomenon that is also not understood. Bursting appears as a spontaneous axial contraction of a trailing vortex, making the vortex thinner in some regions, while generating relatively larger radius pancakes in others. Flow visualization of a bursting vortex shows that the vortex persists through the bursting phase and may burst again. As with collapse, bursting is a phenomenon for which there is no convincing theory nor definitive numerical experiments.

B. Microdynamical Flows

In the new area of MEMS, gases and liquids are forced to flow through micrometer- or even nanometer-sized channels at relatively high densities (atmospheric or higher). Fluidic MEMS systems are being designed for many applications, including drug delivery and sensors for atmospheric contaminants. Often the fluids consist of a background gas or liquid laden with larger particulates, such as water containing biological cells, or air containing spores. In such a flow, the fluid and the particulates may be subjected to electrical forces and undergo chemical reactions. The result is a very complex flow for which the Navier-Stokes equations alone cannot provide an adequate description.

Basic fluid-dynamic problems in MEMS involve developing reliable experimental diagnostics of such small systems. The fundamental issue here is that the sensors need to be smaller than the device being studied! The momentum and energy of the flow is very small, so that only an extremely small amount of momentum and energy exchange can be allowed between the flow and sensor for a good measurement. To date, only a few types of sensors have been developed, including sensors for pressure and temperature. New developments are also needed for micro-flow visualization.

Other problems involve developing the theoretical basis for describing flows of gases or liquids through such small spaces. The problems may be quite different, depending on whether the background fluid is a liquid or a gas. For gases, major theoretical problems hinge around describing very low-speed, high Knudsen number flows with high enough accuracy. Obtaining useful solutions with the statistical methods commonly used for high-speed atmospheric flows requires enormous computational resources. For liquids, the surface effects become increasingly important as the surface-to-volume ratio increases. Electrostatic forces originating at surfaces and imperfections on surfaces are increasingly important. Short- and long-ranged atomic or molecular interactions in the liquid itself become more anisotropic for small systems with large surface areas. Flows with particulates have additional problems, some of which are generic problems in multiphase or non-Newtonian flows, and others are due to interactions that are enhanced due to the increased importance of surface effects.

C. Granular Flows

A granular material is a multiphase material made up of a large collection of closely packed solid particles surrounded by a gas or a liquid. Because the ratio of the volume of solid to fluid phases is very high, the particles are in very close contact with each other. Typical granular materials include sand, a collection of seeds and nuts in a can, a load of cement fragments dumped from a truck, and various types of powders. Depending on local conditions of motion and stress, these particles can act together as either an elastic solid or as a fluid. When it acts as an elastic solid, it can create a very stable and strong supporting structure. Under some circumstances, however, the system begins to "flow," in unusual and interesting ways that are a combination of solid and fluid behavior.

Under certain conditions of applied local stress, imposed motion, or heating, the binding forces between the particles give way and the system begins to flow. At one extreme of the flow, the material moves as chunks of particles that move along shear bands. As the deformation

speed increases, the particles in the material begin to move freely as individual particles instead of chunks. In this fast-flowing regime, the particles behave in a way that is somewhat analogous to a collection of molecules in that they can be described by a bulk velocity and a random thermal motion.

The fast motion in the cascading regime of relatively fast particulate flow can usually be characterized by a "granular temperature," which is related to the flow of internal energy. An important physical fact to note is that the collisions are inelastic for granular flows. Because the interactions are inelastic, these flows are dissipated by collisions between particles, so the temperature always dissipates when there is no external source of energy. For example, shaking a jar of seeds creates a motion that seems to have a temperature, but this motion quickly stops when the shaking stops.

In the past ten years, there has been considerable work on characterizing granular flows. Work has been done to determine material properties, such as equations of state and temperatures, effects of boundary conditions, and the effects of the fluid surrounding the particles. Studies have been performed in which different granular materials have, for example, been shaken, heated, tumbled, melted, squeezed, or dropped. All of these have been done under specified initial conditions conditions and initial variations of material properties.

D. Superfluid Helium

In liquid helium at very low temperatures, quantum mechanical effects allow the onset of a macroscopic "superfluid" state that exhibits ideal, classical fluid dynamic behavior. Flows with zero viscosity (and other strange properties) can be established in which the Reynolds number is effectively infinite. Thus relatively small helium flow tunnels can be set up that exceed the capability of wind tunnels for accurately simulating large-scale aerodynamics and hydrodynamics. The effective Reynolds numbers attained, even at the scale of a few inches, are comparable to or exceed those of a full-scale field trial. This is possible because below 2.712 K helium enters a superfluid state, called Helium II, whose viscosity seems to vanish. Fast flows through very fine capillaries with no measurable pressure drop amply demonstrate this behavior. Between 5.2 and 2.712 K, however, helium under sufficient pressure is in a liquid state called Helium I that behaves as a normal fluid with finite viscosity and boils as the pressure is reduced.

Superfluid helium below 2.7 K is a multiphase fluid in which Helium I and Helium I components are thoroughly intermixed. Bulk viscosity measurements show the Helium I viscosity, but microscopic measurements show a different picture. Using very fine porous media, it stops the normal fluid component while allowing the superfluid component to flow unhindered and even accelerating it. In some situations, counterflows between the two components can be set up and persistent, and other rather counterintuitive fountain effects can also be demonstrated. These phenomena are consistent with the superfluid state having zero entropy.

Continuing, even expanding, research interest in superfluid helium is due to the close connection between quantized vorticity and the macroscopic observed superfluid behavior. Experimental measurements and diagnostics are relatively unsophisticated, once the obstacles to dealing in such a low-temperature regime are overcome. Such experiments are eliciting many types of new information on a quite macroscopic scale. There are thin vortex filaments whose circulation is quantized. There are no experiments showing either fractional or multiple levels of quantization. These thin vortex filaments persist indefinitely and can become lengthened and tangled just as the tube of vorticity on classical high Reynolds number fluid dynamics. Superfluid helium has become a very attractive laboratory for studying classical turbulence.

IV. APPROACHES FOR STUDYING FLUID DYNAMICS

Fluid dynamic systems are studied by (1) analytic approaches that solve the Navier-Stokes equations or other simpler fluid models, (2) computational fluid dynamics approaches that solve the mathematical models of fluids on large digital computers, and (3) experimental observations made on real fluids either as they occur in nature or in idealized laboratory systems in which specific processes and parameter regimes can be isolated more easily. Each of these approaches has inherent strengths and weaknesses, and each makes an important contribution to understanding and predicting the dynamics of a fluid system.

A. Analytic Approaches

The dynamic behavior of most fluids is usually well described by the Navier-Stokes or Euler equations, although additional terms or equations are sometimes necessary to complete the description. A closed-form solution is very valuable because the variation of flow properties with changes in the controlling parameters is explicitly displayed. Thus optimization can be accomplished directly with a minimum of computer time or experimental trial and error. Even though mathematical models of continuum fluids are reasonably complete, their intrinsic nonlinearity means that closed-form analytic solutions de-

scribing the dynamics exist for only a very few special systems. Therefore some approximations are required to find an analytic solution.

Various analytic approaches include analyzing steady-state flows, for which time derivatives can be neglected, and then using these idealized steady flows as a basis for linear perturbation analyses of periodic oscillations and exponential instabilities. For example, using a simplified geometry and reducing problems to one or two dimensions allows the use of complex-variable transformation techniques in the incompressible potential flow limit and similarity techniques in compressible and strong shock limits.

The techniques of singular perturbation theory have been developed to deal with compressible flows developing shocks and rotational flows that become singular through vortex stretching. Indeed, much of the mathematical apparatus for partial differential equations evolved to solve fluid dynamic problems. Asymptotic analytical methods were developed to deal with cases where expansion in one or more small parameters is reasonable. These methods have gone a long way toward providing an understanding of classical boundary-layer theory and problems involving a very thin viscous critical layer. Nonetheless, asymptotic solutions only provide a quantitative description of physical phenomena for limited ranges of parameter values.

Another relatively new methodology currently being explored is wavelets. In this approach, an expansion is used in which there are a number of expansion functions for each region, each representing a different scale length of the solution. A potential advantage of wavelets is the ability to adjust the basis set, since it is not fully determined *a priori*, to satisfy additional constraints or properties of the solution. To date, wavelets have been applied to signal processing, although now there is a notable effort to extend the applications to solutions of partial differential equations.

Turbulence is yet another class of problem for which special statistical analysis tools and expansion procedures have been developed. As described earlier, our views of turbulence have been colored by the mathematical tools we have been able to bring to bear on the problem. The newer tools, such as those mentioned briefly here are having a decided impact. Because turbulence spans a range of space and time scales, the techniques of group renormalization are being investigated. Recently Hamiltonian theory, nonlinear system dynamics (chaos), and fractal theory have revitalized the analysis of complex fluid dynamics and turbulence at a very general level. Generic properties that fluid systems share with much wider classes of dynamical and nonlinear phenomena are being uncovered, although much work remains to convert this understanding to practical use.

B. CFD Approaches

Exact or useful analytic solutions are not available for most fluid problems, even in relatively idealized configurations. To treat fluid nonlinearities, time dependence, and realistic geometries, recourse to numerical solution and experiments is usually required. CFD deals with the solution of fluid dynamic equations on digital computers. Its recent successes are based on innovative numerical algorithms that discretize the continuum equations of fluid dynamics to form a large but finite number of algebraic or ordinary differential equations that can be solved by computers.

The fundamental idea in CFD is that spatial dimensions are divided into discrete contiguous cells, usually called finite volumes or finite elements, and time is discretized into short intervals called time steps. This discretization is forced by conventional computers having finite-sized memories segmented into floating-point words of data. The numerically determined values of the fluid variables in each cell are advanced from one time step to the next using the nearby fluid variables determined at the previous time step. While such marching techniques are developed with time-dependent problems in mind, they are also used to find steady-state solutions where appropriate. For aerodynamic applications, special computational fluid static (CFS) techniques have been developed to accelerate convergence of numerical solutions to the steady state.

The number of possible CFD algorithms is enormous and the correct choice depends on many properties of the problem being solved and the computer resources available. In all cases, however, only a finite number of discrete values comprise the representation, and each value is only specified to finite precision. This means that information is inevitably lost in the computational solution relative to the continuous problem being approximated. The result is uncertainty and errors in the computation arising from the discretization. The source of the uncertainty is the missing information about the detailed solution structure within the discrete spatial cells and time steps. All approaches that use a finite number of values to represent a continuous profile have this problem.

Nevertheless, computers with billions of high-precision floating-point calculations per second are readily available, and this translates into computations that can be more accurate approximations to real fluid flow than modern diagnostic techniques can provide from experiments. Three-dimensional simulations with several millions of cells are possible giving spatial resolution of order one percent. Two-dimensional solutions, where they are valid, can be even more accurate.

Computational fluid dynamics is used for basic studies of fluid dynamics, for engineering design of complex flow configurations in appliances, automobiles, airplanes, or ships, for understanding and predicting the interactions of chemistry with fluid flow in combustion and propulsion, for basic and applied research into the nature of turbulence (as discussed previously), and for extrapolating into parameter regimes that are either inaccessible or very costly experimentally. Relative to analytic approaches, CFD has many fewer restrictive assumptions and provides a more complete description of the flow. Relative to experiments, CFD has fewer Mach number (speed), Reynolds number (viscosity), and diagnostic interference limitations.

Computational fluid dynamics methods are limited intrinsically by the speed and size of the available computers. Today they can be used to simulate flow in complex geometry with simple physics or flows with complex physics in relatively simple geometry, but they cannot do both. Thus research continues on a broad front to develop computers and algorithms that are faster, more robust, more flexible, easier to use, and more accurate. It is thus not surprising that the limits of CFD are most strongly felt in the simulations of turbulent flow. Current work on large-eddy simulations (LES) is exploring many ways to construct physically meaningful subgrid models. One of the more useful and simple of the approaches is called Monotone Integrated Large Eddy Simulation (MILES), which presents an interesting alternative to standard LES approaches.

C. Experimental Approaches

Laboratory experiments and field observations have long been our main approach to study fluid dynamic behavior, but the field is still dynamic and the rate of progress is rapid because of strides made in diagnostic technology. The advent of very localized, highly accurate, and virtually noninvasive diagnostic techniques, generally made possible by the rapid development of laser and computer technology, is the basis of these recent advances.

Two important and distinct classes of experimental approaches are global flow visualization and localized quantitative measurement. Flow visualization is still the best way to see the overall dynamics of a fluid system. Dyes can be used to distinguish different components of a complex fluid system as the flow evolves. Various techniques such as Schlieren photography can be used to record the location and motion of density gradients in a fluid. Very small marker particles or bubbles can be injected into a fluid system to show up local motions. In fluids where the temperature variation is important, as in the study of thermal convection, calibrated temperature-sensitive marker particles can give an instantaneous map of the fluid temperature.

With the exception of acoustic and X-ray techniques, which are considerably more complicated, flow visualizations are generally associated with either gases or liquids that are transparent and bounded in space. Photography, in the form of motion pictures or video tapes, often with lasers as illumination sources, are used to record a sequence of images. Lasers are particularly useful illumination sources because they can deliver narrow pencils or sheets of light to isolate two-dimensional layers in the fluid.

Direct digital recording is also used now to acquire flow visualization information in computer readable form. These data may be analyzed quantitatively and dynamically, although with some difficulty, by digitizing several successive images and using image-processing techniques to identify locations or structures in each frame and deduce their movement.

The second class of experimental approach uses probes to make *in situ* quantitative measurements in the fluid. For example, pitot tubes and flush-mounted transducers are used to measure the pressure time-dependence. Wires, heated by carrying an electrical current, can be used to measure local flow speed because their temperature and hence resistance changes with changes in the speed of the passing fluid. The temperature sensitivity of dyes and other materials can be such that a few tenths of a degree change is enough to cause a complete color change, making possible quantitative measurements of the temperature in the body of a fluid without using an invasive probe. Pressure-sensitive paints can be used to map the pressure over the entire surface of a body.

Laser Doppler velocimetry employs crossed laser beams focused in a tiny volume of a flow to measure the instantaneous velocity of particles passing through the focal region. This is one of the most used of the new techniques because virtually no disturbance of the flow is involved and the space and time resolution can be made extremely fine. Three-dimensional locations can be pinpointed to a small fraction of a millimeter and velocities of a few meters per second can be measured with one or two percent accuracy. This technique for diagnosing fluid dynamics suffers some difficulty in making multiple measurements over a large volume to map out a flow field and so is often used to make separate time traces for statistical analysis at a modest number of points in the flow.

Because lasers can be tuned in a controllable way to very narrow frequency bands, it is possible to use the tightly focused laser light itself to diagnose or even induce reactions. Absorption of the light can be used to measure the density of particular constituents of the flow integrated along the laser line of sight. The flexibility to use the laser photons to diagnose specific short-lived chemical species is particularly important to a wide class of fluid dynamics problems involving combustion and reactive flows. These techniques include laser Rayleigh scattering, laser resonant absorption, a number of Raman spectroscopy techniques, micro-schlieren and shadowgraph methods, and laser-induced fluorescence, and chemiluminescence.

Recent advances in computer technology are tending to blur the distinction between flow visualization and point probe diagnostics. Flow visualizations, as recorded photographically or on video tape, can be digitally encoded and analyzed separately from the experiment by a number of techniques designed to extract major features of a flow in successive frames of the record. Using computers, the location and motion of features such as a passive interface between two distinct fluid components or a dynamic surface such as a flame front or a shock can be calculated accurately and automatically. The motion of specific illuminated markers can be deduced from successive frames, giving more global visualizations of both local fluid velocity and density, assuming that the marker particles or bubbles are small enough that buoyancy and inertial effects can be neglected. These digital records also facilitate the direct comparison of CFD simulations and the experiments.

SEE ALSO THE FOLLOWING ARTICLES

CHEMICAL THERMODYNAMICS • FLIGHT (AERODYNAMICS) • FLUID MIXING • LIQUIDS, STRUCTURE AND DYNAMICS • MOLECULAR HYDRODYNAMICS

BIBLIOGRAPHY

Anderson, J. D., Jr. (1995). "Computational Fluid Dynamic," McGraw-Hill, New York.

Donnelly, R. J. (1993). Quantized vortices and turbulence in helium II. *Ann. Rev. Fluid Mech.* **25,** 325–371.

Landau, L. D., and Lifshitz, E. M. (1959). "Fluid Mechanics," Pergamon, New York.

Lugt, H. J. (1983). "Vortex Flow in Nature and Technology," Wiley, New York.

Liepmann, H., and Roshko, A. (1957). "Elements of Gasdynamics," Wiley, New York.

Oran, E. S., and Boris, J. P. (2000). "Numerical Simulation of Reactive Flow," Cambridge Univ. Press, Cambridge, UK.

Shapiro, A. H. (1954). "The Dynamics and Thermodynamics of Compressible Fluid Flow," Ronald Press, New York.

Schlichting, H. (1951). "Boundary Layer Theory," McGraw-Hill, New York.

Soo, S. L. (1990). "Multiphase Fluid Dynamics," Science Press, Beijing, distributed through Gower Technical, Brookfield (USA).

Shtern, V., and Hussain, F. (1999). Collapse, symmetry breaking, and hysteresis in swirling flows. *Ann. Rev. Fluid Mech.* **31,** 537–566.

Tannehill, J. C., Anderson, D. A., and Pletcher, R. H. (1997). "Computational Fluid Mechanics and Heat Transfer," Taylor & Francis, Washington, D.C.

Van Dyke, M. (1982). "An Album of Fluid Motion," Parabolic Press, Stanford, California.

White, F. M. (1974). "Viscous Fluid Flow," McGraw-Hill, New York.

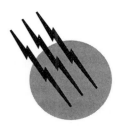

Fluid Dynamics (Chemical Engineering)

Richard W. Hanks
Brigham Young University

GLOSSARY

Field Mathematical representation of a physical quantity; at every point of space the mathematical quantity is defined as continuous for all necessary orders of differentiation.

Ground profile Plot of physical ground elevations along a pipeline route.

Head Any hydraulic energy quantity converted to an equivalent hydrostatic pressure and expressed as a column height of fluid.

Hydraulic grade line Graphic representation of the mechanical energy equation as hydraulic or pressure head against length; slope is frictional head loss per unit length.

Mixing length Mean distance over which a turbulent eddy retains its identity; phenomenological measure of a turbulence length scale in a zero-parameter model.

Physical component Tensor component that has the physical dimensions of the property being represented.

Reynolds stress Nondiagonal element of the correlation dyad for fluctuation velocity components in a turbulent flow; commonly interpreted as a shear component of the extra stress caused by the turbulence.

Tensor Matrix operator that transforms one vector function into another; all tensorial functions and entities must transform properly according to laws of coordinate transformation and retain both formal and operational invariance.

THE MECHANICS OF FLUIDS is a broad subject dealing with all of the phenomena of fluid behavior. Subtended within this subject is the subset of phenomena associated specifically with the kinematic and dynamic behavior of fluids. Kinematics is the study of motion per se, while dynamics includes the response of specific materials to applied forces. This requires one to apply the theory of deformable continuum fields. In its most general form the continuum field theory includes both fluid mechanics and dynamics in all their myriad forms. This article deals specifically with kinematic and dynamic applications.

I. INTRODUCTION

The phenomena of fluid mechanics are myriad and multiform. In the practice of chemical engineering, most applications of fluid mechanics are associated with either flow through a bounded duct or flow around a fixed object in the context of design of processing equipment. The details of such problems may be very simple or extremely complex. The chemical engineer must know how to apply standard theoretical and empirical procedures to solve these problems. In cases where standard methods fail, he or she must also know how to apply fundamental principles and develop an appropriate solution. To this end this article deals with both the fundamentals and the application thereof to bounded duct flows and flows about objects of incompressible liquids of the type commonly encountered by practicing chemical engineers. The phenomena associated with compressible flow, two-phase gas–liquid flow, and flow through porous media are not considered because of space limitations.

II. BASIC FIELD EQUATIONS (DIFFERENTIAL OR MICROSCOPIC)

A. Generic Principle of Balance

The fundamental theory of fluid mechanics is expressed in the mathematical language of continuum tensor field calculus. An exhaustive treatment of this subject is found in the treatise by Truesdell and Toupin (1960). Two fundamental classes of equations are required: (1) the generic equations of balance and (2) the constitutive relations.

The generic equations of balance are statements of truth, which is *a priori* self-evident and which must apply to all continuum materials regardless of their individual characteristics. Constitutive relations relate diffusive flux vectors to concentration gradients through phenomenological parameters called transport coefficients. They describe the detailed response characteristics of specific materials. There are seven generic principles: (1) conservation of mass, (2) balance of linear momentum, (3) balance of ro-

tational momentum, (4) balance of energy, (5) conservation of charge–current, (6) conservation of magnetic flux, and (7) thermodynamic irreversibility.

In the vast majority of situations of importance to chemical engineers, the conservation of charge–current and magnetic flux are of no importance, and therefore, we will not consider them further here. They would be of considerable importance in a magnetohydrodynamic problem.

The four balance or conservation principles can all be represented in terms of a general equation of balance written in integral form as

$$\iiint\limits_V \frac{\partial \psi}{\partial t}\, dV = -\iint\limits_S \psi \mathbf{v} \cdot \mathbf{n}\, ds - \iint\limits_S \mathbf{j}_{D\psi} \cdot \mathbf{n}\, ds$$

$$\underbrace{\phantom{\iiint\limits_V \frac{\partial \psi}{\partial t}\, dV}}_{\substack{\text{Net increase} \\ \text{of } \psi \text{ in } V}} \qquad \underbrace{\phantom{-\iint\limits_S \psi \mathbf{v} \cdot \mathbf{n}\, ds}}_{\substack{\text{Net convective} \\ \text{influx of } \psi}} \qquad \underbrace{\phantom{-\iint\limits_S \mathbf{j}_{D\psi} \cdot \mathbf{n}\, ds}}_{\substack{\text{Net diffusive} \\ \text{influx of } \psi}}$$

$$+ \underbrace{\iiint\limits_V \dot{r}_\psi\, dV}_{\substack{\text{Net production} \\ \text{of } \psi \text{ in } V}} \qquad (1)$$

or in differential form as (**n** is the outward-directed normal vector; hence, $-\psi \mathbf{v} \cdot \mathbf{n}\, ds$ represents influx)

$$\underbrace{\frac{\partial \psi}{\partial t}}_{\substack{\text{Net increase} \\ \text{of } \psi \text{ at point}}} = \underbrace{-\nabla \cdot \psi \mathbf{v}}_{\substack{\text{Net convective} \\ \text{influx of } \psi}} \underbrace{-\nabla \cdot \mathbf{j}_{D\psi}}_{\substack{\text{Net diffusive} \\ \text{influx of } \psi}} + \underbrace{\dot{r}_\psi}_{\substack{\text{Net production} \\ \text{of } \psi \text{ at point}}} \qquad (2)$$

where ψ represents the concentration or density of any transportable property of any tensorial order, $\mathbf{j}_{D\psi}$ represents the diffusive transport flux of property ψ, and $\dot{r}_\psi$ represents the volumetric rate of production or generation of property ψ within the volume V, which is bounded by the surface S.

Equation (2) is expressed in the Eulerian frame of reference, in which the volume element under consideration is fixed in space, and material is allowed to flow in and out of the element. An equivalent representation of very different appearance is the Lagrangian frame of reference, in which the volume element under consideration moves with the fluid and encapsulates a fixed mass of material so that no flow of mass in or out is permitted. In this frame of reference, Eq. (2) becomes

$$D\psi/Dt = -\psi \nabla \cdot \mathbf{v} - \nabla \cdot \mathbf{j}_{D\psi} + \dot{r}_\psi, \qquad (3)$$

where the new differential term $D\psi/Dt$ is called the substantial or material derivative of ψ and is defined by the relation

$$\frac{D\psi}{Dt} = \frac{\partial \psi}{\partial t} + \mathbf{v} \cdot \nabla \psi. \qquad (4)$$

Equations (2) and (3) are related by an obvious vector identity.

B. Equation of Continuity

If the generic property ψ is identified as the mass density ρ of a material, then Eq. (2) represents the generic principle of conservation of mass. The diffusive flux vector $\mathbf{j}_{D\rho}$ is equal to 0 and also $\dot{r}_\rho$ equals 0. Thus, the statement of conservation of mass, or equation of continuity, is

$$\partial\rho/\partial t = -\nabla \cdot \rho\mathbf{v} \tag{5}$$

in the Eulerian frame or

$$D\rho/Dt = -\rho\nabla \cdot \mathbf{v} \tag{6}$$

in the Lagrangian frame. The following are specific expressions for Eq. (5) in the three most commonly used systems:

Cartesian

$$-\frac{\partial\rho}{\partial t} = \frac{\partial}{\partial x}(\rho v_x) + \frac{\partial}{\partial y}(\rho v_y) + \frac{\partial}{\partial z}(\rho v_z) \tag{7}$$

Cylindrical Polar

$$-\frac{\partial\rho}{\partial t} = \frac{1}{r}\frac{\partial}{\partial r}(r\rho v_r) + \frac{1}{r}\frac{\partial}{\partial\theta}(\rho v_\theta) + \frac{\partial}{\partial z}(\rho v_z) \tag{8}$$

Spherical Polar

$$-\frac{\partial\rho}{\partial t} = \frac{1}{r^2}\frac{\partial}{\partial r}(r^2\rho v_r) + \frac{1}{r\sin\theta}\frac{\partial}{\partial\theta}[(\sin\theta)\rho v_\theta]$$
$$+ \frac{1}{r\sin\theta}\frac{\partial}{\partial\phi}(\rho v_\phi) \tag{9}$$

C. Equations of Motion

The vector quantity $\rho\mathbf{v}$ represents both the convective mass flux and the concentration of linear momentum. Its vector product $\mathbf{x} \times \rho\mathbf{v}$ with a position vector $\mathbf{x}$ from some axis of rotation represents the concentration of angular momentum about that axis. If $\mathbf{g} = -\nabla\Phi$ is an external body or action-at-a-distance force per unit mass, where Φ is a potential energy field, then the vector $\rho\mathbf{g}$ represents the volumetric rate of generation or production of linear momentum. The vector $\mathbf{x} \times \rho\mathbf{g}$ is the volmetric production rate of angular momentum.

Surface tractions or contact forces produce a stress field in the fluid element characterized by a stress tensor $\mathbf{T}$. Its negative is interpreted as the diffusive flux of momentum, and $\mathbf{x} \times (-\mathbf{T})$ is the diffusive flux of angular momentum or torque distribution. If stresses and torques are presumed to be in local equilibrium, the tensor $\mathbf{T}$ is easily shown to be symmetric.

When all of these quantities are introduced into Eq. (2), one obtains

$$\frac{\partial}{\partial t}(\rho\mathbf{v}) = -\nabla \cdot \rho\mathbf{v}\mathbf{v} - \nabla \cdot (-\mathbf{T}) + \rho\mathbf{g}, \tag{10}$$

which is known variously as Cauchy's equations of motion, Cauchy's first law of motion, the stress equations of motion, or Newton's second law for continuum fluids. Regardless of the name applied to Eq. (10), Truesdell and Toupin (1960) identify it and the statement of symmetry of $\mathbf{T}$ as the *fundamental equations of continuum mechanics*.

By using the vector identities relating Eulerian and Lagrangian frames together with the equation of continuity, one can convert Eq. (10) to an equivalent form:

$$\rho\frac{D\mathbf{v}}{Dt} = \rho\mathbf{g} - \nabla p - \nabla \cdot \boldsymbol{\tau}. \tag{11}$$

In this equation the stress tensor $\mathbf{T}$ has been partitioned into two parts in accordance with

$$\mathbf{T} = -p\boldsymbol{\delta} + \mathbf{P} = -p\boldsymbol{\delta} - \boldsymbol{\tau}, \tag{12}$$

where $-p$ is the mean normal stress defined by

$$-p = \tfrac{1}{3}(T_{xx} + T_{yy} + T_{zz}) \tag{13}$$

and $\mathbf{P}$ is known variously as the viscous stress tensor, the extra stress tensor, the shear stress tensor, or the stress deviator tensor. It contains both shear stresses (the off-diagonal elements) and normal stresses (the diagonal elements), both of which are related functionally to velocity gradient components by means of constitutive relations. In purely viscous fluids only the shear stresses are important, but the normal stresses become important when elasticity becomes a characteristic of the fluid. In incompressible liquids the mean normal stress is a dynamic parameter that replaces the thermodynamic pressure. It is the gradient of this pressure that is always dealt with in engineering design problems.

If one performs the vector operation $\mathbf{x} \times$ (equations of motion), the balance of rotational momentum or moment of momentum about an axis of rotation is obtained. It is this equation that forms the basis of design of rotating machinery such as centrifugal pumps and turbomachinery.

Equation (11) is written in the form of Newton's second law and states that the mass times acceleration of a fluid particle is equal to the sum of the forces causing that acceleration. In flow problems that are acceleration-less ($D\mathbf{v}/Dt = 0$) it is sometimes possible to solve Eq. (11) for the stress distribution independently of any knowledge of the velocity field in the system. One special case where this useful feature of these equations occurs is the case of rectilinear pipe flow. In this special case the solution of complex fluid flow problems is greatly simplified because the stress distribution can be discovered before the constitutive relation must be introduced. This means that only a first-order differential equation must be solved rather than a second-order (and often nonlinear) one. The following are the components of Eq. (11) in rectangular Cartesian, cylindrical polar, and spherical polar coordinates:

Cartesian:

x Component

$$\rho\left(\frac{\partial v_x}{\partial t} + v_x\frac{\partial v_x}{\partial x} + v_y\frac{\partial v_x}{\partial y} + v_z\frac{\partial v_x}{\partial z}\right) = -\frac{\partial p}{\partial x}$$

$$-\left(\frac{\partial \tau_{xx}}{\partial x} + \frac{\partial \tau_{yx}}{\partial y} + \frac{\partial \tau_{zx}}{\partial z}\right) + \rho g_x \qquad (14)$$

y Component

$$\rho\left(\frac{\partial v_y}{\partial t} + v_x\frac{\partial v_y}{\partial x} + v_y\frac{\partial v_y}{\partial y} + v_z\frac{\partial v_y}{\partial z}\right) = -\frac{\partial p}{\partial y}$$

$$-\left(\frac{\partial \tau_{xy}}{\partial x} + \frac{\partial \tau_{yy}}{\partial y} + \frac{\partial \tau_{zy}}{\partial z}\right) + \rho g_y \qquad (15)$$

z Component

$$\rho\left(\frac{\partial v_z}{\partial t} + v_x\frac{\partial v_z}{\partial z} + v_y\frac{\partial v_z}{\partial y} + v_z\frac{\partial v_z}{\partial z}\right) = -\frac{\partial p}{\partial z}$$

$$-\left(\frac{\partial \tau_{xz}}{\partial x} + \frac{\partial \tau_{yz}}{\partial y} + \frac{\partial \tau_{zz}}{\partial z}\right) + \rho g_z \qquad (16)$$

Cylindrical Polar:

r Component

$$\rho\left(\frac{\partial v_r}{\partial t} + v_r\frac{\partial v_r}{\partial r} + \frac{v_\theta}{r}\frac{\partial v_r}{\partial \theta} - \frac{v_\theta^2}{r} + v_z\frac{\partial v_r}{\partial z}\right) = -\frac{\partial p}{\partial r}$$

$$-\left(\frac{1}{r}\frac{\partial}{\partial r}(r\tau_{rr}) + \frac{1}{r}\frac{\partial \tau_{r\theta}}{\partial \theta} - \frac{\tau_{\theta\theta}}{r} + \frac{\partial \tau_{rz}}{\partial z}\right) + \rho g_r \quad (17)$$

θ Component

$$\rho\left(\frac{\partial v_\theta}{\partial t} + v_r\frac{\partial v_\theta}{\partial r} + \frac{v_\theta}{r}\frac{\partial v_\theta}{\partial \theta} + \frac{v_r v_\theta}{r} + v_z\frac{\partial v_\theta}{\partial z}\right) = -\frac{1}{r}\frac{\partial p}{\partial \theta}$$

$$-\left(\frac{1}{r^2}\frac{\partial}{\partial r}(r^2\tau_{r\theta}) + \frac{1}{r}\frac{\partial \tau_{\theta\theta}}{\partial \theta} + \frac{\partial \tau_{\theta z}}{\partial z}\right) + \rho g_\theta \qquad (18)$$

z Component

$$\rho\left(\frac{\partial v_z}{\partial t} + v_r\frac{\partial v_z}{\partial r} + \frac{v_\theta}{r}\frac{\partial v_z}{\partial \theta} + v_z\frac{\partial v_z}{\partial z}\right) = -\frac{\partial p}{\partial z}$$

$$-\left(\frac{1}{r}\frac{\partial}{\partial r}(r\tau_{rz}) + \frac{1}{r}\frac{\partial \tau_{\theta z}}{\partial \theta} + \frac{\partial \tau_{zz}}{\partial z}\right) + \rho g_z \qquad (19)$$

Spherical Polar:

r Component

$$\rho\left(\frac{\partial v_r}{\partial t} + v_r\frac{\partial v_r}{\partial r} + \frac{v_\theta}{r}\frac{\partial v_r}{\partial \theta} + \frac{v_\phi}{r\sin\theta}\frac{\partial v_r}{\partial \phi} - \frac{v_\theta^2 + v_\phi^2}{r}\right)$$

$$= -\frac{\partial p}{\partial r} - \left(\frac{1}{r^2}\frac{\partial}{\partial r}(r^2\tau_{rr}) + \frac{1}{r\sin\theta}\frac{\partial}{\partial \theta}(\tau_{r\theta}\sin\theta)\right.$$

$$\left.+ \frac{1}{r\sin\theta}\frac{\partial \tau_{r\phi}}{\partial \phi} - \frac{\tau_{\theta\theta} + \tau_{\phi\phi}}{r}\right) + \rho g_r \qquad (20)$$

θ Component

$$\rho\left(\frac{\partial v_\theta}{\partial t} + v_r\frac{\partial v_\theta}{\partial r} + \frac{v_\theta}{r}\frac{\partial v_\theta}{\partial \theta} + \frac{v_\phi}{r\sin\theta}\frac{\partial v_\theta}{\partial \phi} + \frac{v_r v_\theta}{r}\right.$$

$$\left.- \frac{v_\phi^2\cot\theta}{r}\right) = -\frac{1}{r}\frac{\partial p}{\partial \theta} - \left(\frac{1}{r^2}\frac{\partial}{\partial r}(r^2\tau_{r\theta}) + \frac{1}{r\sin\theta}\frac{\partial}{\partial \theta}\right.$$

$$\left.\times(\tau_{\theta\theta}\sin\theta) + \frac{1}{r\sin\theta}\frac{\partial \tau_{\theta\phi}}{\partial \phi} + \frac{\tau_{r\theta}}{r} - \frac{\cot\theta}{r}\tau_{\phi\phi}\right) + \rho g_\theta$$

$$(21)$$

φ Component

$$\rho\left(\frac{\partial v_\phi}{\partial t} + v_r\frac{\partial v_\phi}{\partial r} + \frac{v_\theta}{r}\frac{\partial v_\phi}{\partial \theta} + \frac{v_\phi}{r\sin\theta}\frac{\partial v_\phi}{\partial \phi} + \frac{v_\phi v_r}{r}\right.$$

$$\left.+ \frac{v_\theta v_\phi}{r}\cot\theta\right) = -\frac{1}{r\sin\theta}\frac{\partial p}{\partial \phi} - \left(\frac{1}{r^2}\frac{\partial}{\partial r}(r^2\tau_{r\phi})\right.$$

$$\left.+ \frac{1}{r}\frac{\partial \tau_{\theta\phi}}{\partial \theta} + \frac{1}{r\sin\theta}\frac{\partial \tau_{\phi\phi}}{\partial \phi} + \frac{\tau_{r\phi}}{r} + \frac{2\cot\theta}{r}\tau_{\theta\phi}\right) + \rho g_\phi$$

$$(22)$$

Two terms in Eqs. (17) and (18) are worthy of special note. In Eq. (17) the term $\rho v_\theta^2/r$ is the centrifugal "force." That is, it is the effective force in the r direction arising from fluid motion in the θ direction. Similarly, in Eq. (18) $\rho v_r v_\theta/r$ is the Coriolis force, or effective force in the θ direction due to motion in both the r and θ directions. Both of these forces arise naturally in the transformation of coordinates from the Cartesian frame to the cylindrical polar frame. They are properly part of the acceleration vector and do not need to be added on physical grounds.

D. Total Energy Balance

Two types of energy terms must be considered: (1) thermal and (2) mechanical. The specific internal energy is $u = C_v T$, where C_v is the heat capacity and T is the temperature of the fluid. The specific kinetic energy is $v^2/2$. Thus, the total energy density is $\rho(u + v^2/2)$. Thermal energy diffuses into the fluid by means of a heat flux vector $\mathbf{q}$. Mechanical energy diffuses in by means of work done against the stresses $\mathbf{v} \cdot (-\mathbf{T})$. Energy may be produced internally in the fluid by chemical reactions at a rate $\dot{r}_{CR}$ and by the action of external body forces $\mathbf{v} \cdot \rho\mathbf{g}$. Thus, Eq. (2) can be written as Eulerian-form total energy balance as

$$\frac{\partial}{\partial t}\left[\rho\left(u + \frac{v^2}{2}\right)\right] = -\boldsymbol{\nabla} \cdot \left[\rho\left(u + \frac{v^2}{2}\right)\mathbf{v}\right]$$
$$-\boldsymbol{\nabla} \cdot \mathbf{q} - \boldsymbol{\nabla} \cdot [\mathbf{v} \cdot (-\mathbf{T})]$$
$$+ \mathbf{v} \cdot \rho\mathbf{g} + \dot{r}_{CR}. \qquad (23)$$

By appropriate manipulation as before, this can be written in Lagrangian form as

$$\rho\frac{D}{Dt}\left(u + \frac{v^2}{2}\right) = -\boldsymbol{\nabla} \cdot \mathbf{q} - \boldsymbol{\nabla} \cdot [\mathbf{v} \cdot (-\mathbf{T})]$$
$$+ \mathbf{v} \cdot \rho\mathbf{g} + \dot{r}_{CR}. \qquad (24)$$

By using Eq. (12) the term $-\boldsymbol{\nabla} \cdot [\mathbf{v} \cdot (-\mathbf{T})]$ can be written as

$$-\boldsymbol{\nabla} \cdot [\mathbf{v} \cdot (-\mathbf{T})] = -\boldsymbol{\nabla} \cdot (p\mathbf{v}) - \mathbf{v} \cdot (\boldsymbol{\nabla} \cdot \boldsymbol{\tau}) - \boldsymbol{\tau} : \boldsymbol{\nabla}\mathbf{v}. \qquad (25)$$

In Eq. (25) the term $\mathbf{v} \cdot (\boldsymbol{\nabla} \cdot \boldsymbol{\tau})$ represents reversible stress work, while $\boldsymbol{\tau} : \boldsymbol{\nabla}\mathbf{v}$ represents irreversible or entropy-producing stress work. The following are expressions for the latter quantity in rectangular Cartesian, cylindrical polar, and spherical polar coordinates:

Cartesian

$$(\boldsymbol{\tau} : \boldsymbol{\nabla}\mathbf{v}) = \tau_{xx}\left(\frac{\partial v_x}{\partial x}\right) + \tau_{yy}\left(\frac{\partial v_y}{\partial y}\right) + \tau_{zz}\left(\frac{\partial v_z}{\partial z}\right)$$
$$+ \tau_{xy}\left(\frac{\partial v_x}{\partial y} + \frac{\partial v_y}{\partial x}\right) + \tau_{yz}\left(\frac{\partial v_y}{\partial z} + \frac{\partial v_z}{\partial y}\right)$$
$$+ \tau_{zx}\left(\frac{\partial v_z}{\partial x} + \frac{\partial v_x}{\partial z}\right) \qquad (26)$$

Cylindrical Polar

$$(\boldsymbol{\tau} : \boldsymbol{\nabla}\mathbf{v}) = \tau_{rr}\left(\frac{\partial v_r}{\partial r}\right) + \tau_{\theta\theta}\left(\frac{1}{r}\frac{\partial v_\theta}{\partial \theta} + \frac{v_r}{r}\right) + \tau_{zz}\left(\frac{\partial v_z}{\partial z}\right)$$
$$+ \left[\tau_{r\theta}r\frac{\partial}{\partial r}\left(\frac{v_\theta}{r}\right) + \frac{1}{r}\frac{\partial v_r}{\partial \theta}\right]$$
$$+ \tau_{\theta z}\left(\frac{1}{r}\frac{\partial v_z}{\partial \theta} + \frac{\partial v_\theta}{\partial z}\right) + \tau_{rz}\left(\frac{\partial v_z}{\partial r} + \frac{\partial v_r}{\partial z}\right)$$
$$\qquad (27)$$

Spherical Polar

$$(\boldsymbol{\tau} : \boldsymbol{\nabla}\mathbf{v}) = \tau_{rr}\left(\frac{\partial v_r}{\partial r}\right) + \tau_{\theta\theta}\left(\frac{1}{r}\frac{\partial v_\theta}{\partial \theta} + \frac{v_r}{r}\right)$$
$$+ \tau_{\phi\phi}\left(\frac{1}{r\sin\theta}\frac{\partial v_\phi}{\partial \phi} + \frac{v_r}{r} + \frac{v_\theta\cot\theta}{r}\right)$$
$$+ \tau_{r\theta}\left(\frac{\partial v_\theta}{\partial r} + \frac{1}{r}\frac{\partial v_r}{\partial \theta} - \frac{v_\theta}{r}\right)$$
$$+ \tau_{r\phi}\left(\frac{\partial v_\phi}{\partial r} + \frac{1}{r\sin\theta}\frac{\partial v_r}{\partial \phi} - \frac{v_\phi}{r}\right)$$
$$+ \tau_{\theta\phi}\frac{1}{r}\left(\frac{\partial v_\phi}{\partial \theta} + \frac{1}{r\sin\theta}\frac{\partial v_\theta}{\partial \phi} - \frac{\cot\theta}{r}v_\phi\right)$$
$$\qquad (28)$$

Equation (23) represents the total energy balance or first law of thermodynamics. It includes all forms of energy transport. An independent energy equation, which does not represent a generic balance relation, is obtained by performing the operation $\mathbf{v} \cdot$ (equations of motion) and is

$$\rho\frac{D}{Dt}\frac{v^2}{2} = -\mathbf{v} \cdot \boldsymbol{\nabla}p - \mathbf{v} \cdot (\boldsymbol{\nabla} \cdot \boldsymbol{\tau}) + \mathbf{v} \cdot \rho\mathbf{g}. \qquad (29)$$

This relation, called the mechanical energy equation, describes the rate of increase of kinetic energy in a fluid element as a result of the action of external body forces, pressure, and reversible stress work.

When Eq. (29) is subtracted from Eq. (24), one obtains

$$\rho\frac{Du}{Dt} = -\boldsymbol{\nabla} \cdot \mathbf{q} - p\boldsymbol{\nabla} \cdot \mathbf{v} = \boldsymbol{\tau} : \boldsymbol{\nabla}\mathbf{v} + \dot{r}_{CR}, \qquad (30)$$

which is called the thermal energy equation. It describes the rate of increase of thermal internal energy of a fluid element by the action of heat fluxes, chemical reactions, volumetric expansion of the fluid, and irreversible stress work.

Clearly, only two of the three energy equations are independent, the third being obtained by sum or difference from the first two. The coupling between Eqs. (29) and (30) occurs by means of Eq. (25), which represents the total work done on the fluid element by the stress field. Neither Eq. (29) nor Eq. (30) is a balance relation by itself, but the sum of the two, Eq. (24), is.

E. Entropy Production Principle

We cannot write down *a priori* a generic balance relation for the entropy of a fluid. We can, however, derive a result that can be placed in the same form as Eq. (3) and therefore recognized as a balance relation. By working with the combined first and second laws of thermodynamics, one can show that the rate of increase of specific entropy is given by

$$\rho \frac{Ds}{Dt} = -\nabla \cdot \frac{\mathbf{q}}{T} + \frac{1}{T}\left(-\tau : \nabla \mathbf{v} + \frac{1}{T}\mathbf{q} \cdot \nabla T + \dot{r}_{CR}\right), \tag{31}$$

where s is the specific entropy. Equation (31) is in the Lagrangian form of Eq. (3) with $\psi = \rho s$ and where the eqation of continuity has been invoked. Thus, we recognize the term $-\nabla \cdot (\mathbf{q}/T)$ as the diffusive influx of entropy and the production or generation of entropy as the remaining three terms on the right side of Eq. (31). In the absence of chemical reactions, the principle of entropy production (or "postulate of irreversibility," as Truesdell has called it) states that

$$\frac{1}{T}(-\tau : \nabla \mathbf{v}) + \frac{1}{T^2}\mathbf{q} \cdot \nabla T \geq 0. \tag{32}$$

From Eq. (32) it follows that only the part $-\tau : \nabla \mathbf{v}$ of the stress work contributes to the production of entropy; hence, it is the "irreversible" or nonrecoverable work. The remainder of the stress work, expressed by $\mathbf{v} \cdot (\nabla \cdot \mathbf{T})$, is "reversible" or recoverable, as already described.

F. Constitutive Relations

The generic balance relations and the derived relations presented in the preceding section contain various diffusion flux tensors. Although the equation of continuity as presented does not contain a diffusion flux vector, were it to have been written for a multicomponent mixture, there would have been such a diffusion flux vector. Before any of these equations can be solved for the various field quantities, the diffusion fluxes must be related to gradients in the field potentials ϕ.

In general, the fluxes are related to gradients of the specific concentrations by relations of the form

$$\mathbf{j}_{D\psi} = -\beta \nabla \phi \tag{33}$$

or

$$\mathbf{j}_{d\psi} = -\mathbf{B} \cdot \nabla \phi. \tag{34}$$

In the form of Eq. (33) β is a scalar parameter called a transport coefficient. In the form of Eq. (34) $\mathbf{B}$ is a tensor, the elements of which are the transport coefficients. In either form the transport coefficients may be complex nonlinear functions of the scalar invariants of $\nabla \phi$.

For isotropic fluids the heat flux vector $\mathbf{q}$ takes the form

$$\mathbf{q} = -k_T \nabla T, \tag{35}$$

where k_T is the thermal conductivity. Equation (35) is known as Fourier's law of conduction. The momentum flux tensor τ is expressed in the form

$$\tau = -2\mu_a \mathbf{D}, \tag{36}$$

where μ_a is the apparent viscosity or viscosity function and $\mathbf{D}$ is the symmetric part of $\nabla \mathbf{v}$ given by

$$\mathbf{D} = \tfrac{1}{2}(\nabla \mathbf{v} + \nabla \mathbf{v}^T). \tag{37}$$

In general, μ_a is a complex and often nonlinear function of $\mathrm{II_D}$, the second principal invariant of $\mathbf{D}$; $\mathrm{II_D}$ is given by

$$-\mathrm{II_D} = \tfrac{1}{2}[(\nabla \cdot \mathbf{v})^2 - \mathbf{D}:\mathbf{D}]. \tag{38}$$

In the special case of a Newtonian fluid, $\mu_a = \mu$ is a constant called the viscosity of the fluid and Eq. (36) becomes Newton's "law" of viscosity. In a great many practical cases of interest to chemical engineers, however, the non-Newtonian form of Eq. (36) is encountered.

The formulation of proper constitutive relations is a complex problem and is the basis of the science of rheology, which cannot be covered here. This section presents only four relatively simple constitutive relations that have proved to be practically useful to chemical engineers. Elastic fluid behavior is expressly excluded from consideration. The following equations are a listing of these constitutive relations; many others are possible:

Bingham Plastics

$$\tau = -2\left\{\mu_\infty \pm \frac{\tau_0}{2\left|\sqrt{-2\mathrm{II_D}}\right|}\right\}\mathbf{D}, \quad \tfrac{1}{2}\tau:\tau > \tau_0^2 \tag{39}$$

$$0 = \mathbf{D}, \qquad\qquad\qquad \tfrac{1}{2}\tau:\tau \leq \tau_0^2 \tag{40}$$

Ostwald–DeWael or Power Law

$$t = -2k\left|2\sqrt{-2\mathrm{II_D}}\right|^{n-1}\mathbf{D} \tag{41}$$

Herschel–Bulkley or Yield Power Law

$$\tau = -2\left\{k\left|2\sqrt{-2\mathrm{II_D}}\right|^{n-1} \pm \frac{\tau_0}{2\left|\sqrt{-2\mathrm{II_D}}\right|}\right\}\mathbf{D}$$

$$\tfrac{1}{2}\tau:\tau > \tau_0^2 \tag{42}$$

$$0 = \mathbf{D}, \qquad\qquad \tfrac{1}{2}\tau:\tau \leq \tau_0^2 \tag{43}$$

Casson

$$\frac{\tau}{|2 - 2\mathrm{II}_\tau|^{1/2}} = -2\frac{\pm\tau_0}{|2 - 2\mathrm{II_D}|} + \frac{\mu_\infty}{|2 - 2\mathrm{II_D}|^{1/2}}\mathbf{D}$$

$$\tfrac{1}{2}\tau:\tau > \tau_0^2 \tag{44}$$

$$0 = \mathbf{D}, \qquad\qquad \tfrac{1}{2}\tau:\tau \leq \tau_0^2 \tag{45}$$

When these constitutive relations are coupled with the stress distributions derived from the equations of motion, details of the velocity fields can be calculated, as can the overall relation between pressure drop and volume flow rate.

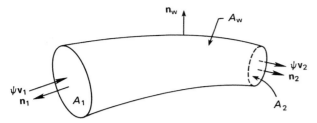

FIGURE 1 Schematic illustration of notation used in developing macroscopic equations.

III. BASIC FIELD EQUATIONS (AVERAGED OR MACROSCOPIC)

While the differential equations presented here are general and can be used to solve all types of fluid mechanics problems, to the average "practical" chemical engineer they are often unintelligible and intimidating. Much more familiar to most engineers are the averaged or macroscopic forms of these equations.

Equation (1) contains the integral form of the general balance relation. In this form it is a Eulerian result. If we take the volume in question to be the entire volume of the pipe located between two planes located at points 1 and 2 separated by some finite distance, as shown in Fig. 1, Eq. (1) can be written in the following a verage or macroscopic form,

$$\frac{\partial \Psi}{\partial t} = -\langle \psi \mathbf{v} \cdot \mathbf{n} \rangle_2 A_2 - \langle \psi \mathbf{v} \cdot \mathbf{n} \rangle_1 A_1 - \langle \mathbf{j}_{D\psi} \cdot \mathbf{n} \rangle_2 A_2$$
$$- \langle \mathbf{j}_{D\psi} \cdot \mathbf{n} \rangle_1 A_1 - \langle \mathbf{j}_{D\psi} \cdot \mathbf{n} \rangle_w A_w + \dot{R}_\psi V, \quad (46)$$

where Ψ is the total content of ψ in volume V and $\dot{R}_\psi$ is the volume average rate of production of ψ in V. In this relation the caret brackets have the significance

$$\langle (\ ) \cdot \mathbf{n} \rangle_k \equiv \frac{1}{A_k} \iint\limits_{A_k} [(\ ) \cdot \mathbf{n}]_k \, ds, \quad (47)$$

which is simply a statement of the mean value theorem of calculus applied to the integral in question. Equation (46) is an averaged or macroscopic form of the general balance relation and can be applied to mass, momentum, and energy.

A. Equation of Continuity

As before, there are no generation or diffusion terms for mass, so Eq. (46) becomes

$$\frac{\partial m}{\partial t} = \rho(\langle v \rangle_1 A_1 - \langle v \rangle_2 A_2). \quad (48)$$

The vast majority of practical chemical engineering problems are in steady-state operation, so that Eq. (48) reduces

simply to the statement that mass flow or volume flow is constant,

$$\langle v \rangle_1 A_1 = \langle v \rangle_2 A_2 = Q, \quad (49)$$

where Q is the volume flow. This relation defines the area mean velocity as Q/A. Equation (49) is the working form most often used.

B. Momentum Balance

Setting ψ equal to $\rho \mathbf{v}$ in Eq. (46) produces the macroscopic momentum balance. The term $\langle \mathbf{j}_{D\psi} \cdot \mathbf{n} \rangle_w$ represents the reaction force of the wall of the pipe on the fluid arising from friction and changes in the direction of flow. The term $\dot{R}_\psi V$ represents the action of the body force $\rho \mathbf{g}$ on the total flow. Thus, Eq. (46) becomes

$$\frac{\partial \mathbf{M}}{\partial t} = -\rho \langle \mathbf{v} \mathbf{v} \cdot \mathbf{n} \rangle_1 A_1 - \rho \langle \mathbf{v} \mathbf{v} \cdot \mathbf{n} \rangle_2 A_2 - \langle p \mathbf{n} \rangle_1 A_1$$
$$- \langle p \mathbf{n} \rangle_2 A_2 + \mathbf{F}_w + \rho V \mathbf{g}, \quad (50)$$

where $\mathbf{M}$ is the total momentum of the flow. Equation (50) can be solved at steady state for the reaction force $\mathbf{F}_w$ as

$$\mathbf{F}_w = \langle p \mathbf{n} \rangle_1 A_1 + \langle p \mathbf{n} \rangle_2 A_2 + \rho \langle \mathbf{v} \mathbf{v} \cdot \mathbf{n} \rangle_1 A_1$$
$$+ \rho \langle \mathbf{v} \mathbf{v} \cdot \mathbf{n} \rangle_2 A_2 - \rho V \mathbf{g}. \quad (51)$$

As an illustration of the use of this result, consider the pipe bend shown schematically in Fig. 2. Presuming the pipe to lie entirely in the x–y plane, we compute $F_{wx} = \mathbf{i} \cdot \mathbf{F}_w$, $F_{wy} = \mathbf{j} \cdot \mathbf{F}_w$, and $F_{wz} = \mathbf{k} \cdot \mathbf{F}_w$ as follows:

$$F_{wx} = -p_1 A_1 \cos \phi_1 + p_2 A_2 \cos \phi_2$$
$$- \rho \langle v \rangle_1^2 A_1 \cos \phi_1 + \rho \langle v x_2^2 A_2 \cos \phi_2 \rangle \quad (52)$$

$$F_{wy} = -p_1 A_1 \sin \phi_1 + p_2 A_2 \sin \phi_2$$
$$- \rho \langle v \rangle_1^2 A_1 \sin \phi_1 + \rho \langle v \rangle_2^2 A_2 \sin \phi_2 \quad (53)$$

$$F_{wz} = \rho V g \quad (54)$$

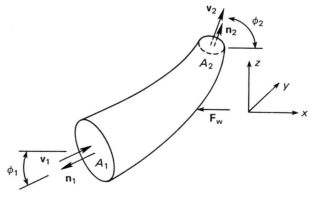

FIGURE 2 Illustration of forces on a pipe bend.

Thus, $(F_{wx}^2 + F_{wy}^2 + F_{wz}^2)^{1/2}$ is the magnitude of the force that would act in a bracing strut applied to the outside of the pipe bend to absorb the forces caused by turning the stream.

C. Energy Equations

When Eq. (46) is applied to energy quantities, a very large number of equivalent representations of the results are possible. Because of space limitations, we include only one commonly used variation here.

1. Total Energy (First Law of Thermodynamics)

When the various energy quantities used in arriving at Eq. (23) are introduced into Eq. (46), we obtain

$$\frac{\partial E}{\partial t} + \langle \rho e' \mathbf{v} \cdot \mathbf{n} \rangle_1 A_1 + \langle \rho e' \mathbf{v} \cdot \mathbf{n} \rangle_2 A_2 = \dot{Q} - \dot{W} + Q'_{CR},$$

(55)

in which $E = u + v^2/2 + \Phi$ is the total energy content of the fluid, $e' = e + p/\rho$, $\dot{Q}$ is the total thermal energy transfer rate,

$$\dot{Q} = - \iint_S \mathbf{q} \cdot \mathbf{n} \, ds,$$

(56)

Q'_{CR} is the total volumetric energy production rate due to chemical reactions or other such sources, and $\dot{W}$ is the total rate of work done or power expended against the viscous stresses,

$$\dot{W} = \iint_S (\mathbf{v} \cdot \mathbf{T}) \cdot \mathbf{n} \, ds.$$

(57)

In common engineering practice Eq. (55) is applied to steady flow in straight pipes and is divided by the mass flow rate $\dot{m} = \rho \langle v \rangle A$ to put it on a per unit mass basis,

$$\Delta u + \Delta \langle v \rangle^2 / 2 + g \Delta z + \Delta p / \rho = \hat{q} - \hat{w} + \hat{q}', \quad (58)$$

where the operator Δ implies average quantities at the downstream point minus the same average quantities at the upstream point. The terms on the right-hand side of Eq. (58) are just those on the right-hand side of Eq. (55) divided by $\rho \langle v \rangle A$. In Eq. (58) z is vertical elevation above an arbitrary datum plane.

2. Mechanical Energy (Bernoulli's Equation)

By considering Cauchy's equations of motion [Eq. (10)], Truesdell derived the theorem of stress means,

$$\iint_S G\mathbf{T} \cdot \mathbf{n} \, ds = \iiint_V \mathbf{T} \cdot \nabla G \, dV + \iiint_V \rho G \frac{D\mathbf{v}}{Dt} \, dV$$
$$- \iiint_V \rho G g \, dV,$$

(59)

where G is a functional operator of any tensorial order and the other terms have the significance already described. In particular, if one sets G equal to $\mathbf{v} \cdot$, Eq. (59) results in

$$\frac{\partial}{\partial t} \frac{v^2}{2} + \langle \rho K' \mathbf{v} \cdot \mathbf{n} \rangle_1 A_1 + \langle \rho K' \mathbf{v} \cdot \mathbf{n} \rangle_2 A_2 = -\dot{W} - \dot{F},$$

(60)

which is the macroscopic form of Eq. (29), the mechanical energy equation. In this expression $K' = e - u$ is the combined kinetic, potential, and pressure energy of the fluid; $\dot{F}$ is the energy dissipated by friction and is given by

$$\dot{F} = - \iiint_V \tau : \nabla \mathbf{v} \, dV.$$

(61)

Consideration of Eqs. (29) and (32) shows that the mechanical energy equation involves only the recoverable or reversible work. In order to calculate this term on the average, however, it is necessary to compute the total work done $\dot{W}$ and subtract from it the part lost due to friction or the irreversible work $\dot{F}$. If Eq. (60) is applied to steady flow in a pipe and divided by the mass flow rate, the following per unit mass form is obtained,

$$\Delta \langle v \rangle^2 / 2 + g \Delta z + \Delta p / \rho = -\hat{w} - \hat{w}_f, \quad (62)$$

where $\hat{w}_f = \dot{F} / \rho \langle v \rangle A$ is the frictional energy loss per unit mass, and all other terms have the same significance as in Eq. (58). In practical engineering problems the key to the use of Eq. (62) is determining a numerical value for $\hat{w}_f$.

As we have seen, the above are variations of the mechanical energy equation. They are variously called the Bernoulli equation, the extended Bernoulli equation, or the engineering Bernoulli equation by writers of elementary fluid mechanics textbooks. Regardless of one's taste in nomenclature, Eq. (62) lies at the heart of nearly all practical engineering design problems.

a. Head concept. If Eq. (62) is divided by g, the gravitational acceleration constant, we obtain

$$\Delta \langle v \rangle^2 / 2g + \Delta z + \Delta p / \rho g = -h_s - h_f. \quad (63)$$

It will be observed that each term in Eq. (63) has physical dimensions of length. For example, if flow ceases, Eq. (63) reduces to

$$\Delta z + \Delta p / \rho g = 0, \quad (64)$$

which is just the equation of hydrostatic equilibrium and shows that the pressure differential existing between points 1 and 2 is simply the hydrostatic pressure due to a column of fluid of height $-\Delta z$. In a general situation each of the terms in Eq. (63) has the physical significance that it is the equivalent hydrostatic pressure "head" or height to which the respective type of energy term could be converted. Thus, $\Delta \langle v \rangle^2 / 2g$ is the velocity head, $\Delta p / \rho g$ is the pressure head, $-h_s$ is the pump or shaft work head, h_f is the friction head, and Δz is the potential or ground head.

b. Friction head.

In order to solve problems using Eq. (63), additional information is required regarding the nature of the friction head loss term $-h_f$. This information can be obtained by empirical correlation of experimental data, by theoretical solution of the field equations, or a combination of both. It is customary to express the friction head loss term as a proportionality with the dimensionless length of the pipe L/D and the velocity head in the pipe $\langle v \rangle^2/2g$,

$$h_f = f\frac{L}{D}\frac{\langle v \rangle^2}{2g}, \qquad (65)$$

where f is called a friction factor. The problem is thus reduced to finding a functional relation between the dimensionless factor f and whatever variables with which it may be found to correlate. In practice, two definitions of the friction factor are in common use. The expression given in Eq. (65) is the Darcy–Weisbach form common to civil and mechanical engineering usage. An alternative form, commonly used by chemical engineers in the older literature, is the Fanning friction factor,

$$f' = f/4. \qquad (66)$$

Care should always be exercised in using friction factors derived from a chart, table, or correlating equation to determine which type of friction factor is being obtained. The Darcy–Weisbach form is gradually supplanting the Fanning form as a consequence of most modern textbooks on fluid mechanics being written by either civil or mechanical engineers. Figure 3 is the widely accepted correlation for f for Newtonian fluids. This is called the Moody diagram. In it f is correlated as a function of the two dimensionless variables ε/D and $\mathrm{Re} = D\langle v \rangle \rho/\mu$, where ε is a relative roughness factor expressed as an average depth of pit or height of protrusion on the wall of a rough pipe and Re is called the Reynolds number. Re is a dynamic similarity parameter. This means that two flows having the same value of Re are dynamically similar to one another. All variables in two pipes therefore scale in similar proportion to their Re values. The Moody diagram does not work for non-Newtonian fluids. In this case other methods, to be discussed below, must be employed.

c. Pump or work head.

The pump head term in Eq. (63) is given by $h_s = \hat{w}_s/g$ and represents the head

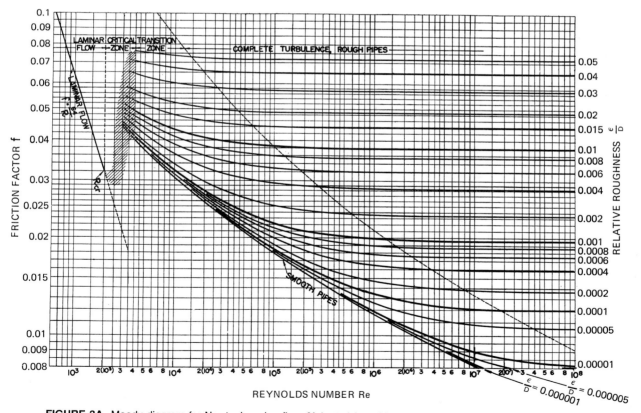

FIGURE 3A Moody diagram for Newtonian pipe flow. [Adapted from Moody, L. F. (1944). *Trans. ASME* **66**, 671–684.]

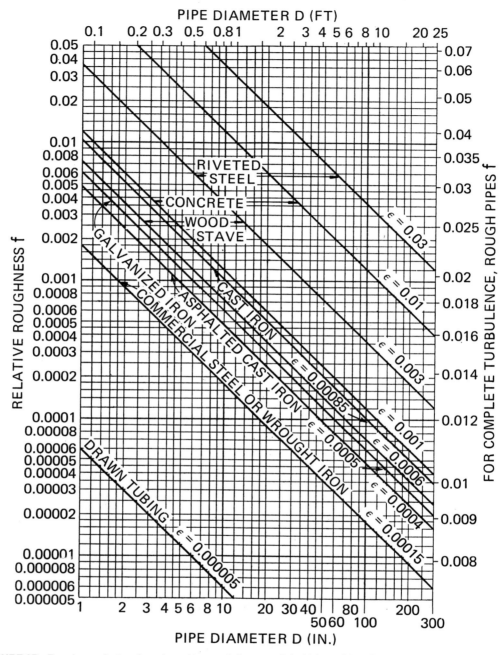

FIGURE 3B Roughness factors for selected types of pipe materials. [Adapted from Moody, L. F. (1944). *Trans. ASME* **66**, 671–684.]

equivalent of the energy input to the fluid by a pump in the system. This is the actual or hydraulic head. In order to obtain the total work that a pump–motor combination performs, one must take into account the efficiency of the pump–motor set.

The efficiency is the ratio of the useful energy delivered to the fluid as work to the total energy consumed by the motor and is always less than unity. This is a figure of merit of the pump that must be determined by experimentation and is supplied by the pump manufacturer. As a rule of thumb, well-designed centrifugal pumps usually operate at about 75–80% efficiency, while well-designed positive displacement pumps generally operate in the 90–95% efficiency range. In the case of positive displacement pumps, the efficiency is determined primarily by the mechanical precision of the moving parts and the motor's electrical efficiency. Centrifugal pumps also depend strongly on the hydraulic conditions inside the pump and are much more variable in efficiency. More is said about this in Section VI.

d. Hydraulic grade line.

Equation (63) is a finite difference equation and applies only to differences in the various energy quantities at two discrete points in the system. It takes no account of any conditions intermediate to these two reference points. If one were to keep point 1 fixed and systematically vary point 2 along the length of the pipe, the values of the various heads calculated would represent the systematic variation of velocity, potential, pressure, pump, and friction head along the pipe route. If all these values were plotted as a function of L, the distance down the pipe from point 1, a plot similar to that shown schematically in Fig. 4, would be obtained.

Figure 4 graphically illustrates the relation between the various heads in Eq. (63). For a pipe of constant cross section, the equation of continuity requires $\Delta \langle v \rangle^2 / 2g = 0$. The pump, located as shown, creates a positive head $h_s = \hat{w}_s / g$, represented by the vertical line of this height at the pump station (PS). The straight line of slope $-h_f/L$ drawn through the point $(h_s, 0)$ is a locus of all values of potential (Δz), pressure $(\Delta p / \rho g)$, and friction (h_f) heads calculated from Eq. (63) for any length of pipe (L). It is called the hydraulic grade line (HGL). The vertical distance between the HGL and the constant value $-h_s$ represents the energy that has been lost to that point due to friction. The height Δz designated as ground profile (GP) is a locus of physical ground elevations along the pipeline route and is also the actual physical location of the pipe itself. The difference between the HGL and the GP is the pressure head $\Delta p / \rho g$ at the length L. The significance of this head is that if one were to poke a hole in the pipe at point L, a fluid jet would spurt upward to a height equal to the HGL at that point. Thus, the HGL shows graphically at each point along the pipeline route the available pressure head to drive the flow through the pipe.

If the HGL intersects and drops below the GP, as in the area of Fig. 4 marked "region of negative pressure," there is not sufficient pressure head in the pipe to provide the potential energy necessary to raise the fluid to the height Δz at that point. Thus, if a hole were poked into the pipe at such a point, rather than a jet spurting out of the pipe, air would be drawn into the pipe. In a closed pipe a negative gauge pressure develops. This negative gauge pressure is the source of operation of a siphon. If, however, the absolute pressure in this part of the pipe drops to the vapor pressure of the liquid, the liquid will boil. This may cause the formation of a vapor bubble at the top of the pipe, or it may result in full vapor locking of the pipe, depending on the pressure conditions. This is called cavitation. Downhill of the negative pressure region where the HGL reemerges above the GP, the pressure rises back above the vapor pressure of the liquid and the vapor recondenses. This can occur with almost explosive violence and can result in physical damage to the pipe. Regions where this cavitation occurs are called "slack flow" regions. The HGL plot provides a simple and easy way to identify potential slack flow regions. In good design, such regions are avoided by the expedient of introducing another pump just upstream of the point where the HGL intersects the GP. Details of such procedures are discussed in Section VI.

3. Thermal Energy (Heat Transfer)

Just as the macroscopic mechanical energy equation is used to determine the relations between the various forms of mechanical energy and the frictional energy losses, so the thermal energy equation, expressed in macroscopic form, is used to determine the relation between the temperature and heat transfer rates for a flow system.

When Eq. (46) is applied to the thermal energy terms, we obtain

$$\frac{\partial U}{\partial t} + \langle \rho u \mathbf{v} \cdot \mathbf{n} \rangle_1 A_1 + \langle \rho u \mathbf{v} \cdot \mathbf{n} \rangle_2 A_2 = \dot{Q} + \dot{F} + Q'_{CR},$$

(67)

where U is the total internal energy of the fluid and the other terms have the significance already discussed.

Equation (67) is the basis of practical heat transfer calculations. In order to use it to solve problems, additional information is required about the total heat transfer rate $\dot{Q}$ and the production rate Q'_{CR}. The first is usually expressed in terms of a heat transfer coefficient analogous to the friction factor,

$$\dot{Q} = U_m A_s \Delta T_m,$$

(68)

where U_m is an "overall" heat transfer coefficient, which is usually related to "local" heat transfer coefficients both

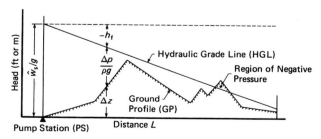

FIGURE 4 Illustration of hydraulic grade line concept.

inside and outside the pipe; A_s is the area of the heated pipe surface; and ΔT_m is some sort of mean or average temperature difference between the fluid and the pipe wall. Depending on the definition of ΔT_m, the definitions of the local heat transfer coefficients vary and so does the definition of U_m. This equation is not discussed further here, as its full discussion properly belongs in a separate article devoted to the subject of heat transfer.

IV. LAMINAR FLOW

In laminar flow the velocity distribution, and hence the frictional energy loss, is governed entirely by the rheological constitutive relation of the fluid. In some cases it is possible to derive theoretical expressions for the friction factor. Where this is possible, a three-step procedure must be followed.

1. Solve the equations of motion for the stress distribution.
2. Couple the stress distribution with the constitutive relation to produce a differential equation for the velocity field. Solve this equation for the velocity distribution.
3. Integrate the velocity distribution over the cross section of the duct to obtain an expression for the average velocity $\langle v \rangle$. Rearrange this expression into a dimensionless form involving a friction factor.

A. Shear Stress Distributions

In some special cases it is possible to solve the equations of motion [Eq. (11)] entirely independently of any knowledge of the constitutive relation and to obtain a universal shear stress distribution that applies to all fluids. In other cases it is not possible to do this because the evaluation of certain integration constants requires knowledge of the specific constitutive relation. Because of space limitations, we illustrate only one case of each type here.

1. Pipes

Equation (11) for the cylindrical geometry appropriate to the circular cross-section pipes so commonly used in practical situations is expressed by Eqs. (17)–(19). For steady, fully developed, incompressible flow, the solution of these equations is

$$\tau_{rz} = \frac{r}{2}\left(-\frac{dp}{dz}\right) + \frac{C}{r}, \tag{69}$$

where C is a constant of integration. Considerations of boundedness at the pipe centerline, $r = 0$, require that $C = 0$. Thus, Eq. (69) reduces to the familiar linear stress distribution,

$$\tau_{rz} = \frac{r}{2}\left(-\frac{dp}{dz}\right) = \xi\tau_w, \tag{70}$$

where $-dp/dz$ is the axial pressure gradient, $\xi = r/R$ is a normalized radial position variable, and τ_w is the wall shear stress given by

$$\tau_w = \frac{D}{4}\left(-\frac{\Delta p}{L}\right). \tag{71}$$

Equation (70) is clearly independent of any constitutive relation and applies universally to all fluids in a pipe of this geometry.

2. Concentric Annulus

Suppose a solid core were placed along the centerline of the pipe described in the preceding section so as to be coaxial and concentric with the pipe. Equation (69) is still valid as the solution of Eqs. (17)–(19). Now, however, the point $r = 0$ is not included in the domain of the solution, so that C is no longer zero. Somewhere between the two boundaries $r = R_i$ and $r = R$ the shear stress will vanish. If this point is called $\xi = \lambda$, then Eq. (69) becomes

$$\tau_{rz} = \tau_R(\xi - \lambda^2/\xi), \tag{72}$$

where τ_R is the shear stress at the outer pipe wall given by

$$\tau_R = \frac{R}{2}\left(-\frac{dp}{dz}\right) \tag{73}$$

and $\xi = r/R$ as before. Note two things: (1) Eq. (72) is now nonlinear in ξ, and (2) we still do not know the value of C. All that has been done is to shift the unknown value of C to the still unknown value of λ. We do, however, know the physical significance of λ. It is the location of the zero-stress surface. Unfortunately, we cannot discover the value of λ until we introduce some specific constitutive relation, integrate the resulting differential equation for the velocity distribution (thus introducing yet another constant of integration), and then invoke the no-slip or zero-velocity boundary conditions at *both* solid boundaries to determine the values of the new integration constant and λ. The value of λ so determined will be different for each different constitutive relation employed.

B. Velocity Distributions

1. Newtonian

When the Newtonian constitutive relation is coupled with Eq. (70) and appropriate integrations are performed, we obtain

$$u = v_z/\langle v \rangle = 2(1 - \xi^2) \tag{74}$$

$$\langle v \rangle = D\tau_w/8\mu \tag{75}$$

which are respectively known as the Poiseuille velocity profile and the Hagen–Poiseuille relation.

When the same operations are performed for the concentric annulus geometry, the results are

$$u = \frac{2}{F(\sigma)}[1 - \xi^2 + (1 - \sigma^2)\ln\xi/\ln(1/\sigma)] \quad (76)$$

$$\langle v \rangle = D\tau_R F(\sigma)/8\mu \quad (77)$$

$$F(\sigma) = 1 + \sigma^2 - (1 - \sigma^2)/\ln(1/\sigma) \quad (78)$$

where $\sigma = R_i/R$ is the "aspect" ratio of the annulus.

2. Non-Newtonian

Because of the extreme complexity of the expressions for the velocity distributions and average velocities in concentric annuli for even simple non-Newtonian fluids, we include here only the results for pipe flow.

a. Bingham Plastic. The pertinent results are

$$u = \frac{2}{F(\xi_0)}[1 - \xi^2 - 2\xi_0(1 - \xi)], \quad \xi > \xi_0 \quad (79)$$

$$u = 2(1 - \xi_0)^2/F(\xi_0), \quad \xi \le \xi_0 \quad (80)$$

$$\langle v \rangle = D\tau_w F(\xi_0)/8\mu_\infty \quad (81)$$

$$F(\xi_0) = 1 - \tfrac{4}{3}\xi_0 + \tfrac{1}{3}\xi_0^4 \quad (82)$$

where $\xi_0 = \tau_0/\tau_w$. Equation (81) is a version of the well-known Buckingham relation and is the Bingham plastic equivalent of the Hagen–Poiseuille result. The parameter ξ_0, because of the linearity of Eq. (70), also represents the dimensionless radius of a "plug" or "core" of unsheared material in the center of the pipe, which moves at the maximum velocity given by Eq. (80). This is a feature of all fluids that possess yield stresses.

b. Power law. The pertinent results are

$$u = \frac{1 + 3n}{1 + n}\left(1 - \xi^{(1+n)/n}\right) \quad (83)$$

$$\langle v \rangle = \frac{D}{2}\left(\frac{n}{1 + 3n}\right)\left(\frac{\tau_w}{k}\right)^{1/n}. \quad (84)$$

Note that these results reduce to the Newtonian results in the limit $n = 1$, $k = \mu$.

c. Herschel–Bulkley. The pertinent results are

$$u = \frac{1 + 3n}{1 + n}\frac{1}{F(\xi_0, n)}\left[(1 - \xi_0)^{(1+n)/n}\right.$$

$$\left. -(\xi - \xi_0)^{(1+n)/n}\right], \quad \xi > \xi_0 \quad (85)$$

$$u = \frac{1 + 3n}{1 + n}\frac{1}{F(\xi_0, n)}(1 - \xi_0)^{(1+n)/n}, \quad \xi \le \xi_0 \quad (86)$$

$$\langle v \rangle = \frac{D}{2}\frac{n}{1 + 3n}\left(\frac{\tau_w}{k}\right)^{1/n}(1 - \xi_0)^{(1+n)/n}F(\xi_0, n) \quad (87)$$

$$F(\xi_0, n) = (1 - \xi_0)^2 + \frac{2(1 + 3n)\xi_0(1 - \xi_0)}{1 + 2n} + \frac{1 + 3n}{1 + n}\xi_0^2 \quad (88)$$

where ξ_0 has the same significance as in the Bingham case.

d. Casson. The pertinent results are

$$u = \frac{2}{G(\xi_0)}\left[1 - \xi^2 + 2\xi_0(1 - \xi) - \tfrac{8}{3}\xi_0^{1/2}(1 - \xi^{3/2})\right],$$
$$\xi > \xi_0 \quad (89)$$

$$u = \frac{2}{G(\xi_0)}\left(1 - \tfrac{8}{3}\xi_0^{1/2} + 2\xi_0 - \tfrac{1}{3}\xi_0^2\right), \quad \xi \le \xi_0 \quad (90)$$

$$\langle v \rangle = D\tau_w G(\xi_0)/8\mu_\infty \quad (91)$$

$$G(\xi_0) = 1 - \frac{16}{7}\xi_0^{1/2} + \frac{4}{3}\xi_0 - \frac{1}{21}\xi_0^4 \quad (92)$$

where ξ_0 has the same significance as in the Bingham case.

It should be observed that in all cases, even the *linear* Bingham plastic case, the resultant average velocity expressions are nonlinear relations between $\langle v \rangle$ and $-dp/dz$. This is true of all non-Newtonian constitutive relations. A direct consequence of this result is that the friction factor relation is also nonlinear.

C. Friction Factors

In Eq. (65) the friction factor was introduced as an empirical factor of proportionality in the calculation of the friction loss head. If Eq. (63) is applied to a length of straight horizontal pipe with no pumps, one finds that

$$-h_f = \Delta p/\rho g. \quad (93)$$

Elimination of h_f between Eqs. (65) and (93) results in

$$f = \frac{8}{\rho\langle v \rangle^2}\left(\frac{-D\,\Delta p}{4L}\right) = \frac{8\tau_w}{\rho\langle v \rangle^2}, \quad (94)$$

which may be looked on as an alternate definition of the friction factor. From Eq. (66) it is evident that Eq. (94) with the numeric factor 8 replaced by 2 defines the Fanning friction factor.

1. Newtonian

Equation (94) provides the means for rearranging all of the theoretical expressions for $\langle v \rangle$ given above into expressions involving the friction factor. For example, when Eq. (75) for Newtonian pipe flow is so rearranged and one eliminates $\langle v \rangle$ in terms of the Reynolds number, $\mathrm{Re} = D\langle v \rangle \rho / \mu$, one obtains

$$f = 64/\mathrm{Re}. \tag{95}$$

Equation (95) is the source of the laminar flow line on the Moody chart (Fig. 3).

In the case of the concentric annulus the problem is somewhat ambiguous, because there are two surfaces of different diameter and hence the specification of a length in Re is not obvious as in the case of the pipe. For example, one could use D_i, D, or $D - D_i$ or a host of other possibilities. Obviously, for each choice a different definition of Re arises. Also, the specification of τ_w in Eq. (94) is ambiguous for the same reason. Here, we list only one of many possible relations,

$$f_R' = 2\tau_R / \rho \langle v \rangle^2 = 16 / F(\sigma) \mathrm{Re}_D, \tag{96}$$

where both f_R' and Re_D are based on τ_w and D for the outer pipe. The function $F(\sigma)$ in Eq. (96) is the same as given by Eq. (78).

2. Non-Newtonian

The ambiguity of definition of Re encountered in the concentric annulus case is compounded here because of the fact that no "viscosity" is definable for non-Newtonian fluids. Thus, in the literature one encounters a bewildering array of definitions of Re-like parameters. We now present friction factor results for the non-Newtonian constitutive relations used above that are common and consistent. Many others are possible.

a. Bingham plastic. The pertinent results are

$$f' = \frac{16}{\mathrm{Re}_{\mathrm{BP}}} + \frac{8}{3}\frac{\mathrm{He}}{\mathrm{Re}_{\mathrm{BP}}^2} - \frac{16}{3}\frac{\mathrm{He}^4}{f'^3\mathrm{Re}_{\mathrm{BP}}^8} \tag{97}$$

$$\mathrm{He} = D^2\rho\tau_0 / \mu_\infty^2 \tag{98}$$

$$\mathrm{Re}_{\mathrm{BP}} = D\langle v \rangle \rho / \mu_\infty \tag{99}$$

Note that a new dimensionless parameter He, called the Hedstrom number, arises because in the constitutive relation there are two independent rheological parameters. Parameter He is essentially a dimensionless τ_0. This multiplicity of dimensionless parameters in addition to the Re parameter is common to all non-Newtonian constitutive relations.

b. Power law. The pertinent results are

$$f' = 16/\mathrm{Re}_{\mathrm{PL}} \tag{100}$$

$$\mathrm{Re}_{\mathrm{PL}} = 2^{3-n}\frac{D^n\langle v \rangle^{2-n}\rho}{k}\left(\frac{n}{1+3n}\right)^n \tag{101}$$

Historically, $\mathrm{Re}_{\mathrm{PL}}$ was invented to force the form of Eq. (100).

c. Herschel–Bulkely. The pertinent results are

$$f' = 16/\left[\mathrm{Re}_{\mathrm{HB}}(1-\xi_0)^{1+n}F(\xi_0,n)^n\right] \tag{102}$$

$$\xi_0 = \frac{2}{f'}\left[\frac{\mathrm{He}_{\mathrm{HB}}^n\left(\dfrac{n}{1+3n}\right)^{2n}(2^{3-n})^2}{\mathrm{Re}_{\mathrm{HB}}^2}\right]^{1/(2-n)} \tag{103}$$

$$\mathrm{He}_{\mathrm{HB}} = \frac{D^2\rho}{\tau_0}(\tau_0/k)^{2/n} \tag{104}$$

and $\mathrm{Re}_{\mathrm{HB}}$ is identical in definition to Eq. (101). Indeed, Eqs. (102)–(104) reduce to Eqs. (100) and (101) for the limit $\tau_0 = 0$. In Eq. (104) $\mathrm{He}_{\mathrm{HB}}$ is the Herschel–Bulkley equivalent of the Bingham plastic Hedstrom number He.

d. Casson. The pertinent results are

$$f' = 16/\mathrm{Re}_{\mathrm{CA}}\,G'(f', \mathrm{Ca}, \mathrm{Re}_{\mathrm{CA}}) \tag{105}$$

$$G'(f', \mathrm{Ca}, \mathrm{Re}_{\mathrm{CA}}) = 1 - \frac{16\sqrt{2}}{7}\frac{(\mathrm{Ca}/f')^{1/2}}{\mathrm{Re}_{\mathrm{CA}}}$$
$$+ \frac{8}{3}\frac{(\mathrm{Ca}/f')}{\mathrm{Re}_{\mathrm{CA}}^2} - \frac{16(\mathrm{Ca}/f')^4}{21\,\mathrm{Re}_{\mathrm{CA}}^8} \tag{106}$$

$$\mathrm{Ca} = D^2\rho\tau_0 / \mu_\infty^2 \tag{107}$$

$$\mathrm{Re}_{\mathrm{CA}} = D\langle v \rangle \rho / \mu_\infty \tag{108}$$

The parameter Ca is called the Casson number and is analogous to the Hedstrom number He for the Bingham plastic and Herschel–Bulkley models.

V. TURBULENT FLOW

A. Transition to Turbulence

As velocity of flow increases, a condition is eventually reached at which rectilinear laminar flow is no longer stable, and a transition occurs to an alternate mode of motion that always involves complex particle paths. This motion may be of a multidimensional secondary laminar form, or it may be a chaotic eddy motion called turbulence. The nature of the motion is governed by both the rheological nature of the fluid and the geometry of the flow boundaries.

1. Newtonian

The most important case of this transition for chemical engineers is the transition from laminar to turbulent flow, which occurs in straight bounded ducts. In the case of Newtonian fluid rheology, this occurs in straight pipes when $Re = 2100$. A similar phenomenon occurs in pipes of other cross sections, as well and also for non-Newtonian fluids. However, just as the friction factor relations for these other cases are more complex than for simple Newtonian pipe flow, so the criteria for transition to turbulence cannot be expressed as a simple critical value of a Reynolds number.

All pressure-driven, rectilinear duct flows, whether Newtonian or non-Newtonian, undergo transition to turbulence when the transition parameter K_H of Hanks, defined by

$$K_H = \frac{\rho|\nabla v^2/2|}{|\rho\mathbf{g} - \nabla p|}, \tag{109}$$

achieves a maximum value of 404 at some point in the duct flow. In this equation v is the *laminar* velocity distribution. In the special limit of Newtonian pipe flow, Eq. (109) reduces the $Re_c = 2100$. For the concentric annulus, it reduces to

$$Re_{DC} = 808F(\sigma)/[(1 - \bar{\xi}^2 + 2\lambda^2 \ln \bar{\xi})|\bar{\xi} - \lambda^2/\bar{\xi}|], \tag{110}$$

where $\bar{\xi}$ is the root of

$$(1 - \bar{\xi}^2 + 2\lambda^2 \ln \bar{\xi})(\lambda^2 + \bar{\xi}^2) - 2(\bar{\xi}^2 - \lambda^2)^2 = 0 \tag{111}$$

with λ defined by

$$\lambda^2 = \tfrac{1}{2}(1 - \sigma^2)/\ln(1/\sigma) \tag{112}$$

and $F(\sigma)$ is given by Eq. (78). There are two roots to Eq. (111), with the result that Eq. (110) predicts two distinct Reynolds numbers of transition, in agreement with experiment.

2. Non-Newtonian

a. Bingham plastic.

The critical value of Re_{BP} is given by

$$Re_{BPc} = He\left(1 - \tfrac{4}{3}\xi_{0c} + \tfrac{1}{3}\xi_{0c}^4\right)/8\xi_{0c}, \tag{113}$$

where He is the Hedstrom number and ξ_{0c} is the root of

$$\xi_{0c}/(1 - \xi_{0c})^3 = He/16,800. \tag{114}$$

The predictions of these equations agree very well with experiental data.

b. Power law.

The pertinent results are

$$Re_{PLc} = \frac{6464n}{(1 + 3n)^2}(2 + n)^{(2+n)/(1+n)}. \tag{115}$$

c. Herschel–Bulkley.

The pertinent results are

$$Re_{HBc} = \frac{6464n}{(1 + 3n)^2}(2 + n)^{(2+n)/(1+n)}\left[\frac{F(\xi_{0c}, n)^{2-n}}{(1 - \xi_{0c})^n}\right], \tag{116}$$

where ξ_{0c} is the root of

$$\left[\frac{\xi_{0c}}{(1 - \xi_{0c})^{1+n}}\right]^{(2-n)/n}\left[\frac{1}{(1 - \xi_{0c})^n}\right]$$
$$= \frac{nHe_{HB}}{3232(2 + n)^{(2+n)/(1+n)}}. \tag{117}$$

He_{HB} is given by Eq. (104) and $F(\xi_{0c}, n)$ is given by Eq. (88) evaluated with $\xi = \xi_{0c}$.

d. Casson.

The pertinent equations are

$$Re_{CAc} = CaG(\xi_{0c})/8\xi_{0c}, \tag{118}$$

where ξ_{0c} must be determined from the simultaneous solution of Eqs. (119) and (120),

$$0 = 1 + 2\xi_{0c} - \tfrac{8}{3}\xi_{0c}^{1/2} + 2\bar{\xi}^{1/2}\xi_{0c}^{3/2} - 8\xi_{0c}\bar{\xi}$$
$$+ \tfrac{26}{3}\xi_{0c}^{1/2}\bar{\xi}^{3/2} - 3\bar{\xi}^2 \tag{119}$$

$$6464\xi_{0c}/Ca = \left[1 - \bar{\xi}^2 + 2\xi_{0c}(1 - \bar{\xi})\right.$$
$$\left. - \tfrac{8}{3}\xi_{0c}^{1/2}(1 - \bar{\xi}^{3/2})\right](\bar{\xi}^{1/2} - \xi_{0c}^{1/2})^2 \tag{120}$$

and $G(\xi_{0c})$ is given by Eq. (92) evaluated with $\xi = \xi_{0c}$.

B. Reynolds Stresses

When full turbulence occurs, the details of the velocity distribution become extremely complicated. While in principle these details could be computed by solving the general field equations given earlier, in practice it is essentially impossible. As an alternative to direct solution it is customary to develop a new set of equations in terms of Reynolds' averages. The model is illustrated schematically in Fig. 5.

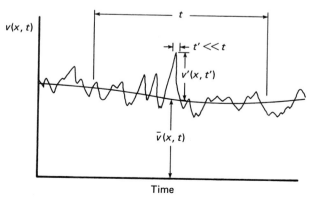

FIGURE 5 Schematic illustration of Reynolds' convention $\mathbf{v} = \bar{\mathbf{v}} + \mathbf{v}'$ for turbulent flow.

The actual velocity field fluctuates wildly. Reynolds modeled it by a superposition of a Eulerian time mean value $\bar{\mathbf{v}}$ defined by

$$\bar{\mathbf{v}}(\mathbf{x}, t) = \frac{1}{t} \int_0^t \mathbf{v}(\mathbf{x}, t') \, dt', \qquad (121)$$

where t' is a time interval of the order of an individual excursion and t is a time interval large in comparison with t' but small enough that gross time variations of the mean field can still be observed and calculated by the basic field equations. In terms of this model then, we write

$$\mathbf{v} = \bar{\mathbf{v}} + \mathbf{v}' \qquad (122)$$

with $\bar{\mathbf{v}}'$ being the instantaneous excursion or "fluctuation" from $\bar{\mathbf{v}}$. After this result is introduced into the field equations and the time-averaging operation defined in Eq. (121) is invoked, we obtain a new set of averaged field equations for the turbulent flow.

For incompressible fluids we obtain the following results:

Equation of Continuity

$$\nabla \cdot \bar{\mathbf{v}} = 0 \qquad (123)$$

Cauchy's Equations of Motion

$$\rho \frac{D\bar{\mathbf{v}}}{Dt} + \nabla \cdot \rho \overline{\mathbf{v}'\mathbf{v}'} = \rho \mathbf{g} - \nabla \bar{p} - \nabla \cdot \bar{\tau} \qquad (124)$$

Thermal Energy Relation

$$\rho \frac{\partial \bar{u}}{\partial t} + \rho \bar{\mathbf{v}} \cdot \nabla \bar{u} + \rho \overline{\mathbf{v}' \cdot \nabla u'}$$
$$= -\bar{\tau} : \nabla \bar{\mathbf{v}} - \overline{\tau' : \nabla \mathbf{v}'} - \nabla \cdot \bar{\mathbf{q}} + \bar{r}_{\mathrm{CR}} \qquad (125)$$

Mechanical Energy Relation

$$\rho(\partial/\partial t)(\bar{v}^2/2) + \rho(\partial/\partial t)\overline{(v'^2/2)} + \rho\bar{\mathbf{v}}\bar{\mathbf{v}} : \nabla\bar{\mathbf{v}}$$
$$+ \rho\bar{\mathbf{v}}\overline{\mathbf{v}'} : \nabla\mathbf{v}' + \rho\overline{\mathbf{v}'\bar{\mathbf{v}}} : \nabla\mathbf{v}' + \rho\overline{\mathbf{v}'\mathbf{v}'} : \nabla\bar{\mathbf{v}}$$
$$+ \rho\overline{\mathbf{v}'\mathbf{v}'} : \nabla\mathbf{v}' = \rho\bar{\mathbf{v}} \cdot \mathbf{g} - \bar{\mathbf{v}} \cdot \nabla\bar{p}$$
$$- \overline{\mathbf{v}' \cdot \nabla p'} - \bar{\mathbf{v}} \cdot (\nabla \cdot \bar{\tau}) - \overline{\mathbf{v}' \cdot (\nabla \cdot \tau')} \qquad (126)$$

Entropy Production Postulate

$$-\bar{\tau} : \nabla\mathbf{v} - \overline{\tau' : \nabla\mathbf{v}'} \geq 0 \qquad (127)$$

All of these relations contains terms involving statistical correlations among various products of fluctuating velocity, pressure, and stress terms. This renders them considerably more complex than their laminar flow counterparts. Reynolds succeeded in partially sol ving this dilemma by the expedient of introducing the turbulent stress tensor $\hat{\tau}$, defined by

$$\hat{\tau} = \bar{\tau} + \rho\overline{\mathbf{v}'\mathbf{v}'}. \qquad (128)$$

With this substitution Eq. (124) becomes identical with Eq. (10), with all terms replaced by their Eulerian time mean values. Thus, any solution of Eq. (10) for the stress distribution also becomes a solution of Eq. (124) for the "turbulent" stress distribution. Howe ver, this small success extracts a dear price. No further progress can be made because a new unknown quantity, $\rho\overline{\mathbf{v}'\mathbf{v}'}$, which has come to be known as the Reynolds' stress tensor, has been introduced with no compensating new equation for its calculation. This is the famous turbulence "closure" problem.

An enormous amount of effort has been expended in attempting to discover new equations for $\rho\overline{\mathbf{v}'\mathbf{v}'}$. Five different levels of approach have been pursued in the literature involving various degrees of mathematical complexity. We cannot discuss all of them here. We outline only two of the most fruitful: (1) the mixing length or zero-equation models and (2) the κ–ε or two-equation models.

C. Mixing Length Models

An early approach to the closure, typified by the work of Prandtl, represented the Reynolds' stress tensor as

$$\rho\overline{\mathbf{v}'\mathbf{v}'} = 2\rho\hat{\epsilon}_\tau \cdot \bar{\mathbf{D}}, \qquad (129)$$

where $\hat{\epsilon}_\tau$ is a second-order eddy diffusivity tensor and $\bar{\mathbf{D}}$ is the symmetric part of $\nabla\bar{\mathbf{v}}$ defined by Eq. (37) for $\mathbf{v} = \bar{\mathbf{v}}$. In this degree of approximation $\hat{\epsilon}_\tau$ is assumed to depend only on the properties of the mean velocity gradient tensor $\bar{\mathbf{D}}$ and is

$$\hat{\epsilon}_\tau = 2L^2\left|\sqrt{-2II_{\bar{\mathbf{D}}}}\right|\delta, \qquad (130)$$

where L is some sort of length measure of the turbulence called a mixing length, and $II_{\bar{\mathbf{D}}}$ is defined by Eq. (38) for $\mathbf{v} = \bar{\mathbf{v}}$.

For the special case of pipe flow, Prandtl modeled L as

$$L = k_t R(1 - \xi), \qquad (131)$$

where k_t, known as the Von Karman constant, is an empirical parameter usually taken to be ~ 0.36. This simple model leads to a rather famous results for the velocity distribution in a pipe:

$$u^+ = v/v^* = \frac{1}{0.36}\ln y^+ + 3.80, \qquad y^+ > 26 \quad (132)$$

$$y^+ = \frac{Rv^*\rho}{\mu}(1 - \xi) \qquad (133)$$

$$v^* = \sqrt{\tau_{\mathrm{w}}/\rho} \qquad (134)$$

The dimensionless variables u^+ and y^+ are called Prandtl's universal velocity profile variables. The parameter v^* is called the friction velocity.

In efforts to increase the range of applicability of the mixing length model, numerous others have modified it.

One of the better versions of the modified mixing length model is

$$L = k_t R(1 - \xi)(1 - E) \qquad (135)$$

$$E = \exp[-\phi^*(1 - \xi)] \qquad (136)$$

$$\phi^* = \left(R^* - R_c^*\right)/2\sqrt{2}B \qquad (137)$$

$$R^* = \mathrm{Re}\sqrt{f'} \qquad (138)$$

The parameter R_c^* is the laminar–turbulent transition value of R^* and has the numerical value 183.3 for Newtonian fluids. For non-Newtonian fluids it would have to be computed from the various results presented above.

The parameter B is called a dampening parameter, as its physical significance is associated with dampening turbulent fluctuations in the vicinity of a wall. For Newtonian pipe flow it has the numerical value 22. For non-Newtonian fluids it has been found to be a function of various rheological parameters as follows:

Bingham Plastic

$$B_{BP} = 22[1 + 0.00352He/(1 + 0.000504He)^2] \qquad (139)$$

Power Law

$$B_{PL} = 22/n \qquad (140)$$

Herschel–Bulkley

$$B_{HB} = B_{BP}/n \qquad (141)$$

No correlation has as yet been developed for the Casson model.

These simple models of turbulent pipe flow for various rheological models do not produce accurate details regarding the structure of the turbulent flow. They do, however, offer the practicing design engineer the opportunity to predict the gross engineering characteristics of interest with reasonable correctness. They are called zero-order equations because no differential equations for the turbulence properties themselves are involved in their solutions. Rather, one must specify some empirical model, such as the mixing length, to close the equations.

D. Other Closure Models

Actually, all methods of closure involve some type of modeling with the introduction of adjustable parameters that must be fixed by comparison with data. The only question is where in the hierarchy of equations the empiricism should be introduced. Many different systems of modeling have been developed. The zero-equation models have already been introduced. In addition there are one-equation and two-equation models, stress-equation models, three-equation models, and large-eddy simulation models. Depending on the complexity of the model and the problem being investigated, one can obtain various degrees of detailed information about the turbulent motions. Most of the more complex formulations require large computing facilities and may result in extreme numerical stability and convergence problems. All of the different methods of computing the turbulent field structure cannot be discussed here. Therefore, only one of these other methods, the so-called κ–ε method, which is a two-equation type of closure model, is outlined. Models such as this have to date been applied only to Newtonian flow problems.

The idea involved in the κ–ε model is to assume that the Reynolds' stress tensor can be written as

$$\rho \overline{\mathbf{v}'\mathbf{v}'} = 2\mu_t \bar{\mathbf{D}} - \tfrac{2}{3}\kappa\delta, \qquad (142)$$

where κ is the turbulent kinetic energy,

$$\kappa = \tfrac{1}{2}\overline{\mathbf{v}' \cdot \mathbf{v}'}, \qquad (143)$$

and μ_t is a turbulent or eddy viscosity function quite analogous to the eddy diffusivity discussed earlier. Just as in the zero-equation modeling situation, one cannot write down a general defining equation for μ_t, but must resort to modeling. In the present case the model used is

$$\mu_t = c_1 \rho \kappa^2 / \varepsilon, \qquad (144)$$

where c_1 is a (possibly) Reynolds number-dependent coefficient that must be determined empirically. The function ε is the turbulent energy dissipation rate function. The functions κ and ε are determined by the pair of simultaneous differential equations

$$\frac{D\kappa}{Dt} = \nabla \cdot \left(c_2 \frac{\mu_t}{\rho} \nabla \kappa\right) + \overline{\tau' : \nabla \mathbf{v}'} - \varepsilon \qquad (145)$$

$$\frac{D\varepsilon}{Dt} = \nabla \cdot \left(c_3 \frac{\mu_t}{\rho} \nabla \varepsilon\right) + c_4 \frac{\varepsilon}{\kappa} \overline{\tau' : \nabla \mathbf{v}'} - c_5 \varepsilon^2 / \kappa \qquad (146)$$

In this model the coefficients c_1 to c_5 are commonly given the numerical values $c_1 = 0.09$, $c_2 = 1.0$, $c_3 = 0.769$, $c_4 = 1.44$, and $c_5 = 1.92$, although these values can be varied at will by the user and are definitely problem specific. They can also be made functions of any variables necessary.

Here ends this article's discussion of this model, but extensive detail is available in numerous books on the subject. Some of these models present very accurate, detailed descriptions of the turbulence in some cases, but may be very much in error in others. Considerable skill and experience are required for their use.

VI. APPLICATIONS

A. Friction Factors

From a practical point of view the chemical engineer is very often interested in obtaining a relation between the overall pressure drop across a pipe, fitting, or piece of processing equipment and the bulk or mean velocity of flow through it. On occasion the details of the velocity, temperature, or concentration profile are important, but most frequently it is the gross pressure drop-flow rate behavior that is important to a chemical engineer.

This is generally obtained by use of the integrated form of the mechanical energy equation with the frictional energy loss calculated by Eq. (65). Thus, the basic problem facing a design engineer is how to obtain numerical values for the friction factor f.

For Newtonian fluids this problem is solved empirically by the introduction of the Moody diagram (Fig. 3). In the case of non-Newtonian fluids, however, this is not appropriate and alternative, semi-theoretical formulations must be developed. The theoretical laminar flow equations for the four rheological models considered here have already been presented, as have the modified mixing length models for turbulent flow of three of these same models. The latter equations can be integrated to obtain velocity distributions, which can in turn be integrated to produce mean velocity–pressure gradient relations. These results can then be algebraically rearranged into the desired friction factor correlations. These results are presented in the following subsections.

1. Bingham Plastic Pipe Flow

When the appropriate integrations are performed using the Bingham model, one obtains

$$\text{Re}_{BP} = \frac{1}{2} R_{BP}^{*2} \int_{\xi_0}^1 \xi^2 g\left(\xi, \xi_0, R_{BP}^*\right) d\xi \tag{147}$$

$$g\left(\xi, \xi_0, R_{BP}^*\right)$$
$$= \frac{\xi - \xi_0}{1 + \left[1 + \frac{1}{2} k_t^2 R_{BP}^{*2}(\xi - \xi_0)(1 - \xi)^2(1 - E)^2\right]^{1/2}}, \tag{148}$$

where E and R_{BP}^* are defined by Eqs. (136) to (138), with Re being replaced by Re_{BP}^* and B being given by Eq. (139). The parameter ξ_0 is related to R_{BP}^* by the relation

$$R_{BP}^{*2} = 2\text{He}/\xi_0. \tag{149}$$

These equations can be used for practical calculations in two ways. One may generate the equivalent of the Moody

diagram from them, or one may solve them iteratively for a specific design case. Both methods are illustrated below.

a. General friction factor plot. The Hedstrom number is the key design parameter. From its definition in Eq. (98) it can be seen that He depends only on the rheological parameters and the pipe diameter. The rheological parameters are obtained from laboratory viscometry data, and the pipe diameter is at the discretion of the designer to specify. Thus, its numerical value is discretionary.

For a given value of He one can compute a complete f'–Re curve as follows:

1. Compute Re_{BPc} from Eqs. (113) and (114).
2. Using ξ_{0c} in Eq. (149) compute R_{BPc}^*.
3. For $\text{Re}_{BP} < \text{Re}_{BPc}$ compute f' from Eq. (97).
4. For $\text{Re}_{BP} > \text{Re}_{BPc}$ choose a sequence of values of $R_{BP}^* > R_{BPc}^*$.
5. For each such value of R_{BP}^* compute ξ_0 from Eq. (149) and Re_{BP} from Eqs. (147) and (148).
6. From the computed value of Re_{BP} and the assumed value of R_{BP}^* compute f' from Eq. (138).
7. Repeat steps 4–6 as many times as desired and plot the pairs of points f', Re_{BP} so computed to create the equivalent Moody plot.

Figure 6 was created in this manner for a series of decade values of He. It may be used in place of the Moody chart for standard pipeline design problems. Because of the manner in which the empirical correlation for B was determined, no correction for pipe relative roughness is needed when one is dealing with commercial grade-steel line pipe.

b. Specific design conditions. A very common design situation involves the specification of a specific throughout and pipe diameter, thus fixing Re_{BP} but not R_{BP}^*. The system of equations presented earlier must therefore be solved iteratively for the value of R_{BP}^*, which produces the design Re_{BP} from Eq. (147).

The procedure to be followed is nearly the same as that already outlined. Steps 1 and 2 are followed exactly to determine R_{BPc}^*. Steps 4 and 5 are repeated iteratively until the Re_{BP} computed from Eq. (147) agrees with the design Re_{BP} to some acceptable convergence criterion. Because of the pinching effect of the curves in Fig. 6 at larger Re_{BP} values, it is best to use slower but more reliable interval halving techniques in searching for the root of the equation rather than a faster but often unstable Newton–Raphson method.

As an alternative to these two techniques, which involve considerable programming and numerical integration,

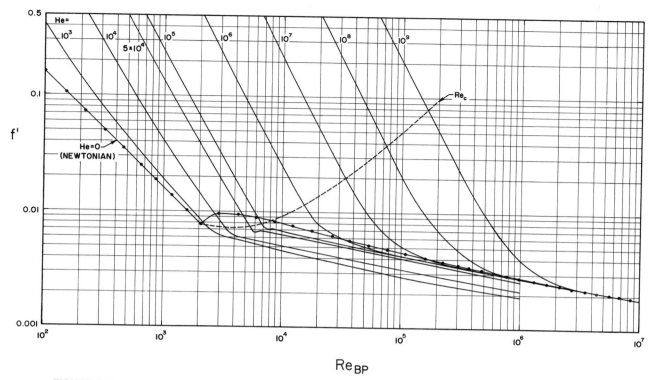

FIGURE 6 Fanning friction factor–Bingham plastic Reynolds number curves for Bingham plastic fluids. [Reproduced from Hanks, R. W. (1981). "Hydraulic Design from Flow of Complex Mixtures," Richard W. Hanks Associates, Inc., Orem, UT.]

the following empirical curve fits of Fig. 6 have been developed:

$$f' = 10^A / \text{Re}_{BP}^{0.193} \tag{150}$$

$$A = -1.378\{1 + 0.146 \exp[-2.9(10^{-5})\text{He}]\} \tag{151}$$

These equations are valid only for turbulent flow.

2. Power Law Model Pipe Flow

The pertinent equations here are

$$\text{Re}_{PL} = \left(\frac{n}{1+3n}\right)^n R_{PL}^{*2} \left[\int_0^1 \xi^2 \zeta\left(\xi, R_{PL}^*\right) d\xi\right]^{2-n} \tag{152}$$

$$R_{PL}^* = \frac{3n+1}{n} \left[\text{Re}_{PL}\left(\frac{f'}{16}\right)^{(2-n)/2}\right]^{1/n} \tag{153}$$

$$\xi = \zeta^n + \tfrac{1}{8} R_{PL}^{*2} L_{PL}^{*2} \zeta^2 \tag{154}$$

As with the Bingham case one first computes Re_{PLc} from Eq. (115) and then uses Eq. (153) to compute R_{PLc}^*. The value of f' to be used in this calculation comes from Eq. (100). Once R_{PLc}^* is known, one then chooses a series of values of $R_{PL}^* > R_{PLc}^*$ and computes Re_{PL} for each

from Eq. (152). These values, together with the specified values of R_{PL}^* and Eq. (153), determine the corresponding values of f'. In Eq. (152) the function $\zeta(\xi, R_{PL}^*)$ is defined implicitly by Eq. (154), where the mixing length L_{PL}^* is equal to L_{PL}/R, with L_{PL} being determined by Eqs. (135)–(137) and (140). The computation of f' for a specific value of Re_{PL} is carried out iteratively using these equations in exactly the same manner as described for the Bingham model.

An approximate value of f' can be computed from the following empirical equation:

$$\sqrt{\frac{1}{f'}} = \frac{4.0}{n^{0.75}} \log\left(\text{Re}_{PL} f'^{(2-n)/2}\right) - \frac{0.4}{n^{1.2}}. \tag{155}$$

3. Herschel–Bulkley Model Pipe Flow

For this model the pertinent equations are

$$\text{Re}_{HB} = (1 - \xi_0)^{(2-n)/n} \left(\frac{n}{1+3n}\right)^n R_{HB}^{*2}$$

$$\times \left[\int_{\xi_0}^1 \xi^2 \zeta\left(\xi, \xi_0, R_{HB}^*\right) d\xi\right]^{2-n} \tag{156}$$

$$\xi = \xi_0 + (1 - \xi_0)\zeta^n + \tfrac{1}{8}R_{HB}^{*2}(1 - \xi_0)^{2/n}L_{HB}^{*2}\zeta^2 \quad (157)$$

$$R_{HB}^{*2} = 2He_{HB}\big/\xi_0^{(2-n)/n} \qquad\qquad (158)$$

Equation (156) is exactly analogous to Eq. (152) for the power law model and to Eq. (147) for the Bingham model. R_{HB}^* is defined in relation to Re_{HB} and f' by Eq. (153), with Re_{PL} being replaced by Re_{HB}. The function $\xi(\xi, \xi_0, R_{HB}^*)$ is defined implicitly by Eq. (157), with $L_{HB}^* = L_{HB}/R$, and L_{HB} is given by Eqs. (135)–(137) and (141). The value of ξ_0 to be used in all of these equations is determined from Eq. (158) for specified values of He and R_{HB}^*. The computational procedures follow exactly the steps outlined for the other models. There are no simple empirical expressions that can be used to bypass the numerical integrations called for by this theory. One must use the above equations.

4. Casson Model Pipe Flow

As of the time of this writing, the corresponding equations for the Casson model have been developed but have not been tested against experimental data. Therefore, we cannot include any results.

5. Other Non-Newtonian Fluids

Thus far we have given exclusive attention to the flow of purely viscous fluids. In practice the chemical engineer often encounters non-Newtonian fluids exhibiting elastic as well as viscous behavior. Such viscoelastic fluids can be extremely complex in their rheological response. The le vel of mathematical complexity associated with these types of fluids is much more sophisticated than that presented here. Within the limits of space allocated for this article, it is not feasible to attempt a summary of this very extensive field. The reader must seek information elsewhere. Here we shall content ourselves with fluids that do not exhibit elastic behavior.

B. Pipeline System Design

1. Hydraulic Grade Line Method

As already indicated, once one has in hand a method for estimating friction factors, the practical engineering problem of designing pumping systems rests on systematic application of the macroscopic or integrated form of the mechanical energy equation [Eq. (63)], with h_f being defined in terms of f by Eq. (65). Section III.C.2.d introduced the concept of the hydraulic grade line, of HGL. This is simply a graphic representation of the locus of all possible solutions of Eq. (63) along a given pipeline for a given flow rate. When coupled with a ground profile (GP) as illustrated schematically in Fig. 4, tis plot provides a

particularly useful and simple means of identifying potential trouble spots in a pipeline. Although in this age of computers graphic techniques have generally fallen into disuse, this method still finds active use in commercial pipeline design practice.

The method is applied as illustrated below for a typical design problem. The conditions of the problem are $Q = 17{,}280$ bbl/day (528 gpm or 0.0333 m^3/sec) of a Newtonian fluid of specific gravity $= 1.18$ and viscosity $= 4.1$ cP (0.0041 Pa $\cdot$ sec) with a reliability factor of 0.9 and a terminal end head of 100 ft (30.48 m). The GP is shown in Fig. 7. The following steps are taken:

1. A pipeline route is selected and a GP is plotted.
2. A series of potential pipe diameters is chosen with a range of sizes such that the average flow velocity of 6 ft/sec (1.83 m/sec) is bracketed for the design throughput of the pipe.
3. For each of these candidate pipes the slope of the HGL, $-h_f/L$, is computed. For the illustrative design problem we chose pipes of schedule 40 size with nominal diameters of 5, 6, 8, and 10 in. The results are shown in Table I.
4. The desired residual head at the terminal end of the pipeline is specified. This is governed by the requirements

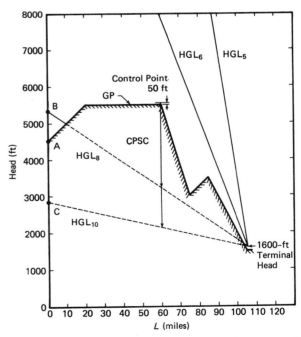

FIGURE 7 Ground profile (GP) plot showing initial hydraulic grade lines (HGLs) for pipes of different diameter. Eight- and 10-in. pipes (HGL$_8$, HGL$_{10}$) require additional control point static correction (CPSC) to clear the control point.

TABLE I   Sample Design Problem Illustrating Hydraulic Grade Line Method

D		V		$-h_f/L$	
(in.)	(m)	(ft/sec)	(m/sec)	(ft/mile)	(m/km)
5	0.1270	9.41	2.87	338	64.0
6	0.1524	6.52	1.99	138	26.1
8	0.2032	3.76	1.15	35.6	6.74
10	0.2540	2.39	0.73	11.9	2.25

of the process to be fed by the pipeline system. For this case 100 ft is used.

5. Once the terminal end pressure head is decided on, it is used as an anchor point through which HGLs for the various pipes are drawn (lines of slope $-h_f/L$ passing through the terminal head point at the end of the line). This is illustrated for the candidate pipes in Fig. 7.

6. From the HGL/GP plot the control points are determined. These are points, such as mp-60 (mp refers to the mileage post along the horizontal axis) in Fig. 7, that must be cleared by the flatter HGLs in order to avoid slack flow conditions. These points, together with the slopes of the HGLs, determine the minimum heights to which the HGL must be raised at mp-0 and thus the pump head requirements for each pipe. Depending on the specific GP, there may be multiple control points.

7. The approximate number and size of pumps required for the job are estimated. This is done by determining the total hydraulic horsepower required for each pipe and dividing by a nominal pump head representative of pump types (of the order of 2000 psi for positive displacement

and 900 psi for centrifugal pumps). For the sample problem the results assuming 900-psi centrifugals are shown in Table II. In making the calculations in Table II a number of factors must be taken into account. The total Δp_f is the HGL $\Delta p/L$ times total length (105 miles). The CPSC is the control point static correction and represents the net head increase that must be added to the HGL at mp-0 to cause it to clear the GP at its critical interior control point by a minimum terrain clearance (taken here to be 50 ft). For the 8-in. pipe it is simply the vertical distance between the GP + 50 ft at the control point (mp-60) and the HGL at that point. This is so because at mp-0 the HGL starts at point B (see Fig. 7), which is above the GP. For the 10-in. pipe, however, the HGL actually starts at point C, which is below GP. Therefore, the CPSC is the vertical distance between GP + 50 ft and the HGL at mp-60 decreased by the negative head at mp-0 (point C minus point A in Fig. 7).

The significance of CPSC is that this is the additional head the pumps must produce in order to get the fluid up over the GP at the control point with a minimum terrain clearance. This, of course, results in the HGL terminating at mp-105 at a much higher head than the specified 1600-ft terminal end head. This excess head, also tabulated in Table II, must be wasted or "burned off" as friction. This can be accomplished in a number of ways, such as introducing an orifice plate, introducing a valve, or decreasing the pipe diameter. Depending on specific pipeline system conditions and economics, any of these alternatives may be desirable.

8. The hydraulic and actual horsepower required for the pumps are determined. The hydraulic horsepower (HHP) is given by

TABLE II   Hydraulic Horsepower Calculations for Candidate Pipes

Nominal D (in.)	CP (miles)[a]	Total Δp_f (psi)[b]	CPSC (psi)[b,c]	Minimum pump pressure (psi)[b]	Approx. number of pump stations	HHP[d,e]	AHP[f]	AHP/PS[g]	Nominal HP/PS[g,h]	Actual head (ft)[i]	Excess head (ft)[j]
5	105	18,144	—	18,144	21	6211	8281	394	400	1714	—
6	105	7,408	—	7,408	9	2536	3381	376	400	1714	—
8	60	1,960	1196	3,156	4	1080	1440	360	400	1714	2348
10	60	639	895	1,534	2	525	700	350	350	767	3415

[a] CP, control point.

[b] 1 psi ≡ 6894.8 Pa.

[c] CPSC, control point static correction.

[d] HHP (hydraulic horsepower) = Δp(psi)Q(gpm)/1714.

[e] 1 hp ≡ 745.7 W = 0.7457 kW.

[f] AHP (actual horsepower) = HHP/Eff; Eff = 0.75 is assumed here.

[g] PS, pump station.

[h] Rounded up to nearest 50 hp.

[i] Based on nominal HHP/PS and 75% efficiency.

[j] Head at mp-105 less terminal head for HGL, which clears interior CP by 50-ft minimum terrain clearance.

$$HHP = \Delta p_f(\text{psi}) Q(\text{gpm})/1714, \qquad (159)$$

while the actual horsepower (AHP) is HHP divided by the pump efficiency (here taken to be 0.75; actual values would be fixed by the vendor in a real case).

9. The nominal horsepower per pump station (HP/PS) is fixed. This is done by rounding the AHP/PS up to the next nearest 50 hp.

10. The actual head required is determined. This is done by taking the nominal HP/PS and computing the pump station pressure rise from Eq. (159).

11. The PS discharge head is determined. This is done by adding to the PS pressure rise just computed the net positive suction head (NPSH) of the pump as specified by the vendor. It is always wise to allow an additional head above this value as a safety factor. Here a 50-ft intake head has been assumed for illustrative purposes.

12. The PSs are located. Figure 8 contains the final results for the 8-in. pipe. The total PS discharge head is plotted above the GP at mp-0 (6264 ft in Fig. 8). From this point the HGL is plotted. When it reaches a point equal to the pump intake head (50 ft in this example) above the GP, the next PS is located (mp-20 in Fig. 8). Here the process repeated, and the PS pressure rise head is plotted above the HGL (7266 ft in Fig. 8). This process is repeated as many times as necessary to cause the HGL to clear

all control points and to terminate on the terminal mp. In this example no more PSs are required, and the HGL terminates at mp-105 at a head of 4240 ft. This is far too much head for the specified conditions of the design. The excess head (4240–1600 ft) must be consumed as friction, as already explained. In Fig. 8 the diameter is decreased to a 6-in, pipe at mp-79.2. This introduces the HGL for the 6-in. pipe, which now terminates at 1600 ft at mp-105 as desired.

13. The system is optimized. Steps 1–12 must be repeated for each candidate pipe. The entire set must then be cost optimized. For example, the design indicated by Fig. 8 will work hydraulically but is not optimum. We see that at mp-60, the interior control point, we have actually cleared GP by 342 ft. This is considerably more than the minimum 50-ft terrain clearance required and is therefore wasteful of pumping energy. The design can obviously be improved by a change in pump specifications and other details. This should be done for each candidate pipe. The final design to be selected is based on an economic minimum-cost evaluation.

The method just outlined and illustrated is route specific. It is very flexible and simple to use. It can also be easily computerized if the GP data can be fed in as numerical values. Here we have illustrated its use in the context of a cross-country pipeline, such as a crude oil, products, or perhaps slurry pipeline, which might be commonly encountered by chemical engineers. The method is completely adaptable to any hydraulic flow problem and could be used equally well for a short in-plant pumping system analysis. It can help the designer of flow systems to avoid sometimes subtle traps for slack flow and siphons that might not be immediately obvious if the mechanical energy equation is applied only once between the initial and final points of the flow system.

2. Pumps

Pumps come in a bewildering array of shapes, sizes, capacities, head characteristics, chemical and corrosion resistance features, materials of construction, and prime mover types. The choice of a specific pump for a specific application is best made in consultation with individual vendors who can provide detailed data about their product. Ultimately all choices are based on a cost optimization.

Pumps come basically in two types: (1) positive displacement and (2) centrifugal. As a rule of thumb, positive displacement pumps operate at high head but relatively low capacity. Centrifugals, on the other hand, operate at low head and high capacity. Typically, positive displacement (PD) pumps may operate at heads from 1 to 10,000 psi and from hundreds of gallons per minute to

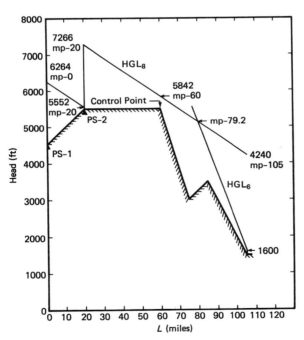

FIGURE 8 Hydraulic grade line (HGL) method design for pipe flow problem showing placement of pumping stations and change of diameter of pipe to handle excess head downstream of control point.

a fraction of a gallon per minute depending on the conditions. Typical centrifugal pumps may operate at heads of a few tens of feet to several hundreds of feet and capacities of several thousands of gallons per minute. It is possible to operate PD pumps in parallel or centrifugal pumps in series to achieve high head and high capacity. Some pump manufacturers also make "staged" centrifugal pumps, which are essentially multiple centrifugal pumps of identical head characteristics mounted on a common shaft and plumbed so as to permit the discharge of one to be the intake of the next stage.

a. Positive displacement pumps.

Positive displacement pumps include gear pumps, piston pumps, plunger pumps, and progressing cavity pumps. All PD pumps have in common the fact that they are volumetric devices in which a fixed volume of fluid is drawn into the pump, pressurized, and discharged at high pressure into the line. As a result, the output is pulsatile, giving rise to a (sometimes violently) fluctuating discharge pressure. This necessitates the installation of pulsation dampeners at the discharge of all PD pumps in a large pumping installation to protect the system against heavy pressure surging.

Another feature of PD pumps is that, if the line for any reason becomes blocked, they simply continue forcing high-pressure fluid into the line and eventually break something if a precautionary rupture system has not been installed. Thus, a PD pump should be protected by a high-pressure shutoff sensor and alarm system and also a bypass line containing a rupture disk or pressure relief valve.

Figure 9 is a schematic illustration of a double-acting PD piston pump. The volumetric capacity of this device per stroke of the piston is given by

$$Q' = \left(\tfrac{1}{4}\pi D_p^2 L_s n - V_R\right) Ne, \qquad (160)$$

where D_p is the diameter of the piston, L_s is the stroke length, $n = 1$ for a single-acting (only one side of the piston drives fluid on one-half of the stroke) or $n = 2$ for a double-acting (the piston drives fluid on both halves of the stroke)

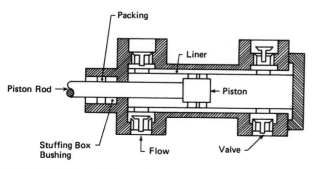

FIGURE 9 Schematic illustration of double-acting piston pump.

pump, V_R is the volume displaced by the rod in the double-acting case, N is the number of cylinders per pump, and e is a volumetric efficiency factor, usually 0.95–0.99.

The total volumetric capacity of the pump is

$$Q = \omega Q', \qquad (161)$$

where ω is the frequency in strokes per time. As an illustration of the use of these equations, suppose that in the previous HGL sample design problem we had elected to use single-acting ($n = 1$, $V_R = 0$), triplex ($N = 3$) piston pumps with a 12-in. piston diameter and a 10-in. stroke. At a total throughput of 587 gpm we calculate $Q' = 3256$ in.^{3}/stroke from Eq. (160) and from Eq. (161) we find $\omega = 41.6$ strokes per/minute. Armed with such information one can now seek a specific vendor. Adjustments in several of the design variables may need to be made to be compatible with vendor specifications.

A useful feature of the PD pump is that for a given power input Eqs. (159)–(161) allow the designer considerable flexibility in adjusting discharge pressure, cylinder capacity, and overall capacity. Positive displacement pumps are favorites on large-scale, high-pressure systems. Details of each of the various types of PD pump are best obtained from individual vendors.

b. Centrifugal pumps.

The operation of centrifugal pumps is entirely different from that of PD pumps. The principle of operation involves spinning a circular vaned disk at high speed inside a casing. The resulting centrifugal force accelerates the fluid to high velocity at the tangential discharge port, where it stagnates against the fluid already in the pipe, creating high pressure as a result of Bernoulli's equation. As a result the discharge pressure of an ideal centrifugal pump is proportional to the square of the velocity of the impeller tip. In actual practice, however, frictional energy losses and turbulence within the pump result in a different relationship, which must be determined experimentally for each pump. This is routinely done by pump manufacturers, and the information is presented in the form of a pump head curve, such as that illustrated in Fig. 10.

Manufacturers' performance curves, such as those in Fig. 10, contain a great deal of useful information. Actual average head–capacity curves are shown for a number of impeller diameters. Also superimposed on these head curves are curves of constant efficiency. A third set of curves superimposed on the head curves are the NPSH requirement curves (dashed line in Fig. 10), which indicate the required NPSH at any given condition of operation. A fourth set of curves sometimes included are the BHP (brake horsepower) curves. BHP is the actual horsepower calculated in the previous HGL method illustration. It is the HHP divided by the effciency.

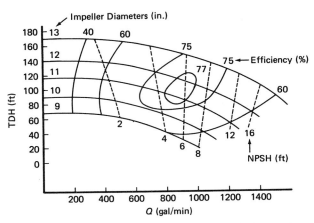

FIGURE 10 Typical centrifugal pump characteristic curves showing efficiency curves and NPSH (net positive suction head) for several impeller diameters.

We have discussed only a very small amount of information about pumps. A great deal more detail and practical operating information is available in books dealing with the selection of pumps. Space limitations preclude the inclusion of this detail here. In any specific application the user should consult with the pump vendors for assistance with details regarding materials of construction, installation, operation and maintenance, bearings, seals, valves, couplings, prime movers, and automatic controls.

3. Fitting Losses

From Eq. (63), the mechanical energy equation in head form, it is seen that, in the absence of a pump head, losses in a pipe system consist of pressure head changes, potential head changes, and velocity head changes. When fittings or changes in pipe geometry are encountered, additional losses occur.

It is customary to account for these losses either as pressure head changes over a length of pipe that produces the same frictional loss (hence an "equivalent length") or in terms of a velocity head equivalent to the actual fitting head loss. In the earlier literature the equivalent length method was popular, with various constant equivalent lengths being tabulated for fittings of various types. More recently, however, it has been realized that flows through fittings may also be flow-rate dependent so that a single equivalent length is not adequate.

In the velocity head method of accounting for fitting losses, a multiplicative coefficient is found empirically by which the velocity head term $\langle v \rangle^2/2g$ is multiplied to obtain the fitting loss. This term is then added to the regular velocity head losses in Eq. (63). Extensive tables and charts of both equivalent lengths and loss coefficients and formulas for the effect of flow rate on loss coefficients

are published by manufacturers of fittings and valves. They are much too extensive to be reproduced here.

C. Noncircular Ducts

The mathematical analysis of flow in ducts of noncircular cross section is vastly more complex in laminar flow than for circular pipes and is impossible for turbulent flow. As a result, relatively little theoretical base has been developed for the flow of fluids in noncircular ducts. In order to deal with such flows practically, empirical methods have been developed.

The conventional method is to utilize the pipe flow relations with pipe diameter replaced by the hydraulic diameter,

$$D_H = 4A_c/P_w, \tag{162}$$

where A_c is the cross-sectional area of the noncircular flow channel and P_w is its wetted perimeter. For Newtonian flows this method produces approximately correct turbulent flow friction factors (although substantial systematic errors may result). It has not been tested for non-Newtonian turbulent flows. It can easily be shown theoretically to be invalid for laminar flow. However, for purposes of engineering estimating of turbulent flow one can obtain rough "ballpark" figures.

D. Drag Coefficients

When fluid flows around the outside of an object, an additional loss occurs separately from the frictional energy loss. This loss, called form drag, arises from Bernoulli's effect pressure changes across the finite body and would occur even in the absence of viscosity. In the simple case of very slow or "creeping" flow around a sphere, it is possible to compute this form drag force theoretically. In all other cases of practical interest, however, this is essentially impossible because of the difficulty of the differential equations involved.

In practice, a loss coefficient, called a drag coefficient, is defined by the relation

$$F_D/A_c = C_D \rho v_\infty^2/2, \tag{163}$$

which is exactly analogous to the definition of f', the Fanning friction factor. In this equation F_D is the total drag force acting on the body, A_c is the "projected" cross-sectional area of the body (a sphere projects as a circle, etc.) normal to the flow direction, ρ is the fluid density, v_∞ is the fluid velocity far removed from the body in the undisturbed fluid, and C_D is the drag coefficient.

In the case of Newtonian fluids, C_D is found to be a function of the particle Reynolds number,

$$Re_p = d_p v_\infty \rho/\mu, \tag{164}$$

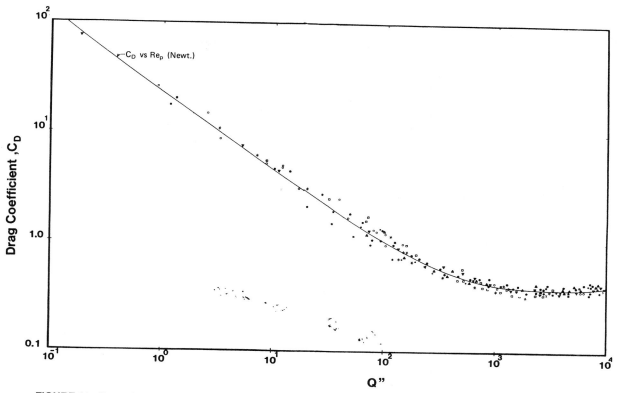

FIGURE 11 Generalized correlation of drag coefficient for Herschel–Bulkley model fluids; Q'' is defined by Eq. (165) and reduces to appropriate parameters for Bingham plastic, power law, and Newtonian fluid limits.

where d_p is the "effective" spherical diameter of the particle, v_∞ and ρ are as defined above, and μ is the viscosity of the fluid. The effective spherical diameter is the diameter of a sphere of equal volume. Also of importance are "shape" factors, which empirically account for the nonsphericity of real particles and for the much more complex flow distributions they engender.

Figure 11 is a plot of C_D as a function of a generalized parameter Q'', defined by

$$Q'' = \frac{\text{Re}_{pHB}^2}{\text{Re}_{pHB} + (7\pi/24)\text{He}_{pHB}}, \qquad (165)$$

where Re_{pHB} and He_{pHB} are the Reynolds number and Hedstrom number, respectively, for the Herschel–Bulkley rheological model defined as in the pipe flow case with D replaced by d_P.

This parameter is defined to accommodate Herschel–Bulkley model fluids. In the limit $\tau_0 = 0$, it reduces to an equivalent power law particle Reynolds number. In the limit $n = 1$, it reduces to a compound parameter involving the Bingham plastic particle Reynolds number and particle Hedstrom number. In both limits it reduces to the Newtonian particle Reynolds number. This correlation permits one to determine drag coefficients for spheres in a wide variety of non-Newtonian fluids.

The curve in Fig. 11 has been represented by the following set of empirical equations to facilitate computerization of the iterative process of determining C_D,

$$C_D = 24/Q'', \qquad Q'' \le 1 \qquad (166)$$

$$C_D = \exp[q(\ln Q'')], \qquad (167)$$

where the function $q(x)$ with $x = \ln(Q'')$ has the form

$$q(x) = 3.178 - 0.7456x - 0.04684x^2$$
$$+ 0.05455x^3 - 0.01796x^4$$
$$+ 2.4619(10^{-3})x^{5x} - 1.1418(10^{-4})x^6. \quad (168)$$

For $Q'' > 1000$, $C_D = 0.43$ is used. In the Newtonian limit, Eq. (166) is Stokes' law.

SEE ALSO THE FOLLOWING ARTICLES

FLUID DYNAMICS • FLUID MIXING • LIQUIDS, STRUCTURE AND DYNAMICS • REACTORS IN PROCESS ENGINEERING • RHEOLOGY OF POLYMERIC LIQUIDS

BIBLIOGRAPHY

Alexandrou, A. N. (2001). "Fundamentals of Fluid Dynamics," Prentice Hall, Englewood Cliffs, NJ.

Batchelor, G. K. (2000). "An Introduction to Fluid Dynamics," Cambridge Univ. Press, Cambridge, U.K.

Darby, R. (1996). "Chemical Engineering Fluid Mechanics," Dekker, New York.

Dixon, S. L. (1998). "Fluid Mechanics and Thermodynamics of Turbomachinery," Butterworth-Heinemann, Stoneham, MA.

Fuhs, A. E., ed. (1996). "Handbook of Fluid Dynamics and Fluid Machinery," 99E, Vols. 1–3, Wiley, New York.

Garg, V. K. (1998). "Applied Computational Fluid Dynamics," Dekker, New York.

Kleinstreuer, C. (1997). "Engineering Fluid Dynamics: An Interdisciplinary Systems Approach," Cambridge Univ. Press, Cambridge, U.K.

Lin, C. A., Ecer, A., and Periaux, J., eds. (1999). "Parallel Computational Fluid Dynamics '98: Development and Applications of Parallel Technology," North-Holland, Amsterdam.

Mc Ketta, J. J. (1992). "Piping Design Handbook," Dekker, New York.

Middleman, S. (1997). "An Introduction to Fluid Dynamics: Principles of Analysis and Design," Wiley, New York.

Sabersky, R. H., and Acosta, A. J. H. (1998). "Fluid Flow: A First Course in Fluid Mechanics," 4th ed., Prentice Hall, Englewood Cliffs, NJ.

Siginer, D. A., De, D., and Kee, R. (1999). "Advances in the Flow and Rheology of Non-Newtonian Fluids," Elsevier, Amsterdam/New York.

Sirignano, W. A. (1999). "Fluid Dynamics and Transport of Droplets and Sprays," Cambridge Univ. Press, Cambridge, U.K.

Smits, A. J. (1999). "A Physical Introduction to Fluid Mechanics," Wiley, New York.

Srivastava, R. C., and Leutloff, D. (1995). "Computational Fluid Dynamics: Selected Topics," Springer-Verlag, Berlin/New York.

Upp, E. L. (1993). "Fluid Flow Measurements: Practical Guide to Accurate Flow Measurement," Gulf Pub., Houston.

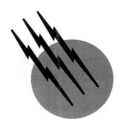

Fluid Inclusions

Edwin Roedder
Harvard University

GLOSSARY

Daughter phase Any new phase, solid, liquid, or gas, formed within an inclusion as a result of the change in pressure and temperature (P–T) conditions from that of trapping to that of observation.

Decrepitation method Procedure for measuring the temperatures at which fluid inclusions explode because of an increase in internal pressure.

Homogenization method Procedure for determining the temperature of trapping of an inclusion, wherein a sample on a special microscope stage is heated until all daughter phases homogenize to a single fluid phase.

Isochore Line in a P–T diagram representing the locus of all points for a fluid of a given composition and density.

Primary inclusion One that is trapped during the growth of the host crystal.

Secondary inclusion One that is trapped by the healing of fractures in the host crystal at some unspecified time after its growth.

FLUID INCLUSIONS are small volumes of fluid trapped within a crystal during its original growth from that fluid, or during subsequent healing of fractures in the presence of fluid. Changes in pressure and temperature from those of trapping to those of observation usually result in the formation of new daughter phases in the inclusion. Thus, water-rich fluid inclusion commonly split to form liquid and vapor phases, and sometimes solid crystals as well. If the original fluid that was trapped was a silicate melt at high temperatures, as in volcanic lavas, the fluid may now be a glass.

The study of fluid inclusions provides a wide range of data, some of which are unavailable from any other source, on the various physical and chemical processes that have yielded the minerals we see in terrestrial, lunar, and meteoritic samples. As a result, they have been of interest and value in prospecting, planetary geology, gemology, volcanology, petroleum exploration, geothermal power, nuclear waste disposal, and even in the siting of atomic power plants.

I. OCCURRENCE AND GENERAL SIGNIFICANCE

Fluid inclusions occur in most natural crystals, because most rocks and minerals, including ores and lunar and meteoritic samples, formed from a fluid or had a fluid present during later fracturing and healing. Inclusions are generally small. Very few are visible to the unaided eye, and usually the abundance in a given sample increases with decrease in size. In most samples, inclusions <10 μm in size are one or two orders of magnitude more abundant than those >10 μm. In many samples, they are exceedingly numerous. Ordinary white minerals such as common quartz or calcite are white because of the refraction of incident light at the many mineral–fluid inclusion interfaces; it is not uncommon to find as many as 10^9 cm^{-3}. Those crystals of hard minerals that are relatively free of inclusions and other defects and have a pleasing color are called gemstones, and the fewer the inclusions, the more valuable the stone.

A geologist is, in effect, a detective, trying to piece together the events of the distant past through a careful study of the many clues present in the rocks as they are found. Fluid inclusions, although small, provide a remarkable array of such clues, in that they can be made to yield quantitative data on the temperature, pressure, density, and composition of the fluids that existed in the past. These data have proven valuable in many aspects of geology and geochemistry, including the search for ore deposits, and some provide striking visual observations of the various phenomena at the critical point.

Crystals of many substances are grown commercially for use in a wide variety of optical and electrical devices. These applications normally require relatively perfect crystals, as free of defects as possible, and hence the growth conditions must be carefully maintained so that fluid inclusions are *not* formed.

II. MECHANISMS OF TRAPPING

With a few very minute exceptions, all crystals are imperfect, and fluid inclusions constitute one kind of imperfection. A cubic water inclusion 10 μm in size contains $\sim10^{-9}$ g of water, or $\sim10^{13}$ molecules. It is probable that something resembling a continuum exists between this size imperfection and single water molecules trapped within a crystal, but the bulk of fluid inclusion studies today are of inclusions in the range of 10–100 μm. Much work has been done on the mechanisms of trapping of such imperfections, and although some mechanisms have been well documented in the laboratory, in theory, or in nature, it is probably true that for many natural inclusions, the exact mechanism of trapping is simply not known. A sudden jump in the growth rate of an otherwise "perfect" crystal can yield a feathery or dendritic zone, which can subsequently be covered by solid growth. Hollow or tubular growth on certain crystal faces can similarly be covered.

Subparallel growth may trap some of the growth fluid between portions of the crystal. Growth spirals may develop central cavities, or cavities may form between adjacent growth spirals. Another common mechanism involves the enclosure of a grain of another solid phase within the growing crystal; often the enclosure traps some fluid along with the solid.

All the above mechanisms pertain to the trapping of primary inclusions, during the growth of the host crystal. However, in most samples, primary inclusions are much less abundant than secondary ones, formed from the healing of a fracture, as shown in Fig. 1. In fact, many samples contain millions of obvious secondary inclusions and not a single recognizable and verifiable primary inclusion. Secondary inclusions represent samples of fluids present at some time after the growth of the host crystal. They may form minutes, years, or millions of years later, and it is not rare to find that there have been several separate periods of fracturing and rehealing, under changing P–T conditions, from fluids of differing composition. A special type of inclusion, called pseudosecondary, forms by the same process of crack healing, but *during* the growth of the host crystal. Distinction among the inclusions from various origins and generations in any given sample may require careful and extensive petrography. Some planes of secondary inclusions are plainly evident in most samples, but unambiguous identification of a primary origin, although crucial to many uses of fluid inclusions, is often difficult and sometimes impossible.

III. COMPOSITION OF THE FLUID TRAPPED

This aspect of fluid inclusion study is the most important of all, as fluid inclusions provide us with the only actual *samples* we have of the fluids responsible for, or at least intimately involved in, many geochemical processes in the past. Other than silicate melts, fluid inclusions most commonly consist of a dilute to concentrated water solution of various solutes, with Na and Cl generally predominant, but with variable amounts of K, Ca, Mg, SO_4, HCO_3, CO_2, etc. Others consist mainly of liquid CO_2 under pressure, with various amounts of N_2, CH_4, and other gases and hydrocarbons. Ever since an 1858 seminal paper by H. C. Sorby, often called the father of fluid inclusions, a wide range of methods have been applied to determine the composition of the fluids. The small size of most inclusions, the extremely wide range of possible composition (at least 40 elements have been found in significant quantities), the occurrence in many different host minerals, and the difficulties of avoiding serious loss or contamination during extraction for analysis present a formidable quantitative

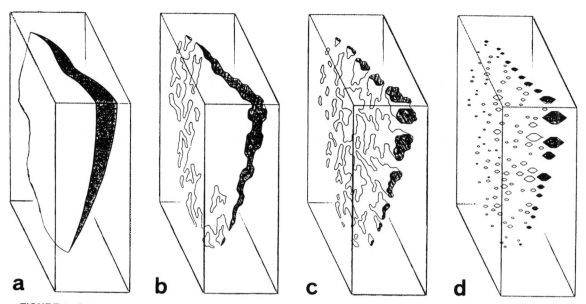

FIGURE 1 Stages in the healing or "necking down" of a crack in a quartz crystal, resulting in secondary inclusions. The solution of the curved surfaces having nonrational indices, and redeposition as dendritic crystal growth on other surfaces, eventually results in the formation of sharply faceted negative crystal inclusions.

analytical problem, for which there is no simple panacea. As a result of these difficulties, semiquantitative and even qualitative methods are still used widely. One of the most important reasons for this is that these methods, using the microscope, are not only quick and nondestructive, but can be easily used on *single,* small inclusions of recognized origin that are far too small for most other analytical procedures. They also show that inclusions from different growth zones of a single zoned crystal may vary widely in composition. For such reasons, most studies of the composition of fluid inclusions begin by the application of a series of nondestructive optical tests, to be followed by quantitative and generally destructive methods.

A. Nondestructive Methods

Any property of the fluid in an inclusion that can be measured under the microscope may provide at least some compositional evidence. Thus, these features of a liquid phase may be determined and interpreted in terms of probable composition: viscosity, color, wetting characteristics (relative to other daughter phases and the host phase), infrared and ultraviolet absorption, Raman spectra, fluorescence, index of refraction, thermal expansion, temperature of homogenization, critical phenomena, phase changes on heating or freezing, bubble movement in thermal gradients, and many others. Unopened inclusions can be bombarded with electrons, protons, X-rays, or other radiation, and the spectra from their constituent elements or polynu-

clear species such as CO_2 or SO_4^{-2} can then be detected. Many of these same features can be used in the identification of solid phases in inclusions, but here one has many more possible definitive properties that can be measured or estimated, particularly crystal habit, interfacial angles, extinction angles, elongation, birefringence, and indices of refraction and their relation to crystal orientation. Some data can even be obtained on opaque daughter phases, such as reflected light color, crystal shape, magnetic properties, and the Curie point. Perhaps the two most commonly used properties are the depression of the freezing point of water-rich fluids and the homogenization temperature and triple point of carbon dioxide. The first provides a measure of the solute content, and the second provides a measure of the density and purity of the CO_2 phase.

B. Destructive Methods

Many of the earlier chemical analyses of fluid inclusions are bulk analyses of the inclusions in large samples (100 or even 1000 g). Very few samples of this size can be found in which only one generation of inclusions is present; most will have a mixture of primary and perhaps more than one generation of secondary inclusions, and hence, the analysis of such a mixture is close to meaningless. Those very specially selected bulk samples that do contain essentially only one generation of inclusions present relatively few problems to the chemist in terms of needed sensitivity, as the total mass of inclusion fluid available is ample, but the most serious problems lie in the step involving the extraction of the fluid from the host mineral without serious loss and/or contamination. Generally, the extraction is done by fine grinding to open the inclusions, followed by leaching with one or more types of solutions and filtration. Major loss of cations to absorption on the host mineral surfaces can be minimized by the addition of a high-valence element that is not expected in the analysis, but the contamination from various sources can be a serious limitation on the usefulness of the results. The leach solutions can be analyzed by inductively coupled plasma, ion chromatograph, ion-specific electrodes, or other procedures.

Such analyses provide the ratios of the various solutes, but not their concentrations in the original fluids. To obtain concentrations, it is necessary to determine the amount of water (and CO_2). For this, somewhat smaller samples may be crushed in a vacuum system that permits the analysis of the amounts of water and other volatile materials, as well as the determination of the isotopic ratios of the oxygen and hydrogen in the water. Following this crushing, the broken mineral grains are leached with water or acids, and the concentrations of released solutes in the original fluids are calculated.

In view of the multiple origins for the inclusions in many samples, much more effort has been expended in recent years on procedures for the quantitative analysis of *individual* inclusions. Although many of the modern analytical procedures are adequately sensitive for samples in the range of 10^{-9} g, the bulk of this 10^{-9} g is generally water; the most valuable information pertains to the solutes in this water. Thus, the total mass of a constituent present in the range of 1000 ppm in the fluid would be only 10^{-12} g, which taxes the limits of sensitivity of some analytical methods. An additional problem is that of obtaining truly duplicate samples for check analyses or for the determinations of other constituents. In spite of these difficulties, a variety of procedures have been developed for quantitative multiple element analysis of individual inclusions. A major development in this direction has been in the methods for quantitative analysis of the gases in individual inclusions, by gas chromatography and mass spectrometry, after release by crushing, or by thermal or laser decrepitation.

IV. APPLICATIONS TO GEOLOGICAL AND GEOCHEMICAL PROBLEMS

A. Temperature Determinations

By far the most common use of fluid inclusions is to obtain a measure of the temperature of past events. H. C. Sorby showed in 1858 that the bubble present in most inclusions represents the differential shrinkage of the host crystal and the fluid that was trapped in it, from the temperature of trapping to that of observation. By reversing this shrinkage on a heating stage on the microscope until homogenization occurs, the minimum temperature of trapping can be determined (Fig. 2). The actual temperature of trapping must be at this temperature or higher, because any and all inclusions trapped along that particular isochore will have the identical homogenization temperature. The dissolution of solid daughter phases on heating also yields a minimum temperature. The true temperature of trapping can only be obtained if some independent estimate of the pressure is available. Several conditions are necessary for this method to be valid, and there are numerous problems in both the measurement itself and in the interpretation of the results, but still, the homogenization method is one of the most accurate and widely applicable geothermometers available.

Temperature determinations on past geologic events are of far more than mere academic value. Many mining companies make use of them in understanding the processes that went on to form known deposits and in the exploration for new deposits; oil companies look at the

FIGURE 2 Serial photomicrographs of a large primary inclusion in sphalerite (S) from Creede, Colorado, after equilibration at the temperatures indicated. The horizontal bars are oscillatory striae on the cavity walls. On heating, the bubble (v) decreased in volume to the homogenization temperature; this inclusion is said to "homogenize in the liquid phase at 210°C."

temperatures of inclusions in core samples from wells to determine, among other things, whether the rocks have been hot enough to generate oil, or so hot that any oil would have been destroyed; geothermal energy companies use them to guide drilling and to determine the potential for a field; gemologists use them to validate the place of origin of gemstones, to distinguish between manmade and natural stones, and to detect laboratory heat treatments made to improve the color of an otherwise less valuable natural stone.

B. Pressure Determinations

Under some conditions, fluid inclusions may provide an estimate of the pressure at the time of trapping, or at least some limits on the possible range of pressure. The various caveats are beyond the scope of this article, and the accuracy of the pressure values obtained differ widely with the specific method used, but they have provided important data on processes that cannot be observed directly. Volcanoes erupt some solid materials along with the molten silicate melt or lava. Pressure determinations on the inclusions in these solids have provided insight into the "plumbing system" of the volcano, and may help eventually in predicting volcanic hazards. Estimates of the pressure and hence depth at which some ore veins have formed have aided in the search for similar veins in otherwise unknown new areas.

Fluid inclusions provide one of the many examples of "pure" scientific research, pursued for what might have been considered purely academic reasons, that turn out to have social value. The siting of nuclear reactors involves careful evaluation of the hazards from various possible geologic processes that might occur after the reactor is in operation. One such process is movement on faults in or near the site. Movement on a fault actually cutting the site could cause major disruption, and seismic action from movement on a fault near the site could also cause damage. Not uncommonly, excavations for major construction projects uncover evidence of faults. When such a fault is found, the geologist is immediately asked to predict the possibility of renewed movement on the fault. This question is not easy to answer. The probability of such renewed movement is assumed to be inversely related to the length of time since the last movement; if fault movement has not occurred for a very long time, the fault can be considered inactive. However, an old inactive fault may look very similar to a recently active one. Fortunately, fluid inclusion geobarometry can sometimes help to place some constraints on the time since the last movements.

In 1974 at the Ginna Project in New York State, fluid inclusions were found in euhedral crystals protruding into vuggy cavities along a fault that cut the proposed site. The occurrence of the crystals suggested that they must have grown since the last movement on the fault. An elevated pressure at trapping, shown by thermometric data on the inclusions in the crystals, established a minimum depth below the surface at the time of formation. (The inclusions can be either primary or secondary for this procedure, and in fact, the latest inclusions, the secondaries, are actually the most useful.) Subsequent erosion had removed this much material from the site, so by estimating the rates of denudation, a minimum time since the last movement was obtained. Similar procedures have since been used at several other reactor sites.

C. Density Determinations

If the compositions (and hence density) of the individual daughter phases in an inclusion can be determined, and the relative volumes of the various phases determined by optical measurement (or other procedures), the bulk density of the fluid that was trapped can be calculated. These densities range, for water-rich inclusions, from very low densities of <0.1 g cm^{-3} (from the trapping of a dense steam phase) to densities >1.3 g cm^{-3} (from the trapping of a strongly saline fluid or hydrosaline melt). Inclusions in many ore deposits have densities near 1.0 g cm^{-3}. Most ores have formed as a result of a hot ore-bearing fluid moving through the earth. One of the problems in understanding the origin of ore deposits (and in the search for others) is in understanding the hydrologic system—what fluid moved which way? Since most rocks near the surface of the earth are saturated with cold, essentially fresh water of density 1.0 g cm^{-3}, it is important to know the density of the ore fluid at the time of ore deposition, and hence which direction convection might have taken. Fluid inclusions provide the only known source of data on the densities of these ore fluids.

D. Compositional Determinations

Two examples from the extensive literature on fluid inclusions will have to suffice to illustrate the application of compositional data to geological and geochemical problems. First, in the study and exploration of a given type of ore deposit, the geologist commonly establishes a model of the process, built on the basis of available data. Ore deposits do not occur randomly; they are the result of transport of the ore metal or metals in a fluid from a dilute source to a site where they are deposited in a more concentrated form, that is, as an ore. The chemical and physical nature of the fluid that transports and deposits the ore is the essential element in any such model. In recent years, there has been a major breakthrough in our understanding of the chemistry of ore metal transport and deposition.

Fluid inclusion compositions have helped to refine these models, as it has become obvious that the chemistry of the ore fluid has been very different in various types of ore deposits. Thus, one well-known type of very large copper deposit, the prophyry copper, has involved fluids that, based on inclusion study, were almost always exceedingly saline (commonly as much as 50% salts in solution), very hot (as much as 700°C), and generally two-phase at the time of trapping (i.e., boiling or condensing). Other ore deposits formed from very different fluids. One class of gold deposit formed from very low salinity aqueous fluids, containing much CO_2, at moderate temperature and low pressure. Such model characteristics are sufficiently typical that they are useful in exploration for deposits. During surface weathering of a prophyry copper deposit, all copper may be leached out, but fluid inclusions in residual quartz in the soil or stream sediments may still preserve the recognizable characteristics. Unfortunately, no such set of characteristics is completely limited to a given class of ore deposits, but they may still be used as possible indicators.

Another example of the use of fluid inclusion compositions is found in the studies that have been made in the search for possible sites to store nuclear waste. The basic question is, where in the earth can nuclear waste be stored for thousands of years so that it will not be a hazard to mankind? One type of geological environment that has been considered for nuclear waste storage is the massive salt deposit. Fluid inclusions in such salt play two roles, one beneficial and one detrimental. On the beneficial side, study of the fluid inclusion composition can provide some clues on what processes have taken place in that salt deposit in the millions of years since it formed. These data on the past are of major concern, since the geologist is being asked to predict what may happen to the waste site in future millennia. For example, has there ever been an incursion of fresh groundwater into these beds in the past (and hence might one occur again in the future)? Fluid inclusions trapped from such an incursion would have grossly different solute compositions and isotopic signatures for the water than those from the original evaporated seawater.

On the detrimental side, fluid inclusions in the salt complicate the engineering problems in designing a nuclear waste site. A nuclear waste canister buried in the ground becomes thermally hot from the intense radioactivity of the waste. Such heat, in flowing out into the surrounding salt, set up a thermal gradient. Fluid inclusions in salt in a gradient tend to dissolve the host salt on the hot side of the inclusion and redeposit it on the cold side, and hence, they migrate up the thermal gradient toward the waste canister. Yet salt beds were originally selected for consideration because they were considered to be an exceedingly *dry* en-

vironment. An additional problem comes from the cation composition of the fluids. Ordinary saturated NaCl–H_2O solutions can be expected to be corrosive to the canister materials, but the fluid inclusions in salt normally contain high concentrations of such divalent cations as Mg (as chloride). These solutions may show metal corrosion rates orders of magnitude higher than simple saturated NaCl brines. As a result, if a salt repository is to be used, it may be necessary to engineer the use of very special and expensive canister materials.

V. SPECIAL PROBLEMS

The small size, ambiguity as to origin, and extremely wide range of possible combinations of inclusion constituents and host materials make measurements difficult and the interpretation sometimes uncertain. In addition to these expectable problems, two others are particularly pertinent and of some scientific interest.

A. Leakage and Volume Change

The term leakage in fluid inclusion studies refers to movement of material either into or out of an inclusion. Many lines of evidence have been used to show that in most geological environments, fluid inclusions have not leaked, at least not enough to have measurable effects with present procedures. One special case, however, does seem to be best explained by significant leakage of hydrogen out of fluid inclusions. In many porphyry copper deposits, the fluid inclusions contain solid crystals of hematite (Fe_2O_3) or anhydrite ($CaSO_4$) or both. Although these might be considered to be daughter minerals, formed on cooling, they do *not* redissolve on heating the inclusions in the laboratory, as a true daughter mineral should. These may represent new phases formed irreversibly in the inclusion after trapping, as a result of leakage of hydrogen. Disproportionation of water at these high temperatures to form hydrogen and oxygen, followed by outward diffusion of the hydrogen, has apparently resulted in auto-oxidation of original ferrous iron ion in solution to precipitate hematite, and of sulfide ion to sulfate, to precipitate anhydrite. Several features of the prophyry copper environment might be expected to lead to the presence of hot but oxygenated surface waters in the ore body during the long cooling period, thus providing a hydrogen sink surrounding the crystal, with enough time at high temperatures to permit significant hydrogen diffusion to that sink.

One of the necessary assumptions implicit in any use of fluid inclusions for geothermometry and geobarometry is a constant cavity volume after trapping. Most geological environments investigated apparently fulfill this requirement within the limitations of modern measurements. However, during metamorphism of large masses of rock in the crust, individual crystals containing inclusions trapped by growth during an early stage may later be subjected to higher pressures and temperatures as the depth of burial increases. In addition, all inclusions trapped at the peak pressures and temperatures must, out of necessity, be subjected to along period of decreasing temperatures and pressures before reaching surface conditions. If the P–T path taken during this cooling follows the isochore for the trapped inclusion fluid, valid data can be obtained from the inclusions. If, however, the path deviates from this isochore, pressure differentials will be set up between the inclusion and the exterior of the host crystal. If the internal pressure is higher, the inclusion may permanently expand (i.e., stretch) and if it is lower, it may permanently collapse; recent studies show that geologically expectable differential pressures can result in such changes, even in quartz, in a few days at high temperatures. The validity of the extrapolation of these results down to the P–T paths encountered most commonly in metamorphic rocks (i.e., possible lower temperatures and pressure differentials, but for much longer times) is still an open question.

B. Metastability

Inclusions are very small systems to consider, even from an atomic viewpoint. Thus, an inclusion 10 μm in diameter containing a 30 wt% solution of a salt may have only 10^{11}–10^{12} "molecules" of that salt; at 10 ppm PbS, it would contain only 25 million "molecules" of PbS, enough to form a crystal only 200 unit cells on an edge. The smaller the system, the more common metastability becomes. Hence, not surprisingly, metastability—resulting from failure to nucleate new but stable phases—becomes a problem that can be observed in (and frequently interferes with) many types of inclusion studies. It is common in samples as found, even though they have had geologic time periods to equilibrate; and it is even more commonly observed in the phase changes during the much shorter time spans of laboratory experiments on inclusions.

Metastability that results from failure to freeze on cooling in the laboratory is almost universal, and can be frustrating, but it normally does not lead to misunderstanding or error. However, one aspect of metastability—the occurrence of metastable superheated ice in liquid-water inclusions at high negative pressures—can lead to serious errors. Although negative pressures are not encountered in many inclusion studies, they are very commonly observed in some, particularly in studies of low-salinity inclusions formed at low to moderate temperatures, in which the vapor phase is eliminated upon freezing and fails to renucleate on warming. In such inclusions, it is not rate to have as much as 1000 atm of negative pressure, resulting in perfectly normal-appearing (but metastable) ice-water equilibria to as high as +6°C, until the spontaneous nucleation of the stable vapor phase results in instantaneous melting of the superheated ice and a return to the stable phase assemblage of liquid + vapor.

SEE ALSO THE FOLLOWING ARTICLES

CRYSTAL GROWTH • CRYSTALLOGRAPHY • EVAPORITES • FLUID DYNAMICS • VOLCANOLOGY

BIBLIOGRAPHY

Gübelin, E. J., and Koivula, J. I. (1986). "Photo Atlas of Inclusions in Gemstones," ABC Edition, Zurich.

Hollister, L. S., and Crawford, M. L., eds. (1981). "Fluid Inclusions: Applications to Petrology," Mineralogical Association of Canada Short Course Handbook, Vol. 6.

Roedder, E. (1972). "Composition of Fluid Inclusions," U.S. Geological Survey Professional Paper 440JJ.

Roedder, E. (1984). "Fluid Inclusions" (P. H. Ribbe, series ed.), Vol. 12 of "Reviews in Mineralogy," Mineralogical Society of America, Washington, D.C.

Roedder, E. (1984). "The fluids in salt," Am. Mineral. 69, 413–439.

Roedder, E., and Bodnar, R. J. (1980). "Geologic pressure determinations from fluid inclusion studies," Annu. Rev. Earth Planet. Sci. 8, 263–301.

Fluid Mixing

J. Y. Oldshue

Mixing Equipment Company, Inc.

GLOSSARY

Axial flow impellers Impellers that pump the fluid primarily in an axial direction when installed in a baffled mixing tank.

Chemical or mass transfer criteria Criteria for fluid mixing evaluation that involves measuring the rate of chemical reactions or rates of mass transfer across liquid, gas, or solid interfaces.

Computational fluid mixing Computer programs that use velocity data to calculate various types of flow patterns and various types of fluid mechanics variables used in analyzing a mixing vessel.

Fluidfoil impellers Axial flow impellers in which the blade shape and profile is patterned after airfoil concepts. The blade normally has camber and has a twist in toward the shaft with a rounded leading edge to pro-

duce a uniform velocity across the entire face width of the axial flow impeller.

Fluid shear rate Velocity gradient at any point in the mixing tank.

Fluid shear stress Product of shear rate and viscosity, which is responsible for many mixing phenomena in the tank.

Macroscale mixing, macroscale shear rates Particles on the order of 500–1000 μm and larger, or fluid elements of this size, respond primarily to average velocities at any point in the tank and are characterized as macroscale shear rate sensitive or related to macroscale mixing. Visual inspection of a tank normally yields information on the macroscale mixing performance.

Microscale shear rates, microscale mixing Any particles or fluid elements on the order of 100 μm or less respond primarily to the fluctuating velocity components in turbulent flow or to shear rate elements on the order of that same size in viscous flow. Measurement of fluid mixing parameters at the microscale level involve the ability to resolve small elements of fluid parameters, as well as understanding the dissipation of energy at the microscale level.

Physical uniformity criteria Criteria for fluid mixing which involves physical sampling of tank contents or estimation of pumping of tank contents or estimation of pumping capacity and/or velocity values.

Radial flow impellers Impellers that pump fluid in essentially a radial direction when installed in a baffled mixing tank.

FLUID MIXING, as an engineering study, is the technology of blending fluid substances, including gases and solids, and is an integral process in most manufacturing operations involving fluid products. An important aspect of fluid mixing is the design and use of equipment. Fluids can be mixed in containers with rotating impellers or by means of jets, or in pipelines by internal baffles and passageways. Fluid mixing can involve primarily a physical suspension or dispersion that can be analyzed by the degree of composition or uniformity. Other operations may involve mass transfer across two-phase interfaces or chemical reactions in one or more phases. Information about microscale and macroscale mixing requirements are needed for process analysis and scaleup.

I. GENERAL PRINCIPLES

The power put into a fluid mixer produces pumping Q and a velocity head H. In fact all the power P which is proportional to QH appears as heat in the fluid and must be dissipated through the mechanism of viscous shear. The pumping capacity of the impeller has been measured for a wide variety of impellers. Correlations are available to predict, in a general way, the pumping capacity of the many impeller types in many types of configurations. The impeller pumping capacity is proportional to the impeller speed N and the cube of the impeller diameter D,

$$Q \propto ND^3$$

The power drawn by an impeller in low- and medium-viscosity fluids is proportional to the cube of impeller speed N and the impeller diameter D to the fifth power,

$$P \propto N^3D^5 \tag{1}$$

At higher viscosities other exponents are involved (discussed later).

If these three relations are combined it is seen that at constant power, one can vary the ratio of flow to impeller velocity head by a choice of D given by Eq. (2)

$$(Q/H)_P \propto D^{8/3} \tag{2}$$

This equation indicates that large-diameter impellers running at low speed give high flow and low shear rates, but small-diameter impellers running at high speed give us high shear rates and low pumping capacities. This important relationship also indicates that impeller velocity head is related in principle to macroscale shear rates. Thus, one has the ability to change the flow to fluid shear ratio.

In addition to the mathematical concepts brought out in Eq. (2), axial flow impellers, often applied as the pitched blade turbine (Fig. 1a), are inherently able to produce more flow at a given horsepower and impeller speed than radial flow impellers, typified by the flat blade disc turbine, shown in Fig. 1b. Some processes, such as blending and solids suspension, are affected primarily by pumping capacity and are not greatly influenced by the fluid shear rate. Therefore, it is typical in practice to use axial flow impellers when dealing with solids suspension and blending. Changes in D/T (where T is the tank diameter) can affect the flow-to-fluid-shear rate ratio relative to the various diameters:

$$(Q/H)_P \propto (D/T)^{8/3} \tag{3}$$

The introduction in recent years of the fluidfoil type of impeller, shown in Fig. 1c, further improves the pumping capacity of axial impellers and reduces the fluid shear rate by the actual design of the impeller blades themselves. Figure 2 illustrates the phenomena of the fluidfoil. The illustration indicates the desired flow pattern over the blade shape to minimize shear rate and maximize flow. For comparison, Fig. 2b shows fluid flow if the angle of the impeller blade in the fluid is not set at this optimum flow position. As shown in Fig. 2b, the turbulence and drag behind the impeller blade will cause increased power and reduced pumping efficiency. However, the turbulence and drag are not always a problem, because some processes require a certain level of turbulence and energy dissipation. In such processes, the use of the fluidfoil impeller type would not be as effective as other types that develop higher internal impeller zone shear rates.

There are now several varieties of fluidfoil impellers in use. The A310 is an effective impeller for the low viscosity region and has a negative response to viscosity at a Reynolds number of approximately 600. As shown in Fig. 3, the angle that the flow stream makes with the vertical starts to become greater than with the A200 impeller, so we can say effectively that the Reynolds number limitation on the A310 is approximately 200.

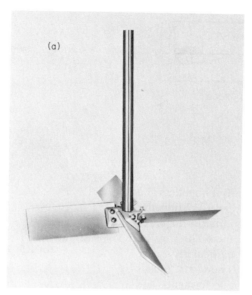

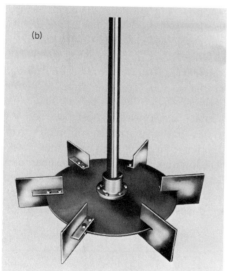

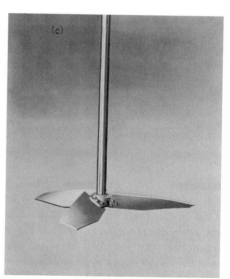

FIGURE 1 Three typical impellers for low and medium viscosity: (a) Axial flow, 45° blade (A200), (b) radial flow, disc turbine (R100), and (c) fluidfoil axial flow impeller (A300).

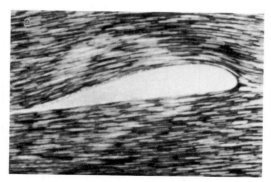

FIGURE 2 Typical air foil profiles. (a) Proper blade angle of attack for minimum drag and maximum flow for a given power. (b) Different blade angle of attack, giving higher drag coefficient and less flow per unit power.

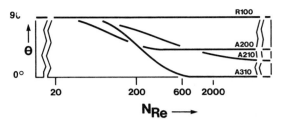

FIGURE 3 Changes in flow discharge angle with Reynold's number for four different impellers.

In order to carry this concept of fluidfoil impellers at a uniform velocity of discharge further, the A312 Impeller (Fig. 4) was developed and is used primarily in paper pulp suspensions. Carrying it further is the A320 Impeller (Fig. 5). The A320 has been studied particularly in the transitional area of traditional Reynolds numbers. This is shown in Fig. 6. This figure shows its performance and Reynolds numbers between 10 and 1,000.

For gas–liquid processes, the A315 impeller (Fig. 7) has been developed. This further increases the blade area and is used for gas–liquid applications.

The family of impellers shown here can be characterized by the solidity ratio, which is the ratio of area to blades to disc area circumscribing the impeller.

As shown in Fig. 8, the solidity ratio goes from 22% with the A310 up to 87% with the A315.

A. Shear Rate

There is a need to distinguish at this point how the shear rate in the impeller zone differs from the shear rate in the tank zone. To do this, however, one must carefully define shear rate and the corresponding concepts of macroscale shear rate and microscale shear rate. When one studies the localized fluid velocity through utilization of a small dimension probe, or as is currently used, a laser Doppler velocity meter device, one sees that at any point in the

FIGURE 5 Photograph of A320 fluidfoil impeller.

stream of the tank there is a fluctuating velocity if we have turbulent flow (Fig. 9). From the curve in Fig. 9, one can calculate the average velocity at any point, as well as the fluctuating velocity above and below the average at this point. Figure 10 is a plot of the average velocity obtained from this curve. If these velocities are plotted at a constant discharge plane from the impeller, then the average impeller zone shear rate can be calculated. This average rate is really a macroscale shear rate, and it only refers to particles that have sizes much greater than $1000 \ \mu m$ that experience an effect from these shear rates. Also note that there is a maximum macroscale shear rate around the impeller. There are a variety of shear rates around the impeller, so that one needs to recognize the effect of each on a given process.

FIGURE 4 Photograph of A312 fluidfoil impeller.

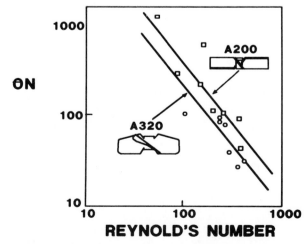

FIGURE 6 Effect of Reynolds number on blend number, θN, for the two impellers shown. θ, blend time; N, impeller rotational speed.

FIGURE 7 Photograph of A315 fluidfoil impeller.

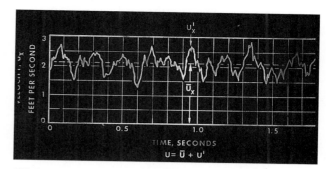

FIGURE 9 Schematic representation of turbulent flow recorded from a velocity probe as a function of time, showing average velocity and fluctuating velocity.

In addition, the turbulent fluctuations set up a microscale type of shear rate. Microscale mixing tends to affect particles that are less than 100 μm in size. The scaleup rules are quite different for macroscale controlled process in comparison to microscale. For example, in microscale processes, the major variables are the power per unit volume dissipated in various points in the vessel and the total average power per unit volume. In macroscale mixing, the energy level is important, as well as the geometry and design of the impeller blades and the way that they set up macroscale shear rates in the tank.

The fluidfoil impeller, shown in Fig. 1c, is often designed to have about the same total pumping capacity as the axial flow turbine (Fig. 1a). However, the flow patterns are somewhat different. The fluidfoil impeller has an axial discharge, while the axial flow turbine discharge tends to deviate from axial flow by 20–45°. Nevertheless at the same total pumping capacity in the tank, the tank shear rates are approximately equal. However, the axial flow fluidfoil turbine requires between 50 and 75% of the power required by the axial flow turbine. This results in a

much smaller energy loss and dissipation in the impeller zone, and much lower microscale mixing in the impeller zone. There is also some difference in microscale mixing in the rest of the tank.

The lower horsepower is an important factor in the efficient design of axial flow or fluidfoil impellers. Such lower horsepower must be considered in the efficient design involving fluid velocity and overall macroscale mixing phenomena. On the other hand, if the process involves a certain amount of microscale mixing, or certain amounts of shear rate, then the fluidfoil impeller may not be the best choice.

Radial flow impellers have a much lower pumping capacity and a much higher macroscale shear rate. Therefore they consume more horsepower for blending or solids suspension requirements. However, when used for mass transfer types of processes, the additional interfacial area produced by these impellers becomes a very important factor in the performance of the overall process. Radial flow turbines are primarily used in gas–liquid, liquid–solid, or liquid–liquid mass transfer systems or any combinations of those.

B. Baffles and Impeller Position

Unbaffled tanks have a tendency to produce a vortex and swirl in the liquid. Such conditions may be wanted. Frequently, however, a good top-to-bottom turnover and the elimination of vortexing is needed. Therefore, baffles

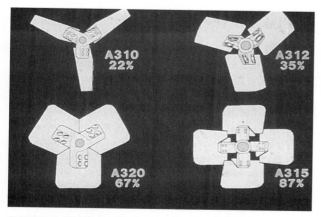

FIGURE 8 Solidity ratio of total blade area ratio to disc area of circumscribed circle at blade tips expressed as a percentage.

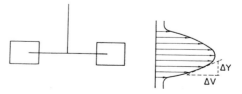

FIGURE 10 Illustration of average velocity from the radial discharge of a radial flow impeller, showing the definition of fluid shear rate ($\Delta V/\Delta Y$).

84

are used more often than not. Wall baffles for low-viscosity systems consist of four baffles, each $\frac{1}{12}$ of the tank diameter in width. Another method is to install an axial flow impeller type in an angualar, off-center position, such that it gives good top-to-bottom turnover, avoids vortexing, and also avoids the use of baffles. Figure 11a shows a typical flow pattern for an unbaffled tank. A baffled tank axial radial flow is shown in Fig. 11b, and the angular off-center position is in Fig. 11c.

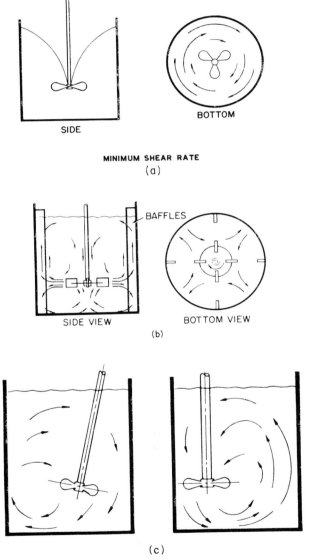

FIGURE 11 Effect of baffles in position on flow pattern. (a) Typical swirling and vortexing flow in a tank without baffles. (b) Typical top-to-bottom flow pattern with radial flow impellers with four wall baffles. (c) Typical angular off-center position for axial flow impellers to give top-to-bottom flow pattern to avoid swirl without the use of wall baffles.

TABLE I Elements of Mixer Design

Process design
  Fluid mechanics of impellers
  Fluid regime required by process
  Scaleup; hydraulic similarity
Impeller power characteristics
  Relate impeller hp, speed, and diameter
Mechanical design
  Impellers
  Shafts
  Drive assembly

The need to use wall baffles to eliminate vortexing decreases as fluids become more viscous (5000–10,000 cP or more). But swirl will still be present if there are no baffles. Accordingly, quite often baffles of about one-half normal width are used in viscous materials. In such cases they are placed about halfway between the impeller and the wall.

C. Power Consumption

Table I shows the three areas of consideration in mixer design. The first area is process design, which will be covered in detail in succeeding pages. Process design entails determining the power and diameter of the impeller to achieve a satisfactory result. The speed is then calculated by referring to the Reynolds number–power number curve, shown in Fig. 12. Such a curve allows trial-and-error calculations of the speed once the fluid properties, P, D, and the impeller design are known.

D. Process Considerations

Table II gives a representation of the various types of mixing processes. The second column lists the nine basic areas of mixing: gas-liquid, liquid-solid, liquid-liquid, miscible liquid, fluid motion, and combinations of those. However, of more importance are the two adjacent columns. The first column includes physical processing, and has mixing criteria which indicate a certain degree of uniformity. The third column has chemical and mass transfer requirements, which involve the concept of turbulence, mass transfer, chemical reactions, and microscale mixing. Thus, there are summarized ten separate mixing technologies, each having its own application principles, scaleup rules, and general effect of process design considerations. In a complex process such as polymerization, there may possibly exist solids suspension, liquid–liquid dispersion, chemical reaction, blending, heat transfer, and other important steps. In general, it is more advantageous to break the process down into the component steps and consider the effect

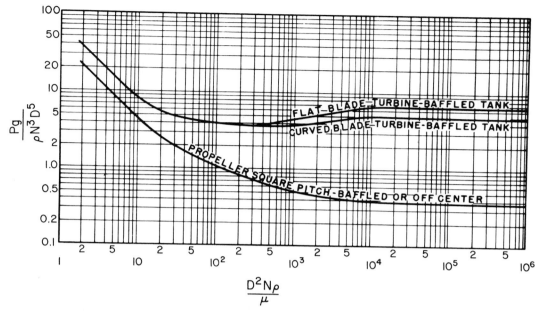

FIGURE 12 Reynolds number–power number curve for several impeller types: D, impeller diameter; N, impeller rotational speed; ρ, liquid density; μ, liquid viscosity; P, power; and g, gravity constant.

of mixing on each of these steps. One can then determine how the process will be affected by making changes in the mixer variables to the various mixing steps in the process. In scaleup, this is normally done by first determining the relative importance of the various steps, such as chemical reaction, mass transfer, blending, and so forth. The next step is to scaleup each of these steps separately to see the change on full-scale mixing. Later sections on scaleup and pilot planting will give some ideas on how scaleup affects typical performance variables.

Generally, heat transfer, blending, and solids suspension are governed primarily by the impeller's pumping capacity and not by fluid shear rates. Solid–liquid mass transfer, liquid–liquid mass transfer, and gas–liquid mass transfer have certain requirements for fluid shear in addi-

tion to pumping capacity: There are optimum ratios for those kinds of processes. There are many different combinations of impeller type and D/T ratios that can be used to get an optimum combination once the optimum flow to fluid shear is achieved. Thus, impeller design is not critical in terms of process performance but is critical in terms of economics of the overall mixer.

It is possible to use mixers as low head pumps by suitably installing them in a draft tube or above the orifice. They can then be used to pump large volumes of liquid at low heads.

The fluid mixing process involves three different areas of viscosity which affect flow patterns and scaleup, and two different scales within the fluid itself, macroscale and microscale. Design questions come up when looking at the design and performance of mixing processes in a given volume. Considerations must be given to proper impeller and tank geometry as well as the proper speed and power for the impeller. Similar considerations come up when it is desired to scaleup or scaledown and this involves another set of mixing considerations.

If the fluid discharge from an impeller is measured with a device that has a high frequency response, one can track the velocity of the fluid as a function of time (Fig. 9). The velocity at a given point in time can then be expressed as an average velocity ($\bar{v}$) plus fluctuating component (v'). Average velocities can be integrated across the discharge of the impeller and the pumping capacity normal to an arbitrary discharge plane can be calculated. This arbitrary discharge plane is often defined as the plane bounded by

TABLE II   Characterization of Various Types of Mixing Processes

Physical processing	Application classes	Chemical process
Suspension	Liquid-Solid	Dissolving
Dispersion	Liquid-Gas	Absorption
Emulsions	Immiscible liquids	Extraction
Blending	Miscible liquids	Reactions
Pumping	Fluid motion	Heat transfers
	Liquid-solid-gas	
	Liquid-liquid-solid	
	Liquid-liquid-gas	
	Liquid-liquid-gas-solid	

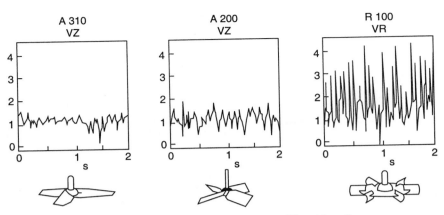

FIGURE 13 Velocity versus time for three different impellers.

the boundaries of the impeller blade diameter and height. Because there is no casing, however, an additional 10–20% of flow typically can be considered as the primary flow of an impeller.

The velocity gradients between the average velocities operate only on larger particles. Typically, these larger size particles are greater than 1000 μm. This is not a proven definition, but it does give a feel for the magnitudes involved. This defines macroscale mixing. In the turbulent region, these macroscale fluctuations can also arise from the finite number of impeller blades passing a finite number of impeller blades passing a finite number of baffles. These set up velocity fluctuations that can also operate on the macroscale.

Smaller particles primarily see only the fluctuating velocity component. When the particle size is much less than 100 μm, the turbulent properties of the fluid become important. This is the definition of the boundary size for microscale mixing.

All of the power applied by a mixer to a fluid through the impeller appears as heat. The conversion of power to heat is through viscous shear and is approximately 2500 Btu/hr/hp. Viscous shear is present in turbulent flow only at the microscale level. As a result, the power per unit volume is a major component of the phenomena of microscale mixing. At a 1-μm level, in fact, it doesn't matter what specific impeller design is used to apply the power.

Numerous experiments show that power per unit volume in the zone of the impeller (which is about 5% of the total tank volume) is about 100 times higher than the power per unit volume in the rest of the vessel. Making some reasonable assumptions about the fluid mechanics parameters, the root-mean-square (rms) velocity fluctuation in the zone of the impeller appears to be approximately 5–10 times higher than in the rest of the vessel. This conclusion has been verified by experimental measurements.

The ratio of the rms velocity fluctuation to the average velocity in the impeller zone is about 50% with many open impellers. If the rms velocity fluctuation is divided by the average velocity in the rest of the vessel, however, the ratio is on the order of 5–10%. This is also the level of rms velocity fluctuation to the mean velocity in pipeline flow. There are phenomena in microscale mixing that can occur in mixing tanks that do not occur in pipeline reactors. Whether this is good or bad depends upon the process requirements.

Figure 13 shows velocity versus time for three different impellers. The differences between the impellers are quite significant and can be important for mixing processes.

All three impellers are calculated for the same impeller flow, Q, and same diameter. The A310 (Fig. 2) draws the least power, and has the least velocity fluctuations. This gives the lowest microscale turbulence and shear rate.

1. The A200 (Fig. 3) shows increased velocity fluctuations and draws more power.
2. The R100 (Fig. 4) draws the most power and has the highest microscale shear rate.
3. The proper impeller should be used for each individual process requirement.

The velocity spectra in the axial direction for an axial flow impeller A200 is shown in Fig. 14. A decibel correlation has been used in Fig. 5 because of its well-known applicability in mathematical modeling as well as the practicality of putting many orders of magnitude of data on a reasonably sized chart. Other spectra of importance are the power spectra (the square of the velocity) and the Reynolds stress (the product of the R and Z velocity components), which is a measure of the momentum at a point.

The ultimate question is this: How do all of these phenomena apply to process design in mixing vessels? No one

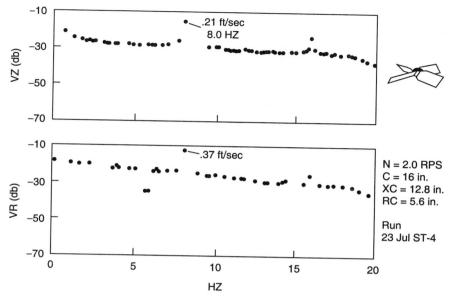

FIGURE 14 The velocity spectra in the axial direction for an axial impeller A200.

today is specifying mixers for industrial processes based on meeting criteria of this type. This is largely because processes are so complex that it is not possible to define the process requirements in terms of these fluid mechanics parameters. If the process results could be defined in terms of these parameters, sufficient information probably exists to permit the calculation of an approximate mixer design. It is important to continue studying fluid mechanics parameters in both mixing and pipeline reactors to establish what is required by different processes in fundamental terms.

Recently, one of the most practical results of these studies has been the ability to design pilot plant experiments (and, in many cases, plant-scale experiments) that can establish the sensitivity of process to macroscale mixing variables (as a function of power, pumping capacity, impeller diameter, impeller tip speeds, and macroscale shear rates) in contrast to microscale mixing variables (which are relative to power per unit volume, rms velocity fluctuations, and some estimation of the size of the microscale eddies).

Another useful and interesting concept is the size of the eddies, L, at which the power of an impeller is eventually dissipated. This concept utilizes the principles of isotropic turbulence developed by Komolgoroff [1]. The calculations assume some reasonable approach to the degree of isotropic turbulence, and the estimates do give some idea as to how far down in the microscale size the power per unit volume can effectively reach

$$L = (\nu^3/e)^{1/4}$$

where ν is the dynamic viscosity and e is the power per unit volume.

II. SCALEUP RELATIONSHIPS

Scaleup involves determining the controlling factors in a process, the role that mixing plays, and the application of a suitable scaleup technique. In this section, the general scaleup relationships will be presented, and the particular types of processes involved will be covered. Section X will cover pilot planting, how runs are made to determine the controlling factor, and how to choose a suitable design relationship for that situation.

Table III is a key for understanding scaleup relationships. In the first column are listed many design variables involved in mixing processes. These include power, power per unit volume, speed, impeller diameter, impeller

TABLE III Properties of a Fluid Mixer on Scaleup

Property	Pilot scale (80 Liters)	Plant scale (17.280 liters)			
P	1.0	216	7776	36	0.16
P/Vol	1.0	1.0	36	0.16	.0007
N	1.0	0.3	1.0	0.16	.03
D	1.0	6.0	6.0	6.0	6.0
Q	1.0	65	216	36	6.0
Q/Vol	1.0	0.3	1.0	0.16	.03
ND	1.0	1.8	6.0	1.0	0.16
$\dfrac{ND^2\rho}{\mu}$	1.0	10.8	36	5.8	1.0

pumping capacity, pumping capacity per unit volume, impeller tip speed, and Reynolds number. In the second column, all these values are given a common value of 1.0, to examine the changes relative to each other on scaleup. In the remaining columns, a specific variable is held constant. When power per unit volume is constant, the speed drops, the flow increases, but the flow per unit volume decreases. The impeller tip speed goes up, and the Reynolds number goes up. It is quite apparent that the ratio of all the variables cannot be maintained as in the pilot plant. In addition, it appears that the maximum impeller zone shear rate will increase, while the circulating time and the impeller Reynolds number increase. This means that the big tank will be much different from the small tank in several potentially key parameters. When the flow per unit volume is held constant, the power per unit volume increases in proportion to the square of the tank diameter ratio. This is possible to do but is normally impractical.

When the impeller tip speed is held constant, the same maximum shear rate is maintained. However, the average impeller shear rate related to impeller speed drops dramatically, and the power per unit volume drops inversely to the tank size ratio. In general, this is a very unconservative scaleup technique and can lead to insufficient process results on full scale.

The final column shows results for a constant Reynolds number, which requires that the total power decrease on scaleup. This is not normally practical, and therefore we must accept an increased Reynolds number on scaleup. To complete this picture refer to Fig. 15, which shows that the maximum impeller zone shear rate increases, while the average impeller zone shear rate decreases during scaleup.

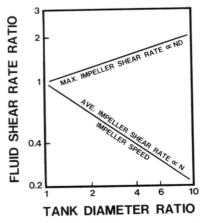

FIGURE 15 Schematic illustration of the increase in maximum impeller zone macroscale shear rate and a decrease of average impeller zone macroscale shear rate as tank size is increased, illustrating a wider distribution of shear rates in a large tank than in a small tank. The figure is based on a constant power/volume ratio and geometric similarity between the two tanks.

Both Table III and Fig. 15 are based on geometric similarity. One way to modify these marked changes in mixing parameters and scaleup is to use nongeometric similarity on scaleup. The problem is that the big tank has a much longer blend time than the small tank. The large tank has a greater variety of shear rates and has a higher Reynolds number than a small tank. These effects can be greatly modified by using nongeometric impellers in the pilot plant.

A. Role of Dynamic and Geometric Similarity

Equations (4) and (5) show the relationship for geometric and dynamic similarity, respectively, and illustrates four basic fluid force ratios.

$$\frac{X_M}{X_P} = X_R \tag{3}$$

$$\frac{(F_I)_M}{(F_I)_P} = \frac{(F_\mu)_M}{(F_\mu)_P} = \frac{(F_g)_M}{(F_g)_P} = \frac{(F_\sigma)_M}{(F_\sigma)_P} = F_R \tag{4}$$

where F is the fluid force, I the inertia force, μ, the viscous force, g the gravitational force, σ the surface tension force, M the model, P the prototype, and R the ratio. Subscript I is the inertia force added by the mixer, and it is desirable that it remain constant between the model M and the prototype P. Three fluid forces oppose the successful completion this process: viscosity, gravity, and fluid interfacial surface tension. It is impossible to keep these force ratios constant in scaleup with the same fluid. Therefore, we must choose two to work with. This, then, has led to the concept of dimensionless numbers, shown below.

$$\frac{F_I}{F_v} = N_{Re} = \frac{ND^2\rho}{\mu}$$

$$\frac{F_I}{F_g} = N_{Fr} = \frac{N^2D}{g}$$

$$\frac{F_I}{F_\sigma} = N_{We} = \frac{N^2D^3\rho}{\sigma}$$

in which the Reynolds number (the ratio of inertia force to viscosus force) is shown, as well as the Froude number and the Weber number. The Reynolds number and power number curve have been discussed, in which the power number is the ratio of inertia force to acceleration. To illustrate the characteristics of dimensionless numbers in mixer scaleup, examine the case of blending. We can express blending performance in terms of blend time multiplied by impeller speed, which gives a dimensionless process group. This is shown in Fig. 16 and gives a good correlation against the Reynolds number. However, for the other thousands of applications that are designed each year, there is normally no good way to write a

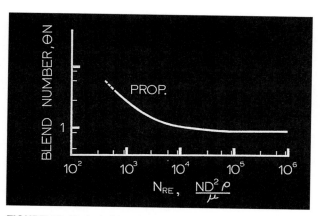

FIGURE 16 Typical dimensionless process correlation of blend number θN versus Reynolds number.

dimensionless group around the process result. For example, including polymerization yield, including productivity of a fermentation process, or incorporating the rate of absorption of flue gas into caustic does not allow a dimensionless type of process grouping. Thus, it is not practical to deal with dimensionless numbers when we do not have the ability to write a dimensionless group around the mixing process result.

There are as many potential scaleup parameters as there are individual process mixing results. However, we can make some generalizations which are very helpful in dealing with actual mixing problems, but for reliable scaleup, some experimental verification of the scaleup method to be used is desirable.

For example, it is found that the mass transfer coefficient, $K_G a$, for gas–liquid processes, is mostly a function of the linear superficial gas velocity and the power per unit volume with the constant D/T ratio for various size tanks. This is because the integrated volumetric mass transfer coefficient over the entire tank can be quite similar in large and small tanks even though the individual bubble size, interfacial area, and mass transfer coefficient can vary at specific points within the small and large tanks.

It has also been observed that suspension and blending of slurries operating in the hindered settling range (such as with particle sizes on the order of 100 mesh or smaller) tend to show a decreasing power per unit volume on scaleup. When this relationship is used, the blend time for the large tank is much longer than it is for the small tank. Blend time is not a major factor in a large slurry holding tank in the minerals processing industry, and therefore, that factor is not an important one to maintain on scaleup.

For homogeneous chemical reactions, most of the effect of the mixer occurs at the microscale level. Microscale mixing is largely a function of the power per unit volume, and maintaining equal power per unit volume gives similar chemical reaction requirements for both small and large tanks.

Some processes are governed by the maximum impeller zone shear rate. For example, the dispersion of a pigment in a paint depends upon the maximum impeller zone shear rate for the ultimate minimum particle size. However, when constant tip speed is used to maintain this, the other geometric variables must be changed to maintain a reasonable blend time, even though process results on full scale will probably take much longer than those on small scale.

Two aspects of scaleup frequently arise. One is building a model based on pilot plant studies that develop an understanding of the process variables for an existing full-scale mixing installation. The other is taking a new process and studying it in the pilot plant in such a way that pertinent scaleup variables are worked out for a new mixing installation.

There are a few principles of scaleup that can indicate what approach to take in either case. Using geometric similarity, the macroscale variables can be summarized as follows:

- Blend and circulation times in the large tank will be much longer than in the small tank.
- Maximum impeller zone shear rate will be higher in the larger tank, but the average impeller zone shear rate will be lower; therefore, there will be a much greater variation in shear rates in a full-scale tank than in a pilot unit.
- Reynolds numbers in the large tank will be higher, typically on the order of 5–25 times higher than those in a small tank.
- Large tanks tend to develop a recirculation pattern from the impeller through the tank pack to the impeller. This results in a behavior similar to that for a number of tanks in a series. The net result is that the mean circulation time is increased over what would be predicted from the impeller pumping capacity. This also increases the standard deviation of the circulation times around the mean.
- Heat transfer is normally much more demanding on a large scale. The introduction of helical coils, vertical tubes, or other heat transfer devices causes an increased tendency for areas of low recirculation to exist.
- In gas-liquid systems, the tendency for an increase in the gas superficial velocity upon scaleup can further increase the overall circulation time.

What about the microscale phenomena? These are dependent primarily on the energy dissipation per unit volume, although they must also be concerned about the energy spectra. In general, the energy dissipation per unit volume around the impeller is approximately 100 times higher than in the rest of the tank. This results in an rms velocity fluctuation ratio to the average velocity on the order of 10:1 between the impeller zone and the rest of the tank.

Because there are thousands of specific processes each year that involve mixing, there will be at least hundreds of different situations requiring a somewhat different pilot plant approach. Unfortunately, no set of rules states how to carry out studies for any specific program, but here are a few guidelines that can help one carry out a pilot plant program.

- For any given process, take a qualitative look at the possible role of fluid shear stresses. Try to consider pathways related to fluid shear stress that may affect the process. If there are none, then this extremely complex phenomena can be dismissed and the process design can be based on such things as uniformity, circulation time, blend time, or velocity specifications. This is often the case in the blending of miscible fluids and the suspension of solids.
- If fluid shear stresses are likely to be involved in obtaining a process result, then one must qualitatively look at the scale at which the shear stresses influence the result. If the particles, bubbles, droplets, or fluid clumps are on the order of 1000 μm or larger, the variables are macroscale and average velocities at a point are the predominant variable.

When macroscale variables are involved, every geometric design variable can affect the role of shear stresses. They can include such items as power, impeller speed, impeller diameter, impeller blade shape, impeller blade width or height, thickness of the material used to make the impeller, number of blades, impeller location, baffle location, and number of impellers.

Microscale variables are involved when the particles, droplets, baffles, or fluid clumps are on the order of 100 μm or less. In this case, the critical parameters usually are power per unit volume, distribution of power per unit volume between the impeller and the rest of the tank, rms velocity fluctuation, energy spectra, dissipation length, the smallest microscale eddy size for the particular power level, and viscosity of the fluid.

- The overall circulating pattern, including the circulation time and the deviation of the circulation times, can never be neglected. No matter what else a mixer does, it must be able to circulate fluid throughout an entire vessel appropriately. If it cannot, then that mixer is not suited for the tank being considered.

Qualitative and, hopefully, quantitative estimates of how the process result will be measured must be made in advance. The evaluations must allow one to establish the importance of the different steps in a process, such as gas–liquid mass transfer, chemical reaction rate, or heat transfer.

- It is seldom possible, either economically or time-wise, to study every potential mixing variable or to compare the performance of many impeller types. In many cases, a process needs a specific fluid regime that is relatively independent of the impeller type used to generate it. Because different impellers may require different geometries to achieve an optimum process combination, a random choice of only one diameter of each of two or more impeller types may not tell what is appropriate for the fluid regime ultimately required.
- Often, a pilot plant will operate in the viscous region while the commercial unit will operate in the transition region, or alternatively, the pilot plant may be in the transition region and the commercial unit in the turbulent region. Some experience is required to estimate the difference in performance to be expected upon scaleup.
- In general, it is not necessary to model Z/T ratios between pilot and commercial units, where Z is the liquid level.
- In order to make the pilot unit more like a commercial unit in macroscale characteristics, the pilot unit impeller must be designed to lengthen the blend time and to increase the low maximum impeller zone shear rate. This will result in a greater range of shear rates than is normally found in a pilot unit.

All of these conditions can be met using smaller D/T ratios and narrower blade heights than are used normally in a pilot unit. If one uses the same impeller type in both the pilot and commercial units, however, it may not be possible to come close to the long blend time that will be obtained in the commercial unit. Radial flow impellers can be excellent models in a pilot plant unit for axial flow impellers in a commercial unit.

III. LIQUID–SOLID CONTACTING

Solids suspension involves producing the required distribution of solids in the tank and is essentially a physical phenomenon. The criterion is normally a physical description of the degree of uniformity required in the suspension. A key variable for solids suspension is the settling velocity of the solids. This is usually measured by timing the fall velocity of individual solid particles in a defined depth of

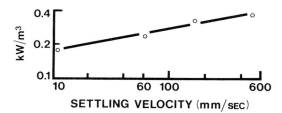

FIGURE 17 Effect of settling velocity to achieve a 60% suspension of particle sizes when there is a mixture of particle sizes.

mother liquor. When there is a wide range of particle sizes, there may well be a wide range of settling velocities.

Much of the literature is based on experimental data with similarly sized particles and observations of the speed required to keep particles in motion with at most 1 or 2 sec of rest on the bottom of the tank. This is done by visually observing solids in a transparent tank. This, of-course, means that relatively small-scale experiments are conducted and that this particular criterion cannot be used for studies in large-sized tanks or in field tests.

Sizing procedures to design a mixer for one closely sized particle settling velocity are modified considerably when there are other solids present. Figure 17 shows the effect of settling velocity on power when there are other solids present in the system. The slope is much less pronounced than it is when a single particle size alone is being suspended.

Much of the literature correlations for solids suspension are based on the so-called critical impeller speed. Attempts to duplicate experiments between various investigators often yield deviations of ± 30–50% from the critical speed shown by other investigators. Because power is proportional to speed cubed, power varies on the order of 2 to 3 times, which is not sufficiently accurate for industrial full-scale design. Therefore, many approximate, conservative estimates have been made in the literature as general guidelines for choosing mixers for solids suspension. Table IV is one such guideline for solid particles of a closely sized nature.

The study of solids suspension in quantitative terms normally involves a method of sampling. Typically, samples are withdrawn from the side of the mixing tank through openings or tubes inserted into the vessel wall. It may also be done by submerging a container and quickly removing

TABLE IV   Motor Horsepower for Estimating Purposes for Solids Suspension[a]

Settling velocity	1'/min	2'/min	4'/min
Off bottom	1	2	5
Uniform	1.5	5	15

[a] 15,000 gal tank; $D/T = 0.33$. $C/D = \frac{1}{2}$; axial flow turbine; 1–20% solids by weight.

the top and replacing it, allowing the slurry to flow into the container. Neither of these methods gives the absolute percentage of solids at the measurement point. In the case of a tube, the withdrawal velocity of the tube can affect the percent of solids that comes out of the discharged slurry as well as the orientation of the tube relative to the flow pattern in the tank. In the case of the sample container, its location, fill rate, and other variables can affect the actual solids composition measured compared to the actual. On the other hand, as long as measurement techniques are consistent, a reliable effect of mixer variables can be determined, which is of value in predicting operating conditions for full-scale units. One such test on a pilot plant scale yields data shown in Fig. 18, which shows the difference in axial and radial flow of solids suspension characteristics and indicates, as we mentioned previously, that axial flow impellers require less horsepower for the same degree of solids suspension.

The use of the new type of fluidfoil impeller has reduced the power required for solids suspension to about one-half to two-thirds of the values formerly used with 45° pitch blade turbines.

In continuous flow, the only point in the tank that must be equal to the feed composition for steady-state operation is the drawoff point. Thus, if the drawoff point is at the bottom, middle, or top of the tank, different average tank compositions can result, even though the composition of the entrance and exit streams are the same. If the mixer is large enough to provide complete uniformity of all the solids, including the coarse particles as well as the fine particles, then the drawoff point does not make any difference in the composition of the tank. However, if the mixer is designed only to just suspend the solids to the drawoff point, then tank compositions vary widely, depending upon the drawoff conditions.

Many times a fillet can be left in a tank, which will reduce the horsepower considerably for what will be required to completely clean out the last corners of a flat bottom tank. Depending upon the value of the solids in the process, they may either be left to form their own fillet, or the tank may be streamlined by using concrete or other materials to give a more streamlined shape.

When solids increase in percentage, the effect is to make the process requirement more difficult, and a curve similar to that in Fig. 19 results, until a point which often occurs around 40–50% by weight solids, at which there may be a discontinuity. At this point, the viscosity of the slurry is becoming a parameter, which reduces the settling velocity and, thus, minimizes its importance as a criterion to one in which we are essentially blending and providing motion through a pseudo-plastic fluid. Then as the solids percentage gets up toward 70 or 80% (and this point can be normalized by relating it to the percentage of the ultimate

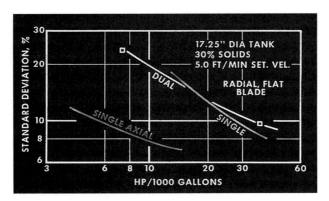

FIGURE 18 Typical comparison of power required for axial flow impeller compared to radial flow impellers in solids suspension.

settled solids), power becomes extremely high and approaches infinity where there is no supernatant liquid left in the tank. To evaluate this effect, a mixing viscosimeter is valuable, in which the slurry is agitated at the same time that the viscosity is measured, so that the measurement gives a reasonable value for the overall slurry.

A. Typical Mass Transfer Processes

Many processes involve criteria other than solids suspension, for example, crystallization, precipitation, and many types of leaching and chemical reactions. In crystallization, the shear rate around the impeller and other mixing variables can affect the rate of nucleation, and can affect the ultimate particle size. In some cases, the shear rate can be such that it can break down forces within the solid particle and can affect the ultimate particle size and shape. There are some very fragile precipitate crystals that are very much affected by the mixer variables.

In leaching, there usually is a very rapid leach rate that occurs when the mineral is on the surface of the particle, but many times the internal diffusion of the solid through the solid particle becomes controlling, and mixer variables do not affect the leaching rate beyond that point. In studying the effect of mixing on leaching processes, it is normally desirable to run separate experiments with

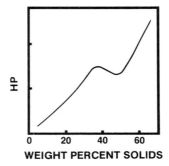

FIGURE 19 Increase of process horsepower versus weight percent solids, showing discontinuity when criteria changes from solids suspension to pseudo-plastic blending.

the fine particle sizes, the average particle sizes, and the coarse particle sizes, since the leaching curves are often quite different. In addition, the suspension of the fine, average, and coarse particles may be different in the leach tank due to the fact that all these particles may not be completely uniform throughout the system.

Typical design of an industrial leaching system looks at the extraction rate versus time and power level and determines the optimum combination of tank size and mixer horsepower in terms of return on the product leached.

B. Industrial Examples

One area for industrial studies is the whole area of slurry pipelines. Coal is by far the most common material in slurry pipelines, but other pipelines, but other pipelines include iron ore and potash. In large volume solid suspension applications, there is a considerable trade-off between volume of a tank, mixer horsepower, shape of a tank, and many other areas of cost consideration that are important in overall design. In addition to the tanks in these sorts of slurry systems, it must be capable of incorporating slurries into water or vice versa to either increase or decrease the solids concentration of a given system.

Another industrial application is mixing of paper pulp and slurries. An entire technology exists for this fluid, which is quite unique compared to other liquid–solid systems. Basically, there is a question of whether to use baffles, comparison of both top-entering and side-entering mixers, as well as the very large effect of type of paper pulp and the consistency of the paper pulp in the vessel. Other examples include fermentation, in which there is a biological solid producing the desired product, and the role of fluid shear rates on the biological solids is a critical consideration as well as the gas–liquid mass transfer (see Section VI).

Another class of applications is the high shear mixers used to break up agglomerates of particles as well as to cause rapid dissolving of solids into solvents. A further type includes the catalytic processes such as hydrogenation, in which there is a basic gas–liquid mass transfer to be satisfied, but in addition, effective mixing and shear rate on the catalyst particle fluid film as well as degradation must be considered.

IV. GAS–LIQUID CONTACTING

Many times a specification calls for a fluid mixer to produce a "good dispersion" of so many computational fluid mixing (CFM), of gas into a given volume of liquid. Actually, there are very few applications in which dispersion of gas–liquid is the ultimate process requirement. Usually there is a mass transfer requirement involved, and the role of a mixer to provide a certain mass transfer coefficient K_Ga can entirely supercede any requirement for a particular type of visual description of the gas–liquid dispersion. In general, linear gas superficial velocity, normally given the symbol F, in feet per second, is based on dividing the tank cross-sectional area by the flow of gas at the temperature and pressure of the gas at the midpoint of the tank. This quantity is very basic both in the scaleup correlation and in predicting the power imparted to the liquid by the gas stream.

It is characteristic that this ratio F increases on scaleup, since if we maintain equal volumes of gas per volume of liquid per time on scaleup, which is necessary to provide the same stoichiometric percentage of gas absorbed from the gas phase, then the linear velocity increases directly proportional to the depth of the large tank.

While the variables are many and complex, in a general concept, if the power in the tank is equal to the energy provided by the gas stream, we will get a gas-controlled flow pattern. This has different characteristic coefficients of the mass transfer rate than the case where the mixer horsepower is three or more times higher than the gas power. For radial flow impellers, this factor of three will provide a mixer-controlled flow pattern, which again, has different exponents on the correlating equation for mass transfer coefficient K_Ga or K_La. To drive the gas down to the bottom of the tank, below the sparge ring, the power level must be on the order of 5–10 times higher than the gas power level.

For axial flow impellers, the ratio of mixer power to gas stream power for a mixer-controlled flow pattern is approximately 8–10. This means that radial flow impellers are more commonly used for gas–liquid dispersion than axial flow impellers.

Figure 20 gives a typical curve for the effect of gas velocity and power level on mass transfer coefficient K_Ga. In a given application, knowledge of the required gas ab-

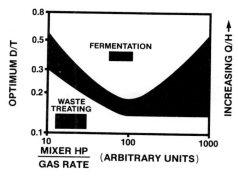

FIGURE 21 Schematic representation of optimum D/T as a function of flow of gas compared to mixer horsepower input. Shaded area is optimum D/T. Two industrial examples, fermentation and aeration of biological waste, are shown.

sorption rate and the partical pressure in the incoming–outgoing gas stream, coupled with an estimate of the equilibrium partial pressure of gas related to the dissolved gas in the liquid, allows the calculation of the average concentration driving force and then the mass transfer coefficient K_Ga when needed to provide that mass trasfer rate. This then allows the mixer to be chosen for that particular combination. It is typical to try different gas rates, different tank shapes, or perhaps different head pressures to see the effects on the mixer design and the cost for process optimization.

Another consideration is the optimum flow to fluid shear ratio involved for gas-liquid dispersion. Figure 21 shows the optimum D/T for different combinations for gas flow and mixer power level in conceptual form. At the left edge of the curve, where gas rates are high and power levels are low, large D/T values are desired to produce high flow and low shear rates. In the middle of the graph, which is more common, where the gas flow pattern is controlled by the mixer, desired D/T values are very small (on the order of 0.15–0.2). At the far right-hand side of the graph, we have a mixer power level greater than 10 times the gas power level, and it makes very little difference what ratio of flow-to-fluid shear rate we have, as shown by the effect of D/T. This relationship shows the difficulty in comparing impellers in gas–liquid mass transfer systems, because the comparison of fluid shear and fluid flow requires a knowledge of the mixer power to gas flow ratio. In addition, in a process such as fermentation, where there are certain maximum shear rates possible without damaging the organism, the D/T chosen for the process may not be the optimum for the gas–liquid mass transfer step, and correlations must be available for the effect of D/T ratio on mass transfer coefficient to complete design of those kinds of processes.

Scaleup is normally based on the fact that the correlation of K_Ga versus power per unit volume and superficial gas velocity is the same for both pilot and full-scale tanks.

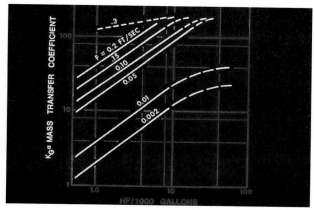

FIGURE 20 Typical correlation of gas–liquid mass transfer coefficient K_Ga as a function of impeller power and superficial gas velocity.

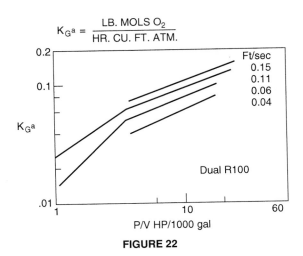

FIGURE 22

This allows the calculation of full-scale mixers when pilot plant data is available in that particular fluid system.

The curve shown in Fig. 22, for an R100 impeller illustrates that there is a break point in the relationship with $K_G a$ versus the power level at the point where the power of the mixer is approximately three times the power in the expanding gas stream. The power per unit volume for an expanding gas stream at pressures from 1 to 100 psi can be expressed by the equation P/V (HP/1000 gal) = 15F (ft/sec). The A315 impeller, Fig. 23, is able to visually disperse gas to a ratio of about 1 to 1 in expanding gas power and mixer power level. It does not have a break point in the curve, although slopes are somewhat different than those in Fig. 22.

A comparison of the curves is such that in some areas the A315 has a somewhat better mass transfer and in other areas the R100 has a better mass transfer performance.

The large difference in the A315, however, is more in its blending ability compared to the R100, so that the blend

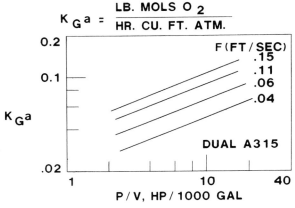

FIGURE 23   Effect of power per unit volume, P/V, and superficial gas velocity, F, on the mass transfer coefficient $K_G a$, for dual A315 impellers.

time in a large mixing vessel equipped with A315 impellers will be about one-third that of the blend time in the same vessel equipped with R100 impellers. Blending is relatively long on full scale compared to pilot scale, so the improvement in blending characteristics on full scale can lead to a much more uniform blending condition. Many fermentations are responsive to improved blending and this is another factor in addition to the requirement of gas–liquid mass transfer that exists in many fermentation systems as well as in other gas–liquid operations.

A. Combination of Gas–Liquid and Solids Systems

As mentioned previously, axial flow impellers are typically used for solids suspension. It is also typical to use radial flow impellers for gas–liquid mass transfer. In combination gas-liquid-solid systems, it is more common to use radial flow impellers because the desired power level for mass transfer normally accomplishes solids suspension as well. The less effective flow pattern of the axial flow impeller is not often used in high-uptake-rate systems for industrial mass transfer problems. There is one exception, and that is in the aeration of waste. The uptake rate in biological oxidation systems is on the order of 30 ppm/hr, which is about $\frac{1}{2}$ to $\frac{1}{10}$ the rate that may be required in industrial processes. In waste treatment, surface aerators typically use axial flow impellers, and there are many types of draft tube aerators that use axial flow impellers in a draft tube. The gas rates are such that the axial flow characteristic of the impeller can drive the gas to whatever depth is required and provide a very effective type of mass transfer unit.

B. Effect of Gas Rate on Power Consumption

At a given mixer speed, there is a reduction in the horsepower of a mixer when gas is added to the system, and normally the horsepower decreases somewhat proportional to the increase in gas velocity. Figure 24 shows a typical curve, but there are many other variables that affect the location of the curve markedly. This brings up a key point for industrial design. If the mixer is to be run both with the gas off and on in the process, then an interlock is used to prevent gas-off operation or change the mixer speed to prevent overloading, or else the mixer must be capable of transmitting power and torque possibly two, three, or four times higher than is needed during the actual process step. This often is solved by a two-speed motor, which allows a lower speed to be used when the gas is off, compared to the normal speed at processing conditions.

A new impeller is now being used for gas-liquid contacting, call the Smith Turbine. It is a radial flow turbine with blades as shown in Fig. 25. It is rotated in the concave

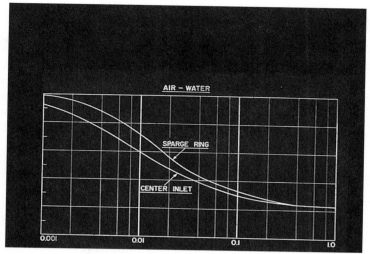

FIGURE 24 Typical plot of *K* factor, ratio of horsepower with gas-on to horsepower with gas-off at constant speed, as a function of superficial gas velocity with two different gas inlets.

direction. It has the characteristic of giving the same mass transfer as the radial flat blade turbine shown in Fig. 1a but does not drop off in power as much as the radial flow turbine does. Figure 25 shows the *K* factor which relates the power drawn when the gas is on to the power drawn in the ungased liquid. This means that variable speed drives do not have to be used many times to keep the desired power in the gased condition.

V. LIQUID–LIQUID CONTACTING

A. Emulsions

There is a large class of processes where the final product is an emulsion. It includes homogenized milk, shampoos, polishing compounds, and some types of medical preparations. A key factor is the chemistry of the product, to ensure that the emulsion remains stable over a desired product shelf-life, when produced properly by the fluid mixer. There are numerous correlations in the literature on drop size in a two-phase liquid–liquid system, relative to fluid properties and mixer variables. However, most

of these have been done with pure liquid components, and do not apply to the complicated chemicals used industrially. Small amounts of surface active agents make dramatic differences in emulsion characteristics, so it is not usually possible to calculate in advance the mixer needed to provide the particular type of emulsion particle size. Therefore, test work is normally required where conditions required in the pilot plant are evaluated for scaleup.

Large tanks have a longer blend time and a much greater variety of shear rates than small tanks, therefore, emulsion characteristics on full scale are difficult to predict. Usually, the pilot plant work is aimed at trying to elaborate the key role of maximum impeller zone shear rate, average impeller zone shear rate, and general circulation rate and velocity out in the main part of the tank. If these can be even qualitatively determined, scaleup to full scale can be done with reliability. There are a variety of mixers used in these processes. For various types of emulsion polymerization, it is typical to use axial flow impellers because the shear rate requirements do not demand the use of radial flow impellers. Getting into the other end of the spectrum, where extremely high shear mixers are needed, various kinds of radial flow blades, usually with very narrow width blades, allow speeds to go up to 1000 or 2000 rpm giving very intense shear rates that are needed for many types of emulsion processes.

B. Liquid Extraction

Figure 26 shows a typical curve relating the performance of many kinds of mixing devices for liquid extraction. The main advantage of using mixing or some type of mechanical energy, compared to packed plate or spray

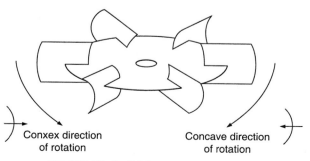

Conxex direction of rotation   Concave direction of rotation

FIGURE 25 Radial flow turbine with blades.

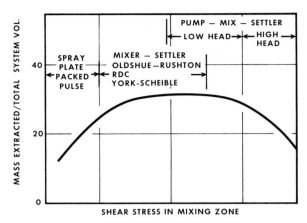

FIGURE 26 Illustration of optimum shear stress in a mixing zone of various types of countercurrent liquid–liquid extraction columns.

columns, is the ability to get a smaller volume for the same degree of extraction. However, if an attempt is made to use too much energy, then problems of settling characteristics are encountered, and this negates the advantages of the mixed system many times. In the mining industry, it is quite typical to use mixer settlers. These usually involve an extraction step, a scrubbing step, and then a stripping step. Usually the requirement is for only one or two stages in each of these areas with the use of very selective ion exchange chemicals in the system. To eliminate interstage pumps a pump–mixer is used in which some of the head component of the impeller is converted to a static head so that fluids can be pumped against small static heads in the mixers and settlers of the whole train. This has worked well in many applications, although there is a potential problem that the conditions required for effective pumping are not optimum for the mixing that is required in the mixing stage, and there may be some design parameters that are difficult to satisfy in the systems.

The other area is the countercurrent liquid–liquid extraction system, shown in Fig. 27, using mixer stages separated by stationary horizontal discs. These have the advantage of only one interface for settling to occur, plus the fact that solids can be handled in one or both phases. Also, all the principals of fluid mixing can be used to design an effective transfer system. The design procedure is also based on the $K_L a$ concept, discussed in Section IV, and allows the calculation of reliable full-scale performance, based on pilot plant work, often done in a laboratory column about 6 in. in diameter.

One of the key variables to be studied in the pilot plant is the effect of turndown ratio, which is the ratio of flow to the design flow through the column, so that predictions can be made of performance during reduced throughput during certain parts of the plant processing startup.

VI. BLENDING

A. Low-Viscosity Blending

Low-viscosity blending involves evaluation of the degree of uniformity required and the operating cycle. There is a difference in performance, depending on whether the materials to be blended are added continuously and

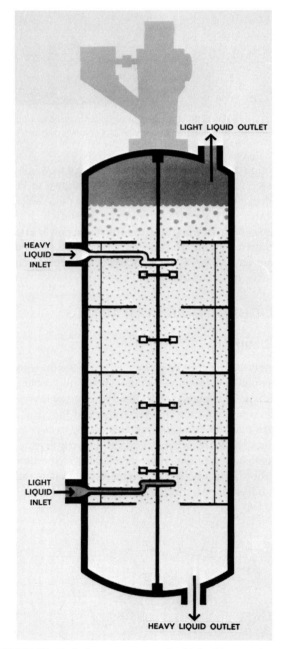

FIGURE 27 Typical countercurrent liquid–liquid extraction column with mixing phases: Oldshue/Rushton column illustrated.

uniformly into the tank, with the tank originally in motion or whether the tank has become stratified during the filling application, and mixing must be accomplished with a stratified liquid level situation. In general, blend time is reduced at constant mixer power with larger D/T ratios. The exponent on D/T with blend time is approximately -1.5, with the range observed experimentally of from 0.5 to 3.0. This leads to the fact that larger impellers running at slow speeds require less power than a small mixer running at high speed for the same blend time. In that case, there is an evaluation needed which relates the capital cost of the equipment, represented by the torque required in the mixer drive which is usually greatest for the big impeller, versus the cost of horsepower, which is usually greatest for the small impeller. This leads to the concept of optimization of the economics of a particular process. In all cases, at least two or three mixers must be selected for the same blend time with different power and impeller diameter to carry out this evaluation.

Table V gives a typical values for estimation purposes of blending horsepower required for various low- and mediumviscosity situations. Mixers may be either top entering or side entering. Again, a side-entering mixer requires more power and less capital dollars, and this must be evaluated in looking at practical equipment. Side-entering mixers have a stuffing box or mechanical seal and are limited for use on materials that are naturally lubricating, noncorrosive, or nonabrasive.

B. High-Viscosity Blending

Blending of high-viscosity materials, which are almost always pseudo-plastic, involves a different concept. The degree of pseudo-plasticity is determined by the exponent n in the equation

$$\text{shear stress} = K(\text{shear rate})^n$$

with value 1 for Newtonian fluids and a value less than 1 representing the degree of decrease of viscosity with an increase in shear rate. For very viscous materials (on the order of 50 m cP and higher), the helical impeller

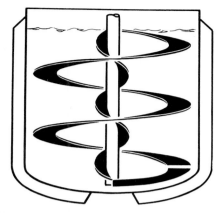

FIGURE 28 Typical helical flow impeller for high-viscosity blending with close clearance to the tank wall.

(Fig. 28) is often used. Many times this is a double helix, in which pumping on the outside is done by the outer flight, while pumping on the inside is done by the inner flight. Reverse rotation, of course, reverses the direction of the flow in the tank. These impellers typically run at about 5–15 rpm and have the unique characteristics that the circulating time and blend time are not a function of the viscosity of the fluid. At a given velocity, there is a certain turnover time for a given Z/T ratio, and changing viscosity does not affect that parameter, nor does the degree of pseudo-plasticity affect it. However, the power is directly proportional to the viscosity at the shear rate of the impeller, and so doubling or tripling the viscosity at the impeller shear rate will cause an increase of power of two or three times, even though circulation time will remain the same. Helical impellers are very effective for macroscale blending, but do not typically have the microscale shear rate required for some types of uniformity requirements or process restraints.

Open impellers, such as the axial flow turbine (Fig. 1a) or the radial flow turbine (Fig. 1b), may also be used in high-viscosity pseudo-plastic fluids. These require a level of power four to five times higher than the helical impeller, but only cost about one-third as much. Another economic comparison is possible to see which is the most effective for a given operation. This higher power level, however, does provide a different level of microscale blending. Occasionally the flow from a blend system with a helical impeller will be passed through a mechanical type of line blender, which imparts a higher level of microscale mixing.

C. Side-Entering Mixers

Figure 29 shows the importance of orientation on side-entering mixers on low-viscosity systems. The mixer must be inclined about 7° from the tank diameter, to ensure a

TABLE V Motor Horsepower for Estimating Purposes for Blending Purposes[a]

Blend time θ (min)	H.P.[b]			
	100	250	500	1000
6	5	7.5	10	15
12	3	5	7.5	10
30	1.5	2	3	5

[a] 15,000 gal tank; axial flow impeller, $D/T = \frac{1}{3}$, $Z/T = 1$; $C/D = 1$.
[b] For viscosities in centipoises.

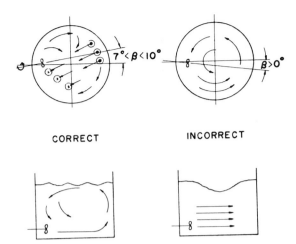

FIGURE 29 Typical orientation of side-entering mixer in large petroleum storage tanks.

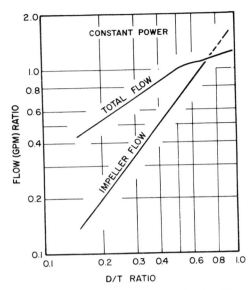

FIGURE 30 Schematic illustration of total flow in mixing tank as compared to impeller flow.

top-to-bottom flow pattern. However, even when this is done, there still are some relatively stagnant areas of the tank, and side-entering mixers are not usually satisfactory when solid suspension is a critical factor. As discussed previously, larger-diameter impellers at slower speeds require less horsepower, and so there can be an economic evaluation of the power versus capital equipment cost for various types of side-entering mixers and a given blending process. Typical applications are crude oil tanks, gasoline tanks, and paper stock. In addition, they are used for various kinds of process applications where the advantages are considerable over the use of a conventional top-entering mixer.

VII. FLUID MOTION

Many times the objective is to provide a pumping action throughout the tank. The pumping capacity of impellers can be measured by photographic techniques, hot wire or hot film velocity meters, or laser Doppler velocity meters. There is no generally agreed upon definition of the discharge areas for impellers, so that the primary pumping capacity of mixing impellers varies somewhat, depending on the definition used for discharge area. There is considerable entrainment of fluid in the tank, due to the jet action of the flow from the impeller. Figure 30 shows the increase in total flow in the tank at various D/T ratios. This also indicates that at about 0.6 D/T ratio further increases in total flow in the tank are difficult to achieve, since there is no more entraining action of the impeller in the total system.

The pumping capacity of a mixing impeller is specified by either the flow from the impeller or the total flow of the tank. Flow varies for any impeller as the speed and diameter cubed. Table VI gives some for constants in the equation $Q = KND^3$ for various impeller types. The radial

flow impeller has essentially less flow and higher shear rates than does the axial flow impeller type.

If the impeller is required to pump against a static head or a friction head within the channel of the mixing tank, then there must be a series of head flow curves developed, (Fig. 31) for the impeller being used. This is a function of the clearance between a radial impeller and a horizontal baffle. The hole in it allows the flow to come into the impeller zone but not circulate back, or the clearance of an axial impeller in a draft tube (Fig. 32). The operating point, then, will be the intersection of the impeller head flow curve and the system head flow curve. Draft tube circulators have the advantage of giving the highest flow in the annulus for a given level of power or requiring the least power to provide a given flow of the annulus. When pumping down the draft tube, the flow in the annulus must equal the settling velocity of the particles, and the total flow can be calculated on that basis. In practice, the flow coming up the annulus is not a uniform flat velocity profile; so that additional total flow is needed because of the nonuniform distribution of the upward axial velocity to the annulus. Pumping down the draft tube allows the tank bottom to be flat or have very small conical fillets at the sidewalls.

Pumping up the draft tube requires that the solids are to be suspended in the draft tube with a much lower total

TABLE VI Constant in Flow versus Speed and Diameter of Various Mixing Impellers

Figure	K
1a	0.8
1b	0.6

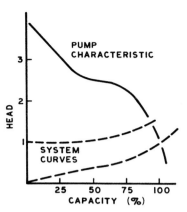

FIGURE 31 Typical head flow curve for mixing impeller and draft tube with corresponding system curves.

flow, and also power, and then make their own way down the outside of the annulus coming into the bottom of the draft tube again. This means that the bottom of the tank must usually have a steep cone, and suitable flares and baffles must be added to the draft tube bottom so that the flow comes up in a uniform fashion for proper efficiency.

When using a draft tube, the back flow possibility in the center of the impeller requires the use of a large-diameter hub. This is not normally desirable in fluidfoil impellers used in open tanks. The system head for a draft tube circulator is a function primarily of the design of the entrance and exit of the draft tube, and considerable work has been

done on the proper design and flaring of these tubes for special applications. The main use of draft tube circulators has been in precipitators and crystallizers. A further requirement is that the liquid level be relatively uniform in depth above the top of the draft tube, which means that variable liquid levels are not practical with draft tube systems. In addition, slots are often provided at the bottom of the draft tube, so that should a power failure occur and solids settle at the bottom of the tank, flow can be passed through these slots and scrub out particles at the bottom of the tank for resuspension.

Sometimes it is desired to have a large working area in a tank where, for example, a conveyor belt containing car bodies can be passed through for electrostatic painting. One way to accomplish this is to put a series of propeller mixers in a side arm of the long side of the tank, so that the flow is directed into the middle zone, but there are no mixer shafts or impellers in the center to impede the flow of the parts through the equipment.

VIII. HEAT TRANSFER

Another area for pumping consideration is heat transfer. The only sources of turbulence provided in heat transfer are flow around the boundary layer of a jacketed tank and around a helical coil or vertical tubes. There are several good heat transfer correlations available, and most of them have fairly common exponents on the correlation of the Nusselt number hD/k. This is correlated with the Reynolds number ND^2p/μ and the Prandtl number $Cp\mu/k$ plus other geometric ratios. The exponential slope on the effect of power on heat transfer coefficient is very low (on the order of 0.2). This means that most heat transfer design involves determining the mixer required for just establishing forced convection through the tank, and usually not going beyond that point if heat transfer is the main requirement. If other requirements are present which indicate a high horsepower level, then advantage can be taken of these higher power levels by use of the 0.2 exponent. However, if it is desired to increase the heat transfer capacity of a mixing tank, it is normally done by increasing or changing the heat transfer surface, since very little can be done by changing the mixer power level. Figure 33 gives a good working correlation for the effect of viscosity on both heating and cooling coefficients for helical coil systems. Jacketed tanks have values about two-thirds of those in Fig. 33. This is the mixer side coefficient only, and it holds for organic materials, The heat transfer coefficient for aqeuous materials is higher than the value shown in Fig. 33. Bear in mind that the overall coefficient is made up of other factors, including the coefficient on the inside of the tube or jacket, as well as the thermal conductivity value of the heat transfer surface.

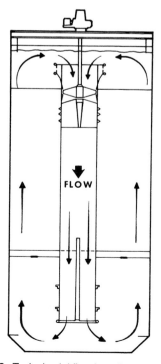

FIGURE 32 Typical axial flow impeller and draft tube.

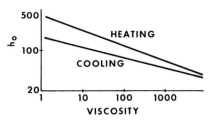

FIGURE 33 Practical heat transfer coefficients for use in estimating with helical coils and vertical tubes.

IX. CONTINUOUS FLOW

A mixing tank has a variety of residence times. The definition of perfect mixing requires that one particle leave in 0 time and one particle stay in forever. Curves shown in Fig. 34; developed by McMullen and Weber, show the percentage of material that is in the tank for various lengths of time. To provide good mixing in a system but avoid the detrimental effect of a variety of residence times, multiple staging can be used. This curve shows, for example, that if the total residence time in a tank were 60 min, then at the end of 30 min, 33% of the material is already gone and 67% of the material is still there. Out at the very long residence time, there is still a small amount of material that stays in an infinitely long length of time. This means that processes involving pharmaceuticals or food products must take into account that small contaminants or mutants may stay in the system for a very long time and can cause problems in yield and productivity.

Another purpose of a mixing tank is to dampen out fluctuations. A mixing tank cannot change the frequency of fluctuations but can dampen the amplitude. As a general principle, a residence time equal to the cycle time of the fluctuations will cause the amplitude to be dampened by about a factor of six.

For any chemical reaction of an order greater than zero, the process takes longer in a continuous flow tank than it

does in a corresponding batch tank. An infinity of mixing stages is equivalent to a batch tank or to a plug flow reactor. Usually, however, 5, 10, or 20 stages are sufficient to give a good efficient reaction time and to possess the advantages of continuous flow compared to the reaction time in a batch system.

A. Inline Mixers

Mixers in a flowing pipeline are of two general types, one utilizing static elements and the other using a rotating impeller.

A static inline mixer is essentially a device that provides transverse uniformity and not longitudinal or time-interval blending. Hence, if a particle in Fig. 35 is ever to catch up with another particle behind it, there must be a tank volume such that the first particle can remain until the latter one catches up with it.

There are two kinds of static mixers. One type has helical elements that twist the fluid, and another set of elements that cut the fluid, divide it, and twist it again. The twisting and cutting is continued until the production and scaleup uniformity is achieved. This is useful in viscous fluids.

Attempts to use these kinds of devices on low-viscosity materials showed that the flows did not twist and curl in quite that same fashion. In the low-viscosity region, pressure drop is a key factor. The second type of static mixer gets pressure drop through controlled channels, different types of static elements, as well as random placement of baffles, blades, orifices, or other devices inside the pipeline.

Mechanical inline mixers have a relatively high-speed impeller, rotating in a small volume, usually on the order of $\frac{1}{4}$ gal to perhaps 50 or 60 gal. Obviously, with a big enough tank, you then have a system that really does not fit in the pipe-line itself. Usually, the flow is directed through two stages, the flow comes in the bottom of the container, flows up through a hole in a static plate into a stage divider, and then flows in the second impeller. The power is such that

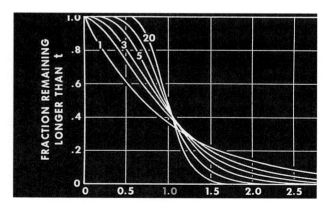

FIGURE 34 Curves based on perfect mixing in each compartment of the multistage compartment system, showing percentage material retained for various lengths of time in continuous flow.

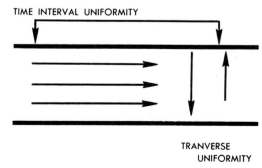

FIGURE 35 Pipeline flow showing that time-interval mixing normally must have a volume for retention time, compared to radial flow with usual static mixer elements.

the flow pattern is completely disrupted, so the pressure drop to these units is at least one velocity head. The rpm can be adjusted to achieve almost any required level of dispersion for contacting.

X. PILOT PLANT PROCEDURES

Pilot planting involves gathering sufficient information from model runs so that the major controlling factors in the process are understood for a suitable scaleup analysis.

The heart of the pilot plant study normally involves varying the speed over two or three steps with a given impeller diameter. The analysis is done on a chart, shown in Fig. 36. The process result is plotted on a log-log curve as a function of the power applied by the impeller. This, of course, implies that a quantitative process result is available, such as a process yield, a mass transfer absorption rate, or some other type of quantitative measure. The slope of the line reveals much information about likely controlling factors. A relatively high slope (0.5–0.8) is most likely caused by a controlling gas–liquid mass transfer step. A slope of 0, is usually caused by a chemical reaction, and a further increase of power is not reflected in the process improvement. Point A indicates where blend time has been satisfied, and further reductions of blend time do not improve the process performance. Intermediate slopes on the order of 0.1–0.4, do not indicate exactly which mechanism is the major one. Possibilities are shear rate factors, blend time requirements, or other types of possibilities.

To further sort out the effect of mixing, it is usually desirable to vary the impeller diameter. For example, if a 100-mm impeller had been used in a 300-mm diameter tank for the original runs, and if it were thought that pumping capacity would be more helpful in fluid shear rate, a series of runs with 125- or 150-mm diameter im-

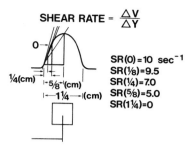

FIGURE 37 Schematic illustration that the macroscale shear rate around the impeller is a function of the size of the fluid element of interest.

peller would be appropriate. On the other hand, if it were thought that fluid shear was more important, then runs with a 50- or 75-mm impeller would be indicated.

If separation of the microscale mixing phenomenon from the macroscale mixing phenomenon is desired, then it is necessary to systematically vary the ratio of blade width to blade diameter.

There is a minimum size pilot tank. Referring now to Fig. 37, the shear rate at the boundary layer of the impeller jet in the tank has approximately a value of 10 in this example. The impeller is approximately 1 cm in blade width. The shear rate across a $\frac{1}{8}$ cm is about 9.5, shear rate across a $\frac{1}{4}$ cm is 7.5, and the shear rate across a $\frac{1}{2}$ centimeter is 5, and is the average shear rate. The shear rate across the entire blade 1 cm wide is 0, since it has the same velocity on both sides of the impeller blade. Thus, a particle of 1 cm size would have a zero shear rate, while a particle having a 1 μm size would have a shear rate of 10. This leads to the general rule that the impeller blade must be at least three times larger in physical dimension than the biggest particle that is desired to disperse, react, or coalesce. In practice, this indicates that most gas–liquid processes should be done in tanks at least 12 in. in diameter, while most viscous and pseudo-plastic materials should probably be handled in tanks from 12 to 18 in. in diameter. Homogenous chemical reactions could be carried out in a thimble, if desired, since there is no problem getting the scale of the molecule to be smaller than the scale of an impeller blade, even a small laboratory size.

It is usually desirable to either measure or calculate horsepower, and there are several methods by which this can be done. One is to have impellers calibrated by the manufacturer, which provides a curve of power versus speed. By using suitable factors for judging viscosity and gas flow, power in the batch can be estimated as a function of the impeller speed. Another possibility is to place the impeller on a trunion bearing mounting, in which the motor is held stationary by a pulley arm, and the force required is measured on a scale. Another method involves the use of strain gauges, which measure either the elongation on the surface of a shaft or the changes in conductivity

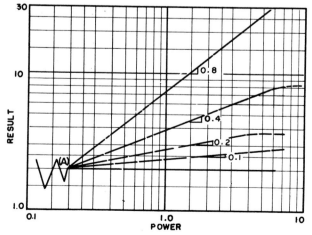

FIGURE 36 Typical plot of a given process result as a function of mixer power level in a pilot plant study.

or reluctance with various kinds of electrical signals. It is possible today to use micro-sized amplifiers that rotate with the shaft and feed a signal through the slip rings with very little loss in accuracy.

In general, a large mixing tank has a much longer circulation time and a much higher maximum macroscale impeller shear rate than does a small tank. In addition, it has a greater variety of shear rates than does a small tank. This means that a small tank can be changed in its performance compared to a big tank by using a nongeometric approach to the design of the mixer. There are usually two extremes of pilot plant objectives. One involves the use of a more-or-less standard impeller geometry in small scale, and attempts to determine the maximum efficiency of the process on that scale. Estimates on a full-scale performance must be modified because the big tank is different in many regards, which may have beneficial or detrimental effects on the process.

The other approach looks at either existing equipment in the plant or a probable design of a full-scale device. How can this be modeled in a pilot plant? This usually involves using narrow-blade impellers and/or small-diameter impellers to more closely decrease the blend time and increase the shear rate over what might usually occur when geometric similarity is used in a pilot plant.

In addition, the variety of shear rates in a big tank means that for bubble or droplet dispersion requirements, the big tank will have a different distribution of bubble sizes than the small tank. This can be very important in such areas as polymerization and particle size analysis.

A. Step #1—What to Do First

First ask yourself' if there is any role for fluid shear stresses in determining and obtaining the desired process result. About half of the time the answer will likely be no. That is the percentage of mixing processes where fluid shear stresses either have no effect or seem to have no effect on the process result. In these cases, mixer design can be based on pumping capacity, blend time, velocities and other matters of that nature. Impeller type location and other geometric variables are major factors in these types of processes.

However, if the answer to this first question is yes; there is an effect of fluid shear stresses on the process, then there needs to have a second question asked. Is it at the micro- or macroscale that the process participants are involved? And, of course, it may be both.

B. Scaleup/Scaledown

Table III shows what happens to many of the variables on scale up. A summary of this is that blend time typically increases and the standard deviation of circulation times

around the mean circulation times also normally increases. The quantitative effect depends somewhat on the degree of uniformity required and the blend time being considered.

As a general rule, the operating speed of the mixer tends to go down, while the peripheral speed of the impeller tends to go up. The speed of the mixer is related to the average impeller zone macroscale shear and thus typically goes down in scaleup while the impeller peripheral speed is often related to the maximum impeller zone macroscale shear rate, see Fig. 5. Out in the rest of the tank (away from the impeller) there another spectrum of shear rates which typically is about a factor of 10 lower than the average impeller zone shear rate. These particular impeller zone shear rates tend to decrease on scaleup.

The microscale environment tends to have a power per unit volume of dissipation around the impeller about 100 times higher than it is in the rest of the tank more or less regardless of the tank size. Thus, the magnitudes of these quantities can be quite similar. This brings up another consideration in the following paragraph.

C. Shear Rate Magnitude and Total Shear Work

Shear stresses and their origin from shear rates (shown in Table VII) gives the magnitude of the shear stress environment that the process participants see. The time they are exposed to that magnitude is a major factor in the process result. For example, it may take a minimum shear stress magnitude to create a certain size particle. However, the ultimate distribution of particle sizes may well relate to the length of time that a particle is exposed to that shear rate. The product of shear stress and time determines what is likely to happen to the process. This obviously is a matter of the spectrum of shear stresses throughout the tank and the statistical distribution of circulation times that particles have going through these zones.

With constant viscosity between the model and the prototype and/or a constant change in viscosity to the process during a batch operation, we can substitute shear rate for shear stress and the product of shear rate times the time is a dimensionless number. Considerable progress is being made toward calculating the velocities, shear rates, and circulating times in mixing vessels, and suitable models and calculations could be made to model these effects in more quantitative detail both on a point-by-point basis and at an overall vessel average. What still is challenging, however,

TABLE VII

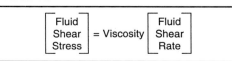

is that it is not usually known what effect these particular properties will have on the process participants in a given process and, thus, it is usually necessary to measure the process result either full scale in the plant, or in smaller size systems in pilot plant or laboratory.

To summarize the situation, geometric similarity controls no mixing variable whatsoever. The question is does that make a difference to the process. In the portion following, we will take a look at the ten basic mixing technology classifications and see what effect these considerations might have. Added to this is the fact that most industrial mixing processes involve two or more of the ten mixing technological classifications and so their interaction between those technology classification parameters must be considered to give the overall performance of the mixing process.

D. What to Do in the Pilot Plant

There are several considerations to bear in mind when planning a pilot plant program.

1. The pilot tank is blending rich while full scale tanks are blending poor. This means that relatively inefficient blending impellers are needed in the pilot

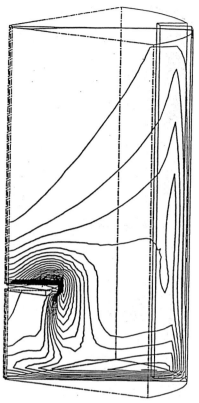

FIGURE 39 Contours of kinetic energy of turbulence.

plant to correspond to the blending efficient impellers used in the plant.

2. One technique to make the pilot plant unit more similar to the plant scale unit is to use impellers of relatively narrow blade width compared to their traditional blade widths used with commercial impellers in the plant. This is purposely reducing the blending performance and improving the shear rate performance in the pilot plant by using impellers of relatively narrow blade width. The blade width cannot be so small that it gets out of proportion to the process participant particles.

3. Always bear in mind the qualitative relationship with viscosity is that the full-scale tank will appear to be less viscous than the pilot plant tank, somewhere in the range of a factor of 10–50.

4. If it appears that upon a qualitative examination that the role of circulation time, blend time, and shear rate may not be important to the process on scaleup, then go ahead and use geometric similarity on the pilot plant study and all of these differences noted above will play no part in the results of the scaleup prediction. It may be that there are compensating effects that while circulation time becomes longer and shear rates become larger, there is a compensating effect that makes the process result satisfactory.

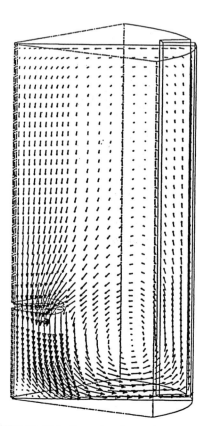

FIGURE 38 Velocity vectors for an A310 impeller.

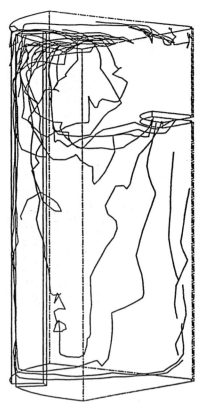

FIGURE 40 A particle trajectory approach with neutral buoyancy particles.

XI. COMPUTATIONAL FLUID DYNAMICS

There are several software programs that are available to model flow patterns of mixing tanks. They allow the prediction of flow patterns based on certain boundary conditions. The most reliable models use accurate fluid mechanics data generated for the impellers in question and a reasonable number of modeling cells to give the overall tank flow pattern. These flow patterns can give velocities, streamlines, and localized kinetic energy values for the systems. Their main use at the present time is to look at the effect of making changes in mixing variables based on doing certain things to the mixing process. These programs can model velocity, shear rates, and kinetic energy, but probably cannot adapt to the actual chemistry of diffusion or mass transfer kinetics of actual industrial process at the present time.

Relatively uncomplicated transparent tank studies with tracer fluids or particles can give a similar feel for the overall flow pattern. It is important that a careful balance be made between the time and expense of calculating these flow patterns with computational fluid dynamics compared to their applicability to an actual industrial process. The future of computational fluid dynamics appears very encouraging and a reasonable amount of time and effort placed in this regard can yield immediate results as well as potential for future process evaluation.

Figures 38–40 show some approaches. Figure 38 shows velocity vectors for an A310 impeller. Figure 39 shows contours of kinetic energy of turbulence. Figure 40 uses a particle trajectory approach with neutral buoyancy particles.

Numerical fluid mechanics can define many of the fluid mechanics parameters for an overall reactor system. Many of the models break the mixing tank up into small microcells. Suitable material and mass transfer balances between these cells throughout the reactor are then made. This can involve long and massive computational requirements. Programs are available that can give reasonably acceptable models of experimental data taken in mixing vessels. Modeling the three-dimensional aspect of a flow pattern in a mixing tank can require a large amount of computing power.

SEE ALSO THE FOLLOWING ARTICLES

FLUID DYNAMICS • FLUID DYNAMICS (CHEMICAL ENGINEERING) • FLUID INCLUSIONS • HEAT TRANSFER • REACTORS IN PROCESS ENGINEERING • SOLVENT EXTRACTION

BIBLIOGRAPHY

Dickey, D. S. (1984). *Chem. Eng.* **91,** 81.

Mcmullen, R., and Weber, M. (1935). *Chem. Metall. Eng.* **42,** 254–257.

Nagata, S. (1975). "Mixing Principles and Applications," Halsted Press, New York.

Nienow, A. W., Hunt, G., and Buckland, B. C. (1994). *Biotech, Bio Eng.* **44,** No. 10, 1177.

Oldshue, J. Y. (1996). *Chem. Eng. Prog.* Vol. **92.**

Oldshue, J. Y. (1980). *Chem. Eng. Prog.* June, pp. 60–64.

Oldshue, J. Y. (1981). *Chemtech.* Sept., pp. 554–561.

Oldshue, J. Y. (1981). *Chem. Eng. Prog.* May, pp. 95–98.

Oldshue, J. Y. (1982). *Chem. Eng. Prog.* May, pp. 68–74.

Oldshue, J. Y. (1983). "Fluid Mixing Technology," McGraw-Hill, New York.

Patwardhan, A. W., Joshi, J. B. (1999). *Ind. Eng. Chem. Pres.* **38,** 49–80.

Tatterson, G. B. (1991). *Fluid Mixing and Gas Dispersion in Agitated Tanks.*

Uhl, V. W., and Grey, J. B. (1966). "Mixing Theory and Practice," Vols. I, II, and III, Academic Press, New York.

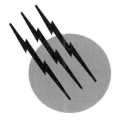

Food Colors

Pericles Markakis
Michigan State University

I. Introduction
II. Natural Food Pigments
III. Food Browning
IV. Color Additives in Foods

GLOSSARY

Anthocyanins Red, blue, and violet water-soluble plant pigments of a phenolic nature.

Browning, food Darkening of foods as a result of enzymatic or nonenzymatic reactions.

Caramel Brown coloring matter made by heating sugars dry or in solution.

Carotenes Chiefly orange-yellow plant and animal pigments; some are provitamins A.

Certification, color Submission of a sample of a listed color additive to the Food and Drug Administration and, after chemical analysis, issuance of a certificate permitting marketing of the batch from which the sample was taken; certain color additives are exempt from certification.

Chlorophyll Green pigment of plants; chemically it is related to the red pigment of blood.

Colorant Substance that colors or modifies the color of another substance.

Excipient Inert substance used as a diluent or vehicle of a colorant.

Heme Color-furnishing portion of the red pigment molecule of blood and meat.

Lakes, color Water-insoluble pigments prepared by precipitating soluble dyes on an insoluble substratum, alumina in the case of food lakes.

Listed color(ant)s Color additives that have been sufficiently evaluated to convince the Food and Drug Administration of their safety for the application intended.

FOOD COLORS are both the sensations evoked when light reflected from foods stimulates the retina of the eye and the particular food components involved in the process. These components, also known as food colorants, may be present in foods naturally, or formed during food processing, or intentionally added to foods, or all of these. This article deals with all groups of food colorants.

I. INTRODUCTION

Color is important for identifying foods, judging their quality, and eliciting aesthetic pleasure in our encounters with them. Because color is usually the first food attribute to strike the senses, its significance in food marketing is obvious ("eating" with the eyes). Thus, all food providers (growers, grocers, homemakers, chefs, and industrial food processors) do their best to present a food with an attractive color. In certain instances, the original color of the food must be preserved, as is the case with most fruits and vegetables. In other instances, culinary art is required to create new, pleasing colors, as when turkey is roasted, bread is baked, or potato chips are fried. In still other instances, colors (colorants) are added to foods, as is done with many beverages and candies.

The coloring matter of foods is discussed under three headings: natural food colors, food browning, and food color additives.

II. NATURAL FOOD PIGMENTS

Approximately 1500 colored compounds, also known as natural food pigments, have been isolated from foodstuffs. On the basis of their chemical structure, these food pigments can be grouped in the following six classes: heme pigments, chlorophylls, carotenoids, flavonoids, betalains, and miscellaneous pigments.

A. Heme Pigments

Heme (from the Greek for blood) is the basic chemical structure (Fig. 1) responsible for the red color of two important animal pigments: hemoglobin, the red pigment of blood, and myoglobin, the red pigment of muscles. Practically all the red color of red meat is due to myoglobin, since the hemoglobin is removed with the bleeding of the slaughtered animal. Other colored muscle compounds (cytochromes, vitamin B_{12}, flavoproteins) do not contribute significantly to the color of red meat.

Myoglobin is a protein that facilitates the transfer of oxygen in muscles. It was the first protein to be fully elucidated with regard to the three-dimensional arrangement of its atoms. Hemoglobin, the oxygen-carrying pigment of blood, is composed of four heme groups attached to four polypeptide chains.

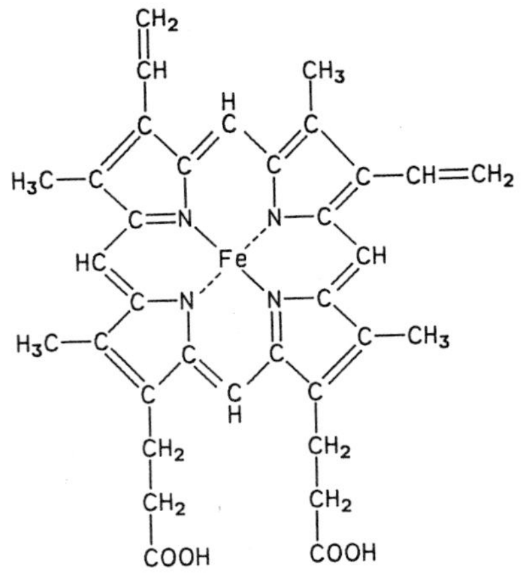

The myoglobin in meat is subject to chemical and color changes. Freshly cut meat looks purplish. On exposure to air, the surface of the meat acquires a more pleasing red hue (blooming of the cut). The color change is due to the oxygenation of myoglobin (an oxygen molecule is attached to the heme group in a fashion parallel to the oxygenation of hemoglobin). The oxygenated myoglobin is called oxymyoglobin. When meat is packed in plastic film, the oxygen permeability of the film should be sufficient to keep the myoglobin oxygenated. In both myoglobin and oxymyoglobin the heme iron is in the Fe^{2+} form. In the presence of oxygen, myoglobin is eventually oxidized to brown metmyoglobin, in which the heme iron is in the Fe^{3+} form. Both the oxygenation and oxidation processes are reversible. Severe oxidative deterioration may result in the formation of green pigments (sulfmyoglobin, cholemyoglobin).

When meat is cooked, the protein moiety (globin) of myoglobin is denatured and the heme is converted chiefly to nicotinamide hemichrome, the entire pigment acquiring a brown hue. These changes are irreversible. Heated meat is also subject to the browning reactions discussed in Section III. A simplified scheme of the red-pigment changes in fresh and heated meat is shown in Fig. 2.

In cured meats, in which nitrite is used, many reactions occur, some of which lead to color changes. Among the established reactions are the following: (1) the nitrite salt is converted to nitric oxide (NO), nitrate, and water; (2) the NO replaces the H_2O attached to the iron of heme and forms nitrosyl myoglobin, which is reddish; (3) on heating, the nitrosyl myoglobin is transformed to nitrosyl hemochrome, which has the familiar pink color of cured meats; and (4) any metmyoglobin present in the cured meat is similarly nitrosylated, reduced, and finally converted to nitrosyl hemochrome.

B. Chlorophylls

Several chlorophylls have been described. Two of them, chlorophyll *a* and chlorophyll *b*, are of particular interest in food coloration because they are common in green plant tissues, in which they are present in the approximate ratio 3 : 1, respectively. Their structures resemble that of heme since they are all derivatives of tetrapyrrole. An important difference is that the central metal atom is iron in heme and magnesium in the chlorophylls. Another difference is that the pyrrole unit IV in the chlorophylls is hydrogenated. In addition, the chlorophylls contain a 20-carbon hydrophobic "tail," the phytyl group (Fig. 3).

The chlorophylls are located in special cellular bodies, the chloroplasts, where they function as photosynthetic

FIGURE 1 Structure of heme.

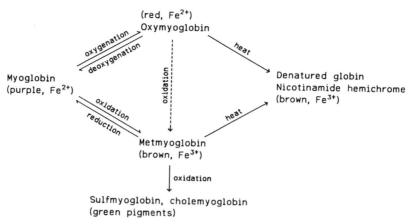

FIGURE 2 Pigment changes in fresh and heated red meat.

agents. As food pigments, chlorophylls impart their green color to many leafy (spinach, lettuce, etc.) and nonleafy (green beans and peas, asparagus, etc.) vegetables and to unripe fruits. They are not very stable pigments, however. Ethylene, a gaseous plant hormone, destroys chlorophylls, and it is occasionally used to degreen fruits. The acids naturally present, formed, or added to plant tissues during food processing convert the bright green chlorophylls to dull olive brown pheophytins by replacing the magnesium of the molecule with hydrogen. Unfortunately, no fail-safe procedure has been proposed for preventing this discoloration in heated and stored green vegetables. Freezing storage is an effective method of preserving the green color of vegetables.

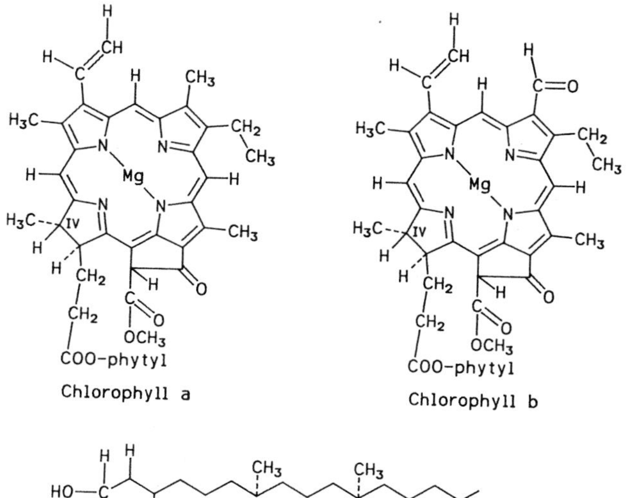

C. Carotenoids

Many of the yellow, orange, and red colors of plants and animals are due to carotenoids, pigments similar to those of carrots. The basic structure of carotenoids is a chain of eight isoprenoid units. Certain isoprenoid derivatives with shorter chains (e.g., vitamin A) are also considered carotenoids. Most of the structural differences among carotenoids exist at the ends of the chain. Some carotenoids are hydrocarbons and are known as carotenes, while others contain oxygen and are called xanthophylls. The structures of several carotenoids, along with the foods or tissues in which they are present, are shown in Table I.

Because of the numerous double bonds in the carotenoid molecule, a large number of cistrans isomers are theoretically possible. The carotenoids of foods, however, are usually in the all-trans form (Table I). Trans to cis transformation is possible and is accelerated by heat, light, and acidity.

Carotenoids occur free or as esters of fatty acids or as complexes with proteins and carbohydrates; for example, in paprika, capsanthin is esterified with lauric acid. In live lobster, astaxanthin is complexed with protein; the astaxanthin–protein complex is blue-gray, the color of live lobster, but on heating, the complex is broken and the freed astaxanthin imparts its red color to the cooked lobster.

Carotenoids are present in a large variety of foods, from yeast and mushrooms, to fruits and vegetables, to eggs, to fats and oils, to fish and shellfish. As fat-soluble substances, carotenoids tend to concentrate in tissues or products rich in lipids, such as egg yolk and skin fat, vegetable oils, and fish oils.

Plants and microorganisms synthesize their own carotenoids, while animals appear to obtain theirs from primary producers. In the development of many fruits (e.g., citrus fruits, apricots, tomatoes) ripening is associated with the accumulation of carotenoids and the

Chlorophyll a

Chlorophyll b

(phytol)

FIGURE 3 Structure of chlorophylls *a* and *b*. [From Aronoff, S. (1966). *In* "The Chlorophylls" (L. P. Vernon and G. R. Seeley, eds.), Academic Press, New York.]

TABLE I   Types of Carotenoids and Their Natural Sources

Structure	Name and source
	β-Carotene (carrot, egg, orange, chicken fat)
	Xanthophyll (vegetables, egg, chicken fat)
	Zeaxanthin (yellow corn, egg, liver)
	Cryptoxanthin (egg, yellow corn, orange)
	Physalien (asparagus, berries)
	Bixin (annatto seeds)
	Lycopene (tomato, pink grapefuit, palm oil)
	Capsanthin (paprika)
	Astaxanthin (lobster, shrimp, salmon)
	Torularhodin (*Rhodotorula* yeast)
	Canthaxanthin (mushrooms)
	β-Apo-8′-carotenal (spinach, orange)

disappearance of chlorophyll. The intensity of the yellow color of certain animal products, such as egg yolk and milk fat or butter, depends on the carotenoid content of the feed the animals ingest. In view of this dependency, the seasonal variation in the color of these products is understandable. A nutritionally important interconversion of carotenoids is the formation of retinol (vitamin A) from

β-carotene and other carotenoids possessing a β-ionone ring and known as provitamins A.

The stability of carotenoids in foods varies greatly, from severe loss to actual gain in carotenoid content during storage. Carotenoid losses amounting to 20 or 30% have been observed in dehydrated vegetables (e.g., carrots, sweet potatoes) stored in air. These losses are minimized when the dry product is stored in vacuum or inert gas (e.g., nitrogen), at low temperatures, and protected from light. The main degradative reaction of carotenoids is oxidation. Oxygen may act either directly on the double bonds or through the hydroperoxides formed during lipid autoxidation. Hydroperoxides formed during enzymatic lipid oxidation can also bleach carotenoids by a coupled lipid–carotenoid oxidation mechanism. On the other hand, certain vegetables, such as squash and sweet potatoes, in which carotenoid biosynthesis continues after harvesting, may manifest an increase in carotenoid content during storage.

D. Flavonoid Pigments

Hundreds of flavone-like pigments are widely distributed among plants. On the basis of their chemical structure, these pigments are grouped in several classes, the most important of which are listed in Table II. The basic structure of all these compounds comprises two benzene rings, A and B, connected by a heterocycle. The classification of flavonoids is based on the nature of the heterocycle (which is open in one class).

Most of these pigments are yellow (Latin, *flavus*). One important exception is the anthocyanins, which display a great variety of red and blue hues. Because of the strong visual impact of anthocyanins on the marketing of fruits and vegetables, these pigments will be discussed in greater detail than other flavonoids.

1. Anthocyanins

The name of these pigments was originally coined to designate the blue (*kyanos*) pigments of flowers (*anthos*). It is now known that not only the blue color, but also the

purple, violet, magenta, and most of the red hues of flowers, fruits, leaves, stems, and roots are attributable to pigments chemically similar to the original "flower blues." Two exceptions are notable: tomatoes owe their red color to lycopene and red beets owe theirs to betanin, pigments not belonging to the anthocyanin group.

Anthocyanins are glycosides of anthocyanidins, the latter being polyhydroxyl and methoxyl derivatives of flavylium. The arrangement of the hydroxyl and methoxyl groups around the flavylium ion in six anthocyanidins common in foods is shown in Fig. 4.

There are at least 10 more anthocyanidins in nature, practically always appearing as glycosides. The number of anthocyanins far exceeds that of anthocyanidins, since monosaccharides, disaccharides, and at times trisaccharides glycosylate the anthocyanidins at various positions (always at 3, occasionally at 5, and seldom at other positions). Eventual acylation with *p*-coumaric, caffeic, and ferulic acids increases the number of natural anthocyanins. An example of acylated anthocyanin is the dark purple eggplant pigment delphinidin, 3-[4-(p-coumaroyl)-L-rhamnosyl-(1 → 6)-D-glycosido] 5-D-glucoside.

The color of anthocyanins is influenced not only by structural features (hydroxylation, methoxylation, glycosylation, acylation), but also by the pH of the solution in which they are present, copigmentation, metal complexation and self-association.

The pH affects both the color and the structure of anthocyanins. In very acidic solution, anthocyanins are red, but as the pH rises the redness diminishes. In freshly prepared alkaline or neutral solution, anthocyanins are blue or violet, but (with the exception of certain multiacylated anthocyanins) they fade within hours or minutes.

In acidic solution four molecular species of anthocyanins exist in equilibrium: a bluish quinoidal (or quinonoidal) base A, a red flavylium cation AH^+, a colorless carbinol pseudo-base B, and a colorless or yellowish chalcone C (Fig. 5).

At very low pH (below 1), the red cation AH^+ dominates, but as the pH rises to 4 or 5, the concentration of the colorless form B increases rapidly at the expense of AH^+, while forms A and C remain scarce. In neutral and alkaline solutions, the concentration of base A rises and its phenolic hydroxyls ionize, yielding unstable blue or violet quinoidal anions A^- (Fig. 6).

Although it is true that the reaction of most plant tissues pigmented with anthocyanins (fruits, flowers, leaves) is slightly acidic, pH alone cannot explain the vivid colors encountered in these tissues. One mechanism leading to the enhancement and stability of anthocyanin coloration is copigmentation, that is, the association of anthocyanins with other organic substances (copigments). This association results in complexes that absorb more

TABLE II   Major Classes of Flavonoids

Class	Structure	Example (source)
Flavones		Apigenin (chamomile)
Flavan-3-ols (catechins)		(−)-Epicatechin (cocoa)
Flavan-3, 4-diols		Leucocyanidin (peanut)
Flavanones		Naringenin, hesperidin (citrus fruits)
Flavonols		Quercetin (apples, grapes)
Flavanonols		Taxifolin (*Prunus*)
Isoflavones		Genistein (soybeans)
Anthocyanidins		Pelargonidin, cyanidin, delphinidin (berries, red apples, red grapes)
Chalcones		Butein (*Butea*)
Aurones		Aureusidin (*Oxalis*)

Pelargonidin
Cyanidin
Delphinidin
Petunidin
Peonidin
Malvidin

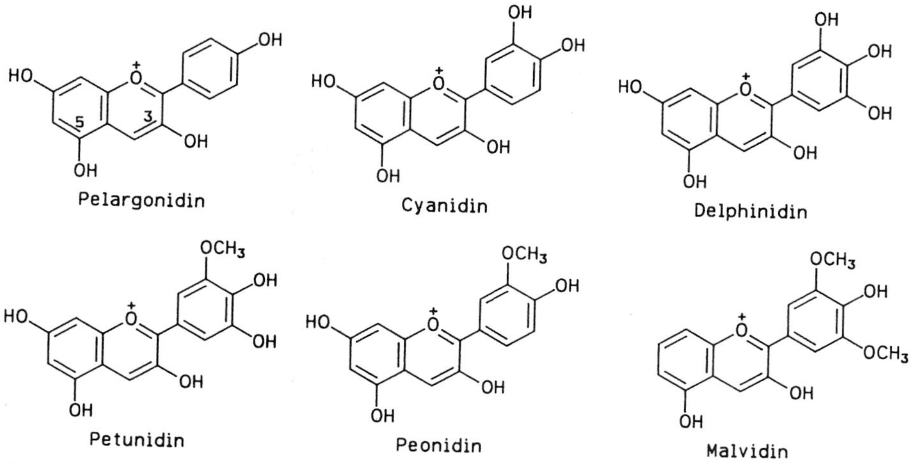

FIGURE 4 Six anthocyanidins common in foods. The electric charge shown at position 1 is delocalized over the entire structure by resonance.

visible light (they are brighter) and light of lower frequency (they look bluer—the bathochromic effect) than the free anthocyanins at tissue pH. Most of these copigments are flavonoids, although compounds belonging to other groups (e.g., alkaloids, amino acids, nucleotides) can function similarly. A stacked molecular complex between an acylated anthocyanin and a copigment (flavocommelin) is shown in Fig. 7.

Self-association is the binding of anthocyanin molecules to one another. It has been observed that the complexes absorb more light than the sum of the single molecules. This explains why a 100-fold increase in the concentration of cyanidin 3,5-glucoside results in a 300-fold rise in absorbance.

A: Quinoidal base
AH⁺: Flavylium cation
C: Chalcone
B: Carbinol pseudo-base

FIGURE 5 Four anthocyanin structures present in aqueous acidic solutions: R is usually H, OH, or OCH_3. Gl is glycosyl. [Adapted from Brouillard, R. (1982). *In* "Anthocyanins as Food Colors" (P. Markakis, ed.), Academic Press, New York.]

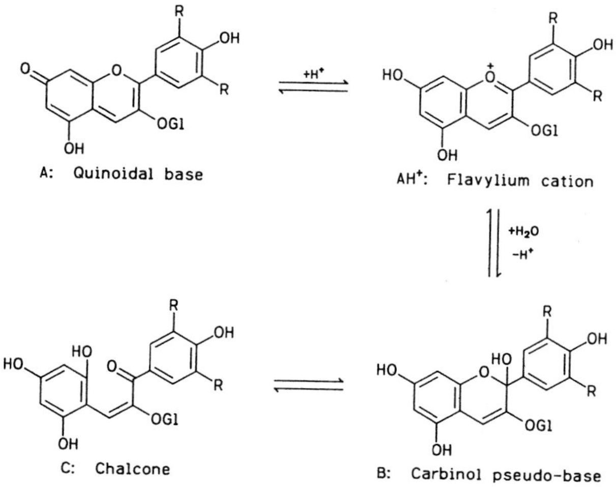

Certain anthocyanins form complexes with metals (e.g., iron, aluminum, magnesium), and the result is an augmentation of the anthocyanin color. At times the complexes involve an anthocyanin, a copigment, and a metal.

A large number of the anthocyanins present in fruits and vegetables have been identified. It is not unusual for a plant tissue to contain several anthocyanins (17 in certain grape varieties), all genetically controlled. Table III shows the anthocyanidin moieties of anthocyanins in common fruits and vegetables.

Generally, the attractive color of anthocyanin-pigmented foods is not very stable. Canning of red cherries or berries results in products with considerable bleaching. Strawberry preserves lose one-half of their anthocyanin content after a few weeks on the shelf, although the browning reaction may mask the loss. And red grape juice is subject to extensive color deterioration during storage.

pH 10
A⁻
pH 2
AH+
pH 6
A
Absorbance
1.0
0.5
0
500
600
700
Wavelength (nm)

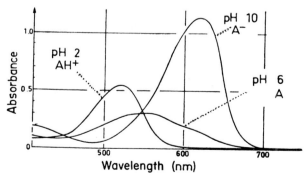

FIGURE 6 Absorption spectra recorded immediately after dissolving an anthocyanin (malvin chloride) in buffers of pH 2, 6, and 10. The absorption peaks at pH 6 and 10 disappeared within 1 to 3 hr. (Adapted from Brouillard, R. (1982). *In* "Anthocyanins as Food Colors" (P. Markakis, ed.), Academic Press, New York.]

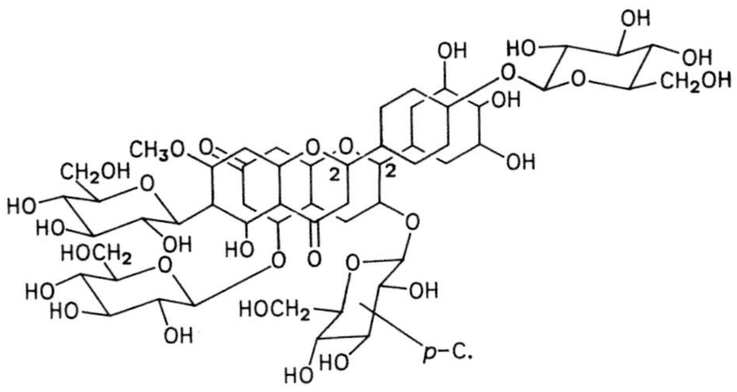

FIGURE 7 Stacked molecular complex of awobanin and flavocommelin; *p*-C. denotes *p*-coumaroyl. [From Osawa, Y. (1982). *In* "Anthocyanins as Food Colors" (P. Markakis, ed.), Academic Press, New York.]

Exposure to high temperatures and contact with the oxygen of the air appear to be two factors affecting anthocyanin stability most adversely. Ascorbic acid accelerates the destruction of anthocyanins, and so does light. Certain oxidizing enzymes, such as phenol oxidase, and a hydrolyzing enzyme known as anthocyanase may contribute to the degradation of anthocyanin pigments. Oxidizing enzymes act on the anthocyanidin moiety, while anthocyanase splits off the sugar residue(s); the freed anthocyanidin is very unstable and loses its color spontaneously. Sulfur dioxide, which is used for the preservation of some fruit products (pulps, musts), bleaches anthocyanin pigments, but on heating of the fruit prduct in vacuum the SO_2 is removed and the anthocyanin color reappears. Large concentrations of SO_2, combined with lime, decolorize anthocyanins irreversibly and are used in the preparation of maraschino cherries. Anthocyanins act as anodic and cathodic depolarizers and thereby accelerate the internal corrosion of tin cans. It is therefore necessary to pack anthocyanin-colored products in cans lined with special enamel. In aging red wines anthocyanins condense with other flavonoids and form polymeric (MW $\leq$ 3000) redbrown pigments (Fig. 8). On continued polymerization these pigments become insoluble and form sediments in bottled red wines.

Anthocyanins possessing more than one acyl group show extraordinary color stability over a wide pH range. One of them, peonidin-3-(dicaffeyl sophoroside) 5-glucoside, isolated from 'Heavenly Blue' morning glory flowers (*Ipomoea tricolor*), has been shown to "produce a wide range of stable colors in foods and beverages which have a pH range of 2.0 to about 8.0." United States patent 4,172,902 covers its use as a colorant in foods.

2. Other Flavonoids

Among flavonoids other than anthocyanins, the catechins, flavonols, and leucoanthocyanidins have the widest dis-

tribution in foodstuffs, while flavonone glycosides are of special interest in citrus fruits.

Catechins, or flavan-3-ols, are present mainly in woody tissues. Among common foods, tea leaves contain at least six catechins representing about 25% of the dry weight of tea leaves. Tea catechins are excellent substrates for the catechol oxidase that is present in tea leaves and participates in the conversion of green tea to black tea. The reddish brown color of tea brew is due to a mixture of

TABLE III Anthocyanidins Present as Anthocyanins in Fruits and Vegetables

Fruit or vegetable	Anthocyanidin
Apple (*Malus pumila*)	Cyanidin
Blackberry (*Rubus fructicosus*)	Cyanidin
Black currant (*Ribes nigrum*)	Cyanidin. delphinidin
Blueberry (lowbush,*Vaccinium angustifolium;* highbush, *V. corymbosum*)	Delphinidin, petunidin, malvidin, peonidin, cyanidin
Cherry (sour, 'Montmorency,' *Prunus cerasus;* sweet, 'Bing,' *P. avium*)	Cyanidin, peonidin
Cranberry (*Vacinnium macrocarpon*)	Cyanidin, peonidin
Elderberry (*Sambucus nigra*)	Cyanidin
Fig (*Ficus carica*)	Cyanidin
Gooseberry (*Ribes grossularia*)	Cyanidin
Grape (red European. *Vitis vinifera*)	Malvidin, peonidin, delphinidin, cyanidin, petunidin, pelargonidin
Grape ('Concord,' *Vitis labrusca*)	Cyanidin, delphinidin, peonidin, malvidin, petunidin
Mango (*Mangifera indica*)	Peonidin
Mulberry (*Morus nigra*)	Cyanidin
Olive (*Olea europea*)	Cyanidin
Orange ('Ruby,' *Citrus sinesis*)	Cyanidin. delphinidin
Passion fruit (*Passiflora edulis*)	Delphinidin
Peach (*Prunus persica*)	Cyanidin
Pear (*Pyrus communis*)	Cyanidin
Plum (*Prunus domestica*)	Cyanidin, peonidin
Pomegranate (*Punica granatum*)	Delphinidin, cyanidin
Raspberry (*Rubus ideaus*)	Cyanidin
Strawberry (*Fragaria chiloensis* and *F. virginiaca*)	Pelargonidin, little cyanidin
Beans (red, black; *Phaseolus vulgaris*)	Pelargonidin, cyanidin, delphinidin
Cabbage (*red, Brassica oleracea*)	Cyanidin
Corn (red, *Zea mays*)	Cyanidin, pelargonidin
Eggplant (*Solanum melongena*)	Delphinidin
Onion (*Alium cepa*)	Cyanidin, peonidin
Potato (*Solanum tuberosum*)	Pelargonidin, cyanidin, delphinidin, petunidin
Radish (*Raphanus sativus*)	Pelargonidin, cyanidin

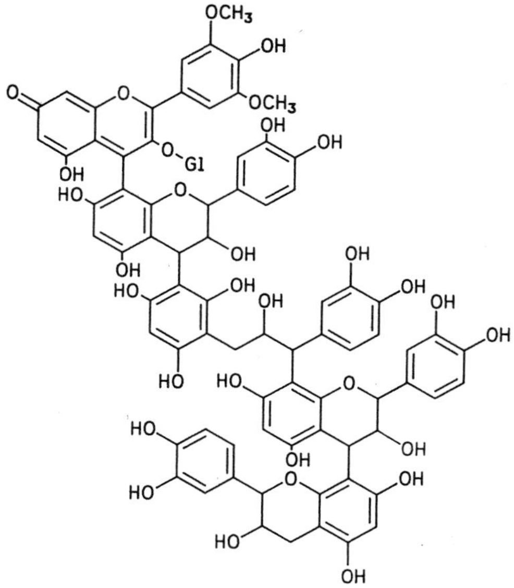

FIGURE 8 Proposed structure for a polymeric red-brown pigment in aging red wine. [From Somers, T. C. (1971). *Phytochemistry* **10,** 2184.]

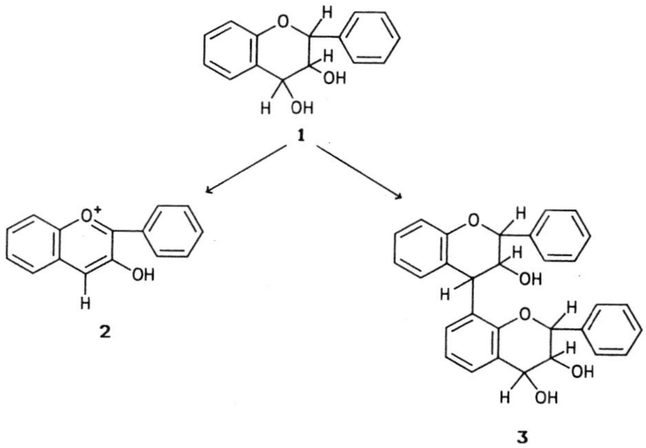

FIGURE 10 Basic structures of leucoanthocyanidins (**1**), anthocyanidins (**2**), and dimeric leucoanthocyanidins (**3**).

pigments known as theaflavins and thearubigins. The structure of one of them is shown in Fig. 9.

Flavonols, like anthocyanidins, exist almost exclusively as glycosides. Three common flavonols are kaempferol, quercetin, and myricetin, resembling pelargonidin, cyanidin, and delphinidin, respectively, in the hydroxylation pattern of the B ring. Flavonol glycosides impart weak yellow hues to apples, apricots, cherries, cranberries, grapes, onions, plums, potatoes, strawberries, tea, tomatoes, and other commodities.

Leucoanthocyanidins are compounds of the general formula **1** shown in Fig. 10. They have no color of their own, but in acidic environments and at elevated temperatures they are converted to colored anthocyanidins (**2**). This reaction is in competition with the condensation to a dimeric leucoanthocyanidin (**3**). Low temperature favors the formation of the dimeric compound, which can polymerize to yield products with pronounced tanning properties.

The most common leucoanthocyanidins are leucopelargonidin, leucocyanidin, and leucodelphinidin, which are converted to the corresponding anthocyanidins. This conversion results in the undesirable "pinking" of certain products such as canned pears, canned banana puree, processed brussels sprouts, and beer. On the other hand, polymerization to tannins leads to astringency and the formation of haze in beer (insolubilization of beer proteins).

E. Betalains

Betalain is a relatively new term used to describe a class of water-soluble plant pigments exemplified by the red-violet betacyanins and yellow betaxanthins. (In a parallel fashion, flavonoids comprise the red-blue anthocyanins and the typical yellow flavonoids that some authors call anthoxanthins.) Betalains owe their name to the red beet (*Beta vulgaris*), from which they were originally extracted, and they are not as widely distributed as flavonoids. Other foods containing betalains include chard, pokeberries, and Indian cactus fruits. The major red pigment of red beets is betanin, and their major yellow pigment is vulgaxanthin (Fig. 11).

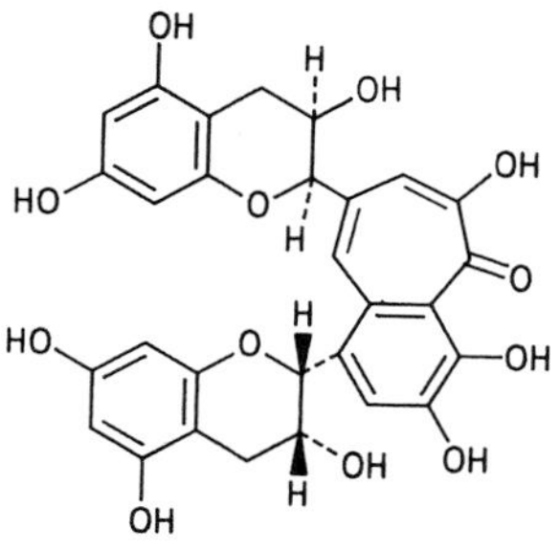

FIGURE 9 Structure of theaflavin.

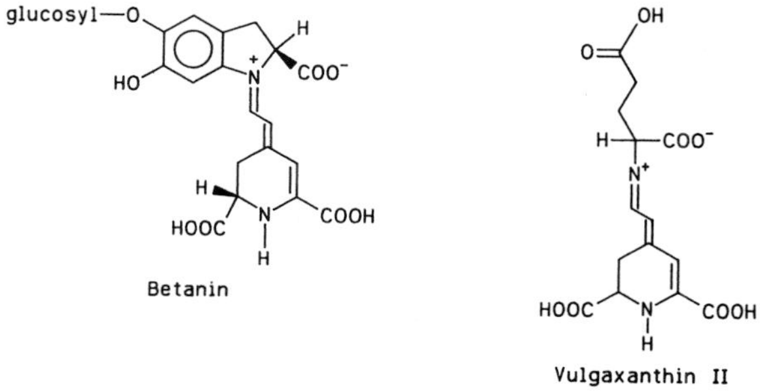

FIGURE 11 Two major pigments of red beets.

Betalains are stable in the pH range 3.5–7.0, which is the pH range of most foods, but they are sensitive to heat, oxidation, and light.

F. Miscellaneous Natural Food Colors

There are several hundred additional natural pigments that are not as widely represented in foods as the previously discussed coloring substances. Among them are the quinones and xanthones, which are yellow pigments. An example of a quinone is juglone, which is present in walnuts and pecans. Mangiferin, a representative of xanthones, is found in mangoes. Tannins include two types of pale yellow to light brown compounds, characterized by their property to convert animal hides to leather. One type consists of condensed tannins, to which reference was made in relation to the leucoanthocyanidins, and the other type comprises hydrolyzable tannins, which are esters of a sugar, usually glucose, with gallic acid, ellagic acid, or both. Corilagin is an example of a gallotannin, in which glucose is esterified with three gallic acid molecules. A yellow pigment that has attracted much attention because of its toxicity to humans and nonruminant animals is gossypol. It is present in cottonseeds, which are used as animal feed and have been considered a potential source of protein for human use. Several biologically very important food constituents are colored, such as phytochrome (yellow), vitamin B_2 (riboflavin, orange-yellow), and vitamin B_{12} (red), although their contribution to food coloration is negligible.

III. FOOD BROWNING

Foods may develop a variety of brown colors, from yellow-brown to red-brown to black-brown, during handling, processing, and storage. These colors are desirable in certain foods (e.g., coffee, beer, bread, maple syrup). In other foods, such as most dehydrated fruits and vegetables, dried eggs, and canned or dried milk, browning is detrimental. Even when desirable, browning should not be excessive, as in potato chips, french fries, and apple juice. Numerous reactions lead to browning in foods. Some of these may also generate flavors and/or alter the nutritional properties of foods. Conventionally, browning is discussed as enzymatic and nonenzymatic browning.

A. Enzymatic Browning

Several enzymes may initiate reactions that eventually produce brown colors in foods. For example, the action of ascorbate oxidase on ascorbic acid or of lipoxidase on lipids leads to carbonyl products that may either polymer-

ize or react with amino compounds and form brown products. Phenolase (or phenol oxidase), however, is the principal browning enzyme. This enzyme oxidizes o-diphenols to o-quinones, which, by nonenzymatic processes, are ultimately converted to brown polymers known as melanins. Melanins are formed in both animal and plant tissues. A typical substrate of phenolase in animals is tyrosine. This amino acid is converted to melanin by a series of reactions, some of which are shown in Fig. 12.

In dark hair, skin, eyes, and other animal tissues, melanin is attached to proteins. Tyrosine is also a phenolase substrate in plant tissues (e.g., potatoes), but o-diphenols and polyphenols are by far the most common substrates of enzymatic browning in foods of plant origin. The following phenolic compounds have been associated with enzymatic browning in some foods: chlorogenic acid, caffeic acid, and catechin in apples, apricots, peaches, and pears; 3,4-dihydroxyphenylethylamine in bananas; (−)-epicatechin, (+)-catechin, (+)-gallocatechin, and (−)-epigallocatechin galate in tea leaves and cocoa beans; catechins in grapes; and tyrosine and chlorogenic acid in potatoes. The structures of four of these phenolics are shown in Fig. 13.

Many fresh fruits and vegetables brown slowly as they senesce. The enzymatic browning of these commodities is more rapid when they are subjected to processing, such as the pressing of apples in making cider or the peeling and cutting of potatoes in preparing potato products. Since enzyme, substrate, and oxygen must all be present for the development of this type of browning, elimination of any one of the three agents will prevent the browning. Heat inactivation of the enzyme, the exclusion of oxygen (by keeping the commodity under water or packaging it under vacuum or inert gas), and the selection of varieties poor in substrate content or enzyme activity are ways of preventing this discoloration. Also, storage at low temperature and the addition of sulfur dioxide, ascorbic acid, citric acid, sodium chloride, or combinations of these compounds will inhibit browning.

B. Nonenzymatic Browning

A number of chemical processes not involving enzymes may result in food browning. Briefly discussed here are the Maillard reaction, caramelization, ascorbic acid browning, and metalpolyphenol browning.

1. The Maillard Reaction (Maillard Browning)

This reaction is actually a series of reactions occurring from the first encounter of a carbonyl compound with an amine compound to the formation of brown pigments. It is also known as the carbonyl–amine reaction, and its brown

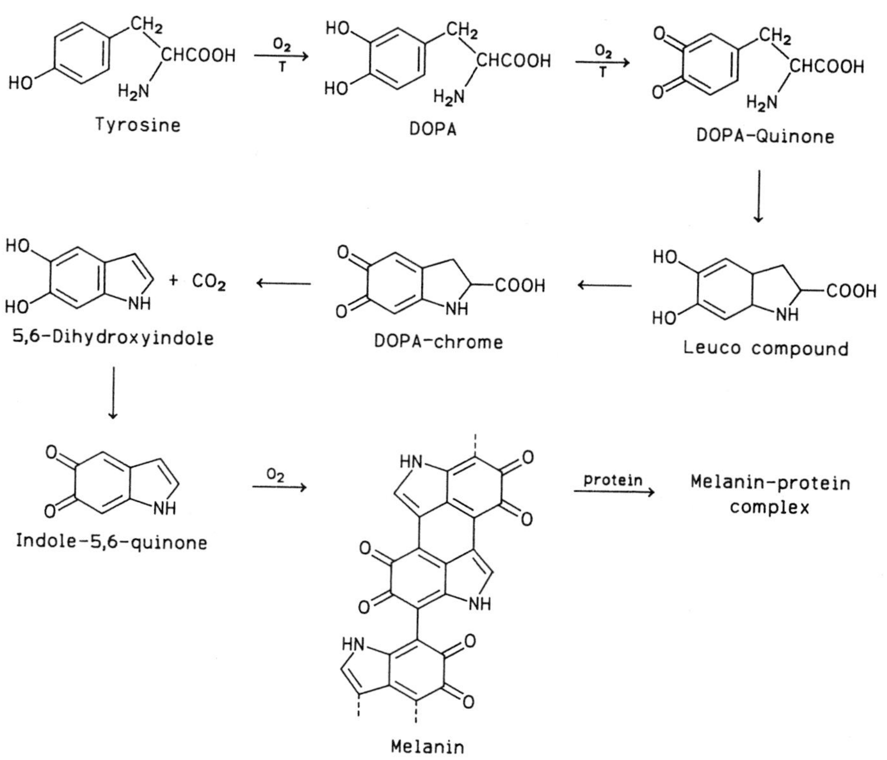

FIGURE 12 Conversion of tyrosine to melanin, catalyzed in part by tyrosinase (T). DOPA, Dihydroxyphenylalanine. Only part of melanin is shown.

products are often called melanoidins, indicating their visual similarity to the melanins of enzymatic browning. The most common carbonyl compounds of foods involved in the Maillard reaction are reducing sugars, and the most common amine compounds are amino acids.

The intermediate reactions and their relative velocities vary with the type of initial reactants and the conditions of

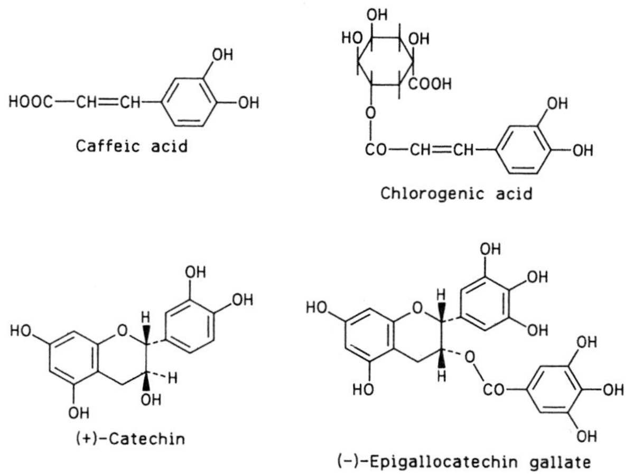

FIGURE 13 Four phenolic compounds involved in enzymatic browning.

the reactions. Among sugars, pentoses are more reactive than hexoses, and hexoses are more reactive than reducing disaccharides. When free amino acids react with sugars, lysine appears to be the most active among them. In peptides and proteins, the N-terminal amino acid is the most reactive, followed by a nonterminal lysine. Raising the temperature and/or the pH accelerates the Maillard reaction. Intermediate water activity appears to maximize this reaction.

Several pathways have been proposed for the formation of melanoidins through the interaction of carbonyl and amine compounds. A simplified scheme is shown in Fig. 14. This scheme involves first the condensation of a carbonyl compound (an aldohexose in this scheme) with an amine to a Schiff base via an intermediate product (not shown). The Schiff base quickly cyclizes to an N-substituted glycosylamine. Up to this step the process is reversible because the glycosylamine can be hydrolyzed back to the initial reactants. The N-substituted glycosylamine is then rearranged to an N-substituted 1-amino-1-deoxy-2-ketose, the Amadori compound, which is in equilibrium with its enol form. If the initial carbonyl compound is a 2-ketose (e.g., fructose), the corresponding N-substituted 2-amino-2-deoxy-1-aldose is formed by the Heyns rearrangement, which is analogous to the Amadori

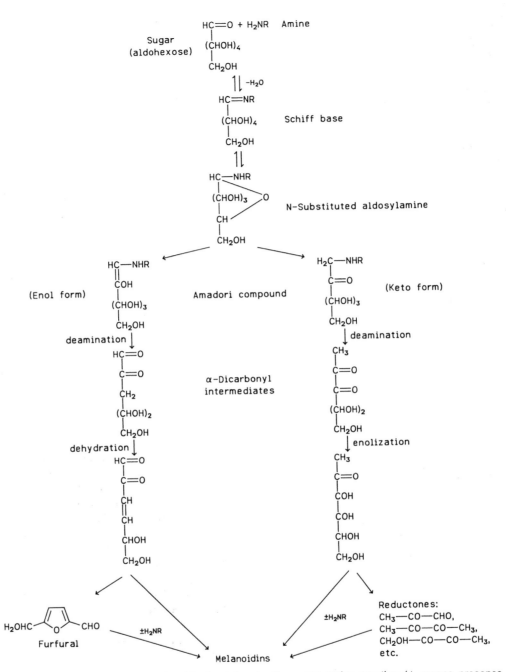

FIGURE 14 Simplified scheme of melanoidin formation by the sugar–amine reaction (± means presence or absence).

rearrangement for aldoses. The Amadori and Heyns compounds are subsequently subjected to a variety of transformations, which may include deamination, dehydration, enolization, cyclization, and degradation. The products of those reactions finally condense or polymerize with or without the participation of an additional amino compound to form dark pigments of a colloidal nature and ill-defined composition—the melanoidins. Some of the

intermediate products are collectively called reductones, because they are strongly reducing compounds that account for the reducing properties of systems undergoing Maillard browning.

It should be mentioned here that a side reaction of the Maillard browning results in the formation of flavorful compounds, such as those associated with the roasting of meat, coffee, or nuts and the baking of bread. This side reaction, known as the Strecker degradation, occurs between α-amino acids and dicarbonyl compounds and leads to the formation of aldehydes possessing one less carbon atom than the corresponding initial amino acids. The newly formed aldehydes are responsible for the pleasing flavors.

A significant consequence of the Maillard reaction is the loss of the nutritional value of the amino acid involved in the reaction. If the participating amino acid is essential, and especially if it is a limiting one, as is lysine in most cereal grains, the Maillard reaction can seriously lower the nutritive value of the food. Toasting, for example, may reduce to one-half the protein efficiency ratio of bread.

2. Caramelization

This reaction leads to brown products when sugars are heated dry or in solution. Certain conditions of caramelization favor the formation of flavor compounds as well. The chemical transformations involved in caramelization are complex and poorly understood. They include dehydration, fragmentation, and polymerization. On the heating of pentoses, furfural is formed which polymerizes to brown products. Heating hexoses results in hydroxymethylfurfural, which polymerizes similarly. The large quantities of industrial caramel color that are added to beverages (cola drinks), baked goods, and confections are made by heating high-conversion corn syrups in the presence of catalysts (acids, alkalis, salts).

3. Ascorbic Acid Browning

When ascorbic acid is heated in the presence of acids, furfural is formed. The latter, either by itself or after reacting with amino compounds, polymerizes to brown products. Citrus juices, especially their concentrates, develop browning, which has been attributed to ascorbic acid degradation.

4. Metal–Polyphenol Browning

Polyphenolic compounds form complexes with certain metals. The polyphenols of fruits and vegetables most commonly chelate iron. The resulting iron complexes are bluish black pigments. Cutting apples with a non-stainless-steel knife results in darkening of both the blade and the cut surface of the apple. This darkening is independent of the enzymatic browning that might develop as a result of cutting. Wine makers avoid contact between the wine and iron implements because of the black iron–tannin precipitate that forms on such contact. Certain varieties of potatoes tend more than others to darken after cooking. This darkening is attributed to a complex between iron and chlorogenic acid. The iron of the tissue must first be oxidized to the ferric state for the blackish complex to appear. The stem end, which contains much less citric acid than the rest of the tuber, displays the deepest darkening. Canned or pickled cauliflower may turn dark due to the interaction of polyphenols in the tissue with iron from external sources.

As already indicated, nonenzymatic browning is desirable in certain instances and undesirable in others. The availability of reactants and the type of conditions (temperature, pH, moisture) will determine the extent of browning. A chemical preservative often used to inhibit nonenzymatic (and enzymatic) browning is sulfur dioxide. An obvious way to prevent metal-polyphenol browning is to eliminate contact between susceptible tissues and reactive metals and use inoffensive equipment (stainless steel, glass-lined tanks, etc.)

IV. COLOR ADDITIVES IN FOODS

Colored substances are added to foods to modify the appearnce of insufficiently colored or discolored foods and to create new foods. The following criteria must be met if a food color (colorant) is to be used; (1) it must be safe at the level and under the conditions of use; (2) it must be stable in the products in which is added; (3) it must not impart any offensive property (flavor, texture) to the product; (4) it must be easy to apply; (5) it must have a high tinctorial power; and (6) it must not be too costly.

There are two classes of color additives, those that must be certified and those that are exempt from certification. Both are strictly controlled in the United States by regulatory statutes (Food Color Additives Amendments), but an official certificate is required for each commercial batch of color of the first group, while no such certificate is necessary for the second group. For certification the manufacturer must submit a sample of the batch to the Food & Drug Administration for chemical analysis. The results of the analysis are compared with the specifications for certified colors published in the Code of Federal Regulations. If the compliance is complete, a certificate is issued for that particular batch of color.

When a new color is to be introduced, the petitioner is expected to provide data proving that the new additive is safe and effective. The safety tests are elaborate and expensive. They include chronic toxicity feeding tests with two animal species over several generations. The effectiveness tests include long-term color stability experiments in the foods to which the color is to be added.

The food color additives subject to certification are listed in Table IV. The initials FD&C stand for the Food, Drug & Cosmetic Act under which these additives are regulated. The food color additives exempt from certification are listed in Table V. Generally, only synthetic organic

TABLE IV Permanently Listed Food Colors Subject to Certification

Color	Structure	Uses[a]
FD&C Blue #1		Used in amounts consistent with GMP
FD&C Green #3		Used in amounts consistent with GMP
FD&C Yellow #5		Used in amounts consistent with GMP
FD&C Red #3[b]		Used in amounts consistent with GMP
FD&C Red #40		Used in amounts consistent with GMP
Orange B		Coloring sausage surfaces or casings (150 ppm max. based on finished product)
FD&C Blue #2		Used in amounts consistent with GMP
FD&C Yellow #6		Used in amounts consistent with GMP
Citrus Red #2		In skins of oranges that are not intended for processing (at 2 ppm max., based on whole fruit)

[a] GMP. Good manufacturing practices.

[b] The FDA has recently banned the use of Red #3 in such products as cake frostings, certain processed foods, cough drops, and lipstick, in which the color is mixed with other additives reacting with it. This dye can still be applied directly to meat, nut products, fruit and fruit juices, candy, confections, and breakfast cereals.

TABLE V Permanently Listed Colors Exempt from Certification

Colorant	Uses[a]	Colorant	Uses[a]
Caramel	In foods, generally consistent with GMP	Cochineal extract; carmine	In foods, generally consistent with GMP
β-Carotene	In foods, generally consistent with GMP	Dehydrated beets	In foods, generally consistent with GMP
Annatto extract	In foods, generally consistent with GMP	Riboflavin	In foods, generally consistent with GMP
Paprika	In foods, generally consistent with GMP	Carrot oil	In foods, generally consistent with GMP
Paprika oleoresin	In foods, generally consistent with GMP	β-Apo-8'-carotenal	In foods, generally not to exceed 25 mg/lb
Turmeric	In foods, generally consistent with GMP	Titanium dioxide	In foods, generally not to exceed 1% by weight
Turmeric oleoresin	In foods, generally consistent with GMP	Grape skin extract	In still and carbonated beverages and alcoholic beverages
Saffron	In foods, generally consistent with GMP	Ferrous gluconate	For coloring ripe olives, consistent with GMP
Fruit juice	In foods, generally consistent with GMP	Canthaxanthin	In foods, generally not to exceed 30 mg/lb
Vegetable juice	In foods, generally consistent with GMP		
Toasted, partially defatted, cooked cottonseed flour	In foods, generally consistent with GMP		

[a] GMP, Good manufacturing practices.

colorants are subject to certification, while natural organic and inorganic colors, such as paprika and titanium oxide are not. The colorant β-carotene is not subject to certification whether it is obtained from a natural source or it is synthetically produced.

While synthetic food dyes are generally water-soluble, food lakes are water-insoluble. Food lakes are prepared by precipitating dyes on alumina. These lakes are useful for coloring water-repelling foods, such as fats and oils, certain gums, as well as packaging materials, e.g., plastic films, lacquers and inks, from which soluble dyes would leach out. Listing of a food dye does not necessarily imply listing the corresponding lake.

Polymeric food dyes have been developed that cannot pass the gastrointestinal wall and are excreted virtually intact in the feces. Toxicity and efficacy tests must be completed before FDA approval is granted to these dyes.

In recent years, plant tissue culture techniques have been applied to the production of food colors. Also the pigments of two fungi: *Monascus anka* and *Monascus purpureus* are being considered for use in foods. These fungal pigments have been used as food colors and medicines in the Far East for hundreds of years.

The regulations regarding color additives can be found in the Code of Federal Regulations, Title 21, Parts 70–82. Changes in these regulations are published in the "Federal Register." Additional information on color additives can be obtained from:

Food & Drug Administration
Division of Color & Cosmetics
200 C Street. S. W. Washington, DC 20204

SOME DOMESTIC SUPPLIERS OF COLOR ADDITIVES

Beatrice Foods Co., 156 W. Grand Ave., Beloit, WI 53511

BIOCON Inc., 518 Codell Dr., Lexington, KY 40509

COLORCON Inc., Moyer Blvd., West Point, PA 19486

Crompton & Knowles Co., 1595 MacArthur Blvd., Mahwah, NJ 07430

Hilton-Davis Co., 2235 Langdon Farm Rd., Cincinnati, OH 45237

H. K. COLOR Group, 155 Helen St., South Plainfield, NJ 07080

Hoffmann-LaRoche Inc., 304 Kingsland St., Nutley, NJ, 07110

Meer Corp., 9500 Railroad Ave., North Bergen, NJ 07047

Pylam Products Co., 1001 Stewart Ave., Garden City, NY 11530

Sethness Products Co., 2367 W. Logan Blvd., Chicago, IL 60647

Sun Chemical Corp., 441 Tompkins Ave., Staten Island, NY 10305

Warner-Jenkinson Co., 2526 Baldwin St., St. Louis, MO 63106

Whittaker, Clark & Daniels Inc., 1000 Coolidge St., South Plainfield, NJ 07080

SEE ALSO THE FOLLOWING ARTICLES

BIOPOLYMERS • NATURAL ANTIOXIDANTS IN FOODS • POLYMERS, SYNTHESIS

BIBLIOGRAPHY

Association of Official Analytical Chemists (1995). "Official Methods of Analysis" 16 ed., Washington, DC.

Food & Drug Administration (1982). "Toxicological Principles for the Safety Assessment of Direct Food Additives and Color Additives in Food," (The Redbook), Washington, DC.

Francis, F. J. (1986). "Handbook of Food Colorant Patents," Food & Nutrition Press (1999). Westport, CT.

Hutchings, J. B. (1999). "Food Color and Appearance," 2nd ed., Aspen Frederick, MD.

Ilker, R. (1987). "In-vitro pigment production," *Food Technol.* **41**(4), 70.

Markakis, P., (ed.). (1982). "Anthocyanins as Food Colors," Academic Press, New York.

Marmion, D. M. (1991). "Handbook of U.S. Colorants for Foods, Drugs and Cosmetics," 3rd ed., Wiley-Interscience, New York.

Walford, J., (ed.). (1984). "Developments in Food Colours," Vols 1 and 2, Elsevier, London.

Fossil Fuel Power Stations— Coal Utilization

L. Douglas Smoot
Larry L. Baxter
Brigham Young University

GLOSSARY

Ash The oxidized product of inorganic material found in coal.

ASTM American Society for Testing and Materials, one of several organizations promulgating standards for fuel analysis and classification.

BACT Best available control technology.

CAAA Clean Air Act amendments of 1990 in the United States, which introduced substantial new controls for nitrogen oxides, sulfur-containing compounds, and, potentially, many metals emissions by power plants.

Char Residual solid from coal particle devolatilization.

Char oxidation Surface reaction of char with oxygen.

Clean Air Act A federal law governing air quality.

Coal devolatilization Thermal decomposition of coal to release volatiles and tars.

Coal rank A systematic method to classify coals based on measurement of volatile matter, fixed carbon and heating value.

Comprehensive code Computerized numerical model for describing flow and combustion processes.

Criteria pollutants Pollutants for which federal law governs emissions; (nitrogen oxides, sulfur-containing compounds, CO, and particulates are the most significant criteria pollutants for coal-based power systems).

daf Dry, ash-free, a commonly cited basis for many coal analyses.

Fouling and slagging Accumulation of residual ash on inner surfaces of power plant boilers, more generically referred to as ash deposition.

Greenhouse gases Gases (e.g., CO_2, CH_4, N_2O, halocarbons) that contribute to atmospheric warming.

Inorganic material The portion of coal that contributes

to ash or is not a fundamental portion of the organic matrix.

ISO International Organization for Standards, an international association of national/regional standards organizations that publishes standards for fuel characterization and classification.

Mineral matter The portion of the inorganic material in coal that occurs in the form of identifiable mineral structures.

NAAQS National Ambient Air Quality Standard.

NSPS New source performance standard; regulatory standards applied to new power stations.

PSD Prevention of Significant Deterioration.

SCR Selective catalytic reduction of nitrogen oxides by NH_3 over a generally vanadium-based catalytic surface.

SNCR Selective noncatalytic conversion of nitrogen oxides by NH_3 via gas-phase reactions.

COAL is the world's most abundant fossil fuel and is the energy source most commonly used for generation of electrical energy. World use of electrical energy represents about 35% of the total world consumption of energy and this share is increasing. Even so, coal presents formidable challenges to the environment. This article outlines the commercial technology used for generation of electricity and mentions developing technologies. Properties of coals and their reaction processes are discussed and advantages and challenges in the use of coal are enumerated. New analysis methods for coal-fired electrical power generators are noted and future directions are indicated.

I. COAL USE IN POWER STATIONS

Most of the world's coal production of over 5 billion short tons a year is used to help generate nearly 14 trillion kWhr of electricity, with especially significant use in the United States, China, Japan, Russia, Canada, and Germany. Coal generates about 56% of the world's electrical power.

World generation of electricity is growing at a more rapid rate than total world energy consumption. The use of fossil fuels (i.e., coal, natural gas, liquified natural gas, petroleum) dominates total energy and electric power production, accounting for 85–86% of the total for both the United States and the world. Petroleum provides the largest fraction of the U.S. (39%) and the world (37%) energy requirements, while coal is the most commonly used fuel for electrical power generation in the United States (52% of total kWhr) and the world. Approximately 87% of all coal consumed in the United States is for gener-

ation of electricity. The U.S. production of coal has increased by 55% in the past 20 years, during the time that U.S. production of natural gas and petroleum has held steady or declined. During this same period, world coal production has increased 39% (with little increase in the past decade). Proven world recoverable reserves of coal represent 72% of all fossil fuel reserves, with petroleum and natural gas each representing about 14% of the total reserve. The world depends increasingly on electrical energy, while coal represents the most common fuel of choice for electric power generation and the world's most abundant fossil fuel reserve. With electricity generation from nuclear energy currently declining and with hydroelectric power increasing more slowly than electric power from fossil fuels, coal may continue to be the world's primary energy source for electricity generation. Yet coal is not without substantial technical challenges, which include mining, pollutant emissions, and global warming concerns.

II. POWER STATION TECHNOLOGIES

A. Power Station Systems and Conversion Cycles

Fuel conversion technologies of current commercial significance include cyclone boilers, pulverized coal boilers, combustion grates, and fluidized beds. Similar technologies, generally operated at less extreme conditions of pressure and temperature, provide steam and hot water in regions with high demand for district or process heating. The most important advanced technologies include gasifiers, high-temperature furnaces and fuel cells. Some of the general characteristics of these technologies are given in Table I, and characteristics of advanced systems are shown in Table II. The advantages and challenges of the various technologies occur in part from the thermodynamic cycles involved in these heat engines.

1. Carnot Cycle

The most efficient means of converting thermal energy into work or electricity is the Carnot cycle. The Carnot cycle represents an idealized heat engine that is unachievable in practical systems but that sets a useful standard for heat engine performance and comparisons. The Carnot efficiency is given by

$$\eta = 1 - \frac{T_{lo}}{T_{hi}}, \tag{1}$$

where T_{hi} and T_{lo} are the peak and exhaust temperatures, respectively, of the working fluid (expressed as absolute

TABLE I   Figures of Merit for Coal-Based Power Generation Systems[a]

	Pulverized coal boilers				Fluid beds			
	Total	Wall-fired	Tangential	Cyclones	Total	Circulating	Bubbling	Grate-fired
Capital cost ($/kW)	950	950	950	850	1350	1450	1250	1250
Peak furnace temperature (K)	2000	2000	2000	2400	1700	1700	1700	1700
Energy/fuel efficiency (%)	38/100	38/100	38/100	38/100	37/99	38/99	36/98	32/95
Fuel flexibility	Moderate	Moderate	Moderate	Excellent	Excellent	Excellent	Excellent	Excellent
Average/max size (MW$_e$)	250/1000	250/1000	250/1000	600/1000	30/250	30/250	30/250	10/150
Pollutant emissions	Moderate	Moderate	Moderate	High	Low	Low	Low	Moderate
Installed capacity (%)								
World	82	43	39	2	8	4	5	1
North America	80	43	38	8	1	<1	<1	<1
Western Europe	80	43	37	0	10	4	6	8
Asia	84	43	41	0	8	4	4	8
Australia	88	43	45	0	3	1	2	2

[a] Values are illustrative of typical trends but vary significantly on case-by-case bases.

temperatures). The strong dependence of efficiency on temperature is clear from this expression.

2. Rankine Cycle

Nearly all steam-based power boilers use the Rankine cycle to generate electricity. A simple Rankine cycle is based on pressurizing water with a pump, heating the pressurized water beyond its boiling point to produce superheated steam in the boiler, expanding the steam through a steam turbine, cooling the residual steam to regenerate water, and returning the water to the pump to start through the cycle again. Commercial boilers operate at both supercritical (>221.2 bar) and subcritical pressures, with supercritical boilers exhibiting higher efficiencies and subcritical boilers encountering less severe pressure requirements.

The efficiency of the Rankine cycle can be improved by returning the steam to the boiler after partial steam expansion in a process referred to as *reheat*. Reheat cycles add significant complexity to the turbine, the boiler, and the controls but, at large scale, the increased complexity and cost can be justified by the increase in efficiency of a few percent. Other significant factors affecting overall

TABLE II   Figures of Merit for Coal-Based Advanced Power Generation Systems

	Gasifiers	Fuel cells	High-temperature furnaces
Capital cost ($/kW)	2000	2500	1100
Energy/fuel efficiency (%)	50/85	45/100	40/100
Fuel flexibility	Moderate	Very poor	Poor
Average/max size (MW$_e$)	10/200	<0.2/15	250/1000

efficiency in the Rankine cycle are parasitic losses (sootblowing, pumps, fans, etc.) and nonidealities in pumps, turbines, etc. In theory, Rankine efficiencies of about 60% are achievable. Actual plant efficiencies for coal-fired systems range from approximately 40% (on a higher heating value or gross calorific value basis) for large-scale systems to less than 20% for small-scale systems. Mathematical expressions for the efficiency of the Rankine cycle are more cumbersome than those for the Carnot efficiency, but the strong reliance on peak temperature is common to all of them.

3. Brayton (or Joule) Cycle

Essentially all gas turbines are based on the Brayton cycle, which is sometimes referred to as a Joule cycle. In this cycle, fuel and air are pressurized, burned, pass through a gas turbine, and exhausted. The exhaust gases are generally used to preheat the fuel or air. Modern gas turbines operate at significantly higher temperatures than steam turbines. The turbine blades present the most critical materials issues, and blades are thus cooled by internal circulation or transpiration. The Brayton cycle is of little current commercial significance in coal-fired power plants but is a major consideration in the most promising proposed schemes for the next generation of coal-driven heat engines. Gas turbine efficiencies can be as high as 60% and gas turbines enjoy a significant turndown ratio with little loss in efficiency.

B. Power Station Technologies

The Rankine cycle forms the basis of power generation from nearly all current coal-fired power stations. The

Brayton cycle offers efficiency and other advantages and is incorporated into many future coal-based technologies, generally in combination with the Rankine cycle (so-called combined cycle systems). The Brayton cycle is also the basis of gas- and oil-fired turbine technologies. Some of the distinguishing characteristics of the major coal conversion technologies and the leading advanced technologies are summarized here.

Most of the energy associated with the efficiency losses compared with ideal systems is available as heat, but not as work or electricity. Somewhat over half of the total energy potentially available from fuel is not converted to electricity/work because of these losses. In many countries (including the United States) the available heat from these losses is rarely used for industrial processes or district heating. Recovering this heat would require collocation of power stations with industrial parks or communities and installation of district heating or process heating systems in the communities or collocated industries. These changes are fraught with social and economic challenges. Nevertheless, these energy losses represent major inefficiencies in the overall energy budgets in many countries.

1. Pulverized-Coal Boilers

The most widely implemented coal-based technology for power generation is the pulverized-coal (pc) boiler. These boilers provide 88% of the total coal-based electric capacity in the United States and similar large fractions of electric capacity abroad. A typical pulverized coal boiler is shown in Fig. 1. Steps in the process are (1) mining of the coal from near-surface or deep mines, (2) transporting crushed coal several centimeters in size to the power sta-

tion, (3) pulverizing the coal to a powder predominantly less than 100 μm in size, (4) pneumatically transporting this coal dust with combustion air into a large combustion chamber with steel walls cooled by water flowing in wall tubes, (5) ignition and combustion of the coal dust and air to create high temperatures, which heats the pressurized cooling water to high temperatures, (6) expansion of the high-pressure water through large steam turbines to create electricity, (7) cleaning of the combustion off-gases to remove gaseous pollutants, (8) marketing or disposal of the noncombustible ash, and (9) quality control, cooling, and recycling of the water.

Fuel preparation for a pc boiler occurs in a mill, where coal is typically reduced to 70% through a 200-mesh (74-μm) screen. Figure 2 shows a particle-size distribution from a bowl-mill grinder typically used in power plants. A typical utility specification for particle size distribution of the coal is that 70% of the mass should pass through a 200-mesh screen, which is equivalent to 70% less than 74 μm. Sometimes the top size is limited to about 120 μm. The pulverized fuel is pneumatically transported to burner levels (commonly one level fed by each mill) with each level made up of a series of burners.

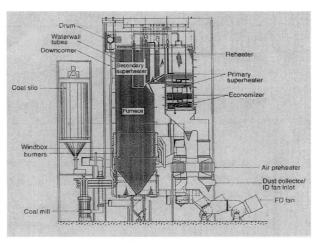

FIGURE 1 Tangentially fired boiler. [Reproduced with permission from Singer, J. G. (ed.), (1981). "Combustion: Fossil Fuel Systems," 3rd Ed., Combustion Engineering, Windsor, CT.

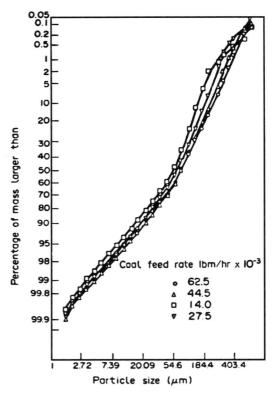

FIGURE 2 Measured pulverized-coal size distribution for a full-scale exhauster-type mill operating at a steady state. [Reproduced with permission from Beér, J. M., Chomiak, J., Smoot, L. D. (1984). *Progress Energy Combustion Sci.* **10**, 229–272.

Two basic types of burners define the two major designs of pc boilers: wall and corner burners. Wall burners are generally configured as a series of annular flows, with the pulverized coal entering near the center and air, often swirled, entering through one or more outer flows. Wall burners are located on either the front or the back wall, or both. Corner burners comprise a stack of individual ducts in or near the boiler corners that individually contain air or a coal suspended in air. The direction of the air and coal streams is biased either clockwise or counterclockwise, creating a large vortex that fills the furnace. The burner tilt and yaw (angle made in the horizontal plane) can be adjusted as operational parameters, although in practice the yaw is generally fixed.

The furnace exit is generally defined as the location at which the flue gas enters in-flow heat exchangers. Such heat exchangers are first encountered either in the form of platens hanging in the top of the furnace or in the form of tube bundles over the furnace nose. Beginning at the in-flow heat exchangers, the gas exits the furnace and enters the convection pass. The highest temperature tube bundles are called superheaters, which are the last heat exchangers through which steam passes before entering the turbine. The steam temperature and pressure entering the turbine are two of the most closely controlled set points in a boiler. The steam temperature range is 500–650°C, with 540°C being a common value. Pressures vary from subcritical to supercritical (>221 bar), with subcritical boilers being most common and generally requiring less maintenance, and supercritical boilers having higher efficiencies. Steam turbines in large-scale power plants are split into two or three sections. Steam exits the high-pressure turbine after a partial expansion, where it flows through another set of heat exchangers called reheaters before returning to the intermediate-pressure turbine.

2. Gasifiers

Coal gasification is a process in which coal is converted to a low-grade gas; it can be regarded in many ways as fuel-rich combustion. Oxygen-blown coal gasifiers operate at overall equivalence ratios (fuel to oxidizer ratio normalized by fuel to oxidizer ratio required to fully oxidize fuel) greater than two. Commercial or near-commercial systems are available that are either air-blown or oxygen-blown. Different designs use suspension firing, slurry firing, or fluid beds. They can be slagging or dry-bottom systems. They operate at either atmospheric or high pressure. The most commercially significant systems for coal are entrained-flow, high-temperature, high-pressure systems. Gasification offers potential environmental and efficiency advantages over conventional coal combustion. Efficiency benefits derive from the potential combustion of the gasi-

fied coal, commonly referred to as producer gas, in a gas turbine. A Brayton cycle for a gas turbine is a more efficient heat engine than the Rankine cycle used in a boiler. If the gases are introduced into the turbine at their peak gasification temperatures and if a conventional Rankine cycle is used to burn the residual char and for the low-grade heat, overall thermal efficiencies as high as 60% (higher-heating-value basis) can be achieved. Gas-phase sulfur forms H_2S in fuel-rich systems, while gas-phase nitrogen forms HCN and NH_3, all of which are much easier to scrub from combustion gases than are their fully oxidized counterparts (SO_2 and nitrogen oxides, NO_x). Cleaning coal-derived gases to the standards demanded by gas turbines (in particular, residual ash, alkali, and sulfur) is a technical challenge.

Other commercial systems include cyclone boilers, fluidized-bed boilers, grate-based furnaces, and fuel cells.

C. Fuel Considerations

A boiler converts chemical energy stored in fuels into steam that is used to produce electricity or for process use. The theoretical and practical limitations of this process determine the ultimate technology and systems making up the overall power plant. Carbon, hydrogen and oxygen are responsible for the majority of the calorific value of the fuel.

A convenient illustration of the relationships between many fuels is based on the carbon, hydrogen, and oxygen content and is referred to as a coalification diagram when restricted to coal (Fig. 3). Natural gas has a hydrogen-to-carbon ratio of about 3.6 and an oxygen-to-carbon ratio of near 0, but is not conveniently illustrated at the scale of this diagram. As is seen, major solid fuels fall into consistent regions of the diagram.

The remaining fuel components are typically nitrogen, sulfur, and inorganic impurities. It is ultimately the

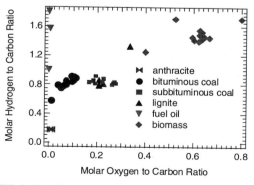

FIGURE 3 Coalification diagram indicating the relationships among a variety of solid fuels in terms of their chemical composition.

impurities in the fuel and process materials that drive engineering decisions regarding power station technologies and operating conditions. The contributions most directly associated with carbon, hydrogen, and oxygen such as peak flame temperature have less influence on performance than idealized theoretical analyses might suggest. Fuel impurities and limitations of current materials place the greatest restrictions on boiler design and operation. With low-grade fuels these restrictions can be substantial.

Coal is commonly classified by rank, with the major classifications being, in order of decreasing rank, anthracite, bituminous coal, subbituminous coal, and lignite. The data in Fig. 3 illustrate how fuel properties vary with rank and how they relate to each other. Coals most commonly used for steam generation have molecular hydrogen-to-carbon ratios ranging from 0.7 to 0.9 and molecular oxygen-to-carbon ratios ranging from near 0 to about 0.2.

While coal rank correlates with oxygen and hydrogen content, coal rank is determined principally by heating value except for the highest rank coals and anthracites, which are distinguished on the basis of fixed carbon contents (according to ASTM standards). Heating value is the amount of energy released from a fuel during combustion and is discussed later in this article.

III. COAL PROPERTIES AND PREPARATION

A. Coal Classification

A common means of characterizing the broad range of coals in use is by rank, moisture, inorganic material, and heating value, which ranges from about 9 to 35 MJ/kg (4000 to about 15,000 Btu/lb). Moisture and inorganic material percentages typically range from 1% to 30% and from 5% to over 50%, respectively. Table III summarizes several properties of various coals. The heating values are conveniently cited for scientific purposes on dry, ash-free bases, but for commercial purposes they are typically moist, inorganic-matter-free values. In the United States, coals are ranked primarily on a heating-value basis up to a value of 33 MJ/kg (14,000 Btu/lb). Coals with heating values higher than 33 MJ/kg are ranked based on volatile matter/fixed carbon contents. Coals range in rank from lignites to low-volatile, bituminous coals and anthracites, with the bulk of the coals representing either subbituminous, low-sulfur coals or high-volatile bituminous coals.

Figure 3 illustrates the relationship of anthracite, bituminous and subbituminous coals, lignite, fuel oil, and biomass in terms of the atomic hydrogen-to-carbon (H:C) and oxygen-to-carbon (O:C) ratios. This type of diagram indicates some chemical structure, combustion, and inorganic aspects of coals and other solid fuels. For example, increasing the H:C or O:C ratio implies decreasing aromaticity of the fuel. Increasing the O:C ratio implies increasing hydroxyl, carboxyl, ether, and ketone functional groups in the fuel. Both the aromaticity and the oxygen-containing functional groups influence the modes of occurrence of inorganic material in fuel and inorganic transformations during combustion.

B. Inorganic Matter

One type of inorganic material in coal is inherent inorganic material that can be volatile at combustion temperatures and includes some of the alkali and alkaline earth metals, most notably sodium and potassium. Anthracites typically lose less than 10% of their mass by pyrolysis. Bituminous and subbituminous coals lose between 5% and 65% of their mass by this process. Lignites, peats, and biomass can lose over 90% of their mass in this first stage of combustion. The large quantities of gases or tars leaving coals, lignites, and biomass fuels can convectively carry inorganic material out of the fuel, even if the inorganic material itself is nonvolatile. The combination of high oxygen content and high organic volatile matter in subbituminous coals indicates a potential for creating large amounts of inorganic vapors during combustion.

The second class of inorganic material in solid fuels includes material that is added to the fuel from extraneous sources. Geologic processes and mining techniques contribute much of this material to coal. This adventitious material is often particulate in nature (in contrast to the atomically dispersed material) and is the dominant contributor to fly ash particles larger than about 10 μm. Examples include silica, pyrite, calcite, kaolinite, illite, and other silicates.

Typically, a boiler designed for combustion of subbituminous coals is about 40% taller and 20% wider and deeper than a similarly rated boiler designed for bituminous coal. This increase in size is required even though the low-rank fuel combusts notably faster than the high-rank fuel. The primary reason for the increase in size is to accommodate differences in the behavior of the inorganic material associated with these two coals, specifically the properties of the ash deposits that they form. The increased surface area in the boiler for the low-rank fuel allows the operator to manage this ash behavior without compromising boiler lifetime or damaging boiler components. In the past, boilers largely burned coals supplied by local or regional mines under long-term contracts. A combination of environmental pressures and deregulation has created much larger variation in the coals used by most current boilers and has led to widespread fuel blending.

TABLE III Major Fuel Properties as Determined by Standardized Analyses for Typical Samples of Each Major Coal Rank

	Beulah	Black Thunder	Pittsburgh #8	Pocahontas #3
Nominal rank	Lignite	Subbituminous	Bituminous	Low Volatile Bituminous
Proximate analysis				
Moisture (as received)	26.89	21.3	1.02	0.62
Ash (dry basis)	13.86	6.46	10.68	4.51
Volatile matter (dry basis)	42.78	54.26	40.16	18.49
Fixed carbon (dry basis)	43.36	39.28	49.16	77
Ultimate analysis (daf basis)				
Carbon	70.49	74.73	80.06	91.65
Hydrogen	4.75	5.4	5.63	4.46
Nitrogen	1.22	1	1.39	1.31
Sulfur	2.14	0.51	5.35	0.79
Oxygen (by difference)	21.36	18.27	7.57	1.62
Chlorine	0.05	0.08	0	0.17
Ash chemistry/elemental ash (% of ash)				
SiO_2	21.23	30.67	41.7	37.92
Al_2O_3	13.97	16.46	20.66	24.16
TiO_2	0.42	1.29	0.9	1.14
Fe_2O_3	12.25	5.1	29.24	17.14
CaO	16.36	21.32	2.08	7.67
MgO	4.46	4.8	0.79	2.4
K_2O	0.22	0.35	1.74	1.84
Na_2O	6.5	1.43	0.4	0.83
SO_3	24.6	17.67	2.35	6.79
P_2O_5	0	0.92	0.15	0.1
Heating value (MJ/kg, daf basis)	23.4	29.8	32.2	35.0
Form of sulfur (% of daf fuel)				
Sulfatic	0.16	0	0.15	0
Pyritic	0.47	0.08	1.81	0.21
Organic	1.18	0.4	2.82	0.54
Ash fusion temperature (°C)				
Reducing atmosphere				
Initial deformation	774	1161	1047	1191
Hemispherical	1636	1178	1082	1249
Spherical	1650	1189	1179	1278
Fluid	1659	1210	1222	1331
Oxidizing atmosphere				
Initial deformation	1714	1179	1337	1297
Hemispherical	1753	1200	1372	1313
Spherical	1756	1209	1381	1337
Fluid	1769	1243	1389	1364

In the past, long-term experience with a limited number of coals partially compensated for an inability to quantitatively anticipate the behavior of the inorganic material in these coals. Current fuel purchasing trends require much more adept boiler operation.

In general, ash behavior depends on fuel properties, boiler design, and boiler operation. Industrial practice has led to the development of a wide variety of indices, nearly all of which are based on fuel properties, for predicting ash behavior. These include arithmetic combinations of elemental mass fractions in the ash, fusion temperatures, and slag viscosity-related measurements. It is on the basis of these parameters that the coal-fired electrical power industry has developed a large, stable, reliable suite of boilers. However, there is a critical need for improved accuracy, given the combination of the effect of ash on boiler performance and current market trends leading to increased variation and generally decreased quality of coals used in most boilers.

C. Standardized Analysis Methods

A large number of coal analyses have been standardized by ASTM, ISO, and similar organizations around the world. The most commonly cited and generally useful among these include energy-content, proximate, ultimate, ash-chemistry, and ash-fusion-temperature analyses. The specific procedures by which these analyses are performed differ slightly from organization to organization, but generally they are more similar than they are different. The following briefly discusses the general objectives of the analyses, some typical results, and a few easily misinterpretable features of these analyses.

1. Coal Energy Content and Flame Temperatures

Energy content represents possibly the single most relevant property of a fuel. The standard measure of a fuel's energy content is the heating value or calorific value, which typically ranges from 23 to 35 MJ/kg (10,000–15,000 Btu/lb) on a dry, ash-free basis. Since some low-rank coals contain large quantities of both ash and moisture, as-received values may be as low as 13 MJ/kg (6000 Btu/lb) or sometimes lower in exceptional cases. The practical importance of this parameter for combustion technology is obvious. The heating value is more commonly referred to as calorific value in countries that use SI units in commerce. Two different heating values/calorific values are commonly cited. The higher heating value (gross calorific value) assumes that water forms a liquid product after combustion, whereas the lower heating value (net calorific value) assumes that steam vapor is the final form of the water, the difference between the two being the heat of vaporization of the water formed during combustion and contained in the fuel. Most computations in the United States and Australia are based on higher heating value or gross calorific value, whereas most of the rest of the world bases calculations on net calorific value. Performance statistics such as efficiencies are consequently not directly comparable among various countries.

All standardized heating value analyses measure the heat released from the fuel when burned with enough air to fully oxidize the fuel (carbon forms carbon dioxide, hydrogen forms water, etc.), normalized by the mass of the fuel, *not the mass of all combustion products*. This value determines the amount of fuel required to release a given amount of heat during combustion. It does not directly indicate the peak temperatures of the resulting flames. These temperatures play an important role in some

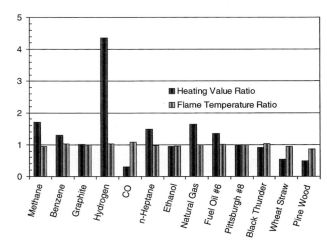

FIGURE 4 Variation of heating value and adiabatic flame temperature with fuel type. All values are normalized by those for bituminous coal.

theoretical analyses of boiler performance such as the heat engine cycles discussed previously.

A useful figure of merit for most fuels is the adiabatic flame temperature. The adiabatic flame temperature defines the temperature of the products of combustion after all chemical reactions have reached equilibrium and when no heat is allowed to escape (or enter) the combustor. Each fuel has a unique adiabatic flame temperature for a given amount of air. As the ratio of fuel to air is varied, the adiabatic flame temperature varies. The highest value generally occurs at an air-to-fuel ratio with slightly less air than is required to convert all of the carbon of CO_2 and hydrogen to H_2O.

The adiabatic flame temperature does not correlate well with heating value over a broad range of fuels. Figure 4 illustrates both heating values and adiabatic flame temperatures for a variety of fuels. All values are normalized to those for a high-volatile bituminous coal and all calculations assume dry fuels. The lack of correlation between heating value and flame temperature is clear. A comparison of hydrogen and carbon monoxide vividly illustrates this point. The heating values for these two fuels differ by a factor of about 14, whereas the adiabatic flame temperatures differ by less than 6%, with that of CO exceeding that of hydrogen. The wide disparity in heating values but similar adiabatic flame temperatures arises primarily from the varying oxygen content of the fuels. Fuels that contain significant oxygen require less air to completely burn, so the heat release is distributed within less mass. The net effect of less heat generation and less mass to absorb that heat is a flame temperature that does not significantly vary from fuel to fuel, presuming dry fuels.

Despite the dominant role that flame temperature plays in theoretical analysis of cycle efficiencies, peak flame temperature has little role in practical systems. The efficiency of virtually all combustion systems is limited ultimately by materials, not flame temperatures. Most dry fuels have peak adiabatic flame temperatures that slightly exceed 2000°C (assuming air as an oxidizer). However, no commercially viable boiler materials can withstand such temperatures as either structural members or pressure parts. Therefore, the peak temperatures achieved in gas and steam turbines are determined by materials constraints, not fuel constraints. Furthermore, simple theoretical cycles such as the Carnot cycle assume ideal heat transfer and fluid flow that are not achievable in practice, further limiting the efficiency of real systems. A consequence of these limitations is that most coal combustion systems convert less than 40% of the energy in the fuel (as measured by gross calorific value or higher heating value) to electricity. The departure from ideal behavior arises, in order of significance, from a lack of suitable high-temperature materials, nonidealities in fluid flow and heat transfer, and process-related losses of energy for auxiliary equipment. Steam turbines typically operate with peak steam temperatures of 500–600°C (900–1100°F). Gas turbines typically operate with peak temperatures as high as 1425°C (2600°F), with the high-temperature limit depending on sophisticated blade-cooling systems. There are many examples of lower performance systems, especially in small systems and those that are optimized for steam rather than power production.

2. Proximate and Ultimate Analyses

The proximate analysis indicates the moisture, ash, and volatiles content of a fuel. The volatile yield is sensitive to heating rate and peak temperature. The yield experienced during the relatively slow and cool proximate analysis is lower than that experienced by coal in many combustors by about 30%. However, the relative volatile yields of many coals are approximately indicated by the proximate analysis.

The ultimate analysis apportions the composition of the organic portion of the fuel among the elements C, H, N, and S. The remaining mass is generally assumed to be oxygen. Some details of this analysis are commonly misunderstood with respect to sulfur, chorine, and ash. Ash typically includes some sulfate. The sulfur content reported in the ultimate analysis is most commonly based on a sample heated to 1350°C in an oxidizing environment, well above the proximate ashing temperature of about 750–815°C. Often, 20% or more of the ash from western coals is in the form of sulfates or other forms represented as SO_3. When proximate analysis ash is included with the ultimate analysis, the sum of ash, carbon, hydrogen, nitrogen, sulfur, and oxygen will exceed 100%. The excess arises from a portion of the sulfur being counted twice, once in the ash

and once in the ultimate analysis for sulfur. Since oxygen is normally determined by difference, this confusion potentially biases both the sulfur and oxygen values. To avoid such confusion, ash should be included with the proximate analysis and the ultimate analysis should be reported on an ash-free basis. A significant fraction of the sulfur in high-rank U.S. coals derives from pyrite; an accurate estimate of the amount of pyrite in the fuel can be obtained from relatively simple, standardized forms of sulfur analyses.

Chlorine analyses generally are performed separately from ultimate analyses. Essentially all of the chlorine in coals is released during the ultimate analysis process. Corrections to the oxygen content should be made to account for the presence of chlorine in the evolved gases if oxygen is determined by difference. U.S. coals typically have less than 0.15% chlorine, with 1% chlorine being an extreme upper bound. Oxygen content, on the other hand, ranges from 4% to 20% depending strongly on rank. Even small amounts of chlorine can have large impacts on inorganic transformation and deposition. Chlorine is much more significant in some international coals and in many fuels fired in conjunction with coal.

3. Ash Chemistry and Ash Fusion Temperatures

The most uniquely suited standardized analyses for ash deposition include ash chemistry and ash fusion temperature. The total ash content from proximate analysis and ash composition provide the fuel information that goes into the majority of common empirical indices of ash behavior, along with ash fusion temperature. The ash chemistry analysis typically reports the ash elemental composition on an oxide basis. This does not mean that all of the species exist as oxides in the fuel (which they do not). It is a convenient method of checking the consistency of the data. The sum of the oxides should be about the same as the total ash content. The analysis is fundamentally an elemental analysis with no distinction of the chemical speciation of the inorganic species.

D. Advanced Analysis Methods

A variety of advanced techniques exists for determining the species composition of the fuel, which is vital to improved predictive methods for coal, and especially ash behavior, in boilers. X-ray diffraction, performed on low-temperature ash, is the most widely used technique for qualitatively identifying the presence of minerals in their crystalline form in concentrations of a few weight percent or greater. Thermal analytical techniques such as differential thermal analysis (DTA) and thermogravimetric analysis (TGA) have been used as a signature analysis based on changes in physical properties with temperature.

Microanalytical techniques, including scanning electron microscopy (SEM) equipped with energy dispersive microscopy (TEM), resolve particles to submicrometer scales. The techniques may be applied to low-temperature ash as well as raw coal and provide visual images of the morphology of the inorganic structure. Extended X-ray analysis (XANE) is another signature analytical technique sometimes applied to coal and petroleum coke. Mössbauer spectroscopy has been widely used for characterizing the inorganic forms of iron in coal as well as in slags and deposits. Computer-controlled scanning electron microscopy (CCSEM) is a commonly used technique that measures the chemical composition and size of individual inorganic grains in coal particles. It forms the basis of several models of ash deposition during coal combustion. Chemical fractionation is a nonstandardized but relatively accessible technique for characterization of the modes of occurrence of both mineral and nonmineral inorganic material in coal. These analyses, particularly the last two, are becoming increasingly important for predicting ash behavior in boilers. Detailed descriptions of inorganic material in coal can be inferred from chemical fractionation, as illustrated in Fig. 5.

IV. COAL REACTION PROCESSES

A. Reaction Types

Coal particle reactions are an essential aspect of all coal combustion processes. These reactions have a direct impact on the formation of fine particles, nitrogen-

TABLE IV Typical Process Characteristics for Various Coal-Related Combustion Technologies[a]

Process	Particle size (μm)	Flame temperature (K)	Typical residence time (sec)
Entrained flow	1–100	1900–2000	<1
Fluidized bed	1500–6000	1000–1450	10–500
Stoker/fixed bed	10,000–50,000	<2000	500–5000

[a] Adapted from Table 3.1, *in* Smoot, L. Douglas, and Smith, Philip J. (1985). "Coal Combustion and Gasification," Plenum Press, New York.

and sulfur-containing species, and other pollutants. They also dominate heat transfer and fluid dynamic processes. Table IV summarizes typical residence times, flame temperatures, and particle sizes for three generic process types: entrained flow, fluidized bed, and fixed bed.

A schematic diagram of a reacting coal particle is given in Fig. 6. The diagram suggests that the particle, at any time in the reaction process, may be composed of moisture, (raw) coal, char, inorganic matter, and ash. Ash represents the condensed-phase product of inorganic matter transformations in a manner similar to char representing the condensed-phase product of coal pyrolysis. The particle may be reacting in either oxidizing or reducing gases, depending on boiler design and operation and the location of the particle. Commonly, the center of the particle is in a reducing environment while the surface is oxidizing. Reactions of the organic portion of the coal proceed by two primary processes: devolatilization and char oxidation.

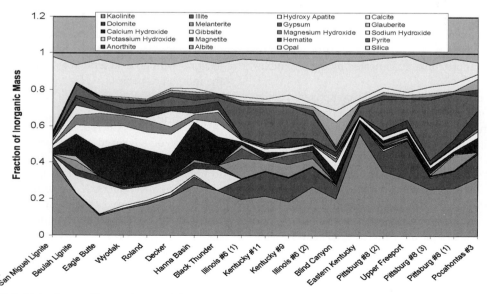

FIGURE 5 Major forms of inorganic material in a suite of U.S. coals, arranged in rank order and by mineral class. Species are presented from bottom to top in the order indicated in the legend.

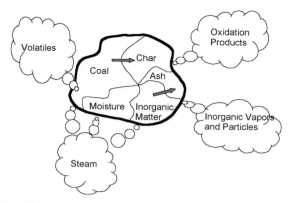

FIGURE 6 Schematic diagram of coal particle reactions during combustion.

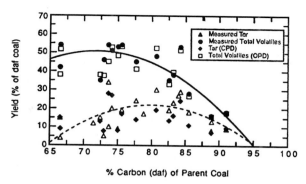

FIGURE 7 Comparison of predicted and measured tar and total volatiles yields for a wide range of coals. Carbon content is used to illustrate coal rank. Predictions were made with a coal devolatilization network model. Solid and dashed lines represent correlation of the measured total volatiles and tar yields, respectively. [Reproduced with permission from Fletcher, T. H., Solum, M. S., Grant, D. M., Pugmire, R. J. (1992). "The Chemical Structure of Chars in Transition from Devolatilization to Combustion," *Energy and Fuels* **6**, 643–650.]

1. Coal Devolatilization

Coal devolatilization or pyrolysis dominates the early stages of combustion. This process involves the homogeneous, solid-phase thermal decomposition of coal chemical structure, producing gaseous and solid products. The gaseous products include light gases (CO, CO_2, CH_4, H_2O, H_2, etc.) and heavy hydrocarbon gases that condense at room temperature. The latter are called *tars*. The general, but not universal, convention is to call this process *devolatilization* when it occurs under unspecified conditions or combinations of reducing and oxidizing conditions. When it occurs under neutral or reducing conditions, it is sometimes called *pyrolysis*. Devolatilization generates the gases that form visible flames in wood (and coal) fires.

This part of the reaction cycle occurs as the raw coal is heated and thermally decomposes. The particle may soften and undergo internal transformation. Moisture present in the coal evolves early as the temperature rises. As the temperature continues to increase, gases and heavy tarry substances are emitted as the coal chemical structure thermally decomposes. The extent of devolatilization can vary from a few percent up to 70–80% of the total particle weight and can take place in a few milliseconds to several minutes or longer, depending on coal size and type and temperature. Devolatilization typically occurs in a few milliseconds and accounts for about 60% of the organic mass loss in pulverized coal processes. The residual mass, enriched in carbon and inorganic material and depleted in oxygen and hydrogen, is called *char*. The char particle often has many cracks or holes made by escaping gases, may have swelled to a larger size than the original coal particle, and can have high porosity. The nature of the char is dependent on the original coal type and size and the conditions of devolatilization.

Five principal devolatilization phases have been identified. The first phase is moisture vaporization near the boiling point of water. The second phase (627–675 K)

is associated with the initial evolution of carbon dioxide and carbon monoxide and a small amount of tar. The third phase (800–1000 K) is the evolution of chemically formed water and carbon oxides. The fourth phase (1000–1200 K) involves the final evolution of tar, hydrogen, and hydrocarbon gases. The fifth phase is the formation of carbon oxides at high temperatures. Structural properties of the coal strongly affect devolatilization mechanisms and rates. Yield of volatiles depends on external factors such as peak temperature, heating rate, pressure, and particle size as well as coal type and chemical structure. Proportions of gases and tars vary widely, as shown in Fig. 7. Rates and amounts of weight loss during devolatilization differ among coal types, with much more liquid and tar products for the bituminous coals and more gaseous products (including H_2O) from lower rank coals.

Weight loss is not significantly size dependent for particles up to about 400 μm in diameter. However, very large particles behave differently from finely pulverized coal. Larger particles heat less rapidly and less uniformly, so that a single temperature cannot be used to characterize the entire particle. The internal char surface provides sites where secondary reactions occur. Devolatilization products generated near the center of a particle must migrate to the outside to escape. During this migration, they may crack, condense, or polymerize with some carbon deposition and smaller yield of volatiles. Devolatilization rates and yields consequently decrease with increasing particle size.

2. Char Oxidation

Char oxidation is an inherently heterogeneous, gas–solid reaction with the gas composition dominating the rates and yields. When this reaction occurs with oxygen it is generally called *oxidation*. When it occurs with CO_2, H_2O, or other gases it can be called *gasification*. Gasification is sometimes used to describe the combination of devolatilization/pyrolysis reactions and heterogeneous reactions with the char under overall reducing conditions.

The residual char particles can be oxidized or burned away by direct contact with oxygen at sufficiently high temperature. This reaction of the char and oxygen is heterogeneous, with gaseous oxygen diffusing toward and into the particle, adsorbing and reacting on the particle surface, and desorbing as either CO or CO_2. For small particles at high temperature, CO is the dominant surface product. This heterogeneous process is often much slower than the devolatilization process, requiring seconds for small particles to several minutes or more for larger particles. These rates vary with coal type, temperature, pressure, char characteristics (e.g., size, surface area, porosity), presence of inorganic impurities, and oxidizer concentration. At atmospheric pressure, gasification reactions of char with CO_2 and H_2O are much slower than oxidation.

These two processes (i.e., devolatilization and char oxidation) may take place simultaneously, especially at very high heating rates. If devolatilization takes place in an oxidizing environment (e.g., air), then the fuel-rich gaseous and tar products react further in the gas phase to produce high temperatures in the vicinity of the coal particles.

B. Rate Processes and Equations

1. Coal Devolatilization

The earliest coal devolatilization models were simple single-step, single-equation expressions in which coal was conceived to form gases and char with a specified yield, as follows:

$$\frac{dv}{dt} = k(v_\infty - v), \qquad (2)$$

$$k = A \exp(-E/RT), \qquad (3)$$

where v is the mass of volatiles emitted, t is time, k is the reaction rate coefficient, v_∞ is the ultimate volatiles yield, T is temperature, and A and E are respectively the preexponential factor and the activation energy that describe the reaction rate coefficient.

This single-step process lacks the flexibility required to describe much of the experimental data and nonisothermal effects. More sophisticated but still essentially empirical models describe devolatilization with two parallel, first-order reactions, by a first-order reaction with a statistical distribution of activation energy, and by various combina-

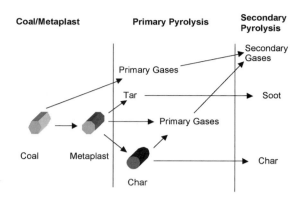

FIGURE 8 A schematic illustration of the major steps during coal devolatilization (pyrolysis).

tions of parallel and sequential first-order reactions. The most successful of these empirical models describe the dependence of volatile yields on temperature, heating rate, and pressure with some amount of differentiation of the species produced by devolatilization (tars, light gases, and chars).

The organic properties and structural characteristics of coal have a large impact on its devolatilization behavior. Fundamental devolatilization models base predictions on measurable organic properties of the coal. For example, devolatilization of higher rank, melting and swelling coals has been described by the following two-step process, as illustrated in Fig. 8. In the first step, coal undergoes a reduction of hydrogen bonding, and the macromolecular network softens to partially form a metaplast containing liquid coal components. Tars are evaporated from the generated metaplast, and the mobile phase is released during this stage. Not all of the macromolecular structure is depolymerized into metaplast. Low-rank coals produce very little metaplast. In the second step, further bond breaking leads to the evolution of tars and gases and repolymerization of coal fragments from char. Competition between bond scission and cross-linking reactions controls the rate of tar evolution and the properties of char. Gas formation is correlated with the composition of the coal. Large amounts of char and primary gases are generated directly from the coal macromolecular phase, especially for low-rank coals. Tar formation is viewed as a combined depolymerization and vaporization process. The structural evolution of char during pyrolysis suggests that functional groups are released from the aromatic clusters, with cross-linking occurring among the aromatic clusters. At the end of devolatilization, the carbon skeletal structures of chars from a variety of coals become very similar chemically. However, the physical structure of chars varies significantly depending on the type and rank of the coal from which they are derived.

Coal network models have more recently been developed to relate the structures of coal to devolatilization processes and rate measurements. Model parameters have been obtained from several optical or spectral measurements of coals and from rate measurements. Initially, physical melting occurs due to weakening of hydrogen bonds in the network, and the mobile phase is released as early tar. Further scission of covalent bonds in the network generates fragments corresponding to the main evolution of aromatic tars. Depending on coal rank, oxygen structures in coal such as carboxyl and hydroxyl groups thermally decompose early in the devolatilization process, generating radical groups that rapidly cross-link the coal network structure. Tar precursors are then not allowed to form and evolve. Thus, early cross-linking is empirically correlated with the evolution of CO_2.

Coal structure is typically modeled with four basic groups: aromatic nuclei, labile bridges, char or stable bridges, and peripheral groups. These structural parameters have been determined from many advanced coal characterization techniques. Competition occurs in the evolving coal structure between the cleavage of labile bonds and the formation of stable char links. With increasing temperature, the labile bonds are rapidly consumed and the decomposition of methyl and methoxy groups forms radicals that rapidly cross-link the structure into char. With further increases in temperature, methane and hydrogen are evolved from the char structure. It has been shown that two moieties, aryl–alkyl ethers and hydroaromatic structures, can be used to correlate weight loss and tar yield.

Network models use lattice statistics to quantitatively describe the thermal breakup of the coal macromolecular structure and interpret the interrelationships in the lattice characteristics of coal, tar, and char. The capability of one network model to predict total volatiles and char release for a large number of coals of various rank without the use of any fitting parameters was illustrated in Fig. 7. Char is formed from charring reactions within the network and from cross-linking of nonvolatile metaplast.

2. Coal Volatiles Ignition and Combustion

Many products are produced during devolatilization of coal, including tars and hydrocarbon liquids, hydrocarbon gases, CO_2, CO, H_2, H_2O, and HCN. These products react with oxygen in the vicinity of the char particles, increasing temperature and depleting oxygen. This complex reaction process is important to control soot formation, flame stability, and char ignition. Rate processes include the following: volatiles release, tar condensation and repolymerization in char pores, hydrocarbon cloud evolution through small pores from the moving particle, cracking of the hydrocarbons to smaller hydrocarbon fragments with local production of soot, condensation of gaseous hydrocarbons and agglomeration of sooty particles, macromixing of the devolatilizing coal particles with oxygen, micromixing of the volatiles cloud and oxygen, oxidation of the gaseous species to combustion products, production of nitrogen oxides and sulfur oxides by reaction of devolatilized products with oxygen, and heat transfer from the reacting fluids to the char particles.

At high combustion temperatures, up to 70% or more of the reactive coal mass can be consumed through this process. Volatiles combustion in practical systems is complicated by turbulent mixing of fuel and oxidizer, soot formation and radiation, and near-burner fluid dynamics. Such systems usually exhibit volatiles cloud combustion, rather than single-particle combustion.

3. Char Oxidation

The rate of char oxidation is determined by a combination of chemical kinetics and mass transport. Heterogeneous char oxidation can proceed simultaneously with or after vaporization and devolatilization, depending on reaction conditions. The time required for char and coke combustion is substantially greater than that required for devolatilization. There is also much emphasis on reducing the amount of carbon in the ash, which means that the burnout of the char or coke needs to be more effective. The physical structure of the char, such as pore structure, surface area, particle size, and inorganic content, controls the reaction processes of the char. However, the structure of char is determined by the devolatilization process as well as by the parent fuel. Models of char combustion can include the following processes: (1) diffusion of gaseous reactants and products through the boundary layer surrounding the particle, (2) diffusion of reactants within the porous structure of the particle, (3) conduction of heat through the particle, (4) adsorption of oxygen, (5) chemical reaction of the oxidizer on the particle surface, (6) desorption of products (e.g., CO) from the solid surface, (7) the homogeneous reaction of carbon monoxide with oxidizer within the pores and in the boundary layer, and (8) the evolution of particle and pores during the reactions.

Global models of heterogeneous reaction typically include only processes 1 and 5 described above as resistances in the reaction rate expression. The resistance due to film diffusion is usually calculated through the use of diffusional coefficients corrected for the mass leaving the particle surface (i.e., blowing factor). The effective diffusivity of the gases into the particle and the portion of the particle that participates in the reaction can be estimated by using methods that have been published.

The char reacts with oxygen to form CO primarily, and CO_2, depending on conditions. The measured oxidation rate is correlated through reaction rate expressions such as

$$r_1 = \eta \xi k_x T_p^m p_{O_2,s}^n = \eta \xi k_x R_g^n T_p^{m-n} C_{O_2}^n, \quad (4)$$

where T_p is the particle temperature, k_x is the kinetic rate coefficient, which varies exponentially with surface temperature, and $p_{O_2,s}$ is the oxygen partial pressure at the solid surface. The parameter ξ is the ratio of active surface area to external surface area and can be used to incorporate intrinsic kinetics in the model. The parameter η is the fraction of the surface area that is available for reaction. As the ash fraction of the char particle increases, rates of reaction of the entire particle based on total surface area generally decrease. The particle diameter changes as the particle burns and is typically modeled as a shrinking sphere with constant density, a constant-diameter sphere with changing porosity, or some intermediate relationship. Some chars fragment during combustion, a process most pronounced in high-rank, initially large coal particles. Since the oxygen partial pressure is not known at the particle surface, the molecular diffusion equation is used in conjunction with the oxidation equation, Eq. (4), to solve for reaction rate, assuming quasi-steady state as the char particle reduces in size and/or mass. Other complications during char oxidation include possible changes in reactivity during reaction because of char annealing and graphitization, effects of impurities on surface reaction rates, and the ratio of CO/CO_2 formed on the char surface. More recent char oxidation models include oxygen adsorption and product desorption steps through the classic Langmuir expression, which differs in form substantially from the empirical expression above.

Descriptions of pore development and structure require microscopic models of the particle. These models include intrinsic kinetics and pore structural changes during burnoff. Three of the most popular microscopic models are a random capillary pore model, one in which the pores are considered spherical vesicles connected by cylindrical micropores, and one in which the pores have a treelike structure. These models allow for pore growth and coalescence in their respective fashions and provide estimates of reactive surface area. Parameters required for these models are obtained from experimental measurements of the various chars.

Other microscopic models have been developed that are based on percolation or other char transformation theories. These microscopic models currently serve more as scientific descriptors than practical tools.

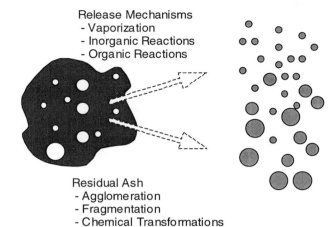

FIGURE 9 Schematic illustration of the fate of inorganic material in coals during combustion.

C. Fate of Inorganic Material

1. Ash Formation

Figure 9 schematically illustrates the fate of inorganic material in coal leading to the production of fly ash. Inorganic grains may be imbedded in the particle or may be extraneous to the particle itself. The fate of this second class of inorganic material differs substantially from that of the inherent material. The minerals undergo chemical reactions and phase changes determined by their thermochemistry as well as interacting with other inorganic components and the organic material. Components of the minerals may be released from the fuel by either thermal decomposition or vaporization during combustion.

The transformations are divided into two types: release mechanisms and the fate of the residual ash. Release mechanisms are indicated as vaporization, thermal or chemical disintegration of the inorganic material (inorganic reaction), or convection during rapid devolatilization or other organic reactions. These mechanisms tend to produce small ($<0.8 \ \mu$m) particles or vapors. The residual ash may undergo fragmentation either as an inorganic grain or in conjunction with fragmentation of burning char particles, may coalesce with some or all of the remaining inorganic material, and may undergo significant chemical or physical transformations. This material tends to produce larger ash particles. Depending on the type of inorganic material and the combustion conditions, the ash produced during combustion is composed of varying amounts of vapor, fume (<1-μm-diameter particles), and larger particles.

Many inorganic materials occur as hydrates in coal, all of which dehydrate through endothermic reactions, generally at temperatures from 100°C to 200°C. Sulfates typically thermally decompose as they are heated, as do carbonates and sulfides. In cases such as most of the alkaline-earth-containing materials, these salts decompose prior to melting and appreciable vaporization, while in cases such as most of the alkali-containing material, the salts melt, partially vaporize, and react further in the gas phase. Chlorides and hydroxides are more stable than the salts with bivalent anions and generally do not completely decompose or vaporize. At flame temperatures, these species represent the most stable form of the alkali metals and of chlorine and hydroxyl. Phosphates and silicates undergo transformations when heated but generally do not completely decompose or vaporize. Their melting behavior depends on their composition and structure, and they can generally be grouped into high- and low-temperature-melting materials. High-temperature-melting materials include phosphates, kaolinite, and their decomposition products. Low-temperature-melting materials include illite, salts, and alkali-containing species. Mathematical models have recently been developed to predict the decomposition and reaction rates of the inorganic species in coal.

2. Ash Deposition

A great deal of information is available on rates and mechanisms of ash deposition. Five major mechanisms of deposit formation include (1) inertial transport including impaction and sticking, (2) eddy impaction, (3) thermophoresis, (4) condensation, and (5) chemical reaction. Rates of inertial impaction in most environments are well established, and inertial impaction commonly accounts for most of the accumulated mass. Rates of eddy impaction are less well known. The capture efficiency, a measure of the propensity of material to stick to a surface upon impaction, is far less well established. The rates of thermophoretic deposition are reasonably well established when local temperature gradients and the functional form of the thermophoretic force on a particle (or the thermophoretic velocity) are known. Condensation rates can be predicted reasonably well, given accurate vapor pressure and concentration data. The accuracy to which rates of chemical reaction are known is often inadequate, especially those involving sulfation and alkali adsorption in silicates.

3. Deposit Properties

Properties of deposits that are important to boiler behavior and management include emittance and absorbance, thermal conductivity, strength, tenacity and thermal shock resistance, viscosity, and porosity. Methods for measurement of each of these properties are available, while techniques for prediction or correlation of these properties are in various states of development. Figure 10 illustrates the

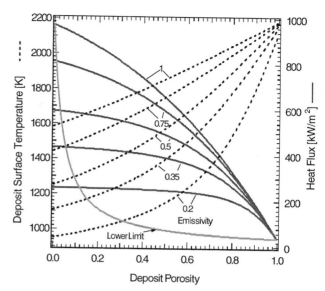

FIGURE 10 Deposit surface temperature and heat flux as a function of porosity and emissivity for a specified set of assumptions.

dramatic impact of deposit thermal conductivity and emissivity on a boiler's deposit surface temperature and heat flux.

V. COAL-GENERATED POLLUTANTS

A. Criteria Pollutants

1. Pollutant Species

The most challenging problem associated with coal consumption is the control of pollutants. Increasing power demand will lead to increased production and use of fossil fuels with the potential for increased pollutant emissions.

In the United States, burning of coal accounts for 93% of the SO_2 emissions from electric generating units and 85% of the nitrogen oxides (NO_x) with total emissions of about 14 million tons of SO_2 and nearly 9 million tons of NO_x per year (figures for 1997). Total SO_2 emissions from electric power generation declined by 16% from 1989 to 1997, while total NO_x emissions increased by about 3%. During the same period, electric power generation increased by 22%.

Concern for the atmospheric quality can be dated to several early episodes when coal began to replace wood as the primary source of domestic heating and industrial fuel. By the mid-20th century, public support and government resolve were strong enough to regulate air pollutant emissions from stationary combustors.

The combustion of coal produces both primary and secondary pollutants. Primary pollutants include all species of the combustor exhaust gases that are considered contaminates to the environment. The major primary pollutants include carbon monoxide (CO), hydrocarbon fragments, sulfur-containing compounds (SO_x), nitrogen oxides (NO_x), particulate material, and various trace metals. Secondary pollutants are defined as environmentally detrimental species that are formed in the atmosphere from precursor combustion emissions. The list of secondary pollutants includes particulate matter and aerosols that accumulate in the size range of 0.1–10 μm in diameter, NO_2, O_3, other photochemical oxidants, and acid vapors. The connection between combustion-generated pollutants and airborne toxins, acid rain, visibility degradation, and stratospheric ozone depletion is well established. The detrimental impact of these contaminants on ecosystems in the biosphere has been the impetus for stricter standards around the world.

Studies have shown that anthropogenic (i.e., human-generated) emissions of sulfur-containing pollutants decreased in the United States during the 1970s and early 1980s, whereas nitrogen oxides continued to increase. In 1985, stationary fuel combustion in electric utilities and industry accounted for approximately 50% of all NO_x emissions and more than 90% of all SO_x emissions. The increase in NO_x emissions can be correlated with the construction of many coal-fired, power-generating facilities throughout the U.S. midwest and west. Coal naturally contains nitrogen in the range 0.5–1.5 wt%, while sulfur typically varies from 0.5 to 10 wt%, with 1–2 wt% being more common.

2. Air-Quality and Emissions Regulations

The National Ambient Air Quality Standards (NAAQS) set by the United States specify the maximum allowable concentrations for the following *criteria pollutants*: CO, hydrocarbons, particulate matter, NO_2, and SO_2 (see Table V). The last three are of most concern in the burning of coal. The NAAQS define Primary Standards, which are intended to protect public health with an adequate margin of safety. Secondary Standards are also specified by the NAAQS, which are intended to protect public welfare (e.g., soils, vegetation, wildlife). Other countries have similar, sometimes more stringent, requirements, as shown in Table V. Regions are classified as either in "attainment" or "nonattainment" depending on whether the ambient level of pollutant concentrations exceeds or does not exceed the standards established by the regulations. While these standards specify the maximum allowable ambient concentrations for key pollutants, they do not effectively control point-source emissions rates.

Two common provisions dictate the type and level of pollutant abatement required by existing and new combustion facilities in the United States. These include the New Source Performance Standards (NSPS) and the Prevention of Significant Deterioration (PSD) regulations. NSPS establish nationwide emission standards for new and upgraded facilities regardless of source location or regional air quality. The NSPS are set at levels that reflect the degree of control achievable through the application of the best system of continuous emission reduction that has been adequately demonstrated for each category of sources. Under PSD regulations, the permitting of most new facilities requires the installation of best available control technology (BACT) for the control of nitrogen and sulfur oxides. BACT is defined as "an emissions limitation based on the maximum degree of reduction for each pollutant ... which the reviewing authority, on a case-by-case basis, taking into account energy, environmental, and economic impacts and other costs, determines is achievable." A BACT analysis must be performed for each new, modified, or reconstructed combustion source.

The U.S. Clean Air Act and its 1990 amendment provide the framework for controlling major sources, including stationary combustors. The regulations establish NSPS which apply to units of more than 73 MW$_e$ (250×10^6 Btu hr^{-1}) of heat input and are directed toward the control of SO_2, particulates, and NO_x.

The 1980 Clean Air Act regulation of SO_2 concentrations for gaseous and liquid fuels is 96 mg/10^6 J (0.20 lb/10^6 Btu). The standard for coal-fired unit emissions is somewhat complicated. A unique maximum emission rate and a unique minimum reduction of potential (uncontrolled) emissions is based on the sulfur content and heating value of the coal. For high-sulfur coals (>1.5% sulfur content), a 90% reduction of uncontrolled SO_2 emissions is required with a maximum allowable SO_2 emission rate of approximately 520 mg/10^6 J (1.2 lb/10^6 Btu). For low-sulfur coals, sometimes called compliance coals, the standard requires at least a 70% reduction in uncontrolled SO_2 emissions and limits the emission rate to 287 mg/10^6 J (0.60 lb/10^6 Btu).

The 1990 Clean Air Act amendment (CAAA) further reduced SO_2 emissions to 50% of their 1980 levels by the year 2000, and caps SO_2 emissions at that level. The new standards not only affect new sources, but also require a constant total emissions allowed from all electric utility generation units of 8.1×10^6 metric tons of SO_2 per year. The Act contains provisions for trading of emission allowances among units in order to achieve the most economical application of control technologies.

In Germany, NO_x emissions limits for new coal-fired boilers are almost half the currently prescribed level in the United States. Japan has placed nearly as stringent limits on NO_x emissions. However, under CAAA, by 2003–04, emissions limits in the United States will drop by more than half. Strict limits on sulfur oxides have also been set in Japan, where owners of SO_2 emitters pay a proportional

TABLE V Ambient Air Quality Standards or Recommendations for Combustion Pollutants in Various Countries[a]

	Country/organization	Concentration	Time
NO_x (as NO_2)	Germany	0.05 ppm	2–12 mo
	Japan	0.04–0.06 ppm	24 hr
	United States	0.05 ppm[b,c,d]	Annual arithmetic mean
	Former USSR	0.05	24 hr
SO_2	Germany	0.06 ppm	24 hr
	United States	0.03 ppm[b,c]	Annual arithmetic mean
		0.14 ppm[b,c]	24 hr
		0.5 ppm[b,d]	3 hr
	Former USSR	0.02 ppm	24 hr
	World Health Organization	38–37 ppb[e]	24 hr
		15–23 ppb[e]	Annual arithmetic mean
CO	Canada	13 ppm	8 hr
	Germany	26 ppm	0.5 hr
	Japan	20 ppm	8 hr
	World Health Organization	25 ppm[f]	24 hr
	United States	9.0 ppm[c]	8 hr
		35 ppm[d]	1 hr
	Former USSR	1.3 ppm	24 hr
NMHC[g]	Canada	0.24 ppm	—
	United States	0.24 ppm[c,d]	Average from 6 to 9 A.M.
Particulate matter	Japan	200 $\mu g\, m^{-3}$	1 hr
	United States	150 $\mu g\, m^{-3}$[h]	24 hr
		50 $\mu g\, m^{-3}$[h]	Annual arithmetic mean
	World Health Organization	60–90 $\mu g\, m^{-3}$	Annual arithmetic mean

[a] From Smoot, L. D., (Ed.) (1993). "Fundamentals of Coal Combustion," Table 6.1, Elsevier Science Publishers, New York.

[b] NAAQS in effect in 1984.

[c] NAAQS primary standard.

[d] NAAQS secondary standard.

[e] Based on observation of community populations exposed simultaneously to mixtures of SO_2 and "smoke," that is, particulate matter.

[f] Based on recommended limit in blood of 4% carboxyhemoglobin (COHb) or less.

[g] Nonmethane hydrocarbons; expressed as ppm carbon (ppmC).

[h] Primary and Secondary Standards effective July 1, 1987. Includes only particles having an equivalent spherical diameter of 10 Φm or less. Referred to as the PM10 standard.

release levy. Some European communities have low limits on nitrogen and sulfur oxide emissions. These strict standards have been the impetus for developing and implementing BACT, which includes flue gas treatment for both nitrogen and sulfur oxides. In fact, Japan and Germany are the first to use catalyst beds in the flue duct system to control nitrogen oxide emissions. Similar technologies are widely expected to be deployed in the United States in the near future as power plants prepare for Phase II reductions of NO_x required by the 1990 CAAA.

In the United States, the 1990 CAAA called on the Environmental Protection Agency (EPA) to analyze the health impacts of over 180 specific chemicals and all chemicals containing any of 12 hazardous metals. Mercury, arsenic, and cadmium are among the metals listed and are also of specific concern for coal power plants. The Act required the EPA to submit a report to Congress on the health hazards of power plant (and other industrial) hazardous air pollutant emissions and to describe possible control strategies. The EPA must then regulate those hazardous air pollutants found to be significant in power plant stack exhausts. Currently, utilities in the United States are not required to control the emission of hazardous toxins. Small particles, mercury, arsenic, and cadmium are among the emissions of greatest concern from power plants.

3. NO_x Formation and Reduction

A detailed understanding of nitrogen oxide formation during the combustion of fossil fuels has been achieved by years of research. Most of the nitrogen oxides emitted to the atmosphere by combustion systems are in the form of nitric oxide (NO), with smaller fractions appearing as nitrogen dioxide (NO_2) and nitrous oxide (N_2O). Other nitrogenous pollutants include ammonia and hydrogen cyanide.

Nitric oxide is formed in flames predominantly by three mechanisms: thermal NO_x (the fixation of atmospheric molecular nitrogen by reaction with oxygen atoms at high temperatures), fuel NO_x (the formation of NO_x from nitrogen contained in the fuel), and prompt NO_x (the attack of a hydrocarbon radical on molecular nitrogen, producing NO_x precursors). N_2O can be formed by a number of paths in gaseous and coal-laden reactors and is more prevalent at lower temperatures. N_2O levels in coal combustors are generally less than 5 ppm (except for fluidized-bed systems). NO_2 is typically significant only in combustion sources such as fuel-lean gas turbines. Neither N_2O nor NO_2 reaches high concentrations in pulverized coal flames or in the unquenched effluent of most gas combustors. Hence, NO is normally the most significant nitrogen oxide species emitted from coal-fired furnaces.

It is generally agreed that the reaction network for thermal NO is qualitatively described by the modified Zel'dovich mechanism:

$$N_2 + O \longleftrightarrow NO + N, \tag{5}$$

$$O_2 + N \longleftrightarrow NO + O, \tag{6}$$

$$N + OH \longleftrightarrow NO + H, \tag{7}$$

and where O atom concentrations are often assumed to be in equilibrium.

The first step is rate-limiting and requires high temperatures to be effective due to the high activation energy barrier with the designation *Thermal* NO. The third step is important in fuel-rich environments.

Prompt NO occurs by the collision and fast reaction of hydrocarbons with molecular nitrogen in fuel-rich flames. This mechanism accounts for rates of NO formation in the early flame region that are much greater than the rates of formation predicted by the thermal NO mechanism just described. This mechanism is much more significant in fuel-rich hydrocarbon flames. Two such reactions are believed to be particularly significant:

$$CH + N_2 \rightarrow HCN + N, \tag{8}$$

$$C + N_2 \rightarrow CN + N. \tag{9}$$

The cyanide species is further oxidized to NO.

Fuel NO typically accounts for 75–95% of the total NO_x accumulation in coal flames. Nitrogen in coal is predominantly found as a heteroatom in aromatic rings. Nitrogen primarily evolves from tars as HCN, with much

lower concentrations of NH_3. The conversion of fuel nitrogen to HCN is independent of the chemical nature of the initial fuel nitrogen (assuming that it is in-ring aromatic nitrogen) and is not rate limiting. Once the fuel nitrogen has been converted to HCN, it rapidly decays to NH_i ($i = 0, 1, 2, 3$), which reacts to form NO and N_2 as illustrated by the following simplified global reaction scheme, though the process is far more complicated:

$$\text{fuel-N} \longrightarrow \text{HCN} \longrightarrow \text{NH}_i \underset{O_2}{\overset{NO}{\diagdown}} \begin{matrix} N_2 \\ NO \end{matrix} \tag{10}$$

In fuel-rich combustion systems, NO can also be reduced by hydrocarbon radicals leading to the formation of HCN and then molecular N_2. Based on such observations an NO_x abatement technology termed *reburning* has been developed by injecting and burning secondary fuel in the postburner flame to convert NO back to HCN. It is generally accepted that light hydrocarbon gases are the most effective staging fuels, although pulverized coal, particularly lignite, has been shown to work.

In a typical coal flame, char retains a significant fraction (30–60%) of the initial nitrogen. NO formation during char reactions often accounts for 20–30% or more of the total NO formed.

Heterogeneous destruction of gas-phase NO is significant during fuel-lean char burnout, but homogeneous decay reactions appear to dominate NO removal in fuel-rich zones of coal combustion. Soot formation and subsequent conversion of soot nitrogen to nitrogen pollutants represents yet another way to produce NO_x from coal burning. The mechanism of NO destruction by soot particles is also well documented, with results indicating that NO removal rates on carbon black particles are much greater than heterogeneous destruction of NO by char.

Commercially significant NO_x abatement technologies include installation of low-NO_x burners, fuel/air staging, reburning, nonselective catalytic reduction (NSCR), and selective catalytic reduction (SCR). These are listed in approximate order of their implementation. NO_x reductions as high as 70% can be expected by incorporating all of the noncatalytic processes, and nearly all power stations currently do so. However, the next phase of NO_x regulations incorporated in the CAAA require many power plants to reduce NO_x emissions further than is practical using burner and simple boiler modifications, and SCR represents one of the few commercially demonstrated technologies capable of such high reduction. Therefore, SCR systems are expected to become much more prevalent in the United States in the near future. Such systems are already installed in Germany and Japan, where NO_x emission limits are more stringent than in the United States.

4. Sulfur Oxides Formation and Capture

Sulfur oxides are formed in stationary combustors from the sulfur entering with the coal. The sulfur content of coal typically varies from 0.5 to 10 wt%, depending on rank and the environmental conditions during the geological coal formation process. Sulfur oxides emissions can be reduced substantially by simply converting from high- to low-sulfur coal. Such coal switching or partial switching (blending) is responsible for the large shift in coal production from the eastern and midwestern coal seams in the United States to the low-sulfur coals predominantly found in Wyoming and Montana. For certain coals, the sulfur content can be reduced by removing some of the sulfur from the coal through washing or magnetic separation before burning it.

There are three forms of sulfur in coal: (1) organic sulfur such as thiophene, sulfides, and thiols, (2) pyritic sulfur (FeS_2), and (3) sulfates such as calcium or iron salts. Both organic and inorganic sulfur phases undergo significant chemical transformations during coal devolatilization and combustion. Sulfur-containing pollutants formed by burning coal include SO_2, SO_3, H_2S, COS, and CS_2. Under normal boiler operating conditions, with overall excess oxygen being used, virtually all of the sulfur is oxidized to SO_2, with small quantities of SO_3. The SO_2 is the thermodynamically favored product at high temperatures. At the stack temperatures, SO_3 is the most stable species, but kinetic limitations prevent significant amounts of SO_3 from forming, leading to SO_x emissions being dominated by SO_2.

Natural minerals such as limestone can be pulverized and added directly to the furnace to capture the sulfur. The limestone calcines to form the calcium oxide with formation of pores in the particle:

$$CaCO_3 \longleftrightarrow CaO + CO_2. \qquad (11)$$

The sulfur dioxide penetrates the pores and reacts with the calcium oxide to form solid calcium sulfate that can be removed with the ash. Dolomite ($CaCO_3 \cdot MgCO_3$) and hydrated lime [$Ca(OH)_2$] are also used as sorbents. Sulfur scrubbers based on a variety of chemical reactions have become more common since 1990. Such systems produce by-products with some commercial value, such as elemental sulfur, sulfuric acid, and gypsum. Scrubbers have added benefits of removing some NO_x, mercury, arsenic, and other pollutants that either currently or in the future may fall under formal regulation. However, scrubbers add about 25% to the capital and operating costs of a power station, leading most power stations to switch to low-sulfur coals rather than build scrubbers.

B. Non-Criteria Pollutants

Coal contains significant amounts of many trace elements. Most trace elements are associated with the inorganic portion of the coal, although some are also chemically and physically bound with the organic material. Trace element concentrations vary significantly among coal seams and even among mines within the same seam. During coal combustion, these trace elements become partitioned in various effluent streams, depending on their chemical behavior. Classification of trace element emissions has been partitioned into three general groups.

- *Group 1*. Elements concentrated in coarse residues, such as bottom ash or slag, or partitioned between coarse residues and particulates that are trapped by particulate control systems (e.g., Ba, Ce, Cs, Mg, Mn, and Th).
- *Group 2*. Elements concentrated in particulates compared with coarse slag or ash and tend to be enriched in fine particulate material that may escape particle control systems (e.g., As, Cd, Cu, Pb, Se, and Zn).
- *Group 3*. Elements that volatilize easily during combustion, are concentrated in the gas phase, and deplete in solid phases (e.g., Br, Hg, and I).

Emission of elements found in group 1 is controlled by standard particle handling systems. Because group 2 elements are concentrated in fine particles, the release of these elements depends largely on the efficiency of the gas-cleaning system. The emission of volatile trace elements, such as those in group 3, can be controlled by combustion conditions or downstream processing. For example, trace element release may be lower from a fluidized-bed coal furnace than from entrained-flow systems because of the lower temperatures in fluid beds. Furthermore, wet scrubbers used for flue gas desulfurization allow volatile elements to condense and therefore provide an effective method of reducing emissions of certain trace elements. Trace elements that are of particular concern from coal combustion are Hg, As, Cd, and Pb.

C. Greenhouse Gas Emissions

Greenhouse gas emissions represent a major and complex concern for the future use of fossil energy in general and coal in particular. Greenhouse gases are atmospheric components that are generally transparent in the visible region but absorb longer wavelength radiation, thereby contributing to a higher atmospheric temperature. They include water vapor (H_2O), carbon dioxide (CO_2), nitrous oxide (N_2O), methane (CH_4), hydrofluorocarbons

(HFCs), perfluorocarbons (PFCs), and sulfur hexafluoride (SF_6). Water vapor plays a large role in the greenhouse effect and, in the form of clouds, introduces much of the uncertainty in global circulation models. However, water is rarely discussed in terms of anthropogenic sources because of its very high background concentration and vital role in weather and precipitation. Most analyses, and this discussion, are presented on a water-vapor-free basis. Because of differences in the absorption efficiency of these gases and wide differences in their existing concentrations, they vary widely in potency as greenhouse gases. For example, methane is about 21 times as potent as carbon dioxide as a greenhouse gas. However, CO_2 is by far the largest potential contributor to emissions due to much higher concentration in both the atmosphere and in net emissions from most sources. In globally averaged numbers, CO_2 accounts for 80–90% of the total anthropogenic greenhouse gas effect, CH_4 between 5% to 10%, and NO slightly less than CH_4, with all other gases making up the remaining few percent. Therefore, greenhouse gas emissions are often almost synonymous with carbon dioxide and methane emissions.

The complexity of the issue can be partially reduced by analyzing its aspects, starting at the most certain and progressing to the less certain but more important. There is no doubt that the average concentration of greenhouse gases in the earth's atmosphere increased during the 20th century. There is also no doubt that much of this increase is anthropogenic as opposed to natural fluctuations. The large increase in fossil energy utilization during the same period is responsible for much of this increase. There is debate among scientists as to whether these global atmospheric chemistry changes are driving a measurable increase in the average world temperature, i.e., global warming, although a consensus seems to be growing that measurable warming is occurring. There is considerable uncertainty regarding the current and long-term magnitude of this change and its impact on global climates, national and international economies, and global and regional environments. The most radical predicted changes from credible sources suggest very large impacts on both environments and economies, with potentially massive destruction of property and loss of life associated with changes in sea levels, regional precipitation and drought, crop yields, and ecosystems. Others predict more modest but still measurable changes. Most scientists acknowledge that such changes are difficult to predict with confidence. However, the possible consequences are such that some proactive measures may be warranted. The difficulty and magnitude of lifestyle change required to affect a decrease or even to significantly slow the increase in global CO_2 concentrations is not widely acknowledged.

Coal figures prominently in this discussion. Coal contributes a disproportionately high amount of CO_2 from power production. For example, in the United States about 54% of the electric power production is derived from coal, whereas about 88% of the CO_2 emissions from electric power production are derived from coal. Furthermore, the largest increases in CO_2 emissions in the future are anticipated to arise in association with increased industrialization of large population centers, specifically China and India. Both countries have large coal reserves and are actively developing them to support their improving economies. If the per capita energy consumption of such countries becomes comparable to that of the United States or western Europe, or even if it becomes a significant fraction of it, the impact on greenhouse gas emissions will increase significantly.

Technologies exist that effectively reduce greenhouse gas emissions. Most are renewable energy options (solar, wind, biomass, and hydrogen-based power, among others), which generally do not produce a net change in greenhouse gases. Another effective and proven, but highly unpopular and nonrenewable option is nuclear power. Those familiar with the practical implementation of renewable systems recognize the enormous change in infrastructure required to implement these at large enough scale to impact greenhouse gas emissions. Most renewable options are more expensive, less proven, and less convenient than fossil energy systems. Many suggest that the increased costs of renewable or sustainable power production are more accurate indicators of the true cost of power production than are current market costs based mainly on fossil energy.

Government programs are actively exploring increased efficiency, renewable energies, energy conservation, and other systems to reduce greenhouse gases. Nevertheless, several prominent nations, including the United States, have yet to commit to specific reduction targets such as those described in the Kyoto Protocol and many nations that have committed to these targets are well behind schedule in meeting these commitments. Greenhouse gas emissions may well become one of the major issues in coal utilization in the coming years.

VI. NUMERICAL SIMULATION OF COAL BOILERS AND FURNACES

A. Foundations

Development and application of comprehensive, multidimensional, computational combustion models for coal-fired power stations is increasing across the world. While once confined to specialized research computer codes, these combustion models are now commercially available. Simulations made with such computer codes offer substantial potential for use in analyzing, designing, retrofitting, and optimizing the performance of coal combustion during power generation.

Development of fossil-fuel combustion technology in the past was largely empirical in nature, being based primarily on years of accumulated experience in the operations of utility furnaces and on data obtained from subscale and full-scale test facilities. Empirically based experience and data have limited applicability, however, when considering changes in process parameters, such as evaluating firing strategies for improving combustion efficiencies or mitigating pollutant formation. Large-scale furnace data are typically limited to effluent measurements. Subscale data provide valuable information and insights into controlling phenomena and are less expensive to obtain. However, the results of subscale tests can be difficult to extrapolate to large-scale systems because of the complex nature of turbulent, reactive flow processes in furnace diffusion flames. Combustion modeling technology bridges the gap between subscale testing, which tends to be phenomenological in nature, and the operation of large-scale furnaces typically used for power generation by providing information about the combustion processes that experimental data alone cannot practically provide.

When modeling pulverized coal combustion, the framework for the gas-phase solution approach is most rigorously based on computational fluid dynamics (CFD) using numerical solutions of multidimensional differential equations describing mass, energy, and momentum transport. Source terms and transport coefficients in the CFD models are derived from particle models, gaseous turbulent combustion models, radiation models, and similar coupled descriptions. Information available from model predictions includes temperature, composition, and velocity for gases and particles, ash/slag accumulation, and so forth. Most models of this type provide spatial variation of such quantities, while some also provide variation with time.

Comprehensive combustion models offer many advantages in characterizing combustion processes that can effectively complement experimental tests. Code predictions are typically less expensive and take less time than experimental programs, and such models often provide additional information that cannot always be measured. Computational modeling thereby becomes a cost-effective, complementary tool to testing in designing, retrofitting, analyzing, and optimizing the performance of coal-fired power stations. Typical objectives of these applications have been to determine key information required for program planning, study the effects of various firing alternatives or different fuels on process performance, or aid in process design and optimization. Other objectives

include trend analysis, retrofitting, identification of test variables, or scaling of measurements. Examples include reducing carbon loss, optimization of low-NO_x burners, definition of regions of particle impact on walls of furnaces, effects of changes in fuel feedstock quality, effects of changes in burner configuration, or impacts of larger and smaller coal particle sizes.

Computer capacity for running these codes is also becoming much more acceptable due to improved numerical methods and more advanced computer hardware. Three-dimensional code solutions for full-scale furnaces utilizing hundreds of thousands of grid nodes are being obtained on advanced, high-powered engineering workstations within a few hours to a few days.

While combustion modeling technology offers great potential, a major challenge in the use of such technology is establishing confidence that such models adequately characterize, both qualitatively and quantitatively, the combustion processes of interest. This is typically accomplished by making comparisons of code predictions with experimental data measured from flames in reactors that embody the pertinent aspects of the turbulent combustion of coal. Consequently, data from a range of different-size facilities are necessary to validate the adequacy of code predictions and establish code reliability in simulating the behavior of industrial furnaces. Such detailed data also provide new insights into combustion processes and strategies.

B. Equations and Solutions

The significant physical and chemical phenomena that are typically required to classify a combustion model as *comprehensive* include (see Fig. 11) (1) gaseous, turbulent fluid mechanics with heat transfer, (2) gaseous, turbulent combustion, (3) radiative energy transport, (4) multiphase, turbulent fluid mechanics, (5) water vaporization from particles, (6) particle devolatilization, (7) particle oxidation, (8) soot formation, (9) NO_x and SO_x pollutant formation and distribution, and (10) fouling/slagging behavior. The equations that constitute the mathematical model of reacting gaseous and particulate turbulent fluid flow with heat transfer make use of the fundamental transport equations for mass, momentum, and energy. For the gas phase, the equations are generally cast into fixed-coordinate (i.e., Eulerian) form:

$$\frac{\partial \bar{\rho} \tilde{u} \phi}{\partial x} + \frac{\partial \bar{\rho} \tilde{v} \phi}{\partial y} + \frac{\partial \bar{\rho} \tilde{w} \phi}{\partial z} - \frac{\partial}{\partial x}\left(\Gamma_\phi \frac{\partial \phi}{\partial x}\right)$$

$$- \frac{\partial}{\partial y}\left(\Gamma_\phi \frac{\partial \phi}{\partial y}\right) - \frac{\partial}{\partial z}\left(\Gamma_\phi \frac{\partial \phi}{\partial z}\right) = S_\phi, \qquad (12)$$

where ϕ refers to the conserved scalar (mass, momentum, enthalpy), x, y, and z represent the three coordinate

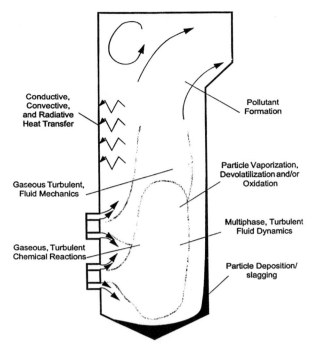

FIGURE 11 Major physical and chemical mechanisms in a pulverized-coal combustion process. [Reproduced with permission from Eaton, A. M., Smoot, L. D., Hill, S. C., Eatough, C. N. (1999). "Components, Formulations, Solutions, Evaluation and Application of Comprehensive Combustion Models," *Progress Energy Combustion Sci.* **25**, 387–436.]

TABLE VI Variables Typically Considered in Comprehensive Simulations of Coal Boilers[a]

Reactor parameters	Independent variables
Configuration (e.g., shape, upflow, downflow)	Physical coordinates (x, y, z) (x, r, θ)
Inlet configurations (e.g., location, presence of quarl)	Time (t)
Inlet locations	Dependent variables
Dimensions	Gas species composition
Wall properties	Gas temperature
Wall thickness	Gas velocity
Wall temperature	Pressure
Boundary/initial conditions	Turbulent kinetic energy
Gas velocity	Turbulent energy dissipation rate
Gas composition	Mixture fraction (mean and variance)
Gas temperature	
Gas mass flow-rate	Particle composition
Pressure	Particle temperature
Particle velocity	Particle velocity
Particle composition	Particle diameter
Particle temperature	Gas density
Particle size distribution	Gas viscosity
Particle loading	
Particle bulk density	

[a] Adapted from Smoot, L. Douglas, and Kramer, Stephen, K. (1995). "Combustion Modeling (Table 2)" *Energy Technology and the Environment*, **2**, (Bisio, A. and Boots, S. eds.) John Wiley and Sons, pp. 863–897.

directions (in a Cartesian coordinate system), u, v, and w are component velocities, ρ is the density, Γ_ϕ is the turbulent transport coefficient, and S_ϕ is the source term. Substitutions of algebraic approximations for the derivatives in such equations produces solvable equations that describe three-dimensional systems. To complete the model formulation, mathematical expressions are required for (1) gaseous reactions, including pollutant formation, (2) effects of turbulence on chemical reaction rates, (3) coal particle reaction rates, (4) radiative heat transfer, (5) temperature and composition dependence of physical properties, (6) particulate flow within the gas continuum, (7) behavior of inorganic matter, and (8) initial and boundary conditions. Recent reviews discuss these components in detail including code solution methods.

A list of typical dependent and independent variables for a furnace simulation is shown in Table VI. Coal particles typically are treated in a procedure that couples the continuum (Eulerian or fixed reference fame) gas-phase grid and the discrete (Lagrangian or moving reference frame) particles. Numerical solutions are repeated until the gas flow field is converged for the computed particle source terms, radiative fluxes, and gaseous reactions. Lagrangian particle trajectories are then calculated. After solving all particle-class trajectories, the new source terms

are compared with the old values and the entire procedure is repeated until convergence is obtained. Use of comprehensive combustion models typically requires a technical specialist with appropriate training and experience.

C. Code Evaluation—Field Tests

Data for validating pulverized coal combustion predictions requires accurate information for the reactor parameters shown in Table VI. Data measured in the combustion chamber typically include (1) locally measured values of the gaseous flow field velocity, temperature, and species composition, (2) coal particle burnout, number density, velocity, temperature, and composition, and (3) wall temperatures and heat fluxes. Evaluation should include comparisons with measurements from a wide variety of combustors and furnaces that range in scale from very small laboratory combustors (0.01–0.5 MW) and industrial furnaces (1–10 MW) to large utility boilers (up to 1000 MW).

Laboratory-scale data are very important to comprehensive model validation for several reasons. First, the nonintrusive, laser-based instrumentation needed to characterize the turbulent behavior in a flame is used more conveniently in smaller laboratory reactors. Second, a wider range and better control of reactor operating conditions are possible for a smaller facility. Third, the entire reactor volume can be traversed with small incremental steps, so that much more detailed datasets can be obtained. Other advantages of small laboratory facilities are relatively low operating cost, flexibility and accessibility, and ability to control and to define carefully the boundary and inlet conditions. They can be large enough to give sufficient spatial resolution and to create a near-burner furnace environment, and small enough to utilize the advanced measurement techniques which are essential to providing accurate and complete data for model evaluation and for detailed understanding of combustion processes.

In large-scale facilities, detailed measurements are difficult to make, inlet conditions are often not well defined, operating costs for obtaining data can be high, and the facility may have instrumentation limitations. For example, detailed model evaluation measurements of species and temperature within or across large flames cannot easily be made, although such spatially resolved profile data are often best for comparing flame characteristics, near-field burner performance, and jet mixing behavior with predictions. Such measurements are essential for the evaluation of comprehensive combustion models. These detailed measurements provide important information concerning flame response to parametric variation. Effluent measurements (measurements at the outlet of the system or process) can be useful, particularly when effects of key system variables, such as excess air percentage, firing rate, or burner tilt angle, are measured. In some facilities, effluent data are the only kind available because of furnace size or access constraints, or because that was the only objective for making the measurements. Three-dimensional data at the utility furnace scale obtained with the specific intent of evaluating and validating model predictions have been collected at large-scale 85-MW$_e$ and 160-MW$_e$ corner-fired boilers of one U.S. utility.

VII. THE FUTURE

Energy production from coal will face at least three major challenges in the future: Legislative reduction of allowable pollutant emissions, effects of deregulation of the power industry, and consideration of greenhouse gas emissions. The coal community has always faced pressure to reduce pollutant production; however, constraints relating to deregulation and greenhouse gas emissions are relatively new.

Emissions from coal-fired power plants have decreased substantially in the last decades due to improved understanding of pollutant formation mechanisms and combustion science. NO$_x$ is routinely reduced by 50–70% in major power stations, well over 99% of the particulate matter is filtered from the stack gases, and low-sulfur coals or sulfur scrubbers reduce SO$_2$ emissions by 40–90% compared to uncontrolled emissions. However, newer and tighter pollutant regulations portend continued pressure to reduce emissions. In particular, NO$_x$ emissions have not been reduced nearly as far as can be accomplished by advanced but proven technologies or as far as is necessary to prevent environmental harm such as acid rain, ozone reactions, or smog. Impending environmental standards will require installation of treatment systems such as selective catalytic reduction (SCR) of NO$_x$. SCR systems are established technology but have not been widely implemented in commercial operation in the United States due to the success of alternative and generally cheaper burner or boiler modifications. Germany and Japan are currently implementing SCR systems and the coming years should see large increases in the number of SCR installations or similarly effective NO$_x$ controls on commercial coal-fired power plants in most countries. Catalyst lifetime extension and catalyst regeneration are not as well understood as are SCR NO$_x$ reduction mechanisms, and these issues may play a prominent role in near-future plant operation. Proposed particle cleanup regulations targeted at particles less than 2.5 μm in size and mercury, arsenic, and other heavy or toxic metal regulations will also be major pollutant-related issues in the near future.

Many world power markets are deregulating, with power producers now able to compete for customers throughout the grid. As one consequence of this deregulation, power production cost is likely to become increasingly more important than production reliability or consistency. This has the potential to fundamentally change the way in which power plants operate and are managed. The intended consequence of deregulation is an overall decrease in the cost of power, but some regions with inadequate local production capacity have experienced rapid increases in cost and dramatic decreases in reliability during their early efforts to deregulate power systems. Regions with adequate supply from competing interests have made smoother transitions to deregulated systems. Issues associated with greenhouse gases have been discussed above. If nuclear power remains a socially or technically unacceptable form of power generation, it is difficult to foresee the availability of sufficient capacity for inexpensive power generation without coal continuing to play a major role.

To accomplish substantial reductions in greenhouse gas emissions, the efficiency of power production from coal will necessarily increase, the cost of power will increase as more renewable energies are developed, significant conservation of energy will be practiced, or methods of controlling/capturing CO$_2$ emissions will be deployed. A combination of these efforts is likely to be most effective. Efficiency increases as high as 50% (on paper) are achievable through advanced cycles such as integrated gasification combined cycles and fuel cells. However, capital costs associated with these technologies may be twice those of conventional power systems and technical risks are very high. Since approximately half of the cost of coal-derived power is associated with capital investment, there is likely no amount of fuel savings that could economically compensate for a doubling of capital costs. Global warming concerns may provide the motivation for conservation, sequestration, and advanced power systems that are otherwise difficult to justify on economic terms alone.

SEE ALSO THE FOLLOWING ARTICLES

BIOMASS UTILIZATION, LIMITS OF • ENERGY FLOWS IN ECOLOGY AND IN THE ECONOMY • ENERGY RESOURCES AND RESERVES • GREENHOUSE EFFECT AND CLIMATE DATA • POLLUTION, AIR • POLLUTION, ENVIRONMENTAL • RENEWABLE ENERGY FROM BIOMASS • THERMODYNAMICS

BIBLIOGRAPHY

Abbas, T., Costen, P. G., and Lockwood, F. C. (1996). "Solid fuel utilization: From coal to biomass," In "Twenty-Sixth Symposium (International) on Combustion," pp. 3041–3058, Combustion Institute, Pittsburgh, PA.

Bartok, W., and Sarofim, A. F. (eds.) (1991). "Fossil Fuel Combustion: A Source Book," Wiley, New York.

Baxter, L. L. (1993). "Ash deposition during biomass and coal combustion: A mechanistic approach," Biomass Bioenergy 4(2), 85–102.

Baxter, L. L., and DeSollar, R. W. (1993). "A mechanistic description of ash deposition during pulverized coal combustion: Predictions compared to observations," Fuel 72(10), 1411–1418.

Baxter, L. L., and DeSollar, R. W. (eds.) (1996). "Application of Advanced Technology to Ash-Related Problems in Boilers," Plenum Press, New York.

Cengel, Y. A., and Boles, M. A. (1998). "Thermodynamics: An Engineering Approach," 3rd ed., McGraw-Hill, New York.

EIA. (1999). "Annual Energy Review 1998," Energy Information Administration, U.S. Department of Energy, Washington, DC.

Gupta, R. P., Wall, T. F., and Baxter, L. L. (eds). (1999). "Impact of Mineral Impurities in Solid Fuel Combustion," Plenum Press, New York.

Hart, R. H. (1998). "Structure, properties, and reactivity of solid fuels," In "Twenty-Seventh Symposium (International) on Combustion," pp. 2887–2904, Combustion Institute, Pittsburgh, PA.

Hinds, W. C. (1999). "Aerosol Technology: Properties, Behavior and Measurements of Airborne Particles," 2nd ed., Wiley, New York.

Leckner, B. (1996). "Fluidized bed combustion: Achievements and problems," In "Twenty-Sixth Symposium on Combustion," pp. 3231–3241, Combustion Institute, Pittsburgh, PA.

Niksa, S. (1996). "Coal Combustion Modeling," IEA PR/31, IEA Coal Research, London.

Patankar, S. V. (1980). "Numerical Heat Transfer and Fluid Flow," Hemisphere, Washington, DC.

Rosner, D. E. (2000). "Transport Processes in Chemically Reacting Flow Systems," Dover, New York.

Seinfeld, J. H., and Pandis, S. N. (1998). "Atmospheric Chemistry and Physics," Wiley, New York.

Smith, K. L., Smoot, L. D., Fletcher, T. H., and Pugmire, R. J. (1994). "The Structure and Reaction Processes of Coal," Plenum Press, New York.

Smoot, L. D. (ed.) (1993). "The Fundamentals of Coal Combustion," Elsevier, Amsterdam.

Smoot, L. D. (1998). "International research centers' activities in coal combustion," Prog. Energy Combustion Sci. 24, 409–501.

Solomon, P. R., and Fletcher, T. H. (1994). "Impact of coal pyrolysis on combustion," In "Twenty-Fifth Symposium (International) on Combustion," pp. 463–474, Combustion Institute, Pittsburgh, PA.

Tree, D. R., Black, D. L., Rigby, J. R., McQuay, M. Q., and Webb, B. W. (1998). "Experimental measurements in the BYU controlled profile reactor," Prog. Energy Combustion Sci. 24, 355–384.

Wall, T. F. (1992). "Mineral matter transformations and ash deposition in pulverised coal combustion," In "Twenty-Fourth Symposium (International) on Combustion," Combustion Institute, Pittsburgh, PA.

Willeke, K., and Baron, P. A. (eds.). (1993). "Aerosol Measurement: Principles, Techniques and Applications," Wiley, New York.

Williams, A., Pourkashanian, M., Jones, J. M., and Skorupska, N. (2000). "Combustion and Gasification of Coal," Taylor and Francis, New York.

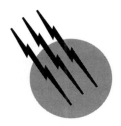

Foundations

Fred H. Kulhawy

Cornell University

Anwar Hirany

Electric Power Research Institute

GLOSSARY

Bearing capacity Capacity of the soil or rock to sustain the applied foundation load. *Ultimate* bearing capacity is the maximum stress sustained by the foundation at shear failure of the soil or rock. *Allowable* bearing capacity is the maximum design stress that can be imposed on the foundation and is equal to the ultimate bearing capacity divided by the *factor of safety*.

Deep foundation Generally, a long, cylindrical foundation having its depth significantly greater than its diameter or width and deriving its capacity from both side and tip resistance. Common types are the *pile*, which is generally between 6 and 18 in. in diameter or width (but can be 5 feet or more for bridges or offshore platforms), and is driven into the ground by mechanical means, and the *drilled shaft*, which is generally between 2 and 5 feet (but can be more than 10 feet) in diameter and is constructed by augering a hole into the ground and backfilling it with reinforced concrete (also called *drilled pier*, *bored pile*, or *drilled caisson*).

Drained and undrained loading Time-dependent loading conditions. *Drained* conditions occur when the excess pore water pressure resulting from the applied load has been dissipated (long-term loading condition in fine-grained soil). *Undrained* conditions occur when excess pore water pressure has developed under the applied load (short-term loading conditions in fine-grained soil).

Effective stress Difference between the total stress and pore water pressure at a point within the soil/rock mass. *Total stress* at a point is defined as the sum of the stress exerted by the total weight (solids plus water) of the soil/rock and the load on the foundation. *Pore water pressure* is the pressure exerted by water in the soil/rock voids. The ratio of the horizontal effective stress to the vertical effective stress at a point is the *coefficient of horizontal soil stress* (K_0).

Excavation support system Generally, a temporary system used to support the walls of the excavation during foundation construction. *Braced* (horizontally supported) vertical timber planks (*sheeting*), steel *sheet* *piles*, or a combination of steel H-sections (*soldier piles*) and horizontal timber planks (*lagging*) may be used to support the excavation walls for shallow and mat (slab) foundations. Steel pipe (*casing*) or *drilling slurry* may be used to support the excavation for a drilled shaft.

Inspection Quality assurance and control performed during foundation construction and installation.

Load test Test performed on full-scale foundation to determine its load–displacement characteristics and ultimate bearing capacity.

Settlement (heave) Downward (upward) displacement of the foundation resulting from elastic, consolidation, and secondary compression (expansion) of the soil and/or rock. *Elastic* response results from the initial elastic displacement to the applied load. *Consolidation* (*swell*) response results from dissipation of excess pore water pressure (i.e., flow of water from the pores). *Secondary* response occurs from soil skeleton readjustment to the new effective stress after dissipation of the excess pore water pressure.

Shallow foundation Generally, a large, rectangular foundation embedded to a depth less than or equal to its width and deriving its capacity predominantly from end bearing (tip resistance). Common types are the *spread footing*, which is constructed of cast-in-place reinforced concrete, and the *grillage*, which is constructed of prefabricated steel or treated lumber placed in a crisscross pattern.

A FOUNDATION can be described simply as the support system that rests on or in the ground and transfers to the ground the weight of the supported structure or equipment and the loads acting on them.

I. BACKGROUND

All man-made structures, from buildings to bridges to towers to dams, require a foundation to support them. Foundations also are necessary to support various equipment such as pumps, motors, turbines, generators, etc. They can range from a simple peg driven into the ground for supporting a camping tent to a complex system of very deep foundations for supporting an offshore drilling platform, which will require elaborate design and often innovative construction techniques. Foundations can ultimately affect or even control the performance, safety, and economics of a structure, and it is almost impossible to design or construct these structures without considering the engineering characteristics of the foundation soils and rocks.

The primary function of a foundation is to distribute the supported load to the soil and rock in a manner that restricts movements (displacement, settlement, heave, strain, rotation, and vibration) to tolerable limits. These movements are associated with the shear failure, compressibility, and/or volume change characteristics of the soil or rock, all of which can be affected adversely by the foundation construction procedures. Therefore, two important requisites for a successful foundation project are adequate engineering knowledge of the natural ground materials—soil and rock—and adequate experience in the art of foundation construction.

Foundations must be designed so that they are stable and do not move excessively. They must resist various loads that could be axial (vertical uplift and compression), lateral (horizontal and moment), inclined, or torsional in direction and can be static (gravity load from a building), transient (wind loads on a high-rise building), cyclic (traffic loads on a bridge), or dynamic (loads from earthquakes or machine vibrations) in nature. For each of these loading combinations, a number of different types of foundations can be used. The final choice commonly is governed by various factors such as the ground conditions at the site (soil or rock characteristics mentioned above), depth of frost penetration, wind and water erosion, type of structure or equipment to be supported, influence on or of adjacent structures, constructibility concerns, and project economics.

Foundation engineering is a very broad subject. Accordingly, in this article, the emphasis will be only on the basic concepts involved. Details can be found in the bibliography.

II. TYPES OF FOUNDATIONS

A wide variety of different types of foundations have evolved in response to the many variables affecting foundation design, and these include (see Fig. 1) spread foundations, mat foundations, driven piles, drilled shafts, and other specialty foundations such as direct embedment and anchor foundations. These different foundation types can be grouped under two major categories: shallow and deep. The difference between these two categories is based on their relative depth below the ground level and how the load is distributed to the soil. Generally, if the foundation is embedded in soil to a depth less than or equal to its width, it is considered a shallow foundation. As described later, the applied compressive loads from the structure are resisted by the base of a shallow foundation and by both the base (tip) and sides of a deep foundation. Table I presents a summary of the more commonly used foundations for various soil and loading conditions.

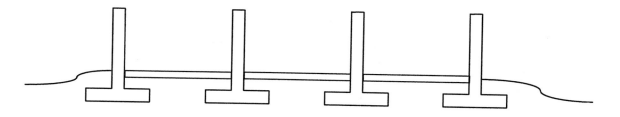

a. Shallow Foundation (spread footing)

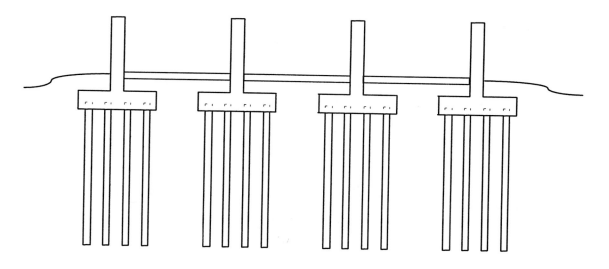

b. Deep Foundation (piles and pile cap)

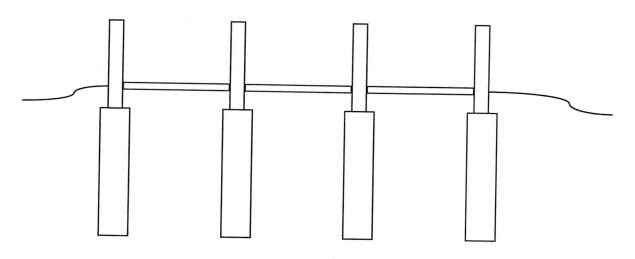

c. Deep Foundation (drilled shaft)

FIGURE 1 Foundation types. (a) Shallow foundation (spread footing). (b) Deep foundation (piles with pile cap). (c) Deep foundation (drilled shaft).

TABLE I   Applicability of Foundation Types

	Applicability for given foundation type[a]					
Parameter	**Spread**	**Grillage**	**Mat**	**Piles**	**Drilled shafts**	**Anchors**
Soil/rock condition						
Competent soil/rock near surface	**MF**	**MF**	SC	SC	SC	SC
Expansive/collapsing soil	SC	SC	**MF**	SC	SC	NR
Competent soil/rock at depth	NR	NR	SC	**MF**	**MF**	SC
Swamps	NA	NA	NA	**MF**	SC	SC
Structure type						
Single-family residential	**MF**	SC	**MF**	SC	SC	SC
Low-rise buildings	**MF**	SC	**MF**	SC	SC	SC
High-rise buildings	SC	SC	SC	**MF**	**MF**	SC
River crossings (bridges)	NA	NA	NA	**MF**	**MF**	SC
Highway crossings (overpasses)	**MF**	**MF**	NA	**MF**	**MF**	SC
Equipment/machinery	**MF**	NA	**MF**	SC	SC	NA
Offshore platforms	NA	NA	NA	**MF**	SC	SC
Transmission towers	SC	**MF**	SC	SC	**MF**	**MF**
Loading type						
Axial uplift	**MF**	**MF**	**MF**	**MF**	**MF**	**MF**
Axial compression	**MF**	**MF**	**MF**	**MF**	**MF**	NA
Horizontal/moment	SC	SC	SC	**MF**	**MF**	SC
Vibratory	**MF**	SC	**MF**	SC	SC	SC
Costs	**Low**	**Low**	Medium to high	High	High	Medium to high

[a] MF, Most feasible; NA, not applicable; NR, not recommended; SC, under special conditions.

A. Spread Foundations

This type of foundation generally is used where competent soil or rock is found at shallow depths and the groundwater table is below the bottom of the foundation. It has large, rectangular plan dimensions, is placed in a shallow excavation, and then is backfilled. Effectively, the load from the structure is spread out over a larger area of the soil, thereby reducing both the stress exerted on the soil and the settlement. The two main variations are the footing, which is constructed of cast-in-place concrete, and the grillage, which is prefabricated steel or treated lumber placed in a crisscross pattern. One footing or grillage commonly is used for each column of the structure. Spread foundations are shallow foundations and are used most commonly to support the columns and walls of lightly-loaded structures, such as single-family dwellings or low-rise buildings.

B. Mat Foundations

Mat or slab foundations are similar to spread foundations in that they have large, rectangular plan dimensions, but they basically cover the entire footprint of the structure. Generally, all of the columns and load-bearing walls of the structure are supported by one large mat foundation. These foundations have been used for high-rise buildings where competent soil or rock is found at shallow depths and for single-family dwellings that are constructed on soil susceptible to moisture-related volume change (e.g., expansive clays and collapsible sands). Mat foundations fall under the category of shallow foundations.

C. Driven Piles

Piles are long, slender foundation elements that are driven into the soil by mechanical means. They may consist of steel H-sections, steel pipe, precast concrete, wood, fiber glass, or a composite of these materials. Piles normally are not used singly to constitute a foundation; instead, two or more are driven to create a closely-spaced group, which is then integrated structurally through a cast-in-place concrete pile cap. Piles fall under the category of deep foundations, and they generally are used for high-rise or low-rise buildings and river crossings where the water table is high and/or competent soil or rock is found at depths where excavation for a spread or mat foundation would be uneconomical.

D. Drilled Shafts

Drilled shafts (also known as drilled piers, bored piles, and drilled caissons) are constructed by augering a cylindrical hole into the ground and backfilling it with reinforced concrete. The bottom of the shaft can be belled in appropriate soil to give a larger bearing surface, although bells are becoming less common. A single shaft normally is used to support a building column, but groups can be used, similar to driven piles. Drilled shafts fall under the category of deep foundations and often are a suitable and economical alternative to pile foundations. They are most commonly used for river and highway crossings, large electricity transmission structures, high-rise and low-rise buildings, and soil-retaining structures.

E. Specialty Foundations

Under certain circumstances, the supporting column of a structure can itself be used as the foundation. A hole is excavated, and the column is placed directly into the hole and then backfilled with the excavated soil. Smaller pole structures for electricity and telephone transmission, sign structures, etc., are supported this way. This type of foundation is called a direct embedment foundation and falls under the category of a deep foundation.

Anchors are long, very slender structural elements, typically of steel, although other options exist. There are several major variations of anchors. Plate anchors are long rods with flat plates at various levels and are placed in excavated holes backfilled with soil or concrete. Screw anchors essentially are augers which are screwed directly into the ground. Grouted anchors are long rods or tendons that are placed in excavated holes and then grouted to fill the annulus. Generally, the tendons and rods are prestressed. Inflated anchors are similar to the grouted type, except that they are expanded to grip the wall of the hole and may or may not be grouted afterward. Anchors generally are used to resist uplift or tension force and fall under the category of deep foundations. They are most commonly used for supporting soil-retaining structures, mobile homes, and structures supported by guy wires.

III. GEOTECHNICAL CONCEPTS FOR FOUNDATION DESIGN

The rate of loading and the geotechnical (or ground engineering) characteristics of the underlying soil govern the foundation response to the applied load, and therefore familiarity with basic geotechnical concepts is necessary to understand the behavior of foundations under load.

A. Consolidation

When a load is applied to any material, it deforms or strains, and a plot of the load (or stress) versus the corresponding deformation defines the load–deformation (load–settlement, stress–strain, or stress–deformation) characteristics of the material. In most elastic materials such as steel, the entire deformation is almost instantaneous, occurring as soon as the load is applied. In saturated particulate materials such as soil, the applied load causes elastic deformation of the soil structure, and it also increases the water pressure in the soil voids.

The soil compresses as the water flows from the loaded region to areas of lower water pressure. This time-dependent compression or deformation of the soil, known as consolidation, is controlled by the rate at which water flows from the pores. As this flow occurs, the soil particles move closer together and the soil mass becomes denser and stronger.

In saturated, fine-grained soils such as clays, this consolidation process takes a long time because of its low coefficient of permeability (hydraulic conductivity), and therefore the deformation can continue to occur for months, years, or even decades after the application of load. On the other hand, coarse-grained soils such as sands and gravels have a relatively high hydraulic conductivity, and the water flow is almost instantaneous. For foundations in competent sand and gravel, most of the deformation occurs during construction, and therefore most of the settlement has taken place by the time the structure is completed.

B. Effective Stress

Soils at the bottom of the ocean or other water body generally are very soft or loose, even though they have been subject to a tremendous amount of water pressure for a long time. This condition implies that the weight or pressure of water does not influence the process of consolidation, otherwise the ocean-bottom soils would be dense or hard. The deeper soils get consolidated under the weight—more specifically, buoyant or submerged weight—of the overlying soils. In soil under water, the reduced buoyant weight means less grain to grain stress in the soil. This grain-to-grain stress is called the effective stress σ' and the water pressure in the pores between the grains is called the pore water pressure u. Mathematically, the effective stress at a point is defined as the difference between the total stress σ and the pore water pressure at that point. Total stress at a given depth below the soil surface is the product of the depth and the total unit weight of the soil.

As stated above, when external stress is applied to saturated soil, the water pressure increases. This increase is above the hydrostatic pressure and is called excess pore water pressure. Theoretically, at the instant the external stress is applied, the excess pore water pressure is equal to the applied stress, and there is no increase in the grain-to-grain (effective) stress. As consolidation occurs and water flows from the pores, the excess pore water pressure dissipates and the effective stress increases.

C. Overconsolidation Ratio

When a soil is loaded, say by the weight of an ice sheet or any other boundary load, it undergoes consolidation and the ground surface settles under the applied load. Once consolidation has occurred and the ice sheet melts away or the boundary load is removed, the unloaded soil tries to expand. This expansion can occur in the vertical direction only because adjacent soil particles restrict expansion in the horizontal direction. However, the ground surface does not return to its original elevation after unloading because the particles have shifted and realigned themselves during consolidation. The soil is now in a denser or stiffer state than it was prior to the loading. The overconsolidation ratio (OCR) is a qualitative indicator of this densification or stiffening of the soil, and it is defined as the ratio of the maximum overburden stress ever experienced by the soil (i.e., with the ice sheet on top) to the present overburden stress (i.e., without the ice sheet). Soils with an OCR = 1 are called normally consolidated (NC) and unaged (which means the soil layer is young and has not experienced more load than what it supports presently), while an OCR from 1 to 1.3 can still indicate NC soil, but usually soil that is older. Soil with an OCR from 1.3 to 4 is called lightly overconsolidated (LOC), soil with an OCR from 4 to 10 is called moderately overconsolidated (MOC), and soil with an OCR greater than 10 is called heavily overconsolidated (HOC).

The OCR governs the response of the soil to loading. Soils with low OCR (typically NC to LOC) tend to contract when subjected to shearing forces, and expel the water during shear. Soils with high OCR (typically MOC to HOC) tend to expand or dilate when subjected to shearing forces, which causes water to be drawn into the soil.

D. Coefficient of Horizontal Soil Stress

The stress acting at a point within a body of water is the same in all directions, and that is why it is called pressure. If this were true for granular soils, sand dunes would not exist, and the soil would flow like water. Unlike water, soil offers resistance to shear (frictional resistance), which causes the stress to be different in different directions within a soil mass. The ratio of the horizontal stress to the vertical stress at a point within the soil mass is defined as the coefficient of horizontal soil stress K_0 at that point. Depending on the soil OCR, K_0 can be less than, equal to, or greater than one (from as low as $1/4$ or $1/3$ to as much as 3 or 4), and therefore the horizontal stress can be less than,

equal to, or greater than the vertical stress, respectively. Generally, the higher the OCR, the higher the K_0.

E. Undrained and Drained Loading

Undrained loading conditions develop when loads are applied relatively rapidly to fine-grained soils and no consolidation takes place because of the low coefficient of hydraulic conductivity. Initially, the pore water supports the entire applied load, and the soil volume, water content, and effective stress remain unchanged. With time, as water flows from the pores, the soil consolidates under the applied load, its volume and water content decrease, and the applied load is supported gradually by the soil particles, thereby increasing the effective stress. This limit state is referred to as the drained loading condition. In coarse-grained soil with high hydraulic conductivity, the drained loading condition occurs almost instantaneously after the application of load. Depending on the OCR, the drained strength may be greater or less than the undrained strength, and this fact must be considered in the design. In moderately to heavily overconsolidated soil, the drained strength generally is less than the undrained strength, which implies that long-term conditions generally will govern for this type of soil. The opposite is true for normally consolidated and lightly overconsolidated soil.

IV. FACTORS AFFECTING FOUNDATION PERFORMANCE

Foundations have to be designed for a variety of loading conditions on the basis of site-specific characteristics. Many factors must be considered to design a foundation that will perform satisfactorily, including geologic environment, loading characteristics, structure/equipment characteristics, foundation characteristics, foundation response, soil or rock response factors, and construction procedure.

A. Geologic Environment

Geologic environment provides the starting point for the foundation selection and design and includes site geology, soil and rock conditions, *in situ* state of stress, and groundwater conditions. Some soils are dense or hard, some are loose and soft, and some are hard and later become soft. When exposed to water, some soils expand and lift the foundation, whereas other soils collapse and lower the foundation. Among other factors that need to be considered are frost susceptibility and chemical composition of the soil and rock, seasonal groundwater fluctuations, erosion, and formation of solution cavities in rock. Details of the local geologic environment are quite important in both design and construction.

B. Loading Characteristics

Loading characteristics include the load combinations (axial, lateral, moment, inclined, torsional) mentioned earlier. The key variables for foundation design are the magnitude, direction, and line of action of the load and how fast (rate of loading) and how often (frequency of loading) it is applied. Depending on the magnitude of the load, soil and rock response associated with a gradually applied static load can be significantly different if the same magnitude load is applied instantaneously or repeatedly.

C. Foundation Characteristics

Foundation characteristics include the weight, size, shape, and material of the foundation. The weight of the foundation helps in resisting uplift and horizontal force but reduces the amount of compression (downward) force that can be applied. The size and shape (geometry) of the foundation determine the volume of soil that will be affected by the applied loads. Larger foundations have a greater zone of influence both horizontally and vertically. Under certain load combinations, the frictional resistance between the foundation material and soil or rock will affect the load-carrying capacity of the foundation. The foundation material can be affected by the chemistry of the soil and rock or by groundwater fluctuations. Certain chemicals present in the ground can attack concrete and steel, and timber can deteriorate with wet and dry cycles caused by groundwater fluctuations.

D. Structure or Equipment Characteristics

The supported structure or equipment dictates the limits of tolerable movement of the foundation. Generally, it is the differential movement between adjacent foundations of a structure or interconnected equipment that is of greater concern than the absolute movement. Differential movement imposes additional load on the structural members (beams and columns) or misaligns the interconnected equipment so that they cannot function properly. The amount of differential movement that can be tolerated depends primarily on the foundation spacing (the greater the spacing, the greater the tolerance) and on the functional requirements of the structure or equipment. However, if all of the individually supported columns of a building settle by the same amount simultaneously, no additional loads will be imposed on the structure. The structure will be safe, but it could lose its functionality if the settlement is large.

E. Foundation Response

Foundation response describes how the foundation reacts to the applied loads, which defines the method of analysis. Foundation response is a function of its relative stiffness (or flexibility) compared to the surrounding soil and rock. Stiffer foundations generally tend to have a rigid-body movement (that is, they do not flex under the applied load) and transfer the load to the soil more evenly than flexible foundations. For example, if a lightly-loaded column is supported concentrically by a thick, heavily-reinforced concrete foundation having an area slightly greater than the cross-sectional area of the column, the load will be distributed evenly below the foundation. If this foundation is replaced by a very thin plate of steel having a significantly large area, more of the load will be transferred to the soil immediately below the column than below the edges of the plate. This flexibility also influences the development of potential failure surfaces and how they vary with depth. These surfaces can take various forms and determine how the load will be transferred to the soil or rock in end bearing (tip resistance) or side resistance. The actual load distribution will determine the deformation response, which is also a function of the soil and foundation material characteristics.

F. Soil and Rock Response Factors

Soil or rock response factors determine how the subsurface material will respond under the applied loads. Fine-grained soils, such as clays and silts, respond differently than coarse-grained soils, such as sands and gravels, when subjected to different types of loading. This response differs mostly because of the very different coefficients of permeability. Also, soil response is governed by its water content, and therefore the same soil will behave differently under dry, moist, and saturated conditions. A typical example of this can be found at the beach. The dry sand near the dunes is very loose and difficult to walk or drive on. The saturated soil close to the water (or under water) is comparatively stronger, yet sinks under weight. However, the moist soil on the main portion of the beach between the dunes and the water generally is strong enough to drive on.

G. Construction Procedures

Construction procedures influence the ultimate performance of the foundation system because they can alter the properties of the soil and rock. Sloppy construction practices will adversely affect the foundation performance. For example, poor backfilling will reduce the uplift capacity of shallow foundations or lateral capacity of direct embedment foundations. Often, a number of different methods can be used to construct a particular type of foundation, and a contractor is given the freedom to choose a construction procedure that will allow a competitive bid for the project. This freedom keeps the project costs down and encourages innovation, but care must be exercised that the procedure used does not adversely affect the soil and rock conditions, foundation performance, and adjacent structures or property.

V. BASIC THEORIES FOR FOUNDATION DESIGN

Soil and rock deform under load, causing the supported structure to move. The foundation will fail if the applied load is greater than the strength of the underlying soil or rock (bearing capacity failure) or if the soil and rock deformation results in excessive damage to the structure. Foundations must be designed to avoid both types of unsatisfactory behavior. Generally, a significant amount of foundation displacement is required to mobilize the ultimate bearing capacity (the maximum load that can be applied to the foundation without causing a shear failure of the soil or rock) of a foundation. This displacement may be much greater than the tolerable limits of the structure or equipment.

The description and evaluation of bearing capacity and displacement given below pertain to axial compression loads only. Procedures for other loading modes such as uplift, horizontal, and dynamic loads can be found in the bibliography.

A. General Description of Bearing Capacity

When you observe your footprint on a moist portion of the beach, you notice that along its edges the sand has moved or bulged upward. The grains immediately below the foot are forced downward because of your weight, resulting in the depression of the footprint. To make room for this downward movement, the adjacent soil grains are pushed outward and upward. The greater the resistance offered by the soil to this outward and upward movement, the greater the bearing capacity of the soil and the shallower the depth of the footprint. The frictional resistance of the sand and the confining or vertical stress around the footprint offer resistance to this outward and upward movement. This type of bearing capacity failure is called general shear failure. It develops in soils that exhibit a brittle-type load-settlement behavior (i.e., the failure occurs suddenly), and it is catastrophic under most circumstances.

In saturated soft clays, the edges of the footprint will not show any bulging because the weight causes the foot to simply "punch" through the soil and there is no outward movement of the soil particles. Only the soil directly below the footprint is involved in resisting the load. The

greater the resistance to punching, the greater the bearing capacity of the soil and the shallower the depth of the footprint. Resistance to punching in clays is mostly offered by the undrained shear strength of the soil. This type of bearing capacity failure is called punching shear failure, and it develops in soils that exhibit very plastic-type load-settlement behavior (i.e., the failure is progressive, with continuous downward movement or punching of the foundation).

For deep foundations in medium-stiff clays or medium-dense sands, the bearing capacity (tip resistance) failure mode often lies between the two extreme cases of general shear and punching shear failure. In this case, there is both a punching of the foundation into the ground and some lateral and upward movement of the soil particles. Because of the high confining stress at the tip, the upward movement of the particles is restricted and does not manifest itself at the ground surface. This type of failure is called local shear failure, and it develops in soils that exhibit somewhat plastic-type load-settlement behavior (i.e., the failure occurs gradually with significant downward movement).

Bearing capacity of the foundation is governed by the hydraulic conductivity and shear strength of the soil and by the foundation geometry and location below the surface. Bearing capacity failure results when the shear strength of the soil is exceeded. For undrained loading, the undrained shear strength of the soil s_u and the total unit weight of the soil are required to evaluate bearing capacity. For drained loading, the effective stress friction angle, submerged unit weight, and coefficient of horizontal soil stress are required to evaluate bearing capacity.

Deep foundations develop their capacity from end bearing or tip resistance (like shallow foundations) and from side resistance. The side resistance results from the frictional force between the foundation material and soil, and therefore it is a function of the normal (horizontal in case of vertical foundations) force on the foundation and the coefficient of friction between the soil and foundation material. The horizontal force on the foundation normally increases with depth and is a function of the buoyant or effective unit weight of the soil and the coefficient of horizontal soil stress.

B. General Description of Settlement

Foundation geometry influences the volume of soil that will be affected by the foundation load. When shallow foundations are loaded, there is an increase in stress (bearing or contact stress) immediately below the foundation. As one moves farther away below the center of the foundation, the load is spread over a larger area (for a square foundation, this increase in area with depth can be visualized approximately as a truncated pyramid), and the stress gradually decreases. At a depth equal to about twice the

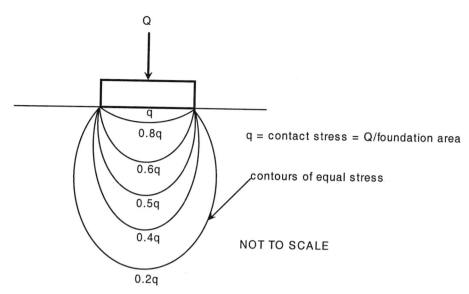

q = contact stress = Q/foundation area

contours of equal stress

NOT TO SCALE

FIGURE 2 Stress bulb.

width of the foundation, the stress decreases to about 10% of the contact stress. Figure 2 shows a contour plot of stress increase below a foundation. These contours also are called pressure bulbs or envelopes.

Foundation settlement results from the magnitude of the stress increase experienced by the soil. Because the zone of influence of wider foundations is greater than that of narrow foundations, they cause a stress increase in deeper layers of soil than do narrow foundations, and therefore, for the same bearing stress, wider foundations will experience more settlement than narrow ones. However, for the same bearing stress, wider foundations will carry more load than narrow ones. If the load is the same on both foundations, the wider one will have a lower contact stress. This contact stress may not be enough to cause significant increase in stress, even at depths equal to half the width of the foundation, and therefore the settlement may be less than for a narrower foundation carrying the same load.

When a load is applied to a foundation, three distinct phases of settlement occur: elastic, consolidation, and secondary settlement. The elastic settlement occurs immediately after load application because of initial stress–displacement response. Concurrently, excess pore water pressure develops in the soil, and there is no increase in effective stress. As this excess pore water pressure dissipates with time, the load is transferred to the soil grain-to-grain skeleton, and the effective stress increases. This transfer process results in consolidation settlement. Further settlement continues after consolidation because the soil skeleton gradually readjusts to the new level of effective stress. This readjustment leads to secondary settlement.

In fine-grained soils such as clays, all three settlement components are distinct. The elastic response occurs as soon as the load is applied; the consolidation settlement is time-dependent and can take years to complete, because of the low hydraulic conductivity of the soil. The secondary response also is time-dependent and continues indefinitely. In coarse-grained soils such as sands, these three components are rarely distinct. Because of the high hydraulic conductivity of sands, the elastic and consolidation settlement occur simultaneously and immediately after load application. The secondary settlements in coarse-grained soils usually are very small and are often disregarded in sands and gravels of sound mineralogy.

In addition to the three settlement components described above, two additional components need to be considered. The first is the elastic deformation of the structural material of the foundation, which normally is minor for shallow foundations but can be significant for deep foundations. The second is the broad category of special cases that has to be evaluated on a case-by-case basis. These include moisture-susceptible soils such as expansive clays and collapsible sands, biochemical decay in landfills and other organic soils, groundwater chemistry that can adversely affect the foundation material, earthquake loading that could lead to liquefaction, and other special local problems.

C. Shallow Foundations

1. Evaluation of Bearing Capacity

The boundaries between the general shear, punching shear, and local shear failure modes described above are not distinct, primarily because the changes in soil behavior modes are gradual. However, for the relatively low stress levels associated with shallow foundations, general shear failure normally occurs in the medium to denser sands and in most clays during undrained loading.

Different methods of analysis are used for the different failure modes. For the general shear mode, a rational approach based on limiting states of equilibrium is employed. The approach is based on theory and has been confirmed in principle by laboratory and field testing. For the local and punching shear modes, a variety of approaches have been suggested, but none is strictly correct. A more rational approach was suggested by Vesic (Winterkorn and Fang, 1975), which incorporates the stress–deformation characteristics of the soil and is therefore applicable over a wide range of soil behavior. This method is described below.

The ultimate bearing capacity for the general shear mode is computed utilizing the shear surface illustrated in Fig. 3, assuming that the foundation is of infinite length. The soil within the shear surface is assumed to behave as a rigid-plastic medium and is idealized as three zones: an active Rankine zone (I), a radial Prandtl zone (II), and a passive Rankine zone (III). The soil above the foundation is modeled as an equivalent surcharge (i.e., additional weight only), as shown in Fig. 3.

The general solution is known as the Buisman–Terzaghi equation and is given as follows:

Ultimate bearing capacity $= q_{\text{ult}}$
$$= cN_c + 0.5B\gamma N_\gamma + qN_q, \tag{1}$$

where c is the soil cohesion, B is the foundation width, q is the surcharge ($=\gamma D$), D is the foundation depth, γ is the soil unit weight, and N_c, N_γ, and N_q are dimensionless bearing capacity factors that depend upon the friction angle of the soil.

The Buisman–Terzaghi equation is used in either of two derivative forms, which depend upon the soil type and rate of loading. For undrained loading, which develops when loads are applied relatively rapidly to fine-grained soils such as clays, the friction angle $\phi = 0$ and $N_c = 5.14$, $N_\gamma = 0$, and $N_q = 1$. The resulting equation is

$$q_{\text{ult}} = 5.14c + q. \tag{2}$$

In this case the cohesion term c is actually the undrained shear strength of the soil, more commonly denoted as s_u.

For drained loading, which develops under most loading conditions in coarse-grained soils such as sands and for long-term sustained loading of fine-grained soils, $c = 0$ and the resulting equation is

$$q_{\text{ult}} = 0.5B\gamma N_\gamma + qN_q. \tag{3}$$

The above two equations apply very well to stiffer or denser soils, where the deformation under load is small, and this corresponds well with the assumed rigid-plastic soil behavior for general shear. For softer or looser soils, the deformation under load becomes large, which is at variance with the assumption of rigid-plastic soil behavior. To account for this variance, Vesic introduced the concept of rigidity index, which accounts for soil deformability. In this method, the rigidity index I_r of the soil is compared to the critical rigidity index I_{rc} of the soil–foundation system. The rigidity index is a function of the shear strength parameters c and ϕ, the elastic modulus, and Poisson's ratio of the soil. Fundamentally, it is defined as the ratio of the shear modulus to strength, as modified for volume changes. The theoretically-derived critical rigidity index is a function of the foundation geometry (aspect ratio) and the friction angle ϕ of the soil. When the rigidity index is greater than the critical rigidity index, the soil behaves as a rigid-plastic material, and therefore the general shear failure equations given above can be used. When the rigidity

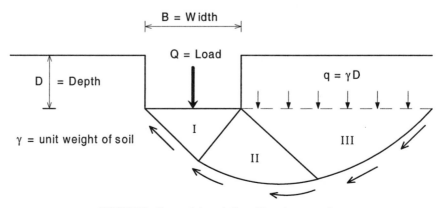

FIGURE 3 General description of bearing capacity.

index is less than the critical, the equation is modifed as follows:

$$\text{Ultimate bearing capacity} = q_{\text{ult}}$$

$$= cN_c\zeta_{cr} + 0.5B\gamma N_\gamma\zeta_{\gamma r} + qN_q\zeta_{qr}. \quad (4)$$

The ζ_{xr} terms are the rigidity factors and are functions of the soil friction angle, soil rigidity index, and foundation aspect ratio.

The bearing capacity equations developed above are based on the following assumptions:

- Horizontal ground surface
- Horizontal, infinitely long strip foundation at ground surface
- Vertical loading, concentrically applied

Taking into account the variations that occur in practice, the basic bearing capacity equation is modified by a number of factors that separately treat each variation:

$$q_{\text{ult}} = cN_c\zeta_{cr}\zeta_{cs}\zeta_{ci}\zeta_{ct}\zeta_{cg}\zeta_{cd} + 0.5B\gamma N_\gamma\zeta_{\gamma r}\zeta_{\gamma s}\zeta_{\gamma i}\zeta_{\gamma t}\zeta_{\gamma g}\zeta_{\gamma d}$$

$$+ qN_q\zeta_{qr}\zeta_{qs}\zeta_{qi}\zeta_{qt}\zeta_{qg}\zeta_{qd}. \quad (5)$$

Each ζ factor is doubly indexed to indicate the term to which it applies (c, γ, or q) and the phenomenon it describes (r for rigidity of the soil, s for shape of the foundation, i for inclination of the load, t for tilt of the foundation base, g for ground surface inclination, and d for depth of foundation). Although this equation seems formidable at first, it really only represents the inclusion of simple modification factors that depend upon the soil shear strength parameters and foundation geometry. The above equation does not account for the influences of the groundwater table and soil layering, but procedures exist to incorporate them in the design of shallow foundations.

2. Evaluation of Displacements

As stated above, when a load is applied to the foundation, three distinct phases of settlement occur: immediate or elastic (ρ_e), consolidation (ρ_c), and secondary settlement (ρ_s):

$$\text{Total settlement} = \rho_t = \rho_e + \rho_c + \rho_s. \quad (6)$$

In coarse-grained soil, the elastic and consolidation settlements basically occur immediately and simultaneously, and the secondary settlement usually is very small.

The elastic settlement of a foundation is determined by assuming that the soil behaves as an elastic material. The basic approach employed is to use the traditional Boussinesq solution for stress changes in an isotropic, elastic medium subjected to a point load applied normal to the surface of a half-space (the ground is a half-space because it does not exist above the ground surface). The elastic settlement of the foundation is given by

$$\text{Elastic settlement} = \rho_e = [\Delta\sigma(1 - \nu^2)A^{0.5}]/(E\beta_z), \quad (7)$$

where $\Delta\sigma$ is the applied stress (Q_c/A, with Q_c the applied compressive load), A is the foundation area, ν is Poisson's ratio of the soil, E is the elastic modulus of the soil, and β_z is the shape and rigidity factor, ranging from about 1.04 for flexible circular foundations to 1.41 for rigid, rectangular foundations with a length-to-width ratio of 10. Real foundations are neither perfectly rigid nor perfectly flexible, but the usual reinforced concrete footings are nearer to rigid, while the usual grillages are nearer to flexible. The above solution is based on a homogeneous soil with constant elastic modulus with depth, both of which are incorrect for real situations. However, procedures exist for modifying the results to account for this variation.

For fine-grained soils, the consolidation settlement often is the major component of the total settlement. As noted previously, consolidation settlement is the time-dependent volume change that occurs as the excess pore water pressure, developed during load application, gradually dissipates and is transferred to the soil skeleton. The consolidation settlement is given by

$$\text{Consolidation settlement} = \rho_c = (\Delta e \cdot H_c)/(1 + e_0), \quad (8)$$

where e_0 is the initial void ratio (=volume of voids/volume of solids) of the soil (which can be determined by laboratory tests on undisturbed soil samples), Δe is the change in void ratio caused by the increase in stress from the foundation, and H_c is the height (thickness) of the compressible soil (which can be determined by soil exploratory borings). Although this is a simple expression, the computation of the consolidation settlement requires detailed knowledge of the stress changes occurring in the soil and both the volume change characteristics and the stress history of the soil. The change in void ratio caused by the increase in stress from the foundation is determined by laboratory consolidation tests, the results of which give the stress–deformation (volume change) characteristics of the soil and the amount of time it would take for that volume change to occur. Details of these tests can be found in many soil mechanics (geotechnical engineering) textbooks.

Secondary settlement continues after consolidation settlement is complete, as the soil structure gradually readjusts to the new level of effective stress. This process continues indefinitely. The prediction of secondary settlement is of considerable importance with clays of high plasticity and with micaceous and organic soils, but it typically is of minor concern with coarse-grained soils and overconsolidated, low plasticity, inorganic clays. Secondary settlement often is computed from the relation

$$\rho_s = C_\alpha H_c \log(t_2/t_1), \qquad (9)$$

where C_α is the coefficient of secondary settlement (determined from laboratory consolidation tests), H_c is the thickness of the soil layer, t_1 is the time when secondary settlement begins to be computed, often defined as the time for 90% consolidation settlement, and t_2 is an arbitrary time to which the secondary settlement is computed. Values for C_α typically are very low for overconsolidated soil, but they can be high for soft, normally consolidated soil.

D. Deep Foundations

As the depth of embedment of a foundation increases, the applied compressive loads are resisted by a combination of tip resistance (end bearing) and side resistance, as shown in Fig. 4. When a deep foundation is loaded, the foundation moves downward with respect to the surrounding soil, mobilizing side resistance in the soil. For small loads, small displacements occur, and most of the load is supported by the side resistance. As the load increases, larger displacements occur, and the full side resistance is reached. From this point on, additional applied loads are resisted by the tip. Typically, the displacement needed to mobilize side resistance is in the range of 0.2–0.6 in. (5–15 mm), and it is relatively independent of the foundation dimensions or soil type. On the other hand, the displacement necessary to mobilize the tip resistance is a function of foundation size and can range from about 4% to 10% of the foundation diameter, depending on the foundation type and construction method.

The side resistance is analyzed as a problem in determining the shearing resistance of the foundation–soil interface acting over the surface of the foundation. The tip resistance is treated as an ultimate bearing capacity problem, similar to that for a shallow foundation.

1. Evaluation of Tip Resistance

The tip resistance of a deep foundation is a bearing capacity problem. The maximum tip resistance is given by

$$Q_{tc} = q_{ult} A_{tip}, \qquad (10)$$

where q_{ult} is the maximum bearing capacity at the tip (given above by the modified equation for shallow foundations) and A_{tip} is the area of the foundation tip.

2. Evaluation of Side Resistance

The side resistance of a deep foundation is given by

$$Q_{sc} = \int_{surface} \tau(z)\,dz, \qquad (11)$$

where $\tau(z)$ is the shearing resistance with depth along the foundation side, and the integral represents a summation with depth of $\tau(z)$ times the foundation surface area.

For drained loading, which develops under most loading conditions in coarse-grained soils such as sands and for long-term sustained loading of fine-grained soils, the cohesion $c = 0$, and the side resistance is evaluated by

$$Q_{sc} = \pi B \int \sigma_h'(z) \tan \delta(z)\,dz, \qquad (12)$$

where $\sigma_h'(z)$ is the horizontal effective stress at depth z, which acts as a normal stress on the soil–foundation interface, and $\delta(z)$ is the effective stress angle of friction for the soil–foundation interface at depth z. Expressing the horizontal stress in terms of the overburden stress $\sigma_v'(=\gamma' z)$, we obtain for the above equation

$$Q_{sc} = \pi B \int \gamma' z K(z) \tan \delta(z)\,dz, \qquad (13)$$

where γ' is the effective unit weight of the soil (submerged below the water table; dry or moist above) and K is the operative coefficient of horizontal soil stress $(=\sigma_h'/\sigma_v')$.

The most complex term to evaluate in the above equation is K, which is a function of the original *in situ* horizontal stress coefficient K_0 and the stress changes caused in response to construction, loading, and time. Values of K can range from as low as 0.1 to over 5.

For undrained loading conditions, which develop when loads are applied relatively rapidly to fine-grained soils such as clays, $\phi = 0$ and the above equation can be expressed as

$$Q_{sc} = \pi B \alpha \int s_u(z)\,dz, \qquad (14)$$

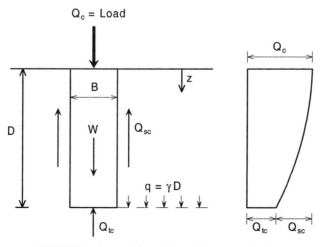

FIGURE 4 General description of deep foundation.

where $s_u(z)$ is the undrained shear strength, which varies with depth z, and α is the adhesion or reduction factor applied to the average s_u over the shaft depth D to account for construction effects, disturbance, etc. The adhesion factor is a purely empirical factor that has been used to correlate the results of full-scale load tests with the average soil undrained shear strength.

3. Evaluation of Displacements

The settlement analysis of deep, shaft-type foundations is somewhat different from that of shallow, spread-type foundations because of two primary reasons. First, the installation process for deep foundations (driving or "pounding" the pile into the ground or excavating the hole and keeping it open for drilled shafts) causes soil disturbance and alters the state of stress in the soil surrounding the foundation. Second, the load is transferred to the soil in both tip resistance and in side resistance along the foundation shaft. This load transfer process is complex and varies with load intensity. The problem is complicated further by the installation of adjacent foundations and possible group action.

A number of procedures exist for evaluating the settlement of deep foundations, and each of them has its advantages and shortcomings. The procedure recommended by Vesic (Winterkorn and Fang, 1975) is a simple method that gives reasonable results. This procedure is described below.

The total settlement ρ_t of deep, shaft-type foundations is given by

$$\rho_t = \rho_f + \rho_{tt} + \rho_{ts}, \tag{15}$$

where which ρ_f is the settlement caused by axial deformation of the foundation structural material, ρ_{tt} is the settlement of the tip from the tip load, and ρ_{ts} is the settlement of the tip from the shaft side resistance load. Both ρ_{tt} and ρ_{ts} include elastic, consolidation, and secondary settlement components. For coarse-grained and overconsolidated fine-grained soils, the elastic settlement normally constitutes at least 80% of the total settlement, and secondary effects usually are minimal. However, if a compressible soil underlies the foundation, the situation is different. For this case, a settlement analysis similar to that for shallow foundations is performed to evaluate the settlement of the underlying stratum.

The value of ρ_f is determined from conventional formulas for the deformation of a structural material and is given by

$$\rho_f = D(Q_t + \alpha_s Q_s)/(A E_m), \tag{16}$$

where Q_t is the load transmitted at the foundation tip, Q_s is the load transmitted in side resistance along the foundation shaft, D is the foundation depth, A is the foundation cross-sectional area, E_m is the foundation elastic modulus, and α_s is a coefficient that depends on the distribution of side resistance along the foundation depth. For a linearly increasing side resistance with depth, $\alpha_s = 2/3$.

The tip settlement equations for the simplified Vesic solutions employ elastic theory and empirical correlations, resulting in

$$\rho_{tt} = (C_t Q_t)/(B q_{ult}), \tag{17}$$

$$\rho_{ts} = (C_s Q_s)/(D q_{ult}), \tag{18}$$

where q_{ult} is the maximum bearing capacity at the tip and C_t and C_s are empirical coefficients, with C_t varying from 0.02 to 0.18 and $C_s = [0.93 + 0.16(D/B)^{0.5}]C_t$.

E. Safety Assessment

1. Factor of Safety

The primary purpose of foundation design is to ensure that the foundation performs satisfactorily during a specified time period. Economic constraints often make it impossible to design for all possible circumstances (e.g., for a major hurricane or nuclear blast). However, the designer does need to consider other factors that can have an adverse effect on the foundation, such as variation in load, unrecognized loads, variation in material strength, unforeseen *in situ* conditions, inaccuracies in design equations, errors from poorly supervised construction, change in system function from original intent, etc.

Traditional design practice does not require a designer to go through the process of considering each of these factors separately and in explicit detail. The designer is required to have a global appreciation of these factors, which ultimately leads to a single factor of safety in design, which is defined as

$$\begin{aligned} \text{FS} = {} & \text{ultimate bearing capacity}/ \\ & \text{allowable bearing capacity}, \end{aligned} \tag{19}$$

in which the ultimate bearing capacity is evaluated from equations similar to those given above, and allowable bearing capacity is equal to the maximum design stress that can be applied on the foundation. In conventional foundation design, factors of safety between two and three generally are considered to be acceptable when site conditions are reasonably uniform and soil and rock properties are known with some confidence.

The single factor of safety used in foundation design suffers from a major flaw in that it is not unique. It depends upon how it is defined, and it can vary significantly over a wide range. For example, in uplift loading, in which the foundation weight contributes to the uplift capacity of the foundation, certain designers choose not to apply the factor of safety to the foundation weight because it can be calculated fairly accurately by the foundation geometry and material density. Others choose to include the foundation weight when defining the factor of safety. Also, since the factor of safety is not defined within a consistent and common framework, engineers cannot communicate and share their experiences effectively. This approach also does not distinguish between model (design equation) and parameter (soil and rock property) uncertainty, and therefore makes it difficult to justify any reduction in the safety level if there is additional information about the soil and rock properties or if there are advances in the state of the art in foundation design equations.

2. Reliability-Based Design

An alternative design philosophy is the concept of reliability-based design (RBD), which is also known as LRFD or load and resistance factor design. This approach is well-advanced for structural design, but it is still largely in the development stage for foundation design. In contrast to the single factor of safety approach, factors greater than or equal to one are applied to the various loads, while factors less than one are applied to the soil or rock resistances. These various factors should be established by rigorous calibrations using reliability theory, with an in-depth knowledge of the uncertainties and their variations for both the loads and resistances. Foundation design is slowly progressing toward RBD.

VI. FOUNDATION CONSTRUCTION

It is important for the foundation designer to be familiar with the various equipment and methods used during foundation construction because they can adversely affect the foundation and soil response, project economics, and overall effectiveness of the design. Generally, the contractor chooses the method of foundation construction, and the designer is unaware of the exact details of the method until the time of construction. There may be occasions when the designer specifies a particular method of construction, for example, to protect existing structures in close proximity to the new foundation, but this will almost certainly add to the cost of construction. Whatever method of construction is used, the key to ensuring that the design methods are appropriate is to ensure that the construction is done properly for that method. This approach requires that the specifications describe clearly the proper construction techniques that result in quality construction and also allow the contractor as much freedom as possible.

A. Shallow Foundations

Construction of spread footings involves excavation of the hole, placement of the required forms, reinforcing rods,

and concrete, and then backfilling of the hole with the excavated soil. The forms are used to contain the concrete within the specified dimensions. Care must be exercised that the hole is not overexcavated and that the bottom of the hole is subjected to the minimum amount of disturbance during the placement of the forms and reinforcing rods. If the hole is overexcavated too deeply beyond that called for in design, it should be backfilled with concrete and not with the excavated soil. This approach will prevent additional settlement from occurring.

Excavations for shallow foundations that are up to 5 ft (1.5 m) deep generally do not require support for the excavation walls. For these shallow depths, vertical cuts can be made in moist silts, clays, and sands containing little amounts of silt or clay. In dry sands and gravels, the walls can be sloped back for stability. Because of worker safety issues, excavations deeper than about 5 ft (1.5 m) either have to be sloped back or require wall support even in clays, where vertical cuts can be made up to 20 ft (6.0 m) or more. The steepness of the slopes depends upon the soil or rock characteristics, depth of the excavation, amount of time the excavation is to remain open, and weather conditions.

When the excavation is close to the property line or adjacent to existing buildings, the sides of the excavation must be made vertical and have to be structurally supported. For individual spread foundations, the excavation generally is less than about 10 ft (3.0 m) deep, but if a single-story or multistory basement (as for an underground parking facility) is to be constructed, the excavation can be more than 15 ft (4.5 m) deep. For shallow (less than 10–12 ft or 3.0–3.7 m in depth) vertical cuts, vertical timber planks (sheeting) can be used to support the excavation walls. The plank is driven with its wider side against the soil face to a few feet (up to 1 m) below the surface. As excavation proceeds, the sheets are driven farther into the ground so that they are always a few feet below the bottom of the excavation. Horizontal beams (wales or walers) are placed close to the top and bottom of the hole to support the sheeting. In narrow excavations, horizontal timber or pipe struts that run from side to side, as shown in Fig. 5a, support the wales. In wider excavations (such as for basements), inclined struts (rakes or rakers) can support the wales, as shown in Fig. 5b.

For excavations deeper than 12–15 ft (3.7–4.5 m), timber sheeting becomes uneconomical, and other alternatives such as soldier pile and lagging or steel sheet piles can be used. If a vertical face of about 30–50 ft^2 (2.78–4.65 m^2) of soil can be left unsupported for a short period without collapsing, a system of soldier piles and lagging can be used to support the vertical cuts. Steel H-sections (soldier piles) are driven into the ground with their flanges (flat part) parallel to the sides of the excavation to a depth a

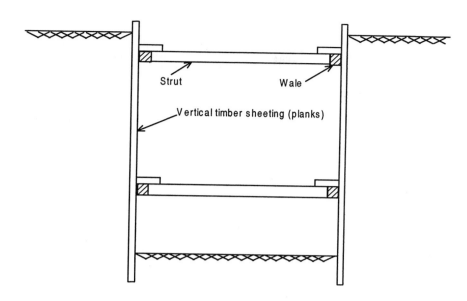

a. Horizontal bracing for narrow shallow excavations

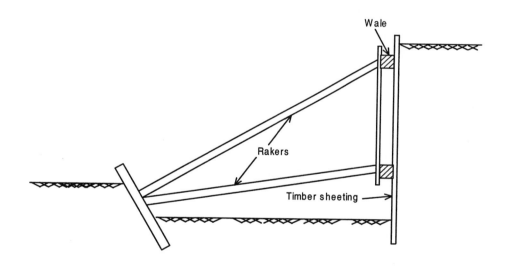

b. Inclined bracing for wide shallow excavations

FIGURE 5 Bracings for shallow excavations. (a) Horizontal bracing for narrow, shallow excavations. (b) Inclined bracing for wide, shallow excavations.

few feet below the maximum depth of the excavation and at a spacing of about 4–8 ft (1.2–2.4 m). As excavation proceeds, horizontal timber boards with their wide sides against the soil are wedged behind the exposed flanges of the soldier piles, as shown in Fig. 6. Wales and struts then are placed at regular intervals to support the soldier piles. For wide excavations where horizontal struts may not be feasible, inclined braces have to be used. Under these cir-

cumstances, it may be possible to excavate the central portion to the maximum depth, leave a bench (berm) of soil around the perimeter to support the soldier piles, and construct a portion of the foundation to support the inclined braces. Once the braces are in place to support the soldier piles, the berms can be excavated. Under certain conditions, it may be feasible to use an anchor system (also called tiebacks) to support the soldier pile wall (see Fig. 7)

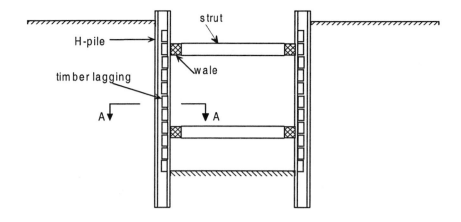

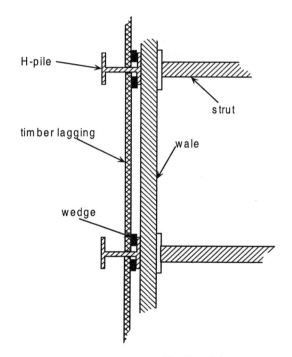

SECTION A-A

FIGURE 6 Soldier pile and lagging.

instead of the inclined braces or rakers. These tiebacks are constructed by drilling an inclined hole into the soil, placing a reinforcing rod in the hole, and then backfilling the hole with concrete. Once the concrete has set, the reinforcing rods generally are prestressed before excavation proceeds.

If the soil conditions are such that a soldier pile system cannot be used (e.g., clean, dry sand that will run as soon as it is excavated or sand below the water table), steel sheet piles can be driven to support the vertical sides of the excavation. Interlocking steel sections are driven into the

ground around the perimeter of the excavation to a depth a few feet below the final depth of the excavation. Wales and struts (or tiebacks) then are placed at regular intervals, as described above for the soldier pile and lagging system.

B. Deep Foundation

1. Pile Foundations

a. General. Wooden pegs for supporting a camping tent generally are driven into the ground by a hand-held hammer. If the ground is hard, the hammer blows can

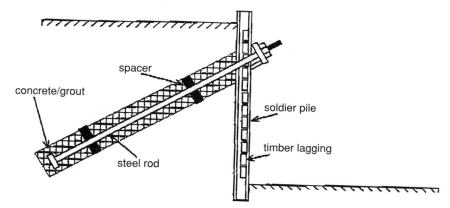

FIGURE 7 Tieback system.

damage the top of the peg, and the bottom of the peg can be crushed or split because of the ground resistance. Similarly, pile foundations are driven into the ground by impact or vibratory-type hammers. If the ends of the pile are not protected adequately, the hammer blows can damage the top or butt of the piles, and the resistance offered by the ground can damage the tip of the pile. Special attention also must be paid to pile storage and handling, placement tolerances, driving, and extraction to ensure a properly constructed pile foundation.

Piles can be damaged during storage and handling, which is especially true for precast concrete piles that can crack or develop a camber (permanent bending) if they are supported or lifted at locations away from the designated pickup points. Excessive camber can cause the pile to be driven out of plumb and be overstressed when loaded. Special care also must be exercised when handling piles with protective coatings, and the damaged areas must be repaired prior to installation. Generally, all piles are visually examined at the driving location by a qualified inspector immediately prior to installation.

Piles have to be driven close to the location and inclination designated on the construction plans; otherwise, they will be unable to support the design load with an adequate margin of safety. However, it is impractical to drive a pile exactly at the given location and inclination, and the engineer must allow for reasonable placement tolerances and account for these deviations in the design. These tolerances need to be designated in the construction specifications.

Perhaps the single most important aspect of pile construction is the process of driving the pile into the ground. Pile driving should not result in permanent deformation of steel H-piles or steel pipe piles, crushing or spalling of concrete piles, or splitting or brooming of wood and fiber glass piles. Pile damage can be avoided if the proper

equipment is utilized for the prevailing subsurface conditions and the type and capacity of pile being used. Even if proper equipment is used, piles can still be damaged if the pile and hammer are misaligned.

Sometimes, when piles are driven through hard layers of dense sand, cobbles, boulders, or other obstructions, it may be necessary to supplement pile driving with the help of pile shoes, jetting, or preboring to prevent damage to the pile from overdriving. These methods usually result in reduced pile capacity and therefore should be used with caution.

Specially designed pile shoes (tip protector/reinforcer) improve driveability and also provide increased cutting ability and pile tip protection. These shoes usually are fabricated from cast steel. Jetting is accomplished by pumping water under high pressure through pipes attached to the pile. The water loosens and removes the soil and thereby allows the pile to be driven more easily. When hard layers of soil are encountered, drilling a pilot hole or preboring can reduce the driving resistance. Preboring can also be used when passing through materials such as cobbles and boulders that can deflect the pile.

When pile damage is suspected during installation, it may be necessary to extract the pile for inspection. Extraction also may be required if excessive drift occurs during driving and the pile location and inclination exceed the specified placement tolerances. Pile extraction is an expensive and difficult process that requires experienced personnel. Care should be taken that the tensile strength of the piles, especially concrete piles, is not exceeded during the extraction process.

b. Pile-driving equipment. Typical pile-driving equipment includes a crawler-mounted crane (see Fig. 8) with a boom, leads, hammer, driving caps, and other accessories all connected to form a unit. This unit must be

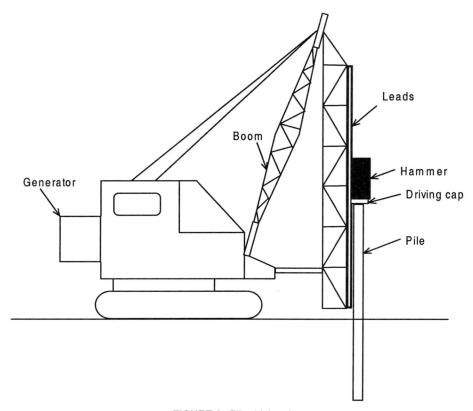

FIGURE 8 Pile-driving rig.

strong enough to guide and drive each pile accurately to its final position and withstand safely all loads imposed during the pile driving process.

Pile hammers can be grouped under impact (drop hammer, steam or air hammer, diesel hammer) or vibratory types. Drop hammers are the simplest and oldest type of impact hammers. They consist of a guide weight or ram that is lifted up to a required height by a hoist line and then released. The drop hammer is best suited for small projects that require relatively low capacity and employ lightweight timber or steel piles. Other types of impact hammers include the steam or air hammer, in which steam or compressed air pressure is used to raise the ram. The ram falls freely and strikes the drive cap when the pressure is released automatically. These hammers can be used for a variety of different pile–soil combinations, but they are best suited for timber and H-piles in sand.

Diesel hammers are self-contained, lightweight, economical, and easy to service. Initially, the ram is raised by mechanical means, and the fuel is injected into the cylinder while the ram drops. When the ram strikes the anvil, the fuel is ignited, explodes, and forces the anvil down against the pile and the ram upward. The sequence repeats itself automatically, and the pile is driven into the ground by the

ram impact and the explosion of the fuel. Diesel hammers are best suited for hard driving conditions and do not work well in soft soil because of the driving resistance required for compression and ignition of the fuel.

Vibratory hammers utilize eccentric weights that are rotated with the help of electric or hydraulic motors to produce vertical vibrations. The vibrations reduce the frictional soil resistance along the sides of the pile and also loosen the soil at the tip . This type of hammer is best suited for steel piles in sand. Clay soils dampen the vibrations of the hammer and therefore reduce pile penetration. These hammers are not effective in penetrating soils with large obstacles and boulders and are not suitable for concrete or timber piles.

Pile-driving leads are used to align concentrically the pile head and hammer, provide lateral support for the pile, and maintain proper pile alignment and position during driving. If the hammer is not aligned properly with the pile head, eccentric loading can cause excessive stresses that result in structural damage of the pile. Also, the driving energy transferred to the pile will be reduced. Leads or leaders are usually fabricated of steel.

Driving caps are placed between the pile head and the hammer and transfer the driving energy from the hammer

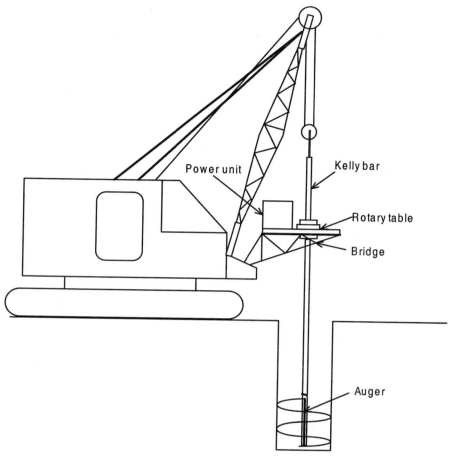

Power unit

Kelly bar

Rotary table

Bridge

Auger

FIGURE 9 Drilling rig for drilled shafts.

to the pile without causing distress to the pile. Driving caps consist of a pile cushion placed directly on top of the pile butt, a helmet placed on top of the pile cushion, and an anvil placed on top of the helmet. Sometimes an additional cushion is placed between the helmet and the hammer ram to protect the hammer from the shock wave reflected back to it. Plywood and oak boards normally are used for pile cushions, and thick blocks of hardwood are used for the hammer cushion.

2. Drilled Shafts

a. General. Drilled shaft construction generally involves excavating a cylindrical hole in the ground, providing a suitable method of keeping the hole from caving in, placing tied steel reinforcing rods (also called a rebar cage) in the excavated hole, and then filling the hole with concrete. Drilled shaft construction can be classified in three broad categories, dry, casing, and wet, according to the method used for keeping the hole open during excavation. The method selected for construction depends

upon the soil in which the shaft is being constructed. The performance of a drilled shaft depends on the construction method, and therefore it is included as part of the design process.

The most common excavation method used for drilled shafts is by rotary drilling, although in some developing countries hand excavation also is used. The rotary drilling machines are either truck-mounted or crane-mounted (see Fig. 9) and can be equipped with a number of different types of drilling tools, such as a drilling bucket, flight auger, rock auger, core barrel, or shot barrel. The type of drilling tool used depends upon the type of soil or rock being excavated.

b. Construction method. Drilled shafts generally are constructed by the dry, casing, or wet method or a combination of these methods, depending upon the type of ground encountered during excavation. The dry method can be used above the water table for soil and rock that will not cave or slump when the hole is drilled. Above the water table, stiffer clay and sands mixed with some cohesive material are best suited for dry construction. This method sometimes can be used for low-permeability soils below the water table if the concrete is placed immediately after the hole has been excavated. A short piece of casing commonly is used near the surface during dry construction to prevent the loose surface soil from caving into the hole. Generally, this casing is corrugated steel pipe and extends some distance above the ground surface. The casing also serves as a guide for drilling tools, safety barrier for the workers, and means for preventing material falling into the hole. After the hole is excavated to the required depth, the bottom of the hole is cleaned to remove the loose material. Hand cleaning sometimes is performed, but this should be avoided as much as possible because of safety considerations. Special regulations have to be followed if a person is lowered into the hole for hand cleaning or inspection. If the bottom of the hole has to be enlarged, or belled, to form a larger bearing area, an underreaming (belling) tool can be used.

The casing method is used for drilled shaft construction when excessive caving and deformation of the soil or rock will occur during excavation. A casing consists of a simple steel pipe or tube that is either driven or vibrated into the ground, either before drilling or as the drilling proceeds. Casings also are used when the drilled shaft extends through water-bearing soil into rock. The casing is driven or vibrated through the soil to form a seal at the soil–rock boundary that will prevent water from flowing into the excavated hole. When large obstructions such as boulders are anticipated, the casings can be telescoped. Telescoping involves placing a casing with a larger diameter than that required for the drilled shaft to a certain depth below the surface. A smaller diameter borehole is then drilled and cased below the bottom of the first casing. This process then can be repeated to the desired depth of the drilled shaft.

Casings can be temporary or permanent, depending on the type of soils involved. Temporary casings are withdrawn as the concrete is placed in the shaft, but they may have to be left in place if it is difficult to extract them. This smooth-wall casing can reduce the side resistance of the shaft and therefore its overall load-carrying capacity. Permanent casings often are used for very soft and loose soil conditions and are left in the ground after the concrete is placed. Construction with permanent casing is more expensive than for temporary casing.

An alternative to the casing method is the wet or slurry-displacement method, which can be used in any of the soil conditions described for the casing method. This method generally involves use of slurry to keep the excavated hole stable for the total depth of the excavation. Either mineral slurry or polymer slurry can be used for this purpose. The wet method is more feasible than the casing method for very deep holes and when it becomes difficult to seal casing in rock or impermeable soil below water-bearing granular soils. This method can be adapted for construction of drilled shafts where the surface of the ground is under water, as in a lake or river.

When minerals such as bentonite are mixed with water, they form a suspension of platelike solids in the water that is called slurry. It also can be formed with certain polymers. When the fluid pressure of the slurry exceeds that of the ground water in permeable soil, the slurry displaces the water, penetrates the soil, and forms a membrane on the walls of the hole. This membrane or mudcake helps to keep the hole stable. The effectiveness of the slurry depends on the proper amount of bentonite or polymer used and the chemistry of the water with which it is mixed. The use of slurry involves a significant additional cost resulting from the materials, handling, mixing, placing, recovering, cleaning, and testing of the slurry. For mineral slurry, there is an additional cost for properly disposing of the slurry because of environmental concerns.

Basically, the wet method either can be a static process or a circulation process. In the static process, the cuttings are lifted up by the drilling tool, and in the circulation process the cuttings are brought to the surface with the circulating slurry. Once excavation has been completed to the required depth, it is important to check that too much soil material is not in suspension within the slurry; otherwise, the particles can settle to the bottom and form a loose bearing surface for the drilled shaft.

c. Drilling rigs and tools. Excavations for drilled shafts are performed most commonly by rotary methods using some form of a rotary drilling unit or rig. These drilling units are mounted on trucks, cranes, crawler tractors, or skids. The type of drilling rig that is best suited for the job depends on the soil and rock conditions at the site and the size of the foundation. The capacity of the rig is determined by the amount of downward force (crowd) and the twisting force (torque) it can apply. A steel drive shaft called a kelly bar is turned by the power unit and transfers the crowd and torque to the drilling tools. The crowd can be provided by the weight of the kelly bar and the drilling tool or by other mechanical or hydraulic means. Although the crowd and the torque are important factors for achieving the best drilling rate, progress can be hindered if the proper drilling tool is not used.

Tools for rotary drilling include drilling buckets and flight augers for soil excavation and rock augers, core barrels, shot barrels, and full-faced excavators for rock excavation. To form an enlarged base, underreamers or belling buckets can be used. Drilling buckets commonly are used for excavation in granular soils and are equipped with cutting teeth at the bottom of the bucket. Two openings at the bottom of the bucket allow the excavated soil to enter the bucket. Flaps inside the bucket prevent the excavated soil from falling out of the bucket. When the bucket is full, it is raised and withdrawn from the hole, and the hinged bottom is opened to remove the excavated soil. These buckets also can be used with drilling slurries.

Flight augers are best suited for cohesive soils and soft rock and look like enlarged versions of a carpenter's drill bit. A cutting edge is fixed to the bottom of the auger that drills through the soil or rock. The cuttings are brought to the surface by traveling up the flights of the auger in much the same fashion as the wood shavings are brought to the surface when a hole is made in wood with a drilling bit. The cutting faces on the augers are aligned to result in a flat base at the bottom of the excavation. The bottoms of the comparable rock augers are equipped with tungsten carbide conical teeth that are used to break up hard rock. The flights of the rock augers are made of thicker metal than soil augers. Augers generally are not well suited for slurry construction, and care must be exercised when lowering or removing the auger from a slurried hole. Otherwise, differential pressures can be developed above and below the auger that can cause instability of the walls of the shaft and result in a cave-in.

Sometimes the rock is too hard to be excavated by a rock auger. Under these conditions, a core barrel can be used for excavating the hole. One of the simplest types of core barrels consists of a cylindrical steel tube equipped with hard metal teeth at the cutting edge. The tube cores into the rock, and the rock core is brought to the surface by lifting the core barrel to the surface. Friction from the cuttings holds the core in place inside the barrel. Rock coring is a slow process. If the hardness of the rock makes rock coring ineffective and only a minor penetration into rock is required, a shot barrel can be used. This barrel is similar to the core barrel but has a plain bottom. Hard steel shot are fed below the base of the barrel and grind the rock as the barrel is rotated.

d. Concrete placement. Concrete can be segregated and lose its strength if it is not placed properly inside the excavated hole. In dry holes, concrete generally is allowed to fall freely inside the hole, but it can be segregated if it strikes an obstruction such as a rebar cage or the sides of the hole. To prevent segregation of free-falling concrete, a flexible hose generally is used to direct it toward the center of the hole. Some studies have shown that the impact of the free-falling concrete at the bottom of the hole has a beneficial effect on the strength of the concrete because it drives air out of the voids, making it denser and therefore stronger.

When the excavated hole is not completely dry and is partially or fully filled with water or slurry, it is not feasi-

ble to place the concrete with the free-fall method. Also, the free-fall method is not recommended when the shafts are not vertical. Under these circumstances, concrete is placed through a tremie pipe. This pipe is made of steel and is equipped with a flap or plug at its bottom end to prevent the concrete from being contaminated by the slurry or water as the tremie is lowered into the hole. The concrete is allowed to flow through the tremie by opening the flap or removing the plug only when it reaches the bottom of the hole. Because of the large difference between the unit weights of the concrete and the slurry, the slurry is displaced by the concrete and floats up above the concrete, as long as the head of concrete is larger than the slurry head (head at a point is the unit weight of the material times its height above the point). As the column of concrete rises in the hole, the tremie pipe is raised gradually to allow reasonable flow of concrete through the tremie. Care must be exercised that the bottom of the tremie stays well below the top of the column of concrete; otherwise, pockets of slurry can get trapped within the concrete and weaken the shaft.

e. Effects of construction method. Construction procedures have a significant influence on the performance of the drilled shaft. Even the length of time the hole is allowed to remain open before concrete is placed can have a negative impact on the load-carrying capacity of the drilled shaft. For dry hole construction, the soil immediately surrounding the shaft expands and gets weaker with time. For slurry construction, if the slurry is allowed to remain standing in the hole without circulation, it can form a cake on the sides of the excavation and reduce the frictional side resistance of the shaft.

Frictional resistance between smooth steel casing and soil is lower than the frictional resistance between rough concrete and soil. Therefore, when temporary steel casings are left in place after concrete placement, the side resistance of the shaft will be decreased. All of these factors need to be considered in the design process.

VII. QUALITY ASSURANCE AND CONTROL

The success of a foundation system depends on the care exercised during its construction, and therefore quality assurance and quality control in the field are extremely important. Qualified personnel (inspectors) should be present on the site for verifying the soil conditions and inspecting the construction materials and procedures. On large projects, it might be feasible to perform full-scale load tests on the foundations to verify the design assumptions.

Verification of subsurface conditions is important because the foundation capacity and construction procedures depend on the soil and rock properties. Soil conditions can be verified easily during construction of shallow foundations because the soils are exposed during construction. For drilled shafts and anchors, soil conditions can be verified by the cuttings brought to the surface. For pile foundations, the soil conditions cannot be determined by visual examination but they can be inferred from the performance of the pile as it is being driven into the ground.

Materials used for foundation construction, such as reinforcing rods, concrete, piles, etc., are inspected prior to use to ensure that they meet the required specifications. The type, number, and size of both the longitudinal and transverse reinforcement are verified prior to placement in the excavated hole. Tests on concrete are performed for verifying the temperature, air content, and slump (amount of water) of the concrete, and cylinder samples are prepared for compression testing in the laboratory. The field tests are performed for verification only and not for control. Control tests on the concrete are performed at the concrete batch plants. Piles are inspected prior to installation to verify the condition, dimensions, and straightness of the pile. The pile is rejected if it does not meet the required specifications.

Full-scale load testing of a foundation is expensive and often is feasible only for large projects or critical structures. The success of the load test depends on the experience and qualifications of both the contractor and the designer and, if proper care is not exercised during the conduct of the load test, the results would be meaningless. Generally, foundation load tests are conducted for construction control or for verifying that the design capacity is comparable to the capacity of the as-built foundation. These tests are called "proof" tests, probably because they are envisioned as the "proof of the pudding." Load tests also can be performed during the design stage to obtain design parameters or for verifying the design assumptions.

A number of different procedures are used for foundation load testing, and the results obtained are influenced by the testing procedure. The simplest form of load test consists in applying a specified load at the top of the foundation and measuring its corresponding displacement. However, load tests have been performed with sophisticated instruments to measure the load (or stress) and displacements (or strains) along the length and at the tip of deep foundations and in the surrounding soil or rock. It should be noted that the results of the load tests are only applicable to the particular foundation being tested and the prevailing soil or rock conditions at the time of the test. Extrapolation of the load test results to other foundations on the project must be done with caution, especially if there is a change in construction procedure or soil and rock conditions. Also, load tests are performed over a short time duration compared to the life of the foundation. Extrapolation to long-term response should be done with care because soil conditions can change with time (e.g., change in ground water elevation) and because long-term response is governed by soil parameters different from those that govern short-term performance.

SEE ALSO THE FOLLOWING ARTICLES

CONCRETE, REINFORCED • CONSTITUTIVE MODELS FOR ENGINEERING MATERIALS • DAMS, DIKES, AND LEVEES • EARTHQUAKE ENGINEERING • GEOENVIRONMENTAL ENGINEERING • MASONRY • MECHANICS OF STRUCTURES • ROCK MECHANICS • SOIL MECHANICS

BIBLIOGRAPHY

Bell, F. G. (1987). "Ground Engineering Reference Book," Butterworths, London.

Broms, B. B. (1998). "Foundation Engineering," www.geoforum.com/knowledge/texts/broms/index.asp.

Canadian Geotechnical Society. (1992). "Canadian Foundation Engineering Manual," 3rd ed., BiTech, Richmond, BC, Canada.

Fang, H.-Y. (1991). "Foundation Engineering Handbook," 2nd ed., Van Nostrand Reinhold, New York.

Handy, R. L. (1995). "The Day the House Fell," ASCE Press, New York.

Macaulay, D. (1976). "Underground," Houghton Mifflin, Boston.

Peck, R. B., Hanson, W. E., and Thornburn, T. H. (1974). "Foundation Engineering," 2nd ed., Wiley, New York.

Sowers, G. F. (1979). "Introductory Soil Mechanics and Foundations; Geotechnical Engineering," 4th ed., Macmillan, New York.

Winterkorn, H. F., and Fang, H.-Y. (1975). "Foundation Engineering Handbook," Van Nostrand Reinhold, New York.

Witherspoon, W. T. (2000). "Residential Foundation Performance," ADSC, Dallas, TX.

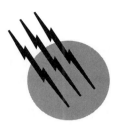

Fourier Series

James S. Walker
University of Wisconsin–Eau Claire

GLOSSARY

Bounded variation A function f has *bounded variation* on a closed interval $[a, b]$ if there exists a positive constant B such that, for all finite sets of points $a = x_0 < x_1 < \cdots < x_N = b$, the inequality $\sum_{i=1}^{N} |f(x_i) - f(x_{i-1})| \leq B$ is satisfied. Jordan proved that a function has bounded variation if and only if it can be expressed as the difference of two nondecreasing functions.

Countably infinite set A set is *countably infinite* if it can be put into one-to-one correspondence with the set of natural numbers $(1, 2, \ldots, n, \ldots)$. Examples: The integers and the rational numbers are countably infinite sets.

Continuous function If $\lim_{x \to c} f(x) = f(c)$, then the function f is *continuous* at the point c. Such a point is called a *continuity point* for f. A function which is continuous at all points is simply referred to as continuous.

Lebesgue measure zero A set S of real numbers is said to have *Lebesgue measure zero* if, for each $\epsilon > 0$, there ex-

ists a collection $\{(a_i, b_i)\}_{i=1}^{\infty}$ of open intervals such that $S \subset \cup_{i=1}^{\infty}(a_i, b_i)$ and $\sum_{i=1}^{\infty}(b_i - a_i) \leq \epsilon$. Examples: All finite sets, and all countably infinite sets, have Lebesgue measure zero.

Odd and even functions A function f is *odd* if $f(-x) = -f(x)$ for all x in its domain. A function f is *even* if $f(-x) = f(x)$ for all x in its domain.

One-sided limits $f(x-)$ and $f(x+)$ denote limits of $f(t)$ as t tends to x from the left and right, respectively.

Periodic function A function f is *periodic*, with *period* $P > 0$, if the identity $f(x + P) = f(x)$ holds for all x. Example: $f(x) = |\sin x|$ is periodic with period π.

FOURIER SERIES has long provided one of the principal methods of analysis for mathematical physics, engineering, and signal processing. It has spurred generalizations and applications that continue to develop right up to the present. While the original theory of Fourier series applies to periodic functions occurring in wave motion, such as with light and sound, its generalizations often

Encyclopedia of Physical Science and Technology, Third Edition, Volume 6

relate to wider settings, such as the time-frequency analysis underlying the recent theories of wavelet analysis and local trigonometric analysis.

I. HISTORICAL BACKGROUND

There are antecedents to the notion of Fourier series in the work of Euler and D. Bernoulli on vibrating strings, but the theory of Fourier series truly began with the profound work of Fourier on heat conduction at the beginning of the 19th century. Fourier deals with the problem of describing the evolution of the temperature $T(x, t)$ of a thin wire of length π, stretched between $x = 0$ and $x = \pi$, with a constant zero temperature at the ends: $T(0, t) = 0$ and $T(\pi, t) = 0$. He proposed that the initial temperature $T(x, 0) = f(x)$ could be expanded in a series of sine functions:

$$f(x) = \sum_{n=1}^{\infty} b_n \sin nx \tag{1}$$

with

$$b_n = \frac{2}{\pi} \int_0^{\pi} f(x) \sin nx \, dx. \tag{2}$$

A. Fourier Series

Although Fourier did not give a convincing proof of convergence of the infinite series in Eq. (1), he did offer the conjecture that convergence holds for an "arbitrary" function f. Subsequent work by Dirichlet, Riemann, Lebesgue, and others, throughout the next two hundred years, was needed to delineate precisely which functions were expandable in such trigonometric series. Part of this work entailed giving a precise definition of function (Dirichlet), and showing that the integrals in Eq. (2) are properly defined (Riemann and Lebesgue). Throughout this article we shall state results that are always true when Riemann integrals are used (except for Section IV where we need to use results from the theory of Lebesgue integrals).

In addition to positing Eqs. (1) and (2), Fourier argued that the temperature $T(x, t)$ is a solution to the following *heat equation with boundary conditions:*

$$\frac{\partial T}{\partial t} = \frac{\partial^2 T}{\partial x^2}, \qquad 0 < x < \pi, \ t > 0$$

$$T(0, t) = T(\pi, t) = 0, \qquad t \geq 0$$

$$T(x, 0) = f(x), \qquad 0 \leq x \leq \pi.$$

Making use of Eq. (1), Fourier showed that the solution $T(x, t)$ satisfies

$$T(x, t) = \sum_{n=1}^{\infty} b_n e^{-n^2 t} \sin nx. \tag{3}$$

This was the first example of the use of Fourier series to solve *boundary value problems* in partial differential equations. To obtain Eq. (3), Fourier made use of D. Bernoulli's method of *separation of variables,* which is now a standard technique for solving boundary value problems.

A good, short introduction to the history of Fourier series can be found in *The Mathematical Experience.* Besides his many mathematical contributions, Fourier has left us with one of the truly great philosophical principles: "The deep study of nature is the most fruitful source of knowledge."

II. DEFINITION OF FOURIER SERIES

The Fourier sine series, defined in Eqs. (1) and (2), is a special case of a more general concept: the Fourier series for a *periodic function.* Periodic functions arise in the study of wave motion, when a basic waveform repeats itself periodically. Such periodic waveforms occur in musical tones, in the plane waves of electromagnetic vibrations, and in the vibration of strings. These are just a few examples. Periodic effects also arise in the motion of the planets, in AC electricity, and (to a degree) in animal heartbeats.

A function f is said to have period P if $f(x + P) = f(x)$ for all x. For notational simplicity, we shall restrict our discussion to functions of period 2π. There is no loss of generality in doing so, since we can always use a simple change of scale $x = (P/2\pi)t$ to convert a function of period P into one of period 2π.

If the function f has period 2π, then its *Fourier series* is

$$c_0 + \sum_{n=1}^{\infty} \{a_n \cos nx + b_n \sin nx\} \tag{4}$$

with *Fourier coefficients* c_0, a_n, and b_n defined by the integrals

$$c_0 = \frac{1}{2\pi} \int_{-\pi}^{\pi} f(x) \, dx \tag{5}$$

$$a_n = \frac{1}{\pi} \int_{-\pi}^{\pi} f(x) \cos nx \, dx, \tag{6}$$

$$b_n = \frac{1}{\pi} \int_{-\pi}^{\pi} f(x) \sin nx \, dx. \tag{7}$$

[*Note:* The sine series defined by Eqs. (1) and (2) is a special instance of Fourier series. If f is initially defined over the interval $[0, \pi]$, then it can be extended to $[-\pi, \pi]$ (as an odd function) by letting $f(-x) = -f(x)$, and then extended periodically with period $P = 2\pi$. The Fourier series for this odd, periodic function reduces to the sine series in Eqs. (1) and (2), because $c_0 = 0$, each $a_n = 0$, and each b_n in Eq. (7) is equal to the b_n in Eq. (2).]

It is more common nowadays to express Fourier series in an algebraically simpler form involving complex exponentials. Following Euler, we use the fact that the complex exponential $e^{i\theta}$ satisfies $e^{i\theta} = \cos\theta + i\sin\theta$. Hence

$$\cos\theta = \frac{1}{2}(e^{i\theta} + e^{i\theta}),$$

$$\sin\theta = \frac{1}{2i}(e^{i\theta} - e^{-i\theta}).$$

From these equations, it follows by elementary algebra that Formulas (5)–(7) can be rewritten (by rewriting each term separately) as

$$c_0 + \sum_{n=1}^{\infty}\left\{c_n e^{inx} + c_{-n} e^{-inx}\right\} \tag{8}$$

with c_n defined for all integers n by

$$c_n = \frac{1}{2\pi}\int_{-\pi}^{\pi} f(x) e^{-inx}\, dx. \tag{9}$$

The series in Eq. (8) is usually written in the form

$$\sum_{n=-\infty}^{\infty} c_n e^{inx}. \tag{10}$$

We now consider a couple of examples. First, let f_1 be defined over $[-\pi, \pi]$ by

$$f_1(x) = \begin{cases} 1 & \text{if } |x| < \pi/2 \\ 0 & \text{if } \pi/2 \le |x| \le \pi \end{cases}$$

and have period 2π. The graph of f_1 is shown in Fig. 1; it is called a *square wave* in electric circuit theory. The constant c_0 is

$$c_0 = \frac{1}{2\pi}\int_{-\pi}^{\pi} f_1(x)\, dx$$

$$= \frac{1}{2\pi}\int_{-\pi/2}^{\pi/2} 1\, dx = \frac{1}{2}.$$

While, for $n \ne 0$,

$$c_n = \frac{1}{2\pi}\int_{-\pi}^{\pi} f_1(x) e^{-inx}\, dx$$

$$= \frac{1}{2\pi}\int_{-\pi/2}^{\pi/2} e^{-inx}\, dx$$

$$= \frac{1}{2\pi}\frac{e^{-in\pi/2} - e^{in\pi/2}}{-in}$$

$$= \frac{\sin(n\pi/2)}{n\pi}.$$

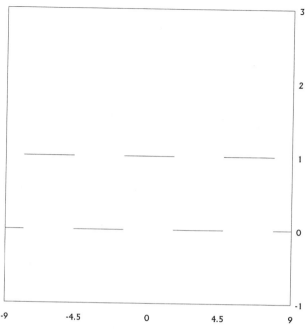

FIGURE 1 Square wave.

Thus, the Fourier series for this square wave is

$$\frac{1}{2} + \sum_{n=1}^{\infty}\frac{\sin(n\pi/2)}{n\pi}(e^{inx} + e^{-inx})$$

$$= \frac{1}{2} + \sum_{n=1}^{\infty}\frac{2\sin(n\pi/2)}{n\pi}\cos nx. \tag{11}$$

Second, let $f_2(x) = x^2$ over $[-\pi, \pi]$ and have period 2π, see Fig. 2. We shall refer to this wave as a *parabolic wave*. This parabolic wave has $c_0 = \pi^2/3$ and c_n, for $n \ne 0$, is

$$c_n = \frac{1}{2\pi}\int_{-\pi}^{\pi} x^2 e^{-inx}\, dx$$

$$= \frac{1}{2\pi}\int_{-\pi}^{\pi} x^2 \cos nx\, dx - \frac{i}{2\pi}\int_{-\pi}^{\pi} x^2 \sin nx\, dx$$

$$= \frac{2(-1)^n}{n^2}$$

after an integration by parts. The Fourier series for this function is then

$$\frac{\pi^2}{3} + \sum_{n=1}^{\infty}\frac{2(-1)^n}{n^2}(e^{inx} + e^{-inx})$$

$$= \frac{\pi^2}{3} + \sum_{n=1}^{\infty}\frac{4(-1)^n}{n^2}\cos nx. \tag{12}$$

We will discuss the convergence of these Fourier series, to f_1 and f_2, respectively, in Section III.

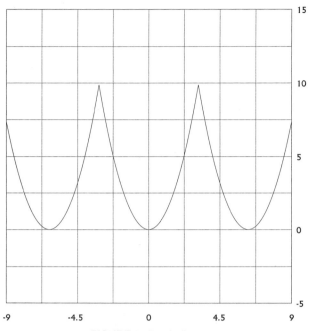

FIGURE 2 Parabolic wave.

Returning to the general Fourier series in Eq. (10), we shall now discuss some ways of interpreting this series. A complex exponential $e^{inx} = \cos nx + i \sin nx$ has a smallest period of $2\pi/n$. Consequently it is said to have a *frequency* of $n/2\pi$, because the form of its graph over the interval $[0, 2\pi/n]$ is repeated $n/2\pi$ times within each unit-length. Therefore, the integral in Eq. (9) that defines the Fourier coefficient c_n can be interpreted as a *correlation* between f and a complex exponential with a precisely located frequency of $n/2\pi$. Thus the whole collection of these integrals, for all integers n, specifies the *frequency content* of f over the set of frequencies $\{n/2\pi\}_{n=-\infty}^{\infty}$. If the series in Eq. (10) converges to f, i.e., if we can write

$$f(x) = \sum_{n=-\infty}^{\infty} c_n e^{inx}, \tag{13}$$

then f is being expressed as a superposition of elementary functions $c_n e^{inx}$ having frequency $n/2\pi$ and amplitude c_n. (The validity of Eq. (13) will be discussed in the next section.) Furthermore, the correlations in Eq. (9) are *independent* of each other in the sense that correlations between distinct exponentials are zero:

$$\frac{1}{2\pi} \int_{-\pi}^{\pi} e^{inx} e^{-imx} \, dx = \begin{cases} 0 & \text{if } m \neq n \\ 1 & \text{if } m = n. \end{cases} \tag{14}$$

This equation is called the *orthogonality property* of complex exponentials.

The orthogonality property of complex exponentials can be used to give a derivation of Eq. (9). Multiplying

Eq. (13) by e^{-imx} and integrating term-by-term from $-\pi$ to π, we obtain

$$\int_{-\pi}^{\pi} f(x) e^{-imx} \, dx = \sum_{n=-\infty}^{\infty} c_n \int_{-\pi}^{\pi} e^{inx} e^{-imx} \, dx.$$

By the orthogonality property, this leads to

$$\int_{-\pi}^{\pi} f(x) e^{-imx} \, dx = 2\pi c_m,$$

which justifies (in a formal, nonrigorous way) the definition of c_n in Eq. (9).

We close this section by discussing two important properties of Fourier coefficients, *Bessel's inequality* and the *Riemann-Lebesgue lemma*.

Theorem 1 (Bessel's Inequality): If $\int_{-\pi}^{\pi} |f(x)|^2 \, dx$ is finite, then

$$\sum_{n=-\infty}^{\infty} |c_n|^2 \leq \frac{1}{2\pi} \int_{-\pi}^{\pi} |f(x)|^2 \, dx. \tag{15}$$

Bessel's inequality can be proved easily. In fact, we have

$$0 \leq \frac{1}{2\pi} \int_{-\pi}^{\pi} \left| f(x) - \sum_{n=-N}^{N} c_n e^{inx} \right|^2 dx$$

$$= \frac{1}{2\pi} \int_{-\pi}^{\pi} \left(f(x) - \sum_{m=-N}^{N} c_m e^{imx} \right) \left(\overline{f(x)} - \sum_{n=-N}^{N} \overline{c_n} e^{-inx} \right) dx.$$

Multiplying out the last integrand above, and making use of Eqs. (9) and (14), we obtain

$$\frac{1}{2\pi} \int_{-\pi}^{\pi} \left| f(x) - \sum_{n=-N}^{N} c_n e^{inx} \right|^2 dx$$

$$= \frac{1}{2\pi} \int_{-\pi}^{\pi} |f(x)|^2 \, dx - \sum_{n=-N}^{N} |c_n|^2. \tag{16}$$

Thus, for all N,

$$\sum_{n=-N}^{N} |c_n|^2 \leq \frac{1}{2\pi} \int_{-\pi}^{\pi} |f(x)|^2 \, dx \tag{17}$$

and Bessel's inequality (15) follows by letting $N \to \infty$.

Bessel's inequality has a physical interpretation. If f has *finite energy,* in the sense that the right side of Eq. (15) is finite, then the sum of the moduli-squared of the Fourier coefficients is also finite. In Section IV, we shall see that the inequality in Eq. (15) is actually an equality, which says that the sum of the moduli-squared of the Fourier coefficients is precisely the same as the energy of f.

Because of Bessel's inequality, it follows that

$$\lim_{|n| \to \infty} c_n = 0 \tag{18}$$

holds whenever $\int_{-\pi}^{\pi} |f(x)|^2 \, dx$ is finite. The Riemann-Lebesgue lemma says that Eq. (18) holds in the following more general case:

Theorem 2 (Riemann-Lebesgue Lemma): If $\int_{-\pi}^{\pi} |f(x)| \, dx$ is finite, then Eq. (18) holds.

One of the most important uses of the Riemann-Lebesgue lemma is in proofs of some basic pointwise convergence theorems, such as the ones described in the next section.

See Krantz and Walker (1998) for further discussions of the definition of Fourier series, Bessel's inequality, and the Riemann-Lebesgue lemma.

III. CONVERGENCE OF FOURIER SERIES

There are many ways to interpret the meaning of Eq. (13). Investigations into the types of functions allowed on the left side of Eq. (13), and the kinds of convergence considered for its right side, have fueled mathematical investigations by such luminaries as Dirichlet, Riemann, Weierstrass, Lipschitz, Lebesgue, Fejér, Gelfand, and Schwartz. In short, convergence questions for Fourier series have helped lay the foundations and much of the superstructure of mathematical analysis.

The three types of convergence that we shall describe here are *pointwise*, *uniform*, and *norm* convergence. We shall discuss the first two types in this section and take up the third type in the next section.

All convergence theorems are concerned with how the *partial sums*

$$S_N(x) := \sum_{n=-N}^{N} c_n e^{inx}$$

converge to f(x). That is, *does* $\lim_{N \to \infty} S_N = f$ *hold in some sense?*

The question of pointwise convergence, for example, concerns whether $\lim_{N \to \infty} S_N(x_0) = f(x_0)$ for each fixed x-value x_0. If $\lim_{N \to \infty} S_N(x_0)$ does equal $f(x_0)$, then we say that the *Fourier series for* f *converges to* f(x_0) *at* x_0.

We shall now state the simplest pointwise convergence theorem for which an elementary proof can be given. This theorem assumes that a function is Lipschitz at each point where convergence occurs. A function is said to be *Lipschitz at a point* x_0 if, for some positive constant A,

$$|f(x) - f(x_0)| \leq A \, |x - x_0| \tag{19}$$

holds for all x near x_0 (i.e., $|x - x_0| < \delta$ for some $\delta > 0$). It is easy to see, for instance, that the square wave function f_1 is Lipschitz at all of its continuity points.

The inequality in Eq. (19) has a simple geometric interpretation. Since both sides are 0 when $x = x_0$, this inequality is equivalent to

$$\left| \frac{f(x) - f(x_0)}{x - x_0} \right| \leq A \tag{20}$$

for all x near x_0 (and $x \neq x_0$). Inequality (20) simply says that the difference quotients of f (i.e., the slopes of its secants) near x_0 are bounded. With this interpretation, it is easy to see that the parabolic wave f_2 is Lipschitz at all points. More generally, if f has a derivative at x_0 (or even just left- and right-hand derivatives), then f is Lipschitz at x_0.

We can now state and prove a simple convergence theorem.

Theorem 3: Suppose f has period 2π, that $\int_{-\pi}^{\pi} |f(x)| \, dx$ is finite, and that f is Lipschitz at x_0. Then the Fourier series for f converges to f(x_0) at x_0.

To prove this theorem, we assume that f(x_0) = 0. There is no loss of generality in doing so, since we can always subtract the constant f(x_0) from f(x). Define the function g by $g(x) = f(x)/(e^{ix} - e^{ix_0})$. This function g has period 2π. Furthermore, $\int_{-\pi}^{\pi} |g(x)| \, dx$ is finite, because the quotient $f(x)/(e^{ix} - e^{ix_0})$ is bounded in magnitude for x near x_0. In fact, for such x,

$$\left| \frac{f(x)}{e^{ix} - e^{x_0}} \right| = \left| \frac{f(x) - f(x_0)}{e^{ix} - e^{x_0}} \right|$$

$$\leq A \left| \frac{x - x_0}{e^{ix} - e^{x_0}} \right|$$

and $(x - x_0)/(e^{ix} - e^{x_0})$ is bounded in magnitude, because it tends to the reciprocal of the derivative of e^{ix} at x_0.

If we let d_n denote the nth Fourier coefficient for $g(x)$, then we have $c_n = d_{n-1} - d_n e^{ix_0}$ because $f(x) = g(x)(e^{ix} - e^{ix_0})$. The partial sum $S_N(x_0)$ then telescopes:

$$S_N(x_0) = \sum_{n=-N}^{N} c_n e^{inx_0}$$

$$= d_{-N-1} e^{-iNx_0} - d_N e^{i(N+1)x_0}.$$

Since $d_n \to 0$ as $|n| \to \infty$, by the Riemann-Lebesgue lemma, we conclude that $S_N(x_0) \to 0$. This completes the proof.

It should be noted that for the square wave f_1 and the parabolic wave f_2, it is not necessary to use the general Riemann-Lebesgue lemma stated above. That is because for those functions it is easy to see that $\int_{-\pi}^{\pi} |g(x)|^2 \, dx$ is finite for the function g defined in the proof of Theorem 3. Consequently, $d_n \to 0$ as $|n| \to \infty$ follows from Bessel's inequality for g.

In any case, Theorem 3 implies that the Fourier series for the square wave f_1 converges to f_1 at all of its points of continuity. It also implies that the Fourier series for the parabolic wave f_2 converges to f_2 at all points. While this may settle matters (more or less) in a pure mathematical sense for these two waves, it is still important to examine specific partial sums in order to learn more about the nature of their convergence to these waves.

For example, in Fig. 3 we show a graph of the partial sum S_{100} superimposed on the square wave. Although Theorem 3 guarantees that $S_N \to f_1$ as $N \to \infty$ at each continuity point, Fig. 3 indicates that this convergence is at a rather slow rate. The partial sum S_{100} differs significantly from f_1. Near the square wave's jump discontinuities, for example, there is a severe spiking behavior called *Gibbs' phenomenon* (see Fig. 4). This spiking behavior does *not* go away as $N \to \infty$, although the width of the spike does tend to zero. In fact, the peaks of the spikes overshoot the square wave's value of 1, tending to a limit of about 1.09. The partial sum also oscillates quite noticeably about the constant value of the square wave at points away from the discontinuities. This is known as *ringing*.

These defects do have practical implications. For instance, oscilloscopes—which generate wave forms as

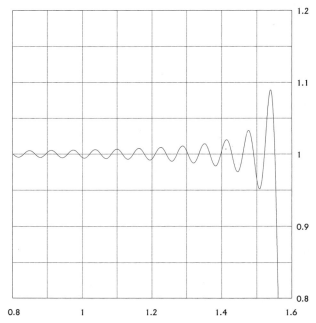

FIGURE 4 Gibbs' phenomenon and ringing for square wave.

combinations of sinusoidal waves over a limited range of frequencies—cannot use S_{100}, or any partial sum S_N, to produce a square wave. We shall see, however, in Section V that a clever modification of a partial sum does produce an acceptable version of a square wave.

The cause of ringing and Gibbs' phenomenon for the square wave is a rather slow convergence to zero of its Fourier coefficients (at a rate comparable to $|n|^{-1}$). In the next section, we shall interpret this in terms of energy and show that a partial sum like S_{100} does not capture a high enough percentage of the energy of the square wave f_1.

In contrast, the Fourier coefficients of the parabolic wave f_2 tend to zero more rapidly (at a rate comparable to n^{-2}). Because of this, the partial sum S_{100} for f_2 is a much better approximation to the parabolic wave (see Fig. 5). In fact, its partial sums S_N exhibit the phenomenon of *uniform convergence*.

We say that the Fourier series for a function f *converges uniformly* to f if

$$\lim_{N \to \infty} \left\{ \max_{x \in [-\pi, \pi]} |f(x) - S_N(x)| \right\} = 0. \qquad (21)$$

This equation says that, for large enough N, we can have the maximum distance between the graphs of f and S_N as small as we wish. Figure 5 is a good illustration of this for the parabolic wave.

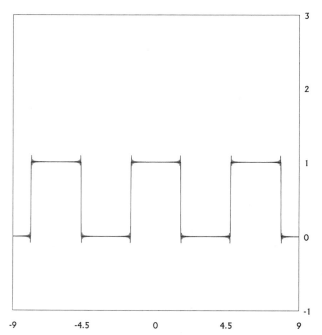

FIGURE 3 Fourier series partial sum S_{100} superimposed on square wave.

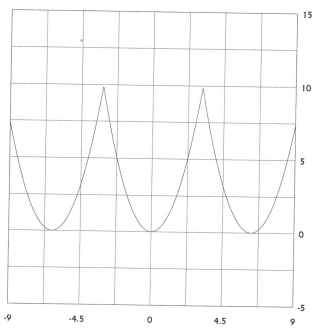

FIGURE 5 Fourier series partial sum S_{100} for parabolic wave.

We can verify Eq. (21) for the parabolic wave as follows. By Eq. (21) we have

$$|f_2(x) - S_N(x)| = \left| \sum_{n=N+1}^{\infty} \frac{4(-1)^n}{n^2} \cos nx \right|$$

$$\leq \sum_{n=N+1}^{\infty} \left| \frac{4(-1)^n}{n^2} \cos nx \right|$$

$$\leq \sum_{n=N+1}^{\infty} \frac{4}{n^2}.$$

Consequently

$$\max_{x \in [-\pi, \pi]} |f_2(x) - S_N(x)| \leq \sum_{n=N+1}^{\infty} \frac{4}{n^2}$$

$$\to 0 \quad \text{as } N \to \infty$$

and thus Eq. (21) holds for the parabolic wave f_2.

Uniform convergence for the parabolic wave is a special case of a more general theorem. We shall say that f is *uniformly Lipschitz* if Eq. (19) holds for all points using the same constant A. For instance, it is not hard to show that a continuously differentiable, periodic function is uniformly Lipschitz.

Theorem 4: Suppose that f has period 2π and is uniformly Lipschitz at all points, then the Fourier series for f converges uniformly to f.

A remarkably simple proof of this theorem is described in Jackson (1941). More general uniform convergence theorems are discussed in Walter (1994).

Theorem 4 applies to the parabolic wave f_2, but it does not apply to the square wave f_1. In fact, the Fourier series for f_1 cannot converge uniformly to f_1. That is because a famous theorem of Weierstrass says that a uniform limit of continuous functions (like the partial sums S_N) must be a continuous function (which f_1 is certainly not). The Gibbs' phenomenon for the square wave is a conspicuous failure of uniform convergence for its Fourier series.

Gibbs' phenomenon and ringing, as well as many other aspects of Fourier series, can be understood via an integral form for partial sums discovered by Dirichlet. This integral form is

$$S_N(x) = \frac{1}{2\pi} \int_{-\pi}^{\pi} f(x - t) D_N(t) \, dt \qquad (22)$$

with *kernel* D_N defined by

$$D_N(t) = \frac{\sin(N + 1/2)t}{\sin(t/2)}. \qquad (23)$$

This formula is proved in almost all books on Fourier series (see, for instance, Krantz (1999), Walker (1988), or Zygmund (1968)). The kernel D_N is called *Dirichlet's kernel*. In Fig. 6 we have graphed D_{20}.

The most important property of Dirichlet's kernel is that, for all N,

$$\frac{1}{2\pi} \int_{-\pi}^{\pi} D_N(t) \, dt = 1.$$

From Eq. (23) we can see that the value of 1 follows from cancellation of signed areas, and also that the contribution

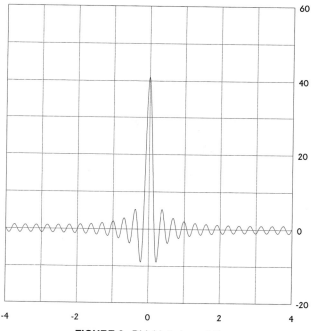

FIGURE 6 Dirichlet's kernel D_{20}.

of the main lobe centered at 0 (see Fig. 6) is significantly greater than 1 (about 1.09 in value).

From the facts just cited, we can explain the origin of ringing and Gibbs' phenomenon for the square wave. For the square wave function f_1, Eq. (22) becomes

$$S_N(x) = \frac{1}{2\pi} \int_{x-\pi/2}^{x+\pi/2} D_N(t) \, dt. \qquad (24)$$

As x ranges from $-\pi$ to π, this formula shows that $S_N(x)$ is proportional to the signed area of D_N over an interval of length π centered at x. By examining Fig. 6, which is a typical graph for D_N, it is then easy to see why there is ringing in the partial sums S_N for the square wave. Gibbs' phenomenon is a bit more subtle, but also results from Eq. (24). When x nears a jump discontinuity, the central lobe of D_N is the dominant contributor to the integral in Eq. (24), resulting in a spike which overshoots the value of 1 for f_1 by about 9%.

Our final pointwise convergence theorem was, in essence, the first to be proved. It was established by Dirichlet using the integral form for partial sums in Eq. (22). We shall state this theorem in a stronger form first proved by Jordan.

Theorem 5: If f has period 2π and has bounded variation on $[0, 2\pi]$, then the Fourier series for f converges at all points. In fact, for all x-values,

$$\lim_{N \to \infty} S_N(x) = \tfrac{1}{2}[f(x+) + f(x-)].$$

This theorem is too difficult to prove in the limited space we have here (see Zygmund, 1968). A simple consequence of Theorem 5 is that the Fourier series for the square wave f_1 converges at its discontinuity points to $1/2$ (although this can also be shown directly by substitution of $x = \pm\pi/2$ into the series in (Eq. (11)).

We close by mentioning that the conditions for convergence, such as Lipschitz or bounded variation, cited in the theorems above cannot be dispensed with entirely. For instance, Kolmogorov gave an example of a period 2π function (for which $\int_{-\pi}^{\pi} |f(x)| \, dx$ is finite) that has a Fourier series which fails to converge at *every* point. More discussion of pointwise convergence can be found in Walker (1998), Walter (1994), or Zygmund (1968).

IV. CONVERGENCE IN NORM

Perhaps the most satisfactory notion of convergence for Fourier series is convergence in L^2-norm (also called L^2-convergence), which we shall define in this section. One of the great triumphs of the Lebesgue theory of integration is that it yields necessary and sufficient conditions

for L^2-convergence. There is also an interpretation of L^2-norm in terms of a generalized Euclidean distance and this gives a satisfying geometric flavor to L^2-convergence of Fourier series. By interpreting the square of L^2-norm as a type of energy, there is an equally satisfying physical interpretation of L^2-convergence. The theory of L^2-convergence has led to fruitful generalizations such as Hilbert space theory and norm convergence in a wide variety of function spaces.

To introduce the idea of L^2-convergence, we first examine a special case. By Theorem 4, the partial sums of a uniformly Lipschitz function f converge uniformly to f. Since that means that the maximum distance between the graphs of S_N and f tends to 0 as $N \to \infty$, it follows that

$$\lim_{N \to \infty} \frac{1}{2\pi} \int_{-\pi}^{\pi} |f(x) - S_N(x)|^2 \, dx = 0. \qquad (25)$$

This result motivates the definition of L^2-convergence.

If g is a function for which $|g|^2$ has a finite Lebesgue integral over $[-\pi, \pi]$, then we say that g is an L^2-function, and we define its L^2-norm $\|g\|_2$ by

$$\|g\|_2 = \sqrt{\frac{1}{2\pi} \int_{-\pi}^{\pi} |g(x)|^2 \, dx} \, .$$

We can then rephrase Eq. (25) as saying that $\|f - S_N\|_2 \to 0$ as $N \to \infty$. In other words, *the Fourier series for f converges to f in L^2-norm.* The following theorem generalizes this result to all L^2-functions (see Rudin (1986) for a proof).

Theorem 6: If f is an L^2-function, then its Fourier series converges to f in L^2-norm.

Theorem 6 says that Eq. (25) holds for every L^2-function f. Combining this with Eq. (16), we obtain *Parseval's equality:*

$$\sum_{n=-\infty}^{\infty} |c_n|^2 = \frac{1}{2\pi} \int_{-\pi}^{\pi} |f(x)|^2 \, dx. \qquad (26)$$

Parseval's equation has a useful interpretation in terms of energy. It says that the energy of the set of Fourier coefficients, defined to be equal to the left side of Eq. (26), is equal to the energy of the function f, defined by the right side of Eq. (26).

The L^2-norm can be interpreted as a generalized Euclidean distance. To see this take square roots of both sides of Eq. (26): $\sqrt{\sum |c_n|^2} = \|f\|_2$. The left side of this equation is interpreted as a Euclidean distance in an (infinite-dimensional) coordinate space, hence the L^2-norm $\|f\|_2$ is equivalent to such a distance.

As examples of these ideas, let's return to the square wave and parabolic wave. For the square wave f_1, we find that

$$\|f_1 - S_{100}\|_2^2 = \sum_{|n| > 100} \frac{\sin^2(n\pi/2)}{(n\pi)^2}$$

$$= 1.0 \times 10^{-3}.$$

Likewise, for the parabolic wave f_2, we have $\|f_2 - S_{100}\|_2^2 = 2.6 \times 10^{-6}$. These facts show that the energy of the parabolic wave is almost entirely contained in the partial sum S_{100}; their energy difference is almost three orders of magnitude smaller than in the square wave case. In terms of generalized Euclidean distance, we have $\|f_2 - S_{100}\|_2 = 1.6 \times 10^{-3}$ and $\|f_1 - S_{100}\|_2 = 3.2 \times 10^{-2}$, showing that the partial sum is an order of magnitude closer for the parabolic wave.

Theorem 6 has a converse, known as the *Riesz-Fischer theorem*.

Theorem 7 (Riesz-Fischer): If $\sum |c_n|^2$ converges, then there exists an L^2-function f having $\{c_n\}$ as its Fourier coefficients.

This theorem is proved in Rudin (1986). Theorem and the Riesz-Fischer theorem combine to give necessary and sufficient conditions for L^2-convergence of Fourier series, conditions which are remarkably easy to apply. This has made L^2-convergence into the most commonly used notion of convergence for Fourier series.

These ideas for L^2-norms partially generalize to the case of L^p-norms. Let p be real number satisfying $p \geq 1$. If g is a function for which $|g|^p$ has a finite Lebesgue integral over $[-\pi, \pi]$, then we say that g is an L^p-function, and we define its L^p-norm $\|g\|_p$ by

$$\|g\|_p = \left[\frac{1}{2\pi} \int_{-\pi}^{\pi} |g(x)|^p \, dx \right]^{1/p}.$$

If $\|f - S_N\|_p \to 0$, then we say that the Fourier series for f converges to f in L^p-norm. The following theorem generalizes Theorem 6 (see Krantz (1999) for a proof).

Theorem 8: If f is an L^p-function for $p > 1$, then its Fourier series converges to f in L^p-norm.

Notice that the case of $p = 1$ is not included in Theorem 8. The example of Kolmogorov cited at the end of Section III shows that there exist L^1-functions whose Fourier series do not converge in L^1-norm. For $p \neq 2$, there are no simple analogs of either Parseval's equality or the Riesz-Fischer theorem (which say that we can characterize L^2-functions by the magnitude of their Fourier coefficients). Some partial analogs of these latter results for L^p-functions, when $p \neq 2$, are discussed in Zygmund (1968) (in the context of *Littlewood-Paley* theory).

We close this section by returning full circle to the notion of pointwise convergence. The following theorem was proved by Carleson for L^2-functions and by Hunt for L^p-functions ($p \neq 2$).

Theorem 9: If f is an L^p-function for $p > 1$, then its Fourier series converges to it at almost all points.

By *almost all points,* we mean that the set of points where divergence occurs has Lebesgue measure zero. References for the proof of Theorem 9 can be found in Krantz (1999) and Zygmund (1968). Its proof is undoubtedly the most difficult one in the theory of Fourier series.

V. SUMMABILITY OF FOURIER SERIES

In the previous sections, we noted some problems with convergence of Fourier series partial sums. Some of these problems include Kolmogorov's example of a Fourier series for an L^1-function that diverges everywhere, and Gibbs' phenomenon and ringing in the Fourier series partial sums for discontinuous functions. Another problem is Du Bois Reymond's example of a continuous function whose Fourier series diverges on a countably infinite set of points, see Walker (1968). It turns out that all of these difficulties simply disappear when new summation methods, based on appropriate modifications of the partial sums, are used.

The simplest modification of partial sums, and one of the first historically to be used, is to take *arithmetic means*. Define the Nth arithmetic mean σ_N by $\sigma_N = (S_0 + S_1 + \cdots + S_{N-1})/N$. From which it follows that

$$\sigma_N(x) = \sum_{n=-N}^{N} \left(1 - \frac{|n|}{N} \right) c_n \, e^{inx}. \qquad (27)$$

The factors $(1 - |n|/N)$ are called *convergence factors*. They modify the Fourier coefficients c_n so that the amplitude of the higher frequency terms (for $|n|$ near N) are damped down toward zero. This produces a great improvement in convergence properties as shown by the following theorem.

Theorem 10: Let f be a periodic function. If f is an L^p-function for $p \geq 1$, then $\sigma_N \to f$ in L^p-norm as $N \to \infty$. If f is a continuous function, then $\sigma_N \to f$ uniformly as $N \to \infty$.

Notice that L^1-convergence is included in Theorem 10. Even for Kolmogorov's function, it is the case that $\|f - \sigma_N\|_1 \to 0$ as $N \to \infty$. It also should be noted that no assumption, other than continuity of the periodic function, is needed in order to ensure uniform convergence of its arithmetic means.

For a proof of Theorem 10, see Krantz (1999). The key to the proof is Fejér's integral form for σ_N:

$$\sigma_N(x) = \frac{1}{2\pi} \int_{-\pi}^{\pi} f(x-t) F_N(t)\, dt \qquad (28)$$

where *Fejér's kernel* F_N is defined by

$$F_N(t) = \frac{1}{N} \left(\frac{\sin Nt/2}{\sin t/2} \right)^2. \qquad (29)$$

In Fig. 7 we show the graph of F_{20}. Compare this graph with the one of Dirichlet's kernel D_{20} in Fig. 6. Unlike Dirichlet's kernel, Fejér's kernel is positive [$F_N(t) \geq 0$], and is close to 0 away from the origin. These two facts are the main reasons that Theorem 10 holds. The fact that Fejér's kernel satisfies

$$\frac{1}{2\pi} \int_{-\pi}^{\pi} F_N(t)\, dt = 1$$

is also used in the proof.

An attractive feature of arithmetic means is that Gibbs' phenomenon and ringing do not occur. For example, in Fig. 8 we show σ_{100} for the square wave and it is plain that these two defects are absent. For the square wave function f_1, Eq. (28) reduces to

$$\sigma_N(x) = \frac{1}{2\pi} \int_{x-\pi/2}^{x+\pi/2} F_N(t)\, dt.$$

As x ranges from $-\pi$ to π, this formula shows that $\sigma_N(x)$ is proportional to the area of the positive function F_N over

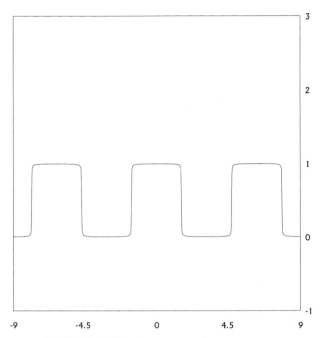

FIGURE 8 Arithmetic mean σ_{100} for square wave.

an interval of length π centered at x. By examining Fig. 7, which is a typical graph for F_N, it is easy to see why ringing and Gibbs' phenomenon do not occur for the arithmetic means of the square wave.

The method of arithmetic means is just one example from a wide range of summation methods for Fourier series. These summation methods are one of the major elements in the area of *finite impulse response filtering* in the fields of electrical engineering and signal processing.

A *summation kernel* K_N is defined by

$$K_N(x) = \sum_{n=-N}^{N} m_n e^{inx}. \qquad (30)$$

The real numbers $\{m_n\}$ are the *convergence factors* for the kernel. We have already seen two examples: Dirichlet's kernel (where $m_n = 1$) and Fejér's kernel (where $m_n = 1 - |n|/N$).

When K_N is a summation kernel, then we define the modified partial sum of f to be $\sum_{n=-N}^{N} m_n c_n e^{inx}$. It then follows from Eqs. (14) and (30) that

$$\sum_{n=-N}^{N} m_n c_n e^{inx} = \frac{1}{2\pi} \int_{-\pi}^{\pi} f(x-t) K_N(t)\, dt. \qquad (31)$$

The function defined by both sides of Eq. (31) is denoted by $K_N * f$. It is usually more convenient to use the left side of Eq. (31) to compute $K_N * f$, while for theoretical purposes (such as proving Theorem 11 below), it is more convenient to use the right side of Eq. (31).

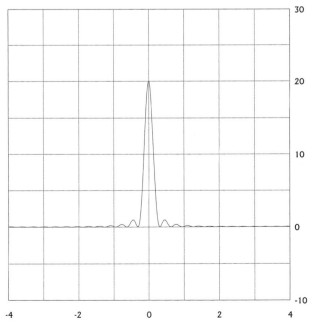

FIGURE 7 Fejér's kernel F_{20}.

We say that a summation kernel K_N is *regular* if it satisfies the following three conditions.

1. For each N,

$$\frac{1}{2\pi} \int_{-\pi}^{\pi} K_N(x)\,dx = 1.$$

2. There is a positive constant C such that

$$\frac{1}{2\pi} \int_{-\pi}^{\pi} |K_N(x)|\,dx \le C.$$

3. For each $0 < \delta < \pi$,

$$\lim_{N \to \infty} \left\{ \max_{\delta \le |x| \le \pi} |K_N(x)| \right\} = 0.$$

There are many examples of regular summation kernels. Fejér's kernel, which has $m_n = 1 - |n|/N$, is regular. Another regular summation kernel is Hann's kernel, which has $m_n = 0.5 + 0.5\cos(n\pi/N)$. A third regular summation kernel is de le Vallée Poussin's kernel, for which $m_n = 1$ when $|n| \le N/2$, and $m_n = 2(1 - |m/N|)$ when $N/2 < |m| \le N$. The proofs that these summation kernels are regular are given in Walker (1996). It should be noted that Dirichlet's kernel is *not* regular, because properties 2 and 3 do not hold.

As with Fejér's kernel, all regular summation kernels significantly improve the convergence of Fourier series. In fact, the following theorem generalizes Theorem 10.

Theorem 11: Let f be a periodic function, and let K_N be a regular summation kernel. If f is an L^p-function for $p \ge 1$, then $K_N * f \to f$ in L^p-norm as $N \to \infty$. If f is a continuous function, then $K_N * f \to f$ uniformly as $N \to \infty$.

For an elegant proof of this theorem, see Krantz (1999).

From Theorem 11 we might be tempted to conclude that the convergence properties of regular summation kernels are all the same. They do differ, however, in the *rates* at which they converge. For example, in Fig. 9 we show $K_{100} * f_1$ where the kernel is Hann's kernel and f_1 is the square wave. Notice that this graph is a much better approximation of a square wave than the arithmetic mean graph in Fig. 8. An oscilloscope, for example, can easily generate the graph in Fig. 9, thereby producing an acceptable version of a square wave.

Summation of Fourier series is discussed further in Krantz (1999), Walker (1996), Walter (1994), and Zygmund (1968).

VI. GENERALIZED FOURIER SERIES

The classical theory of Fourier series has undergone extensive generalizations during the last two hundred years.

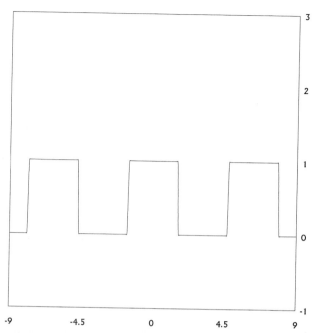

FIGURE 9 Approximate square wave using Hann's kernel.

For example, Fourier series can be viewed as one aspect of a general theory of *orthogonal series expansions*. In this section, we shall discuss a few of the more celebrated orthogonal series, such as Legendre series, Haar series, and wavelet series.

We begin with Legendre series. The first two *Legendre polynomials* are $P_0(x) = 1$, and $P_1(x) = x$. For $n = 2, 3, 4, \ldots$, the nth Legendre polynomial P_n is defined by the recursion relation

$$n P_n(x) = (2n - 1)x P_{n-1}(x) + (n - 1) P_{n-2}(x).$$

These polynomials satisfy the following *orthogonality relation*

$$\int_{-1}^{1} P_n(x)\, P_m(x)\,dx = \begin{cases} 0 & \text{if } m \ne n \\ (2n + 1)/2 & \text{if } m = n. \end{cases} \tag{32}$$

This equation is quite similar to Eq. (14). Because of Eq. (32)—recall how we used Eq. (14) to derive Eq. (9)—the *Legendre series* for a function f over the interval $[-1, 1]$ is defined to be

$$\sum_{n=0}^{\infty} c_n P_n(x) \tag{33}$$

with

$$c_n = \frac{2}{2n + 1} \int_{-1}^{1} f(x) P_n(x)\,dx. \tag{34}$$

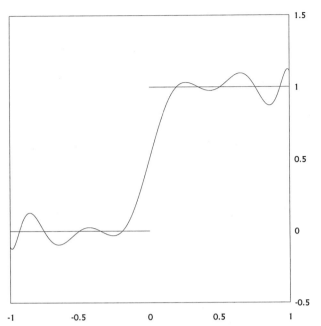

FIGURE 10 Step function and its Legendre series partial sum S_{11}.

The partial sum S_N of the series in Eq. (33) is defined to be

$$S_N(x) = \sum_{n=0}^{N} c_n P_n(x).$$

As an example, let $f(x) = 1$ for $0 \le x \le 1$ and $f(x) = 0$ for $-1 \le x < 0$. The Legendre series for this step function is [see Walker (1988)]:

$$\frac{1}{2} + \sum_{k=0}^{\infty} \frac{(-1)^k (4k+3)(2k)!}{4^{k+1}(k+1)!k!} P_{2k+1}(x).$$

In Fig. 10 we show the partial sum S_{11} for this series. The graph of S_{11} is reminiscent of a Fourier series partial sum for a step function. In fact, the following theorem is true.

Theorem 12: If $\int_{-1}^{1} |f(x)|^2 \, dx$ is finite, then the partial sums S_N for the Legendre series for f satisfy

$$\lim_{N \to \infty} \int_{-1}^{1} |f(x) - S_N(x)|^2 \, dx = 0.$$

Moreover, if f is Lipschitz at a point x_0, then $S_N(x_0) \to f(x_0)$ as $N \to \infty$.

This theorem is proved in Walter (1994) and Jackson (1941). Further details and other examples of *orthogonal polynomial series* can be found in either Davis (1975), Jackson (1941), or Walter (1994). There are many important orthogonal series—such as Hermite, Laguerre, and Tchebysheff—which we cannot examine here because of space limitations.

We now turn to another type of orthogonal series, the Haar series. The defects, such as Gibbs' phenomenon and ringing, that occur with Fourier series expansions can be traced to the unlocalized nature of the functions used for expansions. The complex exponentials used in classical Fourier series, and the polynomials used in Legendre series, are all non-zero (except possibly for a finite number of points) over their domains. In contrast, Haar series make use of localized functions, which are non-zero only over tiny regions within their domains.

In order to define Haar series, we first define the *fundamental Haar wavelet* $H(x)$ by

$$H(x) = \begin{cases} 1 & \text{if } 0 \le x < 1/2 \\ -1 & \text{if } 1/2 \le x \le 1. \end{cases}$$

The *Haar wavelets* $\{H_{j,k}(x)\}$ are then defined by

$$H_{j,k}(x) = 2^{j/2} H(2^j x - k)$$

for $j = 0, 1, 2, \ldots$; $k = 0, 1, \ldots, 2^j - 1$. Notice that $H_{j,k}(x)$ is non-zero only on the interval $[k2^{-j}, (k+1)2^{-j}]$, which for large j is a tiny subinterval of $[0, 1]$. As k ranges between 0 and $2^j - 1$, these subintervals partition the interval $[0, 1]$, and the partition becomes finer (shorter subintervals) with increasing j.

The *Haar series* for a function f is defined by

$$b + \sum_{j=0}^{\infty} \sum_{k=0}^{2^j-1} c_{j,k} H_{j,k}(x) \tag{35}$$

with $b = \int_0^1 f(x) \, dx$ and

$$c_{j,k} = \int_0^1 f(x) H_{j,k}(x) \, dx.$$

The definitions of b and $c_{j,k}$ are justified by orthogonality relations between the Haar functions (similar to the orthogonality relations that we used above to justify Fourier series and Legendre series).

A partial sum S_N for the Haar series in Eq. () is defined by

$$S_N(x) = b + \sum_{\{j,k \,|\, 2^j + k \le N\}} c_{j,k} H_{j,k}(x).$$

For example, let f be the function on $[0, 1]$ defined as follows

$$f(x) = \begin{cases} x - 1/2 & \text{if } 1/4 < x < 3/4 \\ 0 & \text{if } x \le 1/4 \text{ or } 3/4 \le x. \end{cases}$$

In Fig. 11 we show the Haar series partial sum S_{256} for this function. Notice that there is no Gibbs' phenomenon with this partial sum. This contrasts sharply with the Fourier

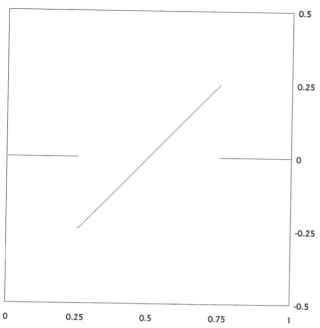

FIGURE 11 Haar series partial sum S_{256}, which has 257 terms.

series partial sum, also using 257 terms, which we show in Fig. 12.

The Haar series partial sums satisfy the following theorem [proved in Daubechies (1992) and in Meyer (1992)].

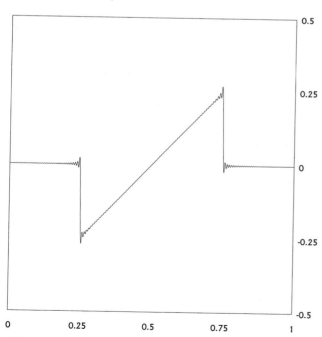

FIGURE 12 Fourier series partial sum S_{128}, which has 257 terms.

Theorem 13: Suppose that $\int_0^1 |f(x)|^p\, dx$ is finite, for $p \geq 1$. Then the Haar series partial sums for f satisfy

$$\lim_{N \to \infty} \left[\int_0^1 |f(x) - S_N(x)|^p\, dx \right]^{1/p} = 0.$$

If f is continuous on [0, 1], then S_N converges uniformly to f on [0, 1].

This theorem is reminiscent of Theorems 10 and 11 for the modified Fourier series partial sums obtained by arithmetic means or by a regular summation kernel. The difference here, however, is that for the Haar series no modifications of the partial sums are needed.

One glaring defect of Haar series is that the partial sums are discontinuous functions. This defect is remedied by the wavelet series discovered by Meyer, Daubechies, and others. The fundamental Haar wavelet is replaced by some new fundamental wavelet Ψ and the set of wavelets $\{\Psi_{j,k}\}$ is then defined by $\Psi_{j,k}(x) = 2^{-j/2}\Psi[2^j x - k]$. (The bracket symbolism $\Psi[2^j x - k]$ means that the value, $2^j x - k$ mod 1, is evaluated by Ψ. This technicality is needed in order to ensure periodicity of $\Psi_{j,k}$.) For example, in Fig. 13, we show graphs of $\Psi_{4,1}$ and $\Psi_{6,46}$ for one of the Daubechies wavelets (a Coif18 wavelet), which is *continuously differentiable*. For a complete discussion of the definition of these wavelet functions, see Daubechies (1992) or Mallat (1998).

The *wavelet series,* generated by the fundamental wavelet Ψ, is defined by

$$b + \sum_{j=0}^{\infty} \sum_{k=0}^{2^j-1} c_{j,k}\Psi_{j,k}(x) \qquad (36)$$

with $b = \int_0^1 f(x)\, dx$ and

$$c_{j,k} = \int_0^1 f(x)\Psi_{j,k}(x)\, dx. \qquad (37)$$

This wavelet series has partial sums S_N defined by

$$S_N(x) = b + \sum_{\{j,k \,|\, 2^j+k \leq N\}} c_{j,k}\Psi_{j,k}(x).$$

Notice that when Ψ is continuously differentiable, then so is each partial sum S_N. These wavelet series partial sums satisfy the following theorem, which generalizes Theorem 13 for Haar series, for a proof, see Daubechies (1992) or Meyer (1992).

Theorem 14: Suppose that $\int_0^1 |f(x)|^p\, dx$ is finite, for $p \geq 1$. Then the Daubechies wavelet series partial sums for f satisfy

$$\lim_{N \to \infty} \left[\int_0^1 |f(x) - S_N(x)|^p\, dx \right]^{1/p} = 0.$$

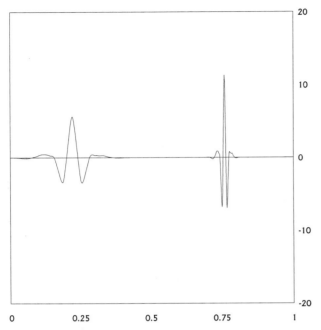

FIGURE 13 Two Daubechies wavelets.

If f is continuous on [0, 1], then S_N converges uniformly to f on [0, 1].

Theorem 14 does not reveal the full power of wavelet series. In almost all cases, it is possible to rearrange the terms in the wavelet series *in any manner whatsoever* and convergence will still hold. One reason for doing a rearrangement is in order to add the terms in the series with coefficients of largest magnitude (thus largest energy) *first* so as to speed up convergence to the function. Here is a convergence theorem for such permuted series.

Theorem 15: Suppose that $\int_0^1 |f(x)|^p \, dx$ is finite, for $p > 1$. If the terms of a Daubechies wavelet series are permuted (in any manner whatsoever), then the partial sums S_N of the permuted series satisfy

$$\lim_{N \to \infty} \left[\int_0^1 |f(x) - S_N(x)|^p \, dx \right]^{1/p} = 0.$$

If f is uniformly Lipschitz, then the partial sums S_N of the permuted series converge uniformly to f.

This theorem is proved in Daubechies (1992) and Meyer (1992). This type of convergence of wavelet series is called *unconditional convergence*. It is known [see Mallat (1998)] that unconditional convergence of wavelet series ensures an optimality of compression of signals. For details about compression of signals and other applications of wavelet series, see Walker (1999) for a simple introduction and Mallat (1998) for a thorough treatment.

VII. DISCRETE FOURIER SERIES

The digital computer has revolutionized the practice of science in the latter half of the twentieth century. The methods of computerized Fourier series, based upon the *fast Fourier transform* algorithms for digital approximation of Fourier series, have completely transformed the application of Fourier series to scientific problems. In this section, we shall briefly outline the main facts in the theory of discrete Fourier series.

The Fourier series coefficients $\{c_n\}$ can be discretely approximated via Riemann sums for the integrals in Eq. (9). For a (large) positive integer M, let $x_k = -\pi + 2\pi k/M$ for $k = 0, 1, 2, \ldots, M - 1$ and let $\Delta x = 2\pi/M$. Then the nth Fourier coefficient c_n for a function f is approximated as follows:

$$c_n \approx \frac{1}{2\pi} \sum_{k=0}^{M-1} f(x_k) \, e^{-i2\pi n x_k} \Delta x$$

$$= \frac{e^{-in\pi}}{M} \sum_{k=0}^{M-1} f(x_k) e^{-i2\pi kn/M}.$$

The last sum above is called the *Discrete Fourier Transform* (DFT) of the finite sequence of numbers $\{f(x_k)\}$. That is, we define the DFT of a sequence $\{g_k\}_{k=0}^{M-1}$ of numbers by

$$G_n = \sum_{k=0}^{M-1} g_k \, e^{-i2\pi kn/M}. \tag{38}$$

The DFT is the set of numbers $\{G_n\}$, and we see from the discussion above that the Fourier coefficients of a function f can be approximated by a DFT (multiplied by the factors $e^{-in\pi}/M$). For example, in Fig. 14 we show a graph of approximations of the Fourier coefficients $\{c_n\}_{n=-50}^{50}$ of the square wave f_1 obtained via a DFT (using $M = 1024$). For all values, these approximate Fourier coefficients differ from the exact coefficients by no more than 10^{-3}. By taking M even larger, the error can be reduced still further.

The two principal properties of DFTs are that they can be inverted and they preserve energy (up to a scale factor). The inversion formula for the DFT is

$$g_k = \sum_{n=0}^{M-1} G_n \, e^{i2\pi kn/M}. \tag{39}$$

And the conservation of energy property is

$$\sum_{k=0}^{M-1} |g_k|^2 = \frac{1}{N} \sum_{n=0}^{M-1} |G_n|^2. \tag{40}$$

Interpreting a sum of squares as energy, Eq. (40) says that, up to multiplication by the factor $1/N$, the energy of the discrete signal $\{g_k\}$ and its DFT $\{G_n\}$ are the same. These

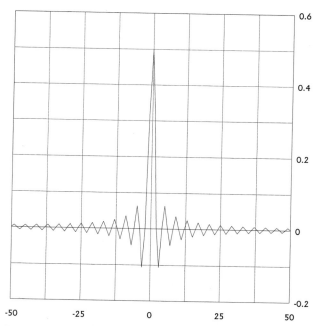

FIGURE 14 Fourier coefficients for square wave, $n = -50$ to 50. Successive values are connected with line segments.

facts are proved in Briggs and Henson (1995) and Walker (1996).

An application of inversion of DFTs is to the calculation of Fourier series partial sums. If we substitute $x_k = -\pi + 2\pi k/M$ into the Fourier series partial sum $S_N(x)$ we obtain (assuming that $N < M/2$ and after making a change of indices $m = n + N$):

$$S_N(x_k) = \sum_{n=-N}^{N} c_n\, e^{in(-\pi + 2\pi k/M)}$$

$$= \sum_{n=-N}^{N} c_n\, (-1)^n\, e^{i2\pi nk/M}$$

$$= \sum_{m=0}^{2N} c_{m-N}\, (-1)^{m-N}\, e^{-i2\pi kN/M}\, e^{i2\pi km/M}.$$

Thus, if we let $g_m = c_{m-N}$ for $m = 0, 1, \ldots, 2N$ and $g_m = 0$ for $m = 2N+1, \ldots, M-1$, we have

$$S_M(x_k) = e^{-i2\pi kN/M} \sum_{m=0}^{M-1} g_m\, (-1)^{m-N}\, e^{i2\pi km/M}.$$

This equation shows that $S_M(x_k)$ can be computed using a DFT inversion (along with multiplications by exponential factors). By combining DFT approximations of Fourier coefficients with this last equation, it is also possible to approximate Fourier series partial sums, or arithmetic means, or other modified partial sums. See Briggs and Henson (1995) or Walker (1996) for further details.

These calculations with DFTs are facilitated on a computer using various algorithms which are all referred to as *fast Fourier transforms* (FFTs). Using FFTs, the process of computing DFTs, and hence Fourier coefficients and Fourier series, is now practically instantaneous. This allows for rapid, so-called *real-time,* calculation of the frequency content of signals. One of the most widely used applications is in calculating *spectrograms.* A spectrogram is calculated by dividing a signal (typically a recorded, digitally sampled, audio signal) into a successive series of short duration subsignals, and performing an FFT on each subsignal. This gives a portrait of the main frequencies present in the signal as time proceeds. For example, in Fig. 15a we analyze discrete samples of the function

$$\sin(2\nu_1 \pi x)e^{-100\pi(x-0.2)^2} + [\sin(2\nu_1 \pi x) + \cos(2\nu_2 \pi x)]$$
$$\times\, e^{-50\pi(x-0.5)^2} + \sin(2\nu_2 \pi x)e^{-100\pi(x-0.8)^2} \quad (41)$$

where the frequencies ν_1 and ν_2 of the sinusoidal factors are 128 and 256, respectively. The signal is graphed at the bottom of Fig. 15a and the magnitudes of the values of its spectrogram are graphed at the top. The more intense spectrogram magnitudes are shaded more darkly, while white regions indicate magnitudes that are essentially zero. The dark blobs in the graph of the spectrogram magnitudes clearly correspond to the regions of highest energy in the signal and are centered on the frequencies 128 and 256, the two frequencies used in Eq. (41).

As a second example, we show in Fig. 15b the spectrogram magnitudes for the signal

$$e^{-5\pi[(x-0.5)/0.4]^{10}}[\sin(400\pi x^2) + \sin(200\pi x^2)]. \quad (42)$$

This signal is a combination of two tones with sharply increasing frequency of oscillations. When run through a sound generator, it produces a sharply rising pitch. Signals like this bear some similarity to certain bird calls, and are also used in radar. The spectrogram magnitudes for this signal are shown in Fig. 15b. We can see two, somewhat blurred, line segments corresponding to the factors $400\pi x$ and $200\pi x$ multiplying x in the two sine factors in Eq. (42).

One important area of application of spectrograms is in *speech coding.* As an example, in Fig. 16 we show spectrogram magnitudes for two audio recordings. The spectrogram magnitudes in Fig. 16a come from a recording of a four-year-old girl singing the phrase "twinkle, twinkle, little star," and the spectrogram magnitudes in Fig. 16b come from a recording of the author of this article singing the same phrase. The main frequencies are seen to be in harmonic progression (integer multiples of a lowest, fundamental frequency) in both cases, but the young girl's main frequencies are higher (higher in pitch) than the adult male's. The slightly curved ribbons of frequency content are known as *formants* in linguistics. For more

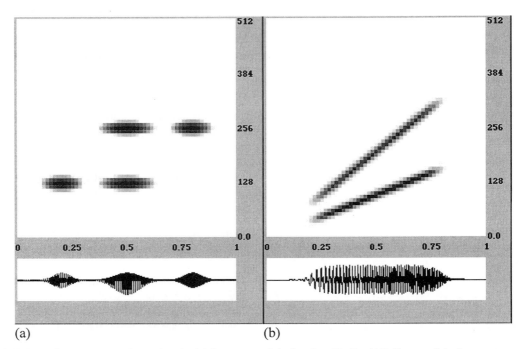

(a)   (b)

FIGURE 15 Spectrograms of test signals. (a) Bottom graph is the signal in Eq. (41). Top graph is the spectrogram magnitudes for this signal. (b) Signal and spectrogram magnitudes for the signal in (42). Horizontal axes are time values (in sec); vertical axes are frequency values (in Hz). Darker pixels denote larger magnitudes, white pixels are near zero in magnitude.

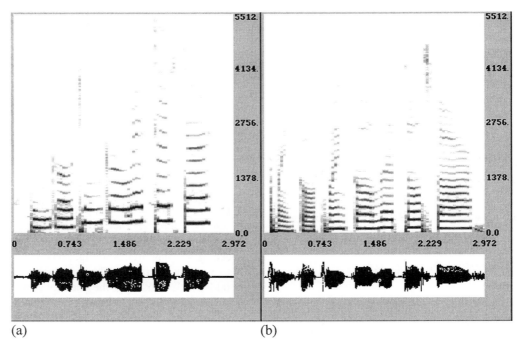

(a)   (b)

FIGURE 16 Spectrograms of audio signals. (a) Bottom graph displays data from a recording of a young girl singing "twinkle, twinkle, little star." Top graph displays the spectrogram magnitudes for this recording. (b) Similar graphs for the author's rendition of "twinkle, twinkle, little star."

details on the use of spectrograms in signal analysis, see Mallat (1998).

It is possible to invert spectrograms. In other words, we can recover the original signal by inverting the succession of DFTs that make up its spectrogram. One application of this inverse procedure is to the *compression of audio signals*. After discarding (setting to zero) all the values in the spectrogram with magnitudes below a threshold value, the inverse procedure creates an approximation to the signal which uses significantly less data than the original signal. For example, by discarding all of the spectrogram values having magnitudes less than 1/320 times the largest magnitude spectrogram value, the young girl's version of "twinkle, twinkle, little star" can be approximated, *without noticeable degradation of quality,* using about one-eighth the amount of data as the original recording. Some of the best results in audio compression are based on sophisticated generalizations of this spectrogram technique— referred to either as *lapped transforms or as local cosine expansions*, see Malvar (1992) and Mallat (1998).

VIII. CONCLUSION

In this article, we have outlined the main features of the theory and application of one-variable Fourier series. Much additional information, however, can be found in the references. In particular, we did not have sufficient space to discuss the intricacies of multivariable Fourier series which, for example, have important applications in crystallography and molecular structure determination. For a mathematical introduction to multivariable Fourier series, see Krantz (1999), and for an introduction to their applications, see Walker (1988).

SEE ALSO THE FOLLOWING ARTICLES

FUNCTIONAL ANALYSIS • GENERALIZED FUNCTIONS • MEASURE AND INTEGRATION • NUMERICAL ANALYSIS • SIGNAL PROCESSING • WAVELETS

BIBLIOGRAPHY

Briggs, W. L., and Henson, V. E. (1995). "The DFT. An Owner's Manual," SIAM, Philadelphia.

Daubechies, I. (1992). "Ten Lectures on Wavelets," SIAM, Philadelphia.

Davis, P. J. (1975). "Interpolation and Approximation," Dover, New York.

Davis, P. J., and Hersh, R. (1982). "The Mathematical Experience," Houghton Mifflin, Boston.

Fourier, J. (1955). "The Analytical Theory of Heat," Dover, New York.

Jackson, D. (1941). "Fourier Series and Orthogonal Polynomials," Math. Assoc. of America, Washington, DC.

Krantz, S. G. (1999). "A Panorama of Harmonic Analysis," Math. Assoc. of America, Washington, DC.

Mallat, S. (1998). "A Wavelet Tour of Signal Processing," Academic Press, New York.

Malvar, H. S. (1992). "Signal Processing with Lapped Transforms," Artech House, Norwood.

Meyer, Y. (1992). "Wavelets and Operators," Cambridge Univ. Press, Cambridge.

Rudin, W. (1986). "Real and Complex Analysis," 3rd edition, McGraw-Hill, New York.

Walker, J. S. (1988). "Fourier Analysis," Oxford Univ. Press, Oxford.

Walker, J. S. (1996). "Fast Fourier Transforms," 2nd edition, CRC Press, Boca Raton.

Walker, J. S. (1999). "A Primer on Wavelets and their Scientific Applications," CRC Press, Boca Raton.

Walter, G. G. (1994). "Wavelets and Other Orthogonal Systems with Applications," CRC Press, Boca Raton.

Zygmund, A. (1968). "Trigonometric Series," Cambridge Univ. Press, Cambridge.

Fractals

Benoit B. Mandelbrot
Michael Frame
Yale University

GLOSSARY

Dimension An exponent characterizing how some aspect—mass, number of boxes in a covering, etc.—of an object scales with the size of the object.

Lacunarity A measure of the distribution of hole sizes of a fractal. The prefactor in the mass–radius scaling is one such measure.

Self-affine fractal A shape consisting of smaller copies of itself, all scaled by affinities, linear transformations with different contraction ratios in different directions.

Self-similar fractal A shape consisting of smaller copies of itself, all scaled by similitudes, linear transformations with the same contraction ratios in every direction.

FRACTALS have a long history: after they became the object of intensive study in 1975, it became clear that they had been used worldwide for millenia as decorative patterns. About a century ago, their appearance in pure math-

ematics had two effects. It led to the development of tools like fractal dimensions, but marked a turn toward abstraction that contributed to a deep and long divide between mathematics and physics. Quite independently from fundamental mathematical physics as presently defined, fractal geometry arose in equal parts from an awareness of past mathematics and a concern for practical, mundane questions long left aside for lack of proper tools.

The mathematical input ran along the lines described by John von Neumann: "A large part of mathematics which became useful developed with absolutely no desire to be useful... This is true for all science. Successes were largely due to... relying on... intellectual elegance. It was by following this rule that one actually got ahead in the long run, much better than any strictly utilitarian course would have permitted... The principle of laissez-faire has led to strange and wonderful results."

The influence of mundane questions grew to take on far more importance than was originally expected, and recently revealed itself as illustrating a theme that is common

Encyclopedia of Physical Science and Technology, Third Edition, Volume 6

in science. Every science started as a way to organize a large collection of messages our brain receives from our senses. The difficulty is that most of these messages are very complex, and a science can take off only after it succeeds in identifying special cases that allow a workable first step. For example, acoustics did not take its first step with chirps or drums but with idealized vibrating strings. These led to sinusoids and constants or other functions invariant under translation in time. For the notion of roughness, no proper measure was available only 20 years ago. The claim put forward forcibly in Mandelbrot (1982) is that a workable entry is provided by rough shapes that are dilation invariant. These are fractals.

Fractal roughness proves to be ubiquitous in the works of nature and man. Those works of man range from mathematics and the arts to the Internet and the financial markets. Those works of nature range from the cosmos to carbon deposits in diesel engines. A sketchy list would be useless and a complete list, overwhelming. The reader is referred to Frame and Mandelbrot (2001) and to a Panorama mentioned therein, available on the web. This essay is organized around the mathematics of fractals, and concrete examples as illustrations of it.

To avoid the need to discuss the same topic twice, mathematical complexity is allowed to fluctuate up and down. The reader who encounters paragraphs of oppressive difficulty is urged to skip ahead until the difficulty becomes manageable.

I. SCALE INVARIANCE

A. On Choosing a "Symmetry" Appropriate to the Study of Roughness

The organization of experimental data into simple theoretical models is one of the central works of every science; invariances and the associated symmetries are powerful tools for uncovering these models. The most common invariances are those under Euclidean motions: translations, rotations, reflections. The corresponding ideal physics is that of uniform or uniformly accelerated motion, uniform or smoothly varying pressure and density, smooth submanifolds of Euclidean physical or phase space. The geometric alphabet is Euclidean, the analytical tool is calculus, the statistics is stationary and Gaussian.

Few aspects of nature or man match these idealizations: turbulent flows are grossly nonuniform; solid rocks are conspicuously cracked and porous; in nature and the stock market, curves are nowhere smooth. One approach to this discrepancy, successful for many problems, is to treat observed objects and processes as "roughened" versions of an underlying smooth ideal. The underlying geometry is

Euclidean or locally Euclidean, and observed nature is written in the language of noisy Euclidean geometry.

Fractal geometry was invented to approach roughness in a very different way. Under magnification, smooth shapes are more and more closely approximated by their tangent spaces. The more they are magnified, the simpler ("better") they look. Over some range of magnifications, looking more closely at a rock or a coastline does not reveal a simpler picture, but rather more of the same kind of detail. Fractal geometry is based on this ubiquitous *scale invariance*. "A fractal is an object that doesn't look any better when you blow it up." Scale invariance is also called "symmetry under magnification."

A manifestation is that fractals are sets (or measures) that can be broken up into pieces, each of which closely resembles the whole, except it is smaller. If the pieces scale isotropically, the shape is called *self-similar*; if different scalings are used in different directions, the shape is called *self-affine*.

There are deep relations between the geometry of fractal sets and the renormalization approach to critical phenomena in statistical physics.

B. Examples of Self-Similar Fractals

1. Exact Linear Self-Similarity

A shape S is called *exactly (linearly) self-similar* if the whole S splits into the union of parts S_i: $S = S_1 \cup S_2 \cup \ldots \cup S_n$. The parts satisfy two restrictions: (a) each part S_i is a copy of the whole S scaled by a linear contraction factor r_i, and (b) the intersections between parts are empty or "small" in the sense of dimension. Anticipating Section II, if $i \neq j$, the fractal dimension of the intersection $S_i \cap S_j$ must be lower than that of S. The roughness of these sets is characterized by the *similarity dimension d*. In the special equiscaling case $r_1 = \cdots = r_n = r$, $d = \log(n)/\log(1/r)$. In general, d is the solution of the *Moran equation*

$$\sum_{i=1}^{n} r_i^d = 1.$$

More details are given in Section II.

Exactly self-similar fractals can be constructed by several elegant mathematical approaches.

a. Initiator and generator. An *initiator* is a starting shape; a *generator* is a juxtaposition of scaled copies of the initiator. Replacing the smaller copies of the initiator in the generator with scaled copies of the generator sets in motion a process whose limit is an exactly self-similar fractal. Stages before reaching the limit are called *protofractals*. Each copy is anchored by a fixed point, and one may have to specify the orientation of each replacement. The

FIGURE 1 Construction of the Sierpinksi gasket. The initiator is a filled-in equilateral triangle, and the generator (on the left) is made of $N = 3$ triangles, each obtained from the initiator by a contraction map T_i of reduction ratio $r = 1/2$. The contractions' fixed points are the vertices of the initiator. The middle shows the second stage, replacing each copy of the initiator with a scaled copy of the generator. On the right is the seventh stage of the construction.

Sierpinski gasket (Fig. 1) is an example. The eye spontaneously splits the whole S into parts. The simplest split yields $N = 3$ parts S_i, each a copy of the whole reduced by a similitude of ratio $1/2$ and with fixed point at a vertex of the initiator. In finer subdivisions, $S_i \cap S_j$ is either empty or a single point, for which $d = 0$. In this example, but not always, it can be made empty by simply erasing the topmost point of every triangle in the construction.

Some of the most familiar fractals were orignally constructed to provide instances of curves that exemplify properties deemed counterintuitive: classical curves may have one multiple point (like Fig. 8) or a few. To the contrary, the Sierpinski gasket (Fig. 2, far left) is a curve with dense multiple points. The Sierpinski carpet (Fig. 2, mid left) is a universal curve in the sense that one can embed in the carpet every plane curve, irrespective of the collection of its multiple points. The Peano curve [initiator the diagonal segment from $(0, 0)$ to $(1, 1)$, generator in Fig. 2 mid right] is actually not a curve but a motion. It is plane-filling: a continuous onto map $[0, 1] \to [0, 1] \times [0, 1]$.

b. Iterated function systems.
Iterated function systems (IFS) are a formalism for generating exactly self-similar fractals based on work of Hutchinson (1981) and Mandelbrot (1982), and popularized by Barnsley (1988). IFS are the foundation of a substantial industry of image compression. The basis is a (usually) finite collection $\{T_1, \ldots, T_n\}$ of contraction maps $T_i: \mathbf{R}^n \to \mathbf{R}^n$ with contraction ratios $r_i < 1$. Each T_i is assigned a probability p_i that serves, at each (discrete) instant of time, to select the next map to be used. An IFS attractor also can be viewed

as the limit set of the orbit $\mathcal{O}^+(x_0)$ of any point x_0 under the action of the semigroup generated by $\{T_1, \ldots, T_n\}$.

The formal definition of IFS, which is delicate and technical, proceeds as follows. Denoting by $\mathcal{K}$ the set of nonempty compact subsets of $\mathbf{R}^n$ and by h the Hausdorff metric on $\mathcal{K}$ [$h(A, B) = \inf\{\delta: A \subset B_\delta$ and $B \subset A_\delta\}$, where $A_\delta = \{x \in \mathbf{R}^n: d(x, y) \leq \delta$ for some $y \in A\}$ is the δ-thickening of A, and d is the Euclidean metric], the T_i together define a transformation $\mathcal{T}: \mathcal{K} \to \mathcal{K}$ by $\mathcal{T}(A) = \bigcup_{i=1}^{n} \{T_i(x): x \in A\}$, a contraction in the Hausdorff metric with contraction ratio $r = \max\{r_1, \ldots, r_n\}$. Because $(\mathcal{K}, h)$ is complete, the contraction mapping principle guarantees there is a unique fixed point C of $\mathcal{T}$. This fixed point is the *attractor* of the IFS $\{T_1, \ldots, T_n\}$. Moreover, for any $K \in \mathcal{K}$, the sequence $K, \mathcal{T}(K), \mathcal{T}^2(K), \ldots$ converges to C in the sense that $\lim_{n \to \infty} h(\mathcal{T}^n(K), C) = 0$.

The IFS *inverse problem* is, for a given compact set A and given tolerance $\delta > 0$, to find a set of transformations $\{T_1, \ldots, T_n\}$ with attractor C satisfying $h(A, C) < \delta$. The search for efficient algorithms to solve the inverse problem is the heart of *fractal image compression*. Detailed discussions can be found in Barnsley and Hurd (1993) and Fischer (1995).

2. Exact Nonlinear Self-Similarity

A broader class of fractals is produced if the decomposition of S into the union $S = S_1 \cup S_2 \cup \ldots \cup S_n$ allows the S_i to be the images of S under nonlinear transformations.

a. Quadratic Julia sets.
For fixed complex number c, the "quadratic orbit" of the starting complex number z is a sequence of numbers that begins with $f_c(z) = z^2 + c$, then $f_c^2(z) = (f_c(z))^2 + c$ and continues by following the rule $f_c^n(z) = f_c(f_c^{n-1}(z))$. The *filled-in (quadratic) Julia set* consists of the starting points that do not iterate to infinity, formally, the points $\{z: f_c^n(z)$ remains bounded as $n \to \infty\}$. The *(quadratic) Julia set J_c* is the boundary of the filled-in Julia set. Figure 3 shows the Julia set J_c for $c = 0.4 + 0 \cdot i$ and the filled-in Julia set for $c = -0.544 + 0.576 \cdot i$. The latter has an attracting 5-cycle, the black region is the basin of attraction of the

FIGURE 2 The Sierpinski gasket, Sierpinski carpet, the Peano curve generator, and the fourth stage of the Peano curve.

FIGURE 3 The Julia set of $z^2 + 0.4$ (left) and the filled-in Julia set for $z^2 - 0.544 + 0.576 \cdot i$ (right).

5-cycle, and the Julia set is the boundary of the black region. Certainly, J_c is invariant under f_c and under the inverses of f_c, $f_{c+}^{-1}(z) = \sqrt{z-c}$ and $f_{c-}^{-1}(z) = -\sqrt{z-c}$. Polynomial functions allow several equivalent characterizations: J_c is the closure of the set of repelling periodic points of $f_c(z)$ and J_c is the attractor of the nonlinear IFS $\{f_{c+}^{-1}, f_{c-}^{-1}\}$.

Much is known about Julia sets of quadratic functions. For example, McMullen proved that at a point whose rotation number has periodic continued-fraction expansion, the J set is asymptotically self-similar about the critical point.

The J sets are defined for functions more general than polynomials. Visually striking and technically interesting examples correspond to the Newton function $N_f(z) = z - f(z)/f'(z)$ for polynomial families $f(z)$, or entire functions like $\lambda \sin z$, $\lambda \cos z$, or $\lambda \exp z$ (see Section VIII.B). Discussions can be found in Blanchard (1994), Curry *et al.* (1983), Devaney (1994), Keen (1994), and Peitgen (1989).

b. The Mandelbrot set.

The quadratic orbit $f_c^n(z)$ always converges to infinity for large enough values of z. Mandelbrot attempted a computer study of the set M^0 of those values of c for which the orbit does *not* converge to infinity, but to a stable cycle. This approach having proved unrewarding, he moved on to a set that promised an easier calculation and proved spectacular. Julia and Fatou, building on fundamental work of Montel, had shown that the Julia set J_c of $f_c(z) = z^2 + c$ must be either connected or totally disconnected. Moreover, J_c is connected if, and only if, the orbit $\mathcal{O}^+(0)$ of the critical point $z = 0$ remains bounded. The set M defined by $\{c: f_c^n(0)$ remains bounded$\}$ is now called the *Mandelbrot set* (see the left side of Fig. 4). Mandelbrot (1980) performed a computer investigation of its structure and reported several observations. As is now well known, small copies of the M set are infinitely numerous and dense in its boundary. The right side of Fig. 4 shows one such small copy, a nonlinearly distorted copy of the whole set. Although the small copy on the right side of Fig. 4 appears to be an isolated "island," Mandelbrot conjectured and Douady and Hubbard (1984) proved that the M set is connected. Sharpening an obser-

FIGURE 4 Left: The Mandelbrot set. Right: A detail of the Mandelbrot set showing a small copy of the whole. Note the nonlinear relation between the whole and the copy.

vation by Mandelbrot, Tan Lei (1984) proved the convergence of appropriate magnifications of Julia sets and the M set at certain points named after Misiurewicz. Shishikura (1994) proved Mandelbrot's (1985) and Milnor's (1989) conjecture that the boundary of the M set has Hausdorff dimension 2. Lyubich proved that the boundary of the M set is asymptotically self-similar about the Feigenbaum point.

Mandelbrot's first conjecture, that the interior of the M set consists entirely of components (called *hyperbolic*) for which there is a stable cycle, remains unproved in general, though McMullen (1994) proved it for all such components that intersect the real axis. Mandelbrot's notion that M may be the closure of M^0 is equivalent to the assertion that the M set is locally connected. Despite intense efforts, that assertion remains a conjecture, though Yoccoz and others have made progress.

Other developments include the theory of quadratic-like maps (Douady and Hubbard, 1985), implying the universality and ubiquity of the M set. This result was presaged by the discovery (Curry *et al.*, 1983) of a Mandelbrot set in the parameter space of Newton's method for a family of cubic polynomials.

The recent book by Tan Lei (2000) surveys current results and attests to the vitality of this field.

c. Circle inversion limit sets.

Inversion I_C in a circle C with center O and radius r transforms a point P into the point P' lying on the ray OP and with $d(O, P) \cdot d(O, P') = r^2$. This is the orientation-reversing involution defined on $\mathbf{R}^2 \cup \{\infty\}$ by $P \to I_C(P) = P'$. Inversion in C leaves C fixed, and interchanges the interior and exterior of C. It contracts the "outer" component not containing O, but the contraction ratio is not bounded by any $r < 1$.

Poincaré generalized from inversion in one circle to a collection of more than one inversion. As an example, consider a collection of circles $C_1, \ldots, C_N$ each of which is external to all the others. That is, for all $j \neq i$, the disks bounded by C_i and C_j have disjoint interiors. The *limit set* $\Lambda(C_1, \ldots, C_N)$ of inversion in these circles is the set of limit points of the orbit $\mathcal{O}^+(P)$ of any point P, external to $C_1, \ldots, C_N$, under the group generated by $I_{C_1}, \ldots, I_{C_N}$. Equivalently, it is the set left invariant by every one of the inversions $I_{C_1}, \ldots, I_{C_N}$.

The limit set Λ is nearly always fractal but the nonlinearity of inversion guarantees that Λ is nonlinearly self-similar. An example is shown in Fig. 5: the part of the limit set inside C_1 is easily seen to be the transform by I_1 of the part of the limit set inside C_2, C_3, C_4, and C_5.

How can one draw the limit set Λ when the arrangement of the circles $C_1, \ldots, C_N$ is more involved? Poincaré's original algorithm converges extraordinarily slowly. The first alternative algorithm was advanced in Mandelbrot

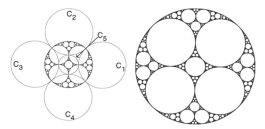

FIGURE 5 Left: The limit set generated by inversion in the five circles $C_1, \ldots, C_5$. Right: A magnification of the limit set.

(1982, Chapter 18); it is intuitive and the large-scale features of Λ appear very rapidly, followed by increasingly fine features down to any level a computer's memory can support.

d. Kleinian group limit sets. A Kleinian group (Beardon, 1983; Maskit, 1988) is a discrete group of Möbius transformations

$$z \to \frac{az + b}{cz + d}$$

acting on the Riemann sphere $\hat{\mathbf{C}}$, the sphere at infinity of hyperbolic 3-space $\mathbf{H}^3$. The isometries of $\mathbf{H}^3$ can be represented by complex matrices

$$\begin{bmatrix} a & b \\ c & d \end{bmatrix}.$$

[More precisely, by their equivalence classes in $PSL_2(\mathbf{C})$.] Sullivan's side-by-side dictionary (Sullivan, 1985) between Kleinian groups and iterates of rational maps is another deep mathematical realm informed, at least in part, by fractal geometry. Thurston's "geometrization program" for 3-manifolds (Thurston, 1997) involves giving many 3-manifolds hyperbolic structures by viewing them as quotients of $\mathbf{H}^3$ by the action of a Kleinian group G (Epstein, 1986). The corresponding action of G on $\hat{\mathbf{C}}$ determines the limit set $\Lambda(G)$, defined as the intersection of all nonempty G-invariant subsets of $\hat{\mathbf{C}}$. For many G, the limit set is a fractal. An example gives the flavor of typical results: the limit set of a finitely generated Kleinian group is either totally disconnected, a circle, or has Hausdorff dimension greater than 1 (Bishop and Jones, 1997). The Hausdorff dimension of the limit set has been studied by Beardon, Bishop, Bowen, Canary, Jones, Keen, Mantica, Maskit, McMullen, Mumford, Parker, Patterson, Sullivan, Tricot, Tukia, and many others. Poincaré exponents, eigenvalues of the Laplacian, and entropy of geodesic flows are among the tools used.

Figure 5 brings forth a relation between some limit sets of inversions or Kleinian groups and Apollonian packings

(Keen *et al.*, 1993). In fact, the limit set of the Kleinian groups that are in the Maskit embedding (Keen and Series, 1993) of the Teichmüller space of any finite-type Riemann surface are Apollonian packings. These correspond to hyperbolic 3-manifolds having totally geodesic boundaries. McShane *et al.* (1994) used automatic group theory to produce efficient pictures of these limit sets, and Parker (1995) showed that in many cases the Hausdorff dimension of the limit set equals the circle packing exponent, easily estimated as the slope of the log–log plot of the number of circles of radius $\geq r$ (*y* axis) versus *r* (*x* axis).

Limit sets of Kleinian group actions are an excellent example of a deep, subtle, and very active area of pure mathematics in which fractals play a central role.

3. Statistical Self-Similarity

A tree's branches are not exact shrunken copies of that tree, inlets in a bay are not exact shrunken copies of that bay, nor is each cloud made up of exact smaller copies of that cloud. To justify the role of fractal geometry as a geometry of nature, one must take a step beyond exact self-similarity (linear or otherwise). Some element of randomness appears to be present in many natural objects and processes. To accommodate this, the notions of self-similarity and self-affinity are made statistical.

a. Wiener brownian motion: its graphs and trails. The first example is classical. It is one-dimensional Brownian motion, the random process $X(t)$ defined by these properties: (1) with probability 1, $X(0) = 0$ and $X(t)$ is continuous, and (2) the increments $X(t + \Delta t) - X(t)$ of $X(t)$ are Gaussian with mean 0 and variance Δt. That is,

$$Pr\{X(t + \Delta t) - X(t) \leq x\} = \frac{1}{\sqrt{2\pi \Delta t}}$$
$$\times \int_{-\infty}^{x} \exp\left(\frac{-u^2}{2\Delta t}\right) du.$$

An immediate consequence is independence of increments over disjoint intervals. A fundamental property of Brownian motion is statistical self-affinity: for all $s > 0$,

$$Pr\{X(s(t + \Delta t)) - X(st) \leq \sqrt{s}x\} = Pr\{X(t + \Delta t)$$
$$- X(t) \leq x\}.$$

That is, rescaling *t* by a factor of *s*, and of *x* by a factor of $\sqrt{s}$, leaves the distribution unchanged. This correct rescaling is shown on the left panel of Fig. 6: *t* (on the horizontal axis) is scaled by 4, *x* (on the vertical axis) is scaled by $2 = 4^{1/2}$. Note that this magnification has about the same degree of roughness as the full picture. In the center panel, *t* is scaled by 4, *x* by 4/3; the magnification is flatter than the original. In the right panel, both *t* and

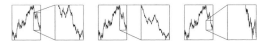

FIGURE 6 Left panel: Correct rescaling illustrating the self-affinity of Brownian motion. Center and right panels: Two incorrect rescalings.

x are scaled by 4; the magnification is steeper than the original.

A sequence of increments of Brownian motion is called Gaussian white noise. Even casual inspection of the graph reveals some fundamental features. The width of an old pen-plotter line being equal to the spacing between successive difference values, the bulk of the difference plot merges into a "band" with the following properties (see Fig. 7):

- The band's width is approximately constant.
- The values beyond that band stay close to it (this is due to the fact that the Gaussian has "short tails").
- The values beyond that band do not cluster.

Positioning E independent one-dimensional Brownian motions along E coordinate axes gives a higher dimensional Brownian motion: $B(t) = \{X_1(t), \ldots, X_E(t)\}$. Plotted as a curve in E-dimensional space, the collection of points that $B(t)$ visits between $t = 0$ and $t = 1$ defines a *Brownian trail*.

When $E > 2$, this is an example of a statistically self-similar fractal. To split the Brownian trail into N reduced-scale parts, pick $N - 1$ arbitrary instants t_n with $0 = t_0 < t_1 < \cdots < t_{N-1} < t_N = 1$. The Brownian trail for $0 \le t \le 1$ splits into N subtrails B_n for the interval $t_{n-1} < t < t_n$. The parts B_n follow the same statistical distribution as the whole, after the size is expanded by $(t_n - t_{n-1})^{-1/2}$ in every direction.

Due to the definition of self-similarity, this example reveals a pesky complication: for $i \ne j$, $B_i \cap B_j$ must be of dimension less than B. This is indeed the case if $E > 2$, but not in the plane $E = 2$. However, the overall idea can be illustrated for $E = 2$. The right side of Fig. 8 shows $B_1(t)$ for $0 \le t \le 1/4$, expanded by a factor of $2 = (1/4 - 0)^{-1/2}$ and with additional points interpolated so the part and the whole exhibit about the same number of turns. The details for $E = 2$ are unexpectedly complex, as shown in Mandelbrot (2001b, Chapter 3).

The dimensions (see Sections II.B and II.E) of some Brownian constructions are well-known, at least in most cases. For example, with probability 1:

FIGURE 7 Plot of 4000 successive Brownian increments.

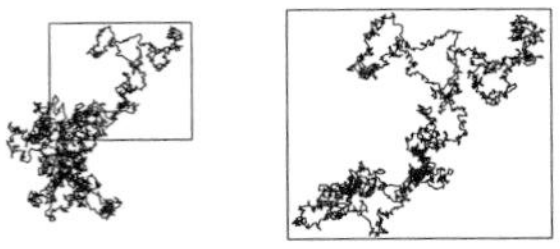

FIGURE 8 A Brownian trail. Right: The first quarter of the left trail, magnified and with additional turns interpolated so the left and right pictures have about the same number of turns.

- For $E \ge 2$ a Brownian trail $B: [0, 1] \to \mathbf{R}^E$ has Hausdorff and box dimensions $d_H = d_{box} = 2$, respectively.
- The graph of one-dimensional Brownian motion $B: [0, 1] \to \mathbf{R}$ has $d_H = d_{box} = 3/2$.

Some related constructions have been more resistant to theoretical analysis. Mandelbrot's *planar Brownian cluster* is the graph of the complex $B(t)$ constrained to satisfy $B(0) = B(1)$. It can be constructed by linearly detrending the x- and y-coordinate functions: $(X(t) - tX(1), Y(t) - tY(1))$. See Fig. 9. The cluster is known to have dimension 2. Visual inspection supported by computer experiments led to the *4/3 conjecture*, which asserts that the boundary of the cluster has dimension 4/3 (Mandelbrot, 1982, p. 243). This has been proved by Lawler *et al.* (2000).

Brownian motion is the unique stationary random process with increments independent over disjoint intervals and with finite variance. For many applications, these conditions are too restrictive, drawing attention to other random processes that retain scaling but abandon either independent increments or finite variance.

b. Fractional Brownian motion. For fixed $0 < H < 1$, *fractional Brownian motion* (FBM) of exponent H is a random process $X(t)$ with increments $X(t + \Delta t) - X(t)$ following the Gaussian distribution with mean 0 and standard deviation $(\Delta t)^H$. Statistical self-affinity is straightforward: for all $s > 0$

$$Pr\{X(s(t + \Delta t)) - X(st) \le s^H x\}$$
$$= Pr\{X(t + \Delta t) - X(t) \le x\}.$$

The correlation is the expected value of the product of successive increments. It equals

FIGURE 9 A Brownian cluster.

FIGURE 12 Lévy flight on the line. Left: the graph as a function of time. Right, the increments.

FIGURE 10 Top: Fractional Brownian motion simulations with $H = 0.25$, $H = 0.5$, and $H = 0.75$. Bottom: Difference plots $X(t + 1) - X(t)$ of the graphs above.

$$E((X(t) - X(0)) \cdot (X(t + h) - X(t)))$$
$$= \tfrac{1}{2}((t + h)^{2H} - t^{2H} - h^{2H}).$$

If $H = 1/2$, this correlation vanishes and the increments are independent. In fact, FBM reduces to Brownian motion. If $H > 1/2$, the correlation is positive, so the increments tend to have the same sign. This is *persistent FBM*. If $H < 1/2$, the correlation is negative, so the increments tend to have opposite signs. This is *antipersistent FBM*. See Fig. 10. The exponent determines the dimension of the graph of FBM: with probability 1, $d_H = d_{box} = 2 - H$. Notice that for $H > 1/2$, the central band of the difference plot moves up and down, a sign of long-range correlation, but the outliers still are small. Figure 11 shows the trails of these three flavors of FBM. FBM is the main topic of Mandelbrot (2001c).

c. Lévy stable processes. While FBM introduces correlations, its increments remain Gaussian and so have small outliers. The Gaussian distribution is characterized by its first two moments (mean and variance), but some natural phenomena appear to have distributions for which these are not useful indicators. For example, at the critical point of percolation there are clusters of all sizes and the expected cluster size diverges.

Paul Lévy studied random walks for which the jump distributions follow the power law $Pr\{X > x\} \approx x^{-\alpha}$. There is a geometrical approach for generating examples of Lévy processes.

The *unit step function* $\xi(t)$ is defined by

$$\xi(t) = \begin{cases} 0 & \text{for } x < 0 \\ 1 & \text{for } x \geq 0 \end{cases}$$

and a (one-dimensional) Lévy stable process is defined as a sum

$$f(t) = \sum_{k=1}^{\infty} \lambda_k \xi(t - t_k),$$

where the pulse times t_n and amplitudes λ_n are chosen according to the following Lévy measure: given t and λ, the probability of choosing (t_i, λ_i) in the rectangle $t < t_i < t + dt, \lambda < \lambda_i < \lambda + d\lambda$ is $C\lambda^{-\alpha-1} d\lambda\, dt$. Figure 12 shows the graph of a Lévy process or flight, and a graph of its increments.

Comparing Figs. 7, 10, and 12 illustrates the power of the increment plot for revealing both global correlations (FBM) and long tails (Lévy processes).

The effect of large excursions in Lévy processess is more visible in the plane. See Fig. 13. These Lévy flights were used in Mandelbrot (1982, Chapter 32) to mimic the statistical properties of galaxy distributions.

Using fractional Brownian motion and Lévy processes, Mandelbrot (in 1965 and 1963) improved upon Bachelier's Brownian model of the stock market. The former corrects the independence of Brownian motion, the latter corrects its short tails. The original and corrected processes in the preceding sentence are statistically self-affine random fractal processes. This demonstrates the power of invariances in financial modeling; see Mandelbrot (1997a,b).

d. Self-affine cartoons with mild to wild randomness. Many natural processes exhibit long tails or global dependence or both, so it was a pleasant surprise that both can be incorporated in an elegant family of simple cartoons. (Mandelbrot, 1997a, Chapter 6; 1999, Chapter N1; 2001a). Like for self-similar curves (Section I.B.1.a), the basic construction of the cartoon involves an initiator and a

FIGURE 11 Top: Fractional Brownian motion simulations with $H = 0.25$, $H = 0.5$, and $H = 0.75$. Bottom: Difference plots $X(t + 1) - X(t)$ of the graphs above.

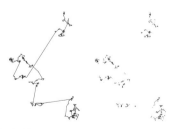

FIGURE 13 Left: Trail of the Lévy flight in the plane. Right: The Lévy dust formed by the turning points.

FIGURE 14 The initiator (left), generator (middle), and first generation (right) of the Brownian cartoon.

generator. The process used to generate the graph consists in replacing each copy of the initiator with an appropriately rescaled copy of the generator. For a Brownian cartoon, the initiator can be the diagonal of the unit square, and the generator, the broken line with vertices $(0, 0)$, $(4/9, 2/3)$, $(5/9, 1/3)$, and $(1, 1)$. Pictured in Fig. 14 are the initiator (left), generator (middle), and first iteration of the process (right).

To get an appreciation for how quickly the local roughness of these pictures increases, the left side of Fig. 15 shows the sixth iterate of the process.

Self-affinity is built in because each piece is an appropriately scaled version of the whole. In Fig. 14, the scaling ratios have been selected to match the "square root" property of Brownian motion: for each segment of the generator we have $|\Delta x_i| = (\Delta t_i)^{1/2}$.

More generally, a cartoon is called *unifractal* if there is a constant H with $|\Delta x_i| = (\Delta t_i)^H$ for each generator segment, where $0 < H < 1$. If different H are needed for different segments, the cartoon is *multifractal*.

The left side of Figure 15 is too symmetric to mimic any real data, but this problem is palliated by shuffling the order in which the three pieces of the generator are put into each scaled copy. The right side of Fig. 15 shows a Brownian cartoon randomized in this way.

Figure 16 illustrates how the statistical properties of the increments can be modified by adjusting the generator in a symmetrical fashion. Keeping fixed the endpoints $(0, 0)$ and $(1, 1)$, the middle turning points are changed into $(a, 2/3)$ and $(1 - a, 1/3)$ for $0 < a \leq 1/2$.

e. Percolation clusters (Stauffer and Aharony, 1992).

Given a square lattice of side length L and a number $p \in [0, 1]$, assign a random number $x \in [0, 1]$ to each lattice cell and fill the cell if $x \leq p$. A *cluster* is a maximal collection of filled cells, connected by sharing common edges. Three examples are shown in Fig. 17. A

spanning cluster connects opposite sides of the lattice. For large L there is a *critical probability* or *percolation threshold* p_c; spanning clusters do not arise for $p < p_c$. Numerical experiments suggest $p_c \approx 0.59275$. In Fig. 17, $p = 0.4, 0.6$, and 0.8. Every lattice has its own p_c.

At $p = p_c$ the masses of the spanning clusters scale with the lattice size L as L^d, independently of the lattices. Experiment yields $d = 1.89 \pm 0.03$, and theory yields $d = 93/49$. This d is the mass dimension of Section II.C. In addition, spanning clusters have holes of all sizes; they are statistically self-similar fractals.

Many fractals are defined as part of a percolation cluster. The *backbone* is the subset of the spanning cluster that remains after removing all parts that can be separated from both spanned sides by removing a single filled cell from the spanning cluster. Numerical estimates suggest the backbone has dimension 1.61. The backbone is the path followed by a fluid diffusing through the lattice.

The *hull* of a spanning cluster is its boundary. It was observed by R. F. Voss in 1984 and proven by B. Duplantier that the hull's dimension is $7/4$.

A more demanding definition of the boundary yields the *perimeter*. It was observed by T. Grossman and proven by B. Duplantier that the perimeter's dimension is $4/3$.

Sapoval *et al.* (1985) examined discrete diffusion and showed that it involves a fractal diffusion front that can be modeled by the hull and the perimeter of a percolation cluster.

f. Diffusion-limited aggregation (DLA; Vicsek, 1992).

DLA was proposed by Witten and Sander (1981, 1983) to simulate the aggregates that carbon particles form in a diesel engine. On a grid of square cells, a cartoon of DLA begins by occupying the center of the grid with a "seed particle." Next, place a particle in a square selected at random on the edge of a large circle centered on the seed square and let it perform a simple random walk. With each tick of the clock, with equal probabilities it will move to an adjacent square, left, right, above, or below. If the moving particle wanders too far from the seed, it falls off the edge of the grid and another wandering particle is started at a randomly chosen edge point. When a wandering particle reaches one of the four squares adjacent to the seed, it sticks to form a cluster of two particles, and another moving particle is released. When a moving particle reaches a square adjacent to the cluster, it sticks there. Continuing in this way builds an arbitrarily large object called a diffusion-limited aggregate (DLA) because the growth of the cluster is governed by the particles' diffusing across the grid. Figure 18 shows a moderate-size DLA cluster.

Early computer experiments on clusters of up to the 10^4 particles showed the mass $M(r)$ of the part of the cluster a distance r from the seed point scales as $M(r) \approx k \cdot r^d$,

FIGURE 15 Left: The sixth iterate of the process of Fig. 14. Right: A sixth iterate of a randomized Brownian cartoon.

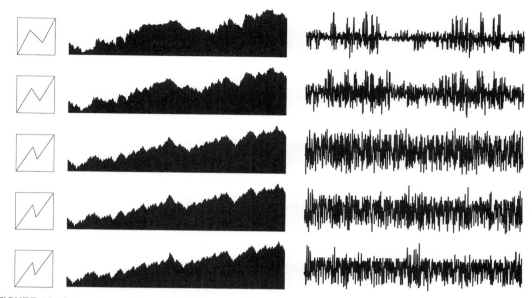

FIGURE 16 Generators, cartoons and difference graphs for symmetric cartoons with turning points $(a, 2/3)$ and $(1 - a, 1/3)$, for $a = 0.333, 0.389, 0.444, 0.456$, and 0.467. The same random number seed is used in all graphs.

with $d \approx 1.71$ for clusters in the plane and $d \approx 2.5$ for clusters in space. This exponent d is the mass dimension of the cluster. (See Sections II.C and V.) These values match measured scalings of physical objects moderately, but not terribly well. A careful examination of much larger clusters revealed discrepancies that added in due time to a very complex picture of DLA. Mandelbrot *et al.* (1995) investigated clusters in the 10^7 range; careful measurement reveals an additional dimension of 1.65 ± 0.01. This suggests the clusters become more compact as they grow. Also, as the cluster grows, more arms develop and the largest gaps decrease in size; i.e., the lacunarity descreases. (See Section VI.)

II. THE GENERIC NOTION OF FRACTAL DIMENSION AND A FEW SPECIFIC IMPLEMENTATIONS

The first, but certainly not the last, step in quantifying fractals is the computation of a dimension. The notion of Euclidean dimension has many aspects and therefore extends in several fashions. The extensions are distinct in the most general cases but coincide for exactly self-similar fractals. Many other dimensions cannot be mentioned here.

A more general approach to quantifying degrees of roughness is found in the article on *Multifractals.*

A. Similarity Dimension

The definition of similarity dimension is rooted in the fact that the unit cube in D-dimensional Euclidean space is self-similar: for any positive integer b the cube can be decomposed into $N = b^D$ cubes, each scaled by the similarity ratio $r = 1/b$, and overlapping at most along $(D - 1)$-dimensional cubes.

The equiscaling or isoscaling case. Provided the pieces do not overlap significantly, the power-law relation $N = (1/r)^D$ between the number N and scaling factor r of the pieces generalizes to all exactly self-similar sets with all pieces scaled by the factor r. The *similarity dimension* d_{sim} is

$$d_{\mathrm{sim}} = \frac{\log(N)}{\log(1/r)}.$$

FIGURE 17 Percolation lattices well below, near, and well above the percolation threshold.

FIGURE 18 A moderate-size DLA cluster.

The pluriscaling case. More generally, for self-similar sets where each piece is scaled by a possibly different factor r_i, the similarity dimension is the unique positive root d of the *Moran equation*

$$\sum_{i=1}^{N} r_i^d = 1.$$

The relation $0 \leq d_{\text{sim}} \leq E$. If the fractal is a subset of E-dimensional Euclidean space, E is called the *embedding dimension*.

So long as the overlap of the parts is not too great (technically, under the *open set condition*), we have $d_{\text{sim}} \leq E$. If at least two of the r_i are positive, we have $d_{\text{sim}} > 0$. However, Section III.F shows that some circumstances introduce a latent dimension d, related indirectly to the similarity dimension, and that can satisfy $d < 0$ or $d > E$.

B. Box Dimension

The similarity dimension is meaningful only for exactly self-similar sets. For more general sets, including experimental data, it is often replaced by the box dimension. For any bounded (nonempty) set A in E-dimensional Euclidean space, and for any $\delta > 0$, a δ-*cover* of A is a collection of sets of diameter δ whose union contains A. Denote by $N_\delta(A)$ the smallest number of sets in a δ-cover of A. Then the *box dimension* d_{box} of A is

$$d_{\text{box}} = \lim_{\delta \to 0} \frac{\log(N_\delta(A))}{\log(1/\delta)}$$

when the limit exists. When the limit does not exist, the replacement of lim with lim sup and lim inf defines the *upper* and *lower box dimensions*:

$$\overline{d_{\text{box}}} = \limsup_{\delta \to 0} \frac{\log(N_\delta(A))}{\log(1/\delta)},$$

$$\underline{d_{\text{box}}} = \liminf_{\delta \to 0} \frac{\log(N_\delta(A))}{\log(1/\delta)}.$$

The box dimension can be thought of as measuring how well a set can be covered with small boxes of equal size, because the limit (or lim sup and lim inf) remain unchanged if $N_\delta(A)$ is replaced by the smallest number of E-dimensional cubes of side δ needed to cover A, or even the number of cubes of a δ lattice that intersect A.

Section V describes methods of measuring the box dimension for physical datasets.

C. Mass Dimension

The mass $M(r)$ of a d-dimensional Euclidean ball of constant density ρ and radius r is given by

$$M(r) = \rho \cdot V(d) \cdot r^d$$

$$\text{with} \quad V(d) = \left(\Gamma\left(\tfrac{1}{2}\right)^d \right) / \Gamma\left(d + \tfrac{1}{2}\right),$$

where $V(d)$ is the volume of the d-dimensional unit sphere. That is, for constant-density Euclidean objects, the ordinary dimension—among many other roles—is the exponent relating mass to size. This role motivated the definition of mass dimension for a fractal. The definition of mass is delicate. For example, the mass of a Sierpinski gasket cannot be defined by starting with a triangle of uniform density and removing middle triangles; this process would converge to a mass reduced to 0. One must, instead, proceed as on the left side of Fig. 1: take as initiator a triangle of mass 1, and as generator three triangles each scaled by $1/2$ and of mass $1/3$. Moreover, two very new facts come up.

Firstly, the $=$ sign in the formula for $M(r)$ must be replaced by $\approx$. That is, $M(r)$ fluctuates around a multiple of r^d. For example, as mentioned in Sections I.B.3.e and I.B.3.f, the masses of spanning percolation clusters and diffusion-limited aggregates scale as a power-law function of size. Consequently, the exponent in the relation $M(r) \approx k \cdot r^{d_{\text{mass}}}$ is called the *mass dimension*.

Second, in the Euclidean case the center is arbitrary but in the fractal case it must belong to the set under consideration. As an example, Fig. 19 illustrates attempts to measure the mass dimension of the Sierpinski gasket. Suppose we take circles centered at the lower left vertex of the initiator, and having radii $1/2$, $1/4, 1/8, \ldots$. We obtain $M(1/2^i) = 1/3^i = (1/2^i)^d$, where $d_{\text{mass}} = \log 3 / \log 2$. See the left side of Fig. 19. That is, the mass dimension agrees with the similarity dimension.

On the other hand, if the circle's center is randomly selected in the interior of the initiator, the passage to the limit $r \to 0$ almost surely eventually stops with circles bounding no part of the gasket. See the middle of Fig. 19.

Taking a family of circles with center c a point of the gasket, the mass–radius relation becomes $M(r) = k(r, c) \cdot r^d$, where the prefactor $k(r, c)$ fluctuates and depends on both r and c. See the right side of Fig. 19. Even in this case, the exponent is the mass dimension. The prefactor is no longer a constant density, but a random variable, depending on the choice of the origin.

FIGURE 19 Attempts at measuring the mass dimension of a Sierpinski gasket using three families of circles.

Section V describes methods of measuring the mass dimension for physical datasets.

D. Minkowski–Bouligand Dimension

Given a set $A \subset \mathbf{R}^E$ and $\delta > 0$, the *Minkowski sausage* of A, also called the δ-*thickening* or δ-*neighborhood* of A, is defined as $A_\delta = \{x \in \mathbf{R}^E: d(x, y) \leq \delta$ for some $y \in A\}$. (See Section I.B.b.) In the Euclidean case when A is a smooth m-dimensional manifold imbedded in $\mathbf{R}^E$, one has $vol(A_\delta) \sim \Lambda \cdot \delta^{E-m}$. That is, the E-dimensional volume of A_δ scales as δ to the codimension of A. This concept extends to fractal sets A: if the limit exists,

$$E - \lim_{\delta \to 0} \frac{\log(vol(A_\delta))}{\log(\delta)}$$

defines the *Minkowski–Bouligand dimension*, $d_{MB}(A)$ (see Mandelbrot, 1982, p. 358). In fact, it is not difficult to see that $d_{MB}(A) = d_{box}(A)$. If the limit does not exist, lim sup gives $\underline{d_{box}}(A)$ and lim inf gives $\overline{d_{box}}(A)$.

In the privileged case when the limit

$$\lim_{\delta \to 0} \frac{vol(A_\delta)}{\delta^{E-m}}$$

exists, it generalizes the notion of Minkowski content for smooth manifolds A. Section VI will use this prefactor to measure lacunarity.

E. Hausdorff–Besicovitch Dimension

For a set A in Euclidean space, given $s \geq 0$ and $\delta > 0$, consider the quantity

$$\mathcal{H}_\delta^s(A) = \inf\left\{ \sum_i |U_i|^s: \{U_i\} \text{ is a } \delta\text{-cover of } A \right\}.$$

A decrease of δ reduces the collection of δ-covers of A, therefore $\mathcal{H}_\delta^s(A)$ increases as $\delta \to 0$ and $\mathcal{H}^s(A) = \lim_{\delta \to 0} \mathcal{H}_\delta^s(A)$ exists. This limit defines the *s-dimensional Hausdorff measure* of A. For $t > s$, $\mathcal{H}_\delta^t(A) \leq \delta^{t-s} \mathcal{H}_\delta^s(A)$. It follows that a unique number d_H has the property that

$$s < d_H \quad \text{implies} \quad \mathcal{H}^s(A) = \infty$$

and

$$s > d_H \quad \text{implies} \quad \mathcal{H}^s(A) = 0.$$

That is,

$$d_H(A) = \inf\{s: \mathcal{H}^s(A) = 0\} = \sup\{s: \mathcal{H}^s(A) = \infty\}.$$

This quantity d_H is the *Hausdorff–Besicovitch dimension* of A. It is of substantial theoretical significance, but in most cases quite challenging to compute, even though it suffices to use coverings by disks. An upper bound often is relatively easy to obtain, but the lower bound can be much

more difficult because the inf is taken over the collection of all δ-covers. Because of the inf that enters in its definition, the Hausdorff–Besicovitch dimension cannot be measured for any physical object.

Note: If A can be covered by $N_\delta(A)$ sets of diameter at most δ, then $\mathcal{H}_\delta^s(A) \leq N_\delta(A) \cdot \delta^s$. From this it follows $d_H(A) \leq \underline{d_{box}}(A)$, so $d_H(A) \leq d_{box}(A)$ if $d_{box}(A)$ exists. This inequality can be strict. For example, if A is any countable set, $d_H(A) = 0$ and yet d_{box}(rationals in $[0, 1]) = 1$.

F. Packing Dimension

Hausdorff dimension measures the efficiency of covering a set by disks of varying radius. Tricot (1982) introduced packing dimension to measure the efficiency of packing a set with disjoint disks of varying radius. Specifically, for $\delta > 0$ a δ-*packing* of A is a countable collection of disjoint disks $\{B_i\}$ with radii $r_i < \delta$ and with centers in A. In analogy with Hausdorff measure, define

$$\mathcal{P}_\delta^s(A) = \sup\left\{ \sum_i |B_i|^s: \{B_i\} \text{ is a } \delta\text{-packing of } A \right\}.$$

As δ decreases, so does the collection of δ-packings of A. Thus $\mathcal{P}_\delta^s(A)$ decreases as δ decreases and the limit

$$\mathcal{P}_0^s(A) = \lim_{\delta \to 0} \mathcal{P}_\delta^s(A)$$

exists. A technical complication requires an additional step. The *s-dimensional packing measure* of A is defined as

$$\mathcal{P}^s(A) = \inf\left\{ \sum_i \mathcal{P}_0^s(A_i): A \subset \bigcup_{i=1}^\infty A_i \right\}.$$

Then the *packing dimension* $d_{pack}(A)$ is

$$d_{pack}(A) = \inf\{s: \mathcal{P}^s(A) = 0\} = \sup\{s: \mathcal{P}^s(A) = \infty\}.$$

Packing, Hausdorff, and box dimensions are related:

$$d_H(A) \leq d_{pack}(A) \leq \overline{d_{box}}(A).$$

For appropriate A, each inequality is strict.

III. ALGEBRA OF DIMENSIONS AND LATENT DIMENSIONS

The dimensions of ordinary Euclidean sets obey several rules of thumb that are widely used, though rarely stated explicitly. For example, the union of two sets of dimension d and d' usually has dimension $\max\{d, d'\}$. The projection of a set of dimension d to a set of dimension d' usually gives a set of dimension $\min\{d, d'\}$. Also, for Cartesian products, the dimensions usually add:

$\dim(A \times B) = \dim(A) + \dim(B)$. For the intersection of subsets A and B of $\mathbf{R}^E$, it is the codimensions that usually add: $E - \dim(A \cap B) = (E - \dim(A)) + (E - \dim(B))$, but only so long as the sum of the codimensions is nonnegative. If this sum is negative, the intersection is empty. Mandelbrot (1984, Part II) generalized those rules to fractals and (see Section III.G) interpreted negative dimensions as measures of "degree of emptiness."

For simplicity, we restrict our attention to generaling these properties to the Hausdorff and box dimensions of fractals.

A. Dimension of Unions and Subsets

Simple applications of the definition of Hausdorff dimension give

$$A \subseteq B \quad \text{implies} \quad d_{\mathrm{H}}(A) \le d_{\mathrm{H}}(B)$$

and

$$d_{\mathrm{H}}(A \cup B) = \max\{d_{\mathrm{H}}(A), d_{\mathrm{H}}(B)\}.$$

Replacing max with sup, this property holds for countable collections of sets. The subset and finite union properties hold for box dimension, but the countable union property fails.

B. Product and Sums of Dimensions

For all subsets A and B of Euclidean space, $d_{\mathrm{H}}(A \times B) \ge d_{\mathrm{H}}(A) + d_{\mathrm{H}}(B)$. Equality holds if one of the sets is sufficiently regular. For example, if $d_{\mathrm{H}}(A) = \overline{d_{\mathrm{H}}}(A)$, then $d_{\mathrm{H}}(A \times B) = d_{\mathrm{H}}(A) + d_{\mathrm{H}}(B)$. Equality does not always hold: Besicovitch and Moran (1945) give an example of subsets A and B of $\mathbf{R}$ with $d_{\mathrm{H}}(A) = d_{\mathrm{H}}(B) = 0$, yet $d_{\mathrm{H}}(A \times B) = 1$.

For upper box dimensions, the inequality is reversed: $\overline{d_{\mathrm{box}}}(A \times B) \le \overline{d_{\mathrm{box}}}(A) + \overline{d_{\mathrm{box}}}(B)$.

C. Projection

Denote by $\mathrm{proj}_P(A)$ the projection of a set $A \subset \mathbf{R}^3$ to a plane $P \subset \mathbf{R}^3$ through the origin. If A is a one-dimensional Euclidean object, then for almost all choices of the plane P, $\mathrm{proj}_P(A)$ is one-dimensional. If A is a two- or three-dimensional Euclidean object, then for almost all choices of the plane P, $\mathrm{proj}_P(A)$ is two-dimensional of positive area. That is, $\dim(\mathrm{proj}_P(A)) = \min\{\dim(A), \dim(P)\}$.

The analogous properties hold for fractal sets A. If $d_{\mathrm{H}}(A) < 2$, then for almost all choices of the plane P, $d_{\mathrm{H}}(\mathrm{proj}_P(A)) = d_{\mathrm{H}}(A)$. If $d_{\mathrm{H}}(A) \ge 2$, then for almost all choices of the plane P, $d_{\mathrm{H}}(\mathrm{proj}_P(A)) = 2$ and $\mathrm{proj}_P(A)$ has positive area. So again, $d_{\mathrm{H}}(\mathrm{proj}_P(A)) = \min\{d_{\mathrm{H}}(A), d_{\mathrm{H}}(P)\}$.

The obvious generalization holds for fractals $A \subset \mathbf{R}^E$ and projections to k-dimensional hyperplanes through the origin.

Projections of fractals can be very complicated. There are fractal sets $A \subset \mathbf{R}^3$ with the surprising property that for almost every plane P through the origin, the projection $\mathrm{proj}_P(A)$ is any prescribed shape, to within a set of area 0. Consequently, as Falconer (1987) points out, in principle we could build a fractal digital sundial.

D. Subordination and Products of Dimension

We have already seen operations realizing the sum, max, and min of dimensions, and in the next subsection we shall examine the sum of codimensions. For certain types of fractals, multiplication of dimensions is achieved through "subordination," a process introduced in Bochner (1955) and elaborated in Mandelbrot (1982). Examples are constructed easily from the Koch curve generator (Fig. 20a). The initiator (the unit interval) is unchanged, but the new generator is a subset of the original generator. Figure 20 shows three examples.

In Fig. 20, generator (b) gives a fractal dust (B) of dimension $\log 3/\log 3 = 1$. Generator (c) gives the standard Cantor dust (C) of dimension $\log 2/\log 3$. Generator (d) gives a fractal dust (D) also of dimension $\log 2/\log 3$. Thinking of the Koch curve K as the graph of a function $f: [0, 1] \to K \subset \mathbf{R}^2$, the fractal (B) can be obtained by restricting f to the Cantor set with initiator $[0, 1]$ and generator the intervals $[0, 1/4]$, $[1/4, 1/2]$, and $[1/2, 3/4]$. In this case, the *subordinand* is a Koch curve, the *subordinator* is a Cantor set, and the *subordinate* is the fractal (B). The identity

$$\frac{\log 3}{\log 3} = \frac{\log 4}{\log 3} \cdot \frac{\log 3}{\log 4}$$

expresses that the dimensions multiply,

$$\dim(\text{subordinate}) = \dim(\text{subordinand})$$

$$\cdot \dim(\text{subordinator}).$$

Figure 20, (C) and (D) give other illustrations of this multiplicative relation. The *seeded universe* model of the distribution of galaxies (Section IX.D.1) uses subordination to obtain fractal dusts; see Mandelbrot (1982, plate 298).

FIGURE 20 The Koch curve (A) and its generator (a); (b), (c), and (d) are subordinators, and the corresponding subordinates of the subordinand (A) are (B), (C), and (D).

E. Intersection and Sums of Codimension

The dimension of the intersection of two sets obviously depends on their relative placement. When $A \cap B = \emptyset$, the dimension vanishes. The following is a typical result. For Borel subsets A and B of $\mathbf{R}^E$, and for almost all $x \in \mathbf{R}^E$,

$$d_{\mathrm{H}}(A \cap (B + x)) \leq \max\{0, d_{\mathrm{H}}(A \times B) - E\}.$$

If $d_{\mathrm{H}}(A \times B) = d_{\mathrm{H}}(A) + d_{\mathrm{H}}(B)$, this reduces to

$$d_{\mathrm{H}}(A \cap (B + x)) \leq \max\{0, d_{\mathrm{H}}(A) + d_{\mathrm{H}}(B) - E\}.$$

This is reminiscent of the transversality relation for intersections of smooth manifolds.

Corresponding lower bounds are known in more restricted circumstances. For example, there is a positive measure set M of similarity transformations of $\mathbf{R}^E$ with

$$d_{\mathrm{H}}(A \cap T(B)) \geq d_H(A) + d_H(B) - E$$

for all $T \in M$. Note $d_{\mathrm{H}}(A \cap T(B)) = d_{\mathrm{H}}(A) + d_{\mathrm{H}}(B) - E$ is equivalent to the addition of codimensions: $E - d_{\mathrm{H}}(A \cap T(B)) = (E - d_{\mathrm{H}}(A)) + (E - d_{\mathrm{H}}(B))$.

F. Latent Dimensions below 0 or above E

A blind application of the rule that codimensions are additive easily yields results that seem nonsensical, yet become useful if they are properly interpreted and the Hausdorff dimension is replaced by a suitable new alternative.

1. Negative Latent Dimensions as Measures of the "Degree of Emptiness"

Section E noted that if the codimension addition rule gives a negative dimension, the actual dimension is 0. This exception is an irritating complication and hides a feature worth underlining.

As background relative to the plane, consider the following intersections of two Euclidean objects: two points, a point and a line, and two lines. Naive intuition tells us that the intersection of two points is emptier than the intersection of a point and a line, and that the latter in turn is emptier than the intersection of two lines (which is almost surely a point). This informal intuition fails to be expressed by either a Euclidean or a Hausdorff dimension. On the other hand, the formal addition of codimensions suggests that the three intersections in question have the respective dimensions -2, -1, and 0. The inequalities between those values conform with the above-mentioned naive intuition. Therefore, they ushered in the search for a new mathematical definition of dimension that can be measured and for which negative values are legitimate and intuitive. This search produced several publications leading to Mandelbrot (1995). Two notions should be mentioned.

Embedding. A problem that concerns $\mathbf{R}^2$ can often be reinterpreted as a problem that really concerns $\mathbf{R}^E$, with $E > 2$, but must be approached within planar intuitions by $\mathbf{R}^2$. Conversely, if a given problem can be embedded into a problem concerning $\mathbf{R}^E$, the question arises, "which is the 'critical' value of $E - 2$, defined as the smallest value for which the intersection ceases to be empty, and precisely reduces to a point?" In the example of a line and a point, the critical $E - 2$ is precisely 1: once embedded in $\mathbf{R}^3$, the problem transforms into the intersection of a plane and a line, which is a point.

Approximation and pre-asymptotics in mathematics and the sciences. Consider a set defined as the limit of a sequence of decreasing approximations. When the limit is not empty, all the usual dimensions are defined as being properties of the limit, but when the limit is empty and all the dimensions vanish, it is possible to consider instead the limits of the properties of the approximations. The Minkowski–Bouligand formal definition of dimension generalizes to fit the naive intuitive values that may be either positive or negative.

2. Latent Dimensions That Exceed That of the Embedding Space

For a strictly self-similar set in $\mathbf{R}^E$, the Moran equation defines a similarity dimension that obeys $d_{\mathrm{sim}} \leq E$. On the other hand, a generator that is a self-avoiding broken line can easily yield $\log(N)/\log(1/r) = d_{\mathrm{sim}} > E$. Recursive application of this generator defines a parametrized motion, but the union of the positions of the motion is neither a self-similar curve nor any other self-similar set. It is, instead, a set whose points are covered infinitely often. Its box dimension is $\leq E$, which *a fortiori* is $< d_{\mathrm{sim}}$. However, one can load a mass on this set by following the route that applies in the absence of multiple points. Mass is distributed on the generator's intervals in proportion to the values of $r_i^{d_{\mathrm{sim}}}$. By infinite recursion, the difference between the times t' and t'' when points P' and P'' are visited is defined as the mass supported by the portion of the curve that links these points.

If so and $d_{\mathrm{sim}} > E$, the similarity dimension acquires a useful role as a latent dimension. For example, consider the multiplication of dimensions in Section III.D. Suppose that our recursively constructed set is not lighted for all instants of time, but only intermittently when time falls within a fractal dust of dimension d''. Then, the rule of thumb is that the latent dimension of the lighted points is $d_{\mathrm{sim}} d''$. When $d_{\mathrm{sim}} d'' < E$, the rule of thumb is that the true dimension is also $d_{\mathrm{sim}} d''$.

Figure 21 shows an example. The generator has $N = 6$ segments, each with scaling ratio $r = 1/2$, hence latent dimension $d_{\mathrm{sim}} = \log 6 / \log 2 > 2$. Taking as subordinator

FIGURE 21 Left: Generator and limiting shape with latent dimension exceeding 2. Right: generator and limiting shape of a subordinate with dimension <2. For comparison, this limiting shape is enclosed in the outline of the left limiting shape.

a Cantor set with generator having $N = 3$ segments, each with scaling ratio $r = 1/2$, yields a self-similar fractal with dimension $\log 3/\log 2$.

G. Mapping

Recall f satisfies the Hölder condition with exponent H if there is a positive constant c for which $|f(x) - f(y)| \le c|x - y|^H$. For such functions, $d_H(f(A)) \le (1/H)d_H(A)$. If $H = 1$, f is called a Lipschitz function; f is bi-Lipschitz if there are constants c_1 and c_2 with $c_1|x - y| \le |f(x) - f(y)| \le c_2|x - y|$. Hausdorff dimension is invariant under bi-Lipschitz maps. The analogous properties hold for box-counting dimension.

IV. METHODS OF COMPUTING DIMENSION IN MATHEMATICAL FRACTALS

Upper bounds for the Hausdorff dimension can be relatively straightforward: it suffices to consider a specific family of coverings of the set. Lower bounds are more delicate. We list and describe briefly some methods for computing dimension.

A. Mass Distribution Methods

A *mass distribution* on a set A is a measure μ with $\mathrm{supp}(\mu) \subset A$ and $0 < \mu(A) < \infty$. The *mass distribution principle* (Falconer, 1990, p. 55) establishes a lower bound for the Hausdorff dimension: Let μ be a mass distribution on A and suppose for some s there are constants $c > 0$ and $\delta > 0$ with $\mu(U) \le c \cdot |U|^s$ for all sets U with $|U| \le \delta$. Then $\delta \le d_H(A)$.

Suitable choice of mass distribution can show that no individual set of a cover can cover too much of A. This can eliminate the problems caused by covers by sets of a wide range of diameters.

B. Potential Theory Methods

Given a mass distribution μ, the *s-potential* is defined by Frostman (1935) as

$$\phi_s(x) = \int \frac{d\mu(x)}{|x - y|^s}.$$

If there is a mass distribution μ on a set A with $\int \phi_s(x)\,d\mu(x) < \infty$, then $d_H(A) \ge s$. Potential theory has been useful for computing dimension of many sets, for example, Brownian paths.

C. Implicit Methods

McLaughlin (1987) introduced a geometrical method, based on local approximate self-similarities, which succeeds in proving that $d_H(A) = \overline{d_{\mathrm{box}}}(A)$, without first determining $d_H(A)$. If small parts of A can be mapped to large parts of A without too much distortion, or if A can be mapped to small parts of A without too much distortion, then $d_H(A) = \overline{d_{\mathrm{box}}}(A) = s$ and $\mathcal{H}^s(A) > 0$ (in the former case) or $\mathcal{H}^s(A) < \infty$ (in the latter case). Details and examples can be found in Falconer (1997, Section 3.1).

D. Thermodynamic Formalism

Sinai (1972), Bowen (1975), and Ruelle (1978) adapted methods of statistical mechanics to determine the dimensions of fractals arising from some nonlinear processes. Roughly, for a fractal defined as the attractor A of a family of nonlinear contractions F_i with an inverse function f defined on A, the *topological pressure* $P(\phi)$ of a Lipschitz function $\phi\colon A \to \mathbf{R}$ is

$$P(\phi) = \lim_{k \to \infty} \frac{1}{k} \log \left\{ \sum_{x \in \mathrm{Fix}(f^k)} \exp[\phi(x) + \phi(f(x)) + \cdots \phi(f^{k-1}(x))] \right\},$$

where $\mathrm{Fix}(f^k)$ denotes the set of fixed points of f^k. The sum plays the role of the partition function in statistical mechanics, part of the motivation for the name "thermodynamic formalism." There is a unique s for which

$P(-s \log |f'|) = 0$, and $s = d_H(A)$. Under these conditions, $0 < \mathcal{H}^s(A) < \infty$, $\mathcal{H}^s(A)$ is a Gibbs measure on A, and many other results can be deduced. Among other places, this method has been applied effectively to the study of Julia sets.

V. METHODS OF MEASURING DIMENSION IN PHYSICAL SYSTEMS

For shapes represented in the plane—for example, coastlines, rivers, mountain profiles, earthquake faultlines, fracture and cracking patterns, viscous fingering, dielectric breakdown, growth of bacteria in stressed environments—box dimension is often relatively easy to compute. Select a sequence $\epsilon_1 > \epsilon_2 > \cdots > \epsilon_n$ of sizes of boxes to be used to cover the shape, and denote by $N(\epsilon_i)$ the number of boxes of size ϵ_i needed to cover the shape. A plot of $\log(N(\epsilon_i))$ against $\log(1/\epsilon_i)$ often reveals a scaling range over which the points fall close to a straight line. In the presence of other evidence (hierarchical visual complexity, for example), this indicates a fractal structure with box dimension given by the slope of the line. Interpreting the box dimension in terms of underlying physical, chemical, and biological processes has yielded productive insights.

For physical objects in three-dimensional space—for example, aggregates, dustballs, physiological branchings (respiratory, circulatory, and neural), soot particles, protein clusters, terrain maps—it is often easier to compute mass dimension. Select a sequence of radii $r_1 > r_2 > \cdots > r_n$ and cover the object with concentric spheres of those radii. Denoting by $M(r_i)$ the mass of the part of the object contained inside the sphere of radius r_i, a plot of $\log(M(r_i))$ against $\log(r_i)$ often reveals a scaling range over which the points fall close to a straight line. In the presence of other evidence (hierarchical arrangements of hole sizes, for example), this indicates a fractal structure with mass dimension given by the slope of the line. Mass dimension is relevant for calculating how density scales with size, and this in turn has implications for how the object is coupled to its environment.

VI. LACUNARITY

Examples abound of fractals sharing the same dimension but looking quite different. For instance, both Sierpinski carpets in Fig. 22 have dimension $\log 40/\log 7$. The holes' distribution is more uniform on the left than on the right. The quantification of this difference was undertaken in Mandelbrot (1982, Chapter 34). It introduced *lacunarity* as one expression of this difference, and took

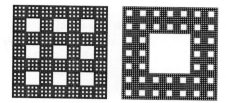

FIGURE 22 Two Sierpinski carpet fractals with the same dimension.

another step in characterizing fractals through associated numbers. How can the distribution of a fractal's holes or gaps ("lacunae") be quantified?

A. The Prefactor

Suppose A is either carpet in Fig. 22, and let A_δ denote the δ-thickening of A. As mentioned in Section II.D, $\text{area}(A_\delta) \sim \Lambda \cdot \delta^{2 - \log 40/\log 7}$. One measure of lacunarity is $1/\Lambda$, if the appropriate limit exists.

It is well known that for the box dimension, the limit as $\epsilon \to 0$ can be replaced by the sequential limit $\epsilon_n \to 0$, for ϵ_n satisfying mild conditions. For these carpets, natural choices are those ϵ_n just filling successive generations of holes. Applied to Fig. 22, these ϵ_n give $1/\Lambda \approx 0.707589$ and 0.793487, agreeing with the notion that higher lacunarity corresponds to a more uneven distribution of holes.

Unfortunately, the prefactor is much more sensitive than the exponent: different sequences of ϵ_n give different limits. Logarithmic averages can be used, but this is work in progress.

B. The Crosscuts Structure

An object is often best studied through its *crosscuts* by straight lines, concentric circles, or spheres. For a fractal of dimension d in the plane, the rule of thumb is that the crosscuts are Cantor-like objects of dimension $d - 1$. The case when the gaps between points in the crosscut are statistically independent was singled out by Mandelbrot as defining "neutral lacunarity." If the crosscut is also self-similar, it is a Lévy dust.

Hovi *et al.* (1996) studied the intersection of lines (linear crosscuts) with two- and three-dimensional critical percolation clusters, and found the gaps are close to being statistically independent, thus a Lévy dust.

In studying very large DLA clusters, Mandelbrot *et al.* (1995) obtained a crosscut dimension of $d_c = 0.65 \pm 0.01$, different from the value 0.71 anticipated if DLA clusters were statistically self-similar objects with mass dimension $d_{\text{mass}} = 1.71$. The difference can be explained by asserting the number of particles $N_c(r/l)$ on a crosscut of radius r scales as $N_c(r/l) = \Lambda(r)(r/l)^{d_c}$. Here l is the scaling length, and the lacunarity prefactor varies with r. Assuming slow variation of $\Lambda(r)$ with r, the observed linear log–log fit requires $\Lambda(r) \sim r^{\delta d}$, where $\delta d = d_{\text{mass}} - 1 - d_c = 0.06 \pm 0.01$. Transverse crosscut analysis reveals lacunarity decreases with r for large DLA clusters.

C. Antipodal Correlations

Select an occupied point p well inside a random fractal cluster, so that the $R \times R$ square centered at p lies within the cluster. Now select two vectors V and W based at p and separated by the angle θ. Finally, denote by x and y the number of occupied sites within the wedges with apexes at p, apex angles ϕ much less than θ, and centered about the vectors V and W. The *angular correlation function* is

$$C(\theta) = \frac{\langle xy \rangle - \langle x \rangle \langle y \rangle}{\langle x^2 \rangle - \langle x \rangle \langle x \rangle},$$

where $\langle \cdots \rangle$ denotes an average over many realizations of the random fractal. *Antipodal correlations* concern $\theta = \pi$. Negative and positive antipodal correlations are interpreted as indicating high and low lacunarity; vanishing correlation is a weakened form of neutral lacunarity.

Mandelbrot and Stauffer (1994) used antipodal correlations to study the lacunarity of critical percolation clusters. On smaller central subclusters, they found the antipodes are uncorrelated.

Trema random fractals. These are formed by removing randomly centered discs, tremas, with radii obeying a power-law scaling. For them, $C(\pi) \to 0$ with ϕ because a circular hole that overlaps a sector cannot overlap the opposite sector. But nonconvex tremas introduce positive antipodal correlations. For θ close to π, needle-shaped tremas, though still convex, yield $C(\theta)$ much higher than for circular trema sets. From this more refined viewpoint, needle tremas' lacunarity is much lower.

VII. FRACTAL GRAPHS AND SELF-AFFINITY

A. Weierstrass Functions

Smooth functions' graphs, as seen under sufficient magnification, are approximated by their tangents. Unless the function itself is linear, the existence of a tangent contradicts the scale invariance that characterizes fractals. The early example of a continuous, nowhere-differentiable function devised in 1834 by Bolzano remained unpublished until the 1920s. The first example to become widely

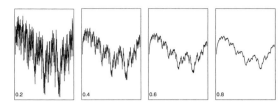

FIGURE 23 The effect of H on Weierstrass graph roughness. In all pictures, $b = 1.5$ and H has the indicated value.

known was constructed by Weierstrass in 1872. The *Weierstrass sine function* is

$$W(t) = \sum_{n=0}^{\infty} b^{-Hn} \sin(2\pi b^n t),$$

and the complex *Weierstrass function* is

$$W_0(t) = \sum_{n=0}^{\infty} b^{-Hn} \exp(2\pi i b^n t).$$

Hardy (1916) showed $W(t)$ is continuous and nowhere-differentiable if and only if $b > 1$ and $0 < H < 1$.

As shown in Fig. 23, the parameter H determines the roughness of the graph. In this case, H is not a perspicuous "roughness exponent." Indeed, as b increases, the amplitudes of the higher frequency terms decrease and the graph is more clearly dominated by the lowest frequency terms. This effect of b is a little-explored aspect of lacunarity.

B. Weierstrass–Mandelbrot Functions

The Weierstrass function revolutionized mathematics but did not enter physics until it was modified in a series of steps described in Mandelbrot (1982, pp. 388–390; (2001d, Chapter H4). The step from $W_0(t)$ to $W_1(t)$ added low frequencies in order to insure self-affinity. The step from $W_1(t)$ to $W_2(t)$ added to each addend a random phase φ_n uniformly distributed on $[0, 1]$. The step from $W_1(t)$ to $W_3(t)$ added a random amplitude $A_n = \sqrt{-2 \log V}$, where V is uniform on $[0, 1]$. A function $W_4(t)$ that need not be written down combines a phase and an amplitude. The latest step leads to another function that need not be written down: it is $W_5(t) = W_4(t) + W_4(-t)$, where the two addends are statistically independent. Contrary to all earlier extensions, $W_5(t)$ is not chiral. We have

$$W_1(t) = \sum_{n=-\infty}^{\infty} b^{-Hn}(\exp(2\pi i b^n t) - 1),$$

$$W_2(t) = \sum_{n=-\infty}^{\infty} b^{-Hn}(\exp(2\pi i b^n t) - 1)\exp(i\varphi_n),$$

$$W_3(t) = \sum_{n=-\infty}^{\infty} A_n b^{-Hn}(\exp(2\pi i b^n t) - 1).$$

C. The Hölder Exponent

A function $f: [a, b] \to \mathbf{R}$ has *Hölder exponent* H if there is a constant $c > 0$ for which

$$|f(x) - f(y)| \leq c \cdot |x - y|^H$$

for all x and y in $[a, b]$ (recall Section III.G). If f is continuous and has Hölder exponent H satisfying $0 < H \leq 1$, then the graph of f has box dimension $d_{\text{box}} \leq 2 - H$.

The Weierstrass function $W(t)$ has Hölder exponent H, hence its graph has $d_{\text{box}} \leq 2 - H$. For large enough b, $d_{\text{box}} = 2 - H$, so one can think of the Hölder exponent as a measure of roughness of the graph.

VIII. FRACTAL ATTRACTORS AND REPELLERS OF DYNAMICAL SYSTEMS

The modern renaissance in dynamical systems is associated most often with chaos theory. Consequently, the relations between fractal geometry and chaotic dynamics, mediated by symbolic dynamics, are relevant to our discussion. In addition, we consider fractal basin boundaries, which generalize Julia sets to much wider contexts including mechanical systems.

A. The Smale Horseshoe

If they exist, intersections of the stable and unstable manifolds of a fixed point are called *homoclinic points*. Poincaré (1890) recognized that homoclinic points cause great complications in dynamics. Yet much can be understood by labeling an appropriate coarse-graining of a neighborhood of a homoclinic point and translating the corresponding dynamics into a string of symbols (the coarse-grain bin labels). The notion of symbolic dynamics first appears in Hadamard (1898), and Birkhoff (1927) proved every neighborhood of a homoclinic point contains infinitely many periodic points.

Motivated by work of Cartwright and Littlewood (1945) and Levinson (1949) on the forced van der Pol oscillator, Smale (1963) constructed the *horseshoe map*. This is a map from the unit square into the plane with completely invariant set a Cantor set Λ, roughly the Cartesian product of two Cantor middle-thirds sets. Restricted to Λ, with the obvious symbolic dynamics encoding, the horseshoe map is conjugate to the shift map on two symbols, the archetype of a chaotic map.

This construction is universal in the sense that it occurs in every transverse homoclinic point to a hyperbolic saddle point. The Conley–Moser theorem (see Wiggins, 1990) establishes the existence of chaos by conjugating the dynamics to a shift map on a Cantor set under general conditions. In this sense, chaos often equivalent to simple dynamics on an underlying fractal.

B. Fractal Basin Boundaries

For any point c belonging to a hyperbolic component of the Mandelbrot set, the Julia set is the boundary of the basins of attraction of the attracting cycle and the attracting fixed point at infinity. See the right side of Fig. 3.

Another example favored by Julia is found in Newton's method for finding the roots of a polynomial $f(z)$ of degree at least 3. It leads to the dynamical system $z_{n+1} = N_f(z_n) = z_n - f(z_n)/f'(z_n)$. The roots of $f(z)$ are attracting fixed points of $N_f(z)$, and the boundary of the basins of attraction of these fixed points is a fractal; an example is shown on the left side of Fig. 24. If contaminated by even small uncertainties, the fate of initial points near the basin boundary cannot be predicted. Sensitive dependence on initial conditions is a signature of chaos, but here we deal with something different. The eventual behavior is completely predictable, except for initial points taken exactly on the basin boundary, usually of two-dimensional Lebesgue measure 0.

The same complication enters mechanical engineering problems for systems with multiple attractors. Moon (1984) exhibited an early example. Extensive theoretical and computer studies by Yorke and coworkers are described in Alligood and Yorke (1992). The driven harmonic oscillator with two-well potential

$$\frac{d^2x}{dt^2} + f\frac{dx}{dt} - \frac{1}{2}x(1 - x^2) = A\cos(\omega t)$$

is a simple example. The undriven system has two equilibria, $x = -1$ and $x = +1$. Initial values (x, x') are painted white if the trajectory from that point eventually stays in the left basin, black if it eventually stays in the right basin. The right side of Fig. 24 shows the initial condition portrait for the system with $f = 0.15$, $\omega = 0.8$, and $A = 0.094$.

IX. FRACTALS AND DIFFERENTIAL OR PARTIAL DIFFERENTIAL EQUATIONS

The daunting task to which a large portion of Mandelbrot (1982) is devoted was to establish that many works of nature and man [as shown in Mandelbrot (1997), the latter includes the stock market!] are fractal. New and often important examples keep being discovered, but the hardest present challenge is to discover the *causes* of fractality. Some cases remain obscure, but others are reasonably clear.

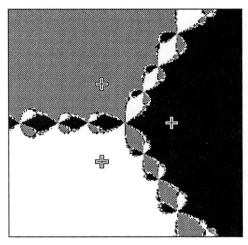 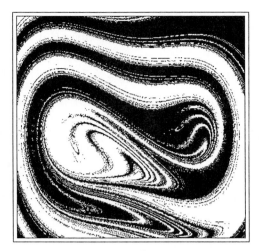

FIGURE 24 Left: The basins of attraction of Newton's method for finding the roots of $z^3 - 1$. Right: The basins of attraction for a damped, driven two-well harmonic oscillator.

Thus, the fractality of the physical percolation clusters (Section I.B.3.e) is the geometric counterpart of scaling and renormalization: the analytic properties of those objects follow a wealth of power-law relations. Many mathematical issues, some of them already mentioned, remain open, but the overall renormalization framework is firmly rooted. Renormalization and the resulting fractality also occur in the structure of attractors and repellers of dynamical systems. Best understood is renormalization for quadratic maps. Feigenbaum and others considered the real case. For the complex case, renormalization establishes that the Mandelbrot set contains infinitely many small copies of itself.

Unfortunately, additional examples of fractality proved to be beyond the scope of the usual renormalization. A notorious case concerns DLA (Section I.B.3.f).

A. Fractal Attractors of Ordinary Differential Equations

The Lorenz equations for fluid convection in a two-dimensional layer heated from below are

$$\frac{dx}{dt} = \sigma(y-x), \quad \frac{dy}{dt} = -xz+rx-y, \quad \frac{dz}{dt} = xy-bz.$$

Here x denotes the rate of convective overturning, y the horizontal temperature difference, and z the departure from a linear vertical temperature gradient. For the parameters $\sigma = 10$, $b = 8/3$, and $r = 28$, Lorenz (1963) suggested that trajectories in a bounded region converge to an attractor that is a fractal, with dimension about 2.06, as estimated by Liapunov exponents. The Lorenz equations are very suggestive but do not represent weather systems very well. However, Haken established a con-

nection with lasers. The sensitivity to initial conditions common to chaotic dynamics is mediated by the intricate fractal interleaving of the multiple layers of the attractor. In addition, Birman and Williams (1983) showed an abundance of knotted periodic orbits embedded in the Lorenz attractor, though Williams (1983) showed all such knots are prime. Grist (1997) constructed a *universal template*, a branched 2-manifold in which all knots are embedded. Note the interesting parallel with the universal aspects of the Sierpinski carpet (Section I.B.1.a). It is not yet known if the attractor of any differential equation contains a universal template. The Poincaré–Bendixson theorem prohibits fractal attractors for differential equations in the plane, but many other classical ordinary differential equations in at least three dimensions exhibit similar fractal attractors in certain parameter ranges.

B. Partial Differential Equations on Domains with Fractal Boundaries ("Can One Hear the Shape of a Fractal Drum?")

Suppose $D \subset \mathbf{R}^n$ is an open region with boundary ∂D. Further, suppose the eigenvalue problem $\nabla^2 u = -\lambda u$ with boundary conditions $u(x) = 0$ for all $x \in \partial D$ has real eigenvalues $0 < \lambda_1 < \lambda_2 < \cdots$. For D with sufficiently smooth boundary, a theorem of Weyl (1912) shows $N(\lambda) \sim \lambda^{n/2}$, where the *eigenvalue counting function* $N(\lambda) = \{$the number of λ_i for which $\lambda_i \le \lambda\}$. If the boundary ∂D is a fractal, Berry (1979, pp. 51–53) postulated that some form of the dimension of ∂D appears in the second term in the expansion of $N(\lambda)$, therefore can be recovered from the eigenvalues. This *could not* be the Hausdorff dimension, but Lapidus (1995) showed that it

FIGURE 25 Perimeter of an extensively studied fractal drum.

is the Minkowski–Bouligard dimension. The analysis is subtle, involving some deep number theory.

For regions with fractal boundaries, the heat equation $\nabla^2 u = (\partial/\partial t)u$ shows heat flow across a fractal boundary is related to the dimension of the boundary. Sapoval (1989) and Sapoval *et al.* (1991) conducted elegant experiments to study the modes of fractal drums. The perimeter of Fig. 25 has dimension $\log 8/\log 4 = 3/2$. A membrane stretched across this fractal curve was excited acoustically and the resulting modes observed by sprinkling powder on the membrane and shining laser light transverse to the surface. Sapoval observed modes localized to bounded regions A, B, C, and D shown in Fig. 25. By carefully displacing the acoustic source, he was able to excite each separately.

Theoretical and computer-graphic analyses of the wave equation on domains with fractal boundaries have been carried out by Lapidus *et al.* (1996), among others.

C. Partial Differential Equations on Fractals

The problem is complicated by the fact that a fractal is not a smooth manifold. How is the Laplacian to be defined on such a space? One promising approach was put forward by physicists in the 1980s and made rigorous in Kigami (1989): approximate the fractal domain by a sequence of graphs representing successive protofractals, and define the fractal Laplacian as the limit of a suitably renormalized sequence of Laplacians on the graphs. Figure 26 shows the first four graphs for the equilateral Sierpinski gasket. The values at the boundary vertices are specified by the boundary conditions at any nonboundary vertex x_0. The mth approximate Laplacian of a function $f(x)$ is the product of a renormalization factor by $\sum (f(y) - f(x_0))$, where the sum is taken over all vertices y in the mth protofractal graph corresponding to the mth-stage reduction of the whole graph.

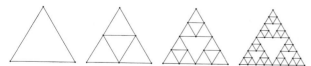

FIGURE 26 Graphs corresponding to protofractal approximations of the equilateral Sierpinski gasket.

With this Laplacian, the heat and wave equations can be defined on fractals. Among other things, the wave equation on domains with fractal boundaries admits localized solutions, as we saw for the wave equation on fractal drums. A major challenge is to extend these ideas to fractals more complicated than the Sierpinski gasket and its relatvies.

D. How Partial Differential Equations Generate Fractals

A quandary: It is universally granted that physics is ruled by diverse partial differential equations, such as those of Laplace, Poisson, and Navier–Stokes. A differential equation necessarily implies a great degree of local smoothness, even though close examination shows isolated singularities or "catastrophes." To the contrary, fractality implies everywhere-dense roughness or fragmentation. This is one of the several reasons that fractal models in diverse fields were initially perceived as being "anomalies" contradicting one of the firmest foundations of science.

A conjecture–challenge responding to the preceding quandry. There is no contradiction at all: fractals arise unavoidably in the long-time behavior of the solution of very familiar and innocuous-looking equations. In particular, many concrete situations where fractals are observed involve equations that allow free and moving boundaries, interfaces, or singularities. As a suggestive "principle," Mandelbrot (1982, Chapter 11) described the following possibility: under broad conditions that largely remain to be specified, these free boundaries, interfaces, and singularities converge to suitable fractals. Many equations have been examined from this viewpoint, but we limit ourselves to two examples of central importance.

1. The Large-Scale Distribution of Galaxies

Chapters 9 and 33–35 of Mandelbrot (1982) conjecture that the distribution of galaxies is fractal. This conjecture results from a search for invariants that was central to every aspect of the construction of fractal geometry. Granted that the distribution of galaxies certainly deviates from homogeneity, one broad approach consists in correcting for local inhomogeneity by using local "patches." The next simplest global assumption is that the distribution is nonhomogeneous but scale invariant, therefore fractal.

Excluding the strict hierarchies, two concrete constructions of random fractal sets were subjected to detailed mathematical and visual investigation. These constructions being random, self-similarity can only be statistical. But a strong counteracting asset is that the self-similarity ratio can be chosen freely, is not restricted to powers of a prescribed r_0. A surprising and noteworthy finding came forth. These constructions exhibited a strong hierarchical structure that is not a deliberate and largely arbitrary input. Details are given in Mandelbrot (1982).

The first construction is the *seeded universe* based on a Lévy flight. Its Hausdorff-dimensional properties were well known. Its correlation properties (Mandelbrot 1975) proved to be nearly identical to those of actual galaxy maps. The second construction is the *parted universe* obtained by subtracting from space a random collection of overlapping tremas. Either construction yields sets that are highly irregular and involve no special center, yet, with no deliberate design, exhibit a clear-cut clustering, "filaments" and "walls." These structures were little known when these constructions were designed.

Conjecture: Could it be that the observed "clusters," "filaments," and "walls" need not be explained separately? They may not result from unidentified features of specific models, but represent unavoidable consequences of a variety of unconstrained forms of random fractality, as interpreted by a human brain.

A problem arose when careful examination of the simulations revealed a clearly incorrect prediction. The simulations in the *seeded universe* proved to be visually far more "lacunar" than the real world. That is, the simulations' holes are larger than in reality. The *parted universe* model fared better, since its lacunarity can be adjusted at will and fit to the actual distribution. A lowered lacunarity is expressed by a positive correlation between masses in antipodal directions. Testing this specific conjecture is a challenge for those who analyze the data.

Does dynamics make us expect the distribution of galaxies to be fractal? Position a large array of point masses in a cubic box in which opposite sides are identified to form a three-dimensional torus. The evolution of this array obeys the Laplace equation, with the novelty that the singularities of the solution are the positions of the points, therefore movable. All simulations we know (starting with those performed at IBM around 1960) suggest that, even when the pattern of the singularities begins by being uniform or Poisson, it gradually creates clusters and a semblance of hierarchy, and appears to tend toward fractality. It is against the preceding background that the limit distribution of galaxies is conjectured to be fractal, and fractality is viewed as compatible with Newton's equations.

2. The Navier–Stokes Equation

The first concrete use of a Cantor dust in real spaces is found in Berger and Mandelbrot (1963), a paper on noise records. This was nearly simultaneous with Kolmogorov's work on the intermittence of turbulence. After numerous experimental tests designed to create an intuitive feeling for this phenomenon (e.g., listening to turbulent velocity records that were made audible), the fractal viewpoint was extended to turbulence, and circa 1964 led to the following conjecture.

Conjecture. The property of being "turbulently dissipative" should not be viewed as attached to domains in a fluid with significant interior points, but as attached to fractal sets. In a first approximation, those sets' intersection with a straight line is a Cantor-like fractal dust having a dimension in the range from 0.5 to 0.6. The corresponding full sets in space should therefore be expected to be fractals with Hausdorff dimension in the range from 2.5 to 2.6.

Actually, Cantor dust and Hausdorff dimension are not the proper notions in the context of viscous fluids because viscosity necessarily erases the fine detail essential to fractals. Hence the following conjecture (Mandelbrot, 1982, Chapter 11; 1976). The dissipation in a viscous fluid occurs in the neighborhood of a singularity of a nonviscous approximation following Euler's equations, and the motion of a nonviscous fluid acquires singularities that are sets of dimension about 2.5–2.6. Several numerical tests agree with this conjecture (e.g., Chorin, 1981).

A related conjecture, that the Navier–Stokes equations have fractal singularities of much smaller dimension, has led to extensive work by V. Scheffer, R. Teman, and C. Foias, and many others. But this topic is not exhausted.

Finally, we mention that fractals in phase space entered the transition from laminar to turbulent flow through the work of Ruelle and Takens (1971) and their followers. The task of unifying the real- and phase-space roles of fractals is challenging and far from being completed.

X. FRACTALS IN THE ARTS AND IN TEACHING

The Greeks asserted art reflects nature, so it is little surprise that the many fractal aspects of nature should find their way into the arts—beyond the fact that a representational painting of a tree exhibits the same fractal branching as a physical tree. Voss and Clarke (1975) found fractal power-law scaling in music, and self-similarity is designed in the music of the composers György Ligeti and Charles Wuorinen. Pollard-Gott (1986) established the presence of fractal repetition patterns in the poetry of Wallace Stevens. Computer artists use fractals to create both abstract aesthetic images and realistic landscapes. Larry Poons' paintings since the 1980s have had rich fractal textures. The "decalcomania" of the 1830s and the 1930s and 1940s used viscous fingering to provide a level of visual complexity. Before that, Giacometti's Alpine wildflower paintings are unquestionably fractal. Earlier still, relatives of the Sierpinski gasket occur as decorative motifs in Islamic and Renaissance art. Fractals abound in architecture, for example, in the cascades of spires in Indian temples, Bramante's plan for St. Peter's, Malevich's Architektonics, and some of Frank Lloyd Wright's designs. Fractals occur in the writing of Clarke, Crichton, Hoag, Powers, Updike, and Wilhelm, among others, and in at least one play, Stoppard's *Arcadia*. Postmodern literary theory has used some concepts informed by fractal geometry, though this application has been criticized for its overly free interpretations of precise scientific language. Some have seen evidence of power-law scaling in historical records, the distribution of the magnitudes of wars and of natural disasters, for example. In popular culture, fractals have appeared on t-shirts, totebags, book covers, MTV logos, been mentioned on public radio's *A Prairie Home Companion*, and been seen on television programs from *Nova* and *Murphy Brown*, through several incarnations of *Star Trek*, to *The X-Files* and *The Simpsons*. While Barnsley's (1988) slogan, "fractals everywhere," is too strong, the degree to which fractals surround us outside of science and engineering is striking.

A corollary of this last point is a good conclusion to this high-speed survey. In our increasingly technological world, science education is very important. Yet all too often humanities students are presented with limited choices: the first course in a standard introductory sequence, or a survey course diluted to the level of journalism. The former builds toward major points not revealed until later courses, the latter discusses results from science without showing how science is done. In addition, many efforts to incorporate computer-aided instruction attempt to replace parts of standard lectures rather than engage students in exploration and discovery.

Basic fractal geometry courses for non-science students provide a radical departure from this mode. The subject of fractal geometry operates at human scale. Though new to most, the notion of self-similarity is easy to grasp, and (once understood) handles familiar objects from a genuinely novel perspective. Students can explore fractals with the aid of readily available software. These instances of computer-aided instruction are perfectly natural because computers are so central to the entire field of fractal geometry. The contemporary nature of the field is revealed by a supply of mathematical problems that are simple to state but remain unsolved. Altogether, many fields of interest to non-science students have surprising examples of fractal structures. Fractal geometry is a powerful tool for imparting to non-science students some of the excitement for science often invisible to them. Several views of this are presented in Frame and Mandelbrot (2001).

The importance of fractals in the practice of science and engineering is undeniable. But fractals are also a proven force in science education. Certainly, the boundaries of fractal geometry have not yet been reached.

SEE ALSO THE FOLLOWING ARTICLES

CHAOS • PERCOLATION • TECTONOPHYSICS

BIBLIOGRAPHY

Alligood, K., and Yorke, J. (1992). *Ergodic Theory Dynam. Syst.* **12,** 377–400.

Alligood, K., Sauer, T., and Yorke, J. (1997). "Chaos. An Introduction to Dynamical Systems," Springer-Verlag, New York.

Barnsley, M. (1988). "Fractals Everywhere," 2nd ed., Academic Press, Orlando, FL.

Barnsley, M., and Demko, S. (1986). "Chaotic Dynamics and Fractals," Academic Press, Orlando, FL.

Barnsley, M., and Hurd, L. (1993). "Fractal Image Compression," Peters, Wellesley, MA.

Batty, M., and Longley, P. (1994). "Fractal Cities," Academic Press, London.

Beardon, A. (1983). "The Geometry of Discrete Groups," Springer-Verlag, New York.

Beck, C., and Schlögl, F. (1993). "Thermodynamics of Chaotic Systems: An Introduction," Cambridge University Press, Cambridge.

Berger, J., and Mandelbrot, B. (1963). *IBM J. Res. Dev.* **7,** 224–236.

Berry, M. (1979). "Structural Stability in Physics," Springer-Verlag, New York.

Bertoin, J. (1996). "Lévy Processes," Cambridge University Press, Cambridge.

Besicovitch, A., and Moran, P. (1945). *J. Lond. Math. Soc.* **20,** 110–120.

Birkhoff, G. (1927). "Dynamical Systems, American Mathematical Society, Providence, RI.

Birman, J., and Williams, R. (1983). *Topology* **22,** 47–82.

Bishop, C., and Jones, P. (1997). *Acta Math.* **179,** 1–39.

Blanchard, P. (1994). In "Complex Dynamical Systems. The Mathematics behind the Mandelbrot and Julia Sets" (Devaney, R., ed.), (pp. 139–154, American Mathematical Society, Providence, RI.

Bochner, S. (1955). "Harmonic Analysis and the Theory of Probability," University of California Press, Berkeley, CA.

Bowen, R. (1975). "Equilibrium States and the Ergodic Theory of Anosov Diffeomorphisms," Springer-Verlag, New York.

Bunde, A., and Havlin, S. (1991). "Fractals and Disordered Systems," Springer-Verlag, New York.

Cartwright, M., and Littlewood, L. (1945). *J. Lond. Math. Soc.* **20,** 180–189.

Cherbit, G. (1987). "Fractals. Non-integral Dimensions and Applications," Wiley, Chichester, UK.

Chorin, J. (1981). *Commun. Pure Appl. Math.* **34,** 853–866.

Crilly, A., Earnshaw, R., and Jones, H. (1991). "Fractals and Chaos," Springer-Verlag, New York.

Crilly, A., Earnshaw, R., and Jones, H. (1993). "Applications of Fractals and Chaos," Springer-Verlag, New York.

Curry, J., Garnett, L., and Sullivan, D. (1983). *Commun. Math. Phys.* **91,** 267–277.

Dekking, F. M. (1982). *Adv. Math.* **44,** 78–104.

Devaney, R. (1989). "An Introduction to Chaotic Dynamical Systems," 2nd ed., Addison-Wesley, Reading, MA.

Devaney, R. (1990). "Chaos, Fractals, and Dynamics. Computer Experiments in Mathematics," Addison-Wesley, Reading, MA.

Devaney, R. (1992). "A First Course in Chaotic Dynamical Systems. Theory and Experiment," Addison-Wesley, Reading, MA.

Devaney, R. (ed.). (1994). "Complex Dynamical Systems. The Mathematics Behind the Mandelbrot and Julia Sets," American Mathematical Society, Providence, RI.

Devaney, R., and Keen, L. (1989). "Chaos and Fractals. The Mathematics Behind the Computer Graphics," American Mathematical Society, Providence, RI.

Douady, A., and Hubbard, J. (1984). "Étude dynamique des polynômes complexes. I, II," Publications Mathematiques d'Orsay, Orsay, France.

Douady, A., and Hubbard, J. (1985). *Ann. Sci. Ecole Norm. Sup.* **18**, 287–343.

Edgar, G. (1990). "Measure, Topology, and Fractal Geometry," Springer-Verlag, New York.

Edgar, G. (1993). "Classics on Fractals," Addison-Wesley, Reading, MA.

Edgar, G. (1998). "Integral, Probability, and Fractal Measures," Springer-Verlag, New York.

Eglash, R. (1999). "African Fractals. Modern Computing and Indigenous Design," Rutgers University Press, New Brunswick, NJ.

Encarncação, J., Peitgen, H.-O., Sakas, G., and Englert, G. (1992). "Fractal Geometry and Computer Graphics," Springer-Verlag, New York.

Epstein, D. (1986). "Low-Dimensional Topology and Kleinian Groups," Cambridge University Press, Cambridge.

Evertsz, C., Peitgen, H.-O., and Voss, R. (eds.). (1996). "Fractal Geometry and Analysis. The Mandelbrot Festschrift, Curaçao 1995," World Scientific, Singapore.

Falconer, K. (1985). "The Geometry of Fractal Sets," Cambridge University Press, Cambridge.

Falconer, K. (1987). *Math. Intelligencer* **9**, 24–27.

Falconer, K. (1990). "Fractal Geometry. Mathematical Foundations and Applications," Wiley, Chichester, UK.

Falconer, K. (1997). "Techniques in Fractal Geometry," Wiley, Chichester, UK.

Family, F., and Vicsek, T. (1991). "Dynamics of Fractal Surfaces," World Scientific, Singapore.

Feder, J. (1988). "Fractals," Plenum Press, New York.

Feder, J., and Aharony, A. (1990). "Fractals in Physics. Essays in Honor of B. B. Mandelbrot," North-Holland, Amsterdam.

Fisher, Y. (1995). "Fractal Image Compression. Theory and Application," Springer-Verlag, New York.

Flake, G. (1998). "The Computational Beauty of Nature. Computer Explorations of Fractals, Chaos, Complex Systems, and Adaptation," MIT Press, Cambridge, MA.

Fleischmann, M., Tildesley, D., and Ball, R. (1990). "Fractals in the Natural Sciences," Princeton University Press, Princeton, NJ.

Frame, M., and Mandelbrot, B. (2001). "Fractals, Graphics, and Mathematics Education," Mathematical Association of America, Washington, DC.

Frostman, O. (1935). *Meddel. Lunds. Univ. Math. Sem.* **3**, 1–118.

Gazalé, M., (1990). "Gnomon. From Pharaohs to Fractals," Princeton University Press, Princeton, NJ.

Grist, R. (1997). *Topology* **36**, 423–448.

Gulick, D. (1992). "Encounters with Chaos," McGraw-Hill, New York.

Hadamard, J. (1898). *J. Mathematiques* **5**, 27–73.

Hardy, G. (1916). *Trans. Am. Math. Soc.* **17**, 322–323.

Hastings, H., and Sugihara, G. (1993). "Fractals. A User's Guide for the Natural Sciences," Oxford University Press, Oxford.

Hovi, J.-P., Aharony, A., Stauffer, D., and Mandelbrot, B. B. (1996). *Phys. Rev. Lett.* **77**, 877–880.

Hutchinson, J. E. (1981). *Ind. Univ. J. Math.* **30**, 713–747.

Keen, L. (1994). *In* "Complex Dynamical Systems. The Mathematics behind the Mandelbrot and Julia Sets" (Devaney, R., ed.), pp. 139–154, American Mathematical Society, Providence, RI.

Keen, L., and Series, C. (1993). *Topology* **32**, 719–749.

Keen, L., Maskit, B., and Series, C. (1993). *J. Reine Angew. Math.* **436**, 209–219.

Kigami, J. (1989). *Jpn. J. Appl. Math.* **8**, 259–290.

Lapidus, M. (1995). *Fractals* **3**, 725–736.

Lapidus, M., Neuberger, J., Renka, R., and Griffith, C. (1996). *Int. J. Bifurcation Chaos* **6**, 1185–1210.

Lasota, A., and Mackey, M. (1994). "Chaos, Fractals, and Noise. Stochastic Aspects of Dynamics," 2nd ed., Springer-Verlag, New York.

Lawler, G., Schramm, O., and Warner, W. (2000). *Acta Math.*, to appear [xxx.lanl.gov./abs/math. PR/0010165].

Lei, T. (2000). "The Mandelbrot Set, Theme and Variations," Cambridge University Press, Cambridge.

Le Méhauté, A. (1990). "Fractal Geometries. Theory and Applications," CRC Press, Boca Raton, FL.

Levinson, N. (1949). *Ann. Math.* **50**, 127–153.

Lorenz, E. (1963). *J. Atmos. Sci.* **20**, 130–141.

Lu, N. (1997). "Fractal Imaging," Academic Press, San Diego.

Lyubich, M. (2001). *Ann. Math.*, to appear.

Mandelbrot, B. (1975). *C. R. Acad. Sci. Paris* **280A**, 1075–1078.

Mandelbrot, B. (1975, 1984, 1989, 1995). "Les objects fractals," Flammarion, Paris.

Mandelbrot, B. (1976). *C. R. Acad. Sci. Paris* **282A**, 119–120.

Mandelbrot, B. (1980). *Ann. N. Y. Acad. Sci.* **357**, 249–259.

Mandelbrot, B. (1982). "The Fractal Geometry of Nature," Freeman, New York.

Mandelbrot, B. (1984). *J. Stat. Phys.* **34**, 895–930.

Mandelbrot, B. (1985). *In* "Chaos, Fractals, and Dynamics" (Fischer, P., and Smith, W., eds.), pp. 235–238, Marcel Dekker, New York.

Mandelbrot, B. (1995). *J. Fourier Anal. Appl.* **1995**, 409–432.

Mandelbrot, B. (1997a). "Fractals and Scaling in Finance. Discontinuity, Concentration, Risk," Springer-Verlag, New York.

Mandelbrot, B. (1997b). "Fractales, Hasard et Finance," Flammarion, Paris.

Mandelbrot, B. (1999). "Multifractals and $1/f$ Noise. Wild Self-Affinity in Physics," Springer-Verlag, New York.

Mandelbrot, B. (2001a). *Quant. Finance* **1**, 113–123, 124–130.

Mandelbrot, B. (2001b). "Gaussian Self-Affinity and Fractals: Globality, the Earth, $1/f$ Noise, & R/S," Springer-Verlag, New York.

Mandelbrot, B. (2001c). "Fractals and Chaos and Statistical Physics," Springer-Verlag, New York.

Mandelbrot, B. (2001d). "Fractals Tools," Springer-Verlag, New York.

Mandelbrot, B. B., and Stauffer, D. (1994). *J. Phys. A* **27**, L237–L242.

Mandelbrot, B. B., Vespignani, A., and Kaufman, H. (1995). *Europhy. Lett.* **32**, 199–204.

Maksit, B. (1988). "Kleinian Groups," Springer-Verlag, New York.

Massopust, P. (1994). "Fractal Functions, Fractal Surfaces, and Wavelets," Academic Press, San Diego, CA.

Mattila, P. (1995). "Geometry of Sets and Measures in Euclidean Space. Fractals and Rectifiability," Cambridge University Press, Cambridge.

McCauley, J. (1993). "Chaos, Dynamics and Fractals. An Algorithmic Approach to Deterministic Chaos," Cambridge University Press, Cambridge.

McLaughlin, J. (1987). *Proc. Am. Math. Soc.* **100**, 183–186.

McMullen, C. (1994). "Complex Dynamics and Renormalization," Princeton University Press, Princeton, NJ.

McShane, G., Parker, J., and Redfern, I. (1994). *Exp. Math.* **3**, 153–170.

Meakin, P. (1996). "Fractals, Scaling and Growth Far from Equilibrium," Cambridge University Press, Cambridge.

Milnor, J. (1989). *In* "Computers in Geometry and Topology" (Tangora, M., ed.), pp. 211–257, Marcel Dekker, New York.

Moon, F. (1984). *Phys. Rev. Lett.* **53**, 962–964.

Moon, F. (1992). "Chaotic and Fractal Dynamics. An Introduction for Applied Scientists and Engineers," Wiley-Interscience, New York.

Parker, J. (1995). *Topology* **34**, 489–496.

Peak, D., and Frame, M. (1994). "Chaos Under Control. The Art and Science of Complexity," Freeman, New York.

Peitgen, H.-O. (1989). "Newton's Method and Dynamical Systems," Kluwer, Dordrecht.

Peitgen, H.-O., and Richter, P. H. (1986). "The Beauty of Fractals," Springer-Verlag, New York.

Peitgen, H.-O., and Saupe, D. (1988). "The Science of Fractal Images," Plenum Press, New York.

Peitgen, H.-O., Jürgens, H., and Saupe, D. (1992). "Chaos and Fractals: New Frontiers of Science," Springer-Verlag, New York.

Peitgen, H.-O., Rodenhausen, A., and Skordev, G. (1998). *Fractals* **6**, 371–394.

Pietronero, L. (1989). "Fractals Physical Origins and Properties," North-Holland, Amsterdam.

Pietronero, L., and Tosatti, E. (1986). "Fractals in Physics," North-Holland, Amsterdam.

Poincaré, H. (1890). *Acta Math.* **13**, 1–271.

Pollard-Gott, L. (1986). *Language Style* **18**, 233–249.

Rogers, C. (1970). "Hausdorff Measures," Cambridge University Press, Cambridge.

Ruelle, D. (1978). "Thermodynamic Formalism: The Mathematical Structures of Classical Equilibrium Statistical Mechanics," Addison-Wesley, Reading, MA.

Ruelle, D., and Takens, F. (1971). *Commun. Math. Phys.* **20**, 167–192.

Samorodnitsky, G., and Taqqu, M. (1994). "Stable Non-Gaussian Random Processes. Stochastic Models with Infinite Variance," Chapman and Hall, New York.

Sapoval, B. (1989). *Physica D* **38**, 296–298.

Sapoval, B., Rosso, M., and Gouyet, J. (1985). *J. Phys. Lett.* **46**, L149–L156.

Sapoval, B., Gobron, T., and Margolina, A. (1991). *Phys. Rev. Lett.* **67**, 2974–2977.

Scholz, C., and Mandelbrot, B. (1989). "Fractals in Geophysics," Birkhäuser, Basel.

Shishikura, M. M. (1994). *Astérisque* **222**, 389–406.

Shlesinger, M., Zaslavsky, G., and Frisch, U. (1995). "Lévy Flights and Related Topics in Physics," Springer-Verlag, New York.

Sinai, Y. (1972). *Russ. Math. Surv.* **27**, 21–70.

Smale, S. (1963). *In* "Differential and Combinatorial Topology" (Cairns, S., ed.), pp. 63–80, Princeton University Press, Princeton, NJ.

Stauffer, D., and Aharony, A. (1992). "Introduction to Percolation Theory," 2nd ed., Taylor and Francis, London.

Strogatz, S. (1994). "Nonlinear Dynamics and Chaos, with Applications to Chemistry, Physics, Biology, Chemistry, and Engineering," Addison-Wesley, Reading, MA.

Sullivan, D. (1985). *Ann. Math.* **122**, 410–418.

Tan, Lei (1984). *In* "Étude dynamique des polynômes complexes" (Douardy, A., and Hubbard, J., eds.), Vol. II, pp. 139–152, Publications Mathematiques d'Orsay, Orsay, France.

Tan, Lei. (2000). "The Mandelbrot Set, Theme and Variations," Cambridge University Press, Cambridge.

Thurston, W. (1997). "Three-Dimensional Geometry and Topology," Princeton University Press, Princeton, NJ.

Tricot, C. (1982). *Math. Proc. Camb. Philos. Soc.* **91**, 54–74.

Vicsek, T. (1992). "Fractal Growth Phenomena," 2nd ed., World Scientific, Singapore.

Voss, R., and Clarke, J. (1975). *Nature* **258**, 317–318.

West, B. (1990). "Fractal Physiology and Chaos in Medicine," World Scientific, Singapore.

Weyl, H. (1912). *Math. Ann.* **71**, 441–479.

Williams, R. (1983). *Ergodic Theory Dynam. Syst.* **4**, 147–163.

Wiggins, S. (1990). "Introduction to Applied Nonlinear Dynamical Systems and Chaos," Springer-Verlag, New York.

Witten, T., and Sander, L. (1981). *Phys. Rev. Lett.* **47**, 1400–1403.

Witten, T., and Sander, L. (1983). *Phys. Rev. B* **27**, 5686–5697.

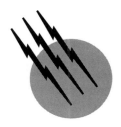

Fracture and Fatigue

K. M. Nikbin
Imperial College of Science Technology and Medicine

I. Introduction
II. High-Temperature Fracture Mechanics
III. Discussions and Conclusions

GLOSSARY

Fatigue Failure of a material through repeated action of cyclic stresses.
Stress intensity factor Scalar amplitude factor that uniquely characterizes the linear elastic crack tip stress and deformation fields.

THIS SECTION considers fatigue assisted crack growth at elevated temperatures. Components with preexisting defects in the creep range which undergo cyclic loading may invariably fail by crack growth due to either creep or fatigue. The mechanism of time-dependent deformation and fracture in creep as well as the time-independent fatigue are analyzed in terms of fracture mechanics parameters K and C^*. Cumulative damage concepts are then used for predicting crack growth under static and cyclic loading conditions.

I. INTRODUCTION

Most engineering components which operate at elevated temperatures are subjected to non-steady loading during service. For example, electric power plants may be required to follow the demand for electricity and equipment used for making chemicals may undergo a sequence of operations during the production process. The power plants have to change their operating temperature and pressure to follow the demands of electricity need and to shut down and re-start for their routine maintenance as depicted in Fig. 1. Also, aircraft experience a variety of loading conditions during take-off, flight, and landing. There may, in addition, be a superimposed high frequency vibration. Similarly, equipment that is subjected to predominantly steady operating conditions may experience transients during start-up and shutdown.

The gradual increase over the years in operating temperatures to achieve improved efficiency and performance from plants is causing materials to be used under increasingly arduous conditions. The first stage blades in aircraft gas-turbines can, for example, experience centrifugal stresses in the region of 150–200 MPa at gas temperatures that can exceed their incipient melting temperatures. Under these circumstances, survival is only possible if cooling is adopted to reduce average blade temperatures and coatings are used to limit erosion and environmental attack. The steep temperature gradients produced by the cooling will, however, introduce thermal stresses which will be regenerated each flight cycle and which can give rise to a mode of failure called thermal fatigue. The same situation can occur during rapid start-ups and shutdowns in thick sections of other components. Thermal fatigue

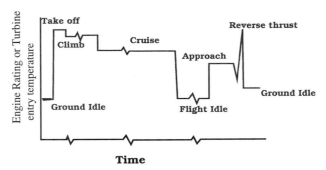

FIGURE 1 Example of operation scheme of a power plant.

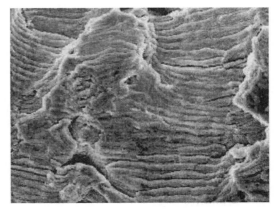

FIGURE 2 Example of striation and transgranular fatigue fracture surface of type 316 LN stainless steel tested at 650°C.

is the type of failure that can occur by the repeated application of predominantly thermal stresses that are produced by the local constraint imposed by the surrounding material.

It is apparent, depending on the material of manufacture and the operating conditions, that creep, fatigue, and environmental processes may contribute to failure. The dominating mode of failure in a particular circumstance will depend on such factors as material composition, heat-treatment, cyclic to mean load ratio, frequency, temperature, and operating environment.

In the preceding section, low temperature fatigue crack growth was discussed. For the prediction of the crack growth in components at elevated temperatures, the interaction of creep and fatigue crack growth must be evaluated appropriately. In this section, the characteristics of fatigue crack growth and creep crack growth are introduced separately. Then a prediction of crack growth under the combination of creep and fatigue will be outlined.

A. Characterisitics of Fatigue and Creep Crack Growth

As fatigue crack growth has been detailed fully in the previous section only elements relevant to fatigue and creep crack growth will be discussed here before considering creep-fatigue crack growth interaction. References [1–11] give various description and views of the problems involved in fatigue and creep/fatigue conditions. Fatigue mechanisms usually dominate at room temperature. Procedures for measuring the fatigue crack propagation properties of materials at room temperature are described in reference [8]. Fatigue crack growth is usually observed as transgranular cracking at low temperature (as shown in Fig. 2 and is characterized by the stress intensity factor range ΔK using the Paris law [1]. At elevated temperature, transgranular cracks are also observed under relatively high frequency cycles ($f > 1$ Hz) and this fatigue crack growth rate can still be characterized by elastic

or elastic-plastic fracture mechanics parameters in most cases. When the effect of plasticity can be neglected, the stress and strain states around a crack tip are characterized by K, and the crack growth per cycle da/dN is often plotted against stress the intensity factor range ΔK which is defined as the difference between the maximum K_{max} and minimum K_{min} stress intensity factors applied each load cycle. ΔK is given by;

$$\Delta K = Y \Delta \sigma \sqrt{a} \qquad (1)$$

where $\Delta \sigma$ is the range of the applied stress, and Y is the nondimensional geometry factor. The crack growth rate da/dN is correlated with ΔK using the power law relation [1] in steady-state fatigue as

$$(da/dN)_f = C \Delta K^m \qquad (2)$$

where C and m are material constants and m is typically around 3. Typically, da/dN is sensitive to the mean stress or the load ratio R defined by;

$$R = \frac{\sigma_{min}}{\sigma_{max}} \qquad (3)$$

where σ_{min} and σ_{max} are the minimum stress and the maximum stress, respectively [89]. To include the effect of the load ratio in Eq. (2), the effective stress intensity factor range ΔK_{eff} is used. Several formulae to define ΔK_{eff} are proposed using ΔK and R [1–12].

B. Types of Loading Cycle in Creep/Fatigue

A range of specimens and loading cycles is used in laboratory experiments to simulate different applications and operating conditions. Mostly, precracked compact tension specimens are employed to determine crack growth behavior, although sometimes thin, single-edge notch samples are used to represent blade aerofoil sections. In other circumstances natural cracks are allowed to develop in round

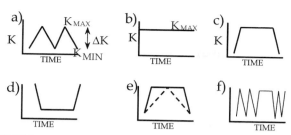

FIGURE 3 Representative loading cycles: (a) continuous loading, (b) static load, (c) hold at peak load, (d) hold at minimum load, (e) wave shape effects, (f) cyclic/static loading.

bar specimens. In these experiments load P, crack length a, number of cycles N, and sometimes displacement Δ are measured continuously throughout a test. Crack length is usually determined by optical, electric potential, compliance, or some combination of these methods [1–12]. Crack growth/cycle da/dN is then normally correlated in terms of linear or nonlinear fracture mechanics parameters.

Either load or displacement can be controlled in these tests. Load is controlled when different applied loading conditions are being represented. It is more usual to perform experiments between fixed displacement limits when simulating thermally induced stresses as these normally result from fixed boundary constraints. Sinusoidal, saw tooth, and trapezoidal wave shapes are commonly employed as indicated in Fig. 3. Crack growth behavior is then investigated over a range of frequencies f, hold times t_h and minimum to maximum load (or displacement) ratios R.

Normally these experiments are performed at constant temperature. The temperature chosen is that which is expected to cause the fastest cracking rate. However, particularly when it is intended to represent thermally induced stresses, both load and temperature may be cycled together. This cycling may be performed in phase or out of phase. For many start-up and shutdown situations compression tends to be produced at high temperature and tension at low temperature. In these cases direct electric current, or induction heating, with forced cooling often have to be employed to achieve the desired rates of change of temperature. These can sometimes be obtained by alternately immersing samples in hot and cold media when only thermally generated stresses are of interest.

C. Elevated Temperature Cyclic Crack Growth

In the previous section cyclic crack growth was considered where high-temperature time-dependent mechanisms were not relevant and where cracking was controlled mainly by fatigue mechanisms. As temperature is increased, time-dependent processes become more signifi-

cant. Creep and environmentally assisted crack growth can take place more readily since they are aided by diffusion and rates of diffusion increase with rise in temperature. The effects of temperature, frequency, mean stress, and environment will be considered in turn [3–12].

Frequency effects can be investigated in continuous cycling tests or by introducing hold periods into a cycle as illustrated in Fig. 3. However, it is found that wave shape is relatively unimportant compared with the temperature and mean load at which the cycling is taking place. Generally the influence of frequency on crack propagation rate is more pronounced with increase in temperature and R ratio. Figure 4 shows a schematic description of cracking rate versus ΔK for cyclic cracking at elevated temperatures showing the effects of frequency, R ratio and temperature. A higher crack propagation rate is observed with increase in load range. In all cases, crack propagation rate increases slowly at first before accelerating rapidly as final fracture is approached. As a consequence, most of the time in these tests is spent in extending the crack a small distance and the number of cycles to failure is not significantly influenced by the crack size at which fracture occurs.

Generally it is found that fatigue crack growth at room temperature is not affected by frequency, cycle wave shape, or specimen size (ignoring thin sheet material), but that stages I and III (Fig. 4) are sensitive to microstructure and mean stress effects whereas region II tends to be little affected by these factors unless crack closure occurs.

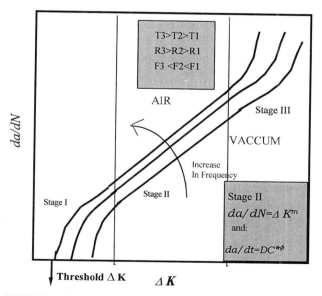

FIGURE 4 Schematic representation of cyclic crack growth at elevated temperatures showing three stages of cracking behavior. T = temperature, R = min/max stress ratio, and F = frequency.

Environment can influence all regions of the fatigue crack growth curve depending on the particular metal-fluid composition combination.

There is evidence of the relative insensitivity of stage II to microstructure [7] as all the crack growth results invariably lie within a reasonably narrow scatter band although different slopes m are obtained for each material. Generally, it is found that m is in the range 2–4 but can be as high as 7 [7].

Transgranular cycle-dependent and intergranular time-dependent controlled cracking processes are identified. Conditions favoring each mechanism are clarified and it is shown that cumulative damage concepts can be applied to predict interaction effects. It is found that cyclic controlled processes are most likely to dominate at high frequencies and low R. Creep and environmental mechanisms, which are favored at low frequencies, high temperatures, and high R, are identified as contributing to the time-dependent component of cracking. It is shown that these processes can enhance crack growth/cycle significantly and reduce component lives.

II. HIGH-TEMPERATURE FRACTURE MECHANICS

The arguments for correlating high temperature crack growth data essentially follow those of elastic-plastic fracture mechanics methods. For creeping situations [13, 14] where elasticity dominates, the stress intensity factor may be sufficient to predict crack growth. However, as creep is a nonlinear, time-dependent mechanism even in situations where small scale creep may exist linear elasticity may not be the answer. By using the J definition to develop the fracture mechanics parameter C^* it is possible to correlate time-dependent crack growth using nonlinear fracture mechanics concepts.

A simplified expression for stress dependence of creep is given by a power law equation which is often called the Norton's creep law and is comparable to the power law hardening material giving

$$\varepsilon = A\sigma^N \qquad (4)$$

and by analogy for a steady state creeping material

$$\dot{\varepsilon} = C\sigma^n \qquad (5)$$

where C and n are material constants and $\dot{\varepsilon}$ and σ are creep strain rate and applied stress, respectively. Equations (4, 5) are used to characterize the steady-state (secondary) creep stage where the hardening by dislocation interaction is balanced by recovery processes. The typical value for n is between 3 and 10 for most metals. When $N = n$ for creep and plasticity the assumption is that the state of stress is characterized in the same manner for the two

conditions. The stress fields characterized by K in elasticity will be modified to the stress field characterized by the J integral in plasticity in the region around the crack tip. In the case of large-scale creep where stress and strain rate determine the crack tip field the C^* parameter is analogous to J. The C^* integral has been widely accepted [13–15] as the fracture mechanics parameter for this purpose.

As the J integral characterizes the stress and strain state, the C^* integral is also expected to characterize the stress and strain rate around a crack. For a nonlinear creeping material, the asymptotic stress and strain rate fields are expressed by equations

$$\sigma_{ij} = \sigma_0 \left(\frac{C^*(t)}{I_n \sigma_0 \dot{\varepsilon}_0 r} \right)^{1/(n+1)} \tilde{\sigma}_{ij} \qquad (6)$$

$$\varepsilon_{ij} = \dot{\varepsilon}_0 \left(\frac{C^*(t)}{I_n \sigma_0 \dot{\varepsilon}_0 r} \right)^{n/(n+1)} \dot{\tilde{\varepsilon}}_{ij} \qquad (7)$$

Therefore the stress and strain rate fields of nonlinear viscous materials are also HRR [15] type fields with I_n being the nondimensional factor of n, and $\tilde{\sigma}_{ij}$ and $\tilde{\varepsilon}_{ij}$ are functions of angle θ and n, and are normalized to make their maximum equivalent stress and strain unity. By analogy also with the energy release rate definition of J, the C^* integral can be obtained from

$$C^* = \frac{1}{B} \frac{dU^*}{da} \qquad (8)$$

where B is the thickness, da is the crack extension, and U^* is the rate of change of the potential energy dU^*/dt. From the view of an energy balance, the C^* integral is the rate of change of potential energy with crack extension. Experimentally C^* can be calculated [16] from the general relationship

$$C^* = (P\dot{\Delta}_c/WB_n)F \qquad (9)$$

where $\dot{\Delta}_c$ is the load-line creep displacement rate, F is a nondimensional factor which can be obtained from limit analysis techniques, B_n is the net thickness of the specimen with side groove, and W is the width. In general, Eq. (8) is used to estimate the values of C^* for tests in the laboratory. Another method available is one based on reference stress concepts [17–18]. Reference stress procedures are employed to evaluate C^* for feature and actual component tests where the load-line deformation rate is not available. By determining

$$C^* = \sigma_{ref} \cdot \dot{\varepsilon}_{ref} \left(\frac{K}{\sigma_{ref}} \right)^2 \qquad (9)$$

where $\dot{\varepsilon}_{ref}$ is the creep strain rate at the reference stress, σ_{ref}, and K is the stress intensity factor. Usually it is most convenient to employ limit analysis to obtain σ_{ref} from

$$\sigma_{ref} = \sigma_y \frac{P}{P_{lc}} \qquad (10)$$

where P_{lc} is the collapse load of a cracked body and σ_y is the yield stress. The value of P_{lc} will depend on the collapse mechanism assumed and whether plane stress or plane strain conditions apply. This procedure will be employed to interpret the results of the tests on the thick-walled cylinder and thin-walled tube specimens.

Models for creep crack growth are available [13, 14] which describe steady-state crack growth rates in creep versus the parameter C^*. In most cases it has been found that testing and analysis of the data have been performed in ASTM E1457 [16] which describes crack growth rate $\dot{a}$ in terms of the creep fracture mechanics parameter C^*

$$\dot{a} = \frac{DC^{*\phi}}{\varepsilon_f^*} \qquad (11)$$

Typically D and ϕ are relatively insensitive to material properties and Eq. (11) is often written as

$$\dot{a} = \frac{D3C^{*0.85}}{\varepsilon_f^*} \qquad (12)$$

with $\dot{a}$ in mm/h and C^* in MPam/h, D and ϕ are materials parameters which can be derived from multiaxial creep data [13, 14]. In Eq. (10–11) ε_f^* is the ductility appropriate to the state of stress at the crack tip; the bounds [14] are usually taken to be the uniaxial failure strain (ε_f) for plane stress conditions and $\varepsilon_f/30$ for plane strain conditions. Several studies have been reported in the literature of the influence of specimen size and geometry on creep crack growth rate. There is evidence to suggest that an increase in size and the introduction of side grooves in relatively creep brittle materials [14] causes an increase in crack propagation rate at constant C^*, consistent with a transition from plane stress toward plane strain conditions. The situation is not so clear for more ductile conditions.

Therefore by using Eq. (2) it is assumed that when the fatigue component is the primary driving mechanism for crack growth, then the crack tip is dominated by the stress intensity factor K. For low frequencies where the creep dominates C^* as in Eq. (11) will best describe the stress field ahead of the crack and the use of Eq. (2) becomes increasingly more inaccurate.

A. Modeling Creep and Creep/Fatigue Crack Growth

When alternating loads are applied to high-temperature structures, the crack growth in the structures will be subject to both creep and fatigue. Interaction between creep

and fatigue is expected under cyclic loading. Some of the causes of creep-fatigue interaction might be the enhancement of fatigue crack growth due to embrittlement of grain boundaries or weakening of the matrix in grains and enhancement of creep crack growth due to acceleration of precipitation or cavitation by cyclic loading [4–5]. The importance of creep-fatigue interaction effects is largely dependent on the material and loading conditions. Nevertheless, the simple linear summation rule for creep-fatigue crack growth defined by the following equation has been successfully applied to predict the total crack growth/cycle (da/dN) for several engineering metals [9–11] is given as

$$\frac{da}{dN} = \left(\frac{da}{dN}\right)_f + \left(\frac{da}{dN}\right)_c \qquad (13)$$

where subscripts f and c define the creep and fatigue mode of crack growth. Equation (13) can be rewritten as;

$$\left(\frac{da}{dN}\right) = C\Delta K^m + \frac{\dot{a}}{f} \qquad (14)$$

where f is frequency and $\dot{a}$ is the creep component of cracking which can be determined from Eq. (12) or any other models of creep crack growth [13], and $(da/dN)_f$ can be derived from Eq. (2). Therefore by determining the crack growth rate according to Eqs. (2, 10) Eq. (13) becomes

$$\frac{da}{dN} = C\Delta K^m + \int_0^{t_h} \dot{a}[C^*(a)]\,dt \qquad (15)$$

where the value of the parameter C^* can be taken at maximum load at the start of the dwell as an upper bound but expressions are available for its determination under displacement-controlled conditions. The two contributions in Eq. (15) are summed over the operating cycles, updating the crack size as necessary. This linear cumulative damage law has been used extensively in the analysis and assumes little or no interaction between the time-dependent creep and the time-independent fatigue component of crack growth.

Effects of the influence of frequency [9–11] on crack growth/cycle in a nickel-base alloy (AP1) at 700°C are shown in Figs. 5 and 6. Figure 5 indicates a dependence of da/dN on frequency at $R = 0.7$. In the steady-state cracking region crack growth can be described by the Paris law [Eq. (2)] with $m \approx 2.5$. This value is within the range expected for room temperature behavior. Correlation of results obtained at 20 MPa$\sqrt{m}$ for specimen thickness ($B = 25$ mm) is depicted in Fig. 5. Furthermore in Fig. 6 the two lines show the relative behavior of the creep and the fatigue components of da/dN.

The slope of -1 indicates time-dependent creep cracking and the horizontal line indicates fatigue control of

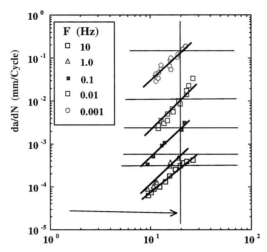

FIGURE 5 Frequency dependence of fatigue cracking at high temperatures for AP1 nickel-base superalloy tested at 700°C and $R = 0.7$.

da/dN. Therefore by adding the two components it is clear that at high-frequency creep is seen to have third-order effect on cracking rate, and conversely at low frequencies fatigue has a third-order effect. From metallurgical and fractographic investigations performed on the alloy tested in the creep and creep/fatigue range similar qualitative conclusions can be reached with respect to the mode of the creep/fatigue interaction. Figure 7 shows the fractographs for the nickel-base superalloy AP1 tested at 700°C. There is a transition from intergranular cracking at $f = 0.001$ Hz to transgranular cracking at 10 Hz.

The intermediate frequencies show a mixture of inter- and transgranular cracking modes. These suggest that the

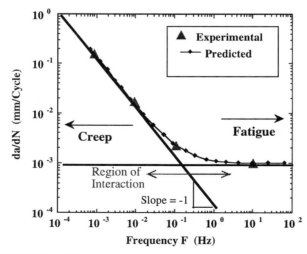

FIGURE 6 Fatigue crack growth sensitivity to frequency at constant $\Delta K = 20$ MPa$\sqrt{m}$ in an AP1 nickel-base superalloy tested at 700°C. [From Winstone, M. R., Nikbin, K. M., and Webster, G. A. (1985). *J. Mater. Sci.* **20**, 2471–2476.]

Surface Micrographs in Creep/Fatigue Interaction

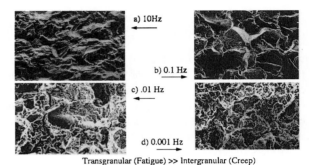

FIGURE 7 Effects of frequency on mode of failure for AP1 astroloy nickel-base superalloy tested at 700°C. [From Winstone, M. R., Nikbin, K. M., and Webster, G. A. (1985). *J. Mater. Sci.* **20**, 2471–2476.]

two mechanisms work in parallel and that cumulative damage concepts proposed above can well describe the total cracking behavior.

The effects of R ratio on creep/fatigue interaction is shown in Figs. 8 and 9. Generally as shown in Fig. 8, an increase in the R ratio reduces the ΔK needed for crack growth per cycle and increase in frequency reduced the da/dN. Figure 9a,b compares the da/dN in terms of frequency in terms of constant ΔK and constant K_{max}. Figure 9a shows a dependence of da/dN versus frequency on R ratio. It is clear that at low frequencies cracking is independent of R ratio when plotted in terms of K_{max} as shown in Fig. 9b. This suggests that creep failure dominates at maximum load for low frequencies.

In both Figs. 6 and 9 it is clear that the interaction region of creep fatigue is contained within a small frequency band of 1 decade in the region of 0.1–1 Hz. From experimental

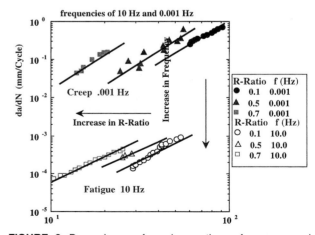

FIGURE 8 Dependence of crack growth on frequency and R ratio for an AP1 nickel-base superalloy tested at 700°C.

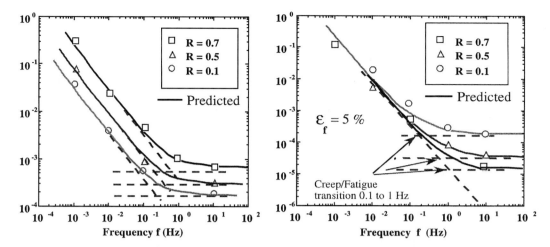

FIGURE 9 Fatigue crack growth sensitivity to frequency and R ratio in an AP1 nickel-base superalloy tested at 700°C at (a) constant $\Delta K = 30$ MPa$\sqrt{m}$ and (b) constant $K_{max} = 30$ MPa$\sqrt{m}$.

and data analysis performed [6] it has been shown that static crack growth data correlates well with low frequency (=0.001 Hz) data. Therefore Eqs. (13–15) can be used to incorporate static data with cyclic data. This method is useful for cases where no cyclic data exists and only static and high-frequency data is available.

The effects of temperature and the predictions of creep/fatigue using static data are shown in Fig. 10. This figure shows for a range of test temperatures both static and cyclic data for a martensitic alloy FV448. While the high-frequency data varies very little with temperature and frequency, the effects of predicted static and low frequency data shift to lower frequencies

indicating that fatigue will dominate failure at low frequencies.

The interpretations of elevated temperature cyclic crack growth behavior have so far been presented mainly in terms of linear elastic-fracture mechanics concepts. It is anticipated that this approach is adequate when fatigue and environmental processes dominate as stress redistribution will not take place at the crack tip. When creep mechanisms control, stress redistribution will occur in the vicinity of the crack tip and use of the creep fracture mechanics parameter C^* becomes more appropriate.

In the region where the Paris law is relevant, substitution of the C^* relation to crack growth into Eq. (15)

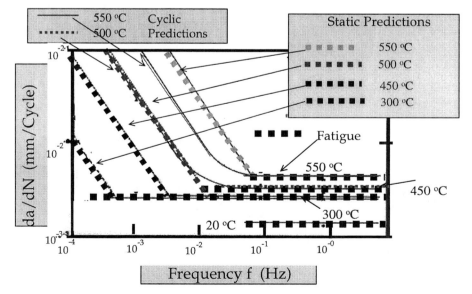

FIGURE 10 Prediction of da/dN for a Martensitic steel from static data over a temperature range of 20–550°C using static and cyclic data at a constant $\Delta K = 50$ MPa$\sqrt{m}$.

allows cyclic crack growth under creep/fatigue loading conditions to be established from

$$da/dN = C\Delta K^m + DC^{*\phi}/f \qquad (16)$$

Alternatively, if the approximate expression in Eq. (12) is employed for the creep component of cracking, crack growth/cycle becomes,

$$da/dN = C\Delta K^m + 3C^{*0.85}/\varepsilon_f^* \cdot 3600 f \qquad (17)$$

and for static loading

$$da/dN = C\Delta K^m + 3C^{*0.85}/\varepsilon_f^* \cdot 7200 f \qquad (18)$$

where da/dN is in mm/cycle with frequency in Hz. To allow for crack closure effects, ΔK in these equations can be replaced by ΔK_{eff}. Similarly ΔJ can be used instead of ΔK when plastic deformation is significant. In order to make predictions of creep/fatigue crack growth in components it is necessary to be able to calculate ΔK and C^* as crack advance occurs. The same procedures that are employed for estimating K and C^* under static loading can be employed.

The effects of constraint due to geometry and specimen size can be described using the cumulative method and over a range of frequencies or using C^* at low frequencies where creep dominates. Figure 11 shows the effects of geometry and specimen size on crack growth at constant ΔK over a range of frequencies for the AP1 alloy at 700°C.

At high frequencies there are no size and geometry effects, however, at low frequencies the corner crack tension (CCT) specimens, shown in Fig. 12a exhibit lower da/dN compared to the thin CT specimens, and this in turn shows a lower da/dN compared to the 25-mm thick CT. Therefore constraint effects are exaggerated under creep conditions and increase in constraint tends to increase cracking rates.

The effects of geometry on crack growth for the static and low frequency data are plotted against C^* in Fig. 13. It is clear that cracking rate is state of stress controlled in the creep range. Figure 12b shows a fractograph exhibiting the effects of creep and fatigue on the actual cracking of the CCT. In this case the frequency was varied between 10 and 0.001 Hz. Where fatigue dominates the crack front profile is approximately a quadrant and the cracking rate is the same in the inner section and the surface. When frequency is reduced the crack leads in the center and where the surface is more in plane stress crack growth rate is reduced. The effect is reversed when the frequency is increased once again.

Finally it has been shown that high-temperature crack growth occurs by either cyclic-controlled or time-dependent processes. Over the limited range where both mechanisms are significant, a simple cumulative damage law can be employed to predict behavior. Interpretations have been developed in terms of linear elastic and non-linear fracture mechanics concepts. Linear elastic fracture mechanics descriptions are expected to be adequate when fatigue and environmental processes dominate. When creep mechanisms control stress redistribution takes place in the vicinity of the crack tip and use of the creep fracture mechanics parameter C^* should be employed for characterizing the creep component of cracking.

III. DISCUSSIONS AND CONCLUSIONS

A. Relevance to Life Assessment Methodology Under Creep Fatigue

In this section the relevance of what has been discussed above is highlighted by describing the practical problems associated with creep and creep/fatigue in plant components. For new plant design procedures are required to avoid excessive creep deformation and fracture in electric power generation equipment, aircraft gas turbine engines, chemical process plant, supersonic transport, and space vehicle applications. There is also a continual trend toward increasing the utilization and efficiency of plant for economic reasons. This can be achieved in a number of ways. One is to fabricate components out of new materials with improved creep properties. Another is to increase the temperature of operation which may itself require the introduction of new materials.

By extending plant life and to avoid premature retirement on the basis of reaching the design life which may have been obtained previously using conservative procedures is an economic reality for most industrial operators.

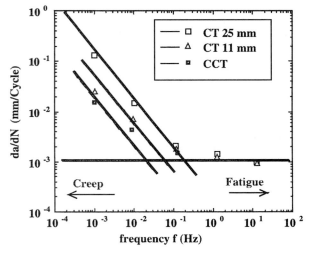

FIGURE 11 Geometry and size effects in the creep/fatigue crack growth of AP1 nickel-base superalloy tested at 700°C.

Corner Cracked Tension CCT

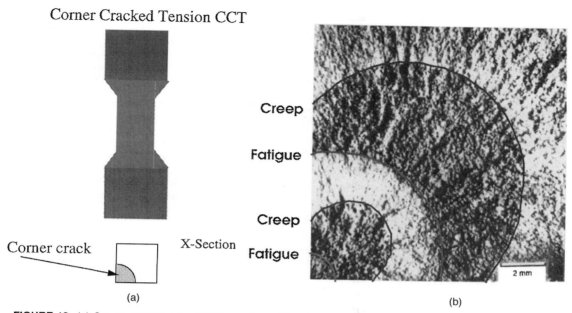

FIGURE 12 (a) Corner crack tension (CCT) specimen AP1 superalloy tested at 700°C, (b) a fractograph showing influence of creep and fatigue on the shape and texture of the crack front.

Many electric power generation and chemical process facilities are designed to last for 30 years or more assuming specific operating conditions. It is possible that if they are used under less severe conditions to those anticipated or a re-evaluation using the modern understanding of creep and fracture is performed, it may justify a further extension of the plant life.

To avoid excessive creep deformation and prevent fracture are the two important high-temperature design considerations. In most industries these objectives are achieved

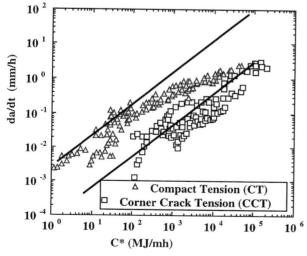

FIGURE 13 Effect of geometry in creep crack growth of AP1 nickel-base superalloy tested at 700°C.

by the application of design codes which incorporate procedures for specifying maximum acceptable operating stresses and temperatures. The maximum values allowed depend upon the type of component and consequences of failure. The magnitudes chosen for the safety factors are determined by experience and are dependent on the type of calculation performed and whether average or minimum properties data are employed. Different safety factors may also be applied to normal operating conditions, frequent, infrequent, or emergency excursions.

Most high-temperature design codes [19, 20] have been developed from those that have been produced for room temperature applications. They are therefore aimed at avoiding failure by plastic collapse, fatigue, and fast fracture as well as creep and it is possible to define temperatures, which vary somewhat between codes because of the different procedures employed, below which creep need not be considered for particular classes of materials. Guidance has been in existence for some to guard against fast fracture and fatigue below the creep range and is now becoming available for elevated temperature situations where creep is of concern.

The high-temperature design codes in current use have been developed principally for application to defect free equipment. There is an increasing trend for critical components to be subjected to nondestructive examination to search for any possible flaws. Flaws can be detected by visual, liquid penetrant, magnetic particle, eddy current, electrical potential, radiographic, and ultrasonic means depending on whether they are surface breaking or buried.

Consequently there is a requirement for establishing tolerable defect sizes. Also the improving sensitivity of these techniques is causing smaller and smaller flaws to be found and the question of whether they must be removed, repaired, or can be left is being encountered more frequently.

Finally during operation different equipment can experience a wide range of types of loading. For example, stress and temperature can be cycled in-phase or out-of-phase, a stress can be applied or removed at different rates and dwell periods can be introduced during which stress or strain are maintained constant. In these circumstances, it is important that the processes governing failure are properly understood so that the appropriate predictive models and equations can be employed.

SEE ALSO THE FOLLOWING ARTICLES

CORROSION • ELASTICITY • EMBRITTLEMENT, ENGINEERING ALLOYS • PLASTICITY (ENGINEERING)

BIBLIOGRAPHY

Paris, P. C. (1977). Fracture Mechanics in the Elastic Plastic Regime, ASTM STP, 631, pp. 3–27, American Society for Testing and Materials, Philadelphia.

Forman, R. G., Kearney, V. E., and Engle, R. M. (1967). Numerical analysis of crack propagation in a cyclic-loaded structures, ASME Transaction, *J. Basic Engi.* **89**(D), 459.

Paris, P. C., Gomez, M. P., and Anderson, W. E. (1961). A Rational analytic theory of fatigue, *Trend Engi.* **13,** 9–14.

Kaneko, H., *et al.* (1962). "Study on Fracture Mechanism and a Life Estimation Method for Low Cycle Creep-Fatigue Fracture of Type 316 Stainless Steels in Low Cycle Fatigue and Elastic-Plastic Behaviour of Materials —3′," Elsevier Applied Science, Berline.

Nakazawa, T., *et al.* (1992). "Study on Metallography of Low Cycle Creep Fatigue Fracture of Type 316 Stainless Steels in Low Cycle Fatigue and Elastic-Plastic Behaviour of Materials —3," Elsevier Applied Science, Berline.

Nikbin, K. M., and Webster, G. A. (1987). Prediction of Crack Growth under Creep-Fatigue Loading Conditions, STP 942, pp. 281–292, American Society for Testing and Materials, Philadelphia.

Kennedy, A. J. (1962). "Processes of Creep and Fatigue in Metals," Wiley, New York.

ASTM E647-86a (1987). Standard Test Method for Measuring Fatigue Crack Growth Rates, Book of Standards, pp. 899–926, American Society for Testing and Materials, Philadelphia.

Nikbin, K. M., and Webster, G. A. (1984). *In* "Creep and Fracture of Engineering Materials and Structures" (B. Wilshire and D. R. J. Owen, eds.), pp. 1091–1103, Pineridge Press, Swansea.

Winstone, M. R., Nikbin, K. M., and Webster, G. A. (1985). Modes of failure under creep/fatigue loading of a nickel-base superalloy, *J. Mater. Sci.* **20,** 2471–2476.

Dimopulos, V., Nikbin, K. M., and Webster, G. A. (1988). Influence of cyclic to mean load ratio on creep/fatigue crack growth, *Met. Trans. A* **19A,** 873–880.

Nikbin, K. M., Smith, D. J., and Webster, G. A. (1984). Prediction of creep crack growth from uniaxial creep data, *Proc. R. Soc. London* **A396,** 183–197.

Nikbin, K. M., Smith, D. J., and Webster, G. A. (1986). An engineering approach to the prediction of creep crack growth, *J. Eng. Mater. Technol.* **108,** 186–191.

Saxena, A. (1984). "Crack Growth under Non Steady-State Condition," in 17th ASTM National Symposium on Fracture Mechanics, Albany, N.Y.

Hutchinson, J. W. (1968). Singular behaviour at the end of a tensile crack in a hardening material, *J. Mech. Phys. Solids* **16,** 13–31.

ASTM (1998). Standard Test Method for Measurement of Creep Crack Growth Rates in Metals, ASTM E1457-98.

Ainsworth, R. A. (1984). The Assessment of defects in structures of strain hardening material, *Eng. Fracture Mech.* **19**(4), 633–642.

Webster, G. A., Nikbin, K., Chorlton, M. R., Celard, N. J. C., and Ober, M. (1998). A comparison of High temperature defect assessment methods, *J. Mater. High Temp.* **15**(3/4), 337–347.

Ainsworth, R. A., Chell, G. G., Coleman, M. C., Goodall, I. W., Gooch, D. J., Haigh, J. R., Kimmins, S. T., and Neate, G. J. (1987). CEGB assessment procedure for defects in plant operating in the creep range, *Fatigue Fract. Eng. Mater. Struct.* **10,** 115–127.

Drubay, B., Moulin, D., Faidy, C., Poette, C., and Bhandari, S. (1993). Defect assessment procedure: a French approach, ASME PVP, Vol. 266, 113–118.

Froth Flotation

Richard R. Klimpel

Dow Chemical Company

GLOSSARY

Activators Chemicals that enhance the collector attachment to an otherwise difficult to float mineral species.

Cleaner bank Group of cells that perform an additional flotation upgrading of concentrate from an earlier bank of cells in the plant.

Collector Chemical reagent that produces the hydrophobic film on the valuable mineral particles.

Contact angle Angle formed by an air bubble on a solid mineral submerged in water.

Depressants Chemicals that prevent an undesired mineral species from floating with a given collector.

Double layer Tendency of opposite electrical charges to line up at a solid–water interface.

Flotation cell Unit piece of equipment that holds the pulp, creates the bubbles by aeration, collects the froth phase, and removes the froth phase.

Froth phase Collection of air bubbles, including the attached mineral particles.

Frother Chemical reagent that influences the collision frequency between particles and bubbles as well as the efficiency of particle–bubble attachment.

Gangue That portion of the feed to flotation that is made up of undesired minerals.

Grade Fraction or percent of the total mass recovered in a flotation operation that is actually valuable material.

Liberation Degree of separation of valuable from gangue, with higher liberation occurring at finer particle sizes.

Middlings That size range of particles that are capable of floating but are often not completely liberated, thus often requiring more grinding.

Modifying agents Series of chemicals that assist in the flotation process by controlling pH or changing the particle surface characteristics or changing the pulp fluidity character, and so on.

Recovery Fraction or percent of total valuable in the feed to flotation that is removed in the froth phase.

Rougher bank Group of cells that perform the initial rough flotation of feed pulp after size reduction.

Scavenger bank Group of cells that operate on the

tailings from other portions of the flotation plant in order to prevent loss of valuable in the upgrading stages.

Selectivity Degree to which one or more valuable minerals can be separated from the remainder of the materials present in the feed ore.

Slime material Particles that are too small in size to be efficiently processed by flotation (usually less than 10–30 μm).

Tailings That material that is rejected (not floated) after a particular bank of cells.

Valuable That portion of the feed to flotation that is desired to end up in the froth phase for removal.

FROTH FLOTATION is a physicochemical process for separating finely ground minerals from their associated gangue material. The process involves chemical treatment of a finely divided ore in a water pulp to create conditions favorable for the attachment of certain of the mineral particles to air bubbles. The air bubbles then carry the selected minerals, called valuable, to the surface of the pulp to form a stabilized froth, which is removed and recovered. The unattached gangue material remains submerged in the pulp and is either discarded or reprocessed. Although flotation was originally developed in the mineral industry, the process has been extended to other applications, such as the recovery of bitumen from tar sands, the removal of solids from white water in paper making, the separation of ink from repulped paper stock, and the removal of oil or organic contaminants from aqueous solutions.

I. APPLICATIONS

It is in the separation of various minerals from naturally occurring ores that flotation demonstrates its greatest utility. The flotation process is the most efficient, most widely used, and most complex of all mineral concentration methods in use by the mining industry. Froth flotation is the principal means of processing copper, molybde-num, nickel, lead, zinc, phosphate, and potash ores. Also amenable to flotation are fine coal and other mineral commodities such as fluorspar, glass sand, barite, pyrite, talc, and iron oxide. Globally, an estimated two billion tons of solids are treated per year by this process. In theory, it can be applied to any mixture of particles that is sufficiently liberated from one another and small enough in size to be lifted by rising gas bubbles.

The flotation process encompasses a series of steps: grinding the ore to a size giving reasonable liberation of valuable and gangue materials; making conditions favorable for the adherence of the desired minerals or valuable to air bubbles, which is usually done by chemical means; creating a rising current of air bubbles in the ore pulp; forming a valuable laden froth on the surface of the ore pulp; and removing the valuable laden froth. Although size reduction of the ore is not, strictly speaking, a part of flotation, grinding does play a major role in flotation. This is true because of the relatively higher cost of size reduction over flotation.

Thus, in industrial practice, a complex optimization problem exists involving flotation and grinding, which often leads to less than optimal liberation and particle sizes for the flotation process itself. Generally, optimal flotation performance requires that particle sizes of the valuable material lie in a range between several hundred micrometers (μm) in diameter on the coarse side to 20 μm or so on the fine side. The efficiency of the flotation process is strongly influenced by particle size, with recovery of valuable dropping off considerably when the valuable is contained in slime particles produced by too much grinding. The recovery of valuable-containing particles that are too coarse in size also drops off but may not be as crucial, as this type of material can be reclassified by size, reduced in size again, and reprocessed. Figure 1 gives a typical recovery response of a copper sulfide ore by the froth flotation process as a function of particle size.

The creation of a rising current of air bubbles is accomplished by a flotation machine (cell), which produces bubbles by mechanical agitation of the ore pulp, the direct introduction of air under pressure, or both. These operations

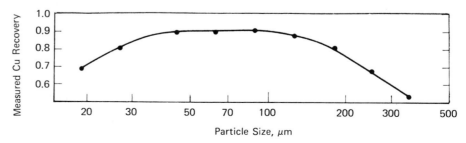

FIGURE 1 The recovery of copper as a function of mineral contained in each particle size.

may be considered the mechanical adjuncts of the flotation process.

To obtain adherence of the desired mineral particles to the air bubbles, two specific steps must occur: a hydrophobic (water-hating) surface film must be formed on the valuable particles to be floated, along with a hydrophilic (water-loving) film on all other gangue particles; and a controlled bubble surface tension interface must be maintained, allowing for high particle–bubble collision frequency and efficient attachment or sticking, of the particle to the bubble once collision has taken place. In most flotation applications, the above two steps are controlled by chemical flotation reagents. The collector is a chemical reagent that produces the hydrophobic film on the valuable mineral particle and is the primary driving force that initiates the flotation process. The frother is a chemical reagent that influences the collision frequency and attachment efficiency of hydrophobic particles and air bubbles. Members of a third class of chemical reagents, called modifying agents, represent chemicals that perform one or more functions, such as controlling pH, changing the particle surface characteristics, altering surface charge, or changing the pulp fluidity character.

The adjustment of the flotation process to selectively remove one or more valuable minerals from an often complex mixture of many minerals in the feed ore is often the most difficult part of the industrial practice of flotation. While flotation-cell equipment and operating parameters such as particle size, pulp density, and pulp temperature are important, the selectivity of the flotation process in any given application is generally dominated by the makeup of the various minerals constituting the feed ore (the ore mineralogy) and by the choice and dosages of the chemical reagents. As chemistry is the primary control mechanism available to a flotation-process operator, the remainder of this article deals chiefly with the chemical reagent aspects of flotation as a function of ore mineralogy. The chemical side of flotation has been extensively researched by chemical reagent supply companies such as the Dow Chemical Company and the American Cyanamid Company (historically, the two major global mining chemical vendors), as well as by numerous academic and independent research institutions.

II. FUNDAMENTALS

The flotation separation of one mineral species from another depends on the relative water-wetting ability of the surfaces. Most of the current chemical collectors in use are heteropolar. That is, they contain both a polar (charged) group and a nonpolar (uncharged) group. When attached to the mineral particle, these surface-active collector molecules are oriented so that the charged end of the molecule is attached (adsorbed) to the mineral surface and the nonpolar or hydrocarbon group is extended outward, giving a hydrophobic film to the particle. Typically, the surface free energy is lowered by the adsorption of heteropolar surface-active reagents so the hydrophobic film then can act as a bridge so that the particle may attach to an air bubble.

The basic principles of surface chemistry have been applied extensively to the understanding of flotation phenomena. While some success has been achieved in fundamental studies, it is generally not possible to quantitatively predict flotation phenomena very well from first principles. This is not surprising in light of the complex nature of flotation, involving several heterogeneous phases acting under dynamic conditions in the presence of complex mass and momentum transfer conditions. Thus, the practice of flotation is one that takes advantage of the qualitative guidelines offered by surface chemistry studies, coupled with actual flotation testing of reagents and other conditions in various sizes of flotation equipment. Some of the surface phenomena that have shown useful correlations with flotation practice will now be briefly reviewed.

The first is the measurement of contact angle, which is the angle θ formed by an air bubble on a solid submerged in water. An air bubble when put into contact with a clean mineral surface does not adhere to the surface. If, however, an appropriate chemical is added, the mineral acquires a hydrophobic film, or coating, and an air bubble may be readily attached, as shown in Fig. 2. The presence of the collector significantly increases the contact angle, indicating enhanced particle–bubble contact.

Another fundamental approach that has shown usefulness in explaining flotation behavior and in the selection of an appropriately charged collector is the concept of an electrical double layer. Mineral particles in aqueous solution possess an electrical charge because an excess of cations or anions exists at the solid surface due to solid defects, dissociation of surface groups, unequal dissolution of lattice ions, and so forth. Figure 3 is a representation of the electrical double layer existing at a negatively charged mineral surface. Attracted to this surface and extending into the solution are counterions—in this case, cations—that maintain electroneutrality. Potential-determining ions can be either part of the mineral structure or an ion from the solution adsorbed onto the surface. The double layer extends into the bulk of the solution for varying distances, depending on the valence and concentration of the counterions.

The potential energy measured in moving a particle through a solution is the potential at the slipping plane and is denoted as the zeta potential, ζ. The zeta potential is not the surface potential ψ, but is the potential at

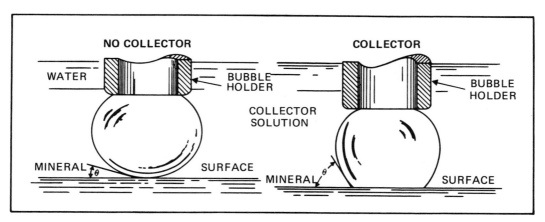

FIGURE 2 Contact angle with and without the presence of a collector.

some distance from the surface. The sign of the charge on the surface, as measured by zeta potential, is very important, and the valence and concentration of potential-determining ions controls the sign and magnitude of the surface potential. A point (or points) of zero charge (isoelectric point) can thus be measured for each mineral, depending on conditions. For example, with the mineral cerargyrite, AgCl, the potential determining ions are Ag^+ and Cl^-. The AgCl mineral is positively charged in solutions containing greater than 10^{-4} mol Ag^+ per liter, requiring an anionic collector, and negatively charged in solutions containing Cl^- in excess of 2×10^{-6} mol/liter, requiring a cationic collector, and the isoelectric point occurs at a pAg of 4.

For insoluble oxide minerals, H^+ and OH^- are the potential-determining ions, so that these minerals are positively charged at pH values below the isoelectric point and negatively charged at higher pH values. Polyvalent cations such as Al^{+3} and Fe^{+3} are also known to alter the surface charge of a mineral in aqueous solutions. This is especially prevalent near the pH of Al^{+3} and Fe^{+3} ion precipitation in dilute solutions, causing specific adsorption on a mineral surface.

In recent years, electrochemical techniques have been successfully used to explain the behavior of thiol collectors with sulfide minerals. Briefly, such work has shown that in this particular type of flotation, the transformation of the sulfide mineral surface to a more hydrophobic state involves simultaneous electrochemical reactions in which the cathodic step, the reduction of oxygen, is coupled with an anodic step involving the oxidation of the collector. Typical anodic reactions are chemisorption of a thiol ion (X^-),

$$X^- = X \text{ (adsorbed)} + e^-;$$

oxidation of a thiol ion to its disulfide,

$$2X^- = X_2 + 2e^-;$$

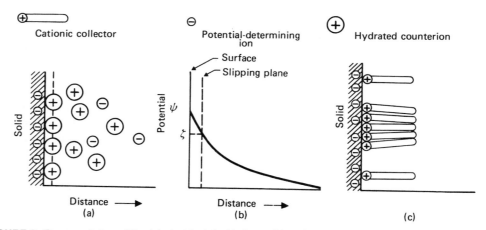

FIGURE 3 Representation of the (a) electrical double layer, (b) surface potential, and (c) collector adsorption.

and formation of a metal thiol compound with a metal portion of the sulfide mineral,

$$MX + 2X^- = MX_2 + S + 2e^-.$$

It is becoming more evident that interfacial electrochemical potential is the governing factor in sulfide mineral systems using thiol collectors. At reducing potentials, the sulfide minerals are depressed, and with increasing oxidizing conditions, the same minerals are floated in the presence of thiol collectors. The location where each sulfide mineral type will start to float as a function of electrochemical potential is relatively easy to determine in an idealized laboratory test using pure minerals and well-controlled water chemistry conditions. For example, Table I gives an example of such a laboratory determination on a series of sulfide minerals using sodium ethyl xanthate collector (see Section III of this article) at a pH of 8.7. In a plant environment, this type of analysis is somewhat more difficult to perform, but nevertheless the general concepts are still valid.

The detailed understanding of the mechanisms involved with the attachment of collectors to various mineral surfaces is far from complete at this time. The term "adsorption" is sometimes used to indicate a multitude of phenomena, ranging from physical adsorption to chemisorption to chelation, depending on the particular system under study. In some practical mixed mineral systems, it is quite apparent that at least several simultaneous mechanisms are occurring with a single given collector. Attachment

by chemisorption is generally desired because it leads to greater selectivity at lower dosage (hence, lower cost). Physical adsorption of the collector may involve van der Waals, hydrogen, or hydrophobic bonding, with electrostatic attraction between an ionic collector and a mineral of opposite charge being the most common mechanism.

Currently, strictly chelation-oriented chemistries functioning as collectors are being evaluated for potential commercial use, but in practice their use at present is limited, primarily because of their higher cost. It is quite likely that many of the existing collector reagents, such as the thiol collectors, also exhibit what may be termed chelation activity under certain conditions. In addition, some reagents undergo rearrangements at the mineral surface, leading to further complications in mechanism identification.

The differences in performance of flotation collectors in idealized systems (one mineral, pure water, air as a bubble former, etc.) compared to actual flotation systems (many minerals, highly impure water—even including sea-water in some processes, nitrogen or oxygen enrichment of air, etc.) are great. Therefore, empirical testing in small laboratory flotation cells, followed by large-scale plant verification, has played a significant role in collector chemistry development and use. The plant use of collector chemistry is extremely complicated and involves collector interactions with dosage, frother type, mineral particle size, time rate of recovery, and so on. This will be discussed further in Section VI.

The role of frothers is also not well understood from a strictly fundamental characterization. Recent work has shown that a major role of a frother is to increase the rate of valuable recovery. This is accomplished by the formation of a froth of consistent character (consistent bubble size and bubble density) under a variety of operating conditions, an increase in the ability to disperse air in the flotation cell, a reduction in the rate of coalescence of individual bubbles in the cell, and a decrease in the rate of bubble rise to the pulp surface. In terms of stability, a successful frother must achieve a delicate balance between allowing sufficient thinning of the liquid film between the colliding bubble and particle so that attachment can take place in the time frame of the collision. At the same time, the frother must provide sufficient stability of the bubble-particle moiety to allow the weakly adhering or mechanically trapped particles of unwanted gangue materials to escape with the draining liquid.

A common feature of most commercial frothers is their heteropolar nature, consisting of nonionic polar group(s) exhibiting hydrophilic character, coupled with a hydrophobic nonpolar character. It is generally accepted that adding a surface-active agent to water lowers the surface tension of the solution as a result of the heteropolar nature of the molecules. This is because the molecules

TABLE I An Example of the Influence of Electrochemical Potential on the Recovery of Various Sulfide Minerals in Presence of a Thiol Collector

Mineral	Potential[a]	% Mineral recovered
Chalcocite (Cu_2S)	−0.4	0.0
	−0.3	3.1
	−0.2	11.0
	−0.1	41.7
	0	85.3
	+0.1	96.1
	+0.2	100.0
Chalcopyrite ($CuFeS_2$)	−0.1	0.0
	0	2.4
	+0.1	21.3
	+0.2	75.6
	+0.3	100.0
Pyrite (FeS_2)	+0.1	0.0
	+0.2	5.3
	+0.3	42.7
	+0.4	82.3

[a] Potential in volts vs a standard reference electrode.

are preferentially adsorbed at the air–water interface, with the polar or hydrophilic group situated in the water phase and the nonpolar or hydrophobic chain in the vapor phase. Thus, the surface tension of a solution containing frother is a partial indication of the activity of a frother, with chemicals that strongly lower the surface tension often producing more stable (persistent) froths.

It will be demonstrated in Section VI that the major influence of changing frother structure appears to be in the effective particle size range recoverable by that frother.

III. CLASSES OF CHEMICAL REAGENTS

Typical groups of flotation collectors are shown in Fig. 4. For industrial sulfide mineral flotation, the xanthates and dithiophosphates represent the major classes of compounds used where M^+ is either Na^+ or K^+ and R is a relatively short hydrocarbon chain, typically C_2 to C_5 in length. The solubility, pH application range, physical state (solid or liquid), and adsorption characteristics for different minerals vary widely for different compounds within Groups 1 and 2 of Fig. 4, indicating the difficulty of selecting an optimal collector from *a priori* reasoning. Typical sulfide collector dosages range from 0.01 to 0.1 kg/metric ton of dry feed to flotation, with some applications ranging up to 0.5 kg/metric ton.

Specifically, the uptake of xanthate by a given sulfide mineral has been shown in idealized laboratory tests to depend on the rest potential of that mineral. The reduction potential for the formation of the disulfide (dixanthogen) is +0.13 V, so that those minerals having a rest potential greater than +0.13 V, such as is the case with chalcopyrite, molybdenite, pyrite, and pyrrhotite, xanthate ions are

oxidized at the mineral surface to dixanthogen, and this is most likely the chemical form imparting hydrophobicity to the mineral. The reaction product at the mineral surface for those minerals having less than +0.13 V, such as is the case with galena, chalcocite, and bornite, is most likely the simple metal xanthate. Obviously, from a practical viewpoint, the effectiveness of the hydrophobic character imparted by the relatively short hydrocarbon chain thiol collectors of Group 1 of Fig. 4 implies that the hydrophobic contribution of the sulfur atoms of the collector and those of the mineral in whatever surface arrangement may be present is extremely important and necessary.

For nonsulfide mineral flotation, the range of collectors used is again variable, depending on ore type and operating conditions, with the hydrocarbon chain length typically being from C_{12} to C_{22}. In the case of nonsulfide flotation, the dosages are somewhat higher than in sulfide mineral flotation, with amounts of 0.1–1.0 kg/metric ton being common. It is interesting to note that in the case of the collectors of Groups 3 and 4 of Fig. 4, the hydrophobicity being imparted to the mineral by the collector requires a collector having a much larger hydrocarbon backbone than in the thiol collectors for sulfide minerals. In almost all cases, those non-sulfide collectors will have at least 10 carbon atoms, with the single most common anionic collector being oleic acid (or its salts) having 18 carbon atoms. Likewise, the industrially used cationic amines typically will also contain 12 to 18 carbon atoms, or even more in a few cases.

Thus, while the theoretical mechanism of attachment of nonsulfide minerals is often more straightforward, involving simple charge neutralization between the mineral surface and the collector, the burden of these collectors to impart the necessary hydrophobicity for flotation to occur

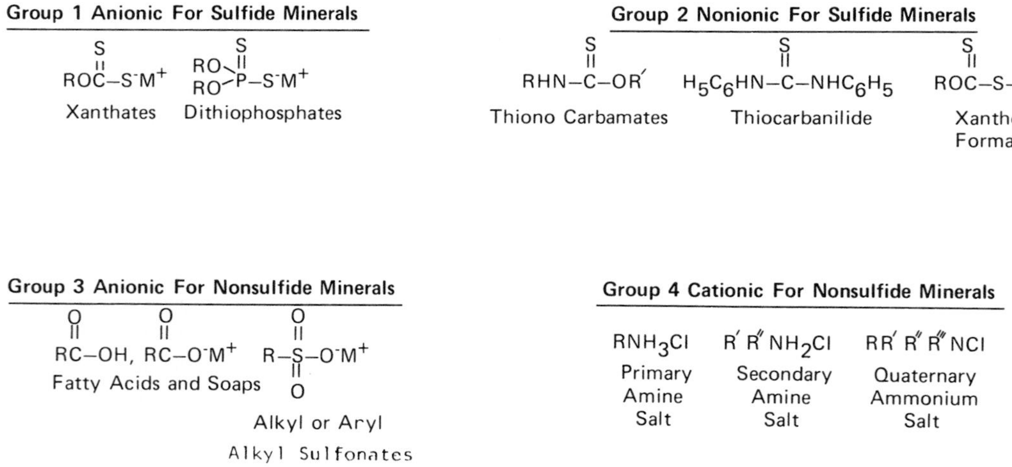

Group 1 Anionic For Sulfide Minerals

$$\underset{\text{Xanthates}}{RO\overset{\overset{\displaystyle S}{\|}}{C}-S^-M^+} \qquad \underset{\text{Dithiophosphates}}{\overset{RO}{\underset{RO}{>}}P\overset{\overset{\displaystyle S}{\|}}{-}S^-M^+}$$

Group 2 Nonionic For Sulfide Minerals

$$\underset{\text{Thiono Carbamates}}{RHN\overset{\overset{\displaystyle S}{\|}}{-}C-OR'} \qquad \underset{\text{Thiocarbanilide}}{H_5C_6HN\overset{\overset{\displaystyle S}{\|}}{-}C-NHC_6H_5} \qquad \underset{\underset{\text{Formates}}{\text{Xanthogen}}}{RO\overset{\overset{\displaystyle S}{\|}}{C}-S\overset{\overset{\displaystyle S}{\|}}{-}COR'}$$

Group 3 Anionic For Nonsulfide Minerals

$$RC\overset{\overset{\displaystyle O}{\|}}{-}OH, \; RC\overset{\overset{\displaystyle O}{\|}}{-}O^-M^+ \qquad R\overset{\overset{\displaystyle O}{\|}}{\underset{\underset{O}{\|}}{-}S}-O^-M^+$$

Fatty Acids and Soaps

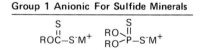

Alkyl or Aryl

Alkyl Sulfonates

Group 4 Cationic For Nonsulfide Minerals

$$\underset{\underset{\underset{\text{Salt}}{\text{Amine}}}{\text{Primary}}}{RNH_3Cl} \qquad \underset{\underset{\underset{\text{Salt}}{\text{Amine}}}{\text{Secondary}}}{R'R''NH_2Cl} \qquad \underset{\underset{\underset{\text{Salt}}{\text{Ammonium}}}{\text{Quaternary}}}{RR'R''R'''NCl}$$

FIGURE 4 Typical types of compounds used as collectors.

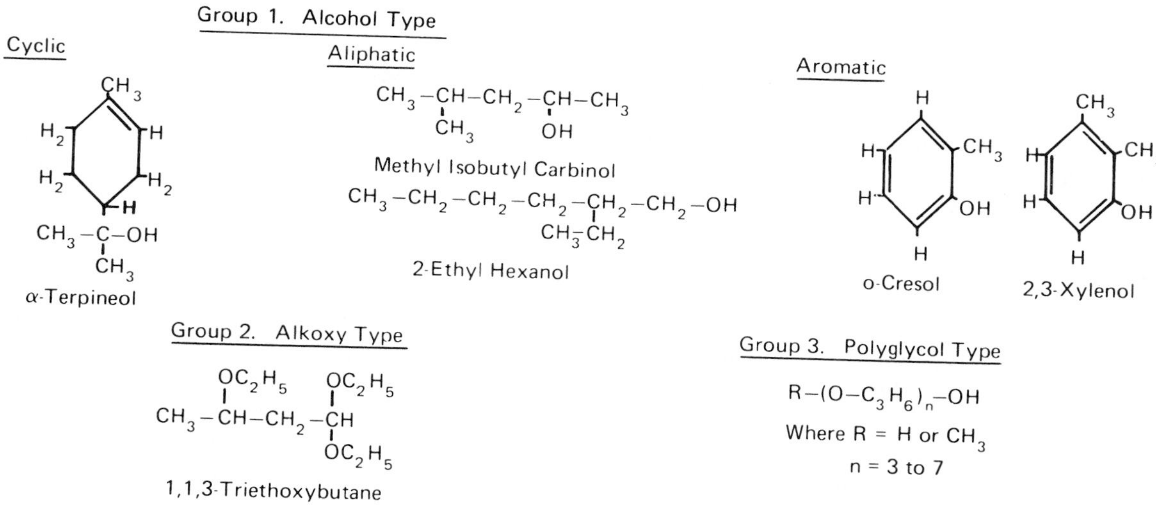

Group 1. Alcohol Type

Cyclic

Aliphatic

$CH_3-\underset{\underset{CH_3}{|}}{CH}-CH_2-\underset{\underset{OH}{|}}{CH}-CH_3$

Methyl Isobutyl Carbinol

$CH_3-CH_2-CH_2-CH_2-\underset{\underset{CH_3CH_2}{|}}{CH_2}-CH_2-OH$

2-Ethyl Hexanol

α-Terpineol

Aromatic

o-Cresol

2,3-Xylenol

Group 2. Alkoxy Type

$CH_3-\underset{\underset{OC_2H_5}{|}}{CH}-CH_2-\underset{\underset{OC_2H_5}{\overset{\overset{OC_2H_5}{|}}{|}}}{CH}$

1,1,3-Triethoxybutane

Group 3. Polyglycol Type

$R-(O-C_3H_6)_n-OH$

Where R = H or CH_3

n = 3 to 7

FIGURE 5 Typical types of compounds used as frothers.

is also greater. This accounts for the greater dosage required of Group 3 and Group 4 collectors than was the case with thiol collectors on sulfide minerals. It also partially explains the often observed poorer selectivity of valuable minerals over gangue in the same nonsulfide mineral systems, as often the nonsulfide mineral desired and the undesired gangue will have the same surface charge. Improving the selectivity of nonsulfide ore systems is a major new area of chemical research.

Typical frothing agents are shown in Fig. 5. As was the case with collectors, a wide range of compounds are used, depending on mineral type and operating conditions. The dosage of frother required in any given application is quite variable, ranging from zero for some systems to 0.3 kg/metric ton in others. Part of the reason for this variability is that many of the collectors presented in Fig. 4 demonstrate some frothing action as well as collecting action. This is particularly true of the nonsulfide collectors.

The third class of reagents used was previously referred to as modifying agents, which can be broken down further into five categories: pH regulators, activators, depressants, dispersants, and flocculants. The major pH regulators used today are lime, soda ash, caustic, sulfuric acid, and hydrochloric acid. The role of pH in flotation is very important. For example, with those minerals having H^+ or OH^- as potential-determining ions, the role of pH is quite direct. In other systems, the influence of pH can be complicated, causing simultaneous changes in many of the variables involved, such as altering the potential-determining ions through changes in solubility of the species involved and controlling the ionization of certain of the collector species, which influences adsorption.

A second important class of modifying agents is known by the name of activators. These are materials that when added to a flotation system cause better collector attachment on the desired mineral to be recovered. A well-known example is the addition of copper sulfate to the flotation of a zinc-containing mineral called sphalerite, ZnS. The cupric ion attaches to the sphalerite by a mechanism involving replacing some of the Zn ions in the sphalerite lattice by Cu ions, which then allows direct flotation using xanthate collectors. Without cupric ion activation, sphalerite flotation with a xanthate collector is not possible. Another mechanism by which activation can occur is the sulfidization of metal-containing ores that have been exposed to an oxygen environment, thus enabling the more efficient use of sulfide ore collectors.

A third and important activation mechanism is the adsorption of the hydroxy complex of a metal ion onto a mineral surface near the pH of precipitation of that metal ion in dilute solution. This then allows an appropriate collector to attach to the mineral surface by means of this hydroxy species activated mineral surface.

Another class of modifying agents is denoted as depressants. The idea of a depressant chemical is to prevent the flotation of an undesired mineral that, without special treatment, will have a tendency to float. A wide variety of depressant mechanisms are employed in practice. One common approach is basically deactivation, where care is taken to remove or modify a situation that normally leads to undesired activation of some mineral species. Sodium cyanide is commonly used for this purpose. Another important depressant, especially in copper–molybdenum flotation, is sodium hydrosulfide. Another mechanism of depression is called surface blocking, where a high-molecular-weight polymer, such as starch or quar, is first adsorbed on a particular mineral, thus blocking the collector attachment to the same mineral so that flotation of that mineral does not occur. A final depression mechanism is the use of oxidizing agents such as sodium hypochlorite to oxidize a mineral surface so as to prevent collector attachment. This latter approach is also commonly employed to remove the residual collector on a mineral species that was previously floated.

The final two classes of modifying agents are called dispersants and flocculants. These materials basically alter the particle–particle interaction (fluidity) of a flotation pulp. Dispersants are used to distinctly separate fine slime particles that normally tend to coagulate or coat each other, thus making separation difficult. This is particularly important for pulps containing fine clays. Common reagents used as dispersants are sodium silicates, lignin sulfonates, and various metaphosphates. Flocculating agents are normally added after flotation to aid in the filtration of recovered valuable concentrate or in the thickening of finely ground gangue materials. Common reagents used for this purpose are high-molecular-weight polyacrylamides, with a wide variety of modifications and various starches.

IV. INFLUENCE OF MINERALOGY ON FLOTATION

The influence of mineralogy, including the crystal structure of minerals, is important in flotation. In addition, the random substitution or adsorption of various ions into a crystal lattice can sometimes dominate the flotation behavior of a given mineral. The mineralogy effects have been categorized into five mineral groups: naturally floatable, sulfides, insoluble oxides and silicates, slightly soluble, and soluble salts. As each group of minerals responds to flotation in a different manner, this categorization offers a convenient method for assessing the potential to float any given mineral.

Some finely divided materials subjected to flotation can exhibit a net hydrophobic character without any collector adsorption. Such a phenomenon is usually explained by the presence of an electrically neutral surface condition. This does not necessarily imply that there are no localized hydrophilic sites on these mineral surfaces which may exhibit a surface charge and hence have an adsorption potential, but rather that the overall surface potential of such materials exhibits a minimum. Some typical minerals that can exhibit natural hydrophobicity are graphite, coal, sulfur, molybdenite (MoS_2), and talc [$Mg_3Si_4O_{10}(OH)_2$]. In addition, some of the more common sulfide minerals, such as galena (PbS), chalcopyrite ($CuFeS_2$), chalcocite (Cu_2S), pentlandite [(Fe, Ni)$_9S_8$], and pyrite (FeS_2), can

exhibit natural hydrophobicity in flotation systems that have only small amounts of oxygen present. There are a number of explanations for this unusual behavior of some sulfide minerals, including the formation of highly hydrophobic elemental sulfur or surface polysulfides under slightly oxidizing conditions. A second explanation is that unlike many of the nonsulfide minerals in the remaining mineralogy groups to be discussed, the sulfide minerals do not exhibit a large amount of hydrogen bonding of water molecules to the surface sulfide atoms because of the large size of the sulfide ion. Such hydrogen bonding of water molecules is responsible for much of the hydrophilic character exhibited by many minerals in water, which accounts for the need of highly hydrophobic collectors to initiate flotation. The full potential of this natural hydrophobicity phenomenon exhibited by certain sulfide minerals has not yet been achieved in industrial practice.

From an industrial viewpoint, coal is one of the more important minerals, exhibiting a broad range of hydrophobicity as a function of its chemical structure and pH. Figure 6 demonstrates the variation of contact angle (or floatability) of coal as a function of the carbon content of the coal. In industrial practice, most coals and the other naturally floatable minerals listed (especially MoS_2) are recovered using simple neutral hydrocarbon promoters such as fuel oil and kerosene to increase the hydrophobic development

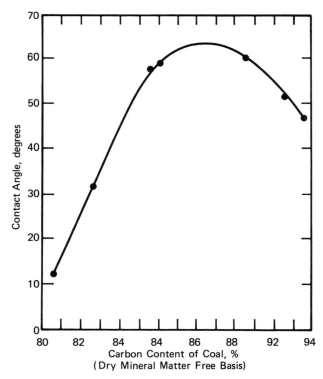

FIGURE 6 Variation of contact angle with carbon content of coal.

required to get sufficient rates of mineral attachment to frother stabilized bubbles. Such promoter materials may be used in relatively high amounts of 0.2–2.0 kg/metric ton. The major as yet unsolved research problem in coal flotation is the ability to selectively separate coal from pyrite, which is a contributor to acid rain when the coal is combusted.

The second category of mineralogy, the sulfide minerals, has been already discussed at some length in this article as it has received the most chemical research investigation. The roles of oxygen and oxidation as well as pH control are particularly important with this family of minerals. In addition, the thiol collectors discussed earlier are industrially used with generally good recoveries and grades. Typical dosages of thiol collectors required a range from 0.01 to 0.1 kg/metric ton. Specific sulfide minerals that have been studied in depth include pyrite (FeS_2), arsenopyrite $(FeAsS)$, pyrrhotite $(Fe_{1-x}S)$, chalcopyrite $(CuFeS_2)$, covellite (CuS), bornite (Cu_5FeS_4), galena (PbS), sphalerite (ZnS), chalcocite (Cu_2S), and pentlandite $[(Fe, Ni)_9S_8]$.

It is with sulfide mineral flotation that the greatest amount of engineering application experience in flotation has also been gained, which will be discussed further in Section VI. The sulfide minerals are an unusual mineralogy group in that they have high specific surface energies when fresh particle surfaces are initially created by the grinding process. This implies that they are quite reactive to a wide variety of chemicals, including oxygen, which does cause some unique problems at the industrial level. The prior time and water chemistry history that the sulfide minerals have experienced markedly affects the nature of the flotation process. Just aging for a few minutes after grinding, for example, can almost destroy the ability to economically recover at any reasonable grade some sulfide minerals. It is common to see the simple timing of where and how chemical reagents and pH regulators are added making a significant difference in recovery. This is part of the challenge of implementing and optimizing froth flotation at the industrial level. Experience has shown that even for the same mineral found in different ore matrices in different physical locations, the best technical and economic engineering practices can be quite different from plant to plant. Identifying the potential determining ions for sulfide minerals in different ores can be complex and includes such ions as HS^-, H^+, OH^-, S_2^-, and M^{n+}. Thus, knowledge of pH alone is not sufficient to predict the surface charge associated with any given sulfide mineral.

In the third mineralogy category, denoted as insoluble oxides and silicates, the number of minerals involved is large and includes such important industrial minerals as beryl $[Be_3Al_2(Si_6O_{18})]$ (3–4), cassiterite (SnO_2) (4–5),

chromite $(FeCr_2O_4)$ (5–8), chrysocolla $[Cu_4H_4Si_4O_{10}(OH)_8]$ (2–4), corundum (Al_2O_3) (9–10), cuprite (Cu_2O) (9–10), goethite $(\alpha FeO \cdot OH)$ (6–7), hematite (Fe_2O_3) (5–7), kaolinite $[Al_2Si_2O_5(OH)_4]$ (3–4), magnetite (Fe_3O_4) (6–7), pyrolusite (MnO_2) (5–8), quartz (SiO_2) (1.8), rutile (TiO_2) (6–7), and zircon $(ZrSiO_4)$ (5–6). Whether or not a given mineral in this category can be treated by froth flotation depends in large part on the electrical properties of the mineral surface and of the collector, along with the molecular weight of the collector and the stability of the metal–collector salts. The basic mechanism prevalent is the adsorption of a charged collector by electrostatic interaction with a surface containing the opposite charge (physical adsorption).

Groups 3 and 4 of Fig. 4 gave the most prevalent anionic and cationic collectors used in the physical adsorption mechanism. With the minerals of this category, the potential-determining ions are generally H^+ and OH^-, so that pH is usually the dominating factor controlling the mineral surface charge of pure mineral particles and, hence, the charge of the collector required. The pH associated with the point of zero charge (pzc) is an important mineral characteristic, with the surface of pH readings below the pzc being positively charged (requiring an anionic collector) and above the pzc being negatively charged (requiring a cationic collector). In the mineral listing of the previous paragraph, the final numbers given after each mineral name and formula indicate the typical range in which the pzc has been shown to occur in pure mineral surface potential tests.

Extensive hydrogen bonding of water molecules occurs on oxide and silicate surfaces, causing them to be highly hydrophilic. This implies that for charged collectors to be effective in flotation with this mineral category, they must have high hydrocarbon content (high hydrophobicity) and must be present in sufficient dosage so that there is some reinforcement of hydrophobic interaction between the hydrocarbon tails of the collectors (denoted as micellization). Typical collector dosages are in the range of 0.3–1.0 kg/metric ton.

By far the most important industrial anionic collectors for nonsulfide minerals are the carboxylic acids (and their salts), with a special preference for the partially unsaturated fatty acids containing 18 carbon atoms (e.g., oleic, linoleic, linolenic acids). Fatty acids dissociate in aqueous solution into negatively charged carboxylate ions,

$$RCOOH \text{ (aqueous)} \leftrightarrows H^+ + RCOO^-$$

with pK_a values ranging from 5 to 6. This then implies that at pHs less than 5 or 6, the acid is undissociated and not effective for attachment to positively charged minerals. Above pH 5 to 6, the acid is ionized, thus defining the range of collector effectiveness, which, of course, must then be matched to the specific surface charge being exhibited by the particular mineral under study at that pH. The advantage of using sulfonic acids (or their salts) is that they are stronger acids than the carboxylic acids and, hence, dissociate at a much lower pH range, giving them a wider window of potential use as collectors.

With regard to industrially viable cationic collectors, the use is almost totally limited to amines. Amines ionize in aqueous solution by protonation.

$$RNH_2 \text{ (aqueous)} + H_2O \leftrightharpoons RNH_3^+ + OH^-.$$

The actual form of the amines used (the makeup and carbon content size of R and the number of R groups attached to the nitrogen atom) is quite variable and includes primary, secondary, tertiary, and quaternary forms. Quaternary amines are strong bases and therefore are ionized over essentially the complete range of pH, while the ionization of the remaining amine forms is pH dependent. Typically for a primary amine, the ionization starts to fall off in the range of pH 10 or greater, thus defining the window of potential use as a collector. The major practical use of amine collectors is in the flotation of silica (quartz) gangue particles away from other materials, a process denoted as reverse flotation. It is also important to point out that in many flotation systems, including the minerals of this category, the actual chemical reagent schemes used industrially may be much more complex in mechanism than simple physical adsorption. Often a mixture of chemisorption and/or surface charge reversal is present, involving activating species such as other inorganic ions and hydroxide formation on particle surfaces. Each specific industrial mineral system needs to be evaluated scientifically as a unique system, starting with the simple mechanisms of physical adsorption, chemisorption, and so on.

The fourth category of mineralogy is the slightly soluble minerals, which include such important industrial minerals as apatite $[Ca_5(PO_4)_3 (F, Cl, OH)]$ (5–8), calcite $(CaCO_3)$ (10–11), dolomite $[CaMg(CO_3)_2]$ (4–7), malachite $[Cu_2CO_3(OH)_2]$ (5–8), and barite $(BaSO_4)$ (3–4). The minerals are described by their moderate solubility in water (typically on the order of 10^{-4} mol/liter) and their surface charge. Thus, as might be expected, the surface charge of these minerals is a function of the ions of which their lattices are made up. In the listing of minerals just given, the typical range of the isoelectric point of typical pure mineral samples is given just after the chemical formula. The isoelectric point of solid (iep) is the point at which the zeta potential is zero. Note that when the zeta potential is measured in the presence of univalent acids or bases, the iep and the pzc of the solid are the same. Anionic collectors such as oleic acid (or its salts) are generally the most frequently used collectors for this category of minerals.

The final mineralogical category is the soluble salt minerals, which include the important example of potash processing involving the upgrading of sylvite (KCl) from halite (NaCl). In this rather unusual system, the zeta potential is near zero, the electrical double layer is approximately one ion in thickness, and the solubilities of collectors are limited. Typically, amine collector chemistry is used at the industrial level at dosages in the range of 0.1–0.5 kg/metric ton.

V. FLOTATION EQUIPMENT AND PLANT CIRCUITRY

Industrial flotation machines can be divided into four classes: (1) mechanical, (2) pneumatic, (3) froth separation, and (4) column. The mechanical machine is clearly the most common type of flotation machine in industrial use today, followed by the rapid growth of the column machine. Mechanical machines consist of a mechanically driven impeller, which disperses air into the agitated pulp. In normal practice, this machine appears as a vessel having a number of impellers in series. Mechanical machines can have open flow of pulp between each impeller or are of cell-to-cell designs which have weirs between each impeller. The procedure by which air is introduced into a mechanical machine falls into two broad categories: self-aerating, where the machine uses the depression created by the impeller to induce air, and supercharged, where air is generated from an external blower. The incoming slurry feed to the mechanical flotation machine is introduced usually in the lower portion of the machine. Figure 7 shows a typical industrial flotation cell of each air delivery type.

The most rapidly growing class of flotation machine is the column machine, which is, as its name implies, a vessel having a large height-to-diameter ratio (from 5 to 20) in contrast to mechanical cells. The mechanism behind this machine to is provide a countercurrent flow of air bubbles and slurry with a long contact time and plenty of wash water. As might be expected, the major advantage of such a machine is the high separation grade that can be achieved, so that column cells are often used as a final concentrate cleaning step. Special care has to be exercised in the generation of fine air bubbles and controlling the feed rate to column cells.

The flotation machine must perform a series of complicated functions. Some of the more obvious are:

1. Good mixing of pulp. To be effective, a flotation machine should maintain all particles uniformly in suspension within the pulp, including those of relatively high density and/or size. Good mixing of pulp is required for maximizing bubble-particle collision frequency.

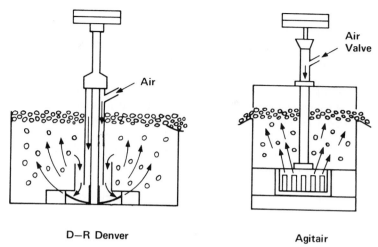

D–R Denver   **Agitair**

FIGURE 7 Illustration of two air delivery systems in common mechanical flotation cells (induced, D-R Denver; supercharged, Agitair).

2. Appropriate aeration and dispersion of fine air bubbles. An important requirement of any flotation machine is the ability to provide uniform aeration throughout as large a volume of the machine as is possible. In addition, the size distribution of the air bubbles generated by the machine is also important, but experience has shown that the proper choice of frother type and dosage generally dominates the bubble size distributions being produced.

3. Sufficient control of pulp agitation in the froth zone. As mentioned earlier, good mixing in the machine is important; however, equally important is that near and in the actual froth bed at the top of the machine, sufficiently smooth or quiescent pulp conditions must be maintained to ensure suspension of hydrophobed (collector coated) particles.

4. Efficient mass flow-mechanisms. It is also necessary in any flotation machine that appropriate provisions be made for feeding pulp into the machine and also for the efficient transport of froth concentrate and tailing slurry out of the machine.

Probably the most significant area of change in mechanical flotation machine design has been the dramatic increase in machine size. This is typified by the data of Fig. 8, which shows the increase in machine (cell) volume size that has occurred with a commonly used cell manufactured by Wemco. The idea behind this approach is that as machine size increases, both plant capital and operating costs per unit of throughput decrease.

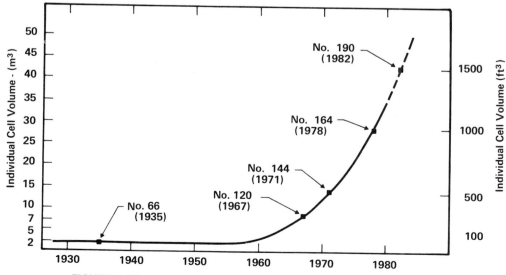

FIGURE 8 Wemco flotation machine size availability vs year of development.

The throughput capabilities of various cell designs will vary with flotation residence time and pulp density. The number of cells required for a given operation is determined from standard engineering mass balance calculations. In the design of a new plant, the characterization of each cell's volume and flotation efficiency is generally calculated from performing a laboratory-scale flotation on the same type of equipment on the ore in question, followed by the application of empirically derived design (scale-up) factors. Research work is currently under way to improve the understanding and performance of commercial flotation cells. Currently, flotation-cell design is primarily a proprietary function of the various cell manufacturers.

Flotation plants are built in multiple cell configurations (called banks), and the flow through various banks is adjusted in order to optimize plant recovery of the valuable as well as the valuable grade of the total recovered mass from flotation. This recovery vs grade trade-off is economically important in flotation, as increased recovery of the valuable is associated with decreased grade. For example, a 95% recovery of copper in the feed ore might give a concentrate grade of 18% Cu in the total recovered mass, while 80% Cu recovery might give a grade of 25% in the concentrate. Obviously, the higher the valuable recovery is, the higher the potential income, but if this higher recovery requires a great deal more grinding and/or expensive downstream processing (including further flotation) in order to upgrade the concentrate for metal refining such as smelting, the increase in potential recovery income may actually cause a net loss of total income. This grade-recovery optimization is generally worked out by individual flotation operators in each plant (and each mineral) and sets the operating philosophy of that plant. Figure 9 shows a typical industrial recovery vs grade trade-off curve for a copper sulfide ore containing pyrite. The higher the copper recovery is, the greater the amount of undesired pyrite contained in the concentrate.

The various banks of flotation cells in an industrial plant are given special names to denote the particular purpose of the banks. The rougher bank is the first group of cells that the pulp sees after size reduction. The goal of the roughers is to produce a concentrate with as high a recovery of valuable as possible with generally low grade of the valuable. The rejected gangue material from any bank of

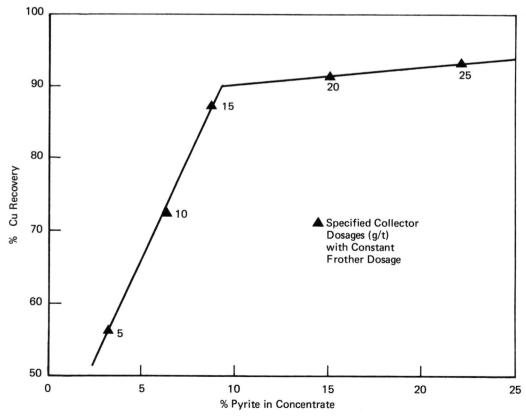

FIGURE 9 Typical trade-off of copper sulfide mineral recovery vs undesired pyrite in concentrate.

cells is commonly denoted as the tails or tailings. Usually, rougher tails are discarded so that valuable mineral not recovered in the rougher bank is lost. The concentrate of the rougher bank can be further concentrated, sometimes after additional grinding, in banks of cells called cleaners or recleaners. The tailings from the cleaners or recleaners can be recirculated to a bank of cells known as scavengers in order not to lose any valuable material in the upgrading process. Various banks of cells are also sometimes known by the particle size of the particular pulps being floated. Coarse particles, fine or slime particles, and middle-sized particles, denoted as middlings, can all be treated in separate banks.

As to overall capacities of flotation plants, the range is quite variable, depending on the type and value of the mineral being processed, the amount of valuable mineral in the feed ore to flotation, the degree and cost of size reduction involved, and the relative response of the valuable(s) to the flotation process. Smaller plants ranging in size from 500 to 5000 metric tons of feed per day are common, with feed materials having high amounts of valuable per ton of feed ore (>40%), such as coal, phosphate, and oxide ores. On the other hand, the sulfide minerals that are typically a small percentage of the ore (<10% and often less than 1%) require much greater capacity in order to achieve a reasonable economic return on investment. Thus, typical copper sulfide plants have capacities in the range of 20,000 to more than 60,000 metric tons of feed ore per day.

VI. THE FLOTATION SYSTEM

The successful industrial practice of froth flotation requires that flotation be analyzed from an interactive systems approach, as illustrated in Fig. 10. The three major components of the flotation system have been discussed in this article and include chemistry, equipment, and operations. Each component has a series of factors that can be manipulated. Essentially, the nature of flotation is that many of the factor settings involved with the various components are either self-compensating or capable of strongly reinforcing desired or undersired system performance.

Thus, any particular economic or technical performance goal can often be achieved by a number of various combinations of factor settings available to flotation operators. For example, poor equipment performance can often be compensated for by appropriate chemical reagent selection, and vice versa. Likewise, equipment and reagent factors can be adjusted to compensate for limitations in operating factors such as the production of excessive coarse or fine feed particles or too high a throughput for a given cell capacity. An important industrial question, therefore, is: how can the system be operated most economically to achieve desired goals?

In recent years, a great deal of work has been done to quantify the plant influences of the various factors of the three major flotation components. This work has required

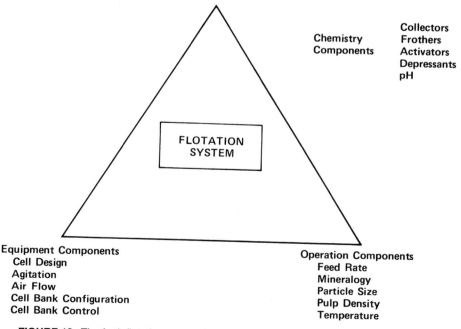

FIGURE 10 The froth flotation system illustrated as a three-cornered interactive system.

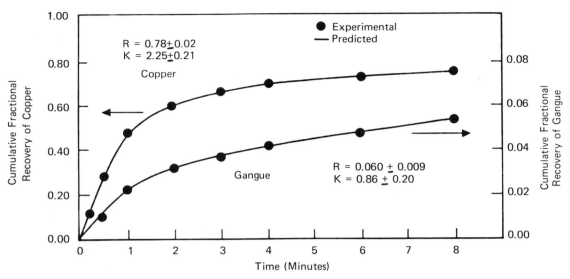

FIGURE 11 Typical time–recovery profiles of valuable and gangue with time of flotation in a batch laboratory-scale cell.

that valuable recovery vs time of flotation be measured in small-scale batch laboratory flotation cells capable of handling feed charges of 500–2000 g. These smaller cells offer many advantages, not the least of which is the ability to accurately determine many sets of experimental data for various factor settings quite cheaply. Once a flotation system has been analyzed in the small laboratory cell, direct inferences can be made on the large-scale industrial plant.

This correspondence of laboratory data to plant prediction can only be done when accurate time–recovery profiles are generated for the valuable and the gangue, as illustrated in Fig. 11. Figure 11 also shows the results of fitting an appropriate rate mathematical model to the data of the form

$$r = R[1 - (1 - e^{-Kt})/Kt],$$

where r is the measured cumulative recovery of either valuable or gangue at time t, R is the calculated equilibrium recovery as $t \rightarrow \infty$, and K is a first-order rate constant of mass removal (units of time^{-1}) that describes the shape of the time–recovery curve at small flotation times. Thus, each time-recovery profile can be characterized by two parameters, R and K, which have statistical confidence limits. When comparing the influence of different factor settings, one can statistically test the differenceof the factor settings by statistically testing the parameters R and K. The ratios of R(valuable) to R(gangue) and K(valuable) to K(gangue) give grade information.

Therefore, an important concept is the identification of an appropriate laboratory time for which to compare to

a particular industrial bank of cells. Historically, it has been assumed that comparison of plant flotation factors should be made in the laboratory at longer flotation times (in the region of R control in Fig. 12). Recent work has shown that often the laboratory time needed for comparison to the plant can be anywhere on the laboratory time scale, depending on the manner in which the plant is being operated (especially with regard to plant feed throughput relative to a given cell bank capacity). Figure 12 shows clearly that the same factor setting that may be negative at longer laboratory flotation times (a poorer R) can actually be better at shorter laboratory flotation times (a higher K). This crossing of time–recovery profiles with any given factor change is common in the flotation system and clearly shows that plant operators must be careful to understand the complex interactions under their control.

The use of R and K parameters, based on laboratory data, is also important, as significant trends in R and K have been identified for controlled changes in each factor setting of the flotation system. Thus, for example, increasing the dosage of any collector consistently causes an increase in R for a given flotation feed but eventually lowers the K at higher dosage (thus, the flotation curve becomes flatter at low times but higher at high times). This effect is shown in Fig. 12, with System 2 being at low collector dosage and System 1 being at higher collector dosage. This is called the R–K trade-off.

Such a phenomenon is important in plants involving a fixed cell capacity, as increased collector dosage, while theoretically giving a higher R, actually can give lower observed recovery at the end of a given cell bank in the

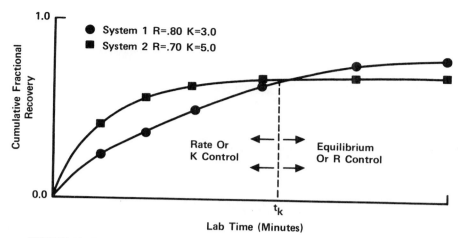

FIGURE 12 Two laboratory time–recovery profiles run under different test conditions.

plant. This is true as one simply runs out of time (cell capacity) to achieve the higher R due to the associated decrease in K. An illustration of this effect on an industrial rougher bank floating a copper mineral is given in Fig. 13. Other common factors that have significant influences on the values of R and K include particle size, slurry density and temperature, air flow rate, frother type and dosage, pH, collector type, and cell design.

Besides the influence of collector dosage on industrial circuit performance from a rate of recovery viewpoint, the effect of both frother and collector dosage on the particle size range of mineral that can be recovered is also quite

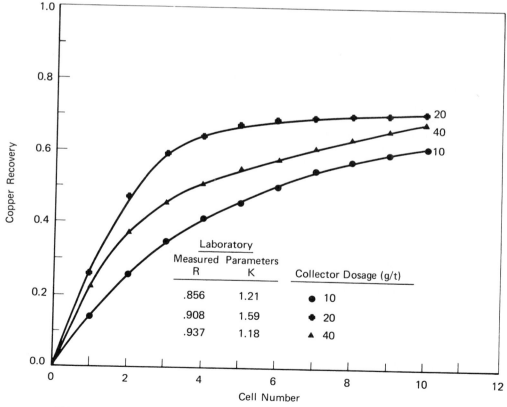

FIGURE 13 The R–K trade-off with collector dosage on an industrial rougher bank.

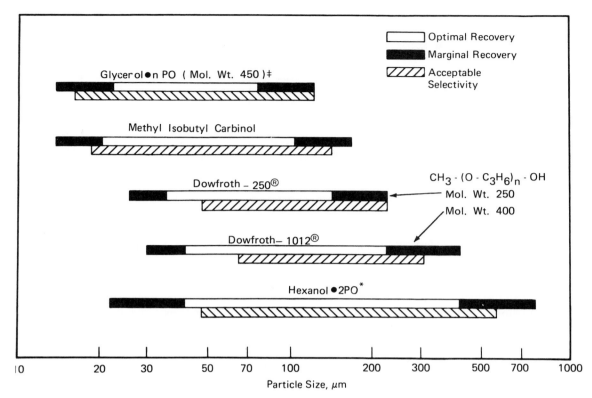

FIGURE 14 The influence of frother structure as a function of particle size on a copper sulfide ore. ‡Experimental coarse particle frother of the Dow Chemical Co. ®Trademark of The Dow Chemical Co.

important. As was described in the introduction to this article, there is a close tie between grinding of the mineral containing ore to achieve liberation of the desired mineral from the undesired gangue and the efficiency of the subsequent flotatation process. As was shown in Fig. 1, this liberation must occur in a relatively narrow window of particle size. Thus, each plant operation must choose the operation conditions to maximize the choice of many variables such as grind size, pH, collector and frother type, as well as dosage, all of which are being superimposed on a naturally occurring ore that often changes its characteristics of hardness, liberation size range, mineral content, and so on, sometimes by the hour.

Another interesting example of how critical the choice of chemical reagents can be at the industrial level is illustrated in Fig. 14. Here, the effective particle size range of copper sulfide mineral being floated by different frother chemical types is illustrated. It is clear that each frother structure has a unique particle size range of maximum recovery effectiveness. In order to properly cover a typical broad feed size range, the plant will require a blending of appropriate frother reagents.

SEE ALSO THE FOLLOWING ARTICLES

COAL PREPARATION • ELECTROCHEMICAL ENGINEERING • MINERAL PROCESSING • MINING ENGINEERING • SEPARATION AND PURIFICATION OF BIOCHEMICALS • SURFACE CHEMISTRY

BIBLIOGRAPHY

Finch, J. A., and Dobby, G. S. (1990). "Column Flotation," Pergamon, Elmsford, NY.

Kawatra, S. K., ed. (1995). "High Efficiency Coal Preparation: An International," Soc. Min. Eng.

Kawatra, S. K., ed. (1997). "Comminution Practices," Soc. Min. Eng.

Laskowski, J. S., and Ralston, J., eds. (1992). "Colloid Chemistry in Mineral Processing," Elsevier, Amsterdam/New York.

Leonard, J. W., III, ed. (1991). "Coal Preparation," 5th ed., Soc. Min. Eng.

Matis, K. A. (1994). "Flotation Science and Engineering," Dekker, New York.

Parekh, B. K., and Miller, J. D., eds. (1999). "Advances in Flotation Technology," Soc. Min. Eng.

Fuel Cells, Applications in Stationary Power Systems

Edward Gillis
Energy Consultant

John O'Sullivan
Energy Consultant

GLOSSARY

Anode The negative fuel cell electrode where electrons are released by electrochemical reaction to the electrode, thus made available to the external electrical circuit; the electrode where the fuel is (1) oxidized by a negatively charged ion or (2) becomes a positively charged ion; either ion completes the electrical circuit by migrating through the fuel cell electrolyte.

Cathode The positive fuel cell electrode where electrons from the external circuit are consumed by the electrochemical reaction; the electrode where the oxidant is reduced to (1) a negatively charged ion or (2) a chemical product by a positively charged ion.

Fuel cell A device that generates electricity by the direct electrochemical reaction of a fuel and oxidant. The reaction can be isothermal and thus not subject to Carnot cycle efficiency limitations.

FUEL CELLS generate electricity by the electrochemical oxidation of a fuel, typically hydrogen, with an oxidant that is typically oxygen from air. This is an isothermal process with several attractive characteristics that include low audible noise, no vibration, low emissions, no visible exhaust plume, excellent efficiency, and possibility of reject heat recovery. These characteristics make fuel cell power plants well suited for installation near the electrical load of an individual consumer or a group of consumers. Reciprocating- and turbine-engine generators are sometimes used for electrical power generation in these applications, generally in conjunction with the electric grid, and most often as emergency backup or peak-shaving systems. The operating characteristics of fuel cell systems are sufficiently different than engine generators that they do not compete in these intermittent duty applications. For example, fuel cells have lengthy start times and high capital cost, hence are better suited to continuous power generation to either supplement or replace the electric grid. Most fuel cell development and commercialization efforts are aimed at providing electricity to a single consumer or to a small region rather than large-scale central station systems. This article will focus on fuel cell power systems for these on-site power generation applications.

Fuel cells for stationary power are made up of four major subsystems: (1) a fuel processor which converts a common fuel (typically natural gas, petroleum, or propane) to a hydrogen-rich product; (2) the fuel cell assembly which electrochemically oxidizes the fuel with air to direct current electricity; (3) a power conditioner that converts the dc electricity to ac at the voltage and frequency desired; and (4) a control system that coordinates the other three systems.

There are four types of fuel cells that can be used for stationary power generation. These are referred to by the type of electrolyte they utilize:

- Polymer exchange membrane (PEM) fuel cells. These operate at a temperature below $100°C$. PEM fuel cells are being developed for transportation systems, and hence may become inexpensive because of high-volume production.
- Phosphoric acid fuel cells (PAFC). These operate at temperature of approximately $175°C$. The PAFC has greater tolerance to carbon monoxide than PEM, which simplifies the fuel processor and provides higher temperature waste heat for possible cogeneration uses.
- Molten carbonate fuel cells (MCFC). These operate at approximately $650°C$, which is similar to the fuel processor subsystem operating temperature. This permits the fuel processor to be integrated with the fuel cell assembly, which can improve system efficiency and reduce system complexity and cost. MCFC have the highest electrochemical efficiency of the four types of fuel cell systems.
- Solid oxide fuel cells (SOFC). These operate at temperatures ranging from $650°C$ to $1000°C$, different manufacturers choosing different temperatures. Again, the fuel processor can be integrated with the fuel cell assembly with benefits similar to MCFC. The SOFC is perhaps better suited than MCFC for low-power systems such as single-family homes.

I. TECHNOLOGY COMMON TO ALL STATIONARY FUEL CELL SYSTEMS

A. Electrochemistry

Hydrogen is the electrochemically active fuel in the four types of fuel cells. It is oxidized electrochemically with oxygen from air by the overall reaction

$$H_2 + \tfrac{1}{2}O_2 = H_2O.$$

This is the same overall reaction as combustion; however, in fuel cells the reactants only make contact ionically. The low temperature fuel cells, i.e., PEM and PAFC, have acid electrolytes with the hydrogen ion H^+ as the ionic conductor. The fuel electrode (anode) reaction is

$$H_2 \rightarrow 2H^+ + 2e^-.$$

The hydrogen molecule gives up two electrons to the electrode to form two hydrogen ions. The hydrogen ions migrate across the electrolyte to the oxygen electrode (cathode) and combine with oxygen by the reaction

$$2H^+ + \tfrac{1}{2}O_2 + 2e^- \rightarrow H_2O.$$

This reaction consumes two electrons from the electrode to maintain electrical neutrality; the path for the electron flow from the anode to the cathode is through the external circuit, i.e., the electrical load. The water produced by the reaction forms on the cathode and is generally allowed to evaporate into the air stream flowing over the cathode, and thus is removed from the fuel cell assembly.

An oxygen-containing species is the ionic conductor in high temperature fuel cells, i.e., CO_3^{2-} in MCFC and O_2^{2-} in SOFC. The cathode reaction for MCFC is

$$CO_2 + \tfrac{1}{2}O_2 + 2e^- \rightarrow CO_3^{2-}.$$

This carbonate ion migrates through the fused salt electrolyte (a mixture of lithium carbonate, potassium carbonate, and/or sodium carbonate) to react with hydrogen by the reaction

$$H_2 + CO_3^{2-} \rightarrow H_2O + CO_2 + 2e^-.$$

The respective reactions for the SOFC are

$$\tfrac{1}{2}O_2 + 2e^- \rightarrow \tfrac{1}{2}O_2^{2-}.$$
$$H_2 + \tfrac{1}{2}O_2^{2-} \rightarrow H_2O + 2e^-.$$

The product water is formed on the anode in these high-temperature fuel cells. The MCFC is unique in that CO_2 is both a reactant and product and must be recycled from the anode to the cathode.

The anode reactions in all fuel cells provide two electrons to the external load for each hydrogen molecule consumed. Thus all fuel cells produce exactly the same electrical current from a given quantity of hydrogen, and there is direct proportionality between the electrical current demanded by the load and hydrogen consumption. If the electrical load drops to zero, the ions cannot migrate, and therefore hydrogen consumption also drops to zero.

The theoretical potential for the reaction $H_2 + \tfrac{1}{2}O_2 \rightarrow H_2O$ is 1.23 V at standard temperature and pressure for water vapor product. This potential decreases slightly at elevated temperature and is also dependent on the partial pressures of the reactants and products; these dependences are described by the Nernst equation (see Hirschenhofer, 1994, or Energy and Environmental Solutions, 2000, for details).

A fuel cell operating voltage is significantly lower than the theoretical value. The principal voltage losses are due to the relatively high ionic resistance of the electrolyte and high overpotential required for the cathode reaction to proceed at a reasonable rate. Other losses of lesser magnitude include the electronic resistance of the electrodes and other components in the assembly and gaseous diffusion losses within and near the electrodes. For thorough discussions of these losses see the references listed in the Bibliography.

The magnitudes of the various losses are different for each type of fuel cell and are to some extent under the control of the fuel cell designer. As a general rule, the full-load operating voltage is 0.6–0.7 V per cell for PEM and SOFC, ~0.75 V for PAFC, and 0.75 to 0.85 V for MCFC. The electric energy (electricity) produced by each of these fuel cells is the product of voltage and current, while the current produced is proportional to the hydrogen consumed in all cases. In principle, the MCFC has the highest electrochemical efficiency. The difference between the Nernst potential and the operating potential times the electrical current quantifies the losses that must be removed as excess heat from the fuel cell assembly. This heat may be utilized in cogeneration applications or in bottoming-cycle engines. At less than full load, the operating voltage for all of the fuel cells increases. Therefore the part-load system efficiency also increases. This is in contrast to the part-load performance of heat engines and provides one of the reasons why fuel cells are attractive in stationary power applications.

B. Fuel Cell Stack Assemblies

An individual fuel cell generates less than 1 V dc. Therefore, numerous fuel cells must be connected electrically in series to reach a voltage level needed for stationary power applications. Most fuel cell developers construct flat (planar) fuel cell assemblies with the electrolyte sandwiched between the two electrodes. A separator plate with grooves formed in each face is placed between neighboring fuel cell assemblies. The grooves permit the flow of the reactants over the electrodes. If the separator plate is an electronic conductor, it serves to short the positive electrode of one cell electrically to the negative of the neighboring cell, thus building voltage. The fuel cell developer selects the number of individual cell to build a stack for the desired total system voltage. The size (area) of the individual cell sandwich is selected to give the desired electrical current. The resultant fuel cell stack assembly resembles a formed plate heat exchanger. Figure 1 is a representative configuration for an SOFC stack assembly.

FIGURE 1 Nine-cell SOFC stack. (Courtesy of Global Thermoelectric, Inc.)

Carbon is a commonly used electrical conducting material for the separator plates in low-temperature fuel cells. Graphitic carbon is used in PAFCs due to corrosion considerations, while carbon-filled plastics (and some metals) are suitable in PEMFCs.

The high-temperature fuel cells commonly use metals for separator plates. The reducing atmosphere in the anode chamber and oxidizing atmosphere in the cathode chamber usually dictate the use of different metals for each chamber, i.e., nickel for the anode and a chromium-rich alloy for the cathode, which often results in the use of a nickel-clad alloy for the separator plate. Some SOFC developers use a semiconducting ceramic for this plate.

Some developers construct stack assemblies using tubular-shaped cells, most notably in the Siemens–Westinghouse SOFC. This requires different kinds of electrical connections between individual cells and these are described in the references listed in the Bibliography.

C. Fuel Cell Performance Decay

The performance of all types of fuel cells decays over time as is typical of electrochemical processes. The performance decay is manifested by a gradual reduction in the operating voltages of individual cells and thus the operating voltages of the cell stack assemblies as well. To maintain constant power output, the electrical current must increase in proportion to the voltage decrease. Consequently, the specific fuel consumption of the power plant will increase over time.

The fuel cell voltage decay is small and gradual, with most fuel cell developers aiming at a decay rate of 2% or less per year for stationary power applications.[1] Thus either the fuel consumption will increase by 10% or the power output will decrease by a similar amount over a 5-year period. At that time, the power plant owner must make the decision to continue operation at lower efficiency/power or to restore efficiency/power by replacing the fuel cell stack assemblies. This is an economic decision as the fuel cell system reliability remains virtually unchanged and is generally capable of longer operating time. Causes of the voltage decay are different for different types of fuel cells. For SOFCs, solid-state diffusion of metallic ions occurs between the anode, electrolyte, and cathode layers and also to and from the semiconducting ceramic separator plate or interconnect if used in the cell assembly. The slight changes in material compositions result in higher internal resistance of these components. The decay rate for SOFCs using ceramic separator plates is the lowest for any type of fuel cell and amounts to much less

---

[1]The decay rate in transportation fuel cells can be much higher because the operating lifetime is much shorter, e.g., 100,000 miles at an average speed of 30 mph requires an operating life of 3,333 hr.

than 2% per year. If metallic separator plates are used, an oxide scale forms on the cathode side of the separator plate, which increases the electrical contact resistance between it and the cathode. Steps must be taken to make this oxide layer electronically conductive, to minimize cell voltage decay.

The principal causes of voltage decay in MCFCs are, first, corrosion of the cathode-side separator plate that increases the contact resistance with the cathode; second, loss of electrolyte by chemical reaction with the separator corrosion scale (which has a beneficial aspect in that it dopes the scale to make it a semiconductor); and third, loss of electrolyte by evaporation. The thickness of the cathode-side oxide scale increases with time, generally in proportion to the square root of time. Since the electrical resistance is proportional to the thickness of this layer, the MCFC decay rate decreases over time. The dopant that makes this layer semiconducting is lithium from the electrolyte. The volatile component of the electrolyte is the heavier alkali metal, either sodium or potassium; therefore the ratio of the two alkali metals in the electrolyte does not change significantly over time. The net loss of electrolyte does, however, increase the resistance of the electrolyte, thus adding to the overall voltage decay. If sufficient electrolyte is lost, the reactants may cross through the cells and react directly. This process results in rapid voltage loss and increased heat rejection, but cells are designed with excess electrolyte so this occurs only long after the manufacturer's planned stack replacement interval. MCFC developers have demonstrated cell stack assemblies with voltage decay rates within the target of 2% per year with durability predicted to be more than 5 years.

The principal causes of voltage decay in PAFC are reduced catalyst activity over time and loss of electrolyte by evaporation. The catalyst is a platinum alloy deposited as very small platelets on high-surface-area carbon black. The platelets do not bond strongly to the carbon and thus migrate over its surface. When platelets touch they tend to merge and form a hemispherical shape which has reduced surface area and reduced chemical activity. This coalescing of platelets decreases with time (the distance between catalyst platelets increases) therefore the rate of performance decay also decreases with time. The electrolyte loss is similar to the situation with MCFCs. The vapor pressure of the phosphoric acid is very low but finite, and less than optimum electrolyte fill decreases the cell ionic conductance. The PC25® power plants, manufactured by International Fuel Cell, LLC, have demonstrated the low decay rates and durability necessary for more than 5 years of operation.

The principal decay mechanism in PEMFCs is contamination of the electrolyte through replacement of the conducting hydrogen ion with a metal ion. The metal ions come from contaminants present in the electrodes and from products of corrosion of the upstream piping and fuel-processor reactors. The electrolyte is a polymer with sulfonic acid grafted to it and the number of sulfonic acid groups per unit area of electrolyte is not large. Over time, replacement of the hydrogen ion by a metal results in the formation of a nonconducting metal salt leading to fewer acid molecules, which, in turn, increases the electrolyte ionic resistance and lowers the cell operating voltage.

Fuel cell developers include the effects of cell decay in their power plant designs by oversizing the balance of plant systems to provide the extra hydrogen and air reactants and cell stack assembly heat rejection required by the degraded fuel cell. These balance-of-plant systems are described in the following discussion.

D. Fuel Processing Subsystems

Stationary fuel cell power plants must utilize common hydrocarbon fuels. The fuel of choice in developed countries is natural gas. In developing countries, easily transported fuels such as distillates are preferred. Research exploring means to oxidize these fuels directly in fuel cells has been extensive but unsuccessful. Because fuel cells use hydrogen readily, fuel cell power plants include a fuel processing subsystem to convert the hydrocarbon fuel to hydrogen. There are two well-known processes suited to fuel cell power plant needs: catalytic steam reforming (CSR) and partial oxidation (POx).

1. Catalytic Steam Reforming

This process involves the reaction of the hydrocarbon fuel with steam in the presence of a catalyst. The steam reacts with the carbon in the fuel to form carbon monoxide, releasing hydrogen from the fuel and steam. Using methane as an example fuel, the reaction is

$$CH_4 + H_2O \rightarrow 3H_2 + CO.$$

This reaction proceeds rapidly at temperatures above 700°C over relatively inexpensive catalysts such as nickel deposited on alumina or magnesia. Excess steam is always added, typically 2.5–4 molecules of H_2O per carbon atom, to promote the gasification of elemental carbon that may form from the dissociation of fuel or CO.

Additional steam can react with the carbon monoxide to produce additional hydrogen,

$$CO + H_2O \rightarrow CO_2 + H_2.$$

This is known as the water–gas shift (or simply "shift") reaction. Chemical equilibrium considerations require much lower temperatures of approximately 250°C for the reaction to approach completion. Hence, the overall reaction with excess steam is

$$CH_4 + 3H_2O \rightarrow CO_2 + 4H_2 + H_2O.$$

The reform reaction is strongly endothermic and the shift reaction moderately exothermic. If fuel is burned to provide the endothermic energy, it requires the equivalent of $0.7\ H_2$, leaving a net $3.3\ H_2$ for both fuel-processing reactions.

2. Partial Oxidation Reactions

The objective is to react the fuel with a small amount of oxygen (air) and steam. This is an exothermic reaction. Heat does not have to be transferred through the walls of the reactor as in catalytic steam reforming. This process allows greater freedom in selecting the shape and size of the POx reactor, more rapid startup, and quicker response to changes in throughput. If only the carbon in the fuel is oxidized to CO, the reaction is

$$CH_4 + \tfrac{1}{2}O_2 \rightarrow CO + 2H_2.$$

There is sufficient heat generated by this reaction to dissociate the fuel and to generate the excess steam required for the shift reaction,

$$CO + 2H_2 + 2H_2O \rightarrow CO_2 + 3H_2 + H_2O.$$

This last reaction produces approximately 10% less net hydrogen than catalytic steam re-forming.

The partial oxidation reaction can take place with or without a catalyst. The reaction can also occur at a higher temperature than catalytic steam reforming because there is no heat transfer materials limitation on the reactor vessel wall. Internally insulated reactor vessels permit reaction temperatures above 1000°C. The higher temperature increases the reaction rate, which shrinks the vessel size substantially and often reduces capital cost as well. The higher reaction temperature also helps with dissociation of heavy hydrocarbons such as distillate fuels. Because of the smaller reactor size, fast startup, and rapid transient response, partial oxidizers are receiving much attention by the developers of fuel cell systems for transportation applications. The shift reactions are the same for both fuel processors and therefore the catalytic shift reactors will have similar requirements for both processes.

If air is used in the reaction, the fuel product is diluted by nitrogen (1.88 moles N_2 in the partial oxidation reaction) yielding a hydrogen concentration of less than 50%. The dilute hydrogen reduces the electrochemical performance of the fuel cell and reduces the amount of hydrogen that can be consumed by the fuel cell. If $3.3\ H_2$ of the $4\ H_2$ from CSR are consumed in a low-temperature fuel cell, the remaining hydrogen concentration is 26%; if $2.7\ H_2$ are consumed of the $3\ H_2$ available from the POx reaction using air, the remaining hydrogen concentration is

7%, which is about the minimum concentration that can sustain a usable current density in the fuel cell assembly. Thus, in this example, the CSR process, which generates 10% more hydrogen than POx, permits 22% more electrical current to be produced ($3.3/2.7 = 1.22$) for the same methane consumption. Therefore, catalytic steam reforming is generally selected for stationary power plants because it yields higher system efficiency. Exceptions occur for very small systems in which some developers have selected POx as a means to reduce system capital cost and for systems that must utilize heavy hydrocarbons such as diesel fuel.

POx systems are favored in transportation systems where total system weight and volume are important.

Both hydrogen generation systems utilize catalysts that are poisoned by sulfur compounds present in the fuel. The sulfur compounds may exist naturally in the fuel or may be added as an odorant. These compounds must be removed to levels of 1 ppmv or below. In small systems, an absorbent such as activated carbon is commonly used. In larger scale systems, it is more economical to react the sulfur compound with hydrogen to form hydrogen sulfide and then react the H_2S with ZnO to form solid ZnS. In both desulfurization processes, the sulfur-absorbing material is replaced periodically.

The water used to generate steam must be of boiler-feed water quality. Obviously, minerals that may be in the water will foul a boiler and chlorine ion will corrode the metals of the piping, heat exchangers, catalyst, and catalytic vessels at the elevated temperatures occurring in the fuel-processing systems. Some fuel cell developers include a water purification system as a standard component of their fuel cell systems.

All of the specified fuel-processing steps have been practiced commercially for many years. Huge quantities of hydrogen are produced, for instance, to synthesize ammonia for fertilizer, upgrade petroleum products, and hydrogenate fats and oils for food products. These are generally large-scale installations, i.e., the equivalent of tens to hundreds of megawatts if the hydrogen were used in fuel cells, and these installations usually operate continuously. The challenge for fuel cell developers has been to scale down these processes to almost miniature scale and to engineer the features of rapid startup and shutdown, fast transient response, and efficient integration with the fuel cell subsystem at low capital cost. This challenge has been comparable in difficulty to that of developing durable, efficient, and low-cost fuel cell assemblies.

E. Power Conditioning Systems

Fuel cells generate direct current (dc) electricity. This must be converted to alternating current (ac) electricity for most consumers at the voltage, frequency, and phasing similar to those in the electric grid. DC to ac power conditioners using solid-state electronics that meet stationary power plant requirements are available commercially, although at higher than desired capital cost. In the past, power conditioners for stationary power plants were generally specially ordered or limited production items and hence expensive. Electric vehicle drives use similar technology and, at the $\sim$100-kW power level of interest to many stationary fuel cell system developers, the capital costs are expected to decrease as manufacturing levels increase.

The stationary power plant should operate when connected to and when isolated from the electric utility grid to be of most benefit to the user. When operating independently of the grid, the power source must be able to follow step changes in load as electrical devices are switched on and off. The electronic power conditioner can react instantaneously to load changes, but the balance of the system and particularly the fuel processor may not be capable of reacting rapidly. Some power plant designers will use batteries in parallel with the fuel cell assembly to accommodate the electrical transients. If battery storage is used, the charging circuitry is included in the power conditioner.

F. System Control Subsystem

A supervisory control system is required to coordinate operations of the fuel processing system, fuel cell assembly, and power conditioning subsystems at all operating conditions, i.e., through startup, steady-state and transient electrical output, standby, and shutdown. As mentioned above, the development of the fuel processing system has been a formidable challenge, as has integrating it with the fuel cell assembly and controlling both to give the desired instantaneous electrical output. Technical details including process flow diagrams and heat and mass balances of systems that overcome these challenges are given in the texts listed in the Bibliography. A brief historical description of the challenges follows, to put them in perspective.

The initial efforts to develop hydrocarbon-fueled fuel cell systems date to about 1960. Most efforts at the time focused on PAFC, as this was the most advanced technology at the time. The catalytic steam reformer, operating at a temperature much higher than the fuel cell, obviously needs to be flame-fired. To minimize system fuel consumption, it is necessary to minimize the fuel consumed in this burner. Therefore, the fuel and steam feeds to the reformer and the fuel and air feeds to the burner need to be preheated. This approach requires gas-to-gas heat exchangers with wall temperatures of 500°C. The high- and low-temperature shift reactors and the sulfur-removal reactors each have heat exchangers upstream to condition the reactant to the proper temperature. The hydrogen-rich fuel and air feeds to the fuel cell assembly need to be brought to their proper temperature and humidity, which adds two heat exchangers. The fuel cell assembly must be cooled, and most PAFC (and PEM) developers opt for circulating liquid coolants that include water, water–steam mixtures, and dielectric fluids. The fuel cell reject heat is often used for cogeneration or within the hydrogen-generating process, but one or more heat exchangers are needed to capture this heat. Most power plant developers also want to recover the product water formed within the fuel cell rather than consume makeup water continuously. Therefore, condensers were installed to recover the excess water in the hydrogen-rich feed upstream of the fuel cell assembly and in the oxygen-depleted air exhaust from the PAFC (or PEM). In summary, the fuel cell systems contained 10 or more heat exchangers in addition to several pumps, blowers, and other motor-operated control devices. Also, the fluid flow rate through the system varies widely because it varies in proportion to the electrical load, which adds to the control complexity of this system. Systems constructed by several developers from the 1960s through the 1980s were functional but suffered from poor reliability and durability and high capital cost. The reliability and durability problems have been overcome, however, as evidenced by the excellent operating records of the PC25® power plants manufactured by In-

ternational Fuel Cells. The high capital cost of the complex system remains a problem.

Developers of high-temperature fuel cells recognized that the overall hydrocarbon–air fuel cell system would be less complex and thus potentially less expensive. MCFC and SOFC developers have the option to use reject heat from the fuel cell assembly to provide the heat to the reformer, thus eliminating the reformer burner. Demonstration model power plants have been constructed using fired and unfired reformers. No shift reactors are needed in these systems (which eliminates associated heat exchangers) because the shift reaction takes place over the fuel electrode. The steam produced from oxidation of the hydrogen fuel drives the shift reaction to completion. The fuel cell reject heat is removed by providing excess air to the air electrode. However, the reformer reactants and the fuel cell reactant air must be preheated to temperatures of approximately 550°C, so that high temperature gas-to-gas heat exchangers are still needed. The heat in the exhaust air can provide this preheat plus raise steam for the reformer and for cogeneration applications. A simplified schematic of an SOFC system developed at Global Thermoelectric Corporation is shown in Fig. 2. In summary, the high-temperature fuel cell systems require only approximately half as many heat exchangers and related control devices as the low-temperature systems. This change reduces the

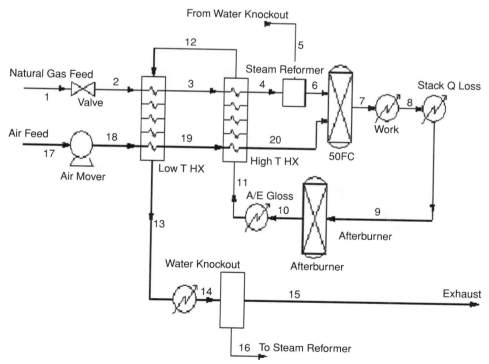

FIGURE 2 Simplified SOFC system schematic. (Courtesy of Global Thermoelectric, Inc.)

plant capital cost but may be offset to some extent by a higher cost for the high-temperature fuel cell assemblies.

G. Hybrid Fuel Cell Systems

All energy conversion technologies have waste streams as a result of conversion inefficiencies. Whether these streams are useful or not depends on their quality (temperature) and quantity. The best example, which leads into a discussion of hybrid cycles, is that involving gas or combustion turbines. These produce a hot exhaust stream whose temperature is a function of the turbine efficiency. The higher the efficiency, the lower the temperature. This exhaust stream can be passed through a heat recovery steam generator (HRSG) to provide high-pressure steam to operate a steam turbine. This approach permits the production of more electricity without the expenditure of any additional fuel, thus increasing the overall fuel efficiency. This marriage is called a combined or hybrid cycle, wherein the waste of one device is utilized to power another device. The higher temperature device is referred to as the topping cycle, while the lower temperature device is the bottoming cycle.

High-temperature fuel cells can be used as the topping cycle in hybrid/combined cycles. Thermodynamically, it is desirable that the most efficient device (which is the fuel cell) be the topping cycle. A number of possible bottoming cycles can be used, for example, Stirling, Rankine (steam or other fluid), and Brayton (gas turbine) cycles. However, most developers are pursuing gas turbine technology because they are available in sizes that are useful for this application.

There are two major combined cycles under development. One is based on SOFCs and the other on MCFCs.

The inherent differences in these two technologies lead to major differences in overall system designs. Each of these is discussed in detail in the following paragraphs.

1. SOFC—Hybrid Cycle

In the spring of 2000, Siemens Westinghouse (S-W) started up a SOFC–CT hybrid unit designed for 250 kW. This was the first fuel cell hybrid cycle power plant constructed. It incorporated a 200-kW S-W SOFC and a 50-kW Northern Research and Engineering combustion turbine. This unit was primarily intended to identify and resolve technical problems relating to the dynamic interaction of the two devices at steady-state and transient conditions. Neither device was optimized for this application because modification of existing hardware was the most cost-effective path to determining what design and operational changes are required for optimization. This unit is under test at the National Fuel Cell Research Center at the University of California, Irvine, financially supported by the U.S. Department of Energy, S-W, Southern California Edison, and others. The S-W unit incorporates the largest tubular cell array they have built (also the largest SOFC of any configuration ever built), but its modular design permits construction of larger stacks even though the tubular cell size is fixed. Operation at pressure has increased the SOFC output by one-third, from 150 to 200 kW.

Although simplified, the schematic in Fig. 3 highlights a number of points: (1) the system is relatively simple, i.e., few heat exchangers (HX) are used, (2) the S-W design, with partial anode recycle, eliminates the need for water recovery and a steam generator, and (3) although shown externally, the reformer is actually thermally integrated into the fuel cell stack assembly. This integration permits

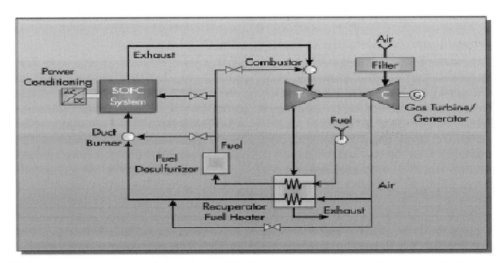

FIGURE 3 Simplified schematic of the Siemens–Westinghouse hybrid system. (Courtesy U.S. Department of Energy.)

significant heat removal from the stack by the endothermic reforming reaction, thus reducing the magnitude of cooling required. Also, it is not necessary for the reformer module to work perfectly, i.e., CH_4 fuel does not have to react completely since it will react at the anode with the product H_2O and CO_2. The nickel–cermet anode electrode is a catalyst for the methane-reforming reaction.

In this hybrid system configuration, the power turbine drives the air compressor as usual, but the compressed air is supplied to the fuel cell. The waste energy (thermal and chemical) from the fuel cell increases the air temperature and drives the power turbine. The fuel cell serves in effect as the gas turbine combustor. The output of the gas turbine is expected to be about 20–30% of the total electrical output. If the SOFC is 50% efficient, then the hybrid cycle will be 60–65% efficient with the energy contributed by the combustion turbine.

The power-generating units in the hybrid cycle are interdependent due to the close coupling of the flow streams. If one unit fails, the control system must respond in such a manner that neither unit is damaged. The fuel cell is generally more sensitive to upsets than the turbine because it does not tolerate large pressure differentials between the anode and cathode streams. The fuel cell anode exhaust mixes with the air exhaust downstream of the fuel cell to combust the residual fuel, which in turn flows into the power turbine. Thus, both streams must operate at the turbine inlet pressure. The pressure difference between the anodes and cathodes must be kept small because pressure surges at one electrode may cause a backflow through the mix point or through cell gas seals from one electrode to the other, which will permanently degrade the performance of the electrode exposed to the wrong reactant and thus reduce the SOFC power output or efficiency in further operation.

2. SOFC—Hybrid Cycle II

The Department of Energy (DOE) also funded an SOFC hybrid design study in which the fuel cell operates at atmospheric pressure. This study by McDermott Technology, Incorporated (MTI), and Northern Research and Engineering targeted a 70% efficient (LHV) system based on planar SOFC and microturbine technologies. The plant design resulted in an ~750-kW power plant with ~70 kW of the output from the turbine. This fuel cell-to-turbine power ratio of ~10:1 is considerably higher than the 4:1 and 5:1 power ratios in the other hybrid designs described. This design also identified the need for an advanced recuperator capable of accepting inlet gases above 950°C.

3. MCFC—Hybrid Cycle

FuelCell Energy (FCE) has completed an initial design for DOE of a hybrid system involving an MCFC topping cycle with the fuel cell operating near atmospheric pressure. As shown in the simplified schematic in Fig. 4, the fuel cell replaces the turbine recuperator and combustor, adding the MCFC reject heat to the turbine power cycle through heat

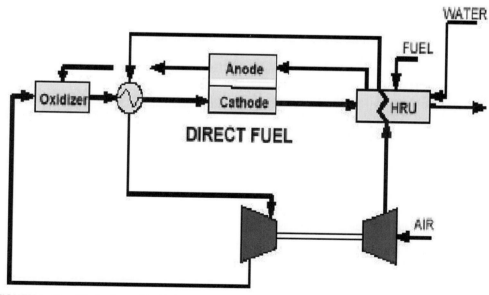

FIGURE 4 Simplified schematic of hybrid MCFC power plant. (Courtesy of FuelCell Energy Corporation.)

exchangers. The incoming air passes through the turbine compressor and is then preheated in a heat exchanger by the cathode exhaust. This air stream is further heated in a high-temperature recuperator and then expanded in the power turbine. The turbine exhaust is combined with the anode exhaust in a combustor that provides the energy for the recuperator. This combustor exhaust is passed through the cathode providing the O_2 and CO_2 for the CO_3^{2-} ion. The unique features of this hybrid cycle are that (1) no combustion is required in the turbine to achieve high power plant efficiency, (2) the fuel cell and the turbine are decoupled (which allows independent operation of the fuel cell and turbine), (3) combustion turbines with any pressure ratio available can be used in the hybrid, and (4) the regenerative heat exchanger operates at approximately 850°C using commercial alloys.

It is expected that the system efficiencies of the MCFC and SOFC hybrid cycles will be similar, in the range of 65–75% LHV, dependent on final design tradeoffs made by the developers. The MCFC operates at higher cell voltage and thus higher efficiency than the SOFC. However, the turbine temperature in the pressurized SOFC hybrid is potentially higher and more efficient, which compensates. Because the MCFC is not operating at the turbine pressure, perturbations in the power turbine have little influence of the fuel cell pressure and hence on the fuel cell durability.

As noted in prior discussions, MCFC systems have the unique requirement of partial recirculation of CO_2 produced at the anode to the cathode, where it reacts with oxygen in the air to form the CO_3^{2-} required for continued operation. In this hybrid, the recirculation loop is avoided by directing all anode exhaust and turbine air to the cathode. This combustion step is an interesting feature that eliminates the need for a feed air preheating heat exchanger. However, the high mass flows through the cathode may accelerate electrolyte loss. Combining some waste streams and doing heat recovery on one rather than two streams can reduce the number of HXs in the system. Unfortunately, it is not always obvious which is best without a detailed analysis of a likely system configuration to determine the probable temperatures and mass flows of the individual streams. In this direct MCFC design, the reformer is thermally integrated into the fuel cell stack and the final reform reaction will be completed within the anode cell chambers in a manner similar to the SOFC cycle. However, without recycle for the fuel processor, this plant requires treatment of local water to achieve boiler quality standards.

The three hybrid systems described above range broadly in degree of technical maturity. The S-W system has been built and operated, the FCE unit is a design based on MCFC technology endurance tested at 250 kW stack size, and the MTI unit is based on the performance of SOFC stacks currently operating at the 2-kW level.

II. FUEL CELL POWER PLANT APPLICATIONS

A. Single-Family Residential Applications

Several fuel cell developers have the goal of installing small fuel cells in single-family residences to provide all electrical energy plus some thermal energy needs. This goal is technically feasible but it may prove difficult to compete economically with the electrical grid. A part of the difficulty stems from the characteristics of the electrical load. Since the average single-family residence in the United States (absent the air conditioning load) uses approximately 600 kWhr per month, the average load is less than 1 kW. Most of the electric loads are steady and either on or off, i.e., lights, refrigerator compressor, etc., with the notable exceptions of television and audio systems and certain motor-driven tools. Thus, the electric load undergoes numerous step changes as electrical devices are turned on and off, and the peak load can reach 3–4 kW for brief periods of time. There are significant periods of time, as much as 10 hr per day, when the electrical load is near zero. Thus, if the power plant is designed to handle the peak load, it will be operating at just a small fraction of the rated capacity most of the time and this oversizing adds to the capital cost of the plant. During the periods of near-zero electrical loads, fuel will still be consumed to keep the power plant systems at the required temperature. Alternatively, the plant can be designed for the average load by including battery energy storage, with the battery recharged during periods of low power demand. The battery also manages electrical load transients so that the fuel cell system is not required to have rapid transient response. The decision on whether to design the plant for peak or average load is based on capital and operating costs of the systems. Fuel cell developers are pursing both approaches. In either case, there is a good match between the fuel cell system reject heat and the hot-water energy requirement of the residence.

The allowable cost for the fuel cell power plant, if owned by the homeowner, can be estimated readily. We assume that the electricity purchased from the grid costs 12 ¢/kWhr, natural gas costs $6/MM Btu, and the power plant has 40% electrical efficiency and provides another 40% as hot water. Thus, purchased electricity at 600 kWhr/month would cost $72 per month and the hot water available from the fuel cell, if provided by burning natural gas, would be an additional $13 per month with a 90% efficient heater. The fuel cell plant would consume

$31 per month of natural gas (at the stated efficiency) to provide the same electrical and thermal energy, thus saving $54 per month or approximately $650 per year. This is the amount available to purchase and install the unit and to pay for any maintenance. If the homeowner requires a simple 3-year payback, the installed plant cost would have to be approximately $2000 or less. To put this in perspective, this is approximately one-half the cost of installing central heating and air conditioning. The fuel cell system has a similar number of interfaces as the heating and air conditioning plant with fuel, water, and electrical facilities and therefore a similar installation labor cost. Hence, the fuel cell system capital cost would have to be less than that of the heating and air conditioning plant, which is a difficult target. At a simple 5-year payback, the allowable cost increases to $3250, which is still a difficult target to achieve.

The allowable cost changes if the power plant is owned by a third party such as an energy service company or utility. As an example, an energy service company can negotiate lower natural gas prices by aggregating large groups of consumers. An electric utility can offset capital and operating costs by replacing or deferring energy purchases from other generators and deferring the construction of other transmission and distribution systems. The magnitude of these offsets is a complex calculation and often unique to each potential third-party owner. Most fuel cell developers pursuing this application are working with or are partners with energy service or utility companies.

Air conditioning, if powered by the fuel cell, adds to the peak power and energy required during the cooling season. The additional electrical energy produced during air conditioner operation also provides additional thermal energy for which there is no practical use and therefore the system efficiency decreases. During the seasons when air conditioning is not required or is used only occasionally, the mismatch between power plant capacity and load is greater than before. The use of gas-fired air conditioning systems, such as absorption cycle or engine-driven systems, permits a smaller and more efficient fuel cell system to provide the other electrical power needs.

There is one niche market where the fuel cell power plant may be the economically preferred choice for a homeowner. This is in remote locations where the electrical grid is not nearby. Homes in remote locations are becoming more popular and practical with the advent of wireless communication systems. In this application, the fuel cell competes with engine-driven generators and photovoltaic or wind-energy systems. These systems have either high capital cost or high operating cost.

Two types of fuel cells, PEMs and SOFCs, are receiving most of the development efforts for single-family residential applications. These cells can be made in small sizes with many cells stacked electrically in series to produce the low power needed for this application at high voltage, which is not easily done with the other fuel cells. Companies that have demonstration model PEM power plants under test include Analytic Power, Ballard, H-Power, and Plug Power. Ceramic Fuel Cells, Global Thermoelectric, SOFCo, Sultzer Brothers, and others are testing demonstration model SOFC power plants. Further information about the progress made by these firms can be found on their web sites.

Figure 1 shows an SOFC stack assembly of ~1 kW fabricated by Global Thermoelectric Corporation and Fig. 2 a simplified process flow diagram representative of their natural gas- or propane-fueled SOFC for the single-family residential application.

B. Multifamily Residential Applications

The electrical load in an individual apartment is similar to that of a single-family residence. But individual lifestyles differ enough that neither the peak loads nor the minimum loads are coincident among the tenants. The multifamily-building peak load is proportionally lower in magnitude and longer in duration than in a single-family dwelling. Thus, there is a leveling of the electrical load that improves the utilization of self-generating equipment.

Apartment buildings are located in urban or suburban areas where electrical power is available from the grid. The most economical application for a fuel cell or other self-generators in apartment buildings is to interconnect it with the grid and size it for the base load of the building. Thermal energy can be utilized to heat water for laundry and other domestic hot water uses. The on-site power plant will operate at maximum capacity most of the time with the peak power provided by the grid. These installations may qualify for special natural gas rates as cogenerators, which reduces operating costs. As is the case for the single-family residence, the owner of the power plant will probably be an energy service or utility company rather than the building owner.

The base load for the apartment building will be less than 1 kW per apartment unit. Hence, most buildings need only relatively low capacity power plants. The developers of PEM and SOFC are aiming at the market for small power plants, i.e., power plants in the tens of kilowatts size. All types of fuel cells are being developed for large multifamily buildings that require plant sizes in the hundreds of kilowatts. Some of the manufacturers pursuing the larger power plant market include Ballard Power Systems, FuelCell Energy, Fuji Electric, International Fuel Cells, Mitsubishi Electric, Toshiba, and Ansaldo. More information about their power plants can be found on their web sites.

FIGURE 5 A 200-kW commercial PAFC power plant. (Courtesy of International Fuel Cells, LLC.)

Figure 5 is a photograph of a 200-kW PC25® cogeneration unit manufactured by International Fuel Cells (IFC) and installed at an office building in Connecticut. IFC has constructed more stationary fuel cell power plants than any other manufacturer and has approximately 200 of these units in operation in North America, Europe, and the Far East. Figure 6 is a block diagram of the PC25 showing the energy products available from the unit.

FUEL CELL POWER PLANT CHEMISTRY
Natural gas fuel and air

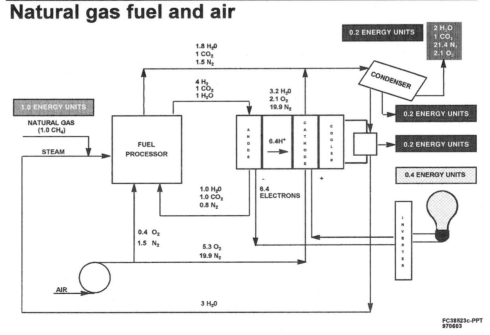

FIGURE 6 The 200-kW PAFC fuel cell system energy balance. (Courtesy of International Fuel Cells, LLC.)

C. Commercial and Industrial Applications

The most economical fuel cell applications in the commercial and industrial sectors will be those that require substantial electric power around the clock, i.e., installations where the ratio of peak- to base-load power is low. This means that the plant will have a high utilization factor that makes the most effective use of the capital cost of the plant and also maximizes the fuel cost savings due to the high generating efficiency of the plant. Some examples of buildings in the commercial sector that have low peak-to-base-load ratios include hospitals, hotels, prisons, and grocery stores. For example, a typical 300-bed hospital will have a minimum load of approximately 1000 kW and a peak load that may approach 2000 kW. In most instances, hotels and motels have lower power requirements but similar peak-to-base-load ratios. These applications also require large quantities of hot water for laundry and bathing, and therefore cogeneration installations are favored.

Grocery stores, due to large refrigeration loads, have an even lower peak-to-base-load ratio (especially those that stay open 24 hr per day) but have little need for hot water. The reject heat from the high-temperature fuel cell systems could be used for absorption refrigeration, however.

There are many industrial applications for fuel cells that span the full breadth of the industrial sector. These range from small installations such as machine shops to very large plants such as petroleum refineries. The common thread between the different industrial installations where fuel cells are economical is multiple-shift operation so that the base-to-peak-load ratio is low. There will usually be one type of fuel cell that is best suited for each installation depending on the desired characteristics of electrical performance, plant efficiency, plant reliability, reject heat characteristics, and fuel options. But all fuel cell types will find favorable applications.

Most of the commercial and industrial installations will be in urban or suburban areas. Thus, an electric grid will be available and this will set the threshold energy costs with which the fuel cell installation must compete. However, the electrical loads that have low peak-to-base-load ratios are also those that are most profitable for the local electrical utility. Thus, the local utility may negotiate a lower energy rate rather than lose the customer. The fuel cell installation must therefore promise significantly lower energy costs than the utility to win the competition.

The energy pricing structure that utility companies apply to commercial and industrial customers is usually made up of two or more components, an energy component measured in kilowatt-hours and a power component (usually referred to as electrical demand) expressed in kilowatts. The latter is usually extended back in time for some extended period, for example, 6 months to 1 year, to account for the generation capacity the utility must have available whether or not it is used regularly. There may be other charges as well, such as a flat monthly connection charge, standby generation charges, and charges for higher quality or higher reliability service. The sum of all these charges, aggregated monthly and annually, is what the fuel cell installation must compete with.

The prospective fuel cell owner must decide whether to size the installation for the base load, leaving the utility to provide the peak-power component, or to handle the peak load as well. In the latter case, questions arise: How will the load be serviced during maintenance of the fuel cell? Will the installation remain connected to the utility grid in order to purchase emergency power from the utility or be independent of the grid with standby emergency generation included in the installation? In all cases, there may be economic (and operational) advantages to installing several small units rather than a single larger capacity unit. In all cases, the units will be shut down periodically for maintenance and repair. If the installation is connected to the grid, the maintenance shutdown will trigger a demand charge that can range up to the capacity (in kilowatts) of the unit being maintained. Clearly planned maintenance will be scheduled for the spring or fall when the peak loads are at their minimum, but breakdowns can occur anytime. If the electrical load is served by several small units rather than a single large unit, it is unlikely that more than one unit would be undergoing repair. Therefore, the demand charge would be assessed only for the smaller unit. For example, if the utility demand charge is $5/kW-month and is assessed on the basis of the highest demand over the past 6 months, the repair outage for a 1000-kW unit could be as much as $5/kW-month $\times$ 1000 kW $\times$ 6 months $=$ $30,000. If the 1000-kW load was serviced by four 250-kW units, the demand charge could be reduced to $5/kW-months $\times$ 250 kW $\times$ 6 months $=$ $7500 for an unplanned outage. Performing planned maintenance on the small units sequentially during periods of low peak power can essentially eliminate the demand charge.

If the prospective fuel cell owner wishes to provide peak power, whether or not the system is connected to the grid, the most economical installation may use a mix of fuel cells and engine generators. A diesel-engine generator will not have the durability of a fuel cell, will cost more per hour to operate because of higher fuel consumption, and many areas will limit its operating hours because of its exhaust emission levels. The engine-generator capital cost, however, is only 10–20% of the cost of a similarly sized fuel cell system. An engine generator of the same electrical capacity as one of the fuel cell units can substitute for a fuel cell during maintenance and contribute to the seasonal peak-power demand. This may aggregate to only several hundred operating hours per year; thus the annual operating cost is low and the engine will last for many years.

All of the commercial and industrial applications described above have the fuel cell installation competing with power purchased from the local utility. There are instances where the fuel cell may provide added value, i.e., where higher power quality or higher reliability of service than that offered by the utility are of value to the consumer. The fuel cell unit, operating in parallel with the electric grid, can be switched to continue powering critical loads during a grid outage. Therefore, the fuel cell serves as an uninterruptible power source. The IFC PC25 shown in Fig. 5 operates in this manner. Another concept has the fuel cell connected as the primary power source for the critical loads with the electrical grid as the backup; this approach provides the benefit to the consumer of being able to check on the availability of the backup power at all times. The fuel cell installation can also protect the electrical load from voltage fluctuations and waveform distortions that are normal grid occurrences. In this service, however, the fuel cell must have load-following capability.

D. Electric Utility Applications

The principal applications for fuel cells on utility grids in the near term are to mitigate local problems with the distribution system. The fuel cells can be dispersed on the distribution system either at customers' sites (possibly with cogeneration) or at utility substations. The fuel cell may support grid voltage at peak power demand, postpone an upgrade of a substation or a distribution line (this is particularly valuable if the distribution line is below ground), or help improve local power quality and reliability. The cost savings for each of these installations will be unique to the installation but can be substantial. In the longer term, as fuel cell capital costs are reduced and fuel prices increase, they may compete with other large-scale generators for bulk-power generation.

The deregulation of investor-owned utilities has reduced the market for fuel cells in this distribution-system-support application. Deregulation has, in essence, split these utilities into three companies, generation, transmission, and distribution companies, the latter being the retail provider to the electricity users. In some states, the distribution company is prevented from owning any generation capacity even if it would be the most economical means to address local service problems, the idea being that the distribution company could pick the most profitable customers to serve with their own generation and thus have an unfair competitive edge. In a vertically inte-

grated utility, the fuel cell installed on the distribution grid would also reduce transmission-line losses, offset (or defer) other generation construction, count against standby reserves needed for reliable service, and provide emission credits to offset those of other generators. These approaches all reduce the utility costs. A company engaged in only distribution service is unable to accrue these cost savings. Some vertically integrated utilities remain, particularly among municipality-owned and cooperative utilities, and these utility sectors are among the strongest advocates of fuel cell development and commercialization activities.

The electric utility owning a fuel cell system can determine the power level at which the unit should operate at different times during the day. Thus, there is no need for fast transient response because the electric grid can provide it. Backup power is also available from the grid so there is little incentive to install multiple units at one location for redundancy. While all types of fuel cells are candidates for distribution system support, the higher electrical efficiency of the high-temperature systems (MCFC and SOFC) may yield the lowest total costs.

Figure 7 shows at 250-kW Direct Fuel Cell® constructed by FuelCell Energy Corporation, using MCFC technology. The other manufacturers of fuel cell power plants that can be considered for these applications are the same as listed under commercial and industrial applications plus firms developing hybrid-cycle systems such as Siemens–Westinghouse.

E. Emission Credits and Low-Cost Fuels

Potential industrial users of fuel cells that have air-emission allocations (i.e., permission to emit a specified quantity of pollutants into the atmosphere) may reduce their total emissions by incorporating fuel cells into their operations. Fuel cells have much lower air emissions than any other fuel-burning generators. The Clean Air Act Amendment of 1990 permits the exchange or sale of emission allowances, thus providing an economic advantage for the use of fuel cells. This is not a large sum versus other clean generators such as combustion turbines using the best available emission-reduction technology, amounting to approximately 1 ¢/kWhr for both SO_2 and NOx credits. However, if an industrial plant expansion is limited by an emissions-allocation ceiling, retrofitting the plant for the use of fuel cells for some of the electrical and thermal needs may be the economically preferred solution.

Fuel cells also have unique advantages in using some low-energy, low-cost fuels. These are fuels that include landfill gas, sewage treatment digester gas, some refinery off-gas products, and some natural gas reserves

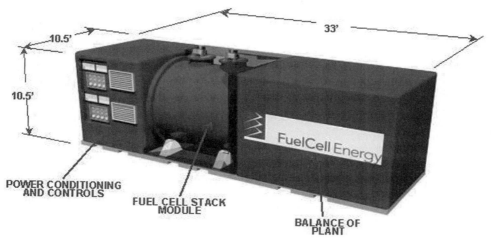

FIGURE 7 Artist's concept of 250-kW direct fuel cell. (Courtesy of FuelCell Energy, Inc.)

that contain nitrogen or carbon dioxide diluents. The landfill and digester gases contain approximately 50% each of methane and CO_2 (dry basis) corresponding to ~500 Btu/scf heating value. The refinery gas is a mixture of H_2, CO, and CO_2 with a typical heating value less than 300 Btu/scf. The natural gas wells that contain diluents generally have higher heating values than landfill gas but vary significantly among wells. All of these gases contain small amounts of contaminants, usually sulfur- or chlorine-containing chemicals, that must be removed, as they will harm the fuel cell system. The contaminants are removed upstream of the fuel processing subsystem as is done with conventional fuels.

The low-Btu fuels are used in engine generators and steam plants. However, the diluents in these fuels displace fuel and oxidant in the engine combustion chamber, thus reducing engine power. The work of compressing the diluents further reduces engine power and efficiency. The diluents have much smaller effects on the fuel cell anode potential, and there are fuel cell systems in the appropriate size that operate at atmosphere pressure which eliminates losses due to compression work. There may be some derating of the fuel cell system output due to the diluent volume that will cause an increase in the pressure drop in the fuel processing subsystem, but this is less important than engine derating. There are negligible changes in the efficiencies of the fuel cell system using these dilute fuels. The fuels containing CO_2 may actually improve the MCFC system performance, as CO_2 is a reactant that improves the cathode potential.

The quantity of these low-Btu fuels is substantial. Landfill gas is the largest of these resources and is estimated to have ~6000-MW base load capacity in the United States if used in fuel cells (Electric Power Research Institute, 1992). Landfills, sewage-treatment plants, and refineries

are all located near population centers where the electricity is needed. In contrast, low-Btu natural gas may be in areas with low population and may require expensive electrical transmission lines.

The net effect of using fuel cells with landfill gas is that much more electricity can be produced than with engine generators, perhaps twice as much as with diesel-engine generators or steam plants for the same gas resource. Emission credits can accrue to these fuel cell plants and will be maximized due to the increase in kilowatt-hours produced. These fuels release their CO_2 to the atmosphere whether or not they are burned and hence are considered CO_2-neutral under the global warming debate. If CO_2 is taxed as a means to reduce greenhouse gas emissions, the CO_2 credits from these sites may have more value than the electricity produced.

F. Hybrid Cycles

The incentive for using fuel cell hybrid cycles is fuel economy. Simply stated, no other power generation technology comes near the 65–70% LHV electrical efficiency that these systems are capable of. What is remarkable is the high efficiency of the hybrids in relatively small (hundreds of kilowatts) packages. For comparison, the highest efficiencies with combined-cycle (gas–steam) turbine systems require power plants in the 250- to 500-MW range to achieve ~58% HHV efficiency. Diesel generators need to be in the tens of megawatt size to operate at +40% efficiency. Power plants of these sizes are simply too large for all but a handful of industrial installations; at hundreds of megawatts they are suited only for electric utility central station power. Thus, the early market for hybrid technologies is believed to be in the 250 kW to 2.5 MW range in commercial and industrial self-generation applications and as dispersed generators for electric utilities. Some of these may be cogeneration installations, although there is not much reject heat available to reduce the plant operating cost.

A potential weak link in fuel cell hybrids is the gas turbine. Although turbine technology, in general, is mature, experience with turbines below 100 kW for continuous duty is very limited. Reliability data for MCFC are now beginning to be generated at a useful size, and SOFC units have operated for periods over 1500 hr with almost imperceptible decay. Should the reliability and capital cost of the hybrid configurations prove to be reasonable, the market will rapidly adopt them. At the 250-kW size, the PAFC and PEMFC units are 35% efficient. A ~60% SOFC–CT unit, if priced competitively, should be welcome, as fuel costs would be reduced by over 40%. The likely continued increase in cost of fuel and/or concern for CO_2 emissions will be a further economic and environmental incentive for eventual adoption of hybrids.

III. COMMERCIALIZATION PROGRAMS

The total program cost to commercialize a fuel cell power plant for any of these stationary power applications is very high. It was estimated to be several hundred million dollars in testimony given to the U.S. Congress by public and private organizations seeking federal support. This amount is more than any single industrial firm has been willing to invest in order to capture a segment of these markets. Therefore, strategic partnerships have developed where the program costs are shared between technology developers, potential users, public R & D funding agencies, component suppliers, and venture capitalists. Many of these partnerships are international in scope. A source for general information concerning partnerships is the Fuel Cells 2000 web site listed in the Bibliography, which is updated frequently with technical highlights, press releases, and other pertinent information.

Government funding plays a key role in many of these programs. For this reason, there is much descriptive information about the programs available in the public domain. However, for cost-shared programs with other partners detailed technical and business information is withheld as proprietary. In the United States, the Department of Energy (DOE), Department of Defense (DOD), and National Aeronautics and Space Administration (NASA) have sponsored government funding for the research, development, demonstration, and, in some instances, market entry of fuel cells for stationary power. Descriptions of the projects sponsored and literature are available from their web sites. The web site for DOE is http://www. metc.gov—and contains links to other agencies. The

Japanese government has sponsored fuel cell development and market entry programs for many years under their New Energy & Industrial Technology Development Organization. Descriptions of the Japanese programs can be found at http://www.enaa.or.jp/WE-NET/contents—e. html. In Europe, government funds from individual countries are funneled through the European Commission, and project descriptions can be found at http://www. hyweb.de/gazette-e/.

Industrial research and/or trade organizations have also made major investments in fuel cell development and commercialization activities. In the United States, the Electric Power Research Institute (EPRI) and the Gas Research Institute (GRI) have sponsored fuel cell R, D, & D programs continuously since the 1970s and reports of these programs are publicly available. These programs often augmented programs sponsored by individual gas and electric companies. A similar situation exists in Japan, with funding provided by the Central Electric Power Council, the Japan Gas Association, Petroleum Energy Center, and many individual gas and electric companies. In Europe and in other countries worldwide, gas and electric utilities and some research organizations have been active participants in R, D, & D activities.

The first fuel cell power plant to be offered for sale under commercial terms is the 200-kW PC25[®] power plant manufactured by IFC. This is an example of a strategic partnership. IFC is a joint venture between United Technologies Corporation (UTC) and Toshiba. Development of this power plant was initiated in the 1960s sponsored by a group of natural gas utility companies and Pratt & Whitney Aircraft (a subsidiary of UTC). DOE and GRI joined the partnership in the 1970s by funding portions of the development and demonstration efforts. Toshiba joined in the late 1970s. More gas utilities, including utilities from Europe and the Far East, and Ansaldo Ricerche joined the program in the 1980s. During this ~25-year period, a series of power plant demonstrations was conducted initially with 12-kW units for residential applications followed by 40-kW units in multifamily and commercial applications. The gas utility partners hosted these demonstrations. The 200-kW commercial units were offered for sale in the mid 1990s and the U.S., Japanese, and some European governments subsidized some installations. IFC has produced approximately 200 PC25 units that are in operation in multifamily residential and commercial applications.

Other manufacturers with similar, internationally funded programs are at the demonstration phase of the commercialization effort with power plants representing all four types of fuel cells. Commercial sale offers will follow soon. Announcements of these offers will certainly appear on the listed web sites.

IV. BUSINESS AND TECHNICAL OPPORTUNITIES

The interest in fuel cells by automobile companies has also given impetus to the stationary power plant sector. Automotive interest helps to legitimize the technology and therefore legitimize investments by the private sector. It is likely, therefore, that numerous fuel cell developers will succeed commercially because (1) no single type of fuel cell technology is best for all stationary power applications, (2) the needs for different applications are diverse, and (3) some developers claim that the most cost-effective approach to entering the automotive market is to enter the stationary power market first, as the system performance and allowable cost goals are easier to achieve.

It is likely that the companies presently involved in fuel cell development will be the initial fuel cell suppliers to the market. The question is, then, are there are situations where new entries can develop a role in the field? Some of the roles for new companies are as component suppliers, for components that fuel cell manufacturers will want to purchase, such as heat exchangers, control devices, power conditioners, and perhaps complete fuel processing systems. There are also sales and service opportunities, leasing opportunities, and other possible business relationships that can be arranged with fuel cell manufacturers.

There are also some applications for which new or improved technology would be desirable. For example, a research area receiving much attention relates to means to improve the tolerance of the PEMFC anode to carbon monoxide. Two approaches receiving attention are to find a CO-tolerant anode catalyst and to develop a membrane electrolyte that will operate at higher temperatures. Also, lower cost membranes, more durable membranes, and membranes that retain water through good hygroscopic properties are of interest.

The developers of all types of fuel cells continually look for improved materials. The following are some desired improvements:

- Carbon is used for electrodes and separator plates in PEMFC and PAFC. High-purity carbon blacks and electrically conductive carbon-filled resins that are corrosion resistant in the electrochemical environment are desired. Also needed are corrosion-resistant and electrically nonconducting gaskets and adhesives.
- Many developers of high-temperature fuel cells use metallic separator plates. These are often bi-metals or ceramic-coated metals for corrosion resistance. An alloy that would function in both the oxidizing and reducing atmospheres in the fuel cell would be

desirable. Alternatively, lower cost cladding or coating processes would be desirable. Furthermore, electronic insulating and corrosion-resistant gasket and seal materials are desired.
- The electrochemically active SOFC components are thin ceramic and cermet layers, the electrodes are porous, and the electrolyte dense. Low-cost means to fabricate and bond these layers, to achieve some small amount of flexibility in the assembly, and to further reduce their operating temperature are all desirable objectives.

Some fuel cell developers believe that there would be cost and performance benefits with fuel cells operating at temperatures intermediate between PAFCs and MCFCs. The objective would be a fuel cell operating at approximately $500°C$ with preferably a proton-conducting electrolyte. The advantages of such a fuel cell would be use of base metals for catalysts, cell structural components, and system heat exchangers. This "ideal" fuel cell could likely be integrated easily with a minimum-complexity fuel processing system, thus promising both high efficiency with low cost.

SEE ALSO THE FOLLOWING ARTICLES

CHEMICAL THERMODYNAMICS • ELECTROCHEMICAL ENGINEERING • ELECTROCHEMISTRY • ELECTROLYTE SOLUTIONS, THERMODYNAMICS • ENERGY EFFICIENCY COMPARISONS AMONG COUNTRIES • ENERGY FLOWS IN ECOLOGY AND IN THE ECONOMY • GEOTHERMAL POWER STATIONS • PHOTOVOLTAIC SYSTEM DESIGN • TRANSPORTATION APPLICATIONS FOR FUEL CELLS • WIND POWER SYSTEMS

BIBLIOGRAPHY

Blomen, L., and Mugerwa, M. (1993). "Fuel Cell Systems," Plenum Press, New York.
Courtesy Associates. (2000). "2000 Fuel Cell Seminar Abstracts," Courtesy Associates, Washington, D.C.
Electric Power Research Institute. (1992). "Survey of Landfill Gas Generation Potential," EPRI Report TR-101068, Electric Power Research Institute, Palo Alto, CA.
Energy and Environmental Solutions. (2000). "Fuel Cell Handbook," 5th ed., U.S. Department of Energy, Morgantown, WV [CD format.]
Hirschenhofer, J (1994). "Fuel Cells. A Handbook, (Revision 3)," U.S. Department of Energy, Morgantown, WV.
Fuel Cells. (2000). Web site, http://www.fuelcells.org, Fuel Cells 2000, Washington D.C.

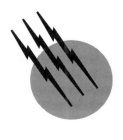

Fuel Chemistry

Sarma V. Pisupati

Pennsylvania State University

GLOSSARY

Atomization A process of breaking a liquid stream into fine droplets.

Combustion Rapid oxidation of fuels generating heat.

Equivalence ratio Ratio of the amount of air required for stoichiometric combustion to the actual amount of air supplied per unit mass of fuel.

Flame A chemical interaction between fuel and oxidant accompanied by the evolution of heat and light.

Fluidization A process of making a bed of particles behave like a fluid by a jet of air.

Fuel Any substance that can be used to produce heat and or light by burning.

Gasification Conversion of solid fuel into a combustible gas.

Liquefaction Conversion of solid fuel into liquid fuel.

Selective catalytic reduction A method used to selectively reduce nitrogen oxides to nitrogen using a catalyst.

ENERGY is the life-blood of a modern society. The quality of life of any society is shown to be directly proportional to the energy consumption of the society. In the past century (1900–2000), the population of the world and the United States has increased by 263 and 258%, respectively. The energy demand increased by a startling 16

TABLE I Total Primary Energy Consumption and Their Sources

1999 (Quadrillion Btus)		
Source	World	U.S.
Petroleum	152.20	37.71
Natural Gas	86.89	22.1
Coal	84.77	21.7
Nuclear	25.25	7.73
Hydro Electric	27.29	3
Renewable	2.83	4.37
Total	381.88	96.6

and 10 times, respectively. A vast majority of this energy (about 85%) comes from fossil fuels. These fuels—coal, oil, natural gas, oil shale, and tar sands were formed over millions years by compression of organic material (plant and animal sources) prevented from decay and buried in the ground. Table I shows the total primary energy use and the sources. Most of the fuels are used to generate heat and/or power. This chapter deals with the fuels origin, properties, and utilization methods and chemistry of these processes.

I. FUELS

Fossil fuels are hydrocarbons comprising of primarily carbon and hydrogen and are classified as solid, liquid, and gaseous based on their physical state. Solid fuels include not only naturally occurring fuels such as wood, peat, lignite, bituminous coal, and anthracites, but also certain waste products by human activities like petroleum coke and municipal solid waste. Approximately 95% of the coal mined in the United States is combusted in boilers and furnaces to produce heat and/or steam. The other 5% is used to produce coke for metallurgical uses. Liquid fuels are mostly produced in a refinery by refining naturally occurring crude oil, which include gasoline, diesel, kerosene, light distillates, and residual fuels oils. Each of these has different boiling range and is obtained from a distillation process. Gaseous fuels include natural gas, blast furnace gas, coke oven gas, refinery gases, liquefied natural gas, producer gas, water gas and coal gas produced from various gasification processes. Except for natural gas, most of the gaseous fuels are manufactured.

A. Origin of Fossil Fuels

Earth is a closed system with respect to carbon, and therefore carbon on this planet has to be used and reused. A total account of carbon in the world would explain fos-

sil fuel formation. Global carbon cycle illustrates the fate carbon in the world. Carbon exists in the world in three major reservoirs: in the atmosphere as CO_2, in the rocks as CO_3^{--}, and in the oceans, which occupy two thirds of the planet's surface, as carbonate (CO_3^{--}) and bicarbonates (HCO_3^-). The CO_2 in the atmosphere has a vital role in the formation of fossil fuels. The CO_2 in the atmosphere reacts with water vapor in the presence of sunlight to form the organic matter and oxygen by *photosynthesis* reaction. The organic matter can be of microscopic plant (phytoplankton) or microscopic animal (zooplankton) or higher plants. The dead organic matter through *decay* reaction combines with oxygen and forms CO_2 and H_2O. This decay reaction is exactly the reverse of the photosynthesis reaction. Fossil fuels have formed by minimization or prevention of the decay reaction by possibly inundating the organic matter by water or covering by sediments. The organic matter, microscopic plants and animals, and higher plants, is chemically comprised of protiens, lipids, carbohydrates, glycosides, resins, and lignin. Of these, lignin is predominantly present in higher plants. Other components are predominantly present in zoo, phytoplankton, and algae (microscopic organic matter). Coal is a complex material composed of microscopically distinguishable, physically distinctive, and chemically different organic substances called macerals and inorganic substances called minerals.

During the transformation of organic matter to coal, there is a significant loss in oxygen and moderate loss of hydrogen with an increase in carbon content by initial aerobic reactions and subsequent anaerobic reactions. This process leads to the formation of kerogen. The organic matter, which is rich in algae, forms alginitic kerogen (type I kerogen); whereas the organic matter, which is rich in fatty acids and long-chain hydrocarbons, leads to the formation of liptinic or Type II kerogen. The organic matter consisting of lignin structure forms Type III kerogen and leads to the formation of coal. Formation of kerogen from the organic matter is by a process called "diagenesis," and the transformation of kerogen to fossil fuels is by a process called "catagenesis." The temperature in the earth's surface increases with depth at a typical rate of 10–30°C/km. As the temperature and pressure due to overburden increase, peat is transformed into lignite in about 30–50 million years. The primary reactions that are believed to occur during this transformation are dehydration, decarboxylation, and condensation leading to a loss of oxygen, hydrogen, and some carbon. Progressive transformation of lignite to subbituminous coal to bituminous coal and then to anthracite occurs in a time period ranging from 50 to 300 million years. The reactions responsible for these changes resulting in rapid loss of hydrogen are dealkylation, aromatization, and condensation. Formation of anthracites from bituminous coals

TABLE II Compositional Analysis of Various Solid Fuels

Fuel component	Wood[a]	Peat[a]	Lignite, Darco seam, TX[b]	Subbituminous coal[b]	Bituminous coal (hvAb), Upper Clarion seam[b]	Anthracite, Primrose seam[b]	Petroleum Coke[a]
Moisture (as rec'd)	48.00	—	32.60	27.12	1.73	3.77	5.58
Volatile matter (d.b.)[c]	72.80	75.00	67.39	47.56	39.41	3.71	10.41
Fixed carbon (d.b.)	24.2	23.1	21.34	38.12	51.68	82.38	88.89
Ash (d.b)	3.00	2.70	11.27	14.32	8.91	13.91	0.71
Heating value (d.b) (Btu/lb)	9,030	8,650	11,375	10,842	13,390	12,562	15,033
Carbon	55.00	8,650	74.90	73.88	83.67	96.65	88.64
Hydrogen	5.77	5.60	4.58	6.28	5.33	1.25	3.56
Nitrogen	0.10	0.70	1.75	1.22	1.46	0.78	1.61
Sulfur	0.10	0.17	0.78	1.78	4.82	0.52	5.89
Oxygen	39.10	40.10	18.78	18.62	9.54	1.32	0.30

[a] C, H, N, S, and O are given on a dry ash free basis.
[b] C, H, N, S, and O are given on a dry mineral matter free basis.
[c] Dry basis.

requires higher pressures and temperatures in excess of 200°C. The higher temperatures are encountered when there is a magma nearby in the ground, and high lateral pressures are encountered where land masses collide leading to the formation of mountains. Both of these scenarios lead to the formation of anthracites. The nature of the constituents in coal is related to the degree of coalification, the measurement of which is termed rank. Coals may be classified according to (1) rank, based on the degree of coalification; (2) type, based on megascopic and microscopic observations (physical appearance) that recognize differences in the proportion and distribution of various macerals and minerals; and (3) grade, based on value for a specific use. Typical properties of these solid fuels are shown in Table II.

II. PROPERTIES FOR UTILIZATION

Coals vary widely from place and to place, and sometimes even within a few feet in a particular seam because of the nature of the precursor material and the depositional environment. Therefore, coals and other solid fuels are analyzed for certain important properties for utilization (summarized in Table III).

Coal rank is usually determined from an empirical analysis called the proximate analysis and calorific value or optical reflectance of vitrinite. The proximate analysis consists of determination of moisture, volatile matter, and ash contents, and, by difference from 100%, the fixed carbon content of a coal. The American Society for Testing and Materials (ASTM) method of classification of coals

uses the dry, mineral matter-free volatile matter to classify coals above the rank of medium volatile bituminous. For coals with greater than 69% volatile matter, the method uses the moist (containing natural bed moisture but not surface moisture), mineral matter-free calorific value to classify coals below the rank of high volatile bituminous. The moisture content is obtained by heating an air-dried coal sample at 105–110°C under specified conditions until a constant weight is obtained. The moisture content, in general, increases with decreasing rank and ranges from 1 to 40% for the various ranks of coal. The moisture content is an important factor in both the storage and the utilization behavior of coals. The presence of moisture adds unnecessary weight during transportation, reduces the available heat consuming latent heat of vaporization, and poses

TABLE III Important Properties for Utilization

Property	Factors affecting
Compositional analysis	Proximate analysis
	Ultimate analysis
Heating value	
Grindability	Coal rank
	Moisture
	Ash
Combustibility	Proximate analysis
	Surface area
	Porosity
	Petrographic Analysis
Inorganic constituents	Associated with the organic structure
	Discrete inorganic minerals

some handling problems. Volatile matter is the material driven off when coal is heated to 950°C in the absence of air under specified conditions, and is determined practically by measuring the loss of weight. It consists of a mixture of gases, low-boiling-point organic compounds that condense into oils upon cooling, and tars. Volatile matter increases with decreasing rank. In general, coals with high volatile matter are known to ignite easily and are highly reactive in combustion applications. The calorific value is the amount of chemical energy stored in the coal, which is released as thermal energy upon combustion and is directly related to the rank. The calorific value determines in part the value of a coal as a fuel for combustion applications. The ultimate analysis includes elemental analysis for carbon, hydrogen, nitrogen, and sulfur and oxygen on a dry ash free basis. Oxygen content in the fuel is obtained by subtracting the sum of the percentages of C, H, N, and S from 100.

The grindability of a coal is a measure of its resistance to crushing. The ball-mill and Hardgrove grindability tests are the two commonly used methods for assessing grindability. The Hardgrove method is often preferred to the ball-mill test because the former is faster to conduct. The test consists of grinding a specially prepared coal sample in a laboratory mill of standard design. The percentage by weight of the coal that passes through a 200-mesh sieve (screen with openings of three-thousandths of an inch or 74 μm) is used to calculate the Hardgrove Grindability Index (HGI). Two factors affecting the HGI are the moisture and ash contents of the coal. A correlation between the HGI and rank indicates that, in general, lignites and anthracite coals are more resistant to grinding (low indices) than are bituminous coals. The index is used as a guideline for sizing grinding equipment in the coal processing industry.

Inorganic constituents in coal upon oxidation produce ash. The inorganic constituents are present in the form of discrete mineral particles or bonded to the organic coal structure or sometimes water associated. The manner in which the inorganic constituents are present in raw coals determines the intermediate and final form of ash. Lower-rank coals (lignites and subbituminous) tend to have more organically associated inorganics compared to high-rank coals.

Combustibility is measured by proximate analysis and some empirical laboratory tests. Volatile matter is an indicator of ease of ignition and the fraction of coal that is burnt in gas phase. The higher the volatile matter the less is left as char that needs to be burnt. The combustion of char takes a longer time because of the heterogeneous reaction between carbon and oxygen. Combustion of high volatile coals therefore is easier to ignite and releases most of the heat closer to the burner. This requires an increase in the velocity of the fuel and air mixture jet to keep the flame at a distance from the burner tip. However, coals with lower volatile matter have a delay in ignition and therefore cause flame stability problems. These coals require additional design measures such as high swirl or a "bluff body" to promote recirculation of hot gases to stabilize the flame. Combustibility of coal is also characterized by bench-scale laboratory techniques such as drop tube reactor, entrained flow reactor, or thermogravimetric analyzer to determine the reactivity of fuels. One such method is used to determine a burning profile. This is widely used to determine the combustion behavior of an unknown fuel relative to a reference fuel. This method originally was developed by Babcock and Wilcox. A burning profile is a plot of the rate at which a solid fuel sample loses its weight as a function of temperature, when heated at constant rate in air. The burning profile is used to determine the temperatures at which the onset of devolatilization and char combustion, peak reation rate, and the completion of char oxidation occur. The temperatures range over which these take place also indicates the heat release rates and zones.

For a better understanding of behavior of coal in a furnace, pilot-scale combustion tests are performed to characterize ignition and flame stability, combustion efficiency, gaseous and particulate emissions, slagging and fouling, and the erosion and corrosion aspects of a fuel.

III. COMBUSTION OF COAL

Combustion is rapid oxidation of fuels generating heat. Combustion of fuels is a complex process the understanding of which involves chemistry (structural features of the fuels), thermodynamics (feasibility and energetics of the reactions), mass transfer (diffusion of fuel and oxidant molecules for a reaction to take place, and products away from the surface), reaction kinetics (rates of reactions), and the fluid dynamics (bulk movement of the fuel and combustion gases) of the process.

Combustion of solid fuels generally involves three steps—drying or evaporation of moisture, thermal decomposition and devolatilization, and oxidation of solid residue or char. Since solid fuels differ widely in terms of physical and chemical properties, a general qualitative discussion on the combustion behavior of a coal particle is given here. The duration and chemistry of each of these processes depend on the type of fuel burnt and the size of the particles, heating rate, furnace temperature, and particle density. For example, wood, peat, and lignite fuels contain a large percentage of moisture and the drying time is quite long. Also the volatile matter is quite high and therefore the time required for heterogeneous combustion of solid residue (char) is shorter. The drying time of a small particle is the time required to heat the particle to the boiling point of the water and evaporate the water. This depends on the amount of water in the particle, particle size, and the heating rate of the particle. The rate of heat transfer to the particle depends on the temperature difference between the boiler furnace and the particle temperature.

Devolatilization of a fuel particle involves thermal decomposition of the organic structure, thereby releasing the resulting fragments and products. The rate of devolatilization depends on the rate of heat transfer to the particle to break weak bonds resulting in the primary decomposition of products. The primary products of decomposition travel from the interior of the particle to the surface through the pores of the solid. While doing so, the primary products, depending on their nature, could react with each other or with the char surface and result in secondary products or deposits on the walls of the pores. The devolatilization process begins at about 350°C and is a strong function of temperature. The products consist of water, hydrogen-rich gases and vapors (hydrocarbons), carbon oxides, tars, light oils, and ammonia. The product distribution for bituminous coal and lignite is shown in Fig. 1. These volatiles may be released as jets and play an important role in ignition and char oxidation steps. A carbon-rich solid product called char/coke also is produced. The composition of the products of decomposition depends on the type of the fuel, peak temperature, and rate of heating of the particle.

Combustion of char is a slower process compared to volatile combustion and is critical in determining the total time for combustion of a coal particle. The steps that are involved in the oxidation of char are

1. Diffusion of oxygen from the bulk to the char surface
2. Diffusion of oxygen from the surface to the interior of pores of the char
3. Chemisorption of reactant gas on the surface of the char
4. Reaction of oxygen and carbon
5. Desorption of CO and CO_2 from the char surface
6. Diffusion of CO and CO_2 to the surface through the pores
7. Diffusion of products from the exterior surface to the bulk of fluid

The properties of the char are important for its oxidation. Pulverized coal particles, during devolatilization, are known to swell by 5–15% depending of the coal type and heating rate. Various types of chars with varying physical structures are produced. The volatile matter when released

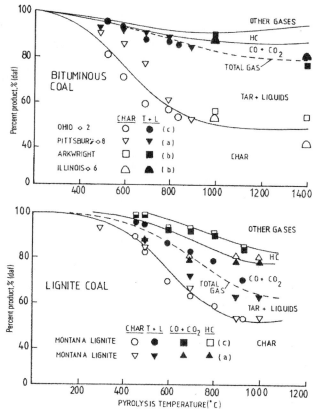

FIGURE 1 The product distribution for bituminous coal and lignite. From Lawn, C. W. (1987). "Principles of Combustion," Academic Press.

from a coal particle produces void space or porosity. The porosity of the char gradually increases as the conversion progresses. Ultimately the char particle becomes so porous that the particle fragments, as shown in Fig. 2. The combustion char that is produced can take place as surface reaction a (constant density and shrinking diameter) or a volumetric reaction (constant diameter and changing density). The surface reaction generally occurs when the temperatures are high and chemical reaction is fast and diffusion is the rate-limiting step, whereas volumetric reaction dominates when the temperatures are low and the oxidant has enough time to diffuse into the interior of the particle. The concentration profile of the reactant is shown in Fig. 2.

Ash formation occurs predominantly by two mechanisms. Volatile inorganic species are vaporized during char combustion and subsequently condense when the temperature is low downstream. The ash particles formed by this mechanism tend to be submicron in size; whereas, mineral inclusions come into contact with one another, and since the temperature in the pulverized coal combustion

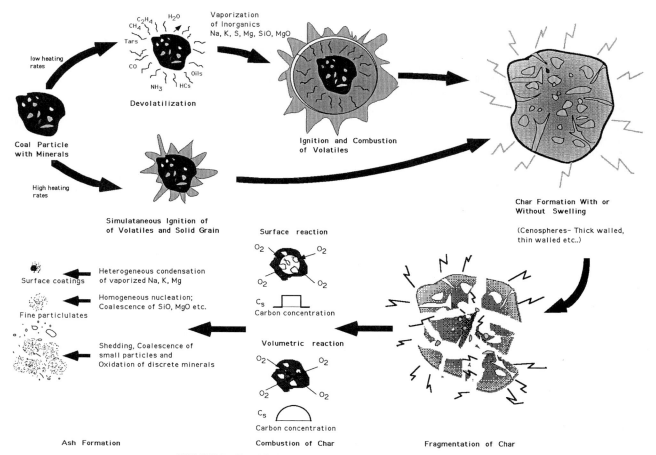

FIGURE 2 Simplified schematic of combustion of coal particles.

units is high enough to melt the molten ash, particles coalesce to form larger particles. Hence, the size distribution of pulerized coal ash tends to be bimodal.

A. Major Combustion Processes

Combustion methods of solid fuels are classified into four categories. These are fixed-bed combustion, fluidized-bed combustion, pulverized or suspension firing, and cyclone firing. These methods differ in the fuel size used and the dynamics of the fuel particles.

IV. FIXED BED COMBUSTION

Combustion of coal in fixed beds (e.g., in stokers) is the oldest method of coal use and it used to be the most common. Large-size coal pieces, usually size graded between 1/8 and 2 in., are supported on a grate and air is supplied from the bottom through the grate. The coal particles are stationary, hence the term "fixed" bed. Fixed-bed combus-

tion systems can be further classified based on the coal feed system—overfeed, underfeed, spreader stoker, or traveling grate. Large size limits the rate of heating of the particles to about 1°C/sec and requires about 45 to 60 min for combustion. A schematic of a traveling grate combustion furnace is shown in Fig. 3. Coal particles initially devolatilize and the combustible volatile gases are burnt above the bed by the overfire air. The overfire air is crucial in obtaining complete combustion of the volatiles. In most of the stoker systems, coals which exhibit caking properties can be a problem due to clinker formation. Clinkers prevent proper air distribution combustion, the air leading to high unburnt carbon and higher carbon monoxide emissions.

When the fuel is drawn on to the grate from the coal hopper by the moving grate, the bed is approximately 10–15 cm. The coal particles initially require less air for drying and combustion. As the coal bed ignites and the plane of ignition travels down from the top to the grate, combustion air demand increases to a maximum. As the coal bed burns and reaches the other side, the demand for air decreases

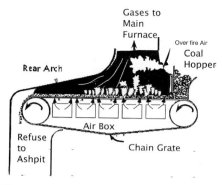

FIGURE 3 Schematic diagram of a traveling grate stroker.

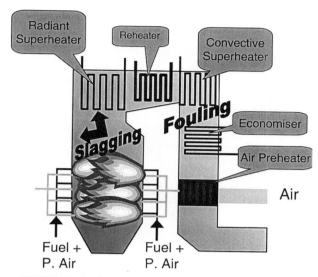

FIGURE 5 A schematic of a pulverized coal firing system.

to a minimum. Ash particles produced are dropped into a refuse pit on the other side of the combustion chamber. As shown in Fig. 4, the oxidant and the products of combustion have to travel through the layer of ash produced. This combustion method tends to produce a high amount of unburnt carbon and high(er) CO levels. Gaseous emissions from stoker systems are hard to control.

Since the coal bed height is constant, an increase in the throughput (larger size) can only be achieved by increasing the size of the grate. Due to this physical limitation of the grate size, stoker combustion units have an upper size limit of about 100 MMBtu/hr or 30 MW(thermal). Therefore, the application of this type of combustion method is limited to industrial and small on-site power generation units.

V. PULVERIZED COAL COMBUSTION

In the pulverized coal combustion system, coal ground to a very fine size (70–80% passing through a 200-mesh screen or 75 μm) is blown into a furnace. Approximately about 10% of the total air required for combustion is used to transport the coal. This primary air, preheated in the airheater, is used to drive out the moisture in the coal and transport the coal to the furnace. A majority of the combustion air is admitted into the furnace as secondary air, which is also preheated in the air preheater. The particles are heated at 10^5–10^{6}°C/sec, and it takes about 1–1.5 sec for complete combustion. This increases the throughput to furnace and, hence, it is the most preferred form of com-

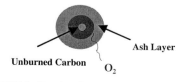

FIGURE 4 Combustion of a large coal particle.

bustion method by electric utilities. Based on the burner configuration, pulverized coal combustion systems can be divided into wall-fired, tangentially fired, and down-fired systems. In wall-fired units the burners are mounted on a single wall or two opposite walls. Wall-fired systems with single wall burners are easy to design. However, due to uneven heat distribution, the flame stability and combustion efficiency are poor compared to the opposed wall-fired system. Because of the possibility of flame impingement on the opposite wall, this system may lead to ash slagging problems. An opposite wall-fired boiler avoids most of these problems by providing even heat distribution and better flame stability. A diagram of the opposed firing system is shown in Fig. 5. However, the heat release rate can be higher than single wall burner systems and could lead to slagging and fouling problems. A tangential firing system has burners installed in all four corners of the boiler (as shown in Fig. 6). The coal and primary air jets are issued at an angle, which is tangent to an imaginary circle at the center of the furnace. The four jets from four corners create a "fireball" at the center. A top view of the furnace is provided in Fig. 6. The swirling flame then travels up to the furnace outlet. This method provides the highest even heating flame stability. The temperature in the "fireball" can reach as high as 1700–1800°C. The burner is a key component in the design of the combustion system. It also determines the flow and mixing patterns, ignition of volatiles, flame stability, temperature, and generation of pollutants. Burners can be classified into two types—swirl and nonswirl. Primary air transports coal at velocities exceeding 15 meters/sec (the flame velocity) of the fuel. The flame velocity is a function of volatile matter. Secondary air ignites the fuel, determines the mixing pattern, and

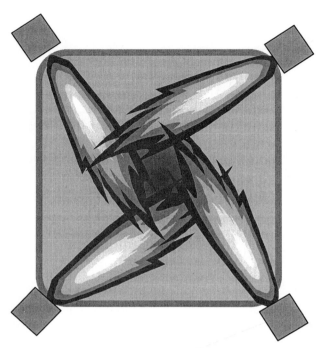

FIGURE 6 A plan view of a tangentially fired furnace.

also determines the NO_x formation. High volatile bituminous coals tend to have higher flame velocities compared to low volatile coals. Swirling flows promote mixing and esatablish recirculation zones. In swirling flows, the axial flux of angular momentum G and the axial flux of axial momentum Gx are conserved.

$$G_x = \int_{r_1}^{r_2} 2\pi r(p + \rho u^2)\, dr = \text{constant}$$

$$G_\phi = \int_{r_1}^{r_2} 2\pi \rho u w r^2 \, dr = \text{constant},$$

where u and w are the axial and tangential components of the velocity at radius r, p is the static gage pressure, and r_1 and r_2 are radial limits of the burner. Swirl number is a measure of swirl intensity

$$S = \frac{G_\phi}{G_x r_b},$$

where r_b is the radius of the burner. A strong swirl usually has a value of $S > 0.6$. Stronger swirl is necessary when burning is difficult to ignite coals.

VI. FLUIDIZED BED COMBUSTION

Fluidization is a process of making solids behave like a "fluid." When a jet of air is passed though a bed of solids from underneath, the force due to drag pushes the particles up while the gravity pulls the particles down. When the velolity of the air increases above a certain point, the drag force exceeds the weight of the particles and the particles will start to fluidize. Under this condition, the solid bed material is made to exhibit properties such as static pressure per unit cross section, bed surface level and flow of solids through a drain etc., similar to that of a fluid in the bed. This fluidization behavior promotes rapid mixing of the bed material, thus promoting gas–solid and solid–solid heat transfer.

In a typical fluidized bed combustion process, solid, liquid, and or/gaseous fuels, together with noncombustible bed material such as sand, ash, and/or sorbent are fluidized. The primary advantage of this mode of combustion over pulverized and/or fixed-bed combustion is its operating temperature (800–900°C). At this temperature, sulfur dioxide formed by combustion of sulfur can be captured by naturally occurring calcium-based sorbents (limestone or dolostone). The reactions between the sorbent and SO_2 are thermodynamically and kinetically balanced in this range of temperatures. The operating temperature range of FBC boilers is low enough to minimize thermal NO_x production (which is highly temperature dependent) compared to other modes of combustion and yet high enough to achieve good combustion efficiency. In addition to these advantages, FBC boilers can be adapted to burn a wide range of fuels such as low-grade, low-calorific-value coal wastes, and high-ash and/or high-sulfur coals in an environmentally acceptable manner. Fluidized beds can be further subdivided into bubbling or circulating fluidized beds based on the operating characteristics, and into atmospheric and pressurized fluid beds based on the operating pressure.

Bubbling fluidized beds characteristically operate with a mean bed particle size between 1000 and 1200 μm (1.0 to 1.2 mm). The operating velocity in a BFBC boiler is above the minimum fluidizing velocity and less than a third of the terminal velocity (1–4 min/sec). Under the conditions of low operating velocity and large particle size, the bed operates in the bubbling mode, with a defined bed surface separating the densely loaded bed and the low solids freeboard region. In bubbling fluidized beds the goal is to prevent solids from carrying over, or elutriating, from the bed into the convective pass(es). Therefore, particles smaller than 500–1500 μm are not used in BFBCs so as to avoid excessive entrainment.

Circulating fluidized beds operate with a mean particle size between 50 and 1000 μm (0.05 to 1 mm). The hydrodynamics of the fast fluidized beds allow CFBC boilers to use sorbent particles as fine as 50 μm. CFBC boilers operate typically at about twice the terminal velocity of the mean size particles, ranging from 3 to 10 min/sec. This high operating gas velocity, recirculation rate, solids

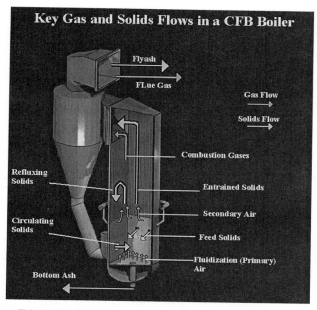

FIGURE 7 A schematic of a circulating fluidized bed boiler.

characteristics, volume of solids, and the geometry of the system create a hydrodynamic condition known as "fast bed" or "dilute phase refluxing." As a result of this condition solid agglomerates are continuously formed, dispersed, and reformed again. This motion also increases the slip velocity between the gas and solid phases, which increases the attrition or fragmentation of char and sorbent particles. The air required for the combustion is usually admitted in two streams: as primary and secondary air. In CFBC boilers, as shown in Fig. 7, the furnace can be conceptually divided into three sections: (1) the lower furnace zone, below the secondary air level where the solids flow regime is similar to a bubbling or turbulent bed and the particles range in size from approximately 1000 to 500 μm; (2) the upper furnace zone, above the secondary air level where fast fluidization occurs and the flow pattern approaches almost plug flow and particles from 500 to 150 μm are distributed in the circulating zone; and the (3) cyclone, where the material coarser than the cut point of the cyclone is recycled back into the combustor. The larger particles remain in the system for an extended period of time, ranging from hours for the dense phase in the lower zone to several minutes for the circulating phase in the upper zone of the furnace. Some of the fines make only one pass through the combustor and exit the cyclone, so that their residence time is similar to that of the gas.

The heating rate of a coal particle in an FBC boiler is approximately 10^3–$10^4 °C$/sec depending on the particle size and the total average residence time of a particle in the combustion chamber is estimated to be 20 min.

Some of the advantages of CFB boilers include fuel flexibility (any fuel such as coal, peat, wood chips, agricultural waste, Refuse Derived Fuel (RDF), and oil, with sufficient heating value to rise the fuel and the combustion air to above the ignition temperature of the fuel can be successfully burnt in a CFB boiler), high combustion efficiency, smaller furnace cross section: good load following capabilities, higher turn down, and more environmentally friendly.

A. Behavior of Ash in Boiler Furnaces

The mineral matter in coal probably is the factor most responsible in considering coal a "dirty" fuel as compared to oil or gas. Because of the dominant role played by coal mineral matter in boiler performance, the need for a better understanding by R&D on the behavior of this impurity is now more important than ever. The most commonly observed problems associated with mineral matter in coal are slagging and fouling. Slagging is the most serious problem with pulverized-coal-fired steam generators, particularly wall slagging, the buildup of slag deposits on water wall tubes. This has several undesirable aspects:

1. It decreases heat transfer appreciably.
2. It can interfere with gas flow when massive deposits accumulate near burners.
3. It results in masses of slag that can cause physical damage to the furnace bottom when they become dislodged.
4. It plugs normal ash handling facilities with excessively large masses of bottom ash.
5. It provides a cover to encourage tube wastage by external corrosion.

The standard methods that are used to predict the slagging and fouling behavior depend on the ash analysis, which is reported as oxides of various inorganic elements. Slagging characteristics of coal ash can be predicted to some extent using the silica percentage. Generally, ash with a low silica percentage will cause the most problems with slagging, and coals with a high silica percentage will be the least troublesome. Thus coal benefeciation to control silica percentage provides one way of modifying slagging behavior, although the effectiveness of coal washing will depend on the chemical composition of the ash in the product. However, there are still a lot of factors that cause slagging (boiler operating parameters) that are not well understood.

Fouling of superheaters and reheaters causes problems such as plugging of gas passages, interference with heat transfer, and erosion of metal tubes. Of the factors leading to such deposits, the physical ones involving particle motion, molecular diffusion, and inertial impaction are, to some extent, controllable by the design of the unit. The chemical factors associated with the ash, particularly the alkalies in the flue gas which are condensed on the surface of fly ash particles and the sintering characteristics of these particles depend upon the mineral species in the coal. Two alkalies, sodium and potassium, are generally held responsible for superheater and reheater fouling, with sodium given the most attention.

A fouling index has been developed to correlate sintering strength more closely with total ash composition:

$$Fouling\ Index = Base/Acid \times Na_2O,$$

where the base is computed as $Fe_2O_3 + CaO + MgO + Na_2O + K_2O$, and the acid as $SiO_2 + Al_2O_3 + TiO_2$ is expressed as weight percentage of the coal ash.

The fuel ash properties which boiler manufacturers generally consider important for designing and establishing the size of coal-fired furnaces include:

- The ash fusibility temperatures
- The ratio of basic to acidic ash constituents
- The iron/calcium ratio
- The fuel–ash content (mg/MJ)
- The ash firability

In addition to the furnace design, the ash content and characteristics also affect the superheater, reheater, and convective pass heat transfer surface design features. These characteristics along with a few others translate into the relative furnace sizes. It is recognized that the ash deposition is much better understood if the inorganic constituents in the coal are known rather than the ash composition after oxidation of the minerals. With the advancement in the computing power and image analysis in the late 1980s and early 1990s, Computer Controlled Scanning Electron Microscopy (CCSEM) has been a very useful tool in predicting ash behavior in furnaces. The determination of organically associated inorganic constituents is being made by chemical fractionation. These advanced characterization methods are now being more effectively used to help understand and solve the ash depositional problems in furnaces.

VII. ENVIRONMENTAL ISSUES IN COAL COMBUSTION

A. Sulfur Dioxide Emissions

Coal contains sulfur and nitrogen which vary typically between 0.5–5% and 0.5–2%, respectively. Sulfur upon combustion forms sulfur dioxide. About 65% of the total sulfur dioxide emissions are from electric utilities burning fossil fuels. Sulfur, 97–99%, in the fuel forms SO_2 and a small fraction of it is oxidized to SO_3. The equilibrium level of SO_3 in fuel lean combustion products is determined by the overall reaction

$$SO_2 + 1/2\,O_2 <=> SO_3.$$

The equilibrium constant ($Kp = 1.53 \times 10 - 5\,e\,11{,}760/T$ atm $-1/2$) indicates that the SO_3 formation is not favorable at combustion temperatures. SO_3 concentration in the flue gas is typically 1–5% of the SO_2.

The SO_2 control techniques can be classified into precombustion, during combustion, and postcombustion methods.

1. Precombustion Control Methods

Precombustion control technologies can be fuel switching or fuel desulfurization. Since sulfur dioxide emissions are directly proportional to the amount of sulfur in the fuel, switching to a low sulfur fuel is a choice when permitted. However, fuel switching may not be an alternative if the regulations require a certain percentage of reduction in SO_2 emissions regardless of the fuel sulfur content. Fuel desulfurization involves reduction of sulfur from the fuels prior to combustion. Sulfur in coals is present in three forms: pyritic, organic, and sulfatic. Pyritic sulfur is the sulfur that is present as a mineral pyrite (FeS_2). The density of pyrite is higher than coal. Therefore, when coal is crushed and washed in water, the particles that are rich in the pyrite sink and the particles that are rich in carbon (organic) float. The sinks are rejected and the floats with less sulfur are used as clean coal. The degree of desulfurization depends on the distribution of sulfur in the coal. The finer the pyrite particles are, the more difficult the coal is to clean. These characteristics are of a coal that can be evaluated by washability curves based on a standard float and sink analysis. The organic sulfur is present in coal bonded to the organic structure of the coal. This form of sulfur is finely distributed throughout the coal matrix, and therefore is not possible to remove by physical coal cleaning.

2. During the Combustion Control Method

In an FBC system, limestones or dolostones are introduced into the combustion chamber along with the fuel. In the combustion chamber, limestones and dolostones undergo thermal decomposition, a process commonly known as calcination. The decomposition of calcium carbonate, the principal constituent of limestone, proceeds according to the following equation:

$$CaCO_3 \; CaO + CO_2$$

$$MgCa(CO_3)_2 \; MgO + CaO + 2CO_2. \tag{1}$$

Calcination, an endothermic reaction, occurs at temperatures above 760°C. Some degree of calcination is thought to be necessary before the limestone can react with gaseous sulfur dioxide. Calcined limestone is porous in nature due to the voidage (pores) created by the expulsion of carbon dioxide.

Capture of the gaseous sulfur dioxide is accomplished via the following reaction, which produces a solid product, calcium sulfate:

$$CaO + SO_2 + 1/2\,O_2 \; CaSO_4. \tag{2}$$

The reaction of porous calcium oxide with sulfur dioxide produces a continuous variation in the physical structure of the reacting solid as the conversion proceeds. Because of the relatively high molar volume of $CaSO_4$ of CaO, the pore network within the reactant can be progressively blocked as conversion increases. For pure CaO prepared by the calcination of reagent grade $CaCO_3$, the theoretical maximum conversion of $CaCO_3$ to $CaSO_4$ has been calculated to be 57%. In practice, the actual conversion obtained using natural limestones is much lower due to the nature of the porosity formed upon calcination. Calcium utilizations as low as 15–20 mol% have been reported in some cases, although utilizations of about 30–40 mol% are typical. MgO will not react with sulfur dioxide at temperatures above 760°C; therefore, the sulfation reaction of dolomite is basically the reaction of sulfur dioxide with calcium oxide.

In pressurized fluidized bed combustion, however, the partial pressure of CO_2 is so high that calcination does not proceed because of thermodynamic restrictions. For example, at 850°C, calcium carbonate does not calcine if the CO_2 partial pressure exceeds 0.5 atmosphere. Under these conditions, the sulfation reaction is

$$CaCO_3 \,(s) + SO_2 + 1/2\,O_2 \; CaSO_4 + CO_2.$$

Increasing the pressure from one to five atmospheres significantly increases the sulfation rate and calcium utilization.

3. Flue Gas Desulfurization

Dry sorbent use in the case of pulverized coal combustion units is not efficient because of the operating temperature of the combustion chamber. At higher temperatures (>1100°C), CaO is known to sinter with a loss in the porosity and, therefore, the conversion. Also, at high temperatures, $CaSO_4$ is not stable and decomposes to CaO and SO_2 limiting this technology to FBC units.

Flue gas desulfurization systems in principle use alkaline reagents to neutralize the SO_2 and are classified as *throwaway* or *regenerative* types. This classification is based on the product fate. While, in a *throwaway* process the product produced by absorbing medium is thrown away (discarded), in a *regenerative* process, the SO_2 is regenerated from the product.

Most of the flue gas desulfurization systems operating in the United States use limestone or lime slurry scrubbing. In this system, limestone is finely ground (90% passing though a 325-mesh screen or 45 μm) and made into slurry. This slurry is finely sprayed in the absorption (scrubber) column. This slurry absorbs SO_2 in water as shown here

$$SO_2\,(g) + H_2O <=> SO_2 \cdot H_2O \text{ (dissolving gaseous } SO_2)$$

$$SO_2 \cdot H_2O <=> H^+ + HSO^{3-} \text{ (hydrolysis of } SO_2).$$

This slurry with absorbed SO_2 is sent to a retention tank where the precipitation of $CaSO_3$, $CaSO_4$ and unreacted $CaCO_3$ occurs. Calcium carbonate has low solubility in water. Low pH promotes dissolution of $CaCO_3$ but low pH also lowers solubility of SO_2 in the scrubber. Therefore, a careful balance of pH is needed for this system. The following reactions take place in the retention tank, where

$$H^+ + CaCO_3 <=> Ca^{2+} + HCO^{3-1}$$
$$\text{(dissolution of limestone)}$$

$$Ca^{2+} + HSO^{3-} + 2H_2O <=> CaSO_3 \cdot 2H_2O + H^+$$
$$\text{(precipitation of calcium sulfate)}$$

$$H^+ + HCO^{3-1} <=> CO_2 \cdot H_2O$$
$$\text{(acid-base neutralization)}$$

$$CO_2 \cdot H_2O <=> CO_2\,(g) + H_2O \text{ (} CO_2 \text{ stripping)},$$

the overall reaction being

$$CaCO_3 + SO_2 + 2H_2O \; CaSO_3 \cdot 2H_2O + CO_2.$$

B. Nitrogen Oxides

Nitrogen oxides, NO and NO_2, collectively known as NO_x, are formed during combustion in three ways. About 85–90% of the NO_x emitted from the combustion chamber is NO and 5 or 10% as NO_2. The three types of NO_x that form are by thermal, fuel, and prompt mechanisms.

Nitrogen oxide emissions from coal combustion can occur from three sources. Thermal NO_x primarily forms from the reaction of nitrogen and oxygen in the combustion air. The Fuel NO_x is a component that forms mainly from the conversion of nitrogen in the fuel to nitrogen oxides. Prompt NO_x is formed when hydrocarbon radical fragments in the flame zone react with nitrogen to form nitrogen atoms, which then form NO.

No definite rules exist to determine which nitrogen oxide formation mechanism dominates for a given stationary combustor configuration because of the complex interactions between burner aerodynamics and both fuel oxidation and nitrogen species chemistry. But in general, fuel nitrogen has been shown to dominate pulverized coal fired boilers, although thermal NO is also important in the postflame regions where over-fire air is used. Thermal NO contributions only become significant at temperatures above 2500°F in coal flames. Prompt NO formation is not typically an important mechanism during coal combustion. In the absence of fuel nitrogen, fuel NO is not a problem for natural gas flames. Prompt NO, however, is important in the vicinity of the inlet burners where reacting fuel fragments mix with the oxidizing air. Thus, in natural gas burners, both prompt and thermal NO contribute to the formation of nitrogen oxides.

C. Thermal NO

The principal reactions governing the formation of thermal NO are

$$N_2 + O \quad\quad NO + N$$

$$N + O_2 \quad\quad NO + O.$$

These two reactions are usually referred to as the thermal-NO formation mechanism or the Zeldovich mechanism. In fuel-rich environments, it has been suggested that at least one additional step should be included in this mechanism:

$$N + OH \quad\quad NO + H$$

The reactions (1), (2), and (3) are usually referred to as the extended Zeldovich mechanism. Experiments have shown that as the complexity of the reactors and fuels increase, it is difficult to evaluate the rate forms. The general conclusion is that although the nonequilibrium effects are important in describing the initial rate of thermal-NO formation, the accelerated rates are still sufficiently low that very little thermal NO is formed in the combustion zone and that the majority is formed in the postflame region where the residence time is longer.

D. Prompt NO

Prompt NO occurs by the fast reaction of hydrocarbons with molecular nitrogen in fuel-rich flames. This Fennimore mechanism accounts for rates of NO formation in the primary zone of the reactor, which are much greater than the expected rates of formation predicted by the thermal NO mechanism alone. NO measurements in both hydrocarbon and nonhydrocarbon flames have been interpreted

to show that prompt NO is formed only when hydrocarbon radicals are available. It has been shown that the amount of NO formed in the fuel-rich systems is proportional to the concentration of N_2 and also to the number of carbons present in the gas phase. Hence, this mechanism is much more significant in fuel-rich hydrocarbon flames or in the reburning flames. Two reactions are believed to be the most significant to this mechanism:

$$CH + N_2 \quad\quad HCN + N \quad\quad (3)$$

$$C + N_2 \quad\quad CN + N. \quad\quad (4)$$

Reaction (4) was originally suggested by Fenimore in 1971. Reaction (5) is considered to be only a minor, but nonnegligible, contributor to prompt NO with its importance increasing with increasing temperature.

E. Fuel NO

Fuel NO is by far the most significant source of nitric oxide formed during the combustion of nitrogen-containing fossil fuels. Fuel NO accounts for 75 to 95% of the total NO_x accumulation in coal flames and greater than 50% in fuel oil combustors. The reason for fuel NO dominance in coal systems is because of the moderate temperatures (1500–2000 K) and the locally fuel-rich nature of most coal flames. Fuel NO is formed more readily than thermal NO because the N–H and N–C bonds common in fuel-bound nitrogen are weaker than the triple bond in molecular nitrogen which must be dissociated to produce thermal NO.

The main step, at typical combustion temperatures, consists of conversion of fuel nitrogen into HCN, step which according to some investigators (Fenimore, 1976; Rees *et al.*, 1981) is independent of the chemical nature of the initial fuel nitrogen. Once the fuel nitrogen has been converted to HCN, it rapidly decays to NH_i ($i = 1, 2, 3$) which reacts to form NO and N_2.

F. NO_x Control Technologies

In order to comply with the regulations for nitrogen oxides emissions, various abatement strategies have been developed. The most common methods used for NO_x control are air staging, fuel staging, flue gas recirculation, selective noncatalytic reduction (SNCR), or selective catalytic reduction (SCR). A determination of the most effective and least expensive abatement technique depends on specific boiler firing conditions and the emission standards. The principle of air staging is mainly reduction the level of available oxygen in zones where it is critical for NO_x formation. By doing so, the amount of fuel burnt at the peak temperature is also reduced. Air staging is accomplished in the furnace by splitting the air stream for combustion. A part of the combustion air is introduced downstream as over-fire air (OFA), intermediate air (IA), or over-burner air (OBA). The principle of fuel staging or reburning involves reduction of the NO_x already formed in the flame zone by reducing it back to nitrogen during combustion. This is usually accomplished by injecting fuel into a second substoichiometric combustion zone in order to let the hydrocarbon radicals from the secondary fuel reduce the NO_x produced in the primary zone. To complete the combustion process of the reburn fuel, additional air is introduced downstream. However, the main drawback of this system is the short residence time available for the reburn fuel for complete oxidation. Therefore, this method uses mostly natural gas or any other highly reactive fuel so that combustion can be completed within the residence available.

In SNCR, chemicals are injected into the boiler, which then react with NO_x and reduce it to N_2.

Most commonly used chemicals are ammonia or urea. Other chemicals used in research work include amines, amides and amine salts, and cyanuric acid. Good mixing is essential for the success of this process and the optimum temperature window at which the reactions take place is 900–1100°C. This happens to be in a region where the heat transfer surfaces are present. Achieving good mixing is difficult.

When the limits of NO_x cannot be met by combustion control or by SNCR, SCR methods are used. NO_x concentration in the flue gas is reduced by injection of ammonia in the presence of a catalyst. The use of a catalyst reduces the optimum temperature window. The reaction products are water and nitrogen and this reaction is accomplished at much lower temperatures than SNCR (between 300 and 400°C). At lower temperatures, the unreacted ammonia can react with sulfur trioxide in the presence of water to form ammonium bisulfate (NH_4HSO_4), a sticky compound, which can cause corrosion, fouling, and blocking of downstream equipment.

This temperature is usually high enough to prevent condensation of (NH_4HSO_4). This system can reduce NO_x emissions by about 90–95%.

Most of the electric utilities are installing SCR systems to meet the current and future NO_x emission regulations.

VIII. COAL GASIFICATION

Gaseous fuels are easy to handle and use. They produce less by-products. As naturally occurring gaseous and liquid fuel resources deplete, synthetic gases and liquids produced from coal can act as suitable replacements. Coal gasification is a process of conversion of solid coal into combustible gases such as synthetic gas mixture (CO and H_2) or by methanation reaction conversion to synthetic natural gas (SNG). This is usually done by thermal decomposition, partial combustion with air or oxygen, and reaction with steam or hydrogen or carbon dioxide. Coal contains impurities such as sulfur, nitrogen, and minerals, which generate pollutants when burned. Most of these impurities and pollutant gases can be removed during the process of converting the coal into synthetic fuels. Therefore, gasification of coal is desirable from an environmental viewpoint and to overcome the transportation constraints associated with solid fuels. While the goal of combustion is to produce the maximum amount of heat possible by oxidizing all of the combustible material, the goal of gasification is to convert most of the combustible solids into combustible gases (such as CO, H_2, and CH_4) with the desired composition and heating value.

A. Gasification Reactions

During gasification, coal initially undergoes devolatilization (thermal decomposition) and the residual char undergoes some or all of the reactions listed here.

solid–gas

$$2C + O_2 = 2CO \text{ (partial combustion)}$$

$$C + O_2 = CO_2 \text{ (combustion)}$$

$$C + CO_2 = 2CO \text{ (Boudward reaction)}$$

$$C + H_2O = CO + H_2 \text{ (water gas)}$$

$$C + 2H_2 = CH_4 \text{ (Hydrogasification)}$$

gas–gas

$$CO + H_2O = CO_2 + H_2 \text{ (Shift)}$$

$$CO + 3H_2 = CH_4 + H_2O$$

For the thermodynamic and kinetic considerations, coal char is normally considered to be 100% carbon. The heats of reactions of the combustion or partial combustion reactions are exothermic (and fast), whereas most other gasification reactions (boudward, water gas, and hydrogasification) are endothermic and relatively slower. Usually, the heat requirement for the endothermic gasification reactions is met by partial combustion of some of the coal. It is also important to note that gasification reactions are sensitive to the temperature and pressure in the system. Equilibrium considerations for the reversible gasification reactions show that high temperature and low pressure are suitable for the formation of most of the gasification products, except for methane. Methane formation is favored at low temperature and high pressure. High temperatures tend to reduce methane, carbon dioxide, and water vapor in the products. Coal char is considered to be a mixture of pure carbon and some inorganic impurities and structural defects. Certain impurities and structural defects are known to be catalytic; the absolute reaction rate depends on the amount and nature of the impurities and structural defects and also on physical characteristics such as surface area and pore structure. The physical structure controls the accessibility of gasification medium to the interior surface. These physical structural characteristics depend on the feed coal and the devolatilization conditions (heating rate and peak temperature).

There has been considerable interest in gasifying coals with several catalysts in order to minimize secondary reactions and produce the desired product gas distribution. However, the major problem with using catalysts is poisoning by various hydrocarbons and other impurities in coal. A major research effort in this area has been directed toward the gasification of carbon using alkali metals compounds. Potassium chloride and carbonate have been found to increase the amount of carbon gasified to CO and H_2. The same catalysts are found to reduce methane yield.

B. Classification of Gasification Processes

Gasification processes are primarily classified according to the operating temperature, pressure, reactant gas, and the mode of contact between the reactant gases and the coal/char. The operating temperature of a gasifier, usually dictates the nature of the ash removal system. Operating temperatures below 1000°C allow dry ash removal, whereas temperatures between 1000 and 1200°C cause the ash to partially melt, become "sticky," and form agglomerates. Temperatures above 1200°C result in melting of the ash and it is removed mostly in the form of liquid slag. Gasifiers may operate at either atmospheric or elevated pressure. Both temperature and pressure affect the composition of the final product gases. Gasification processes use one or a combination of three reactant gases: oxygen, steam, and hydrogen. The heat required for the endothermic reactions (heat absorbing) is supplied by combustion reactions between the coal and oxygen. Table IV illustrates the effect of gasification medium on the product species and the calorific value. Methods of contacting the solid feed and the gaseous reactants in a gasifier are of four main types: fixed bed, fluidized bed, entrained flow, and molten bath. The operating principle of fixed bed, fluidized bed, and entrained flow systems is similar to that discussed for combustion systems (see previous section). The molten bath approach is similar to the fluidized bed concept in that reactions take place in a molten medium (either slag or salt) with high thermal inertia and the medium both disperses the coal and acts as a heat sink, with high heat transfer rates, for distributing

TABLE IV Effect of Gasification Medium on Products and Calorific Value

Gasification medium	Products	Calorific value of the products (MJ/m^3)
Air and steam	N_2, CO, H_2 and CO_2	5.6–11.2
Oxygen and steam	CO, H_2	11.2–14.5
Hydrogen	CH_4	35–38

the heat of combustion. Table V summarizes some important gasification processes, conditions, and product gas compositions.

C. Major Gasification Processes

The most important fixed-bed gasifier available commercially is the Lurgi Gasifier. It is a dry-bottom, fixed-bed system usually operated between 30 and 35 atmospheres pressure. Since it is a pressurized system, coarse-sized coal (25–37 mm) is fed into the gasifier through a lock hopper from the top. The steam–oxygen mixture (gasifying medium) is introduced through the grate located in the bottom of the gasifier. The coal charge and the gasifying medium move in opposite directions (counter-currently). The gasifier is operated at about 980°C and the oxygen reacts with coal to form carbon dioxide, thereby producing heat to sustain the endothermic steam–carbon and carbon dioxide–carbon reactions. The raw product gas consisting mainly of carbon monoxide, hydrogen, and methane leaves the gasifier for further cleanup. Besides participating in the gasification reactions, steam prevents high temperatures at the bottom of the gasifier so as not to sinter or melt the ash. Therefore, this gasification system is most suitable for highly reactive coals. Hot ash is periodically removed through a lock hopper at the bottom. Large commercial gasifiers measuring about 4 m in diameter and 6.3–8.0 m in height are capable of gasifying about 50 tons of coal per hour. Improved versions of the Lurgi Gasifier have been developed but not yet commercialized.

The Winkler gasifier is a fluidized-bed gasification system, which operates at atmospheric pressure. In this gasifier, crushed coal is fed using a screw feeder and is fluidized by the gasifying medium (steam–air or steam–oxygen mixture depending on the desired calorific value of the product gas) entering through a grate at the bottom. The coal charge and the gasification medium move cocurrently (in the same direction). In addition to the main gasification reactions taking place in the bed, some may also take place in the freeboard above the bed. The temperature of the bed is usually maintained at 980°C (1800°F) and the product gas consists primarily of carbon monoxide and hydrogen. The low operating temperature and pressure limit the throughput of the gasifier. Because of the low operating

TABLE V Summary of Some Gasification Processes

Process	Temperature (°C)	Pressure (Psia)	Gasification medium	Type of coal feed	Product gas composition[b] (dry basis)						Product gas calorific value[b] (Btu/scf)	Comments
					CO	CO_2	H_2	CH_4	C_2H_6	Others		
Lurgi[a]	980	400	O_2 + steam	Noncaking	68.5	29.5	40.4	9.4	—	1.2	302	
Winkler[a]	815	15	O_2 + steam	Any coal	33.4	20.5	41.9	3.1	—	1.1	275	
Synthane	980	1000	O_2 + steam	Any coal	16.7	28.9	27.8	24.5	0.8	1.3	405	
Hygas steam–oxygen	980	1000	O_2 + steam	Any coal	23.8	24.5	30.2	18.6	0.6	2.3	374	
Koppers–Totzek[a]	1510	20	O_2 + steam	Any coal	55.8	6.2	36.6	—	—	1.4	298	
Bi-gas	1480	1200	O_2 + steam	Any coal	21.5	29.3	32.1	15.6	—	1.5	367	
Texaco	1480	600	O_2	Any coal in the form of coal–water slurry	46.6	11.5	38.7	0.7	—	2.7	300	
Cogas	980	75	Steam, and air as heat source	Any coal	7.4	9.3	26.2	34.0	8.3	14.8	726	
Producer	980	15	Air + steam	Any coal	29.0	4.0	12.0	3.0	—	52.0	130	
Water gas	980	15	Air + steam	Any coal	41	5.0	49.0	0.5	—	4.5	300	
Westinghouse	1930	20	Air + steam	Any coal	19.2	9.4	14.4	2.8	—	54.2	140	Oxygen can also be used
U-gas	1040	350	Air + steam	Any coal	9.8	12.0	10.3	—	—	67.9	75	
Combustion engineering	1930	20	Air	Any coal	24.4 / 60.6	4.1	10.7	—	—	—	125	Oxygen can also be used
CO_2 acceptor	815	150	Air + steam	Low rank	17.0 / 1.0	6.6	53.8	—	20.9	0.4	440	
Hydrane	815	>1000	H_2	Any coal	3.9	—	22.0	73.2	—	—	826	
Kellog salt	930	1200	O_2 + steam	Any coal	33.5	13.3	45.0	7.5	—	0.7	348	Oxygen can also be used

[a] Commercially available processes.
[b] Composition varies with coal used.

temperatures, lignites and subbituminous coals that have high ash fusion temperatures are ideal feedstocks for this type of gasifier. Units capable of gasifying 40–45 tons per hour are commercially available.

The Koppers–Totzek gasifier has been the most successful entrained-flow gasifier. This process uses pulverized coal (usually less than 74 μm) entrained (blown) into the gasifier by a mixture of steam and oxygen. The gasifier is operated at atmospheric pressure and high temperatures of about 1600–1900°C. The coal dust and gasification medium flow cocurrent (in the same direction) in the gasifier and because of the small coal particle size, the particle residence time is approximately 1 sec. Although this residence time is relatively short, high temperatures enhance the reaction rates, and therefore almost any coal can be gasified in this system. Tars and oils are evolved at moderate temperatures and crack at higher temperatures, and therefore, there is no condensible tarry material in the products and the ash melts and flows as slag in the K–T gasifier. Gasifiers with two diametrically opposite nozzles (also called heads) are most commonly used. However, the use of four nozzles doubles the throughput (over the two, nozzle design) and such configurations are also in use. The product gas is mainly synthesis gas (a mixture of CO and H_2) and is primarily used for ammonia manufacture. Since no heavy-duty moving parts are involved in this system, maintenance is minimal and availability is expected to be high.

D. Advanced Coal Gasification Systems

Many attempts have been made to improve the first-generation commercial gasifiers. British Gas Corporation has converted the Lurgi gasifier into a slagging type by increasing the operating temperature and thereby accommodating higher-rank coals which require higher temperatures for complete gasification. Another version of the Lurgi gasifier is the Ruhr-100 process with operating pressures about three times higher than that of the basic Lurgi process. Developmental work on the Winkler gasification process has lead to a pressurized version called the pressurized Winkler Process with an aim of increasing the yield of methane to produce Synthetic Natural Gas (SNG). Other processes in the developmental stages are the U-gas, Hy-Gas, Cogas, Westinghouse, and Synthane processes. The Texaco gasification system appears to be most promising entrained bed gasification system that has been developed. In this system coal is fed into the gasifier in the form of coal–water slurry and the water in the slurry serves as both a transport and a gasification medium. This system operates at 1500°C and the ash is removed as molten slag. Experience on demonstration units has indicated that the process has a potential to be used with combined-cycle plants for power generation. Another entrained bed gasification process under development is a pressurized version of the K–T process called the Shell–Koppers system.

These developing processes are primarily aimed at either increasing the operating pressure to increase the throughput and provide pressurized product gas for advanced power systems or increasing the operating temperatures to accommodate a variety of fuels, or both.

IX. COAL LIQUEFACTION

Similar to coal gasification, coal liquefaction is the process of converting solid coal into liquid fuels. The main difference between naturally occurring petroleum fuels and coal is the hydrogen to carbon molar ratio. Petroleum has approximately a hydrogen to carbon ratio of 2 where as coal has about 0.8. Coal also contains higher concentrations of oxygen, nitrogen, and significantly higher amounts of mineral matter than petroleum. Therefore, conversion of coal into liquid fuels involves hydrogenation (addition of hydrogen either directly or indirectly). Direct hydrogenation either from gaseous hydrogen or from a hydrogen donor solvent is termed direct liquefaction. If the hydrogen is added indirectly through an intermediate series of compounds, the process is called indirect liquefaction. In direct liquefaction processing, the macromolecular structure of the coal is broken down ensuring that the yield of the correct size of molecules is maximized and that the production of the very small molecules that constitute fuel gases is minimized. In contrast, indirect liquefaction methods break down the coal structure all the way to a synthesis gas mixture (CO and H_2), and these molecules are used to rebuild the desired liquid hydrocarbon molecules. This can be achieved by a variety of gasification techniques as discussed in the Coal Gasification section.

A. Liquefaction Reactions

Since coal in a complex substance it is often represented by an average composition and the reactions occurring during direct and indirect liquefaction can be illustrated as follows.

1. Direct Liquefaction

$$CH_{0.8}S_{0.2}O_{0.1}N_{0.01} + H_2 \longrightarrow (RCH_x)$$
$$+ CO_2, H_2S, NH_3, H_2O.$$

2. Indirect Liquefaction

$$CH_{0.8}S_{0.2}O_{0.1}N_{0.01} \longrightarrow CO + H_2$$
$$+ H_2S, NH_3, H_2O \text{ (Gasification)}$$

$$CO + 2H_2 \longrightarrow CH_3OH \text{ (Methanol Synthesis)}$$

$$2CH_3OH \longrightarrow CH_3OCH_3 + H_2O$$
$$\text{(Methanol-to-Gasoline)}$$

$$nCO + 2nH_2 \longrightarrow (-CH_2-)n + H_2O$$
$$\text{(Fischer–Tropsch Synthesis)}$$

$$nCO + (2n+1)H_2 \longrightarrow (-CH_2-)_{n+1} + H_2O.$$

Direct liquefaction of coal can be achieved with and without catalysts using pressures of 250 to 700 atmospheres and temperatures ranging between 425 and 480°C. In the indirect liquefaction process, coal is gasified to produce synthesis gas and cleaned to remove impurity gases and solids. The processes used to clean the gases depend on the impurities.

The principal variables that affect the yield and distribution of products in direct liquefaction are the solvent properties such as stability and hydrogen transfer capability, coal rank and maceral composition, reaction conditions, and the presence or absence of catalytic effects. Bituminous coals are the most suitable feed-stock for direct liquefaction as they produce the highest yields of desirable liquids, although most coals (except anthracites) can be converted into liquid products. Medium rank coals are the most reactive (react fast) under liquefaction conditions. Among the various petrographic components, the sum of the vitrinite and liptinite maceral contents correlates well with the total yield of liquid products.

3. Liquefaction Processes

The Bergius process was the first commercially available liquefaction process. It was developed during the First World War and involves dissolving coal in a recycled solvent oil and reacting with hydrogen under high pressures ranging from 200 to 700 atmospheres. An iron oxide catalyst is also employed. The temperatures in the reactor were in the range of 425–480°C. Light and heavy liquid fractions are separated from the ash to produce gasoline and recycle oil, respectively. In general, 1 ton of coal produces about 150–170 liters of gasoline, 190 liters of diesel fuel, and 130 liters of fuel oil. Separation of ash and heavy liquids, and erosion due to cyclic pressurization, posed difficulties which caused the process to be taken out of use after the war.

In the first generation, in a commercially operated, indirect liquefaction process called Fischer–Tropsch synthesis, coal is gasified first using the high-pressure Lurgi gasifier and the synthesis gas is reacted over an iron-based catalyst either in a fixed-bed Arge reactor or a fluidized-bed Synthol reactor. Depending on the reaction conditions, the products obtained consist of a wide range of hydrocarbons. Although this process was developed and used widely in Germany during the Second World War, due to poor economics it was discontinued after the war. This process is used in South Africa (Sasol) for reasons other than economics.

4. Developments in Liquefaction Processes

Lower operating temperatures are desirable in direct liquefaction processes since higher temperatures tend to promote cracking of molecules producing more gaseous and solid products at the expense of liquids. Similarly, lower pressures are desirable from an ease and cost-of-operation point of view. Recent research efforts in the area of direct liquefaction have concentrated on reducing the operating pressure, improving the separation process by using a hydrogen donor solvent (Consol Synthetic Fuels Process), operation without catalysts (Solvent Refined Coal (SRC)), and by using a solvent without catalysts (Exxon Donor Solvent) but using external catalytic rehydrogenation of the solvent. Catalytic effects in liquefaction are due to the inherent mineral matter in the coal and to added catalysts. Recent research efforts have focused on the area of multistage liquefaction to minimize hydrogen consumption and maximize overall process yields.

Later versions of Sasol plant (Sasol 2 and 3 units which also use indirect liquefaction process) have used only synthol reactors to increase the yield of gasoline and have reacted excess methane with steam to produce more CO and H_2.

Recent developments include producing liquid fuels from synthesis gas through an intermediate step of converting the synthesis gas into methanol at relatively low operating pressures (750–1500 psi) and temperatures (205–300°C). Methanol is then converted into a range of liquid hydrocarbons. The use of zeolite catalysts (as developed by Mobil) has enabled the direct production of gasoline from methanol with high efficiency.

X. LIQUID FUELS

Liquid fuels are obtained by refining naturally occurring crude oil. Like coal, crude oil from different places can differ in composition because of the precursor materials and the conditions for transformation organic matter to crude oil. Crude oil is a complex naturally occurring liquid containing mostly hydrocarbons and some compounds containing N, S, and O atoms. Crude oil consists of paraffins (straight-chain and branched-chain compounds), naphthenes (cyclo paraffins), and aromatics (benzene and its derivatives).

The average composition of crude oils from various parts of the world does not vary significantly. However, because of the variations in viscosity, density, sulfur, and

TABLE VI Typical Yield from a Barrel of Gasoline

Product	Yield (gallons)
Gasoline	19.5
Distillate fuel oil	9.2
Kerosene	4.1
Residual fuel oil	2.3
Lubricating oil, asphalt, wax	2
Chemicals for use in manufacturing (petrochemicals)	2

boiling points, these are separated into different fractions in a refinery. The most common refining operations are distillation, cracking, reforming alkylation, and coking. The demand for various products changes with the season and the lifestyle of the society. Typical yield from a barrel of crude oil is shown in Table VI.

XI. PROPERTIES FOR UTILIZATION

The majority of products (first four) are burnt in various devices. Gasoline is a mixture of light distillate hydrocarbons with a boiling range of 25–225°C consisting of paraffins, olefins, naphthenes, aromatics, oxygenates, lead, sulfur, and water. The exact composition varies with the season and geographic location. During summer months low volatile components are added to reduce the vapor pressure, whereas in winter months low boiling components are added to make it more volatile. Under the 1990 Clean Air Act Amendments (CAAA), the U.S. Environmental Protection Agency (EPA) developed reformulated gasoline (RFG) to significantly reduce vehicle emissions of ozone-forming and toxic air pollutants. RFG is required to be used in the nine major cities with the worst ozone air pollution problems. Similar to normal gasoline, RFG will contain oxygnates. Oxygenates increase the combustion efficieny and reduce emission of carbon monoxide. Table VII provides a comparison of properties for various gasolines.

Diesel fuel is also a mixture of light distillates but with higher boiling point components with a boiling point range of 185–345°C consisting of lower volatile and more viscous compounds. The average molecular weight is approximately 200.

Kerosene fuels are used in jet engines. Kerosene fuels have a wide range of boiling points. The aromatics in kerosene are limited due to their tendency to form soot.

Residual fuel oils are classified into five categoroies. Some of the important properties are listed here for good atomization.

1. Specific gravity—ratio between the weight of any volume of oil at 60°F to the weight of equal volume of water at 60°F
2. Viscosity—a measure of resistance to motion of a fluid
3. Flash point—The temperature at which the vapors generated "flash" when ignited by external ignition source
4. Pour point—The temperature at which the oil ceases to flow when cooled under prescribed conditions.

A. Combustion of Liquid Fuels

Fuel oil-fired furnaces, diesel engines, and distillate fuel-fired gas turbines utilize fine liquid sprays to increase the rate of evaporation and combustion rate of the fuel. In general the combustion of a liquid fuel takes place in a series of stages: atomization, vaporization, mixing of the vapor with air, ignition, and maintenance of combustion (flame stabilization). Recent advances have shown the atomization step to be one of the most important stages of liquid fuel combustion. The main purpose of atomization is to increase the surface area to volume ratio of the mixture. For example, breaking up of a 3-mm droplet into 30-μm drops results in 10^6 droplets. This increases the burning rate by 10,000 times. The finer the atomization spray the greater the subsequent benefits are in terms of mixing, evaporation, and ignition. The function of an atomizer is twofold: atomizing the oil and matching the momentum of the issuing jet with the aerodynamic flow in the furnace.

The atomizers for larger boiler burners are usually of the swirl pressure jet or internally mixed two fluid types, producing hollow conical sprays. Less common are the externally mixed two fluid types. The principal considerations in selecting an atomizer for a given application are turn-down performance and auxiliary costs.

There are differences in the structures of the sprays between atomizer types which may affect the rate of mixing of fuel droplets with the combustion air and hence the initial development of a flame.

For distillate fuels of moderate viscosity, (30 mm^2 sec^{-1}) at ordinary temperatures, a simple pressure atomization with some type of spray nozzle is most commonly used. Operating typically with a fuel pressure of 700–1000 kPa (7–10 atm) such a nozzle produces a distribution of droplet diameters from 10 to 150 μm. They range in design capacity of 0.5–10 or more, cm^3 sec^{-1}. A typical domestic oil burner nozzle uses about 0.8 cm^3 sec^{-1} of No. 2 fuel oil at the design pressure. Although pressure-atomizing nozzles are usually equipped with filters, the very small internal passages and orifices of the smallest tend to be easily plugged, even with clean fuels. With decreasing fuel pressure the atomization becomes

TABLE VII Properties of Various Types of Gasoline

| | Fuel parameter values (national basis) | | | | |
| | Conventional gasoline | | Gasohol (2.7 wt% oxygen) | Oxyfuel | Phase I RFG |
	Avg[a]	Range[b]	Avg	Avg	Avg
RVP3 (psi)	8.7-S 11.5-W[c]	6.9-15.1	9.7-S 11.5-W	8.7-S 11.5-W	7.2/8.1-S 11.5-W
T50 (°F)	207	141–251	202	205	202
T90 (°F)	332	286–369	316	318	316
Aromatics (vol%)	28.6	6.1–52.2	23.9	25.8	23.4
Olefins (vol%)	10.8	0.4–29.9	8.7	8.5	8.2
Benzene (vol%)	1.60	0.1–5.18	1.60	1.60 (1.3 max)	1.0
Sulfur (ppm)	338	10–1170	305	313 (500 max)	302
MTBE4[d] (vol%)	—	0.1–13.8	—	15 (7.8–15)	11
EtOH4 (vol%)	—	0.1–10.4	10	7.7 (4.3–10)	5.7

[a] As defined in the Clean Air Act.
[b] 1990 MVMA survey.
[c] Winter (W) higher than Summer (S) to maintain vehicle performance.
[d] Oxygenate concentrations shown are for separate batches of fuel; combinations of both MTBE and ethanol in the same blend can never be above 15 vol% total.

progressively less satisfactory. Much higher pressures often are used, especially in engine applications, to produce a higher velocity of liquid relative to the surrounding air and accordingly smaller droplets and evaporation times. Other mechanical atomization techniques for production of more monodisperse sprays or smaller average droplet size (spinning disk, ultrasonic atomizers, etc.) are sometimes useful in burners for special purposes and may eventually have more general application, especially for small flows.

Conventional spray nozzles are relatively ineffective for atomizing of fuels of high viscosity such as No. 6 or residual oil (Bunker C) and other viscous dirty fuels. In order to transfer and pump No. 6 oil, it must usually be heated to about 373 K, at which its viscosity is typically 40 mm² sec⁻¹. Relatively large nozzle passages and orifices are necessary for the possible suspended solids. Dry steam may also be used in a similar way, as is common practice in the furnaces of power plant boilers using residual oil.

Combustion of fuel oil takes place through a series of steps, namely, vaporization, gasification, ignition, dissociation, and finally attaining the flame temperature. Vaporization or gasification of the fine spray of fuel droplets takes place as a physicochemical process in the combustion chamber. The temperature of vaporization for fuel oil is in the range of 100–500°F, depending on the grade of the fuel. Gasification takes place at about 800°F. The final flame temperature attained is between 2000 and 3000°F. The combustion of an oil droplet takes place in 2–20 msec depending on the size of the droplet. A typical characteristic of an oil flame is its bright luminous nature, which is due to incandescent carbon particles in the fuel-rich zone.

$$\tau = \frac{d_{p_p}^2}{\beta}.$$

Figure 8 illustrates the combustion of a single liquid droplet. Evaporation of liquid supplies the gaseous fuel

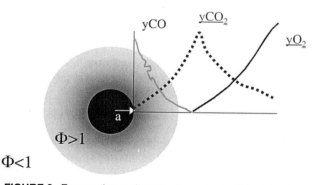

FIGURE 8 Evaporation and combustion of a liquid fuel droplet.

that burns in the gas phase. Evaporation is caused by heat transfer to the surface of the droplet. The time required for complete evaporation is given by

$$\beta = \frac{8\lambda}{\rho_l c_p} \ln(1 + B_T),$$

where, B_T is transfer coefficient, lambda is thermal Conductivity, $\rho_l = $ liquid density, and c_p is the heat capacity.

XII. MAJOR COMBUSTION METHODS

Internal combustion engines are devices that produce work using the hot combustion gases directly rather than steam. Three major types of IC engines are commonly used today using the top three products from a refinery. They are (1) spark engine (gasoline engine), (2) compression engine (diesel engine), and (3) gas turbine (aircraft engines).

A. Spark or Gasoline Engines

The most common engine is a 4-stroke engine. During the *intake* stroke, the fuel and air mixture is drawn into the cylinder with the exhaust valve closed. Then the air and fuel mixture is compressed in a *compression* stroke. At the top of the stroke, the spark plug ignites the mixture. During the *expansion* or power stroke, the high-pressure combustion gases expand moving the piston down and delivering the power. The gases expand completely, the exhaust valve opens, and the gases are expelled out during the *exhaust* stroke. The fuel and air are atomized and premixed in a carburetor. The higher the compression ratio the higher the efficiency of the engine. However, higher compression ratios also require higher octane number fuels. The octane number of a fuel is indicative of its antiknock properties. At equivalence ratios below 0.7 and above 1.4, the mixtures are generally not combustible. The equivalence ratio changes as the power requirement changes. For example, as the vehicle accelerates, high torque and power are required for which a fuel-rich mixture is used, whereas when the vehicle is cruising at high speeds the vehicle needs fuel lean mixtures. Therefore, in IC engines it is difficult to maintain the air to fuel ratio constant. Combustion in IC engines takes place in both oxygen-deficient and oxygen-rich environments, and the air and fuel mixtures are preheated by compression. Every time a fresh batch of fuel comes in flame is produced and quenched resulting in unsteady combustion. This results in continuous changes in the pollutant generation and emission.

B. Diesel Engines

Diesel engines are also IC engines. However, in Diesel engines, there is no carburetor. Only air is compressed to much higher pressures and the fuel is injected into the compressed air. As the fuel and air are mixed, the fuel evaporates and ignites (hence called compression ignition). The pressures used in the engines are almost twice those of the gasoline engines. Rate of injection and mixing of fuel and air determine the rate of combustion. Diesel engines are classified based on fuel injection, direct injection (DI), and indirect injection (IDI). Fuel quality is measured by cetane number (CN).

C. Gas Turbines

Another class of internal combustion engine is the gas turbine. Air is compressed to high pressures (10–30 atm) in a centrifugal compressor. Fuel is sprayed into the primary combustion zone where the fuel burns and increases the temerpature of the gases. The gas volume increases with combustion and the gases expand though a turbine. The power generated exceeds that required for the compressor. This drives the shaft to run an electric generator. In the aircraft applications, the gases are released at high velocity to provide the thrust. These systems are light weight compared to land-based systems. Land-based systems use either distillate oil or natural gas. Gas turbine-based power generation is used commonly to meet the peak power requirements rather than for base load operation.

D. Environmental Challenges for Liquid Fuel Utilization

Carbon monoxide is present in any combustion gas from any carbon containing fuel. The main factor that leads to its formation is incomplete combustion and in the IC engines continuous change from fuel-lean to fuel-rich conditions results in large emissions of CO. More than 70% of the CO emitted in the United States is from the transportation sector. CO emissions are also a function of vehicle speed. At lower speeds the emissions are higher. Cold starts also to contribute to higher emissions of CO. Oxygenates in the fuel aid complete combustion and result in a decrease of CO emissions. Catalytic converters placed at the end of the exhaust pipe oxidize the CO catalytically at lower temperatures.

Most of the hydrocarbon emissions are emitted through the exhaust. However, methane, ethane, acetylene, propylene, and aldehydes were found in the exhaust but were not present in the fuel. It can be deduced that these were formed during combustion. A significant amount of hydrocarbon emissions also come from the combustion chamber wall crevices and solid deposits. These hydrocarbon emissions reduce the NO_x emissions. However, with the increase in the air to fuel ratio, CO and hydrocarbon emissions are reduced but NO_x emissions increase. To reduce all the emissions, three-way catalysts are being used. They not only oxidize CO and HCs but also reduce NO_x to N_2.

XIII. GASEOUS FUELS

A. Combustion of Gaseous Fuels

In any gas burner some mechanism or device (flame holder or pilot) must be provided to stabilize the flame against the flow of unburned mixture. This device should fix the position of the flame at the burner port. Although gas burners vary greatly in form and complexity, the distribution mechanism is fundamentally the same in most. By keeping the linear velocity of a small fraction of the mixture flow equal or less than the burning velocity, a steady flame is formed. From this pilot flame, the main flame spreads to consume the main flow at a much higher velocity. The area of the steady flame is related to the volume flow rate of the mixture:

$$\dot{V}_{mix} = A_f \times S_u$$

where,

$\dot{V}_{mix}$ = volumetric flow rate
A_f = area of the steady flame
S_u = burning velocity

The volume flow rate of the mixture is, in turn, proportional to the rate of heat input:

$$\dot{Q} = \dot{V}_{mix} \times HHV$$

where,

$\dot{V}_{mix}$ = volumetric flow rate
HHV = higher heating value of the fuel
$\dot{Q}$ = rate of heat input

In the simple Bunsen flame on a tube of circular cross section, the stabilization depends on the velocity variation in the flow emerging from the tube. For laminar flow (parabolic velocity profile) in a tube, the velocity at a radius, r, is given by:

$$v = const(R^2 - r^2)$$

where,

v = laminar flow velocity
R = tube radius
r = flame radius
$const$ = experimental constant

Most of the commercial gas–air premixed burners are basically laminar-flow Bunsen burners and operate at atmospheric pressure—i.e., the primary air is induced from the atmosphere by the fuel flow with which it mixes in the burner passage leading to the burner ports where the mixture is ignited and the flame stabilized. The induced air flow is determined by the fuel flow-through momentum exchange and by the position of a shutter or throttle at the air inlet. Hence, the air flow is a function of fuel velocity as it issues from its orifice or nozzle, or of the fuel supply pressure at the orifice. With a fixed fuel flow, the equivalence ratio is adjusted by the shutter, and the resulting induced air flow also determines the total mixture flow, since desired air–fuel volume ratio is usually 7 or more, depending on the stoichiometry. Burners of this general type with many multiple ports are common for domestic furnaces, heaters, stoves, and for industrial use. The flame stabilizing ports in such burners are not always round and maybe slots of various shapes conform to the heating task.

Atmospheric industrial burners are made for a heat release capacity of up to 50 kJ/sec^{-1} and even despite their varied designs their principle of stabilization is basically the same as that for the Bunsen burner. In some the mixture is fed through a fairly thick-walled pipe or casting of appropriate shape for the application and the desired distribution of flame. The mixture issues from many small and closely spaced drilled holes, typically 1–2 mm in diameter, and burns, as rows of small pilot flame, spark or heated wire, usually located near the first holes, to avoid accumulation of the unburned mixture before ignition. The rate of total heat release for a given fuel–air mixture can be scaled with the size and number of holes—e.g., for 2-mm-diameter holes it would be 10–100 J/(hole) or in general 0.3–3 kJ cm^2 sec^{-1} of port area, depending on the fuel. The ports may also be narrow slots, sometimes packed with corrugated metal strips, to improve the flow distribution and lessen the tendency to flashback.

Gas burners that operate at high pressures are usually intended for much higher mixture velocity or heating intensity and the stabilization against blowoff must therefore be enhanced. This can be achieved by a number of methods such as (1) surrounding the main port with a number of pilot ports and (2) using a porous diaphragm screen.

In order to achieve high local heat flow the port velocity of the mixture should be increased considerably. In burners that achieve stabilization by causing pilot ports, most of the mixture can be burned at a port velocity as high as 100 S_u to produce a long pencil-like flame, suitable for operations requiring a high heat flux.

SEE ALSO THE FOLLOWING ARTICLES

UTILIZATION • FUELS • INTEGRATED GASIFICATION COMBINED-CYCLE POWER PLANTS • INTERNAL COMBUSTION ENGINES • PETROLEUM REFINING • POLLUTION, AIR

BIBLIOGRAPHY

Elliot, M. A. (ed.) (1980). "Chemistry of Coal Utilization," 2nd Supplementary Volume, Wiley, New York.

Borman, G. L., and Ragland, K. W. (1998). "Combustion Engineering," WCB/McGraw-Hill, Boston, MA.

Lawn, C. J. (ed.), (1987). "Principles of Combustion Engineering for Boilers" Academic Press, Harcourt Brace Jovanovich, New York.

Turns, S. R. (2000). "An Introduction to Combustion: Concepts and Applications," 2nd ed., Mc Graw-Hill Inc., New York.

Wen, C. Y., and Lee, E. S. (eds.) (1979). "Coal Conversion Technology," Addison-Wesley, MA.

Heinsohn, R. J., and Kabel, R. L. (1999). "Sources and Control of Air Pollution," Prentice-Hall, Upper Saddle River, NJ.

Williams, A., Pourkashanian, M., Jones, J. M., and Skorupska, N. (2000). "Combustion and Gasification of Coal," Applied Energy Technology Series, Taylor and Francis, New York.

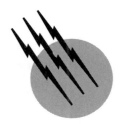

Fuels

E. Lois
National Technical University

A. K. Gupta
University of Maryland

E. L. Keating
Geo-Centers

GLOSSARY

BP Boiling point
ASTM American Society of Testing Materials
CF Watson characterization factor
CI Correlation index, Cetane index
DR Relative Density at 15°C
SG Specific gravity
T Temperature
ν kinematic viscosity
η dynamic viscosity
d density

IN THIS CHAPTER only the petroleum based liquid fuels are considered. Special attention is given to crude oil, gasoline, diesel, biodiesel, and aviation fuels since they provide a major source of energy for almost all kinds of propulsion and power devices worldwide.

I. FUEL COMPOSITION

Petroleum is a mixture of organic compounds and primary hydrocarbons, coming from underground rock formations ranging in age from ten to several hundred million years. The name petroleum comes from the Greek words "petra" (rock) and "elaion" (oil) and literally means "rock oil." Studies indicate that petroleum was formed mainly from microscopic-sized marine animals and plants. When these organisms died in water of low oxygen content, they did not readily decompose. Thus their remains sank to the bottom and were buried under accumulations of sediment. Their conversion to petroleum remains a subject of research today. The theory held generally is that bacteria converted the fats of the marine life into fatty acids. These in turn were changed, by mechanisms still unknown, to the asphaltic material called kerogen. Then, over millions of years, under heat and pressure, plus probably catalytic agents in the rock, the kerogen changed to crude oil and gas.

Encyclopedia of Physical Science and Technology, Third Edition, Volume 6

TABLE I Elemental Analysis of Crude Oils

Elements	Content (wt%)
Carbon	83.90–86.80
Hydrogen	11.40–14.00
Sulfur	0.06–8.00
Nitrogen	0.11–1.70
Oxygen	0.50
Metals (Fe, V, Ni, etc.)	0.03

Principal elements in crude petroleum are carbon and hydrogen atoms, usually in a carbon-hydrogen wt.% ratio between 6 and 8. The hydrocarbons are mainly liquids and gases, with some solids in dispersion or solution. Among the many other materials usually present are small amounts of sulfur, nitrogen, and oxygen in the form of hydrocarbon derivatives; traces of such metals as nickel, vanadium, and iron; along with water (emulsified in the oil by as much as 30% by weight). The water is generally in the form of saturated solutions of calcium and magnesium sulfates and sodium and magnesium chlorides. The typical analysis of crude oil is given in Table I.

A. Hydrocarbons

Hydrocarbons are organic compounds composed entirely of carbon and hydrogen atoms. Each of the four major classes of hydrocarbons—paraffins, olefins, naphthenes and aromatics—represents a family of individual hydrocarbons that have similar structure. The classes differ in the ratio of hydrogen to carbon atoms and how the atoms are arranged in the structure.

The hydrocarbon constituents found in crude petroleum are of three major types: paraffins, naphthenes, and aromatics. The very large range in their proportions and the ratios in which specific series of hydrocarbons appear, determine the *physical* operations to be used in separating these components, such as *by* distillation. The presence and extent of some of the other materials—say sulfur and oxygen—determine the *chemical* operations to be used, i.e., catalytic conversions, hydrogen processing, etc.

Paraffins have the general formula C_nH_{2n+2}, where n is the number of carbon atoms. The carbon atoms in paraffins are joined by single bonds. It is possible for paraffins with four or more carbon atoms to exist as two or more distinct compounds, which have the same number of carbon and hydrogen atoms. These compounds, called structural isomers, differ in the arrangement of the carbon atoms. Normal octane (n-octane) and isooctane are two examples of eight carbon structural isomers. Isooctane is the common name for 2,2,4-trimethylpentane, which specifies the branching pattern of the three-methyl groups on a pentane backbone.

Olefins (C_nH_{2n}) are similar to paraffins, but have two fewer hydrogen atoms and contain at least one double bond between carbon atoms. They rarely occur naturally in crude oil, but are formed during refining processes (especially cracking). Like paraffins, olefins with four or more carbons can exist as structural isomers.

In *naphthenes*, also called *cycloparaffins*, some of the carbon atoms are arranged in a ring. The naphthenes in gasoline have rings of five- and six-carbon atoms. Naphthenes have the same general formula as olefins, C_nH_{2n}. The carbon atoms are joined by single bonds, so naphthenes are saturated compounds.

Like naphthenes, some of the carbon atoms in *aromatics* are arranged in a ring, but aromatic bonds, not single bonds, join them. The structure can be envisioned as an average of two discrete cyclohexatriene molecules. However, the aromatic bond character is distributed evenly around the ring. The shorthand representation for a monocyclic aromatic ring is a hexagon with a circle representing the aromatic bonds. Aromatic rings always contain six carbons; in polycyclic aromatics, like naphthalene, two or more rings share some of the carbons.

Natural gas, the gaseous component of crude petroleum, may also be found some distance away from an oil pool, in separate wells, having been separated from the liquid by natural processes under the ground. It is composed mainly of light paraffins—methane, ethane, propane, butane—plus some higher boiling paraffins, nitrogen, carbon dioxide, and hydrogen sulfide.

B. Crude Oil Classifications

Crudes are commonly classified according to the residue from their distillation, this depending on their relative contents of three basic hydrocarbons: paraffins, naphthenes, and aromatics.

About 85% of all crude oils fall into the following three classifications:

1. Asphalt-base, contains very little paraffin wax and a residue primarily asphaltic (predominantly condensed aromatics). Sulfur, oxygen, and nitrogen contents are often relatively high. Light and intermediate fractions have high percentages of naphthenes. These crude oils are particularly suitable for making high-quality gasoline, machine lubricating oils, and asphalt.
2. Paraffin-base, containing little or no asphaltic materials, is a good source of paraffin wax, quality motor lube oil, and high-grade kerosene. They usually have lower nonhydrocarbon content than asphalt-base crudes.
3. Mixed-base, contains considerable amounts of both wax and asphalt. Virtually all products can be

TABLE II General Properties of Crude Oils

Property	Paraffin base	Asphalt base
API gravity	High	Low
Naphtha content	High	Low
Naphtha octane number	Low	High
Naphtha odor	Sweet	Sour
Kerosene smoking tendency	Low	High
Diesel-fuel knocking tendency	Low	High
Lube-oil pour point	High	Low
Lube-oil content	High	Low
Lube-oil viscosity index	High	Low

obtained, although at lower yields than from the other two classes.

Because of the variations in hydrocarbon fractions in different crude oils, there are several differences in general properties, as indicated in Table II.

Typical parameters used for the characterization and classification of crude oils are

Watson characterization factor

$$CF = \frac{\sqrt[3]{T}}{SG_{60/60°F}}$$

where: T = mean boiling point (R) $(R = °F + 460)$
 $SG_{60/60°F}$ = specific gravity 60/60F°
 CF = 12.9 − 12.15 means paraffin-base crude
 CF = 12.1 − 11.5 means mixed-base crude
 CF = 11.45 − 10.5 means asphalt-base crude

Correlation Index, CI

$$CI = \frac{48640}{K} + 473.7 \cdot SG_{60/60°F} - 456.8$$

where: K = boiling point in (K) $(K = °C + 273.15)$
 $SG_{60/60°F}$ = specific gravity 60/60°F
 CI ≤ 15 means paraffin-base
 CI = 15–50 means naphthenic or mixed-base
 CI > 50 means aromatic base

C. Crude Oil Properties

Crude petroleum is primarily a liquid of widely varying physical and chemical properties. Common colors are green, brown, and black and occasionally almost white or straw color. Specific gravity can range from 0.73 to 1.02; however, most crudes are between 0.80 and 0.95. Viscosity varies, too. Data for a large number of crudes indicate kinematic viscosities range from 0.007 to 13 stokes at 100°F, though most of them fall in the range of 0.023–0.23 stoke.

1. Density

The density (d) of a fuel is the mass of a unit volume of material at a selected temperature. In the case of petroleum products, the reference temperature is 15°C. Relative density (RD), or specific gravity, is the ratio of the density of the material at a selected temperature to the density of a reference material at a selected temperature. For the relative density of petroleum crudes and products in the United States, the reference material is water and both temperatures are 60°F.

$$RD(60/60°F) = \frac{d_{sample}(60°F)}{d_{water}(60°F)}$$

The United States petroleum industry often uses API gravity instead of relative density. API gravity is calculated as:

$$°API = \frac{141.5}{RD(60/60°F)} - 131.5$$

While API gravity measurements may be made on liquids at temperatures other than 60°F, the results are always converted to the values at 60°F, the standard temperature.

2. Color

The color of crude oils is found by optical vision. Crudes have color varying from light yellow to opaque black. The color of the crude depends on their hydrocarbon content and especially on the presence of nitrogen and sulfur compounds. Most crude oils have a dark, almost black color.

3. Vapor Pressure

The composition of the light hydrocarbons of the crude characterizes the vapor pressure of the crude. Vapor pressure is measured with Reid method according to ASTM D323 at 100°F. Because of their light hydrocarbons content, the crude oils are classified in the most volatile and flammable category (Group 1).

4. Viscosity

The viscosity of a fluid is a measure of its resistance to flow. The less the viscosity of the fluid, the easier it flows. The viscosity can be defined as dynamic or kinematic. These two are correlated as follows:

$$\nu = \frac{\eta}{d}$$

where: ν = kinematic viscosity
 η = dynamic viscosity
 d = density

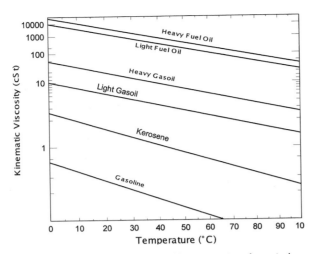

FIGURE 1 Variation of viscosity with temperature for petroleum products.

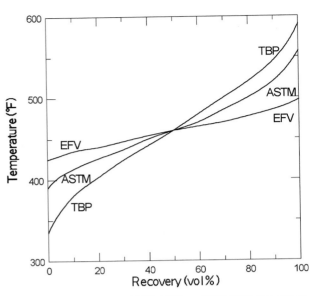

FIGURE 2 Correlation of ASTM, TBP, and EFV distillation curves.

For crude oils and petroleum products viscosity is measured by ASTM D445 as kinematic viscosity, and expressed as cSt at specific temperature.

Depending on the hydrocarbon content, the viscosity of crude oils can vary from very low to very high. Figure 1 gives in graphical form the variation of viscosity with temperature for various petroleum products.

D. Distillation

Distillation is the main process used for the separation of crude oil into valuable products. True boiling point (TBP) method is used to test crude oils using ASTM D2892. Crude oil is distilled in a fractionation column with a large number of theoretical plates and large reflux ratio. True boiling point distillation simulates the conditions of a refinery distillation column. TBP data are used to estimate the quantity and quality of the products that a crude oil can yield. TBP distillation is performed at atmospheric pressure, and beyond a critical point under vacuum, in order to eliminate pyrolysis of the heavier hydrocarbons of the crude. Crude oil is separated in distillates and residue. Distillates are gases, naphtha, kerosene, and gas oil.

Except TBP, there are simpler distillation methods, such as ASTM and EFV (equilibrium flash vaporization). In these distillation methods there is no reflux, and the separation is very poor. The three distillation curves (TBP, ASTM, EFV) are correlated to each other with empirical formulas. Figure 2 gives the correlation of the three distillation curves for a petroleum product.

Data from TBP distillation is the basic parameter that determines the yield of each type of crude in specific products. Figure 3 gives the distillation ranges of the various products on the TBP curve of the crude oil.

E. Sulfur Content

Sulfur content is one of the most critical parameters for the characterization of crude oil. Sulfur content is measured by ASTM D 4222 method, using X-ray fluorescence. Crude oils are characterized as low sulfur (sweet) and high sulfur (sour). Sulfur content is important for environmental (SO_2 and acid rain), operational (poisoning of catalysts), and safety reasons (H_2S, corrosion). Sweet crudes cost more than sour crudes.

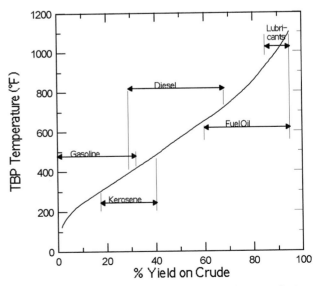

FIGURE 3 Distillation ranges of various petroleum products.

TABLE III Boiling Points of Petroleum Products

Product	Boiling range (°F)	Uses
Light gases, CH_4, C_2H_6	−259–44	Fuel gas, petrochemicals
Propane C_3H_8	−44	LPG, petrochemicals
Butane C_4H_{10}	11–31	LPG, petrochemicals, gasoline
Light naphta, light straight Run	30–300	Gasoline, solvents
Heavy naphtha	300–400	Gasoline, solvents jet fuels
Kerosene	400–500	Jet fuel, heating oil, illuminant
Light gasoil	400–600	Automotive diesel, heating oil
Heavy gasoil	600–800	Automotive diesel, heating oil, marine diesel
Vacuum gasoil	800–1100	Catalytic cracking feed, lubricants
Atmospheric residue	+800	Fuel oil, vacuum distillation feed
Vacuum residue	+1100	Fuel oil, asphalt

F. Crude Oil Refining

Crude oil is not used as is for three reasons:

1. Modern engines require specific products with well-defined properties as fuels and lubricants. Crude oils cannot satisfy these requirements.
2. Even in the hypothetical case in which the engines could operate smooth with a specific type of crude oil, it would be almost impossible to operate with another type of crude oil, because they have many differences.
3. For safety reasons, the use of volatile fuels where less volatile products are required is prohibited.

Refining of crude oil into products is based on the distillation of the crude into fractions, which have specific boiling regions. The usual fractions and their major uses are tabulated in Table III. Figure 4 gives typical distillation curves of crude oil products.

Table IV gives an indication on the effect of crude oil type on the chemical composition by major fractions.

The yield of each type in products from the crude oil is different and depends on the type of crude. Table V gives typical yield in the products from some crudes.

It is clear from Figure 5 that atmospheric distillation is inadequate to cover the local or international demand for distillate products. Conversion processes are used to improve residues into valuable "white" products. Figure 6 shows the possibilities of yielding more distillates by using various configurations of processes.

1. Refining

Petroleum refining begins with the distillation of crude oil into fractions of different boiling ranges. The crude oil is heated to 700°F, pumped into a fractioning tower, and sep-

arated into specific boiling points fractions. Because they are naturally occurring fractions of crude oil, the naphtha fractions obtained by distillation are called virgin naphtha, or straight run gasoline. Both the amounts of these naphthas and their hydrocarbon compositions depend on the type of crude oil being distilled. Thus, straight run gasolines differ widely in such properties as specific gravity, vaporization characteristics, and antiknock quality. The light naphtha fraction is usually not used as a component of finished gasoline without any more refining to remove

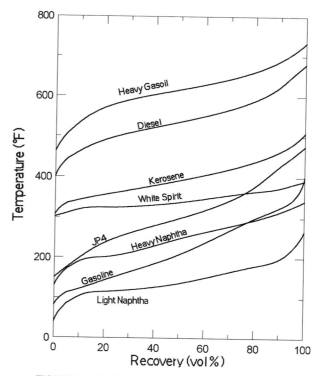

FIGURE 4 Distillation curves of petroleum products.

TABLE IV   Chemical Composition of Petroleum Fractions

Fraction	Paraffin-base crude			Asphalt-base crude		
	Paraffins	Naphthenes	Aromatics	Paraffins	Naphthenes	Aromatics
Naphtha	65	30	5	35	55	10
Kerosene	60	30	10	25	50	25
Gas oil	35	55	15	15	45	40
Heavy Distillate	20	65	15	—	55	45

undesirable impurities. The heavy naphtha is catalytically reformed to higher octane blending stock. The kerosene and light gas-oil fractions, referred to collectively as middle distillates, are used in the production of kerosene, jet fuel, diesel fuel, and furnace oils. The heavy gas-oil fraction may be used in heavy diesel fuel, industrial fuel oil, and bunker oil. However, in the United States, gasoline demand is so high that much of the heavy gas oil and other heavy oils recovered from the reduced crude are cracked into gasolines by the processes given below.

Since natural gasoline is a highly volatile mixture of light hydrocarbons, only a limited amount of it can be included in finished gasoline without exceeding the desired volatility characteristics. Moreover, antiknock quality of natural gasoline is relatively low since it contains a large amount of straight-chain paraffinic hydrocarbons.

Early in the history of petroleum refining, it was found that higher boiling hydrocarbons could be broken down, or cracked, into lower-boiling ones by subjecting them to high temperatures for an appreciable length of time. In fact, a form of cracking was used as early as 1869 to convert heavier crude-oil fractions into kerosene. Use of cracking to produce gasoline began in 1913 and grew rapidly as the motor-vehicle population multiplied.

Cracking in the presence of hydrogen has recently been gaining popularity as a highly flexible refining process because it permits wide variations in yields of gasoline and furnace oils to meet seasonal demand changes. With hydrocracking, refiners can convert heavy fractions into maximum amounts of gasoline and only small amounts of furnace oil, or vice versa, as needed. Moreover, hydrocracking is particularly effective in processing highly re-

fractory (hard-to-crack) stocks. Thus, the process enables refiners to convert practically worthless heavy cracked stocks into gasoline, fuel oil, or other highly sellable products.

All reforming processes have the same general purpose: to convert low-octane gasoline-range hydrocarbons into higher octane ones. Reforming is largely limited to the upgrading of heavier gasoline fractions, such as straight run stocks boiling from 90 to 200°C, because lighter fractions do not contain substantial amounts of hydrocarbons suitable for reforming. Only about 80% of the feedstock volume processed is recovered as full- boiling-range gasoline, because the cracking reactions produce substantial amounts of gases.

Unlike cracking, alkylation makes larger hydrocarbons from smaller ones; it produces gasoline-range liquids from refinery gases. Because of its cost, however, its major value stems from the exceptionally high antiknock quality of its product, rather than from the fact that it offers another way to make gasoline. Extensive use of alkylation began during World War II to produce high-octane (100+) gasolines required by military aircraft. After the war, some alkylation units not required to meet the demand for aviation gasoline were operated to make highly desirable blending stocks for passenger-car gasoline. More recently, a number of refiners have installed new units exclusively for such use.

Like alkylation, polymerization is a way of making gasoline from refinery gases. But in polymerization only the olefinic gases in the feed react, linking together to form olefinic liquids. Any paraffinic gases in the feed pass through the process unchanged. A typical polymerization reaction is that of two molecules of isobutylene (C_4H_8)

TABLE V   Yields of Various Crudes in Products

Type	Arabian light	Iranian light	Zarzaitine	Sarir	Nigerian	Brent	Maya
Gases	0.7	1.8	1.8	2.0	0.6	2.1	1.0
Naphtha	17.8	15.3	24.4	16.4	12.9	17.8	11.7
Middle Distillates	33.1	32.6	41.1	33.7	47.2	35.5	23.1
Residue	48.4	46.0	30.6	47.9	39.3	44.6	64.2

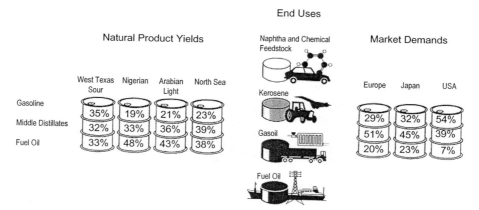

FIGURE 5 The refiner's challenge.

combining into one molecule of a branched-chain octylene (C_8H_{16}). This reaction was used during World War II to produce 2,4,4-trimethyl-2-pentene, which was subsequently hydrogenated to make "isooctane" for aviation gasoline. In polymerization processes for making motor-gasoline components, refinery gases rich in propylene and butylenes are subjected to temperatures of 150–°C and pressures of 150–1200 psi in the presence of a catalyst.

2. Distillation

The distillation process is based on the different boiling points of the hydrocarbons that are present in the crude oil. Each product is assigned a temperature range and the product is obtained by condensing the vapor that boils off in this range at atmospheric pressure (atmospheric distillation). In a distillation column, the vapor of the lowest boiling hydrocarbons—propane and butane—rises to the top. The straight run gasoline, kerosene, and diesel fuel cuts

are drawn off at successively lower positions in the column. Hydrocarbons with boiling points higher than diesel fuel are not vaporized; they remain in liquid form and fall to the bottom of the column (atmospheric residue).

Initially, atmospheric residue was used for paving and sealing. Later it was found that it could yield higher value products like lubricating oil and paraffin wax when it was distilled in a vacuum. Vacuum distillation requires more sophisticated control systems and sturdier stills to withstand the pressure differential and more sophisticated control systems.

Figure 7 gives a typical diagram for an atmospheric distillation unit in combination with a vacuum distillation unit.

G. Cracking Processes

The discovery that hydrocarbons with higher boiling points (the larger ones left in the distillation residue)

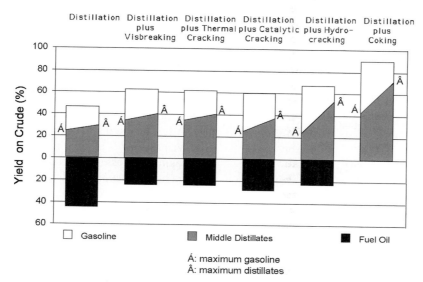

FIGURE 6 Refinery processing options to meet demand.

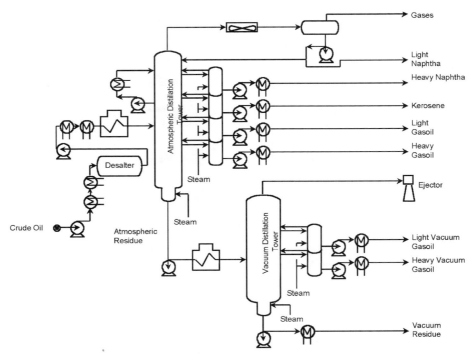

FIGURE 7 Atmospheric and vacuum distillation.

could be cracked into lower boiling hydrocarbons by subjecting them to very high temperatures offered a means to increase the yield of gasoline from the crude. This process, called thermal cracking, was used to increase gasoline production starting in 1913. This process produces a lot of olefins, which have higher octane, but they are unstable and may cause engine deposits (gums). By today's standards, the quality and performance of this early cracked gasoline was low, but it was sufficient for the engines of the day. Typical cracking reactions are as follows:

Molecule crack

$$CH_3(CH_2)_5CH_3 \longrightarrow CH_3(CH_2)_3CH_3 + CH_2{=}CH_2$$

Dehydrogenation of paraffins and naphthenes

$$RCH_2CH_3 \rightleftharpoons RCH{=}CH_2 + H_2$$

Olefins isomerization

$$CH_3CH_2CH{=}CH_2 \rightleftharpoons CH_3CH{=}CHCH_3$$

Olefins polymerization

$$2CH_3CH_2CH{=}CH_2 \rightleftharpoons CH_3\overset{\overset{\displaystyle CH_3}{|}}{\underset{\underset{\displaystyle CH_3}{|}}{C}}CH_2\overset{\overset{\displaystyle CH_3}{|}}{CH}{=}CH_3$$

Dehydrogenation of olefins

$$CH_3CH_2CH{=}CH_2 \rightleftharpoons CH_2{=}CHCH{=}CH_2 + H_2$$

Cracking processes produce some amounts of gasoil. The properties of these gasoil streams are not very good (especially the cetane number), but they are used as blending components in order for the refiner to help meet the market demand.

A typical thermal cracking unit is given in Fig. 8.

Visbreaking is another form of thermal cracking. Visbreaking processes are used to reduce the viscosity of extra heavy vacuum residues using heat under mild pressure. The main product of visbreaking is a lower viscosity residue. Naphtha and gasoil are produced also as side products. Figure 9 shows a typical visbreaking unit.

Delayed coking is also a thermal cracking process. Coking is used to convert vacuum residues and petroleum tars into lighter products and coke. A typical delayed coking unit is shown in Fig. 10.

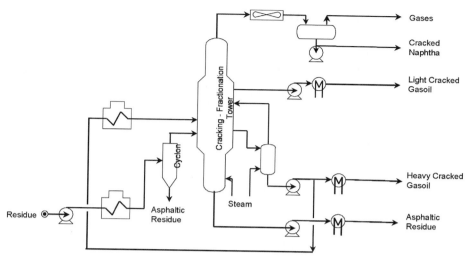

FIGURE 8 Thermal cracking unit.

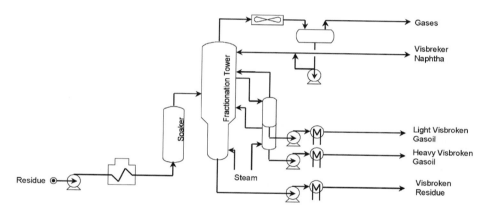

FIGURE 9 Visbreaking unit.

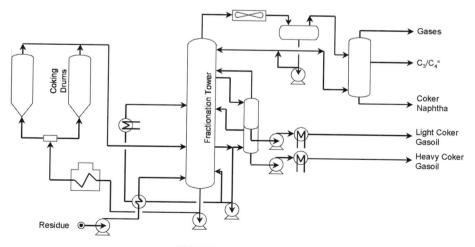

FIGURE 10 Coking unit.

Eventually a catalyst, transforming thermal cracking into catalytic cracking, supplemented heat. Catalytic cracking produces gasoline of higher quality than thermal cracking. There are many variations on catalytic cracking, but *fluid catalytic cracking* (FCC) is the main process for gasoline production in most modern refineries. The term comes from the practice of fluidizing the solid catalyst so that it can be continuously cycled from the reaction section of the reactor to the catalyst regeneration section and back again. The FCC process also produces building blocks (C_4 olefins) for other essential refinery processes, like alkylation.

The reactions of catalytic cracking are the same as in the case of thermal cracking, adding the following:

Hydrogenation of olefins

$$CH_3CH_2CH{=}CH_2 + H_2 \longrightarrow CH_3(CH_2)_2CH$$

Paraffins dehydrocyclation

$$CH_3(CH_2)_5CH_3 \longrightarrow \underset{}{\bigcirc}{\text{-}CH_3} + H_2 \longrightarrow \underset{}{\bigcirc}{\text{-}CH_3} + 3H_2$$

Figure 11 shows a typical fluid catalytic cracking unit.

Hydrocracking is similar to catalytic cracking in that it utilizes a catalyst, but the catalyst operates in a hydrogen atmosphere. Hydrocracking can break down hydrocarbons that are resistant to catalytic cracking alone. The presence of hydrogen saturates the olefins produced from the cracking processes. It is more commonly used to produce diesel fuel than gasoline. Hydrocrackers are operating at very high pressures (up to 3000 psi) and temperatures (800°F). Hydrocracking is used to upgrade heavy vacuum distillates mainly to very good quality gasoil streams. A typical hydrocracking unit is shown in Fig. 12.

H. Chemical Transformation Processes

Another group of processes, conversion processes, is used to increase a refinery's octane pool. While these processes predate the regulation of antiknock additives, they became more important as lead was phased out of gasoline for health and environmental reasons. Without antiknock additives, the only way to produce high octane number gasolines is to use inherently high-octane hydrocarbons or to use oxygenates, which also have high-octane values.

The *reforming* process literally reforms the feed, converting straight chain paraffins into aromatics. For example, reforming cyclizes normal heptane (Research Octane Number, RON = 0) and then abstract hydrogen to produce toluene (RON = 120). The hydrogen by-product is almost as important as the octane upgrade.

Typical reforming processes are:

Dehydrogenation of naphthenes to aromatics and hydrogen

$$\underset{}{\bigcirc}{\text{-}CH_3} \longrightarrow \underset{}{\bigcirc}{\text{-}CH_3} + 3H_2$$

Isomerization of *n*-paraffins and naphthenes

$$CH_3(CH_2)_4CH_3 \longrightarrow \underset{}{\bigcirc}{\text{-}CH_3}$$

$$\underset{}{\bigcirc}{\text{-}CH_3} \longrightarrow \underset{}{\bigcirc}{}$$

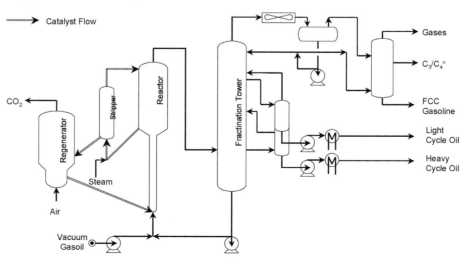

FIGURE 11 Typical catalytic cracking unit.

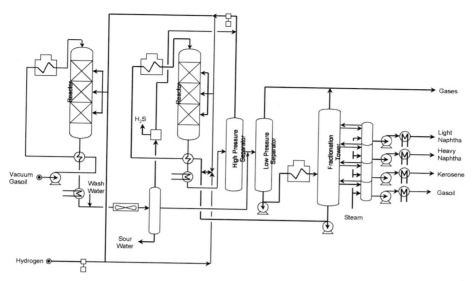

FIGURE 12 A typical hydrocracking unit.

Dehydrocyclation of paraffins to aromatics

$$CH_3(CH_2)_5CH_3 \longrightarrow \text{(methylcyclohexane)} + H_2 \longrightarrow \text{(toluene)} + 3H_2$$

Hydrocracking

$$CH_3(CH_2)_8CH_3 + H_2 \longrightarrow$$
$$CH_3(CH_2)_4CH_3 + CH_3(CH_2)_2CH_3$$

Hydrogen is an essential ingredient for processes like hydrocracking and hydrofining. Refineries often have a hydrogen deficit, which has to be made up by making hydrogen from natural gas (methane) or other light hydrocarbons (such as LPG or naphtha) using steam reforming. A typical naphtha reforming unit is given in Fig. 13.

Alkylation combines small, gaseous hydrocarbons with boiling points too low for use in gasoline to produce liquid hydrocarbons in the boiling range of gasoline. The feed, which comes from the FCC unit, includes C_4 hydrocarbons like isobutanes and butylenes and sometimes C_3 and C_5 paraffins and olefins. The principal products are high-octane isomers of trimethylpentane, like isooctane (RON = 100). Alkylation is a key process for producing reformulated gasolines because the contents of the other classes of high-octane hydrocarbons—olefins and aromatics—are limited by regulation.

Typical alkylation reactions are

$$\underset{\underset{CH_3}{|}}{\overset{\overset{CH_3}{|}}{CH_3CH}} + CH_2{=}CHCH_2CH_3 \longrightarrow \underset{\underset{CH_3}{|}}{\overset{\overset{CH_3}{|}}{CH_3C(CH_2)_3CH_3}}$$

$$\underset{\underset{CH_3}{|}}{\overset{\overset{CH_3}{|}}{CH_3CH}} + \underset{\underset{CH_3}{|}}{CH_2{=}CCH_3} \longrightarrow \underset{\underset{CH_3}{|}\ \underset{CH_3}{|}}{\overset{\overset{CH_3}{|}}{CH_3CCH_2CHCH_3}}$$

Figure 14 gives a diagram of an alkylation unit

Another combination process is the *polymerization* of olefins, typically the C_3 olefin, propylene, into a series of larger olefins differing in molecular weight by three carbon atoms—C_6, C_9, C_{12}, etc. Polymerization is a less favored process than alkylation because the products are also olefins, which may have to be converted to paraffins before they are blended into gasoline due to the high tendency of olefins for gums formation.

A typical polymerization reaction is:

$$\underset{\underset{CH_3}{|}}{\overset{\overset{CH_3}{|}}{CH_2{=}CH}} + \underset{\underset{CH_2}{||}}{\overset{\overset{CH_3}{|}}{CH_3C}} \longrightarrow \underset{\underset{CH_3}{|}}{\overset{\overset{CH_3}{|}}{CH_3CH}}\ \underset{\underset{CH_2}{||}}{\overset{\overset{CH_3}{|}}{CH_2C}}$$

A polymerization unit is shown in Fig. 15.

Isomerization increases a refinery's octane pool by converting straight chain (typically C_5 and C_6) paraffins into their branched isomers. For a given carbon number, branched isomers have higher octane values than the

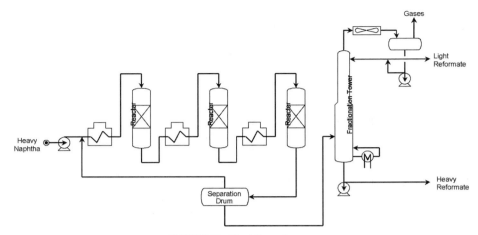

FIGURE 13 Naphtha reformer.

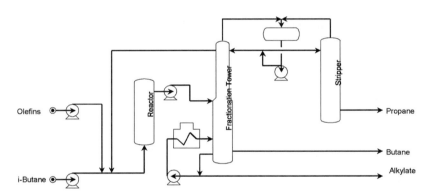

FIGURE 14 Alkylation unit.

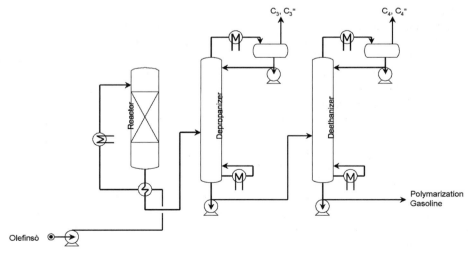

FIGURE 15 Polymerization unit.

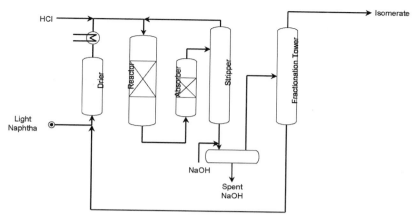

FIGURE 16 Isomerization unit.

corresponding straight chain isomer. Isomerization is used to upgrade the octane of light naphthta that cannot be upgraded with reforming processes. Isomerization—like alkylation—is one important process for the production of reformulated gasoline.

Typical isomerization reactions are

$$CH_3(CH_2)_2CH_3 \longrightarrow CH_3\overset{\overset{\displaystyle CH_3}{|}}{C}HCH_3$$

$$CH_3(CH_2)_3CH_3 \longrightarrow CH_3\overset{\overset{\displaystyle CH_3}{|}}{C}HCH_2CH_3$$

A typical isomerization unit is given in Fig. 16.

Hydrotreating is a generic term for a range of processes that use hydrogen with an appropriate catalyst to remove impurities from a refinery stream. The processes run the gamut from mild selective hydrotreating for removing highly reactive olefins, to heavy hydrotreating for converting aromatics to naphthenes.

Sulfur removal, or desulfurization, is an example of a hydrotreating process. The lower sulfur limits for reformulated gasoline may require the desulfurization of a significant proportion of FCC gasoline. There are also processing reasons to desulfurize refinery streams. In reforming, excess sulfur in the feed deactivates the catalyst, so heavy naphtha has to be *hydrodesulfurized* before being fed to a reformer. In FCC, excess sulfur in the feed results in high levels of sulfur in the FCC gasoline and greater production of sulfur dioxide during catalyst regeneration.

Hydrodesulfurization is a common procedure for diesel fuels. Straight run gasoil has sulfur content much higher than the maximum set by regulations. For this reason, gasoil undergoes a treatment procedure at high pressure in excess hydrogen environment over a catalyst. Under these conditions, hydrogen reacts with sulfur to form hydrogen sulfide (H_2S). H_2S is removed by amine wash.

Typical hydrodesulfurization reactions are

$$RSH + H_2 \longrightarrow RH + H_2S$$

$$R_1SR_2 + 2H_2 \longrightarrow R_1H + R_2H + H_2S$$

$$R_1SSR_2 + 2H_2 \longrightarrow R_1H + R_2H + 2H_2S$$

A hydrodesulfurization unit is shown in Fig. 17.

I. The Modern Refinery

Today's refinery is a complex combination of processing units, which use the advantages of chemistry, engineering, and metallurgy for the separation of crude oil into valuable products.

The schematic layout of a modern refinery is shown in Fig. 18. Crude oil is fed to the distillation column where straight-run light and heavy naphtha, jet, and diesel are separated at atmospheric pressure. Straight-run jet, and diesel are usually acceptable as they are, but the straight-run naphthas require more processing to convert them into gasoline blending components. The straight-run light naphtha may be isomerized to increase octane, or hydrotreated to convert benzene into cyclohexane so that the final gasoline blend will meet a benzene specification limit, or both. The straight-run heavy naphtha is hydrotreated to remove sulfur and then reformed to improve octane and generate hydrogen for the hydrotreaters.

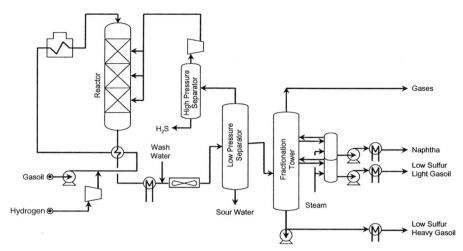

FIGURE 17 Hydrodesulfurization unit.

The atmospheric distillation residue is distilled at reduced pressure (vacuum distillation) to obtain gasoils for FCC or hydrocracker feed. The gasoils are hydrotreated to reduce sulfur and nitrogen to levels that will not interfere with the FCC process. Even though the feed was substantially desulfurized, the FCC product must be sweetened to convert reactive sulfur compounds (mercaptans) to more neutral ones, otherwise the gasoline blend will be odorous and unstable. Alternatively, in California and other areas with tight restrictions on the sulfur content of finished gasoline, the FCC product must be further desulfurized.

Previously, the vacuum residue might have been used as low value, high sulfur fuel oil for power generation or marine fuel. Now, to remain competitive, refiners squeeze

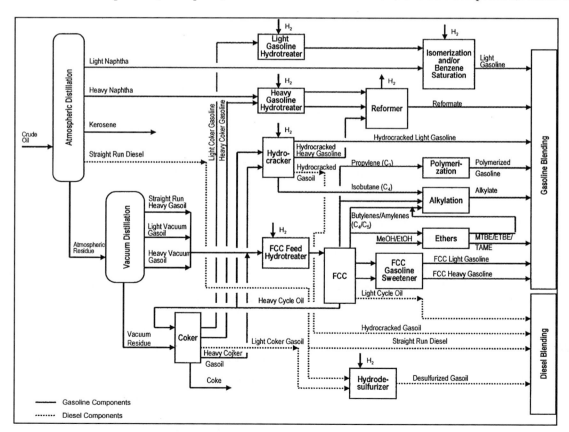

FIGURE 18 Complex refinery scheme.

as much high value product as possible from every barrel of crude. As a result, the vacuum residue is sent to a residue conversion unit, such as a residue cracker, solvent extraction unit or coker. These units produce more transportation fuel or gasoil, leaving an irreducible minimum of residue or coke. The residue-derived streams require further processing and/or treating before they can be blended into light fuels like gasoline or diesel.

II. GASOLINE

The main use of motor gasoline is fuel for automobiles and light trucks for highway use. Smaller quantities are used for off-highway driving, boats, recreational vehicles, and various farms and other equipment.

The fuels characteristics should comply with the fuel requirements of the engine in order to obtain the desired performance. As a result, the gasoline and engine are mutually dependent partners. An engine was not designed without considering the gasolines available in the market place and vice versa. The partnership became a triumvirate in the last decades of the 20th century as environmental considerations began to change both engine design and gasoline characteristics.

The natural gasoline or *naphtha*, has low octane number so it must be upgraded by reformation methods. In more complex steps, nongasoline components of crude are converted into gasoline (cracking processes) and gasoline molecules are rearranged to improve their characteristics.

A. Composition

Gasoline is a complex mixture of hundreds of hydrocarbons. The hydrocarbons vary by class—paraffins, olefins, naphthenes, and aromatics—and, within each class, by size. The mixture of hydrocarbons (and oxygenates) in a gasoline determines its physical properties and engine performance characteristics.

Gasoline is manufactured to meet the property limits of the specifications and regulations, not to achieve a specific distribution of hydrocarbons by class and size. But, to varying degrees, the property limits define chemical composition. For example, gasoline volatility is expressed by its distillation curve. Each individual hydrocarbon boils at a specific temperature called its boiling point and, in general, the boiling point increases with molecular size. Consequently, requiring a distillation curve is equivalent to requiring a certain distribution of hydrocarbons with a range of sizes.

The most common way to characterize the size of a molecule is molecular weight. For a hydrocarbon, an alternate way is carbon number—the number of carbons in its molecular structure. Butane, for example, has a molec-

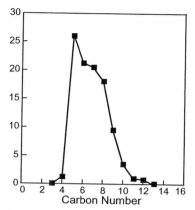

FIGURE 19 Distribution of carbon number in gasoline.

ular weight of 58 g/mole and a carbon number of 4 (C_4). Figure 19 shows the carbon number distribution of a typical gasoline. Note that the range of sizes runs from C_4 to C_{12} with the most prevalent size, C_5, and the average size, $C_{6.8}$. Octane number is another example of how property limits define chemical limits. The RON of hydrocarbons for the same carbon number in molecule is

$$\text{aromatics} > \text{isoparaffins} > \text{naphthenes}$$
$$> \text{olefins} > \text{normal paraffins}$$

The RON of isooctane (2,2,4-trimethylpentane) is 100 by definition, while the RON of normal octane is less than zero. Other properties, such as volatility, also are influenced by isomer structure.

Air pollution regulations and property specifications have been supplemented by some composition specifications. The first gasoline-related air pollution regulation limited the amount of olefins in gasoline sold in Southern California by establishing a *bromine number* maximum specification. More recent regulations limit the amounts of both olefins and aromatics (and more specific benzene), in reformulated gasolines.

Gasolines contain small amounts—less than 0.1% by volume—of compounds with sulfur, nitrogen, and oxygen atoms in their structures (excluding added oxygenates). These compounds either exist in the crude or are formed by the refining processes. Refining processes destroy many nitrogen, in particular the sulfur compounds, but some remain in the final fuel.

B. Gasoline Additives

Gasoline-soluble chemicals are mixed with the gasoline to enhance certain performance characteristics or to provide characteristics not inherent in the gasoline. Typically, they are derived from petroleum-based raw materials and their function and chemistry are highly specialized. They produce the desired effect at the ppm concentration range.

Oxidation inhibitors, also called *antioxidants*, are aromatic amines and hindered phenols. They prevent gasoline components from reacting with oxygen in the air to form *peroxides* or *gums*. They are needed especially for gasolines with high olefin content. Peroxides can degrade antiknock quality and attack plastic or elastomeric fuel system parts, soluble gums can lead to engine deposits, and insoluble gums can plug fuel filters. Inhibiting oxidation is particularly important for fuels used in modern fuel-injected vehicles, as their fuel recirculation design may subject the fuel to higher temperatures and oxygen-exposure stress.

Corrosion inhibitors are carboxylic acids and carboxylates. The facilities—tanks and pipelines—of the gasoline distribution and marketing system are constructed primarily of uncoated steel. Corrosion inhibitors prevent free water in the gasoline from rusting or corroding these facilities. Corrosion inhibitors are less important once the gasoline is in the vehicle. The metal parts in the fuel systems of today's vehicles are made of corrosion-resistant alloys or of steel coated with corrosion-resistant coatings.

Metal deactivators are chelating agents—chemical compounds, which capture specific metal ions. The more active metals, like copper and zinc, effectively catalyze the oxidation of gasoline. These metals are not used in most gasoline distribution and vehicle fuel systems. However, when they are present, metal deactivators inhibit their catalytic activity.

Demulsifiers are polyglycol derivatives. An emulsion is a stable mixture of two mutually insoluble materials. A gasoline-water emulsion can be formed when gasoline passes through the high-shear field of a centrifugal pump if the gasoline is contaminated with free water. Demulsifiers improve the water separating characteristics of gasoline by preventing the formation of stable emulsions.

Antiknock compounds are lead alkyls—tetraethyl lead (TEL) and tetramethyl lead (TML)—and methylcyclopentadienyl manganese tricarbonyl (MMT). Antiknock compounds increase the antiknock quality of gasoline. Because the amount of additive needed is small, they are a low-cost method of increasing octane than changing gasoline chemistry.

The transition from leaded to unleaded gasoline leads to certain problems with the exhaust valve seats of older noncatalytic vehicles. To combat this problem valve seat recession (VSR) additives can be added to unleaded gasoline; these additives usually contain compounds of potassium, phosphorous, or manganese, which have proven effective as lead replacements in protecting exhaust valves of older vehicles. Of the VSR additives, only the ones based on manganese also act as octane improvers.

Deposit control (DC) additives are the first additive of this class. They were introduced in 1970 and were based on polybutene amine chemistry and used in combination with carrier oil. While they have to be used at higher concentrations than detergent-dispersants, DC additives provide benefits throughout the engine intake system. They clean up—and keep clean—the throttle body and upper areas of the carburetor, fuel injectors, intake manifold, intake ports, and intake valves.

Anti-icing additives are surfactants, alcohols, and glycols. They prevent ice formation in the carburetor and fuel system. The need for this additive is disappearing as vehicles with fuel-injection systems replacing older model vehicles with carburetors.

Dyes are oil-soluble solids and liquids used to visually distinguish batches, grades, or applications of gasoline products. For example, gasoline for general aviation, which is manufactured to different and more exacting requirements, is dyed blue to distinguish it from motor gasoline for safety reasons.

Markers are a means of distinguishing specific batches of gasoline without providing an obvious visual clue. A refiner may add a marker to their gasoline so it can be identified as it moves through the distribution system.

Drag reducers are high-molecular-weight polymers that improve the fluid flow characteristics of low-viscosity petroleum products. As energy costs have increased, pipelines have sought more efficient ways to ship products. Drag reducers lower pumping costs by reducing friction between the flowing gasoline and the walls of the pipe.

Gasoline fuel octane number is measured by the following two methods, research and motor:

- ASTM D 2699—Standard Test Method for RON of Spark-Ignition Engine Fuel.
- ASTM D 2700—Standard Test Method for Motor Octane Number (MON) of Spark-Ignition Engine Fuel.

The octane number for the gasoline fuel is the average value of RON and MON.

C. Oxygenated Gasoline

The oxygenated gasoline is a mixture of conventional hydrocarbon-based gasoline and one or more oxygenates. Oxygenates are combustible liquids made up of carbon, hydrogen, and oxygen. The current oxygenates belong to one of two classes of organic molecules: alcohols and ethers. In alcohols, a hydrocarbon group and a hydrogen atom are bonded to an oxygen atom: R—O—H, where "R" represents the hydrocarbon group. All alcohols contain the "OH" atom pair. In ethers, two hydrocarbon groups are bonded to an oxygen atom; the groups may be the same or different: R—O—R or R—O—R'.

Oxygenated gasolines have lower heating values because the heating values of the oxygenate components are lower than those of the hydrocarbons they displace. The percent decrease in heating value is close to the mass percent oxygen in the gasoline. Federal reformulated gasoline and California Phase 2 reformulated gasoline must be oxygenated year round to average oxygen content of about 2% by mass. As a result, their heating values are about 2% lower than conventional gasoline. In addition, California Phase 2 reformulated gasoline sets some limits on distillation temperatures and aromatics content, which have the secondary effect of lowering the density of the fuel. This reduces the heating value by about another 1%.

The oxygenate is regulated by EPA in the United States. The most widely used oxygenates are ethanol, methyl tertiary-butyl ether (MTBE) and tertiary-amyl methyl ether (TAME). Ethyl tertiary-butyl ether (ETBE) may be used more in the future. Methanol has been tested as an alternative oxygenate, but is not preferred because of its toxicity and its high vapor pressure.

The presence of water and acidic compounds may rust or corrode some metal components in the fuel system. The additional water dissolved in oxygenated gasolines does not cause rusting or corrosion, but water from the phase separation of gasoline oxygenated with ethanol will, given time.

Oxygenates can swell and soften natural and some synthetic rubbers (elastomers). Oxygenated gasolines affect elastomers less; the extent of which also depends on the hydrocarbon chemistry of the gasoline, particularly the aromatics content. The effect is of potential concern because fuel systems contain elastomers in hoses, connectors ("O"-rings), valves, and diaphragms. The elastomeric materials used in today's vehicles have been selected to be compatible with oxygenated gasolines. Owner's manuals approve the use of gasoline oxygenated with 10% by volume of ethanol or 15% by volume of MTBE.

D. Reformulated Gasoline

In order to reduce emissions from spark ignition engines, the Environmental Protection Agency (EPA) and the California Air Resources Board (CARB) have established a number of regulations in the last 35 years to control gasoline properties to reduce emissions from gasoline-fueled vehicles. Figure 20 shows the actions in brief.

The most significant changes occurred in the 1990s. In 1992, the EPA required the decrease of the maximum vapor pressure of summertime gasoline to reduce evaporative VOC emissions. They placed an upper limit on vapor pressure at 7.8 psi in ozone nonattainment areas in the southern states, where average summer temperatures are high, and at 9.0 psi elsewhere.

In 1992 California Phase 1 RFG was required throughout California. The Phase 1 RFG regulations set maximum summertime vapor pressure at 7.8 psi for the entire state,

not just for ozone nonattainment areas, and forbade the use of lead-containing additives. They also made the use of deposit control additives mandatory, on the basis that engine intake system deposits increase emissions.

In 1992 the EPA started the winter oxygenate program. This program requires the addition of oxygenates to gasoline sold in the 39 areas of the country that have not attained the National Ambient Air Quality Standard for CO. Gasoline in these areas must contain 2.7 wt% oxygen minimum, averaged over the high CO months.

The Clean Air Act Amendments of 1990 mandated Federal RFG. Federal Phase I RFG was introduced in 1995. It must be used in the nine extreme or severe ozone nonattainment areas across the country. Less severe nonattainment areas may decide on the program. A few characteristics of Federal Phase I RFG are fixed. The average benzene content must be less than 1 vol.% and the average year-round oxygen content must be greater than 2.1 wt%. Otherwise the general approach is to set vehicle emission reduction targets, rather than property or composition limits. The EPA provided refiners with two equations, which relate gasoline composition to vehicle emissions—a *Simple Model* and a *Complex Model*. The Simple Model involves fewer gasoline characteristics than the Complex Model. The Simple Model was in use from 1995 to 1997 only. It requires the refiner to adjust gasoline composition to reduce average toxics by 16.5%, relative to 1990 baseline gasoline. In lieu of a VOC target, it limits average summertime vapor pressure to 8.1 psi in the northern states and to 7.2 psi in the southern states. The Complex Model was optional from 1995 to 1997 and mandatory beginning in 1998. It requires the refiner to adjust gasoline composition to meet VOC, toxics, and NO_x limits. Federal Phase II RFG, to be introduced in 2000, continues the Phase I limits on benzene and oxygen content and the use of the Complex Model, but requires greater reductions in VOCs, toxics, and NO_x. Table VI summarizes the emissions reductions, which must be achieved for gasolines formulated under the Phase I and Phase II programs. Decreasing vapor pressure, benzene content, and sulfur content are the primary strategies refiners are expected to use to meet the Phase I Complex Model and Phase II emission limits.

The California Air Resources Board (CARB) predicts Phase 2 RFG will reduce VOC emissions by 17%, CO and NO_x emissions by 11%, and organic toxics by 44%, relative to Phase 1 RFG. This is equivalent to removing 3.5 million cars from California's roads.

E. Gasoline Properties and Trend

During the 1990s gasoline and diesel fuel were "reformulated" many times to meet requirements included in the Clean Air Act Amendments of 1990 (CAAA90) and

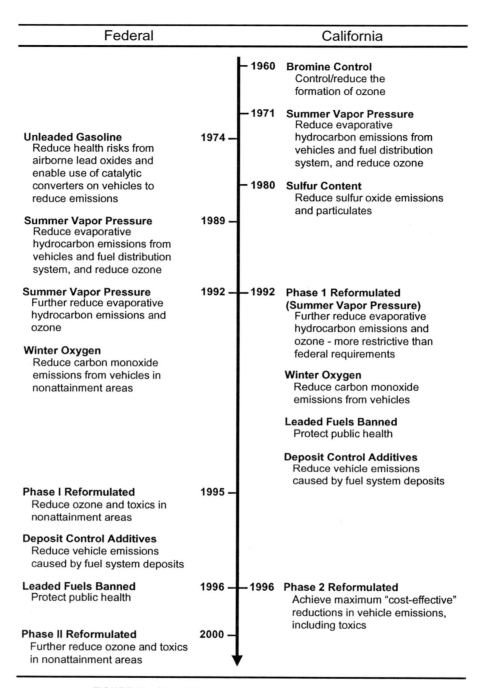

Federal		California
	1960	**Bromine Control** Control/reduce the formation of ozone
	1971	**Summer Vapor Pressure** Reduce evaporative hydrocarbon emissions from vehicles and fuel distribution system, and reduce ozone
Unleaded Gasoline Reduce health risks from airborne lead oxides and enable use of catalytic converters on vehicles to reduce emissions	1974	
	1980	**Sulfur Content** Reduce sulfur oxide emissions and particulates
Summer Vapor Pressure Reduce evaporative hydrocarbon emissions from vehicles and fuel distribution system, and reduce ozone	1989	
Summer Vapor Pressure Further reduce evaporative hydrocarbon emissions and ozone **Winter Oxygen** Reduce carbon monoxide emissions from vehicles in nonattainment areas	1992 — 1992	**Phase 1 Reformulated (Summer Vapor Pressure)** Further reduce evaporative hydrocarbon emissions and ozone - more restrictive than federal requirements **Winter Oxygen** Reduce carbon monoxide emissions from vehicles **Leaded Fuels Banned** Protect public health **Deposit Control Additives** Reduce vehicle emissions caused by fuel system deposits
Phase I Reformulated Reduce ozone and toxics in nonattainment areas **Deposit Control Additives** Reduce vehicle emissions caused by fuel system deposits	1995	
Leaded Fuels Banned Protect public health	1996 — 1996	**Phase 2 Reformulated** Achieve maximum "cost-effective" reductions in vehicle emissions, including toxics
Phase II Reformulated Further reduce ozone and toxics in nonattainment areas	2000	

FIGURE 20 Chronology of United States gasoline regulations.

other State-initiated requirements (Table VI). Although the changes went unnoticed by most motorists, they required many adjustments at refineries and in fuel distribution systems. Refineries changed existing processes and invested in new ones, and storage and distribution systems were modified to handle additional products.

"Phase II" reformulated gasoline, which was required by 2000, is the last fuel quality change specified by the CAAA90, but further changes are on the horizon. Two widely publicized fuel quality issues—sulfur removal and the reduction of the widely used gasoline additive MTBE—point to new challenges for the refining industry. The U.S. EPA is in the process of finalizing regulations that would severely restrict the sulfur content of gasoline (and also diesel). The State of California is already phasing MTBE out of gasoline, and there have been numerous

TABLE VI Reductions in Vehicle Emissions for Federal Phase I and Phase II Reformulated Gasoline Programs

	Effective date	Reduction in emissions, % (Averaged standard, compared to 1990 baseline refinery gasoline)		
		VOC	Toxics	NO$_x$
Phase I				
Simple model	1995	Vapor pressure limits	≥16.5	No increase
Complex model	1998	≥17.1[a], ≥36.6[b]	≥16.5	≥1.5
Phase II				
Complex model	2000	≥27.4[a], ≥29.0[b]	≥21.5	≥6.8

[a] Northern states.
[b] Southern states.

proposals to restrict its use at the national level. Because it is the current law, the California ban on MTBE is reflected in AEO2000. Major recent quality changes, as well as those proposed are given in Table VII.

Cleaner-burning gasoline is fuel that meets requirements established by the Air Resources Board (ARB). All gasoline sold in California for use in motor vehicles must meet these requirements, which have been in effect since spring 1996. Cleaner-burning gasoline reduces smog-forming emissions from motor vehicles by 15% and reduces cancer risk from exposure to motor vehicle toxics by about 40%.

The basic specifications for cleaner-burning gasoline are

1. Reduced sulfur content—Sulfur inhibits the effectiveness of catalytic converters. Cleaner burning

TABLE VII Major Fuel Quality Changes, Past and Future

Current

1975	Gasoline lead phaseout begins
1989–1990	Phase I summer gasoline volatility
1992	Oxygenated gasoline, wintertime
	Phase II summer gasoline volatility
	California gasoline Phase I
1995	Phase I reformulated gasoline: Simple Model
1996	California cleaner gasoline Phase II
1998	Phase I reformulated gasoline: Complex Model
2000	Phase II reformulated gasoline
2002	California ban on MTBE

Proposed

2000–2003	Removal of oxygen requirement on reformulated gasoline
	Reduction of MTBE blended in gasoline
2002	California cleaner gasoline Phase III, proposed
2004–2007	Reduced-sulfur gasoline, proposed 30 ppm

gasoline enables catalytic converters to work more effectively and further reduce tailpipe emissions.

2. Reduced benzene content—Benzene is known to cause cancer in humans. Cleaner burning gasoline has about one-half the benzene of earlier gasoline, thus reducing cancer risks.

3. Reduced levels of aromatic hydrocarbons, which react readily with other pollutants to form smog.

4. Reduced levels of olefins, which also react readily with other pollutants to form smog.

5. Reduced vapor pressure, which ensures that gasoline evaporates less readily.

6. Two specifications for reduced distillation temperatures, which ensure the gasoline burns more completely.

7. Use of an oxygen-containing additive, such as MTBE or ethanol, which also helps the gasoline to burn more cleanly.

The approach in Europe is not similar, even though Europe also focuses on reduction of emitted pollutants. The European Union sets limits on specific properties, and does not use a model for emissions calculations like CAA. This has resulted in less flexibility for European refiners than for Americans.

The wide use of MTBE faces a serious problem. MTBE moves more quickly into water than do other gasoline components and has made its way from leaking pipes and underground storage tanks to water sources. MTBE has not been classified as a carcinogen, but it has been shown to cause cancer in animals. For the most part, MTBE found in water supplies has been well below levels of health concern, but it has become a big water quality issue because only trace amounts cause water to smell and taste bad. In 1999, water quality concerns resulted in the announcement by the Governor of California of a statewide phaseout of MTBE, as well as numerous legislative proposals at both the state and federal levels aimed at reducing or eliminating the use of MTBE in gasoline. The future of MTBE in Europe is currently under discussion.

Legislation that would ban MTBE at either the national or state level without waiving the CAAA90 requirement for oxygen in RFG would force the refining industry to find an alternative source of oxygen. Other EPA-approved oxygenates, including ETBE and TAME, would be suitable replacements; however, those ethers are similar to MTBE in some respects and could raise some of the same groundwater contamination concerns. Ethanol, which is currently used chiefly as an octane enhancer and volume extender in traditional gasoline, would be the leading candidate to replace MTBE. Ethanol is thought to be less toxic than ethers, has a high-octane value, and enjoys a fair amount of political support at both the state and federal levels.

Because automotive emissions and fuel sulfur are linked, tighter standards on the sulfur content of gasoline will be issued. Sulfur reduces the catalyst effectiveness used in emission control systems, increasing their emissions of hydrocarbons, CO, and NO_x. As a result, gasoline with significantly reduced sulfur levels will be required for the control systems to work properly and to meet the new Tier 2 standards. The Notice of Proposed Rulemaking of EPA sets the average annual sulfur content of gasoline at 30 ppm, compared with the current standard of 1000 ppm.

III. DIESEL

Diesel fuel is a very important fuel for transportation and for electric power generation. It is the main fuel for off-road diesel engines. Compression ignition engines are known as diesel engines from the name of Dr. Rudolf Diesel, the German engineer who patented this type of engine. The main uses of diesel fuel are

- On-road transportation
- Rail transportation
- Marine shipping
- Off-road uses (mainly mining, construction, and logging)
- Electric power generation
- Military transportation

The diesel fuel refers to distillate fuels with well-determined distillation profile for use in compression ignition engines. In the United States, this is primarily Grade No. 2-D diesel fuel. However, two other grades, Grade No. 1-D and Grade No. 4-D, are also in commercial use. The American Society for Testing and Materials (ASTM) establishes these grade designations. The grades are numbered in order of increasing density and viscosity, with No. 1-D the lightest and No. 4-D the heaviest. In Europe, diesel fuel is available as automotive diesel and heating gasoil. These two grades differ mainly in their ignition quality characteristics and their sulfur content. The term gasoil, the main component for diesel fuel production, comes from the early days of petroleum industry where the middle distillates were used for the production of town gas. Later, it was found that these fuels were capable for use as fuels in compression ignition engines. The development of diesel fuels followed the development of diesel engine. Fuels with more specific properties were necessary for the operation of high-speed and high-power diesel engines. These engines required fuels with low viscosity and good ignition quality. The ignition quality was initially expressed by the diesel index, an index based on API gravity and aniline point, and later by the Cetane number.

Viscosity was another critical parameter in order to ensure good atomization of the fuel. The distillation curve of the fuels was used mainly to classify the fuel into categories according to their volatility and not as a quality index. Additional specifications were introduced for sulfur content, water content, and ash content.

Another problem that led to the establishment of new specifications for diesel fuel was the separation of paraffinic wax at low temperatures. Cloud point and pour point were the first methods used to predict the cold flow properties of diesel fuels. Different specifications were set for winter and summer periods and for different places, according to their geographic latitude. Both cloud point and pour point are static tests and were unable to predict the behavior of the fuel in real situations. Thus new dynamic tests, like the cold filter plugging point (CFPP) and the low-temperature flow test (LTFT), were introduced.

Marine diesel fuel is a category of diesel fuels and contains more sulfur than on-road diesel fuel.

A. Diesel Fuel Production

Diesel fuel is made from crude petroleum. Small amounts of sulfur, nitrogen, and oxygen (so called heteroatoms) are present in this fuel. When heteroatoms are bound into molecular structures with carbon and hydrogen, the resulting compounds are not characterized as hydrocarbons. Typical examples of nonhydrocarbon compounds found in diesel include dibenzothiophene (sulfur compound) and carbazole (nitrogen compound).

The distillates heavier than kerosene derived by atmospheric distillation of crude oil are the gasoil streams. Gasoils are the main blending components for diesel fuel production. Additional types of gasoil can be derived from cracking processes. Gasoils from hydrocrackers have very good ignition quality properties, while gasoils from thermal and catalytic cracking have poor ignition quality properties. Depending on the type of crude oil used, the properties of the produced gasoil are different. Table VIII gives some relevant information for specific types of crude oil on diesel fuel properties.

B. Diesel Composition

Diesel fuels are complex hydrocarbon mixtures, containing all the classes of hydrocarbons: paraffins, naphthenes, aromatics, and in small concentrations, olefins. They are manufactured to meet the property limits of the specifications and regulations, and not to achieve a specific hydrocarbon distribution by class and size. But, to varying degrees, the property limits define the chemical composition of the fuel. The volatility of a diesel fuel is expressed by its distillation curve. Each individual hydrocarbon boils

TABLE VIII Impact of Crude Oil Type on Diesel Fuel Properties

Crude oil origin	Hydrocarbon type	Cetane number	Sulfur content	Cloud point	Heating value
U.K./Norway	Paraffinic	High	Low/Moderate	High	Low
Denmark	Naphthenic	Moderate	Low	Moderate	Moderate
Middle East	Paraffinic	High	High	High	Low
Nigeria	Naphthenic	Low	Low	Moderate	Moderate
Venezuela/Mexico	Naphthenic/Aromatic	Very Low	Low/Moderate	Moderate	High
Australia	Paraffinic	High	Low	High	Low
Indonesia	Paraffinic	High	Low	High	Low

at a specific temperature called its boiling point and, in general, the boiling point increases with molecular size. Consequently, defining a distillation curve is equivalent to defining a certain distribution of hydrocarbons with a range of sizes. The temperature limits of the distillation curve exclude smaller hydrocarbons with lower boiling points and larger hydrocarbons with higher boiling points.

The most common way to characterize the size of a molecule is molecular weight. For a hydrocarbon, an alternative way is carbon number—the number of carbons in its molecular structure. Hexadecane, for example, has a molecular weight of 226 g/mole and a carbon number of 16 (C16); naphthalene has a molecular weight of 128 g/mole and a carbon number of 10 (C10). Figure 21 shows the carbon number distribution of a typical diesel fuel. Note that the range of sizes runs from C9 to C23 with the most prevalent size as C16 and the average size as C16.1.

Cetane number is another example of how property limits define chemical limits. The cetane number ranking of

hydrocarbons for the same carbon number in molecule is:

$$normal\ paraffins > olefins > isoparaffins$$
$$> naphthenes > aromatics$$

So, in order to achieve a specific cetane number limit, the diesel fuel must contain greater amounts of some classes and less of others. The properties of hydrocarbons are not affected by carbon number only, but also by the types of isomers for the same molecular type. Other properties, such as volatility, viscosity, and cold flow properties also are influenced by isomer structure. Table IX gives the impact of the various hydrocarbon classes on basic diesel fuel properties.

Air pollution regulations and property specifications have been supplemented by some composition specifications. The first diesel fuel properties related to air pollution regulation limited the sulfur content to reduce particulate matter (PM) emissions and to help phase-out acid rain in several areas. More recent regulations limit the backend volatility of the fuel and the polyaromatics content.

C. Air Quality Standards

Air pollutants are natural and artificial airborne substances that are introduced into the environment in a concentration sufficient to have a measurable effect on humans,

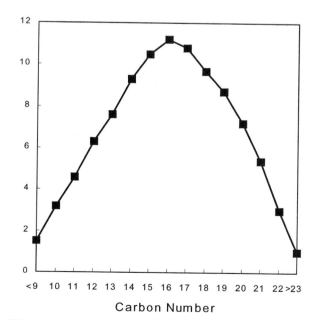

FIGURE 21 Carbon number distribution for a typical diesel fuel.

TABLE IX Effects of Hydrocarbon Class Properties to Some Fuel Properties

Fuel property	Normal paraffin	Isoparaffin	Naphthene	Aromatic
Cetane number	++	0/+	0/+	0/−
Cold flow properties	—	0/+	+	+
Volumetric heating value	—	—	0	+

+   indicates a positive or beneficial effect on the fuel property.
0   indicates a neutral or minor effect.
—   indicates a negative or detrimental effect.

animals, vegetation, or building materials. From a regulatory standpoint of EPA these substances become air pollutants. As part of the regulatory process, the Clean Air Act requires the EPA to issue a criteria document for each pollutant documenting its adverse effects. Regulated pollutants are therefore referred to as criteria pollutants. The EPA uses the information in the criteria documents to set National Ambient Air Quality Standards (NAAQS) at levels that protect public health and welfare. Table X lists the criteria pollutants and the federal and California standards. In most cases, the California standards are more stringent. Some of the criteria pollutants, like carbon monoxide, are primary pollutants, which are emitted directly by identifiable sources. Others, like ozone, are secondary pollutants, which are formed by reactions in the atmosphere. And others, like particulates, are of mixed origin.

Table XI lists the EPA standards for heavy-duty highway engines. The EPA also has engine emission standards for diesel buses, off-road diesels, marine diesels, and railroad diesels. California sets its own limits on diesel emissions, which are generally the same as the federal standards, although sometimes slightly more restrictive. The European standards for heavy trucks are listed in Table XII.

Exhaust emissions are very dependent on how a vehicle is operated. To standardize the test conditions, specific test cycles have been established and each engine is tested according to a specified speed-time cycle on an engine dynamometer.

TABLE XI Federal Heavy-Duty Highway Diesel Engine Emission Standards

Year	CO (g/bhp-hr)	HC (g/bhp-hr)	NO_x (g/bhp-hr)	PM (g/bhp-hr)
1990	15.5	1.3	6.0	0.60
1991–1993	15.5	1.3	5.0	0.25
1994–1997	15.5	1.3	5.0	0.10
1998+	15.5	1.3	4.0[a]	0.10[b]

[a] This standard had to be met by 1996 in California.
[b] Urban buses must meet a 0.05 g/bhp-hr PM standard.

D. Vehicle Emissions: Future Limits

The EPA issued a rule setting emission standards for 2004 model year engines, where a new category of a single nonmethane hydrocarbons (NMHC) + NO_x is introduced. The NMHC and NO_x are of concern because they participate in the reactions that generate ozone. The limit for the combined category will be lowered to 2.4 g/bhp-hr, or 2.5 g/bhp-hr with a limit of 0.5 g/bhp-hr on NMHC. This will reduce NO_x emissions from these engines by 50%, which the EPA expects will lead to lower ambient ozone concentrations.

European regulations tend also to separate HC and NO_x emissions to NMHC + NO_x and HC and set fuel consumption or CO_2 emission limits.

Concern about carbon dioxide emissions and their relationship to global warming is leading to serious consideration of using diesel engines in light-duty trucks, vans, sport utility vehicles, and United States passenger cars. The inherent fuel efficiency of the diesel engine, relative to the gasoline engine results in substantially lower emissions of CO_2 per mile. However, emissions control technology will need to advance for a new generation of diesel vehicles to meet EPA standards.

E. Reformulated Diesel Fuel

California was the first place in which environmental regulations were applied to diesel fuel properties. In 1985

TABLE X Ambient Air Quality Standards

Criteria pollutant	Maximum average concentration		
	Averaging time	Federal standard	California standard
Ozone (O_3), ppm	1-hr	0.12	0.09
	8-hr	0.08	—
Carbon monoxide (CO), ppm	1-hr	35	20
	8-hr	9	9.0
Nitrogen dioxide (NO_2), ppm	1-hr	—	0.025
	annual	0.053	—
Sulfur dioxide (SO_2), ppm	1-hr	—	0.25
	24-hr	0.14	0.05
	Annual	0.03	—
Suspended particulate Matter (PM_{10}), $\mu g/m^3$	24-hr	150	50
	Annual	50	30
Suspended particulate Matter ($PM_{2.5}$), $\mu g/m^3$	24-hr	65	—
	Annual	15	—
Lead, $\mu g/m^3$	30-day	—	1.5
	Quarterly	1.5	—
Sulfates, $\mu g/m^3$	24-hr	—	25

TABLE XII European Heavy Diesel Trucks Emission Standards

Year	CO (g/kWh)	HC (g/kWh)	NO_x (g/kWh)	PM (g/kWh)			
1983	14.0	3.5	18.0	—			
1990	11.2	2.4	14.4	—			
1992	4.5	1.1	8.0	>85 kW	0.36	≤85 kW	0.63
1996	4.0	1.1	7.0	0.15			0.30

CARB limited the sulfur content to 0.05% mass in Southern California because of that region's severe air quality problems. In October 1993 the EPA set a maximum sulfur content of 0.05% mass for on-road diesel fuel for all the United States, while CARB applied this same limit to both off- and on-road diesel fuel. In the other states, the limit on the sulfur content of off-road fuel is the ASTM D 975 limit of 0.5% mass for high sulfur No. 2-D diesel fuel.

Also for environmental reasons, CARB limited the aromatics content of on-road diesel fuel to 10% volume maximum. Alternative formulations with higher aromatics contents are allowed if they have been demonstrated to achieve the same or lower emissions as a 10% aromatics reference fuel in a standardized engine test. The critical properties for alternative formulations of on-road diesel fuel are sulfur content, nitrogen content, aromatics content, polycyclic aromatics content, and cetane number. If the formulation passes the emissions test and receives CARB approval, any fuel manufactured under this alternative formulation must not exceed the sulfur, nitrogen, aromatics, and polycyclic aromatics contents of the candidate formulation and must not have a lower cetane number than the candidate formulation.

CARB estimates that the use of reformulated diesel fuel has reduced SO_2 emissions by 82%, PM emissions by 25%, and NO_x emissions by 7%, relative to the emissions that would have been generated by the continued use of pre-1993 high sulfur diesel fuel.

F. Diesel Fuel Specifications

In the United States, Committee D-2 of ASTM sets the specifications. The committee members bring to the D-2 forum the viewpoints of the large number of groups who are interested in and/or are affected by diesel fuel specifications. These groups include:

- Individual refiners
- Petroleum marketing organizations
- Additive suppliers
- Vehicle and engine manufacturers
- Governmental regulatory agencies, like the EPA and state regulatory agencies
- General interest groups, consumer groups, and consultants.

Table XIII lists the diesel fuel properties that have been identified as important and indicates how they affect performance. The figure also notes whether the property is determined by the bulk composition of the fuel or by the presence or absence of minor components.

Table XIV gives the properties of the five grades of diesel fuel defined by the specification ASTM D 975. This

TABLE XIII   Relationship of Diesel Fuel Properties to Composition and Performance

Property	Property type[a]	Effect of property on performance	Time frame of effect
Flash point	Minor	Safety in handling and use—not directly related to engine performance	—
Water and sediment	Minor	Affects fuel filters and injectors	Long-term
Volatility	Bulk	Affects ease of starting and smoke	Immediate
Viscosity	Bulk	Affects fuel spray atomization and fuel system lubrication	Immediate and long-term
Ash	Minor	Can damage fuel injection system and cause combustion chamber deposits	Long-term
Sulfur	Minor	Affects particulate emissions, cylinder wear, and deposits	Particulates: immediate; wear: long-term
Copper strip corrosion	Minor	Indicates potential for corrosive attack on metal parts	Long-term
Cetane number	Bulk	Measure of ignition quality—affects cold starting, combustion, and emissions	Immediate
Cloud and pour point	Minor	Affects low temperature operability	Immediate
Carbon residue	Minor	Measures coking tendency of fuel, may relate to engine deposits	Long-term
Heating value	Bulk	Affects fuel economy	Immediate
Density	Bulk	Affects heating value	Immediate
Stability	Minor	Indicates potential to form insolubles during use and/or in storage	Long-term
Lubricity	Minor	Affects fuel pump and injector wear	Long-term (typically)
Water separability	Minor	Affects ability to produce dry fuel	—

[a] A bulk property is one that is determined by the composition of the fuel as a whole. A minor property is one that is determined by the presence or absence of small amounts of particular compounds.

TABLE XIV ASTM D 975 Requirements for Diesel Fuels

Property	Test method	No. 1-D Low sulfur	No. 1-D	No. 2-D Low sulfur	No. 2-D	No. 4-D
Flash point (°C), min.	D 93	38	38	52	52	55
Water and sediment (vol.%), max.	D 2709	0.05	0.05	0.05	0.05	
	D 1796					0.50
Distillation temperature (°C)	D 86					
90% vol. recovered						
Min.				282	282	
Max.		288	288	338	338	
Kinematic viscosity (cSt @ 40°C)	D 445					
Min.		1.3	1.3	1.9	1.9	5.5
Max.		2.4	2.4	4.1	4.1	24.0
Ash (wt%), max.	D 482	0.01	0.01	0.01	0.01	0.10
Sulfur (wt%), max.	D 2622	0.05	0.50	0.05	0.50	2.00
Copper strip corrosion, max.	D 130	No. 3	No. 3	No. 3		
Cetane number, min.	D 613	40	40	40	40	30
One of the following:						
Cetane index, min.	D 976	40		40		
Aromatic content, %vol., max.	D 1319	35		35		
Ramsbottom carbon residue (wt%), max. on 10% distillation residue	D 524	0.15	0.15	0.35	0.35	

specification also indicates the standard test methods that are to be used to measure the values of the properties. Most of the requirements of D 975 are the minimum ones needed to guarantee acceptable performance for the majority of users. In addition, the specification recognizes some requirements established by the EPA to reduce emissions.

In Europe, specifications for diesel fuel are set by the Commission of the European Union, who has founded the Auto Oil Program, a series of research project which correlate the properties of the fuels with exhaust emissions from internal combustion engines. Table XV gives the specification for diesel fuels in Europe. These specifications are based on the results of the research projects of the Auto Oil Program, and reflect the opinion of petroleum companies, engine and car manufacturers, and environmental authorities of the European Union and the member states.

G. Diesel Fuel Properties and Test Methods

1. Density and Gravity

The density of the fuel is determined by its content in the various classes of hydrocarbons. For compounds of the same class, density increases with carbon number. For compounds with the same carbon number, the order of increasing density is paraffin, naphthene and aromatic. Figure 22 depicts density for some representative diesel fuel hydrocarbons.

The methods employed are

- ASTM D 1298—Standard Practice for Density, Relative Density (Specific Gravity), or API Gravity of Crude Petroleum and Liquid Petroleum Products by Hydrometer Method
- ASTM D 4052—Test Method for Density and Relative Density of Liquids by Digital Density Meter

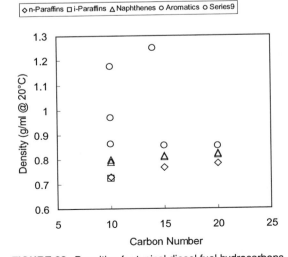

FIGURE 22 Densities for typical diesel fuel hydrocarbons.

TABLE XV European Diesel Fuel Specifications

Properties	Automotive	Heating	Test method
Density (g/ml @ 15°C)	0.820–0.845	Report	EN ISO 3675, 22185
Flash Point (°C), min.	55	55	EN 22179
Sulfur (wt.%), max.	0.035	0.20	PrEN ISO 14596
Distillation (vol.%)			ISO 3405
Recovery at 250°C max.	65	—	
Recovery at 350°C min.	85	85	
Recovery at 360°C min.	95	—	
Cold filter plugging point (°C), max.			EN 116
Kinematic viscosity (cSt @ 40°C)	2.0–4.5	6 max.	EN ISO 3104
Water (mg/kg), max.	200	—	PrEN ISO 12937
Water and sediment (vol.%), max.	—	0.10	
Total contamination (mg/kg), max.	24	—	PrEN 12662
Carbon Conradson Residue (wt%), max.	0.30	0.30	EN ISO 10370
Ash (wt%), max.	0.01	0.02	EN ISO 6245
Copper strip corrosion, max.	3	3	EN ISO 2160
Cetane number, min.	51	—	EN ISO 5165
Cetane index, min.	46	40	EN ISO 4264
Polycyclic aromatic hydrocarbons (wt%), max.	11	—	IP 391
Lubricity HFRR (WSD 1.4 @ 60°C), max.	460	—	ISO 12156-1
Color	—	3.0–5.0 (Red)	
Oxidation stability (g/m^3), max.	25	—	EN ISO 12205

For D 1298, the sample is placed in a cylinder and the appropriate hydrometer is lowered into the sample. After temperature equilibrium has been reached, the sample temperature and hydrometer scale reading are recorded. The Petroleum Measurement Tables are used to convert the recorded value to the value at 60°F.

For D 4052, about 0.7 ml of liquid sample is introduced into an oscillating sample tube and the change in oscillating frequency caused by the change in the mass of the tube is used in conjunction with calibration data to determine the density of the sample.

2. Distillation

The distillation curve is one of the most significant diesel fuel properties. Diesel fuels are mixtures of hundreds of hydrocarbons, which have different boiling points. Thus diesel boils or distills over a range of temperatures, and not at a single temperature, like a pure compound.

The distillation curve of a diesel fuel is the set of temperatures at which gasoline evaporates for a fixed series of increasing volume percentages—5%, 10%, 20%, 30%, etc.—when it is heated under the specific conditions of the test method. Figure 23 gives distillation curves of typical diesel fuels.

Distillation curve of a diesel fuel is related to the hydrocarbons present in the fuel and is determined according to ASTM D 86. For compounds in the same class, boiling point increases with carbon number. For compounds of the same carbon number, the order of increasing boiling point by class is isoparaffin, n-paraffin, naphthene, and aromatic. The compounds with boiling at about 500°F might be C_{12} aromatics, C_{13} naphthenes, C_{14} n-paraffins, and C_{15} isoparaffins. Figure 24 gives normal boiling points of common diesel fuel hydrocarbons.

3. Ignition Quality

The ignition quality in diesel fuels is expressed through the cetane number or the cetane index.

4. Cetane Number

The cetane number is the most significant property of diesel fuels affecting engine performance and emissions. Cetane number varies with hydrocarbon types present in diesel fuels. Normal paraffins have high cetane numbers that increase with carbon number. Isoparaffins have a wide range of cetane numbers, from about 10 to 80. Molecules with many short side chains have low cetane

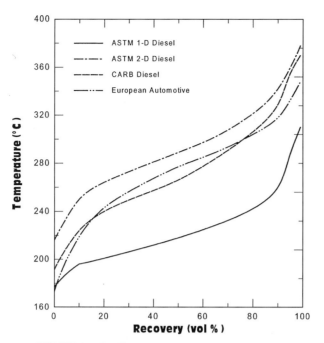

FIGURE 23 Distillation curves of typical diesel fuels.

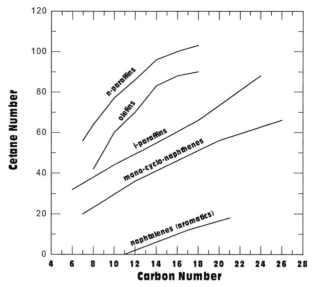

FIGURE 25 Cetane number of diesel fuel hydrocarbons.

numbers. Those with one side chain of four or more carbons have high cetane numbers. Naphthenes generally have cetane numbers from 40 to 70. Higher molecular weight molecules with one long side chain have high cetane numbers; lower molecular weight molecules with short side chains have low cetane numbers. Aromatics have cetane numbers ranging from zero to 60. A molecule with a single aromatic ring with a long side chain will be in the upper part of this range; a molecule with a single

ring with several short side chains will be in the lower part. Molecules with two or three aromatic rings fused together have cetane numbers below 20. Figure 25 shows cetane numbers for various hydrocarbon types.

The cetane number of diesel fuels is measured according to ASTM D 613—Standard Test Method for Cetane Number of Diesel Fuel Oil.

A prototype CFR single cylinder diesel engine is used, running under variable compression ratio until maximum knock is measured. The fuel is rated against mixtures of standard hydrocarbons with known composition. The two standard hydrocarbons used are *n*-hexadecane (cetane), with cetane number 100 (good fuel) and 2,2,4,4,6,8,8-heptamethyl-nonane with cetane number 15 (poor fuel). Heptamethylnonane was replaced in 1962 by *a*-methylnaphthalene, the unstable hydrocarbon with cetane number 0.

The cetane number of a fuel is defined as the volume percent of *n*-hexadecane in a blend of *n*-hexadecane and 1-methylnaphthalene that gives the same ignition delay period as the test sample. For example, a fuel with a cetane number of 50 will have the same performance in the engine as a blend of 50% *n*-hexadecane and 50% 1-methylnaphthalene or 41% *n*-hexadecane and 59% heptamethylnonane.

5. Cetane Index

Cetane number measurement requires the use of the specific CFR engine, which is costly and needs continuous maintenance. Cetane index is an alternative method to estimate the approximate cetane number of the fuel, from

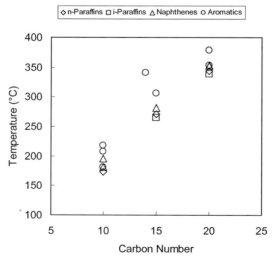

FIGURE 24 Normal boiling points of typical diesel fuel hydrocarbons.

other, easy to measure properties, i.e., density and specific points of the distillation curve, and is used when the CFR engine is not available for measurement.

There are two alternative methods:

1. ASTM D 976—Calculated Cetane Index of Distillate Fuels
2. ASTM D 4737—Calculated Cetane Index by Four Variable Equation

D 976 uses the density of the fuel and 50% distillation temperature to estimate the cetane number. The equation used is:

$$CI = 454.74 - 1641.416 \cdot D + 774.74 \cdot D^2$$
$$- 0.554 \cdot T_{50} + 97.803 \cdot (\log T_{50})^2$$

where

CI = Cetane Index
D = Density at 15°C
T_{50} = 50% recovery temperature (°C)

D 4737 is an improved method that uses the density of the fuel and the distillation temperatures at 10% vol., 50% vol., and 90% vol. recovery to estimate the cetane number. The equation used is:

$$CCI = 45.2 + (0.0892) \cdot (T_{10N}) + [0.131$$
$$+ (0.901) \cdot (B)] \cdot [T_{50N}] + [0.0523$$
$$- (0.420) \cdot (B)] \cdot [T_{90N}] + [0.00049] \cdot [(T_{10N})^2$$
$$- (T_{90N})^2] + (107) \cdot (B) + (60) \cdot (B^2)$$

where

CCI = Calculated Cetane Index
D = Density at 15°C
DN = D-0.85
B = $[e^{(-3.5) \cdot (DN)}] - 1$
T_{10} = 10% recovery temperature (°C)
T_{10N} = $T_{10} - 215$
T_{50} = 50% recovery temperature (°C)
T_{50N} = $T_{50} - 260$
T_{90} = 90% recovery temperature (°C)
T_{90N} = $T_{90} - 310$

Cetane indices are very good tools for the estimation of the natural cetane number of the fuel, but cannot predict the cetane number of fuels with the cetane number improver. Also, these correlations fail to predict cetane number for ultra low aromatic content fuels used in California, according to the CARB requirements.

6. Viscosity

Viscosity is a critical property affecting the fluidity of the fuel. Viscosity is correlated to the shape and the quality of the spray created at the nozzles of the injectors. Viscosity is more related to the carbon number of the compounds and less to the hydrocarbon class. For a given carbon number, naphthenes generally have slightly higher viscosities than paraffins or aromatics. Kinematic viscosity is most commonly used in the petroleum industry.

Viscosity is measured according to ASTM D 445—Kinematic Viscosity of Transparent and Opaque Liquids. The sample is placed in a calibrated capillary glass viscometer tube and held at a controlled temperature bath. The time required for a specific volume of the sample to flow through the capillary under gravity is measured. This time is proportional to the kinematic viscosity of the sample.

7. Flash Point

Flash point is a significant property not for the operability of a diesel fuel, but for its storage and handling. Diesel fuels are classified as nonvolatile fuels, and their storage does not need specific precautions. Flash point is an excellent indication of diesel fuel contamination with more volatile products.

Flash point of diesel fuels is measured according to ASTM D 93—Flash-Point by Pensky-Martens Closed Cup Tester. The sample is stirred and heated at a slow, constant rate in a closed cup. At specific temperature intervals (1 or 2°C), the cup is opened and an ignition source is moved over the top of the cup. The flash point is the lowest temperature at which the application of the ignition source causes the vapors above the liquid to ignite.

8. Cold Flow Properties

Cold flow properties (see Fig. 26) affect the fluidity of diesel fuels at low ambient temperatures. Low temperature operability depends on the size and shape of wax crystals. Dynamic tests that simulate flow through a filter in the fuel system are more reliable for the prediction of cold flow properties than static ones.

9. Cloud Point

Cloud point is the temperature in which wax crystals are observed. It is measured according to ASTM D 2500—Cloud Point of Petroleum Products. A clean clear sample is cooled at a specified rate and examined at 1°C (2°F) intervals. The temperature at which a haze is first observed is the cloud point. Cloud point gives very pessimistic results on the operability of the fuel.

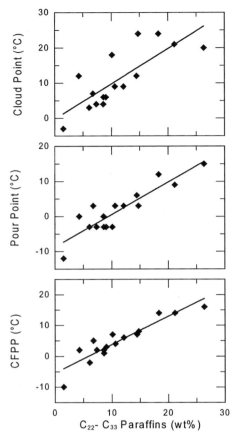

FIGURE 26 Cold flow properties versus C₂₂-C₃₃ paraffins.

10. Pour Point

Pour point is the temperature at which the fuel can flow when cooled under the test conditions. Test method used is ASTM D 97—Pour Point of Petroleum Products. A clean sample is first warmed and then cooled at a specified rate and observed at intervals of 5°F (3°C). The lowest temperature at which sample movement is observed when the sample container is tilted is the pour point. In contrast to cloud point, pour point gives very optimistic results.

11. Low-Temperature Flow Test

Low-temperature flow test (LTFT) is dynamically similar to CFPP, used in the United States. It is measured according to ASTM D 4539—Filterability of Diesel Fuels by Low-Temperature Flow Test (LTFT). A sample is cooled at a rate of 1.8°F/hr (1°C/hr) and filtered through a 17-micron screen under 20-kPa vacuum. The minimum temperature at which 180 ml can be filtered in one minute is recorded.

In contrast to CFPP, LTFT uses a slow constant cooling rate of 1°C/hr. This rate was chosen to simulate the behavior of fuel in the tank of a diesel truck left overnight in a cold environment with its engine turned off. LTFT has been found to correlate well with low-temperature operability field tests. The disadvantage is the long duration of the test (12–24 hr), due to the slow cooling rate. This makes the method impractical for routine fuel testing.

12. Lubricity

Lubricity is a very important diesel fuel property affecting engine performance. Using a fuel with very poor lubricity can destroy a fuel pump. It is very difficult to set a requirement to avoid excessive fuel system wear.

There are two laboratory tests used for the determination of diesel fuel lubricity: scuffing load ball-on-cylinder lubricity evaluator method (SLBOCLE) and high-frequency reciprocating rig method (HFRR). These tests are relatively quick, inexpensive, and easy to perform.

13. Sulfur

Sulfur content is a very critical environmental property for all the types of diesel fuel. The regulations are setting very low sulfur content for automotive diesel fuels, and the trend is to reach zero sulfur fuels. Sulfur content is related to the type of crude oil used for the production of the fuel, as shown in Fig. 27. Hydrotreatment is widely used to decrease the sulfur content of diesel fuel.

Test methods used for the determination of sulfur content are

- ASTM D 2622—Sulfur in Petroleum Products by X-Ray Spectrometry
- ASTM D 4294—Sulfur in Petroleum Products by Energy-Dispersive X-Ray Fluorescence Spectroscopy

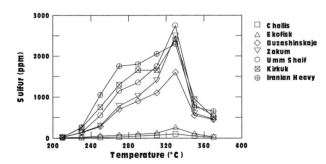

FIGURE 27 Sulfur content of hydrocarbon fractions in diesel fuel range.

• ASTM D 5453—Sulfur in Light Hydrocarbons, Motor Fuels, and Oils by Ultraviolet Fluorescence

X-rays and the measurement of the fluorescence radiation at specific wavelengths base the first two methods on the excitement of sulfur atoms present in the fuel. The third method oxidizes sulfur to SO_2, which is exposed to ultraviolet light and excited. The fluorescence of the excited SO_2^* is proportional to the sulfur content of the fuel.

14. Heating Value

Heating value is a very significant property of diesel fuels, because it gives the energy content of the fuel. The heating value is expressed as gross and net calorific value, depending on the status of water present in the exhaust. If water is present as liquid, then heating value is called gross calorific value. If water is present as vapor, then the heating value is called net calorific value. In real operating situations, water in exhaust gases is present as vapor, so net calorific value is more important for energy efficiency calculations. For compounds with the same carbon number, the order of increasing heating value by class is aromatic, naphthene, and paraffin on a weight basis. However, the order is reversed for a comparison on a volume basis, with aromatic highest and paraffin lowest. Figure 28 gives net heating values of typical diesel fuel hydrocarbons.

15. Carbon Residue

Carbon residue is an indication of the fuel to decompose and form carbonaceous material that can plug diesel fuel injection nozzles. It is determined according to ASTM D

524—Ramsbottom Carbon Residue of Petroleum Products. The sample is first distilled (D 86) until 90% of the sample has been recovered. The residue is weighed into a special glass bulb and heated in a furnace to 1022°F (550°C). Most of the sample evaporates or decomposes under these conditions. The bulb is cooled and the residue is weighed.

16. Additives

Diesel fuel additives are used for a wide variety of purposes, however, they can be grouped into four major categories:

1. Engine performance
2. Fuel handling
3. Fuel stability
4. Contamination control

Engine performance additives are used to improve the performance of diesel engine. They can be cetane number improvers, detergent additives, and lubricity additives.

Cetane number improvers are used to increase the cetane number of the fuel and thus improve the performance of the engine, because it will operate without knock. Cetane number improvers are compounds, which decompose easily with compression, so they help the fuel to ignite, minimizing the ignition delay period. Alkyl nitrates are the most widely used compounds as cetane number improvers, with 2-ethylhexyl nitrate (or isooctyl nitrate) to be the best known.

Injector cleanliness additives are used to remove deposits formed on the nozzles of the injectors. These deposits when formed disturb the fuel spray pattern and tend to create poor air/fuel mixing condition and thus poor combustion. These types of additives are usually polymeric detergent additives, with a polar group that grabs the deposits and nonpolar group that dissolves in the fuel. These additives redissolve the deposits and remove deposit precursors.

Lubricity additives are used to improve the poor lubrication properties of severely hydrotreated diesel fuels. They contain a polar group that is attracted to metal surfaces and creates a thin surface film. The film acts as a boundary lubricant when two metal surfaces come in contact. The types of lubricity additives include fatty acids, esters, and aliphatic amines. The fatty acid type is typically used in the concentration range of 10–50 ppm. Since esters are less polar they require a higher concentration range of 50–250 ppm.

Fuel handling additives are used to improve the properties of the fuel before it is introduced in the combustion chamber.

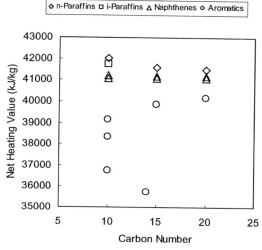

FIGURE 28 Calorific values of typical diesel fuel hydrocarbons.

Low-temperature operability additives are used to improve cold flow properties of diesel fuels. These additives can lower the diesel fuel's pour point, or improve CFPP and LTFT. Most of these additives are polymers that interact with the wax crystals, which form in diesel fuel when it is cooled below the cloud point. The polymers are modifying the size, shape, and degree of agglomeration of the wax crystals when they are formed. The polymer-wax interactions are fairly specific, so a particular additive generally will not perform equally well in all fuels. In order to be effective, cold flow improving additives must be blended into the fuel before any wax has formed, i.e., when the fuel is above its cloud point.

Antifoam additives are used to prevent foam formation that tends to appear during loading some diesel fuels into vehicle tanks. The foaming can interfere with filling the tank completely, or result in a spill. Most antifoam additives are organosilicone compounds and are typically used at concentrations of 10 ppm or lower.

Fuel stability additives can lead to unstable fuel when gums are formed and precipitate from the fuel. These gums can lead to injector deposits or particulates that can plug fuel filters or the fuel injection system. Diesel fuel stability correlates with the processes used for the production of the fuel and the type of crude oil used. Stability additives typically work by blocking one step in a multistep reaction pathway.

Antioxidants are used to prevent oxidation of the fuel, in which oxygen of the dissolved air attacks reactive compounds in the fuel. This initial oxidation begins a series of complex chain reactions. Antioxidants with their active groups quickly terminate these chain reactions. Specific phenols and amines are the most commonly used antioxidants. They typically are used in the concentration range of 10–80 ppm.

Stabilizers are used to prevent acid-base reactions that form insoluble substances. Stabilizers are usually strongly basic amines used in the concentration range of 50–150 ppm. They react with weakly acidic compounds to form products that remain dissolved in the fuel.

Metal deactivators are used to prevent the catalytic action of certain metals, especially copper and iron for producing insoluble substances. Metal deactivators are dissolved in diesel fuel, form chelate complexes with these metals, thus neutralizing their catalytic effect. They typically are used in the concentration range of 1–15 ppm.

Dispersants are used to disperse the particulates that have formed in the fuel, preventing the formation of large size aggregates that can plug fuel filters or injectors' nozzles.

Contaminant control is an additive used mainly for dealing with housekeeping and storage problems.

Biocides are used to prevent the growth of microorganisms like bacteria and fungi, which can form a biofilm at the fuel/water interfaces, which may foul filters and injectors. These microorganisms can be both aerobic and anaerobic. The problem of microorganism growth becomes critical when they have produced enough acidic by-product to accelerate tank corrosion or enough biomass (microbial slime) to plug filters. Although growth can occur in working fuel tanks, static tanks—where fuel is being stored for an extended period of time—are a much better growth environment when water is present.

Biocides can be used when microorganisms reach problem levels. The best choice is an additive that dissolves in both the fuel and the water so it can attack the microbes in both phases. Biocides typically are used in the concentration range of 200–600 ppm. A biocide may not work if a heavy biofilm has accumulated on the surface of the tank or other equipment, because then it doesn't reach the organisms living deep within the film. In such cases, the tank must be drained and mechanically cleaned.

Demulsifiers are surfactants that break up emulsions and allow the fuel and water phases to separate. Emulsions of water in oil are formed when the fuel contains polar compounds that act like surfactants and free water is present. Any operation, which subjects the mixture to high shear forces, like pumping the fuel, can stabilize the emulsion. Demulsifiers typically are used in the concentration range of 5–30 ppm.

Corrosion inhibitors are compounds which form a protective film over metal surfaces and prevent attack by corrosive agents. Except pipe protection, they protect the fuel from contamination caused by rust particles from the corroded surfaces. They typically are used in the concentration range of 5–15 ppm.

IV. BIODIESEL

Chemically, vegetable oils are very different from the conventional diesel fuels, because they consist of glycerol fatty acid esters (triglycerides). These compounds are bigger in size than the ones found in the petroleum diesel.

The operation of diesel engine with vegetable oils for a long period of time causes problems such as deposits in the engine, filter clogging, carbon residues in the injection system, and agglomeration in the lubricant oil, due to the high viscosity of vegetable oils and their polymerization tendency.

In order to avoid the above problems, scientists tried to create a renewable fuel with lower viscosity properties than the vegetable oils. Thus, they turned to fatty acid esters with lower viscosity and properties close to petroleum diesel. This was the birth of biodiesel.

A. Biodiesel Production

Biodiesel production consists of the following basic steps:

1. Collection and preparation of the raw materials
2. Chemical production of biodiesel
3. Separation of final products—distribution in the market

The raw materials include seeds of various oil mill plants, animal fats, or even used fried oils or bad quality oils unsuitable for edible use. There are two ways for the oil extraction from the raw material. The removal of oil could be achieved with the use of a solvent (e.g., hexane), or with mechanical means (sometimes a combination of the above two methods is used). The mechanical removal of oil is mainly used in raw materials with high oil content.

Hexane is separated from the oil, recovered, and then recycled as much as possible. Free fatty acids that may exist in small quantities in the oil should also be removed because these acids are hydrolized forms of fats and cannot transform to biodiesel.

B. Chemical Production of Biodiesel

There are three methods for methylester production by oils and fats:

1. Catalytic transesterification of oil with methanol in basic environment
2. Direct catalytic esterification of oil with methanol in acid environment
3. Oil conversion to fatty acids and then to methylesters through acid catalysis

The majority of methylesters is produced with the first method because it is the most economic for the following reasons:

1. Low temperature ($150°F$) and pressure (20 psi)
2. High reaction yield (98%) with minimum parallel reactions and short time reaction
3. Direct transformation to methylesters without intermediate steps
4. Common materials and reagents

In most of the biodiesel production methods, the removal of free fatty acids takes place before the transesterification reaction. Besides the fact that fatty acids hinder the reaction, they react also with the catalyst forming soaps, as indicated in the following reaction:

$$RCOOH + CH_3ONa \rightarrow RCOONa + CH_3OH$$

Furthermore, the soap formation creates problems to the separation of the final products of transesterification, i.e., methylester and glycerine. To avoid this, caustic sodium or potassium and water are added into the oil. This method is called alkaline cleaning:

$$C_3H_5(OOCR)_3 + 3H_2O \rightarrow 3RCOOH + C_3H_5(OH)_3$$

The chemistry of the transesterification reaction (with catalyst) is as follows:

$$C_3H_5(OOCR)_3 + 3CH_3OH \rightarrow 3RCOOCH_3 + C_3H_5(OH)_3$$

The reaction stoichiometry demands three molecules of methanol for each molecule of triglyceride. However, in order to achieve high yields (greater than 98%), methanol is added in excess and is recovered at the end of the reaction. This happens because the transesterification reaction is an equilibrium reaction, which terminates when two-thirds of ester has been formed. A typical biodiesel production plant is shown in Fig. 29.

Biodiesel can be produced from any plant or animal product, which can give oil or fat after treatment. This means a variety of raw materials available, but not all of them are suitable for oil production. The criteria for selecting the suitable raw materials are availability, cost, yield, and quality. The human factor can successfully intervene and improve all the above criteria. There are about 320 different kinds of plants whose seeds contain fats up to 15% for methylester production. Biodiesel can be produced from used cooking oils, animal fats, and oils (e.g., fish oil).

Rapeseed oil was the first raw material ever used for biodiesel production and has the domain position in Europe, with a share of up to 80%. Others include soybean oil, sunflower oil, and cottonseed oil. Sunflower (~13% share) is mainly used in Italy and the south of France followed by soybean oil, which is preferred in the United States. Other raw materials are palm oil in Malaysia, flax oil and olive oil in Spain, and cottonseed oil in Greece. There are some exotic plants such as the tree Guang-Pi (kind of mulberry tree) in China, the plant Camelina sativa, which can give high yield in low quality soil, the plant Ricinus or the tropical plant Jatropha in Nicaragua. The plant Pogianus, also abudant in Nicaragua, can be used only for biodiesel production because it is toxic and therefore harmful for edible use.

Other raw materials used are the beef tallow in Ireland, pork fat and used oils in Austria, and other used oils and fats in the United States. Used oils and fats are a raw material watched with growing interest worldwide.

The future of biodiesel depends on the availability of cheap raw materials with acceptable quality standards. The major disadvantage of methylesters is the high price, which results in restrictions in the use as fuel during

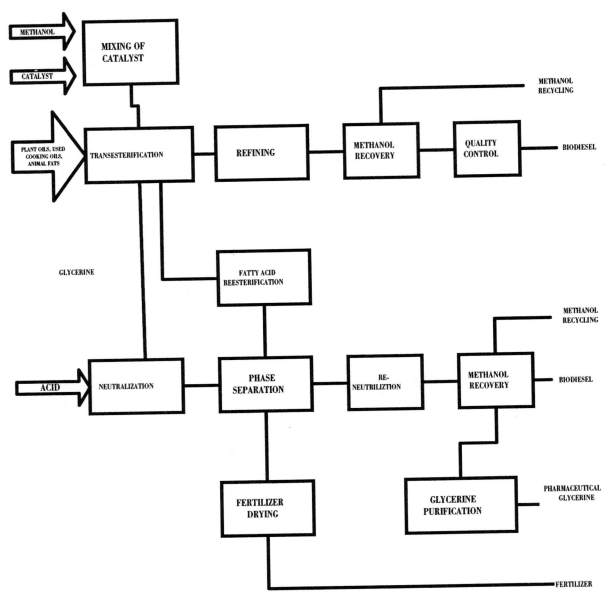

FIGURE 29 Flow chart of biodiesel production.

overproduction periods of cheap petroleum. Some seeds can be bought at very low prices.

C. World Capacity and Production of Biodiesel

In total, 85 industrial plants are recorded in the world for the production of biodiesel. Some of them are pilot plants, over 40 are low capacity (mainly agricultural co-operatives) from 500–3000 tones, and the rest are high capacity, producing 5000–120,000 tones annually.

The 85 biodiesel production plants are distributed in the following areas.

- 44 units in Western Europe, Italy first with 11 industrial plants
- 29 units in Eastern Europe, Czech Republic first with 16 industrial plants
- 8 units in North America
- 4 units in the rest of the world

The oxidation number may affect thermal stability of biodiesel. High oxidation number shows the presence of acids in the biodiesel. The influence of the oxidation number (**TAN**—Total Acid Number) is not fully understood. It is believed that the oxidation number affects the material compatibility, storage, handling, and smell of biodiesel. The oxidation number may cause filter clogging and contributes to thermal instability of biodiesel and its mixtures. High oxidation number may also increase the cetane number of the biodiesel.

The Iodine number reflects the degree of unsaturation in the fuel and depends solely on the origin of the vegetable oil. The restriction of unsaturated fatty acids may be necessary, due to the fact that heating of these compounds leads to the polymerization of glycerides. This results in the formation of deposits and storage stability problems. Soybean and sunflower methylesters have iodine numbers of approximately 130 and 125, which are considered high, because of excessive carbon deposits. A biodiesel generally considered stable, is rapeseed oil, with an iodine number of about 100. A limiting value for the iodine number is generally considered to be 115.

Some properties, like density, cetane number, and sulfur, strongly depend on the selection of plant oil and are not affected by the different production methods or cleaning steps. On the other hand, flash point is strongly related to methanol and viscosity to the nonreacted triglycerides.

The inorganic part in biodiesel, e.g., impurities of the remaining catalyst, is determined by the ash value. High amounts of phosphorous in the fuel may result in higher particulate emissions, which may affect the operation of the catalytic converter. The phosporus in the oil mainly depends on the percentage of the refined oil. Thus, fully refined oils have phosphorus up to some ppm, when unrefined oils or water-degummed oils (rapeseed) have a phosphorus content over 100 ppm. However, during the alkali transesterification reaction this amount could be reduced to 30–40 ppm. This results in an ash content of 0.04%. Further reduction of phosphorus may be achieved by further cleaning steps.

The differences in the specifications among the biodiesel producing countries in the European Union are given in Table XVI.

The European Union, through the European Standardization Organization (CEN), has set European specifications for the biodiesel, see Table XVII.

D. Impact of Biodiesel on Engine Performance and Emissions

Biodiesel seems to emit less carbon monoxide, particulates, sulfur oxides and sulfates, aldehydes, hydrocarbons, and aromatic compounds. Tests showed that biodiesel is less toxic even from salt and clearly less toxic than diesel fuel and crude oil. Diesel fuel, without the addition of biodiesel, is more mutagenic when bacteria do not have metabolic compounds. The existing health surveys for biodiesel and diesel reports that the addition of 20% biodiesel seems to reduce the mutagenic effects of exhaust emissions, although in a lower degree than pure biodiesel. Furthermore, biodiesel particulates are less mutagenic than the particulates of diesel fuel. The impact of the use of biodiesel on engine performance and emissions is summarized below.

- The addition of biodiesel reduces the emissions of unburned hydrocarbons and carbon monoxide emissions. Almost in all cases the black smoke and particulate emissions are reduced (see Fig. 30). In some cases, an increase is observed mainly in low loads.
- The nitrogen oxide problem has preoccupied many researchers. Although most of them observe an increase in NO_x emissions, some of them also note a reduction of these pollutants. There are cases where NO_x in the same test showed both trends. High cetane fuels tend to reduce NO_x emissions, while oxygenated fuels tend to increase the NO_x emissions. Biodiesel belongs in both of these categories of fuels. The reduction or increase of NO_x emissions depends on the driving cycle, the engine, and the presence of exhaust treatment devices. Typical NO_x emissions when biodiesel is added into traditional diesel fuel are shown in Fig. 31.
- The use of biodiesel reduces the PAHs emissions and increases the acrolein emissions.
- The increase or reduction of exhaust emissions depends on the amount of the biodiesel in the mixtures, mainly the emissions of particulates, smoke, and NO_x. The rest of the emitted pollutants have low concentrations even with the use of conventional diesel fuel.
- The continuous reduction of sulfur in the traditional diesel fuel due to environmental impacts results in the reduction of fuel lubricity and the resistance of pumps. The biodiesel addition, even in small quantities, 1–2%, increases the fuel lubricity. Biodiesel reduces the friction deterioration in various parts of the engine, like pumps, nozzles, injection valves and O-rings, pistons, and ring pistons.
- Problems arise by the use of diesel/biodiesel mixtures, where more than 10 or 20% of biodiesel is added, mainly in the fuel injection pumps. Other problems reported are corrosion and deposits on injectors and engine pistons. In other cases, no deposits were formed beyond the usual limits. However, none of the above problems is irreversible.
- In small quantities, biodiesel does not affect the engine performance and power. When the biodiesel percentage in mixtures increases, a small reduction in power and engine performance is usually observed.

TABLE XVI Specifications of Biodiesel in Some European Union Countries

Property	Condition	Unit	Austria ON C 1191	France Journal office	Germany DIN E 51606	Italy UNI 10635	Sweden SS 15 54 36
Day of imposition			1-7-97 FAME	14-9-97 VOME	9-97 FAME	21-4-97 VOME	27-11-97 VOME
Density	15°C	g/cm^3	0.850–0.890	0.870–0.900	0.875–0.900	0.860–0.900	0.870–0.900
Viscosity	40°C	mm^2/s	3.5–5.0	3.5–5.0	3.5–5.0	3.5–5.0	3.5–5.0
Distillation	A.Σ.Z.	°C	—	—	—	>300	—
Distillation	95%	°C	—	<360	—	<360	—
Flash Point		°C	>100	>100	>110	>100	>100
CFPP		°C	<0/−15	—	<0/−10/−20	<0/−15	<−5
Pour point	Summer	°C	—	<−10	—	—	—
Total sulfur		wt.%	<0.02	—	<0.01	<0.01	<0.001
Conradson carbon residue	100%	wt.%	<0.05	—	<0.05		
Conradson carbon residue	10%	wt.%	—	<0.3	—	<0.5	—
Sulfated ash		wt.%	<0.02	—	<0.03	—	—
Ash		wt.%	—	—	—	<0.01	<0.01
Water		mg/kg	—	<200	<300	<700	<300
Total impurities		mg/kg	—	—	<20	—	<20
Cetane number		—	>49	>49	>49	—	>48
Neutralization number		mgKOH/g	<0.8	<0.5	<0.5	<0.5	<0.6
Methanol		wt.%	<0.2	<0.1	<0.3	<0.2	<0.2
Esters		wt.%	—	>96.5	—	>98	>98
Monoglycerides		wt.%	—	<0.8	<0.8	<0.8	<0.8
Diglycerides		wt.%	—	<0.2	<0.4	<0.2	<0.1
Triglycerides		wt.%	—	<0.2	<0.4	<0.1	<0.1
Free glycerol		wt.%	<0.02	<0.02	<0.02	<0.05	<0.02
Total glycerol		wt.%	<0.24	<0.25	<0.25	—	—
Iodine number			<120	<115	<115	—	<125
Phosphorous		mg/kg	<20	<10	<10	<10	<10
Alkalies	Na/K	mg/kg	—	<5/5	<5	—	<10/10

Note: FAME: Fatty acid methylesters; VOME: Vegetable oil methylesters.

- In low temperatures, biodiesel may cause filter and fuel line clogging. In small quantities the problem does not exist.
- Biodiesel blends have shown no increase in engine wear or impact on engine life.

V. AVIATION FUELS

Today aviation fuels are classified into two basic groups used in the aviation field, the gasoline fuels used in reciprocating piston engines and the kerosene type for gas turbine engines. These two main classes of aircraft engines are so fundamentally different that aviation fuels are classified into the above two basic groups. There are currently two main grades of turbine fuel in use in civil commercial aviation: Jet A-1 and Jet A, both are kerosene-type fuels.

There is another grade of jet fuel (Jet B), which is wide cut kerosene (a blend of gasoline and kerosene) but it is rarely used except in very cold climates. For military jets, the main fuel is JP-8, which is the military equivalent of Jet A-1 with the addition of corrosion inhibitor and anti-icing additives. Aviation gasoline is very volatile and thus extremely flammable at normal operating temperatures, so procedures and equipment for safe handling are of the greatest concern.

Aviation gasoline grades are defined primarily by their lean and rich mixture octane ratings. This leads to a multiple numbering system. For example, grade 100/130 corresponds to lean mixture performance rating of 100 and rich mixture performance rating of 130.

Before World War II there was little or no cooperation between engine manufacturers and the authorities, and as a result, there were many different grades of aviation

TABLE XVII   Biodiesel Specifications Set by CEN

Property	Unit	Value		Testing method
		Low	Up	
Esters	%, (m/m)	96.5		ISO 5508
Density at 15°C	Kg/m^3	860	900	EN ISO 3675
Viscosity at 40°C	mm^3/s	3.5	5.0	EN ISO 3104
Flash point (min.)	°C	110		EN22719
CFPP[a]	°C			EN 116
Sulfur	%, (m/m)		0.005	EN 24260/EN
				ISO 8754/PR
				EN ISO 14596
Conradson carbon residue	%		0.3	EN ISO 10370
Distillation:				
5%	°C			
95%	°C			
Cetane number		51		PR EN ISO
				5165
Sulfated ash	%, (m/m)		0.02	ISO 3987 (96)
Water	%, (m/m)		0.05	PR EN ISO
				12937
Sediment	mg/kg		24	PR EN ISO
				12662
Copper corrosion (3 hr at 50°C)				EN ISO 2160
Thermal stability	g/m^3			ISO 6886
Acid Number			0.5	ISO 660
Iodine Number				ISO 3961
Ethanol	%, (m/m)		0.2	E DIN 51608/NF T60 701
Niglycerides	%, (m/m)		0.8	NF T 60 704
Diglycerides	%, (m/m)		0.2	NF T 60 704
Riglycerides	%, (m/m)		0.2	NF T 60 704
Free glycerol	%, (m/m)		0.02	NF T 60 704/ UNI 22054
Total glycerol	%, (m/m)		0.25	NF T 60 704
Alkalis (Na/K)	mg/kg		5	NF T 60 706
Phosphorus	mg/kg		10	NF T 60 705

[a] If CFPP value is −20°C or lower, viscosity should not be greater than 48 mm^2/s.

gasoline in general use e.g., 80/87, 91/96, 100/130, 108/135, 115/145. However, with decreasing demand these have been rationalized down to one principle grade, Avgas 100/130 with a lead content of 1.28 g/liter. (To avoid confusion and to minimize errors in handling aviation gasoline, it is common practice to designate this grade by just the lean mixture performance, i.e., Avgas 100 for Avgas 100/130.)

More recently, an additional grade, called 100LL, where LL represents "low lead," was introduced. This was to allow one fuel to be used in engines originally designed for grades with lower lead contents. The lead content of this type of fuel is 0.56 g/liter. The reason the major oil companies introduced this gasoline was that some of the older 80/87 octane rated engines encountered plug fouling and exhaust valve deterioration when operating on Grade 100/130 with 1.28 g/liter lead.

All equipment and facilities for aviation gasoline handling and storage are color-coded and display prominently the API markings denoting the actual grade carried. In addition to that, the fuels are dyed in order to simplify identification. Currently the two major grades in use internationally are 100LL, which is colored blue, and 100, which is colored green.

The aviation gasoline nozzles for aircraft refueling are painted red. In order to prevent the possibility of jet fuel being supplied to a piston engine aircraft, the nozzle of an aviation gasoline refueling system is limited to a maximum diameter of 40 mm (49 mm in the United States.) and the aperture on an aircraft aviation gasoline tank to a maximum of 60 mm diameter. Nozzles for jet fuels are larger than 60 mm and thus cannot be placed into an aircraft's aviation gasoline tank.

Aviation fuels must consider several other significant properties in addition to the combustion characteristics. They include volatility (ease of evaporation in air), vapor locking tendency (boiling in the fuel lines), ease of engine starting, solvent and corrosion properties (which have adverse effects on fuel systems), and the storage ability of the fuel (resistance to deterioration during prolonged storage).

The specifications for aircraft reciprocating engine fuels used in the past are given in Table XVIII. The low lead, 100LL, is identical in lead content to the 100/130 military grade.

As mentioned above gasolines are graded in terms of knock ratings and when the grade designation includes two numbers such as in grade 100/130, the first number gives the rating at lean mixture and the second at rich mixture. Where the number is 100 or less, it refers to the octane scale, and for 100 or above to the performance scale.

For ratings below 100, octane numbers are used in specifications and grades attributions. However, octane number can be converted into performance number according to the following relationship:

$$[\text{Performance Number, PN}] = \frac{2800}{128 - [\text{Octane Number, ON}]}$$

For example, the required performance number of the fuel for a varying compression ratio in any given engine design will be roughly proportional to the compression ratios used. As a result, if a 70 performance number fuel

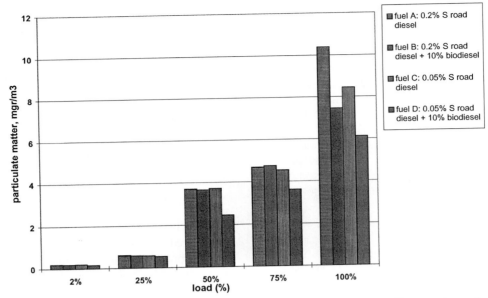

FIGURE 30 Particulate matter emissions under different loads, emitted from a stationary Petter engine, using diesel base fuels and 10% sunflower oil.

were required for a 7:1 compression ratio, a 90 performance number would be theoretically necessary for a 9:1 compression ratio. In practice, however, it would be more than 90 performance number.

Similarly, use of higher grade fuels permits more supercharge. Since increase in compression ratio and manifold pressure both act the same way as fuel grade requirements, it follows that if all conditions except the abovementioned are kept constant, then knock-free power output on a given fuel will be inversely proportional to compression ratio. For example, an engine developing 1600 hp on a compression ratio of 7:1 could only develop $1600 \times \frac{7}{8} = 1400$ hp on an 8:1 compression ratio. These figures are only approximate and serve to illustrate the relationship between the various factors, since the subject is sufficiently complex to justify some degree of oversimplification.

The grading of fuels in this way is of paramount importance because the knocking tendency of a fuel has a

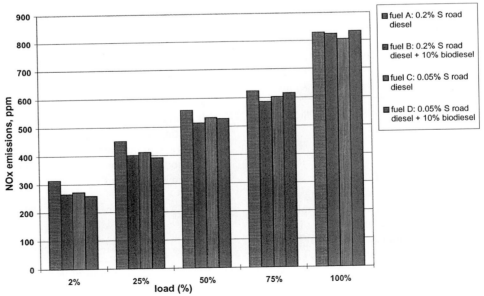

FIGURE 31 NO$_x$ emissions under different loads, emitted from a stationary Petter engine, using diesel base fuels and 10% sunflower oil.

TABLE XVIII U.S. Military and Commercial Aviation Gasoline Specifications

			U.S. Navy MIL-G-5572F-Amd. 1 and ASTM D 910			
	Issuing agency: Specification:					
	Revision date: Grade designation: Fuel type: Color:		1979 80/87 Av. gasoline Red	1981 100/130 Av. gasoline Blue	115/145 Av. gasoline Purple	ASTM test method
Composition	Sulfur (wt.%)	max.	0.05	0.05	0.05	D-1266/D-2622
	Aromatics (vol.%)	min.		5.0	5.0	D-936, D-131 or D-2267
Volatility	Distillation					
	Temp. 10% Rec. (°C)	min.	75	75	75	D-86
	Temp. 40% Rec. (°C)	max.	75	75	75	
	Temp. 50% Rec. (°C)	min.	105	105	105	
	Temp. 90% Rec. (°C)	min.	135	135	135	
	End Point (°C)	max.	170	170	170	
	Sum. of 10% and 50%					
	Evaporated Temp.	min.	135	135	135	
	Residue (vol%)	max.	1.5	1.5	1.5	
	Distillation Loss (vol%)	max.	1.5	1.5	1.5	
	Gravity, °API		Report	Report	Report	D-287
	Reid vapor pressure at 37.8°C, kPa		38.5–49.0	38.5–49.0	38.5–49.0	D-323/D-2551
Fluidity	Freezing point, °C	max.	−60	−60	−60	D-2386
Combustion	Net heat of combustion,					
	MJ/kg	min.	43.5	43.5	44.0	D-240/D-2382
	or					
	Aniline-gravity product	min.	7.500	7.500	9.800	D-611 or D-287
	Knock rating, lean					
	mixture aviation rating	min.	80	100	115	D-2700
	Knock rating, rich					
	mixture supercharge rating	min.	87	130	145	D-909
Corrosion	Copper strip corrosion (2 hr at 100°C)	max.	1	1	1	D-130
Stability	Potential gum, 16-hr aging (mg/100 ml)	max.	6.0	6.0	6.0	D-873
	Precipitate (mg/100 ml)	max.	2.0	2.0	2.0	D-873
Contaminants	Existent gum (mg/100 ml)	max.	3.0	3.0	3.0	D-381
	Water reaction					
	Interface rating	max.	2	2	2	D-1094
	Vol. change (ml)	max.	2	2	2	D-1094
Additives	Tetraethyl lead Content.					D-3341
	g/liter	max.	0.14	0.56	1.28	D-2599/D-2547
	Dye Content					
	Blue dye (mg/liter)		0.131 max.	0.80–1.51	0.713–1.24	D-2392
	Red dye (mg/liter)		1.83–2.29		0.50–0.864	
	Yellow dye (mg/liter)		—	1.4	—	
Other	* NATO Code No.		F-12	F-18	F-22	

pronounced effect on the power obtainable from a given engine. Thus one designed to give say 1450 hp on 115/145-grade fuel, could be operated safely at only about one-third of this power on grade 73.

In a given engine, the use of a higher grade fuel than that for which it was designed will not give any increased performance, since it is already knock-free at all power settings. At best it can have no effect; at worst it can cause lead fouling troubles in engines not designed for high lead fuel.

A. Jet Fuels

In contrast to piston engines, the jet engines use almost exclusively illuminating kerosene as a fuel. The first jet fuel ever was JP-1 (Jet Propellant-1, in 1944). It was kerosene having 60.5°C freezing point and 43°C flash point. Its availability was limited to 3% of the average crude oil. JP-2 (1945) was rejected due to unsatisfactory properties concerning viscosity and combustion characteristics. JP-3 (1947–1951) had high vapor pressure, comparable to that of aviation gasolines. This, combined with the fact that turbine-engine aircraft fly at higher altitudes than piston-engine aircraft, led to fuel losses due to boiling off and vapor locking. JP-4 (1951–1995) referred to as Jet B or with NATO code F-40, was a kerosene-gasoline mixture, with maximum vapor pressure of 2–3 psi in order to reduce fuel boiling off and vapor locking. It has a pour point of −60.5°C and a flash point of −18°C (flash point is not included in specifications for this fuel). In the Mid-1980s an antistatic additive was used for safety reasons. JP-4 was the main operational fuel for all NATO nations for many years, but it was recently abandoned due to its high volatility. JP-5 (1952 to-today), referred to with NATO code F-44, is used by the U.S. Naval air force. It has a minimum flash point of 60°C for safety reasons. Its pour point is −46°C and it has no antistatic additives. JP-6 (1956) was developed for the XB-70 aircraft; it is similar to JP-5 but has a lower pour point (−54°C) and increased thermal stability. There is no specification concerning the flash point for this fuel. JP-7 (1960) was developed for SR-71 aircraft; it has a low vapor pressure and exquisite thermal stability at high altitudes and velocities higher than Mach 3. It has a pour point of −44°C and minimum flash point of 60°C.

JP-8 was first used in 1978 and is referred to with NATO code F-34. JP-8 is the same fuel as Jet A1, but is enhanced by the use of icing inhibitors, lubricity improvers, and antistatic additives. Conversion of aviation fuels to JP-8 was initiated mostly for safety reasons and was completed in 1995.

JPTS (Jet Propellant Thermally Stable, 1956) was developed for use in U-2 aircraft and is kerosene with a pour point of −54°C. The fuel is enhanced with thermal stability additives and has a minimum flash point of 43°C.

Typical specifications for jet fuels, used either in military or civilian airlines, are given in Table XIX.

Subjecting fuels to high temperature thermal-oxidation conditions leads to the formation of particulates, varnishes, and gums, which clog valves and filters, and degrade injector nozzle performance. In the extreme, coking can cause serious fouling in injectors and combustors, leading, for example, to re-light problems. Improvements in the thermal stability of "conventional," e.g., JP8-type kerosene fuels of 100°F through the use of low-cost additive packages have been the subject of the U.S. Air Force's "JP8+100" program since 1989, on the basis that advances in military fighter aircraft systems would require fuels with over 50% improvement in heat sink capability over conventional JP-8 fuel.

During this program, hundreds of commercial additives were tested for thermal stability-enhancing or deposition-reducing characteristics. The program demonstrated that the thermal stability of jet fuels (particularly JP-8) could indeed be enhanced through the use of particular additives and additive blends used in relatively low concentrations. Additionally, flight-testing has highlighted a significant reduction in fuel-related maintenance costs, arising from cleaner combustion.

However, one aspect of the incorporation of the preferred thermal stability additives that has given some cause for concern is the associated effect on the water and solids separation from "JP-8+100" fuel, although, operationally, this is minimized through the introduction of the "+100" additive as close to the skin of the aircraft as possible. As a result, new multifunctional additives that enhance the thermal stability of jet fuels, without compromising other required essentials of jet fuel product quality, namely water and solids separation, are under evaluation by the major oil companies. Moreover, JP-8+100 will be the basis for future jet fuels for both military and civilian aircraft applications requiring thermal stability to 482°C.

This type of additive may, in the future, be used in civil applications and will allow designers of future gas turbines to utilize the increased heat sink availability in the fuel and to produce more powerful and more efficient propulsion systems. Following on from "+100" additives, there are possibilities for the use of additives to improve the performance characteristics by modifying the freezing point/flow point and also smoke/soot emissions.

In the near future pipeline drag reducer additives may be introduced for use in jet fuel distribution. These additives are very high molecular weight polymers, which are already used in the pipelines for transportation sector for some other petroleum products. They act by reducing the turbulence and hence energy loss between the inner

TABLE XIX Typical Aviation Fuel Properties

			USAF MIL-T-56241-Amd. 1		USAF MIL-T-83133A-Amd. 1	
	Issuing agency: **Specification:**					
	Revision date: **Grade designation:** **Fuel-type:**		1980 JP-4 Wide-cut	1980 JP-5 Kerosene	1980 JP-8 Kerosene	**ASTM test method**
Composition	Sulfur, mercaptan (wt.%)	max.	0.001	0.001	0.001	D-1323
	Sulfur, total (wt.%)	max.	0.4	0.4	0.3	D-1266
	Aromatics (vol.%)	max.	25	25	25	D-1319
Volatility	Distillation					
	Temp. 10% Rec. (°C)	max.	Report	205	205	D-86
	Temp. 20% Rec. (°C)	max.	145	Report	Report	
	Temp. 50% Rec. (°C)	max.	190	Report	Report	
	Temp. 90% Rec. (°C)	max.	245	Report	Report	
	End Point (°C)	max.	270	290	300	
	Residue (vol.%)	max.	1.5		1.5	
	Distillation Loss (vol.%)	max.	1.5		1.5	
	Density, 15°C (kg/m^3)	max.	751–802	788–845	775–840	D-1298
	Vapor pressure at 37.8°C, kPa		14–21			D-323/D-2551
Fluidity	Freezing point, °C	max.	−58	−46	−50	D-2386
	Viscosity at −20°C (cSt)	max.	—	8.5	8.0	D-445
Combustion	Net Heat of Combustion, MJ/kg	min.	42.8	42.6	42.8	D-240
	Aniline-gravity product	min.	5250	4500		D-1405
	Smoke point	min.	20.0	19.0	19.0	D-1322
Corrosion	Copper strip corrosion (2 hr at 100°C)	max.	1b	1b	1b	D-130
Stability	JFTOT AP (mmHg)	max.	25	25	25	D-3241
	JFTOT Tube Color Code	max.	<3	<3	<3	
Contaminants	Existent gum (mg/100 ml)	max.	7	7	7	D-381
	Particulates	max.	1	1	1	D-2276
	Water reaction interface	max.	1b	1b	1b	D-1094
	Water separation index					
	Modified	min.	70	85	70	D-2550
	Filtration time (minutes)	max.	15	—	—	
Additives	Anti-icing (vol.%)		0.10–0.15	0.10–0.15	0.10–0.15	5330,5340
	Antioxidant		Required	Required	Option	3527 FED STD 791
	Corrosion inhibitor		Required	Required	Required	
	Metal deactivator		Option	Option	Option	
	Antistatic		Required		Required	
Other	*NATO code No.		F-40	F-44	F-34; F35	

surface of the pipeline and the fluid flowing through. For a given pump size they allow more fluid to be pumped. Interest has been stimulated in the use of these additives because it is cheaper to use them than lay bigger pipes or put in more pumps when supply to an airport has reached its limit.

New products such as synthesized aviation kerosene are coming into the market with increasing frequency and more refining plants are likely to be built where natural gas supplies exist which are remote from this market. "Biofuels" will also emerge on the market eventually, with Scandinavian countries a major driving force in their market penetration and acceptance. While renewable and sustainable fuels may be the answer to many environmental problems facing the world today, they may not necessarily be the best fuels for the engines, both current and future. Increasing pressure to bring these fuels onto the market will certainly come about, but its source and timing are difficult to predict.

SEE ALSO THE FOLLOWING ARTICLES

AIRCRAFT PERFORMANCE AND DESIGN • COMBUSTION • ELECTRIC AND HYBRID-ELECTRIC VEHICLES • ENERGY RESOURCES AND RESERVES • FUEL CHEMISTRY • GREENHOUSE EFFECT AND CLIMATE DATA • HEAT TRANSFER • INTERNAL COMBUSTION ENGINES • JET AND GAS TURBINE ENGINES • PETROLEUM REFINING • POLLUTION, AIR • TRANSPORTATION APPLICATIONS FOR FUEL CELLS

BIBLIOGRAPHY

Amoco, B. P. (1999). Statistical Review of World Energy.

Anasotpoulos, G., Lois, E., Karonis, D., Zanikos, F., Stournas, S., and Kalligeros, S. (2000). "Impact of Aliphatic Mono-Amines and ethyl Esters of Fatty Acids on the Lubrication Properties of Ultra-Low Sulfur Diesel Fuels, Fuels International," vol. 11, no. 11, October.

Aviation Fuel Properties, Coordinating Research Council, SAE, 3rd Printing.

Biodiesel and Biolubricants Project Summaries, Pacific Northwest and Alaska Regional Bioenergy Program, October 23, 1996, pp. 15–26, See also Biodiesel Emissions, <http://www.biodiesel.org/transit>

Bland, W. F., and Davidson, R. L., eds. (1967). "Petroleum Processing Handbook," McGraw-Hill, New York.

DeNevers, N. (2000). "Air Pollution Control Engineering," McGraw-Hill, New York.

El-Gamal, I. M., Khidr, T. T., and Ghuiba, F. M. (1999). "Polar Polymeric Structures as Wax Dipersant Flow Improvers for Paraffinic Distillate Fuels, Fuels 1999," 2nd International Colloquium, pp. 287–299.

Garry, J. H., and Handwerk, G. E. (1984). Petroleum Refining, Technology and Economics, 2nd ed., Marcel Dekker, New York.

Gupta, A. K., Lilley, D. G., and Syred, N. (1984). "Swirl Flows," Abacus Press, U.K.

Keating, E. L. (1993). "Applied Combustion," Marcel Dekker.

Lois, E., Stournas, S., and Karnois, D. (1991). Mathematical Expressions of some Nonadditive Properties of Gasoil-Residual Fuel Blends, "Energy and Fuels," Vol. 5, pp. 855–860.

Lowry, A. (1990). "Alternative Fuels for Automotive and Stationary Engines in developing countries, Fuels for Automotive and Industrial Diesel Engines," The Institution of Mechanical Engineers, London.

Maurice, L. Q., Corporan, E., Harrison, W. E., Minus, D., Mantz, R., Striebich, R.C., Graham, S., Hitch, B., Wickham, D., and Karpuk, M. (1999). "Controlled Chemically Reacting Fuels: A New Beginning," XIV ISABE Conference, September.

Mittebach, M. (1996). Diesel fuel derived from vegetable oils, VI: Specifications and quality control of biodiesel. *Biores. Technol.*, **56**, 7–11.

Odgers, J., and Kretschmer, D. (1986). *In* "Energy and Engineering Science Series" (A. K. Gupta and D. G. Lilley, eds.), Abacus Press, U.K.

Ogston, A. R. (1981). "A short History of Aviation Gasoline Development 1903–1980," SAE Paper No. 810848.

Owen, K., and Coley, T. eds. (1990). Automotive Fuels Handbook, SAE, Warrendale.

Schafer, A. (1996). Biodiesel as an Alternative Fuel for Commercial Vehicle Engines, Proceedings 2nd European Motor Biofuels Forum, Graz, Austria, pp. 233–244.

Schmidt, G. K., and Forster, E. J. (1986). Modern Refining for Today's Fuels and Lubricants, SAE Paper 861176.

Serdari, A., Fragioudakis, K., Teas, C., Zannikos, F., Stournas, S., and Lois, E. (1999). Effect of biodiesel addition to road diesel, on engine performance and emissions. *J. Propulsion Power*, **15(2)**, 224–231.

Yergin, D. (1992). "The Prize: The Epic for Oil, Money and Power," Touchstone, New York.

Ziejewski, M., and Goettler, H. (1992). "Comparative Analysis of the Exhaust Emissions for Vegetable Oil Based Alternative Fuels, Alternative Fuels for CI and SI Engines," SAE, SP-900, February.

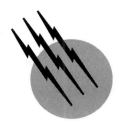

Fullerenes and Carbon Nanotubes

Ado Jorio
Gene Dresselhaus
Massachusetts Institute of Technology

GLOSSARY

Antibonding states Electronic states for which the electron charge distribution is decreased through interactions with neighboring atoms.

Bonding states Electronic states where the energy is lowered by the excess charge distribution coupling adjacent atoms.

Dopant Foreign atom or molecule intentionally added to a host material.

Icosahedron A polyhedron having 20 faces and 12 vertices.

Infrared-active modes Vibrational modes that are excited by infrared radiation.

Intercalation The insertion of guest species in layers or in specially designated sites of the host material.

Raman-active modes Vibrational modes that are symmetry-allowed to participate in the inelastic (Raman) scattering of light.

Space group Ensemble of symmetries that describe a three-dimensional crystal.

Unit cell Unit basis that, repeated in three-dimensional space, forms an entire crystal.

Van der Waals bond Weak dipolar bond between neutral atoms.

THE CARBON ATOM is the basis for many materials important for life and civilization. Due to its unique properties arising from its various possible electronic

Encyclopedia of Physical Science and Technology, Third Edition, Volume 6

configurations (sp^n hybridization, $n = 1, 2, 3$), carbon is not only the cornerstone for all living organisms, but it has been important for the technological development of society. Carbon was first used as a source for world energy supplies, and, in more recent times, carbon materials such as carbon blacks and carbon fibers have been central to our industrial society. Recently, the appearance of carbon nanotubes and fullerenes has brought carbon to the forefront of scientific and technological development.

In basic science, carbon is interesting as a prototype low-dimensional system, with graphite being a two-dimensional (2D) material, nanotubes being a one-dimensional (1D) structure, and fullerenes being a molecular zero-dimensional (0D) structure. Nanotubes and fullerenes show remarkable properties, as outlined in this article, giving rise to significant potential for applications utilizing these special properties. For example, carbon nanotubes can be metallic or semiconducting depending on their geometrical structure, while fullerenes are the basis for the formation of several molecular solids that can be semiconducting, metallic, or superconducting.

This article presents a brief review on the historical appearance of carbon nanotubes and fullerenes (Section I), their structure (Section II), the methods of synthesis (Section III), their electronic structure (Section IV), and the various physical properties (Sections V–VIII). Finally, we discuss some possible applications of carbon nanotubes and fullerenes (Section IX).

I. HISTORY

The experimental observation and identification of carbon nanotubes and fullerenes are relatively recent, with fullerenes appearing in 1985 and carbon nanotubes appearing in 1991. However, the history leading to these discoveries is long and interesting.

The discovery of carbon nanotubes is closely related to research on carbon fibers. Carbon fibers were first studied by Thomas A. Edison, when he prepared a carbon filament for use in an early model of an electric light bulb. Research on carbon fibers was strongly stimulated after World War II because of needs for materials with special properties by the space and aircraft industries. Because of the desire, in the 1960s and early 1970s, to synthesize crystalline filamentous carbons under more controlled conditions, the synthesis of carbon fibers by a catalytic chemical vapor deposition (CVD) process was developed, laying the scientific basis for the mechanism and thermodynamics for the vapor phase growth of carbon fibers. In parallel, other research studies focused on control of the process for the synthesis of vapor grown carbon fibers for various appli-

cations. As research on vapor grown carbon fibers on the micrometer scale proceeded, the growth of very small diameter filaments was occasionally observed and reported, but no detailed systematic studies of such thin filaments were carried out. However, the experimental observation of fullerenes gave direct stimulus to study carbon filaments of very small diameters more systematically.

The roots of fullerene molecules can be traced to early work by Tisza, in 1933, who considered the symmetry for icosahedral molecules similar to the C_{60} fullerene, although the picture of an icosahedral object was already known to Leonardo da Vinci in the 16th century. In 1970, Osawa suggested that an icosahedral C_{60} molecule might be stable chemically, while early Russian workers showed theoretically that C_{60} should have a large electronic gap between the highest occupied molecular orbital (HOMO) and the lowest unoccupied molecular orbital (LUMO). These early theoretical suggestions for icosahedral C_{60} were not widely recognized and were only rediscovered after the experimental work of Kroto and co-workers in 1985 established the stability of the C_{60} molecule in the gas phase.

The experimental identification of the C_{60} molecule as a regular truncated icosahedron occurred because of the coalescence of research activities in two seemingly independent areas. Astrophysicists were working in collaboration with spectroscopists to identify some unusual infrared (IR) emission from large carbon clusters which had been shown to be streaming out of red giant carbon stars. The development of the laser vaporization technique for synthesizing clusters, by Smalley and co-workers at Rice University, suggested the possibility of creating unusual carbon-based molecules or clusters that would yield the same IR spectrum on earth that is seen in red giant carbon stars. This motivation led to a collaboration, in 1985, between Kroto and Smalley and co-workers to use the laser vaporization of a graphite target to synthesize and study cyanopolyynes. It was during these studies that the icosahedral 60 carbon-atom cluster with unusually high stability was discovered, consistent with the results of the EXXON group, who had previously shown mass spectra with peaks at C_{60} and C_{70}, and carbon clusters with only even numbers of carbon atoms. Kroto and Smalley gave the name "buckminster fullerene" to this family of molecules because of their resemblance to the geodesic domes designed and built by R. Buckminster Fuller. With time and usage, the nomenclature for this family of carbon-based molecules has been shortened to fullerenes. The 1996 Nobel Prize for Chemistry was awarded to Richard E. Smalley, Robert F. Curl, Jr., and Sir Harold W. Kroto for their discovery of fullerenes. In the fall of 1990 a new type of condensed matter, based on the C_{60} molecule, was synthesized, and soon thereafter the superconductors

M_3C_{60} (where M = K, Rb, Cs) were discovered, thereby spurring a great deal of interest in C_{60}-related materials.

As pointed out before, the direct stimulus to study carbon filaments of very small diameters more systematically came from the discovery of fullerenes. Although speculations about the existence of carbon nanotubes of sizes comparable to C_{60} were made at various conferences, the real breakthrough in nanotube research came with the first report of the observation of carbon tubules of nanometer dimensions in 1991 by Iijima, soon followed by a method for the synthesis of large quantities of multi-wall nanotubes. The next important advance was the synthesis in 1993 of single wall nanotubes and their preparation in 1996 in crystalline bundles.

II. STRUCTURE

A. Molecular Structure

The structures of carbon nanotubes and fullerenes are strongly related to graphite. Graphite has an *ABAB* planar stacking arrangement shown in Fig. 1a, with an in-plane nearest neighbor distance of 1.421 Å, an in-plane lattice constant of 2.462 Å, a *c*-axis lattice constant of 6.708 Å, and an interplanar distance of 3.354 Å. This structure is described by the $D_{6h}^4 (P6_3/mmc)$ space group and has four atoms per unit cell (see Fig. 1a). The in-plane carbon atoms are strongly bonded in an sp^2 electronic configuration (see Section IV), and the layers are weakly bound through van der Waals forces. Carbon nano-

tubes and fullerenes can be considered as a "rolled up" graphene sheet (a single layer of crystalline graphite), with a graphite-like sp^2-derived nearest neighbor bonding configuration and an average nearest neighbor carbon–carbon (C–C) distance $a_{C-C} = 1.42$ Å, as in graphite. Both nanotube and fullerene structures can be described by a vector $n\mathbf{a_1} + m\mathbf{a_2}$, or simply by (n, m), that determines the nanotube or fullerene structure in relation to the graphene unit cell vectors $\mathbf{a_1}$ and $\mathbf{a_2}$ (see Fig. 1a).

A single wall carbon nanotube (SWNT) is a graphene sheet rolled up into a seamless cylinder (see Figs. 2a–2c). The nanotube structure is determined by its diameter (d_t) and its chiral angle (θ), which, as shown in Fig. 1b, relates the direction of the tube axis according to the C–C atomic bondings. Figure 1b shows the vectors $\mathbf{C_h}$ and $\mathbf{T}$ used to determine the nanotube unit cell formed from a graphene sheet. The chiral vector $\mathbf{C_h}$ connects two crystallographically equivalent sites O and A on a graphene sheet, while the translation vector $\mathbf{T}$ connects the two crystallographically equivalent sites O and B perpendicular to $\mathbf{C_h}$. The nanotube cylinder is formed by superimposing the two ends OA and BB' in Fig. 1b. The chiral vector can be written as $\mathbf{C_h} = n\mathbf{a_1} + m\mathbf{a_2} = (n, m)$ and it determines the nanotube structure, where $d_t = C_h/\pi = \sqrt{3}a_{C-C}(m^2 + mn + n^2)^{1/2}/\pi$ and $\theta = \tan^{-1}[\sqrt{3}/(m + 2n)]$. If the chiral vector is directed along one of the graphene unit cell vectors [e.g., $\mathbf{C_h} = n\mathbf{a_1} = (n, 0), \theta = 0°$] the resulting nanotube is called a *zigzag* tube (see Fig. 2b). If $\theta = 30°$, the resulting tube is called an *armchair* tube (see Fig. 2a). For any $0° < \theta < 30°$, the result is a *chiral* tube (see Fig. 2c).

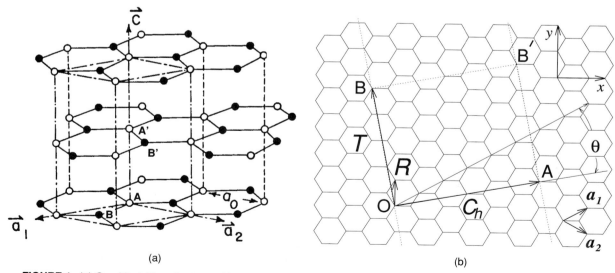

(a)

(b)

FIGURE 1 (a) Graphite lattice structure with the unit cell (dot-dashed lines) and unit cell vectors: a_1, a_2, and c. The projection to form (b) a (4, 2) carbon nanotube, and (c, following page) three icosahedral fullerenes [$(p, q) = (1, 1)$ for C_{60}, (2, 0) for C_{80}, and (2, 1) for C_{140}] on a 2D honeycomb hexagonal lattice by joining like numbers.

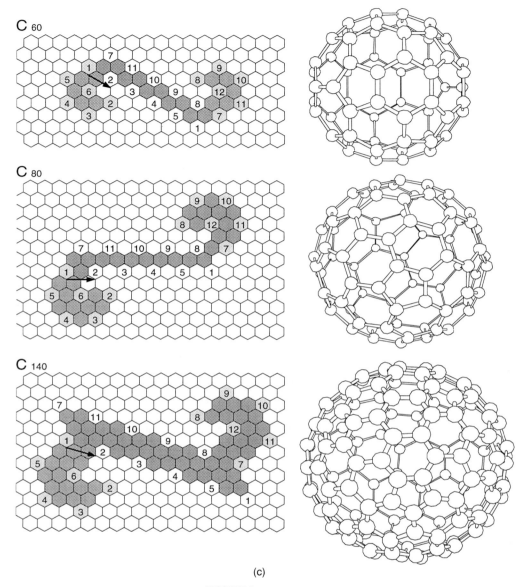

(c)

FIGURE 1 (*continued*)

While carbon nanotubes are graphene sheets rolled up into a seamless cylinder, fullerenes are graphene sheets rolled up into a closed cage structure. The most common fullerene, formed by 60 carbon atoms, is C_{60}, shown in Fig. 2d, and is similar to a soccer ball. Therefore, fullerenes need the substitution of 12 hexagonal carbon rings by pentagons to produce closed (convex) surfaces that lead to the closed cage structures. Figure 1c shows the planar projection of three icosahedral fullerenes a C_{60} molecule, a C_{80} molecule, and a C_{140} molecule on a 2D honeycomb hexagonal lattice. The hexagons are shaded, while the positions of the pentagons are numbered. A closed cage is obtained by substituting the numbered hexagons by pentagons (which leads to a curvature on the graphite layer) and then by joining equally numbered pentagon positions.

From Euler's theorem for polyhedra, it is concluded that all fullerenes must have 12 pentagonal faces to form a closed cage, and the number of hexagonal faces is arbitrary, as shown in Fig. 1c. Therefore, the fullerenes are distinct from each other by the relative positions between two pentagons, as determined by a vector $\mathbf{d_{pq}} = p\mathbf{a_1} + q\mathbf{a_2} = (p, q)$ that relates the fullerene structure with the graphene lattice. The vectors $\mathbf{d_{pq}}$ are shown in

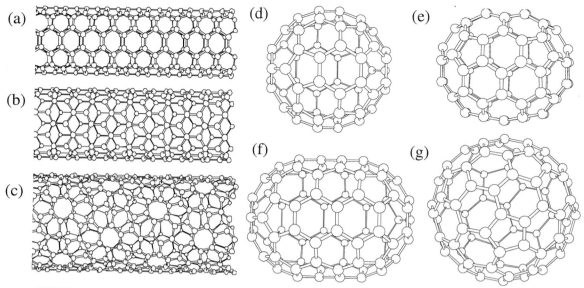

FIGURE 2 Schematic pictures of (a) armchair, (b) zigzag, and (c) chiral carbon nanotubes; (d) icosahedral C_{60}; (e) rugby ball-shaped C_{70} and the two isomers of C_{80}; (f) the extended rugby ball; and (g) the icosahedral fullerene.

Fig. 1c for three icosahedral fullerenes, and the projection method shown here can also be used to construct polyhedra that do not have an icosahedral shape, as shown in Figs. 2e and 2f.

Considering symmetry, while a carbon nanotube is a 1D graphite-like structure formed by the repetition along **T** of a cylindrical unit cell, fullerenes are molecules (zero dimension) for which the unit cell is the whole structure. Many symmetries are possible for different fullerene-related molecules. C_{70} is formed by adding a ring of five hexagons around the equatorial plane of the C_{60} molecule, normal to one of the fivefold axes (see Fig. 2e). If a second ring of five hexagons is added along the equator, we then obtain the C_{80} molecule shown in Fig. 2f. Furthermore, fullerenes generally form isomers, which are polyhedra with a given number n of carbon atoms C_n arranged in different geometrical structures. C_{80} molecules can be formed in the shape of an elongated rugby ball, as shown in Fig. 2f, but also in an icosahedral form, as shown in Fig. 2g. Another example of fullerene isomers is C_{78}, which has five distinct isomers which follow the isolated pentagon rule, such that pentagonal faces are never adjacent to one another.

As shown in Fig. 3, halves of fullerenes can cap the ends of carbon nanotubes with similar diameters. The capping of a nanotube lowers the energy per carbon atom by tying up all the high-energy dangling bonds of the open-ended nanotubes. For example, the (5,5) armchair nanotube ($d_t = 6.83$ Å) can be capped by the C_{60} (1,1) fullerene ($d_i = 6.83$ Å) bisected normal to a fivefold axis (see Fig. 3a), while the (9,0) zigzag nanotube

($d_t = 7.05$ Å) is capped by the C_{60} (1,1) fullerene bisected normal to a threefold axis (see Fig. 3b). Nanotubes down to $d_t = 0.4$ nm, corresponding to the C_{20} fullerene cap, have been identified experimentally. C_{20} is the smallest possible fullerene, consisting of only 12 pentagons.

Many of the experimentally observed carbon nanotubes are multi-layered, consisting of capped concentric cylinders separated by ~3.4 Å (see Fig. 3c), and are called multi-walled carbon nanotubes (MWNTs). In a formal sense, each of the constituent cylinders can be specified by the chiral vector C_h in terms of the indices (n, m). Because of the different numbers of carbon atoms around the various concentric tubules, it is not possible to achieve the ABAB interlayer stacking of graphite in carbon nanotubes. Thus, an interlayer spacing closer to that of turbostratic graphite (3.44 Å) is expected, subject to the quantized nature of the (n, m) integers.

B. Formation of Crystalline Solids

Like the graphene sheets in crystalline graphite, carbon nanotubes and fullerenes can form crystalline solids through the weak van der Waals bonding between fullerenes and nanotubes. These crystalline structures are important for forming intercalated (doped) materials, in analogy to graphite intercalation compounds (GICs).

High-resolution tunneling electron microscopy (TEM) and X-ray diffraction studies on nanotubes have shown that they are organized in bundles containing on the order of 100 well-aligned nanotubes in a close-packed triangular lattice. From theoretical calculations for a (10, 10)

(a)   (b)   (c)

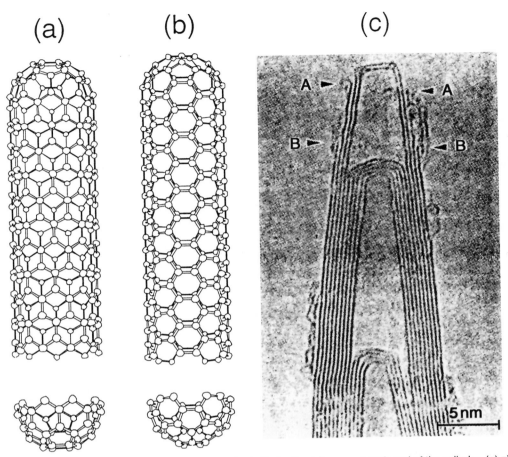

FIGURE 3 Two examples of SWNTs being capped by half of a C_{60} fullerene at each end of the cylinder, (a) zigzag and (b) armchair, and a (c) tunneling electron microscopy (TEM) image of multi-walled carbon nanotubes (MWNTs), consisting of capped concentric cylinders.

nanotube, the shortest C–C intertube distance is found to be relatively small (between 3.17 and 3.26 Å) compared to the interplanar separation in graphite (3.35 Å), and this small distance in nanotubes is attributed to the curved nanotube surface. The variation in the intertube distance is attributed to the breakdown in the sixfold symmetry of an individual nanotube in the crystalline lattice. Furthermore, these bundles can organize themselves into ropes of yet larger diameters. It is assumed that the dopants reside as ions (and also possibly as neutral atoms) primarily in the interstitial channels between the nanotubes in the triangular nanotube lattice, and to some extent in decorating the nanotube surface. Because carbon nanotubes are 1D solids, much scientific effort has been directed toward studying their single nanotube 1D characteristics, which are also promising for technological applications.

Fullerenes, however, can be structurally considered as 0D quantum dots, which also form molecular crystalline solids. For example, the C_{60} molecules crystallize into

a cubic structure (O_h^5 or $Fm3m$ space group) with a lattice constant of 14.17 Å and a density of 1.72 g/cm³ (corresponding to 1.44×10^{21} C_{60} molecules per cubic centimeter). At room temperature the molecules are rotating, and the crystal structure is face-centered cubic (FCC) with four equivalent spinning C_{60} molecules per unit cell (see Fig. 4a). The equivalence of each carbon site on a C_{60} molecule is confirmed by a single sharp line in the nuclear magnetic resonance (NMR) spectrum. Relative to the other allotropic forms of carbon, solid C_{60} is relatively more compressible, with an isothermal volume compressibility of 6.9×10^{-12} cm²/dyn, because the van der Waals charge cloud around the C_{60} molecules can easily be compressed in three dimensions. Below about 250 K, the C_{60} molecules become oriented, and the structure becomes simple cubic (space group T_h^6 or $Pa\bar{3}$) with four inequivalent C_{60} molecules. This orientation is stabilized by bringing the electron-rich double bond hexagonal edges adjacent to the electron-poor pentagonal faces. Supporting evidence for this structural phase transition has been

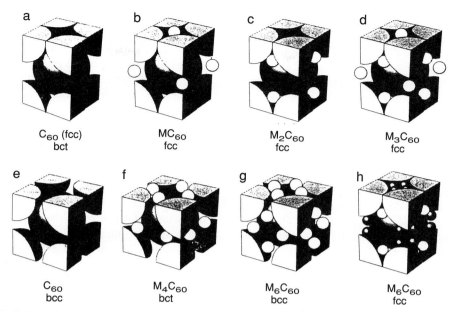

FIGURE 4 Schematic unit cell of the crystal structure for the alkali-metal fullerides (a) undoped fcc C_{60}; (b) MC_{60}; (c) M_2C_{60}; (d) M_3C_{60}; (e) undoped hypothetical bcc C_{60}; (f) M_4C_{60}, and two structures for M_6C_{60}; (g) M_6C_{60} (bcc) for M = K, Rb, Cs; and (h) M_6C_{60} (fcc), which is appropriate for M = Na. The large balls denote C_{60}, and the small balls are alkali-metal ions.

provided by different experimental techniques, such as differential scanning calorimetry, resistivity, NMR, ultrasonic attenuation, Raman spectroscopy, and thermal conductivity. An additional phase transition has also been reported to occur at still lower temperatures. Because of the larger size of the higher mass fullerenes, they crystallize in similar structures with somewhat larger lattice constants.

There are three distinct ways to introduce foreign atoms into a C_{60}-based solid. One method is "endohedral" doping, which involves the addition of a rare earth, an alkaline earth, or an alkali-metal ion into the interior of the fullerene ball. Smalley (1991) has proposed that this endohedral configuration be denoted by La@C_{60} for one endohedral lanthanum in C_{60} or Y_2@C_{82} for two Y atoms inside a C_{82} fullerene. For the larger fullerenes, the addition of up to three metal ions endohedrally has been demonstrated by mass spectroscopy techniques. It has been shown that when La is inside the fullerene cage, the La has a valence of +3, indicating the presence of three delocalized electrons on the C_{60} cage in antibonding states, which could be available for conduction in a solid of endohedrally doped C_{60}.

A second method is the substitutional doping of an impurity atom with a different valence state for a carbon atom on the surface of a fullerene cage. Since a carbon atom is so small, and since the average nearest neighbor C–C distance a_{C-C} on the C_{60} surface is only 1.42 Å, the only species that can be substituted for a carbon atom on the C_{60} ball surface is boron, making the charged ball p-type. Also for graphite, the only substitutional dopant

is boron, for the same reasons as for C_{60}. Smalley and co-workers (1991) have demonstrated that it is possible to replace more than one carbon atom by boron on a given ball. It has also been reported that it is possible to place a potassium atom endohedrally inside the C_{60} ball while at the same time substituting a carbon for a boron atom on the surface of the ball.

In the third method of doping fullerene solids (also called intercalation), the dopant (e.g., an alkali metal or an alkaline earth, M) is introduced into the interstitial positions between adjacent balls (exohedral locations). In this doping method, charge transfer can take place between the M atoms and the balls. This method of doping forms exohedral fullerene solids and closely parallels the process of intercalating an alkali-metal ion between carbon layers in graphite.

Several stable crystalline phases for doped C_{60} have been identified. The high electronegativity of pristine C_{60} relative to graphite hinders acceptor doping. Consequently, relatively few C_{60}-based structural studies have thus far been made with acceptors. Most widely studied are the crystalline phases formed with alkali metals, though some structural work has also been done with the alkaline earth and halogen intercalants. When C_{60} is doped with the alkali metals (M = K, Rb, Cs), stable crystalline phases are formed for the compositions M_1C_{60}, M_3C_{60}, M_4C_{60}, and M_6C_{60} (see Fig. 4). One electron is transferred to the C_{60} molecule per M atom dopant, resulting in M^+ ions at the tetrahedral and/or octahedral interstices of the C_{60} host structure (see Fig. 4a). The M_1C_{60} alkali phase is stable at elevated temperatures only for a limited temperature range (410–460 K), where the M ion is in an octahedral site, thereby forming a rock-salt (NaCl) crystal structure (see Fig. 4b). For the composition M_3C_{60}, the resulting metallic crystal exhibits a fcc arrangement for the C_{60} molecules (see Fig. 4d), and the alkali-metal ions lie adjacent to a C=C double bond. The two other stable phases for M_xC_{60} (namely, M_4C_{60} and M_6C_{60}) each has different crystalline structures. The M_4C_{60} phase, which is more difficult to prepare than M_3C_{60}, or M_6C_{60}, has a body-centered tetragonal structure (bct) (see Fig. 4f). The compound M_6C_{60} has a body-centered cubic structure (bcc) and corresponds to the space group T_h^5 or $Im3$. The compounds K_6C_{60}, Rb_6C_{60}, and Cs_6C_{60} all denote the alkali-metal saturated compound, which is a semiconductor.

Since the ionic radii of the alkali-metal ions Na^+ and Li^+ are much smaller than those of the K^+, Rb^+, and Cs^+ ions, it is possible to fit more than one Na^+ ion in an octahedral site, so that the fcc structure can be preserved upon adding six Na atoms per C_{60} with two Na ions going into tetrahedral sites, and up to four on octahedral sites to form Na_6C_{60} (see Fig. 4h). There is no stable Na_3C_{60} crystal, and materials prepared with this stoichiometry tend to phase separate (disproportionate) into Na_2C_{60} and Na_6C_{60}. Also, because of the small ionic radius of Na, the Na ions tend to fill the smaller tetrahedral sites when alloys of Na with heavier alkali metals are used as a dopant, while the heavier alkali metals tend to be on octahedral sites.

III. SYNTHESIS

Carbon nanotubes and fullerenes are synthesized in the laboratory by three principal techniques: the arc-discharge and the laser vaporization methods, which produce both carbon nanotubes and fullerenes, and the chemical vapor deposition method, which has been largely used to produce carbon fibers and now it is being used to produce carbon nanotubes.

In an arc-discharge method, carbon atoms are evaporated by an energetic plasma of helium gas that is ignited by passing high currents through opposing carbon anode and cathode electrodes. Typical operating conditions for a carbon arc used for the synthesis of carbon nanotubes include the use of carbon rod electrodes of 5–20 mm diameter separated by ~1 mm with a voltage of 20–25 V across the electrodes and a continuous (DC) electric current of 50–120 A flowing between the electrodes. The arc is typically operated in ~500 Torr He with a flow rate of 5–15 ml/sec for cooling purposes. For the MWNT synthesis, no catalyst needs to be used, and the nanotubes are found in bundles in the inner region

of the cathode deposit, where the temperature is maximum (2500–3000°C). For the production of SWNTs, catalysts are used, such as transition metals (e.g., Co, Ni, Fe) and rare earths (e.g., Y, Gd). Mixed catalysts (e.g., Fe/Ni, Co/Ni, Co/Pt) have been used to synthesize ropes of single wall nanotubes. For the production of fullerenes, an alternating current (AC) discharge between the graphite electrodes in ~200 Torr of He is used. This discharge produces a carbon soot which can contain up to ~15% of the C_{60} and C_{70} fullerenes.

In the laser vaporization method, intense laser pulses are used to ablate a carbon target. The target is placed in a tube furnace heated to 1200°C, and a flow of inert gas passes through the growth chamber to propel the grown nanotubes and fullerenes downstream to be collected on a cold finger. The production of high-quality SWNTs at the 1–10 g scale was achieved by using a carbon target containing 0.5 at.% of Ni and Co. This method was found to produce bundles of single wall nanotubes with a narrow diameter distribution, with the average nanotube diameter and diameter distribution being controlled by varying the growth temperature, the catalyst composition, and other growth parameters. For the production of fullerenes, a typical apparatus with a pulsed Nd:YAG laser operating at 532 nm and 250 mJ of power is normally used.

In CVD, a catalyst material is heated up to high temperatures in a furnace with a flow of hydrocarbon gas through the tube reactor. The general nanotube growth mechanism involves the dissociation of hydrocarbon molecules (which is generally catalyzed by a transition metal) and the dissolution and the subsequent saturation of carbon atoms in the metal nanoparticle. The key parameters in nanotube CVD growth are the types of hydrocarbons, catalysts, and growth temperatures that are used. For MWNT growth, most of the CVD methods employ ethylene or acetylene as the carbon feedstock, and the growth temperature is typically in the range of 550–750°C. As in the arc-discharge and laser vaporization methods, Fe, Ni, or Co are often used as catalysts. For SWNTs, it was found that by using methane as the carbon feedstock, reaction temperatures in the range of 850–1000°C, and suitable catalyst materials, high-quality SWNT materials can be produced. High temperatures are necessary to form small diameter nanotubes. One big advantage of the CVD method is that carbon nanotubes can be grown in large quantities under relative well-controlled conditions (see Fig. 5). Therefore, many presently identified applications of carbon nanotubes can be met with the CVD growth method.

Furthermore, the formation of crystalline solids with fullerenes has become a very active research field in synthetic chemistry, largely because of the novelty of the C_{60} molecule and the variety of chemical modifications to the C_{60} shell that appear to be possible. Most of the novel

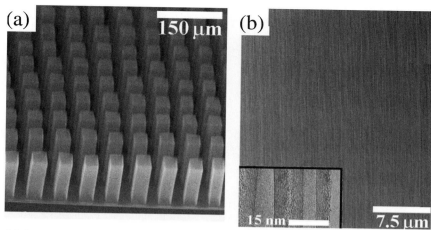

FIGURE 5 (a) Scanning electron microscopy (SEM) image of self-oriented MWNT arrays grown by CVD on a catalytically patterned porous silicon substrate. (b) High-magnification SEM image showing aligned nanotubes in a tower seen in (a).

fullerene synthetic chemistry has been done in solution. The doping of C_{60} with alkali metals can be achieved in a two-zone furnace, similar to the apparatus used to prepare alkali-metal graphite intercalation compounds.

IV. ELECTRONIC STRUCTURE

Carbon is the sixth element of the periodic table, and it has the lowest atomic number of any element in column IV of the periodic table. Each carbon atom has six electrons which occupy $1s^2$, $2s^2$, and $2p^2$ atomic orbitals. The $1s^2$ orbitals contain two strongly bonded core electrons. Four more weakly bound electrons occupy the $2s^2 2p^6$ valence orbitals. In the crystalline phase, the valence electrons give rise to $2s$, $2p_x$, $2p_y$, and $2p_z$ orbitals which are important in forming covalent bonds in carbon materials. Since the energy difference between the upper $2p$ energy levels and the lower $2s$ levels in carbon is small compared with the binding energy of the chemical bonds, the electronic wave functions for these four electrons can readily mix with each other, thereby changing the occupation of the $2s$ and the three $2p$ atomic orbitals, so as to enhance the binding energy of the carbon atom with its neighboring atoms. The general mixing of $2s$ and $2p$ atomic orbitals is called hybridization, whereas the mixing of a single $2s$ electron with one, two, or three $2p$ electrons is called sp^n hybridization with $n = 1, 2, 3$, respectively.

The graphene sheet exhibits an sp^2-derived bonding configuration, where the coplanar interaction between the $2s$, $2p_x$, $2p_y$ atomic orbitals form strongly coupled bonding and antibonding trigonal orbitals. These trigonal orbitals give rise to three bonding and three antibonding electronic bands, called σ-bands, separated by ~ 10 eV. The weakly coupled p_z atomic wave functions correspond to two bands, called π-bands, that are responsible for the

conduction properties of graphite and for the weak van der Waals bonding between adjacent graphene sheets. The dotted curves in Figs. 6a and 6b show the density of electronic states in the valence (negative values) and conduction (positive values) bands. Since the valence and conduction π-bands meet at the Fermi level (E = 0 in Figs. 6a and 6b), and the electronic density of states is zero at the crossing point, the graphene sheet is a zero-gap semiconductor.

Both carbon nanotubes and fullerenes exhibit remarkable electronic properties, first because they are strongly related to those of the zero-gap semiconductor graphene sheet and second because of their low dimensionality. Due to the changes to the graphene sheet structure on forming either carbon nanotubes or fullerenes, different electronic characteristics are obtained for these two cases. However, while the 1D electronic energy band structure for carbon nanotubes is closely related to the energy band structure calculated for the 2D graphene honeycomb sheet used to form the nanotube, the presence of pentagons in the case of fullerenes strongly influences their electronic structure, as discussed below.

A. Unidimensional Electronic Structure in Carbon Nanotubes

In SWNTs, confinement of the structure in the radial direction is provided by the monolayer thickness of the nanotube in the radial direction. Circumferentially, periodic boundary conditions apply to the enlarged unit cell that is formed. These periodic boundary conditions permit only a few wave vectors to exist in the circumferential direction for the 1D SWNTs of small diameter, and these wave vectors k satisfy the relation $n\lambda = \pi d_t$, where $\lambda = 2\pi/k$ is the de Broglie wavelength. This 1D quantization leads

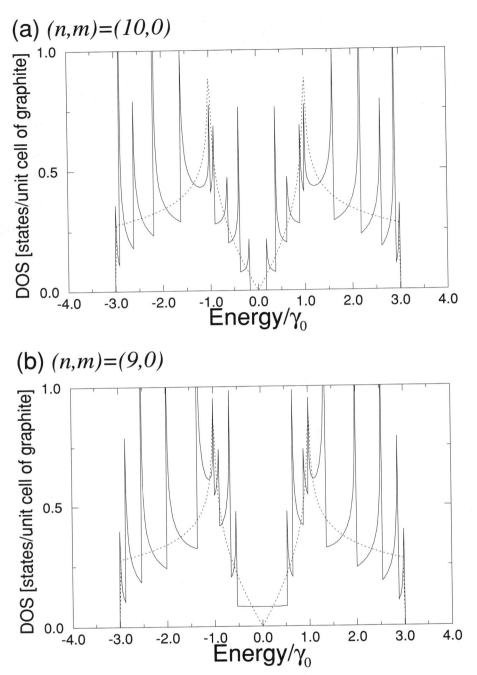

(a) *(n,m)=(10,0)*

(b) *(n,m)=(9,0)*

FIGURE 6 Electronic 1D density of states per unit cell of (a) a (10, 0) semiconducting nanotube and (b) a (9, 0) metallic nanotube obtained by zone folding of the density of states of the 2D graphene Brillouin zone (shown with dotted lines). (c) Derivative of the current-voltage (dI/dV) curves, which are proportional to the 1D DOS, obtained by scanning tunneling spectroscopy on various isolated SWNTs. The nanotubes for traces #1–4 are semiconducting and the nanotubes for traces #5–7 are metallic.

to surprising results and unique features in the electronic structure of SWNTs.

Theoretical calculations show that, depending on the choice of (n, m), the electronic structure of nanotubes can

be either metallic ($n - m = 3q$, where q is an integer) or semiconducting ($n - m = 3q \pm 1$). Solid lines in Figs. 6a and 6b show the 1D electronic density of electronic states (DOS) for the semiconducting (10, 0) and the metallic

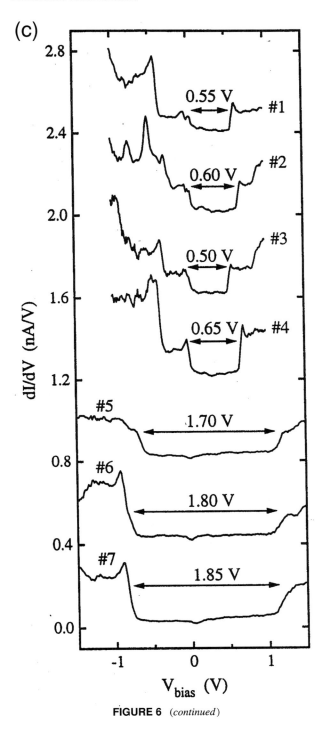

(c)

FIGURE 6 (*continued*)

(see Fig. 6b), and their band gap energy E_g is equal to the energy difference $E_{11}^{S}(d_t)$ between the two first van Hove singularities of the valence and conduction bands. As the nanotube diameter increases, more wave vectors become allowed for the circumferential direction, so that the nanotubes become two dimensional as the semiconducting band gap becomes comparable to the thermal energy. The band gap for isolated semiconducting carbon nanotubes is proportional to the reciprocal nanotube diameter $1/d_t$. At a nanotube diameter of $d_t \sim 3$ nm, the band gap becomes comparable to thermal energies at room temperature, showing that small diameter nanotubes are needed to observe 1D quantum effects.

The general characteristics that are predicted for the 1D electronic DOS of SWNTs have been confirmed by low-temperature STM/STS (scanning tunneling microscopy/spectroscopy) studies carried out on isolated SWNTs. Figure 6c shows the derivative of the current-voltage (dI/dV) curves, which is proportional to the 1D DOS, as obtained by low-temperature STS measurements on various isolated SWNTs. Nanotubes #1–4 are semiconducting and show that the band gap or energy separation is $E_{11}^{S}(d_t) \sim 0.6$ eV; nanotubes #5–7 are metallic and show energy separations of $E_{11}^{M}(d_t) \sim 1.8$ eV. The STS experiments confirm that the density of electronic states near the Fermi level is zero for semiconducting nanotubes and nonzero for metallic nanotubes. The combined STM/STS studies are consistent with the following predictions: (1) about two-thirds of the SWNTs being semiconducting and one-third being metallic; (2) the density of states exhibiting van Hove singularities, characteristic of metallic or semiconducting nanotubes; (3) energy gaps for the semiconducting nanotubes that are proportional to $1/d_t$.

Another important result is that the energy of the van Hove singularities for SWNTs is mainly diameter dependent, but with a small dependence on chiral angle that results in a unique relation between every (n, m) value and a specific E_{ii} value. Figure 7 plots the energy for the transition between the van Hove singularities in the valence and conduction bands $E_{ii}(d_t)$ for all possible (n, m) SWNTs as a function of d_t. Because of these singularities in the DOS, high optical absorbance is expected when the photon energy matches the energy separation between an occupied peak in the electron DOS and one that is empty, as we will discuss further in Section V.A.

B. Molecular Electronic Structure of Fullerenes

Turning now to the electronic structure of fullerenes, the three-dimensional (3D) confinement of this 0D quantum dot structure leads us to the solution of the molecular orbital problem, as illustrated in Fig. 8a. As already pointed

(9, 0) SWNTs, respectively. The quantization of the allowed k vectors results in sharp singularities, called van Hove singularities. Metallic SWNTs have a small, but nonvanishing 1D DOS at the Fermi level. The DOS for semiconducting 1D SWNTs is zero throughout the band gap

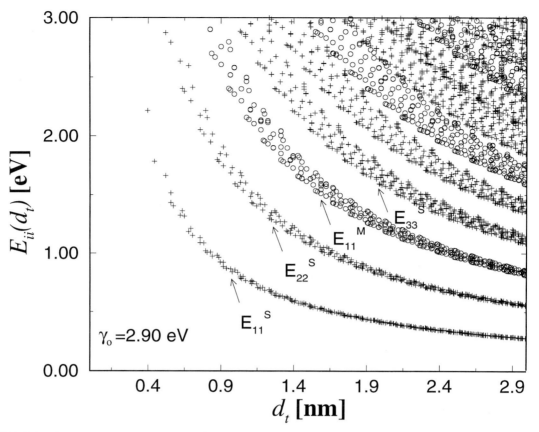

FIGURE 7 Calculation for the energy separations $E_{ii}(d_t)$ for all (m, n) values versus nanotube diameter in the range $0.4 < d_t < 3.0$ nm. Semiconducting and metallic nanotubes are indicated by crosses and open circles, respectively, and the four lowest energy translations are labeled by $E_{11}^S(d_t)$, $E_{22}^S(d_t)$, $E_{11}^M(d_t)$, $E_{33}^S(d_t)$, where S and M, respectively, refer to semiconducting and metallic SWNTs.

out, carbon atoms in fullerenes have single bonds along the sides of the pentagons and double bonds between two adjoining hexagons. If these bonds were coplanar, they would be very similar to the sp^2 trigonal bonding in graphite. However, the curvature of fullerenes causes the planar-derived trigonal orbitals to hybridize with the remaining p-orbitals, thereby giving rise to different electronic states than in graphite. The slight shortening of the double bonds and the slight lengthening of the single bonds in the truncated icosahedron of the C_{60} molecule compared to the C–C bond length in graphite also strongly influence the electronic structure.

The most extensive calculations of the electronic structure of fullerenes thus far have been done for C_{60}. Representative results for C_{60} are shown in Fig. 8b for the free molecule based on a one-electron approximation. Because of the molecular nature of solid C_{60}, the electronic structure for the solid phase would be expected to be closely related to that of the free molecule, and some authors have followed this approach. More attention has, until recently, been given to a band structure approach, as, for example,

shown in Fig. 8c where we see results for the FCC solid of C_{60} molecules. The various band calculations for C_{60} yield a narrow band (~ 0.4–0.6 eV bandwidth) solid, with a HOMO–LUMO-derived direct band gap of $E_{HL} \sim 1.5$ eV separating the LUMO from the HOMO. The allowed electric dipole transitions for the free C_{60} molecule are indicated by arrows in Fig. 8b.

Doping C_{60} with an alkali metal transfers electrons to the LUMO levels which, because of their T_{1u} symmetry (Fig. 8), can accommodate three spin up and three spin down electrons. Total energy calculations on $K_x C_{60}$ show that the tetrahedral sites fill up before the octahedral sites. Assuming one electron to be transferred to the C_{60} molecule per alkali-metal atom dopant, the LUMO levels are expected to be half occupied at the alkali-metal stoichiometry $M_3 C_{60}$ and totally full at $M_6 C_{60}$. Thus, $M_6 C_{60}$ would be expected to be semiconducting with a band gap between the T_{1u} and T_{2g} levels (see Fig. 8b), while $M_3 C_{60}$ should be metallic provided that no band gap is introduced at the Fermi level by a Peierls distortion or by curvature-related effects. Because of the weak

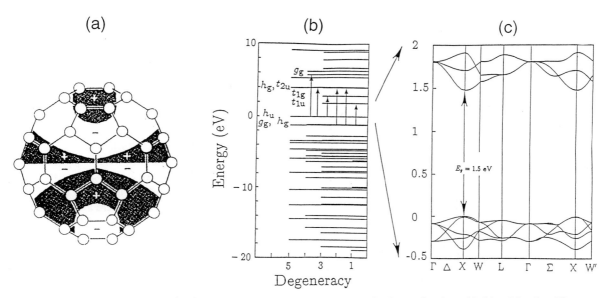

FIGURE 8 (a) Schematic figure for the t_{1u} (f_{1u}) triply degenerate conduction molecular orbital level for C_{60}. The shading on the icosahedron pertains to whether the wave function is negative or positive, showing three nodes and odd parity for the wave function. (b) Calculated electronic structure of an isolated C_{60} molecule and (c) a FCC solid C_{60}, where the direct band gap at the X point is calculated to be 1.5 eV.

interaction of the fullerene molecules with each other and with the alkali-metal dopants, solid M_3C_{60} should also be viewed as a molecular solid having energy levels with little dispersion, giving rise to a very high density of states near the Fermi level, which is important for understanding superconductivity in M_3C_{60} (see Section VII.B).

V. OPTICAL PROPERTIES

A. Optical Properties in Carbon Nanotubes

Due to the presence of the sharp van Hove singularities in the electronic density of states (see Section IV.A), SWNTs absorb light when the photon energy matches the energy separation between an occupied peak in the electron DOS and one that is empty. Therefore, measurements of the absorption of light by carbon nanotubes reflect their 1D electronic structure. Due to the diameter dispersion of the van Hove singularities ($E_{ii} \propto 1/d_t$; see Section IV.A), optical absorption spectra in SWNT samples with different diameter distributions exhibit optical absorption peaks at different energies. Figure 7 gives theoretical predictions for the absorption energies for SWNTs with different diameters.

For example, SWNTs grown by the laser vaporization method with a Ni/Y (4.2/1 at.%) catalyst exhibit a diameter distribution of $1.24 < d_t < 1.58$ nm. Optical absorption spectra performed with these samples show three large absorption peaks at 0.68, 1.2, and 1.7 eV. Considering

the diameter distribution and the energy dispersion with diameter displayed in Fig. 7, these absorption peaks are related, respectively, to the E_{11}^S, E_{22}^S, and E_{11}^M electronic transitions in semiconducting (S superscript) and metallic (M superscript) SWNTs (see Section IV.A).

As another example, SWNTs grown by the laser vaporization method with the Rh/Pd (1.2/1.2 at.%) catalyst exhibit a diameter distribution of $0.68 < d_t < 1.00$ nm. Optical absorption spectra performed with these samples show the first absorption feature at about 1 eV and a second feature at about 1.75 eV. These absorption features are related, respectively, to the E_{11}^S and E_{22}^S electronic transitions (see Fig. 7). Although the diameter distribution is about the same in both samples, the absorption peaks for the SWNTs grown with the Rh/Pd catalyst are broader than the peaks in the optical SWNT absorption spectra with the Ni/Y catalyst. This result reflects the dispersion of the transition energies with diameter, and this dispersion is larger for smaller nanotube diameters (since $E_{ii} \propto 1/d_t$).

B. Optical Properties in Fullerenes and Fullerene-Based Materials

During early research on the separation of C_{60} from impurities and higher fullerenes using liquid chromatography, it became apparent that fullerenes in solution (organic solvents) exhibited characteristic colors which would be useful for identifying and purifying these fullerenes. For example, C_{60} and C_{70} appear magenta and reddish-orange, respectively, in toluene and benzene solutions.

More detailed studies show that the optical properties of fullerenes are complex and provide important insights into the electronic structure of the fullerenes. Since the HOMO–LUMO optical transitions in C_{60} and C_{70} are symmetry-forbidden, a great deal of attention has focused upon understanding the optical behavior near the fundamental absorption edge, which is very sensitively studied by optical density spectra. The strong absorption bands in the solution spectra are identified with electric dipole-allowed transitions between occupied (bonding) and empty (antibonding) sp^2-hybridized molecular orbitals.

Further insight into the electronic structure is provided by pulsed laser studies of C_{60} and C_{70}, which probe the photodynamics of the optical excitation spectra. The importance of these dynamic studies is to show that photoexcitation in the long-wavelength portion of the ultra-violet (UV)-visible spectrum leads to the promotion of the system from the singlet ground state S_0 into a first excited singlet state S_1, which decays quickly with a nearly 100% efficiency via an intersystem crossing to the lowest excited triplet state T_1. In the presence of oxygen, the lifetime of the triplet state becomes very short, and the decay to the ground state becomes very rapid. Furthermore, once C_{60} reaches the T_1 triplet state, efficient electronic dipole excitation to other excited triplet states can occur with high efficiency, thereby greatly increasing the absorption coefficient and providing a mechanism for achieving efficient optical limiting (see Section IX).

To date, most of the optical studies on pristine fullerene solids have been carried out on transmission through thin solid films deposited on various substrates (such as quartz, Si, KBr). The UV-visible transmission spectra for C_{60} and C_{70} solid films are observed to be remarkably similar to their respective solution spectra, thus providing further evidence for the molecular nature of fullerene solids.

Photoemission and inverse photoemission experiments have been especially useful in providing information on the density of states within a few electron volts of the Fermi level for both undoped and doped fullerenes. Typical photoemission spectra ($E < E_F$, where E_F is the Fermi energy) and inverse photoemission spectra ($E > E_F$) for C_{60} and $K_x C_{60}$ ($0 \leq x \leq 6$) show intensity maxima corresponding to peaks in the electronic density of states. Photoemission and inverse photoemission spectra (PES and IPES) provide convincing evidence for charge transfer and for band filling as x in $K_x C_{60}$ increases. The related peak in the DOS moves closer to E_F as potassium is added and some of the $K_3 C_{60}$ metallic phase is introduced. Photoemission and inverse photoemission studies

have confirmed strong similarities between the electronic structure of $K_x C_{60}$ and of both $Rb_x C_{60}$ and $Cs_x C_{60}$.

VI. VIBRATIONAL MODES

The way the atoms vibrate in a material, called vibrational modes or phonons (in analogy to photons), provides important structural information about the material, such as its symmetry, atomic bonding, the presence of impurities, and disorder in its crystalline structure, etc. Two techniques are largely used to study these vibrational modes: Raman spectroscopy, where a high energy photon (generally with a frequency in the visible) is inelastically scattered by a phonon, and IR spectroscopy, where the absorption and reflection of light by a material is studied using photons with the same energy as the atomic vibrations in the material.

A. Phonons in Carbon Nanotubes

Since a graphene sheet (2 carbon atoms per unit cell) has 6 phonon branches in 2D reciprocal space, nanotubes exhibit $2N$ phonon branches, where N is the number of hexagons per nanotube unit cell. Since the propagation of the vibrational waves (phonons) can occur only along the axis direction, the 1D wave vector k is also along the tube axis. Along the circumferential direction, the allowed phonon wave vectors are quantized due to the periodic boundary conditions of the chiral vector C_h, and the vibrational characteristics are dependent on the (n, m) indices.

The dominant experimental technique for studying phonons in SWNTs has been Raman scattering (see Fig. 9). These experiments have been especially informative because of the strong coupling between electrons and phonons in this 1D system. When the excitation incident photon matches the energy for one of the electronic transitions in carbon nanotubes, a highly unusual resonance Raman scattering process becomes dominant and increases the Raman signal by many orders of magnitude, making possible the resonant selection of certain SWNT diameters or even the observation of a Raman spectrum from an individual isolated SWNT. For each laser excitation energy, a diameter-selective resonant Raman spectra is observed, which makes it possible to study different diameter nanotubes selectively and also to differentiate the contribution between metallic and semiconducting tubes, depending on whether the laser energy matches a metallic (E_{ii}^M) or a semiconducting (E_{ii}^S) electronic transition. This selective resonant effect makes Raman scattering an important technique for the study of carbon nanotubes, since the present synthesis techniques produce both

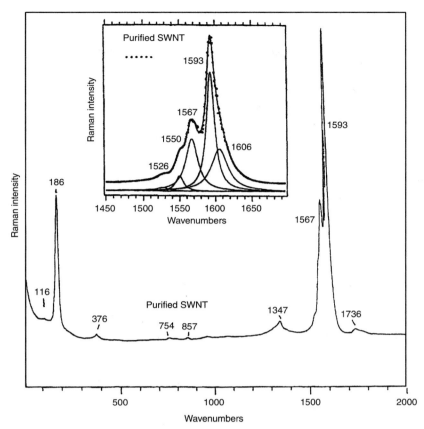

FIGURE 9 Raman spectra taken with 514.5 nm (2.41 eV) laser excitation from a sample consisting primarily of SWNT bundles with diameters $d_t = 1.38 \pm 0.20$ nm. The strongest peaks are the radial breathing mode (peak at 186 cm^{-1}) and the tangential modes (G-band, around 1580 cm^{-1}). The inset shows a Lorentzian analysis of the G-band spectrum.

semiconducting and metallic nanotubes with a range of diameters and all possible chiralities.

Although up to 16 vibrational modes are predicted to be Raman active for all nanotubes, particular attention has been given to the two strongest spectral features, the radial breathing mode (RBM) and the tangential stretching G-band modes, as shown Fig. 9. Theoretical calculations have shown that the frequency ω_{RBM}^0 of the RBM, where the superscript denotes an isolated SWNTs, has a particularly simple $1/d_t$ dependence on the nanotube diameter. This fact, coupled with the large resonantly enhanced Raman cross section for the RBM, the sensitivity of ω_{RBM} to charge transfer and to tube–tube interactions, makes the RBM a valuable probe for characterizing the structure and properties of SWNTs and SWNT-based materials. Raman studies on individual isolated SWNTs show that it is possible to assign the (n, m) structural indices of an isolated SWNT, or the diameter distribution of the SWNTs in a sample, by measuring the RBM frequencies and intensities.

Due to the close connection between the 2D graphene sheet structure and the 1D nanotube structure, the in-plane C–C stretching vibrational mode with frequency equal to 1582 cm^{-1} in graphite (called the E_{2g_2} mode) gives rise to several tangential modes with different symmetries (called the G-band), with carbon nanotubes frequencies occurring between 1510 to 1610 cm^{-1} (see Fig. 9). Based on polarization studies, group theory analysis, and theoretical calculations, six peaks can be present in the G-band spectra, and their relative Raman intensities depend on the SWNT chiral angle θ. Therefore, the analysis of the relative Raman intensity of the G-band modes can also give information relevant to the (n, m) assignment for isolated SWNTs.

In the case of metallic SWNTs, the presence of phonon–plasmon coupling changes the Raman spectra considerably in the region of the G-band. In general, the G-band in metallic SWNTs is downshifted and broader relative to semiconducting SWNTs, and the lower frequency $A(A_{1g})$ mode associated with circumferential displacements exhibits a Breit–Wigner–Fano lineshape, while the upper

$A(A_{1g})$ mode associated with longitudinal displacements has a Lorentzian lineshape. The differences between the Raman spectra of semiconducting and metallic SWNTs, coupled with the diameter-selective resonant Raman effect, make it possible to use Raman scattering to study metallic and semiconducting SWNTs separately.

B. Phonons in Fullerenes

Because of the high symmetry of the C_{60} molecule, only 4 vibrational modes are infrared (IR) active and 10 are Raman active. Raman and IR spectroscopy provide sensitive methods for distinguishing C_{60} from higher fullerenes, since most of the higher molecular weight fullerenes have lower symmetry as well as more degrees of freedom, having many more IR- and Raman-active modes. The IR spectrum of solid C_{60} remains almost unchanged relative to the isolated C_{60} molecule, with only the addition of a few weak features. This is indicative of the highly molecular nature of solid C_{60}.

Likewise, Raman spectroscopy provides valuable information about the intra- and intermolecular bonding in solid C_{60} and C_{60}-related compounds. The Raman spectrum for solid C_{60} shows 10 strong Raman lines, the number of Raman-allowed vibrational modes expected for the free molecule. These spectral lines are therefore assigned to intramolecular modes. The normal modes in molecular C_{60} at high frequency (above about 1000 cm^{-1}) involve carbon atom displacements that are predominantly tangential to the C_{60} ball surface, while the modes at lower frequency (below ~800 cm^{-1}) involve predominantly radial motion. The breathing mode (1469 cm^{-1}), corresponding to breathing type tangential displacements of the 5 carbon atoms around each of the 12 pentagons, downshifts irreversibly to 1458 cm^{-1} if the C_{60} is irradiated with high laser flux. This result has been interpreted as a signature of a photo-induced transformation.

In the solid phase, cubic crystal field interactions are expected to lower the symmetry of the lattice modes. The study of spectral line splittings and their symmetry in solid C_{60} requires the use of single crystals, polarization studies, and studies of the temperature dependence of the Raman spectra. These techniques are also needed to monitor the effect of the 250 K phase transition (associated with C_{60} molecular rotations) on the spectral line shifts, linewidths, line splittings, and their symmetries.

The addition of alkali-metal dopants to the saturation stoichiometry M_6C_{60} (where M = K, Rb, Cs) perturbs the Raman spectra only slightly, with very little change from one alkali-metal dopant to another. For metallic K_3C_{60} and Rb_3C_{60}, the coupling between the phonons and a low-energy electronic continuum in these metallic systems strongly broadens some vibra-

tional modes and gives rise to modifications in the Raman lineshape. Furthermore, as a result of alkali-metal doping, electrons are transferred to the π-electron orbitals on the surface of the C_{60} molecules, elongating the C–C bonds and downshifting the intramolecular tangential modes. The softening of the 1469 cm^{-1} vibrational mode by alkali-metal doping is used as a convenient method to characterize the stoichiometry x of K_xC_{60} samples.

Models for the intramolecular fullerene modes have been used to study the electron–phonon interaction, with particular emphasis on its connection to the observed superconductivity in the M_3C_{60} compounds.

VII. TRANSPORT

A. Electrical and Thermal Conductivity

Transport measurements of nanotubes show different aspects depending on the sample type, such as an isolated SWNT, a single rope of SWNTs, a single MWNT, or a single MWNT bundle. In general, the resistance (R) measurements of various nanotube samples show that there are metallic and semiconducting nanotubes, which was first predicted theoretically, as discussed in Section IV. The resistivity is about 10^{-4} to 10^{-3} Ω-cm for metallic nanotubes, while the room temperature resistivity is about 10^1 Ω-cm in semiconducting nanotubes. The semiconducting tubes exhibit a slope in a plot of $\log R$ versus $1/T$, which indicates an energy gap in the range 0.1–0.3 eV, which is consistent with the theoretical values of the energy gap for the corresponding nanotube diameters.

In a macroscopic conductor, the resistivity ρ does not generally depend on either the length of the wire L or the applied voltage to the sample, but only on the material itself. However, when the size of the wire becomes small compared with the lengths L characteristic of electronic motion, then ρ will depend on the length L through quantum effects. In the quantum regime, the electrons act like waves that show interference effects, depending on the boundary conditions and the impurities and defects present in the nanotube.

There are three possible transport regimes in carbon nanotubes: ballistic, diffusive, and classical transport. In an ideal ballistic conductor, there is no electron scattering, and the wavefunction of the electron is determined over the sample by solution of the Schrödinger equation. In diffusive motion, many elastic scattering events occur. If the scattering of an electron is elastic, a phase shift of the wavefunction can be expected after the elastic scattering event, and thus, the wave remains coherent. Therefore, in a diffusive motion, the length over which the electron retains its coherence (called the phase-relation length L_ϕ) is much longer than the average length that an electron travels before it is scattered by a scattering center (called the mean free path L_m). The electrons can still be considered waves, but the elastic scattering events bring about the localization of the wavefunction. Classical conductance is the conductance which satisfies Ohm's law, where both momentum and phase relaxation occurs frequently, and thus, an electron can be considered as a particle. The appropriate transport regime can be determined by changing L_ϕ and L_m. These characteristic lengths depend on the scattering mechanisms in the material. The scattering can be elastic (with a static potential), by magnetic impurities, by electron–electron interactions, or by phonons. These parameters can be controlled by introducing ionic and magnetic impurities, application of magnetic and electric fields, changes in temperature and pressure, and changes in the carrier concentration by chemical doping.

In the diffusive motion, there is a regime called the weak localization regime ($L_m \ll L_\phi < L_c$, where L_c is the localization length), where interesting effects occur. One effect is that of universal conductance fluctuations. When the phase delocalization L_ϕ is close to the sample size L, interference effects associated with the wavefunction become important. If we prepare many samples of the same size, the conductance will fluctuate from sample to sample because of differences in the inhomogeneous distribution of scattering centers. Another interesting phenomena observed in the weak localization regime is the negative magnetoresistance, which refers to a decrease in the resistance with increasing magnetic field.

Figure 10a shows a single SWNT placed on top of a Si/SiO_2 substrate with Pt electrodes, that are used to perform electrical transport measurements. Figure 10b shows the current-voltage curves for an individual SWNT. The plateaus of nonzero current clearly show ballistic transport effects associated with the discrete levels imposed by the finite length of the 1D nanotube conductor. The observed phenomena can be understood in terms of steps in the quantum conductance.

Let's consider now the electrical transport in fullerene-based crystalline solids. Undoped C_{60} exhibits high electrical resistivity (ρ). The high electrical resistivity and the magnitude of the optical band gap of C_{60} can be reduced by the application of high pressure, with decreases in resistivity of about one order of magnitude observed per 10 GPa pressure. However, at a pressure of ~20 GPa, a phase transition to a more insulating phase has been reported. As the interball C–C distance decreases upon application of pressure and becomes comparable to the on-ball C–C distance, an electronic transition might be expected to occur.

The doping of C_{60} with alkali metals introduces carriers at the Fermi level and decreases the electrical resistivity ρ of C_{60} by several orders of magnitude. As x

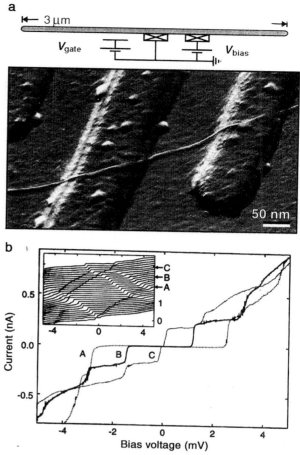

FIGURE 10 (a) Atomic force microscopy image of a carbon nanotube on top of a Si/SiO_2 substrate with Pt electrodes. A schematic circuit diagram is presented at the top of this figure. (b) I-V_{bias} curves of the carbon nanotube shown in (a). The different curves are related to different constant voltages (A = 88.2 mV, B = 104.1 mV, C = 120.0 mV) applied to the third electrode (gate voltage, V_{gate}). The gate voltage changes the position of the energy levels of the nanotube relative to the chemical potentials of the electrodes.

in M_xC_{60} increases, the resistivity ρ approaches a minimum at $x = 3.0 \pm 0.05$, corresponding to a half-filled t_{1u}-derived conduction band. Then, upon further increase in x from 3 to 6, ρ again increases. It should be noted that stable M_xC_{60} compounds only occur for $x = 0, 1, 3, 4$, and 6. Since $x = 4$ is less stable than the others, samples with $x \neq 3$ often phase separate into the more stable compounds with $x = 0, 3$, and 6. The compounds corresponding to filled molecular levels (C_{60} and M_6C_{60}) are the most stable structurally and correspond to maxima in the resistivity. Furthermore, even at the minimum resistivity in M_xC_{60}, a high value of ρ is found for K_3C_{60} ($\rho_{min} = 2.5 \times 10^{-3}$ Ω-cm), typical of a high-resistivity metal. No structure is observed in the $\rho(x)$ data near $x = 4$, indicating that K_4C_{60} is not a metallic phase. Studies of the temperature dependence of the resistivity of polycrystalline M_xC_{60} samples in the normal state show that conduction is by a thermally activated hopping process, except for a small range of x near 3 where the conduction is metallic. The activation energy E_a increases as x deviates further and further from the resistivity minimum at $x = 3$, and E_a has, for example, a magnitude of 0.12 eV for $x = 1$. In the metallic regime, $\rho(T)$ measurements on a superconducting single crystal K_3C_{60} sample show a linear T dependence of $\rho(T)$ above T_c, though the linear T dependence is much weaker than that for the high T_c cuprates. In contrast, many thin film superconducting M_3C_{60} samples have been shown to have a negative slope for $\rho(T)$ versus T for $T \geq T_c$. The observed temperature dependence of the resistivity of films of M_xC_{60} (for $0 < x \leq 6$) has been interpreted in terms of a granular conductor with grain sizes in the 60–80 Å range. This granularity strongly affects the superconducting properties of the M_3C_{60} films, as well as their transport properties in the normal state. Superconductivity in fullerenes is discussed in Section VII.B.

With regard to the thermal conductivity, diamond and graphite (in-plane) display the highest measured thermal conductivity of any known material. The thermal conductivity is entirely by phonons in the case of diamond. In the case of the basal plane graphite, the thermal conductivity is almost totally by phonons, but there is also a very small contribution from electrons. The apparent long-range crystallinity of nanotubes suggests that the thermal conductivity of nanotubes along the nanotube axis should also be high. Thermal conductivity measurements to date indicate that the phonon contribution is dominant in MWNTs and SWNTs at all temperatures. In analogy to graphite, where the in-plane thermal conductivity can be closely approximated by neglecting the interplanar coupling, it is expected that for SWNTs bundles the thermal conductivity is basically in the axial direction of single tubes.

Thermal conductivity measurements from 30 to 300 K on single crystal C_{60} show a low value of the thermal conductivity κ. This small magnitude of κ likely arises because the thermal conduction path is between weakly coupled fullerenes. C_{60} is also expected to have a low Debye temperature. In addition, a discontinuity is observed in the thermal conductivity at about 260 K, and this discontinuity is associated with a first-order structural phase transition in crystalline C_{60}.

B. Superconductivity

Perhaps the most striking transport property of C_{60}-related materials is the observation of high-temperature super-

conductivity. The first observation of superconductivity in an alkali-metal-doped carbon material dates back to 1965 when superconductivity was observed in the first stage alkali-metal graphite intercalation compounds (GIC) C_8K. Except for the novelty of observing superconductivity in a compound having no superconducting constituents, this observation did not attract a great deal of attention, since the T_c was very low ($\sim$140 mK). Later, higher T_c's were observed in other GICs (e.g., $KHgC_8$, $T_c = 1.9$ K).

The early observation of superconductivity at 18 K in K_3C_{60} was soon followed by observations of superconductivity at even higher temperatures: in Rb_3C_{60} ($T_c = 30$ K) and $Rb_xCs_yC_{60}$ ($T_c = 33$ K). A further increase in T_c was achieved by going to compounds with larger intercalate atoms, resulting in unit cells of larger size with larger lattice constants. As the lattice constant increases, the ball–ball coupling decreases, narrowing the width of the LUMO level and thereby increasing the corresponding density of states. Several experiments and calculations provide supporting evidence for an increased density of states at the Fermi level resulting from an increase in lattice constant. T_c values up to 52 K have been reported for electron injected C_{60}.

The reason why the T_c is so much higher in the M_3C_{60} materials relative to other carbon-based superconductors such as C_8K appears to be closely related to the higher density of states expected in a molecular solid. This is achieved in the M_3C_{60} compounds at the Fermi level when the T_{1u} LUMO molecular level is half filled with carriers. In addition, several authors have found an enhancement in the electron–phonon coupling in C_{60}-related materials. If the electron–phonon interaction provides the pairing mechanism for superconductivity in C_{60}-based materials, it is believed that it is the H_g-derived optical phonons that play the dominant role in the coupling. The observation of broad H_g-derived Raman lines in M_3C_{60} is consistent with a strong electron–phonon coupling. Calculations of the contributions of the various phonon modes to the electron–phonon interaction strength show that the high-frequency intraball phonons cannot account for the high T_c values that are observed experimentally.

A small amount of work has also been done to study superconductivity in carbon nanotubes. Preliminary transport measurements at low temperature show evidence for superconductivity in carbon nanotubes, but only in the milli Kelvin (mK) range, as is observed also for alkali-metal intercalated sp^2 carbon materials. This area is now under active investigation and it is expected that superconductivity may someday be found in nanotubes intercalated with specially selected dopants, including fullerene molecules.

VIII. MAGNETIC PROPERTIES

The study of the magnetic properties of sp^2 carbons has been widely pursued, largely because graphite is a highly diamagnetic material, which is also the case for carbon nanotubes.

Electron spin resonance (ESR) has been intensively used to study the properties of the spins in graphite and other sp^2 carbons. Surprisingly, no conduction ESR signal has been observed in SWNTs. One reason for the absence of an ESR line in SWNTs might be the strong electron–electron correlation, giving rise to the so-called Luttinger liquid (LL) character of the 1D electronic system, which would be ESR silent. Another reason is the presence of magnetic particles, which serve as catalysts for the SWNT synthesis, remaining afterwards as impurities in actual SWNT samples, and these particles are difficult to remove completely. Even a small concentration of magnetic particles can cause the ESR line to be unobservably broad. Therefore, only the ESR signal from MWNTs has been studied.

The spin (Pauli) susceptibility χ_s for MWNTs is derived from the ESR signal strength by numerical integration. For free electrons, the Pauli susceptibility is given by $\chi_s = \mu_B^2 N(E_F)$, where μ_B is a constant (Bohr magneton) and $N(E_F)$ is the density of states at the Fermi level. In general, the functional χ_s for MWNTs is similar to that of graphite. Above 40 K, χ_s is temperature independent in MWNTs, where $\chi_s = 7 \times 10^{-9}$ emu/g, which is a little smaller than χ_s for powdered graphite, where $\chi_s = 2 \times 10^{-8}$ emu/g. This Pauli behavior indicates that the aligned nanotubes are at least in part metallic or semimetallic. The measured susceptibility for MWNTs gives an electronic density of states $N(E_F) = 2.5 \times 10^{-3}$ states per electron volt per atom, which is comparable to that in graphite. Estimating the carrier concentration n given by $n = E_F N(E_F)$ and taking the Fermi energy E_F for the MWNTs to be equal to that of graphite (200 K) gives $n \sim 10^{18}/\text{cm}^3$. This low carrier density is consistent with the MWNTs being semimetallic.

ESR measurements are also suitable for characterizing the defects in carbon nanotubes. It is suspected that arc-grown MWNTs contain a high defect density that modifies their electronic properties. Among these defects, vacancies and interstitials can generate paramagnetic centers that can be sensitively studied by ESR. Pentagon–heptagon pair defects can also be present in the lattice. However, ESR measurements show that arc-grown MWNTs contain a very low density of paramagnetic defects. This fact does not apply to catalytically grown MWNT samples, which have a higher defect density, and the observation of an ESR signal is generally precluded by the magnetic impurities left behind in these samples even after purification.

Evidence for low-dimensional behavior can be recognized by comparing the ESR signals from MWNT soot with different MWNT densities. The ESR experiment shows that the ESR linewidth for a dense thick MWNT film is four times larger (32 G) than the ESR linewidth for MWNTs dispersed in paraffin (8 G). This behavior is specific to nanotubes and it is in contrast to what happens in graphitized carbon blacks. The increase in the linewidth with increasing MWNT density is similar to the behavior in a quasi-1D conductor when hydrostatic pressure is applied.

The magnetic properties of fullerenes are interesting and surprising. While the hexagonal rings contribute a diamagnetic term to the total magnetic susceptibility, the pentagonal rings contribute a paramagnetic term of almost equal magnitude. Because of this unusual cancellation of ring currents, the C_{60} molecule exhibits an unusually small diamagnetic susceptibility compared to other fullerenes or to graphite itself, which has only hexagonal rings. From ESR measurements, the temperature independent diamagnetic susceptibility for C_{60} and C_{70} is, respectively, -0.35×10^{-6} and -0.59×10^{-6} emu/g, consistent with the larger number of hexagons in the C_{70} molecule, but much smaller in magnitude than -21×10^{-6} emu/g in graphite for the H field $\parallel$ c-axis.

In the formation of fullerene-based solids, however, the presence of dopants easily introduces paramagnetic behavior into fullerenes. The introduction of dopants with unfilled d or f shells into fullerene host materials (see Section II.B) results in paramagnetic materials. Interestingly, the three ions that have been most commonly used as endohedral dopants into fullerenes (La^{+3}, Y^{+3}, and Sc^{+3}) have no magnetic moments. However, the charge transfer to the C_{60} shell results in paramagnetic behavior. In considering the solid state, a magnetic phase could also be achieved by the exohedral doping of magnetic ions into tetrahedral or octahedral sites (see Section II.B).

IX. APPLICATIONS

Graphite is a material with extreme properties (modulus, strength, thermal conductivity, anisotropy, melting point, etc.), and, as discussed elsewhere in this article, carbon nanotubes and fullerenes are strongly related to graphite. Therefore, we might also expect carbon nanotubes and fullerenes to exhibit extreme properties. Although research on carbon nanotubes, solid C_{60}, and related materials is still at a relatively early stage, these materials are already beginning to show many exceptional properties, some of which may lead to practical applications.

Due to its high stiffness and low density (and high melting point), carbon fibers are commonly used in industrial composites, such as for the aerospace industry. If carbon nanotubes were available in quantity at a price comparable to carbon fibers, it is conceivable that the small diameter and high length to diameter ratio would allow strong processable composites to be developed with practical applications. The high strength and small size of the nanotubes may allow such composites to be processed through an extruder, without fracturing the nanotubes.

Hydrogen (H_2) has attracted a great deal of attention as an energy source. Once it is generated, its use as a fuel creates neither air pollution nor greenhouse gas emissions. A practical problem is how to store H_2 easily and cheaply. SWNTs are promising for H_2 absorption, possibly even at room temperature. H_2 storage capacity of SWNTs synthesized by a hydrogen arc-discharge method, with a relatively large sample quantity (about 500 mg) at ambient temperature under modestly high pressure (about 10 MP), was measured. A H_2 uptake with a 0.52 H/C atom ratio was obtained with an estimated purity of 50 wt% of the sample. In this experiment, about 80% of the absorbed H_2 could be released at room temperature.

In the past several decades, there has been a nearly constant exponential growth in the capabilities of silicon-based microelectronics. However, it is unlikely that these advances will continue far into the new millennium, because of fundamental physical limitations. Besides their nanometric scale, SWNTs exhibit unique electronic, mechanical, and chemical properties. Nanotube-based prototypes, such as a random access memory for molecular computing, have already been built, showing that carbon nanotubes are strong candidates for some applications to post-Si nanotechnology. Other examples of the use of carbon nanotubes are as field emitters for a flat display, indicating that a computer screen with high (nanometric) definition could be built with SWNTs, and as atomic force microscopy (AFM) tips, since the size of the tips define the precision of the microscopy.

Turning now to fullerenes, one promising application for C_{60} is as an optical limiter. Optical limiters are used to protect materials from damage by high light intensities via a rapid saturation of the transmitted light intensity with increasing incident intensity. Outstanding performance for C_{60} relative to presently used optical limiting materials has been observed at an optical wavelength of 532 nm for 8 ns pulses using solutions of C_{60} in toluene and in CH_3Cl. Although C_{70} in similar solutions also showed optical limiting action, the performance of C_{60} was superior. The proposed mechanism for the optical limiting is that C_{60} and C_{70} are more absorptive once the triplet ex-

cited state is populated, so that the initial largely forbidden absorption of a photon leads to an electronic excitation in an excitonic state, with a subsequent intersystem crossing to a triplet state, which can only decay very slowly to the ground state. Stronger absorption of photons can occur as allowed excitations from the triplet state to higher lying states take place.

The possible use of C_{60} for the fabrication of industrial diamonds offers another area for possible applications. It is found that when rapid nonhydrostatic pressures (in the range of 20 GPa) are applied at room temperature to C_{60}, the material is transformed instantaneously into bulk polycrystalline diamond at high efficiency. It is believed that the presence of pentagons in the C_{60} structure promotes the formation of sp^3 bonds during the application of high anisotropic stress. Enhanced nucleation of a high density of small diamond crystallites on a Si substrate has been achieved through the deposition of a 500–100 Å C_{70} film, followed by positive ion bombardment, after which the CVD growth of small diamond can occur in a microwave plasma reactor. A base layer of ion-activated C_{70} was found to be more effective in promoting sp^3 bonding than a similarly treated C_{60} buffer layer. Normally, for diamond films to grow from a mixture of gaseous CH_3 and H_2, the substrate surface (usually Si) must be retreated with diamond grit polish or must contain small diamond seeds.

A photoconducting device can be fabricated using fullerene-doped polymers. Polyvinylcarbazole (PVK) doped with a mixture of C_{60} and C_{70} has been reported to exhibit exceptionally good photoconductive properties, which may lead to the development of future polymeric photoconductive materials. The effects of the fullerenes (∼2.7% by weight) on the charge generation and transport still need to be understood.

As further research on these materials is carried out, it is expected, because of the extreme properties of carbon-based materials generally, that other interesting physics and promising applications will be found for carbon nanotubes, fullerenes, and related materials.

ACKNOWLEDGMENTS

We acknowledge Professors Mildred S. Dresselhaus, Riichiro Saito, and Hui-Ming Cheng for their help and encouragement in the preparation of this article.

SEE ALSO THE FOLLOWING ARTICLES

CARBON FIBERS • CHEMICAL VAPOR DEPOSITION • DIAMOND FILMS, ELECTRICAL PROPERTIES • ELECTRON SPIN RESONANCE • MOLECULAR ELECTRONICS • POLYMERS, ELECTRONIC PROPERTIES • RAMAN SPECTROSCOPY • SUPERCONDUCTIVITY

BIBLIOGRAPHY

Curl, R. F., and Smalley, R. E. (1991). "Fullerenes," *Sci. Am.* **256,** 54–63.

Dresselhaus, M. S., and Eklund, P. C. (2000). "Phonons in Carbon Nanotubes," *Adv. Phys.* **49**(6), 705.

Dresselhaus, M. S., and Endo, M. (2001). "Carbon Nanotubes: Synthesis, Structure, Properties, and Applications," Springer-Verlag, Berlin/Heidelberg.

Dresselhaus, M. S., Dresselhaus, G., and Eklund, P. C. (1996). "Science of Fullerenes and Carbon Nanotubes," Academic Press, New York.

Iijima, S. (1991). *Nature (London)* **354,** 56.

Kroto, H. W., *et al.* (1985). *Nature* **318,** 162–163.

Saito, R., Dresselhaus, G., and Dresselhaus, M. S. (1998). "Physical Properties of Carbon Nanotubes," Imperial College Press, London.

Yakobson, B. I., and Smalley, R. E. (1997). "Fullerene nanotubes: $C_{1,000,000}$ and beyond," *Am. Sci.* **85,** 324.

Functional Analysis

C. W. Groetsch
University of Cincinnati

I. Linear Spaces
II. Linear Operators
III. Contractions
IV. Some Principles and Techniques
V. A Few Applications

GLOSSARY

Banach space A complete normed linear space.

Bounded linear operator A continuous linear operator acting between normed linear spaces.

Compact linear operator A linear operator that maps weakly convergent sequences into convergent sequences.

Decomposition theorem A Hilbert space is the sum of any closed subspace and its orthogonal complement.

Fredholm alternative A theorem that characterizes the solvability of a compact linear operator equation of the second kind.

Hilbert space A complete inner product space.

Inner product A positive definite (conjugate) symmetric bilinear form defined on a linear space.

Linear operator A mapping acting between linear spaces that preserves linear combinations.

Linear functional A linear operator whose range is the field of scalars.

Norm A real-valued function on a linear space having the properties of length.

Orthogonal complement All vectors which are orthogonal to a given set.

Orthogonal vectors Vectors whose inner product is zero; a generalization of perpendicularity.

Riesz representation theorem A theorem that identifies bounded linear functionals on a Hilbert space with vectors in the space.

Spectral theorem Characterizes the range of a compact self-adjoint linear operator as the orthogonal sum of eigenspaces.

FUNCTIONAL ANALYSIS strives to bring the techniques and insights of geometry to bear on the study of functions by treating classes of functions as "spaces." The interplay between such spaces and transformations between them is the central theme of this *geometrization* of mathematical analysis. The spaces in question are *linear* spaces, that is, collections of functions (or more general "vectors") endowed with operations of addition and scalar multiplication that satisfy the well-known axioms of a vector space. The subject is rooted in the vector analysis of J. Willard Gibbs and the investigations of Vito Volterra and David Hilbert in the theory linear integral equations; it enjoyed youthful vigor, spurred by applications in potential theory and quantum mechanics, in the works of Stefan Banach, John von Neumann, and Marshal Stone on the theory of linear operators, and settled into a dignified maturity with the contributions of M. A. Naimark, S. L. Sobolev, and L. V. Kantorovich to normed rings, partial differential equations, and numerical analysis, respectively. Functional analysis provides a general context, that is rich in geometric and algebraic nuances, for organizing and illuminating the structure of many areas of mathematical analysis and mathematical physics. Further, the subject enables rigorous justification of mathematical principles by use of general techniques of great power and wide applicability. In recent decades functional analysis has become the lingua franca of applied mathematics, mathematical physics and the computational sciences, and it has made inroads into statistics and the engineering sciences. In this survey we will concentrate on functional analysis in normed linear spaces, with special emphasis on Hilbert space.

I. LINEAR SPACES

Functional analysis is a geometrization of analysis accomplished via the notion of a linear space. Linear spaces provide a natural algebraic and geometric framework for the description and study of many problems in analysis.

Definition. A *linear space* (vector space) is a collection $\mathcal{V}$ of elements, called *vectors*, along with an associated field of *scalars*, F, and two operations, vector addition (denoted "+") and multiplication of a vector by a scalar (the scalar product of $\alpha \in F$ with $x \in \mathcal{V}$ is denoted αx). The set $\mathcal{V}$ is assumed to form a commutative group under vector addition (the additive identity is called the *zero vector*, denoted θ, and the additive inverse of a vector x is denoted as $-x$). In addition, the scalar product respects the usual algebraic customs, namely:

$$1x = x,$$
$$\alpha(\beta x) = (\alpha\beta)x,$$
$$(\alpha + \beta)x = \alpha x + \beta x,$$
$$\alpha(x + y) = \alpha x + \alpha y,$$

for any $\alpha, \beta \in F$ and any $x, y \in \mathcal{V}$, where 1 denotes the multiplicative identity in F.

The vector $x + (-y)$ is normally written $x - y$. We always take the field F to be the field of real numbers, $\mathbf{R}$, or (less frequently) the field of complex numbers, $\mathbf{C}$. The prototype of all linear spaces is Euclidean n-space, that is, the vector space $\mathbf{R}^n$ consisting of all ordered n-tuples of real numbers with addition and scalar multiplication defined componentwise. That is, for

$$x = [x_1, x_2, \ldots, x_n]$$
$$y = [y_1, y_2, \ldots, y_n]$$

in $\mathbf{R}^n$, and $\alpha \in \mathbf{R}$, the vector operations are defined by

$$x + y = [x_1 + y_1, x_2 + y_2, \ldots, x_n + y_n]$$
$$\alpha x = [\alpha x_1, \alpha x_2, \ldots, \alpha x_n]$$

The most useful linear spaces are *function spaces*. For example, the space, $\mathcal{F}(\Omega)$, of all real-valued functions defined on a set Ω is a linear space when fitted out with the operations

$$(f + g)(s) = f(s) + g(s)$$
$$(\alpha f)(s) = \alpha f(s).$$

Taking $\Omega = \{1, 2, \ldots, n\}$ we see that, with obvious notational conventions, Euclidean n-space is a special case of the function space $\mathcal{F}(\Omega)$. A more useful space is the space $C[a, b]$ consisting of all real-valued functions defined on the interval $[a, b]$ that are continuous at each point of $[a, b]$.

Definition. A subset $\mathcal{S}$ of a linear space $\mathcal{V}$ is called a *subspace* of $\mathcal{V}$ if $\alpha x + \beta y \in \mathcal{S}$ for all $x, y \in \mathcal{S}$ and all $\alpha, \beta \in F$.

A subspace is therefore a subset of a linear space which is a linear space in its own right. Applying the definition iteratively, one sees that a subspace $\mathcal{S}$ contains all *linear combinations* of the form

$$\alpha_1 x_1 + \alpha_2 x_2 + \cdots + \alpha_n x_n$$

where n is a positive integer, $\alpha_1, \alpha_2, \ldots, \alpha_n$ are arbitrary scalars and $x_1, x_2, \ldots, x_n$ are arbitrary vectors in $\mathcal{S}$. A set of vectors $\mathcal{W}$ is called *linearly independent* if no vector in $\mathcal{W}$ can be written as a linear combination of other vectors in the set $\mathcal{W}$. One might then say that a linearly independent set contains no "linear redundancies." A *basis* for a subspace $\mathcal{S}$ is a set of linearly independent vectors $\mathcal{B} \subset \mathcal{S}$ with the property that every vector in $\mathcal{S}$ can be written as a linear combination of vectors in $\mathcal{B}$. If a subspace has a basis consisting of finitely many vectors, then the subspace is called *finite-dimensional*. For example, the set of all real polynomials of degree not more than 7 is an eight-dimensional subspace of $C[a, b]$; the set of polynomials $\{1, t, t^2, \ldots, t^7\}$ is a basis for this subspace. On the other hand, the set of all real polynomials is an infinite dimensional subspace of $C[a, b]$.

Subspaces allow a layering of sets of increasingly regular vectors while preserving the linear structure of the space. For example, the set $C^1[a, b]$ of all real-valued functions on $[a, b]$ which also have a continuous derivative is a subspace of $C[a, b]$, while $C_0^2[a, b]$, the space of all twice continuously differentiable real-valued functions on $[a, b]$ that vanish at the endpoints of the interval, is a subspace of $C^1[a, b]$.

The space $AC[a, b]$ of *absolutely continuous* functions is another important subspace of $C[a, b]$. A function f is called absolutely continuous if the sum $\sum |f(b_i) - f(a_i)|$ can be made arbitrarily small for all finite collections of subintervals $\{[a_i, b_i]\}$ whose total length $\sum(b_i - a_i)$ is sufficiently small. The importance of absolutely continuous functions resides in the fact that they are precisely those functions which have integrable (but not necessarily continuous) derivatives on $[a, b]$. Functions in $C^1[a, b]$ are absolutely continuous because such functions have uniformly bounded derivatives. We therefore have the following relationships for these subspaces of $C[a, b]$:

$$C^1[a, b] \subset AC[a, b] \subset C[a, b].$$

A somewhat less specialized notion than a subspace, that of a convex set, is crucial in various applications of functional analysis.

Definition. A subset K of a linear space is called *convex* if $(1 - t)x + ty \in K$ for all $x, y \in K$ and all real numbers t with $0 \le t \le 1$.

Convex sets have a natural geometric interpretation. Given two vectors x and y, the set of vectors of the form $\{(1 - t)x + ty : t \in [0, 1]\}$ is called, in analogy with the corresponding situation in Euclidean space, the *segment* between x and y. A convex set is then a subset of a linear space which contains the segment between any two of its vectors. Of course, every subspace is a convex set.

A. Normed Linear Spaces

Vector spaces become more serviceable when a metric fabric is stretched over the linear framework. Such a metric structure is provided by a function defined on the space which generalizes the concept of length in Euclidean spaces.

Definition. A nonnegative real-valued function $\|\cdot\|$ defined on a linear space $\mathcal{V}$ is called a *norm* if:

(i) $\|x\| = 0$ if and only if $x = \theta$ (the zero vector in $\mathcal{V}$),

(ii) $\|x + y\| \le \|x\| + \|y\|$, for all $x, y \in \mathcal{V}$,

(iii) $\|tx\| = |t| \|x\|$, for all scalars t and all $x \in \mathcal{V}$.

A linear space that comes bundled with a norm is called a *normed linear space*. For specificity, we will sometimes indicate a linear space $\mathcal{V}$ equipped with a norm $\|\cdot\|$ by the pair $(\mathcal{V}, \|\cdot\|)$. For example, $(C[a, b], \|\cdot\|_\infty)$ is a normed linear space under the so-called *uniform norm*

$$\|f\|_\infty = \max\{|f(t)| : t \in [a, b]\}.$$

The space $C[a, b]$ is also a normed linear space when endowed with the norm

$$\|f\|_1 = \int_a^b |f(t)| \, dt$$

and hence $(C[a, b], \|\cdot\|_1)$ is another distinct normed space consisting of the same set of vectors as the normed linear space $(C[a, b], \|\cdot\|_\infty)$. As another example, $C_0^1[a, b]$ is a normed linear space when equipped with the norm $\|f\| = \|f'\|_\infty$, where f' is the derivative of f.

A norm allows the possibility of measuring the distance between two vectors and it also gives meaning to "equality in the limit," that is, to *convergence*.

Definition. A sequence $\{x_n\}$ of vectors is said to *converge* to a vector x, with respect to the norm $\|\cdot\|$, if $\|x_n - x\| \to 0$ as $n \to \infty$.

The notion of convergence is norm-dependent. For example, let l_n denote the "spike" function in $C[0, 1]$ whose graph is obtained by connecting the points $(0, 0)$, $(\frac{1}{2n}, 1)$, $(\frac{1}{n}, 0)$, $(1, 0)$ with straight line segments. The sequence $\{l_n\}$ converges with respect to the $\|\cdot\|_1$ norm to the zero function since $\|l_n\|_1 = \frac{1}{2n}$. However, this sequence does not converge to the zero function in the uniform norm since $\|l_n\|_\infty = 1$. But note that every sequence that converges with respect to the uniform norm also converges with respect to the norm $\|\cdot\|_1$ since $\|f\|_1 \le \|f\|_\infty$ for all $f \in C[0, 1]$. As another example, consider the space $C_0^1[a, b]$ with the norm $\|f\| = \|f'\|_2$. Wirtinger's inequality asserts that if $f \in C_0^1[a, b]$, then

$$\pi^2 \int_a^b |f(t)|^2 \, dt \le (b - a)^2 \int_a^b |f'(t)|^2 \, dt$$

Hence, if $\|\cdot\|_2$ is the norm on $C_0^1[a, b]$ defined by

$$\|f\|_2 = \sqrt{\int_a^b |f(t)|^2 \, dt}$$

then $m\|f\|_2 \le \|f\|$, where $m = \pi/(b - a)$. Therefore, convergence in the norm $\|\cdot\|$ implies convergence in the norm $\|\cdot\|_2$. Wirtinger's inequality can be generalized to functions defined on domains in $\mathbf{R}^n$; the resulting inequality is known as *Poincaré's inequality*.

Two norms $\|\cdot\|_a$ and $\|\cdot\|_b$ on a linear space $\mathcal{V}$ are called *equivalent* if there are positive constants c_1 and c_2 such that

$$c_1\|x\|_a \le \|x\|_b \le c_2\|x\|_a$$

for all $x \in \mathcal{V}$. Equivalence therefore means equivalent in the sense of convergence: a sequence converges with respect to the norm $\|\cdot\|_a$ if and only if it converges with respect to the norm $\|\cdot\|_b$. It can be shown that any two norms on a finite-dimensional linear space are equivalent and hence when speaking of convergence in a finite-dimensional space one can dispense with mentioning the norm. However, as we have seen above, convergence is norm-dependent in infinite dimensional spaces.

Definition. The *closure*, $\overline{W}$, of a subset W of a normed linear space $(\mathcal{V}, \|\cdot\|)$ is the set consisting of elements of W along with all limits of convergent sequences in W. A set which is its own closure is called *closed*.

For example, the closure of the set of all polynomials in the space $(C[a, b], \|\cdot\|_\infty)$ is, by virtue of the *Weierstrass approximation theorem*, the space $C[a, b]$ itself. Also, for example, the set of nonnegative functions in $C[a, b]$ is a closed subset of $C[a, b]$.

B. Banach Spaces

Certain sequences of vectors in a normed linear space tend to "bunch up" in a way that imitates convergent sequences. Such sequences are named after the nineteenth century mathematician Augustin Cauchy.

Definition. A sequence of vectors $\{x_n\}$ in a normed linear space $(\mathcal{V}, \|\cdot\|)$ is called a *Cauchy* sequence if $\lim_{n,m\to\infty} \|x_n - x_m\| = 0$.

It is easy to see that every convergent sequence is Cauchy, however, it is not necessarily the case that a Cauchy sequence is convergent. Consider, for example, the "ramp" function h_n in $C[-1, 1]$ whose graph consists of the straight line segments connecting the points $(-1, 0), (-\frac{1}{n}, 0), (0, 1), (1, 1)$. The sequence of ramp functions is Cauchy with respect to the norm $\|\cdot\|_1$ since

$$\|h_n - h_m\|_1 = \frac{1}{2}\left|\frac{1}{n} - \frac{1}{m}\right| \to 0, \quad \text{as} \quad n, m \to \infty.$$

However, $\{h_n\}$ does not converge, with respect to the norm $\|\cdot\|_1$, to a function in $C[-1, 1]$ (the $\|\cdot\|_1$ limit of this sequence of ramp functions is the discontinuous Heaviside function). Spaces, such as $(C[-1, 1], \|\cdot\|_1)$, which do not accommodate limits of Cauchy sequences are in some sense "incomplete." A normed linear space is called *complete* if every Cauchy sequence in the space converges to some vector in the space.

Definition. A *Banach space* is a complete normed linear space.

We have just seen that $C[a, b]$ with the norm $\|\cdot\|_1$ is not a Banach space. However, $C[a, b]$ with the uniform norm is a Banach space. The Lebesgue spaces are particularly important Banach spaces. For $1 \le p < \infty$ the space $L^p[a, b]$ consists of measurable real-valued functions f defined on $[a, b]$ such that $|f|^p$ has a finite Lebesgue integral. The norm $\|\cdot\|_p$ on $L^p[a, b]$ is defined by

$$\|f\|_p = \left\{\int_a^b |f(t)|^p \, dt\right\}^{\frac{1}{p}}.$$

Every normed linear space can be imbedded as a dense subspace of a Banach space by an abstract process of "completion." Essentially, the completion process involves adjoining to the original space, for each Cauchy sequence in the original space, a vector in an extended space (the *completion*) which is the limit in the extended space of the Cauchy sequence. For example, it can be shown that the completion of the space $(C[a, b], \|\cdot\|_1)$ is the space $(L^1[a, b], \|\cdot\|_1)$.

C. Hilbert Spaces

Perpendicularity, the key ingredient in the Pythagorean theorem, is one of the fundamental concepts in geometry. A successful geometrization of analysis requires the incorporation of this concept. The essential aspects of perpendicularity are captured in special normed linear spaces called inner product spaces.

Definition. An *inner product* on a real linear space $\mathcal{V}$ is a function $\langle \cdot, \cdot \rangle : \mathcal{V} \times \mathcal{V} \to \mathbf{R}$ which satisfies:

(i) $\langle x, x \rangle \ge 0$ for all $x \in \mathcal{V}$,

(ii) $\langle x, x \rangle = 0$ if and only if $x = \theta$ (the zero vector),

(iii) $\langle x, y \rangle = \langle y, x \rangle$ for all $x, y \in \mathcal{V}$,

(iv) $\langle tx, y \rangle = t \langle x, y \rangle$ for all $x, y \in \mathcal{V}$ and all $t \in \mathbf{R}$,

(v) $\langle x + y, z \rangle = \langle x, z \rangle + \langle y, z \rangle$, for all $x, y, z \in \mathcal{V}$.

Properties (iii)–(v) are summarized by saying that $\langle \cdot, \cdot \rangle$ is a symmetric bilinear form; properties (i) and (ii) say that $\langle \cdot, \cdot \rangle$ is nonnegative and definite. A linear space endowed with an inner product is called an *inner product space*.

The most familiar inner product space is the Euclidean space $\mathbf{R}^n$ with the inner product

$$\langle x, y \rangle = x_1 y_1 + x_2 y_2 + \cdots + x_n y_n.$$

Of course other inner products may be used on the same underlying linear space. For example, in statistics, the Euclidean space is often used with a *weighted* inner product

$$\langle x, y \rangle = w_1 x_1 y_1 + w_2 x_2 y_2 + \cdots + w_n x_n y_n$$

where $w_1, w_2, \ldots, w_n$ are fixed positive weights.

Function spaces serve up a particularly rich stew of inner product spaces. For example, the theory of Fourier series can be developed in the Lebesgue space $L^2[a, b]$ with the inner product

$$\langle f, g \rangle = \int_a^b f(t)g(t) \, dt$$

and many variations are possible. For instance, the space $C^1[0, 1]$ with

$$\langle f, g \rangle = \int_0^1 f'(t)g'(t)\,dt + f(0)g(0)$$

is an inner product space.

In the case of a linear space over the field $\mathbf{C}$ of complex scalars the definition of an inner product requires some modification. Here the inner product is a complex valued function that satisfies the properties above, save that (iii) is replaced by

$$\langle x, y \rangle = \overline{\langle y, x \rangle},$$

where the bar indicates complex conjugation, and (iv) is required to hold for all $t \in \mathbf{C}$. Note that this implies that $\langle x, ty \rangle = \bar{t}\langle x, y \rangle$.

Every inner product on a linear space generates a corresponding norm by way of the definition $\|x\| = \sqrt{\langle x, x \rangle}$. The proof that this defines a norm relies on the *Cauchy-Schwarz inequality*:

$$|\langle x, y \rangle|^2 \le \langle x, x \rangle \langle y, y \rangle.$$

Any norm induced by an inner product must satisfy, by virtue of the bilinearity of the inner product, the *Parallelogram Law*:

$$\|x + y\|^2 + \|x - y\|^2 = 2\|x\|^2 + 2\|y\|^2.$$

Therefore, any norm which does not satisfy the parallelogram law is not induced by an inner product. For example, the space $(C[-1, 1], \|\cdot\|_\infty)$ is not an inner product space since the parallelogram law for $\|\cdot\|_\infty$ fails for the continuous ramp functions h_1 and h_2 introduced above.

Definition. Two vectors x and y in an inner product space are called *orthogonal*, denoted $x \perp y$, if $\langle x, y \rangle = 0$. A subset S of an inner product space is called an orthogonal set if $x \perp y$ for each distinct pair of vectors $x, y \in S$.

From the properties of the inner product it follows immediately that

$$x \perp y \quad \text{if and only if} \quad \|x + y\|^2 = \|x\|^2 + \|y\|^2.$$

This extension of the classical Pythagorean theorem to inner product spaces is what justifies our association of orthogonality, a purely *algebraic* concept, with the geometrical notion of perpendicularity.

Given a subset S of an inner product space $\mathcal{V}$, the *orthogonal complement* of S is the closed subspace

$$S^\perp = \{y \in \mathcal{V} : \langle x, y \rangle = 0 \quad \text{for all} \quad x \in S\}.$$

For example, consider the space $C^1[0, 1]$ of continuously differentiable real functions on $[0, 1]$ equipped with the inner product

$$\langle f, g \rangle = \int_0^1 f'(t)g'(t)\,dt + f(0)g(0).$$

Let S be the set of linear functions. Then $g \in S^\perp$ if and only if

$$0 = \langle 1, g \rangle = g(0) \quad \text{and} \quad 0 = \langle t, g \rangle = \int_0^1 g'(t)\,dt$$
$$= g(1) - g(0).$$

Therefore, $S^\perp = C_0^1[0, 1]$, the space of continuously differentiable functions which vanish at the end points of $[0, 1]$. If $\mathcal{W}$ is a subspace, then its second orthogonal complement is its closure: $\mathcal{W}^{\perp\perp} = \overline{W}$.

Definition. A *Hilbert space* is a complete inner product space.

The best known Hilbert space is the space $L^2[a, b]$. The fact that this space is complete is known as the *Riesz-Fischer Theorem*. The *Sobolev spaces* are important Hilbert spaces used in the theory of differential equations. Let Ω be a bounded domain in $\mathbf{R}^n$. $C^m(\Omega)$ denotes the space of all continuous real-valued functions on $\overline{\Omega}$ having bounded continuous partial derivatives through order m. An inner product $\langle \cdot, \cdot \rangle_m$ is defined on $C^m(\Omega)$ by

$$\langle f, g \rangle_m = \sum_{|\alpha| \le m} \langle D^\alpha f, D^\alpha g \rangle$$

where $\langle \cdot, \cdot \rangle$ denotes the usual L^2 inner product, $\alpha = (\alpha_1, \alpha_2, \ldots, \alpha_n)$ is a multi-index of nonnegative integers, $|\alpha| = \alpha_1 + \cdots + \alpha_n$, and D^α is the differential operator

$$D^\alpha f = \frac{\partial^{|\alpha|}}{\partial x_n^{\alpha_n} \cdots \partial x_1^{\alpha_1}} f.$$

The Sobolev space $H^m(\Omega)$ is the completion of $C^m(\Omega)$ with respect to the norm $\|\cdot\|_m$ generated by the inner product $\langle \cdot, \cdot \rangle_m$. A closely associated Sobolev space is the space $H_0^m(\Omega)$ which is formed in the same way, but using as base space the space $C_0^m(\Omega)$ of all functions in $C^m(\Omega)$ which vanish off of some closed bounded subset of Ω (that is, the space of all *compactly supported* functions in $C^m(\Omega)$). The advantage of using such spaces in the study of differential equations lies in the fact that in these spaces ordinary functions are very smooth, convergence is very strict (convergence in $H^m(\Omega)$ implies uniform convergence on compact subsets of all derivatives through order $m - 1$), and Sobolev spaces are *complete*, allowing the deployment of all the weapons of Hilbert space theory. In particular, Hilbert space theory can be used to develop a rich theory of *weak solutions* of partial differential equations and a corresponding theory of finite element approximations to weak solutions.

An orthogonal set of unit vectors (i.e., vectors of norm one) is called an *orthonormal set*. If S is an orthonormal set in a Hilbert space H, then for any $y \in H$ at most countably many of the numbers $\langle y, x \rangle$, where $x \in S$, are nonzero and

$$\sum_{x \in S} |\langle y, x \rangle|^2 \leq \|y\|^2$$

(this is called *Bessel's inequality*). The numbers $\{\langle y, x \rangle : x \in S\}$ are called the *Fourier coefficients* of y with respect to the orthonormal set S. If the orthonormal set S has the property that the Fourier coefficients of a vector completely determine the vector in the sense that if, for any vector $y \in H$

$$\langle y, x \rangle = 0 \quad \text{for all } x \in S \quad \text{implies} \quad y = \theta$$

then S is called a *complete orthonormal set*.

Definition. A Hilbert space is called *separable* if it contains a sequence which is a complete orthonormal set. Any such complete orthonormal sequence is called a *basis* for the Hilbert space.

It is not hard to see that an orthonormal sequence $\{\phi_n\}$ in a Hilbert space H is complete if and only if

$$\sum_n |\langle y, \phi_n \rangle|^2 = \|y\|^2$$

for each $y \in H$ (*Parseval's identity*). This is equivalent to the assertion

$$\sum_n \langle y, \phi_n \rangle \phi_n = y$$

where the convergence of the series is understood in the sense of the norm generated by the inner product $\langle \cdot, \cdot \rangle$.

For example, the complex space $L^2[-\pi, \pi]$ with inner product

$$\langle f, g \rangle = \int_{-\pi}^{\pi} f(t)\overline{g(t)}\, dt$$

is a separable Hilbert space with complete orthonormal sequence

$$\phi_n(t) = \frac{1}{\sqrt{2\pi}} e^{int}, \qquad , n = 0, \pm 1, \pm 2, \ldots.$$

Note that these orthonormal vectors are integer dilates of a single complex wave, that is,

$$\phi_n(t) = \phi(nt),$$

where

$$\phi(t) = \frac{1}{\sqrt{2\pi}} e^{it} = \frac{1}{\sqrt{2\pi}} (\cos t + i \sin t).$$

None of the functions ϕ_n just defined lies in the space $L^2(-\infty, \infty)$ because the waves do not "die out" as $|t| \to \infty$. Special bases for $L^2(-\infty, \infty)$ that have particularly attractive decay properties, called *wavelet* bases,

have been studied and applied extensively in recent years. We now show how to construct the simplest wavelet basis, the *Haar basis*. As in the previous example, we seek a single function to generate all the basis vectors, but instead of a wave, we choose a function that decays quickly, in fact a function that vanishes off an interval. Our choice is the Haar "wavelet"

$$\psi(t) = \begin{cases} 1 & \text{for} & 0 \leq t < \frac{1}{2} \\ -1 & \text{for} & \frac{1}{2} \leq t < 1 \\ 0 & \text{otherwise} \end{cases}$$

Taking dilates of this wavelet will serve the same purpose as the higher frequency waves in the previous example, that is, the dilates of the fundamental wave ϕ. However, this by itself will not serve to represent functions defined over the entire line. To do that we must also shift the dilates about. If this is done properly, then all the needed time and frequency attributes are captured and the resulting wavelets are orthonormal. Specifically, it can be shown that the functions $\{\psi_{j,k} : j, k = 0, \pm 1, \pm 2, \ldots\}$ defined by

$$\psi_{j,k}(t) = 2^{\frac{j}{2}} \psi(2^j t - k)$$

form an orthonormal basis for $L^2(-\infty, \infty)$ called the *Haar basis*.

II. LINEAR OPERATORS

Mappings between linear spaces that preserve the linear structure are called linear operators.

Definition. A mapping T defined on a linear space $\mathcal{V}$ and taking values in a linear space $\mathcal{W}$ is called *linear* if

$$T(x + y) = Tx + Ty, \quad \text{for all} \quad x, y \in \mathcal{V}$$

and

$$T(tx) = tTx, \quad \text{for all} \quad x \in \mathcal{V} \quad \text{and all scalars} \quad t.$$

The space $\mathcal{V}$ on which a linear operator T is defined is called the *domain* of T and denoted $\mathcal{D}(T)$. In finite-dimensional spaces linear operators have matrix representations relative to specific ordered bases. For example, the differentiation operator acting on the space of polynomials of degree not greater that n has, relative to the basis $\{1, t, \ldots, t^n\}$, the $(n+1) \times (n+1)$ matrix representation

$$\begin{bmatrix} 0 & 1 & 0 & \ldots & 0 \\ 0 & 0 & 2 & \ldots & 0 \\ \cdot & & & & \cdot \\ \cdot & & & & \cdot \\ \cdot & & & & \cdot \\ 0 & 0 & 0 & \ldots & n \\ 0 & 0 & 0 & \ldots & 0 \end{bmatrix}.$$

In functional analysis the primary interest is the study of linear operators on infinite dimensional function spaces that are defined intrinsically, that is, without regard to a specific basis. For example, given a real-valued function $k(\cdot, \cdot)$ which is continuous on the square $[0, 1] \times [0, 1]$ one can define the linear integral operator T on $C[0, 1]$, taking values in $C[0, 1]$, by $Tf = g$ where

$$g(s) = \int_0^1 k(s, t) f(t) \, dt.$$

Such integral operators may be viewed as natural generalizations of matrices to the case of continuous, rather than discrete, variables.

Two important subspaces, the *nullspace* and *range*, are associated with each linear operator. The nullspace, $N(T)$, of a linear operator T consists of all vectors that T maps to the zero vector: $N(T) = \{x \in \mathcal{D}(T): Tx = \theta\}$. For example, the differentiation operator acting on the space $C^1[a, b]$ has nullspace consisting of all constant functions. The range, $R(T)$, of a linear operator T is the set of all images of vectors under T, that is, $R(T) = \{Tx: x \in \mathcal{D}(T)\}$. For example, the range of the integral operator on $C[0, 1]$ generated by the kernel $k(s, t) = \exp(s + t)$ consists of all scalar multiples of the exponential function $f(s) = \exp(s)$.

A. Bounded Operators

Linear operators acting between normed linear spaces that are continuous with respect to the norms are called bounded.

Definition. A linear operator T defined on a normed linear space $(\mathcal{V}, \|\cdot\|_v)$ and taking values in a normed linear space $(\mathcal{W}, \|\cdot\|_w)$ is called *bounded* if there is a constant L such that $\|Tx\|_w \leq L \|x\|_v$ for all $x \in \mathcal{V}$.

The notational specification of norms on various linear spaces is tiresome and annoying; we will therefore often dispense with it and rely on the reader to understand appropriate norms from the context. Hence we will say that a linear operator T is bounded if $\|Tx\| \leq L \|x\|$ for all $x \in \mathcal{D}(T)$ and some constant L. The smallest constant L for which this inequality holds is called the *norm* of the operator T and is denoted $\|T\|$. Equivalently,

$$\|T\| = \sup_{x \neq \theta} \frac{\|Tx\|}{\|x\|}$$

where "sup" stands for supremum, or least upper bound. Using linearity of T one sees that

$$\|Tx - Ty\| \leq \|T\| \|x - y\|$$

and hence $\|T\|$ gives the smallest universal bound for the relative "spread" that a bounded linear operator T can accomplish. From this it also follows that every bounded linear operator is continuous (relative to the norms in question) and, in fact, every continuous linear operator is bounded. Indeed, if T is linear and continuous, then there is a $\delta > 0$ such that $\|z\| \leq \delta$ implies $\|Tz\| \leq 1$. For any $x \neq \theta$ one then has $\|T(\delta x / \|x\|)\| \leq 1$, that is, $\|Tx\| \leq \frac{1}{\delta} \|x\|$ for all x. Therefore, T is bounded and $\|T\| \leq 1/\delta$.

The set $L(\mathcal{V}, \mathcal{W})$ of all bounded linear operators from a normed linear space $\mathcal{V}$ to a normed linear space $\mathcal{W}$ is itself a normed linear space under the natural notions of operator sum and scalar multiplication:

$$(T + S)(x) = Tx + Sx$$

$$(tT)(x) = t(Tx)$$

and with the norm on bounded linear operators as defined above. Further, if $\mathcal{W}$ is a Banach space, then so is $L(\mathcal{V}, \mathcal{W})$. If the image space $\mathcal{W}$ is the scalar field (**R**, or **C**), then $L(\mathcal{V}, \mathcal{W})$ is called the *dual space*, $\mathcal{V}^*$, of $\mathcal{V}$. An operator in $\mathcal{V}^*$ is called a *bounded linear functional* on $\mathcal{V}$. For example, given a fixed $t_0 \in [a, b]$, the point evaluation operator E_{t_0}, defined by

$$E_{t_0}(f) = f(t_0)$$

is a bounded linear functional on the space $(C[a, b], \|\cdot\|_\infty)$. The dual space of $C[a, b]$ may be identified with the space of all functions of bounded variation on $[a, b]$ in the sense that any bounded linear functional ℓ on $C[a, b]$ has a unique representation of the form

$$\ell(f) = \int_a^b f(t) \, dg(t)$$

for some function g of bounded variation (the integral is a Riemann-Stieljes integral). In the case of point evaluation the representative function g is the Heaviside function at t_0:

$$g(t) = \begin{cases} 0 & , \quad t < t_0 \\ 1 & , \quad t_0 \leq t. \end{cases}$$

It can be shown that for $1 < p < \infty$,

$$(L^p[a, b])^* = L^q[a, b]$$

where $\frac{1}{p} + \frac{1}{q} = 1$, in the sense that every bounded linear functional ℓ on $L^p[a, b]$ has the form

$$\ell(f) = \int_a^b f(t) g(t) \, dt$$

for some $g \in L^q[a, b]$ that is uniquely determined by ℓ. With this understanding the space $L^2[a, b]$ is self-dual.

In fact, by the *Riesz Representation Theorem*, all Hilbert spaces are self-dual for the following reason: a bounded linear functional ℓ on a Hilbert space H has the form $\ell(x) = \langle x, z \rangle$, for some vector $z \in H$ which is uniquely determined by ℓ. Further, $\|\ell\| = \|z\|$, where the first norm is an operator norm and the other is the underlying Hilbert space norm. The association of the bounded linear functional ℓ with its Riesz representor z provides the identification of H^* with H.

Bounded linear functionals may be thought of as linear measurements on a Hilbert space in that a bounded linear functional gives a numerical measure on vectors in the space which is continuous with respect to the norm. Such measures distinguish vectors in the space because $x \neq y$ if and only if there is a vector z with $\langle x, z \rangle \neq \langle y, z \rangle$. However, it may happen that no such linear measure can ultimately distinguish the vectors in a sequence from a certain fixed vector. This gives rise to the notion of weak convergence.

Definition. A sequence $\{x_n\}$ in a Hilbert space H is said to converge *weakly* to a vector x, denoted $x_n \rightharpoonup x$, if $\langle x_n, z \rangle \to \langle x, z \rangle$ for all $z \in H$.

It is a consequence of the Cauchy-Schwarz inequality that every convergent sequence is weakly convergent to the same limit. Also, Bessel's inequality shows that every orthonormal sequence of vectors converges weakly to the zero vector.

B. Adjoint Operators

Adjoint operators mimic the behavior of the transpose matrix on real Euclidean space. Recall that the transpose A^T of a real $m \times n$ matrix A satisfies

$$\langle Ax, y \rangle = \langle x, A^T y \rangle$$

for all $x \in \mathbf{R}^n$ and $y \in \mathbf{R}^m$, where $\langle \cdot, \cdot \rangle$ is the Euclidean inner product. If T is a bounded linear operator from a Hilbert space H_1 into a Hilbert space H_2, i.e., $T : H_1 \to H_2$, then for fixed $y \in H_2$ the linear functional ℓ defined on H_1 by

$$\ell(x) = \langle Tx, y \rangle$$

is bounded and hence by the Riesz Representation Theorem

$$\langle Tx, y \rangle = \langle x, z \rangle$$

for some $z \in H_1$. This z is uniquely determined by y, via T, and we denote it by T^*y, that is,

$$\langle Tx, y \rangle = \langle x, T^*y \rangle.$$

The operator $T^* : H_2 \to H_1$ is a bounded linear operator called the *adjoint* of T. If T is a bounded linear operator, then $\|T\| = \|T^*\|$ and $T^{**} = T$.

Suppose, for example, the linear operator $T : L^2[a, b] \to L^2[c, d]$ is generated by the kernel $k(\cdot, \cdot) \in C([c, d] \times [a, b])$, that is,

$$(Tf)(s) = \int_a^b k(s, t) f(t) \, dt, \qquad s \in [c, d],$$

then

$$\langle Tf, g \rangle = \int_c^d \int_a^b k(s, t) f(t) \, dt \, g(s) \, ds$$
$$= \int_a^b \int_c^d k(s, t) g(s) \, ds \, f(t) \, dt = \langle f, T^*g \rangle$$

and hence T^* is the integral operator generated by the kernel $k^*(\cdot, \cdot)$ defined by $k^*(t, s) = k(s, t)$. In particular, if the kernel k is symmetric and $[a, b] = [c, d]$, then the operator T is self-adjoint.

C. Compact Operators

A linear operator $T : H_1 \to H_2$ is called an operator of *finite rank* if its range is spanned by finitely many vectors in H_2. In other words, T has finite rank if there are linearly independent vectors $\{v_1, \ldots, v_m\}$ in H_2 such that

$$Tx = \sum_{k=1}^m a_k(x) v_k$$

where the coefficients a_k are bounded linear functionals on H_1. Therefore, by the Riesz Representation Theorem,

$$Tx = \sum_{k=1}^m \langle x, u_k \rangle v_k$$

for certain vectors $\{u_1, \ldots, u_m\} \subset H_1$. Note that every finite rank operator is continuous, but more is true. If $\{x_n\}$ is a weakly convergent sequence in H_1, say $x_n \rightharpoonup w$, then

$$Tx_n = \sum_{k=1}^m \langle x_n, u_k \rangle v_k \to \sum_{k=1}^m \langle w, u_k \rangle v_k = Tw,$$

and hence a finite rank operator T maps weakly convergent sequences into strongly convergent sequences. Finite rank operators are a special case of an important class of linear operators that enjoy this weak-to-strong continuity property.

Definition. A linear operator $T : H_1 \to H_2$ is called *compact* (also called *completely continuous*) if $x_n \rightharpoonup w$ implies $Tx_n \to Tw$.

Note that every finite rank operator is completely continuous and every completely continuous operator is *a fortiori* continuous. Also, limits of finite rank operators are compact. More precisely, if $\{T_n\}$ is a sequence of finite rank operators converging in operator norm to an operator T, then T is compact.

Completely continuous operators acting on a Hilbert space have a particularly simple structure expressed in terms of certain characteristic subpaces known as eigenspaces. The *eigenspace* of a linear operator $T : H \to H$ associated with a scalar λ is the subspace

$$N(T - \lambda I) = \{x \in H : Tx = \lambda x\}.$$

In general, an eigenspace may be trivial, that is, it may consist only of the zero vector. If $N(T - \lambda I) \neq \{\theta\}$, we say that λ is an *eigenvalue* of T. If T is self-adjoint, then the eigenvalues of T are real numbers and vectors in distinct eigenspaces are orthogonal to each other. If T is self-adjoint, compact, and of infinite rank, then the eigenvalues of T form a sequence of real numbers $\{\lambda_n\}$. This sequence converges to zero, for taking an orthonormal sequence $\{x_n\}$ with $x_n \in N(T - \lambda_n I)$ we have $\lambda_n x_n = Tx_n \to \theta$ since $x_n \rightharpoonup \theta$ (a consequence of Bessel's inequality). Since $\|x_n\| = 1$, it follows that $\lambda_n \to 0$. The fact that $Tx_n \to 0$ for a sequence of unit vectors $\{x_n\}$ is an abstract version of the *Riemann-Lebesgue Theorem*.

The prime exemplar of a compact self-adjoint operator is the integral operator on the real space $L^2[a, b]$ generated by a symmetric kernel $k(\cdot, \cdot) \in L^2([a, b] \times [a, b])$:

$$(Tf)(s) = \int_a^b k(s, t) f(t) \, dt.$$

This operator is compact because it is the limit in operator norm of the finite rank operators

$$T_N f = \sum_{n,m \leq N} c_{n,m} \int_a^b f(t) \phi_m(t) \, dt \, \phi_n$$

where $\{\phi_n\}_1^\infty$ is a complete orthonormal sequence in $L^2[a, b]$ and

$$k(s, t) = \sum_{n,m=1}^\infty c_{n,m} \phi_n(s) \phi_m(t)$$

is the Fourier expansion of $k(\cdot, \cdot)$ relative to the orthonormal basis $\{\phi_n(s)\phi_m(t)\}$ for $L^2([a, b] \times [a, b])$.

D. Unbounded Operators

It is not the case that every interesting linear operator is bounded. For example, the differentiation operator acting on the space $C^1[0, \pi]$ and taking values in the space $C[0, \pi]$, both with the uniform norm, is unbounded. Indeed, for the functions $\phi_n(t) = \sin(nt)$ we find that the quotient

$$\frac{\|\phi_n'\|_\infty}{\|\phi_n\|_\infty} = n$$

is unbounded.

Multiplication by the variable in the space $L^2(-\infty, \infty)$ is another famous example of an unbounded operator. Let

$$\mathcal{D}(M) = \left\{ f \in L^2(-\infty, \infty) : \int_{-\infty}^\infty t^2 |f(t)|^2 \, dt < \infty \right\}.$$

Then $\mathcal{D}(M)$ is a dense subspace of $L^2(-\infty, \infty)$ and one can define the linear operator $M : \mathcal{D}(M) \to L^2(-\infty, \infty)$ by

$$Mf = g \quad \text{where} \quad g(t) = tf(t).$$

Let $\phi_n(t) = 1$ for $t \in [n, n+1]$ and $\phi_n(t) = 0$ otherwise. Then $\|\phi_n\|_2 = 1$ and $\|M\phi_n\|_2 > n$, and hence M is unbounded. The multiplication operator has an important interpretation in quantum mechanics.

The adjoint operator may also be defined for unbounded linear operators with dense domains. Given a linear operator $T : \mathcal{D}(T) \subset H_1 \to H_2$ with dense domain $\mathcal{D}(T)$, let $\mathcal{D}(T^*)$ be the subspace of all vectors $y \in H_2$ satisfying

$$\langle Tx, y \rangle = \langle x, y^* \rangle$$

for some vector $y^* \in H_1$ and all $x \in \mathcal{D}(T)$. The vector y^* is then uniquely defined and we set $T^* y = y^*$. Then the operator $T^* : \mathcal{D}(T^*) \subset H_2 \to H_1$ is densely defined and linear.

As an example, let $\mathcal{D}(T)$ be the space of absolutely continuous complex-valued functions f defined on $[0, 1]$ with $f' \in L^2[0, 1]$ satisfying the periodic boundary condition $f(0) = f(1)$. Then $\mathcal{D}(T)$ is dense in $L^2[0, 1]$. Define $T : \mathcal{D}(T) \to L^2[0, 1]$ by $Tf = if'$. For $g \in \mathcal{D}(T)$ we have

$$\langle Tf, g \rangle = \int_0^1 if'(t)\overline{g(t)} \, dt$$

$$= i\overline{g(t)}f(t) \Big|_0^1 - i \int_0^1 f(t)\overline{g'(t)} \, dt$$

$$= \int_0^1 f(t)\overline{ig'(t)} \, dt = \langle f, ig' \rangle$$

for all $f \in \mathcal{D}(T)$. Therefore, $\mathcal{D}(T) \subset \mathcal{D}(T^*)$ and, in fact, it can be shown that $\mathcal{D}(T^*) = \mathcal{D}(T)$. This calculation shows that $T^* g = Tg$, that is, T is self-adjoint.

A linear operator $T : \mathcal{D}(T) \subseteq H_1 \to H_2$ is called *closed* if its graph $\mathcal{G}(T) = \{(x, Tx) : x \in \mathcal{D}(T)\}$ is a closed subspace of the product Hilbert space $H_1 \times H_2$. This means that if $\{x_n\} \subset \mathcal{D}(T)$, $x_n \to x \in H_1$, and $Tx_n \to y \in H_2$, then $(x, y) \in \mathcal{G}(T)$, that is, $x \in \mathcal{D}(T)$ and $Tx = y$. For example, the differentiation operator defined in the previous paragraph is closed. In fact, the adjoint of any densely defined linear operator is closed.

A densely defined linear operator $T : \mathcal{D}(T) \subseteq H \to H$ is called *symmetric* if

$$\langle Tx, y \rangle = \langle x, Ty \rangle \quad \text{for all} \quad x, y \in \mathcal{D}(T).$$

Every self-adjoint transformation is, of course, symmetric; however, a symmetric transformation is not necessarily self-adjoint. Consider, for instance, a slight modification of the previous example. Let $\mathcal{D}(T)$ be the space of absolutely continuous complex-valued functions on $[0, 1]$ which vanish at the end points, and let $Tf = if'$. For

$f, g \in \mathcal{D}(T)$, integration by parts gives $\langle Tf, g \rangle = \langle f, Tg \rangle$, and hence T is symmetric, and the adjoint of T satisfies $T^*g = if'$. However, $\mathcal{D}(T^*)$ is a proper extension of $\mathcal{D}(T)$, in that no boundary conditions are imposed on functions in $\mathcal{D}(T^*)$, and hence T is not self-adjoint. The examples just given show that a symmetric linear operator is not necessarily bounded. The *Hellinger-Toeplitz* Theorem gives sufficient conditions for a symmetric operator to be bounded: a symmetric linear operator whose domain is the entire space is bounded.

If a linear operator $T : \mathcal{D}(T) \subseteq H_1 \to H_2$ is closed, then $\mathcal{D}(T)$ is a Hilbert space when endowed with the *graph inner product*:

$$\langle (x, Tx), (y, Ty) \rangle = \langle x, y \rangle + \langle Tx, Ty \rangle.$$

If T is closed and everywhere defined, i.e., $\mathcal{D}(T) = H_1$, then since the graph norm dominates the norm on H_1, we find, by the corollary to Banach's theorem (see the *inversion* section), that the norm in H_1 is equivalent to the graph norm. In particular, the operator T is then bounded. This is the *closed graph theorem*: a closed everywhere defined linear operator is bounded.

III. CONTRACTIONS

Suppose X is a Banach space and $D \subseteq X$. A vector $x \in D$ is called a *fixed point* of the mapping $T : D \to X$ if $Tx = x$. Every linear operator has a fixed point, namely the zero vector. But a nonlinear mapping may be free of fixed points. However, a mapping that draws points together in a uniform relative sense (a condition called contractivity) is guaranteed to have a fixed point.

Definition. A mapping $T : D \subseteq X \to X$ is called a *contraction* (relative to the norm $\|\cdot\|$ on X) if, $\|Tx - Ty\| \le \alpha \|x - y\|$, for all $x, y \in D$ and some positive constant $\alpha < 1$.

For example, if $L : X \to X$ is a bounded linear operator with $\|L\| < 1$, and $g \in X$, then the (*affine*) mapping $T : X \to X$ defined by $Tx = Lx + g$ is a contraction with contraction constant $\alpha = \|L\|$. As another example, consider the nonlinear integral operator $T : C[0, 1] \to C[0, 1]$ defined by

$$(Tu)(s) = \int_0^1 k(s, u(t)) \, dt$$

where $k(\cdot, \cdot) \in C^1([0, 1] \times \mathbf{R})$ is a given kernel. If the kernel satisfies

$$\left| \frac{\partial}{\partial t} k(s, t) \right| \le \alpha, \quad \text{for all} \quad s, t$$

then

$$|(Tw)(s) - (Tu)(s)| = \left| \int_0^1 k(s, w(t)) - k(s, u(t)) \, dt \right|$$

$$\le \alpha \int_0^1 |w(t) - u(t)| \, dt \le \alpha \|w - u\|.$$

Therefore, $\|Tw - Tu\| \le \alpha \|w - u\|$ and hence T is a contraction for the uniform norm if $\alpha < 1$.

The *Contraction Mapping Theorem* (elucidated by Banach in 1922) is a constructive existence and uniqueness theorem for fixed points. If D is a closed subset of a Banach space and $T : D \to D$ is a contraction, then the theorem guarantees the existence of a unique fixed point $x \in D$. This fixed point is the limit of any sequence constructed iteratively by $x_{n+1} = Tx_n$, where x_0 is an arbitrary vector in D. If α is a contraction constant for T, then there is an a priori error bound

$$\|x_n - x\| \le \frac{\alpha^n}{1 - \alpha} \|x_1 - x_0\|$$

and an *a posteriori* error bound

$$\|x_n - x\| \le \frac{1}{1 - \alpha} \|x_{n+1} - x_n\|$$

for x_n as an approximation to the unique fixed point x.

The contraction mapping theorem is often used to establish the existence and uniqueness of solutions of problems in function spaces. One such application is a simple *implicit function theorem*. Suppose $f \in C([a, b] \times \mathbf{R})$ satisfies

$$0 < m \le \left| \frac{\partial f}{\partial s}(t, s) \right| \le M$$

for some constants m and M. Then one can show that the equation $f(t, x) = 0$ implicitly defines a continuous function x on $[a, b]$. Indeed, the nonlinear operator $T : C[a, b] \to C[a, b]$ defined by

$$(Tx)(t) = x(t) - \frac{2}{m + M} f(t, x(t))$$

is, under the stated conditions, a contraction mapping on $C[a, b]$ with contraction constant $\alpha = (M - m)/(M + m)$. Therefore, T has a unique fixed point $x \in C[a, b]$. That is, there is a unique function $x \in C[a, b]$ satisfying

$$x(t) = x(t) - \frac{2}{m + M} f(t, x(t))$$

or, equivalently $f(t, x(t)) = 0$.

IV. SOME PRINCIPLES AND TECHNIQUES

A. Projection and Decomposition

The *projection property* is a key feature of the geometry of Hilbert space: a closed convex subset of a Hilbert space contains a unique vector of smallest norm. By shifting the origin to a vector x, one sees that this is equivalent to saying that given a closed convex subset S of a Hilbert space H and a vector $x \in H$, there is a unique vector $Px \in S$ satisfying

$$\|x - Px\| = \min_{y \in S} \|x - y\|.$$

This purely geometric projection property has important applications in optimization theory. For example, consider the following simple example of optimal control of a one-dimensional dynamical system. Suppose a unit point mass is steered from the origin with initial velocity 1 by a control (external force) u. We are interested in a control that will return the particle to a "soft landing" at the origin in unit time while expending minimal effort, where the measure of effort is

$$\int_0^1 |u(t)|^2 \, dt.$$

We may formulate this problem in the Hilbert space $L^2[0, 1]$. The dynamics of the system are governed by the equations

$$\ddot{x} = u, \quad x(0) = 0, \quad \dot{x}(0) = 1, \quad x(1) = 0, \quad \dot{x}(1) = 0.$$

Suppose C is the set of all vectors u in $L^2[0, 1]$ for which the equations above are satisfied for some vector $x \in H^2[0, 1]$. It may be routinely verified that C is a closed convex subset of $L^2[0, 1]$ and hence C contains a unique vector of smallest L^2-norm, i.e., there is a unique minimal effort control that steers the system in the specified manner.

The (generally nonlinear) operator P defined above is called the (metric) *projection* of H onto S. If S is a closed subspace of H, then P is a bounded self-adjoint linear operator and $I - P$, where I is the identity operator, is the projection of H onto $S^\perp$. Since $x = Px + (I - P)x$, this provides a Cartesian decomposition of H, written $H = S \oplus S^\perp$, meaning that each vector in H can be written uniquely as a sum of a vector in S and a vector in $S^\perp$.

For example, suppose H is the completion of the space $C^1[0, 1]$ with respect to the inner product

$$\langle f, g \rangle = \int_0^1 f'(t)g'(t) \, dt + f(0)g(0),$$

and let S be the subspace of linear functions. Then $H = S \oplus S^\perp$ and we have seen that $S^\perp$ consists of those functions in H which vanish at 0 and 1. So in this instance the decomposition theorem expresses the fact that each function in H can be uniquely decomposed into the sum of a linear function and a function in H that vanishes at both end points of $[0, 1]$.

If S is a separable closed subspace of a Hilbert space H, then a representation of the projection operator P of H onto the subspace S can be given in terms of a complete orthonormal sequence $\{\phi_n\}$ for S. Indeed, if $x \in H$, then

$$\sum_n \langle x, \phi_n \rangle \phi_n - x \in S^\perp$$

since this vector is orthogonal to each member of the basis $\{\phi_n\}$ for S. Therefore,

$$Px = Px + P\left(\sum_n \langle x, \phi_n \rangle \phi_n - x \right) = \sum_n \langle x, \phi_n \rangle \phi_n.$$

There is an important relationship involving the nullspace, range, and adjoint of a bounded linear operator acting between Hilbert spaces. If $T : H_1 \to H_2$ is a bounded linear operator, then $N(T^*) = R(T)^\perp$. To see this, note that $w \in R(T)^\perp$ if and only if

$$0 = \langle Tx, w \rangle = \langle x, T^*w \rangle$$

for all $x \in H_1$, that is, if and only if $w \in N(T^*)$. By a previously discussed result on the second orthogonal complement we get the related result that $\overline{R(T)} = N(T^*)^\perp$. In particular, if T is a bounded linear operator with closed range, then the equation $Tf = g$ has a solution if and only if g is orthogonal to all solutions x of the homogeneous adjoint equation $T^*x = \theta$.

Replacing T with T^* and noting that $T^{**} = T$, we obtain two additional relationships between the nullspace, range, and adjoint. Taken together these relationships, namely

$$N(T^*) = R(T)^\perp, \qquad N(T^*)^\perp = \overline{R(T)},$$
$$N(T) = R(T^*)^\perp, \qquad N(T)^\perp = \overline{R(T^*)}$$

are sometimes collectively called the theorem on the *four fundamental subspaces*.

The Riesz Representation Theorem is a simple consequence of the decomposition theorem. If ℓ is a nonzero bounded linear functional on a Hilbert space H, then $N(\ell)^\perp = R(\ell^*)$ is one-dimensional and $H = N(\ell) \oplus N(\ell)^\perp$. Let $y \in N(\ell)^\perp$ be a unit vector. Then $x = Px + \langle x, y \rangle y$, where P is the projection operator of H onto $N(\ell)$. Therefore,

$$\ell(x) = \ell(\langle x, y \rangle y) = \langle x, y \rangle \ell(y) = \langle x, z \rangle$$

where $z = \overline{\ell(y)} y$.

B. The Spectral Theorem

We limit our discussion of the spectral theorem to the case of a compact self-adjoint operator. The spectral theorem gives a particularly simple characterization of the range of such an operator. We have seen that the nonzero eigenvalues of a compact self-adjoint operator T of infinite rank form a sequence of real numbers $\{\lambda_n\}$ with $\lambda_n \to 0$ as $n \to \infty$. The corresponding eigenspaces $N(T - \lambda_n I)$ are all finite-dimensional and there is a sequence $\{v_n\}$ of orthonormal eigenvectors, that is, vectors satisfying

$$\|v_j\| = 1, \quad \langle v_i, v_j \rangle = 0, \quad \text{for} \quad i \neq j,$$
$$\text{and,} \quad T v_j = \lambda_j v_j.$$

The closure of the range of T is the closure of the span of this sequence of eigenvectors. Since T is self-adjoint, $N(T)^\perp = \overline{R(T)}$; the decomposition theorem then gives the representation

$$w = Pw + \sum_{j=1}^{\infty} \langle w, v_j \rangle v_j$$

for all $w \in H$, where P is the projection operator from H onto $N(T)$. The range of T then has the form

$$Tw = \sum_{j=1}^{\infty} \lambda_j \langle w, v_j \rangle v_j.$$

This result is known as the spectral theorem. It can be extended (in terms of a Stieljes integral with respect to a projection valued measure on the real line) to bounded self-adjoint operators (and beyond).

C. The Singular Value Decomposition

The decomposition of a Hilbert space into the nullspace and eigenspaces of a compact self-adjoint operator can be simply extended to obtain a similar decomposition, called the *singular value decomposition* (SVD), for compact operators which are not necessarily self-adjoint. If $T : H_1 \to H_2$ is a compact linear operator from a Hilbert space H_1 into a Hilbert space H_2, then the operators T^*T and TT^* are both self-adjoint compact linear operators with nonnegative eigenvalues. By the spectral theorem, T^*T has an orthonormal sequence of eigenvectors, $\{u_j\}$, associated with its positive eigenvalues $\{\lambda_j\}$, that is complete in the subspace

$$\overline{R(T^*T)} = N(T^*T)^\perp = N(T)^\perp.$$

The numbers $\mu_j = \sqrt{\lambda_j}$ are called the *singular values* of T. If the vectors $\{v_j\}$ are defined by $v_j = \mu_j^{-1} T u_j$, then $\{v_j\}$ is a complete orthonormal set for $N(T^*)^\perp$ and the following relations hold:

$$T u_j = \mu_j v_j \quad \text{and} \quad T^* v_j = \mu_j u_j.$$

The system $\{u_j, v_j; \mu_j\}$ is called a *singular system* for the operator T, and any $f \in H_1$ has, by the decomposition theorem, a representation in the form

$$f = Pf + \sum_{j=1}^{\infty} \langle f, u_j \rangle u_j$$

where P is the projection of H_1 onto $N(T)$ (the sum is finite if T has finite rank). It then follows that

$$Tf = \sum_{j=1}^{\infty} \mu_j \langle f, u_j \rangle v_j.$$

This is called the SVD of the compact linear operator T. The SVD may be viewed as a nonsymmetric spectral representation.

D. Operator Equations

Suppose $T : H_1 \to H_2$ is a compact linear operator and $\lambda \in \mathbf{C}$. If $\lambda \neq 0$, then it can be shown that $R(T - \lambda I)$ is closed. The *Fredholm Alternative* Theorem completely characterizes the solubility of equations of the type

$$Tf - \lambda f = g$$

where $\lambda \neq 0$ and $g \in H_2$. Such equations are called linear operator equations of the *second kind*. The "alternative" is this: either the operator $(T - \lambda I)$ has a bounded inverse, or λ is an eigenvalue of T. If the first alternative holds, then the equation has the unique solution $f = (T - \lambda I)^{-1} g$, and this solution depends continuously on g. In this case we say that the equation of the second kind is *well-posed*. On the other hand, if λ is an eigenvalue of T, then $\bar{\lambda}$ is an eigenvalue of T^* and the equation has a solution only if

$$g \in R(T - \lambda I) = N(T^* - \lambda I)^\perp.$$

That is, the equation has a solution only if g is orthogonal to the finite-dimensional eigenspace $N(T^* - \bar{\lambda} I)$.

The Fredholm Alternative is particularly simple if $T : H \to H$ is self-adjoint. In this case the eigenvalues are real and $N(T^* - \bar{\lambda} I) = N(T - \lambda I)$. Therefore, if λ is not an eigenvalue of T, that is, if the equation $Tf - \lambda f = g$ has no more than one solution, then

$$H = \{\theta\}^\perp = N(T - \lambda I)^\perp = R(T - \lambda I)$$

and hence the equation has a solution for any $g \in H$. Simply put, the Fredholm Alternative for a self-adjoint compact operator says that uniqueness of solutions ($N(T - \lambda I) = \{\theta\}$) implies existence of solutions for any right hand side ($R(T - \lambda I) = H$).

The contraction mapping theorem can be applied to establish the existence and uniqueness of the solutions of certain *nonlinear* operator equations of the second kind, that is, equations of the form $Tf - \lambda f = g$. For example, suppose $T : X \to Y$ is a mapping from a Banach space X to a Banach space Y, satisfying $\|Tu - Tv\| \leq \mu\|u - v\|$. For nonzero $\lambda \in \mathbf{C}$ the equation $Tf - \lambda f = g$ then has a unique solution $f \in X$ for each $g \in Y$, if $\mu < |\lambda|$. Indeed, a solution of this equation is the unique fixed point of the contractive mapping $Af = \frac{1}{\lambda}(Tf - g)$. Further, the inverse mapping $g \mapsto f$ is continuous, that is, the original nonlinear equation of the second kind is well-posed.

Monotone operators form another general class of nonlinear operators for which unique solutions of certain operator equations of the second kind can be assured. Suppose H is a real Hilbert space. An operator $T : H \to H$ is called monotone if $\langle Tx - Ty, x - y \rangle \geq 0$ for all $x, y \in H$. If T is a continuous monotone operator and $\lambda < 0$, then a theorem of Minty insures the existence of a unique solution f of the operator equation of the second kind $Tf - \lambda f = g$, for each $g \in H$. Further, the inverse operator $J = (T - \lambda I)^{-1}$ is not just continuous, but *nonexpansive*, i.e., $\|Jh - Jg\| \leq \|h - g\|$.

When $\lambda = 0$, the linear operator equation equation treated above becomes an operator equation of the *first kind*:

$$Tf = g.$$

In this case, the role of the Fredholm Alternative Theorem is to some extent played by *Picard's Theorem*. Let $\{u_j, v_j; \mu_j\}$ be a singular system for the compact linear operator T. Then $\{v_j\}$ is a complete orthonormal system for $\overline{R(T)}$. Picard's Theorem gives a necessary and sufficient condition for a vector $g \in \overline{R(T)} = N(T^*)^\perp$ to lie in $R(T)$, that is, for the equation of the first kind to have a solution. The condition, *Picard's criterion*, is that the singular coefficients of g decay sufficiently quickly, specifically, that

$$\sum_{j=1}^{\infty} \mu_j^{-2}|\langle g, v_j \rangle|^2 < \infty.$$

If Picard's criterion is satisfied, then the series

$$\sum_{j=1}^{\infty} \frac{\langle g, v_j \rangle}{\mu_j} u_j$$

converges to some vector $f \in H_1$ and

$$Tf = \sum_{j=1}^{\infty} \langle g, v_j \rangle v_j = g,$$

and hence the equation has a solution, namely f. The solution is unique only if $N(T) = \{\theta\}$.

E. Inversion

An invertible bounded linear operator need not have a bounded inverse. For example, the operator $T : L^2[0, 1] \to C[0, 1]$ defined by $(Tf)(s) = \int_0^s f(t)\,dt$ has an inverse defined on the subspace of absolutely continuous functions which vanish at 0. But the inverse operator $T^{-1}g = g'$ is an unbounded operator.

If $T \in L(X, Y)$, where X and Y are Banach spaces, then the existence of a bounded inverse for T is equivalent to the condition $\|Tx\| \geq m\|x\|$ for some $m > 0$. Indeed, if this condition is satisfied, then $N(T) = \{\theta\}$, $R(T)$ is closed in Y, and $T^{-1} : R(T) \to X$ satisfies $\|T^{-1}\| \leq m^{-1}$. On the other hand, if T^{-1} is bounded, then the condition holds with $m = \|T^{-1}\|^{-1}$. If T is bijective, i.e., if $N(T) = \{\theta\}$ and $R(T) = Y$, then *Banach's Theorem* insures that the inverse of T is also bounded, i.e., $T^{-1} \in L(Y, X)$. One consequence of this fact is that if a normed linear space X is a Banach space under two norms $\|\cdot\|_1$ and $\|\cdot\|_2$, and if these norms satisfy $\|x\|_2 \leq M\|x\|_1$ for some $M > 0$, then the norms are equivalent (apply Banach's Theorem to the identity operator $I : (X, \|\cdot\|_1) \to (X, \|\cdot\|_2)$).

The familiar geometric series has an operator analog. If X is a Banach space and $T : X \to X$ is a bounded linear operator with $\|T\| < 1$, then $(I - T)^{-1} \in L(X, X)$ and, in fact,

$$(I - T)^{-1} = I + T + T^2 + \cdots$$

where the series, called the *Neumann series*, converges in the operator norm. This result can be used immediately to guarantee the existence of a unique solution of the linear operator equation of the second kind $Tf - \lambda f = g$, if $|\lambda|$ is sufficiently large ($|\lambda| > \|T\|$ does the trick).

The existence of an inverse of $T \in L(X, Y)$ is equivalent to the unique solvability of the equation $Tf = g$ for each $g \in Y$. This, in turn, is equivalent to $R(T) = Y$ and $N(T) = \{\theta\}$. If either of these conditions is violated, then there is no inverse, but all is not lost. In certain circumstances a *generalized inverse* can be defined; we limit our discussion of the generalized inverse to the Hilbert space context. Suppose $T : H_1 \to H_2$ is a bounded linear operator from a Hilbert space H_1 to a Hilbert space H_2. Suppose we wish to solve the equation $Tf = g$. If $g \notin R(T)$, then the equation has no solution and we might settle for finding an $f \in H_1$ whose image under T is a close to g as possible, that is, we seek $f \in H_1$ satisfying

$$\|Tf - g\| = \inf\{\|Tx - g\| : x \in H_1\}.$$

Such an f is called a *least-squares solution* of $Tf = g$. A least-squares solution exists if and only if the projection of g onto $\overline{R(T)}$ actually lies in $R(T)$. This is equivalent to requiring g to lie in the dense subspace $R(T) + R(T)^\perp$ of H_2. The condition for a least-squares solution may also be phrased as requiring that $Tf - g \in R(T)^\perp = N(T^*)$ (see the theorem on the four fundamental subspaces). This gives another characterization of least-squares solutions: u is a least-squares solution of $Tf = g$ if and only if u

satisfies the *normal equation* $T^*Tu = T^*g$. Since T is bounded, the set of solutions of the normal equation is closed and convex, and hence, by the projection theorem, if there is a least-squares solution, then there is a least-squares solution of smallest norm. These ideas enable us to define a generalized inverse (the *Moore-Penrose generalized inverse*) for T. Let $\mathcal{D}(T^\dagger) = R(T) + R(T)^\perp$ and for $g \in \mathcal{D}(T^\dagger)$, define $T^\dagger g$ to be the least-squares solution having smallest norm. The Moore-Penrose generalized inverse, $T^\dagger : \mathcal{D}(T^\dagger) \subseteq H_2 \to H_1$, is a closed densely defined linear operator. However, $T^\dagger$ is bounded if and only if $R(T)$ is closed. If T is compact, then $R(T)$ is closed if and only if T has finite rank. In particular, a linear integral operator generated by a square integrable kernel has a bounded Moore-Penrose generalized inverse if and only if the kernel is degenerate. If T is compact with singular system $\{u_j, v_j; \mu_j\}$, the the Moore-Penrose generalized inverse has the explicit representation

$$T^\dagger g = \sum_j \frac{\langle g, v_j \rangle}{\mu_j} u_j.$$

F. Variational Inequalities

Suppose H is a real Hilbert space with inner product $\langle \cdot, \cdot \rangle$ and corresponding norm $\|\cdot\|$. A bilinear form $a(\cdot, \cdot) : H \times H \to \mathbf{R}$ is called *bounded* if there is a constant C such that

$$|a(u, v)| \le C \|u\| \|v\| \quad \text{for all} \quad u, v \in H$$

and *coercive* if there is a constant $m > 0$ such that

$$m \|u\|^2 \le a(u, u)$$

for all $u \in H$. A fundamental result of Stampacchia asserts that if $a(\cdot, \cdot)$ is a bounded, coercive bilinear form (which *need not be symmetric*), and if $f \in H$ and K is a closed convex subset of H, then there is a unique $u \in K$ satisfying

$$a(u, v - u) \ge \langle f, v - u \rangle \quad \text{for all} \quad v \in K.$$

This is called a *variational inequality* for the form $a(\cdot, \cdot)$, the closed convex set K, and the vector $f \in H$.

The fundamental nature of this result becomes apparent when one notices that the projection property for Hilbert space is a special case of this result on variational inequalities. Indeed, if $a(x, y) = \langle x, y \rangle$, then the theorem insures the existence of a unique $u \in K$ satisfying

$$\langle u, v - u \rangle \ge \langle f, v - u \rangle$$

or, equivalently

$$\langle f - u, v - u \rangle \le 0 \quad \text{for all} \quad v \in K.$$

Geometrically, this says that the angle between the vectors $f - u$ and $v - u$ is obtuse for all vectors $v \in K$. From this

it follows that $u \in K$ is the vector in K that is nearest to f, that is, $u = Pf$, the metric projection of f onto K.

A nonsymmetric version of the Riesz Representation Theorem also follows from the theorem on variational inequalities. Suppose ℓ is a bounded linear functional on H. Then, by the Riesz Representation Theorem, there is a $f \in H$ such that $\ell(w) = \langle f, w \rangle$ for all $w \in H$. Let the closed convex set K be the entire Hilbert space H, then there is a unique $u \in H$ satisfying

$$a(u, v - u) \ge \langle f, v - u \rangle \quad \text{for all} \quad v \in H$$

and hence $a(u, w) \ge \langle f, w \rangle$ for all $w \in H$. Replacing w by $-w$, we also get $a(u, -w) \ge \langle f, -w \rangle$ for all $w \in H$. Therefore, $a(u, w) = \langle f, w \rangle$ for all $w \in H$. That is, the functional ℓ has the representation $\ell(w) = a(u, w)$ for a unique $u \in H$. This representation of bounded linear functional in terms of a possibly nonsymmetric bilinear form is known as the *Lax-Milgram lemma*. The Lax-Milgram lemma can be used to establish the existence of a unique weak solution for certain nonsymmetric elliptic boundary value problems in the same way that the Riesz Representation Theorem is used to prove the existence of a unique weak solution of the Poisson problem.

As a simple application of the Lax-Milgram lemma, consider the two-point boundary value problem

$$-u'' + u' + u = f, \qquad u'(0) = u'(1) = 0.$$

Integration by parts yields $a(u, v) = \langle f, v \rangle$, where $\langle \cdot, \cdot \rangle$ is the $L^2[0, 1]$ inner product and $a(u, v)$ is the nonsymmetric, bounded, coercive, bilinear form defined on $H^1[0, 1]$ by

$$a(u, v) = \int_0^1 (u'v' + u'v + uv)(s)\, ds.$$

The Lax-Milgram lemma then ensures the existence of a unique weak solution $u \in H^1[0, 1]$ of the boundary value problem, that is, a unique vector $u \in H^1[0, 1]$ satisfying

$$a(u, v) = \langle f, v \rangle \quad \text{for all} \quad v \in H^1[0, 1].$$

V. A FEW APPLICATIONS

A. Weak Solutions of Poisson's Equation

Suppose Ω is a bounded domain in $\mathbf{R}^2$ with smooth boundary $\partial\Omega$. Given $f \in C(\Omega)$ a classical solution of Poisson's equation is a function $u \in C^2(\Omega)$ satisfying

$$-\Delta u = f \qquad \text{in} \qquad \Omega$$
$$u = 0 \qquad \text{on} \qquad \partial\Omega$$

where Δ is the Laplacian operator. If the Poisson equation is multiplied by $v \in C_0^2(\Omega)$ and integrated over Ω, then Green's identity gives

$$\langle \nabla u, \nabla v \rangle = \langle f, v \rangle$$

where ∇ is the gradient operator and $\langle \cdot, \cdot \rangle$ is the $L^2(\Omega)$ inner product. There are two things to notice about this equation: $\langle f, v \rangle$ is defined for $f \in L^2(\Omega)$, allowing consideration of "rougher" data f, and on the left-hand side only first derivatives are required rather than second derivatives, allowing less smooth "solutions" u. These observations permit us to propose a *weaker* formulation of the Poisson problem.

The bilinear form $a(\cdot, \cdot)$ defined by $a(u, v) = \langle \nabla u, \nabla v \rangle$ is an inner product (sometimes called the *energy* inner product) on the space $C_0^1(\Omega)$ (this is a consequence of Poincaré's inequality: $a(u, u) \geq C \|u\|_2^2$, where $\|\cdot\|_2$ is the L^2 norm). The norm generated by this inner product is equivalent to the Sobolev $H_0^1(\Omega)$ norm. Therefore, the completion of $C_0^1(\Omega)$ relative to this inner product is the Sobolev space $H_0^1(\Omega)$, and we define a *weak solution* of the Poisson problem to be a vector $u \in H_0^1(\Omega)$ satisfying

$$a(u, v) = \langle f, v \rangle$$

for all $v \in H_0^1(\Omega)$. The linear functional $\ell : H_0^1(\Omega) \to \mathbf{R}$ defined by $\ell(v) = \langle f, v \rangle$ is (again, as a consequence of Poincaré's inequality) a bounded linear functional on $H_0^1(\Omega)$, and hence, by the Riesz Representation Theorem, there is a unique $u \in H_0^1(\Omega)$ satisfying

$$a(u, v) = \langle f, v \rangle \quad \text{for all} \quad v \in H_0^1(\Omega).$$

In other words, Poisson's problem has a unique weak solution for each $f \in L^2(\Omega)$.

B. A Finite Element Method

A finite element method is a constructive precedure for approximating a weak solution by a linear combination of "basis" functions. As a simple illustration we treat a piecewise linear finite element method for the Poisson problem in the plane. A finite dimensional subspace U_N of $H_0^1(\Omega)$ chosen and a basis (whose members are called *finite elements*) is selected for U_N. The finite element approximation to the weak solution will be a certain linear combination of these basis functions, i.e., a member of U_N. First the region Ω is triangulated; the vertices of the resulting triangles are called *nodes* of the triangulation. The functions in U_N will be continuous on Ω, linear on each triangle, and zero on the boundary of Ω. With each interior node of the triangulation we associate a basis function which is 1 at the node and zero at all other nodes of the triangulation (these basis functions are called the linear *Lagrange* elements).The dimension of the finite element subspace is therefore equal to the number of interior nodes in the triangulation and each basis function has a pyramidal shape with a peak at the associated nodal point of the triangulation.

A finite element solution of the Poisson problem is defined by restricting the conditions for a weak solution to the subspace of finite elements, that is, $u_N \in U_N$ is a finite element solution of the Poisson problem if

$$a(u_N, v) = \langle f, v \rangle \quad \text{for all} \quad v \in U_N.$$

When this condition is expressed in terms of the finite element basis, the resulting coefficient matrix is positive definite and hence there is a unique finite element solution.

If u is the weak solution of the Poisson problem, then

$$a(u - u_N, v) = \langle f, v \rangle - \langle f, v \rangle = 0$$

for all $v \in U_N$. Geometrically, this says that the finite element solution is the projection (relative to the energy inner product) of the weak solution onto the finite element subspace and hence,

$$\|u - u_N\| \leq \|u - v\|$$

for all $v \in U_N$, where $\|\cdot\|$ is the energy norm. That is, the finite element solution is the best approximation to the weak solution, with respect to the energy norm, in the finite element subspace.

C. Two-Point Boundary Value Problems

We briefly treat a simple class of Sturm-Liouville problems. The goal is to find $x \in C^2[a, b]$ satisfying the differential equation

$$\frac{d}{ds}\left[p(s) \frac{dx}{ds} \right] - [\mu + q(s)] x(s) = f(s)$$

where $q, f \in C[a, b]$ and $p \in C^1[a, b]$ are given functions and $\mu \neq 0$ is a given scalar. Define $T : C_0^2[a, b] \to C[a, b]$ by

$$Tx = [px']' - qx.$$

We suppose the this differential operator is *nonsingular*, that is, $N(T) = \{\theta\}$ (such is the case, for example, if $p(s) < 0$ and $q(s) \leq 0$). Then there is a symmetric *Green's function* $k(\cdot, \cdot) \in C[a, b] \times C[a, b]$ for T. That is, T^{-1} is the integral operator generated by the kernel $k(\cdot, \cdot)$. In other words, $Tx = h$ if and only if

$$x(s) = \int_a^b k(s, t) h(t) \, dt.$$

The original problem may then be expressed as

$$Tx = \mu x + f$$

or, in terms of the Green's function $k(\cdot, \cdot)$:

$$x(s) = \int_a^b k(s, t)[\mu x(t) + f(t)] \, dt.$$

Equivalently,

$$Kx - \lambda x = g$$

where $\lambda = 1/\mu$ and $g = -\lambda K f$, where K is the compact self-adjoint operator on $L^2[a, b]$ generated by the kernel $k(\cdot, \cdot)$. If λ is not an eigenvalue of K, then this integral equation of the second kind has, by Fredholm's Alternative, a solution which is expressable, via the spectral theorem, as a series of eigenfunctions of K. Specifically, $\overline{R(K)}$ has an orthonormal basis of eigenfunctions $\{v_j\}$ of K and therefore,

$$x = (K - \lambda I)^{-1} g = -\lambda (K - \lambda I)^{-1} K f$$
$$= \sum_j \frac{\lambda \lambda_j}{\lambda - \lambda_j} \langle f, v_j \rangle v_j.$$

D. Inverse Problems

Many inverse problems in mathematical physics can be modeled as operator equations of the first kind, $Kx = y$, where K is a compact linear operator acting between Hilbert spaces. Consider, for example, the simple model problem in gravimetry in which the vertical component of gravity, $y(s)$, along a horizontal segment $0 \le s \le 1$ is engendered by a mass distribution $x(p)$, $0 \le p \le 1$ on a parallel segment one unit distant from the first. The relationship between y and x is given by

$$y(s) = \gamma \int_0^1 ((s - p)^2 + 1)^{-\frac{3}{2}} x(p) \, dp$$

where γ is a constant. The inverse problem consists of determining the mass distribution from observations of the vertical force y. The model may be phrased abstractly as $y = Kx$, where $K : H \to H$ is a compact linear operator acting on the Hilbert space $H = L^2[0, 1]$.

Formally, the inverse problem may be solved by using the Moore-Penrose generalized inverse: $x = K^\dagger y$. However, the representation of $K^\dagger$ in terms of the SVD of K

$$x = K^\dagger y = \sum_{j=1}^\infty \frac{\langle y, v_j \rangle}{\mu_j} u_j$$

points to a serious problem. The singular values $\{\mu_j\}$ converge to zero leading to instability in the solution process. The vector y is a measured entity and hence is subject to error. Suppose, for example, the measured data consists of a vector $y^\delta \in H$ satisfying $\|y - y^\delta\| \le \delta$, where δ is a known bound for the measurement error. While $y^\delta \to y$ as $\delta \to 0$, generally $\|K^\dagger y - K^\dagger y^\delta\| \to \infty$ as $\delta \to 0$. For example, if $\delta = \sqrt{\mu_n}$, where $n = 1, 2, \ldots$, and $y^\delta = y + \sqrt{\mu_n} v_n$, then $\|y - y^\delta\| = \sqrt{\mu_n} \to 0$, while $\|K^\dagger y - K^\dagger y^\delta\| = 1/\sqrt{\mu_n} \to \infty$ as $n \to \infty$. The lesson is

that even very accurate measurements can lead to wildly unstable computations.

Stability can be restored (at the expense of accuracy) by the method of *regularization*. In its simplest form this method replaces the normal equation with an augmented (or *regularized*) normal equation

$$K^* K x_\alpha + \alpha x_\alpha = K * y,$$

where $\alpha > 0$ is a *regularization parameter*. That is, an ill-posed equation of the first kind is replaced by an approximating well-posed equation of the second kind. The solution x_α of the regularized equation is stable with respect to perturbations in the data y since $-\alpha < 0$ is not an eigenvalue of $K^* K$ and hence, by the Fredholm Alternative, $(K^* K + \alpha I)^{-1}$ is bounded.

The regularized normal equation must be solved using the available data y^δ and hence the choice of the regularization parameter, in terms of the available data, is a matter of considerable importance. One general method for choosing the regularization parameter is known as Morozov's discrepancy principle. According to this principle, if the signal-to-noise ratio is greater than one, i.e., $\|y^\delta\| > \delta \ge \|y - y^\delta\|$ and $x_\alpha^\delta = (K^* K + \alpha I)^{-1} K^* y^\delta$, then the equation

$$\left\| K x_\alpha^\delta - y^\delta \right\| = \delta$$

has a unique positive solution $\alpha = \alpha(\delta)$ and $x_{\alpha(\delta)}^\delta \to K^\dagger y$ as $\delta \to 0$.

E. Heisenberg's Principle

Heisenberg's uncertainty principle in quantum mechanics is a consequence of an inequality involving unbounded self-adjoint operators. We will consider only a very simple case. Suppose a particle moves in one dimension, its position at a given time t denoted by $x \in (-\infty, \infty)$. In the quantum mechanical formalism the position of the particle is understood to be a random variable and it is the *state*, rather than the position, that is at issue. The state is a function ψ which is a unit vector in the complex Hilbert space $H = L^2(-\infty, \infty)$; the interpretation being that

$$\int_a^b \psi(s) \overline{\psi(s)} \, ds$$

represents the probability that at the time in question the particle is positioned between a and b. In other words, the integrable real-valued function $\psi \bar{\psi}$ is a probability density for the position. In general, a unit vector in H is called a *state*. A self-adjoint linear operator $T : \mathcal{D}(T) \subseteq H \to H$ is called an *observable*. The *expected value* of an observable T when the system is in state ψ is

$$E(T) = \langle T\psi, \psi \rangle = \int_{-\infty}^\infty T\psi(s) \overline{\psi(s)} \, ds$$

For example, the expected value of the multiplication by the independent variable operator, $(M\psi)(x) = x\psi(x)$,

$$E(M) = \int_{-\infty}^{\infty} x\psi(x)\overline{\psi(x)}\, dx$$

is the mean position of the particle (slight modifications of the arguments given in the section on unbounded operators show that M is unbounded and self-adjoint). For this reason the observable M is called the *position* operator. The *variance* of an observable T when the system is in state ψ is defined by

$$Var(T) = \|(T - E(T)I)\psi\|^2.$$

That is, the variance gives a measure of the dispersion of an observable from its expected value.

Definition. The *commutator* of two observables S and T is the observable $[S, T] : \mathcal{D}(ST) \cap \mathcal{D}(TS) \to H$ defined by $[S, T] = ST - TS$.

S and T are said to *commute* if $ST = TS$, i.e., $\mathcal{D}(ST) = \mathcal{D}(TS)$ and $[S, T] = 0$. In abstract form the Heisenberg principle says that for any state $\psi \in \mathcal{D}([S, T])$

$$\frac{1}{4}|E([S, T])|^2 \le Var(S) \times Var(T).$$

Suppose $\mathcal{D}(\mathcal{P})$ is the subspace of all absolutely continuous functions in H whose first derivative is also in H. Define the operator $\mathcal{P} : \mathcal{D}(\mathcal{P}) \to H$ by

$$\mathcal{P}\psi(x) = \frac{h}{2\pi i}\psi'(x)$$

where h is *Planck's constant*. Then $\mathcal{P}$ is self-adjoint, that is, an observable. A physical argument shows that $E(\mathcal{P})$ is the expected value of the momentum of the system and hence $\mathcal{P}$ is called the *momentum operator*. The commutator of the momentum and position operators can be found from the relation

$$(\mathcal{P}M\psi)(x) = \frac{h}{2\pi i}\frac{d}{dx}[x\psi(x)] = \frac{h}{2\pi i}[\psi(x) + x\psi'(x)]$$

$$= \frac{h}{2\pi i}\psi(x) + (M\mathcal{P}\psi)(x)$$

i.e., for all $\psi \in \mathcal{D}([\mathcal{P}, M])$:

$$[\mathcal{P}, M]\psi = \frac{h}{2\pi i}\psi.$$

The general Heisenberg principle then gives

$$Var(\mathcal{P}) \times Var(M) \ge \left(\frac{h}{4\pi}\right)^2.$$

This is an expression of the physical uncertainty principle: no matter the state ψ, the position and momentum can not both be determined with arbitrary certainty.

SEE ALSO THE FOLLOWING ARTICLES

CONVEX SETS • DATA MINING AND KNOWLEDGE DISCOVERY • DIFFERENTIAL EQUATIONS, ORDINARY • FOURIER SERIES • GENERALIZED FUNCTIONS • TOPOLOGY, GENERAL

BIBLIOGRAPHY

Brenner, S. C., and Scott, L. R. (1994). "The Mathematical Theory of Finite Element Methods," Springer-Verlag, New York.

Groetsch, C. W. (1980). "Elements of Applicable Functional Analysis," Dekker, New York.

Kantorovich, L. V., and Akilov, G. P. (1964). "Functional Analysis in Normed Spaces," Pergamon, New York.

Kirsch, A. (1996). "An Introduction to the Mathematical Theory of Inverse Problems," Springer-Verlag, New York.

Kreyszig, E. (1978). "Introductory Functional Analysis with Applications," Wiley, New York.

Lebedev, L. P., Vorovich, I. I., and Gladwell, G. M. L. (1996). "Functional Analysis: Applications in Mechanics and Inverse Problems," Kluwer, Dordrecht.

Naylor, A. W., and Sell, G. R. (1971). "Linear Operator Theory in Engineering and Science," Holt, Rinehart and Winston, New York.

Riesz, F., and Sz.-Nagy, B. (1955). "Functional Analysis," Ungar, New York.

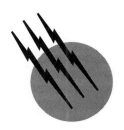

Fuzzy Sets, Fuzzy Logic, and Fuzzy Systems

M. M. Gupta
J. B. Kiszka
University of Saskatchewan

I. Introduction
II. Elements of Fuzzy Set Theory, Fuzzy Logic, and Fuzzy Systems
III. Linguistics and Fuzzy Modeling and Control
IV. Applications of Fuzzy Logic to Systems Modeling and Control
V. Concluding Remarks

GLOSSARY

Approximate reasoning Process or processes by which a possibility imprecise conclusions are deduced from a collection of imprecise premises.

Compositional rule of inference Let the fuzzy conditional inference be:

$$
\begin{array}{ll}
\text{Ant 1:} & \text{if } x \text{ is } A \text{ then } y \text{ is } B. \\
\underline{\text{Ant 2:}} & \underline{\text{if } x \text{ is } A'} \\
\text{Cons:} & \qquad\qquad y \text{ is } B'
\end{array}
$$

The compositional rule of inference is applied to generate a fuzzy subset B', given the antecedent A' and the fuzzy relation $R = A \rightarrow B$, $B' = A' \circ R$.

Expert control systems Computer-controlled systems that use knowledge and reasoning techniques to solve control problems normally requiring the abilities of human operator–experts.

Expert systems Computer programs that perform a specialized, usually difficult, professional task at the level of (or sometimes beyond the level of) a human expert. Because their functioning relies so heavily on large bodies of knowledge, expert systems are sometimes known as knowledge-based systems. Since they are often used to assist the human expert, they are also known as intelligent systems.

Fuzzy algorithm Ordered set of fuzzy instructions that, upon execution, yield an approximate solution to a specified problem. The instructions in a fuzzy algorithm belong to one of three categories: (1) assignment statements: a possibly fuzzy value is assigned to a variable; (2) fuzzy conditional statements: a possible fuzzy value is assigned to a variable, or an action is executed, provided that a fuzzy condition holds; or (3) unconditional action statements: a possible fuzzy mathematical operation or an action to be executed.

Fuzzy controller The purpose of a fuzzy controller is to compute values of action variables from observations of the state variables of the process under control. The relation between state variables and action variables is given as a set of fuzzy implications or fuzzy relations. The composition rule of inference is applied to calculate the action variables.

Fuzzy logic Logic of approximate reasoning, bearing the same relation to approximate reasoning that two-valued does to precise reasoning.

Fuzzy model Finite set of fuzzy relations that, together, form an algorithm for determining the outputs of the process from some finite number of the past inputs and outputs.

Fuzzy relation Fuzzy subset of the Cartesian product $X \times Y$, denoted as a relation R from a set X to a set Y.

Fuzzy relational equation Let A and B be two fuzzy sets, and R be a fuzzy relation. The problem of finding all A's such that $A \circ R = B$, is called a fuzzy relational equation, where $\circ$ stands for the composition of A with R.

Fuzzy set Function with more than two values, usually with values in the unit interval. This function allows a continuum of possible choices and can be used to describe imprecise terms.

Knowledge base Facts, assumptions, beliefs, and heuristics; "expertise"; methods of dealing with the database to achieve desired results such as a diagnosis, or an interpretation, or a solution to a problem.

Rule A pair, composed of an antecedent condition and a consequent proposition, that can support deductive processes.

A NEW APPROACH to the analysis and synthesis of complex engineering systems, originated by Lotfi A. Zadeh in 1965, is based on the premise that the thinking process is vague rather than exact. This new approach emphasized the human ability to extract information from masses of inexact data. The only information extracted is that relevant to the task at hand. Human experience is a very important source of such a thinking process. The theory of the fuzzy logic helps to transfer a linguistic model of the human thinking process to a fuzzy algorithm. During recent years the fuzzy approach has made tremendous progress in many branches of engineering and nonengineering problems.

I. INTRODUCTION

In the design of many engineering systems, there is no precise method of design for a number of reasons. First, it is difficult to understand the complexity of the real world. Second, we have a subjective perception of the real world and can perform only inexact reasoning. However, one fact is very important. By virtue of our knowledge and experience, we can build increasingly better systems in spite of all our inexact reasonings.

Since 1965, when L. A. Zadeh first introduced the fuzzy set theory, our approximate thinking process seems to provide better realizations of the topics. The approach proposed by L. A. Zadeh is based on the premise that the key elements in human thinking are not numbers, but labels of fuzzy sets: that is, classes of objects in which the transition from membership to nonmembership is gradual rather than abrupt. The degree of membership is specified by a number between 1 (full membership) and 0 (full nonmembership). The grade of membership is subjective: it is a matter of perception rather than measurement. By virtue of fuzzy sets, human concepts like "small," "big," "high," "more or less," "most," "few," and "several" can be translated into a form usable by computers. These kinds of linguistic values and modifiers are inherent characteristic of human reasoning.

Generally, the theory of fuzzy set simplifies the task of translating between human reasoning, which is inherently elastic, and the rigid operations of digital computers.

In everyday life, humans tend to use words and sentences rather than numbers of describe how systems behave. Many complex industrial processes are too complicated to be understood fully in terms of exact mathematical relations, but they may be successfully described by natural languages, and can be controlled using some rules or thumb. In order to transfer this human ability to a computer program, we need everyday logic, common-sense logic, or, as L. A. Zadeh said, fuzzy logic. Fuzzy logic is a kind of logic that uses graded qualified statements rather than ones that are strictly true or false. The results of fuzzy reasoning are not definite as are those derived by strict logic, but they cover a large field of discourse.

The first application of fuzzy logic theory was in control systems engineering. E. H. Mamdani and colleagues at the Queen Mary College in London created a set of control rules for a steam engine in fuzzy terms. They proposed a fuzzy self-organizing controller that can modify the control rules by learning from the operator's actions. Their pioneering work led to other research in fuzzy logic control and its application in such industrial processes as the automatic kiln at the Oregon Portland Cement Company.

Expert systems have now emerged as one of the most important applications of artificial intelligence. Reflecting on human expertise, much of the information in the knowledge base of a typical expert system is imprecise, incomplete, or not totally reliable. Fuzzy logic has the potential of becoming an effective tool for the management of uncertainty in expert systems.

Fuzzy theory is now being applied in many humanistic type of process, like meteorology, medicine, psychology, and control engineering.

Some introductory definitions of fuzzy set and fuzzy logic are given in Section II. A general idea of fuzzy model building is given in Section III. The most typical applications of fuzzy systems in control engineering are shown in Section IV. At the end, some simulation results are summarized in Section V.

II. ELEMENTS OF FUZZY SET THEORY, FUZZY LOGIC, AND FUZZY SYSTEMS

A central notion of fuzzy set theory is that it is permissible for elements to be only partly elements of a set.

A. Calculus of Fuzzy Logic

Let $X = \{x\}$ denote a conventional set. A fuzzy set A in the universe X is a set of ordered pairs $A = \{x, \mu_A(x)\}$, $x \in X$, where $\mu_A(x)$ is the grade of membership of x in A:

$$\mu_A : X \to [0, 1]. \tag{1}$$

A fuzzy set is said to be normal if and only if $\max_{x \in X} \mu_A(x) = 1$. The calculus of the fuzzy sets is based on the following important logical notions.

Inclusion. A fuzzy set A is said to be included in a fuzzy set B iff (if and only if) $\mu_A(x) \leq \mu_B(x)$, $\forall x \in X$.

Intersection. A fuzzy set $A \cap B$ is said to be the intersection of the fuzzy sets A and B if

$$\mu_{A \cap B}(x) = \min[\mu_A(x), \mu_B(x)]$$
$$= \mu_A(x) \wedge \mu_B(x), \qquad \forall x \in X. \tag{2}$$

Union. A fuzzy set $A \cup B$ is said to be union of fuzzy sets A and B if

$$\mu_{A \cup B}(x) = \max[\mu_A(x), \mu_B(x)]$$
$$= \mu_A(x) \vee \mu_B(x), \qquad \forall x \in X. \tag{3}$$

Complement. A fuzzy set $\bar{A}$ is said to be complement of a fuzzy set A if

$$\mu_{\bar{A}} = 1 - \mu_A(x), \qquad \forall x \in X. \tag{4}$$

The intersection of A and B corresponds to the connective "**and**" Thus $A \cap B = A$ **and** B.

The union of fuzzy sets A and B corresponds to the connective "**or**." Thus $A \cup B = A$ **or** B. The operation of complementation corresponds to negation **not**. Thus $\bar{A} = $ **not** A.

The product of two fuzzy sets A and B, written $A \cdot B$, is defined as

$$\mu_{A \cdot B}(x) = \mu_A(x) \cdot \mu_B(x), \qquad \forall x \in X. \tag{5}$$

The algebraic sum of two fuzzy sets A, B, written $A \oplus B$, is defined as

$$\mu_{A \oplus B}(x) = \mu_A(x) + \mu_B(x) - \mu_A(x) \cdot \mu_B(x),$$
$$\forall x \in X.$$

The product of a scalar $a \in R$ and a fuzzy set A, written as $a \cdot A$, is defined as $\mu_{a \cdot A}(x) = a \cdot \mu_A(x)$, $\forall x \in X$, where R is the set of nonnegative real numbers, and $0 \leq a \leq 1$.

The kth power of a fuzzy set A, written A^k, is defined as $\mu_{A^k}(x) = \mu_A^k(x)$, $\forall x \in X$, $\forall k \in R$.

These are some of the basic operations on fuzzy sets. A more general approach to operations on fuzzy sets is given by t- or s-norms.

A triangular t-norm is a two-place function $t: [0, 1]^2 \to [0, 1]$ that fulfills the following monotonicity, commutativity, and associativity with boundary conditions

$$t(x, 0) = t(0, x) = 0 \qquad \text{and}$$
$$t(x, 1) = t(1, x) = x, \qquad \text{for} \quad x \in [0, 1]. \tag{6}$$

The t-conorm, called also s-norm, can be introduced via De'Morgen's law:

$$s(x, y) = 1 - t(1 - x, 1 - y) = \overline{t(\bar{x}, \bar{y})}$$
$$\text{for} \quad x, y \in [0, 1]. \tag{7}$$

Some t-norms are:

$$x t_1 y = 1 - \min(1, \sqrt[p]{(1 - x)^p + (1 - y)^p}) \qquad p \geq 1$$
$$x t_2 y = \min(x, y)$$
$$x t_3 y = \max(0, x + y - 1) \tag{8}$$
$$x t_4 y = x \cdot y.$$

And, corresponding s norms are

$$x s_1 y = \min(1, \sqrt[p]{(x^p + y^p)}) \qquad p \geq 1$$
$$x s_2 y = \max(x, y)$$
$$x s_3 y = \min(1, x + y) \tag{9}$$
$$x s_4 y = x + y - x \cdot y$$

The connectives *and*, *or*, *but*, *not*, etc. have local properties and are context-dependent. The mathematical model of these operators should be chosen with respect to the problem under consideration.

The α-cut of a fuzzy set $A \subset X$, written as A_α, is the set

$$A_\alpha = \{x \in X : \mu_A(x) \geq \alpha\}. \qquad (10)$$

This relation is one of the key concepts in fuzzy mathematics.

A fuzzy relation R between the two (nonfuzzy) sets X and Y is a fuzzy set in the Cartesian product $X \times Y$, i.e., $R \subset X \times Y$. Hence it is defined as

$$R = \{\mu_R(x, y), (x, y)\} = \{\mu_R(x, y)/(x, y)\}$$
$$\forall (x, y) \in X \times Y. \qquad (11)$$

The max–min composition of two fuzzy relations $R \subset X \times Y$ and $S \subset Y \times Z$, written as $R \circ S$, is defined as a fuzzy relation $R \circ S \subset X \times Z$ such that

$$\mu_{R \circ S}(x, z) = \max_{y \in Y}(\mu_R(x, y) \wedge \mu_S(y, z)) \qquad (12)$$

for each $x \in X$, $z \in Z$ where $\wedge = \min$.

The Cartesian product of two fuzzy sets $A \subset X$ and $B \subset Y$, written as $A \times B$, is defined as a fuzzy set in $X \times Y$, such that

$$\mu_{A \times B}(x, y) = \mu_A(x) \wedge \mu_B(y) \qquad (13)$$

for each $x \in X$, $y \in Y$.

B. Fuzzy Logic

Most human reasoning is approximate rather than exact. L. A. Zadeh has suggested methods for such fuzzy reasoning in which the antecedent involves a fuzzy conditional proposition, "if x is A then y is B," where A and B are fuzzy concepts.

Consider the following form of inference mechanism in which a fuzzy conditional proposition is contained.

$$\begin{array}{ll} \text{Ant 1:} & \text{if } x \text{ is } A \text{ then } y \text{ is } B \\ \underline{\text{Ant 2:}} & \underline{\text{if } x \text{ is } A'} \\ \text{Cons:} & \quad\quad y \text{ is } B' \end{array}, \qquad (14)$$

where x and y are the names of objects, and A, A', B, and B' are the labels of fuzzy sets in the universes of discourse U, U', V, and V' respectively. This form of inference is called fuzzy conditional inference.

The compositional rule of inference is applied to generate a fuzzy subset B', given the antecedent A' and the implication $A \rightarrow B$.

L. A. Zadeh has proposed a translation rule for translating the fuzzy conditional proposition "if x is A then y is B" into a fuzzy relation $U \times V$.

Let A and B be fuzzy sets in the universe U and V, respectively; then the antecedents Ant 1 and Ant 2 are

translated, respectively, into a binary fuzzy relation, which is expressed as

$$R = 1 \wedge (1 - \mu_A(u) + \mu_B(v)), \qquad \forall u, v \in U \times V. \quad (15)$$

The consequence B' in Cons can be deduced from Ant 1 and Ant 2 by using the max–min composition $\circ$ of the fuzzy set A' and the fuzzy relation R (the compositional rule of inference); that is,

$$B' = A' \circ R, \qquad (16)$$

where the max–min composition $\circ$ of A' and R is defined as

$$\mu_{A' \circ R}(v) = \bigvee_u \{\mu_{A'}(u) \wedge \mu_R(u, v)\}. \qquad (17)$$

Thus, the membership function B' is given by

$$\mu_{B'}(v) = \bigvee_u \{\mu_{A'}(u) \wedge [1 \wedge (1 - \mu_A(u) + \mu_B(v))]\}, \quad (18)$$

where $\vee = \max$, and $\wedge = \min$.

There are many other definitions of the fuzzy implication $A \rightarrow B$ and the definitions of composition $\circ$. For example,

$$\mu_{B'}(u) = \bigvee_u \{\mu_{A'}(u) \wedge [\mu_A(u) \wedge \mu_B(v)]\}$$
$$\mu_{B'}(u) = \bigvee_u \{\mu_{A'}(u) \wedge [\mu_B(v)^{\mu_A(u)}]\}. \qquad (19)$$

The compositional rule of inference has local properties, and a mathematical model should be chosen with respect to the problem under consideration.

The extension principle introduced by L. A. Zadeh is one of the most basic ideas of fuzzy set theory. It provides a general method for extending nonfuzzy mathematical concepts in order to deal with fuzzy quantities.

Let X be a Cartesian product of universes, $X = X_1 \times X_2 \times \cdots \times X_r$, and $A_1, \ldots, A_r$ be r fuzzy sets in $X_1, \ldots, X_r$, respectively.

Let f be a mapping from $X_1 \times \cdots \times X_r$ to a universe Y such that $y = f(x_1, \ldots, x_r)$. The extension principle allows us to induce from r fuzzy sets A_i a fuzzy set B on Y through f such that

$$\mu_B(y) = \sup_{x_1 \cdots x_r} \{\min(\mu_{A_1}(x_1), \ldots, \mu_{A_r}(x_r))\}$$
$$y = f(x_1, \ldots, x_r) \qquad (20)$$
$$\mu_B(y) = 0, \qquad \text{if } f^{-1}(y) = \emptyset,$$

where $f^{-1}(y)$ is the inverse image of y, and $\mu_B(y)$ is the greatest among the membership values $\mu_{A_1 \times \cdots \times A_r}(x_1, \ldots, x_r)$ of the realizations of y using r-tuples $(x_1, \ldots, x_r)$.

C. Fuzzy Dynamic Systems

The basic equation of the fuzzy dynamic system has the following form

$$X_{k+1} = X_k \circ U_k \circ R \qquad k = 0, 1, 2, \ldots, \quad (21)$$

where X_k and X_{k+1} stand for fuzzy sets of the states at the kth and $(k+1)$th time instants, respectively, and U_k the input at the kth instant. The states and control are expressed by means of membership function given by

$$X_k, X_{k+1} \in F(X) \qquad \mu_{X_k}, \mu_{X_{k+1}} \colon X \to [0, 1]$$

$$U_k \in F(U) \qquad \mu_{U_k} \colon U \to [0, 1], \tag{22}$$

where $F(\cdot)$ denotes a family of fuzzy sets defined on a proper space $F(X) = \{X \mid \mu_X \colon X \to [0, 1]\}$. Then R is a fuzzy relation describing the fuzzy dynamical system and is defined on the Cartesian product of $X \times U \times X$

$$R \in F(X \times U \times X), \quad \mu_R \colon X \times U \times X \to [0, 1]. \tag{23}$$

The operator $\circ$ stands for max–min composition. Therefore, the time-evolution of the membership function of a fuzzy dynamic system can be expressed as follows.

$$\mu_{X_{k+1}}(x) = \max_{u \in U} \max_{x \in X} \left\{ \min\left[\mu_{U_k}(u), \mu_{X_k}(x), \mu_R(x, u, x) \right] \right\}. \tag{24}$$

Generally, if X_k and U_k are interactive fuzzy sets, the fuzzy system has the form

$$\begin{aligned} X_{k+1} &= (X_k U_k) \circ R \\ (X_k U_k) &\in F(X \times U) \\ R &\in F(X \times U \times X) \\ X_{k+1} &\in F(X) \end{aligned} \tag{25}$$

and

$$\mu_{X_{k+1}}(x) = \max_{u \in U} \max_{x \in X} \left\{ \min\left[\mu_{X_k U_k}(x, u), \mu_R(x, u, x) \right] \right\}. \tag{26}$$

The above is a fuzzy equation of the first order. A fuzzy equation of higher order has the following form.

$$X_{k+p} = X_k \circ X_{k+1} \circ \cdots \circ X_{k+p-1} \circ R \qquad p \geq 1. \tag{27}$$

Very often, other types of compositions are used in fuzzy systems. Special interest plays sup-prod composition,

$$X_{k+1} = X_k * U_k * R.$$

$$\mu_{X_{k+1}}(x) = \sup_{u \in U} \sup_{x \in X} \left\{ \mu_{X_k}(x) \mu_{U_k}(u) \mu_R(x, u, x) \right\} \tag{28}$$

D. Fuzzy Inverse Relations

For resolution of the fuzzy relational equations of the types $Y = X \circ T$ and $Y = X * T$, consider the following two problems: (1) find T if X and Y are given, and (2) find X if Y and T are given.

If $Y = X \circ T$ is true, then the greatest fuzzy relations $\hat{T}$ satisfying the equation $Y = X \circ \hat{T}$ is equal to $\hat{T} = X \text{\textcircled{α}} Y$,

where $\text{\textcircled{$\alpha$}}$ stands for the α-composition of fuzzy set X and Y. The membership function of $\hat{T}$ is equal to

$$\mu_{\hat{T}}(x, y) = \mu_{X \text{\textcircled{α}} Y}(x, y) = \mu_X(x) \alpha \mu_Y(y)$$

$$= \begin{cases} 1 & \text{if} \quad \mu_X(x) \leq \mu_Y(y) \\ \mu_Y(y) & \text{if} \quad \mu_X(x) > \mu_y(y). \end{cases} \tag{29}$$

If $Y = X \circ T$ is satisfied, the greatest fuzzy set $\hat{X}$ such that $Y = \hat{X} \circ T$ holds true and is equal to

$$\hat{Y} =$$

$$\hat{Y} = T \text{\textcircled{α}} Y \qquad \mu_{\hat{X}}(x) = \inf_{y \in Y} \{ \mu_T(x, y) \alpha \mu_Y(Y) \}. \tag{30}$$

If $Y = X * T$ holds true, then the greater fuzzy relation $\hat{T}$, such that $Y = X * \hat{T}$ is satisfied, is equal to

$$\hat{T} = X \text{\textcircled{ψ}} Y, \tag{31}$$

where $\text{\textcircled{$\psi$}}$ denotes the ψ-composition of fuzzy sets X and Y

$$\mu_{\hat{T}}(x, y) = \mu_{X \text{\textcircled{ψ}} Y} = \mu_X \psi \mu_Y(y)$$

$$= \min(1, \mu_Y(y) / \mu_X(X)). \tag{32}$$

If $Y = X * T$ is satisfied, then the greatest fuzzy set $\hat{X}$ such that $Y = \hat{X} * T$ is satisfied is equal to

$$\hat{X} = T \text{\textcircled{ψ}} Y \qquad \mu_{\hat{X}} = \prod_{y \in Y} \{ \mu_T(x, y) \psi \mu_Y(y) \}. \tag{33}$$

E. Fuzzy Numbers

Fuzzy numbers play the same role in building fuzzy models as real or integer number in the conventional models.

A fuzzy number is defined as a fuzzy set $A \subset R$, where R is the real line, with a convex membership function. It is usually assumed that the fuzzy numbers are normal fuzzy sets.

The operations on the fuzzy numbers may be obtained by applying the extension principle to the respective nonfuzzy operations on the real numbers.

Let $A, B \subset R$ be two fuzzy numbers characterized by $\mu_A(x)$ and $\mu_B(x)$, respectively.

The four basic extended arithmetic operations on them are as follows.

Addition.

$$\mu_{A+B}(z) = \max_{x+y=z} [\mu_A(x) \wedge \mu_B(y)] \qquad \forall z \in R. \tag{34}$$

Subtraction.

$$\mu_{A-B}(z) = \max_{x-y=z} [\mu_A(x) \wedge \mu_B(y)] \qquad \forall z \in R. \tag{35}$$

Multiplication.

$$\mu_{A \cdot B}(z) = \max_{x \cdot y=z} [\mu_A(x) \wedge \mu_B(y)] \qquad \forall z \in R. \tag{36}$$

Division.

$$\mu_{A/B}(z) = \max_{x/y=x,\, y \neq 0} [\mu_A(x) \wedge \mu_B(y)] \qquad \forall z \in R. \quad (37)$$

F. Probability of a Fuzzy Event

Fuzziness and randomness, although they both deal with uncertainty, are different concepts. Randomness deals with uncertainty arising from a physical phenomenon, whereas fuzziness arises from the human thought process. A joint occurence of fuzziness and randomness is a common phenomenon. To formally deal with such problems, the concept of probability of a fuzzy event will be introduced now.

The *fuzzy event* is defined as a fuzzy set A in $X = \{x_1, \ldots, x_n\}$, $A \subset X$, with a Borel measurable membership function. We assume that the probabilities $p(x_1), \ldots, p(x_n)$ are known and $p(x_1) + \cdots + p(x_n) = 1$.

The *(nonfuzzy) probability of a fuzzy event* $A \subset X = \{x_1, \ldots, x_n\}$, written $p(A)$, is defined as

$$p(A) = \sum_{i=1}^{n} \mu_A(x_i) \cdot p(x_i), \quad (38)$$

that is, as the expected value of $\mu_A(x)$.

The *(fuzzy) probability of a fuzzy event* $A \subset X = \{x_1, \ldots, x_n\}$, written $P(A)$, is defined as the following fuzzy set in the interval $[0, 1]$

$$p(A) = \sum_{\alpha \in [0,1]} \alpha \Big/ p(A_\alpha), \quad (39a)$$

or in terms of membership function,

$$\mu_{P(A)}(p(A_\alpha)) = \alpha, \qquad \forall \alpha \in [0, 1], \quad (39b)$$

where A_α is the α-cut of A.

G. Possibility Distribution

Let Y be a variable taking values in X. Then a possibility distribution Π_Y associated with Y may be viewed as a fuzzy constraint on the values that may be assigned to Y. Such a distribution is characterized by a possibility distribution function $\pi_Y : X \to [0, 1]$, which associates with each $x \in X$ the "degree of ease" or the possibility that Y may take x as a value.

If F is a fuzzy subset of X characterized by its membership function $\chi_F : X \to [0, 1]$, then the proposition "Y is F" induces a possibility distribution Π_Y that is equal of F. Equivalently, "Y is F" translates into the possibility assignment equation $\Pi_Y = F$, that is,

$$Y \text{ is } F \to \Pi_Y = F, \quad (40)$$

which signifies that the proposition "Y is F" has the effect of constraining the values that may be assumed by Y, with the possibility distribution Π_Y identified with F.

If Y_1 and Y_2 are two variables taking values in X_1 and X_2, respectively, then the joint and conditional possibility distributions are defined through their respective distribution functions

$$\pi_{(Y_1, Y_2)}(x_1, x_2) = \text{poss}\{Y_1 = x_1, Y_2 = x_2\}$$
$$x_1 \in X_1, x_2 \in X_2 \quad (41)$$

and

$$\pi_{(Y_1 | Y_2)}(x_1 \mid x_2) = \text{poss}\{Y_1 = x_1 \mid Y_2 = x_2\}, \quad (42)$$

where the last equation represents the conditional distribution function of Y_1 given Y_2.

Let π be a possibility distribution induced by a fuzzy set F in X. Let G be a nonfuzzy set of X. The possibility that x belongs to G is $\Pi(G)$ where

$$\Pi(G) = \sup_{x \in G} \chi_F(x) = \sup_{x \in G} \pi(G). \quad (43)$$

H. Modifier Rule

If

$$Y \text{ is } F \to \Pi_Y = F \quad (44)$$

then

$$Y \text{ is } mF \to \Pi_Y = F^+, \quad (45)$$

where m is a modifier such as not, very, or more or less, and F^+ is a modification of F induced by m.

I. Conjunctive, Disjunctive, and Implicational Rules

If

$$Y \text{ is } F \to \Pi_Y = F \quad \text{and} \quad Z \text{ is } G \to \Pi_Z = G, \quad (46)$$

where F and G are fuzzy subset X_1 and X_2, respectively, then

(a) Y is F and Z is $G \to \Pi_{(Y,Z)} = F \times G$ where

$$\chi_{F \times G}(X_1, X_2) = \chi_F(X_1) \wedge \chi_G(X_2) \quad (47)$$

(b) Y is F or Z is $G \to \Pi_{(Y,Z)} = \bar{F} \cup \bar{G}$, where $\bar{F} = F \times X_2$, $\bar{G} = X_1 \times G$, and

$$\chi_{\bar{F} \cup \bar{G}}(x_1, x_2) = \chi_F(X_1) \wedge \chi_G(x_2) \quad (48)$$

(c) If Y is F, then Z is $G \to \Pi_{(Z|Y)} = \bar{F} \oplus \bar{G}$, where $\Pi_{(Z|Y)}$ denotes the conditional possibility distribution of Z given Y, and the bounded sum $\oplus$ is defined by

$$\chi_{\bar{F} \oplus \bar{G}}(x_1, x_2) = 1 \wedge (1 - \chi_F(X_1) + \chi_G(X_2)). \quad (49)$$

J. Truth Qualification Rule

Let τ be a fuzzy truth-value, for example, very true, quite true, or more or less true. Such a truth-value may be regarded as a fuzzy subset of the unit interval that is characterized by a membership function $\chi_\tau : [0, 1] \to [0, 1]$.

A truth-qualified proposition can be expressed as "Y is F is τ." The translation rule for such propositions can be given by

$$Y \text{ is } F \text{ is } \tau \to \Pi_Y = F^+, \tag{50}$$

where

$$\chi_{F^+}(x) = \chi_\tau(\chi_F(x)).$$

K. Projection Rule

Consider a fuzzy proposition with a translation that is expressed as

$$p \to \Pi_{(Y_1,\dots,Y_n)} = F \tag{51}$$

and let $Y(s)$ denote a subvariable of the variable $Y = (Y_1, \dots, Y_n)$. Let $\Pi_{Y(s)}$ denote the marginal possibility distribution of $Y(s)$; that is,

$$\Pi_{Y(s)} = \text{proj}_{X(s)} F, \tag{52}$$

where X_i, $i = 1, \dots, n$, is the universe of discourse associated with Y_i: $X(s) = X_{i_1} \times \cdots \times X_{i_k}$ and the projection of F on $X_{(s)}$ is defined by the possibility distribution function

$$\pi_{Y_s}(x_{i_1}, \dots, x_{i_k}) = \sup_{x_{j1},\dots,x_{jm}} \chi_F(x_1, \dots, x_n). \tag{53}$$

Let q be a retranslation of the possibility assignment equation

$$\Pi_{Y(s)} = \text{proj}_{X(s)} F. \tag{54}$$

Then the projection rule asserts that q may be inferred from p

$$p \to \Pi_{(Y_1,\dots,Y_n)} = F \qquad q \leftarrow \Pi_{Y(s)} = \text{proj}_{x(s)} F. \tag{55}$$

L. Conjunction Rule

Consider a proposition p, which is an assertion concerning the possible values of, say, two variables X and Y that take values in U and V, respectively. Similarly, let q be an assertion concerning the possible values of the variables Y and Z, taking values in V and W. With these assumptions, the translation of p and q may be expressed as

$$p \to \Pi^p_{(X,Y)} = F \quad \text{and} \quad q \to \Pi^p_{(Y,Z)} = G. \tag{56}$$

Let $\bar{F}$ and $\bar{G}$ be, respectively, the cylindrical extensions of F and G in $U \times V \times W$. Thus,

$$\bar{F} = F \times W$$

and

$$\bar{G} = U \times G.$$

Using the conjunction rule, one can infer from p and q a proposition that is defined by the following scheme.

$$r \to \Pi^p_{(X,Y)} = F$$
$$\underline{q \to \Pi^p_{(Y,Z)} = G} \tag{57}$$
$$r \leftarrow \Pi_{(X,Y,Z)} = \bar{F} \cap \bar{G}$$

M. Compositional Rule of Inference

Applying the projection rule, we obtain the following inference scheme.

$$p \to \Pi^p_{(X,Y)} = F$$
$$\underline{q \to \Pi^p_{(Y,Z)} = G} \tag{58}$$
$$r \leftarrow \Pi^r_{(X,Z)} = F \circ G$$

where the composition of F and G is defined by

$$\chi_{F \circ G}(u, w) = \sup_v (\chi_F(u, v) \wedge \chi_G(v, w)).$$

In particular, if p is a proposition of the form "X is F" and q is a proposition of the form "is X is G, then Y is H," then

$$p \to \Pi_X = F$$
$$\underline{q \to \Pi_{(Y|X)} = \bar{G}' + \bar{H}} \tag{59}$$
$$r \leftarrow \Pi_{(Y)} = F \circ (\bar{G}' + \bar{H})$$

A concept that plays an important role in the representation of imprecise data is that of disposition. Informally, a disposition is a proposition that is preponderantly true, but not necessarily always true. More specifically, a proposition such as "most doctors are not very tall" may be regarded as a dispositional proposition, in the sense that it describes a disposition of doctors to be not very tall. Dispositions play a central role in human reasoning, since much of human knowledge and, especially, common-sense knowledge may be viewed as a collection of dispositions.

III. LINGUISTICS AND FUZZY MODELING AND CONTROL

A process operator, having an assigned control goal, observes the process state, control, and process output, and intuitively assesses the variables and parameters of the process. From a subjective assessment of these quantities, the operator makes a decision and performs a manual alternation of the system control value so as to achieve an assigned control goal. In this way, he or she makes a description of control strategy, a description of process behavior, and finally a subjective operational model of the operator's procedure. It should be noted that such a control algorithm is very flexible (elastic) and adequate to the actual process situation, and, as a rule, it is much better than control algorithms obtained using modern control theory. The linguistic algorithm comprises all the "meta physical" skills of the operator, such as intuition, experience, intelligence, learning, adaptation, and memory, which can not be dealt with by the conventional mathematics. The operator's "mind" and his or her knowledge of the controlled technological process can be formalized mathematically, to some extent, using the theory of fuzzy sets and fuzzy logic.

Suppose that in observing an error and error change, the process operator has made a manual control of the process according to his or her own hypothetical verbal description:

If E = big and ΔE = small, then U = zero.

also

If E = medium and ΔE = zero, then U = big.

also

.

.

.

also

If E = zero and ΔE = zero, then U = zero.

Let

$$\mathscr{E} = \{e_i\} \subset (-\infty, \infty), \qquad \Delta\mathscr{E} = \{s_j\} \subset (-\infty, \infty)$$

$$\mathscr{U} = \{u_k\} \subset (-\infty, \infty)$$

be the finite space of error, error change, and control. Further, let

$$\mathscr{F}(E) = \{E \mid \mathscr{E} \to [0, 1]\}$$

$$\mathscr{F}(\Delta E) = \{S \mid \Delta\mathscr{E} \to [0, 1]\} \tag{60}$$

$$\mathscr{F}(U) = \{U \mid \mathscr{U} \to [0, 1]\}$$

be the finite family of fuzzy set of error change, and control; then the above-presented verbal description can be formalized in the fuzzy relation

$$R: \mathscr{F}(E) \times \mathscr{F}(\Delta E) \times \mathscr{F}(U) \to [0, 1],$$

where x is a fuzzy Cartesian product.

In order to calculate control U, when given fuzzy relation R, error E, and error change ΔE, the compositional rule of inference is to be used,

$$U = E \circ \Delta E \circ R,$$

where $\circ$ is the max–min composition given in Eq. (16).

Since in the control system an on-line nonfuzzy control action is necessary, the fuzzy set U must undergo defuzzification. Defuzzification is accomplished using either a "center-of-area" or "mean-of-maxima" algorithm.

Center-of-area algorithm:

$$u_0 = \frac{\sum_{i=1}^{n} \mu_U(u_i) \cdot u_i}{\sum_{i=1}^{n} \mu_u(u_i)} \qquad u_i \in U \tag{61}$$

$$U = \{u_1, u_2, u_3, \ldots, u_n\}.$$

Mean-of-maxima algorithm:

$$u_0 = \frac{\sum_{i-1}^{n} u_i}{I} \tag{62}$$

$$\mu_U(u_i) = \max_u \mu_U(U) \qquad i = 1, 2, 3, \ldots, I.$$

There are two fundamental problems in fuzzy control engineering (1) designing of control rules, and (2) correctness of control rules.

There are four ways to obtain control rules.

1. They may be taken directly from the operator's experience.

2. They may be derived from a fuzzy model of the process, where the process characteristics are expressed in a form similar to control rules.

3. The operator's actions may be monitored and a control derived using a fuzzy model of the operator's behavior.

4. The rule maybe learned by the controller.

The last option is the most promising one and has been used at the Queen Mary College in London for self-organizing controllers for robot arms.

A sophisticated learning algorithm that automatically deduces the fuzzy control rules through a computer analysis of the operator's control actions has been reported by M. Sugeno of the Tokyo Institute of Technology in Japan. The correctness of control rules is improving by evaluating some properties of the fuzzy relation R. The stability, observability, accuracy, good mapping property, and reproducibility of fuzzy control systems are investigated, and some improvements from the operating point of view are obtained.

Another approach to the correctness of the control rules are realized by identification procedures of fuzzy relations. An intelligent automatic controller and an expert fuzzy controller were proposed by R. Tong and E. Mamdani.

IV. APPLICATIONS OF FUZZY LOGIC TO SYSTEMS MODELING AND CONTROL

A. Fuzzy Model of a dc Motor

A dc series motor (see Fig. 1) is investigated. The ratings of the series motor are as follows.

Power: $P_N = 0.7 \text{ kW}$

Voltage: $U_N = 110 \text{ V}$

Rated current: $I_N = 8.84 \text{ A}$

Rated speed: $n_N = 1500 \text{ rpm}$

From the measurement of the current value i and the rotating speed value n in steady states, the real static

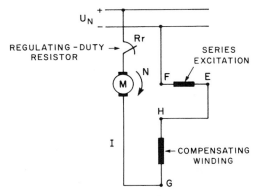

FIGURE 1 DC series motor: *I*, motor current *N*, rpm; U_N, supply voltage; *M*, series motor.

characteristic of the motor was determined, $n = f(i)$ (see Fig. 2).

The process operator observes the influence of current change *I* on the rotating speed *N* of the series motor in steady states and formulates the following verbal description, which is the linguistic static characteristic of the motor.

> If *I* = null then *N* = very big.
> also
> If *I* = zero then *N* = big.
> also
> If *I* = small then *N* = medium.
> also   (63)
> If *I* = medium then *N* = small.
> also
> If *I* = big then *N* = zero.
> also
> If *I* = very big then *N* = zero.

Here, *I* is the linguistic variable of the motor current and *N* the linguistic variable of the motor speed.

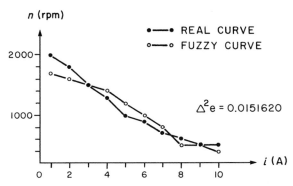

FIGURE 2 Real and fuzzy curves for relation R_{2*}.

TABLE I Fuzzy Sets of Motor Current *I*

A (amp)	Fuzzy sets[a] (Universe)					
	N	**Z**	**S**	**M**	**B**	**SB**
0.00	1.00	0.00	0.00	0.00	0.00	0.00
0.50	0.75	0.25	0.00	0.00	0.00	0.00
1.00	0.50	0.50	0.00	0.00	0.00	0.00
1.50	0.25	0.75	0.00	0.00	0.00	0.00
2.00	0.00	1.00	0.00	0.00	0.00	0.00
2.50	0.00	0.75	0.25	0.00	0.00	0.00
3.00	0.00	0.50	0.50	0.00	0.00	0.00
3.50	0.00	0.25	0.75	0.00	0.00	0.00
4.00	0.00	0.00	1.00	0.00	0.00	0.00
4.50	0.00	0.00	0.75	0.25	0.00	0.00
5.00	0.00	0.00	0.50	0.50	0.00	0.00
5.50	0.00	0.00	0.25	0.75	0.00	0.00
6.00	0.00	0.00	0.00	1.00	0.00	0.00
6.50	0.00	0.00	0.00	0.75	0.25	0.00
7.00	0.00	0.00	0.00	0.50	0.50	0.00
7.50	0.00	0.00	0.00	0.25	0.75	0.00
8.00	0.00	0.00	0.00	0.00	1.00	0.00
8.50	0.00	0.00	0.00	0.00	0.75	0.25
9.00	0.00	0.00	0.00	0.00	0.50	0.50
9.50	0.00	0.00	0.00	0.00	0.75	0.75
10.00	0.00	0.00	0.00	0.00	0.00	1.00

[a] N, null; Z, zero; S, small; M, medium; B, big; SB, super big.

The subjective definitions of fuzzy sets for the motor current *I* and the rotating speed *N* are shown in Tables I and II, respectively.

The verbal description can be formalized mathematically as

$$N = I \circ R, \tag{64}$$

where *N* is the fuzzy variable of the motor rotations, *I* is the fuzzy variable of the motor current, $\circ$ is the max–min composition, and *R* is the fuzzy relation.

The deterministic value of the rotating speed *n* is calculated according to the following defuzzification operator (mean of maximum support values)

$$n = \sum_{k=1}^{n} n_k \Big/ l, \tag{65}$$

where n_k are the support values in which the function $\mu_N(n)$ reaches its maximum, and *l* is the number of support elements in which the membership function $\mu_N(n)$ reaches the maximum value.

The root-mean-square error is assumed to be the criterion for the estimation of the applicability of the fuzzy model,

TABLE II   Fuzzy Sets of Motor Speed N

| n(RPM) | Fuzzy sets[a] (Universe) | | | | |
	Z	S	M	B	SB
400	1.00	0.00	0.00	0.00	0.00
500	0.75	0.25	0.00	0.00	0.00
600	0.50	0.50	0.00	0.00	0.00
700	0.25	0.75	0.00	0.00	0.00
800	0.00	1.00	0.00	0.00	0.00
900	0.00	0.75	0.25	0.00	0.00
1000	0.00	0.50	0.50	0.00	0.00
1100	0.00	0.25	0.75	0.00	0.00
1200	0.00	0.00	1.00	0.00	0.00
1300	0.00	0.00	0.75	0.25	0.00
1400	0.00	0.00	0.50	0.50	0.00
1500	0.00	0.00	0.25	0.75	0.00
1600	0.00	0.00	0.00	1.00	0.00
1700	0.00	0.00	0.00	0.75	0.25
1800	0.00	0.00	0.00	0.50	0.50
1900	0.00	0.00	0.00	0.25	0.75
2000	0.00	0.00	0.00	0.00	1.00

[a] See symbols in Table I.

$$\Delta^2 e = \sum_{l=1}^{v}(n_{rl} - n_{ml})^2 \Big/ \sum_{i=1}^{u} n_{rl}^2, \qquad (66)$$

where n_{rl} is the answer of the real system at point l, n_{ml} is the answer of fuzzy model at point l, and v is the number of discretization intervals of the rotating axis.

The following definitions of fuzzy relations are adopted.

$$\mu_{R_2^*}(i_t, n_j) = \bigwedge_{s=1}^{6} \begin{cases} 1, & \mu_I(i_t) \leq \mu_N(n_j) \\ 0, & \text{otherwise} \end{cases}$$

$$\mu_{R_5^*}(i_t, n_j) = \bigwedge_{s=1}^{6} \begin{cases} 1, & \mu_I(i_t) = \mu_N(n_j) \\ \mu_N(n_j), & \text{otherwise} \end{cases}$$

$$\mu_{R_7^*}(i_t, n_j) = \bigwedge_{s=1}^{6}\{(\mu_I(i_t) \wedge \mu_N(n_j)) \qquad (67)$$
$$\vee (1 - \mu_I(i_t))\}$$

$$\mu_{R_{30}}(i_t, n_j) = \bigwedge_{s=1}^{6}\{1 \wedge (\mu_N(n_j) + \mu_I(i_t))\},$$

where μ_R is the value of the membership function in relation R, t is the number of support elements of variable I, j is the number of support elements of variable N, s is the number of sentences in verbal description, μ_I is the membership function of fuzzy set of variable I, μ_N is the membership function of fuzzy set of variable N, $\vee$ is the max operator, and $\wedge$ is the min operator. The results of the experimental investigation are shown in Figs. 2, 3, 4, and 5.

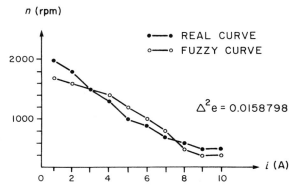

FIGURE 3   Real and fuzzy curves for relation R_{3^*}.

The results of the experiment allow us to state that both the definition of fuzzy implication "If $\cdots$ then" and the definition of the connective "also" exert a significant influence on the accuracy of a fuzzy model.

We suggest that a fuzzy relation of the R_{2^*} type should be used in the fuzzy modeling of electromechanical system because this type gives the smallest error $\Delta^2 e = 0.151620 \times 10^{-1}$ and requires the fewest computational operations.

B. Fuzzy Controller of a dc Motor

A fuzzy control system of dc motor (see Fig. 6) is investigated. The task of the fuzzy controller is to maintain a given constant rotating speed of the motor w_z despite the effect of the disturbing moment $M(t)$.

The process operator formulated the following verbal description of control actions using the variables S (error-sum), I (motor current), and U (control action), where nv is negative and sp is super.

If $S = $ null and $I = $ small, then $U = $ zero.
also
If $S = $ null and $I = $ big, then $U = $ big.
also

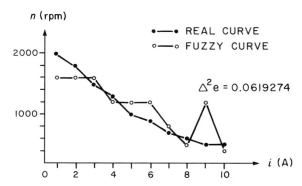

FIGURE 4   Real and fuzzy curves for relation R_{7^*}.

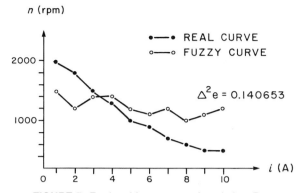

FIGURE 5 Real and fuzzy curves for relation R_{30}.

If S = zero and I = small, then U = zero.
also
If S = zero and I = big, then U = medium.
also
If S = small and I = zero, then U = null.
also
If S = small and I = small, then U = null.
also
If S = small and I = medium, then U = zero.
also
If S = small and I = big, then U = zero.
also
If S = medium and I = small, then U = null.
also
If S = small and I = big, then U = small.
also
If S = big and I = medium, then U = zero.
also
If S = big and I = big, then U = small.
also
If S = sp-big and I = medium, then U = zero.
also

If S = nv-null and I = zero, then U = null.
also
If S = nv-null and I = medium, then U = zero.
also
If S = nv-zero and I = zero, then U = null.
also
If S = nv-zero and I = small, then U = zero.
also
If S = nv-zero and I = medium, then U = big.
also
If S = nv-small and I = zero, then U = zero.
also
If S = nv-small and I = small, then U = big.
also
If S = nv-medium and I = zero, then U = null.
also
If S = nv-medium and I = small, then U = big.
also
If S = nv-medium and I = medium, then U = big.
also
If S = nv-big and I = zero, then U = zero.
also
If S = nv-big and I = small, then U = big.
also
If S = nv-sp-big and I = zero, then U = small.
also
If S = nv-sp-big and I = small, then U = medium.

The fuzzy relation was determined according to the formula

$$\mu_R(s, i, u) = \vee\{\mu_S(s) \wedge \mu_I(i) \wedge \mu_U(u)\}.$$

The result of this experiment is given in Fig. 7. The result was satisfactory.

By virtue of the excellent practical qualities of fuzzy control theory, many applications of it are found in steam

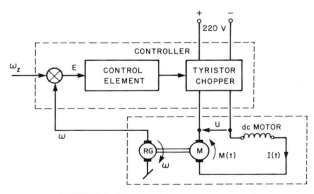

FIGURE 6 Elective-drive process control.

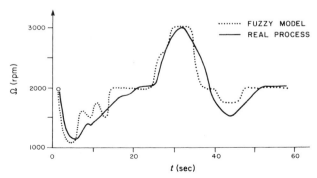

FIGURE 7 Motor vehicles for real process and fuzzy model.

engine fuzzy control, heat exchanger fuzzy control, sinter plant fuzzy control, traffic fuzzy control, fuzzy control of pressurized tank, motor fuzzy control, ship fuzzy autopilot, automobile speed fuzzy control, water cleaning process fuzzy control, room-temperature fuzzy control, pump operations fuzzy control, fuzzy robot control, prosthetic devices fuzzy control, warm-water plant fuzzy control, pressurized tank liquid fuzzy control, batch chemical reactor fuzzy control, nuclear fuel extraction process fuzzy control, cement kiln fuzzy control, activated-sludge wastewater treatment fuzzy control, fuzzy control for maintenance scheduling in transportation system, fuzzy control of stirred tank, fuzzy control of the process of changes of methane concentration, canvas production fuzzy control, fuzzy control of fermentation process of antibiotic synthesis, hydraulic servo-system fuzzy control, pulp-plant fuzzy control, electric network fuzzy control, converter steel-making process fuzzy control, water-tank system fuzzy control, robot arc-welder fuzzy control, predictive fuzzy logic controller of automatic operation of a train, self-learning fuzzy controller of a car, fuzzy controller of aircraft flight, fuzzy algorithm for path selection in autonomous vehicle navigation, and fuzzy adaptive control of continuous casting plant.

V. CONCLUDING REMARKS

Some basic concepts and basic operations of fuzzy set theory are introduced in this article. It is believed that any activity of the human being where knowledge, experience, and approximate reasoning play a dominant role is a potential area of applications for fuzzy expert systems.

However, there are many problems that need to be solved. These include how many rules are needed to guarantee the correctness of linguistic description; how to obtain control rules; how to modify rules under the influence of environmental conditions change; and software and hardware packages for inexact knowledge processing.

SEE ALSO THE FOLLOWING ARTICLES

ARTIFICIAL INTELLIGENCE • ARTIFICIAL NEURAL NETWORKS • HUMAN–COMPUTER INTERACTION • MATHEMATICAL MODELING • PROCESS CONTROL SYSTEMS • SET THEORY

BIBLIOGRAPHY

Ayyub, B. M., Gupta, M. M., and Kanal, L. N. (eds.) (1992). "Analysis and Management of Uncertainty: Theory and Application," Kluwer Academic, Dordrecht.

Bezdek, J. C. (1991). "Pattern Recognition with Fuzzy Objective Function Algorithms," Plenum Press, New York.

Carpenter, G. A., Grossberg, S., Markuzon, N., Reynolds, J. H., and Rosen, D. B. (1992). "Fuzzy ARTMAP: A Neural Network Architecture for Incremental Supervised Learning of Analog Multidimensional Maps," *IEEE Trans. Neural Networks*, Vol. 3, No. 5, September, 698–713.

Dubois, D., and Prade, H. (1980). "Fuzzy Sets and Systems: Theory and Applications," Academic, New York.

Gupta, M. M. (1988). "On the cognitive computing: perspectives," *In* M. M. Gupta and T. Yamakawa (eds.), Fuzzy Computing: Theory, Hardware and Applications. North-Holland, New York.

Gupta, M. M. (1991). "Uncertainty and Information: The Emerging Paradigms," *Int. J. Neuro Mass-Parallel Computing and Information Systems*, Vol. 2, 65–70.

Gupta, M. M. (1992). "Fuzzy Neural Computing Systems," *Neural Network World*, Vol. 2, No. 6, 629–648.

Gupta, M. M., Kandel, A., and Bandler, W. (eds.) (1985). "Approximate Reasoning in Expert Systems," North-Holland, New York.

Gupta, M. M., and Kandel, A., *et al.* (eds.) (1985). "Approximate Reasoning in Expert Systems," North-Holland, Amsterdam.

Gupta, M. M., and Knopf, G. K. (1990). "Fuzzy Neural Network Approach to Control Systems," in *Proc. First Int. Symp. Uncertainty Modeling and Analysis*, Maryland, 483–488.

Gupta, M. M., and Qi, J. (1991). "On Fuzzy Neuron Models," in *Proc. Int. Joint Conf. Neural Networks (IJCNN)*, Seattle, 431–456.

Gupta, M. M., and Rao, D. H. (1993). "Virtual Cognitive Systems (VCS): Neural-Fuzzy Logic Approach," in *Proc. IFAC Conf.*, Sydney, Australia, Vol. 8, 323–330.

Gupta, M. M., Ragade, R. K., and Yager, R. R. (eds.) (1979). "Advances in Fuzzy Set Theory and Applications," North-Holland, Amsterdam.

Gupta, M. M., and Sanchez, E. (eds.) (1982). "Approximate Reasoning in Decision Analysis," North-Holland, Amsterdam.

Gupta, M. M., and Sanchez, E. (eds.) (1982). "Fuzzy Information and Decision Processes," North-Holland, Amsterdam.

Gupta, M. M., and Sanchez, E. (eds.) (1982). "Fuzzy Information and Decision Processes," North-Holland, New York.

Gupta, M. M., and Sanchez, E. (eds.) (1982). "Approximate Reasoning in Decision Analysis," North-Holland, New York.

Gupta, M. M., Saridis, G. N., and Gaines, B. R. (eds.) (1977). "Fuzzy Automata and Decision Processes," North-Holland, New York.

Gupta, M. M., and Singh, M. (eds.) (1987). "International Encyclopedia of Control Systems," Pergamon Press, Oxford.

Gupta, M. M., and Sinha, N. K. (eds.) (1996). "Intelligent Control Systems: Theory and Applications," IEEE Press, New York.

Gupta, M. M., and Yamakawa, T. (eds.) (1988). "Fuzzy Computing: Theory, Hardware and Applications," North Holland, Amsterdam.

Gupta, M. M., and Yamakawa, T. (eds.) (1988). "Fuzzy Logic in Knowledge-Based Systems: Decision and Control," North Holland, Amsterdam.

Gupta, M. M., and Yamakawa, T. (eds.) (1988). "Fuzzy Computing: Theory, Hardware and Applications," North-Holland, New York.

Gupta, M. M., and Yamakawa, T. (eds.) (1988). "Fuzzy Logic in Knowledge-Based Systems, Decision and Control," North-Holland, New York.

Hayashi, I., Nomura, H., and Wakami, N. (1989). "Artificial Neural Network Driven Fuzzy Control and Its Application to Learning of Inverted Pendulum System," in *Proc. Third IFSA Congress*, Seattle, WA, 610–116.

Kandel, A. (1982). "Fuzzy Techniques in Pattern Recognition," Wiley, New York.

Kaufmann, A. (1975). "Introduction to the Theory of Fuzzy Subsets," Vol. 1. Academic, New York.

Kaufmann, A., and Gupta, M. M. (1991). "Introduction to Fuzzy Arithmetic," 2nd ed. Van Nostrand, New York.

Kaufmann, A., and Gupta, M. M. (1988). "Fuzzy Mathematical Models in Engineering and Management Science," North Holland, Amsterdam.

Kaufmann, A., and Gupta, M. M. (1991). "Introduction to Fuzzy Arithmetic: Theory and Applications," 2nd ed. Van Nostrand Reinhold, New York.

Kaufmann, A., and Gupta, M. M. (1992). "Fuzzy Mathematical Models in Engineering and Management Science," North-Holland, New York.

Kiska, J. B., and Gupta, M. M. (1990). "Fuzzy Logic Neural Network," BUSE-FAL, No. 4, 104–109.

Kiszka, J., Kochanska, M., and Sliwinska, D. (1985). "Fuzzy Sets and Systems **15** and **16**," 111–128 and 223–240.

Kosko, B. (1992). "Neural Networks and Fuzzy Systems," Prentice-Hall, Englewood Cliffs, NJ.

Kuncicky, D. C., and Kandel, A. (1989). "A Fuzzy Interpretation of Neural Networks," in *Proc. Third IFSA Congress*, Seattle, WA, 113–116.

Mamdani, E. H. (1983). *In* "Designing for Human–Computer Communication" (M. E. Sime and M. J. Coombs, eds.), Academic, New York.

Nakanishi, S., Takagi, T., Uehara, K., and Gotoh, Y. (1990). "Self-Organizing Fuzzy Controllers by Neural Networks," in *Proc. Int. Conf. Fuzzy Logic and Neural Networks IIZUKA '90*, Japan, 187–192.

Paul, S. K., and Mitra, S. (1992). "Multilayer Perceptron, Fuzzy Sets, and Classification," *IEEE Trans. Neural Networks*, Vol. 3, No. 5, September, 683–697.

Simpson, P. K. (1992). "Fuzzy Min-Max Neural Networks—Part 1: Classification," *IEEE Trans. Neural Networks*, Vol. 3, No. 5, September, 776–786.

Sinha, N. K., Gupta M. M., and Zadeh, L. A. (2000). "Soft-Computing and Intelligent Systems: Theory and Applications," Academic Press, New York.

(1992). "Special Issue on Fuzzy Logic and Neural Networks," *IEEE Trans. Neural Networks*, Vol. 3, No. 5, September.

Tsoukalas, L. H. and Uhrig, R. E. (1997). "Fuzzy and Neural Approachs in Engineering," John Wiley, New York.

Yen, J., Langari, R., and Zadeh, L. A. (eds.) (1995). "Industrial Applications of Fuzzy Logic and Intelligent Systems," IEEE Press, New York.

Zadeh, L. A. (1965). *Inf. Control.* **8,** 338–353.

Zadeh, L. A. (1973). *IEEE Trans. Syst. Man, Cyber.* **SMC-1,** 28–44.

Zadeh, L. A. (1983). *Fuzzy Sets and Systems* **11,** 197–227.

Zadeh, L. A., Fu, K. S., and Shimura, M. (eds.) (1975). "Fuzzy Sets and their Applications to Cognitive and Decision Process," Academic, New York.

Zimmermann, H. J., Zadeh, L. A., and Gaines, B. R. (eds.) (1984). "Fuzzy Sets and Decision Analysis," North-Holland, Amsterdam.

Galactic Structure and Evolution

John P. Huchra

Harvard-Smithsonian Center for Astrophysics

GLOSSARY

Globular cluster A dense, symmetrical cluster of 10^5 to 10^6 stars generally found in the halo of the galaxy. Globular clusters are thought to represent remnants of the formation of galaxies.

Halo The extended, generally low surface brightness, spherical outer region of a galaxy, usually populated by globular clusters and population II stars. Sometimes also containing very hot gas.

H II region A region of ionized hydrogen surrounding hot, usually young, stars. These regions are distinguished spectroscopically by their very strong emission lines.

Hubble constant The constant in the linear relation between velocity and distance in simple cosmolical models. $D = V/H_0$, where H_0 is given in km sec^{-1} Mpc^{-1}. Distance and velocity are measured in Megaparsecs and in kilometers per second, respectively. Current values for H_0 range between 50 and 100 in those units.

Magnitude Logarithmic unit of relative brightness or luminosity. The magnitude scale is defined as $-2.5^* \log(flux) + Constant$ so that brighter objects have smaller magnitudes. Apparent magnitude is defined relative to the brightness of the A0 star Vega (given *magnitude* $= 0$) and absolute magnitude is defined as the magnitude of an object if it were placed at a distance of 10 parsecs.

Metallicity The average abundance of elements heavier than H and He in astronomical objects. This is usually measured relative to the metal abundance of the Sun and quoted as "Fe/H," since iron is a major source of line in optical spectra.

Missing mass The excess mass found from dynamical studies of galaxies and systems of galaxies over and above the mass calculated for these systems from their stellar content.

Parsec, kiloparsec, Megaparsec 1 parsec is the distance at which 1 Astronomical Unit (the Earth orbit radius) subtends 1 arcsec. 1 pc $= 3.086 \times 10^{18}$ cm $= 3.26$ light years.

Population I, II Stars in the galaxy are classified into two general categories. Population II stars were formed at the time of the galaxy's formation, are of low metallicity, and are usually found in the halo or in globular clusters. Population I stars, like the sun, are younger, more metal rich and form the disk of the galaxy.

Solar mass, luminosity

$$L_\odot = 3.826 \times 10^{33} \text{ ergs sec}^{-1};$$
$$M_\odot = 1.989 \times 10^{33} \text{ gm}.$$

Surface brightness Luminosity per unit area on the sky, usually given in magnitudes per square arcsecond. In Euclidean space, surface brightness is distance independent since both the apparent luminosity and area decrease as the square of the distance.

THE OBJECT of the study of galaxies as individual objects is twofold—(1) understand the present morphological appearances of galaxies and their internal dynamics, and (2) to understand the integrated luminosity and energy distributions of galaxies—in both the framework and the timescale of a cosmological model and evolution. The morphology of a galaxy is determined by its formation and dynamical evolution, mass, angular momentum content, and the pattern of star formation that has occurred in it. The Brightness and spectrum of a galaxy are determined by its present stellar population which is in turn a function of the galaxy's star formation history and the evolution of those stars which is also related to the dynamical history of the galaxy and its environment. Significant advances have been made in detailed quantitative modeling of galaxy morphology and internal dynamics. Galaxy form and luminosity evolution are much less well understood despite their great importance for the use of galaxies as cosmological probes.

The study of galactic structure and evolution began only in the early 20th century with the advent of large reflectors and new photographic and spectroscopic techniques that allowed astronomers to determine crude distances to external galaxies and place them in the model universes being developed by Einstein, de Sitter, Friedmann, Lemaître and others. The key discoveries were Harlow Shapley's of a relation between the period and luminosity of Cepheid variable stars, and Edwin Hubble's use of that relation to determine distances to the nearest bright galaxies. Hubble then was able to calibrate other distance indicators (such as the brightest stars in galaxies, which are 10–100 times brighter than Cepheids) to estimate distances to further galaxies. This led to his discovery in 1929 of the redshift-distance relation which not only established the expansion of the universe as predicted by Einstein but also established that the age or timescale of the universe was much greater than 100 million years. Current observational estimates for the age of the universe range between 14 and 17 billion years in the basic Hot Big Bang model, the Friedmann–Lemaître cosmological model which has been favored for the last few decades.

In the past few years, the parameters of the cosmological model have been narrowed down by astronomers and physicists; the current view is that the Einstein–DeSitter globally flat model is most likley correct, that the Universe will probably expand forever, and that matter makes up about one third of the "content" of the Universe with one sixth of that (or about 4–5% of the total mass–energy density of the Universe) in ordinary baryonic matter—the stuff we are made of. The work of astronomers who concentrate on the study of galaxies is to understand the formation and evolution of galaxies and larger structures such as galaxy clusters from the time ∼15 billion years ago when, as seen in the Cosmic Microwave Background, the Universe was uniform and homogeneous to one part in 100,000.

I. GALAXY MORPHOLOGY—THE HUBBLE SEQUENCE

The initial steps in the study of galactic structure were the development of classification schemes that could be tied to physical properties such as angular momentum content, gas content, mass, and age. Another of Hubble's major contributions to extragalactic astronomy was the introduction of such a scheme based only on the galaxy's visual (or, more correctly, blue light) appearance. This morphological classification scheme, known as the Hubble Sequence, has been modified and expanded by Allan Sandage, Gerard de Vaucouleurs, and Sidney van den Bergh. The Hubble Sequence now forms the basis for the study of galactic structure. There are other classification schemes for galaxies based on such properties as the appearance of their spectra or the existence of star like nuclei or diffuse halos; these are of more specialized use.

Hubble's basic scheme can be described as a "tuning fork" with elliptical galaxies along the handle, the two families of spirals, normal and barred, along the tines, and irregular galaxies at the end (see Fig. 1). Elliptical galaxies

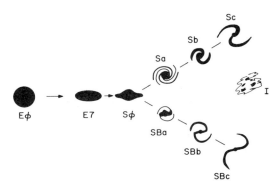

FIGURE 1 Hubble's "tuning fork" diagram (adapted from Hubble, E. 1936.) The Realm of the Nebulae.

FIGURE 2 The EO galaxy NGC 3379 and its comparison SBO galaxy, NGC 3384. This image was obtained with a Charge Coupled Device (CCD) camera at the F. L. Whipple Observatory by S. Kent. Slight defects in the image are due to cosmetic defects in the CCD.

are very regular and smooth in appearance and contain little or no dust or young stars. They are subclassified by ellipticity which is a measure of their apparent flattening. Ellipticity is computed as $e = 10(a-b)/a$, where a and b are the major and minor axis diameters of the galaxy. A typical elliptical galaxy is shown in Fig. 2. A galaxy with a circular appearance has $e = 0$ and is thus classified E0. These are at the tip of the tuning fork's handle. The flattest elliptical galaxies that have been found have $e \approx 7$, and are designated E7. Both normal and barred spiral galaxies range in form from "early" type, designated Sa, through type Sb, to "late" type Sc. Spiral classification is based on three criteria: (1) the ratio of luminosity in the central bulge to that in the disk, (2) the winding of the spiral arms, and (3) the contrast or degree of resolution of the arms into stars and H II regions. Early type galaxies have tightly wound arms of low contrast and large bulge-to-disk ratios. A typical spiral galaxy is shown in Fig. 3. The important transition region between elliptical and spiral galaxies is occupied by lenticular galaxies that are designated S0. It was originally hypothesized that spiral galaxies evolved into S0 galaxies and then ellipticals by winding up their arms and using up their available supply of gas in star formation. This is now known to be false.

Irregular galaxies are split into two types. Type I or Magellanic irregulars are characterized by an almost complete lack of symmetry, no nucleus, and are usually resolved into stars and H II regions. The Magellanic Clouds, the nearest galaxies to our own, are examples of this type. Type II irregular galaxies are objects that are not easily classified—galaxies that have undergone violent dynam-

ical interactions, star formation events, "explosions," or have features uncharacteristic of their underlying class such as a strong dust absorption lane in an elliptical galaxy.

The modifications and extensions to Hubble's scheme added by Sandage and de Vaucouleurs include the addition of later classes for spirals, Sd and Sm, between

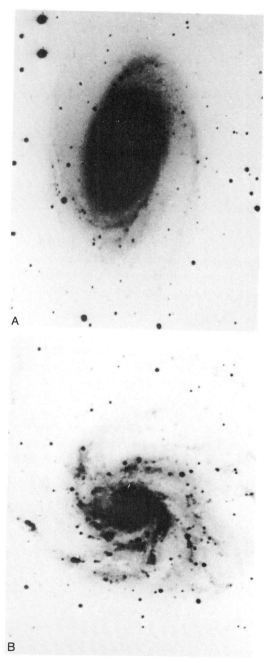

FIGURE 3 (a) The Sb spiral galaxy Messier 81 (=NGC 3031), and (b) the Sc spiral M101 (=NGC 5457 a.k.a. the Pinwheel). The CCD images courtesy of S. Kent.

TABLE I De Vaucouleurs' *T* Classification

T Type	Description
−6	Compact elliptical
−5	Elliptical, dwarf elliptical E, dE
−4	Elliptical E
−3	Lenticular L−, SO−
−2	Lenticular L, SO
−1	Lenticular L+, SO+
0	SO/a, SO-a
1	Sa
2	Sab
3	Sb
4	Sbc
5	Sc
6	Scd
7	Sd
8	Sdm
9	Sm, Magellanic Spiral
10	Im, Irr I, Magellanic Irregular, Dwarf Irregular
11	Compact Irregular, Extragalactic HII Region

classes Sc and Irr I, the further subclassification of spirals into intermediate types such as S0/a, Sab, and Scd, and the inclusion of information about inner and outer ring structures in spirals. S0 galaxies have also been subdivided into classes based either on the evidence for dust absorption in their disks, or, for SB0's, the intensity of their bar components. Van den Bergh discovered that the contrast and developement of spiral arms were correlated with the galaxy's luminosity. Spirals and irregulars can be broken into nine luminosity classes (I, I–II, II, . . . V), but there is considerable scatter ($\approx$0.6–1.0 magnitude) and thus considerable overlap in the luminosities associated with each class. De Vaucouleurs also devised a numerical scaling for morphological type, called the T type and illustrated in Table I, for his *Second Reference Catalog of Bright Galaxies.*

The most recent "modification" to the Hubble sequence is van den Bergh's recognition of an additional sequence between lenticular galaxies and normal spirals. This sequence of spirals, dubbed *Anemic*, has objects designated Aa, ABa, Ab, etc. These Anemic spirals have diffuse spiral features, are usually of low surface brightness, and are gas poor relative to normal spirals of the same form.

II. GALACTIC STRUCTURE

The quantitative understanding of galactic structure is based on only two observables. These are the surface brightness distribution (including the structure of the arms in spiral galaxies) and the line-of-sight (radial) velocity

field. The surface brightness at any point in a galaxy's image is the integral along the line-of-sight through the galaxy of the light produced by stars and hot gas. Measurements of the velocity field are made spectroscopically (either optically or in the radio at the 21-cm emission line of neutral H), and represent the integral along the line-of-sight of the velocities of individual objects (stars, gas clouds) times their luminosity. Dust along the line-of-sight in a galaxy obscures the light from objects behind it; in dusty, edge-on spiral galaxies only the near side of the disk can be observed in visible light. Extinction by dust is a scattering process and thus is a function of wavelength; all galaxies are optically thin at radio wavelengths. Elliptical galxies are considered optically thin at all wavelengths.

A. Elliptical Galaxies

Most galaxies can be readily decomposed into two main structural components, disk and bulge. Elliptical galaxies are all bulge. The radial brightness profile of the bulge component is usually parameterized by one of three laws. The earliest and simplest is an empirical relation called the Hubble law,

$$\mu(r) = \mu_o(1 + r/r_o)^{-2},$$

and is parametrized in terms of a central surface brightness, μ_o, and a scale length, r_o, at which the brightness falls to half its central value. At large radius, the profiles are falling as $1/r^2$. At radii less than the scale length, the profile flattens to μ_o. Giant elliptical galaxies typically have $\mu_o \approx 16$ mag/arcsec^2. A better relation is the empirical $r^{1/4}$ law proposed by de Vaucouleurs

$$\mu(r) = \mu_e \exp\left[-7.67\left((r/r_e)^{1/4} - 1\right)\right],$$

where r_e is the effective radius and corresponds to the radius that encloses $1/2$ of the total integrated luminosity of the galaxy, and μ_e is the surface brightness at that radius, approximately $1/2000$ of the central surface brightness. The third relation is semiempirical and was derived from dynamical models calculated by King to fit the brightness profiles of globular star clusters. These models can be parametrized as

$$\mu(r) = \mu_K\left[\left(1 + r^2/r_c^2\right)^{-1/2} - \left(1 + r_t^2/r_c^2\right)^{-1/2}\right]^2,$$

where r_c again represents the core radius where the surface brightness falls to $\approx$0.5, r_t is the truncation or tidal radius beyond which the surface brightness rapidly decreases, and μ_K is approximately the central surface brightness. Isolated elliptical galaxies are best fit by models with $r_t/r_c \approx 100$–200. Small elliptical galaxies residing in the gravitational potential wells of more luminous galaxies (like the dwarf neighbors of our own galaxy) are tidally stripped and have $r_t/r_c \approx 10$. Figure 4 shows examples of the Hubble and de Vaucouleurs laws and the King models.

Dwarf elliptical galaxies, designated dE, are low luminosity, very low surface brightness objects. Because of their low surface brightness, they are difficult to identify against the brightness of the night sky (airglow). The nearest dwarf elliptical galaxies, Ursa Minor, Draco, Sculptor, and Fornax, are satellites of our own galaxy, and are resolved into individual stars with large telescopes. Although dE galaxies do not contribute significantly to the total luminosity of our own Local Group of galaxies, they dominate its numbers (Table II). As mentioned earlier, dwarf galaxies are often tidally truncated by the gravitational field of neighboring massive galaxies. If M is the mass of the large galaxy, m is the mass of the dwarf, and R is their separation, then the tidal radius, r_t, is given by

$$r_t = R\left(\frac{m}{3M}\right)^{1/3}.$$

Frequently the central brightest galaxy in a galaxy cluster exhibits a visibly extended halo to large radii. These objects were first noted by W. W. Morgan and are designated "cD" galaxies in his classification scheme. Unlike ordinary giant E galaxies, whose brightness profiles exhibit the truncation characteristic of de Vaucouleurs or King profiles at radii of 50 to 100 kpc, cD galaxies have profiles which fall as $1/r^2$ or shallower to radii in excess of 100 kpc. "D" galaxies are slightly less luminous and

TABLE II Presently Known Local Group Members

Name	Type	B_T[a]	Distance (kpc)	Luminosity ($10^9 L_\odot$)
Andromeda	Sb	4.38	730	14.7
Milky Way	Sbc	—	—	(10.0)[b]
M33	Scd	6.26	900	3.9
LMC	SBm	0.63	50	2.2
SMC	Im	2.79	60	0.4
NGC 205	E5	8.83	730	0.24
M32	E2	9.01	730	0.20
IC 1613	Im	10.00	740	0.09
NGC 6822	Im	9.35	520	0.08
NGC 185	dE3	10.13	730	0.07
NGC 147	dE5	10.37	730	0.06
Fornax	dE	9.1	130	0.006
And I	dE	13.5	730	0.003
And II	dE	13.5	730	0.003
And III	dE	13.5	730	0.003
Leo I	dE	11.27	230	0.0026
Sculptor	dE	10.5	85	0.007
Leo II	dE	12.85	230	0.0006
Draco	dE	12.0	80	0.00016
LGS 3	?	(17.5)[b]	730	0.00008
Ursa Minor	dE	13.2	75	0.00005
Carina	dE	—	170	—

[a] B_T is the total, integrated blue apparent magnitude.
[b] Quantities in parentheses are estimated.

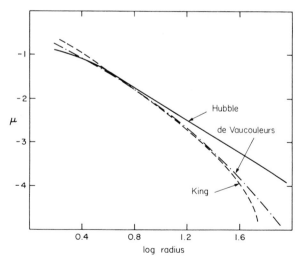

FIGURE 4 The three commonly fit surface brightness profiles for elliptical galaxies and the bulge component of spiral galaxies. The Hubble law (solid line) has core radius $r_0 = 1.0$ unit, the de Vaucouleurs law (dash-dot line) has an effective radius $r_e = 5.0$ units, and the King model (dashed line) has a core radius $r_c = 1.0$, and a tidal radius $r_t = 80.0$ units. All three profiles have been scaled vertically in surface brightness to approximately agree at $r = 5.0$ units ($\log r = 0.7$).

have weaker halos; D galaxies are found at maxima in the density distribution of galaxies. The extended halos of these objects are considered to be the result of dynamical processes that occur during either the formation or during the subsequent evolution of galaxies in dense regions.

The internal dynamics of spheroidal systems (E galaxies and the bulges of spirals) is understood in terms of a self-gravitating, essentially collisionless gas of stars. These systems appear dynamically relaxed; they are basically in thermodynamic equilibrium as isothermal spheres with Maxwellian velocity distributions. Faber and Jackson noted in 1976 that the luminosities of E galaxies were well correlated with their internal velocity dispersions.

$$L \approx \sigma^4.$$

Here, the velocity dispersion, σ, is a measure of the random velocities of stars along the line-of-sight.

The collisional or two-body relaxation time, t_r, for stars is approximately

$$t_r \sim 2 \times 10^8 \frac{V^3}{M^2 \rho} \text{ years,}$$

where V is the mean velocity in km sec^{-1}, ρ is the density of stars per cubic parseconds and M is the mean stellar mass in $M_\odot$. This relaxation time in galaxies is 10^{14} to 10^{18} years—much longer than the age of the universe. Two-body relaxation is generally not important in the internal dynamics of galaxies, but is significant in globular clusters.

The relaxed appearance of galactic spheroids is probably due to the process called violent relaxation. This is a statistical mechanical process described by Lynden-Bell, where individual stars primarily feel the mean gravitational potential of the system. If this potential fluctuates rapidly with time, as in the initial collapse of a galaxy, then the energy of *individual* stars is not conserved. The results of numerical experiments are similar to galaxy spheroids.

For the past decade and a half, the determination and modeling of the true shapes of elliptical galaxies has been a major problem in galaxian dynamics. Most elliptical galaxies are somehat flattened. Early workers assumed that this flattening was rotationally supported as in disk galaxies (see following). Bertola and others observed, however, that the rotational velocites of E galaxies are insufficient to support their shapes (Fig. 5). The velocity dispersion is a measure of the random kinetic energy in the system.

To resolve this problem, Binney, Schwarzschild, and others suggested that E galaxies might be prolate (cigar shaped) or even triaxial systems instead of oblate (disklike) spheroids. Current work favors the view that most flattened elliptical galaxies are triaxial and have internal stellar velocity distributions which are anisotropic.

A small fraction do rotate fast enough to support their shapes.

B. S0 Galaxies

Spiral and S0 galaxies can be decomposed into two surface brightness components, the bulge and disk. Disks have brightness profiles that fall exponentially

$$\mu(r) = \mu_0 \exp(-r/r_s).$$

Bulges have been described earlier. Disks are rotating; their rotational velocity at any radius is presumed to balance the gravitational attraction of the material inside. A typical rotation curve (velocity of rotation as a function of radius) is shown in Fig. 6.

Disks are not infinitely thin. The thickness of the disk depends on the balance between the surface mass density in the disk (gravitational potential) and the kinetic energy in motion perpendicular to the disk. This can be a function of both the initial formation of the system and its later interaction with other galaxies, etc. Tidal interaction with other galaxies will "puff up" a galactic disk of stars. Disks of S0 galaxies are composed of old stars and do not exhibit any indications of recent star formation and associated gas or dust—that is part of the definition of an S0.

Bulges of S0 galaxies and spirals **do** rotate and appear to be simply rotationally flattened oblate spheroids. We will return to this point later when we discuss galaxy formation.

C. Spiral Galaxies

Unlike the smooth and uncomplicated disks of S0 galaxies, the disks of spiral galaxies generally exhibit significant

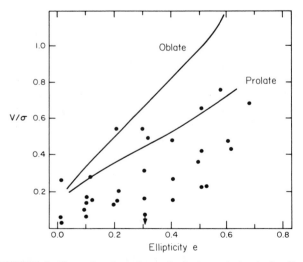

FIGURE 5 The ratio of rotation velocity to central velocity dispersion *versus* the ellipticity, e, for elliptical galaxies. Solid lines are oblate and prolate spheroid models with isotropic velocity distributions from Binney. Data points are from Bertola, Cappacioli, Illingworth, Kormendy and others.

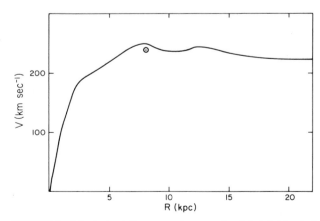

FIGURE 6 A typical rotation curve for a moderately bright spiral galaxy like our own (adapted from the work of V. Rubin *et al.* and M. Roberts). For reference, the rotational velocity of our own galaxy at the approximate radial distance of the sun, 220 km sec^{-1} at 8 pc, is marked with an $\odot$.

amounts of interstellar gas and dust and distinct spiral patterns. The disk rotates differentially (see Fig. 6) and the spiral pattern traces the distribution of recent star formation in the galaxy. Regions of recent star formation have a higher surface brightness than the background disk. Although the spiral is a result of the rotation of the material in the disk, the pattern need not (and generally does not) rotate with the same speed as the material.

The rotation of a spiral galaxy can be described in terms of a velocity rotation curve, $v(r)$, or the angular rotation rate, $\Omega(r)$, where $v = r\Omega$. In 1927, Oort first measured the local differential rotation of our galaxy by studying the motions of nearby stars. He described that rotation in terms of two constants, now known as Oort's Constants, which are measures of the local shear (A) and vorticity (B):

$$A = -\frac{r}{2}\frac{d\Omega}{dr}, \qquad B = -\frac{1}{2r}\frac{d}{dr}(r^2\Omega).$$

The values adopted by the IAU (International Astronomical Union) in 1964 are $A = 15$ km sec^{-1} kpc^{-1} and $B = -10$ km sec^{-1} kpc^{-1}, with the Sun at a distance of $r_o = 10$ kpc from the center of the galaxy, and the rotation rate at the sun $V_\odot = 250$ km sec^{-1}. Current estimates support a smaller Sun-Galactic-center distance ($\sim$8 kpc) and a smaller rotation velocity ($\sim$230 km sec^{-1}) as well as slightly different values for A and B.

Studies of rotation in spiral galaxies are undertaken by long-slit spectroscopy at differenct position angles in the optical or by either integrated (total intensity, big beam) measurements or interferometric mapping in the 21-cm line of neutral hydrogen. The neutral hydrogen (HI) in spiral galaxies is primarily found outside their central regions, reaching a maximum in surface density several kiloparseconds from the center. The gas at the center is mostly in the form of molecular hydrogen, H_2, as deduced from carbon monoxide (CO) maps. From such detailed studies by Rubin, Roberts, and others we know that the rotation curves for spirals generally rise very steeply within a few kiloparseconds of their centers then flatten and stay at an almost constant velocity as far as they can be measured. This result is rather startling because the luminosity in galaxies is falling rapidly at large radii. If the light and mass were distributed similarly, then the rotational velocity should fall off as $1/R$ at large radii as predicted by Kepler's laws. Only a small number of galaxies show the expected Keplerian falloff, leading to the conclusion that the mass in spiral galaxies is **not** distributed as the light.

As in elliptical galaxies, Fisher and Tully noted in 1976 that the luminosities of spiral galaxies are correlated with their internal motions, in particular rotational velocity. The best form of this relation is seen in the near infrared

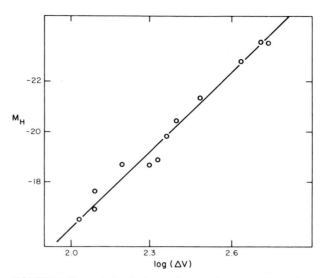

FIGURE 7 The relation between the maximum rotation velocity of spiral galaxies and their absolute infrared (H band = 1.6 μ) magnitude [Adapted from Aaronson, M., Mould, J., and Huchra, J. P. (1980) *Astrophys. J.* **237,** 655.]

where the effects of internal extinction by dust on the luminosities are minimized (Fig. 7). There the relation is approximately

$$L = (\Delta V)^4,$$

where ΔV is the full width of the H I profile measured at either 20 or 50% of the peak. As the H I distribution peaks outside the region where the rotational velocity has flattened, the H I profile is sharp sided and is double peaked for galaxies inclined to the line of sight.

The regularity of the spiral structure seen in these galaxies is exceptional. If the spiral pattern was merely tied to the matter distribution, differential rotation would "erase" it in a few rotation periods. A typical rotation time is a few hundred million years, or $\approx 1/100$th the age of the universe. To explain the persistence of spiral structure, Lin and Shu introduced the Density Wave theory in 1964. In this model, the spiral pattern is the star formation produced in a shock wave induced by a density wave propagating in the galaxy disk. The spiral pattern is in solid body rotation with a pattern speed Ω_p. The main features of such a density wave are the corotation radius, where the pattern and rotation angular speeds are the same, and the inner and outer Lindblad resonances, where

$$\Omega_p = \Omega \pm \kappa/m;$$

m is the mode of oscillation and κ is the epicyclic frequency in the disk,

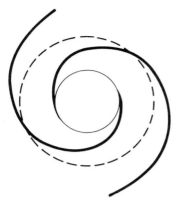

FIGURE 8 Typical two-armed spiral density wave pattern. Solid spiral represents the shock front in the gas. This follows closely behind the gravitational potential perturbation. The outer dashed circle is the corotation radius, where the matter in the disk is rotating at the same speed as the spiral pattern. The inner, thin circle represents the inner Lindblad resonance.

$$\kappa^2 = r^{-3}\frac{d(r^4\Omega^2)}{dr},$$

or

$$\kappa = \frac{2\Omega}{(1 - A/B)^{1/2}}.$$

Figure 8 depicts the features of a density wave. The spiral pattern only exists between the Lindblad resonances. Galaxies in which a single mode dominates are called "Grand Design" spirals.

Two alternative models have been proposed to account for spiral patterns, the Stochastic Star Formation (SSF) model of Seiden and Gerola, and the tidal encounter model. In the first, regions of star formation induce star formation in neighboring regions. With proper adjustment of the rotation and propagation timescales, reasonable spiral patterns result. In the second, spiral structure is the result of galaxy interactions. A possible example of this process is shown in Fig. 9. It is likely that all three processes operate in nature, with density waves producing the most regular spirals, SSF producing "flocculent" spirals, and tidal interactions producing systems like the Whirlpool. A significant fraction (~10%) of all galaxies show some form of interaction with their neighbors. Not all such systems contain spiral galaxies, however.

D. Irregular Galaxies

Galaxies are classified as Irregular for several different reasons. In the morphological progression of the Hubble sequence the true Irregulars are the Magellanic Irregulars, the Irr I's, which are galaxies with no developed spiral structure that are usually dominated by large numbers

of star forming regions. Irregular II's and other peculiar objects have been extensively cataloged by Arp and his coworkers, by Vorontsov–Velyaminov and by Zwicky. These objects are usually given labels to describe their peculiar properties such as "compact," which usually indicates abnormally high surface brightness or a very steep brightness profile, "posteruptive," which usually indicates the existence of jets or filaments of material near the galaxy, "interacting," or "patchy."

There are also galaxies in the form of rings which are thought to be produced by a slow, head-on collision of two galaxies, one of which must be a gas-rich spiral. The collision removes the nucleus of the spiral leaving a nearly round ripple of star formation similar to the ripples produced when a rock is dropped into a lake. Such ring galaxies almost always have compact companion galaxies which are the likely culprits.

The Magellanic irregulars are almost always dwarf galaxies (low luminosity), are very rich in neutral hydrogen, and have relatively young stellar populations. Their internal kinematics may show evidence for regular structure or may be chaotic in nature. These galaxies are generally of low mass; the largest such systems have internal velocity dispersions (usually measured by the width of their 21-cm hydrogen line) less than 100 km sec^{-1}. Their detailed internal dynamics have only been poorly studied until now. There are some indications that star formation proceeds in these galaxies as in the SSF theory mentioned earlier; however evidence also exists for H II regions aligned with possible shock fronts. Although they do not contribute significantly to the total luminosity density of the universe, these galaxies and the dwarf ellipticals dominate the total number of galaxies.

III. INTEGRATED PROPERTIES OF GALAXIES

A. Luminosity Function

The luminosity function or space density of galaxies, $\phi(L)$ is the number of galaxies in a given luminosity range per unit volume. This function is usually calculated from magnituded limited samples of galaxies with distance information. Distances to all but the nearest galaxies are determined from their radial velocities and the Hubble constant. (Note that in the Local Supercluster where the velocity field is disturbed by the gravity of large mass concentrations it is necessary to apply additional corrections to distances measured this way). Figure 10 is the differential luminosity function for field galaxies derived from a recent large survey of galaxy redshifts. The luminosity function is nearly flat at faint magnitudes and falls exponentially

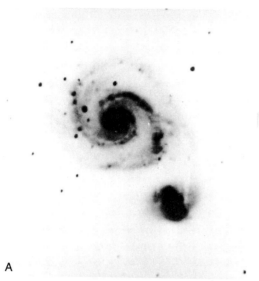

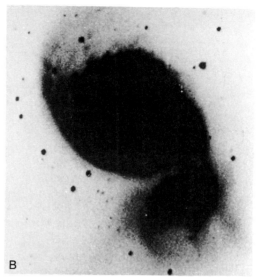

A

B

FIGURE 9 The interacting galaxy pair NGC 5194 + 5195, also known as M51 or the Whirlpool Nebula. This galaxy pair is also known as Arp 85, VV 1, and Ho 526 from the catalogs of peculiar, interacting and binary galaxies of Arp, Vorontsov–Velyaminov and Holmberg. The bright spiral, NGC 5194, is classified Sbc; its companion, NGC 5195, is classified SB0 P. (a) is a low contrast display of the image to show the interior structure of the galaxies; (b) is a high contrast display of the same image to show the effects of interaction on the outer parts of the galaxies. (CCD photo courtesy of S. Kent.)

at the bright end. A useful parametrization of the $\phi(L)$, derived by Schechter, is

$$\phi(L) = \phi_0 L^{-1} (L/L^*)^\alpha \exp(-L/L^*).$$

L^* is the characteristic luminosity near the knee, ϕ_0 is the normalization and α is the slope at the faint end. For a Hubble constant of 100 km sec^{-1} Mpc^{-1} and with blue (B) magnitudes from the Zwicky Catalog of Galaxies and of Clusters of Galaxies

$$L_B^* \approx 8.6 \times 10^9 L_\odot, \quad M_B^* \approx -19.37,$$
$$\phi_0 \approx 0.015 \text{ gal Mpc}^{-3},$$

and

$$\alpha \approx -1.25.$$

M_B^* is the characteristic blue absolute magnitude. The Schechter function has the interesting property that the integral luminosity density, useful in cosmology, is just given by

$$L_{int} = \phi_0 L^* \Gamma(\alpha + 2)$$

where Γ is the incomplete gamma function.

Figure 11 shows the luminosity function of galaxies as a function of morphological type.

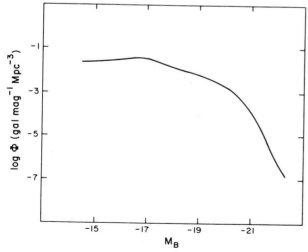

FIGURE 10 The differential galaxy luminosity function, $\phi(L)$, in units of galaxies per magnitude interval per cubic megaparsec, derived from the Center for Astrophysics Redshift Survey.

B. Spectral Energy Distributions

The observed integrated spectra of normal galaxies are functions of stellar population, star formation rate, mean metal content, gas content, and dust content. These properties correlate with, and in some cases are causually related to, galaxy morphology. In our own galaxy there are two relatively distinct populations of stars as discovered by Baade in 1944. Population II stars are the old metal poor stars that form the halo of the galaxy. Globular clusters are population II objects. Population I stars

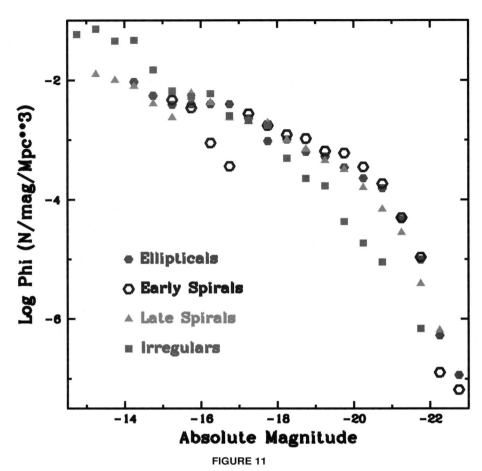

FIGURE 11

are younger and more metal rich. They form the disk of the galaxy. The sun is an intermediate age population I star.

Figure 12a is an example of the optical spectral energy distribution of an old, gasless stellar population. This type of population is typical of the bulge and old disk populations in galaxies. It is dominated by the light of G and K giant stars. The strongest spectral features are absorption lines of (originating in the atmospheres of the giant stars) calcium, iron, magnesium, and sodium, usually in low ionization states, as well as molecular bands of cyanogen, magnesium hydride and, in the red, titanium oxide. The very strong, factor of 2, break in the apparent continuum shortward of 4000 Å is the primary distinguishing characteristic of high redshift normal galaxies. The strengths of the absorption features in old bulge populations is primarily a function of metallicity. In populations where star formation has taken place recently (on timescales of $\leq 10^9$ years), most absorption features in the integrated spectra will appear weaker due to the contribution to the continuum emission from hotter, weaker lined A and F stars. The Balmer lines of hydrogen, which are strength in A and F stars, increase in strength. Table III lists some of the common and strong absorption and emission lines seen in normal galaxy spectra.

Figure 12b is the spectrum of a galaxy who's light is dominated by very young stars and hot gas. The spectrum resembles that of an H II region. This is an irregular galaxy that is undergoing an intense "burst" of star formation. Its optical spectrum is dominated by strong, sharp emission lines of H, He, and the light elements N, O, and S from the photoionized gas. The continuum is primarily from hot O and B stars with a small contribution from the free–free, two-photon, and Paschen and Balmer continuum emission from the gas.

Population synthesis is the attempt to reproduce the observed spectral energy distributions of galaxies by the summation of spectra of well-observed galactic stars, model stars, and models for the emission from the gas photoionized by hot stars in the synthesis. An important parameter in such studies is the Initial Mass Function (IMF), which is the differential distribution of stars as a function of mass in star forming regions. The simplest approximation for the IMF is a powerlaw in the mass, M,

$$N(M)\,dM = (M/M_\odot)^{-\alpha}.$$

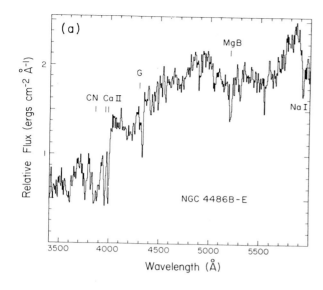

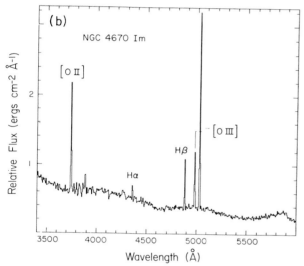

FIGURE 12 Galaxy Spectral energy distributions. (a) NGC 4486B, typical of an old, evolved stellar population. The prominent absorption features of Ca, Na, Mg, the CN molecule and the G band blend of Fe and Cr are marked. (b) NGC 4670, typical of a strong emission line galaxy or extragalactic H II region. The O and H emission lines are marked.

TABLE III   Common Spectral Features in Galaxies

λ_0	Element	Comments[a]
Absorption lines		
3810+	CN	(molecular band)
3933.68	Ca II	K
3968.49, 3970.08	Ca II + He	H
4101.75	$H\delta$	
4165+	CN	(molecular band)
4226.73	Ca I	
4305.5	Fe + Cr	G band
4340.48	$H\delta$	
4383.55	Fe I	
4861.34	$H\beta$	F
5167.3, 5172.7, 5183.6	Mg I	b
5208.0	MgH	(molecular band)
5270.28	Fe I	
5889.98, 5895.94	Na I	D
8542.0	Ca II	
Emission lines[b]		
3726.0	[O II]	
3729.0	[O II]	
3868.74	[Ne III]	
3889.05	$H\zeta$	
3970.07	$H\varepsilon$	
4101.74	$H\delta$	
4340.47	$H\gamma$	
4363.21	[O III]	
4861.34	$H\beta$	F
4958.91	[O III]	
5006.84	[O III]	
5875.65	He I	
6548.10	[N II]	
6562.82	$H\alpha$	C
6583.60	[N II]	
6717.10	[S II]	
6731.30	[S II]	

[a] Letters in comments refer to Fraunhoffer's original designation for lines in the solar spectrum.

[b] Brackets around species indicate that the transition is forbidden.

The value of α found by Salpeter for the solar neighborhood is ≈ 2.35. The real IMF is slightly steeper at the high mass end ($M > 10\ M_\odot$), and slightly flatter at the low mass end ($M > 1\ M_\odot$). Results from population synthesis indicate that the mix of stars in most luminous galaxies is very similar to our own in age and metallicity.

C. Galaxy Masses

The masses of galaxies are measured by a variety of techniques based on measurements of system sizes and relative motion, and the assumption that the system under study is gravitationally bound. Masses are sometimes inferred from studies of stellar populations, but these are extremely uncertain as the lower mass limit of the IMF is essentially unknown. It is customary to quote the mass-to-light ratios for galaxies rather than masses. This is both because galaxy masses span several orders of magnitude so it is easier to compare M/L since mass and luminosity are reasonably well correlated for individual morphological classes, and because the average value of M/L is key in the determination of the mean matter density in the universe. In specifying masses or mass-to-light ratios for galaxies, it is necessary to specify the Hubble constant, as scale lengths vary as the distance while luminosities vary inversely as the square of the distance. It is also useful to specify the magnitude system used for determining luminosity both because different systems measure light to different radii and because galaxies have a wide range of colors and are rarely the same color as the Sun.

Masses of individual spiral galaxies are determined from their rotation curves. Spiral galaxies are circularly symmetric, so the apparent radial velocities relative to their centers can be corrected for inclination. Then the velocity of the outermost point of the rotation curve is a measure of the enclosed mass. If the mass were distributed spherically, the problem would reduce to the classical circular orbit of radius, R, around point mass, M,

$$\frac{1}{2}mV^2 = \frac{GmM}{R}$$

where m is the test particle mass, and V is its orbital velocity, implying

$$M = \frac{1}{2}\frac{V^2 R}{G}.$$

Because the actual distribution is flattened, a small correction must be applied. If the mass distributions of spiral galaxies fall with radius as their optical light, then rotation curves would turnover to a Keplerian falloff, $V \propto R^{-1/2}$. Observations of neutral hydrogen rotation curves made at 21 cm now extend beyond the optical diameters of many galaxies and are still flat. This implies that the masses of spirals are increasing linearly with radius, which in turn, implies that the local mass-to-light ratio in the outer parts of spirals is increasing exponentially.

Masses for individual E galaxies are determined from their velocity dispersions and measures of their core radii or effective radii. For a system in equilibrium, the Virial Theorem states that, averaged over time,

$$\Omega + 2T = 0,$$

where T is the system kinetic energy and Ω is the system gravitational potential energy. For a simple spherical system and assuming that the line-of-sight velocity dispersion, σ, is a measure of the mass weighted velocities of stars relative to the center of mass

$$M\sigma^2 = G\int_0^R \frac{M(r)\,dM}{r}.$$

If the galaxy is well approximated by a de Vaucouleurs Law, then $\Omega = -0.33GM^2 r_e$. More accurate mass-to-light ratios can be determined for the cores of elliptical galaxies alone by comparison with models.

Masses for binary galaxies are also derived from simple orbital calculations, but, unlike rotation curve masses, are subject to two significant uncertainties. The first is the lack of knowledge of the orbital eccentricity (orbital angular momentum). The second is the lack of information about the projection angle on the sky. The combination of these two problems make the determination of the mass for an individual binary impossible. The formula $M = (V^2 R)/(2G)$ produces a minimum mass, the "true" mass is

$$M_T = \frac{V^2 R}{2G}\frac{1}{\cos^3 i \cos^2 \phi},$$

where i is the angle between the galaxies and the plane of the sky and ϕ is the angle between the true orbital velocity and the plane defined by the two galaxies and the observer. The correct determination of the masses of galaxies in binary systems requires a large statistical sample to average over projections as well as a model for the selection effects that define the sample (e.g., very wide binaries are missed; the effect of missing wide pairs depends on the distribution of orbital eccentricities).

Mass-to-light ratios for galaxies in groups or clusters are also determined from the dimensions of the system and its velocity dispersion via the Virial Theorem or a variant called the Projected Mass method. The Virial Theorem mass for a cluster of N galaxies with measured velocities is

$$M_{VT} = \frac{3\pi N}{2G}\frac{\sum_i^N V_i^2}{\sum_{i<j} 1/r_{ij}},$$

where r_{ij} is the separation between the ith and jth galaxy, and V_i is the velocity difference between the ith galaxy and the mean cluster velocity. The projected mass is

$$M_P = \frac{f_p}{GN}\left(\sum_i^N V_i^2 r_i\right),$$

where r_i is the separation of the ith galaxy from the centroid. The quantity f_p depends on the distribution of orbital eccentricities for the galaxies and is equal to $64/\pi$ for radial orbits and $32/\pi$ for isotropic orbits.

Table IV summarizes the current state of mass-to-light determinations for individual galaxies and galaxies in systems via the techniques discussed earlier. These are scaled to a Hubble Constant of 100 km sec^{-1}, and the Zwicky

TABLE IV Mass-to-Light Ratios for Galaxies[a]

Type	Method	M/L
Spiral	Rotation Curves	12
Elliptical	Dispersions	20
All	Binaries	100
All	Galaxy Groups	350
All	Galaxy Clusters	400

[a] In solar units, $M_\odot/L_\odot$.

catalog magnitude system used earlier for the Luminosity Function.

In almost all galaxies, the expected mass-to-light ratio from population synthesis is very small, on the order of 1 ore lower in solar units, because the light of the galaxy is dominated by either old giant stars, with luminosities several hundred Suns and masses less than the Sun, or by young, hot main sequence stars with even lower mass-to-light ratios. The large mass-to-light ratios that result from the dynamical studies have given rise to what is called the "missing mass" problem. Astronomers as yet have not been able to determine what constitutes the mass that binds clusters and groups of galaxies and forms the halos of large galaxies. Possibilities include extreme red dwarf (low luminosity) stars, massive stellar remnants (black holes), exotic elementary particles (axions, neutrinos, etc.), and the possibilty that the dynamical state of clusters is much more complex than the existing simple models.

D. The Fundamental Plane

Much of the work on global properties of galaxies over the last 20 years has centered on finding global relationships like the Faber–Jackson relation and Tully–Fisher relation between velocity dispersion or rotation velocity and galaxy luminosity. The drivers for this are twofold: first to search for redshift independent distance estimators, and second to discover if such relations hold clues for the study of galaxy formation. The most useful of such relations discovered so far is the fundamental plane for early type galaxies, particularly elliptical galaxies. Ellipticals form a multiparameter family. The parameters that describe the global properties of elliptical galaxies as discovered by principal component analysis are the galaxy's size, surface brightness (related to stellar density), and velocity dispersion (related to mass). A typical example of such a relation is shown in Fig. 13, from Jorgensen *et al.* (1996). The equation describing this relation is

$$\log r_e = 1.35 \log \sigma - 0.82 \log \langle I \rangle_e,$$

where r_e is the effective (or half-light) radius, σ is the line-of-sight velocity dispersion and $\langle I \rangle_e$ is the mean surface brightness (luminosity inside the effective radius divided by the area). There are other variants of this relation such as the $D_n - \sigma$ relation which has been used by several groups to study the motions of galaxies relative to the uniform expansion of the Universe and thus derive estimates of the mean mass density of the Universe relative to the critical density.

E. Gas Content

Neutral hydrogen (H I) was first detected in galaxies in 1953 by Kerr and coworkers in the 21-cm (1420.40575 MHz) radio emission line. Since then it has been found that almost every spiral and irregular galaxy contains considerable neutral gas. In spiral galaxies, the H I emission distribution usually shows a central minimum and peaks in a ring which covers the prominent spiral arms. The fractional gas mass ranges from a few percent for early type spirals (Sa galaxies) to more than 50% for some Magellanic irregulars. Neutral hydrogen is only rarely detected in elliptical an S0 galaxies. The gas mass fraction in these objects is usually less than 0.1%. In spiral galaxies, H I can usually be detected at 21 cm out to two or three times the radius of the galaxy's optical image on the Palomar Schmidt Sky Survey plates.

Ionized gas is found in galaxies in H II regions (H II refers to singly ionized hydrogen), in nuclear emission regions, and in the diffuse interstellar medium. H II regions are regions of photoionized gas around hot, usually young stars. The gas temperature is of the order of 10^4 K, and depends on the surface temperatures of the ionizing stars and cooling processes in the gas which depend strongly on its element abundances. Low metallicity H II regions are hotter than those with high metallicity. Ionized gas is often seen in the nuclei of galaxies and is either the result of star formation (as in H II regions) or the result of photoionization by a central nonthermal energy source. In our galaxy, there is a diffuse interstellar medium composed of gas and dust. Much of the volume (although not much mass) of the galaxy is in this state with the gas ionized by the diffuse stellar radiatation field. Because its density is so low, ionized gas in this state takes a long time to recombine. (To a first approximation, the recombination time for diffuse, ionized hydrogen is

$$\tau \sim \frac{10^5}{n_e},$$

where n_e is the electron density in cm^{-3}, and τ is in years.) Ionized gas is also found in supernova remnants.

Carbon monoxide (CO), which is found in cool molecular clouds in the galaxy, has been detected in several other

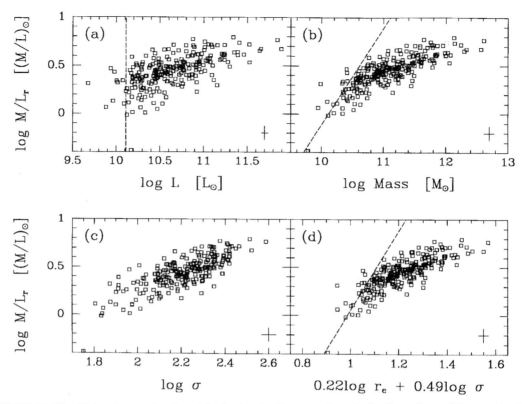

FIGURE 13 The M/L ratio as a function of the luminosity, the mass, the velocity dispersion and the combination $0.22 \log r_c + 0.49 \log \sigma$. $\log M/L = 2 \log \sigma - \log \langle I \rangle_e - \log r_c + \text{constant}$. r_c is in kpc; M/L, mass and luminosity in solar units. For $H_0 = 50$ km sec^{-1} Mpc^{-1} and Mass $= 5\sigma^2 r_c/G$ the constant is -0.73. The dashed lines in (a), (b) and (d) show the selection effect due to a limiting magnitude of -20.45 mag in Gunn r. This is the limit for the Coma cluster sample. (From the Royal Astronomical Society, **280**, 173, 1996.)

nearby spiral and irregular galaxies and even in higher redshift starburst galaxies. CO clouds are associated with regions of both massive and low mass star formation. In sprial galaxies, the CO emission often fills in the central H I minimum. There is an excellent correlation between the gas content of a galaxy and its current star formation rate as measured by integrated colors or the strength of emission from H II regions.

F. Radio Emission

All galaxies that are actively forming stars emit radio radiation. This emission is from a combination of free–free emission from hot, ionized gas in H II regions, and from supernova remnants produced when massive young stars reach the endpoint of their evolution. In addition, certain galaxies, usually ellipticals, have strong sources of nonthermal (i.e., not from stars, hot gas, or dust) emission in their nuclei. These galaxies, called radio galaxies, output the greatest part of their total emission at radio and far-infrared wavelengths. The radio emission is usu-

ally synchrotron radiation and is characterized by large polarizations and self-absorption at long wavelengths. A more detailed treatment of central energy sources can be found in another chapter.

G. X-Ray Emission

As noted earlier all galaxies emit X-ray radiation from their stellar components—X-ray binaries, stellar chromospheres, young supernova remnants, neutron stars, etc. More massive objects, particulary elliptical galaxies, have recently been found by Forman and Jones with the Einstein X-ray Observatory to have X-ray halos, probably of hot gas. A small class of the most massive elliptical galaxies which usually reside at the centers of rich clusters of galaxies also appear to be accreting gas from the surrounding galaxy cluster. This has been seen as cooler X-ray emission centered on the brightest cluster galaxy which sits in the middle of the hot cluster gas. This phenomenon is called a "cooling flow," and results when the hot cluster gas collapses on a central massive object and becomes dense enough to cool efficiently. This process is evidenced by strong optical emission lines as well as radio emission. Cooling flows may be sites of low mass star formation at the centers of galaxy clusters.

Active galactic nuclei—Seyfert 1 and 2 galaxies (discoverd by C. Seyfert in 1943), and quasars are also usually strong X-ray emitters, although the majority are **not** strong radio sources. The X-ray emission in these galaxies is also nonthermal and is probably either direct synchrotron emission or synchrotron-self-Compton emission.

IV. GALAXY FORMATION AND EVOLUTION

A. Galaxy Formation

The problem of galaxy formation is one that remains as yet unsolved. The fundamental observation is that galaxies exist and take many forms. The interiors of most galaxies are many orders of magnitude more dense than the surrounding intergalactic medium.

In the hot Big Bang model, the simplest description of galaxy formation is the gravitational collapse of density fluctuations (perturbations) that are large enough to be bound after the matter in the universe recombines. At early times, the atoms in the universe—mostly hydrogen and helium—are still ionized so electron scattering is important and radiation pressure inhibits the growth or formation of perturbations. Before recombination, the universe is said to be in the *Radiation Era*, because of the dominance of radiation pressure. The period of recombination, that is, hydrogen and helium atoms are formed from protons, α particles, and electrons, is called the *Decoupling Era*, because the universe becomes essentially transparent to radiation. After that the universe is *Matter Dominated* as gravitational forces dominate the formation of structure. The cosmic microwave background radiation, postulated by Gamow and colleagues in the 1950s and detected by Penzias and Wilson in 1965, is the relic radiation field from the primeval fireball and represents a snapshot of the universe at decoupling.

In this simple picture, matter is distributed homogeneously and uniformly before decoupling because the radiation field will tend to smooth out perturbations in the matter. After decoupling statistical fluctuations will form in either the matter density or the velocity field (turbulence). The amplitude of fluctuations which are large enough will grow, and the fluctuations can fragment and, if gravitationally bound, collapse to form globular clusters, galaxies, or larger structures. A simple criterion for the growth of fluctuations in a gaseous medium was derived by Jeans in 1928. The Jeans wavelength, λ_J, is given by

$$\lambda_J = c_s \left(\frac{\pi}{G\rho} \right)^{1/2},$$

where c_s is the sound speed in the medium, and ρ is its density. The sound speed is

$$c_s \sim c/\sqrt{3}$$

in the radiation dominated era, and

$$c_s = \left(\frac{5kT}{3m_P} \right)^{1/2}$$

after recomination. If a fluctuation is larger than that in the Jeans length, gravitational forces can overcome internal pressure. The mass enclosed in such a perturbation, the "Jeans mass," is just

$$M_J \approx \rho \lambda_J^3 \approx \frac{c_s^3}{G^{3/2}\rho^{1/2}}.$$

Before recombination, this mass is a few times $10^{15}\ M_\odot$, which is comparable to the mass of a cluster of galaxies. After recombination, the Jeans mass plunges to $\approx 10^6\ M_\odot$, or about the mass of a globular star cluster. The amplitude $(\delta\rho/\rho)$ required for fluctuations to become gravitationally bound and collapse out of the expanding universe is dependent on the mean mass density of the universe. The ratio of the actual mass density to the density required to close the universe is usually denoted Ω.

$$\Omega = \frac{8\pi G\rho}{3H_o^2},$$

where H_o is the Hubble Constant. If Ω is large, i.e., near unity, small perturbations can collapse.

In the 1970s work on a variety of problems dealing with the existence and form of large-scale structures in the universe made it clear that the formation of galaxies and larger structures had to be considered together. Galaxies cluster on very large scales. Peebles and collaborators introduced the description of clustering in terms of low-order correlation functions. The two-point correlation function, $\xi(r)$, is defined in terms of

$$\delta P = N \left[1 + \xi(r) \right] \delta V,$$

where δP is the excess probability of finding a galaxy in volume δV, at radius r from a galaxy. N is the mean number density of galaxies. Current best measurements of the galaxy 2-point correlation function indicate that it can be approximated as a power law of form

$$\xi(r) = (r/r_o)^{-\gamma},$$

with and index and correlation length (amplitude) of

$$\gamma = 1.8, \quad \text{and} \quad r_o = 5h^{-1}\ \text{Mpc},$$

(where h is the Hubble Constant in units of 100 km sec^{-1} Mpc^{-1}). Power has been found in galaxy clustering on the largest scales observed to date ($\sim$100 Mpc), and the amplitude of clustering of clusters of galaxies is 5 to 10 times that of individual galaxies. In addition, work in the last decade has shown that there are large, almost empty regions of space, called *Voids*, and that most galaxies are usually found in large extended structures that appear filamentary or shell like, with the remainder in denser clusters.

There are currently two major theories for the formation of galaxies and large-scale structures in the universe. The oldest is the gravitational instability plus heirarchical clustering picture primarily championed by Peebles and coworkers. This is a "bottom-up" model where the smallest structures, galaxies, form first by gravitational collapse and then are clustered by gravity. This model has difficulty in explaining the largest structures we see, essentially because gravity does not have time in the age of the universe to significantly affect structure on very large scales. A more recent theory, often called the *pancake* theory, is based on the assumption that the initial perturbations grow adiabatically, as opposed to isothermally, so that smaller, galaxy-sized perturbations are initially damped. In this model the larger structures form first with galaxies fragmenting out later. If dissipation-less material is present, such as significant amounts of cold or hot dark matter (e.g., massive neutrinos), collapse will usually be in one direction first, which produces flattened or pancake-like systems. Models of this kind are somewhat better matches to the spatial observations of structure, but the hot dark matter models fail to produce the observed relative velocities of galaxies. The Hubble flow, or general expansion of the universe, is fairly cool, with galaxies outside of rich, collapsed clusters, moving only slowly ($\sigma < 350$ km sec^{-1}) with respect to the flow.

There are alternative models of galaxy formation, for example, the explosive hypothesis of Ostriker and Cowie. In this model, a generation of extremely massive, pregalactic stars is formed and goes supernova, producing spherical shocks that sweep and compress material soon after recombination. These shells then fragment and the fragments collapse to form galaxies. There is some recent evidence that favors this model.

The starting point for all of the aforementioned models are the observations of small-scale fluctuations in the microwave background radiation. Perturbations produced either before or after recombination appear as perturbations in the microwave background on scales of a few arc minutes. One of the major cosmological results of the 1990s was the discovery of these fluctuations at amplitides of one part in a hundred-thousand, $\Delta T/T < 10^{-5}$, with the COBE (Cosmic Background Explorer) satellite. More recently, higher spatial resolution observations with ground-based telescopes at the South Pole and with balloon-borne telescopes have measured the power spectrum of these fluctuations and appear to strongly confirm that the Universe is geometrically flat.

This is exactly as predicted by an amplification of the Big Bang model called *Inflation*, which was introduced in the early 1980s by A. Guth and later P. Steinhardt and A. Linde. In inflationary models, the dynamics of the early universe is dominated by processes described in th Grand Unified Theories (GUTS). In these models, the universe undergoes a period of tremendous inflation (expansion) at a time near 10^{-35} sec after the Big Bang. If inflation is correct, galaxy formation becomes easier because fluctuations in the very early universe are inflated to scales which are not damped out and can exist *before* decoupling. The problem of galaxy formation and the formation of large-scale structure is then the problem of following the growth of the perturbations we see in the early Universe until they become the objects we see today.

B. Population Evolution

The appearance of a galaxy changes with time (evolves) because its stellar population changes. This occurs because the characteristics of individual stars change with time. and because new stars are being formed in most galaxies. A galaxy's appearance thus reflects its integrated star formation history and the evolution of its gaseous content. The population evolution of galaxies is relatively well understood although detailed models exist for our galaxy and only a few others.

After the initial "formation" of the galaxy, the higher mass stars in the first generation evolve more rapidly than the lower mass stars. For example, the evolutionary timescale for a 100 $M_\odot$ star is only a few million years, while that for a 1 $M_\odot$ star is nearly 10 billion years. Elements heavier than hydrogen and helium are produced in the cores of these stars and are then ejected into the interstellar medium, either by stellar mass loss or supernova explosions, thus enriching the metal content of the gas. Stars formed later and later have increasing metallicities—the oldest and most metal poor stars in our own galaxy have heavy element abundances only 10^{-3} to 10^{-4} solar. The youngest stars have abundances a few times solar. It is possible for the average metallicity of a galaxy's gas to decrease with time if it accretes primordial low metallicity gas from the intergalactic medium faster than its high mass stars eject material.

Elliptical galaxies are objects in which most of the gas was turned into stars in the first few percent of the age of the universe. Only a few exceptional ellipticals—galaxies with cooling flows, for example—show any sign of current star formation. These are objects in which the initial star formation episode was extremely efficient, with almost all of the galaxy's gas being turned into stars. At present, elliptical galaxies, probably get very slightly fainter and redder as a function of time. Their light is dominated by red giant stars that have just evolved off the main sequence, and the number of these stars is a slowly decreasing function of time for an initial mass function with the Salpeter slope. These stars are between 0.5 and 1 $M_\odot$. A competing process, main sequence brightening, can occur in systems where stars still on the main sequence contribute significantly to the light—that is, for systems with steep initial mass functions. Stars on the main sequence evolve up it slightly, becoming brighter and hotter, just before evolving into red giants.

Spiral and irregular galaxies have integrated star formation rates that are more nearly constant as a function of time. In these objects, the gas is used up slowly or replenished by infall. It is probable that in many of these galaxies, star formation is episodic, occurring in bursts caused by either passage of spiral density waves through dense regions, significant infall of additional gas, or interaction with another galaxy. The photometric properties of these galaxies are usually determined by the ratio of the amount of current star formation to the integrated star formation history. The optical light in objects with as little as 1% of their mass involved in recent, $<10^8$-year-old, star formation will be dominated by newly formed stars. Galaxies with constant star formation rates get brighter as a function of time for any reasonable assumption about the initial mass function.

It is common to model the evolutionary history of galaxies by assuming an unchanging initial mass function and parameterizing the star formation rate in terms of the available gas mass or density, or in terms of a simple functional form, usually and exponential. Examples of the possible color and luminosity evolution of an elliptical and a spiral galaxy are shown in Fig. 14. These models assume an exponentially decreasing star formation rate,

$$\Psi(t) \sim A e^{-\beta t},$$

where β is the inverse of the decay time ($\beta = 0$ is a constant star formation rate). Properties of galaxies along the Hubble sequence are well approximated by a continuous distribution of exponentials, from constant star formation rates (Sd and Im galaxies) to initial bursts with little subsequent star formation (E and S0 galaxies).

C. Dynamical Evolution

There are several processes responsible for the dynamical evolution of galaxies, tidal encounters and collisions, mergers, and dynamical friction. If galaxies were uniformly distributed in space, the probability, P_i, that a

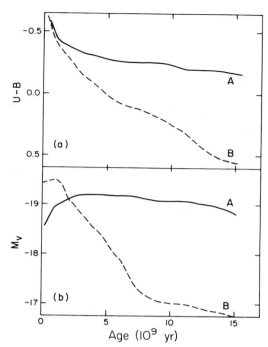

FIGURE 14 Luminosity and color evolution for two model galaxies. (a) The color U–B represents the logarithmic difference (in magnitudes) between two broad bandpasses approximately 800 Å wide centered on 3600 Å(U) and 4400 Å(B). More negative U–B's are bluer. (b) The variable M_v is the absolute visual magnitude. Model A has a star formation rate that is nearly constant as a function of time. This might be typical of a spiral galaxy. Model B has an exponentially decreasing star formation rate—15 e-folds in 15 billion years. This would be typical of an E or S0 galaxy with a small amount of present day star formation.

galaxy would have undergone a close interaction or merger in time t is

$$P_i = \pi R^2 \langle v_{rel} \rangle N t,$$

where R is the size of a galaxy, $\langle v_{rel} \rangle$ is the mean relative velocity, and N is the number density of galaxies. For the average bright galaxy R is about 10 kpc (for a Hubble Constant of 100 km sec^{-1} Mpc^{-1}, $\langle v_{rel} \rangle$ is about 300 km sec^{-1} and P_i is thus less than 10^{-4} in a Hubble time. Galaxies are clustered, however, and this significantly increases the probability that any individual object has undergone an interaction. As stated earlier, on the order of 10%, all galaxies show some evidence of dynamical interaction.

Tidal encounters and collisions that produce observable results are still relatively rare events in the field. Some examples of such events were discussed earlier, e.g., the Whirlpool and ring galaxies. Detailed models of individual events have been made by Toomre and others and the models compare well with observed structures and velocity fields. Encounters at large relative velocity usually produce small effects because the interaction time is short and the stellar components of galaxies can easily pass through one another. Encounters have large effects when the relative velocities are comparable or smaller than the internal velocity fields of the galaxies involved. The effects of encounters between spiral galaxies are also enhanced if the spin and orbital angular momenta of the galaxies are aligned. Fast collisions, however, may be possibly responsible for sweeping gas from early type galaxies in galaxy clusters, although other mechanisms (ablation by the hot intracluster medium, for example) probably dominate. Tidal encounters between protogalaxies were originally thought to produce most of the observed angular momentum in the universe; however, numerical simulations fail to produce the observed amount. The origin of angular momentum remains a mystery.

It was similary proposed that mergers of spiral galaxies might produce elliptical galaxies, a process which could explain the larger fraction of early type galaxies seen in rich clusters, where interactions are more likely to take place. Unfortunately, the photometric properties of elliptical galaxies as well as their relatively large globular cluster populations strongly argue against this hypothesis. As in tidal encounters, the "efficiency" of mergers depends on the relative encounter velocity; encounters at low relative velocities are much more likely to produce mergers than fast encounters. In particular, the line-of-sight velocity dispersion of galaxies in rich clusters is on the order of 1000 km sec^{-1}, which is much higher than the stellar dispersions internal to the individual galaxies.

Dynamical friction is a specialized case of "encounter" whereby a satellite object slowly loses its orbital energy when orbiting inside the halo of a more massive galaxy. An object moving in the halo of a galaxy produces a gravitational wake which itself can exert drag on the moving object. This effect was first described by Chandrasekhar in 1960. An object of mass M moving through a uniform halo of stars of volume density n with velocity v will suffer a drag of

$$\frac{dv}{dt} = -4\pi G^2 M n v^{-2} \left[\phi(x) - x\phi'(x) \right] \ln \Lambda,$$

where ϕ is the error function, $x = 2^{-1/2} v/\sigma$, and Λ is the ratio of the maximum and minimum impact parameters considered. σ is the velocity dispersion of the stars in the halo.

FIGURE 15

Mergers via dynamical friction as well as direct mergers of more massive galaxies almost certainly occur at the very centers of rich clusters of galaxies where the central cD galaxy is often accompanied by a host of satellite galaxies. Such processes probably account for the cD galaxy's extended halo, depressed central surface brightness, and excess luminosity relative to unperturbed bright ellipticals.

D. The High Redshift Universe

The launch of the Hubble Space Telescope in 1990 and the advent of a large number of new and powerful 8-m class ground-based optical telescopes has significantly improved our ability to see the Universe as it was a billion years or so after the Big Bang. Perhaps the best example of our view is the Hubble Deep field (Fig. 15). In this image, we see many galaxies at refshifts of 3 to 4, or as they would appear when the Universe is only a little older than a billion years. These objects are generally morphologically peculiar and show signs of both strong star formation and intense dynamical interaction. This and other observations indicate that the rate of formation of stars in the Universe probably peaked when the Universe was about 1/5th its present age and has declined significantly in the last 5–10 billion years. They also indicate that merging of galaxies and parts of galaxies is an important aspect of the evolutionalry scenario at redshifts greater than 1.

V. SUMMARY

As of this writing, no individual objects have been observed at redshifts greater than 6, so there is a large unexplored region of time and space between the time of recombination (the formation of the Cosmic microwave background at a redshift of ~1000 or an age of a few 100,000 years) and the first observable objects. These "dark ages" will be explored with a new set of space-borne telescopes such as SIRTF (the Space InfraRed Telescope Facility) and the NGST (Next generation Space Telescope).

SEE ALSO THE FOLLOWING ARTICLES

COSMIC INFLATION • COSMOLOGY • INTERSTELLAR MATTER • QUASARS • SOLAR PHYSICS • STAR CLUSTERS • STELLAR SPECTROSCOPY • STELLAR STRUCTURE AND EVOLUTION • SUPERNOVAE

BIBLIOGRAPHY

Bertin, G. (2000). "Dynamics of Galaxies," Cambridge University Press, Cambridge, UK.

Binggeli, B., and Buser, R. (eds.) (1995). "The Deep Universe," Springer, Berlin, Germany.

Binney, J., and Merrifield, M. (1998). "Galactic Astronomy," Princeton, Princeton, NJ.

Binney, J., and Tremaine, S. (1987). "Galactic Dynamics," Princeton, Princeton, NJ.

Bok, B. J., and Bok, P. (1974). "The Milky Way," Harvard University Press, Cambridge.

Bothun, G. (1998). "Modern Cosmological Observations and Problems," Taylor and Francis, London, UK.

Corwin, H., and Bottinelli, L. (eds.) (1989). "The World of Galaxies," Springer-Verlag, New York.

Elmegreen, D. M. (1998). "Galaxies and Galactic Structure," Prentice Hall, Upper Saddle River, NJ.

Ferris, T. (1982). "Galaxies," Stewart, Tabori and Chang, New York.

Hodge, P. (1986). "Galaxies and Cosmology," Harvard University Press, Cambridge.

Hodge, P. (ed.), (1984). "The Universe of Galaxies," Freeman and Co., San Francisco, CA.

Peebeles, P. J. E. (1971). "Physical Cosmology," Princeton University Press, Princeton NJ.

Sandage, A. (1961). "The Hubble Atlas of Galaxies," Carnegie Institution of Washington, Washington, DC.

Sandage, A., Sandage, M., and Kristian, J. (eds.) (1975). "Galaxies and the Universe," University of Chicago Press, Chicago.

Shu, F. (1982). "The Physical Universe," University Science Books, Mill Valley, CA.

Silk, J. (1989). "The Big Bang," Freeman and Co., San Francisco, CA.

Sparke, L. S., and Gallagher, J. S. (2000). "Galaxies in the Universe," Cambridge University Press, Cambridge, UK.

Game Theory

Guillermo Owen

Naval Postgraduate School

GLOSSARY

Complete information A situation in which a player knows all the rules of the game which she is playing.
Maximin Strategy to maximize the minimum payoff.
Mixed strategy A randomization scheme for choosing among a player's pure strategies.
NTU game One in which utility is not freely transferable.
Optimal strategy A strategy which guarantees the value.
Payoff Reward for winning a game.
Perfect information A situation in which a player knows all moves that have been made up to that point.
Strategy Set of instructions telling a player what to do under any possible circumstances.
Zero-sum game One in which the sum of payoffs to all the players is always zero.

GAME THEORY is the mathematical study of situations of conflict of interest. As such, it is applicable to parlor games (hence its name) but also to military, economic, and political situations.

I. CLASSIFICATION OF GAMES

Games are generally classified as being (a) dual or plural, (b) finite or infinite, and (c) cooperative or noncooperative. The game can also be in extensive, normal, or characteristic function form.

A. Extensive and Normal Forms

The *extensive form* of a game shows the logical sequence of moves, the information (or lack thereof) available to players as they move, and the payoff following each play of the game. As an example, consider the following (very rudimentary) form of poker. Each of two players antes $1. Player 2 is given a King; Player 1 is given a card from a deck consisting of one Ace and four Queens. (It is assumed that each card has probability 0.2 of being chosen.) At this

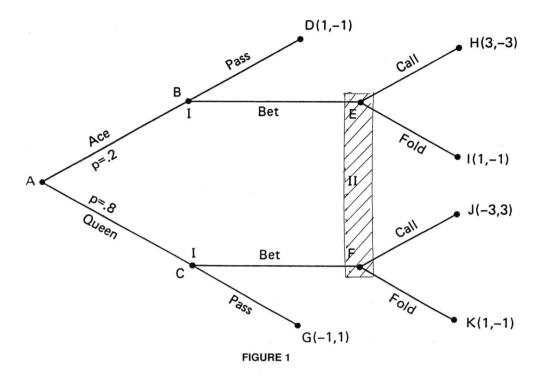

FIGURE 1

point Player 1, seeing her card, has a choice of betting $2 or passing. If Player 1 passes, the game ends immediately. If player 1 bets, then 2 has the choice of folding or calling the bet. If 2 folds, then 1 wins the pot; otherwise there is a showdown for the pot, where Ace beats King and King beats Queen.

The extensive form of this game is shown in Fig. 1. The game starts at node A, with a random move (the shuffle). Player 1 must then bet or pass (nodes B and C). If Player 1 bets, then it is player 2's turn to call or fold (nodes E and F). The remaining four nodes (D, G, H, and J) are terminal nodes, with payoffs of either 1 or 3 from one player to the other. The reader should take note of the shaded area joining E and F: this is supposed to denote the fact that, at that move, Player 2 is unsure as to his position (i.e., he does not know whether Player 1 has an Ace or a Queen). Contrast this with the situation at nodes B and C, where Player 1 knows which card he has.

It has been found that games are best analyzed by reducing them to their strategies. A *strategy*, as the word is used here, is a set of instructions telling a given player what to do in each conceivable situation. Thus, in Fig. 1, Player 1 has four strategies, since he has two possible choices in each of two possible situations. These strategies are BB (always bet), BP (bet on an Ace, pass on a Queen), PB (pass on an Ace, bet on a Queen), and PP (always pass). Player 2 has only the two strategies, C (call) and F (fold). This is due to the fact that he cannot distinguish between nodes E and F.

It may be noticed that, in a game with no chance moves, the several players' strategies will determine the outcome. In a game with chance moves, the strategies do not entirely determine the outcome. Nevertheless, an expected payoff can be calculated. The *normal form* of a game is a listing of all the players' strategies, together with the corresponding (expected) payoffs.

In the poker game of Fig. 1, suppose Player 1's strategy is BB, while Player 2's strategy is F. In that case, Player 1 will always win the antes, so the payoff is +1 (Player 1 wins the dollar that Player 2 loses). If Player 1 plays BB while Player 2 chooses C, then Player 1 has a 0.8 probability of losing $3, and a 0.2 probability of winning $3. Thus 1's expected payoff is $0.8(-3) + 0.2(3) = -1.8$. Other payoffs are calculated similarly, giving rise to the 4×2 matrix shown in Fig. 2: the four rows are Player 1's strategies, while the columns are Player 2's strategies. This matrix is the normal form of the game.

II. TWO-PERSON ZERO-SUM GAMES

It is for dual, or two-person zero-sum games, that the most satisfactory theory has been developed. For these games, the sum of the two players' payoffs is always zero; hence, a single number (the amount won by the first player, and therefore lost by the second) determines the payoff. In the finite case, the normal form for such a game is a matrix (as in Fig. 2), with each row representing one of Player 1's

$$
\begin{array}{c}
& \begin{array}{cc} \mathbf{C} & \mathbf{F} \end{array} \\
\begin{array}{c} \mathbf{BB} \\ \mathbf{BF} \\ \mathbf{FB} \\ \mathbf{FF} \end{array}
\left[
\begin{array}{cc}
-1.8 & 1 \\
-0.2 & -0.6 \\
-2.2 & 1 \\
-0.6 & -0.6
\end{array}
\right]
\end{array}
$$

FIGURE 2 A normal form for the rudimentary poker game.

strategies, and each column, one of player 2's strategies. Such games are called matrix games.

Consider the matrix game shown in Fig. 3. It can be seen that, if Player 1 chooses row E, he can be certain of winning at least 2 units. On the other hand, with row D, he might win as little as 1 unit, and, with row F, he might even lose 3 units, depending on what Player 2 does. Thus row E is his *maximin* strategy (it maximizes his minimum winnings). In a similar way, column A represents Player 2's *minimax* strategy; i.e., the maximum entry, 2, in this column, is the minimum of the column maxima.

In Fig. 3, the maximin and minimax are both equal to 2. By choosing row E, Player 1 is sure of winning at least 2; by choosing column A, Player 2 is sure of losing no more than 2. It is then suggested that both players should choose the maximin/minimax strategies, which are known as *optimal* strategies.

III. MIXED STRATEGIES

In general, there is no guarantee that the maximin and minimax payoffs are equal. In Fig. 2, the maximin is -0.6, in row BP, while the minimax is -0.2, in column C. The difference, some 40¢ per play of the game, represents an indeterminacy.

In cases such as this, it is generally agreed that the players should use mixed strategies—that is, the row and the column should be chosen according to a randomization scheme. By randomizing, each player can prevent the other from guessing his strategy.

Returning to the game in Fig. 2, Player 1 is advised to use row BB with probability 0.125, and BP with probability 0.875. This is represented by the vector $\mathbf{x} = (0.125, 0.875, 0, 0)$ and will give player 1 winnings of -0.4 (i.e., he expects to lose no more than 40¢ per game) whatever Player 2 may do. In turn, Player 2 is advised to choose either column with probability 0.5. This is represented by the vector $\mathbf{y} = (0.5, 0.5)$ and it

$$
\begin{array}{c}
& \begin{array}{ccc} \mathbf{A} & \mathbf{B} & \mathbf{C} \end{array} \\
\begin{array}{c} \mathbf{D} \\ \mathbf{E} \\ \mathbf{F} \end{array}
\left[
\begin{array}{ccc}
1 & 3 & 5 \\
2 & 4 & 3 \\
0 & 1 & -3
\end{array}
\right]
\end{array}
$$

FIGURE 3 A game with optimal pure strategies.

may be seen that the expected payoff will then be at most -0.4 (i.e., Player 2 expects to win at least 40¢ per game) no matter what Player 1 may do. Since Player 1 can guarantee to lose no more than 40¢, and Player 2 can guarantee to win at least this amount, we say that this quantity, -0.4, is the *value* of the game.

In general, in an $m \times n$ matrix game, a mixed strategy for player 1 is a vector $\mathbf{x} = (x_1, x_2, \ldots, x_m)$ with nonnegative components whose sum is 1. (The interpretation is that row i will be chosen with probability x_i, $i = 1, \ldots, m$.) Similarly, a mixed strategy for Player 2 is a vector $\mathbf{y} = (y_1, y_2, \ldots, y_n)$ with nonnegative components whose sum is 1.

If Player 1 uses the mixed strategy $\mathbf{x}$, while 2 uses $\mathbf{y}$, then the expected payoff will be

$$
\sum_{i=1}^{m} \sum_{j=1}^{n} x_i a_{ij} y_j
$$

or, in matrix notation, $\mathbf{x}^{T}A\mathbf{y}$.

It is now possible to define maximin and minimax in terms of pure strategies:

$$
V_{\mathrm{I}} = \max_{\mathbf{x}} \ \min_{\mathbf{y}} \ \mathbf{x}^{T}A\mathbf{y}.
$$

$$
V_{\mathrm{II}} = \min_{\mathbf{y}} \ \max_{\mathbf{x}} \ \mathbf{x}^{T}A\mathbf{y}.
$$

where, in each case, the maximization and minimization are taken over the sets of all mixed strategies. The *minimax* theorem then states that $V_{\mathrm{I}} = V_{\mathrm{II}}$ for all matrix games. The common value of these is the value of the game, and the maximizing $\mathbf{x}$ and minimizing $\mathbf{y}$ are known as *optimal strategies*.

Several techniques exist for computation of the optimal strategies in a matrix game. Most common is linear programming. Also in use is an iterative technique known as *fictitious play*. Apart from this, there exist a few techniques useful for special types of games.

IV. INFINITE GAMES

When each player has an infinite number of pure strategies, it is still possible to define mixed strategies and optimality. Unfortunately, the minimax theorem will not always hold in such games, and where it holds, computation of the optimal strategies can be quite complicated. Nevertheless, some types of infinite games have been analyzed in detail of which we will give a very brief description.

A. Games on the Square

In these games, each player has a continuum of pure strategies—essentially, an interval of these. A value and optimal strategies will exist if the payoff function is continuous. Computation, however, is easiest in the case of some discontinuous games.

B. Stochastic Games

In these games, it is possible to repeat positions, so that, in theory at least, infinitely long play might occur. Solutions occur if either (a) the probability of infinitely long play is zero, or (b) future payoffs can be discounted.

C. Differential Games

These games are played in continuous time; each player is to make a decision at each moment in time. (The typical example would be a pursuit game, in which a pursuer and an evader must choose a speed and direction of motion at each instant in some interval of time.) Differential equation solutions have been obtained under some conditions.

V. NON-ZERO-SUM GAMES

In the non-zero-sum two-person games, the two players' interests are not directly opposed. In such case, it is important to distinguish between *cooperative* games, where communication, binding contracts, side payments, and correlated strategies are allowed, and *noncooperative* games, where all these are forbidden (though in practice there may be intermediate cases where some but not all of these are allowed).

A. Noncooperative Games

1. Equilibrium Points

In the noncooperative two-person games, the search has generally been for equilibrium points. A Nash equilibrium is a pair of strategies such that neither player can gain by a unilateral change of strategies. (This is a generalization of the Cournot equilibrium of classical oligopoly theory.) In Fig. 4, for example, the second row, first column is an equilibrium point: Player 1, by switching unilaterally, would lose three units; Player 2 by a unilateral switch would lose one unit. (This game is a variant of a well-known game called the Battle of the Sexes.)

It may be noticed that in Fig. 4, there is another equilibrium pair of strategies, namely first row, second column. One of the difficulties with the equilibrium concept lies in

$$\begin{bmatrix} (2,3) & (4,4)^* \\ (5,2)^* & (3,1) \end{bmatrix}$$

FIGURE 4 The Battle of the Sexes. The two starred entries are equilibria.

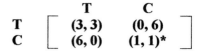

FIGURE 5 The Prisoner's Dilemma.

the possibility that a game may have several nonequivalent equilibria, so that players would then find it difficult to choose among them. (This cannot happen with zero-sum games.) Another difficulty we encounter is that, even when unique, an equilibrium may be unsatisfactory.

Figure 5, usually known as the Prisoner's Dilemma, illustrates this difficulty. In this game, each of two players (prisoners) has a choice between confessing (turning state's evidence, second row and column) and remaining true (thief's honor, first row and column). This game has the unique equilibrium (C, C). The reason for this is that, for each player, C is always better than T, *whatever the other player does*. Hence both will, presumably, confess. (This although both will be better off if neither confesses.) This game has been the subject of serious study by game theorists and psychologists for many years.

Because of the pathologies which some equilibrium points display, Selten and others have sought to refine the concept. These refinements include (but are not limited to) the *undominated equilibrium* and two types of *perfect equilibrium*. Readers are invited to consult the literature.

2. Applications to Biology

An interesting application of game theory to biological evolution has arisen in recent years. It may be noticed that, in many animal species, some members are aggressive fighters while others are peaceful and tend to avoid aggression. For example, males of a certain species of birds may be classified into "hawks" and "doves." Whenever two of these males meet in a conflict situation (say, in the presence of an empty nest which they would both like to appropriate), the "hawks" will fight while the "doves" will yield. It is not difficult to see that, if the species have mainly hawks, it will wear itself out fighting. If, on the other hand, it is mainly doves, an intruding band will be able to dislodge the community from its habitat. Thus, it seems reasonable that there should be some optimal mixture of the two types.

In fact, it is not so much that there is an optimal mixture, but, rather, that certain mixtures of the types are in stable equilibrium in the sense that they are best adapted to the environment (which includes both the habitat and the species itself). This gives us the concept of an *evolutionary stable system* (ESS). Briefly, an ESS consists of a square symmetric matrix, A, together with a (possibly mixed) strategy **x**, such that

(a) $\mathbf{x}^T A \mathbf{x} \geq \mathbf{x}^T A \mathbf{y}$ for all $\mathbf{y}$

(b)   If $\mathbf{x}^T A \mathbf{x} = \mathbf{x}^T A \mathbf{y}$,   then $\mathbf{x}^T A \mathbf{y} > \mathbf{y}^T A \mathbf{y}$.

In essence, condition (a) says that no intruding band ($\mathbf{y}$) will do better against the established population ($\mathbf{x}$) than $\mathbf{x}$ does against itself. Condition (b) says that if the intruding $\mathbf{y}$ does as well against $\mathbf{x}$ as $\mathbf{x}$ itself, then this will create a new environment (a mixture of $\mathbf{x}$ with a small quantity of $\mathbf{y}$), and then $\mathbf{x}$ will do better in this environment than $\mathbf{y}$ does. In either case, the established population will thrive (or reproduce) more successfully than the intruders, thus (eventually) eliminating the latter.

B. Cooperative Games

1. The Axiomatic Approach

For the cooperative games, it is usually more important to focus on the *bargaining* between the two players. The most commonly accepted theory represents such a game as a compact convex subset S of the Euclidean plane (the set of *feasible alternatives*) together with a distinguished point (u^*, v^*) of that set (the *conflict point*). The two coordinates represent payoffs to Players 1 and 2, respectively; it is understood that, by cooperation, the players may obtain any point of S, whereas, if they fail to come to an agreement, they must receive the conflict point (u^*, v^*). As illustrated in Fig. 6, the problem is to choose a "fair" point, $(\bar{u}, \bar{v})$ in set S.

Several axiomatic approaches have been made in this problem. The best known axioms, due to Nash, state that

$(\bar{u}, \bar{v})$ must (1) be feasible (lie in S); (2) be Pareto-optimal (there can be no point in S that is better for both players); (3) be independent of irrelevant alternatives (elimination of a less desirable alternative cannot change the solution); (4) be covariant with linear changes of utility scale; and (5) be at least as symmetric as (S, u^*, v^*) is. Given these axioms, it can be shown that $(\bar{u}, \bar{v})$ is necessarily that feasible point (u, v) which maximizes the product $(u - u^*)(v - v^*)$ of the increments of utilities (subject to $u \geq u^*$, $v \geq v^*$).

2. The Strategic Approach

Apart from alternative axiomatic approaches, a strategic approach has been suggested by Rubinstein. In this approach, the two players alternate in making proposals (i.e., offering points of S). An accepted proposal goes immediately into effect; a rejected proposal causes a delay. By discounting future proposals, an equilibrium point is found which consists of possible offers to be made by the two players. Under certain conditions, these two proposals will converge to Nash's bargaining solution.

VI. *n*-PERSON GAMES

A. Noncooperative Games

The main difference between *n*-person and 2-person games lies in the fact that, for $n \geq 3$, players generally have a choice of coalitions to join. In the noncooperative case, where coalitions are not allowed, there is very little difference between *n*-person and 2-person non-zero-sum games. In general, the search is for equilibrium *n*-tuples of strategies. A generalization of the minimax theorem states that, for finite games, at least one equilibrium *n*-tuple of mixed strategies will exist.

B. Cooperative Games

Where cooperation is allowed, the interest is mainly on the coalitions formed and the bargaining within coalitions, rather than on the available strategies. Thus games are usually studied in their *characteristic function* form.

Let $N = \{1, 2, \ldots, n\}$ be the set of all players in a game. Each nonempty subset S of N is a *coalition*. A game in characteristic function form is a function, v, which assigns, to each coalition S, the set of outcomes that the members of that coalition, acting together, can obtain for themselves (even against the concerted action of the remaining players). In the simplest of cases, it is assumed that utility is freely transferable among members of a coalition, and for each S, $v(S)$ is a real number: the maximum amount of utility that S can obtain, and then distribute (arbitrarily) among its members.

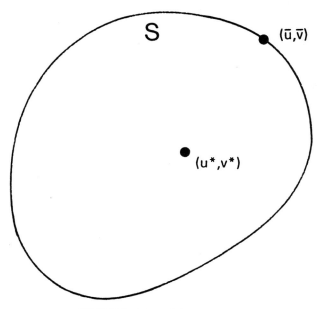

FIGURE 6 A bargaining problem.

As an example, consider the following three-person game. There is a trunk holding jewels worth 100 units of utility. The trunk is too heavy for a single player, but any two of the three players can lift it and thus obtain 100 units. This is represented by the function v, where

$$v(S) = \begin{cases} 0 & \text{if } S \text{ has 0 or 1 element} \\ 100 & \text{if } S \text{ has 2 or 3 elements} \end{cases}$$

As a second example, consider another three-person game. Player 1 owns a horse which she values at \$50. Players 2 and 3 would both like to buy the horse; player 2 values it at \$70, and 3 feels it is worth \$100. In this case players 2 and 3 (the two buyers), alone or together, can obtain no utility. Any other coalition can maximize its utility by giving the horse to the player who values it most; that player can then give money to other members of the coalition so as to reach a bargain.

In this case we have a game u given by:

$$u(\{2\}) = u(\{3\}) = u(\{2, 3\}) = 0$$
$$u(\{1\}) = 50$$
$$u(\{1, 2\}) = 70$$
$$u(\{1, 3\}) = u(\{1, 2, 3\}) = 100$$

We will refer to these two games as the *trunk* game and the *horse-trade* game, for future reference.

C. Imputations: Domination

Let v be the characteristic function of a game with player set N. An imputation is a vector $\mathbf{x} = (x_1, x_2, \ldots, x_n)$ such that $x_i \geq v(\{i\})$ for all i, and $\Sigma x_i = v(N)$. Thus an imputation is an individually rational way of dividing the utility $v(N)$. For the trunk game described previously, the imputations are nonnegative vectors (x_1, x_2, x_3) with components adding to 100. For the horse-trade game, the imputations must additionally satisfy $x_1 \geq 50$. Given two imputations $\mathbf{x}$ and $\mathbf{y}$, we say $\mathbf{x}$ *dominates* $\mathbf{y}$ if there is some coalition which prefers $\mathbf{x}$ to $\mathbf{y}$ and is strong enough to enforce $\mathbf{x}$. Mathematically, $\mathbf{x}$ dominates $\mathbf{y}$ if there is some nonempty $S \subset N$ such that (1) $x_i > y_i$ for all $i \in N$, and (2) $\Sigma_S x_i \leq v(S)$. Thus, for the horse-trade game, imputation $\mathbf{x} = (60, 10, 30)$ dominates $\mathbf{y} = (50, 5, 45)$ through the coalition $\{1, 2\}$. On the other hand, $\mathbf{z} = (65, 10, 25)$ does not dominate $\mathbf{y}$ because $z_1 + z_2 = 75$ which is greater than $v(\{1, 2\})$, that is, $\{1, 2\}$ prefers $\mathbf{z}$ to $\mathbf{y}$ but is not strong enough to enforce $\mathbf{z}$.

D. Solution Concepts

The central problem in n-person game theory lies in choosing some reasonably small set of outcomes (possibly, but not necessarily, a unique outcome) from the set of all imputations. Choices can be made on the basis of either stability or fairness.

1. The Core

Perhaps the most obvious idea is to look for the set of all undominated imputations in a game. This set, the *core*, corresponds more or less to the competitive equilibrium of classical economic theory. In the horse-trade game described earlier, the core consists of all vectors of the form $(t, 0, 100 - t)$, where $70 \leq t \leq 100$. In effect, it suggests that player 3 will buy the horse for some price above 70 but below 100. Player 2 (the low bidder) will be eliminated and receive nothing, but he has, nevertheless, an effect on the game, and that is that he pushes the price up to at least his maximum bid.

It is true that points in the core have a very strong type of stability. Unfortunately, the core is often empty. For the trunk game, there are no core points—it is always possible for two of the players to do better than in any suggested outcome.

2. Stable Sets

Perhaps because the core is so frequently empty, von Neumann and Morgenstern devised the concept of a *stable set* (also known as a *solution*). A set of imputations is *internally stable* if no imputation in the set dominates another. It is *externally stable* if any imputation not in the set is dominated by at some imputation in the set. A *stable set solution* is any set which is both internally and externally stable.

In some cases, stable sets seem very reasonable. As an example, for the trunk game, the three imputations (50, 50, 0), (50, 0, 50), and (0, 50, 50) form a stable set. This is not, however, the only stable set for this game. There are many others, for example, the set of all vectors of the form $(t, 40, 60 - t)$, where $0 \leq t \leq 60$. In fact, most games seem to have a bewildering multiplicity of stable sets, and it is difficult to see how to distinguish among them. Worse yet, some games have no stable sets, though the smallest such game known has 10 players.

3. The Power Indices

An alternative approach consists in looking for some sort of "expected" value for an n-person game. Axiomatically, Shapley defines his power index (the *Shapley value*) as a mapping which assigns, to each n-person game v, an n-vector (imputation) $\Phi[v]$ satisfying (1) *efficiency* (the set of all essential players receives as much as it can enforce), (2) *symmetry* (the vector $\Phi[v]$ is at least as symmetric as the game v, (3) *additivity* ($\Phi[v + w] = \Phi[v] + \Phi[w]$ for two games v and w). It can be shown that there is a unique rule satisfying these conditions, given by

$$\Phi_i[v] = \sum_{S \subset N} \frac{s!(n - s - 1)!}{n!}[v(S \cup \{i\}) - v(S)]$$

where s is the number of elements in S. The trunk game has value (33.3, 33.3, 33.3) as is obvious from symmetry; whereas the horse-trade game has value (78.3, 3.3, 18.3).

The Shapley value has, in particular, been used to analyze power in voting situations; a somewhat similar power index, developed independently by Coleman and Banzhaf, has also been used in this context.

4. The Bargaining Sets

Yet another approach, due to Aumann and Maschler, seeks to reproduce the bargaining that goes on within a coalition.

Let $\Im = \{T_1, T_2, \ldots, T_m\}$ be a partition of the set N of players; we will call it a *coalition structure*. If $\Im$ is the coalition structure, then each of the sets $T_j (j = 1, \ldots, m)$ will have an amount $v(T_j)$ available; we let $\chi(\Im)$ be the set of all vectors $(x_1, \ldots, x_n)$ such that, for each $T_j \in \Im$,

$$\sum_{i \in T_j} x_i = v(T_j)$$

and, for each i, $x_i \geq v(\{i\})$. An *individually rational payoff configuration* (irpc) is a pair $\langle \mathbf{x}, \Im \rangle$, where $\Im$ is a coalition structure and $\mathbf{x} \in X(\Im)$.

If two players, i and k, belong to the same T_j, it is possible for one of them, say k, to object against i in $\langle \mathbf{x}; \Im \rangle$ if he feels that he (k) can obtain more without i's help. This is an *objection*. It may be possible for i to have a *counter-objection*, in which she protects her share, x_i, without k's help.

The *bargaining set*, then is the set of all irpc's $\langle \mathbf{x}; \Im \rangle$ such that, for every objection, there is a counter-objection. (There are, in fact, several bargaining sets, because objections and counter-objections can be defined in several ways.) An important theorem states that, for one particular definition, the corresponding bargaining set M_1^i is nonempty in the strong sense that, for any coalition structure $\Im$, there is at least one $\mathbf{x} \in X(\Im)$ such that $\langle \mathbf{x}; \Im \rangle \in M_1^i$.

For the trunk game described earlier, it is interesting to note that, if $\Im = \{\{1, 2\}, 3\}$, then only corresponding $\mathbf{x}$ is (50, 50, 0). If $\Im = \{\{1, 2, 3\}\}$, the only $\mathbf{x}$ is (33.3, 33.3, 33.3).

5. Related Concepts

Closely related to the bargaining sets are certain other concepts, such as the *kernel* and the *nucleolus*. Interested readers should consult the literature for these.

VII. GAMES WITH A CONTINUUM OF PLAYERS

Essentially, cooperative game theory developed out of a desire to introduce, for situations with a small number of participants, some of the concepts of economics. On the other hand, it is of interest to see whether game-theoretic concepts can be used in situations of "perfect competition," that is, cases with a very large number of players. A very interesting theory of *non-atomic games* has been developed by Aumann and Shapley. In these games, the set of players is isomorphic to the unit interval $I = [0, 1]$. Not all subsets of I are admissible as coalitions; instead there is some σ algebra of coalitions. Otherwise, the characteristic function is defined much as for finite-player games. It has been shown in particular that, for such games, the Shapley value and the core frequently coincide with free-market equilibria.

VIII. GAMES WITH NONTRANSFERABLE UTILITY

A. The Basic Model

A more complicated theory exists when the free transferability of utility is not postulated. In this case, $v(S)$ is not a single number; rather, it is a set in s-dimensional space, where s is the cardinality of S. Attempts have been made to generalize the several solution concepts mentioned above, such as the core, stable sets, power indices, and bargaining sets, to such games. The extensions have proved difficult and many theorems cannot be generalized of which we give an example.

In a three-player game, Player 1 owns a coffeepot. Player 2 has a pound of coffee, while player 3 has a pound of sugar. Thus, players 1 and 2 together can produce sugarless coffee. Player 1 enjoys her coffee without sugar, but Player 2 prefers sugar with his coffee. A coalition with Player 3 will remedy this lack.

Nothing can be produced without the concurrence of players 1 and 2; thus the 1-player sets can give 0 to their members; this happens also with the sets $\{1, 3\}$ and $\{2, 3\}$. Let us assume that the whole pound of coffee will give any one of the players 100 units of utility, but that Player 2 only derives 25 units of utility if there is no sugar available. Then $v(\{1, 2\})$ consists of all vectors $\mathbf{x}$ with $x_1 + 4x_2 \leq 100$. Finally, $v(\{1, 2, 3\})$ will consist of all $\mathbf{x}$ with $x_1 + x_2 + x_3 \leq 100$.

For this game, the core is easily obtained: it is a convex triangle with vertices (100, 0, 0), (0, 100, 0), and (0, 25, 75). As for the Shapley value, several generalizations have been suggested, giving diverse results. One generalization gives

(50, 50, 0), another gives (40, 40, 20), a third yields (50, 37.5, 12.5), and yet another gives (51.8, 47.6, 0.6). No fully satisfactory theory has been given.

B. Spatial Games

Spatial games represent an interesting application to political science, as well as to situations with public goods. In these, outcomes are points in some Euclidean space, Ω, of low dimension. Each player, i, is assumed to have an *ideal point* (preferred outcome) $\mathbf{P}^i$, in Ω. The payoff to player i of outcome α is a decreasing function of the distance from α to $\mathbf{P}^i$. The game is further defined by giving a collection of *winning* coalitions: these coalitions can enforce any point of Ω, while other coalitions can enforce nothing at all. For these games, domination and the core can be defined much as described above. It turns out that, when Ω has dimension 1, the core is nonempty. For higher dimensions, the core is usually empty, so other solution concepts have been sought.

IX. GAMES WITH INCOMPLETE INFORMATION

One problem with the original theory, as developed by von Neumann and Morgenstern, lies in the fact that it assumes all players are aware of all the rules, payoffs, relevant probability distributions, etc., in the game. In real life, things are not so simple; in particular we may well be unaware of the other players' utility functions, let alone what they really know about the game. Now, a player will frequently have private information about a game situation. Use of this information will increase her utility. The problem is that use of this information will reveal it to the other players, who will then be in a position to use it. The player must then decide whether the gains from using this information are outweighed by the losses due to disclosure.

As an example, suppose that (in a non-zero-sum game) Players 1 and 2 come to an agreement. This calls for 1 to carry out some action on Player 2's behalf, after which 2 will pay her back for her help. Now, an honest 2 will pay as promised, because he assigns high utility to keeping his word. A dishonest 2, on the contrary, has no qualms about breaking his word, and will feel free to double-cross 1. Thus Player 1 will come to an agreement only if she feels that 2 is honest. But how is Player 1 to know whether 2 is honest?

Generally speaking, an individual who frequently breaks promises may merely be forgetful, or may be dishonest. Player 1, on the basis of 2's previous actions, can make deductions about his honesty. This being so, a clever thief may act honestly on repeated occasions so as to lull another player, in preparation for a big scam.

In a model due to Harsanyi (on one hand), and to Aumann and Maschler (on the other), each of the players may be of any one of several types (honest, dishonest, brave, cowardly, etc.); this type will determine the player's payoff from the game's possible outcomes. Each one knows his or her own type, but has only a probability distribution (*a priori*) as to the other player's type. Repeated play of a game may then help either to determine the other's type. In other words, we can expect that a player will carry out actions that tend to give her high utility. Since this utility depends on the player's type, the actions are an indication of her type. Hence information about past actions will allow a player to update his information about another player's type, thus obtaining an *a posteriori* distribution. The theory can be extremely complicated. For details, the reader should consult the literature.

SEE ALSO THE FOLLOWING ARTICLES

COMPUTER ALGORITHMS • CYBERNETICS AND SECOND ORDER CYBERNETICS • STOCHASTIC PROCESSES

BIBLIOGRAPHY

Aumann, R. J., and Maschler, M. B., with the collaboration of Stearns, R. E. (1995). "Repeated Games with Incomplete Information," MIT Press, Cambridge, MA.
Aumann, R. J., and Shapley, L. S. (1974). "Values of Non-Atomic Games," Princeton Univ. Press, NJ.
Enelow, J., and Hinich, M. (1984). "The Spatial Theory of Voting: An Introduction," Cambridge Univ. Press, New York.
Ichiishi, T. (1983). "Game Theory for Economic Analysis," Academic Press, New York.
Ordeshook, P., ed. (1978). "Game Theory and Political Science," New York Univ. Press, New York.
Osborne, M. J., and Rubinstein, A. (1990). "Bargaining and Markets," Academic Press, New York.
Owen, G. (1995). "Game Theory," Academic Press, New York.
Shubik, M. (1982). "Game Theory and the Social Sciences: Concepts and Solutions," MIT Press, Cambridge, MA.
van Damme, E. (1987). "Stability and Perfection of Nash Equilibria," Springer-Verlag, New York.

Gamma-Ray Astronomy

J. Gregory Stacy
Louisiana State University and Southern University

W. Thomas Vestrand
Los Alamos National Laboratory

GLOSSARY

Accretion disk A flattened, circulating disk of material drawn in and heated to high temperatures under the influence of the intense gravitational field associated with a black hole or other compact object (such as a neutron star or white dwarf).

Active galactic nuclei (AGN) A collective term for active galaxies whose emission is observed to come predominantly from the central nuclear region of the galaxy. Of these, blazars form a subclass that is observed to emit gamma radiation. It is likely that the viewing angle toward these latter objects is directed along jets of relativistic material ejected from the nucleus of the galaxy by supermassive black holes.

Bremsstrahlung The "braking" radiation given off by free electrons that are deflected (i.e., accelerated) in the electric fields of charged particles and the nuclei of atoms.

Cherenkov light Radiation produced by a charged particle whose velocity is greater than the velocity of light in the medium through which it travels. Cherenkov light is strongly directed along the line of travel of the particle.

Compton scattering The dominant process by which a medium-energy gamma ray interacts with matter by scattering and transferring a part of its energy to an electron. "Inverse" Compton scattering refers to the same process, but where a lower-energy photon is scattered to higher energy after interaction with a relativistic electron.

Cosmic rays High-energy charged particles, such as electrons, protons, alpha-particles (helium nuclei), and heavier nuclei that propagate through interstellar space.

Diffuse emission Radiation that is extended in angular size on the sky, such as the gamma-ray emission arising from the decay of radioactive nuclei dispersed throughout the interstellar medium. A diffuse source is distinguished from a pointlike or point source of emission that is not resolvable into further individual components given the limited angular resolution of a telescope.

Electromagnetic cascades A phenomenon that occurs in the upper atmosphere of the Earth when a very high-energy gamma ray interacts by the pair-production process, followed by further interactions resulting in extensive air showers (or EAS) of particles and photons. The relativistic cascade particles emit optical Cherenkov light that is observable from the ground. Electromagnetic cascades are distinguished from the nucleonic (or hadronic) cascades produced by high-energy cosmic-ray particles in the upper atmosphere.

Electron volt (eV, and keV, MeV, GeV, TeV, PeV) The electron volt (eV) is a fundamental unit of energy commonly used in high-energy astrophysics. It is defined to be the energy acquired by an electron when accelerated through a potential difference of one volt. One eV is equal to 1.602×10^{-19} J or 1.602×10^{-12} ergs.

Gamma rays The highest-energy form of electromagnetic radiation, above the X-ray portion of the spectrum. Gamma rays have energies measured in millions of electron volts and higher.

Pair production and annihilation The process by which the most energetic gamma rays (of MeV energies and above) interact with matter, producing an electron–positron pair (the positron, with positive charge, is the antiparticle to the electron). The inverse process is pair annihilation, in which an electron and positron mutually annihilate and produce a pair of high-energy gamma rays.

Parsec (pc, and kpc, Mpc, Gpc) A unit of distance commonly used in astronomy (an abbreviation for "parallax-arcsecond"). One parsec is equal to 3.086×10^{16} m, or 3.26 light-years.

Point source A source of emission that is not further resolvable into individual components given the limited angular resolution of a telescope. In gamma-ray astronomy angular resolution is relatively poor compared to other branches of astronomy. Thus, in some instances a gamma-ray point source may in fact consist of a number of individual sources whose summed emission is measured by the gamma-ray telescope.

Supernova An endpoint of stellar evolution for the most massive stars, an explosion triggered by the gravitational collapse of the stellar core following the exhaustion of fuel for nuclear burning. The collapsed stellar core, depending on its final mass, can become either a black hole or a neutron star (and some of the latter may be observable as pulsars).

Synchrotron radiation The emission produced by charged particles as they spiral (i.e., accelerate) around magnetic fields.

THE GAMMA-RAY regime constitutes one of the last regions of the electromagnetic spectrum to be opened to detailed astrophysical investigation. Only within the past decade has the field of gamma-ray astronomy become firmly established as a productive and dynamic discipline of modern observational astrophysics. This has been largely due to the successful operation during the 1990s of the Compton Gamma Ray Observatory, whose telescopes carried out the first comprehensive surveys of the sky at gamma-ray energies (as shown in Fig. 1). Gamma rays form the highest-energy portion of the electromagnetic (E-M) spectrum with individual photon energies extending from millions of electron volts (MeV) to values in excess of 10^{16} eV (optical photons in contrast carry energies of only a few electron volts). Observations of these highest-energy photons provide the means of investigating the largest transfers of energy occurring in the Universe and offer the key to understanding a host of challenging cosmic phenomena occurring in a wide variety of astrophysical settings. The environments in and around gamma-ray sources are among the most extreme to be found in the Universe, permitting the testing of models and hypotheses regarding high-energy phenomena under conditions impossible to achieve on the Earth. Further, the Universe is essentially transparent to the propagation of gamma radiation, and since, like all electromagnetic radiation, gamma rays are electrically neutral they are not deviated from their trajectories under the influence of magnetic fields. Gamma rays arriving at the Earth therefore serve as direct messengers from high-energy celestial sources within our own Milky Way galaxy and beyond, extending to the most distant reaches and earliest epochs of the cosmos. Gamma-ray astronomy quite literally provides a new window into space that extends our view out to the edge of the observable Universe.

I. INTRODUCTION AND HISTORICAL OVERVIEW

A. Fundamental Concepts and Terminology

To place the discipline of gamma-ray astronomy in context, we review some basic concepts and nomenclature. The electromagnetic spectrum encompasses the entire range of radiation from the radio through gamma rays, and includes the subregions of radio, infrared (IR), optical (or visible), ultraviolet (UV), X-rays, and gamma rays, in order of increasing energy or frequency of radiation. The fundamental relation between photon energy, frequency, and wavelength is the well-known expression due to Planck,

$$E = h\nu = hc/\lambda,$$

where E is the photon energy, ν the frequency, λ the wavelength of the radiation, c the speed of light ($=3 \times 10^8$ m/s), and h is Planck's constant ($=6.626 \times 10^{-34}$ J s $= 4.135 \times 10^{-15}$ eV s). The speed of an electromagnetic wave in vacuum is the speed of light, thus the product of

EGRET
Skymap E > 100 MeV
Phase 1 - 4

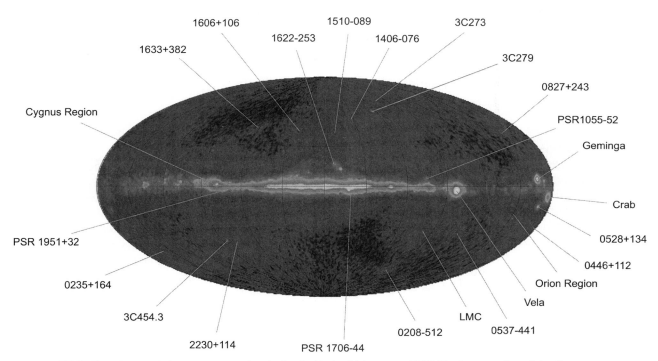

FIGURE 1 A map of the gamma-ray sky obtained with the high-energy EGRET telescope aboard the Compton Gamma Ray Observatory. The map is an all-sky Aitoff equal-area projection in Galactic coordinates with the direction of the Galactic Center at the middle of the map. The bright horizontal band is predominantly diffuse gamma-ray emission from the disk of the Milky Way. Prominent gamma-ray sources are indicated and labeled around the periphery of the figure. (Courtesy NASA, the Max Planck Institute, and the CGRO EGRET Instrument Team.)

frequency and wavelength equals c (or, $\nu\lambda = c$). At energies below or near the optical, sub-bands of the E-M spectrum are traditionally referred to either in terms of their wave properties (in the radio, for example, there are the meter, centimeter, millimeter, submillimeter, and microwave bands), or in terms of their relation to the central optical band (the *near* versus *far* infrared bands, for example, of shorter and longer wavelength, respectively, and the *extreme* ultraviolet beyond the optical and UV). Once in the X-ray band, however, one enters the realm of high-energy astrophysics and tends to abandon the wavelength/frequency nomenclature in favor of particle and energy terminology that is better suited to the description of photon interactions at these energies. Thus X-rays are generally referred to in broad terms as either *soft* or *hard* in energy content (with a loosely defined boundary in the keV energy range). Similarly, gamma rays themselves have been subdivided into soft and hard regimes, or otherwise characterized as of low, medium, or high gamma-ray energy. Given the very broad range of gamma-ray energies

that are now observable (spanning more than 10 orders of magnitude in energy) one now usually refers to gamma rays by their energy designation alone, as either MeV, GeV, TeV, or even PeV, gamma rays, where standard order-of-magnitude prefixes apply (for the above, 10^6, 10^9, 10^{12}, and 10^{15} eV, respectively).

Much of the radiation with which we are familiar in everyday life is of *thermal* origin, arising by definition from matter in thermal equilibrium. In an ideal atomic gas in thermal equilibrium, for example, the upward versus downward transitions of bound electrons between energy levels in individual atoms are in close balance due to the exchange of energy between particles via collisions and the absorption and emission of radiation. The velocities of particles in an ideal thermal gas follow the well-known Maxwellian distribution, and the collective continuous spectrum of the radiating particles is described by the familiar Planck black-body radiation curve with its characteristic temperature-dependent profile and maximum.

Gamma rays, in contrast to other forms of electromagnetic radiation, are most often of *nonthermal* origin, arising usually from interactions involving high-energy, relativistic particles in an ionized plasma whose constituents are not in thermal equilibrium with their surroundings. A *relativistic* particle is one whose kinetic energy is comparable to or exceeds its rest-mass energy given by Einstein's famous relation, $E = mc^2$. Thus,

$$\text{total particle energy} = (\text{rest-mass} + \text{kinetic})\text{energy}$$
$$= \gamma mc^2,$$

where $\gamma = (1 - v^2/c^2)^{-1/2}$ is the relativistic Lorentz factor for a particle traveling at velocity v. Interactions involving relativistic particles are properly treated within the framework of relativity. For sufficiently low velocities the familiar *nonrelativistic* case ($v/c \ll 1$, $\gamma \rightarrow 1$) is re-obtained in which the relations of classical Newtonian physics apply. In the standard relativistic regime the particle velocity approaches c, $v/c \rightarrow 1$ and $\gamma \geq 1$ (for example, a particle traveling at 90% of the speed of light will have $v/c = 0.9$ and $\gamma \cong 2.3$), whereas in the most extreme *ultrarelativistic* case ($v/c \sim 1$, $\gamma \gg 1$) the total energy of the particle is dominated by its kinetic energy. Photons of sufficiently high energy in cosmic sources can be diminished in intensity via the photon–photon pair-production process ($\gamma\gamma \rightarrow e^+e^-$) which is most likely to occur just above the reaction threshold when the product of the two photon energies is equal to the product of the electron–positron rest–mass energies, $E_{\gamma 1}E_{\gamma 2} \sim 2(m_e c^2)^2 \sim 0.52$ (MeV)^2. In astrophysical sources, then, the energy density of gamma rays may be sufficiently high to prevent their escape due to their greater likelihood of producing pairs. Such media are said to have a high pair-production opacity.

Given the relativistic nature of the major gamma-ray production mechanisms we adopt in this review the energy corresponding to the rest–mass energy of the electron, $m_e c^2 \sim 511$ keV ~ 0.511 MeV, as a natural reference energy defining the lower boundary of the gamma-ray regime. (We note that, historically, among physicists of a certain age, gamma rays were defined simply as the radiation resulting from nuclear transitions, independent of the energies involved, in recognition of the predominant nuclear origins of gamma radiation. We adopt here the more current view of a specific energy regime.)

Gamma-ray astronomy is closely linked to several other branches of modern observational astrophysics. A number of these subdisciplines of astronomy are described elsewhere in these volumes. Gamma-rays, by virtue of their high energy, also play a particularly key role in *broadband multiwavelength astronomy*, by which is meant the study of a celestial object or phenomenon over as broad a portion of the electromagnetic spectrum as possible.

B. The Production of Cosmic Gamma Rays

A wide variety of production mechanisms give rise to gamma radiation, resulting in either continuum or spectral-line emission. Gamma rays most often result from high-energy collisions between nuclei, particles, and other photons, or from the interactions of charged particles with magnetic fields. Line emission can arise from the deexcitation of nuclear states, from radioactive decay, or from matter–antimatter annihilation. The primary production mechanisms of interest to gamma-ray astronomy are summarized in the following.

1. Particle–Nucleon Interactions

High-energy nuclear collisions frequently yield charged and neutral mesons as unstable reaction products. Charged pions (π^+ and π^-) decay into positive and negative muons that decay in turn into relativistic electrons and positrons. Neutral pions (π^0) decay almost immediately ($t_{1/2} \sim 10^{-16}$ s) into two gamma rays of total energy equal to approximately 68 MeV in the rest frame of the decaying meson. The resulting gamma-ray spectrum depends on the distribution of particle energies of the original emitted pions, and is generally a broad continuum centered and peaked at $E_\gamma \sim m_\pi c^2/2 \sim 68$ MeV. Nucleon–nucleon interactions play an important role in the production of diffuse high-energy gamma rays in the disk of the Galaxy following the collision of high-energy cosmic rays (primarily protons) with the nuclei of the atoms and molecules of the interstellar gas.

Collisions between high-energy particles and ambient matter can also result in the copious production of numerous other secondary particles, including neutrons. High-energy neutrons are capable of exciting nuclei in secondary collisions, leading to gamma ray line emission (see following). Further, neutrons can be slowed in the interacting medium to thermal energies whereupon they can be quickly captured by nuclei, again giving rise to gamma-ray lines. Neutron processes play an important role in solar flares, and in the production of gamma rays on planetary surfaces after cosmic-ray bombardment.

2. Nuclear Gamma-Ray Lines

Nuclear deexcitation following energetic collisions or radioactive decay gives rise to spectral line radiation whose specific energies are characteristic of the emitting nuclides. Nuclear gamma-ray line radiation extends up to ~ 9 MeV in energy for the most commonly abundant elements and likely interaction processes. The intensities and ratios of observed gamma-ray lines can provide detailed information on elemental composition and relative abundances (for example, in solar flares).

3. Relativistic Electron Interactions

Relativistic electrons can interact with charged particles via the bremsstrahlung process, with photons through Compton scattering, and with magnetic fields by emitting synchrotron radiation. These processes dominate many of the energy regimes in the field of high-energy astrophysics. They give rise to continuum gamma radiation, whose spectral characteristics can be used to deduce the physical conditions at the astrophysical source.

(a) Bremsstrahlung. Bremsstrahlung (or "braking radiation") is the radiation given off by free electrons that are deflected (i.e., accelerated) in the electric fields of charged particles and the nuclei of atoms. Thermal bremsstrahlung is the emission given off by an ionized gas of plasma in thermal equilibrium at a particular temperature, where the distribution of electron velocities follows the well-known Maxwellian distribution. Relativistic electrons, whose distribution of energies often follows a power-law shape in astrophysical settings, give rise to relativistic bremsstrahlung radiation that is also of power-law shape with the same spectral index as the emitting electrons.

(b) Compton scattering. Another major source of cosmic gamma radiation is the Compton scattering of lower-energy photons to gamma-ray energies by relativistic electrons. This process is often referred to as "inverse" Compton scattering since it is the low-energy photon that gains energy from the high-energy electron, in contrast to the more standard view of the Compton mechanism. In the ultrarelativistic case, it can be shown that the energy of the photons scattered by high-energy electrons is $E \sim \gamma^2 E_e$ (in the Thomson limit when the energy of the photon in the center-of-momentum frame of reference is much less than $m_e c^2$), where γ is the relativistic Lorentz factor of the electrons. For relativistic electrons with $\gamma \sim 100$–1000, as observed in many astrophysical sources, this implies that low-energy photons can be up-scattered to very high energies indeed, well into the gamma-ray regime.

(c) Synchrotron radiation. Synchrotron emission results when an electron gyrates around a magnetic field. For electrons of sufficiently high energy, or for magnetic fields of sufficiently high strength, high-energy photon emission readily results. Again, for a power-law distribution of electron energies, a power-law synchrotron emission spectrum follows. The relation between the observed intensity (I) of the synchrotron radiation as a function of frequency (ν), the magnetic field strength (B), and the power-law index (p) of the electron particle distribution is given by $I(\nu) \propto B^{(p+1)/2} \nu^{-(p-1)/2}$.

Collectively, these emission processes represent the primary energy-loss (or "cooling") mechanisms for relativistic electrons in astrophysical sources, the other major process being "ionization" losses via particle collisions. Observations of high-energy emission from celestial sources that can be decomposed into synchrotron, bremsstrahlung, and Compton components from characteristic spectral signatures therefore provides a wealth of information on the physical conditions within the emitting regions (such as particle densities, and the strengths of radiation and magnetic fields).

4. Electron–Positron Annihilation

A free electron and its antiparticle, the positron, may interact to produce annihilation radiation yielding two gamma rays ($e^+ e^- \rightarrow \gamma\gamma$). The total energy of the two photons in the center-of-momentum frame of reference is equal to the combined rest–mass energy of the electron–positron pair, $2m_e c^2 \sim 1.022$ MeV. (Three-photon annihilation can also occur for free electrons and positrons, but is much less likely.) If an electron and positron are essentially at rest upon annihilation then two gamma rays of equal energy (0.511 MeV) are produced. In the more general astrophysical case, however, one or both particles are at relativistic velocities, and a more complicated emergent gamma-ray spectrum usually results.

An electron and positron of sufficiently low energy (typically thermal, ≤ 5 eV) may combine to briefly form a hydrogen-like state of matter referred to as *positronium*. Positronium almost immediately self-annihilates yielding either a two- or three-photon decay into gamma rays ($\tau_{2\gamma} \sim 10^{-10}$ s, $\tau_{3\gamma} \sim 10^{-7}$ s).

C. A Brief History of Gamma-Ray Astronomy

It was quickly recognized at the dawn of the nuclear age that the potential existed for the detection of celestial gamma rays from high-energy sources in the cosmos. In the early 1950s discussions were already underway on the likelihood of gamma-ray production via cosmic-ray interactions in interstellar space (cosmic rays are high-energy, relativistic particles and nuclei of celestial origin). In now-classic papers Burbridge, Burbridge, Fowler, and Hoyle, in 1957, laid out the principles governing the synthesis of heavy elements in stellar nuclear burning and during explosive nucleosynthesis in supernovae, and Morrison in 1958 similarly described many of the fundamental mechanisms and sources for the production of cosmic gamma rays.

Through the 1960s a number of balloon and early spacecraft observations (e.g., the Ranger spacecraft missions to the Moon) provided intriguing but inconclusive evidence for the existence of cosmic gamma-rays. The first positive detection of celestial gamma radiation was made with an instrument aboard the third Orbiting Solar Observatory (OSO-3), whose investigators reported in 1972 the detection of gamma rays from the Galactic disk, with a peak intensity observed toward the Galactic Center. In the early 1970s several high-altitude balloon experiments were also beginning to report positive results, including detection of the Crab pulsar and of diffuse gamma radiation from the disk and central region of the Galaxy. Nuclear gamma-ray lines from the Sun were detected from large solar flares on August 4 and 7, 1972, with a spectrometer aboard OSO-7. The first reported detection of likely positron annihilation radiation (at 0.511 MeV) from the direction of the Galactic Center was based on balloon measurements from 1971, and later confirmed by other investigators with a detector of higher spectral resolution in 1977. As described elsewhere in this review, gamma-ray spectrometers carried to the Moon by both U.S. and Russian spacecraft in the late 1960s and 1970s provided extensive orbital and in situ measurements of the elemental composition of the lunar surface. Similar experiments in the 1970s were carried to Mars aboard the U.S. Viking landers, and to Venus on the Russian Venera landers. Most intriguing was the announcement in 1973 of the discovery of the mysterious cosmic gamma-ray bursts with instruments aboard the Vela series of nuclear surveillance satellites (launched originally to verify compliance with the 1963 Nuclear Test Ban Treaty).

A major advance in the field of gamma-ray astronomy came with the launch in 1972 of the second Small Astronomy Satellite (SAS-2). Over its 7-month lifetime SAS-2 carried out a survey of high-energy gamma-ray emission (>50 MeV) from the Galactic plane, and provided a first measure of the extragalactic diffuse gamma-ray background. This pioneering mission was followed with the launch of the COS-B gamma-ray satellite by the European Space Agency in 1975. Over its 7-year lifetime COS-B greatly extended our knowledge of the gamma-ray sky, providing detailed maps of the diffuse gamma radiation arising from the Galactic plane, as well as cataloging a number of point sources of high-energy gamma rays, including the first detected extragalactic source, the quasar 3C 273.

Two gamma-ray instruments were carried into space in the late 1970s as part of NASA's High Energy Astronomical Observatory (HEAO) series of satellites. HEAO-1 conducted a survey of the sky from 10 keV to 10 MeV in energy, and identified a number of active galaxies and characterized their spectra in the 10- to 100-keV range. The HEAO-3 experiment discovered the first nonsolar nuclear gamma-ray line of celestial origin, the 1.809-MeV spectral line emitted by radioactive ^{26}Al that is produced in

massive stars and is a tracer of recent star formation in the Galaxy. In 1980, NASA launched the Solar Maximum Mission (SMM) satellite which carried a gamma-ray spectrometer among its suite of instruments. Over its extended 10-year lifetime SMM provided a wealth of new information on gamma-ray processes occurring during flares on the Sun, and also made fundamental contributions to nonsolar gamma-ray astronomy. These included the discovery of greater-than-MeV emission from gamma-ray bursts, confirmation of the diffuse ^{26}Al emission detected by HEAO-3, and further observation of the positron annihilation radiation coming from the central region of the Galaxy. Particularly notable was the SMM detection of radioactive ^{56}Co line emission from the Type II supernova SN 1987A in the Small Magellanic Cloud, providing a long-awaited first direct measure of explosive nucleosynthesis in supernovae. The French coded-aperture SIGMA telescope was carried into space aboard the Soviet GRANAT satellite in 1989. Among other observations, this lower-energy (up to $\sim$1.3 MeV) instrument identified a number of black-hole candidate sources in the central region of the Galaxy based on the observed spectral and temporal behavior.

The realm of ground-based gamma-ray astronomy, where different observational challenges present themselves, begins at photon energies of $\sim$50 GeV (sometimes termed the very-high-energy, or VHE, gamma-ray regime). Gamma-rays approaching TeV energies become increasingly rare in number, and cannot be well sampled by existing spacecraft-borne instrumentation. Further, the absorbing medium of the Earth's atmosphere precludes their direct observation from the ground. Their presence, however, can be inferred indirectly from the electromagnetic cascades of electrons and positrons that they produce upon interaction in the upper atmosphere. (These cascades are also referred to as extensive air showers, or EAS.) The relativistic cascade particles emit Cherenkov light over a wide area that can be detected with optical telescopes on the ground. A particular difficulty, however, is distinguishing such photon-induced events from the nucleonic (or hadronic) cascades produced by high-energy cosmic-ray particles in the atmosphere, which exhibit very similar observable effects. Atmospheric Cherenkov imaging telescopes were originally proposed in the 1970s, but only in the 1990s did the techniques and instrumentation become sufficiently developed to achieve breakthrough detections of several high-energy gamma-ray sources. In the late 1980s the Whipple Observatory first detected TeV emission from the Crab pulsar and nebula, and this has been followed in recent years with detections by several groups of TeV emission from a small number of both Galactic and extragalactic sources (outlined in later sections).

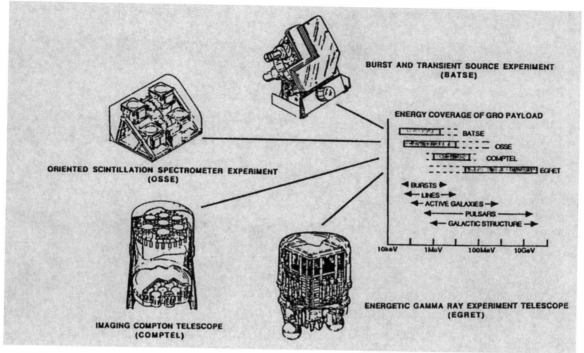

FIGURE 2 Schematic of the four gamma-ray telescopes aboard the Compton Gamma Ray Observatory and their corresponding overlapping energy ranges. Also indicated are classes of prominent gamma-ray sources by energy band. (Courtesy NASA.)

D. The Compton Gamma-Ray Observatory

The great potential of gamma-ray astronomy as a viable, productive branch of observational astrophysics was not fully realized until the launch of the Compton Gamma Ray Observatory (or CGRO) in April 1991. The Compton Observatory was the second of NASA's four planned Great Observatory missions that were designed to study the sky from space in different key regions of the electromagnetic spectrum. The first of these was the Hubble Space Telescope (launched in 1990), the second the CGRO (launched in 1991), and the third the Chandra X-ray Observatory (launched in 1999). At the time of writing, the Space Infrared Telescope Facility (SIRTF) is awaiting an anticipated launch in 2002. The CGRO was named in honor of the American physicist Arthur Holly Compton (1892–1962), whose pioneering investigations into the scattering of X-rays and gamma rays by charged particles earned him the Nobel prize for physics in 1927. After more than 9 years of successful operation, the CGRO mission was terminated in June 2000, when the spacecraft was deorbited for safety reasons.

The CGRO carried four separate, complementary gamma-ray telescopes with overlapping energy ranges,

each designed for specific scientific objectives and developed by an international collaboration of scientists. The combined energy coverage of the CGRO detectors extended over 6 orders of magnitude from ∼30 keV to 30 GeV (see Fig. 2). The key characteristics of each of the four CGRO instruments are summarized in the following.

The *Burst and Transient Source Experiment (BATSE)* was an all-sky monitor consisting of eight separate detectors mounted on the corners of the main platform of the CGRO spacecraft. Its primary objective was to detect and measure rapid brightness variations in gamma-ray bursts and solar flares down to microsecond time scales over the energy range from 30 keV to 1.9 MeV. BATSE continually monitored the sky for transient phenomena, searching for variable emission from both known and new sources.

The *Oriented Scintillation Spectroscopy Experiment (OSSE)* was designed to carry out pointed spectral observations of gamma-ray sources in the range from 0.05 to 10 MeV, with capability above 10 MeV for solar gamma-ray and neutron observations. The four OSSE detectors were collimated scintillators (with a $4° \times 11°$ field of view) that were movable over a single axis, allowing a rapid response to targets of opportunity such as solar flares,

COMPTEL and EGRET Gamma-Ray Sources

MeV < E < GeV

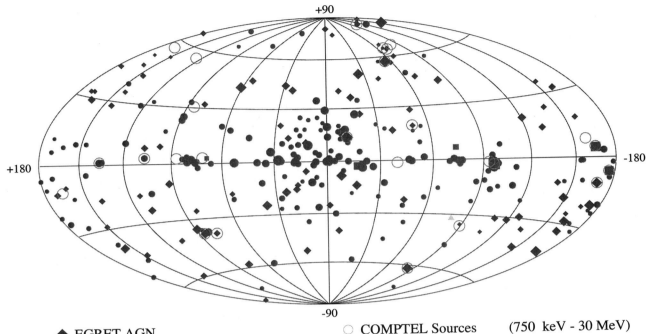

◆ EGRET AGN

■ EGRET Pulsars

▲ LMC

● EGRET Unidentified Sources

○ COMPTEL Sources   (750 keV - 30 MeV)

FIGURE 3 An all-sky map of gamma-ray sources detected with the COMPTEL and EGRET telescopes aboard the Compton Gamma Ray Observatory. Classes of sources are indicated by symbol, with increasing symbol size representing higher source intensity. The map is an Aitoff equal-area projection in Galactic coordinates (as in Fig. 1). (Courtesy NASA, the Max Planck Institute, and the CGRO COMPTEL and EGRET Instrument Teams).

transient X-ray sources, and other explosive astrophysical phenomena.

The *Imaging Compton Telescope* (*COMPTEL*) detected gamma-rays by means of a double-scatter technique whereby an incident gamma photon Compton scattered once in an upper detector module, and then was totally absorbed in a lower detector module. The COMPTEL instrument was sensitive over the energy range from approximately 0.75 to 30 MeV, and was also capable of detecting neutrons from solar flares. With its large field of view (∼1 steradian) COMPTEL carried out the first survey of the gamma-ray sky at MeV energies.

The *Energetic Gamma Ray Experiment Telescope* (*EGRET*) covered the broadest energy range of the CGRO instruments, from ∼20 MeV to 30 GeV. These high-energy photons interact primarily via the pair-production process, and the EGRET spark chamber was designed to de-

tect the electron–positron pairs produced by high-energy gamma rays. EGRET also had a relatively wide field of view (∼0.6 sr), good angular resolution, and very low background.

The coaligned COMPTEL and EGRET instruments operated as wide-field imaging telescopes and together carried out a comprehensive survey of the gamma-ray sky from MeV to GeV energies (see Figs. 1 and 3). Taken together the four CGRO telescopes represented a major improvement in sensitivity, energy coverage, and spectral and angular resolution, compared to previous generations of gamma-ray instruments. It can be said without exaggeration that observations carried out with the CGRO have completely revolutionized our view of the high-energy Universe. The bulk of the scientific results discussed in the sections to follow are based on observations obtained with the four CGRO telescopes.

II. SOURCES OF COSMIC GAMMA RAYS

Our view of the gamma-ray sky has changed dramatically in recent years, and has been particularly influenced by the results obtained with the instruments aboard the Compton Gamma Ray Observatory. The energetic and variable cosmos revealed by gamma-ray telescopes stands in marked contrast to the quiescent night sky viewed in visible light on a placid summer evening. The Universe in the light of gamma rays is a dynamic, diverse, and constantly changing place.

A. The Sun

The Sun is a powerful site for the acceleration of energetic particles. The source of energy for particle acceleration is believed to be the tangled magnetic field in the solar atmosphere. However, our understanding of both the properties of the accelerated particles and the nature of their acceleration during the explosive release of energy in a solar flare is still emerging. Gamma-ray measurements have proved to be an essential tool for studying particle acceleration during flares.

The rich energy spectrum of gamma-ray emission from solar flares is quite complex and shows the signatures of many radiation processes. Below ~1 MeV the observed emission is dominated by a strong line at 0.511 MeV from positron annihilation and a smooth continuum of bremsstrahlung radiation from mildly relativistic electrons. In the energy band from 1 MeV to 10 MeV the emission results predominantly from the deexcitation of nuclear levels following the bombardment of nuclei in the solar atmosphere by energetic particles. This nuclear deexcitation emission is composed of four components: (1) promptly emitted narrow lines from the excitation of a heavy atmospheric nucleus by an energetic proton or alpha particle, (2) broad lines from the excitation of an accelerated heavy ion by collision with an atmospheric hydrogen or helium nucleus, (3) delayed line emission such as the strong line at 2.22 MeV from the capture of secondary neutrons by atmospheric hydrogen to form deuterium, and (4) a quasi-continuum produced by the blending of lines from high-level transitions excited in both the accelerated and target nuclei. Above 10 MeV the emission is dominated by two mechanisms: bremsstrahlung from both ultra-relativistic primary electrons and secondary electrons/positrons from meson decay, and gamma rays from the direct decay of neutral pions. The complexity of the gamma-ray spectra of flares provides many diagnostics for probing the properties of flare-accelerated electrons and ions (see Fig. 4).

Measurements of the bremsstrahlung continuum during solar flares indicate that relativistic electrons are a common product of energy release in flares. The gamma-ray bremsstrahlung generated by relativistic electrons indicates that the yield in relativistic electrons scales roughly with the total energy released in thermal X-rays by the flare. Further, increasingly sensitive searches for gamma-ray bremsstrahlung over the last 2 decades have found no evidence of a flare-size threshold for relativistic electron acceleration. The gamma-ray evidence therefore suggests that relativistic electron acceleration is a property of all flares. The gamma-ray observations also show that the relative amount of high-energy bremsstrahlung increases as the position of the flare approaches the solar limb. Since high-energy bremsstrahlung is directed more strongly along the electron's velocity vector than bremsstrahlung at lower energies, the limb brightening can be explained by a distribution of emitting electrons that increases in directions away from the surface normal at the flare site. The nature of this electron distribution is regulated by the complex magnetic field structure in flaring regions. Future techniques that can measure the angular distribution of gamma-ray bremsstrahlung will allow us to explore the nature of relativistic electron transport in flaring regions.

Nuclear deexcitation emission during flares indicates that energetic ion acceleration is also a common property of solar flares. Gamma rays from nuclear deexcitations were first detected from two giant flares that occurred in August 1972. The enormous size of those flares and the fact that they were the only ones detected during that solar cycle led to an initial suspicion that ion acceleration might only occur when the flare energy surpasses a relatively high threshold. Sensitive detectors aboard the Solar Maximum Mission (SMM) satellite and the Compton Observatory, however, showed that nuclear line emission is present even in relatively small flares. While the relative importance of accelerated ions and electrons is observed to vary by approximately an order of magnitude, existing measurements are consistent with the hypothesis that both components are accelerated in all solar flares.

The temporal structure of variations in gamma-ray flux during flares can be quite rich. Gamma-ray flares can range from a single spiked pulse of 10-s duration to a complex series of pulses with total duration of more than 1000 s. Typically flares are composed of two or more pulses. An interesting property of the pulse structure is that the time of peak intensity is often energy dependent. When this energy dependence is present, the peak at higher energies tends to lag the peak intensity at lower energies by as much as 45 s. At one time, these delays were interpreted as reflecting the timescale needed for particle acceleration during flares. However, we now know that there are many flares where the peaks at X-ray through gamma-ray energies show time coincidence to better than 2 s and that,

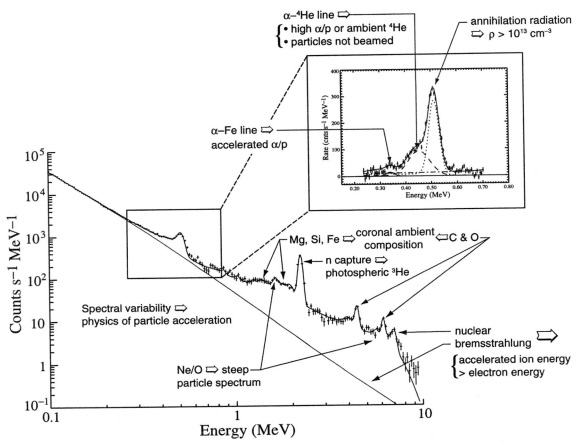

FIGURE 4 The gamma-ray spectrum of the June 4, 1991, solar flare obtained with the OSSE instrument aboard the Compton Gamma Ray Observatory. Prominent gamma-ray emission lines are identified, along with the primary processes responsible for both the observed line and continuum emission. (Courtesy NASA and the CGRO OSSE Instrument Team (G. H. Share and R. J. Murphy.))

even when significant peak delays are present, the pulse starting times are simultaneous to within 2 s. Those observations show that both electrons and nuclei can be rapidly accelerated to relativistic energies within seconds during solar flares. We now believe that the delays are largely generated by propagation and interaction effects as particles move from a low-density acceleration region high in the solar atmosphere to a higher-density interaction region deeper in the atmosphere where the high-energy emission is generated.

B. The Solar System

The question of the origin and evolution of the solar system is one of the most fundamental in astronomy. It bears directly on such related issues as stellar evolution, the formation of planetary systems, and on the existence of life itself. Gamma-ray observations from spacecraft, either via remote sensing from orbit or through *in situ* measurements

from landers, contribute directly to the testing of evolutionary models of solar-system formation. Specifically, they provide a means of directly determining the elemental chemical composition of planetary surfaces, thus providing clues important to reconstructing the geochemical history of the solar system. Related complementary observational techniques include X-ray fluorescence measurements, and the detection of albedo neutrons and charged particles from planetary surfaces.

The gamma-ray observations relevant to planetary studies are spectroscopic in nature, aimed at identifying specific key elements present in planetary surfaces via their characteristic emission energies. The abundances of elements with different condensation temperatures and geochemical behavior relate directly to the origin and evolution of planetary bodies. For example, the K/U ratio provides a measure of the remelting of primordial condensates, while the K/Th ratio indicates the relative abundance of volatile to refractory elements.

Gamma-ray spectral lines originate in nuclear processes, either from radioactive decay or from nuclear de-excitation following particle collisions. Natural radioactivity results from the decay of the primordial radioactive elements ^{40}K, ^{138}La, ^{176}Lu, and those in the uranium and thorium decay sequences. Collisions of primary Galactic cosmic rays (of which ~90% are protons) with planetary material can give rise to numerous interactions and secondary particles, leading to gamma rays. Extensive model calculations of cosmic ray-induced gamma-ray emission from planetary bodies have been carried out and can be readily compared to the available observations.

Since the 1960s measurements of X-rays, gamma rays, alpha particles, and neutrons from the Moon, Mars, and Venus have been undertaken successfully with a variety of instruments aboard both U.S. and Russian spacecraft. The two U.S. Viking landers on Mars, for example, carried out X-ray fluorescence measurements of the Martian surface, while the Russian Venera 8, 9, and 10 spacecraft measured the natural radioactivities of potassium, uranium, and thorium at three landing sites on Venus.

Until very recently the most detailed and extensive remote-sensing observations of a planetary body were carried out during the Apollo 15 and 16 flights to the Moon (1971, 1972) when instruments aboard the orbiting command modules mapped approximately 20% of the lunar surface in X-rays, gamma rays, and alpha particles. The Apollo missions were also unique in that a detailed comparison could be made between the results of the remote mapping and follow-up compositional analysis of actual returned samples of lunar material.

The Apollo measurements clearly demonstrated that the Moon's crust is chemically differentiated, with a pronounced distinction between maria and highland regions, with the maria primarily basaltic in nature. On a more localized scale, material around craters tends to exhibit a significant chemical contrast relative to surrounding regions, suggesting an excavation of material from the subsurface due to impacts from asteroids and comets. Distribution patterns seem to favor an impact rather than a volcanic dispersal. The observed K/Th ratio has provided a measure of the volatile-to-refractory material variation over the lunar surface, which is found to be consistently lower than the terrestrial value, reflecting a global depletion of volatiles on the Moon compared to the Earth.

More recently the U.S. Lunar Prospector mission successfully obtained (1998–1999) global maps of the lunar surface using gamma-ray and neutron spectrometers. The results have generally confirmed the earlier Apollo findings. As a follow-up to the Apollo measurements, there is a particular interest in determining the distribution of "KREEP"-rich material on the Moon. KREEP refers to an unusual mixture of elements containing potassium (K), rare-earth elements (REE), and phosphorous (P) that is believed to have formed at the lunar crust-mantle boundary as the final product of the initial differentiation of the Moon. Understanding the composition and distribution of KREEP-rich material is thus considered key to reconstructing the evolution of the lunar crust. The Lunar Prospector data have demonstrated that KREEP-rich rocks tend to be found on the rims and boundaries of major lunar impact basins where there are surmised to have been exposed, dredged up, and dispersed as a result of these cataclysmic impact events. The most intriguing of the recent lunar neutron observations point to the possible presence of subsurface water ice at the lunar poles (whose existence was first indicated by radar measurements carried out with the Clementine spacecraft in orbit around the Moon in 1994).

In 2001, in an engineering tour de force, the Shoemaker-NEAR (Near Earth Asteroid Rendezvous) spacecraft completed its successful year-long orbital study of the asteroid 433 Eros with a spectacular unplanned landing on the asteroid's surface, transforming its on-board X-ray and gamma-ray spectrometers from remote-sensing to *in situ* instruments. Initial analysis of the NEAR spectrometer data suggests that Eros may remain in an undifferentiated state, unaltered by melting, constituting some of the most primitive material in the solar system yet studied. Also in 2001, the NASA's Mars Odyssey spacecraft is scheduled for launch, carrying a gamma-ray spectrometer among its suite of instruments. If the goals of these missions are fully realized complete global maps of elemental distribution for the Moon, Eros, and Mars, representing three bodies with distinct evolutionary histories, will be available for detailed comparative studies.

C. Galactic Sources

As anyone who has gazed at the night sky from a dark location knows, the distribution of visible stars is not random. Rather, stars tend to cluster in a bright band called the Milky Way that delineates the plane of the flat spiral galaxy in which we live. Like the stars, there is also a population of gamma-ray sources that cluster in the plane of the Galaxy and that are believed to reside within the Milky Way. However, our ability to associate them with visible counterparts in the crowded Galactic plane is hampered by the fact that the angular resolution of the best gamma-ray telescopes is still a hundred times coarser than even small backyard optical telescopes. As a consequence, many Galactic "point" gamma-ray sources have multiple counterpart candidates and, even worse, in many directions in the Galactic plane we know that the gamma emission from several "point" sources is actually blended together and gamma-ray telescopes are "source confused." Nevertheless, by studying the temporal variations of the gamma-ray emission and correlating it with intensity variations measured by higher-resolution X-ray, optical, or radio telescopes, we have been able to identify with certainty several classes of Galactic gamma-ray sources.

1. Isolated Pulsars

Among the best clocks known, natural or man-made, are the spinning, magnetized neutron stars called pulsars. Pulsars emit short bursts of electromagnetic radiation at intervals from one every few seconds to thousands of times a second with a regularity that exceeds that of watches of the highest precision. In some cases this electromagnetic pulse extends across the entire spectrum from radio to gamma-ray wavelengths, thereby allowing us to unambiguously identify these Galactic gamma-ray sources.

The best-known gamma-ray emitting pulsar is the so-called Crab pulsar. It is embedded in and is the source of power for the famous Crab nebula in the constellation Taurus, the first object (M1) listed in the renowned catalog of diffuse, nebular objects compiled by the French astronomer Charles Messier in the late 18th century. From medieval Chinese records describing the temporary appearance of a "guest star" near the star we now call Zeta Tauri, we know that the Crab pulsar was the product of a supernova explosion that occurred in the year 1054 AD. Every 33 ms, the Crab pulsar emits a pair of radiation pulses that are detectable from the radio band all the way up to TeV gamma-ray energies. While pulsar physics still has many open questions, it is generally agreed that the emission from isolated pulsars is generated by energetic particles that are accelerated by electric fields induced by the spinning magnetic field of a rapidly rotating, magnetized, neutron star. The energy reservoir that ultimately powers all the observed emission is therefore the rotational energy of the rapidly spinning pulsar.

At least six other isolated gamma-ray pulsars are currently known to exist and from that small sample a few general patterns are apparent. First, for all known isolated gamma-ray pulsars the gamma emission represents the largest observable fraction of the total power emitted by the pulsar. As a consequence, observational study of the gamma rays provides important diagnostics on the overall efficiency for particle acceleration and interactions in the extreme pulsar environment. Second, the gamma-ray visibility increases with the spin-down luminosity (or, the ratio of the magnetic field strength to the square of the spin period). Finally, all of the isolated gamma-ray pulsars appear to be unvarying point sources when their emission is averaged over the spin period.

Perhaps the most remarkable object in the sample of known gamma-ray-emitting pulsars is Geminga. For nearly 20 years this source, which is the second brightest source in the gamma-ray sky, was a puzzle because it did not appear to emit radiation in any other energy band. The unusual name, Geminga, was coined by Italian astronomers and is derived both from the source's location in the constellation Gemini and from a play on words, geminga meaning "is not there" in the Milanese dialect. The breakthrough in our understanding of Geminga occurred in 1991 when pulsating soft X-ray emission with a period of 0.237 s (=237 ms) was detected from the direction of Geminga by the ROSAT X-ray satellite. Identification of Geminga as pulsar was therefore clinched when a phase analysis revealed that the gamma-ray emission was also modulated at the same 0.237-s period (see Fig. 5). A particularly interesting property of Geminga is that, unlike other rotation-powered pulsars, it is not detectable as a radio pulsar even though it is thought to be only 100 parsecs from the Sun. Since our census of the population of pulsars is based on radio observations, it is possible that Geminga is just the nearest member of a large population of previously unknown pulsars that are only visible at X-ray and gamma-ray energies. Many of the steady, unidentified gamma-ray sources could therefore be Geminga-type pulsars with still unknown spin periods.

2. Accreting Pulsars

Not all pulsars are isolated. Some reside in binary stellar systems and some of these pulsars are X-ray sources that are powered by the gravitational energy released when gas from the companion star is accreted onto the neutron star. Those accreting pulsars are also likely sources of gamma-ray emission. Indeed, several groups using ground-based Cherenkov telescopes have reported the detections of TeV gamma-ray emission from systems containing an accreting pulsar. Unfortunately, most of the reported detections have been of low statistical significance and/or not confirmed with subsequent more sensitive observations. If real, they indicate that the gamma-ray emission from accreting pulsars is sporadic. On theoretical grounds, such behavior is plausible because accretion flows are often unstable and shocks in the flow could efficiently accelerate gamma-ray emitting particles.

Support for the idea that accreting pulsars sporadically emit gamma-ray emission was also found with the EGRET telescope aboard the Compton Gamma Ray Observatory. In October 1994 a week-long outburst of GeV gamma-ray emission was detected from the direction of the massive X-ray binary system Centaurus X-3. During the outburst, the accreting pulsar in Cen X-3 underwent an interval of rapid spin-down. Phase analysis of the gamma-ray emission showed evidence for spin modulation in step with the rapidly drifting X-ray period.

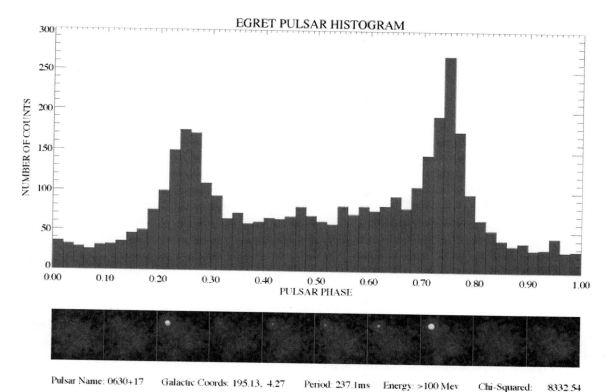

FIGURE 5 Pulsed gamma-ray emission from the Geminga pulsar. *Above*: The light curve of the gamma-ray emission from Geminga, with the gamma rays binned according to the pulsar period of 237 milliseconds. *Below*: Gamma-ray images of the region of the sky containing the Geminga pulsar, showing the brightening of the source (in the upper left of the image plots) in phase with the pulsed emission. The fainter object at the lower right is the Crab pulsar (with a different pulse period of 33 ms) which appears as a steady source in this representation. (Courtesy NASA and the CGRO EGRET Instrument Team (P. Sreekumar)).

3. Black-Hole Binary Systems

Some X-ray binary systems are believed to contain stellar-mass black holes that are embedded in disks of inwardly spiraling accreted gas. These accretion disks are known sources of both soft and hard X-ray emission and, in some cases, are also sources of soft gamma-ray emission. Perhaps the best-known example is Cygnus X-1, the brightest X-ray source in the constellation Cygnus. The X-ray luminosity of this high-mass X-ray binary (in which the normal companion is an O or B star many times the mass of the Sun) is variable and is correlated with different high-energy spectral "states." These states of activity are composed of different admixtures of two principal components: a soft thermal blackbody-like component and a hard nonthermal power-law component. It is this power-law component that can, at times, extend up to gamma-ray energies of at least an MeV.

The origin of the MeV gamma ray emission from Cygnus X-1 is not well understood. Most models assume that the gamma rays are X-rays that were Compton scattered by mildly relativistic electrons. The precise origin of those energetic electrons is still unclear. Some modelers have speculated that reconnecting magnetic fields or shocks in disk outflows accelerate the electrons. However, a particularly attractive idea is that the scattering electrons arise naturally in the convergent accretion flow from the innermost stable orbit of the accretion disk. If the accretion rate is high, Compton emission from the bulk flow could generate the observed gamma ray emission.

Finally, Galactic sources that generate intense outbursts of X-ray emission that can persist for months before fading are termed X-ray novae. They are known to occur in low-mass binary systems in which the normal companion star is of approximately solar mass in close orbit around a compact object. Such binary systems can undergo recurrent outbursts that for a brief period can make them the brightest high-energy sources in the sky. Optical observations of these low-mass X-ray binary systems during their quiescent states indicate that the compact object is typically more massive than 3.0 solar masses and is therefore most likely a black hole. Some outbursts of X-ray novae are also accompanied by low-energy gamma-ray emission. At least one such nova outburst, that of Nova Muscae in 1991 observed with the SIGMA instrument, displayed a variable positron annihilation line. Later observations with telescopes aboard the Compton Observatory of Nova Persei 1992 confirmed that X-ray novae can give rise to gamma-ray emission extending up to photon energies of at least 2 MeV.

D. Gamma-Ray Lines of Galactic and Extragalactic Origin

One of the great triumphs of modern astrophysics has been an increasingly detailed understanding of the nuclear reaction processes that govern the production of energy in stars and that determine the course of stellar evolution. It is now clear that nucleosynthesis, or the production of elements from the primordial building blocks of hydrogen and helium, occurs almost exclusively in the cores of stars, or in the cataclysmic explosive events, supernovae and novae, that mark the end of a star's lifetime. The study of gamma-ray spectral lines provides one of the few direct means of verifying the predictions of the various models of stellar nucleosynthesis, and of the explosive events that disperse this material back into the interstellar medium, out of which new generations of stars are formed.

Gamma-ray lines result predominantly from nuclear processes, either from the decay of radioactive nuclides and the deexcitation of excited nuclei (see Table I), or from the collisions of high-energy particles. The advantages of gamma-ray line spectroscopy are manifest. Spectral line transitions occur at specific characteristic energies that provide immediate identification of the isotopic species that produced them. The comparison of line strengths or intensities between different elements and isotopes can be translated into isotopic abundances, densities, and temperatures in the emitting region. Similarly, the presence of broad, narrow, or Doppler-shifted lines provides a measure of gas motions and velocities, all of which can be interpreted in terms of specific models of production.

1. Nucleosynthesis in Stars and Supernovae

Gamma-ray line radiation in the Galaxy results primarily from two broad categories of production: either *steady-state thermal* or *explosive nucleosynthesis*. In the former case, stars in hydrostatic equilibrium throughout the bulk of their lives generate energy via thermonuclear fusion reactions in their high-temperature cores. Depending on the stellar mass (which determines core temperature), these fusion reactions may progress over time through successive stages of nuclear burning, leading to a buildup of different layers of nuclear reaction products at the center of the star. For low- and intermediate-mass stars such as the Sun, nuclear burning ceases with the fusion of helium into carbon. For the most massive stars (10–100 times the mass of the Sun), however, with much higher core temperatures, nuclear burning leads ultimately to an iron core surrounded by layers of silicon, magnesium, neon, oxygen, and carbon, along with remnant amounts of helium and hydrogen in the outer envelope of the star. Toward the end of its life, as it runs out of nuclear fuel, a star becomes increasingly unstable, and the heavier elements at the center of the star are brought to the surface via mixing and convective processes, and are ultimately dispersed back into the interstellar medium through high-speed stellar winds, flares, outbursts, and variable pulsations of the outer envelope and atmosphere of the star. Mixed in with all of the reaction products are long-lived radioisotopes that act as tracers of the various stages of nuclear burning.

Elements beyond iron in the periodic table are formed in the course of supernovae explosions that result either from the cataclysmic collapse of a white dwarf into a neutron star due to the accretion of matter from a binary companion (a Type I supernova), or from the catastrophic core-collapse of the most massive stars at the end of their lives when nuclear fuel is exhausted (a Type II supernovae). Subcategories of each type also exist. In the former case, the entire star is disrupted, liberating approximately 10^{51} ergs in the explosion and about 0.5 to 1.0 solar masses in synthesized radioactive material, while in the latter class of event slighter higher energies may be liberated (up to ~10^{53} ergs) but thick layers of ejecta partially mask for a time the 0.1 solar masses of radioactive material synthesized in the explosion. Supernovae are among the most violent and luminous events known in the Universe, with ejected material attaining speeds of thousands of kilometers per second. A supernova at peak light may completely outshine its host galaxy (containing billions of stars), and its light curve decays at the exponential rates

TABLE I Primary Radioactive Nuclear Decay Lines from Nucleosynthesis

Decay process	Mean halflife	Line energies (MeV)
$^{56}Ni \rightarrow {}^{56}Co \rightarrow {}^{56}Fe$	111 d	0.511, 0.847, 1.238
$^{57}Co \rightarrow {}^{57}Fe$	272 d	0.014, 0.122
$^{22}Na \rightarrow {}^{22}Ne$	2.6 y	0.511, 1.275
$^{44}Ti \rightarrow {}^{44}Sc \rightarrow {}^{44}Ca$	~60 y	0.068, 0.078, 0.511, 1.157
$^{26}Al \rightarrow {}^{26}Mg$	7×10^5 y	0.511, 1.809
$^{60}Fe \rightarrow {}^{60}Co \rightarrow {}^{60}Ni$	2×10^6 y	0.059, 1.173, 1.332

characteristic of the primary radioactive species produced in the explosion.

The gamma-ray lines of primary observational interest associated with nucleosynthesis are of MeV energies, and are listed in Table I. The radioisotopes ^{56}Ni, ^{57}Ni, ^{44}Ti, and ^{26}Al are particularly important since they span a range of half-lives, from days to millions of years, thus together providing a measure of both the "prompt" emission from individual events, as well as of the "delayed" or integrated cumulative emission dispersed throughout the interstellar medium of the Galaxy arising from generations of star formation. The lines originally predicted to be the most luminous from individual events are the ^{56}Ni and ^{56}Co lines from Type Ia supernovae. The long-anticipated breakthrough in gamma-ray line detection, however, occurred for the now-famous SN 1987A, a type II supernova which occurred at a distance of 55 kpc in our neighbor galaxy, the Large Magellanic Cloud (the LMC). The gamma-ray spectrometer aboard the Solar Maximum Mission (SMM) satellite detected and studied the ^{56}Co lines from this event. The gamma-ray lines were detected earlier after the event than predicted, implying that the inner layers of material containing the radioactive ^{56}Co were more thoroughly mixed than expected, or indicating perhaps that the ejecta were clumpier, allowing clearer lines of sight to the inner regions of the exploding star through which the gamma rays could escape. After the launch of the Compton Observatory in 1991, several years after the event, the OSSE instrument detected the longer-lived ^{57}Co line ($t_{1/2} \sim 272$ days) at 122 keV in energy. Its intensity implied a ratio of synthesized ^{57}Ni to ^{56}Ni (from ^{56}Fe and ^{57}Fe, respectively) of about 1.5 times the solar value, providing constraints on models of the progenitor star's evolution. In other supernova observations, there were tantalizing hints of gamma-ray detections with COMPTEL of the 0.847- and 1.239-MeV lines of ^{56}Co from the Type Ia supernova SN 1991T in the galaxy NGC 4527 at a distance of about 13–17 Mpc, at the limit of this telescope's range of detectability.

Of particular interest in gamma-ray line astronomy is the detection of ^{44}Ti ($t_{1/2} \sim 60$ years) at 1.157 MeV from young, distant obscured supernovae in the Galaxy. About two to three supernovae per century are predicted to occur in the Milky Way, but most have remained undetected because they are believed to remain hidden behind intervening clouds of interstellar gas and dust in the spiral arms. Gamma rays, in contrast to optical light, easily penetrate the interstellar material and provide a means of detecting directly these interesting objects and confirming predictions regarding star formation rates in the Galaxy. Further, ^{44}Ti is formed in the deepest layers of the supernova ejecta, and its predicted line strength is sensitive to the details of the explosion models and to the likelihood of fall-back onto the collapsed, compact stellar core. The COMPTEL instrument aboard the CGRO has reported the detection of ^{44}Ti from the Cas A supernova remnant (a relatively close remnant about 3 kpc distant, believed to have exploded about 250 years ago) and from the Vela region of the Galaxy. The next generation of gamma-ray spectrometer and imager aboard the INTEGRAL spacecraft will provide critical confirmation of these first detections and should also be able to map out in detail the spatial distribution of the emitting radioisotope, and, from a determination of the exact shape of the gamma-ray line profiles, provide a measure of the symmetry of the initial explosion and subsequent expansion of the ejected material.

The longer-lived radioisotope ^{26}Al with a decay line at 1.809 MeV ($t_{1/2} \sim 710,000$ years) traces the sites of massive star formation and nucleosynthesis in the Galaxy over the past million years. This interstellar line was originally detected by the HEAO-3 and SMM spacecraft, followed by a number of balloon instruments. In a long-awaited result, the first all-sky map in the light of this radioisotope was produced by the COMPTEL instrument team following extensive observations and analysis. Clearly evident is the disk of the inner Galaxy, with enhancements of emission in particular regions (see Fig. 6). These tend to coincide in direction to spiral arms in the Galaxy, where recent star formation and supernova activity are most likely to occur (e.g., the Cygnus, Vela, and Carina regions). Further measurements of the spectral line of ^{26}Al with high-resolution spectrometers and imagers are expected to yield detailed identification of specific sources for this emission.

2. Electron–Positron Annihilation Radiation

A gamma-ray line at 0.511 MeV results from the mutual annihilation of an electron and a positron, a particle-antiparticle pair. A number of radioactive decay chains (see Table I) result in the emission of a positron as a decay product, which will annihilate upon first encounter with an electron. Also of astrophysical importance is the production of electrons and positrons via the photon–photon pair-creation process. Such pair plasmas are found in the vicinity of compact objects, such as neutron stars and black holes, that are associated with heated accretion disks and relativistic flows and jets, within which particle acceleration is known to occur. Thus, relatively narrow lines of 0.511-MeV annihilation radiation are expected to arise in the interstellar medium through the decay of dispersed, nucleosynthetic radionuclides, while broadened, Doppler-shifted, and possibly time-variable lines may occur in the high-energy and dense environments associated with compact objects.

Direct annihilation of an electron–positron pair leads to the emission of two photons. If the particles are at

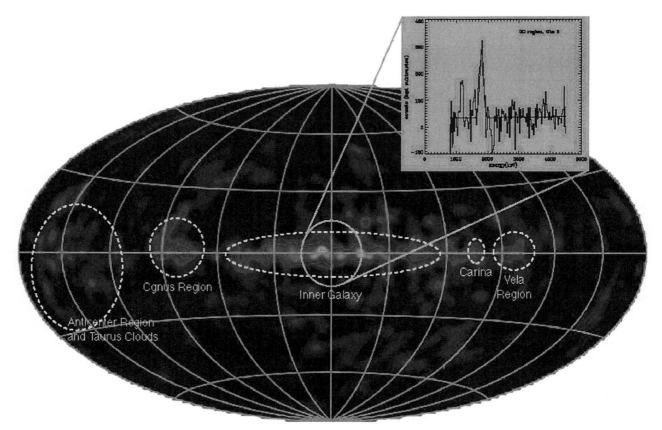

FIGURE 6 An all-sky map of the gamma-ray line emission at 1.809 MeV due to the decay of radioactive ^{26}Al in the Galaxy, obtained with the COMPTEL instrument aboard the Compton Gamma Ray Observatory. The map is an Aitoff equal-area projection in Galactic coordinates (as in Fig. 1). The gamma-ray line from ^{26}Al traces the sites of massive star formation and nucleosynthesis in the Milky Way. Regions with enhanced emission are identified. *Inset*: A gamma-ray spectrum showing the emission line from ^{26}Al at 1.809 MeV observed in the inner Galaxy. (Courtesy NASA, the Max Planck Institute, and the CGRO COMPTEL Instrument Team.)

rest two gamma rays of equal energy, 0.511 MeV, are produced, while in-flight collisions will result in a range of possible photon energies (whose sum must equal the kinetic plus the rest–mass energies of the two particles, $2m_ec^2$). If the velocities of the two particles are small, a bound state consisting of a positron and an electron is possible, the *positronium* atom. Positronium decays after a very short lifetime ($<10^{-7}$ s) via one of two channels. Two-photon decay results again in the emission of two gamma photons of energy 0.511 MeV, while three-photon decay leads to a continuum of emission below the main 0.511-MeV spectral peak. The observed *positronium fraction*, or continuum-to-line ratio, provides a unique diagnostic measure of the physical conditions of the emitting region.

Prior to the launch of the CGRO, from the 1970s on, a number of balloon and early satellite instruments detected the presence of apparently variable annihilation radiation from the direction of the Galactic Center. This led to the widespread supposition that the observed emission arose from two types of sources: a steady, dispersed diffuse component of nucleosynthetic origin in the disk of the Galaxy, and a time-variable point source (or sources) near the Galactic Center. Extensive observations of the 0.511-MeV line from the inner Galaxy with the CGRO OSSE instrument (see Fig. 7), however, have revealed a somewhat different picture: a central bulge, along with diffuse emission in the Galactic plane, and an apparent extension of emission above the Galactic disk in the direction of the Galactic Center. No temporal variability of the annihilation radiation is evident in the OSSE observations. The annihilation emission observed by OSSE can be explained entirely in terms of radioactive isotopes from supernovae and similar sources. The apparent enhancement of 0.511-MeV radiation above the Galactic plane has been variously interpreted as a "positron fountain" arising from an asymmetric outflow of positrons from the region of the Galactic Center following a period of enhanced star

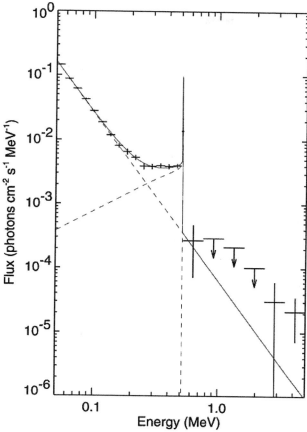

FIGURE 7 The gamma-ray spectrum of the positron annihilation line at 0.511 MeV and associated positronium continuum toward the region of the Galactic Center, obtained with the OSSE instrument aboard the Compton Gamma Ray Observatory. The data were fit with a model consisting of a narrow annihilation line, a positronium component, and an underlying power-law continuum, indicated by the dashed lines. (Adapted from W. R. Purcell *et al.*, "OSSE mapping of galactic 511 keV positron annihilation line emission," *Astrophys. J.* **491**, 725–748, Copyright 1997, reproduced with permission of the AAS.)

formation and supernovae activity, or the result of jet activity from one or more black-hole sources in or near the center of the Galaxy, or even the result of a single cataclysmic gamma-ray burst-like event occurring near the Galactic Center about a million years ago.

E. Diffuse Galactic Gamma-Ray Emission

The most conspicuous feature in the gamma-ray sky is the diffuse continuum radiation arising from the Galaxy itself, originating in the interstellar clouds of gas and dust that reside within the spiral arms, disk, and bulge of the Milky Way. A narrow band of diffuse gamma-ray emission is observed along the Galactic plane, with enhancements toward the inner Galaxy and in directions that are tangent

to the spiral arms (where the column density of interstellar material is greatest along the line of sight) and with hot spots of intensity in particular directions that may be due to unresolved point sources of emission (see Fig. 1). First detected by the OSO 3 satellite, this diffuse radiation was also observed and mapped at medium resolution by the SAS 2 and COS B satellites. The two wide-field gamma-ray telescopes aboard the CGRO, COMPTEL and EGRET, were specifically designed to carry out as one of their primary scientific objectives a complete mapping of the entire sky in gamma rays. A key result of this all-sky survey was the first complete map of the Galactic diffuse radiation from approximately 1 MeV to 30 GeV in energy. The OSSE instrument aboard the CGRO also derived a measure of the Galactic diffuse radiation toward the center of the Galaxy up to an energy of approximately 10 MeV.

The Galactic diffuse emission arises from the interaction of energetic cosmic-ray particles, predominantly electrons and protons, with ambient material and low-energy photons in the interstellar medium. The primary interaction mechanisms are (1) nucleon–nucleon collisions of cosmic-ray protons that result in the creation of neutral pions that decay into gamma-rays, (2) bremsstrahlung from high-energy electrons, and (3) Compton up-scattering by cosmic-ray electrons of the low-energy photons (IR, optical, and UV) that comprise the interstellar radiation field (or ISRF). Each of these gamma-ray production mechanisms is dominant over a particular energy range. Below about 100 MeV the bremsstrahlung component is important, while above that energy the decay of neutral pions from nucleon–nucleon collisions is key. The relative importance of the Compton component is dependent on the spectral index (or "hardness") of the cosmic-ray electron population, and on the details of the ISRF, neither of which is well determined, hence the added interest in studies of the diffuse gamma-ray emission to help fix these contributing physical quantities. Finally, there is also an isotropic component of the diffuse gamma radiation that is believed to be extragalactic in origin (described in the next section).

Since the interstellar medium is essentially transparent to the propagation of gamma rays, the observed gamma-ray intensity in a particular direction represents the total, cumulative emission from all particle interactions and sources along that line of sight. Thus, the diffuse gamma radiation provides a simultaneous measure of both the cosmic-ray and matter distribution throughout the Galaxy. The challenge is to disentangle the cosmic rays (the "projectile" particles) from the interstellar matter (the "target" particles) in the observed gamma-ray signal. The study of the Galactic diffuse gamma radiation is therefore intimately related to Galactic radio astronomy. The distribution of matter in the Galaxy is derived primarily from radio surveys, in particular, those at 21-cm wavelength that map the distribution of atomic hydrogen, and the millimeter CO surveys that serve as tracers of molecular hydrogen. A major advantage of the spectral-line radio observations is that Doppler shifts in the observed line emission can be interpreted in terms of a kinematic model of differential Galactic rotation, which allows a determination of the distance to the emitting gas based on its measured radial velocity. Continuum radio surveys of the synchrotron emission from electrons interacting with Galactic magnetic fields also provide important observational constraints on the distribution and properties of the cosmic-ray electron population. In turn, studies of the diffuse gamma-ray emission complement the radio observations in that they serve to constrain a critical parameter in the molecular radio surveys, namely, the CO-to-H_2 conversion factor (the so-called "X-value," for which the EGRET-determined average over the whole Galaxy is $(1.56 \pm 0.05) \times 10^{20}$ H-molecules cm^{-2} (K km s^{-1})$^{-1}$).

The diffuse gamma-ray studies, when combined with the radio data, provide the best measure to date of the distribution of high-energy cosmic rays within the Galaxy. Cosmic rays presumably arise following particle acceleration via shocks in supernovae and their remnants in the interstellar medium. As charged particles, however, their trajectories are influenced by magnetic fields and thus their exact sites of origin and acceleration, as well as their composition, modes of propagation, and overall lifetime in the Galaxy are still not well understood. This situation is most severe for cosmic-ray electrons, since these less-massive particles "cool" rapidly due to energy losses via synchrotron, bremsstrahlung, and Compton emission, and consequently have relatively little time to diffuse far from their place of origin in the Galaxy to be detected.

A number of approaches have been followed to model the diffuse gamma-ray observations. They can be characterized as either *parametric* models that are fit to the data to study intensity and spectral variations as a function of Galactic radius, *dynamic balance* models that seek to balance the gravitational attraction of interstellar matter against the expansive pressures due to cosmic rays, matter and magnetic fields, and *cosmic-ray propagation* models that mimic the propagation of cosmic rays through the Galaxy by including such processes as diffusion, convection, reacceleration, fragmentation, the production of secondary particles, and the generation of gamma-ray and synchrotron radiation, as constrained by all available observations.

Analysis of the results obtained with the COMPTEL and EGRET instruments aboard the Compton Observatory suggest that the inverse Compton process may be a more important contributor to the Galactic diffuse emission than previously thought. The bremsstrahlung component of the medium-energy (1- to 50-MeV) diffuse emission, though comparatively small, remains of interest in that it can be

combined with surveys of the Galactic radio synchrotron emission (which traces electrons with energies in the range 100 MeV to 10 GeV) to derive the shape of the cosmic-ray electron spectrum. Historically, the electron spectrum below ~10 GeV has been difficult to determine since these particles are excluded from direct detection near the Earth due to the periodic modulation of the solar wind. At higher energies in the EGRET sensitive range (30 MeV to 30 GeV) the observed gamma-ray spectrum is softer in the direction of the outer Galaxy compared to the inner Galaxy, suggestive of a corresponding change in the spectrum of cosmic-ray nuclei with Galactic radius. This could be explained if cosmic rays are accelerated preferentially in the inner Galaxy and then propagate to the outer Galaxy, or if the high-energy cosmic rays are less well-confined in the outer Galaxy. While overall agreement between model predictions and the gamma-ray data is quite good, one surprising finding is an excess in the diffuse emission observed above ~1 GeV (see Fig. 8).

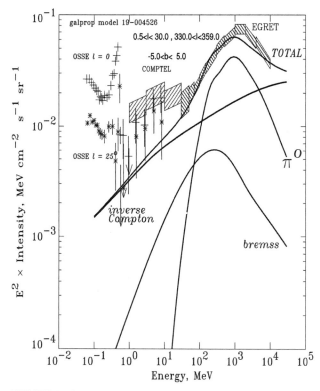

FIGURE 8 Spectrum of the diffuse gamma-ray emission from the inner Galaxy compared with calculations based on cosmic-ray propagation models. The data were obtained with the OSSE, COMPTEL, and EGRET instruments aboard the Compton Gamma Ray Observatory. Curves show the calculated contributions to the observed gamma-ray emission due to inverse Compton, bremsstrahlung, and π^0-decay processes, and the summed total. (Adapted from Strong, A. W., Moskalenko, I. V., and Reimer, O. (2000). "Diffuse continuum gamma rays from the Galaxy," *Astrophys. J.* **537,** 763–784, Copyright 2000, reproduced with permission of the AAS.)

This has been the subject of some debate and may reflect uncertainties in the neutral pion production function used in the model calculations, or perhaps is due to variations in the cosmic-ray spectrum with Galactic radius. Another possible explanation is enhanced Compton emission from a harder interstellar electron spectrum that may also give rise to an electron/inverse-Compton gamma-ray halo surrounding the disk of the Galaxy. Finally, at the very low end of the gamma-ray regime, it has been suggested that a population of unresolved Galactic sources may be responsible for the upturn in emission observed at low gamma-ray energies.

Future space missions (such as GLAST) will be able to confirm any spectral variations in the diffuse emission on sufficiently small angular scales (a few degrees) to permit a serious test of the various production models. Further observations with the next generation of ground-based air-Cherenkov GeV and TeV telescopes (where to date only tantalizing upper limits have been obtained) should also provide important confirmation of the GeV excess observed by EGRET.

F. Active Galaxies

Active galaxies, though they comprise only a small percentage of all known galaxies, rank among the most energetic and exotic objects in the Universe. They are exceptionally luminous (up to 10,000 times brighter than normal galaxies), and their emission is typically broadband and nonthermal in nature, spanning the entire electromagnetic spectrum. Their luminosity is concentrated in the central nucleus, which completely outshines the rest of the galaxy by factors of 100 or more, hence the alternate designation of "active galactic nuclei" (or AGNs). The emission from AGNs is highly variable and fluctuates rapidly enough ($\leq$day) to indicate that the source, or central engine, that drives the active behavior must occupy an extremely small region at the very unresolved core of the galaxy. AGNs are also associated with collimated bipolar jets and lobes of relativistic material emanating from the nucleus of the galaxy, within which ejected blobs of plasma can appear to travel at superluminal velocities, in apparent violation of the laws of physics (though this latter effect is now known to be a consequence of highly relativistic motion viewed along the direction of travel). For decades, as discoveries mounted, sub-categories of AGN proliferated, usually derived from some observational characteristic by which a particular class was originally identified. Thus, AGNs are categorized as radio-loud or radio-quiet, flat-spectrum or steep-spectrum sources, core-dominant or lobe-dominant in their emission, with broad spectral lines (indicative of high-velocity gas motions), narrow lines, highly polarized lines, or *no* lines, Seyfert galaxies of type I or II, Fanaroff-Riley galaxies of type I or II, and, at the most energetic extreme, the quasars, optically violent variables, and BL Lacertae objects (or BL Lacs) which collectively make up the class known as blazars (with luminosities in excess of 10^{48} ergs s^{-1}).

With the successful launch and operation of the Compton Gamma Ray Observatory (CGRO), a remarkable new class of active galaxy was discovered, the gamma-ray blazars. The high-energy EGRET experiment aboard the CGRO detected over 90 definite or probable gamma-ray AGNs (see Figs. 1 and 3), all but a few associated with active galaxies of the blazar class. These discoveries comprise the single largest category of gamma-ray source detected with EGRET and are all the more noteworthy in that prior to the launch of the CGRO only one such gamma-ray AGN was known, the quasar 3C 273 (discovered with the COS-B satellite). Gamma-ray blazars exhibit strong flaring behavior, and during active periods their gamma-ray luminosity can exceed that in other wavebands by factors of 100 or more. The gamma-ray blazars also span a large range in redshift ($z \sim 0.03$ to 2.3) and thus serve as cosmological probes of the intervening intergalactic medium. The discovery and study of the gamma-ray blazars has greatly enhanced our understanding of the blazar phenomenon in general, and has also provided strong confirmation of the relativistic jet model for active galaxies, since energy considerations require that the observed gamma-rays must be beamed in our direction via relativistic Doppler boosting.

In recent years, it has become clear that the vast majority of AGNs can be explained in terms of a single "unified model" for active galaxies. According to this concept the various subclasses of AGN can be explained as arising from geometric effects due to different viewing angles of the observer with respect to the central source. In the now-standard picture, the central engine of an AGN is a supermassive black hole (of $\sim 10^6$–10^{10} solar masses) that is powered by the accretion of surrounding infalling material. It is the partial conversion via accretion of the huge reservoir of gravitational potential energy into thermal energy due to the presence of the black hole that drives the energetic behavior of AGNs. In the standard scenario the central supermassive black hole is surrounded by an accretion disk and a thicker outer obscuring torus of material in the equatorial plane of the rotating black hole (see Fig. 9). Highly collimated bipolar jets of relativistic plasma are formed and ejected along the rotational axis, and intervening clouds of material of varying velocity, subject to bombardment from accelerated particles and radiation from the accretion disk and jets, give rise to the broad and narrow spectral lines observed. Predominantly thermal emission, extending up to X-rays, arises from the heated accretion disk and surrounding torus, while broadband nonthermal synchrotron and Compton emission, including gamma radiation, is given off by the relativistic

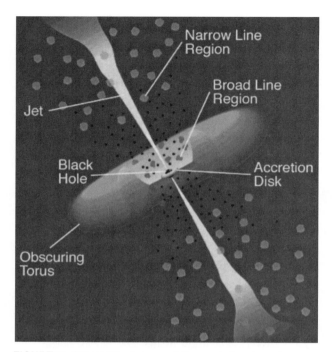

FIGURE 9 Schematic diagram of the central region of an active galaxy illustrating the main components of the "unified model" for active galactic nuclei (AGN). According to this model the various subclasses of AGN can be interpreted as arising from the observer's viewing angle with respect to the central source (a supermassive black hole) and the surrounding accretion disk, torus, intervening clouds, and relativistic jets. Gamma-ray blazars are presumed to arise when the observer's viewing angle is very closely aligned with the relativistic jets. (Adapted from Urry, C. M., and Padovani, P. (1995). "Unified schemes for radio-loud active galactic nuclei," *Publication of the Astronomical Society of the Pacific* **107**, 803–845, Copyright PASP 1995, reproduced with permission of the authors.)

particles in the jets. The appearance of an active galaxy thus depends critically on one's observing angle with respect to the central source and the surrounding accretion disk, torus, and relativistic jets. The most extreme variable behavior will be noted when the observer's viewing angle is very closely aligned to one of the jets, when relativistic effects reach their maximum.

The rapid variability observed in gamma-rays ($\leq$ days) indicates that the high-energy emission arises from regions very near the central engine of the galaxy, where jet formation and the acceleration of relativistic plasma occurs, phenomena that are not yet well understood. The gamma rays therefore constitute a new direct probe of the inner-jet region, heretofore unobservable, even by the techniques of very long baseline radio interferometry. Relativistic beaming plays a critical role in the explanation of the luminous time-variable gamma-ray emission from blazars, allowing rapid variations in the observed high-

energy emission, without the penalty of severe attenuation of the radiation due to the pair-production opacity. Shocks and instabilities in the bulk flow of relativistic particles can lead to variable nonthermal emission over a wide range of energies. Given the broadband nature of the emission from blazars it has become increasingly clear that coordinated multiwavelength observations of flares from blazars offers the prospect of determining the physical structure and properties of the inner-jet region. A key ingredient of such multiwavelength campaigns is the ability to measure time delays, between wavebands, of the brightness variations occurring during a flare. The relative order and delay of these frequency-dependent variations differ according to the predictions of the various models.

The emission mechanism most often employed to model gamma-ray production in AGNs is the Compton scattering of lower-energy photons by high-energy electrons. The scattered "seed" photons are typically either synchrotron photons generated within the jet by the electrons themselves (the so-called Synchrotron Self-Compton, or SSC, model) or photons propagated directly or scattered into the jet from the accretion disk surrounding the central engine (labeled "external" Compton scattering). The broadband spectra predicted by these models present a characteristic double-peaked appearance (see Fig. 10). The low-energy peak, broadly centered in the millimeter radio to UV bands, is representative of the low-energy photons (usually of synchrotron origin) that are then Compton-scattered to higher gamma-ray energies via the relativistic electrons. Of particular interest in this regard are the recent measurements of TeV gamma rays from relatively nearby BL Lac objects (most notably MRK 421 and MRK 501, both at $z \sim 0.03$) with ground-based Cherenkov detectors. These detections have revealed high-energy emission extending up to 30 TeV, combined with the most rapid variability (flux-doubling times of less than 15 min!) observed to date at gamma-ray energies. Further confirming TeV measurements of gamma-ray AGN are eagerly anticipated to place more precise observational constraints on the models described above.

Finally, it is important to note the utility of gamma-ray blazars as cosmological probes of the extragalactic diffuse background radiation, sometimes referred to as the extragalactic background light (or EBL). The propagation of high-energy gamma rays through intergalactic space is fundamentally limited by the pair-production opacity of the intervening medium. That is, the likelihood of interaction of the gamma radiation with the lower-energy photons that make up the universal background radiation. Though the cosmic microwave background radiation, relic of the Big Bang, has been well measured, background

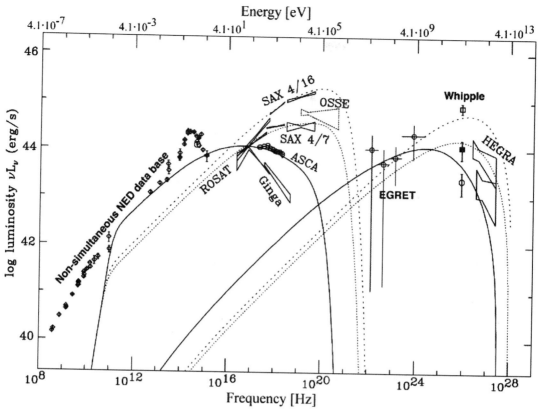

FIGURE 10 Broadband multiwavelength spectrum of the gamma-ray blazar MRK 501 from radio to TeV gamma-ray energies. The broadband spectrum shows the double-peaked structure characteristic of gamma-ray blazars in which lower energy photons are Compton scattered to gamma-ray energies by relativistic electrons. (Adapted from Kataoka, J. *et al.* (1999). "High-energy emission from the TeV blazar Markarian 501 during multiwavelength observations in 1996," *Astrophys. J.* **514**, 138–147, Copyright 1999, reproduced with permission of the AAS.)

radiation fields at other wavelengths are not nearly as well known, and yet are of considerable interest since they provide information about earlier epochs in the history of the Universe. TeV gamma rays, for example, are most likely to interact with infrared and optical photons. The optical/IR background radiation fields therefore limit the distance to which TeV gamma-ray sources can be detected. Observations of gamma-ray blazars as a function of redshift, then, provide a means of estimating the intensity of the EBL at the lower optical and IR wavelengths that provide information on star and galaxy formation in the early Universe.

G. Gamma-Ray Bursts

As their name implies, cosmic gamma-ray bursts (GRBs) are intense bursts of gamma radiation, lasting from fractions of a second to minutes, which emit the bulk of their energy in the gamma-ray regime (above ~0.1 MeV). Unpredictable in occurrence, these transient events form one of the most long-standing and challenging

puzzles in modern astrophysics, dating back to their accidental discovery over thirty years ago with the Vela series of nuclear-testing surveillance satellites. The Burst and Transient Source Experiment (BATSE) aboard the CGRO was specifically designed to serve as an all-sky monitor to detect these mysterious events. Over its nine-year lifetime BATSE detected a total of 2704 GRBs, many times the number recorded previously, and amassed a formidable collection of data on their properties. During their brief appearance GRBs are the brightest objects in the sky, outshining all other gamma-ray sources combined. Indeed, gamma-ray bursts may be the most distant and explosive events (with energies greater than 10^{53} ergs) ever observed in Nature. GRBs occur at random intervals (~1/day); they do not seem to repeat (implying likely destruction of the source); and they are isotropic in their distribution, coming from every direction in the sky. Each burst is different in its temporal structure, which can vary dramatically in duration and complexity from burst to burst (see Fig. 11). Further, the rapid variability of the emission (on the order of

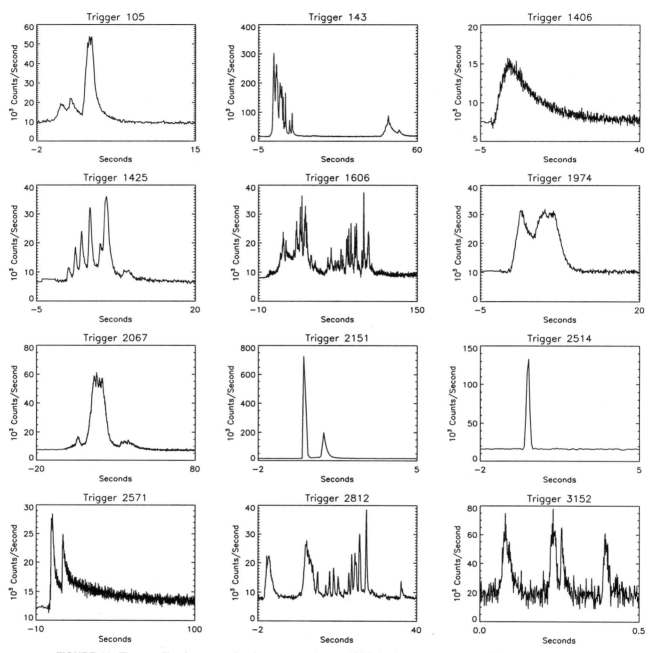

FIGURE 11 Time profiles for a sample of gamma-ray bursts (GRBs), obtained with the BATSE instrument aboard the Compton Gamma Ray Observatory. These profiles illustrate the rich diversity in temporal structure, intensities, and durations for gamma-ray bursts, no two of which are exactly alike in all respects. (Courtesy NASA and the CGRO BATSE Instrument Team.)

milliseconds) measured during a given burst implies that the observed radiation arises from an extremely compact source, requiring the relativistic expansion of the emitting particles to avoid the photon–photon pair-creation opacity that would otherwise quench the observed gamma radiation. This rapid expansion of emitting material could take

the form of a relativistic fireball resulting from an initial explosion or from collimated beams of relativistic jets emanating from a central source.

For many years the mystery of the origin of gamma-ray bursts was compounded due to their lack of detection in any frequency band below the hard X-rays, a fact difficult to reconcile with such an apparently catastrophic release of energy. Further complicating the situation was that few, if any, observational constraints could be placed on the distances to the sources of bursts, nor could GRBs be associated with any known class of object. Consequently, the distance scale to bursts was studied indirectly through the angular distribution on the sky and the intensity distribution of the overall burst ensemble. The observed deficit in the number of weak bursts detected with BATSE (compared to the number expected for a uniform, homogeneous distribution of burst sources in flat three-dimensional Euclidean space) implied, for example, that we were seeing the far "edge" of the burst source population. Still, this allowed for burst sources to be either "local" to our own Galaxy, or "cosmological" at the edge of the observable Universe, leading to endless variety in proposed burst models and widespread debate and controversy within the field of burst studies.

It had long been recognized that the key to unraveling the GRB mystery was the identification of burst counterparts at other wavelengths. The short duration of gamma-ray bursts, however, combined with the unpredictability of their occurrence anywhere in the sky, and the relatively poor location-determination capabilities of gamma-ray instruments (compared to detectors operating in other wavebands) severely limited the effectiveness of coordinated follow-up searches for GRB counterparts. Despite repeated and valiant attempts at many wavelengths to detect either the remnant afterglow of a burst event, or of a quiescent counterpart within a gamma-ray burst error box, none was detected for many frustrating years. An observational breakthrough occurred in early 1997, however, shortly after the launch of the Italian-Dutch BeppoSAX X-ray satellite. On a number of occasions BeppoSAX observed with one of its Wide-Field Cameras (WFCs) the unmistakable X-ray afterglow associated with a gamma-ray burst. The occurrence of a burst was simultaneously registered with the cesium–iodide (CsI) detectors that served primarily as shields for the satellite's main X-ray telescope, but were also configured to operate as a separate gamma-ray burst monitor. Following a suspected burst event the spacecraft was reoriented to allow its high-resolution X-ray instruments to observe the same field, confirm the fading X-ray emission, and to exactly fix its point of origin (see Fig. 12). After rapid communication of precise coordinates, followup observations were immediately initiated by observatories around the world, resulting in the first detections of fading afterglow emission, lasting from hours to weeks, from a gamma-ray burst. In a few instances, further observations of the fading optical source revealed the presence of faint, extended underlying emission, suggestive of very distant host galaxies. Measurement of a host galaxy's redshift has provided the first

incontrovertible evidence that GRBs occur at cosmological distances. In the 3 years following the initial detection of a burst counterpart, over thirty GRBs have been rapidly localized by BeppoSAX, with over a dozen events yielding counterpart detections in other wavebands, including X-ray, optical, infrared, millimeter, and radio, as well as redshift measurements to likely host galaxies (of characteristic redshift $z \sim 1$). These results electrified investigators in the field of gamma-ray burst research, and have completely revolutionized the study of GRBs.

The detection of both prompt and afterglow counterpart emission has permitted a serious revaluation of the many theories put forward to explain the origin of GRBs. In particular favor is the "relativistic fireball" model that posits an enormous, instantaneous energy release within a small volume. In this scenario a relativistic fireball consisting of an electron-positron pair plasma with a Lorentz factor of 100–1000 propagates outward following the initial energetic event. The temporal structure observed in the gamma-ray burst itself results from the collision of shocks with somewhat different Lorentz factors within the relativistic outflow (so-called "internal" shocks). Prompt burst emission (e.g., the dramatic optical flare seen by the robotic ROTSE telescope during GRB 990123) is attributed to "reverse" shocks traveling backward through the dense ejecta. The longer-term afterglow emission arises from forward-moving "external" shocks plowing through the surrounding medium. The relativistic particles accelerated in these shocks radiate both synchrotron and Compton emission. The afterglow spectrum observed hours to weeks following a burst typically consists of a series of synchrotron power laws whose breaks are dependent on the energy distribution in the population of radiating nonthermal electrons. With time, the observed synchrotron spectrum shifts to lower energies as a result of the cooling expansion and the decreasing Lorentz factor of the bulk flow. The delayed onset and characteristic smooth decay observed for the counterpart emission at X-ray, optical, and radio wavelengths can be naturally explained as arising from the expanding pair plasma that becomes progressively more optically thin to lower-frequency radiation during the later stages of the cooling fireball. The fireball model agrees remarkably well with the observations, though the exact details of the broadband emission depend critically on the physical properties of the interstellar or intergalactic environment into which the fireball expands.

A fundamental question is the nature of the energetic event that sparks the detonation of the fireball itself. The enormous energies ($> 10^{53}$ ergs) implied by the cosmological distances inferred from the most recent observations of GRBs greatly restricts the number of possibilities. Only the mergers of the components of evolved binary systems

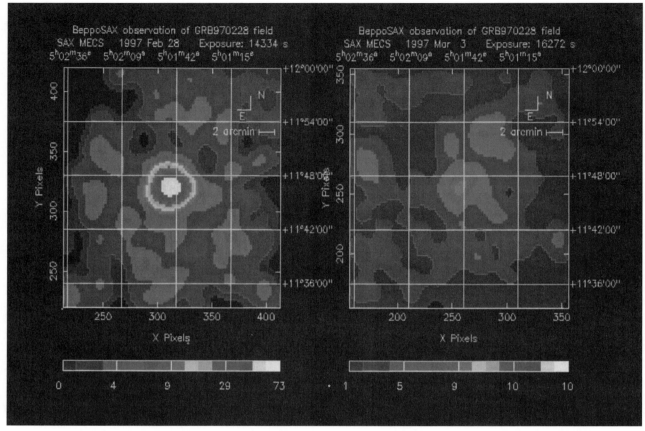

FIGURE 12 The discovery images of the first X-ray afterglow detected from a gamma-ray burst (GRB 970228), observed with the BeppoSAX satellite. The left panel shows bright X-ray emission from the location of the GRB shortly after burst occurrence on February 28, 1997, while the right panel shows the fading X-ray afterglow seen several days later on March 3, 1997. (Adapted from Costa, E. *et al.* (1997). "Discovery of an X-ray afterglow associated with the gamma-ray burst of 28 February 1997," *Nature* **387,** 783–785, Copyright 1997, by permission.)

containing pairs of compact objects (neutron stars or black holes), or the collapse of the most massive stars at the ends of their lifetimes (as described in the "hypernova" and "collapsar" models), appear to meet the gigantic energy requirements. The recent detection of iron-line emission at X-ray wavelengths in GRB afterglows with instruments aboard BeppoSAX and the Chandra Observatory favors the latter class of models, since the quantity of iron estimated from the observations seems most likely to have originated in an evolved massive star that exploded in a supernova-like event.

Virtually all models for the ultimate energy source of GRBs involve an endpoint of stellar evolution, particularly of the most massive stars. Thus it has been proposed that the burst rate must be proportional to the overall cosmic star formation rate. This view is supported by the fact that the typical redshifts ($z \sim 1$) associated with GRB host galaxies correspond to an epoch of early active star formation in the Universe. Burst counterparts also tend to be

found in the outer regions of blue galaxies undergoing recent star formation, or in irregular galaxies that may have undergone recent collisions or mergers, promoting a burst of star-forming activity at an early epoch.

H. The Extragalactic Diffuse Gamma-Ray Emission

Studies of the Galactic diffuse gamma-ray emission revealed the presence of an isotropic component of diffuse emission that is now considered to be extragalactic in origin. The existence of extragalactic diffuse gamma rays was first demonstrated with the SAS-2 satellite and confirmed on a large scale with the all-sky mapping carried out with the COMPTEL and EGRET instruments aboard the CGRO. Sometimes referred to as the cosmic diffuse gamma-ray (CDG) background, this radiation has been measured from ~ 1 MeV to ~ 100 GeV in energy. Since it is by definition the constant, isotropic "background" radiation against which all other sources are measured, it is not characterized by any spatial or temporal signature.

The determination of the extragalactic diffuse gamma-ray background presents a particular observational challenge. This stems from the fact that the basic procedure for determining it involves the subtraction from the accumulated gamma-ray signal of all *other* known sources of emission, including the contribution of point sources, the Galactic diffuse emission, and the effects of the instrumental background. In other words, the extragalactic diffuse emission is assumed to be what is "left over" after all other sources of emission are accounted for and removed from the observations. Thus, by necessity it requires a profound understanding of the instrumental response, and a detailed knowledge of the sources and structure of the gamma-ray sky. Consequently, it is usually the last major result obtained with a gamma-ray telescope, often requiring years of observation, analysis, and study.

In this regard, the COMPTEL measurements of the extragalactic diffuse emission represent a major achievement. The MeV energy band over which COMPTEL was sensitive has long been recognized as a notoriously difficult one in which to operate, due to the high levels of instrumental background to which experiments are susceptible. As the first MeV telescope launched into space for extended observations, COMPTEL was in a unique position to provide a definitive measurement of the CDG at medium gamma-ray energies. The COMPTEL results were particularly anticipated in light of a series of earlier, shorter measurements that had indicated the existence of a puzzling excess, or "MeV bump," in the diffuse spectrum between ~2 and 9 MeV that could not be readily explained and which had provoked widespread discussion as to its possible origin. Initial COMPTEL results were derived from observations of the Virgo region, well removed from the Galactic plane. In contrast with the earlier observations, the COMPTEL measurements provided *no* hint of the MeV bump previously reported. Later confirming observations were obtained in the direction of the South Galactic Pole, and these also did not indicate any excess emission, or any temporal variation or anisotropy in the diffuse gamma radiation. Most significantly, then, COMPTEL disproved the earlier reports of an MeV excess, and provided a first definitive measurement of the cosmic diffuse radiation between ~1 and 30 MeV. At 30 MeV the COMPTEL results join smoothly with an extrapolation of the EGRET measurements at higher energies (see Fig. 13).

The EGRET results were obtained following the analysis of data from 36 distinct regions of the sky well removed from the contaminating influence of the Galactic disk and bulge. From 30 MeV to 100 GeV the extragalactic diffuse emission is well described by a strikingly smooth power-law photon spectrum of index-2.1, in remarkable agreement with the average spectrum derived from the large pool (>90) of gamma-ray blazars detected with EGRET over its lifetime. This immediately suggests that the extragalactic emission at EGRET energies most likely arises from numerous unresolved sources of the blazar class.

The origin of the extragalactic diffuse emission has been the subject of much theoretical speculation for many years. Theories of truly diffuse processes include matter–antimatter annihilation in a baryon-symmetric cosmology, the evaporation of primordial black holes, supermassive black holes ($\sim 10^6$ solar masses) that collapsed at high redshifts ($z \sim 100$) at very early epochs, and the annihilation of exotic supersymmetric particles. Unfortunately, many of these hypotheses cannot be tested realistically at the sensitivity of the current observations. The general consensus at present is that the extragalactic diffuse radiation most likely results from the superposition of a number of classes of unresolved point sources. The CDG, since it represents the integrated emission from sources extending back to the earliest cosmological epochs, provides an important observational constraint for theoretical models describing source evolution with time.

In the COMPTEL energy range around ~1 MeV it has been proposed that a significant fraction of the diffuse emission could arise from gamma-ray lines due to the decay of radionuclides produced in the course of supernovae of types Ia and II in distant, unresolved galaxies. Other possible sources at low MeV energies include active galaxies such as those of the Seyfert I and II class. At the higher gamma-ray energies observed with COMPTEL and EGRET, blazars provide the most likely explanation of the observed diffuse emission. Of note in the EGRET band is the fact that the diffuse spectrum clearly extends up to 100 GeV without significant deviation, implying that the quiescent emission from gamma-ray blazars, assuming that they are responsible for the observed emission, must necessarily extend up to this energy as well. Consequently, the relativistic particles that give rise to gamma-ray emission in blazars must extend to even higher energies. It is anticipated that the next generation of gamma-ray telescopes with higher resolution and sensitivity will begin to detect and map out the individual sources responsible for the extragalactic diffuse gamma-ray emission.

I. Unidentified Gamma-Ray Sources

When any region of the electromagnetic spectrum is opened to new investigation, or detectors suddenly achieve a marked improvement in sensitivity, an exciting era of discovery inevitably ensues. This is particularly true in the gamma-ray regime, where the launch of each new satellite mission has resulted in the discovery of previously

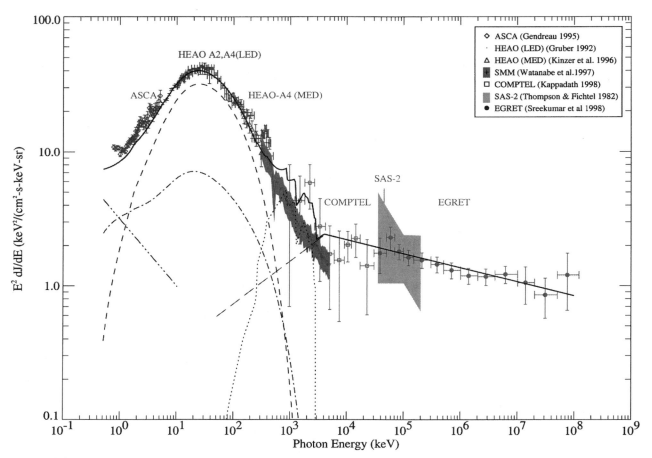

FIGURE 13 Broadband multiwavelength spectrum of the extragalactic diffuse emission from X-ray to gamma-ray energies combining the observations of several high-energy satellite missions, most recently those of the COMPTEL and EGRET telescopes aboard the Compton Gamma Ray Observatory. The dashed and dotted lines indicate the estimated contribution of several classes of unresolved point sources to the observed emission. (Adapted from Sreekumar, P. *et al.* (1998). "EGRET observations of the extragalactic gamma-ray emission," *Astrophys. J.* **494,** 523–534, Copyright 1998, reproduced with permission of the AAS.)

unknown sources of gamma radiation. A large fraction of the gamma-ray sources discovered to date, however, remain unidentified. Indeed, one of the great continuing mysteries of gamma-ray astronomy is the nature of the unidentified sources, which in the Third EGRET Catalog outnumber all other known gamma-ray sources combined (~170 of the 271 cataloged 3EG sources are unidentified, see Fig. 3). These unidentified sources have no obvious counterpart at other wavelengths, and cannot be clearly associated with any other known class of object based on their high-energy emission properties.

The unidentified gamma-ray sources can be broadly categorized by their location in relation to the plane of the Galaxy, and by their spectral and temporal properties. The sources at high Galactic latitudes, well removed from the disk of the Galaxy, are most likely to be extragalactic objects, presumably galaxies. This supposition is

supported by the variable emission often observed from these objects, characteristic of flaring active galaxies. The positional error boxes of these high-latitude unidentified sources, however, do not contain the bright, radio-loud galaxies of the blazar type that are typically associated with extragalactic gamma-ray sources. This suggests that some previously unrecognized class of radio-quiet or active galaxy can give rise to gamma radiation. The detection of gamma rays from the nearby radio galaxy Centaurus A lends support to this hypothesis.

The vast majority of the unidentified gamma-ray sources (>120) appear to belong to a Galactic population. Many of these are likely to be obscured objects whose exact properties are masked by the Galactic diffuse radiation, or cannot be accurately distinguished from those of other nearby sources in the crowded disk of the Milky Way. Gamma rays from the unidentified sources are presumed to arise from the same objects and processes that generate gamma radiation elsewhere in the Galaxy. The objects and phenomena most closely linked to the production of high-energy gamma rays in the Milky Way include (1) molecular clouds within the spiral arms and disk (where high-energy cosmic rays interact to produce gamma-rays); (2) supernova remnants (whose shock waves are a presumed site of particle acceleration and cosmic-ray production, leading to gamma-ray emission in their vicinity); (3) flares, winds, and outflows (within which gamma-rays can be produced) from massive and evolved stars; (4) relativistic jets and accretion disks associated with compact binary systems containing neutron stars or black holes; and (5) radio-quiet pulsars (such as Geminga) in whose intense magnetic fields particles are likely to be accelerated, leading to gamma-ray production. Efforts have been made to study the statistical properties of the unidentified sources and to correlate them with the possible source classes outlined above, though with limited success to date.

It has been noted that the unidentified Galactic sources can be separated into two apparently distinct populations: brighter sources appear confined more closely to the Galactic disk, while fainter sources are found to lie at medium Galactic latitudes (greater than about 5°). The bright sources in the disk are interpreted as more distant luminous objects, while the fainter sources may be much closer and more local to the Sun. Several investigators have proposed that these weaker, medium-latitude sources may be associated with Gould's Belt in the Galaxy. Known for well over a century from optical observations, Gould's Belt is a ring of bright massive O and B stars that define a great circle on the sky inclined by about 20° to the Galactic plane, tilted toward negative Galactic latitudes in the direction of Orion, and toward positive latitudes in Ophiucus. Observations over the years at optical, radio and other wavelengths have established that a slowly expanding ring of material (with an expansion age of about 30 million years) is interacting with and compressing the ambient interstellar gas along the periphery of Gould's Belt. This expansion has likely contributed to periods of enhanced star formation along the boundary of the Belt. The ring of expanding material is presumed to have resulted from a supernova explosion, thus explaining the origin of the local "superbubble" or evacuated region in the interstellar medium within which the Sun currently resides. In the context of this scenario, the most prominent grouping of weak, unidentified gamma-ray sources at medium latitudes lie on the boundary of Gould's Belt that is closest to the Sun, at a distance of about 100–400 pc in the direction of Ophiucus. The exact physical nature of individual sources remains to be determined, but is expected to be found among the list of candidates cited above.

A definitive determination and classification of the unidentified gamma-ray sources discovered to date must await the launch of the next generation of gamma-ray telescopes whose development is currently underway. While investigators are confident that the mystery of the unidentified sources will ultimately be resolved, there still remains the exciting prospect that some of the unidentified gamma-ray sources may in fact represent truly new types of high-energy cosmic sources previously unknown to science. The possibility that such fundamental discoveries remain to be made provides heightened motivation for the continued study of these mysterious objects.

III. THE CHALLENGE OF OBSERVATION AND TECHNIQUES OF DETECTION

As an observational discipline gamma-ray astronomy has always been extremely challenging. The very low fluxes of cosmic gamma-rays, combined with their high energy, necessitate the construction of large complex detectors. Since cosmic gamma rays are severely attenuated by the Earth's atmosphere, the experimental apparatus must be lifted above as much of this absorbing medium as possible by high-altitude scientific balloon or placed in low-Earth orbit via satellite. Either of these possibilities places severe constraints on the size and weight of a gamma-ray experiment. Further, since the detector must be designed to conduct observations quasi-autonomously, and since sophisticated analytical techniques must be employed to accurately identify true cosmic gamma-ray events from the multitude of background interactions that can masquerade as such, the successful operation of a gamma-ray telescope presents unique challenges to the experimental astrophysicist.

A. The Interaction of Gamma Rays with Matter

To observe a cosmic gamma ray, one must first devise a means to stop it (or at least "to slow it down" in some measurable way) within a detecting medium. In the gamma-ray regime, the primary interaction mechanisms of photons with matter are the photoelectric effect, the Compton effect, and pair production. Extensive analyses are available of these fundamental interaction processes. Here, we only briefly review their basic physical characteristics.

1. The Photoelectric Effect

Photoelectric absorption occurs when an incident photon is completely absorbed in an atomic collision with practically all of its energy transferred to an atomic electron, which is ejected. For photoionization to occur, the photon must necessarily have enough energy to overcome the binding energy of the liberated electron. A small, usually minimal, fraction of the initial photon energy is converted into the recoil of the atomic nucleus, required for the conservation of momentum. As a consequence of this conservation requirement, it is the most tightly bound innermost (K-shell) electrons that have the greatest probability of interacting with photons of energies approaching those of gamma rays. The cross section for photoelectric absorption has a strong dependence on atomic number (proportional to Z^5) and an inverse dependence on photon frequency (proportional to $\nu^{-7/2}$). This is valid for photon energies somewhat removed from the so-called "absorption edges," those energies at which strong rises in the absorption cross section occur as photon energies approach the binding energies of individual electron shells.

2. The Compton Effect

As mentioned previously, the Compton scattering process consists of interactions between photons and free electrons. In "standard" Compton scattering, an incident high-energy photon loses a fraction of its energy to an electron, while in the "inverse" Compton process it is the photon that gains energy at the expense of the electron. By applying the standard criteria of conservation of momentum and energy, it is a straightforward exercise to derive the Compton wavelength shift for the scattered photon,

$$\Delta\lambda = \lambda' - \lambda = h/m_e c (1 - \cos\theta) = \lambda_c(1 - \cos\theta),$$

where $\lambda_c = h/m_e c = 2.426 \times 10^{-10}$ cm is referred to as the Compton wavelength. Related expressions are easily obtained for the resultant electron and photon energies. The cross section for Compton scattering by an electron is given by the Klein–Nishina formula, and declines with increasing photon energy. There is a geometric dependence to the scattering probability, and at very high energies the cross section rapidly becomes strongly peaked in the forward direction.

3. Pair Production

In the pair production process, an incident gamma ray of sufficiently high energy is annihilated in the Coulomb field of a nearby charged particle, resulting in the creation of an electron–positron pair. As with the photoelectric effect, conservation of momentum requires the presence of a third body which takes up the balance of the momentum in the form of a recoil. For a recoil nucleus of mass much greater than the mass of the electron, the threshold energy for pair production is simply, $E_{th} \sim 2m_e c^2 = 1.022$ MeV. The total cross section for pair production rises rapidly above

threshold and approaches at strongly relativistic energies a limit that is roughly independent of energy.

4. Attenuation and Absorption of Gamma Radiation

The relative importance of the interaction mechanisms described above as a function of photon energy is represented graphically in Fig. 14. The cumulative effect of these primary interaction processes is usually expressed in terms of attenuation coefficients. The probability that an individual photon will traverse a given amount of absorbing material without interaction is simply the product of the probabilities of survival for each type of possible interaction. Given a collimated beam of monoenergetic gamma rays of initial intensity I_0, the intensity observed after passage through a thickness x of absorber is given by

$$I(x) = I_0 e^{-\mu x},$$

where $\mu = \kappa_{PE} + \kappa_C + \kappa_{PP}$ is the total linear attenuation coefficient (units: cm^{-1}), and κ_{PE}, κ_C, and κ_{PP} are the attenuation coefficients for the photoelectric, Compton and pair-production processes, respectively. The linear attenuation coefficient is related to the cross section σ by an expression of the form $\kappa = \sigma N$ cm^{-1}, where N is the number density of atoms in the target medium. Of more fundamental usefulness in terms of detector design is the *mass attenuation coefficient*, which is simply the linear coefficient divided by the density of the given material (units: cm^2/g), permitting straightforward comparison of the attenuation properties of various types of materials. Numerous tables and graphs of mass attenuation coefficients as a function of energy for the most commonly used detector materials are readily available.

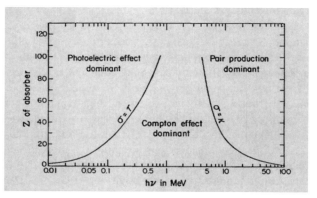

FIGURE 14 The relative importance of the major processes by which gamma rays interact with matter as a function of photon energy for elements of varying atomic number Z. (Adapted from Evans, R. D. (1995). "The Atomic Nucleus," Copyright 1955, McGraw-Hill, by permission.)

Some common detector materials include plastic organic scintillators (i.e., hydrocarbon compounds) typically used in charged-particle anticoincidence shielding, inorganic crystal scintillators such as sodium iodide (NaI), cesium iodide (CsI), and bismuth germanate oxide (BGO), and solid-state semiconductors such as germanium (Ge), and cadmium zinc telluride (CdZnTe) often used in high-resolution spectroscopic applications. As their name implies, scintillators are transparent materials that scintillate in visible light when high-energy photons or particles interact within them. The scintillation light, proportional to the amount of energy deposited, is collected by attached photomultiplier tubes and converted to an electronic pulse for further signal processing and recording. In semiconductor detectors interacting gamma rays deposit energy in the detector material, creating electron-hole pairs that are collected following application of an electric field to the material. Semiconductors are far superior spectroscopically to scintillators, due to their relatively low threshold for electron-hole creation ($\sim$3 eV for Ge). CdZnTe has the advantage that is a room-temperature semiconductor, while germanium, due to its narrow band gap, must be operated at cryogenic temperatures for optimal performance. The technology associated with the detection of cosmic gamma rays is very often derived from experimental techniques developed originally for application in the fields of high-energy and nuclear physics.

B. Sources of Instrumental Background

Gamma-ray telescopes, due to their physical size and mass, are particularly susceptible to background radiation and particle interactions. Below a few MeV in energy, there is a high background due to continuum emission and nuclear decay lines, arising primarily from cosmic rays interactions in and around the detector. Above a few MeV the background is somewhat less severe, but the photon fluxes are much weaker, still presenting a sizable signal-to-noise problem. Gamma-ray telescopes of necessity must combine high sensitivity with superior background-suppression capability. The major sources of experimental background in gamma-ray detectors are summarized below.

1. Atmospheric Gamma Rays

Gamma rays are produced in copious quantities in the upper atmosphere of the Earth as a consequence of cosmic-ray interactions. Balloon experiments rely typically on the "growth curve" technique to estimate the contribution of atmospheric gamma rays to the observed event count rate. In this method, the total count rate of the detector is determined as a function of the residual atmosphere remaining above the balloon-borne instrument as it rises to float altitude. Since the downward vertical atmospheric gamma-ray flux is assumed to be zero at the top of the atmosphere, all remaining event counts are assumed to be truly cosmic in nature (or locally produced within the experiment itself, see following). Both Monte Carlo calculations and semiempirical models are employed to test the reliability of such measurements.

2. Intrinsic Radioactivity

An ubiquitous source of background gamma radiation within a detector is the decay of naturally occurring radioisotopes contained within the structure of the instrument itself. The two most common naturally occurring background lines result from the presence of the long-lived isotopes ^{40}K and ^{232}Th in many detectors and structural materials. ^{40}K, with a half-life of 1.26×10^9 years, undergoes decay to ^{40}Ar giving rise to a gamma-ray line at 1.46 MeV, while a 2.62 MeV gamma ray results from an excited state of ^{208}Pb, a daughter product of ^{232}Th ($t_{1/2} = 1.4 \times 10^{10}$ years).

3. Cosmic-Ray and Secondary Charged-Particle Interactions

Background counts can arise in a gamma-ray telescope due to high-energy cosmic-ray collisions (of protons predominantly) with material in or around the detector, often accompanied by secondary particle production. Fortunately, active anticoincidence shielding has proven to be an effective means of screening out direct charged particle interactions within a detector. Of particular concern, however, are high energy ($E \sim 100$ MeV) proton collisions that result in radioactive spallation products which decay with longer characteristic lifetimes, yielding secondary "activation" gamma rays that can interact within the detector long after the original charged-particle event has been vetoed. The gradual build-up of activation species within detectors in low-Earth orbit, for example, must be monitored with care given the long-term exposure of such instruments to high-energy particles from space or trapped in the Earth's radiation belts. Extensive tables have been compiled detailing the dominant induced radioactive states with corresponding lifetimes and energies for common detector materials.

4. Neutron-Induced Background

Atmospheric and albedo neutrons can interact within detectors at balloon altitudes and in low-Earth orbit. Target nuclei can emit secondary particles or gamma rays that contribute to the background counting rate of the instrument. Three basic types of neutron interactions must be taken into account.

(a) Elastic scattering. Neutrons with energies below the first excited level of target nuclei are elastically scattered without excitation of the recoil nucleus. The amount of energy transferred to the recoil nucleus for the intermediate and heavy elements typical of gamma-ray detectors is relatively small, implying that elastic scattering is unlikely to generate a significant background in most cases. Over a series of such collisions, however, a neutron may lose sufficient energy to permit thermal neutron capture, giving rise to secondary gamma rays as outlined in (c) below.

(b) Inelastic scattering. The inelastic scattering of energetic "fast" neutrons (in interactions of the type $(n, n'\gamma)$ or $(n, x\gamma)$, where the "x" particle is a proton or alpha particle) can lead to significant background counts in a gamma-ray detector. The secondary gamma radiation results from the deexcitation of residual product nuclei. Such gamma-ray emission is usually prompt due to the relatively short lifetimes of low-lying nuclear excited states, though delayed activation gamma rays may result if the product nuclide is also radioactive. An important example of activation gamma rays resulting from fast neutron interactions is one involving aluminum, a commonly used structural material. The reaction $^{27}\text{Al}(n, \alpha)^{24}\text{Na}$ yields *two* gamma rays of energies 1.37 and 2.75 MeV following the radioactive decay via cascade of ^{24}Na ($t_{1/2} = 14.96$ h).

(c) Thermal neutron capture. Secondary gamma radiation can also result from the capture of thermal neutrons by nuclei within a detector. In this type of interaction, the secondary gamma emission is prompt and of energy comparable to the binding energy of the nucleus. A prominent example of a neutron-capture background line occurs for organic scintillator detectors, where the capture of thermal neutrons by the proton nuclei of hydrogen atoms leads to the production of 2.22 MeV photons from the deuterium recoil product.

C. Background Rejection Techniques

In order to optimize the cosmic gamma-ray signal measured with a detector, a number of background-suppression techniques are usually employed. The most common are outlined in the following.

1. Passive Shielding

One of the most fundamental techniques for background rejection is the use of passive high-Z materials (usually lead) to shield the central detector elements from unwanted photons and particles. Typically, several radiation lengths of the chosen material are employed to sharply attenuate the background flux. While the concept is simple in principle, weight constraints tend to limit the total amount of material that can be utilized. Additionally, since it is a passive technique, quantitative information as to the effectiveness of the shield during the course of actual measurements is often difficult to obtain.

2. Active Shielding

Active shielding, as the name implies, provides additional information over the passive approach. The active shield, commonly a plastic organic scintillator, has the advantage of being able to trigger a signal upon passage of a charged particle. Thus, active shielding is most commonly used in an anticoincidence mode. In other words, for a detector placed within charged-particle anticoincidence shielding, only those events which trigger the main detector element and *not* the surrounding active shield will be registered as valid events.

3. Pulse-Shape Discrimination (PSD)

One limitation of most active shields is that they are not efficiently triggered by the passage of neutral particles. Since neutron-induced reactions contribute a significant background to gamma-ray telescopes, it is imperative to devise a scheme for eliminating the neutron component of the detector signal. This neutron rejection is most efficiently achieved by means of pulse-shape discrimination (or PSD). With this technique, one exploits the fact that neutrons which interact in a scintillator crystal give rise to optical pulses that have a fundamentally different time profile compared to signals resulting from photon interactions. Electronic means can then be employed to separate one from the other.

4. Time-of-Flight (TOF) Discrimination

Finally, in telescopes containing multiple detectors one may take advantage of the physical separation between the detector elements to veto certain background events. By precisely timing interactions as they occur in two detectors, one may discriminate between upward-moving (i.e., atmospheric) and downward-moving (celestial) events. By defining a time-of-flight window for allowable events, one may also eliminate "slowly moving" downward particle transitions which could not have resulted from speed-of-light photon interactions. The rejection capability of a TOF system is, of course, limited by the detector-separation distance and the temporal resolution attainable with the available electronics.

TABLE II Gamma-Ray Telescopes by Energy Band

Name	Energy band	Energy (eV)	Telescope type	Instruments
Low energy	keV to MeV	10^5–10^6	Collimated scintillator, Coded-aperture telescopes	HEAO, SMM, GRANAT/SIGMA, CGRO/BATSE, HETE-2, Swift, CGRO/OSSE, INTEGRAL, HESSI
Medium energy	MeV	10^6	Compton telescopes	CGRO/COMPTEL
High energy	MeV to GeV	10^6–10^9	Pair-production telescopes	SAS-2, COS-B, CGRO/EGRET, GLAST
Very high energy (VHE)	GeV to TeV	10^9–10^{12}	Ground-based optical Cherenkov	Whipple, CAT, CANGAROO, HEGRA, Durham, MILAGRO, STACEE, CELESTE, MAGIC, VERITAS
Ultra high energy (UHE)	TeV to PeV	10^{12}–10^{15}	Ground-based optical Cherenkov and particle detection	CASA-MIA, CYGNUS, MILAGRO, HEGRA

D. Types of Gamma-Ray Telescopes

In the descriptions given in the following, refer to Table II for a listing of types of gamma-ray telescopes by energy band.

1. Low-Energy (keV to MeV)

In the low-energy portion of the gamma-ray spectrum photons interact primarily via the photoelectric and Compton processes. Historically, gamma-ray detectors sensitive to this energy range have been collimated, well-type instruments in which the primary detector (often an inorganic scintillator such as NaI or CsI) is surrounded by an active or passive shielding structure. The field of view of the telescope is determined typically by the opening angle of the shields or by passive collimators of some high-Z material, such as lead. The OSSE instrument aboard the CGRO is an example of such a configuration.

Another experimental approach used in the hard X-ray/low gamma-ray regime is that of coded-aperture imaging. The coded-aperture telescope operates essentially as a "multi-pinhole" camera. The telescope employs an absorbing mask to cast a shadow pattern of incident radiation on a position-sensitive detector plane below. Sky images and the positions of sources within the field of view are determined after applying standard deconvolution techniques to the data. To minimize artifacts in the reconstruction process the pattern of the coded mask is usually of a type referred to as a "uniformly redundant array" (or URA) of elements. The angular resolution and source-location accuracy of a coded-aperture telescope ($\sim 10'$) is superior to that of more traditional gamma-ray instruments, and is determined by three factors: the size of the individual mask elements, the mask-detector separation distance, and the spatial resolution attainable in the determination of photon-interaction locations in the central detector. While coded-aperture telescopes have excellent angular resolution, there is a practical upper limit to their

sensitive energy range, determined by the thickness of the mask required to fully attenuate the gamma rays of interest. The French SIGMA experiment aboard the Russian GRANAT spacecraft was a coded-aperture telescope, as are both the imaging and spectroscopic instruments aboard the INTEGRAL spacecraft (see Fig. 15).

2. Medium-Energy (MeV)

The medium-energy gamma-ray regime extends from approximately 1 to 30 MeV and is one of the most difficult in which to observe. The background over this energy range is particularly severe, due to nuclear excitation and decay lines and continuum radiation resulting from particle interactions in and around the detector. The Compton process is the most likely interaction mechanism for incident gamma photons in this energy range, and double-scatter Compton telescopes were developed to exploit this fact.

A Compton telescope consists of two separate detector assemblies which register in coincidence the Compton scattering of an incident cosmic gamma ray in an upper detector array followed by the subsequent total absorption of the scattered photon in a lower detector. Hence, *two* simultaneous interactions (for which interaction positions and energies are recorded) are required for the registration of a valid event. The energy and direction of an incident cosmic photon are determined after event reconstruction. Assuming total absorption of the scattered gamma ray, a straightforward application of the Compton scattering formula can be used to determine the Compton scattering angle from the upper to lower detectors of the telescope. The scattering angle and recorded interaction positions within the detectors define an event circle on the sky on which the source of the incident gamma ray must lie. The position of the gamma-ray source therefore is not unambiguously determined from a single event. Rather, numerous events must be recorded, and event circles computed, whose

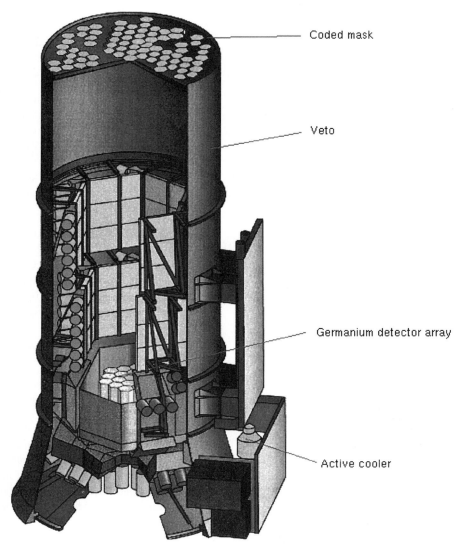

FIGURE 15 Schematic of the INTEGRAL SPI spectrometer showing the main components of this coded-aperture instrument. (Courtesy INTEGRAL SPI Instrument Team.)

common point of intersection reveals the true source location. The partial, rather than total, absorption of incident gamma rays, combined with other measurement errors, introduces uncertainties in source identification. In practice, detailed modeling of the instrumental response to incident radiation is required to accurately identify and characterize cosmic sources.

The COMPTEL instrument aboard the CGRO was a Compton telescope (see Fig. 16). The next generation of Compton telescopes currently under development will seek to remove the scattering angle ambiguity inherent in the traditional design by also recording the energy and direction of the recoil electron in the upper detector from the initial Compton scatter. This will greatly enhance

instrument performance and aid in reducing the instrumental background.

3. High Energy (MeV to GeV)

Above approximately 30 MeV in energy, gamma rays interact predominantly via the pair-production process. The traditional method of detecting such photons is through the use of high-voltage spark-chamber grids, interleaved with sheets of a high-density converter material (such as tungsten). In this technique an incident cosmic gamma ray is first converted into an electron–positron pair. As the pair traverses the layers of the gas-filled spark chamber the trajectories of the charged particles are detected via the

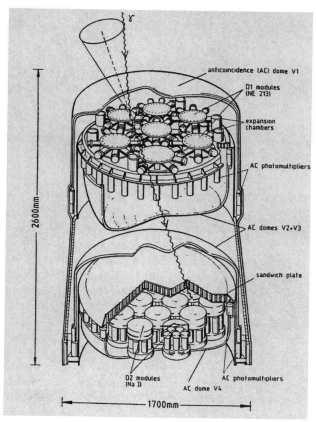

FIGURE 16 Schematic of the Imaging Compton Telescope (COMPTEL) on board the Compton Gamma Ray Observatory illustrating the double-scatter technique employed in Compton telescopes. (Courtesy NASA, the Max Planck Institute, and the CGRO COMPTEL Instrument Team.)

sparks they produce, the positions of which are recorded. A large scintillator below the spark chamber is typically used as a calorimeter to absorb any remaining particle energy. The diverging tracks recorded for the electron and positron are used to determine the incident direction of the original gamma ray. Spark chambers are normally filled with a gas (usually some mixture of neon) that acts as a spark-quenching agent, and which slowly degrades with use. The gas is thus a consumable that must be periodically flushed from the spark chamber and replenished for best performance.

The SAS-2, COS-B, and EGRET experiments all employed the same basic spark-chamber design, albeit on different scales of size and complexity (see Fig. 17). The next generation of pair-production gamma-ray telescope will be the Gamma-Ray Large-Area Space Telescope (GLAST) in which the spark-chamber grids will be replaced by semiconductor silicon strip detectors. GLAST is designed to be sensitive to gamma photons up to ~300 GeV in energy.

4. Very High Energy (GeV to TeV, and above)

Gamma rays in the energy range above ~30 GeV have not been well sampled from space to date, given their low fluxes, and ground-based techniques have only recently matured to the point where efficient observations are feasible. As outlined previously, cosmic gamma rays are absorbed in the Earth's atmosphere and are not directly observable from the ground. Photons of energy greater than ~30 GeV are detectable in principle, however, from the extensive air showers they produce upon interaction in the upper atmosphere. A particular signature of such interactions is the visible pulse of Cherenkov light emitted collectively by the relativistic electrons and positrons produced in these electromagnetic cascades. Ground-based optical telescopes are used to detect this Cherenkov light (see Fig. 18), from which energies and directions of incident cosmic gamma rays can be inferred.

To date the atmospheric Cherenkov technique has been employed with greatest success for photon energies above ~300 GeV by several groups. The U.S.-based Whipple collaboration (USA–UK–Ireland) has pioneered many of the techniques now commonly applied to TeV

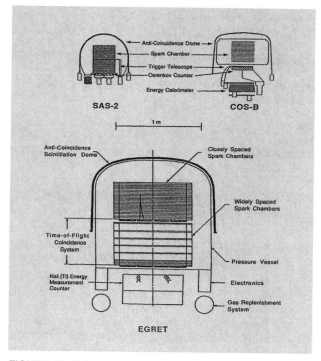

FIGURE 17 Schematic of the EGRET instrument on board the Compton Gamma Ray Observatory illustrating the spark-chamber design characteristic of pair-production telescopes. For comparison, schematics of the earlier SAS-2 and COS-B instruments are shown to scale. (Courtesy NASA and the CGRO EGRET Instrument Team.)

FIGURE 18 Photograph of the Whipple Observatory 10-m imaging atmospheric Cherenkov telescope. (Courtesy of the Whipple Collaboration.)

measurements. Other collaborations actively operating air-Cherenkov telescopes include CAT (in France), the CANGAROO (Japan–Australia) and Durham, UK groups (with telescopes in Australia), and HEGRA (Germany–Armenia–Spain, in the Canary Islands). Among the several TeV sources detected in recent years are the Crab and Vela pulsars, and AGNs such as MRK 421 and MRK 501.

The slightly lower energy domain from approximately 30 to 300 GeV still remains largely unexplored. A lower detection threshold requires a much larger collecting area for the atmospheric Cherenkov light. Two groups, STACEE and CELESTE, are in the process of converting large-area heliostats from experimental solar-power stations to this purpose. A German-Spanish project (MAGIC) is also underway to build a 17-m air-Cherenkov telescope.

At even higher gamma-ray energies, in the PeV range (10^{15} eV, the "ultra-high energy," or UHE, regime), particles associated with the photon-induced electromagnetic cascades have sufficient energy to reach the ground, where arrays of particle detectors can be used to detect them. Very large collecting areas are required, however, given the extremely low gamma fluxes expected. Early attempts to detect photons of PeV energy with particle-detector arrays included the CASA-MIA and CYGNUS efforts, among others. Ongoing attempts are being carried out by the HEGRA and MILAGRO collaborations. Early reports of positive PeV detections in the 1980s proved to be optimistic and have been unconfirmed by later more sensitive measurements.

IV. FUTURE MISSIONS AND PROSPECTS

The development of any new telescope has as its primary goal an increase in sensitivity, combined with improved angular and spectral resolution. In the gamma-ray regime this translates invariably into improved determination of photon-interaction positions and energy-depositions within the detecting medium. More accurate determination of the properties of interacting gamma-rays leads directly to a reduced background rate, since true celestial events are less likely to be confused with background interactions. Virtually every gamma-ray telescope under current development seeks to improve these interaction measurements by exploiting new detector technologies. Spatial and energy resolution within detector materials are greatly enhanced, for example, with the use of newly developed semiconductor strip and pixel detectors (such as silicon, germanium, and CdZnTe). The continuing challenge is to fabricate such sensitive small-scale devices in sufficiently large and reliable quantities to be incorporated into new large-area instrumentation, at costs that can be reasonably afforded. Another common characteristic of high-energy telescopes is the large number of data signals that must be processed and recorded in multi-channel detector systems. Increased use of custom Application-Specific Integrated Circuit (ASICs) designs employing Very Large Scale Integration (VLSI) techniques is imperative for the efficient operation of high-energy instruments. Fortunately, computational speed and data-storage capacities continue to rise at a steady pace, and experimentalists are quick to exploit these new capabilities in their instrument designs.

At the time of writing (2001), a number of gamma-ray missions are scheduled for launch in the near future (see Table II). Key among these is the International Gamma-Ray Astrophysics Laboratory (INTEGRAL), a mission of the European Space Agency (ESA) with participation from Russia and NASA. INTEGRAL is to be launched in 2002 and will be dedicated to high-resolution spectroscopy ($E/\Delta E \sim 500$) and imaging ($\sim 12'$ FWHM) over the energy range from 15 keV to 10 MeV. INTEGRAL carries two gamma-ray instruments, the SPI spectrometer and the IBIS imager, both operated as coded-aperture telescopes for accurate source identification. The SPI employs high-purity germanium detectors, while the IBIS uses two detector planes, a front layer of CdTe elements and a second layer composed of CsI pixels. In recognition of the need for broadband coverage INTEGRAL also carries two coded-aperture X-ray monitors (JEM-X), as well as an optical monitoring camera (the OMC). The primary scientific objective of the INTEGRAL instruments is to carry out high-resolution spectroscopic studies of sources over the nuclear-line region of the spectrum.

The Gamma-Ray Large Area Space Telescope (GLAST), scheduled for launch by NASA in 2005, will be the follow-on mission to the highly successful CGRO EGRET experiment. The sensitivity of GLAST, from 20 MeV to 300 GeV, will extend well beyond the EGRET range, providing much-need coverage in the poorly observed GeV region of the spectrum. A more modern particle-tracking technology (silicon strip detectors) will be employed in GLAST in place of the spark-chambers grids used in earlier pair-production telescopes. GLAST will have a large field of view (~ 2 sr) and achieve a factor of 30 improvement in flux sensitivity and a factor of 10 improvement in point-source-location capability compared to EGRET. GLAST will also carry a gamma-ray burst monitor.

Missions designed specifically for gamma-ray burst studies include HETE-2 and Swift. The High-Energy Transient Experiment-2 (HETE-2) was launched in 2000, and became operational in early 2001. This satellite carries three science instruments: a near-omnidirectional gamma-ray spectrometer, a wide-field X-ray monitor, and a set of soft X-ray cameras. A major goal of the HETE-2 mission is the rapid identification and accurate localization of gamma-ray bursts, whose coordinates will be relayed within seconds to ground-based observatories for deep counterpart searches. The recently selected Swift mission (scheduled for launch in 2003) will also carry out multi-wavelength studies of gamma-ray bursts, in the manner of BeppoSAX and HETE-2. Like its avian namesake Swift will "feed on the fly" by rapidly localizing gamma-ray bursts to ~ 1–$4'$ precision, and transmitting coordinates to the ground within ~ 15 s for follow-up counterpart searches. Swift can also be rapidly reoriented to carry out observations with its X-ray and ultraviolet/optical telescopes that will be used to study afterglow properties, fix positions to arcsecond levels, and determine distances via redshift spectral measurements.

The High-Energy Solar Spectroscopic Imager (HESSI) is a NASA-funded mission to study the characteristics of particle acceleration in solar flares via the X-ray and gamma-ray emission produced in these energetic events.

HESSI, scheduled for launch in 2001 at the peak of the solar cycle, will carry out high-resolution spectroscopic measurements of nuclear lines and underlying bremsstrahlung continuum over the energy range from 3 keV to 20 MeV with a set of cooled high-purity germanium detectors. HESSI will carry out Fourier-transform imaging of the full Sun at $\sim 2''$–$36''$ resolution over its sensitive range by using rotating modulating collimators. Since HESSI is unshielded it can also carry out other non-solar observations, including measurement of the Galactic diffuse lines due to radioactive ^{26}Al (at 1.809 MeV) and positron annihilation (at 0.511 MeV).

In the area of planetary studies, NASA's Mars Odyssey mission is also scheduled for launch in 2001. Among its suite of instruments are a gamma-ray spectrometer and two neutron detectors. These will be used to fully map the Martian surface and determine its elemental composition. The neutron and gamma-ray measurements in combination will also be used to obtain an estimate of the water content of the Martian near-surface.

Other gamma-ray experiments and missions have been identified as a high-priority by the Gamma-Ray Astronomy Program Working Group, an advisory panel to NASA composed of scientists from the high-energy community. Among their recommendations for future development is an advanced Compton telescope employing the latest detector technologies for application in the MeV region of the spectrum.

High-altitude scientific ballooning has long served as a test-bed for new instrumentation. Gamma-ray telescopes require long exposures, due to comparatively low source fluxes and high instrumental backgrounds, while the duration of a typical balloon flight, unfortunately, can often be rather limited (a few days at most). To counter this drawback NASA has recently initiated the Ultra-Long Duration Balloon (ULDB) project whose planned 100-day around-the-world balloon flights will greatly extend the time aloft for scientific instruments. The ULDB program will provide much-needed opportunities for longer-exposure balloon flights, as well as an attractive low-cost alternative to full-scale space missions.

Among the collaborations actively engaged in ground-based air-Cherenkov studies of TeV gamma rays there are also a number of efforts underway to upgrade existing facilities, primarily through an increase in optical collecting area. Perhaps the most ambitious are those of the VERITAS collaboration, with a planned array of seven 10-m telescopes in the USA, the German-French-Italian HESS group with 4 to 16 12-m class telescopes to be built in Namibia, the German-Spanish MAGIC project with a telescope of 17-m aperture, and the Japanese SuperCANGAROO array of four 10-m telescopes in Australia. In a related effort, the MILAGRO collaboration is constructing a water-Cherenkov detector with a wide field of view in New Mexico in the USA for TeV measurements. As a covered light-tight detector MILAGRO has the added advantage that it can remain operational for 24 h a day.

V. CONCLUDING REMARKS

The future of gamma-ray astronomy looks very bright indeed, as the next generation of gamma-ray telescopes and missions stands poised to extend our knowledge of the high-energy sky. Over the past decade gamma-ray astronomy has become firmly established as a productive and dynamic discipline of modern observational astrophysics. With the completion of comprehensive surveys with the telescopes aboard the Compton Gamma Ray Observatory the pioneering phase of gamma-ray exploration has now been achieved and the promise of gamma-ray astronomy confirmed and realized. New questions have been raised and remain to be addressed, and ongoing technological advances will certainly continue to spur the development of future successor missions to build upon the present findings. The results obtained with the Compton Observatory and other spacecraft have revolutionized our view of the high-energy Universe, and have underlined the fundamental importance of the gamma-ray region of the spectrum to a fuller and more complete understanding of the high-energy processes that dominate the cosmos.

ACKNOWLEDGMENT

This review is dedicated to the memory of Frank J. Kerr, one of the founding fathers of the field of Galactic radio astronomy, and an early advisor to both of the authors. His gentlemanly manner and professional example will be long remembered by those who knew him. He would be delighted and amazed at the many recent discoveries in gamma-ray astronomy, and fully appreciative of their relevance to much of his own work.

SEE ALSO THE FOLLOWING ARTICLES

COSMIC RADIATION • GRAVITATIONAL WAVE ASTRONOMY • INFRARED ASTRONOMY • NEUTRINO ASTRONOMY • PULSARS • SOLAR PHYSICS • SUPERNOVAE • ULTRAVIOLET SPACE ASTRONOMY • X-RAY ASTRONOMY

BIBLIOGRAPHY

Catanese, M., and Weekes, T. C. (1999). Very high gamma-ray astronomy. *Publications of the Astronomical Society of the Pacific* **111**, 1193–1222.

Diehl, R., and Timmes, F. X. (1998). Gamma-ray line emission from radioactive isotopes in stars and galaxies. *Publications of the Astronomical Society of the Pacific* **110**, 637–659.

Fichtel, C. E., and Trombka, J. I. (1997). "Gamma-Ray Astrophysics: New Insights into the Universe, Second Edition," NASA Reference Publication 1386. NASA, Greenbelt, MD.

Fishman, G. J., and Meegan, C. A. (1995). Gamma-ray bursts. *Annu. Rev. Astr. Astrophys.* **33**, 415–458.

Gehrels, N., and Paul, J. (1998). The new gamma-ray astronomy. *Physics Today* **51**, 26–32.

Hoffman, C. M., Sinnis, C., Fleury, P., and Punch, M. (1999). Gamma-ray astronomy at high energies. *Rev. Mod. Phys.* **71**, 897–936.

Longair, M. S. (1997). "High Energy Astrophysics, Volume 1: Particles, Photons and their Detection," Second Edition, Cambridge University Press, Cambridge.

Longair, M. S. (1994). "High Energy Astrophysics, Volume 2: Stars, the Galaxy and the Interstellar Medium," Second Edition. Cambridge University Press, Cambridge.

van Paradijs, J., Kouveliotou, C., and Wijers, R. A. M. J. (2000). Gamma ray burst afterglows. *Annu. Rev. Astr. Astrophys.* **38**, 379–425.

Strong, K. T., Saba, J. L. R., Haisch, B. M., and Schmelz, J. T. (eds.). (1998). "The Many Faces of the Sun: A Summary of the Results from NASA's Solar Maximum Mission," Springer-Verlag, Berlin.

Urry, C. M., and Padovani, P. (1995). Unified schemes for radio-loud active galactic nuclei. *Publications of the Astronomical Society of the Pacific* **107**, 803–845.

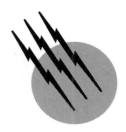

Gamma-Ray Spectroscopy

R. F. Casten
C. W. Beausang
WNSL, Yale University

I. Introduction
II. Gamma-Ray Detection
III. Gamma-Ray Spectroscopy and Nuclear Structure
IV. Conclusions

GLOSSARY

Detector efficiency Loosely defined as the probability of detecting the full energy of a gamma-ray.

Detector resolution Refers to the width of the full energy gamma-ray peak measured in the detector. Typically the resolution is ~2 keV for semiconductor and 10–20 keV for scintillator detectors for a 1000-keV gamma-ray.

Doppler shift The shift in frequency or energy of waves emitted from a moving source.

Nuclear level scheme A graph of the excited energy levels of a nucleus and their connecting gamma-ray transitions. The levels are usually labeled by their angular momentum and parity quantum numbers.

Nucleons The protons and neutrons that make up the nucleus.

Pauli exclusion principle Fundamental principle of quantum mechanics. It states that for certain types of elementary particles, including electrons, protons, and neutrons, no two identical particles can be in the same quantum state.

Potential energy surface A contour plot of the potential energy of the nucleus as a function of deformation. Stable deformations correspond to minima in the potential energy.

Scintillator detector A material, liquid or solid, that converts the energy lost by a gamma-ray into pulses of light.

Semiconductor detector Essentially a large diode usually constructed out of either silicon or germanium.

Spin Angular momentum.

I. INTRODUCTION

THE NUCLEUS is a unique, strongly interacting, quantum mechanical system. Consisting of a few to a few hundred protons and neutrons, its structure combines the macroscopic features expected of bulk nuclear matter (shape, size, etc.) with the microscopic properties associated with the motion of a finite number of nucleons in a potential.

Atomic nuclei studied in the laboratory (whether they are produced in reactions or populated via radioactive decay) are often found in excited states. Since any physical

system seeks its lowest possible energy level, such excited nuclear configurations are unstable. They generally
de-excite to the nuclear ground state in time scales of
10^{-15} to 10^{-6} sec by the emission of one or more gamma-
rays. Hence, the study of the gamma-rays emitted from
excited nuclei provides a means of studying the levels,
de-excitation rates, and structure of these objects.

The study of the decay properties of the atomic nucleus
has provided an enormous quantity of information on the
behavior of such systems when stressed by the application of high temperature, high angular momentum, large
deformation or by large isospin values (isospin is a quantum number which basically counts the difference in the
numbers of protons and neutrons in a nucleus).

In this article we will touch upon some of these topics
while attempting to give a flavor of the field of nuclear
gamma-ray spectroscopy and charting some possible future directions. We begin by introducing some of the detector types used to detect gamma-rays and briefly discuss some of the design criteria for modern gamma-ray
spectrometers. This is followed by a discussion of some
features found in excited nuclear states, broadly separated
into low-spin and high-spin properties, and chosen to illustrate the variety of macroscopic and microscopic features of the nuclear system. To discuss or even list the
enormous number of practical applications of gamma-ray
spectroscopy in medicine, in industry (e.g., the oil industry), in other sciences such as archeology and astronomy,
and in the areas of security and defense is far beyond the
possible scope of this article.

II. GAMMA-RAY DETECTION

In this section we discuss the mechanisms by which
gamma-rays interact with matter (i.e., detectors), the different types of detectors and detector systems, and the
criteria that go into the design and choice of particular
systems. To facilitate this discussion, a simplified example of a nuclear level scheme is shown in Fig. 1.

A. Interaction Mechanisms

For energies ranging from a few kilo-electron volts to a few
mega-electron volts, gamma-rays interact with matter via
one of three principal mechanisms: the photoelectric effect, Compton scattering, or for energies above ~1 MeV,
the electron-positron pair production. Most gamma-ray
detectors exploit one or more of these effects both to detect the gamma-ray and to measure its energy. Of course,
gamma-rays are electromagnetic waves and sometimes
their wave properties are also used in their measurement,
for example, with diffraction techniques. Before we dis-

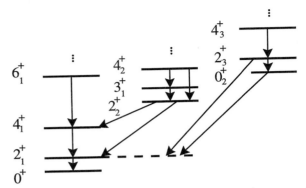

FIGURE 1 A simplified nuclear level scheme showing some of
the levels and gamma-ray transitions that might be observed in
a typical heavy-ion fusion-evaporation reaction. The levels are labeled by their angular momentum and parity quantum numbers.

cuss the design of gamma-ray detectors and spectrometers,
we first briefly describe these interaction mechanisms. The
relative probability for each mechanism is shown schematically in Fig. 2 as a function of gamma-ray energy.

The photoelectric effect is the dominant interaction
mechanism for low gamma-ray energies, below a few hundred kilo-electron volts. In this case, the gamma-ray interacts with an atomic electron somewhere in the bulk of the
detector material. The gamma-ray energy is transferred to
the electron, which is ejected from the atom with energy
$E_e = E_\gamma - E_{BE}$, where E_{BE} is the electron binding energy.
The probability for the photoelectric effect interaction increases very rapidly with the atomic number (Z) of the
material. This is why high-Z materials are favored both
for gamma-ray detectors and for absorbers and shields.

The Compton scattering mechanism is similar to the
photoelectric effect in that the gamma-ray also interacts
with an atomic electron in the detector material. In this
case, however, the initial gamma-ray energy is shared
between the electron and a scattered (lower energy)

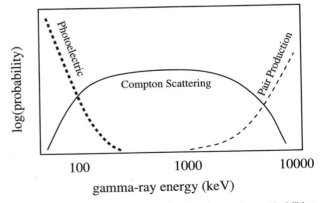

FIGURE 2 Schematic diagram showing the relative probabilities
for photoelectric, Compton scattering, and pair production as a
function of gamma-ray energy.

gamma-ray. Compton scattering is the dominant interaction mechanism for gamma-ray energies in the range from a few hundred kilo-electron volts up to a few mega-electron volts. It is important to realize that this is the energy range of most gamma-rays produced in typical nuclear structure experiments.

For gamma-ray energies greater than $2m_ec^2$ (m_ec^2 being the rest mass of the electron $\sim$511 keV), electron-positron pair production is possible, and, for energies significantly higher than this threshold, pair production begins to dominate the interaction cross section. In this case, some of the incident gamma-ray energy is used to create the electron-positron pair while the remainder (in excess of $2m_ec^2$) is shared as kinetic energy between the electron and positron. Eventually, the positron annihilates with another electron in the detector medium, producing two photons each of energy 511 keV emitted back to back.

As one can see, all three of these interaction mechanisms result in the production of a single energetic electron (or, in the case of pair production, of an electron-positron pair) with a kinetic energy less than or equal to that of the incident gamma-ray energy. These energetic electrons recoil through the bulk material of the detector, their range being typically less than a millimeter or two. They rapidly slow down, losing energy through many collisions with other atomic electrons. In an ideal detector all of the incident gamma-ray energy is eventually absorbed in the detector material by a combination of photoelectric, Compton scattering, and (for high enough gamma-ray energies) pair-production processes.

It is intuitively obvious that the number of collisions, and hence the number of secondary electronic excitations produced, is proportional to the primary electron energy and, hence, is directly related to the incident gamma-ray energy. For example, in semiconductor detectors, which are essentially diodes made out of germanium (Ge) or silicon (Si), the electron-hole pairs produced following electron–electron collisions are extracted by a high voltage placed across the detector and produce a current pulse which is proportional to the deposited gamma-ray energy. For scintillation detectors, such as sodium iodide (NaI(Tl)) detectors, the collisions of the primary electron produce excited atomic or molecular states. The subsequent decay of these states produces scintillation photons (typically in the UV range). These photons are converted into a current pulse using a photocathode and photomultiplier tube. The size of the current pulse is again proportional to the deposited gamma-ray energy.

B. Gamma-Ray Detectors

Different types of detectors have quite different efficiencies and energy resolutions. Indeed, generally speaking,

gamma-ray spectroscopy is a constant trade-off between these two properties. For example, scintillation detectors, such as NaI(Tl) detectors, typically have high efficiencies but poor energy resolution compared to Ge detectors. Detectors based on the technique of crystal diffraction (see below) have superb energy resolution but very small efficiency.

1. Scintillation Detectors

The detection of gamma-rays (or other types of ionizing radiation) by the scintillation light produced in certain materials is one of the oldest techniques on record, and it is still one of the most useful and common techniques today.

A scintillator, either a solid or a liquid, is a material which converts the energy lost by the gamma-ray into pulses of light. The scintillation light is detected in turn by a light-sensitive material which usually forms the cathode of a photomultiplier tube. The light pulses are converted into electrons in the photocathode. These electrons are then accelerated and their number vastly (and linearly) amplified in the photomultiplier. The resulting current pulse is proportional to the energy of the absorbed gamma-ray.

The energy required to produce a light pulse is fairly large, on the order of 30–50 eV. Thus, the average number of light pulses produced when, say, a 500-keV gamma-ray is absorbed is on the order of 10,000. Fluctuations in this number and in the light collection process limit the resolution obtainable with scintillation detectors.

2. Semiconductor Detectors

A semiconductor detector is essentially a large diode constructed out of either Si or Ge. For gamma-ray spectroscopy, Ge detectors are preferred, as Ge has a larger stopping power. The diode is operated under a reverse bias and is normally fully depleted (i.e., with no free charge carriers). The gamma-ray interaction produces electron-hole pairs in the depletion region, which are collected because of the detector bias voltage and which produce a current pulse proportional to the absorbed gamma-ray energy.

In contrast to scintillator detectors, the average energy required to produce a single electron-hole pair is only about 2–3 eV. Therefore, a 500-keV gamma-ray can produce around 250,000 primary charge carriers, much larger than the corresponding number for scintillation detectors with a corresponding decrease in the statistical fluctuations and improvement in detector resolution. Figure 3a shows a typical spectrum of a ^{60}Co source obtained using a modern Ge detector. The energy resolution obtained is about 2 keV for gamma-ray energies of about 1000 keV. Using a NaI(Tl) scintillation detector, the two peaks in

Fig. 3a (having energies of 1174 and 1332 keV) would be barely resolved. Following a brief discussion of crystal diffraction detectors, we will focus most of the remainder of this article on gamma-ray spectroscopy using Ge detectors.

3. Ultra-High Energy Resolution Spectroscopy: Crystal Diffraction

Gamma-rays are, after all, electromagnetic photons, and under certain circumstances their wave properties may be used in their detection and measurement. Indeed, the ultimate in current gamma-ray energy resolution and measurement precision is obtained by the use of crystal diffraction. With such techniques it is routine to measure a 1-MeV gamma-ray with an energy resolution of ~3 eV and an energy precision of better than 1 eV. The cost, however, is low efficiency and the need to scan the energy spectrum one small energy bite at a time.

The technique uses Bragg diffraction from a nearly perfect crystal, usually of Si. As for optical and X-ray transitions, the gamma-ray wavelength λ and diffraction angle θ are related by the Bragg law: $n\lambda = 2d \sin \theta$, where the lattice spacing d is known to an accuracy of 1 part in 10^{10} and

n is the order of diffraction. Clearly, the resolution scales with n. Higher order diffraction gives greater dispersion and, hence, energy precision, although the efficiency generally falls off with n. The accuracy depends on the precision of the angle measurement. In the realization of this technique at the Institut Laue Langevin in Grenoble, in the GAMS (GAMma-ray Spectrometer) family of instruments, accuracies of the latter are typically in the milli-arc seconds range (Koch *et al.*, 1980). Given the nature of the technique, the gamma-ray energy spectrum is stored energy interval by interval rather than sampled fully at each point in time. An example of a crystal diffraction spectrum compared to the corresponding Ge detector spectrum is shown in Fig. 3b. Generally speaking, crystal spectrometers offer the greatest advantages over Ge detectors for gamma-ray energies below ~1 MeV. At higher energies their efficiency drops quickly, and hence, lower orders of diffraction are used with poorer energy resolution.

C. Level Scheme Construction

One might ask why the extraordinary energy precision of crystal diffraction techniques can be useful since nuclear models seldom predict nuclear states to accuracies better

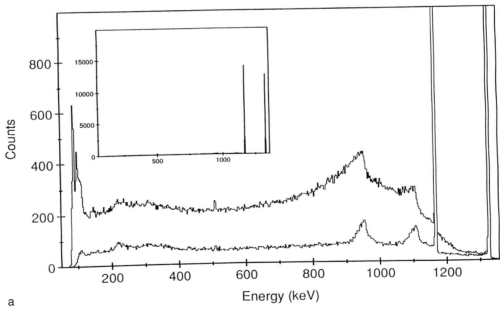

FIGURE 3 (a) Typical spectrum of a ^{60}Co source obtained using a modern Ge detector with and without escape suppression. The vertical scale has been greatly expanded in order to show the Compton background. The dramatic reduction in the height of the background when using a Compton suppression shield is obvious. The insert shows the same spectra but now with the full vertical scale to illustrate the height of the photopeaks compared to the background. (b) An example of the very high energy resolution obtainable using a crystal diffraction system. The top spectrum, obtained using a Ge detector, shows seven peaks, some of which are not resolved. The lower spectrum, obtained using the GAMS crystal diffraction system, shows the same portion of the spectrum, but with a dramatic improvement in resolution. In both spectra, the gamma-rays are labeled with their energies in kilo-electron volts. *Continued.*

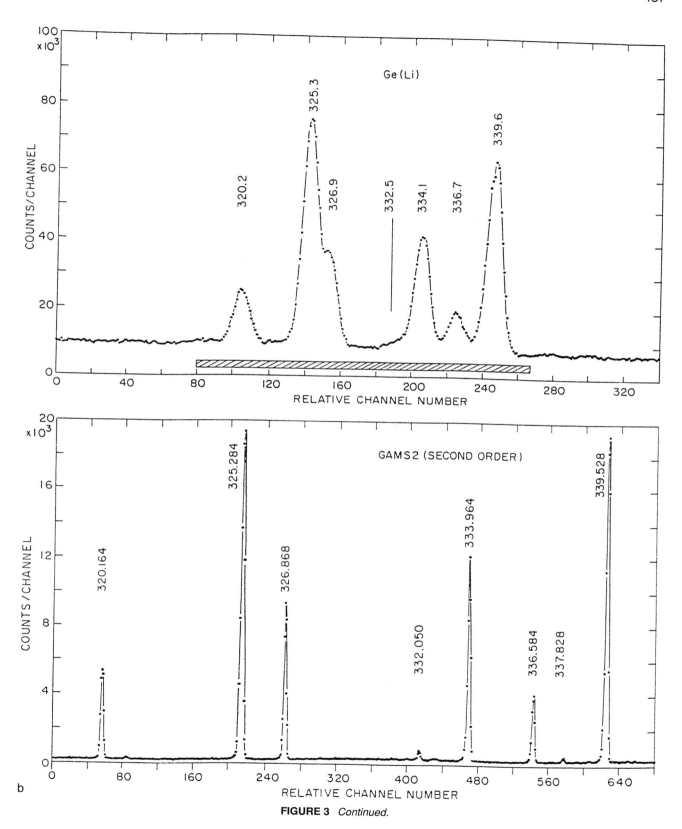

FIGURE 3 *Continued.*

than many kilo-electron volts? The principal reason relates to the construction of reliable level schemes. One method of constructing level schemes utilizes the Ritz Combination Principle. Here, nuclear level energies are determined by demanding that their energy differences be equal to the energy of the gamma-ray transition connecting them (see Fig. 1). A brief discussion of this technique enables us to see both the role of the ultra-high energy resolution in level scheme construction and the complementary and much more commonly used technique of coincidence spectroscopy with Ge detectors.

To see the point, imagine constructing a level scheme which consists of 50 levels spanning the excitation energy range from zero up to 2.0 MeV simply by using the Ritz Combination Principle in a case where one has detected, say, 500 γ-rays with energies below 1.2 MeV, with Ge detector energy accuracy of ± 0.1 keV. This is a typical situation encountered in the spectroscopy of low-spin states of heavy nuclei. In such a case the probability of an accidental Ritz Combination, that is, a level energy difference that inadvertently coincides within uncertainties with a gamma-ray energy, is about 10%. Even with other experimental input, such as information on the angular momenta of the levels to rule out certain transition placements from conservation of angular momentum, it is clear that a large number of incorrect placements and, hence, incorrect physics will result. There are two ways of resolving this situation: either using time coincidence relations between successive gamma-rays to place them correctly in the level scheme of a nucleus or improving the energy resolution significantly. For the latter, the crystal diffraction approach is ideal. With an energy precision of, say, ± 5 eV, the probability of an accidental sum drops to negligible levels. Indeed, data from the GAMS spectrometers at the ILL have often shown that existing Ge detector results (usually data taken without coincidences) are in error.

However, the usual solution to this problem is the use of coincidence spectroscopy. This technique exploits the fact that nuclear levels are generally short lived, with typical half-lives in the pico- to nanosecond range, so that successive gamma-ray de-excitations effectively occur simultaneously, on the time scale of standard pulse analysis electronics. Therefore, if two or more gamma-rays are observed in separate detectors, in time coincidence (within say a few nenoseconds) then they must occur in a cascade in the nuclear level scheme. For example, in Fig. 1 the $6_1^+ \rightarrow 4_1^+$ and $4_1^+ \rightarrow 2_1^+$ transition would be in coincidence, as would the $3_1^+ \rightarrow 2_2^+$ and $2_2^+ \rightarrow 4_1^+$ or $2_2^+ \rightarrow 2_1^+$ transitions. However, the $3_1^+ \rightarrow 2_2^+$ transition is not in coincidence with the $6_1^+ \rightarrow 4_1^+$ transition. Such coincidence relations are of inestimable help in constructing complex nuclear level schemes.

While coincidence spectroscopy is a powerful technique, it also has limitations. For example, it does not help place ground state transitions or very weak transitions, and, in some cases, experimental constraints preclude its use. Nevertheless, it is, by far, the most common approach to sorting out the plethora of gamma-rays observed in nuclear de-excitation. We will further discuss coincidence spectroscopy below when we introduce advanced multidetector arrays of Ge detectors.

We briefly mention that another application of ultrahigh resolution crystal spectroscopy is in the determination of fundamental constants such as the accepted standard for length measurements (the definition of the meter) through the precise measurement of gamma-ray wavelengths. These applications fall outside the scope of this article.

D. The Evolution of Detector Arrays

To illustrate the increasing power and sophistication of gamma-ray detectors, particularly of Ge detector arrays, it is useful to consider the nuclear reactions by which the nuclei to be studied are formed. One of the most common reaction mechanisms used to populate high-spin states in atomic nuclei is the heavy-ion fusion-evaporation reaction. This type of reaction has the advantage of bringing large quantities of angular momentum into the product nucleus (often up to the limit allowed by fission), while at the same time populating only a few product nuclei with significant probability.

In such reactions, a heavy-ion beam is incident on a target at an energy just above the Coulomb barrier. A typical reaction might involve a ^{48}Ca ($Z = 20$, $N = 28$) beam incident on a ^{108}Pd ($Z = 46$, $N = 62$) target at a beam energy of 200 MeV. Following the collision, the beam and target nuclei fuse to form a compound nucleus, in this case ^{156}Dy ($Z = 66 = 20 + 46$, $N = 90 = 28 + 62$). The compound nuclear system will be produced in a highly excited and rapidly rotating state, with typically 60 MeV of excitation energy and about $70\hbar$ of angular momentum.

The initial decay of the compound system is via the emission of a few (3–5) particles, usually neutrons and less frequently protons or alpha particles. This first stage of the decay process typically removes about 40 MeV of excitation energy and about $10\hbar$ of angular momentum. The remainder of the excitation energy and most of the angular momentum is subsequently removed by gamma-ray emission. Each gamma-ray photon removes either one or two units of angular momentum. Thus, we can expect the emission of cascades of up to 30 gamma-rays following each reaction. Because of this multiplicity of gamma-rays, the study of transitions following production of the compound nucleus imposes stringent requirements on detector systems. Other reactions used, such as Coulomb excitation, (n, γ), or β-γ decay, generally present simpler experimental challenges. The driving force in the impressive developments in gamma-ray detector systems in the last 40 years has been the requirements imposed by the fusion evaporation reaction studies.

Information about the changes in nuclear structure during the decay, as the nucleus loses energy and angular momentum, is obtained by measuring the properties of the gamma-rays in these cascades, such as the gamma-ray energy, angular distribution, linear polarization, emission sequence, etc.

The evolution of gamma-ray spectroscopy with time over the past 40 years or so is illustrated in Fig. 4, which plots the population intensity of various nuclear states as a function of angular momentum or spin of the state. Pioneering experiments in the early 1960s were carried out with one or a few NaI(Tl) scintillation detectors (Morinaga and Gugelot, 1963). The sensitivity of these experiments was limited, both by the poor energy resolution of NaI(Tl) detectors (about 80 keV at 1000 keV) and by the small number and size of the detectors, to spins up to about

spin 8–10 $\hbar$ and to states populated with about 10% the intensity of the strongest transition.

The introduction of reversed bias, lithium drifted, Ge detectors in the mid 1960s led to a major increase in sensitivity and major breakthroughs in our physics knowledge. Germanium detectors have very good energy resolution, about 1 keV for $E_\gamma \sim 100$ keV and 2 keV for $E_\gamma \sim 1000$ keV. On the other hand, the detection efficiency of early Ge detectors was often much lower than NaI(Tl) detectors. To compensate for the lower efficiency, and also to measure the time coincidence relationships of successive gamma-rays in a cascade, experiments with more than one Ge detector were soon commonplace. The phenomenon of backbending, at spin ~ 15 $\hbar$ (see below), was discovered by Johnson, Ryde, and Sztarkier (1971) using just two Ge(Li) detectors, while the structure of ^{160,161}Yb was investigated by Riedinger et al. (1980). up to spin ~ 30 $\hbar$ using only four Ge detectors. In the last three decades, the study of the properties of the atomic nucleus through gamma-ray spectroscopy has evolved through the development of larger and more efficient Ge detector arrays and, indeed, has driven the development of these arrays.

Starting in the 1980s and continuing to today, large arrays of Ge detectors such as TESSA, GASP, Eurogam, Gammasphere (Lee, 1990) and Euroball (Gerl and Lieder, 1992) further revolutionized gamma-ray spectroscopy. Future arrays such as the proposed GRETA (Gamma-Ray Energy Tracing Array) spectrometer (Deleplanque et al., 1999), which promise very large increases in sensitivity resulting from modern manufacturing techniques, electronics, and digital data processing, are in the planning stages.

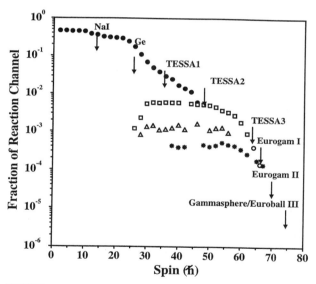

FIGURE 4 A schematic diagram illustrating the evolution of gamma-ray spectroscopy. The various symbols plot the measured intensity of various nuclear states vs angular momentum, giving an indication of the sensitivity of various detector systems. Early experiments using NaI(Tl) and a few Ge detectors were sensitive to excited states which were populated with intensities down to about one-tenth of the reaction channel (solid symbols). As time went on, more sensitive arrays were developed. The current generation of arrays, the Gammasphere and Euroball arrays, are capable of observing excited states populated with a fraction as small as 10^{-6} of the reaction channel. The open symbols and stars plot the intensity of various superdeformed bands as a function of angular momentum.

E. Germanium Detector Performance

1. Peak-to-Total Ratio and Escape Suppression

A major problem encountered with early Ge detectors, and still a problem today, is the poor peak-to-background ratio in the spectrum. The background, clearly seen in Fig. 3a, is caused by incomplete energy collection in the Ge detector occurring when a Compton scattered gamma-ray leaves the active bulk of the detector before being absorbed.

Even with today's large volume Ge detectors, irradiation with a standard ^{60}Co source (which emits two gamma-rays with energies of 1174 and 1332 keV) yields a spectrum where only $\sim 25\%$ of the events lie in the full energy photopeaks. This number, the ratio of the number of counts in the photopeak(s) to the total number of counts in the spectrum, is termed the peak-to-total ratio (PT). For PT = 0.25, the remaining 75% of the events in the detector form a continuous background extending to lower energies.

The preferred solution is to detect these scattered photons in a second, surrounding detector, termed an escape suppression or an anti-Compton shield, and to reject, using fast electronics, coincidence events between the Ge detector and the shield detector. The combination of Ge detector and suppression shield is termed an escape suppressed spectrometer (ESS). The material commonly used in the anti-Compton shield is bismuth germanate (BGO), a dense, high-efficiency scintillator material.

After suppression, typically about 65% of the remaining events are in the photopeaks (PT = 0.65). A typical ESS configuration showing a Ge detector and shield is shown in Fig. 5, while the improvement in the background, and hence spectrum quality, is illustrated in Fig. 3a.

The PT ratio is of prime importance for coincidence spectroscopy. For example, when requiring a coincidence between two Ge detectors, a PT ratio of 0.25 implies that only $(0.25)^2$ or ~6% of the events will be photopeak-photopeak coincidences. The remaining 94% will be background events. Using an ESS, however, the photopeak-photopeak coincidence fraction increases to $(0.65)^2$ or 42%, an improvement of a factor of 7. Even larger improvements are obtained when three- or higher fold coincidence events are recorded. For example, the improvement is a factor of 17 for triples coincidences, 45 for quadruples, and 120 for quintuples. Today's largest gamma-ray spectrometers, the Gammasphere array in the United States and the Euroball array in Europe, regularly record even higher fold coincidence events (Lee, 1990; Gerl and Lieder, 1992).

The efficiency and sensitivity of ESS arrays improved rapidly, so that by the mid 1980s arrays with more than 20 ESS having total absolute peak efficiencies of up to 1% were constructed. By convention, the total photopeak efficiency is defined as the probability of measuring the full energy of the 1332-keV ^{60}Co gamma-ray when the source is placed at the center of the array. These ESS arrays enabled nuclear phenomena that occur at an intensity of about 1% of the total intensity of the nucleus to be studied. Worldwide there were about a dozen arrays with this level of sensitivity. One of the earliest of these, the TESSA3 array (Nolan, Gifford, and Twin, 1985), located at Daresbury Laboratory in the United Kingdom, was used in the discovery of the classic discrete line superdeformed band in ^{152}Dy. Superdeformation will be discussed further below.

In the mid 1990s the latest generation of gamma-ray spectrometers with total photopeak efficiencies of up to ~10% came on line. These spectrometers, namely the Gammasphere (Lee, 1990) and Eurogam/Euroball (Gerl and Lieder, 1992; Beausang et al., 1992) arrays, contain up to 240 individual Ge elements and have sensitivities of better than 0.001% of the production cross section. Some of the detectors in these arrays are composites formed by closely packing several Ge detectors together as a unit. Two varieties of such units, called clover (Duchene et al., 1999) or cluster (Eberth et al., 1996) detectors, are nowadays the backbone of advanced arrays such as the YRAST Ball array at Yale University (Beausang et al., 2000) or the planned Exogam and Miniball arrays in Europe (Simpson et al., 2000). Even more powerful detectors, termed tracking detectors, are under development. These will be discussed below.

2. Counting Rates

It is informative to look at some of the numbers involved in a typical nuclear physics reaction carried out in the laboratory. Once again, we consider the example of the ^{48}Ca + ^{108}Pd reaction, which was used in the experiment in which the first superdeformed band in ^{152}Dy was discovered (Twin et al., 1986).

Typically, the beam intensity from an accelerator is about 10^{10}–10^{11} particles per second incident on a target. This corresponds to an electric current on the order of a few nano-Amperes (nA). About one beam particle in a million will actually strike a target nucleus and induce a nuclear reaction. Therefore, we expect about 100,000 reactions per second. About 20% of the reactions produce ^{152}Dy and about 1% of these will populate the nucleus in the superdeformed state, corresponding to about 200 such events per second.

The array used in the original discovery of the superdeformed band in ^{152}Dy, the TESSA3 array (Nolan, Gifford, and Twin, 1985), had a total photopeak efficiency of about 0.5%. Assuming that each superdeformed nucleus decays by emitting a cascade of ~25 gamma-rays, and that we require a coincidence between two detectors (γ^2) before accepting an event, we might expect to detect about 1 gamma-ray coincidence event per second originating from a superdeformed cascade. Since each cascade is ~25 transitions long, we expect about 1 count per gamma-ray transition every 20 sec or so. The background rate from other processes is many hundreds of times greater.

3. Doppler Effects and Segmentation

The lifetimes of the highest spin states populated via heavy-ion fusion-evaporation reactions are often comparable to, or shorter than, the stopping time of the recoiling nucleus (recoiling due to the momentum imparted by the incident beam nucleus that initiates the reaction) in the target material. Typical recoil velocities are on the order of a few percent the speed of light. Therefore, Doppler effects play a major role.

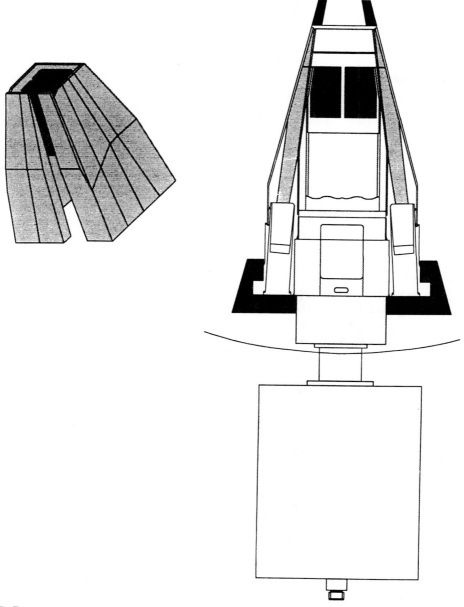

FIGURE 5 Escape suppression spectrometer showing Ge detector and shield. This is the type of ESS used for clover Ge detectors in the Eurogam/Euroball array. The clover Ge detector position is indicated inside the suppression shield. The liquid nitrogen storage dewar is also shown.

The Doppler shifted energy of a gamma-ray emitted from a nucleus in flight is given by

$$E_\gamma = E_0 \left[1 + \frac{v}{c} \cos\theta \right],$$

where E_0 is the unshifted energy and θ is the detector angle. If a detector records the same gamma-ray emitted from different nuclei having a wide range of velocities or traveling at different angles with respect to the beam direction, the resulting energy resolution can be very poor.

One solution is to use very thin targets in order to minimize slowing down effects and detectors that subtend only a small range of angles. Knowing the detector angles, one can correct for the Doppler shift and recover most of the resolution. The limit on detector resolution now becomes the finite opening angle of the Ge detector itself, in other words the uncertainty in knowing in which part of the Ge detector the gamma-ray actually interacted. For a constant recoil velocity, the Doppler broadening is given by

$$dE = E_0 \frac{v}{c} \sin \theta \, d\theta,$$

where $d\theta$ is the opening angle of the Ge detector, typically about 5–10°. For experiments on very high spin nuclear states, the energy resolution is dominated by Doppler broadening effects and is often a factor of two or more worse than the intrinsic resolution of the Ge detector. For very high recoil velocities the problem is much worse. Because of the $\sin \theta$ dependence, the Doppler broadening is worst for detectors placed at $\theta = 90°$ to the beam direction (even though the Doppler shift is zero at 90°).

Various methods have been developed to minimize Doppler broadening effects. Most involve the concept of detector segmentation in which one determines in what part of a detector a given photon was detected. The development of the clover detector, for example, with four separate Ge detectors closely packaged in a single vacuum vessel, was driven by such concerns (Duchene *et al.*, 1999). A schematic diagram of a clover Ge detector is shown in Fig. 6. The idea is that by using four small detectors, one effectively has a much larger detector while preserving the smaller opening angle for each individual segment. Gamma-rays may interact in only a single element of the clover detector. In this case one takes the angle θ to be the

center of this element and $d\theta$ is the opening angle of this segment. A gamma-ray may also scatter between two elements of the clover detector. In this case simulations and measurements have shown that the gamma-ray interaction usually takes place close to the boundary between the two crystals. Thus, one is justified in taking θ as the angle of the boundary. The other enormous advantage of the clover detector is that the energy is measured accurately, even for such scattering events. The energies measured in each separate crystal may be added together while preserving the good energy resolution of the individual crystals. Because of this add-back feature, the efficiency of a clover detector, consisting of four individual Ge crystals, is actually about six times the efficiency of the individual detector crystals.

4. Tracking Detectors

Recently, further advances in detector manufacturing technology allow the electronic segmentation of a single crystal into smaller elements, thus further localizing the interaction site within the volume of the detector. The ultimate goal of these developments is the development of a tracking detector array, which actually allows one to follow the trajectory of each individual gamma-ray as it traverses a detector, even if it undergoes multiple scattering events en route.

Ideally, such an array needs to cover all the available solid angle and localize each gamma-ray interaction to within 1–2 mm in three dimensions. A variety of tracking detectors are under development worldwide, including the Gamma-Ray Energy Tracking Array (GRETA for short) in the United States (Deleplanque *et al.*, 1999).

Such an array would be very efficient. Simulations for the proposed GRETA array indicate that it may be up to a thousand times more sensitive than the best of today's spectrometers. This sensitivity comes about because of the high-count rate capability (the relatively low-count rate in each segment is the limiting factor, rather than the high rate in the entire detector), excellent PT ratio, resolution, and efficiency.

A prototype detector for the GRETA array has already been extensively tested in Lawrence Berkeley National Laboratory, Berkeley, CA. One key test involved the determination of the gamma-ray interaction position by use of a closely collimated source. The interaction positions are determined by detailed measurements of pulse shapes on an event-by-event basis. A comparison of measured pulse shapes, with calculations show excellent agreement, which is a major first step in a proof of principle for the detector. The next step in this project is to purchase a mini-array of such detectors. These multiple detectors, assembled into a closely packed array, allow one both to

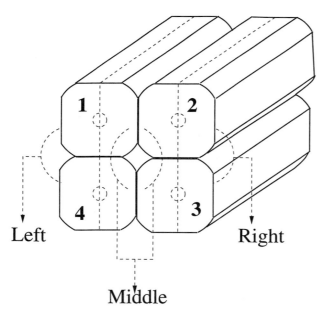

FIGURE 6 Schematic diagram showing the four Ge crystals of a segmented clover Ge detector. In this type of clover detector, to further improve position information, signals are taken from the center contacts of each crystal (labeled 1–4) and also from the left, right, and middle parts of the outer electrical contacts.

do physics and to prove the principle of practical gamma-ray tracking for the first time. The proposal to construct this array is currently awaiting funding.

5. Pair Spectrometers

High-energy gamma-rays (with energies, say, in the 1–10 MeV range) can interact with matter to produce positron-electron pairs. When the positron annihilates, two photons, each of 511 keV, are emitted at an angle of 180° to each other. When such a pair-production process occurs in a Ge detector, one or both of the 511-keV photons may escape from the detector without being detected. Hence, each gamma-ray transition leads to three peaks in the spectrum, the full energy peak plus the so-called single and double "escape" peaks. This proliferation of peaks can significantly complicate spectral analysis and adversely affect nuclear level scheme construction.

To improve such spectra one often uses a pair spectrometer, which, in essence, is the inverse of the anti-Compton shield spectrometer discussed earlier. Whereas in an anti-Compton shield any event detected in the shield is used to veto the coincident event in the Ge detector, in a pair spectrometer the simultaneous detection of 511-keV γ-rays on opposite sides of the Ge detector is used to positively trigger (i.e., to select) the double escape peak in the Ge detector spectrum.

F. Measurements of Nuclear Level Lifetimes

Aside from the measurement of gamma-ray energies and intensities, and the determination of gamma-ray transition placements in nuclear level schemes by coincidence and Ritz Combination techniques, gamma-ray detectors can be used to measure another extremely important observable, namely, nuclear level lifetimes. These lifetimes are proportional to squares of quantum mechanical quantities called transition matrix elements and therefore can directly reveal insights into nuclear structure and the properties of nuclear excitations.

Two general classes of techniques are used: those involved in directly measuring the time difference between successive gamma de-excitations in a nucleus and those based on Doppler effects. The former has traditionally been limited by the rise time of voltage pulses from detectors to the nanosecond range, but advances using faster scintillation detectors have pushed the frontiers of electronic time measurements farther down to nearly the picosecond range. Doppler techniques are typically used to measure lifetimes from hundreds of picoseconds down to the few femtoseconds range. These techniques cover a range of lifetimes characteristic of a wide variety of nuclear decays.

1. Recoil Distance and Doppler Shift Attenuation Methods

Typical recoil velocities following heavy-ion fusion-evaporation reactions are a few percent the speed of light. The associated Doppler shifts of emitted gamma-rays can be used to obtain level lifetimes. The recoil velocity corresponds to an easily measured, maximum Doppler shift of about 20–30 keV in a 1000-keV gamma-ray. The fraction of the gamma-ray intensity which lies in the Doppler shifted peak can be proportional to the lifetime of the nuclear state. Two Doppler-based techniques are commonly used. The first, termed the Recoil Decay Method (RDM), utilizes two parallel foils separated by a distance d. The nuclei of interest are produced in the first, thin foil and recoil out of the foil with a well-defined recoil velocity. Having flown a distance d, they are rapidly stopped in the second, thicker stopper foil. If the nuclear state of interest decays while the nucleus is flying between the two foils, then the gamma-ray will be emitted with the appropriate Doppler shift. On the other hand, if the lifetime is long enough that the nucleus reaches the stopper foil and is stopped, the gamma-ray will be emitted from a nucleus at rest, without a Doppler shift. Changing the distance d between the foils can access different lifetime ranges. Typically, the RDM technique is used to probe lifetimes in the nanosecond to picosecond range.

The Doppler Shift Attenuation Method (DSAM) is similar to the RDM in that two foils are used. However, in this case, the foils are placed in intimate contact with each other. Now the recoiling nuclei immediately enter the second foil and begin to slow down and stop. If the nuclear lifetime of interest is of the same order of magnitude as the slowing down time of the nuclei in the foil, around 1–2 ps, then the gamma-ray transitions will be emitted with a range of Doppler shifts, ranging from the maximum shift down to zero. Level lifetimes may be extracted by carefully analyzing the resulting, complicated peak shapes and comparing them to model calculations. Of course the calculations also have to include the slowing down process itself. The DSAM method is sensitive to level lifetimes on the order of picoseconds to femtoseconds, i.e., somewhat shorter than those accessible with the RDM method.

2. The GRID Technique

Another Doppler-based method is used at the ILL in Grenoble, referred to earlier (Koch *et al.*, 1980), using the ultra-high resolution crystal diffraction instruments GAMS4 and GAMS5. In this approach, a thermal neutron from a reactor is captured by a target nucleus which then emits a series of gamma-rays from the capture state (typically lying at an excitation energy of about 6 MeV) to lower lying states. Each emitted gamma-ray carries a small linear momentum, $p = E/c$. Hence, the emitting nucleus recoils in a direction opposite to that of the gamma-ray. If the same nucleus then emits a subsequent gamma-ray prior to stopping in the target material, the second gamma-ray will be Doppler shifted.

The shifts are exceptionally small. Typical recoil energies are a few electron volts, and therefore, the technique relies of measuring Doppler broadening effects of this order using crystal diffraction techniques. Note that one measures a Doppler broadening rather than a shift because the gamma-ray emission of the ensemble of nuclei is effectively isotropic. The technique is known as the GRID (Gamma-Ray Induced Doppler) technique (Borner and Jolie, 1993) and, like the DSAM, is useful for nuclear level lifetimes shorter than or on the order of the stopping time, typically ~ 1 ps.

3. Fast Timing Spectroscopy

The coincidence measurements with Ge detectors described above are typically used to establish nuclear level schemes. Such measurements utilize coincidence resolving times on the nanosecond time scale (10^{-9} sec). However, the time response of BaF_2 scintillation detectors is much faster than that of Ge detectors and, with special care, can be reduced to the few picosecond range. Hence, coincidence timing can also be used to directly measure nuclear level lifetimes in the few tens of picoseconds range, which is typical of the lifetimes of many collective excitations in medium mass and heavy nuclei. In practice, the technique, called FEST (Fast Electron Scintillation Timing) [see Buescher et al. (1990) for a simplified discussion and for references to more technical literature], is most commonly used in β-decay experiments where the time is measured between the emission of a β-ray (detected in a thin, fast plastic scintillator) and the subsequent gamma-ray emission in the daughter nucleus.

The technique must be used with great care. One problem is that the BaF_2 detectors have very poor energy resolution ($\sim 10\%$). Additional gamma-ray selection, by coincidence with cascade gamma-rays using Ge detectors (with "normal" nanosecond timing), is normally needed to simplify the BaF_2 spectra to one of two gamma-rays at most. Therefore, most applications are in low multiplicity experiments. Another serious problem relates to the energy dependence of the timing. A gamma-ray moves at the speed of light and in 3 ps travels ~ 1 mm. Since typical BaF_2 detectors have sizes on the order of centimeters, it is clear that the timing is sensitive to the exact position in the crystal where the gamma-ray absorption occurs. Hence, the time properties of such detectors are energy dependent and must be carefully calibrated. Nevertheless, the tech-

nique has proven to be quite useful in studies of nuclei off the line of nuclear stability in β-decay experiments.

III. GAMMA-RAY SPECTROSCOPY AND NUCLEAR STRUCTURE

The atomic nucleus is a unique, many-body quantum mechanical system. When describing nuclei, numbers of the order of 100 seem to occur frequently. For example, the depth of the potential holding the protons and neutrons, collectively known as nucleons, together is about 50 MeV. The maximum angular momentum the nucleus can hold before centrifugal forces break it apart is about 100 $\hbar$, which occurs for nuclei around mass 100.

Typical nuclei have a few hundred constituent nucleons. This number implies that the nucleus occupies a unique position in the plethora of quantum systems found in nature. A few hundred particles grouped together is sufficient to allow one to contemplate macroscopic nuclear properties such as shape and surface area and thickness. One the other hand, it is few enough that the addition or subtraction of a single proton or neutron can radically change the behavior of the whole system. Indeed, one of the appealing features of the nucleus is that it is a many-body quantal system in which the number of interacting bodies can be precisely controlled, measured, and varied. We will see a stunning example of the microscopic nature of the nucleus below when we discuss the phenomenon of backbending. This mixture of macroscopic and microscopic behavior in a strongly interacting system (the nucleons are after all bound together in the nucleus by the effects of the strong force) is nearly unique in nature.

The behavior of the nucleons inside the nucleus can be likened to the behavior of a herd of wild animals. The herd clusters together for protection, defining a shape and form. (The Hungarian word for such a herd is gulyas, so the nucleus is a bit like a goulash soup of nucleons.) However, the behavior of a single animal can have dramatic effects on the collective motion of the whole system. In the following sections, we will describe some of the features of the excited atomic nucleus and attempt to describe a few of the many manifestations of its macroscopic and microscopic behavior.

Generally speaking, atomic nuclei can be excited from the "bottom up" using reactions such as Coulomb excitation, inelastic scattering, or direct reactions, or from the "top down" using β-decay, neutron capture, and heavy-ion fusion-evaporation reactions. The former approach most often excites states selectively, while the latter approach is much less selective, tending to populate most states along a myriad of possible de-excitation routes, subject only to constraints due to angular momentum selection rules or phase space considerations. Gamma-ray spectroscopy is most often used in this second approach. When gamma-ray spectrometry is used in "bottom up" techniques, such as Coulomb excitation, it is exploited primarily as an indicator of the excitation probability of particular levels rather than as a study of de-excitation modes per se.

In this section, we will discuss a number of aspects of gamma-ray spectrometry. Although the distinction is a bit artificial, it is convenient, and historically pertinent, to break the discussion up into the study of low- and high-spin states.

A. Low-Spin States

The study of the low-spin nuclear states dates back to the beginning of nuclear structure and is the basis for our understanding of the equilibrium structure of nuclei and its evolution with nucleon number. Low-spin states are typically populated following β-decay, neutron capture, Coulomb excitation, or photon scattering reactions.

1. Beta-Decay

Nuclei formed off the valley of stability decay back toward stable nuclei via β-decay (which includes the processes of β^-, β^+, and electron capture decay). Typically, β-decay populates several excited levels in the daughter nucleus. Half-lives near stability range from seconds to days. Pro-

duction of β-decay parent nuclei can be achieved by simple reactions such as (p, n) or by heavy-ion reactions. The simpler, lower energy reactions tend to form only one or a couple of parent nuclei, whereas heavy-ion reactions may form many times more, and, in that case, selection techniques are needed to select the decay products of interest.

A popular technique in β-decay is the use of moving tape collectors in which the activity is collected on a tape (e.g., movie reel tape or aluminized Mylar) for some period of time (typically ~ 1.8 times the half-life for the desired β-decay). The tape is then moved to a low background area for detection of gamma-rays following decay. Collection of a new activity at another spot on the tape proceeds simultaneously.

Gamma-ray spectroscopy following β-decay was for many years in the 1950s–1970s a standard technique used to elucidate nuclear structure. Since β-decay itself carries off little or no angular momentum, the spin states accessible with this technique are generally those within ± 2–$3\,\hbar$ of the parent (ground or isomeric) state.

In recent years the technique has enjoyed a renaissance with the use of arrays of much higher efficiency Ge detectors (e.g., clover or cluster detectors). Since the gamma-ray multiplicity following β-decay is low and there is no Doppler effect, the detectors can often be mounted in close geometry to maximize count rates and achieve considerable coincidence efficiencies.

One current setup for such studies is the Yale moving tape collector (Casten, 2000). Illustrated in Fig. 7,

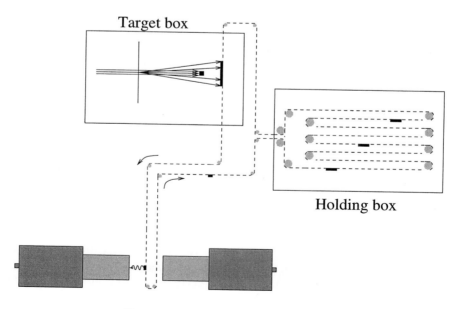

Target box

Holding box

Detector area

FIGURE 7 Diagram of the Yale moving tape collector showing the target box, counting area, and tape holding box. Activity is deposited on the tape in the target box, with the beam entering from the left. It is then transported to the counting area. The holding box provides a delay to let unwanted extraneous activity decay away before the tape once more returns to the target box.

it uses up to four Compton suppressed clover detectors that can be positioned at any angle in a horizontal plane. Studies with this instrument have included searches for possible multi-phonon states in ^{162}Dy and ^{164}Er. Nuclei with ellipsoidal shapes can undergo vibrational oscillations (called phonons) of these shapes about their equilibrium position. In principle, it is possible to superpose two or more identical vibrations. However, the effects of the Pauli Principle acting on the particles in the nucleus may destroy such states. One test of their intact character is to study their gamma decay. If they have predominantly a two-phonon character, then they should decay to the one-phonon state. Experimental searches for weak gamma-ray decay branches to the single phonon excitation are being sought in these two nuclei.

Another application of β-decay exploits the fourfold segmentation of the clover detectors. In the Yale arrangement, four such detectors allow simultaneous coincidence measurements at a large number of different relative angles of emission between the two detected gamma-rays. These angular correlation measurements can be used to constrain spin arguments for levels in the gamma-ray cascade. With clover detectors situated at appropriate angles, it is also possible to exploit their segmentation to measure the linear polarization of the gamma-ray and thereby to deduce the parity relations of the nuclear levels involved (Duchene *et al.*, 1999).

Finally, β-decay measurements are also an important tool in mass measurements, since, often, the daughter or granddaughter mass is known but not that of the parent. Nuclear masses (that is, in effect, binding energies) are of importance in a number of contexts. The binding energy reflects the sum of all the nucleonic interactions. Differences of binding energies for neighboring nuclei give the separation energy of the last nucleon and are therefore sensitive to single particle energies of nucleons in a mean field nuclear potential, as well as to shape and structure changes from one nucleus to the next. Mass measurements are also important for understanding the astrophysical processes occurring in the interiors of stars that lead to nucleosynthesis. Recent studies of nuclei in the mass $A \sim 70$ region, for example, are helping to set constraints on the termination of the rapid proton capture process in certain classes of stars.

Nuclear mass measurements are carried out by measuring gamma-ray spectra in coincidence with β-particle detection in order to deduce the β-decay end point, that is, the maximum β-decay energy (where energy sharing with the simultaneously emitted anti-neutrino is insignificant). The end point energy directly gives the mass of the parent nucleus if the daughter mass is known. The gamma-ray coincidence is used to cleanly select the product nucleus of interest.

2. Coulomb Excitation

When a beam particle passes close to a target nucleus, one or both nuclei may be excited by the changing electromagnetic Coulomb field between them (without any nuclear reaction occurring). Usually, a series of low-spin levels of the target nuclei are excited. The excitation probabilities are deduced by observing the subsequent de-excitation gamma-rays. A typical Coulomb excitation experiment involves bombarding a target of the (stable) isotope to be studied with beams of particles (the beams used range from protons to very heavy ions) at beam energies of roughly 80% of the Coulomb barrier.

Coulomb excitation is a powerful technique to study nuclear structure. Since the excitation mechanism is purely electromagnetic, it is known and calculable. Therefore, one can extract nuclear information from the excitation probabilities. This is in contrast, for example, to inelastic scattering processes at beam energies above the Coulomb barrier where nuclear effects enter in both the excitation mechanism and the nuclear structure itself and must therefore be disentangled.

In typical Coulomb excitation experiments, to correctly account for Doppler effects, the gamma-rays are detected in coincidence with the scattered beam particle. As noted, the excitation probability is enhanced by smaller impact parameters, which often result in scattering at backward angles in the laboratory frame of reference. Hence, often, annular particle detectors are placed at back angles (say, $140° \leq \theta \leq 170°$). These detectors allow the beam to pass through and then selectively identify those scattering events most likely to have resulted in nuclear excitations.

3. (n, γ) Reactions

Historically, an immense amount of critical data on medium and heavy mass nuclei came from the study of radiative neutron capture, or (n, γ), reactions with reactor neutrons. Like other reactions, such as heavy-ion fusion-evaporation reactions or β-decay that populate nuclear levels from the top down, the process is non-selective and, therefore, gives access to a wide variety of nuclear states. Indeed, when used in the average resonance capture (ARC) mode, the technique can actually guarantee that all states in a given angular momentum and excitation energy range can be identified, thus providing very sensitive tests of models (Caston *et al.*, 1980). Such states can be directly observed from the so-called primary transitions that de-excite the capture state. The use of pair spectrometers is important here. When low-energy gamma-ray spectra are studied, one typically observes hundreds of transitions. Therefore, gamma-gamma coincidence techniques are crucial. Alternatively, many of the most important

(n, γ) studies have used the ultra-high energy resolution GAMS crystal diffraction detectors (see Fig. 3b). Studies of nuclei such as ^{196}Pt (Cizewski *et al.*, 1978) and ^{168}Er (Davidson *et al.*, 1981) with (n, γ) have provided some of the most comprehensive and complete level schemes ever produced and have provided key tests of nuclear models, such as the Interacting Boson Model. Today, most fore-front (n, γ) work is carried out using the GRID technique to measure lifetimes with GAMS detectors.

B. High-Spin States

1. Backbending and the Pauli Principle

One of the fundamental questions to ask about a nucleus is: What is the shape? Is the nucleus spherical, like a soccer ball, or deformed, stretched out like an American foot-ball or perhaps flattened like a Frisbee? It turns out that some nuclei are spherical; some are deformed like foot-balls; and some are deformed like Frisbees. The excitation spectrum of deformed nuclei is particularly easy to un-derstand. A deformed system has a defined orientation in space (it is not isotropic), and rotations of this shape can be observed. A quantum mechanical rotor has an excitation energy given by

$$E(I) = \frac{\hbar^2}{2J} I(I + 1),$$

where I is quantum number counting the angular mo-mentum of the state ($I = 0, 2, 4, \ldots$ for the ground state rotational band, the odd spins are missing for symmetry reasons which are not relevant here) and J is the moment of inertia of the nucleus. The gamma-ray energy (which is measured in the experiment) is just the energy difference between adjacent states.

$$E_\gamma(I \rightarrow I - 2) = \frac{\hbar^2}{2J}[I(I + 1) - (I - 2)(I - 1)]$$

$$= \frac{\hbar^2}{2J}[4I - 2].$$

Thus, the gamma-ray energy increases linearly with angu-lar momentum. For gamma-rays linking adjacent levels, the energy difference is given by

$$\Delta E_\gamma = E_\gamma(I \rightarrow I - 2) - E_\gamma(I - 2 \rightarrow I - 4)$$

$$= \frac{\hbar^2}{2J}[(4I - 2) - 4((I - 2) - 2)]$$

$$= \frac{4\hbar^2}{J}.$$

If the moment of inertia, J, does not change, then ΔE_γ is a constant, independent of spin. Usually, this is not the case in nuclei. A rare example of a nearly ideal rotational

band, where the spacing between adjacent transitions is constant, is shown in Fig. 8 (the most intensely popu-lated superdeformed band in ^{150}Gd). However, usually, dramatic changes in structure occur (e.g., due to centrifu-gal forces or quenching of pairing) as a nucleus rotates faster and faster. These are manifest as deviations from the simple linear dependence outlined above. For exam-ple, a spectrum of the ground state rotational band of ^{158}Er is illustrated in Fig. 9, where the lines indicate transitions linking states with increasing spin. Notice that at gamma-ray energies of about 400 keV the transitions double back on themselves. This phenomenon is called backbending and corresponds to a dramatic change in the internal struc-ture of the nucleus.

The origin of this structural change lies in the effects of the familiar Coriolis force on the microscopic structure of the nucleus. As we have stressed, the nucleus is not a rigid body, but instead is made up of only a few hundred protons and neutrons that orbit the center of mass in orbits char-acterized by particular angular momenta. We know that many medium mass and heavy nuclei exhibit properties similar to those of a superconductor. In the ground state of an even-even nucleus, all of the protons are coupled pairwise, in identical but time-reversed orbits, so that the total angular momentum of each pair is zero. Similarly, the neutrons are also paired. Hence, the total angular momen-tum of the ground state of any even-even nucleus is zero. As an interesting aside, it follows that in an odd-proton or odd-neutron nucleus, the ground state spin and parity is usually determined by the quantum numbers of the final unpaired proton or neutron.

The question, therefore, becomes, what happens to these pairs of protons and neutrons as the nucleus as a whole begins to rotate? Just as a person walking on a merry-go-round experiences a force on the moving plat-form, the so-called Coriolis force, the nucleons in the nu-cleus also experience the effect of the rotating bulk. Just as with the merry-go-round, the Coriolis force increases the faster the nuclear rotation or the orbital velocity. In-deed, the size of the Coriolis force is such that at moderate nuclear rotational frequencies it perturbs the orbits of the particles sufficiently that the pairs of nucleons will begin to break apart. This has the effect of dramatically chang-ing the excitation energies of the states and the gamma-ray energies for transitions between them. It is this breaking of the superconducting pairs that is responsible for the backbending observed in Fig. 9.

An illustration of the effects of the Pauli exclusion principle can be seen in the rotational spectra of odd-even nuclei. Figure 10 is a plot of the angular momentum as a function of rotational frequency for ^{133}Pr, which has 59 protons and 74 neutrons. The two curves shown in Fig. 10 correspond to rotational bands in which the final

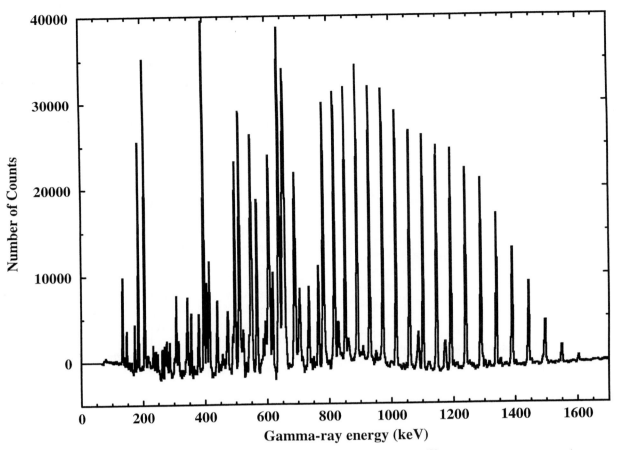

FIGURE 8 Spectrum of the most intensely populated superdeformed band in ^{150}Gd. In addition to the regular picket-fence pattern of gamma-rays associated with decays of superdeformed states, the spectrum also shows, at lower energies, the complex pattern of transitions depopulating the nearly spherical normal deformed states in ^{150}Gd.

unpaired proton is in different orbits about the nucleus. In one of these cases the odd-proton acts like a spectator to the underlying even-even nucleus, and in this case the above backbending phenomenon occurs as before. In the other band, however, the odd-proton occupies one of the orbits of the pair of aligning nucleons. The pair breaking is therefore prohibited by the Pauli exclusion principle, and the backbending is delayed until higher rotational frequencies when it becomes possible to occupy higher lying orbits.

2. Superdeformation

One of the forefront areas of research in high-spin nuclear structure physics over the last decade has been the study of superdeformed (SD) nuclei. These states exist in a second minimum in the nuclear potential energy surface in which the nucleus takes on an ellipsoidally deformed shape which roughly corresponds to an integer ratio of

major to minor axes, typically 2:1 or 3:2. The observation of the first high-spin SD bands in ^{152}Dy and ^{132}Ce, by the Liverpool University groups of Peter Twin and Paul Nolan, respectively (Twin *et al.*, 1986; Nolan *et al.*, 1985), sparked an enormous worldwide effort to discover additional examples of highly deformed nuclei and to characterize the properties of such highly stressed systems (stressed both by the application of very high angular momenta and by extreme values of deformation). Today, about 40 nuclei in four main mass regions, or islands, have been shown to exhibit SD behavior. Most of these nuclei have more than one known SD band.

Superdeformed rotational bands are generally characterized by extremely regular gamma-ray energy spacing. The energy spacing from one transition to the next in the rotational band is either constant or varying slowly and regularly from one transition to the next. For example, the strongest SD band in ^{150}Gd is illustrated in Fig. 8. The regular picket-fencelike pattern of SD transitions is

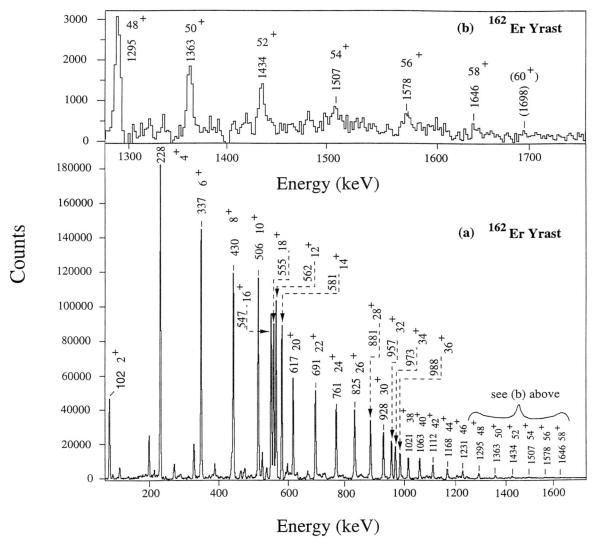

FIGURE 9 Spectrum illustrating the ground state rotational band of ^{162}Er illustrating the backbending phenomenon. (Figure courtesy of Mark Riley).

unmistakable in this spectrum as is the irregular pattern of transitions de-exciting lower lying normal deformed states (^{150}Gd is nearly spherical in its ground state).

Due to this regularity, which is the rule rather than the exception for SD bands, one can feel confident in predicting where transitions in a given band should occur. However, detailed measurements of the strongest SD band in ^{149}Gd, using the Eurogam Ge detector array, revealed a very small deviation from this smooth behavior (Flibotte et al., 1993), Indeed, it was found that every second energy spacing was larger/smaller than the average. The deviation, illustrated in Fig. 11, is very small, only about 0.25 keV, and is measurable only due to the very high quality spectra available from the Eurogam array. It is believed that the deviation is caused by alternate states in

the rotational band being perturbed up and down in energy by very small amounts, on the order of 60 eV. This staggering essentially separates the rotational band into two $\Delta I = 4\,\hbar$ sequences. The origin of the perturbation, which affects states differing in spin by 4 $\hbar$, is still unclear. Several theoretical models have been proposed to explain this phenomena, none of which, however, can reliably predict which SD bands should exhibit staggering and which should not.

3. Magnetic Rotation and Chiral Symmetry

Interesting effects have also emerged from the study of near-spherical nuclei. One of the consequences of quantum mechanics is that the rotation of a spherical shape

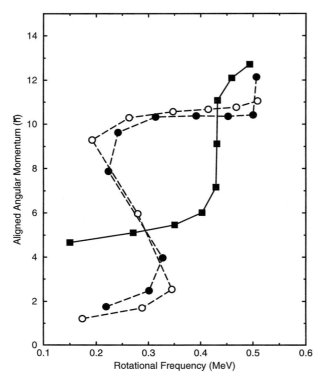

FIGURE 10 Component of the angular momentum along the rotation axis vs rotational frequency for different rotational bands in ^{133}Pr. Notice that the backbending observed at a frequency of ~0.25 MeV is completely missing in one of the bands. The absence of a crossing in this band is a dramatic illustration of the Pauli exclusion principle.

cannot be observed. How then does a spherical or near-spherical nucleus generate angular momentum? Rather than a collective rotation of the whole shape, it does so by rearranging the orbits of its constituent protons and neutrons, by single particle excitations to higher lying excited states with high values of angular momentum. Typically, such excitations have irregularly spaced energies resulting in a gamma-decay spectrum with many irregularly spaced peaks (see the lower energy portion of Fig. 8).

It was a surprise, therefore, when regularly spaced sequences of gamma-rays were observed in some almost spherical light Pb nuclei, near the doubly closed shell ^{208}Pb. Furthermore, these apparently rotational-like cascades were found to consist of very strong magnetic dipole (M1) transitions which change the angular momentum by $\Delta I = 1\,\hbar$, with very weak, or unobserved, $\Delta I = 2\,\hbar$ electric quadrupole transitions (E2). In contrast, a rotational band in a well-deformed nucleus consists of a sequence of strong $\Delta I = 2\,\hbar$ E2 transitions. The absence of E2 transitions in these new bands is an indication of the near-spherical nuclear shape. However, the regularity of the new band structure implies a type of collective behavior.

The tilted axis-cranking model provides an explanation for these bands (Frauendorf, 1993). For certain near-spherical nuclei with proton and neutron particle numbers close to magic numbers, the angular momentum vectors of the unpaired proton particles and neutron holes prefer to align perpendicular to each other, with one vector pointing along the rotation axis and the other perpendicular to the rotation axis. The vector sum of these two angular momenta then lies at an angle to the nuclear symmetry axis. Furthermore, the vector accounts for almost all of the nuclear angular momentum, since the collective rotation of the near-spherical shape is small. Higher angular momentum states are generated by slowly closing these two angular momentum blades, or shears, pushing against the repulsive particle-hole nuclear interaction. The enhanced M1 transitions arrise because the magnetic dipole moment is proportional to the component of the individual proton and neutron angular momenta perpendicular to the total angular momentum.

An interesting extension of the idea of tilted axis cranking comes when we consider the possibilities in doubly odd deformed, triaxial nuclei. As for shears bands, for certain favorable particle numbers the angular momenta of the final unpaired proton and neutron align preferentially perpendicular to each other, along the nuclear rotation (short) and symmetry (long) axes. For a triaxial nuclear shape, considerations of irrotational flow indicate that the collective angular momentum should align preferentially with the intermediate length nuclear axis. Thus, the three angular momentum vectors can form either a

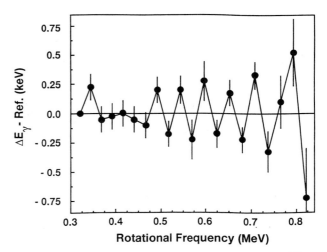

FIGURE 11 Energy staggering in the strongest SD band in ^{149}Gd (Flibotte *et al.*, 1993). The figure shows the deviation of the measured gamma-ray energy from a smooth reference as a function of rotational frequency. Notice that the deviation is extremely small, usually less than 0.25 keV. This deviation corresponds to a tiny perturbation in the nuclear energy levels of only about 60 eV.

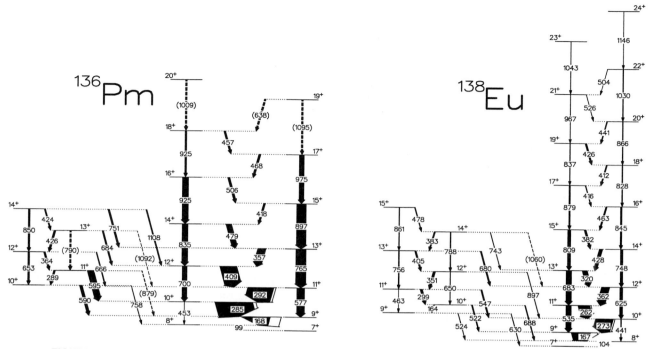

FIGURE 12 A partial level scheme of the odd-odd nuclei ^{136}Pm (61 protons and 75 neutrons) and ^{138}Eu (63 protons and 75 neutrons). The proposed chiral twin bands are shown on the left of each level scheme.

right- or a left-handed coordinate system. The so-called 3d-tilted axis-cranking model, developed by S. Frauendorf and J. Meng (1997), addresses such a system and predicts a doubling of energy levels, one corresponding to each chirality or handedness. For complete symmetry, the levels of the same spin and parity would be degenerate. If the solutions for different chiralities mix, then the degeneracy will be broken, and one set of states, corresponding to a $\Delta I = 1\,\hbar$ rotational band, will be lifted with respect to the second band. Indeed, two $\Delta I = 1\,\hbar$ bands in the doubly-odd nucleus ^{134}Pr have been proposed as a possible chiral candidate (Frauendorf and Meng, 1997). Following on this suggestion, several other candidate bands have been observed in nearby nuclei (Starosta *et al.*, 2001; Beausang *et al.*, 2001; Hecht *et al.*, 2001), while candidate bands have also recently been reported in doubly-odd ^{188}Ir (Balabanski *et al.*, unpublished). The proposed chiral twin bands in ^{136}Pm and ^{138}Eu are shown in Fig. 12.

C. Spectroscopy in Coincidence with Separators

A great deal of exciting new spectroscopy of nuclei far from stability or with very large Z has been achieved over the last several years when large Ge detector arrays have

been coupled to high-transmission magnetic separators. A magnetic separator is a device placed behind the target position which will selectively transport nuclei, produced in a reaction, to its focal plane where they can be detected and identified using a variety of different detectors. Residual nuclei that are not of interest, or scattered beam particles, will not be transmitted through the separator. Very small fractions of the total reaction cross section can be selected using this method. Nuclear structure information is obtained by detecting gamma-rays produced at the target position, in coincidence with recoils detected at the focal plane. One example of the use of this technique is illustrated here.

One of the goals of nuclear physics is to understand the limits of nuclear existence as functions, for example, of angular momentum, isospin, or indeed mass. For example, what are the heaviest nuclei that can exist? For many years now, various models have predicted that an island of superheavy nuclei should exist. However, most models disagree as to the exact proton and neutron numbers categorizing this island and indeed on the extent of the island. Recently, models have predicted that these superheavy nuclei might indeed be deformed. Therefore, it is very relevant to inquire as to what is the structure of the heaviest nuclei accessible to gamma-ray spectroscopy and to ask the simplest type of questions about them, for example, are they spherical or deformed? Unfortunately, the production cross sections for superheavy nuclei are such that, even using very intense beams, only one or two nuclei are produced per week or two. These small numbers are clearly beyond what we can measure with existing gamma-ray facilities. Therefore, we cannot address the spectroscopy of the superheavy elements (yet). However, we can look at the structure of very heavy nuclei lying just below these unattainable regions.

Recently, groups at Argonne National Laboratory in the United States and at the University of Jyvaskyla in Finland carried out tour de force experiments to study the excitation spectrum of ^{254}No (Leino *et al.*, 1999). With $Z = 102$, No is the heaviest nucleus for which gamma-ray spectroscopy has ever been carried out. The gamma-ray spectrum of transitions de-exciting states in ^{254}No is shown in Fig. 13. A rotational band structure is clearly visible, indicating that ^{254}No is in fact a deformed nucleus. A very surprising feature of the spectrum is that the rotational band is observed up to very high spins $\sim 18 \, \hbar$, (an amazing number for such a heavy, fissile nucleus). The existence of a rotational cascade up to spin $\sim 18 \, \hbar$, well beyond the classical fission barrier limit, indicates that ^{254}No is held together primarily by microscopic shell effects, rather than macroscopic liquid drop binding, as in normal nuclei. Shell effects, for certain favorable proton and neutron numbers and for favorable deformation, can provide an additional 1–2 MeV of binding energy. It is this binding energy, which does not depend strongly on angular momentum, which holds ^{254}No together to such high spin.

D. Experiments with Radioactive Beams

Today, a new era in nuclear structure physics is opening up with access to a much wider selection of nuclei, extending far beyond the valley of stability and encompassing nuclei that are expected to be exotic in both proton/neutron composition and structure. The physics opportunities with such beams have been discussed elsewhere (RIA Physics White Paper, 2000) and need not be repeated here. What are relevant are the particular methods of carrying out gamma-ray spectroscopy on exotic nuclei. Basically, the techniques to be used will be familiar ones, such as β-decay, Coulomb excitation, and fusion-evaporation reactions. High-, medium-, and low-spin states will all present topics of interest.

Experiments with radioactive beams differ primarily in two critical respects from their stable beam siblings. First, beam intensities will often be much lower than with stable beams. Instead of beams of 10^{11} particles per second, many experiments will need to be carried out with intensities that are less than 10^6 particles per second and, at

the limits of accessibility, down to 1 particle per second or even less. Therefore, detectors will have to be correspondingly more efficient. Second, because the nucleus to be studied is sometimes the one produced as a beam by the radioactive beam facility, most experiments will be done in inverse kinematics in which the roles of beam and target are interchanged.

In inverse kinematics, $m_b > m_t$ where m_b and m_t are the masses of the beam and target nuclei. Therefore, the reaction products all go forward in the laboratory system. For $m_b \gg m_t$, this forward focusing results in a quite narrow cone of reaction products. For example, for elastic scattering of ^{62}Ni on ^{12}C, the maximum allowed scattering angle is $\sim 10°$. This has two principal effects. First, measuring angular distributions of reaction products is much more difficult. Second, on the other hand, it is possible to capture much larger percentages of the reaction products in the acceptance angles of various types of charged particle spectrometers and mass separators, thereby enhancing counting rates. These considerations impose design constraints on gamma-ray detectors surrounding the target. First of all, ultra-high efficiency is needed. Second, generally, a forward angled cone needs to be left free of detectors.

The requirement of maximal gamma-ray counting efficiency generally means a close geometry and detectors that subtend large solid angles. However, Doppler effects can then be very large, especially when using inverse kinematics, and high detector granularity will generally be critical. This granularity can currently be achieved in two ways and considerable development in both directions is needed. One is the use of highly segmented tracking arrays such as GRETA discussed earlier. The other is the use of position-sensitive Ge detectors of the type developed by Glasmacher and colleagues for use in intermediate energy Coulomb excitation experiments at MSU (Muller *et al.*, in press). In these detectors a resistive readout at the two ends of a linear Ge crystal allows the localization of the γ-ray interaction to an accuracy of ~ 2 mm. This detector is capable of measurements even with beam intensities of ~ 1 particle per second or even less. An example of a gamma-ray spectrum (corrected for Doppler effects) from intermediate energy Coulomb excitation taken with an early generation detector system (using NaI(Tl) detectors) is shown in Fig. 14. This data was taken on ^{40}S in order to test predictions of the underlying particle motion in exotic nuclei with a high excess of neutrons over protons (the heaviest stable isotope of sulfur has 20 neutrons, ^{40}S has 24 neutrons). The Coulomb excitation was accomplished in this case using a ^{197}Au target.

For specialized experiments, such as low-energy Coulomb excitation in inverse kinematics designed explicitly to excite only the lowest one or two states,

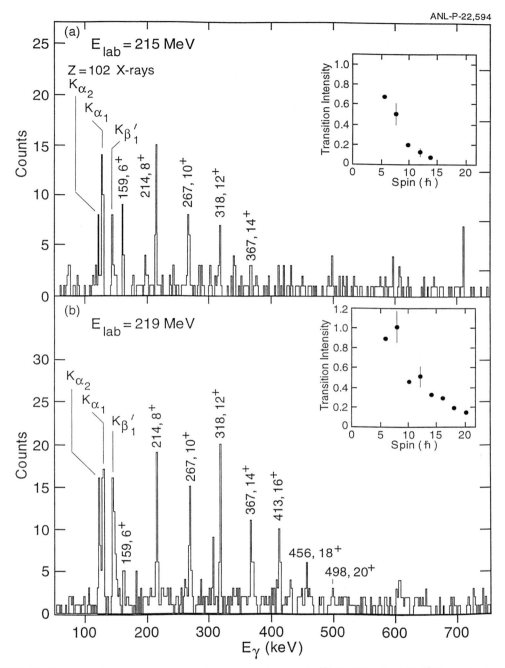

ANL-P-22,594

FIGURE 13 Spectrum of gamma-rays depopulating excited states in ^{254}No. With $Z = 102$, ^{254}No is the heaviest nucleus for which gamma-ray spectroscopy has ever been accomplished. The gamma-rays are labeled by their transition energies in kilo-electron volts and also by the spin of the state they depopulate. The inserts show the population intensity as a function of spin for the two beam energies, 215 (top) and 219 MeV (bottom). More angular momentum is brought into the system at higher beam energies, and this is reflected in the stronger population of higher spin states in the lower spectrum (Leino *et al.*, 1999).

the gamma-ray spectra are particularly simple. Therefore, energy resolution is not a problem and high efficiency can be obtained with, for example, low-resolution NaI(Tl) detectors placed in close geometry. Here, Doppler effects are both unimportant and undetected. One typical design, the GRAFIK detectors (Sheit *et al.*, 1996) actually incorporates the target inside an annular hole in the detector, achieving ~80% of 4π solid angle coverage.

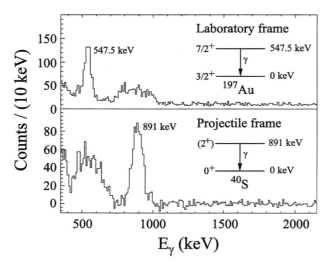

FIGURE 14 Gamma-ray spectra following intermediate energy Coulomb excitation of a radioactive ^{40}S beam on a ^{197}Au target. The spectra are corrected for Doppler effects for gamma-rays emitted from nuclei at rest in the laboratory frame (top) or from nuclei moving at the beam velocity (bottom). The gamma-ray de-exciting the first excited state of ^{40}S is clearly visible in the lower spectrum. [From Muller *et al.* (in press). *Nucl. Instrum. Methods.*]

IV. CONCLUSIONS

Although the dominant interaction binding nucleons into nuclei is the strong force, the electromagnetic interaction, as manifested primarily in gamma-ray spectroscopy, provides an ideal probe of the structure and excitations of the nucleus. Indeed, gamma-ray spectroscopy in nuclear and astrophysics research is a broad and diverse field, utilizing a variety of detector systems that vary greatly (according to the needs of particular experiments) in resolution, efficiency, gamma-ray energy range, and other properties. These detectors are often used alone or in conjunction with auxiliary devices such as charged particle or neutron detectors. The areas of nuclear structure addressed with such instrumentation cover the whole gamut of nuclei spanning the entire nuclear chart and physics problems,, ranging from the motion of individual nucleons to collective flows (e.g., vibrations or rotations) of the nucleus as a whole.

ACKNOWLEDGMENT

We are grateful to many colleagues for advice and discussions and, in particular, to Mark Caprio, Thomas Glasmacher, Hans Borner, and Mark Riley for providing the figures we have used. This work was supported in part by the U.S. DOE grant number DE-FG02-91ER-40609.

SEE ALSO THE FOLLOWING ARTICLES

GAMMA-RAY ASTRONOMY • ION BEAMS FOR MATERIAL ANALYSIS • NUCLEAR PHYSICS • POTENTIAL ENERGY SURFACES

BIBLIOGRAPHY

Barton, C. J., et al. (1997). *Nucl. Instrum. Methods* **A391,** 289.

Balabanski, D., et al., to be published.

Beausang, C. W., et al. (1992). *Nucl. Instrum. Methods* **A313,** 37; Beck, F. A., et al. (1992). *Prog. Part. Nucl. Phys.* **28,** 443; Nolan, P. J. (1990). *Nucl. Phys.* **A520,** 657c.

Beausang, C. W., et al. (2000). *Nucl. Instrum. Methods* **A452,** 431.

Beausang, C. W., et al. (2001). *Nucl. Phys.* **A682,** 394c.

Buescher, M., et al. (1990). *Phys. Rev.* **C41,** 1115; Mach, H., et al. (1990). *Phys. Rev.* **C41,** 1141.

Borner, H. G., and Jolie, J. (1993). *J. Phys.* **G19,** 217.

Casten, R. F., et al. (1980). *Phys. Rev. Lett.* **45,** 1077.

Casten, R. F. (2000). *Nucl. Phys. News Int.* **10,** 4.

Casten, R. F., and Nazarewicz, W. (2000). "White Paper for the RIA Workshop, Raleigh-Durham, North Carolina, July 24–26, 2000."

Cizewski, J. A., et al. (1978). *Phys. Rev. Lett.* **40,** 167.

Davidson, W. F., et al. (1981). *J. Phys.* **G7,** 443, 455.

Deleplanque, M. A., et al. (1999). *Nucl. Instrum. Methods* **A430,** 292.

Duchene, G., et al. (1999). *Nucl. Instrum. Methods* **A432,** 90.

Eberth, J., et al. (1996). *Nucl. Instrum. Methods* **A369,** 135.

Frauendorf, S. (1993). *Nucl. Phys.* **A557,** 259c.

Frauendorf, S., and Meng, J. (1997). *Nucl. Phys.* **A617,** 131.

Flibotte, S., et al. (1993). *Phys. Rev. Lett.* **71,** 4299.

Gerl, J., and Lieder, R. (1992). "Euroball III," GSI Darmstadt Report. Darmstadt, Germany.

Hecht, A., et al. (2001). *Phys. Rev.* **C63,** 051302(R).

Johnson, A., Ryde, H., and Sztarkier, J. (1971) *Phys. Lett.* **B34,** 605.

Koch, H. R., et al. (1980). *Nucl. Instrum. Methods* **175,** 401; Kessler, E. G., et al. (2001). *Nucl. Instrum. Methods* **A457,** 187.

Lee, I. Y. (1990). *Nucl. Phys.* **A520,** 641c; Deleplanque, M. A., and Diamond, R. M., eds. (March 1988). "The Gammasphere Proposal: A National Gamma-Ray Facility," LBL, Berkeley, CA.

Leino, M., et al. (1999). *Eur. Phys. J.* **A6,** 63; Reiter, P., et al. (1999). *Phys. Rev. Lett.* **82,** 509.

Morinaga, H., and Gugelot, P. C. (1963). *Nucl. Phys.* **46,** 210.

Muller, W. F., et al. (in press). *Nucl. Instrum. Methods.*

Nolan, P. J., Gifford, D. W., and Twin, P. J. (1985). *Nuicl. Instrum. Methods* **A236,** 95.

Nolan, P. J., et al. (1985). *J. Phys.* **G11,** L17.

Riedinger, L. L., et al. (1980). *Phys. Rev. Lett.* **44,** 568.

Sheit, H., et al. (1996). *Phys. Rev. Lett.* **77,** 3967.

Simpson, J., et al. (2000). *Heavy Ion Phys.* **11,** 159.

Starosta, K., et al. (2001). *Phys. Rev. Lett.* **86,** 971.

Twin, P. J., et al. (1986). *Phys. Rev. Lett.* **57,** 811.

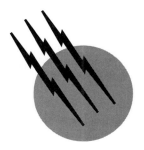

Gas Chromatography

Milos Novotny
Indiana University

GLOSSARY

Mobile phase Gas containing the compounds to be separated while migrating through the stationary phase.
Stationary phase Solid or liquid whose physical properties affect the separation of the mobile-phase compound.

GAS CHROMATOGRAPHY (GC) is a physical separation method used extensively in scientific investigations, chemical practice, petroleum technology, environmental pollution control, and modern biology and medicine. Its primary role is to separate various chemical compounds that were introduced into the system as a mixture and to determine quantitatively their relative proportions. When combined with other physicochemical methods, GC can also provide qualitative (structural) information on the separated substances. The method is limited to relatively volatile (low-molecular-weight) compounds. The principle of separation is a relative affinity of various mixture components to the stationary phase (a solid or a liquid), while the mobile phase (a gas) migrates them through the system. GC is a dynamic separation method, where the separation of components occurs in a heterogeneous phase system.

I. GAS CHROMATOGRAPHIC MOLECULAR SEPARATION

Separating chemical substances from each other has been extremely important to various branches of science and technology for many years. Simple separation procedures such as distillation, crystallization, precipitation, and solvent extraction have been used by humankind from time immemorial. More refined forms of separation, such as chromatography and electrophoresis, have been among the major causes of scientific revolution during the last several decades of this century.

Gas chromatography is one of the several chromatographic methods. The scientific principles of

Encyclopedia of Physical Science and Technology, Third Edition, Volume 6

TABLE I Comparison of Types of Chromatography

Mobile phase	Stationary phase	Types of chromatography[a]	Abbreviation	Separation
Liquid	Solid	Liquid–Solid	LSC[b]	Adsorption
Liquid	Liquid (immiscible)	Liquid–Liquid	LLC	Solubility (partition)
Gas	Solid	Gas–Solid	GSC	Adsorption
Gas	Liquid	Gas–Liquid	GLC	Solubility (partition)

[a] Gas–gas and solid–solid equilibria do occur in nature; however, they are impractical for chromatographic separations.

[b] C = Chromatography.

chromatography were discovered by a Russian botanist M. S. Tswett (1872–1919) but hardly developed into useful chemical separation procedures until the 1930s. The name *chromatography* was originated by Tswett who primarily investigated plant pigments (*chromatos* is the Greek name for color). However, any method that utilizes a distribution of the molecules to be separated between the mobile phase (a gas or a liquid) and the stationary phase (a solid or a liquid that is immiscible with the mobile phase) now qualifies as chromatography. The physical state of the mobile phase determines whether we deal with gas or liquid chromatography.

Variation in the type of stationary phase is important as well: if a solid is used as the stationary phase, the interaction of the molecules under separation with it is due to adsorption forces; if a liquid is used in the same capacity, the molecules under separation interact with it based on their solubilities. According to this type of interaction, we distinguish between adsorption chromatography and partition chromatography. This classification is further evident in Table I.

Tswett's original work pertained to liquid adsorption chromatography, while the first experiments on liquid partition chromatography were described in the early 1940s by A. J. P. Martin and his co-workers in Great Britain

(a decade later, recognized by a Nobel Prize in Chemistry). Several investigations pertaining to the use of gas as the mobile phase in gas/adsorption systems were reported in Austria, Czechoslovakia, Russia, and Sweden during the 1940s. However, the development of gas–liquid chromatography, reported in 1952 by A. T. James and A. J. P. Martin, is widely considered the beginning of GC as a powerful analytical method.

Today, GC is complementary to other separation methods. It can be practiced on either a small (analytical) scale or a large (preparative or industrial) scale. The preparative uses of GC are relatively uncommon. While typical amounts of chemical substances analyzed by the modern GC are between the microgram (10^{-6} g) and nanogram (10^{-9} g), samples as small as a femtogram (10^{-15} g) can be measured in special circumstances. Importantly, contemporary GC can often simultaneously recognize up to several hundred chemical substances.

II. PHYSICAL PRINCIPLES

The apparatus designated to separate compounds by GC is called the gas chromatograph. Its essential parts are shown in Fig. 1. At the heart of the system is the separation

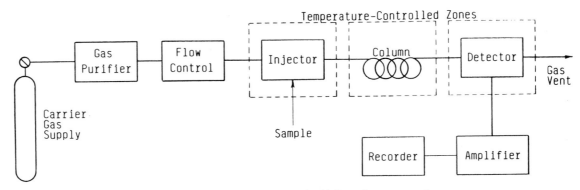

FIGURE 1 A gas chromatograph with its main components.

column, at which the crucial physicochemical processes of the actual compound separation occur.

The separation column contains the stationary phase, while the mobile phase (frequently referred to as the carrier gas) is permitted to flow through this column from a pressurized gas cylinder (source of the mobile phase). The rate of mobile-phase delivery is controlled by a pressure and/or flow-regulating unit. An exclusive separation mode for the analytical GC is elution chromatography, in which the sample (a mixture of chemicals to be separated) is introduced at once, as a sharp concentration impulse (band), into the mobile-phase stream. The unit where sample introduction is performed is called the injector. The unfractionated sample is transferred from the injector into the chromatographic column, where it is subjected to a continuous redistribution between the mobile phase and the stationary phase. Due to their different affinities for the stationary phase, the individual sample components eventually form their own concentration bands, which reach the column's end at different times. A detector is situated at the column's end to sense and quantitatively measure the relative amounts of these sample components.

The detector, together with auxiliary electronic and recording devices, is instrumental in generating the chromatogram, shown in Fig. 2. Such a chromatogram is, basically, a plot of the sample concentration versus time. It represents the individual component bands, separated by the chromatographic column and modified by a variety of physical processes into a peak shape. The position of a peak on the time scale of the total chromatogram bears some qualitative information, since each chromatographic peak represents at least one chemical substance. The areas under the peaks are, however, related to the amounts of individual substances separated in time and space.

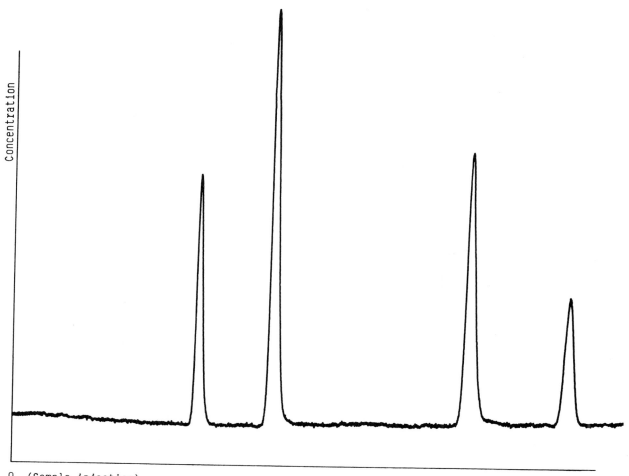

FIGURE 2 Chromatogram.

A typical gas chromatograph has three independently controlled thermal zones: proper temperature of the injector zone ensures rapid volatilization of the introduced sample; the column temperature is controlled to optimize the actual separation process; and the detector must also be at temperatures where the individual sample components are measured in the vapor phase. For certain GC separations, it is advisable to program the temperature of the chromatographic column.

As shown in Fig. 2, different sample components appear at the column's end at different times. The retention time t_R is the time elapsed between injection and the maximum of a chromatographic peak. It is defined as

$$t_R = t_0(1 + k), \tag{1}$$

where t_0 is the retention time of a mixture component that has no interaction with the stationary phase (occasionally referred to as dead time), and k is the capacity factor. The capacity factor is further defined as

$$k = K \frac{V_s}{V_M}, \tag{2}$$

where K is the solute's distribution coefficient (pertaining to a distribution between the stationary phase and the mobile phase), V_s is the volume of the stationary phase, and V_M is the volume of the mobile phase in a chromatographic column. The distribution coefficient $K = C_s/C_M$ (where C_s is the solute concentration in the stationary phase and C_M is the solute concentration in the mobile phase) is a thermodynamic quantity that depends on temperature as do all equilibrium constants. The molecular interactions between the phases and the solutes under separation are strongly temperature-dependent. If, for example, a solid adsorbent (column material) is brought into contact with a permanent (inorganic) gas and a defined concentration of organic (solute) molecules in the gas phase at a certain temperature, some solute molecules become adsorbed on the solid, and others remain in the permanent gas. When we elevate the system temperature, less solute molecules are adsorbed, and more of them join the permanent gas; the distribution (adsorption) coefficient, as defined above, changes correspondingly. Likewise, if the stationary phase happens to be a liquid, the solute's solubility in it decreases with increasing temperature, according to Henry's law, resulting in a decrease of the distribution (partition) coefficient.

According to Eqs. (1) and (2), the retention time in GC depends on several variables: (a) the chemical nature of the phase system and its temperature, as reflected by the distribution coefficient; (b) the ratio of the phase volumes in the column V_s/V_M; and (c) the value of t_0. In the practice of chromatography, these variables are used to maximize the component separation and the speed of analysis.

Unlike some other chromatographic processes, the physical interactions between the mobile phase and solute molecules in GC are, for all practical purposes, negligible. Thus, the carrier gas serves only as means of molecular (solute) transport from the beginning to the end of a chromatographic column. The component separation is then primarily due to the interaction of solute molecules with those of the stationary phase. Since a variety of column materials are available, various molecular intertactions can now be utilized to enhance the component separation. Moreover, these interactions are temperature-dependent.

For the mixture component with no affinity for the stationary phase, the retention time t_0 serves merely as the marker of gas linear velocity u (in cm/s) and is actually defined as

$$t_0 = \frac{L}{u}, \tag{3}$$

where L is the column length. The gas velocity is, in turn, related to the volumetric flow rate F since

$$u = \frac{F}{s}, \tag{4}$$

where s is the column cross-sectional area. The gas-flow rate is chiefly regulated by the inlet pressure value; the higher the inlet pressure the greater the gas-flow rate (and linear velocity) becomes, and consequently, the shorter t_0 is. The retention time t_R of a retained solute is also modified accordingly. Correspondingly, fast GC separations are performed at high gas-inlet pressures. The so-called retention volume V_R is a product of the retention time and volumetric gas-flow rate:

$$V_R = t_R F. \tag{5}$$

Since the retention times are somewhat indicative of the solute's nature, a means of their comparison must be available. Within a given chemical laboratory, the relative retention times (the values relative to an arbitrarily chosen chromatographic peak) are frequently used:

$$\alpha_{2,1} = \frac{t_{R_2}}{t_{R_1}} = \frac{V_{R_2}}{V_{R_1}} = \frac{K_2}{K_1}. \tag{6}$$

This equation is also a straightforward consequence of Eqs. (1) and (2). Because the relative retention represents the ratio of distribution coefficients for two different solutes, it is frequently utilized (for the solutes of selected chemical structures) as a means to judge selectivity of the solute–column interactions.

For interlaboratory comparisons, the retention index appears to provide the best method for documenting the GC properties of any compound. The retention index system compares retention of a given solute (on a logarithmic scale) with the retention characteristics of a set of standard solutes that are the members of a homologous series:

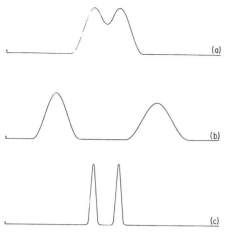

FIGURE 3 Enhancement of component resolution as based on the selectivity and efficiency of the separation process: (a) two unresolved components, (b) resolution based on the column selectivity, and (c) resolution based on the column kinetic efficiency.

$$I = 100z + 100\frac{\log t_{R(x)} - \log t_{R(z)}}{\log t_{R(z+1)} - \log t_{R(z)}}. \qquad (7)$$

The subscript z represents the number of carbon atoms within a homologous series, while x relates to the unknown. For example, a series of n-alkanes can be used in this direction; each member of a homologous series (differing in a single methylene group) is assigned an incremental value of 100 (e.g., 100 for methane, 200 for ethane, and 300 for propane, etc.) and if a given solute happens to elute from the column exactly half-way between ethane and propane, its retention index value is 250). Retention indices are relatively independent of the many variables of a chromatographic process.

The success of GC as a separation method is primarily dependent on maximizing the differences in retention times of the individual mixture components. An additional variable of such a separation process is the width of the corresponding chromatographic peak. Whereas the retention times are primarily dependent on the thermodynamic properties of the separaton column, the peak width is largely a function of the efficiency of the solute mass transport from one phase to the other and of the kinetics of sorption and desorption processes. Figure 3 is important to understanding the relative importance of both types of processes.

In Fig. 3, (a) depicts a situation where two sample components are eluted too closely together, so that the resolution of their respective solute zones is incomplete; (b) represents a situation where the two components are resolved from each other through choosing a (chemically) different stationary phase that retains the second component more strongly than the first one; and (c), which shows the same component retention but much narrower chromatographic peaks, thus represents the most "efficient" handling of the two components. This efficiency, represented by narrow chromatographic zones, can actually be attained in GC practice by a proper design in physical dimensions of a chromatographic column. Width of a chromatographic peak is determined by various column processes such as diffusion of solute molecules, their dispersion in flow streamlines of the carrier gas, and the speeds by which these molecules are transferred from one phase to another.

An arbitrary, but the most widely used, criterion of the column efficiency is the number of theoretical plates, N. Figure 4 demonstrates its determination from a

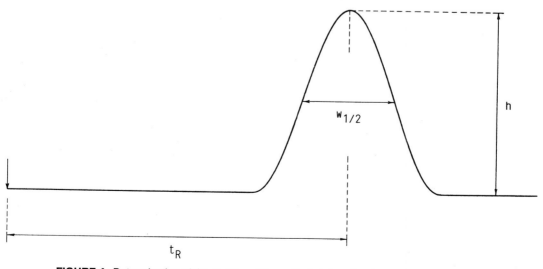

FIGURE 4 Determination of the number of theoretical plates of a chromatographic column.

chromatographic peak. This number is simply calculated from the measured retention distance t_R (in length units) and the peak width at the peak half-height $W_{1/2}$:

$$N = 5.54\left(\frac{t_R}{w_{1/2}}\right)^2. \qquad (8)$$

The length of a chromatogrphic column L is viewed as divided into imaginary volume units (plates) in which a complete equilibrium of the solute between the two phases is attained. Obviously, for a given value of t_R, narrower peaks provide greater numbers of theoretical plates than broader peaks. Turning once again to Fig. 3, we see that cases (a) and (b) represent low column efficiencies (plate numbers), while case (c) demonstrates a high-efficiency separation.

Equation (8), used to determine the number of theoretical plates, relates to a perfectly symmetrical peak (Gaussian distribution). While good GC practice results in peaks that are nearly Gaussian, departures from peak symmetry occasionally occur. In Fig. 5, (a) is usually caused by a slow desorption process and undesirable interactions of the solute molecules with the column material, and (b) is associated with the phenomenon of column overloading (if the amount of solute is too large, exceeding saturation of the stationary phase, a fraction of the solute molecules is eluted with a shorter retention time than the average). When feasible, GC should be carried out at the solute concentrations that give a linear distribution between the two phases.

The length element of a chromatographic column occupied by a theoretical plate is the plate height (H):

$$H = \frac{L}{N}. \qquad (9)$$

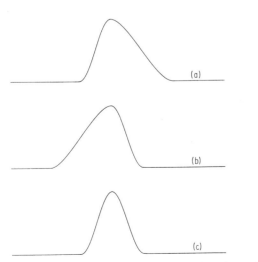

FIGURE 5 Departures from peak symmetry: (a) slow desorption process and (b) column overloading. (c) Gaussian distribution.

The column efficiency N can be dependent on a number of variables. Most importantly, the plate height is shown to be a function of the linear gas velocity u according to the van Deemter equation:

$$H = A + \frac{B}{u} + Cu, \qquad (10)$$

where the constant A describes the chromatographic band dispersion caused by the gas-flow irregularities in the column. The B-term represents the peak dispersion due to the diffusion processes occurring longitudinally inside the column, and the C-term is due to a flow-dependent lack of the instantaneous equilibrium of solute molecules between the gas and the stationary phase. The mass transfer between the two phases occurs due to a radial diffusion of the solute molecules.

Equation (10) is represented graphically by a hyperbolic plot, the van Deemter curve, in Fig. 6. The curve shows the existence of an optimum velocity at which a given column exhibits its highest number of theoretical plates. Shapes of the van Deemter curves are further dependent on a number of variables: solute diffusion rates in both phases, column dimensions and various geometrical constants, the phase ratio, and retention times. Highly effective GC separations often depend on thorough understanding and optimization of such variables.

III. SEPARATION COLUMNS

Since the introduction of GC in the early 1950s, many different column types have been developed, as is widely documented by numerous column technology studies reported in the chemical literature. The column design is extremely important to the analytical performance and utility for different sample types and applications. The most important features include (a) type of column sorption material (in both physical and chemical terms), (b) column diameter, (c) column length, and (d) surface characteristics of a column tubing material. A proper combination of these column design features can often be crucial to a particular chemical separation.

Based on their constructional features, GC columns can be divided into three main groups: packed columns, capillary (open tubular) columns, and porous-layer open tubular columns. Their basic geometrical characteristics are shown in Fig. 7.

A packed column is basically a tube, made from glass or metal, that is filled with a granular column material. The material is usually held in place by small plugs of a glass wool situated at each column's end. During a GC run, such a column is attached to the instrument through a gas-tight connection; the carrier gas is forced through the

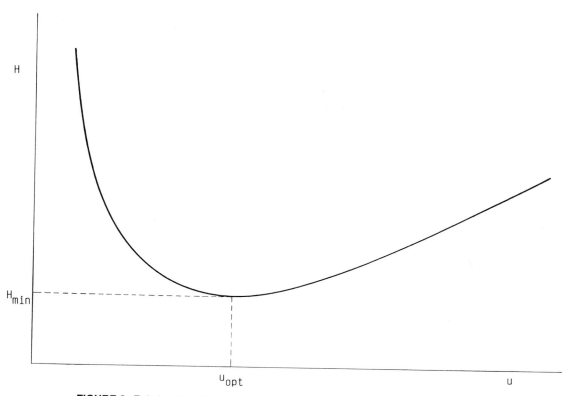

FIGURE 6 Relationship of the plate height and linear gas velocity (van Deemter curve).

free channels between the individual particles, while the sample molecules are allowed to interact with the particles. Typical column inner diameters are 1–4 mm, and the lengths are 1–3 m, although departures from these dimensions may exist for special applications. The inner diameters of preparative columns can be considerably larger.

The granular packing can be either an adsorbent (if the method of choice is gas–solid chromatography) or an inert solid support that is impregnated with a defined amount of a liquid stationary phase (for gas–liquid chromatography). In either case, packing materials with uniformly small particles are sought, as the column performance is strongly dependent on the particle size. In fact, a distinct advantage of small particles is their closer contact with diffusing sample molecules and a greater number of the mentioned equilibrium units (i.e., theoretical plates). Because extremely small particles present a great hindrance to gas flow, materials with a particle size between 100 and 150 μm are typically used as a sensible compromise between the column efficiency and technological limitations of high gas pressure at the column inlet.

In gas–solid chromatography, the solute molecules interact with the surface of solid adsorbents through relatively weak physical adsorption forces. Such weak forces are desirable, because the adsorption process must be reversible, preserving the chemical integrity of the solutes (unlike in some forms of contact catalysis, where a strong compound adsorption precedes chemical conversion). Consequently, not all adsorbing solids qualify as suitable column packings in GC. Examples of suitable GC adsorbents are silica gel, alumina, zeolites, carbonaceous adsorbents, and certain porous organic polymers. Surface porosity and a relatively large surface area are among the characteristic features of GC adsorbents. For example, certain synthetic zeolites, molecular sieves, may have a specific surface area as high as 700–800 m^2/g.

Specificity of certain solute-adsorbent interactions is a major advantage of gas–solid chromatography. Various adsorbents readily discriminate between different molecular geometries of otherwise similar solutes (e.g., geometrical isomers). At present, however, major difficulties exist as well: (1) large distribution coefficients (compared with partitioning liquids) result in long retention times; (2) the separaton process can often be strongly dependent on sample size, which is a serious problem for analytical determinations; (3) the physical processes in adsorption chromatography are less amenable to a rigorous theoretical description compared with gas-partition chromatography; and (4) the current adsorbent technology does not permit an effective suppression of minor catalytic

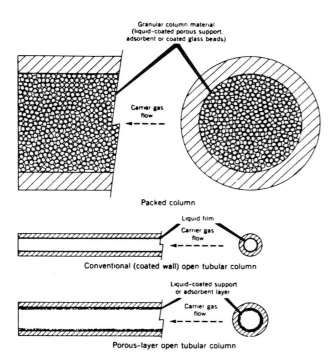

FIGURE 7 The three major types of GC columns. [From Horvath, C. (1967). *In* "The Practice of Gas Chromatography" (L. S. Ettre and A. Zlatkis, eds.), p. 133. Wiley (Interscience), New York.]

effects. The current applications of gas–solid chromatography are largely confined to the separation of relatively small molecules (such as permanent gases and lower alkanes). They represent a relatively small fraction of all GC applications.

Gas–liquid chromatography has found considerable use in chemical analysis. The packing materials (solid supports) utilized in this method are macroscopically similar to the described adsorbents. Yet their function is entirely different: They serve only as a supporting medium for the liquid stationary phase and do not participate directly in the separation process. The specific surface area of such solid supports is considerably less than that of adsorbents (i.e., their microstructure is considerably less developed).

The most commonly used solid supports are the diatomaceous earths. They are fossil-originated minerals found in abundance in various parts of the world. Prior to their use in chromatography, the diatomaceous earths are washed, thermally treated, chemically modified, and sieved, in a large manufacturing process. The diatomaceous earths are basically siliceous materials that contain reactive surface structures, the silanol groups. Since such groups could adversely affect the chromatographic analyses, causing "tailing" of certain polar sample components,

they are effectively blocked (deactivated) by a silylation reaction, an example of which is given below:

(a part of original surface structure)   (dimethyldichlorosilane. a deactivation agent)   (deactivated surface)

The solid support is subsequently impregnated by a liquid stationary phase. While many solid supports can carry up to 25% by weight of a liquid phase before becoming visibly wet, much lower phase loadings (a few percent) are used in practice. Both the amount and the chemical type of a stationary phase are crucial to the separation characteristics (efficiency and sample capacity) of a chromatographic column. Packed columns are considered to be low-efficiency, high-capacity GC columns. While their best efficiencies amount to no more than a few thousand theoretical plates, packed columns can tolerate microgram amounts of samples. Only carefully and totally packed columns yield the expected efficiencies.

The concept of the open tubular (capillary) column was introduced in 1956 by a Swiss scientist, M. J. E. Golay. Due to their extremely high separation efficiencies, open tubular columns have recently revolutionized analytical separations. As seen in Fig. 7, there is no granular packing inside the capillary column. The stationary liquid phase is uniformly deposited as a thin film on the surface of the inner wall, along the entire length of a long column. Typical lengths of capillary columns range from 10–100 m, with 0.2–0.5 mm inner diameters. The columns of smaller diameters (50–100 μm) have also been prepared for extremely efficient separations. Capillary columns with inner diameters larger than 0.5 mm are uncommon.

Column efficiencies between 10^5 and 10^6 theoretical plates have been achieved in capillary GC. Very narrow chromatographic peaks elute from such columns, allowing a high degree of resolution among the individual components of complex mixtures. The resolution advantage of a capillary column over a packed column is clearly indicated by Fig. 8, where numerous constituents of Calmus oil are separated from each other using a capillary column (a). A packed column (b) shows considerably less component resolution. The high separation efficiency of capillary columns is due to their high permeability to the carrier gas (the absence of column packing); consequently, long columns, featuring a great number of theoretical plates, can be prepared.

Another outstanding feature of GC capillary columns is their geometrical simplicity and consequent accessibility

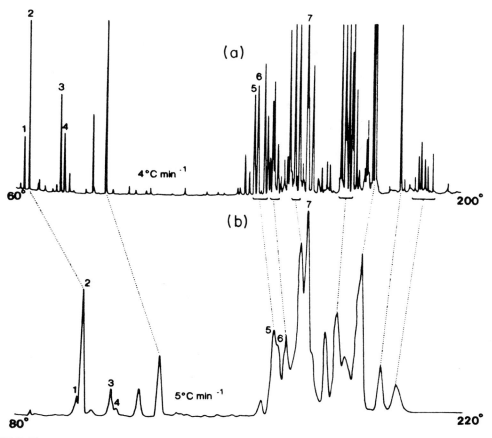

FIGURE 8 The resolution advantage of a capillary column (a) over a packed column (b), in Calmus oil analysis. [From Grob, K., and Grob, G. (1979). *J. High Resolut. Chromatogr.* **3,** 109.]

to theoretical description. An example is a description of the physical processes that occur inside such capillary columns. Equation (10), the van Deemter equation, can be translated, for the capillary column into

$$H = \frac{2D_G}{u} + \frac{(1 + 6k + 11k^2)}{(1 + k)^2} \frac{r^2}{24D_G} u, \quad (11)$$

where D_G is the solute diffusion coefficient in the mobile (gas) phase and r the capillary inner radius. The equation shows explicitly how the plate height is dependent on the diffusion processes and the column radius. At low gas velocities, molecular diffusion significantly increases the plate height. At higher velocities, the opposite is true (D_G is in the denominator), reflecting the fact that the solute mass transfer from one phase to another is primarily diffusion controlled. Reducing the column radius is a powerful way to increase the column performance.

Note that Eq. (11) is an accurate description of the column processes because the column geometry is well defined. Although somewhat similar equations exist for the packed columns, various (less accurate) empirical constants must be used.

Equation (11) is strictly valid only for the cases where the stationary phase film thickness amounts to no more than a few tenths of a micrometer (thin-film columns). Columns with film thicknesses up to several micrometers can also be prepared. Although their efficiencies are lower than those obtained for the thin-film columns, due to the impaired solute mass transfer, they can tolerate larger sample amounts without signs of overloading.

Refined aspects of column technology have been crucial to the success of GC capillary columns. Early in the development of such columns, metal or plastic tubes were used exclusively. Highly efficient glass capillary columns were developed at a later stage, and the problems of glass fragility were successfully overcome through the technology of fused-silica flexible tubes. Production of fused-silica capillaries is reminiscent of the fabrication of optical fibers: thin-walled sillica tubes, drawn from a hot zone are immediately protected by an overcoat of a stable organic polymer (Fig. 9).

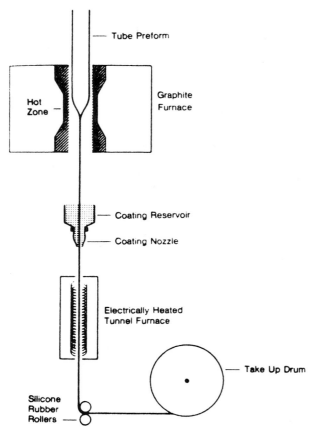

FIGURE 9 Column drawing apparatus for the preparation of fused silica capillaries. [From Lipsky, S. R., McMurray, W. J., Hernandez, M., Purcell, J. E., and Billeb, K. A. (1980). *J. Chromatogr. Sci.* **18**, 1.]

To secure a uniform film deposition from the solution of a stationary phase, the inner column's surface is first treated by an organic compound, the task of which is to improve the surface wettability and to mask potentially adsorptive column sites. A variety of stationary phases can now be successfully coated for capillary GC with a controlled film thickness. Refined procedures now exist even for the chemical immobilization of some stationary liquids.

The third type of a GC column (Fig. 7) is a porous-layer, open tubular column. While such a column has an inner diameter and a length comparable to the wall-coated columns, its inner wall is modified through a chemical treatment or deposition of finely dispersed particles. The porous layer can be either an adsorbent or a thin layer of the solid support impregnated with a liquid stationary phase. Efficiencies of the porous-layer, open tubular columns are not as high as those of "true" capillary columns, but their greater sample capacity is an advantage to some separations. The enhanced sample capacity is obtained because

of a greater surface area of such columns and, consequently, a relatively larger amount of stationary phase.

A proper choice of the liquid stationary phase is exceedingly important to a successful chromatographic separation. A great number of chemically different stationary phases have been described in the scientific and commercial literature. Several requirements govern the choice of a chemical substance as a GC stationary phase. First of all, it should have adequate selectivity for the substances to be resolved. It must be chemically stable at the column temperatures used in a given separation problem. The stationary phase must easily adhere as a uniform film to the column support without running off the column; if such mechanical instability occurs, the phase contaminates the detector and, naturally, the columns function properly for only a limited time. Finally, the stationary phase should be a well-defined chemical compound, so that the column preparation as well as the chromatographic process itself are reproducible.

In spite of the above strict requirements, many chemical substances can adequately perform as the stationary phase. The thermal stability requirement has made various synthetic polymers (silicones, polyglycols, polyesters, polyimides, etc.) most popular. Since the stabilities vary according to chemical stucture, nonpolar polymers are more stable than the polar column substrates. Column temperatures above $300°C$ are seldom used in the practice of GC.

The general solution rules roughly determine the suitability of a stationary phase for a given separation task: Polar substances (solutes) are readily dissolved and chromatographically retained by the polar stationary phases, while the nonpolar column materials retain the chromatographed sample components according to their boiling points, without any particular regard to the presence of unique functional groups in the sample molecules. Although the rules appear relatively straightforward, the stationary phases for many practical separations are still selected empirically.

The main solute–column interactions can be classified as dispersion forces and dipole–dipole interactions. The dispersion forces are present in any solute–solvent system, a hydrocarbon solute interacting with a nonpolar paraffin being often shown as an example. The polar solute molecules have permanent dipoles that can interact with those of the polar phases; on occasions, the dipole moments can also be induced in certain solute molecules in the presence of highly polar column materials. Dipole–dipole interactions are clearly evident in the separations of alcohols, esters, amines, aldehydes, and so on, on the polyglycol, polyamide, polyester, or cyanoalkylsilicone stationary phases.

Some extremely selective GC separations have been accomplished. In a number of cases, the hydrogen-bonding mechanism has been utilized. Synthesized optically active polymers are highly effective in resolving various racemic mixtures. Certain metal chelates, used as additives to the common stationary phases, can retain selected solutes through the formation of reversible complexes. Finally, highly organized liquids (such as various liquid crystals) tend to retain more strongly the molecules of elongated rather than bulky structures.

Through advances in synthetic chemistry and polymer research, new GC stationary phases will become available. Additional column selectivities can also be achieved by mixing the existing stationary phases with each other, in suitable proportions.

IV. DETECTORS AND ANCILLARY TECHNIQUES

The detector has an extremely important role in the overall process of GC analysis. The current popularity and success of GC as an analytical method is attributable in great part to the early development of highly sensitive and reliable means of detection. In sensing the vapor concentration at the column outlet, the detector provides information on the distribution of individual peaks within a chromatogram (which compound?) as well as their relative amounts (how much?). The area measured under a chromatographic peak is generally related to the quality of the compound.

Many detection principles in GC were investigated over the years, but only a few pass the criteria of reliability needed for precise analytical measurements. Detectors can broadly be classified as universal or selective. Universal detectors measure all (or nearly all) components of a mixture, although their response to the same quantities of different compounds is seldom similar. Selective detectors respond only to mixture components that possess a unique structural feature in their molecules. For example, a typical gasoline sample contains a number of organic components which, after being separated by an appropriate chromatographic column, are all detected by a universal detector. However, if a lead-selective detector is used instead, only a few peaks are recorded, those due to the lead-containing additives in gasoline, while the remaining mixture constituents are ignored. The so-called ancillary techniques go a step further as highly selective detectors, because they actually characterize the individual GC peaks qualitatively.

A. DETECTORS

The most important analytical properties of a GC detector are sensitivity, linearity over an extensive concentration range, long-term stability, and ease of operation.

While most GC determinations are performed with solute quantities between 10^{-6} and 10^{-9}, certain selective detectors can reach down to the 10^{-15}-g levels, representing some of the most sensitive measurement techniques available to the chemist. Some GC detection principles are based on the measurement of certain transport properties of the solutes (e.g., thermal conductivity or optical properties), while other detectors are transducers, measuring ultimately some product of a solute molecule (e.g., gas-phase ionization products). The latter detectors are destructive to the solutes.

1. The Thermal-Conductivity Detector

This detector, occasionally referred to as the hot-wire detector or katharometer, operates on the basis of measuring the difference in thermal conductivity of pure carrier gas and the carrier gas plus a solute. Typically, the column effluent is passed through a thermostatted cavity (measuring cell) that contains a resistor element heated by passage of a constant current. Various changes in the thermal conductivity of the surrounding gas causes the element temperature (and, consequently, its electric resistance) to decrease or increase. Pure carrier gas is passed, under the same conditions, through a reference cell of identical design. The resistor elements of both cells are parts of a Wheatstone bridge circuit that records any imbalance caused by the passage of individual solutes.

In thermal-conductivity measurements, it is advisable to choose a carrier gas that differs maximally from the organic solutes (e.g., hydrogen or helium). The detector is a truly universal and simple device, but its sensitivity is marginal; at best, submicrogram amounts are detected. The thermal-conductivity detector is most typically employed for the analysis of permanent gases and light hydrocarbons.

2. The Flame-Ionization Detector

This detector is the workhorse of GC. It operates on the basis of decomposing the solute-neutral molecules in a flame into charged species and electrically measuring the resultant changes of conductivity. A cross-sectional view of a flame-ionization detector is shown in Fig. 10. A small flame is sustained at the jet tip by a steady stream of pure hydrogen, while the necessary air (oxidant) is supplied through the diffuser. At the detector base, the column effluent is continuously introduced, mixed with hydrogen, and passed into the flame. Conductivity changes between the electrodes are monitored, electronically amplified, and recorded. A conventional carrier gas contributes little to the flame conductivity; however, when organic solute molecules enter the flame, they are rapidly ionized,

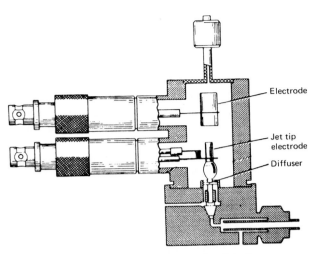

FIGURE 10 The flame ionization detector.

increasing the current in accordance with the solute concentration. With most flame-ionization detectors, this current increase is linear with the solute concentration up to six orders of magnitude.

The flame-ionization detector is a carbon counter; each carbon atom in the solute molecule that is capable of hydrogenation is believed to contribute to the signal (compounds with C—C and C—H bonds), while the presence of nitrogen, oxygen, sulfur, and halogen atoms tends to reduce the response. The detector is most sensitive for hydrocarbons. Practically, no response is obtained for inorganic gases, carbon monoxide, carbon dioxide, and water.

Because of its high sensitivity (the minimum detectable amounts are of the order of 10^{-12} g/s), linearity, and ease of operation, this detector is most popular, in spite of the somewhat incomplete understanding of the physical (ionization) processes involved.

3. The Electron-Capture Detector

This detector is a device based on certain gas-phase ionization phenomena within the ionization chamber. Its schematic diagram is given in Fig. 11. The carrier gas

molecules, flowing through the ionization chamber, are bombarded by the radioactive rays from the source of radiation (usually a foil containing ^{63}Ni or ^{3}H) incorporated into the detector body. In a rather complicated process, radicals, positive ions, and low-energy electrons are created. Application of electric potential between the electrodes permits the easily collected electrons to be continuously monitored as the so-called standing current (typically, around 10^{-9} A). This steady current provides a baseline value for the measurement of substances with a strong affinity to such low-energy electrons. When an electron-capturing solute enters the detector, it decreases the population of electrons by an electron attachment process. A decrease of standing current thus occurs during the passage of a solute band, resulting in a negative chromatographic peak.

The decrease of standing current due to the electron-capture process is proportional to the solute concentration in a process reminiscent and formally similar to Beer's law of optical absorption, except that thermal-energy electrons rather than photons are involved:

$$E = E_0 \exp(-Kxc), \qquad (12)$$

where E is the number of electrons reaching the anode per second, E_0 is the initial number of electrons, K is the electron-capture coefficient (a function of molecular parameters), x is the detector geometrical constant, and c is the solute concentration.

The electron-capture detector is a selective measurement device since only certain compounds exhibit appreciable affinities toward the low-energy electrons. Among the structures exhibiting strong electron affinities are various halogenated compounds, nitrated aromatics, highly conjugated systems, and metal chelates. The detector is extremely sensitive (amounts between 10^{-12} and 10^{-15} g can be detected) to various pesticides, herbicides, dioxins, freons, and other substances of great environmental concern. To achieve this extremely high sensitivity for normally noncapturing types of molecules (e.g., hormones and drug metabolites), various electron-capturing moieties can be introduced via chemical derivatization (a controlled sample alteration).

4. Other Detection Techniques

Several additional detectors were developed for GC. A major aim of such measurement devices is selectivity together with high sensitivity. Selective detectors should be blind to compounds in a mixture that do not possess certain unique structural features (i.e., chromophores or heteroatoms). In practice, some detectors qualify for such selectivity; in other cases, certain substances merely enhance the detector response.

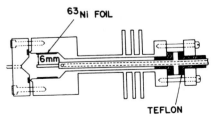

FIGURE 11 Electron capture detector. [From Fenimore, D. C., Loy, P. R., and Zlatkis, A. (1971). *Anal. Chem.* **39,** 1972.]

TABLE II Properties of Some Gas Chromatography Selective Detectors

Detector	Selectivity mode	Approximate sensitivity (g)
Electron-capture	Affinity to low-energy electrons	10^{-13}–10^{-14}
Thermionic	Nitrogen	10^{-12}
	Phosphorus	10^{-13}
Flame-photometric	Sulfur	10^{-9}
	Phosphorus	10^{-11}
Electrolytic-conductivity	Halogen compounds	10^{-10}
Ultraviolet	Aromatics	10^{-9}
Photoionization	Partially enhanced response to certain organic molecules as compared with flame ionization (not truly selective)	10^{-11}–10^{-12}

The most common GC selective detectors are listed in Table II together with their analytically important features. These selective detectors have been finding an increasing utilization in the analysis of environmental and biological mixtures. As seen from Table II, sensitivities at the low nanogram level are very common, while some detectors reach levels even below picogram amounts. Parallel uses of a nonselective and a selective detector are quite popular in chemical identification efforts.

B. Ancillary Techniques

While GC is a powerful separation method, it provides only limited information on the chemical nature of the substances it so effectively separates. Consequently, it has to be combined with ancillary techniques. These are certain sample manipulative techniques that are coupled in either a precolumn or a postcolumn arrangement to GC. Their purpose is to enhance qualitative information about the sample, to characterize it chemically, or ideally, to determine unequivocally its structure. Some of these ancillary techniques chemically alter the sample during the process; others measure only its physical parameters, such as optical spectra. On occasion, ancillary tools may represent instruments that are considerably more sophisticated and expensive than the GC instrumentation itself. The three GC ancillary techniques discussed below are among the most powerful and illustrative of this direction.

1. Pyrolysis/GC

This combination is an example of the precolumn arrangement. Pyrolysis/GC combines a controlled thermal degradation of a sample with the subsequent separation of neu-

tral thermal fragments. Most typically, the samples under investigation are large and nonvolatile compounds, such as synthetic or natural polymers. A reproducible pyrolysis/GC process results in the formation of pyrograms that are often highly indicative of some structural details of the original substance; both the presence of certain chromatographic peaks and their areas are judged. As small as submicrogram samples have been successfully analyzed by this combination.

Design of a precolumn pyrolysis unit and the method of thermal degradation are crucial to the acquisition of diagnostically useful pyrograms. Sample size and the pyrolysis temperature must also be carefully controlled. The three most common pyrolysis techniques use (a) filament (ohmical) heating, (b) rapid warp-up of a ferromagnetic conductor in a high-frequency field (Curie-point pyrolysis), and (c) direct thermal degradation in a heated quartz tube. In each case, the sample is deposited from its solution onto a suitable matrix, and the solvent is dried off prior to pyrolysis. Alternatively, small pieces of solids are directly pyrolyzed.

Pyrolysis/GC is used extensively in the analysis of polymers, paints, textile fibers, and even whole microorganisms. Certain materials of forensic interest have been characterized by this approach. A unique pyrolysis/GC system was aboard the Viking 1975 Mission spacecraft to investigate the possible occurrence of organic compounds in the martian soil.

2. GC/Mass Spectrometry

Mass spectrometers are sophisticated instruments that work on the principles of compound ionization and fragmentation (typically through the bombardment by electrons or selected ions), the physical separation of the charged fragments, and their detection. The information obtained by mass spectrometry is a mass spectrum (ion intensity versus mass) that is highly indicative of the sample's original structure, virtually a fingerprint of a molecule. Consequently, the method provides a powerful means to identify various organic compounds but works more effectively with pure substances than with substance mixtures. The combination of GC with mass spectrometry provides an ideal analytical system, in which the complex mixtures are first separated, and the mass spectrometer is permitted to analyze the substances, one at a time.

Commercial instruments that combine the two techniques vary in several respects. The low-resolution instruments provide the designation of nominal molecular weights, while the high-resolution instruments can work up to the precision of a small fraction of such nominal masses. For example, a low-resolution masses. mass spectrometer "sees" the proton (^1H) as the mass 1 and oxygen (^{16}O) as 16; a high-resolution instrument can measure the same species as 1.0078 and 15.9949, respectively. Consequently, the high-resolution instruments are capable of providing measurements of exact elemental composition for various compounds. Different physical principles of mass separation are involved with these instruments. Importantly, both the low- and high-resolution mass spectrometers can be combined with GC. The methods also strongly overlap with respect to the amounts necessary for analysis.

At first, a coupling of GC and mass spectrometry encountered technological difficulties because the gas chromatograph operates at gas pressures above atmospheric pressure, while most mass spectrometers operate at high vacuum. To overcome these difficulties, molecule separators were developed. These devices, working on principles such as molecular effusion, the jet separation effect, and preferential adsorption on a membrane, selectively remove most carrier gas, reduce pressure in the interface, and allow most sample molecules to pass into an evacuated mass spectrometer. The process of coupling GC to mass spectrometry is further aided by modern pumping technology. In fact, modern combination instruments need no molecule separators for capillary columns (typical flow rates around 1 ml/min).

Contemporary GC/mass spectrometry instruments are greatly aided by computers, which can control various instrumental parameters, provide data reduction, and compare acquired mass spectra with the extensive libraries of many thousands of previously recorded spectra.

3. GC/Infrared Spectroscopy

Infrared (IR) spectra of organic compounds are characteristic of various functional groups in the molecules. IR spectral information is somewhat complementary to mass spectral information. Therefore, the combination of GC with IR spectroscopy is, after GC/mass spectrometry, the second most important structural identification tool. Since conventional IR spectroscopy is less sensitive than most GC detectors, the necessary sensitivity enhancement is achieved through the use of Fourier transform techniques. With the advent of refined optical systems and fast computational techniques, the combination of GC with Fourier-transform IR spectrometry is becoming widely used, although its sensitivity is currently less than that of mass spectrometry. Special optical cells were designed for the purposes of this combination.

V. INSTRUMENTATION

The variety of GC analytical applications, columns, and specialized techniques make the modern gas chromatographs quite sophisticated instruments with precise electronic and pneumatic controls.

The carrier gas and the auxiliary gases for detectors are controlled by a set of pneumatic devices (pressure regulators, flow-controllers, and restrictors) to assure (a) reproducibility of the column flow rate, and thus retention times, in multiple analyses; (b) adjustment of the gas linear velocity for optimal column efficiencies; and (c) reproducibility of detector response for reliable quantitative measurements. In addition, filtering devices are inserted in the gas lines to purify all gases mechanically and chemically.

Type and design of the injection port are crucial to performing separations with different types of chromatographic columns. Different physical dimensions of the packed and capillary columns cause substantial differences in the optimum volumetric flow rates. While typical values for conventional capillary columns range around 1 ml/min, various packed columns pass one to two orders of magnitude greater gas flows. The volumes of injected samples must be adjusted accordingly. In a typical sampling procedure with a packed column, liquid samples of up to a few microliters are injected by a miniature syringe, through a rubber septum, into the hot zone of the injection port. Rapid sample evaporation and transfer into the first section of the column are feasible because of a sufficiently high flow rate of the carrier gas.

Considerably smaller samples are necessary for the much narrower capillary columns. Since small fractions of a microliter can be neither reproducibly measured nor easily introduced into the capillary GC system, indirect sampling techniques are employed. In a commonly used sampling method, a sample volume of approximately $1 \mu l$, or slightly less, is injected into a heated T-piece, where an uneven separation of the vaporized sample stream occurs. While the major part of the sample is allowed to escape from the system, a small fraction (typically, less than 1%) enters the first section of a capillary column. Sampling devices based on this principle are called splitting injectors or splitters. They are generally adequate in situations where samples with high concentrations of the analyzed substances are encountered.

Other ways of indirect sampling onto a capillary column involve the injections of (relatively nonvolatile) samples diluted in a sufficiently large (measurable) volume of a volatile solvent (which serves as a sample "vehicle"). With the column inlet kept at a sufficiently low temperature, the nonvolatile sample trace is trapped at the inlet and focused into a narrow zone, while the volatile solvent is allowed to pass through the column and widely separate from the sample. A subsequent increase of temperature permits the sample zone to desorb from its inlet position and enter the usual separation process.

Most sample introduction techniques in GC have now been automated. Process automation permits repeatable analysis and unattended operation of the instrument. Moreover, reproducibility of the sample injection is improved considerably.

A precise column temperature control is now required for all commercial gas chromatographs. In practice, the GC ovens are designed to have low thermal mass. Resistance spirals situated inside the oven are proportionally heated, while the air circulation throughout the oven is provided by a fan. For adequate analytical work, the column temperature should be reproducible within at least ±0.1°C.

Retention times in GC are affected by temperature. In accordance with Eqs. (1) and (2), the retention time decreases with increasing temperature because the partition coefficient is decreased. Various solutes, depending on their structures and the chemical nature of a particular phase system, have different dependencies on temperature. Consequently, temperature optimization is necessary for the maximum resolution of the analyzed components. For the mixtures of components with very different values of partition coefficient, column temperature programming is often employed. Commercial gas chromatographs are equipped with convective heating systems that facilitate linear temperature programs. As the column temperature is being gradually raised from a certain initial value to the maximum permissible temperature for an analysis, the sample components with increasingly higher boiling points are eluted from the column. According to the needs of analysis, programming rates are adjustable from as slow as 0.5°C/min up to 30°C/min. Nonlinear and multistep temperature programs are also feasible for special applications.

Modern GC utilizes sensitive detectors. As the measured detector signals (changes in current, voltage, etc.) are quite small, electronic signal amplification is necessary. Since the gas chromatographs are further provided with integrating devices and small computers to calculate exact retention times and peak areas for quantitation, the signals are converted to their digital forms. In addition to displaying a chromatogram on a recorder, modern GC instruments are capable of performing some advanced tasks, such as computing relative retention values, adjusting the detector baseline, and performing certain forms of data reduction.

VI. APPLICATIONS

Gas chromatography is a highly developed analytical method. It has found great use in the routine analysis of various mixtures of organic compounds. Quantitative GC measurements can frequently be carried out with a remarkable degree of reproducibility (analytical error within a few percent). For accurate determinations, it is advisable

to use appropriate standard compounds. Some automated analyzers based on GC principles are also used in the process control and continuous analysis of industrial streams. Specialized GC techniques find their place in scientific research. The extremely high sensitivities of some GC detectors are unparalleled.

The GC method is employed for a variety of mixtures, ranging from permanent gases up to molecules that are almost as large as 1000 Da of molecular weight. The variety of chromatographic columns and detectors available to GC continues to expand its applicability to various analytical problems. Several representative examples will now be described to demonstrate the method's versatility, resolving power, selectivity, and sensitivity. These examples have been chosen from the areas of industrial analysis, occupational hygiene, and biochemical research. Other major areas, not covered here, are geochemistry, food and aroma analysis, various agricultural and environmental analyses, atmospheric measurements, and forensic investigations.

The analysis of light gases (permanent gases, gaseous oxides, and C_1–C_5 hydrocarbons) has been traditionally performed in gas–solid chromatographic systems. Various porous adsorbents possess the capability to adsorb and separate these relatively small molecules. An example is shown in Fig. 12, where the carbon molecular sieve column (6 ft × 1/8 in. i.d.) rapidly resolved a mixture consisting of air, methane, carbon dioxide, acetylene, ethylene,

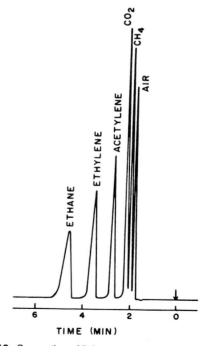

FIGURE 12 Separation of light gases on a carbonaceous adsorbent. [From Zlatkis, A., Kaufman, H. R., and Durbin, D. E. (1970). *In* "Advances in Chromatography 1970" A. Zlatkis, ed. Chromatography Symposium, University of Houston, Texas, p. 120.]

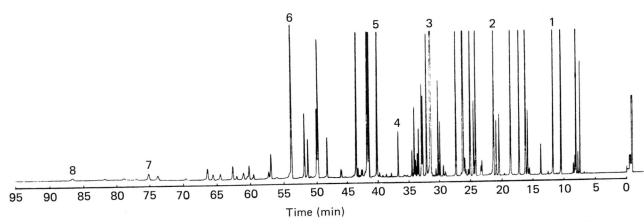

FIGURE 13 Capillary chromatogram of a gasoline sample. [From Adlard, E.R., Bowen, A.W., and Salmon, D. G. (1979). *J. Chromatogr.* **186**, 207.]

and ethane at 150°C (thermal conductivity detection was used). In particular, the separation of acetylene and ethylene is industrially important. While air (a mixture of two major components, oxygen and nitrogen) is eluted here as a single peak, there exist other GC adsorbents that can separate oxygen from nitrogen.

The petrochemical industries have long utilized GC as the analytical method for characterization of various fossil fuels, in monitoring the efficiency of distillation procedures, cracking processes, various chemical conversions, identification of oil spills, and so on. Most samples of

petrochemical interest are very complex, so the highly efficient capillary columns are frequently utilized. An example of major-component analysis is shown in Fig. 13, where a full-range gasoline sample has been resolved into a substantial number of components. A 70-m long capillary column was employed, the column temperature was programmed from 0 to 95°C, and the flame-ionization principle was used in the peak detection. A neat gasoline sample was injected (using a sample-splitting technique).

Environmental pollution is among the chief concerns of our industrial society. Highly sensitive analytical methods

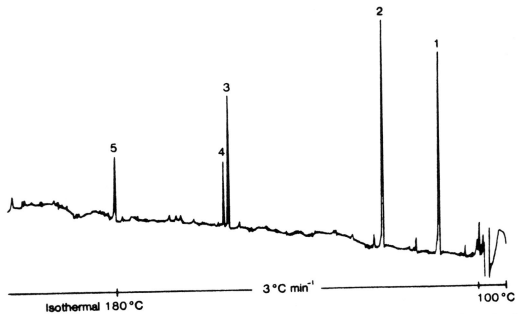

FIGURE 14 Chromatogram of trace aromatic amines (after preconcentration) from the atmosphere of a film-processing laboratory. [From Becher, G. (1981). *J. Chromatogr.* **211**, 103.]

have been developed over the years to identify the sources of water and air pollution, to study biodegradation and transformation of various pollutants in the environment, and to monitor their levels on a continual basis. A great majority of such methods involve GC measurement principles: both packed and capillary columns, selective detectors, the gas chromatography/mass spectrometry combination, and so on. For example, in film-processing laboratories air analysis must periodically be carried out to measure the levels of toxic aromatic amines. With a capillary column (Fig. 14) and the nitrogen-sensitive flame-based detector, five different aromatic amines can be quantitated at the airborne levels of 3 to 13 $\mu g/m^3$.

Prior to the GC analysis, the air sample is first concentrated by passing it through a small adsorbent column. Such a preconcentration step is common if trace organics are to be analyzed in dilute media (air, water, soil, etc.).

Gas chromatography has been applied to analyze numerous biologically important substances such as fatty acids, amino acids and peptides, steroids, carbohydrates, and prostaglandins. Since these compounds are mostly polar and nonvolatile, chemical modifications (sample derivatization) are necessary to block the polar groups and thus enhance volatility of such compounds. To ensure the necessary reliability of GC analyses, such chemical modifications must have highly reproducible yields. For compounds with diverse functional groups, multiple derivatizations (through more than one reaction) are needed. Examples of these are the various steroid hormone metabolites that feature ketonic and hydroxy functional groups in their molecules. Prior to their GC analysis, these compounds are first subjected to treatment with methoxylamine hydrochloride (to form methoximes from ketones) and then to reaction with a trimethylsilyl donor reagent (to form trimethylsilylethers from alcohols). An example of a fully derivatized steroid is a methoxime-trimethylsilyl product of the glucocorticoid hormone, cortisol.

Other steroids (i.e., compounds structurally related to cortisol) can also be derivatized in a similar manner and subjected to GC analysis. If a high-resolution (capillary) column is employed for their separation, entire profiles of

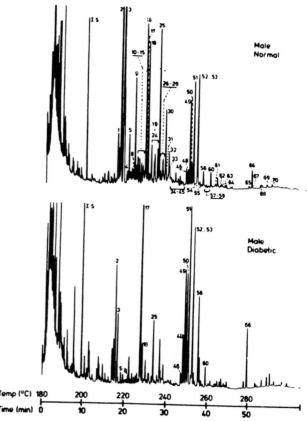

FIGURE 15 Urinary steroid profiles (after sample derivation) from a normal and a diabetic patient, as recorded by capillary chromatography. [From Alasandro, M., Wiesler, D., Rhodes, G., and Novotny, M. (1982). *Clin. Chim. Acta* **126**, 243.]

closely related substances can be monitored under different circumstances of health and disease (Fig. 15). While this demonstrated case has been related to an effort to improve our understanding of hormonal alterations in human diabetes, similar analytical GC techniques have been employed to detect abnormalities in adrenal function and reproductive processes.

Gas-chromatographic methods are widely used to analyze amino acids in the hydrolyzates of small protein samples. The method's sensitivity is the major reason for these applications. In addition, GC-based techniques provide opportunities to distinguish and quantitate amino acids (and several other compound types) as different optical isomers. The most popular procedure to separate R and S isomers employs an optically active (chiral) stationary phase. Because of the zwitterionic nature of amino acids, a two-step derivatization is necessary prior to GC. As the first step, the acid (carboxy) function is blocked through esterification. During the second treatment, the amino groups are acylated. Figure 16 demonstrates

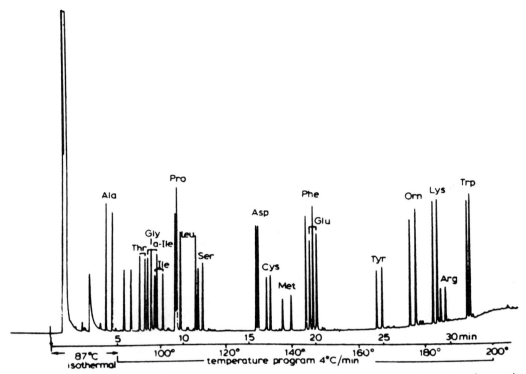

FIGURE 16 Capillary GC separators of a racemic mixture of 19 amino acids on an optically active stationary phase. [From Frank, H., Nicholson, G. J., and Bayer, E. (1978). *J. Chromatogr.* **167**, 187.]

a chromatogram of the 19 naturally occuring amino acids, separated in the form of their volatile derivatives into their respective enantiomers.

SEE ALSO THE FOLLOWING ARTICLES

ELECTROPHORESIS • LIQUID CHROMATOGRAPHY • OR-GANIC CHEMISTRY, COMPOUND DETECTION

BIBLIOGRAPHY

Ettre, L. S., and Zlatkis, A. (eds.). (1984). "The Practice of Gas Chromatography," Wiley (Interscience), New York.

Lee, M. L., Young, F. J., and Bartle, K. D. (1984). "Open Tubular Column Gas Chromatography," Wiley, New York.

Novotny, M., and Wiesler, D. (1984). *In* "New Comprehensive Biochemistry," (Z. Deyl, ed.), Vol. 8, p. 41. Elsevier, Amsterdam.

Poole, C. F., and Schute, S. A. (1984). "Contemporary Practice of Chromatography," Elsevier, Amsterdam.

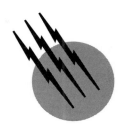

Gas Hydrate in the Ocean Environment

William P. Dillon

U.S. Geological Survey

I. The Nature of Gas Hydrate
II. Environment of Stability
III. Presence in Nature
IV. Effect on Seafloor Slides and Slumps
V. Effect on Climate
VI. Energy Resource

GLOSSARY

Biogenic methane Methane formed by microbial processes.

BSR An abbreviation for the so-called "bottom simulating reflection." A reflection recorded in seismic reflection profiles that results from an acoustic velocity contrast produced by the decrease in sound speed caused primarily by the presence of gas trapped beneath the gas hydrate stability zone. BSRs provide a remotely sensed indication of the presence of gas hydrate.

Dissociation The breakdown of gas hydrate into its components, water and gas, commonly resulting from some combination of pressure reduction and temperature increase.

Gas hydrate or clathrate A solid crystalline structure formed of cages of water molecules; a cage commonly encloses and is supported by a gas molecule.

Gas hydrate stability zone The region within deep ocean water and seafloor sediments where gas hydrate is stable. Gas hydrate commonly occurs in the zone from the seafloor down to some depth, perhaps several hundred meters below the seafloor, where temperature/pressure changes result in conditions where gas hydrate is unstable.

Geothermal gradient The rate at which temperature increases downward into the solid Earth.

Phase boundary The interface in temperature/pressure space where a phase change occurs. In the case of gas hydrate, the phase change is from stability of solid gas hydrate to stability of gas + water.

Seismic reflection profile An image of the structure below the seafloor along a ship trackline that is created by using sound pulses and recording the time of return from various reflecting layers.

Thermogenic methane Methane formed from organic matter by heating at sediment depths exceeding about 1 km.

Encyclopedia of Physical Science and Technology, Third Edition, Volume 6

A GAS HYDRATE, also known as a gas clathrate, is a gas-bearing, icelike material. It occurs in abundance in marine sediments and stores immense amounts of methane, with major implications for future energy resources and global climate change. Furthermore, gas hydrate controls some of the physical properties of sedimentary deposits and thereby influences seafloor stability.

I. THE NATURE OF GAS HYDRATE

Gas hydrate or gas clathrate is a crystalline solid; its building blocks consist of a gas molecule surrounded by a cage of water molecules. The water molecules are connected by hydrogen bonds, but the gas molecule is not bonded; rather it is simply trapped within the cage and held by Van der Waal forces (the term "clathrate" refers to a cage or lattice-like structure). Because the material is formed of hydrogen-bonded water molecules, it looks like ice and is very similar to ice, although of different crystallographic form, having its crystal structure stabilized by the guest gas molecules. A conceptual model of a gas hydrate cage is shown in Fig. 1, containing a methane molecule. This is the most common gas hydrate structure, which is classified as structure I. Structure I gas hydrate has the smallest cavity sizes and is comfortable holding methane molecules or other gas molecules of similar size, such as nitrogen, oxygen, carbon dioxide, and hydrogen sulfide. The structure of the clathrate cages varies depending on the size of the guest molecule, and some molecules are too small (smaller than argon) or too big to exist in any of the gas hydrate structures. If larger gas molecules are available, at least two other gas hydrate structures are known. These are structure II, holding molecules up to the size of propane and isobutane, and structure H, holding even larger molecules like isopentane and neohexane. Actually, more than one cavity size exists in a structure, and thus, a mixture of gases can be accommodated. Because the gas molecules are held within a crystal lattice, they may be held closer together than in the gaseous state, at least at lower pressures, so gas hydrate can act as a concentrator of gas. For example, a unit volume of methane hydrate at one atmosphere (and 0°C) can hold 163 volumes of methane gas at the same conditions.

Many gas hydrates are stable in the deep ocean conditions, but methane hydrate is by far the dominant type, making up >99% of gas hydrate in the ocean floor. The methane is almost entirely derived from microbial methanogenesis, predominantly through the process of carbon dioxide reduction. In some areas, such as the Gulf of Mexico, gas hydrates are also created by other thermogenically formed hydrocarbon gases and other clathrate-forming gases such as hydrogen sulfide and carbon dioxide. Such gases escape from sediments at depth, rise along faults, and form gas hydrate at or just below the seafloor, but on a worldwide basis these are of minor volumetric importance compared to methane hydrate. Methane hydrate exists in several forms in marine sediments. In coarse-grained sediments it often forms as disseminated grains and pore fillings, whereas in finer silt/clay deposits it commonly appears as nodules and veins. Gas hydrate also is observed as surface crusts on the seafloor.

II. ENVIRONMENT OF STABILITY

Gas hydrate forms wherever appropriate physical conditions exist—moderately low temperature and moderately high pressure—and the materials are present—gas near saturation and water. These conditions are found in the deep sea commonly at water depths greater than about 500 m or somewhat shallower depths (about 300 m) in the Arctic, where bottom-water temperature is colder. Gas hydrate also occurs beneath permafrost on land in arctic conditions, but, by far, most natural gas hydrate is stored in ocean floor deposits. A simplified phase diagram is shown in Fig. 2A, in which pressure has been converted to water depth in the ocean (thus, pressure increases downward in the diagram). The heavy line in Fig. 2A is the phase boundary, separating conditions in the temperature/pressure field where methane hydrate is stable to the left of the curve (hatched area) from conditions where it is not. In Fig. 2B, some typical conditions of pressure and temperature in the deep ocean were chosen to define the region where methane hydrate is stable. The phase boundary indicated is the same as in Fig. 2A, so methane hydrate is stable

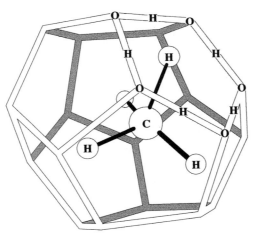

FIGURE 1 Diagrammatic concept of a structure I gas hydrate cage holding a methane molecule. H, O, and C represent hydrogen, oxygen, and carbon atoms. Double lines, filled or unfilled, represent chemical bonds.

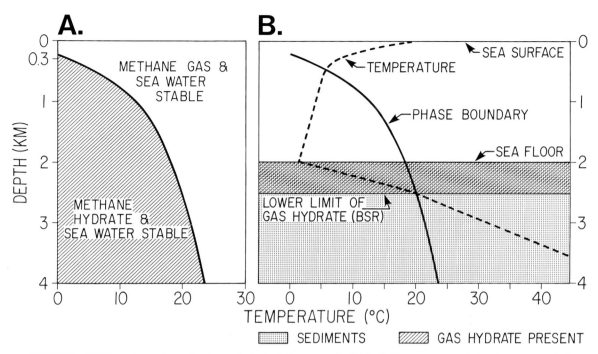

FIGURE 2 (A) Phase boundary of methane hydrate in the ocean (solid line). The pressure axis has been converted to depth into the ocean, so pressure increases downward. (B) The same phase boundary as shown in (A) with a seafloor inserted at 2 km depth and a typical temperature curve (dashed line). Hatched region shows the vertical extent of the gas hydrate stability zone under these assumed conditions.

at locations in the pressure/temperature field where conditions plot to the left of the phase boundary. The dashed line shows how temperature conditions typically vary with depth in the deep ocean and underlying sediments. In this case typical western North Atlantic Ocean thermal conditions were chosen and a seafloor at 2 km water depth (Fig. 2B) was selected. Near the ocean surface, temperatures are too warm and pressures too low for methane hydrate to be stable. Moving down through the water column, temperature decreases and an inflection in the temperature curve is reached, known as the main thermocline, which separates the warm surface water from the deeper cold waters. At about 500 m, the temperature and phase boundary curves cross; from there downward, temperatures are cold enough and pressures high enough for methane hydrate to be stable in the ocean. This intersection would occur at a shallower depth in colder, arctic waters.

If methane is sufficiently concentrated (near saturation), gas hydrate will form. However, like ice, the density of crystalline methane hydrate is less than that of water (about 0.9), so if such hydrate formed in the water (e.g., at methane seeps) it would float upward and would dissociate when it crossed the depth where the curves intersect. However, if the gas hydrate formed within sediments, it would be bound in place. Minimum temperature occurs

at the seafloor (Fig. 2). Downward through the sediments, the temperature rises along the geothermal gradient toward the hot center of the Earth. At the point where the curve of conditions in the sediments (dashed line) crosses the phase boundary, one reaches the bottom of the zone where methane hydrate is stable.

The precise location of the base of the gas hydrate stability zone under known pressure/temperature conditions varies somewhat depending on several factors, most important of which is gas chemistry. In places where the gas is not pure methane, for example, in the Gulf of Mexico, at a pressure equivalent to 2.5 km water depth, the base of the gas hydrate stability zone will occur at about 21°C for pure methane, but at 23°C for a typical mixture of approximately 93% methane, 4% ethane, 1% propane, and some smaller amounts of higher hydrocarbons. At the same pressure (2.5 km water depth) but for a possible mixture of about 62% methane, 9% ethane, 23% propane, plus some higher hydrocarbons, the phase limit will be at 28°C. These differences will cause major shifts in depth to the base of the gas hydrate stability zone as would be implied by Fig. 2B. Such mixtures of gases essentially make the formation of gas hydrate easier and therefore can result in the formation of gas hydrate near the seafloor at shallower depths (lower pressures) than for methane hydrate

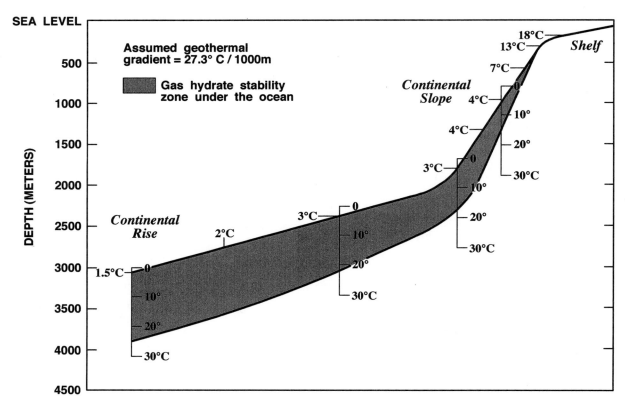

FIGURE 3 Assumed thickness of the gas hydrate stability zone for changing water depth (pressure) and a constant geothermal gradient. Factors that change the chemistry (changes in gas composition, salinity variations, etc.) or that affect the thermal structure (landslides, salt diapirs) can complicate this simple picture. [From Kvenvolden, K. A., and Barnard, L. A. (1982). Hydrates of natural gas in continental margins, *In* "Studies in Continental Margin Geology" (J. S. Watkins, and C. L. Drake, eds.), Am Assoc. Petroleum Geologists Memoir No. 34, pp. 631–640.]

at equal temperatures. Below the base of the gas hydrate stability zone (500 m in our example in Fig. 2B), methane and water will be stable and methane hydrate will not be found.

The geothermal gradient tends to be quite uniform across broad regions where sediments do not vary. Thus, for a given water depth, the sub-bottom depth to the base of the gas hydrate stability zone will be quite constant. However, because a change in water depth causes change in pressure, we anticipate that the base of the gas hydrate stability zone will extend further below the seafloor as water depth increases (Fig. 3).

III. PRESENCE IN NATURE

Two basic issues that concern us about gas hydrate are: (1) where does it exist and (2) how much is present? These questions have turned out to be very difficult to answer in a precise manner because gas hydrate exists beneath the ocean floor in conditions of temperature and pres-

sure where human beings cannot survive, and if gas hydrate is transported from the sea bottom to normal Earth-surface conditions, it dissociates. Thus, drilled samples cannot be depended upon to provide accurate estimates of the amount of gas hydrate present, as would be the case with most minerals. Even the heat and changes in chemistry (methane saturation, salinity, etc.) introduced by the drilling process, including the effect of circulating drilling fluids, affect the gas hydrate, independent of the changes brought about by moving a sample to the surface. Gas hydrate has been identified in nature generally from drilling data or by using remotely sensed indications from seismic reflection profiles.

A. Identifying Gas Hydrate in Nature

1. Measuring Gas Hydrate in Wells

Drilled samples of gas hydrate have survived the trip to the surface from hydrate accumulations below the seafloor, despite the transfer out of the stability field, just because the dissociation of gas hydrate is fairly slow. Such samples have been preserved, at least temporarily, by returning them to the pressure/temperature conditions where they formed or, more commonly, by keeping them at surface pressure, but at the ultra-cold temperature of liquid nitrogen. Such preserved samples are valuable for many studies, but certainly they do not provide quantitatively accurate indications of the amount or distribution of gas hydrate or its relation to sediment fabric that existed in the undisturbed seafloor sediments.

Several indirect approaches have been used to gain indication of the amount and/or distribution of gas hydrate in the sediments. Some are as obvious as making temperature measurements along a core to indicate where gas hydrate existed, because the dissociation of gas hydrate, being an endothermic reaction, will leave cold spots.

The dissociation of gas hydrate leaves another indirect marker of its former existence, because when gas hydrate forms it extracts pure water to form the clathrate structure, excluding all salts as brine. Therefore, when hydrate dissociates in a core, the interstitial water becomes much fresher, and the amount of gas hydrate present before dissociation can be calculated. An example is shown in Fig. 4 from Ocean Drilling Program hole 997 off the South Carolina coast. Measured chloride content is shown in the left panel. The values near zero depth (depth at the seafloor) represent seawater chloride concentration (chlorinity). It is assumed that a smooth curve of chlorinity vs depth, following the main trend of data points (solid curve), rep-

resents the undisturbed (predrilling) chlorinity and that spikes of low chlorinity to the left of the curve represent the result of hydrate dissociation. The base of hydrate stability here is at 450 m below the seafloor, and the top of significant gas hydrate concentrations is apparently about 200 m. The estimation of the base curve is still a controversial issue in using this method. The second panel shows the base curve straightened out (dashed line) so that the plot is considered to represent the chlorinity anomaly that results from dissociation of previously existing hydrate. The third panel shows the calculated results in terms of proportion of sediment that had been occupied by gas hydrate.

A third approach that is important for identifying and quantifying gas hydrate in wells is the downhole logging methods in which sensors are pulled through the hole and measurements are made. These measurements can give porosity information about the sediments, using gamma-rays or neutron flux, and provide indications about gas hydrate concentration, using electrical resistivity or acoustic velocity measurements.

2. Sensing Gas Hydrate in Seismic Reflection Profiles

Beyond the issue of precisely quantifying gas hydrate in the few spots on the seafloor where wells have been drilled, we would like to have a technique of remotely sensing gas hydrate without drilling a hole and by some means that

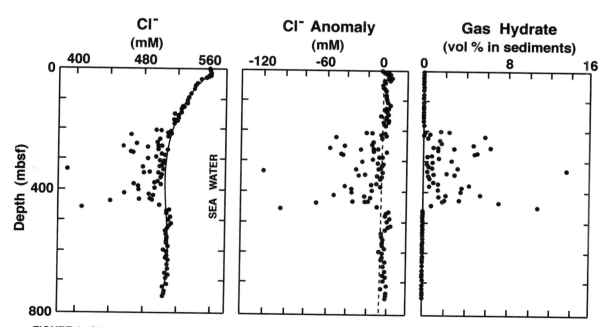

FIGURE 4 Chloride content, chloride anomaly, and volume of gas hydrate calculated from this information for Ocean Drilling Program hole 997, Blake Ridge, off the southeastern United States. [From Paull, C. K., Matsumoto, R., and Wallace, P. J. *et al.* (1996). "Proceedings of the ODP Initial Reports," Vol. 164, Figure 47, p. 317, College Station, TX.]

would allow us to survey large areas. The only effective approaches that seem to fulfill these requirements are the use of acoustic imaging of the near-bottom sediments. Most commonly, this means using seismic reflection profiling to image the sediments in the zone of gas hydrate stability just below the seafloor. A seismic reflection profile is created by steaming a ship along a line and triggering a sound source in the water every few seconds. The sound source for gas hydrate studies can be a fairly small one, as the need is to provide penetration only down to the base of the gas hydrate stability zone, which is unlikely to be more than 1000 m below the seafloor. Use of a small seismic source, which provides higher frequency sound, is also consistent with obtaining the highest possible resolution. The sound source is often an "airgun," which is a small high-pressure air chamber (commonly 100–160 in.[3] volume, which is equivalent to 1.6–2.6 liters), from which the air can be abruptly dumped by an electrical signal. Hydrophones trailed in the water behind the ship receive the echoes of each sound pulse from the seafloor and below, and the echoes are recorded on the ship for computer processing into a profile. In a sense, any seismic profile represents a cross section through the seabed, but keep in mind that the image is created by reflections of sound from reflectors formed by density and acoustic velocity changes that are inherent to sedimentary layers and gas hydrate accumulations. The vertical axis is imaged in the travel time of sound.

Fortunately, the base of the the gas hydrate stability zone is often easy to detect in seismic reflection profiles. Free

gas bubbles commonly accumulate, trapped just beneath the base of the gas hydrate stability zone, where free gas is stable and gas hydrate will not exist. Presence of bubbles in intergranular spaces reduces the acoustic velocity of the sediment markedly. Conversely, in the gas hydrate stability zone the velocity is increased slightly by the presence of gas hydrate, which in the pure state has twice the velocity of typical deep sea sediments, and furthermore, bubbles generally cannot be present because any free gas would convert to gas hydrate as long as water is present. A large velocity contrast generates a strong echo when an acoustic pulse impinges on it. Thus, we can receive an image of the base of the gas hydrate stability zone in a seismic reflection profile. The base of gas hydrate stability, as disclosed by this reflection, generally occurs at an approximately uniform sub-bottom depth throughout a restricted area because it is controlled by the temperature. Thermal gradients across an area tend to be consistent, so isothermal surfaces have consistent depth. Hence, the reflection from the base of the gas hydrate stability zone roughly parallels the seafloor in seismic profiles and has become known as the "Bottom Simulating Reflection" (BSR) (Fig. 5). The BSR is a sure sign that gas exists trapped beneath the base of the gas hydrate stability zone and strongly implies that gas hydrate is present, because free gas, which has a tendency to rise, exists just below and in contact with the zone where gas would be converted to gas hydrate. The coincidence in depth of the BSR to the theoretical, extrapolated pressure/temperature conditions that define the base of gas hydrate stability and

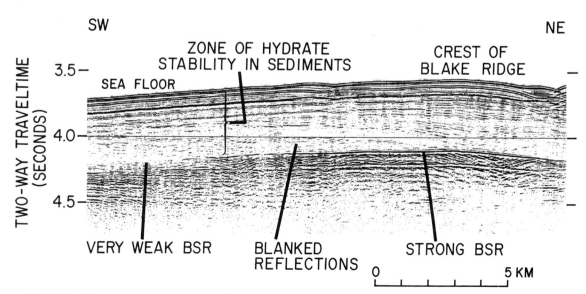

FIGURE 5 Seismic reflection profile near the crest of the Blake Ridge off South Carolina in the southeastern United States. Note the strong reflection marked BSR that defines the base of the gas hydrate stability zone. The vertical axis is in two-way travel time of sound, which varies with sound velocity in the medium; thus, this axis is a variable scale with respect to distance.

TABLE I   Global Organic Carbon Distribution in Giga-tons (Gt = 10^{15} g)[a]

Gas hydrate (methane)	10,000	Land	
		Biota	830
		Detritus	60
Fossil fuels	5,000	Soil	1400
		Peat	500
Atmosphere (methane)	3.6		
		Ocean	
		Marine biota	3
Dispersed carbon	20,000,000	Dissolved	980

[a] Data from Kvenvolden (1988).

the sampling of gas hydrate above BSRs and gas below give confidence that this seismic indication of the base of gas hydrates is dependable.

A second significant seismic characteristic of gas hydrate cementation, called "blanking," is also displayed in Fig. 5; blanking is the reduction of the amplitude (weakening) of seismic reflections, which appears to be caused by the presence of gas hydrate. Many observations of blanking in nature have been associated with gas hydrate accumulations, but the reality of blanking has been questioned because an undisputed physical model for it has not yet been identified.

B. Estimated Quantities of Gas Hydrate Worldwide

Many estimates of the global amount of gas hydrate have been made. Keith Kvenvolden of the U.S. Geological

Survey recently did a survey of these and looked at their variability over time. In 1980 the estimates covered a very broad range, but since 1988 the range of disagreement has been considerably reduced, even though the methods used by ten different sets of workers have varied widely. From 1988 to 1997 the estimates of worldwide methane content of gas hydrate range from 1 to 46×10^{15} m^3, with a "consensus value" of 21×10^{15} m^3.

How does this amount of methane compare to other stores of carbon in the natural environment? Table I attempts to answer that question by comparing the amount of organic carbon in various reservoirs on Earth. Obviously, by far, the largest amount of organic carbon exists dispersed in the rocks and sediments. However, consider the first three numbers (Table I). The organic carbon in gas hydrate is estimated to be twice as much as in all fossil fuels on Earth (including coal). Organic carbon in gas hydrate is also estimated to be about 3000 times the amount in the atmosphere, and here, we are directly comparing methane to methane. As methane is a very powerful greenhouse gas, this reservoir of methane, which might be released to the atmosphere, may have significant implications.

C. Environments of Gas Hydrate Concentration

The presence of gas hydrate, identified on the basis of drilling and seismic reflection profiling, has been reported all around the world, but most commonly around the edges of the continents (Fig. 6). Methane accumulates in

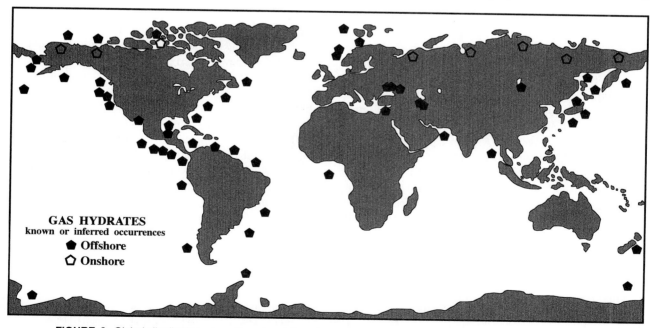

FIGURE 6 Global distribution of confirmed or inferred gas hydrate sites, 1997 (courtesy of James Booth, U.S. Geological Survey). This information represents our very limited knowledge. Gas hydrate probably is present in essentially all continental margins.

continental margin sediments probably for two reasons. First, the margins of the oceans are where the flux of organic carbon to the seafloor is greatest because oceanic biological productivity is highest there and organic detritus from the continents also collects to some extent. Second, the continental margins are where sedimentation rates are fastest. The rapid accumulation of sediment serves to cover and seal the organic material before it is oxidized, allowing the microorganisms in the sediments to use it as food and form the methane that becomes incorporated into gas hydrate. In addition to sites in the marine environment, the map (Fig. 6) also shows some sites where gas hydrate has been reported associated with permafrost in the Arctic, the one location where it has been found in fresh water in Lake Baikal, Siberia, and in intermediate salinity (between fresh and oceanic salinities) in the Caspian Sea.

Gas hydrate concentration commonly increases downward through the gas hydrate stability zone in marine sediments, based on the evidence of well and seismic reflection profiling data. For example, note the gas hydrate distribution described by chlorinity in Fig. 4 and the increase in blanking downward toward the BSR (base of the gas hydrate stability zone) in Fig. 5. The increase in concentration toward the base of the gas hydrate stability zone suggests that methane is being supplied below the base of the gas hydrate stability zone by some process and that the gas is probably being trapped there before entering the lower part of the gas hydrate stability zone. The presence of trapped gas just beneath the base of the gas hydrate stability zone is demonstrated by the common occurrence of BSRs and has been observed in drilling. This methane may rise from below, either because it is being generated by microbes at greater depth in the sediment or because of solubility considerations. Alternatively, it may be recycled from above. Many continental margin settings have ongoing sediment deposition. As the seafloor builds up, the geothermal gradient tends to remain constant, so the isothermal surfaces must rise with the accreting seafloor. A sediment grain or bit of gas hydrate in the shallow sediments effectively sees the gas hydrate stability zone migrate upward past it as the seafloor builds up. Eventually, that bit of gas hydrate ends up far enough below the seafloor that it is beneath the base of the gas hydrate stability zone and thus outside the range of gas hydrate stability. Then the gas hydrate dissociates and releases its methane, which will tend to rise through the sediments because of the low density of gas and accumulate at the base of the gas hydrate stability zone, ultimately to work its way up into the gas hydrate stability zone, where it forms more gas hydrate.

IV. EFFECT ON SEAFLOOR SLIDES AND SLUMPS

When gas hydrate forms or dissociates within seafloor sediments, this has major effects on sediment strength. Dissociation converts a solid material, which sometimes at high concentrations acts as a cement, to gas and water. Only one triaxial shear strength test of a natural gas hydrate-bearing sediment has been made at *in situ* conditions (at the U.S. Geological Survey, Gas Hydrate And Sediment Test Laboratory Instrument—GHASTLI; Booth, Winters, and Dillon, 1999), but this confirmed the tremendous increase in strength that was expected from gas hydrate in high concentrations. However, in most locations in natural seafloor sediments, hydrate concentrations are fairly low (<5%), so the cementing effect is minimal. The effect of dissociation of gas hydrate on sediment strength can still be immense, though, because when hydrate breaks down at shallower depths, the products of breakdown, gas + water, occupy greater volume than the gas hydrate they were derived from, and thus, the dissociation will increase the internal pressure in the pore space. Such pressures are called "overpressures," which are pressures greater than the column of water plus sediment above the spot. Generation of overpressures weakens the sediment significantly and is likely to initiate sediment slides and slumps on continental slopes and rises.

Dissociation of gas hydrate that leads to weakening of sediments can result from pressure reduction or warming (Fig. 2). Obviously, warming will occur if ocean bottom waters warm up. However, gas hydrate will only dissociate at its phase boundary. In Fig. 2B, for example, the gas hydrate at the seafloor exists at pressure/temperature conditions some considerable distance within its zone of stability, so a few degrees of warming will not cause it to dissociate. The deep limit of gas hydrate stability at conditions shown in Fig. 2B is about 500 m below the seafloor. If bottom water became abruptly warmer, the warming front would have to propagate downward through the sediment to the depth where the gas hydrate is at the stability limit, which might take hundreds or thousands of years. The change would have to occur as a conductive heat flow, as downward flow of water that could transfer (advect) heat is extremely limited in ocean sediments. Obviously, the place where warming of bottom water will have a rapid influence is where the base of the gas hydrate stability zone is very close to the seafloor (see Fig. 3).

Atmospheric warming is presently occurring. Global surface air temperature has probably increased by roughly 0.8°C over the last century. This warming is probably being transferred to the ocean in a manner comparable to chemical tracers that have been observed, which means that warmed surface water can be expected to circulate down to depths of the shallower gas hydrates in several tens of years. In specific cases in the Gulf of Mexico, where warm bottom currents sometimes sweep through the region, and off northern California, active dissociation of gas hydrate at the seafloor has been observed (evidence is the absence of previously observed gas hydrate and release of methane bubbles). Some of this activity has been related to identifiable water temperature changes, so present atmospheric warming may be leading to hydrate dissociation.

Dissociation of oceanic gas hydrate by pressure reduction clearly must also occur because a lowering of sea level must cause an instantaneous reduction of pressure at the seafloor and down into the sediments. Pressure is dependent on the weight of a column of water and sediment above a spot, and if that changes there is no delay in changing pressure at all depths, as there is in changing temperature. Thus, a lowering of sea level, as occurred during buildup of continental icecaps that removed water from the ocean, will immediately reduce the pressure at the base of the gas hydrate stability zone and cause gas hydrate to dissociate. The most recent glacial sea level lowering of about 120 m ended about 15,000 years ago and must have caused significant dissociation of gas hydrate.

The result of hydrate dissociation on a continental slope is diagrammed in Fig. 7. The initial hydrate breakdown releases a slide mass. When the slide takes place, the removal of sediment reduces the load on the sediment that was be-

low it and, thus, creates another pressure reduction that may cause further gas hydrate dissociation, resulting in cascading submarine slides. Circumstantial evidence that gas hydrate processes and submarine landslides are related is provided by a map of the shallow limit of gas hydrate stability compared to the tops of known landslide scars on the U.S. Atlantic margin that is shown in Fig. 8. Clustering of slope failures within the zone of gas hydrate stability, just below its upper limit, is suggestive of a relationship comparable to that diagrammed in Fig. 7. Even on relatively flat slopes where slides are not triggered, evidence for buildup of pressure due to gas hydrate dissociation; mobilization of gas + water-rich sediments; and escape of methane, water, and sediment has been interpreted. This evidence consists of apparent blowout structures (craters with chaotically disturbed sediments beneath, sometimes with associated ejecta) in the Gulf of Mexico and the U.S. Atlantic margin.

The dissociation of gas hydrate, resulting in gas and water release and sediment deformation, can also be a safety issue in petroleum exploration and production. Throughout the world, oil drilling is moving to deeper water where gas hydrate can be anticipated. In the marine environment, gas hydrate commonly is concentrated near the base of the gas hydrate stability zone. If the gas hydrate is dissociated by warm drilling fluid, gas can enter the wellbore and circulating fluid. This reduces the density of drilling muds, which are placed in the hole to maintain high pressure at depth, and the reduced pressure encourages further dissociation of hydrate. Muds have foamed

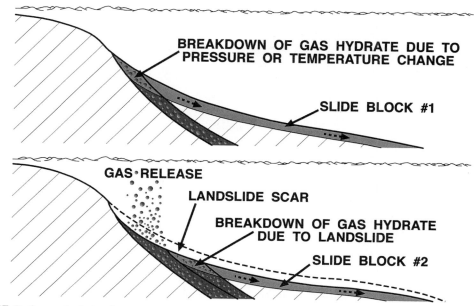

FIGURE 7 Conceptual drawing of the triggering of a landslide and release of methane due to gas hydrate dissociation.

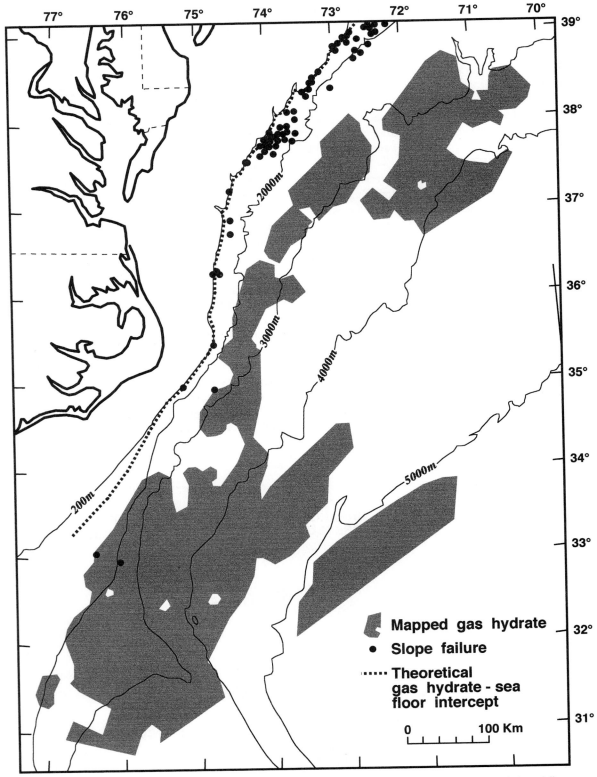

FIGURE 8 Theoretical shallow limit of gas hydrate stability (dashed line) and distribution of mapped slope failures on the Atlantic continental margin of the United States. [From Booth, J. S., Winters, W. J., and Dillon, W. P. (1994). "Circumstantial evidence of gas hydrate and slope failure associations on the United States Atlantic continental margin." *Ann. N.Y. Acad. Sci.* **715,** 487–489].

and even been completely lost from the hole in gas hydrate areas. However, anticipation of the problems allows proper engineering of the drilling, and such issues can be avoided by controlling temperatures and pressures. During the production phase, flow of warm fluids from below the gas hydrate stability zone upward through the conductor pipe could warm adjacent sediments, which could create an expanding dissociation front around the pipe, possibly resulting in such problems as subsidence around the conductor pipe, escape of gas, loss of support for the platform foundation, or erosion by rafting away of sediment with newly formed gas hydrate at the sediment surface. The same considerations of dissociation by warming and loss of support can affect pipelines. Again, control is possible as long as the situation is recognized in advance.

V. EFFECT ON CLIMATE

The gas hydrate reservoir in the ocean sediments has significant implications for climate because of the vast amount of methane situated there and the strong greenhouse warming potential of methane in the atmosphere. It has been suggested that ancient climate shifts, such as the Late Paleocene Thermal Maximum, might have been caused by the release of methane associated with gas hydrate. A reasonably conservative estimate (see Table I) suggests that there is roughly 3000 times as much methane in the gas hydrate reservoir as there is in the present atmosphere. Methane absorbs energy at wavelengths that are different from other greenhouse gases, so a small addition of methane can have important effects. If a mass of methane is released into the atmosphere, it will have an immediate greenhouse impact that will slowly decrease as the methane is oxidized to carbon dioxide in the air. The global warming potential of methane is calculated to be 56 times by weight greater than carbon dioxide over a 20-year period. That is, a unit mass of methane introduced into the atmosphere would have 56 times the warming effect of an identical mass of carbon dioxide over that time period. Because of chemical reactions (oxidation) in the atmosphere, this factor decreases over time; for example, the global warming potential factor is 21 for a 100-year time period. Oxidation of methane can also occur in the ocean water when methane or methane hydrate is released from the seafloor. The methane that reaches the atmosphere can be gas released by dissociation of gas hydrate and/or gas that escapes from traps beneath the gas hydrate seal. The slides and collapses, which were discussed in the previous section, are probably the mechanisms that allow gas to escape from the gas hydrate/sediment system to the ocean/atmosphere system.

In the long term, methane from the gas hydrate reservoir might have had a stabilizing influence on global climate. When the Earth cools at the beginning of an ice age, expansion of continental glaciers binds ocean water in vast continental ice sheets and thus causes sea level to drop. Lowering of sea level would reduce pressure on seafloor hydrates, which would cause hydrate dissociation and gas release, possibly in association with seafloor slides and collapses. Such release of methane would increase the greenhouse effect, and so global cooling might trigger a response that would result in general warming. Thus, speculatively, gas hydrate might be part of a great negative feedback mechanism leading to stabilization of Earth temperatures during glacial periods. Currently, researchers are just beginning to analyze the potential effects of this huge reservoir of methane on the global environment and to consider many hypotheses.

VI. ENERGY RESOURCE

The very large amount of gas hydrate that probably exists in nature suggests the idea that methane hydrate may represent a significant energy resource. However, most of the reports of gas hydrate in marine sediments are only indications that gas hydrate exists at some place (Fig. 6). Almost every natural resource that we extract for human use, including petroleum, is taken from the unusual sites where there are natural high concentrations. Much of the large volume of gas hydrate may be dispersed material and, therefore, may have little significance for the extraction of methane from hydrate as an energy resource. But even if only a small fraction of the estimated gas hydrate exists in extractable concentrations, the resource could be extremely important. The goals for the future use of methane hydrate as an energy resource are to be able to predict where methane hydrate concentrations exist and develop methods to safely extract the gas.

A. Concentration

Most of the gas hydrate in the world (commonly stated as 99%) is formed of biogenic methane. Therefore, understanding the processes that can cause this gas, which is formed by bacteria dispersed through the sediment, to become concentrated into a hydrate accumulation is a critical issue. A small amount of gas hydrate is formed from thermogenic gas rising from great depths along fault channelways that apparently are kept warm (thus outside the gas hydrate stability field) as a result of circulation of warm fluids. The gas-bearing warm fluids reach the ocean floor, where temperature abruptly drops and gas hydrate is formed. This process is common in the Gulf of Mexico.

The accumulations are likely to be small, geometrically complex, and very near the seafloor. Thus, they are likely to be difficult to extract, although, at this point, categorical statements are premature.

In normal, microbially formed, gas hydrate deposits, the common increase in gas hydrate concentration downward through the gas hydrate stability zone toward its base and probable causes were discussed in Section III.C. Direct evidence from drilling and seismic reflection profiling (presence of BSRs) indicates that free gas bubbles typically accumulate just beneath the base of the gas hydrate stability zone, and, presumably, this gas is feeding the gas hydrate accumulation directly above. The concentration of gas beneath the gas hydrate stability zone can vary significantly because the gas can migrate laterally, rising along the sloping sealing surface formed by the base of the gas hydrate stability zone. When it does so, the gas will often reach a site where it becomes trapped at a culmination (shallow spot) and then probably will produce high concentrations of gas hydrate above that spot. Such gas traps can take several forms (Fig. 9). The simplest is formed at a hill on the seafloor, where the base of the gas hydrate stability zone parallels the seafloor and forms a broad arch or dome that acts as a seal to form a gas trap (Fig. 9, top panel). Such a hill can be a sedimentary buildup, such as at the Blake Ridge off South Carolina shown in Fig. 5, where the BSR appears much stronger near the crest, presumably indicating higher concentrations of trapped gas beneath the gas hydrate stability zone there. Alternatively, the seafloor hill could be a fold in an active tectonic setting.

In some cases, a culmination that traps gas at the base of the gas hydrate stability zone can form independent of the morphology of the seafloor. This happens over salt diapirs (plastically rising bodies of lower density salt) as a result of control of the depth of the base of the gas hydrate stability zone by two parameters. First, salt has a higher heat conductivity than sediment, so a warm zone will exist above a salt diapir. Second, the ions that are dissolved out of the salt act as inhibitors (antifreeze) to gas hydrate, just as salt lowers the freezing temperature of water ice. This double effect of chemical inhibition and disturbance of the thermal structure causes the base of the gas hydrate stability zone to be warped upward above a salt diapir, creating a gas trap (middle panel in Fig. 9). A seismic reflection profile across an isolated salt diapir is shown in Fig. 10. Notice that the deep strata appear to be bent up sharply by the rising salt. The base of the gas hydrate stability zone, as indicated by the BSR, also rises over the diapir, and the BSR cuts through reflections that represent sedimentary layers. This does not actually represent a physical bending of the base of the gas hydrate stability zone, but rather a thermal/chemical inhibition that prevents gas hydrate from forming as deeply as it would if the salt diapir

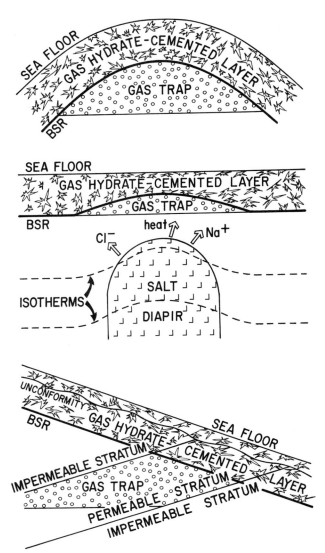

FIGURE 9 Conceptual diagrams showing cross sections through three of the many types of traps that confine gas beneath gas hydrate seals.

were not present. In this case, the BSR certainly does not fulfill its name by "simulating" the seafloor because of the strong influence of variable temperature and chemistry, but it does show us where the base of gas hydrate is located. The extremely strong reflections just below the BSR to the left of the diapir indicate that considerable gas is trapped there and is likely to cause concentration of gas hydrate above the trap. Of course, there are innumerable ways in which gas can be trapped beneath the gas hydrate-bearing zone. A common, simple trap (Fig. 9, bottom panel) is formed where dipping strata of alternating permeability are sealed at their updip ends by the gas hydrate-bearing layer, forming traps in the more permeable layers.

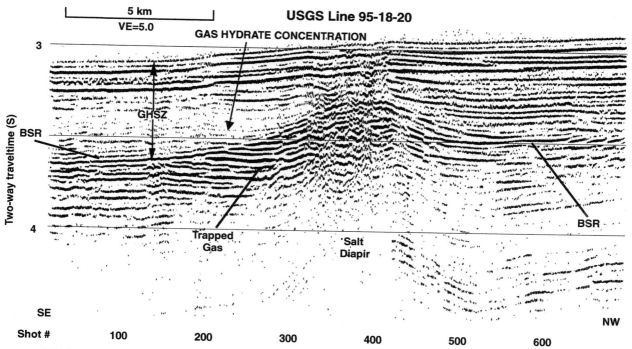

FIGURE 10 Seismic reflection profile across a salt diapir beneath the continental slope off North Carolina in the southeastern United States. GHSZ indicates gas hydrate stability zone within the sediments. The BSR, denoting the base of the gas hydrate stability zone, rises markedly over the diapir because of the thermal and chemical effects created by the diapir. This doming of the base of gas hydrate stability forms a trap for gas. Compare this to the middle diagram of Figure 9.

B. Extraction

The best places to look for gas hydrate deposits for methane extraction are the continental slopes, which, as discussed in Section III.C, seem to have the greatest concentrations of gas hydrate because they contain higher concentrations of organic matter for methane generation by microbes and have greater rates of sediment accumulation, which buries the organic matter and allows microbial decomposition to form gas. Furthermore, the shallower depths of the upper slopes allow the greatest concentrations of methane compared to conventional gas traps. From the minimum depth of hydrate stability down to total depths of about 1600 m, gas is held in hydrate in concentrations greater than in conventional traps. This concentration effect happens because gas molecules are held closer together in the crystal lattice than they would be by simple pressurization of free methane. However, the maximum concentration of methane in hydrate is essentially fixed, in contrast to free gas in reservoirs, where pressure increase will cause increased concentration, so at greater depths, free gas reservoirs will be more concentrated as a result of the higher pressure. The fixed concentration of gas in gas hydrate also means that shallower gas hydrate deposits will release gas that will expand more and fill a greater proportion of the pore volume than comparable deposits at greater depths (at greater pressures). For example, a 5% pore concentration of gas hydrate at 1000 m will provide the same gas expansion as a 15% concentration at 3000 m. Therefore, there is greater likelihood at shallow depths of exceeding the minimum limit of gas volume needed to generate spontaneous gas flow. Both the concentration factor at shallow depths and the expansion considerations suggest that the shallower part of the gas hydrate range may be more favorable for gas extraction—perhaps in the 1000–2000 m range of water depths. Methane probably will be extracted from hydrate by using a combination of depressurization (perhaps primarily), warming (perhaps using warm fluids from deeper in the sediments), and selective use of chemical inhibitors to initiate dissociation at critical locations.

C. Environmental Concerns

As previously discussed, methane is a powerful greenhouse gas, so why would we wish to extract it, perhaps risking its escape to the atmosphere, for purposes of using it as a fuel? The primary environmental reason is that methane is considerably less polluting than other fossil fuels. Methane has the highest hydrogen-to-carbon ratio of all fossil fuels (4:1), and therefore, its combustion produces a minimum amount of carbon dioxide per energy unit—about 34% less than fuel oil and 43% less than coal. Furthermore, methane does not produce any other pollutants such as particulate matter or sulfur compounds, so conversion to methane fuel would significantly clean up our present polluting emissions. The extraction of methane from gas hydrate probably will entail reducing pressure at the base of the gas hydrate stability zone (see Section VI.B). Therefore, overpressures would be reduced, and the likelihood of slides and slumps with release of gas to the ocean/atmosphere system might actually be diminished.

D. The Future

The future of gas hydrate as an energy resource depends on the evolution of (1) geology (to identify concentration sites and settings where methane can be effectively extracted from gas hydrate); (2) engineering (to determine the most efficient means of dissociating gas hydrate in place and extracting the gas safely); (3) economics (the costs are likely to be higher than for extraction of gas from conventional reservoirs); and (4) politics (increased energy security and independence for energy-poor nations are significant driving forces). We are using other energy resources, which are nonrenewable, at significant rates, and the use of methane as fuel is preferable from a greenhouse-warming point of view because of its favorable hydrogen-to-carbon ratio and other pollution concerns. Therefore, methane hydrate holds promise as a future energy resource.

SEE ALSO THE FOLLOWING ARTICLES

ENERGY RESOURCES AND RESERVES • FUELS • GEO-CHEMISTRY, LOW-TEMPERATURE • GEOCHEMISTRY, OR-GANIC • GREENHOUSE EFFECT AND CLIMATE DATA • OCEANIC CRUST • OCEANOGRAPHY, CHEMICAL • OCEAN THERMAL ENERGY CONVERSION • PETROLEUM GEOLOGY

BIBLIOGRAPHY

Booth, J. S., Winters, W. J., and Dillon, W. P. (1999). "Apparatus investigates geological aspects of gas hydrates," *Oil Gas J.* **Oct. 4,** 63–69.

Buffett, B. A. (2000). "Clathrate hydrates," *Annu. Rev. Earth Planet. Sci.* **28,** 477–507.

Henriet, J. P., and Mienert, J., eds. (1998). "Gas Hydrates: Relevance to World Margin Stability and Climate Change," Spec. Publication 137, 338 p., Geological Society, London.

Holder, G. D., and Bishnoi, P. R. eds. (2000). "Gas hydrates—Challenges for the future," *Ann. NY Acad. Sci.* **912,** 1044 p.

Kvenvolden, K. A. (1988). "Methane hydrate—A major reservoir of carbon in the shallow geosphere?" *Chem. Geol.* **71,** 41–51.

Kvenvolden, K. A. (1993). "Gas hydrates—Geological perspective and global change," *Rev. Geophy.* **31,** 173–187.

Max, M. D., ed. (2000). "Natural Gas Hydrates in Oceanic and Permafrost Environments," Kluwer Science, New York.

Paull, C. K., and Dillon, W. P., eds. (2001). "Natural Gas Hydrates: Occurrence, Distribution, and Dynamics," Monograph GM 124, 316 pp., Am. Geophys. Union, Washington, DC.

Paull, C. K., Matsumoto, R., Wallace, P. J., and Dillon, W. P. (2000). "Proceedings of the Ocean Drilling Program," Scientific Results Volume 164, 459 p., College Station, TX (Ocean Drilling Program).

Sloan, E. D., Jr. (1998). "Clathrate Hydrates of Natural Gases," 2nd ed., 705 p., Dekker, New York.

Gas-Turbine Power Plants

Albert C. Dolbec and Arthur Cohn
Electric Power Research Institute

GLOSSARY

Brayton cycle Ideal constant-pressure heating cycle used as a model for the actual gas-turbine simple cycle.

Brayton–Lorentz cycle Ideal cycle used as a model for the combined cycle.

Combined cycle Power plant consisting primarily of gas turbine, heat-recovery steam generator (HRSG), and steam-turbine system. It has a very high efficiency and is used for base and intermediate load.

Combustor(s) Can be one or two silos, many can-type components, or an annular chamber around the shaft. The purpose is to raise the temperature of the compressed air at nearly constant pressure by combustion with injected fuel. Usually there is a primary zone that is approximately stoichiometric, followed by secondary combustion zones and further dilution zones to reduce the temperature to the turbine inlet temperature.

Combustor, dry-low NOx (DLN) To reduce NOx emissions, dry low-NOx (DLN) combustors may be used. These premix the fuel and air plus schedule burning zones to promote more homogeneous combustion avoiding hot zones during starting and loading.

Compressor Component that raises the pressure of the airflow by the action of rotating blades that increase the rotational velocity of the flow followed by stationary vanes that straighten and diffuse the flow. Typically, there are 10–25 stages of compression in axial flow compressors.

Gas-turbine engine Engine consisting of (in simplest form) a compressor section, combustion section, and turbine section. The turbine provides the power to turn the compressor, plus the net output of the engine.

Gas-turbine power plant Consists of gas-turbine engine and electrical generator, plus the controls and accessories that allow controlled power and frequency

Encyclopedia of Physical Science and Technology, Third Edition, Volume 6

output. For the combined cycle it includes the heat-recovery steam generator (HRSG), steam turbine, condenser and cooling tower, and other equipment.

Heat-recovery steam generator (HRSG) Component that transfers enthalpy from the gas turbine exhaust gas to the steam of the steam turbine cycle.

Intercooler Component that removes enthalpy between compressor spools, lowering the power required to run the compressor.

Recuperator Component for the recuperative cycle that transfers enthalpy from the gas turbine exhaust gas to the compressed air exiting the compressor.

Simple cycle Power plant consisting of a gas turbine engine plus controls, accessories, and electrical equipment. It is not high in efficiency, but is low in cost and is used for peaking power.

Turbine (expander) Component that absorbs the power of the hot compressed gas by lowering the pressure. It consists of typically one to five stages, each stage containing a row of stationary vanes (also termed nozzles) to rotate and expand the flow, followed by a row of rotating blades (also termed buckets) that straightens the flow by absorbing the rotational momentum. The turbine is sometimes termed the expander.

Turbine-firing temperature Average total temperature of combustion gas between the first stationary and the first rotating stage of the turbine with all first stationary stage cooling flows mixed into the gas stream.

Turbine inlet temperature Average total temperature of combustion gas upstream of the first stationary stage of the turbine.

GAS-TURBINE POWER PLANTS are those power plants in which gas-turbine engine(s) provide the major portion of the power output. The most common type is the simple cycle, where the comparatively inexpensive gas-turbine engine provides all of the power but the enthalpy remaining in the exhaust gas is *not* utilized. The combined-cycle power plant utilizes the exhaust enthalpy to provide superheated steam to a steam turbine that generates additional power. This type of plant provides the highest efficiency now available, which is expected to increase further as utility gas-turbine engines incorporate advances being developed for aviation applications. This type of power plant is used as a basic unit of many power systems. Of more limited application, to date, are recuperated and steam cooling/injection cycles, where the exhaust enthalpy is utilized back into the gas turbine itself. Also in use for larger units is a reheat or sequential combustion cycle described later.

I. INTRODUCTION

Gas-turbine power plants have evolved continuously since the first land-based installation by Brown Boveri Corporation at Neuchatel, Switzerland, in 1939, and the first aircraft engine turbines developed in Great Britain by Whittle and, in Germany, by von Ohain. Continuing applications of gas turbines for power generation did not occur until the 1950s, and large-scale deployment of this type of power plant was not popularized until the 1960s. Today there are over 115,000 MW of gas-turbine-type plants in U.S. commercial service, representing more than 2700 separate units. Unit sizes range from a few hundred kilowatts for special power-generation applications to the more commonly considered 20- to 270-MW units operated by electric utilities. Unit sizes continue to grow as higher temperatures, greater airflow, and innovative cycle technologies are incorporated into new power generators.

The power-generation application uses two types of design: aeroderivative units modified from aircraft engines, and those especially designed for power-generation service. This distinction exists because gas turbines are available in numerous sizes that have been modified from existing aircraft engine types so that they will operate satisfactorily at constant speed for power generation. Contrasted to this type are gas turbines that have been mechanically designed for power generation to resemble steam turbines as "heavy-duty" designs, used mostly to generate electrical power, but sometimes to operate as mechanical drives, in the latter case frequently at variable speeds. The main applications prompting gas-turbine use are, first of all, aircraft engines and, a distant second, the power-generation application. Other uses for the gas turbine, such as marine propulsion and compressor or other mechanical load drive, use modifications of engines designed for the prime applications mentioned. Generally advances in technology to the state of the art are made in aircraft engines, then later transferred to heavy-duty designs.

During the early development phases of the 1940s and 1950s, gas turbines were largely inefficient compared to other prime movers, but in the case of aircraft had an advantageous output per unit weight and better high speed propulsive efficiency, while in the case of power generation had better unit cost and physical size per unit output. Technology was driven for many years to increase unit sizes and reduce unit costs. During the 1970s, gas turbines operating in combined cycles with steam turbines achieved superior efficiencies to those of other available power generation cycles and thus have broadened their application phases. The types of components employed in gas turbines fall into the general categories of compressors, turbines, and combustors, with particular emphasis on the packaging for power generation application and the materials in the hot gas path compatible with utility service.

II. GENERAL DESIGN FEATURES

A gas-turbine power plant has certain design features and components, regardless of whether it is aircraft derivative or the customized, heavy-duty type design. To illustrate these features, a sectional view of a heavy duty design is shown in Fig. 1. The airflow moves from left to right on this figure, entering the turbine at an aerodynamically designed bell mouth to minimize losses entering the first stage of the compressor. In larger power-generation turbines, the compressor is always an axial flow design, having many stages, exceeding 20 in the largest and newest designs. In Fig. 1, the airflow is reversed after the compressor and reversed again as it flows through the combustion system. There are many different designs in this area, some employing silo-type, single or double combustors, resembling small, pressurized combustion systems. In others, as in Fig. 1, can type or annular type combustors are used. In aircraft derivative units, the combustor is generally annular and in line with the compressor, so airflow continues in the same direction from the compressor through the combustion system. Regardless of how the combustion system is designed, the compressed air is used for primary combustion air, secondary combustion air, and cooling. The primary combustion air is that used for the process of combustion, and it is metered into the flame zone or in the case of DLN combustors, the premix zone. Cooling air is needed to keep the metal parts of the combustion system within their expected temperatures, and all other air is used to dilute the products of combustion and shape the temperature pattern leaving the combustion system. Downstream of the combustion system is the turbine. Roughly two-thirds of the power developed by the turbine is used to drive the compressor, and the remainder of the power is then used

to drive the load. Many configurations exist connecting compressors and turbines together, and the type shown is a single-shaft unit with three stages for the turbine. More or fewer stages can be used, depending on the design and the size, as well as a separate shaft for the power turbine and the load alone. After the turbine, the hot gas exhausts, and in order to minimize the leaving loss, a diffuser is used with appropriate turning vanes as necessary. Outside of the section shown in Fig. 1, exhaust equipment for the turbine generally includes silencers or heat-recovery equipment that can utilize the exhaust heat to improve station efficiency.

The fuel used in the gas turbine can have a serious effect on the design and performance of the power station. Aircraft-derivative turbines generally used only the lightest and cleanest of fuels, including natural gas, various forms of aviation fuel, and distillate oil. The heavy-duty engines occasionally are configured to burn heavy residual oil, but not without significantly increasing maintenance and affecting performance. No commercially available gas turbine can burn directly any of the solid fuels commonly used in electric utility fossil power plants. In the 1940s and 1950s, considerable experimental work was devoted to a direct coal-burning turbine, but no practical design resulted. Gasifying coal or other solid fuels is a practical means for utilizing coal while providing a clean gas to existing turbines. Experimental work continues today toward the utilization of coal in a more direct fashion as fuel for the gas-turbine power plant.

III. COMPRESSOR

Heavy-duty gas-turbine compressors in turbines above 10 MW are multistage axial flow type with pressure ratios between 9 and 30. The large number of stages, up to more than 20, keeps each stage lightly loaded. More heavily loaded stages would permit smaller, fewer numbers of stages and are available on aircraft derivative units. The design techniques are well established, so that these types

FIGURE 1 Sectional view of a heavy-duty gas turbine. (Courtesy of General Electric Co.)

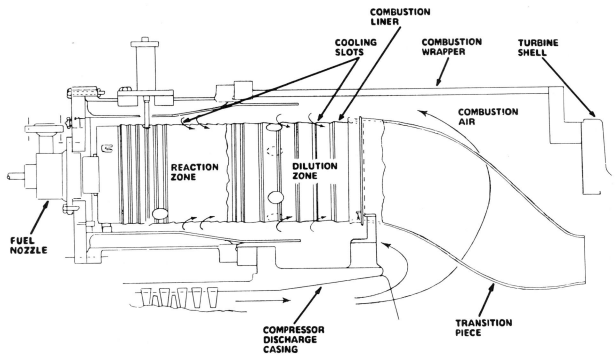

FIGURE 2 Gas-turbine combustion system using can-type combustors.

could be included in heavy-duty designs if the economics warrant. Since weight and size are not significant factors in land-based gas turbines, the compressor design can be optimized for efficiency.

Some aspects of axial-flow heavy-duty compressor design are required to minimize the possibility of stalling or surging, either during startup when speed is a variable, or during low-ambient-temperature operation when mass flow, and hence pressure ratio, increase dramatically. Bleed valves are often located part way through the compressor to unload the airflow during operating conditions near the surge line. In addition, inlet guide vanes to the compressor are variable, so that airflow can be modulated over a small range. Additional variable stationary vane stages are sometimes used to minimize destructive surges during operation and improve part load efficiency.

IV. COMBUSTION

Three distinct types of combustion systems occur in both aircraft and land-based gas turbines. The annular combustor, mainly used in aircraft derivatives, has a multitude of nozzles to inject fuel into a continuous annular chamber located circumferentially around the turbine shaft. Air is metered into the chamber for premix combustion, cooling of the metal, and dilution, with the hot-gas exit annulus connecting to the first stationary stage of the turbine.

Most heavy-duty turbines have either a can-type of combustion system, or a silo-type combustion system. In the former, various cylindrical cans, as many as 16, are located circumferentially around the turbine, and compressor air exits to the outer shell of each can (see Fig. 2). Each can has its own method of metering the air into the primary air for combustion, the cooling air for the metal parts in each can, and the secondary, or dilution, air. Each can that contains the flame is connected to an additional duct called a transition piece. Typically, no additional air is introduced into the system after the combustion can, although in high-temperature units special cooling of the transition pieces or hot-gas ducting is common. The transition piece connects each can into a section of the first-stage stationary vanes. The effect is to have a uniform circumferential flow of hot gas into the first stationary turbine stage: thus the turbine ends of the transition pieces are mated closely together.

The silo-type combustor system can be either single or double pressurized furnaces to which the compressor air is ducted along the outer wall of the silo combustor and hot gas is ducted back to the turbine, entering through a hot-gas casing connected to the first stationary vanes. Silo-type combustors use either ceramic or metal tiles mounted in the flame zone to reduce structural metal temperatures due to flame radiation. An example of a single silo-type combustor mounted directly on the gas turbine is shown in Fig. 3.

FIGURE 3 Gas-turbine combustion system using a single silo-type combustor. (Courtesy of Brown Boveri Corp.)

Reducing the oxides of nitrogen in the gas turbine exhaust can be accomplished by steam or water injection, or by the design of a DLN combustor. Dry systems utilize a premixed combustion arrangement to uniformly lower the flame temperature.

Many factors influence a design of a combustion system, including the type of fuel and environmental performance, but a major design consideration is to present a uniform hot-gas temperature to the first-stage stationary vane of the turbine. Since length and size considerations for heavy-duty turbines are not as significant as for the aeroengines, the pattern of temperatures presented to the first stationary turbine vanes is considerably more uniform for the heavy-duty machines than for the aeroderivatives.

V. TURBINE (EXPANDER)

The turbine (also termed the expander) driving the load equipment plus its own compressor can be mechanically arranged on more than one shaft and have any number of stages appropriate for the size and design practice. Figure 4 illustrates a four-stage turbine that is directly coupled to its own compressor. Single-shaft units with both the compressor and the turbine on the same shaft as the load equipment or generator are most popular in the largest size. Large modern turbine designs provide an efficiency in excess of 90%. They present some unique

and difficult design requirements, not the least of which is cooling the metal to less than 1600°F with gas temperatures entering the first stationary stage greater than 2700°F. Historically, compressor air extracted at the appropriate pressure is used to cool turbine stationary and rotating parts. Sometimes this air is artificially cooled, but in all cases the cooling air is inducted through channels internal to the stationary vanes and rotating blades of the turbine so that metal temperatures may be kept cool in the high-temperature gas environment. Some of the largest units employ steam generated by the turbine exhaust to cool hot gas path parts, thus releasing compressed air to increase unit output and improve efficiency.

The internal passages of the stationary vanes and rotating blades of the turbine are designed in great detail to provide proper cooling. These are generally referred to as one of the three following types. Convection cooling is the direct cooling of the metal by a channel flowing cooling fluid through the part. Impingement cooling is defined as cooling fluid normal to the metal surface, such as a jet onto a flat plate. The third method employs film cooling used as angled holes through the metal to interpose a thin film of coolant between the hot gas and the metal surface. In some designs only one of the cooling methods is employed, yet in others all three are used in varying degrees on different parts of the vanes and blades.

VI. THERMODYNAMIC CYCLES

There is a spectrum of gas-turbine power-plant types. The type most commonly used commercially is the "simple

FIGURE 4 Turbine section of a gas-turbine power plant. (Courtesy of Westinghouse Electric Corp.)

cycle," where an unadorned gas turbine engine provides all of the power. Less common, but currently coming into more prominence, is the "combined cycle" where the considerable energy remaining in the exhaust of the gas turbine engine is used to generate steam, which is then utilized in a steam turbine, providing additional power. Another arrangement being utilized is the steam-injected power plant, in which the steam generated using the exhaust energy is injected back into the gas-turbine engine itself to provide additional power. Also used commercially is the recuperated cycle, where the exhaust energy is utilized to preheat the air prior to the combustor, thus reducing the fuel consumption. In recent years the reheat or sequential combustion cycle has been successfully commercialized.

A. Ideal Brayton Cycle

The thermodynamics of the simple cycle can be understood by first considering the ideal Brayton cycle. The ideal Brayton cycle consists of the following processes for the working gas: (1) an adiabatic (no heat exchange) compression process, which is reversible and thus is also isentropic (no entropy change), (2) a constant-pressure heat-addition process to the maximum temperature, (3) an isentropic expansion process through a turbine back to the original pressure, and (4) a constant-pressure cooling back to the original pressure (exhaust).

On a temperature–entropy diagram, the Brayton cycle has the shape plotted in Fig. 5. For an ideal working gas of constant specific heat at constant pressure (C_p), the enthalpy is proportional to the temperature, so that Fig. 5 can also be considered an enthalpy-entropy diagram with proper scaling of the ordinate. Element (1–2) is isentropic compression. Element (2–3) is constant-pressure heating. Element (3–4) is isentropic expansion, while element (4–1) is constant-pressure cooling. The power, per unit mass flow per second, required for compression is

$h_2 - h_1$, where h stands for the enthalpy per unit mass, while the power produced in the expansion is $h_3 - h_4$. The important feature of Fig. 5 is that the slopes of the constant-pressure elements are proportional to the temperature, so that, at a given entropy, element (2–3) has a greater slope than does element (1–4). Therefore, the lines diverge and the expansion power ($h_3 - h_4$) is greater than the required compressor power ($h_2 - h_1$) and there is net positive power output equal to $(h_3 - h_4) - (h_2 - h_1)$ per unit mass flow.

The power output and efficiency for the ideal Brayton cycle can be evaluated from the thermodynamics of the individual elements. For the isentropic compression process,

$$h_2/h_1 = (p_2/p_1)^{(\gamma-1)/\gamma} \tag{1}$$

where p is the pressure and γ is the ratio of the specific heat at constant pressure to the specific heat at constant volume (for air, γ equals about 1.4). Thus, the required compressor power (per unit mass flow) is

$$h_2 - h_1 = h_1\left[(p_2/p_1)^{(\gamma-1)/\gamma} - 1\right] \tag{2}$$

while for the expansion process the power out is

$$h_3 - h_4 = h_4\left[(p_3/p_4)^{(\gamma-1)/\gamma} - 1\right] \tag{3}$$

Because of the nature of the constant-pressure processes, $p_4 = p_1$ and $p_3 = p_2$; therefore the net power is

$$(h_3 - h_4) - (h_2 - h_1) = (h_4 - h_1)\left(p_r^{(\gamma-1)/\gamma} - 1\right) \tag{4a}$$

$$= (h_3 - h_2)\left(1 - p_r^{-(\gamma-1)/\gamma}\right) \tag{4b}$$

where p_r is the pressure ratio, defined as

$$p_r = p_2/p_1 = p_3/p_4$$

But the energy input rate to the cycle is $(h_3 - h_2)$. Therefore, the efficiency η_B of the Brayton cycle is

$$\eta_B = 1 - p_r^{-(\gamma-1)/\gamma} \tag{5}$$

Since the temperature ratios for both the isentropic compression and expansion elements are given by

$$T_2/T_1 = T_3/T_4 = p_r^{(\gamma-1)/\gamma} \tag{6}$$

then

$$\eta_B = 1 - T_1/T_2 = 1 - T_4/T_3 \tag{7}$$

This differs from, and is lower than, the efficiency η_C of the Carnot cycle for the same temperature limits, which would be

$$\eta_C = 1 - T_1/T_3 \tag{8}$$

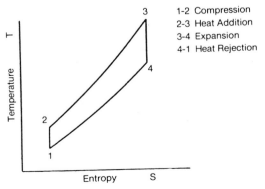

1-2	Compression
2-3	Heat Addition
3-4	Expansion
4-1	Heat Rejection

FIGURE 5 Ideal Brayton-cycle thermodynamic diagram.

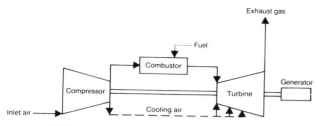

FIGURE 6 Simple-cycle equipment schematic.

B. Simple Cycle

The actual simple open-cycle gas-turbine engine, whose schematic is shown in Fig. 6, differs from the ideal Brayton cycle in a number of ways:

1. The cooling process (4–1) is not directly carried out, but the hot gas at 4 is exhausted and new air at 1 is ingested from the atmosphere. However, this has no effect on the analysis.

2. The compression process, while adiabatic, is not isentropic, having internal friction enthalpy losses. This is accounted for by introducing a compressor efficiency η_c that is typically around 85%.

3. Air may be extracted from various locations in the compressor, mainly for use in turbine cooling.

4. The heat addition process (for all except the very few indirect heated engines) takes place by internal combustion of fuel. The combustion products contain virtually all the original nitrogen in the air, about 70% of the original oxygen, depending on the combustor temperature addition, but the remaining approximately 30% of the oxygen forms CO_2 and HO_2 combustion products. This changes the mass flow, the specific heat C_p, and the ratio of the specific heats γ, so that these cannot be taken as constant through the cycle as in the idealized analysis above. There is also the formation of trace compounds such as NO, NO_2, CO, and unburned hydrocarbons, whose emissions have to be limited in order to meet ecological standards. The unburned hydrocarbons and the CO formation are associated with incomplete combustion, so that the combustion efficiency is not exactly 100%, but it usually is close enough for most practical purposes.

5. The combustion process (or the heat-addition process for indirect cycles) has a pressure drop usually in the neighborhood of about 5%.

6. The turbine (expander) is not isentropic but typically has a turbine efficiency of about 90%.

7. The turbine blades and vanes are cooled with air derived from the compressor or with steam generated from the turbine exhaust gases. This fluid is introduced at various points into the turbine, complicating the cycle analysis.

8. There are also the pressure drops associated with the inlet and the exhaust, as well as the shaft, gear, and generator efficiencies.

A key parameter in determining the performance of the Brayton cycle and thus of the actual gas turbine cycles is the temperature at point 3, which is the average total temperature ahead of the first-stage rotor blades (buckets) but after the first-stage stator vanes (nozzles). This is termed the turbine firing temperature.

Material and cooling constraints limit the firing temperature. However, it should be kept in mind that the turbine firing temperature is the average of the temperature pattern that varies radially and circumferentially. It is the peak temperature of the pattern that tends to be directly constrained by the material and cooling limitations.

In order to minimize capital cost, the simple-cycle gas-turbine engines are usually designed toward maximization of the specific power, the power per unit air flow, rather than the efficiency. The specific power goes up rapidly with the firing temperature but maximizes at a moderate pressure ratio, as shown in Fig. 7. Thus, even though the efficiency (which is not much dependent on the firing temperature) would peak at pressure ratios of about 30 : 1, typical simple-cycle gas-turbine power plants have pressure ratios of between 10 : 1 and 18 : 1, thus optimizing the power out.

VII. OTHER GAS-TURBINE CYCLES

While higher-pressure-ratio machines would have higher efficiency in a simple cycle, if high efficiency is desired, it is better to utilize the considerable enthalpy remaining in the exhaust of the turbine at point 4. This is done in combined cycles, recuperated cycles, and in steaminjected-cycle power plants. These types of power plants rank as having the highest efficiency now available (and with newer gas-turbine engines should improve even further), while having moderate capital cost.

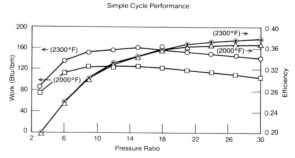

FIGURE 7 Simple-cycle efficiency and specific power screening curves for firing temperatures of 2000°F (1093°C) and 2300°F (1260°C).

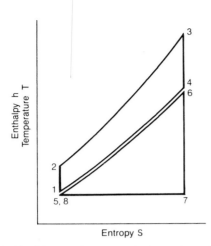

FIGURE 8 Ideal Brayton–Lorentz cycle thermodynamic diagram.

A. Ideal Brayton—Lorentz Cycle

An idealized cycle for the combined cycle is the Brayton–Lorentz cycle, shown in Fig. 8. The gas-turbine part is the Brayton cycle as in Fig. 5. However, now the exhaust of the gas turbine is cooled from point 4 down to ambient at point 1 by heating another high pressure working fluid from liquid at ambient temperature T_5, (which is equal to T_1) up to T_6 (which is equal to T_4) where it is a gas, yet assuming no latent heat of vaporation. This working fluid is then expanded isentropically back down to ambient temperature at point 7 at a pressure that has no relationship whatsoever to ambient pressure or to the pressure ratio of the Brayton part of the cycle. The working fluid is then condensed at constant temperature and pressure to the liquid state at point 8. This liquid is then pumped with negligible change in the enthalpy and temperature up to the high pressure of point 5 to start the Lorentz component of the cycle again.

The efficiency η_L of the Lorentz component of the cycle (obtained by breaking it up into differential Carnot cycles and then integrating) is

$$\eta_L = 1 - \frac{T_1 \ln(T_4/T_1)}{T_4(1 - T_1/T_4)} \qquad (9)$$

where the symbol ln stands for the natural logarithm. The efficiency η_{BL} of the overall Brayton–Lorentz cycle, obtained from the relationship $\eta_{BL} = \eta_B + (1 - \eta_B)\eta_L$, is

$$\eta_{BL} = 1 - \left(\frac{T_1}{T_3}\right)\frac{\ln(T_4/T_1)}{(1 - T_1/T_4)}$$

$$= 1 - \left(\frac{T_1}{T_3}\right)\frac{\ln(T_3/T_2)}{(1 - T_2/T_3)} \qquad (10)$$

Since for given temperature and pressure ratios the Brayton and the Brayton–Lorentz cycle have the same heat input rates $h_3 - h_2$, the power outputs of the two cycles are in exactly the same proportion as are their efficiencies:

$$P_B = (T_3/T_1)(1 - T_2/T_3)(1 - T_1/T_2) \qquad (11)$$

$$P_{BL} = (T_3/T_1)(1 - T_2/T_3)[1 - (T_1/T_3)]\frac{\ln(T_3/T_2)}{(1 - T_2/T_3)} \qquad (12)$$

where P_B is the power output (per unit inlet enthalpy flow) for the Brayton cycle and P_{BL} is for the Brayton–Lorentz cycle. In Fig. 9 are plotted the efficiencies of these two cycles as a function of their power outputs for various values of T_3/T_1. The Brayton cycle has two branches—the low-efficiency, low-pressure ratio branch up to the maximum power [where $T_2/T_1 = (T_3/T_1)^{1/2}$], and the high-efficiency, high-pressure ratio branch back to zero power at the Carnot efficiency. The Brayton–Lorentz has only one branch going from zero power (where the efficiency equals the Carnot) to the maximum power point where the Brayton pressure ratio equals 1 and the cycle reduces to the pure Lorentz. In all nonzero cases, the Brayton–Lorentz has both higher power and efficiency than the Brayton (for the same T_3/T_1 ratio).

B. Combined Cycle

In the actual combined plant, the enthalpy remaining in the turbine exit gas is utilized to raise high-pressure

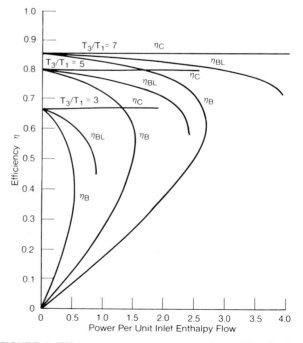

FIGURE 9 Efficiency versus power comparison of the Brayton and Brayton–Lorentz cycles.

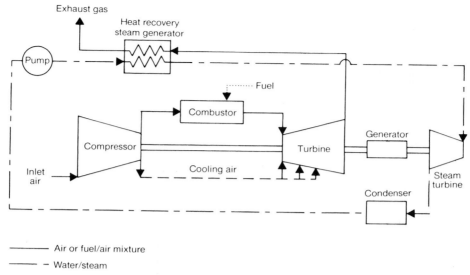

FIGURE 10 Combined-cycle equipment schematic. Solid lines show air or fuel–air mixture, and dashed lines indicate water–steam.

superheated steam, which is then expanded in a steam turbine to a high vacuum to make additional power. The steam is then condensed to liquid water, pumped back to high pressure, and then goes back through the cycle to be heated by the gas-turbine exhaust. One mode is for there to be one steam turbine per gas turbine sharing the same shaft with a generator as shown in the schematic of Fig. 10, but just as common are installations where a number of gas turbines' exhausts produce steam for a single steam turbine, each machine having a separate generator.

The transfer of enthalpy from the gas-turbine exhaust to the steam occurs in what is termed the heat-recovery steam generator (HRSG). This enthalpy transfer may be improved by judicious selection of steam conditions and by multiple steam pressure and flows. However, only the single-pressure steam system will be analyzed.

As the gas-turbine exhaust transfers enthalpy to the steam via heat transfer to the steam flowing in tubes imbedded in the exhaust, it loses temperature. The specific heat at constant pressure, C_p of the exhaust products is reasonably constant, and thus the loss in temperature is almost linear with the enthalpy transferred. Ideally (as in the Brayton–Lorentz Cycle), the enthalpy pickup by the steam should also be linear (and at a temperature as close to that of the exhaust gas, as feasible). However, because of the large latent heat of vaporization of the steam, this is impossible.

The process can be visualized by the temperature–enthalpy transfer diagram of Fig. 11. The heat-transfer process is countercurrent. The enthalpy-transfer process starts (for the water) with the pressurized water at point L being heated to boiling at point B_L, then boiled at virtually

constant temperature until it is fully vaporized, and then superheated to the maximum temperature at point M.

The critical points of this process are the boiling temperature T_B and the maximum temperature T_M. The maximum steam temperature must be below the gas turbine exit temperature T_4. The difference $T_4 - T_M$ is termed the superheat approach temperature and is typically about 50°F (28°C). If the gas-turbine exit temperature is above about 1050°F (565°C), the steam final temperature may be further limited by material constraints to be not above about 1000°F (537°C).

The boiling point must be below the value of the exhaust gas temperature T_p at the point of the HRSG where the boiling begins. The difference $T_p - T_B$ is termed the pinch temperature difference. Typically, the pinch temperature difference is also about 50°F (28°C). However,

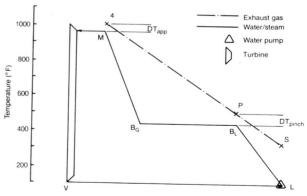

FIGURE 11 Temperature–enthalpy transfer diagram for the combined cycle.

decreasing it improves the efficiency, although increasing capital costs. Lower values of about 30°F (17°C) may be optimum economically, depending on the relative costs of fuel and capital.

The steam flow that can be made per unit exhaust gas flow is determined by these temperatures. The enthalpy transferred by exhaust gas going from T_4 down to T_p equals the enthalpy pickup of the steam from the beginning of boiling to the final superheat point; thus,

$$C_p(T_4 - T_P) = (h_M - h_{BL})m_s \qquad (13)$$

where m_s is the ratio of the mass flow of the steam to the mass flow of the gas turbine exhaust, h_{BL} is the enthalpy of liquid water at the boiling temperature and pressure, and C_p is the average specific heat of the exhaust gas flow. For a choice of steam pressure and thus the boiling temperature, for given gas-turbine exit temperature and approach and pinch-temperature differences, Eq. (13) determines the steam flow rate m_s.

The power of the steam turbine is then determined by

$$P_{ST} = m_s(h_M - h_V)\eta_{ST} \qquad (14)$$

where h_V is the enthalpy per unit mass flow of the steam at the vacuum conditions of the steam turbine exit and η_{ST} is the steam-turbine efficiency.

The steam-turbine power is typically about 30–50% of the gas turbine power and is obtained without any additional fuel consumption. This raises the efficiency of the combined cycle plant typically into the 40–50% range. The amount of steam produced and the final superheat enthalpy both decrease for a lower gas-turbine exit temperature that is associated with a high-pressure-ratio gas-turbine cycle. Thus, the efficiency of the combined cycle optimizes at a lower pressure ratio than for the simple cycle, typically at about 15 : 1, which is in the same range for which the simple cycle optimizes on specific power.

The combined-cycle efficiency may be improved significantly by the use of more than one steam pressure and by the use of steam reheat. This allows more of the exhaust energy to be utilized, and less loss of thermodynamic availability in the heat transfer process. The improvement in efficiency in going from a single pressure nonreheat to a three-pressure reheat steam-bottoming system can be 2 points. The steam turbine power can be increased, but at a cost to the efficiency, by firing further fuel at the beginning of HRSG.

Some of the newest gas turbines have firing temperatures approaching 2700°F. If this is the case, and a nonfired, two-pressure, reheat steam-bottoming cycle is applied, thermodynamic efficiencies approaching 60% (lower heating value) are realized. While this efficiency is very attractive, the application of these types of power plants was initially limited because of their requirement for premium fuels such as natural gas and distillate liquid fuels. Combined-cycle use has become quite common as the cost, availability, and environmental intrusion of these fuels has been favorable. The development of the integrated coal gasification combined cycle (IGCC) plant brings the option of utilizing low-cost coal, albeit at an increase in capital cost.

C. Steam-Injected Gas Turbine Cycle (STIG)

The steam produced in the HRSG may be utilized in the gas turbine rather than being expanded in a separate steam turbine. This is sometimes partially done in order to reduce the NO_x formation in the gas-turbine combustor. However, the purposeful use of all the HRSG steam in the gas turbine is considered a separate cycle, termed the steam-injected gas turbine or the dual-fluid cycle. The efficiency (based on the higher heating value) and specific power for a steam-injected cycle as a function of the steam to compressor inlet airflow ratio (for a firing temperature of 2200°F) are shown in Fig. 12. In this calculation, all of the steam is injected into the combustor region ahead of the first turbine stage. From Fig. 12 it can be seen that the peak efficiency occurs in the pressure-ratio range of 20 : 1 to 30 : 1, almost as high as that for the simple cycle. This is because the steam expansion is restricted to the expansion ratio of the gas turbine engine. Therefore a higher pressure ratio allows more expansion down to a lower exit enthalpy. Aeroderivative engines, which tend to have higher pressure ratios, thus are good candidates for conversion to the steam-injected cycle.

The improvement in both power and efficiency compared to the simple cycle is dramatic. Much of the enthalpy that would have been wasted at the turbine exit is now put to use. However, the efficiency of the optimum steam-injected cycle still tends to be slightly lower than that of the optimum combined cycle. This is because of the limitation on the expansion of the steam in the gas turbine down to the minimum pressure of 1 atm, while in the steam turbine it is expanded to a high vacuum. However, the specific power of the steam-injected cycle is even higher than that of the combined cycle. This is because the steam, after its superheating in the HRSG, is further very highly superheated up to the level of the gas-turbine firing temperature. More fuel is utilized in combustion to heat this steam than would be used for the same airflow and firing temperature in the combined cycle, thus providing more power, although at a small decrease in efficiency compared to the combined cycle.

Also, the steam pressures used in the HRSG are now limited by compatibility with the pressure in the location of the gas turbine where the steam is introduced. In order for the steam to be compatible, the steam pressure should

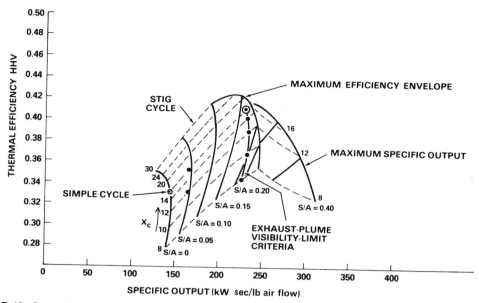

FIGURE 12 Steam-injected gas-turbine efficiency and specific power screening curve for a firing temperature of 2200°F (1240°C). S/A, steam/air flow; X_c, compressor pressure, ratio.

be about 50 psi greater than the pressure in the gas turbine at the injection location. It is desirable to introduce as much steam as is feasible into the combustor and other areas between the compressor exit and the entrance to the turbine (expander) in order to obtain the most expansion ratio of the steam. However, similarly to the multipressure HRSG for the combined cycle, additional steam can be made using a second, or even a third, lower pressure level. This lower pressure steam can be injected downstream between turbine stages or, more practically, between turbine spools for multispool gas turbines. Since the expansion ratio is lower for this type of injection location, the added power is proportionately less.

D. Recuperative Gas-Turbine Cycle

The fourth gas-turbine type with some widespread commercial use is the recuperative gas turbine. This is intermediate in efficiency between the combined and simple-cycle power plants and is also intermediate in capital costs. Thus, it finds application where a reasonably high efficiency is desired but the higher capital cost, water usage, and/or the complications of running a steam plant are not desired.

The recuperative cycle utilizes the exhaust enthalpy, by heat transfer in a recuperator, directly back into the compressor exit air. It ideally has the same power output compared to the simple cycle but improves the efficiency by lowering the need for fuel use (or external heat addition). This is in contrast to the combined cycle, which ideally has the same fuel use as the simple cycle (for the same engine parameters), but has a higher power output.

The cycle of the ideal recuperative–Brayton cycle is shown in Fig. 13. From point 4 to point 6, the turbine exhaust gases transfer enthalpy to the compressor exit air, which gains enthalpy from point 2 to point 5. This enthalpy transfer takes place by heat transfer in what is termed the recuperator. The effectiveness ε of the recuperator is defined as

$$\varepsilon = \frac{T_5 - T_2}{T_4 - T_2} = \frac{T_4 - T_6}{T_4 - T_2} \qquad (15)$$

In this ideal cycle, as for the previous ideal cycles, it is assumed that the mass flow and the specific heat at constant pressure are constant throughout the cycle.

Even with an ideal recuperator with an effectiveness equal to 1, the proportion of exhaust enthalpy usable in

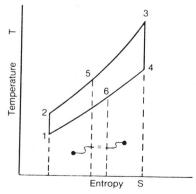

FIGURE 13 Ideal recuperative–Brayton cycle thermodynamic diagram: 1–2 compression, 2–3 head addition, 3–4 expansion, 4–1 heat rejection, 4–6 recuperation, 2–5 recuperation.

the recuperative–Brayton cycle is limited to $(T_4 - T_2)/(T_4 - T_1)$. The efficiency of the recuperative cycle, η_R, is

$$\eta_R = \frac{(h_3 - h_4) - (h_2 - h_1)}{(h_3 - h_5)} \quad (16)$$

As noted above, this is the same power out (numerator), but with reduced heat input (denominator) as the pure Brayton cycle. Assuming the effectiveness ε is 1 and using the transformations of Eqs. (1), (4), and (6) (for constant specific heat),

$$\eta_R = 1 - \frac{T_2}{T_3} \quad (17)$$

This is lower than the Carnot efficiency if $T_2 > T_1$. For the limit $T_2 = T_1$, the power output of the recuperative–Brayton cycle is zero; there would be no compression and no heat input. Conversely, for $T_2 > T_4$ the recuperator is ineffective and no heat is transferrable.

The power of the cycle per unit mass flow is, of course, the same as the pure Brayton cycle. The efficiency at a given power level is exactly the same as for the upper branch of the efficiency–power curve for the pure Brayton cycle of Fig. 9, except that now it occurs at the same low-pressure ratio as for the lower branch of the curve.

The actual recuperated gas-turbine power plant differs from the ideal recuperated–Brayton by the recuperator having an effectiveness of less than 1 and in having pressure drops on both the cold and hot sides of the recuperator. (This is in addition to the other nonideal features listed above in connection with the simple cycle.) These cause the efficiency to be lower than the ideal, lower the power output below that of the simple cycle, and optimize the efficiency at a pressure ratio typically in the range of 6 : 1 to 10 : 1. Typically, recuperator effectiveness is in the range of 0.80 to 0.90, while total pressure drop through both sides of the recuperator is approximately 5–10%.

E. Closed Cycle

The recuperative cycles are in limited use as open cycle machines. However, it is the standard arrangement for the closed-cycle power plants. In the closed cycle, the same working fluid is reused through the cycle requiring a cooler between point 6 and point 1 of Fig. 13. The heating from point 5 to point 3 is now done by heat transfer from an external heater. For the closed cycle, point 1 is not generally at ambient conditions. In particular, the pressure at point 1 may be quite higher than 1 atm. The mass flow, and thus the power per unit volume flow, may therefore be greatly increased compared to the open cycle. However, because of the material limitations of the heat exchanger, the temperature at point 3 is restricted to well below that of open-cycle combustion-fired turbines, typically to 1300–1400°F (700–750°C).

F. Intercooled Cycle

Another cycle in limited use is the intercooled cycle. The intercooler is inserted between two spools of the compressor, thus lowering the inlet temperature of the second spool. This decreases the power requirement of the second spool of the compressor. Therefore, the net power of the cycle is larger than that of the simple cycle. However, even though the power is improved the efficiency is little changed, since the enthalpy at the combustor inlet is now lower than for the simple cycle. Thus, more heat addition is required. More than one stage of intercooling is feasible, but all known existing installations contain only one stage.

A small number of power plants have been built that combine the intercooling and recuperative components. This raises the efficiency significantly compared to having either feature separately and also raises the specific power toward that of intercooling alone. However, the efficiency and the overall economics have, to date, still not measured up to that of the combined cycle and have not found widespread use.

Recent studies have further indicated that the provision of after-cooling the compressor exit air followed by saturation with hot water ahead of the recuperator inlet would significantly increase the efficiencies of these cycles.

VIII. REHEAT (SEQUENTIAL COMBUSTION)

The combustor exit gas in the gas turbine engine usually contains enough oxygen (typically about 15%) to sustain further combustion. In a reheat engine, after expansion partially through the turbine the gas is introduced into a second-stage reheat combustor where additional fuel is burned, raising the gas temperature again. The gas is then fully expanded through the rest of the turbine. High pressure ratios are favored with reheat. Whereas reheat cycles have been utilized occasionally for more than 50 years, a large unit (265 MW, 50 Hz) has been successfully commercialized in the 1990s. Applications have been made that also include intercooling and recuperation. The intercooled–reheat combined cycle (for given cycle restraints) tends to have a somewhat higher efficiency and specific power than the standard combined cycle. However, it is at the cost of additional equipment, and these multicomponent cycles have not yet had more than occasional use.

IX. MATERIALS LIMITATIONS

The evolution of gas-turbine technology had to await the development of high-temperature metals before

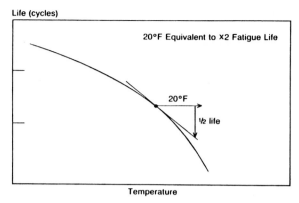

FIGURE 14 Typical metal temperature versus fatigue life for superalloy material.

competitive efficiencies and even a net power output could be achieved. During the 1940s and 1950s, various materials called superalloys were developed based on nickel or cobalt but containing chronomium, aluminum, and other metals in varying percentages. This class of materials enabled metal temperatures to approach 1600°F currently for heavy duty engines, and even higher, for aircraft applications. Figure 14 shows a typical fatigue life in cycles versus temperature for modern super alloys. At modern temperatures, the creep life of these materials can be cut in half by as small as a 20°F increase in temperature.

The materials problem in modern gas turbines for power generation is complicated by the fact that corrodents existing in the fuels and in the ambient air can subject hot-gas path materials to high-temperature oxidation and also to what is termed sulfidation. The term sulfidation is a corrosion of the metal parts that occurs when there is sulfur in the fuel or air. Both sulfidation and oxidation are corrosion phenomena, and in addition, if residual fuel is used, a vanadium attack can occur. Corrosion effects require the use of coatings for most stationary gas-turbine applications. These coatings are another metallic compound generally applied to the surface of the metal by processes that permit adherence over many thousands of operating hours.

Considerable research is directed toward materials to operate at higher temperatures. Currently available techniques involve solidifying the superalloy metal casting to produce several single crystals in the direction of the stress, or even a complete single crystal part. Materials composed of one or several crystals along the direction of the stress have considerably more strength at high temperatures than the normal polycrystalline materials. Additional techniques to form metals for high-temperature applications are being researched, as well as ceramic parts that may some day be appropriate in large gas turbines.

X. PACKAGING, CONTROLS, AND ACCESSORIES

The design of a power plant employing gas turbines is approached quite differently than other forms of utility power plants. The gas turbine provides a considerably higher output in a small volume than other plants and is of such a compact design that the basic engine can be shipped virtually complete from a factory. Thus a gas-turbine power plant consists of a group of compact modules, one of which is the gas turbine, and others are a control package and various mechanical accessory systems, such as lubricating oil, fuel, or water. These are arranged in an overall plant that is remarkably similar from one site to another. A typical site for a simple-cycle power plant is shown on Fig. 15. Gas-turbine plants are characterized by very low profiles and minimum required acreage. In addition, a simple-cycle unit requires no water supply, except for services, and these plants have been located in many areas generally not considered suitable for power-generating equipment.

The mechanical accessories required for the gas turbine can be either mechanically driven off the main turbine shaft or electrically driven. The latter arrangement permits more flexibility of location, and in addition allows redundancy to improve the reliability of the unit. Owing to the compact nature of the plant, plus the fast start and rapid load-change features of gas turbines, the control system is considerably more automated than usually found in other types of power plants. Originally, mechanical controls were used, but since the 1960s gas turbine controls have been electrical and of an integrated design to accomplish the functions of control, automation, and protection for the entire power plant. In general, gas turbines are controlled by speed or load for power-generation use, but with limits on maximum gas temperature that prevent units from being overfired.

FIGURE 15 Simple-cycle gas-turbine packaged power plant. (Courtesy of Westinghouse Electric Corp.)

XI. POWER-GENERATION APPLICATION

There are two distinct applications of gas turbines for power-generation purposes on electric utility systems. These are generally referred to as the peaking application and the baseload application.

A look at Fig. 16 will help to understand the nature of these applications on a power system. This figure shows the load duration curve of an electric utility system describing the percentage of the year each percent of the yearly maximum load is required. Note that there is a percentage of maximum load that is required for 100% of the entire year. This is usually referred to as the base load of the power system and is required day and night, weekend and weekdays, for all seasons of the year. There are just a few hours during the year that require a certain amount of capacity, perhaps about 10%. Generally these will be a few hours in the late afternoon and early evening on weekdays and are the peak load time of the day. Often this occurs during the summer because of air-conditioning uses, but also in northern areas it can occur during the winter months. It is economic for power systems to install generation serving only this peak load. Generally these peak loads are needed for less than 1000 h per year, but equipment serving the load must be started and stopped frequently as each daily "peak" may only be 2–5 h. Usage is generally these 2 to 3 months in summer and/or winter. Characteristics of the peak load application are that the generation required should be low cost, not necessarily the highest efficiency, and suitable for frequent fast starts and stops. The application is well suited to gas turbines, and it is for this reason that much of the gas-turbine generation in the United States has been installed by electric utilities.

The baseload portion of the power system requirement, as well as load duration cycles between base and peak,

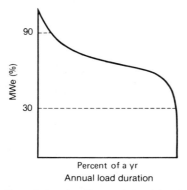

FIGURE 16 Typical electric utility yearly load duration curve, percent electrical megawatts versus percent of year.

can also be served by gas turbines, but since the operating hours per year are longer, more efficient units are generally required than for simple-cycle turbines. In these areas combined cycles, reheat cycles, etc., have been applied and used where oil- or gas-fired capacity is needed for power generation.

Between peaking application and baseload application are loads that must be served from perhaps 80% of the time to maybe 20% of the time, depending on the power-system characteristics. This area under the load duration curve is sometimes called intermediate load and often is served by generation capacity less efficient than baseload equipment, but more efficient than peaking generation.

SEE ALSO THE FOLLOWING ARTICLES

COMBUSTION • ENERGY RESOURCES AND RESERVES • FUELS • INTEGRATED GASIFICATION COMBINED-CYCLE POWER PLANTS • JET AND GAS TURBINE ENGINES • TURBINE GENERATORS

BIBLIOGRAPHY

Becker, B. (1990). "Development of Gas Turbines in Germany," Section 7, Proceedings: 1989 Conference on Technologies for Producing Electricity in the Twenty-First Century. EPRI GS-6991, Palo Alto, CA.

Brown, D. H., and Cohn, A. (1979). "Steam Injected Gas Turbine Study," EPRI AF-1186, Palo Alto, CA.

Boyce, M. P. (1982). "Gas Turbine Engineering Hand-book," Gulf Publishing, Houston, TX.

Boyen, J. L. (1980). "Thermal Energy Recovery," 2nd ed., Wiley, New York.

Eldrid, R. et al. (1999). "Continuing Evolution of the F Gas Turbine," Presented at Power Gen International, New Orleans, Dec. 2, 1999.

Foster-Pegg, R. W. (1978). "Steam bottoming plants for combined cycles," J. Eng. Power **100**(2).

Joos, F., Brunner, P., Stalder, M., and Tschirren, S., "Field Experience with the Sequential Combustion System of the GT24/GT26 Gas Turbine Family," ABB Review May 1998.

Lucas, H., and Cohn, A. (1986). "Survey of Alternative Gas Turbine Engine and Cycle Design," EPRI AP-4450, Palo Alto, CA.

Morton, T. R., and Rao, A. D. (1990). "Perspective for Advanced High Efficiency Cycles Using Gas Turbines," Section 7, Proceedings: 1989 Conference on Technologies for Producing Electricity in the Twenty-First Century. EPRI GS-6991, Palo Alto, CA.

Sawyer, J. W., and Japikse, D. (1986). "Sawyers Gas Turbine Engineering Handbook," 3rd ed., 3 vols. Turbomachinery International Publishers, Norwalk, CT.

Southall, L., and McQuiggan, G. (1995), "New 200MW Class 501G Combustion Turbine," ASME Paper No. 95-GT-215, American Society of Mechanical Engineers, New York.

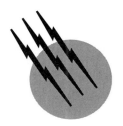

Gene Expression, Regulation of

Göran Akusjärvi

Uppsala University

GLOSSARY

Exon In eukaryotes, the part of the precursor-RNA that reaches the cytoplasm as part of a mRNA, rRNA, or tRNA (see also intron).

Intron The part of the precursor-RNA that is removed during RNA splicing, before the mature mRNA, rRNA, or tRNA is transported to the cytoplasm (see also exon).

Nucleosome The basic structural unit used to condense DNA in a cell. The nucleosome consists of a disc-shaped core of histone proteins (H2A, H2B, H3, and H4) around which an approximately 146-base-pair segment of DNA is wrapped.

Precursor-RNA The nuclear RNA transcript produced by transcription of the DNA. The precursor-RNA contains both exonic and intronic sequences. Introns are removed by RNA splicing before the mature RNA is transported to the cytoplasm.

Promoter The DNA sequence element that determines the site for transcription initiation for an RNA polymerase.

RNA splicing The nuclear process by which introns are removed from the precursor-RNA before the mature mRNA, rRNA, or tRNA is transported to the cytoplasm.

U snRNP An RNA protein complex consisting of uridine-rich small nuclear RNAs (U snRNAs) complexed with a common set of proteins, the Sm proteins, and U snRNA-specific proteins. The U1, U2, U4, U5, and U6 snRNPs are involved in RNA splicing. Other U snRNPs serve other functions in the cell.

OUR GENES are stored in a stable DNA molecule that is faithfully replicated and transmitted to new daughter cells. Each cell in our body contains the same genetic information. The development of complex multicellular organisms such as humans with highly specialized organs and cell types then arises through a complex regulation of expression of the genetic material. Recent advances in whole-genome sequencing have suggested that the difference between a simple organism like a bacteria and a much more complex organism like a human may only result from as little as a 5- to 10-fold difference in the number of genes. Thus, a prototypical bacterial genome encodes for 2000–4000 genes, whereas the recently completed sequencing of the human genome suggests a total gene number slightly more than 20,000. Although this estimate may be too low, it appears unlikely, based on other measurements, that the number of genes in humans will exceed 50,000. At a first, and even a second, glance this small difference in the number of genes makes it difficult to understand why humans and bacteria are so different from each other.

The past decade has seen an explosive increase in information about regulation of gene expression. This review summarizes some of the general themes that have emerged. It is focused on expression of protein-encoding genes in higher eukaryotes. At appropriate places a comparison with gene expression in prokaryotes is made in an attempt to highlight similarities and to show differences that might provide some answers to how mechanistic differences in the regulation of genes may provide at least part of the solution of to how a complex organism like humans may have arisen without an enormous increase in the number of genes compared to prokaryotes.

I. INTRODUCTION

Expression of the genetic information has been summarized in the so-called central dogma, which postulates that the genetic information in a cell is transmitted from the DNA to an RNA intermediate to protein. A major difference between simple and complex organisms is the existence of a cell nucleus. Thus, prokaryotes, which include the bacteria and the blue-green algae, do not have a nucleus, whereas eukaryotes, which include animals, plants, and fungi, have cells with a nucleus that encapsulates the DNA. The basic mechanisms to regulate gene expression in eukaryotes and prokaryotes are very similar, although eukaryotes generally use more sophisticated methods to squeeze out more information from the DNA sequence. In prokaryotes on–off switches of transcription appear to be the key mechanism to control gene activity, although other mechanisms also contribute to the control of gene expression: transcriptional attenuation, transcriptional terminations, and posttranscriptional effects. In eukaryotes similar mechanisms are in operation. However, a key difference between prokaryotes and eukaryotes is the extensive use of RNA processing to generate a mature mRNA. Thus, eukaryotic genes are encoded by discontinuous DNA segments that require a posttranscriptional maturation to produce a functional mRNA. As will be discussed later in this review, the requirement for RNA splicing may be a key to the development of a highly differentiated organism like humans. The general postulate that one gene makes one protein was derived from genetic studies of bacteriophages and does not apply to higher eukaryotes. Because of alternative RNA processing events a large fraction of eukaryotic genes encode for multiple proteins (see Section V).

II. DEFINITION OF A TRANSCRIPTION UNIT

A transcription unit represents the combination of regulatory and coding DNA sequences that together make up an expressible unit, whose expression leads to synthesis of a gene product that often is a protein but also may be an RNA molecule. In prokaryotes, proteins in a specific metabolic pathway are often encoded by genes that are clustered and transcribed into one polycistronic mRNA. A polycistronic mRNA encodes for multiple proteins. In such mRNAs, ribosomes are recruited to internal translational initiation sites through an interaction between the 16S ribosomal RNA and the so-called Shine–Delgarno sequence located immediately upstream of the translational start codon that is used to initiate protein synthesis.

In eukaryotes, in contrast, the primary transcription product is a precursor-RNA that undergoes several posttranscriptional maturation steps before it is transported to the cytoplasm and presented to the ribosomes. Thus, the 5′ end of the pre-mRNA is capped early after transcription initiation by addition of an inverted methylated guanosine nucleotide (the m7G-cap), the pre-mRNA is cleaved at its 3′ terminus, and an approximately 250-nucleotide poly(A) tail is added posttranscriptionally; finally, the pre-mRNA is spliced to remove the intervening intron sequences, and thus form the spliced mRNA which is transported to the cytoplasm. These posttranscriptional processing events give eukaryotes a unique, very important level to control gene expression (see Section V.). Furthermore, a eukaryotic mRNA usually is functionally monocistronic. This means that even if the mRNA encode for multiple open translational reading frames, the open reading frame closest to the 5′ end of the mRNA is typically the only one translated into protein. This results from the fact that the eukaryotic ribosome recognizes the mRNA by binding to the modified 5′ end of a mRNA (recognizing the cap nucleotide), whereas prokaryotic ribosomes recognizes internal Shine–Delgarno sequences in the polycistronic mRNA.

Transcription involves synthesis of an RNA chain that is identical in sequence to one of the two complementary DNA strands. DNA sequence elements upstream of the initiation site for transcription make up the promoter that binds the RNA polymerase responsible for synthesis of the precursor-RNA. Transcription can be subdivided into at least three stages: (1) initiation, which begins by RNA polymerase binding to the double-stranded DNA molecule and incorporation of the first nucleotide(s); (2)

elongation, during which the RNA polymerase, by a processive mechanism, moves along the DNA template in a 5′-to-3′ direction and extends the growing RNA chain by copying one nucleotide at a time; and (3) termination, where RNA synthesis ends and the RNA polymerase complex disassembles from the transcription unit.

Because of space limitation this review will mostly cover RNA synthesis and maturation of protein-encoding messenger RNAs (mRNAs). However, similar mechanisms are used to regulate synthesis of other types of RNA molecules.

III. REGULATION OF TRANSCRIPTION IN EUKARYOTES

Eukaryotic cells contain three DNA-dependent RNA polymerases which are responsible for synthesis of specific RNA molecules in the cell. RNA polymerase I is responsible for the synthesis of ribosomal RNA, RNA polymerase II is responsible for the synthesis of protein-coding mRNA and four small stable RNAs involved in splicing (U1, U2, U4, and U5), and RNA polymerase III is responsible for the synthesis of small RNAs such as transfer RNA (tRNA) and 5S RNA and a whole array of small RNAs including U6, which is involved in splicing. The three polymerases are large enzymes consisting of approximately 15 subunits each. Many subunits are shared among the different polymerases, whereas others are unique and determine the promoter specificity during the transcription process. All three eukaryotic RNA polymerases contain core subunits that show a great homology with the *Escherichia coli* RNA polymerase, suggesting that the basic mechanism of RNA synthesis evolved early during evolution and is conserved.

As will become important later in this review, the largest subunit of RNA polymerase II contains a 52-times-repeated stretch of seven amino acids at the carboxy-terminus. This heptapeptide repeat is refereed to as the carboxy-terminal domain (CTD) and contains serines and a tyrosine that contribute to transcriptional regulation via reversible phosphorylation. The RNA polymerase that assembles at the promoter contains an unphosphorylated CTD tail. This CTD tail anchors the polymerase to the promoter by making interactions with TFIID bound at the TATA box. The release of the RNA polymerase from the promoter (i.e., start of elongation) is associated with a phosphorylation of the CTD tail.

A. Structure of a Eukaryotic Promoter

The transcriptional activity of a prototypical RNA polymerase II gene is regulated by a series of DNA sequence

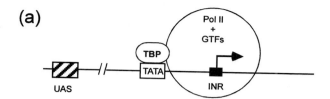

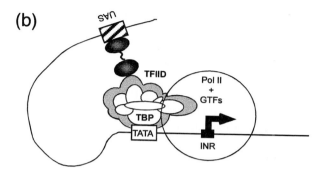

FIGURE 1 A schematic model for preinitiation complex formation on a core promoter (a) and a promoter regulated by enhancer-binding transcription factors (b). The figure is meant to illustrate that the TATA-binding factor TBP is sufficient for basal transcription, whereas enhancer-dependent transcription requires TFIID, which consists of TBP plus the TBP associated factors (TAFs). Pol II, RNA polymerase II; GTFs, general transcription factors; INR, initiator region.

elements that can be subdivided into the core promoter element, which consists of the transcriptional start site and the TATA element, and upstream regulatory elements, which are needed for regulated transcription (Fig. 1).

1. The Core Promoter

The TATA element located 25–30 base pairs (bp) upstream of the transcription initiation site is critical for formation of the preinitiation complex by functioning as the binding site for the TATA-binding protein (TBP). The core promoter is sufficient to direct basal (unregulated) transcription. In some genes the transcriptional start site includes an initiator (Inr) element that binds specific factors that may substitute for the TATA box in recruiting the basal transcriptional machinery to the promoter. Although the core promoter is of fundamental importance for binding of the general transcription apparatus, the composition of elements may influence regulation of promoter activity.

2. Upstream Regulatory Factors

Upstream activating sequences (UAS), or transcriptional enhancer elements, are binding sites for transcription factors that stimulate RNA synthesis. The term UAS is used to describe DNA sequence elements that are located close to the core promoter, so-called promoter proximal elements. UAS elements are typically located within 200 base pairs upstream of the transcription initiation site. Enhancer sequences are DNA segments containing binding sites for multiple transcription factors that activate transcription independent of their orientation and at a great distance [up to 85 kilobases (kb)] from the start site of transcription. Enhancer elements can be located either upstream or downstream of the transcription initiation site. Enhancer sequences activate transcription in a position-independent manner because they become spatially positioned close to the core promoter through bending of the DNA molecule (Fig. 1).

In addition to enhancer elements, eukaryotic promoters contain upstream repressor elements, which block RNA synthesis by various mechanisms by recruiting factors that interefere with enhancer factors or directly block RNA polymerase II recruitment. A third class of DNA sequence elements regulating transcription are the transcriptional silencers. A classical silencer represses transcription in a position- and orientation-independent fashion. The silencer element is thought to block transcription by functioning as the nucleation site for binding of histones or silencing proteins that coat the region, thereby making the promoter inaccessible for RNA polymerase recruitment.

The human genome encodes for several thousand different transcription factors. Promoters that contain combinations of binding sites for different transcription factors regulate different genes. Thus, for example, a gene specifically expressed in the liver or the brain uses liver- or brain-specific enhancer binding transcription factors, respectively, to achieve a tissue-specific gene expression. The basal transcriptional machinery appears to a large extent to be the same in all cell types.

B. Regulation of Promoter Activity

From a regulatory point of view it is important to note that TBP is sufficient to recruit RNA polymerase II and direct basal transcription from the core promoter. However, the basal transcription factor TFIID has been shown to play a central role in activated transcription by binding to the TATA element in the core promoter and facilitating the recruitment of the RNA polymerase holoenzyme to the promoter (Fig. 1). TFIID is a multiprotein complex consisting of TBP and approximately 11 TBP-associated factors (TAFs). The TAFs have been shown to be essential for regulated transcription by mediating contact with enhancer binding factors. Thus, TBP is sufficient for constitutive transcription but TAFs are necessary for regulated transcription (Fig. 1). *In vitro* studies suggest that assembly of an initiation-competent RNA polymerase at a promoter can be subdivided into several steps where different

basal transcription factors are sequentially recruited to the promoter. However, fractionation experiments have shown that on certain promoters the RNA polymerase and most or all of the general transcription factors may be recruited as a single complex. *In vivo* activation of the thousands of promoters present in the human genome may use a large spectrum of mechanistic possibilities.

An important finding was the observation that UAS-binding transcription factors are modular in structure with a DNA-binding domain and an effector domain that could be exchanged without losing their predicted biological activity. The effector or activation domains in different transcription factors perform the same task but have different properties, for example, consisting of acidic blobs, proline-rich, glutamine-rich, or serine/threonine-rich sequences. Different classes of UAS-binding transcription factors may transmit a signal to the basal promoter complex by making specific contacts with different TAFs. For example, an interaction between the UAS-binding transcription factor SP1 and TAF-110 has been shown to be necessary for SP1-mediated activation of transcription. Collectively stabilized protein–protein interactions between UAS-binding factors and the general transcriptional factor TFIID are likely to facilitate recruitment of the RNA polymerase to the core promoter element, and as a consequence increase the transcriptional activity of the promoter (Fig. 1).

Transcription factors can be subdivided into families based on the structural feature of the DNA-binding domain. Thus, the DNA-binding domain may interact with the DNA through structural types like the helix-turn-helix motif found in homeodomain proteins, zinc fingers, or leucine-zipper-basic DNA-binding domain motifs. Heterodimerization between members of UAS-activating transcription factors belonging to such structural types is not uncommon and has been shown to increase the repertoire by which transcription factors can interact with different promoter sequences. For example, the prototypical AP1 transcription factor, which belong to the leucine-zipper family of transcription factors, consists of a heterodimer of c-*jun* and c-*fos*. It binds to its cognate DNA motif with a higher affinity than, for example, a c-*jun*– c-*jun* homodimer or a *JunB*–c-*fos* heterodimer. The combinatorial complexity is further increased by the fact that c-*jun* may form heterodimers with members of the ATF family of transcription factors. Thus, heterodimerization between different members of a transcription factor family is an important mechanism to generate factors with alternative DNA-binding specificity.

When the RNA polymerase leaves the promoter, TFIID remains bound at the TATA element and is ready to help a second RNA polymerase to bind and initiate transcription at the same promoter. The activity of TFIID appears also to be regulated by inhibitory proteins that interact with TBP. Such TBP–inhibitory protein complexes may serve an important regulatory role by keeping genes which have been removed from inactive chromatin in a repressed but rapidly inducible state.

1. Regulation of Transcription by Chromatin Remodeling

During the last decade a wealth of information has demonstrated the significance of the chromatin template for the transcriptional activity of a promoter. It has been known for decades that the DNA in our cells is wrapped around a protein core called the nucleosome. The nucleosome consists of two copies each of four histones: H2A, H2B, H3, and H4. The DNA is packaged into either a loose structure called euchromatin or a more highly ordered structure called heterochromatin. The heterochromatin fraction is transcriptionally inactive, whereas active genes are found in the euchromatin fraction of the DNA.

Histones are modified by acetylation, phosphorylation, methylation, and ubiquitnation. During recent years an impressive amount of work has demonstrated the significance of reversible histone acetylation as a regulatory mechanism controlling gene expression. Several lysines on the amino-terminal tail of each core histone can be acetylated. Lysines are negatively charged and make strong interaction with the phosphate backbone of DNA, thereby preventing basal transcription factors like TBP from interacting with DNA. Acetylation of lysines neutralizes this negative charge and reduces the electrostatic interaction of the histones with the DNA, thereby making the promoter region accessible for interaction with the basal transcription machinery (Fig. 2). A considerable amount of work shows that a general theme in transcriptional regulation is that acetylation of core histones results in looser nucleosomal structure, which makes the DNA more accessible for binding of transcription factors, and hence a gene more transcriptionally active. In contrast, histone deacetylation has the opposite effect and functions as a signal to repress transcription (Fig. 2).

Several transcriptional enhancer proteins have been shown to activate transcription by binding so-called coactivator proteins which have histone acyltransferase (HAT) activity. The best characterized are Gcn5 (yeast), TAF$_{II}$250, CBP, and p300. TAF$_{II}$250, which is a component of the TATA-binding basal transcription factor TFIID, may activate transcription by inducing acetylation of histones located in the vicinity of the TATA box. On the other hand, transcriptional repressor proteins have often been shown to inhibit RNA synthesis by recruiting histone deacetyltransfereases (HDACs), which cause a condensation of nucleosomes to a more compact, transcriptionally

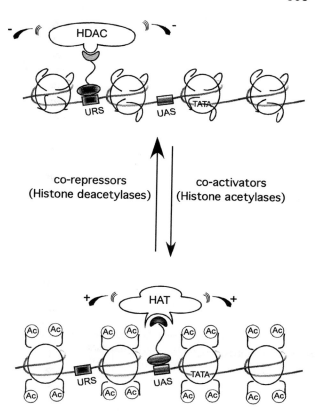

FIGURE 2 Role of the nucleosome in gene expression. Recruitment of histone deacetylases (HDACs) to a promoter inhibits binding of general transcription factors to the TATA element, thereby blocking transcription. Recruitment of histone acetylases (HATs) to the promoter results in acetylation of the amino-terminal tails of the core histones, thereby facilitating binding of the general transcription factors required for initiation of transcription. URS, upstream repressor sequence; UAS, upstream activating sequence.

inactive structure (Fig. 2). In yeast, HATs and HDACs are found in multiprotein complexes such as SAGA and Sin3 complexes, respectively. The equivalent, and additional, multienzyme complexes are also found in higher eukaryotes.

In addition, the nucleus contains so-called chromatin remodeling factors, such as the Swi/Snf complex, which has the capacity to reposition nucleosomes and transiently dissociate the DNA from the surface of the nucleosome. Depending on the promoter context, chromatin remodeling factors may cause an activation or repression of transcription.

It is likely that other histone modifications, such as phosphorylation, ubiquitinilation, and methylation, also play a significant regulatory role in transcriptional control of promoter activity, although the importance of these modifications has not yet been characterized to the same extent as has that of reversible acetylation. The main conclusion from these studies is that a linear assessment of the

DNA sequence elements capable of binding transcriptional activator or repressor proteins only tells us part of the story, namely which factors have the capacity to control promoter activity. However, actual RNA synthesis requires a complex interplay between UAS-binding factors and the chromatin or the chromatin remodeling factors that have positive or negative effects on promoter activity.

2. Regulation of Transcription Factor Activity

The activity of a UAS-binding transcription factor is subjected to a posttranslational regulation. There are in principle three ways that the activity of an UAS-binding transcription factor may be tuned (Fig. 3); covalent (like phosphorylation) or noncovalent (like hormone binding) modification of the UAS-binding factor, or variation of the subunit composition (like binding of an inhibitory protein). These mechanisms may be used individually or in combination with other mechanisms to regulate transcription. To illustrate the flexibility of transcriptional control in eukaryotic cells two examples are presented. The first example concerns the activation of steroid hormone-dependent gene transcription (Fig. 4). Steroid hormones are a group of substances derived from cholesterol which exert a wide range of effects on processes such as growth, metabolism, and sexual differentiation. A prototypical member of a steroid hormone-inducible transcription factor is the glucocorticoid receptor (GR). In the absence of hormone this receptor is found as a monomer in the cy-

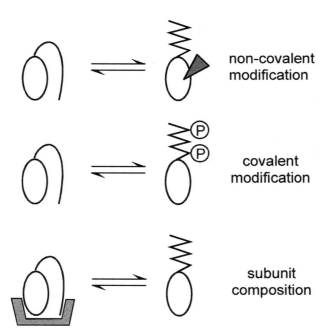

FIGURE 3 Three common mechanisms to regulate transcription factor activity.

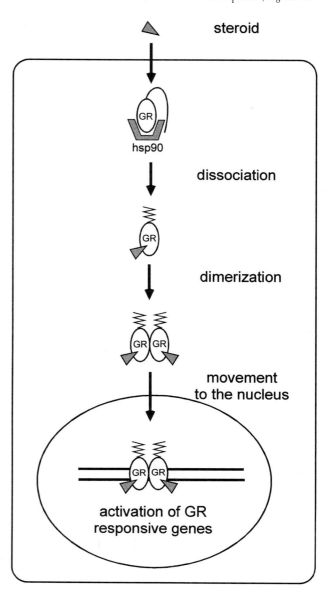

FIGURE 4 A schematic drawing showing the activation of glucocorticoid receptor (GR) by steroid hormone binding, which results in the dissociation of the cytoplasmic GR–hsp90 complex, followed by GR dimerization and translocation of GR to the nucleus.

toplasm complexed to the heat-inducible hsp90 protein. Treatment with steroid hormones results in the release of GR from hsp90, which renders GR free to dimerize and move to the nucleus, where it binds to its cognate DNA sequence element and activates transcription (Fig. 4). In addition to inducing a dissociation of the receptor from hsp90, ligand binding also induces a conformational change in the activation domain of GR, such that the activation domain binds transcriptional coactivator proteins that stimulate transcription. Interestingly,

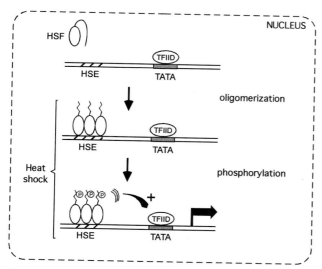

FIGURE 5 Activation of transcription of heat-shock genes in mammalian cells. The heat shock factor (HSF) is present as a monomer in normal cells. An increase in temperature results in a trimerization of HSF, which binds to the heat shock element (HSE). HSF is activated as a transcriptional enhancer protein by phosphorylation.

heat-shock proteins do not regulate other nuclear hormone receptors, such as the retinoic acid and thyroid hormone receptors. Thus, these receptors bind DNA in the absence of the ligand. In this case ligand binding results in a conformational change of the activation domain permitting binding of coactivator proteins.

The second example concerns heat-shock activation of transcription in mammalian cells (Fig. 5). When cells are subjected to an elevated temperature (heat shock) they respond by activating synthesis of a small number of genes encoding for so-called heat-shock proteins. These proteins serve an important function during heat shock by binding to cellular proteins, which become denatured by the increase in temperature. Subsequently, the heat-shock proteins help to renature the proteins to their native conformation. Transcription of heat-shock genes is controlled by the heat-shock transcription factor (HSF), which binds to the heat-shock element (HSE) found in the promoter of all genes regulated by heat shock. HSF is activated by two mechanisms (Fig. 5). Thus, in normal cells HSF exists as a monomer. An increase in temperature results in unfolding of HSF, which exposes the DNA-binding domain and allows it to bind to other HSFs and form a trimer that binds to the HSE. However, binding of HSF to DNA is not enough to activate transcription. Thus, HSF needs to be modified by phosphorylation before it activates transcription of the heat-shock genes. Interestingly, TFIID is bound to the TATA element in heat-shock genes also in uninduced cells. Thus, binding of an active HSF to the

HSE is needed for recruitment of the RNA polymerase to the heat-shock promoter. The binding of TFIID to the uninduced promoter may help heat-shock genes respond more rapidly to an increase in temperature.

Cell type and differentiation-specific gene expression is often regulated by the availability of specific transcription factors. Genes that are expressed in specific organs contain binding sites for cell type-specific transcription factors. Thus, tissue-specific transcription is often regulated by the precise arrangement of regulatory UAS motifs in the promoter, the availability of the cognate transcription factors, and the way these transcription factors influence the activity of the promoter. Thus, for example, the liver and the brain encode for respectively liver- and brain-specific transcription factors that ensure a tissue-specific expression gene expression. Since transcription factors are typically dimeric proteins, the exact composition of the two partners may vary among cell types and have different transcription regulatory properties.

3. Regulation of Transcription Elongation

Although transcription initiation has only been discussed, RNA polymerase elongation is also an important step in regulating gene expression in eukaryotes. Thus, there are several examples where the RNA polymerase halts at specific pause sites during elongation. To be able to complete the synthesis of the precursor-RNA the polymerase has to be able to override this attenuation of transcription. The best-characterized example is the human immunodeficiency virus (HIV) Tat protein, which binds to a stem-loop structure at the 5′ end of the HIV transcript, the TAR sequence. In the absence of the Tat protein, HIV transcription terminates approximaqtely 50 nucleotides downstream of the initiation site. When Tat is present it binds to the TAR sequence and recruits a cyclin T/Cdk9 complex which is responsible for phosphorylation of the CTD tail of RNA polymerase II, thereby alleviating termination and permitting the RNA polymerase to synthesize the full-length HIV genomic RNA.

IV. REGULATION OF TRANSCRIPTION IN PROKARYOTES

A. Introduction

The mechanisms to initiate transcription in eukaryotes and prokaryotes are similar. As a comparison to control of transcription in eukaryotes some key features in transcriptional control in bacteria will be given. Prokaryotic cells contain only one type of RNA polymerase, which is responsible for synthesis of all types of RNA: mRNA, rRNA, and tRNA. The core polymerase is a four-subunit enzyme consisting of two a, one b, and one b′ subunit. However, the holoenzyme, which is the complete enzyme, contains the core polymerase plus the sigma factor, which may regarded as the prokaryotic equivalent of the general transcription factors found in eukaryotes. The sigma factor is required for proper RNA polymerase binding to a prokaryotic promoter. After initiation of transcription the sigma factor leaves the polymerase complex and elongation is taken care of by the core polymerase.

Similar to a eukaryotic promoter a prototypical prokaryotic core promoter contains a conserved TATAAT located at position −10 relative to the transcription start site and resembling the eukaryotic TATA element. In addition, prokaryotic promoters contain a conserved TTGACA located at position −35. The spacing between the two elements is of critical importance for the efficiency by which the RNA polymerase binds to the promoter. The exact sequences at the −10 and −35 positions vary slightly for different transcription units. Usually promoters that have a better homology to the consensus sequences also initiate transcription more efficiently. An important mechanism to regulate the transcriptional activity of a prokaryotic promoter is to provide the core polymerase with different sigma factors. Thus, different sigma factors determine promoter specificity by recognizing −10 and −35 elements with different base sequences. This strategy mediates the heat shock response and the regulated expression of genes during developmental processes. For example, sporulation in *Bacillus subtilis* uses a cascade of different sigma factors to cause the transformation of a vegetative bacterium to a spore.

The existence of transcriptional enhancers similar to those found in eukaryotic cells has also been described in prokaryotes. For example, the enhancer-binding protein nitrogen regulatory protein C (NTRC) in the *glnA* promoter from *Salmonella typhimurium* activates transcription from a distance by means of DNA looping. NTRC stimulates transcription by a transient contact between the activator and the polymerase. This catalyzes an unwinding of the DNA at the promoter, which then allows the RNA polymerase to initiate transcription.

B. The *Lac* Operon

Transcriptional repression is a key mechanism to control the activity of prokaryotic promoters. Enzymes used in a specific metabolic pathway are often organized into an operon that is transcribed into a single polycistronic mRNA. Specific repressor proteins then control the transcriptional activity of the operon by regulating RNA polymerase binding to the promoter. Repressor proteins are DNA-binding proteins that typically block RNA polymerase access to the −10 and/or −35 regions in the pro-

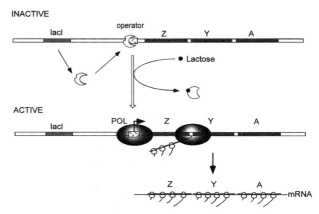

FIGURE 6 Regulation of the *lac* operon in *E. coli*. The *lac I* gene encodes for a transcriptional repressor protein that binds to an operator sequence in the *lac* operon, thereby preventing synthesis of the structural genes required for metabolism of lactose. If *E. coli* is grown on lactose as the sole carbon source, lactose binds to the *lac I* repressor protein and inactivates it as a repressor of *lac* operon transcription. As a consequence, the β-galactosidase (*lac Z*), the permease (*lac Y*), and the β-galactosidase transacetylase (*lac A*) enzymes are synthesized.

moter or transcription elongation by associating with an operator sequence that is positioned downstream of the start site of transcription. Usually these regulatory proteins undergo allosteric changes in response to binding of a specific ligand. The paradigm of a prokaryotic operon regulated by a specific repressor protein is the *lac* operon in *E. coli*. In this system synthesis of proteins necessary for usage of lactose as a carbon source is repressed by the *lac* repressor protein if cells have the possibility to use glucose for growth. Thus, in the presence of glucose the *lac* repressor binds to its operator sequence, which overlaps the transcription start site in the *lac* operon (Fig. 6), and blocks RNA polymerase binding to the *lac* promoter. If cells are grown on lactose as the carbon source, lactose functions as an inducer of *lac* operon transcription by binding to the *lac* repressor and converting it to an inactive form that does not bind DNA (Fig. 6) and therefore is unable to inhibit transcription of the *lac* operon. The polycistronic *lac* mRNA encodes for the specific proteins necessary for metabolism of lactose. The *lac* operon represents an example of an inducible system where an inducer activates transcription. However, inducers can also have the opposite effect and repress transcription of an operon, like the *trp* operon in *E. coli*.

V. POSTTRANSCRIPTIONAL REGULATION OF GENE EXPRESSION

Expression of eukaryotic genes is not only controlled at the level of initiation of RNA synthesis. Thus, the precursor-RNA synthesized by RNA polymerase II undergoes several posttranscriptional modifications before a mature mRNA is formed. For example, the transcript is capped at its 5′ end, the 3′ end is generated by a specific cleavage polyadenylation reaction, and intronic sequences are removed by RNA splicing.

Early after initiation of transcription, when the nascent RNA chain is 25–30 nucleotides long, the 5′ end is modified by addition of an inverted 7-methylguanosine, the cap nucleotide. The capping enzymes are brought to the transcribing polymerase by specific association with the hyperphosphorylated form of the CTD tail on RNA polymerase II. As mentioned above, the CTD tail becomes phosphorylated when RNA polymerase II progress from the initiation to the elongation phase of RNA synthesis. Since RNA polymerases I and III do not have a CTD tail, only RNA polymerase II transcripts are capped. The cap plays a crucial role in initiation of translation by binding the translational initiation factor eIF4F required for the recruitment of the small subunit of the ribosome to the mRNA. The translational start site is then identified by a scanning mechanism where the ribosome usually selects the first AUG triplet as the start codon for protein synthesis. The selective addition of a cap to RNA polymerase II transcripts therefore provides a logical explanation to why this class of RNAs is used for translation. Polyadenylation and RNA splicing are key mechanisms to regulate eukaryotic gene expression and are therefore described in more detail below.

A. Exons and Introns: General Considerations

Virtually all prokaryotic genes are encoded by a collinear DNA sequence: the concept one gene, one mRNA. In contrast, most eukaryotic genes are discontinuous, with the coding sequences (exons) interrupted by stretches of noncoding sequences (introns). Introns are present at the DNA level and in the primary transcription product of the gene (the precursor-RNA), and are removed by RNA splicing before the mature mRNA is transported to the cytoplasm. Recent experiments suggest that splicing is necessary for efficient transport of intron-containing precursor-RNAs. Introns have been found in all types of eukaryotic RNA—mRNA, rRNA, and tRNA. Because of space limitation, only introns in protein-encoding genes will be described.

The number of introns in mRNA-encoding genes varies considerably among genes. For example, c-*jun*, histone, heat-shock, and the α-interferon genes have no introns, whereas the gene for dystrophin has more than 70 introns. Also, the size of introns can vary from less than 100 nucleotides to several million nucleotides in length. The extreme example is the *Drosophila Dhc*7 gene, which contains a 3.6 million-nucleotide-long intron. This intron is approximately double the size of most bacterial genomes and takes days to transcribe. In contrast, exons are typically short, usually less than 350 nucleotides. This comes from the fact that splice sites used to define the borders of the splicing reaction are defined across the exon, not the intron—the so-called exon definition model (see below). Some eukaryotic genes are remarkably large. For example, the human gene for dystrophin covers approximately 2.4 million base pairs. The RNA polymerase that initiates transcription requires approximately 20 hr to synthesize the full-length precursor-RNA. Subsequently, more than 99.5% of the transcript is removed by RNA splicing. Thus, the final mRNA that is transported to the cytoplasm is only around 14,000 nucleotides. The extreme lengths of eukaryotic genes place a high demand on the stability of the transcription complex. Thus, an RNA polymerase that binds to a promoter must stay attached for days with the DNA template to be able to complete synthesis of the longest genes.

It is interesting to note that introns in eukaryotic genes almost always interrupt the protein-coding portion of the precursor-RNA. Thus, introns are rarely found after the translational stop codon, within the 3′ noncoding portion of the mRNA. This organization is significant since the presence of an intron downstream of the translational stop codon in a reading frame is sensed as a signal that the precursor-RNA has been incorrectly spliced or for other reasons is defective, and will not produce the correct protein after translation in the cytoplasm. Such nuclear transcripts are sent for destruction by a mechanism that is collectively called the non-sense-mediated mRNA decay mechanism. How the translational reading frame is read already in the nucleus is not known. The easiest explanation would be that there exists a nuclear ribosome-like structure that scans the spliced mRNA for a full-length translational reading frame before the mRNA is transported to the cytoplasm. However, this question is controversial and has not been proven.

1. Mechanism of RNA Splice Site Choice During Spliceosome Assembly

The sequence elements used to specify the splice sites are remarkably short and degenerate in a eukaryotic precursor-RNA. Thus, short conserved sequence motifs at the beginning (5′ end) and the end (3′ end) of the intron guide the assembly of a large RNA protein particle, the spliceosome (Fig. 7), which catalyzes the cleavage and ligation reactions necessary to produce the mature cytoplasmic mRNA. The nucleus of eukaryotic cells contains several abundant low-molecular-weight RNAs, so-called U snRNAs. The U snRNAs derive their name from the fact that they were initially characterized as RNAs rich in uridines. Five of these U snRNAs (U1,

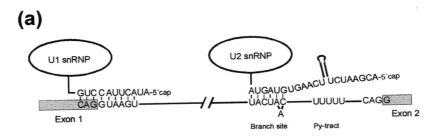

(a)

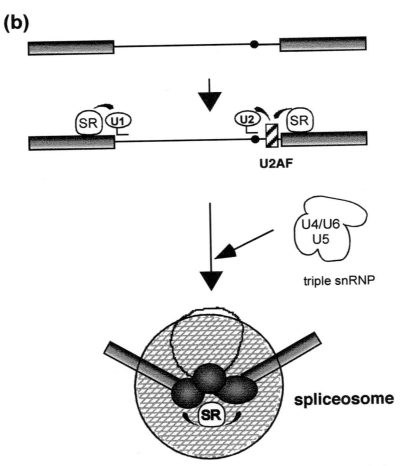

(b)

FIGURE 7 A simplistic model for spliceosome assembly. (a) The 5′ splice site and the branch site are defined via a direct base pairing between the RNA components of the U1 and U2 snRNPs and the precursor-RNA, respectively. (b) Efficient recruitment of the U snRNPs to the spliceosome is aided by non-snRNP proteins. Thus, SR proteins facilitate U1 snRNP binding to the 5′ splice site, whereas U2AF binds to the polypyrimidine tract at the 3′ splice site and helps U2 snRNP binding to the branch site. The U4/U6–U5 triple snRNP is recruited to form the mature spliceosome. SR proteins bring the 5′ and 3′ splice sites in close proximity for the catalytic steps of splicing by making simultaneous contact with splicing factors binding to the 5′ and 3′ splice sites, respectively (see also Fig. 8).

U2, U4, U5, and U6), ranging in size from 107 to 210 nucleotides, have been shown to participate in splicing. *In vivo* the snRNAs are found complexed to 6–10 proteins, generating the so-called small nuclear ribonucleoprotein particles (snRNPs). Some snRNP proteins are shared among different U snRNPs, whereas other snRNP proteins are unique to each U snRNP. During spliceosome assembly the ends of the introns are in part identified by RNA–RNA base pairing between the precursor-RNA and a U snRNP (Fig. 7). For example, the 5′ splice site is recognized through a short base pairing between the U1 snRNA and the precursor-RNA. Similarly, a base pairing between U2 snRNA and the branch point defines the 3′ splice site. Later during spliceosome formation the U5–U4/U6 triple snRNP is recruited. In the triple snRNP, U4 and U6 snRNP form an extensive base pairing. The catalytically active spliceosome is generated by conformational changes, which results in a breakage of the base pairing between U4 and U6 snRNP and formation of new U–U snRNA and U snRNA–precursor-RNA base pairings. It is generally believed, although not proven, that the U snRNAs in the spliceosome are the enzymes that catalyze the two transesterification reactions required to excise the intron.

2. Non-snRNP Proteins Required for Splicing

The spliceosome, which is a large RNA–protein complex, with a size similar to a cytoplasmic ribosome, also contains numerous non-snRNP proteins which are important for correct splice site recognition. Assembly of the spliceosome proceeds over several stable intermediates (Fig. 7).

Efficient recruitment of U2 and U1 snRNP to the 3′ and 5′ splice sites also requires specific proteins. Here only two factors will be described. The first is U2 snRNP auxiliary factor (U2AF), which binds to the pyrimidine tract located between the branch site and the 3′ splice site in the precursor-RNA. U2AF stabilizes U2 snRNP binding to the branch site. The second factor is not one protein, but a family of proteins, designated SR proteins. SR proteins contain one or two amino-terminal RNA-binding domains and a carboxy-terminus rich in arginine (R) and serine (S) dipeptide repeats (the RS domain); hence the name SR proteins. Mechanistically, SR proteins appear to perform the same function in RNA splicing that transcriptional enhancer proteins do in transcription initiation. Thus, SR proteins bind to splicing enhancer sequences through their RNA-binding domains and stimulate spliceosome assembly by facilitating protein–protein interaction (Fig. 7). The RS domain functions as a protein interaction surface that makes contact with other SR proteins and so-called SR-related proteins. Thus, many proteins involved in RNA splicing contain RS domains. For example, SR proteins aid in efficient U1 snRNP binding to a 5′ splice site by interacting with the U1-70K protein, which is an RS-domain containing protein. However, in contrast to transcriptional enhancer proteins, which active transcription irrespective of the position where they bind, SR protein function is position dependent. Thus, in general, SR proteins function as splicing-enhancer proteins if they bind to the exon and function as splicing-repressor proteins if they bind to the intron in the precursor-RNA.

The number of SR proteins found in mammalian cells is surprisingly few considering the multitude of regulated splicing events for which they are required. Thus, only around 12 "true" SR proteins have been identified. Even more surprising, gene knockout experiments suggest that only one of the SR proteins is essential in *Caenorhabditis elegans*. Thus, disrupting the expression of the SR protein ASF/SF2 resulted in early embryonic lethality, whereas gene knockout of other SR proteins resulted in no change in phenotype. Probably, SR proteins show a large extent of functional redundancy, and disruption of one is compensated for by another SR protein. The essential role of SR proteins in spliceosome assembly makes them prime targets for regulation of gene expression.

3. The Exon Definition Model

The conserved sequences at the 5′ and 3′ ends of the intron are surprisingly short considering the precision by which very large introns are excised during splicing. The answer to this puzzle appears to be resolved by the fact that the 5′ and 3′ splice sites that are joined in the splicing reaction are not recognized over the intron. Instead splice sites are recognized across the exons—the so-called exon definition model. Thus, whereas introns can vary in length from less than 100 to more than 1 million nucleotides, in ternal exons in a precursor-RNA have a constant length and rarely exceed 350 nucleotides. The exon definition model postulates that U2 snRNP binding to a 3′ splice site makes contact with U1 snRNP binding to the downstream 5′ splice site (Fig. 8). If the 3′ and 5′ splice sites are too far away the model postulates that the intervening sequence is not recognized as an exon because U2 and U1 snRNP binding to respective splice sites cannot interact with each other. Once the exons have been defined in the precursor-RNA, adjacent exons are aligned for the splicing reaction.

B. Alternative RNA Splicing Is an Important Mechanism to Generate Protein Diversity

A major difference in gene regulation between a prokaryotic and a eukaryotic cell is the existence of mechanisms in eukaryotic cells that permit one gene to express multiple gene products. In bacteria a protein is encoded by a collinear DNA sequence. In contrast, in eukaryotes a single gene may encode for thousands of proteins. Thus, the discontinuous arrangement of eukaryotic genes, with introns interrupting the coding segments of the precursor-RNA, permit production of multiple, alternatively spliced mRNAs from a single gene. Examples of how a precursor-RNA can be alternatively spliced are shown in Fig. 9. This, of course, means that multiple proteins with different primary amino acid sequence and biological activity can be produced from a single eukaryotic gene. Of specific interest is that the production of alternatively spliced mRNAs in many cases is a regulated process, either in a temporal,

Exon definition

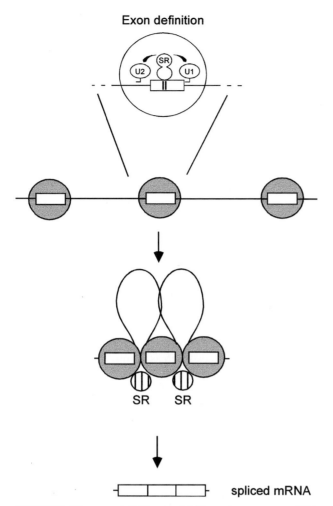

FIGURE 8 The exon definition model. Exons in a precursor-RNA are recognized as units by U2 snRNP (U2) binding to the 3′ splice site and U1 snRNP (U1) binding to the downstream 5′ splice site. Subsequently adjacent exons are defined across the intron. In both recognition steps SR proteins function as bridging proteins.

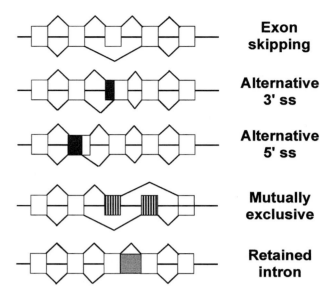

Exon skipping

Alternative 3′ ss

Alternative 5′ ss

Mutually exclusive

Retained intron

FIGURE 9 Examples of different patterns of alternative RNA splicing.

developmental, or tissue-specific manner. Changes in splicing have been shown to determine the ligand-binding specificity of growth factor receptors and cell adhesion molecules and to alter the activation domains of transcription factors. For example, the fibronectin precursor-RNA is alternatively spliced in hepatocytes and fibroblasts. In fibroblasts two exons which are skipped in hepatocytes are included during the splicing reaction. These two exons encode for protein domains that make fibroblast fibronectin adhere to many cell surface receptors. Fibronectin produced in hepatocytes lacks these two exons and therefore is translated to a hepatocyte-specific fibronectin protein that does not adhere to cells, allowing it to circulate in the serum.

The impact of alternative splicing on the coding capacity of a eukaryotic gene is mind-boggling. For example,

the *Drosophila DSCAM* gene, which encodes for an axon guidance receptor, has been estimated to produce 38,016 DSCAM protein isoforms by alternative splicing. This figure is remarkable since the total gene number calculated from the *Drosophila* DNA sequence suggests a total of only approximately 14,000 genes. Thus, a single *Drosphoila* gene produces almost three times the number of proteins compared to the number of genes in *Drosophila*. The *DSCAM* gene is not unique. There are many examples of human genes, like those for neurexins, *n*-cadherins, and calcium-activated potassium channels, that are known to produce thousands of functionally divergent mRNAs. A low estimate suggests that approximately 35% of all human genes produce alternatively spliced mRNAs. Thus, the estimate of 20,000–50,000 genes in the human genome could easily produce several hundred thousand, or million, proteins. Such differences in numbers are comforting because they make it easier to explain how a complex organism like humans with highly differentiated organs have evolved without an enormous increase in the number of genes compared to bacteria.

1. Regulation of Alternative RNA Splicing by Changes in SR Protein Activity

With a few exceptions little is known about the mechanistic details of how production of alternatively spliced mRNAs is regulated. However, it appears clear that the SR family of splicing factors partake in many regulated splicing events. SR proteins are highly phosphorylated, primarily within the RS domain. Thus, reversible RS domain phosphorylation has been shown to regulate SR protein

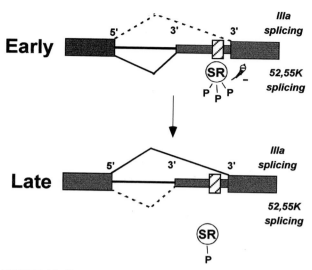

Early

IIIa
splicing

52,55K
splicing

Late

IIIa
splicing

52,55K
splicing

FIGURE 10 Regulation of alternative RNA splicing by SR protein phosphorylation. Hyperphosphorylated SR proteins present in normal cells and early virus-infected cell bind to a specific repressor element in the adenovirus L1 precursor-RNA and block spliceosome assembly at the IIIa 3′ splice site. This results in an exclusive production of the 52,55 K mRNA early after infection. Adenovirus induces a dephosphorylation of SR proteins late during infection, which alleviates the repressive effect of SR proteins on IIIa splicing, hence a shift to IIIa mRNA splicing.

interaction with other splicing factors and control alternative RNA splicing.

One of the best-characterized examples is the human adenovirus L1 unit (Fig. 10). The L1 unit produces two mRNAs, the 52,55K and the IIIa mRNAs, which are generated by alternative 3′ splice site selection. Splicing during an adenovirus infection is temporally regulated such that the IIIa mRNA is produced exclusively late during virus infection. It has been shown that highly phosphorylated SR proteins bind to an intronic repressor element and inhibit IIIa splicing during the early phase of infection. At late times of infection IIIa splicing is activated by a virus-induced dephosphorylation of SR proteins. This change in the phosphorylated status of SR proteins reduces their binding capacity to the repressor element and hence results in an alleviation of their repressive effect on IIIa 3′ splice site usage.

2. Maintenance of Sex in *Drosophila*: Sxl Regulation of Splicing

One of the most spectacular and best-characterized examples where alternative RNA splice site choice is used to regulate gene expression is the somatic sex-determination pathway in *Drosophila melanogaster* (Fig. 11). In this system sex determination has been shown to involve a cascade of regulatory events taking place at the level of alternative

RNA splice site choice. The X chromosome encodes for transcription factors that control *Sex-lethal* (*Sxl*) transcription. In females, which contain two X chromosomes, the double dose of these transcription factors results in an activation of an early promoter of the *Sxl* gene. This promoter is inactive in males, which contain one X chromosome. The female-specific Sxl protein is an RNA-binding protein that binds to certain pyrimidine tracts and outcompetes U2AF binding to that site. Since the Sxl protein lacks the splicing activator function of U2AF, Sxl binding to a pyrimidine tract prohibits spliceosome formation at the 3′ splice site. Thus, once made, the female-specific Sxl protein autoregulates its own expression by ensuring that exon 3 in the Sxl pre-mRNA is efficiently skipped, thereby establishing a female-specific splicing of the Sxl pre-mRNA (Fig. 11). In male flies exon 3 is incorporated during splicing, resulting in the translation of a functionally inactive Sxl protein. This results from the fact that the third exon in the Sxl pre-mRNA contains a translational stop codon that causes a premature termination of translation. In addition, Sxl controls the splicing of downstream targets in the sex determination pathway. Thus, Sxl regulates splicing of the *transformer* (*Tra*) precursor-RNA by repressing usage of the male-specific 3′ splice site. Subseqently the female-specific Tra protein complexes with the Tra2 protein and activates a female-specific 3′ splice site in the *double-sex* (*Dsx*) precursor-RNA (Fig. 11). In males where a biologically inactive Sxl protein is expressed, the *Sxl*, *Tra*, and *Dsx* precursor-RNAs are processed by a default-splicing pathway, resulting in the development of male flies. The Dsx protein, the final protein in the cascade, is a transcription factor. The male- and female-specific Dsx proteins regulate development of flies along the male- or female-specific pathways.

It is widely accepted that alternative splicing is an important mechanism to regulate gene expression during growth and development in eukaryotic cells. A large number of eukaryotic genes have been shown to mature alternatively spliced mRNAs, examples include growth factors, growth factor receptors, intracellular messengers, transcription factors, oncogenes, and muscle proteins. The number of examples is constantly increasing.

C. The Basic Mechanism of 3′ End Formation

1. Introduction

In eukaryotes all mRNAs except the histone mRNAs have a 200- to 250-long 3′ poly(A) tail. It is noteworthy that the 3′ end of a eukaryotic mRNA is not generated by termination of transcription. Thus, the RNA polymerase continues to synthesize RNA beyond the actual 3′ end of the mature mRNA. Sequence analysis of a number of RNA polymerase II genes has reveled two elements that

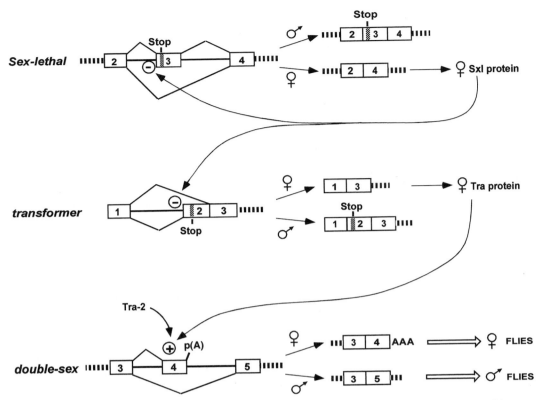

FIGURE 11 The cascade of regulated alternative splicing events controlling *Sxl, transformer,* and *double-sex* expression in *Drosophila melanogaster.* The positions of translational stop codons that will cause premature termination of protein synthesis are indicated by Stop. See text for further details.

by biochemical assays have been shown to specify the position of the 3′ end of a mRNA (Fig. 12). Thus, an almost invariable AAUAAA sequence located 25–30 nucleotides upstream of the cleavage site and a GU-rich sequence located within 50 nucleotides downstream of the cleavage–polyadenylation site are critical for 3′ end formation. The AAUAAA sequence, which resembles the AT-rich TATA box important for transcription initiation, binds the essential cleavage–polyadenylation specificity factor (CPSF). The GU-rich sequence binds the cleavage–stimulatory factor (CStF). In addition, two cleavage factors, CFI and CFII, and the poly(A) polymerase (PAP) assemble to form an active enzyme complex that cleaves the growing RNA chain and catalyzes the addition of the 200- to 250-nucleotide poly(A) tail. The transcribing RNA polymerase terminates RNA synthesis in an ill-defined sequence downstream of the cleavage–polyadenylation site. However, the cleavage–polyadenylation reaction has been shown to be required for transcription termination, suggesting that breakage of the primary transcript is coupled to termination of transcription, possibly through the action of 5′ exonucleases that degrade the downstream RNA chain generated by the cleavage reaction. Mechanistically,

this may be analogous to Rho-dependent transcription termination in bacteria.

2. Regulation of Gene Expression at the Level of Alternative Poly(A) Site Usage

In addition to control of alternative RNA splicing, eukaryotic cells frequently use alternative poly(A) site usage to further increase the coding capacity of a gene. As described above, the 3′ end of a eukaryotic mRNA is generated by an endonucleolytic cleavage of the primary transcript rather than termination of transcription. This means that the RNA polymerase transcribing a gene may pass by several potential poly(A) sites before terminating transcription. Thus, the selection of different poly(A) signals in a precursor-RNA can be used to regulate gene expression. For example, if the first poly(A) signal is ignored, a second poly(A) signal further downstream in a transcription unit may be used to incorporate novel exons into the mRNA. As an example of this type of regulation, the production of secreted or membrane-bound immunoglobulin M (IgM) is described. When a hematopoietic stem cell differentiates to a pre-B lymphocyte it produces IgM as an

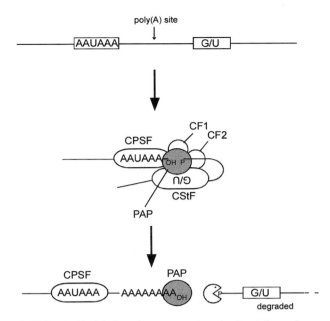

FIGURE 12 Model for cleavage and polyadenylation of a precursor-RNA in mammalian cells. The cleavage and polyadenylation specificity factor (CPSF) binds to the conserved AAUAAA sequence located 10–35 nucleotides upstream of the poly(A) site. The cleavage-stimulatory factor (CStF) binds to the GU-rich element located downstream of the poly(A) site. Binding of two cleavage factors (CF1 and CF II) and the poly(A) polymerase (PAP) then stimulates cleavage of the precursor-RNA. The PAP synthesizes the 200- to 250-nucleotide-long poly(A) tail, while the downstream RNA fragment is rapidly degraded.

antibody anchored to the plasma membrane. After binding to an antigen the lymphocyte undergoes differentiation and produces the same IgM molecule as a secreted protein. The shift from making a membrane-bound or secreted antibody is regulated at the level of alternative poly(A) site usage (Fig. 13). Thus, in the unstimulated B cell the first poly(A) signal is ignored and the downstream poly(A) signal is used as the default poly(A) site. The last two exons incorporated in the mRNA encode for a membrane-binding domain that anchors the antibody to the plasma membrane. After stimulation the upstream poly(A) signal is activated resulting in the processing of a mRNA that is translated to the same antibody but lacking the membrane-binding domain. Hence this antibody is secreted.

Also, a new translational reading frame, encoding for a completely different protein, may be produced by alternative poly(A) site usage. For example, production of calcitonin, which occurs in thyroid cells, and the calcitonin-related protein, which is produced in the brain, is regulated by tissue-specific poly(A) site usage.

D. Transcription and RNA Processing Are Coordinately Regulated Events

Recent experiments suggest that transcription and RNA processing are tightly coupled events. Thus, the RNA polymerase that assembles at the promoter has been shown to associate with factors required for polyadenylation and SR-related proteins that may partake in RNA splicing. The CTD tail of RNA polymerase II appears to function as a platform that recruits the RNA processing factors to the preinitiation complex. Therefore, the elongating RNA polymerase appears to have been loaded with the factors required for polyadenylation, and probably

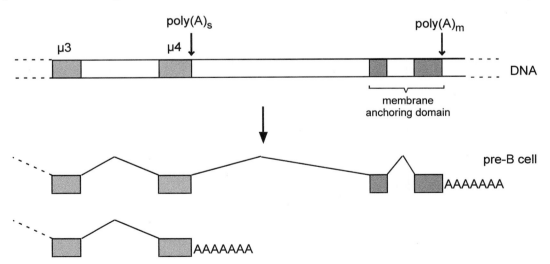

FIGURE 13 Alternative polyadenylation signals in the constant region of IgM yields heavy chains that are either membrane bound (pre-B cells) or secreted (plasma cells). Note that only the exon structure of the 3′ part of the IgM transcription unit is shown. The upstream region, which encodes for the antigen-binding domain, is identical in the two forms of IgM.

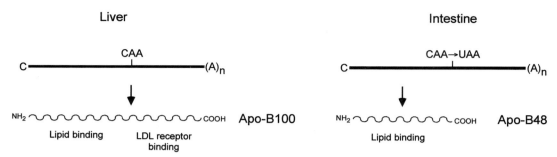

FIGURE 14 Editing of the Apo-B mRNA in intestinal cells. A CAA codon is edited to a UAA translational stop signal resulting in the production of a shorter protein (Apo-B48) corresponding to the amino-terminal half of the Apo-B100 protein expressed in liver cells.

deposits them at the polyadenylation signal used for 3′ end formation. Even more surprising, evidence has been presented suggesting that the composition of a promoter may specify alternative RNA splicing. Thus, the same gene under the transcriptional control of different promoters produces different types of alternatively spliced mRNAs. This finding suggests that enhancer binding transcription factors, in addition to stimulating recruitment of the RNA polymerase to the promoter, also may partake in the recruitment of selective RNA processing factors to the CTD tail. The future will tell whether alternative RNA splicing is regulated already at the level of transcription initiation.

E. Other Mechanisms of Posttranscriptional Regulation

Although this review has focused on control of gene expression at the level of synthesis and processing of mRNA, there are additional mechanisms that make significant contributions to gene expression. For example, a few genes in vertebrates have been shown to use RNA editing to produce different protein isoforms. The serum protein apolipoprotein B (Apo-B) is expressed in two forms. The Apo-B100 is expressed in hepatocytes, whereas a shorter polypeptide, Apo-B48, is expressed in intestinal epithelial cells. The cell-type-specific expression of Apo-B results from a posttranscriptional editing of the Apo-B mRNA in intestinal epithelial cells. Thus, a CAA codon encoding for the amino acid glutamate is converted to a UAA stop codon by cytosine deamination. As a result the Apo-B48 protein translated from the edited mRNA in the intestine differs from Apo-B100 by lacking the carboxy-terminus (Fig. 14). Both proteins bind to lipids. However, only the liver-specific Apo-B100 contains the carboxy-terminal domain required for binding to the low-density lipoprotein receptor, necessary for delivery of cholesterol to body tissues.

In nuclear genes RNA editing appears to be rare. Also, editing in such genes is restricted to modification of single nucleotides. In contrast, RNA editing in genes expressed in the mitochondria of protozoa, plants, and chloroplasts results in a more dramatic change of the mRNA sequence. Thus, a precursor-RNA may be edited such that more than 50% of the sequence in the mature mRNA is altered compared to the primary transcription product.

Gene expression is also regulated at other levels, such as nuclear to cytoplasmic transport of mRNA, translational efficiency of mRNA, RNA and protein stability, or protein modification. As is the case for transcriptional regulation, control of gene expression at the level of translation often occurs at the initiation step of the decoding process. Thus, not all mRNAs that reach the cytoplasm are used directly to synthesize protein. In fact, as much as 10% of genes in a eukaryotic cell may be regulated at the level of translation.

SEE ALSO THE FOLLOWING ARTICLES

CHROMATIN STRUCTURE AND MODIFICATION • DNA TESTING IN FORENSIC SCIENCE • ENZYME MECHANISMS • HYBRIDOMAS, GENETIC ENGINEERING OF • NUCLEIC ACID SYNTHESIS • RIBOZYMES • TRANSLATION OF RNA TO PROTEIN

BIBLIOGRAPHY

Lodish, H., Berk, A., Zipursky, S. L., Matsudaria, P., Baltimore, D., and Darnell, J. (2000). "Molecular Cell Biology," Freeman and New York.
Latchman, D. S. (1998). "Eukaryotic Transcription Factors," Academic Press, San Diego, CA.
Gesteland, R. F., Cech, T. R., and Atkins, J. F. (1999). "The RNA World," 2nd ed., Cold Spring Harbor Laboratory Press, Cold Spring Harbor, New York.

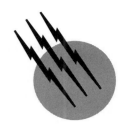

Generalized Functions

Ram P. Kanwal
Pennsylvania State University

GLOSSARY

Convolution The convolution $f*g$ of two functions $f(x)$ and $g(x)$ is defined as $\int_{-\infty}^{\infty} f(y)\,g(x-y)\,dy$. This concept carries over to the generalized functions also.

Dirac delta function $\delta(x)$ This function is intuitively defined to be zero when $x \neq 0$ and infinite at $x=0$, in such a way that the area under it is unity.

Distribution A linear continuous functional.

Distributional (generalized) derivatives Nonclassical derivatives of the generalized functions which do not have classical derivatives at their singularities.

Functional A rule which assigns a number $\langle t, \phi \rangle$ to a test function $\phi(x)$ through a generalized function $t(x)$.

Hadamard finite part Finite difference of two infinite terms for defining divergent integrals.

Impulse response The particular solution of a differential equation with $\delta(x)$ (impulse) as its forcing term.

Pseudo-function Singular functions such as $1/x^k$ which are regularized at their singularity with the help of concepts such as the Hadamard finite part.

Regular distributions Distributions based on locally integrable functions.

Singular distributions (generalized functions) Distributions which are not regular.

Test function space A space of suitable smooth functions $\phi(x)$ which is instrumental in defining a generalized function such as $\delta(x)$ as a distribution.

Encyclopedia of Physical Science and Technology, Third Edition, Volume 6

GENERALIZED FUNCTIONS are objects, such as the Dirac delta function, with such inherent singularities that they cannot be integrated or differentiated in the classical sense. By defining them as functionals (distributions) which carry smooth functions to numbers, we can overcome the difficulty. Then they possess remarkable properties that extend the capabilities of the classical mathematics. Indeed, the techniques based on the generalized functions not only solve the classical problems in a simple fashion but also produce many new concepts. Accordingly, they have influenced many topics in mathematical and physical sciences.

I. INTRODUCTION

Functions such as Dirac's delta function and Heaviside function have been used by scientists even though the former is not a function and the latter is not a differentiable function in the classical sense. The theory of distributions has provided us not only with mathematical foundations for these functions but also for various other nonclassical functions. The rudiments of the theory of distributions can be found in the concepts of the Hadamard finite part of the divergent integrals and sequences which in the limit behave like delta function. However, it was Sobolev and Schwartz who established the modern theory of distributions.

The Dirac delta function $\delta(x - \xi)$, also called the impulse function, is usually defined as a function which is zero everywhere except at $x = \xi$, where it has a spike such that $\int_{-\infty}^{\infty} \delta(x - \xi)\, dx = 1$. More generally, it is defined by its sifting property,

$$\int_{-\infty}^{\infty} f(x)\delta(x - \xi)\, dx = f(\xi), \tag{1}$$

for all continuous functions $f(x)$. However, for any acceptable definition of integration there can be no such function. This difficulty was subsequently overcome by two approaches. The first is to define the delta function as the limit of delta sequences, while the second is to define it as a distribution. The reasoning behind a delta sequence is that although the delta function cannot be justified mathematically, there are sequences $\{s_m(x)\}$ which in the limit $m \to \infty$ satisfy the relation (1). An interesting sequence is

$$s_m(x) = \frac{1}{\pi}\frac{m}{1 + m^2 x^2}. \tag{2}$$

It is instructive to imagine (2) as a continuous charge distribution on a line, so that the total charge $s_m(x)$ to the left of x is $r_m(x) = \int_{-\infty}^{x} s_m(u)\, du = \frac{1}{2} + (1/\pi)\tan^{-1} mx$.

Because $\int_{-\infty}^{\infty} s_m(x)\, dx = 1$, the total charge on the line is equal to unity. Furthermore, it can be readily proved that

$$\lim_{m \to \infty} \int_{-\infty}^{\infty} f(x)s_m(x)\, dx = f(0).$$

Accordingly, it satisfies the sifting property (1).

Defining $\delta(x)$ as a distribution is much more convenient and useful. Moreover, the theory of distributions helps us define many more functions which are singular in nature.

II. DISTRIBUTIONS

To make $\delta(x)$ meaningful, we appeal to various spaces of smooth functions. We start with the real-valued functions $\phi(x)$, where $x = (x_1, \ldots, x_n) \in \mathbb{R}^n$, the n-dimensional Euclidean space. To describe various properties of these functions we introduce the multi-index notation. Let k be an n-tuple of non-negative integers: $k = (k_1 \ldots, k_n)$. Then we define

$$|k| = k_1 + \cdots + k_n; \qquad x^k = x_1^{k_1} \cdots x_n^{k_n};$$

$$k! = k_1! \cdots k_n!;$$

$$\binom{k}{m} = \frac{k!}{m!(k - m)!}; \qquad D_j = \frac{\partial}{\partial x_j}; \tag{3}$$

$$D^k = \frac{\partial^{|k|}}{\partial x_1^{k_1} \cdots \partial x_n^{k_n}} = \frac{\partial^{k_1 + \cdots + k_n}}{\partial x_1^{k_1} \cdots \partial x_n^{k_n}} = D_1^{k_1} \ldots D_n^{k_n}.$$

We are now ready to state the properties of the functions $\phi(x)$. They are (1) $D^k\phi(x)$ exists for all multi-indices k. (2) There exists a number A such that $\phi(x)$ vanishes for $r > A$, where r is the radial distance $r = (x_1^2 + \cdots + x_n^2)^{1/2}$. This means that $\phi(x)$ has a compact support. These two properties are written symbolically as $\phi(x) \in C_0^\infty$, where the superscript stands for infinite differentiability, the subscript for the compact support. This space of functions is denoted by $\mathcal{D}$, and $\phi(x)$ are called the test functions. The prototype example in $\mathbb{R}$ is

$$\phi(x) = \begin{cases} \exp\left(-\dfrac{a^2}{a^2 - x^2}\right), & |x| < a, \\ 0, & |x| > a. \end{cases} \tag{4}$$

Note that the definition of $\mathcal{D}$ does not demand that all $\phi(x)$ have the same support. Furthermore, we observe that (a) $\mathcal{D}$ is a linear (vector) space, because if ϕ_1 and ϕ_2 are in $\mathcal{D}$, then so is $c_1\phi_1 + c_2\phi_2$ for arbitrary real numbers. (b) If $\phi \in \mathcal{D}$, then so is $D^k\phi$. (c) For a C^∞ function of $f(x)$ and a $\phi(x) \in \mathcal{D}$, $f\phi \in \mathcal{D}$. (d) If $\phi(x_1, \ldots, x_m)$ is an m-dimensional test function and $\psi(x_{m+1,\ldots,x_n})$ is an $(n - m)$-dimensional test function, then $\phi\psi$ is an n-dimensional test function in the variables $x_1, \ldots, x_n$.

With the following four definitions we are able to grasp the concept of distributions:

1. A sequence $\{\phi_m(x)\}$, $m = 1, 2, \ldots$, for all $\phi_m \in \mathcal{D}$, converges to ϕ_0 if the following two conditions are satisfied. (a) All ϕ_m as well as ϕ_0 vanish outside a common region. (b) $\mathcal{D}^k \phi_m \to \mathcal{D}^k \phi_0$ uniformly over $\mathbb{R}^n$ as $m \to \infty$, for all multi-indices k. For the special case $\phi_0 = 0$, the sequence $\{\phi_m\}$ is called a null sequence.

2. A linear functional $t(x)$ on the space $\mathcal{D}$ is an operation by which we assign to every test function $\phi(x)$ a real number $\langle t(x), \phi(x) \rangle$ such that $\langle t, c_1 \phi_1 + c_2 \phi_2 \rangle = c_1 \langle t, \phi_1 \rangle + c_2 \langle t, \phi_2 \rangle$ for arbitrary test functions $\phi_1(x)$ and $\phi_2(x)$ and real numbers c_1 and c_2. Then it follows that $\langle t, \sum_{j=1}^m c_j \langle \phi_j \rangle = \sum_{j=1}^m c_j \langle t, \phi_j \rangle$, where c_j are arbitrary real numbers.

3. A linear functional $t(x)$ on $\mathcal{D}$ is called *continuous* if and only if the sequence of numbers $\langle t, \phi_m \rangle \to \langle t, \phi \rangle$, as $m \to \infty$, when the sequence $\{\phi_m\}$ of test functions converges to $\phi \in \mathcal{D}$, that is,

$$\lim_{m \to \infty} \langle t, \phi_m \rangle = \langle t, \lim_{m \to \infty} \phi_m \rangle.$$

4. A continuous linear functional on the space $\mathcal{D}$ is called a distribution.

The set of distributions that is most useful is the ones generated by locally integrable functions $f(x)$ (that is, $\int_\Omega [f(x)] \, dx$ exists for every bounded region Ω of $\mathbb{R}^n$). Indeed, every locally integrable function $f(x)$ generates a distribution through the relation $\langle f, \phi \rangle = \int_\Omega f(x) \phi(x) \, dx$. The linearity and continuity of this functional can be readily proved. The distributions produced by locally integrable functions are called *regular distributions* or distributions of order zero. All the other ones are called *singular distributions*. We present two examples in the one-dimensional space $\mathbb{R}$.

1. Heaviside distribution:

$$H(x) = \begin{cases} 0, & x < 0, \\ 1, & x > 0. \end{cases}$$

The Heaviside distribution defines the regular distribution $\langle H, \phi \rangle = \int_0^\infty \phi(x) \, dx$. It is clearly a linear functional. It is also continuous because $\lim_{m \to \infty} \langle H(x), \phi_m(x) \rangle = \lim_{m \to \infty} \int_0^\infty \phi_m(x) = \int_0^\infty \phi(x) \, dx$, where $\phi(x)$ is the limit of the sequence $\{\phi_m(x)\}$ as $m \to \infty$. Thus,

$$\lim_{m \to \infty} \langle H(x), \phi_m(x) \rangle = \langle H(x), \lim_{m \to \infty} \phi_m(x) \rangle.$$

2. The Dirac delta function $\delta(x - \xi)$. By the sifting property (1) we have $\langle \delta(x - \xi), \phi(x) \rangle = \int_{-\infty}^\infty \delta(x - \xi) \times \phi(x) \, dx = \phi(\xi)$, a number. The linearity also follows from the sifting property. This functional is continuous because $\lim_{m \to \infty} \langle \delta(x - \xi), \phi_m(x) \rangle = \lim_{m \to \infty} \phi_m(\xi) = \langle \delta(x - \xi), \lim_{m \to \infty} \phi_m(x) \rangle$. Since $\delta(x - \xi)$ is not a locally integrable function, it produces a singular distribution. These results hold in $\mathbb{R}^n$ as well. The functions which generate singular distributions are called generalized functions. We shall use these words interchangeably. The definition of a distribution can be extended to include complex-valued functions.

The dual space $\mathcal{D}'$. The space of all distributions on $\mathcal{D}$ is called the *dual space* of $\mathcal{D}$ and is denoted as $\mathcal{D}'$. It is also a linear space.

There are many other interesting spaces. For example, the test function space $\mathcal{E}$ consists of all the functions $\phi(x)$, so there is no limit on their growth at infinity. Accordingly, $\mathcal{E} \supset \mathcal{D}$. The corresponding space $\mathcal{E}'$ consists of the distributions which have compact support, so that $\mathcal{E}' \supset \mathcal{D}'$. The distributions of slow growth are defined in Sect. IX.

III. ALGEBRAIC OPERATIONS ON DISTRIBUTIONS

Let $\langle t(x), \phi(x) \rangle$ be a regular distribution generated by a locally integrable function $t(x) \times \in \mathbb{R}$. Let $x = ay - b$, where a and b are constants. Then we have

$$\langle t(ay - b), \phi(y) \rangle = \int_{-\infty}^\infty t(ay - b) \phi(y) \, dy$$

$$= \frac{1}{|a|} \int_{-\infty}^\infty t(x) \phi\left(\frac{x + b}{a}\right) dx$$

$$= \frac{1}{|a|} \left\langle t(x), \phi\left(\frac{x + b}{a}\right) \right\rangle. \quad (5)$$

For the special case $a = 1$, relation (5) becomes $\langle t(y - b), \phi(y) \rangle = \langle t(x), \phi(x + b) \rangle$. As another special case, (5) yields $\langle \delta(-y), \phi(y) \rangle \langle \delta(x), \phi(-x) \rangle = \phi(0)$. Thus $\delta(x)$ is an even function.

A distribution is called homogeneous of degree λ if $t(ax) = a^\lambda t(x)$, $a > 0$. In view of relation (5) with $b = 0$, we find that for a homogeneous distribution we have $\langle t(ay), \phi(y) \rangle = (1/a)\langle t(x), \phi(x/a) \rangle = a^\lambda \langle t(x), \phi(y) \rangle$, so that $\langle t(x), \phi(x/a) \rangle = a^{\lambda+1} \langle t(x), \phi(x) \rangle$.

All these relations hold also for distributions $t(x)$, $x \in \mathbb{R}^n$. In that case we set $x = Ay - B$, where A is a nonsingular $n \times n$ matrix and B is a constant vector. Then we have $\langle t(Ay - B), \phi(y) \rangle = (1/\det A)\langle t(x), \phi[A^{-1}(x + B)] \rangle$, where A^{-1} is the inverse of the matrix A.

Product of a distribution and a fucntion: In general it is difficult to define the product of two distributions, even for two regular distributions. Indeed, if we take $f(x) = g(x) = 1/\sqrt{x}$, which is a regular distribution, then their product $1/x$ is not a locally integrable function in the neighborhood of $x = 0$ and as such does not define a regular distribution (we shall discuss the function $1/x$ in Sect. V). However, we can multiply a distribution $t(x)$ with a C^∞ function $\psi(x)$ so that we have $\langle \psi t, \phi \rangle = \langle t, \psi \phi \rangle$. Because $\psi \phi \in \mathcal{D}$, it follows that ψt is a distribution.

IV. ANALYTIC OPERATIONS ON DISTRIBUTIONS

Let us start with a regular distribution generated by C^1 function $t(x)$, $x \in \mathbb{R}$, so that $\langle t(x), \phi(x) \rangle = \int_{-\infty}^{\infty} t(x) \phi(x) \, dx$. When we integrate the quantity $\int_{-\infty}^{\infty} t'(x)$, $\phi(x) \, dx$ by parts, we get $\int_{-\infty}^{\infty} t'(x) \times \phi(x) \, dx = -\int_{-\infty}^{\infty} t(x) \phi'(x) \, dx$, where we have used the fact that $\phi(x)$ has a compact support. Thus, $\langle t'(x), \phi(x) \rangle = -\langle t(x), \phi'(x) \rangle$. Because $\phi'(x)$ is also in $\mathcal{D}$, this relation helps us in defining the distributional derivative $t'(x)$ of a distribution $t(x)$ (regular or singular) as above. Continuing this process, we have $\langle t^{(n)}(x), \phi(x) \rangle = (-1)^n \langle t(x), \phi^{(n)}(x) \rangle$, where the superscript (n) stands for nth-order differentiation. The corresponding formula in $\mathbb{R}^n$ is

$$\langle D^k t(x), \phi(x) \rangle = (-1)^k \langle t(x), D^k \phi(x) \rangle. \tag{6}$$

Thus a generalized function is infinitely differentiable. This result has tremendous ramifications, as we shall soon discover.

The primitive of a distribution $t(x)$ is a solution of $ds(x)/dx = t(x)$. This means that we seek $s(x) \in \mathcal{D}'$ such that $\langle s, \phi' \rangle = -\langle t, \phi \rangle$, $\phi \in \mathcal{D}$.

Let us illustrate the concept of distributional differentiation with the help of a few applications.

1. Recall that the Heaviside function is defined as a distribution by the relation (6) $\langle H(x), \phi(x) \rangle = \int_0^\infty \phi(x) \, dx$. According to definition (6), we have $\langle H'(x), \phi(x) \rangle = -\langle H(x), \phi'(x) \rangle = -\int_0^\infty \phi'(x) = \phi(0) = \langle \delta(x), \phi(x) \rangle$, where we have use the fact that $\phi(x)$ vanishes at ∞. Thus,

$$\frac{dH}{dx} = \delta(x). \tag{7}$$

Because $H(x)$ is a distribution of order zero, $\delta(x)$ is a distribution of order 1. We can continue this process and find that $\langle \delta^{(n)}(x), \phi(x) \rangle = (-1)^n \langle \delta(x), \phi^{(n)}(x) \rangle$. Just as $\delta(x)$ stands for an impulse or a pole at $x = 0$, $\delta'(x)$ stands for a dipole and $\delta^{(n)}(x)$ is a multiple of order $(n + 1)$ as well as a distribution of order $(n + 1)$.

2. *Impulse response.* With the help of the formula (7) we can find the derivative of the signum function sgn x, which is 1 for $x > 0$ and -1 for $x < 0$. Thus, sgn $x = 2H(x) - 1$. Then it follows from (7) that $(\text{sgn } x)' = 2\delta(x)$. This result, in turn, helps us in finding the derivative of the function $|x|$. Indeed, $\langle |x|, \phi(x) \rangle = \int_{-\infty}^{\infty} |x| \phi(x) \, dx = \int_0^\infty x \phi(x) \, dx - \int_{-\infty}^0 x \phi(x) \, dx$. Thus, $\langle |x|', \phi(x) \rangle = -\langle |x|, \phi'(x) \rangle = -\int_0^\infty x \phi'(x) \, dx + \int_\infty^0 x \phi' \times (x) \, dx$, which, when integrated by parts, yields the formula $\langle |x|', \phi(x) \rangle = \langle \text{sgn } x, \phi(x) \rangle$. Thus, $|x|' = \text{sgn } x$. The second differentiation yields $(d^2/dx^2)(|x|) = 2\delta(x)$.

3. A distribution $E(x)$ is said to be a fundamental solution or a free-space Green's function or an impulse response if it satisfies the relation $LE(x) = \delta(x)$, where L is a differential operator. Then from the previous paragraph we find that $\frac{1}{2}|x|$ is the impulse response for the differential operator $L = (d^2/dx^2)$.

Relation (6) for differentiation helps us in deriving many interesting formulas such as

$$\begin{aligned} t(x)\delta^{(n)}(x) =\ & (-1)^n t^{(n)}(0)\delta(x) \\ & + (-1)^{(n-1)} t^{(n-1)}(0)\delta'(x) \\ & + (-1)^{n-2} \frac{n(n-1)}{2!} t^{(n-2)}(0)\delta''(x) + \cdots \\ & + f(0)\delta^{(n)}(x) \end{aligned} \tag{8}$$

for a distribution $t(x)$. The proof follows by evaluating the quantity $\langle t(x)\delta^{(n)}(x), \phi(x) \rangle = \langle \delta^{(n)}(x), t(x)\phi(x) \rangle = (-1)^n \langle \delta(x), (t(x)\phi(x))^{(n)} \rangle$. For $n = 0$ and 1, formula (8) becomes

$$\begin{aligned} t(x)\delta(x) &= t(0)\delta(x); \\ t(x)\delta'(x) &= -t'(0)\delta(x) + t(0)\delta'(x). \end{aligned} \tag{9}$$

One of the most important consequences of (8) is

$$x^m \delta^{(n)}(x) = \begin{cases} (-1)^m \dfrac{n!}{(n-m)!} \delta^{(n-m)}(x), & n \geq m, \\[2mm] 0, & n < m. \end{cases} \tag{10}$$

4. Next, we attempt to evaluate $\delta[f(x)]$. Let us first assume that $f(x)$ has a simple zero at x_1 such that $f(x_1) = 0$ but $f'(x_1) > 0$ [the case $f'(x)_1 < 0$ follows in a similar fashion]. Thus $f(x)$ increases monotonically in the neighborhood of x_1 so that $H[f(x)] = H(x - x_1)$, where $H(x)$ is the Heaviside function. Then we use (7) and find that $(d/dx)H[f(x)] = \delta(x - x_1)$ or $\delta[f(x)] = |(f'(x_1))|^{-1}\delta(x - x_1)$. If there are n zeros of $f(x)$, then the above result yields

$$\delta[f(x)] = \sum_{m=1}^{n} \frac{\delta(x - x_m)}{f'(x_m)}. \tag{11}$$

This result has many applications.

5. The formula $(\text{sgn}\, x)' = 2\delta(x)$, derived in Example 2, is also instrumental in deriving an integral representation for the delta function. Indeed, if we appeal to the relations

$$\int_{-\infty}^{\infty} \frac{\sin tx}{x}\, dx = \begin{cases} \pi, & t > 0 \\ -\pi, & t < 0; \end{cases} \qquad \int_{-\infty}^{\infty} \frac{\cos tx}{x} = 0,$$

from calculus, we find that $(1/2\pi)\int_{-\infty}^{\infty}(e^{itx}/ix)\, dx = \frac{1}{2}\,\text{sgn}\, t$. When we differentiate this relation with respect to t we get the important formula

$$\delta(t) = \frac{1}{2\pi}\int_{-\infty}^{\infty} e^{itx}\, dx = \frac{1}{2\pi}\int_{-\infty}^{\infty} e^{-itx}\, dx, \qquad (12)$$

where we have used the fact that $\delta(t)$ is an even function. This formula can be generalized to n dimensions if we observe that $\delta(x_1, \ldots, x_n) = \delta(x_1)\cdots\delta(x_n)$. Accordingly, in the four-dimensional space (x_1, x_2, x_3, t), relation (12) yields the planewave expansion of the delta function,

$$\delta(\boldsymbol{x}, t) = \frac{1}{(2\pi)^4}\int_{-\infty}^{\infty} e^{-i(\boldsymbol{k}\cdot\boldsymbol{x} - \omega t)}\, d^3\boldsymbol{k}\, d\omega, \qquad (13)$$

where $\boldsymbol{k} = (k_1, k_2, k_3)$ and we have a fourfold integral.

V. PSEUDO-FUNCTION, HADAMARD FINITE PART AND REGULARIZATION

The functions $1/x^m$ and $H(x)/x^m$, where m is an integer, are not locally integrable at $x = 0$ and, as such, do not define distributions. However, we can appeal to the concepts of Cauchy principal value and the Hadamard finite part and define these functions as distributions. The simplest example is the function $1/x$. For a test function $\phi(x)$ the integral $\int[\phi(x)/x]\, dx$ is not absolutely convergent unless $\phi(0) = 0$. However, it has an interpretation as a principal value integral, namely,

$$\left\langle Pv\left(\frac{1}{x}\right), \phi(x)\right\rangle = \lim_{\epsilon \to 0}\int_{|x| > \epsilon} \frac{\phi(x)}{x}\, dx. \qquad (14)$$

Writing $\phi(x) = \phi(0) + [\phi(x) - \phi(0)]$, relation (14) takes the form

$$\left\langle Pv\left(\frac{1}{x}\right), \phi(x)\right\rangle = \lim_{\epsilon \to 0}\int_{|x| > \epsilon} \frac{\phi(0)}{x}\, dx$$
$$+ \int_{|x| > \epsilon} \frac{[\phi(x) - \phi(0)]}{x}\, dx. \qquad (15)$$

Since the function $1/x$ is odd, the first term on the right side of (15) vanishes. The integrand in the second integral on the right side of (15) approaches $\phi'(0)$ as $x \to 0$. Accordingly, we have

$$\left\langle Pv\left(\frac{1}{x}\right), \phi(x)\right\rangle = \lim_{\epsilon \to 0}\int_{|x| > \epsilon} \frac{\phi(x) - \phi(0)}{x}\, dx.$$

Because $\epsilon > 0$ is arbitrary, the previous relation is also written as

$$\left\langle Pv\left(\frac{1}{x}\right), \phi(x)\right\rangle = \lim_{\epsilon \to 0}\int_{|x| > 1} \frac{\phi(x) - \phi(0)}{x}\, dx. \qquad (16)$$

In this form the function $Pv(1/x)$ is easily proved to be a linear continuous function. As such, it is a distribution and is written as $Pf(1/x)$, where Pf stands for pseudo-function.

We come across many divergent integrals whose principal values do not exist. Consider, for instance, the integral $\int_a^b dx/x^2$, $a < 0 < b$. Because

$$\int_{[a,b]/[-\epsilon,\epsilon]} \frac{dx}{x^2} = \lim_{\epsilon \to 0}\left(-\frac{1}{b} + \frac{1}{a} + \frac{2}{\epsilon}\right), \qquad (17)$$

we cannot get the principal value. In these situations the concept of the Hadamard finite part becomes helpful. Indeed, in relation (17), the value $(-1/b + 1/a)$ is the finite part and $(2/\epsilon)$ is the infinite part. Thus, we write

$$Fp\int_a^b \frac{dx}{x^2} = -\frac{1}{b} + \frac{1}{a}. \qquad (18)$$

As another example, we consider the integral $\int_0^1 dx/x^\alpha$. It is divergent if $\alpha \geq 1$. Indeed, for $\alpha > 1$, we obtain

$$\int_\epsilon^1 \frac{dx}{x^\alpha} = \frac{1}{1 - \alpha} - \frac{\epsilon^{1-\alpha}}{1 - \alpha} \qquad (19)$$

so that we have

$$Fp\int_0^1 \frac{dx}{x^\alpha} = \frac{1}{1 - \alpha}, \qquad \alpha > 1. \qquad (20)$$

When $\alpha = 1$, $\int_\epsilon^1 dx/x = -\ln \epsilon$, which is infinite, so that

$$Fp\int_0^1 \frac{dx}{x} = 0. \qquad (21)$$

The process of finding the principal value and the finite part of divergent integrals is called the regularization of these integrals. When both of them exist, they are equal.

Let us use the definition of the Hadamard finite part to regularize the function $H(x)/x$. The action of $H(x)/x$ on a test function $\phi(x)$ is $\int_0^\infty(\phi(x)/x)\, dx$. Thus, we consider

$$\int_\epsilon^\infty \frac{\phi(x)}{x}\, dx = \int_\epsilon^1 \frac{\phi(x)}{x}\, dx + \int_1^\infty \phi(x)\, dx. \qquad (22)$$

In the numerator of the first term on the right side of this equation we add and subtract $\phi(0)$ to get

$$\int_\epsilon^\infty \frac{\phi(x)}{x}\,dx = \int_\epsilon^1 \frac{\phi(0)}{x}\,dx + \int_\epsilon^1 \frac{\phi(x)-\phi(x)}{x}$$
$$+ \int_1^\infty \phi(x)\,dx$$

$$= -\phi(0)\ln\epsilon + \int_\epsilon^1 \frac{\phi(x)-\phi(0)}{x}\,dx$$
$$+ \int_1^\infty \frac{\phi(x)}{x}\,dx.$$

The first term on the right side of this equation is infinite as $\epsilon \to 0$, while the other two terms are finite. Accordingly, we define

$$\int \frac{H(x)\phi(x)}{x}\,dx = \int_\epsilon^1 \frac{\phi(x)-\phi(0)}{x}\,dx + \int_1^\infty \frac{\phi(x)}{x}\,dx. \tag{23}$$

The function $Pf(1/|x|)$ can be regularized in the same way. Indeed,

$$\left\langle Pf\left(\frac{1}{|x|}\right), \phi(x)\right\rangle = \int_{|x|\le 1} \frac{\phi(x)-\phi(0)}{|x|}\,dx$$
$$+ \int_{|x|>1} \frac{\phi(x)}{|x|}\,dx. \tag{24}$$

These concepts can be used to regularize the functions $1/x^m$ and $H(x)/x^m$ to yield the generalized functions $Pf(1/x^m)$ and $Pf(H(x)/x^m)$. The combination of $Pf(1/x^m)$ and $\delta^{(m)}(x)$ yield the Heisenberg distributions

$$\delta^{\pm(m)}(x) = \frac{1}{2}\delta^{(m)}(x) \mp \frac{1}{2\pi i}\,Pf\left(\frac{1}{x^m}\right).$$

They arise in quantum mechanics.

Let us end this section by solving the equation $xt(x) = g(x)$. The homogeneous part $xt(x) = 0$ has the solution $t(x) = \delta(x)$ because $x\delta(x) = 0$. Thus, the complete solution is

$$t(x) = \delta(x) + g(x)Pf\left(\frac{1}{x}\right). \tag{25}$$

VI. DISTRIBUTIONAL DERIVATIVES OF DISCONTINUOUSFUNCTIONS

Let us start with $x \in \mathbb{R}$ and consider a function $F(x)$ that has a jump discontinuity at $x = \xi$ of magnitude a but that

has the drivative $F'(x)$ in the intervals $x < \xi$ and $x > \xi$. The derivative is undefined at $x = \xi$. To find the distributional derivative of $F(x)$ in the entire interval we define the function $f(x) = F(x) - aH(x-\xi)$, where H is the Heaviside function. This function is continuous at $x = \xi$ and has a derivative which coincides with $F'(x)$ on both sides of ξ. Accordingly, we differentiate both sides of this equation and get $F'(x) = \bar{F}'(x) - a\delta(x-\xi)$, where the bar over F stands for the generalized (distributional) derivative of F. Thus,

$$\bar{F}'(x) = F'(x) + [F]\delta(x-\xi), \tag{26}$$

where $[F] = a$ is the value of the jump of F at $x = \xi$. Before we present the corresponding n-dimensional theory, we give an interesting application of (26) to the Sturm-Liouville differential equation,

$$\frac{d}{dx}\left[p(x)\frac{dE(x;\xi)}{dx}\right] = q(x)E(x,\xi) - \delta(x-\xi), \tag{27}$$

where $E(x,\xi)$ stands for the impulse response and $p(x)$ and $q(x)$ are continuous at $x = \xi$. When we compare (22) and (23) and use the continuity of $p(x)$ at $x = \xi$, we find that the jump of $[dE/dx]_{x=\xi}$ is

$$[dE/dx]_{x=\xi} = -1/p(\xi).$$

These concepts easily generalize to higher-order derivatives and to the surfaces of discontinuity and, therefore, have applications in the theory of wave fronts. Accordingly, we include time t in our discussion and consider a function $F(\boldsymbol{x}, t)$, $\boldsymbol{x} \in \mathbb{R}^n$, which has a jump discontinuity across a moving surface $\Sigma(\boldsymbol{x}, t)$. Such a surface can be represented locally either as an implicit equation of the form $u(x_1, \ldots, x_n, t) = 0$, or in terms of the curvilinear Gaussian coordinates $v_1, \ldots, v_{n-1}$ on the surface: $x_i = x_i(v_1, \ldots, v_{n-1}, t)$. The surface Σ is regular, so the above-mentioned functions have derivatives of all orders with respect to each of their arguments, and for all values of t, the corresponding Jacobian matrices of transformation have appropriate ranks, that is, grad $u \ne 0$, and the rank of the matrix $(\partial x_i/\partial v_j) = n - 1$. Furthermore, Σ divides the space into two parts, which we shall call positive and negative.

The basic distribution concentrated on a moving and a deforming surface $\Sigma(\boldsymbol{x}, t)$ is the delta function $\delta[\Sigma(\boldsymbol{x}, t)]$ whose action on a test function $\phi(\boldsymbol{x}, t) \in \mathcal{D}$ is

$$\langle \delta(\Sigma), \phi \rangle = \int_{-\infty}^\infty \int_\Sigma \phi(\boldsymbol{x}, t)\,dS(\boldsymbol{x})\,dt, \tag{28}$$

where dS is the surface element on Σ. This is a simple layer. The second surface distribution is the normal derivative operator, given as

$$\langle d_n \delta(\Sigma), \phi \rangle = -\int_{-\infty}^{\infty} \int_{\Sigma} \frac{d\phi}{dn} \, dS(\mathbf{x}) \, dt, \qquad (29)$$

where $d\phi/dn$ stands for the differentiation along the normal to the surface. This is a dipole layer. Another surface distribution that we need in our discussion is $\delta'(\Sigma)$, defined as

$$\delta'(\Sigma) = n_i \frac{\bar{\partial}}{\partial x_i} [\delta(\Sigma)], \qquad (30)$$

where n_i are the components of the unit normal vector $\mathbf{n}$ to the surface. These three distributions are connected by the relation

$$d_n \delta(\Sigma) = \delta'(\Sigma) - 2\Omega \delta(\Sigma), \qquad (31)$$

where Ω is the mean curvature of Σ.

Let a function $F(\mathbf{x}, t)$ have a jump discontinuity across $\Sigma(\mathbf{x}, t)$. Then the formulas for the time derivative, gradient, and curl of a discontinuous function $F(\mathbf{x}, t)$ can be derived in a manner similar to (26). Indeed,

$$\frac{\bar{\partial} \mathbf{F}}{\partial t} = \frac{\partial \mathbf{F}}{\partial t} - G[\mathbf{F}]\delta(\Sigma), \qquad (32)$$

$$\overline{\text{grad}} \, \mathbf{F} = \text{grad} \, \mathbf{F} + \mathbf{n}[\mathbf{F}]\delta(\Sigma), \qquad (33)$$

$$\overline{\text{div}} \, \mathbf{F} = \text{div} \, \mathbf{F} + \mathbf{n}[\mathbf{F}]\delta(\Sigma), \qquad (34)$$

$$\overline{\text{curl}} \, \mathbf{F} = \text{curl} \, \mathbf{F} + \mathbf{n} \times [\mathbf{F}]\delta(\Sigma), \qquad (35)$$

where G is the normal speed of the front Σ and $[\mathbf{F}] = \mathbf{F}_+ - \mathbf{F}_-$ is the jump of $\mathbf{F}$ across Σ.

Before we extend these concepts, we apply relation (33) to a scalar function F and derive the distributional derivative of $1/r$, where r is the radial distance. In mathematical physics the function $1/r$ is important because it describes the gravitational and the Coulomb potentials. In the modern theories of small particles the singularity of $1/r$ at $r = 0$ makes a significant contribution. Accordingly, its distributional derivatives are needed to get the complete picture of the singularity at $r = 0$ and we present it as follows. For the sake of computational simplicity, we restrict ourselves to $\mathbb{R}^3$, so that $r = (x_1^2 + x_2^2 + x_3^2)^{1/2}$. The function $1/r$ corresponds to the $Pf[H(x)/x]$ that we considered in the previous section. Accordingly, we introduce the function $F(\mathbf{x}) = H(r - \epsilon)/r$, where $H(r - \epsilon)$ is the Heaviside function, which is unity for $r > \epsilon$ and is zero for $r < \epsilon$. This, in turn, helps us in defining the distribution $1/r$ as

$$\left\langle \frac{1}{r}, \phi(\mathbf{x}) \right\rangle = \lim_{\epsilon \to 0} \int_{-\infty}^{\infty} \frac{1}{r} H(r - \epsilon) \phi(\mathbf{x}) \, dx. \qquad (36)$$

Our aim is to differentiate $1/r$ by taking into account the singularity at $r = 0$. For this we appeal to formula (33) and observe that $F(\mathbf{x}) = H(r - \epsilon)/r$ so that the jump

of this function at the spherical surface of discontinuity $\Sigma : r = \epsilon$, is $1/\epsilon$. Thus, formula (33) becomes

$$\frac{\bar{\partial}}{\partial x_j} \left(\frac{H(r - \epsilon)}{r} \right) = -\frac{x_j}{r^3} H(r - \epsilon) + \hat{r}_j \frac{1}{\epsilon} \delta(\Sigma), \qquad (37)$$

where $n_j = \hat{r}_j - x_j/r$. To evaluate the limit of the second term on the right side of (37) we evaluate $\lim_{\epsilon \to 0} \int_{\Sigma} \phi(\mathbf{x})(1/\epsilon)\hat{r} \, dS = \lim_{\epsilon \to 0}(1/\epsilon) \int_{\Sigma(1)} \phi(\mathbf{x})\hat{r} \, \epsilon^2 \, d\omega = 0$, where $\Sigma(1)$ is the unit sphere and ω is the solid angle. Thus $\bar{\partial}/\partial x_j (1/r) = -x_j/r^3$, which is the same as the classical derivative. In order to compute the second-order distributional derivatives we apply formula (33) to the function $(\partial/\partial x_j)[H(r - \epsilon)/r]$ and get

$$\frac{\bar{\partial}^2}{\partial x_i \, \partial x_j} \left[\frac{H(r - \epsilon)}{r} \right] = \left[\frac{\partial^2}{\partial x_i \, \partial x_j} \left(\frac{1}{r} \right) \right] H(r - \epsilon)$$
$$+ \left(\frac{x_i}{r} \right) \left(\frac{-x_j}{r^3} \right) \delta(\Sigma)$$
$$= \left(\frac{3x_i x_j - r^2 \delta_{ij}}{r^5} \right) H(r - \epsilon)$$
$$- \frac{x_i x_j}{r^4} \delta(\Sigma), \qquad (38)$$

where δ_{ij} is the Kronecker delta, which is 1 when $i = j$ and 0 when $i \neq j$. Because $\lim_{\epsilon \to 0} \int_{\Sigma} (x_i x_j/r^4)\phi(\mathbf{x}) \, dS = (4\pi/3)\delta_{ij}\phi(\mathbf{0})$, relation (38) becomes

$$\frac{\bar{\partial}^2}{\partial x_i \, \partial x_j} \left(\frac{1}{r} \right) = \frac{3x_i x_j - r^2 \delta_{ij}}{r^5} - \frac{4\pi}{3} \delta_{ij} \delta(\mathbf{x}). \qquad (39)$$

From this formula we can derive the impulse response for the Laplace operator. Indeed, if we set $i = j$ and sum on j, we get $\nabla^2(1/r) = -4\pi\delta(\mathbf{x})$. Thus, the impulse response of $-\nabla^2$ is $1/4\pi r$. We can continue this process and drive the nth-order distributional derivative of the function $1/r^k$.

With the help of the foregoing analysis we can obtain the results that correspond to (32)–(35) for singular surfaces which carry infinite singularities, such as charge sheets and vortex sheets. Suppose that the surface of discontinuity $\Sigma(\mathbf{x}, t)$ carries a single layer of strength $f(\mathbf{x}, t)\delta(\Sigma)$. Then

$$\frac{\bar{\partial}}{\partial t} [f\delta(\Sigma)] = \frac{\bar{\delta} f}{\delta t} \delta(\Sigma) - Gf\delta'(\Sigma), \qquad (40)$$

where $\bar{\delta}/\delta t = \bar{\partial} f/\partial t + G \, df/dn$ is the distributional time derivative as apparent to an observer moving with the front Σ. Similarly, $\bar{\delta} f/\delta x_i = \bar{\partial} f/\partial x_i - n_i \, df/dn$ is the distributional surface derivative with respect to the Cartesian coordinates of the surrounding space. Thereby,

$$\frac{\bar{\partial}}{\partial x_i} [f\delta(\Sigma)] = \frac{\bar{\delta} f}{\delta x_i} \delta(\Sigma) + n_i f\delta'(\Sigma). \qquad (41)$$

If we use relation (31) between $d_n \delta(\Sigma)$ and $\delta'(\Sigma)$, we can rewrite formulas (40) and (41) which contain the mean curvature Ω.

VII. CONVERGENCE OF DISTRIBUTIONS AND FOURIER SERIES

A sequence $\{t_m(x)\}$ of distributions with $t_m(x) \in \mathcal{D}'$, $m = 1, 2, \ldots$, is said to converge to a distribution $t(x) \in \mathcal{D}'$ if $\lim_{m \to \infty} \langle t_m, \phi \rangle = \langle t, \phi \rangle$ for all $\phi \in \mathcal{D}$. This is called distributional (or weak) convergence. An important consequence is that if the sequence $\{t_m(x)\}$ converges to $t(x)$, then the sequence $\{D^k t_m\}$ converges to $\{D^k t\}$ because $\lim_{m \to \infty} \langle D^k t_m, \phi \rangle = \lim_{m \to \infty} (-1)^{|k|} \langle t_m, D^k \phi \rangle = (-1)^{|k|} \langle t, D^k \phi \rangle = \langle D^k t, \phi \rangle$. As an example, consider the sequence $\{\cos mx/m\}$, $x \in \mathbb{R}$, which is a sequence of regular distributions and converges to zero pointwise. Then the sequence $\{-\sin mx\}$ which arises by differentiating $\{\cos mx/m\}$, also converges to zero. It is remarkable because the sequence $\{\sin mx\}$ does not have a pointwise limit, as $m \to \infty$, in the classical sense.

A series of distributions $\sum_{p=1}^{\infty} s_p(x)$ converges distributionally to $t(x)$ if the sequence of the partial sums $\{t_m(x)\} = \{\sum_{p=1}^{m} s_p(x)\}$ converges to t distributionally. Thus, we observe from the above analysis that term-by-term differentiation of a convergent series is always possible, provided the resultant series is interpreted in the sense of distributions.

It is known that the Fourier series $\sum_{m=-\infty}^{\infty} c_m e^{imx}$ converges uniformly if for a large m, $|c_m| \le M/m^k$ where m is a constant and k is an integer greater than 2. The series may diverge for other values of k. However, the above analysis assures us that the series converges distributionally for any integer because it can be obtained from the uniformly convergent series $\sum_{m=-\infty}^{\infty} (im)^{k-2} c_m e^{imx}$ by $(k+2)$ successive differentiations. For example, the series $\sum_{m=-\infty}^{\infty} e^{imx}$ has no meaning in the classical sense, but it can be written as $1 + (d^2/dx^2)\{\sum_{m=-\infty}^{\infty} (1/m^2) e^{imx}\}$. Accordingly, the series $\sum_{m=-\infty}^{\infty} e^{imx}$ is distributionally convergent.

VIII. DIRECT PRODUCT AND CONVOLUTION OF DISTRIBUTIONS

The direct product of distributions is defined by using the space of test functions in two variables x and y. Let us denote the direct product of the distributions $s(x) \in \mathcal{D}'(x)$ and $t(y) \in \mathcal{D}'(y)$ as $s(x) \otimes t(y)$. Then

$$\langle s(x) \otimes t(y), \phi(x, y) \rangle = \langle s(x), \langle t(y), \phi(x, y) \rangle \rangle \quad (42)$$

where $\phi(x, y)$ is a test function in $\mathcal{D}(x, y)$. This makes sense because the function $\psi(x) = \langle t(y), \phi(x, y) \rangle$ is a test function in $\mathcal{D}(x)$ and $D^k \psi(x) = \langle t(y), D_x^k \phi(x, y) \rangle$, where D_x^k implies kth-order derivative with respect to x. Thus, $s(x) \otimes t(y)$ is a functional in $\mathcal{D}'(x, y)$. Indeed, it is a linear and continuous functional. Let us mention some of the properties of the direct product:

1. Linearity: $s \otimes (\alpha t + \beta u) = \alpha s \otimes t + \beta s \otimes u$.
2. Commutativity: $s \otimes t = t \otimes s$.
3. Continuity: if $s_m(x) \to s(x)$, in $\mathcal{D}'(x)$ as $m \to \infty$, then $s_m(x) \otimes t(y) \to s(x) \otimes t(y)$ in $\mathcal{D}'(x, y)$ as $m \to \infty$.
4. Associativity: when $s(x) \in \mathcal{D}'(x)$, $t(y) \in \mathcal{D}'(y)$ and $u(z) \in \mathcal{D}'(z)$, then $s(x) \otimes [t(y) \otimes u(z) = s(x) \otimes t(y)] \otimes u(z)$.
5. Support: $\text{supp} (s \times t) = \text{supp } s \times \text{supp } t$, where $\times$ stands for the Cartesian product.
6. Differentiation: $D_x^k [s(x) \otimes t(y)] = D^k[s(x)] \otimes t(y)$.
7. Translation: $(s \otimes t)(x + h, y) = s(x + h) \otimes t(y)$.

An interesting example of the direct product is $\delta(x) \otimes \delta(y) = \delta(x, y)$, where $\delta(x, y)$ is the two-dimensional delta function. This follows by observing that $\langle \delta(x) \otimes \delta(y), \phi(x, y) \rangle = \langle \delta(x), \langle \delta(y), \phi(x, y) \rangle \rangle = \langle \delta(y), \phi(0, y) \rangle = \phi(0, 0) = \langle \delta(x, y), \phi(x, y) \rangle$.

The *convolution* $s * t$ of two functions $s(x)$ and $t(x)$ is $(s * t)(x) = \int_{-\infty}^{\infty} s(y)t(x - y) \, dy = \int_{-\infty}^{\infty} t(y)s(x - y) \, dy = t * s(x)$. When the functions $s(x)$ and $t(x)$ are locally integrable, then $s(x) * t(x)$ is also locally integrable, and as such, defines a regular distribution.

Next, let us examine $\langle s * t, \phi \rangle$:

$$
\begin{aligned}
\langle s * t, \phi \rangle &= \int_{-\infty}^{\infty} (s * t)\phi(z) \, dz \\
&= \int_{-\infty}^{\infty} \left[\int_{-\infty}^{\infty} s(z - y)t(y) \, dy \right] \phi(z) \, dz \\
&= \int_{-\infty}^{\infty} t(y) \left[\int_{-\infty}^{\infty} s(z - y)\phi(z) \right] dy \\
&= \int_{-\infty}^{\infty} t(y) \int_{-\infty}^{\infty} [s(x)\phi(x + y) \, dx] \, dy \\
&= \int_{-\infty}^{\infty} \int_{-\infty}^{\infty} s(x)t(y)\phi(x + y) \, dx \, dy.
\end{aligned}
$$

Thus,

$$\langle s(x) * t(y), \phi(x, y) \rangle = \langle s(x) \otimes t(y), \phi(x + y) \rangle. \quad (43)$$

There is, however, one difficulty with this definition. Even though the function $\phi(x + y)$ is infinitely differentiable, its support is not bounded in the (x, y) plane because x and y may be unboundly large while $x + y$ remains finite. This is remedied by ensuring that the intersection of the supports of $s(x) \otimes t(y)$ and $\phi(x + y)$ is a bounded set. This happens if (1) either s or t has a bounded support or (2) both s and t have support bounded on the same side.

As an example, we find that $(\delta * t)(x), = \int \delta(y)$ $t(x - y)\, dx = t(x)$, that is, $\delta(x)$ is the unit element in the convolution algebra. The operation of the convolution has the following properties:

1. Linearity: $s * (\alpha t + \beta u) = \alpha s * t + \beta s * u$.
2. Commutivity: $s * t = t * s$.
3. Associativity: $(s * t) * u = s * (t * u)$.
4. Differentiation: $(D^k s) * t = D^k(s * t) = s * D^k t$.
5. For instance, $[H(x) * t(x)]' = \delta(x) * t(x) = H(x) * t'(x)$.

IX. FOURIER TRANSFORM

Let us write formula (12) as

$$\frac{1}{2\pi} \int\limits_{-\infty}^{\infty} e^{iu(x-\xi)}\, du = \delta(x - \xi),$$

and multiply both sides of this formula by $f(x)$, integrate with respect to x from $-\infty$ to ∞, use the sifting property, and interchange the order of integration. The result is the celebrated Fourier integral theorem,

$$f(\xi) = \frac{1}{2\pi} \int\limits_{-\infty}^{\infty} du \int\limits_{-\infty}^{\infty} f(x)e^{iu(x-\xi)}\, dx, \qquad (44)$$

which splits into the pair

$$\hat{f}(u) = \int\limits_{-\infty}^{\infty} f(x)e^{iux}\, dx, \qquad (45)$$

$$f(x) = \frac{1}{2\pi} \int\limits_{-\infty}^{\infty} \hat{f}(u)e^{-iux}\, du, \qquad (46)$$

where we have relabeled ξ with x in (46). The quantity $\hat{f}(u)$ is the Fourier transform of $f(x)$, while formula (46) gives the inverse Fourier transform $F^{-1}[f(u)]$. We shall now examine formula (45) in the context of test functions and distributions. If we attempt to define the Fourier transform of a distribution $t(x)$ as in (45), then we get $\hat{t}(u) = \langle t(x),\ e^{iux} \rangle$, but we are in trouble because e^{iux} is not in $\mathcal{D}$. We could try Parseval's theorem, $\int_{-\infty}^{\infty} \hat{f}(x)g(x)\, dx = \int_{-\infty}^{\infty} f(x)\hat{g}(x)\, dx$, which follows from (45) and (46) and connects the Fourier transforms of two functions. Then we have $\langle \hat{t}, \phi \rangle = \langle t, \hat{\phi} \rangle, \phi \in \mathcal{D}$. We again run into trouble because $\hat{\phi}$ may not be in $\mathcal{D}$ even if ϕ is in it. These difficulties are overcome by enlarging the class of test functions.

Test functions of rapid decay. The space $\mathcal{S}$ of test functions of rapid decay contains the complex-valued func-

tions $\phi(x)$ with the properties: (1) $\phi(x) \in C^{\infty}$, (2) $\phi(x)$ and its derivatives of all orders vanish at infinity faster than the reciprocal of any polynomial, i.e., $|x^p D^k \phi(x)| < C_{pq}$, where $p = (p_1, \ldots, p_n), k = (k_1, \ldots, k_n)$ are n-tuples and C_{pq} is a constant depending on p, k and $\phi(x)$. Another way of saying this is that after multiplication by any polynomial $P(x)$, the function $P(x)\phi(x)$ still tends to zero as $x \to \infty$. Clearly, $S \supset \mathcal{D}$. Convergence in this space is defined as follows. The sequence of functions $\{\phi_m(x)\} \to \phi(x)$ if and only if $|x^k(D^k \phi_m - D^k \phi)| < C_{pq}$ for all multi-indices p and k and all m.

Functions of slow growth. A function $f(x)$ in $\mathbb{R}^n$ is of slow growth if $f(x)$, together with all its derivatives, grows at infinity more slowly than some polynomial, i.e., there exist constants C, m, and A such that $|D^k f(x)| \langle C|x|^m, |x| \rangle A$.

Tempered distributions. A linear continuous functional $t(x)$ over the space S of test functions is called a *distribution of slow growth* or *tempered distribution.* According to each $\phi \in S$, there is assigned a complex number $\langle t, \phi \rangle$ with the properties: (1) $\langle t, c_1\phi_1 + c_2\phi_2 \rangle = c_1 \langle t, \phi_1 \rangle + c_2 \langle t, \phi_2 \rangle$; (2) $\lim_{m \to 0} \langle t, \phi_m \rangle = 0$, for every null sequence $\{\phi_m(x)\} \in S$. This yields us the dual space S'. If follows from the definitions of convergence in $\mathcal{D}$ and S that a sequence $\{\phi_m(\xi)\}$ that converges in the sense of $\mathcal{D}$ also converges in the sense of S, so that $S' \subset \mathcal{D}'$. Fortunately, most of the distributions in $\mathcal{D}'$ encountered previously are also in S'. However, some locally integrable functions in D' are not in S'. The regular distributions in S' are those generated by functions $f(x)$ of slow growth through the formula $\langle f, \phi \rangle = \int_{\infty}^{\infty} f(x)\phi(x)\, dx, \phi \in S$. It is a linear continuous functional.

Fourier transform of the test functions. We shall first give the analysis in $\mathbb{R}$ and then state the corresponding results in $\mathbb{R}^n$. For $\phi(x) \in S$, we can use the definition (45) so that $\hat{\phi}(u) = \int_{\infty}^{\infty} \phi(x)e^{iux}\, dx$. The inverse follows from (46). Since $\hat{\phi}(u)$ is also in S, as can be easily proved, the difficulty encountered earlier has disappeared. Moreover, it follows by the inversion formula (46) that

$$[\hat{\phi}]\hat{}(x) = 2\pi\phi(-x), \qquad (47)$$

which shows that every function $\phi(x) \in S$ is a Fourier transform of some function in S. Indeed, the Fourier transform and its inverse are linear, continuous, and one-to-one mapping of S on to itself.

In order to obtain the transform of tempered distributions, we need some specific formulas of the transform of the test functions $\phi(x)$. Let us list them and prove some of them. The transform of $d^k\phi/dx^k$ is equal to $\int_{-\infty}^{\infty} (d^k\phi/dx^k)e^{iux}\, dx = (-iu)^k \int_{-\infty}^{\infty} \phi(x)e^{iux}\, dx$ so that $[d^k\phi/dx^k]\hat{}(u) = (-iu)^k \hat{\phi}(u)$. Thus, if $P(\lambda)$ is an arbitrary polynomial with constant coefficients, we find that

$[P(d/dx)\phi]\hat{} = P(-iu)\hat{\phi}(u)$. Similarly, from the relation $\int_{-\infty}^{\infty}(ix)^k\phi(x)e^{iux}\,dx = (d^k/du^k)\int_{-\infty}^{\infty}\phi(x)e^{iux}\,dx$, we derive the relation $[x^k\phi]\hat{}(u) = P(-d/du)\hat{\phi}(u)$. Because $\int_{-\infty}^{\infty}\phi(x-a)e^{iux}\,dx = e^{iua}\int_{-\infty}^{\infty}\phi(y)e^{iuy}\,dy$, we have $[\phi(x-a)]\hat{}(u) = e^{iau}\hat{\phi}(u)$. Similarly, $[\phi(x)]\hat{}(u+a) = [e^{ax}(x)]\hat{}(u)$ and $[\phi(ax]\hat{}(u) = (1/|a|\hat{\phi}(u/a))$. The corresponding n-dimensional formulas are listed below.

$$\hat{\phi}(u) = \int_{-\infty}^{\infty} e^{iu\cdot x}\phi(x)\,dx, \tag{48}$$

$$\phi(x) = \frac{1}{(2\pi)^n}\int_{-\infty}^{\infty} e^{iu\cdot x}\hat{\phi}(u)\,du, \tag{49}$$

$$[\hat{\phi}]\hat{}(u) = (2\pi)^n\phi(-u), \tag{50}$$

$$[D^k\phi]\hat{}(u) = (-iu)^k\hat{\phi}(u), \tag{51}$$

$$\left[P\left(\frac{\partial}{\partial x_1},\ldots,\frac{\partial}{\partial x_n}\right)\phi(x)\right]\hat{}(u) = P(-iu_1,.,-u_n)\hat{\phi}(x), \tag{52}$$

$$[x^k\phi]\hat{}(x) = (-iD)^k x\hat{\phi}(u), \tag{53}$$

$$[P(x_1,\ldots,x_n)\phi]\hat{}(u) = P\left(-i\frac{\partial}{\partial u_1},\ldots,-i\frac{\partial}{\partial u_n}\right)\hat{\phi}(u), \tag{54}$$

$$[\phi(x-a)]\hat{}(u) = e^{ia\cdot u}\hat{\phi}(u), \tag{55}$$

$$[\hat{\phi}(x)](u+a) = [e^{ia\cdot x}]\hat{}\phi(u), \tag{56}$$

$$[\phi(Ax)]\hat{}(u) = |\det A|^{-1}\hat{\phi}(A^T)u), \tag{57}$$

where $u\cdot x = u_1x_1+\cdots+u_nx_n$, A is a nonsingular matrix, and A^T is its transpose. We shall refer to the numbered formulas above for $n=1$ also.

As a simple example we consider the function $\exp(-x^2/2)$, which is clearly a member of S. To find its transform we first observe that it satisfies the equation $\phi'(x)+x\phi(x)=0$. Taking Fourier transforms of both sides of this equation and using (52) and (53), we find that $(d/du)[\hat{\phi}(u)e^{u^2/2}]=0$. Thus, $\hat{\phi}(u)=Ce^{-u^2/2}$, where C is a constant. To evaluate C, we observe that $\hat{\phi}(u)=\int_{-\infty}^{\infty}\exp(-x^2/2)\,dx=\sqrt{2\pi}$, so that $C=\sqrt{2\pi}$ and we have $\hat{\phi}(u)=\sqrt{2\pi}\phi(u)$. Thus we have found a function which is its own inverse. [The multiplicative $\sqrt{2\pi}$ disappear if we use the factors $1/\sqrt{2\pi}$ in the definition of the transform pairs (48) and (49)].

Fourier transform of tempered distributions. Having discovered that $\hat{\phi}\in S$ when ϕ is, we can apply the relation $\langle\hat{t},\phi\rangle = \langle t,\hat{\phi}\rangle$ to define the Fourier transform of the tempered distributions $t(x)$. Then all the formulas given above for $\hat{\phi}$ carry over for $\hat{t}$. For instance, $\langle|(d^k/dx^k)t(x)]\hat{}u$,

$\phi(u)] = \langle d^k t/dx^k, \hat{\phi}(x)\rangle = (-1)^k\langle t(x),[(d^k/dx^k)\phi(x)]\hat{}\rangle = \langle t(x),(-iu)^k\phi(u)]\hat{}(x)\rangle = \langle\hat{t}(u),(-iu)^k\phi(u)\rangle = \langle(-iu)^k\hat{t}(u),\phi(u)\rangle$, which shows that $[d^kt/dx^k]\hat{}(u) = (-iu)^k\hat{t}(u)$, which agrees with (51). Instead of writing the formulas (50)–(57) all over again for a distribution $t(x)$, we shall merely refer to them with ϕ replaced by t. Let us now give some important applications of these formulas.

1. *Delta function.* $\langle[\delta(x)]\hat{},\phi(u)\rangle = \langle\delta(x),\int_{-\infty}^{\infty}\phi(u)e^{iux}\,du\rangle = \int_{-\infty}^{\infty}\phi(u)\,du = \langle1,\phi(u)\rangle$. Thus $\hat{\delta}(x)=1$ so that $\hat{1}=2\pi\delta(x)$. In the n-dimensional case, the corresponding formulas are $\hat{\delta}(x)=1$ and $\hat{1}=(2\pi)^n\delta(x)$. Incidentally, we recover formula (12) from this relation.

2. *Heaviside function.* We use formula (51), so that $[t'(x)]\hat{}(u)=-iu\hat{t}(u)$. Because $t(x)=H(x)$, we have $[\delta(x)]\hat{}(u)=-iu[H(x)]\hat{}(u)$ or $u(H)(x)]\hat{}(u)=i$, whose solution follows from (25) as $[H(x)]\hat{}(u)=c\delta(u)+iPf(1/u)$. Similarly, from (57) and the above result we have $[H(-x)]\hat{}(u)=c\delta(u)-iPf(1/u)$. To find the constant c we observe that $H(x)+H(-x)=1$, whose Fourier transform is $2c\delta(u)=2\pi\delta(u)$, so that $c=1$. Thus $[H(\pm x)]\hat{}(u)=\pi\delta(u)\pm iPf(1/u)$.

3. *Signum function.* Because $\text{sgn}\,x = H(x)-H(-x)$, we take Fourier transforms of both sides and use Example 2 above to get $[\text{sgn}\,x]\hat{}(u)=2iPf(1/u)$.

4. $Pf(1/x)$. We use formulas (47) for $t(x)$ and the example above. The result is $[(\text{sgn}\,x)\hat{}(u)](x) = [2iPf(-1/u)]\hat{}(x)$, so that

$$\left[Pf\left(\frac{1}{x}\right)\right]\hat{}(u) = i\pi\,\text{sgn}\,u. \tag{58}$$

5. *The function* $|x|$. We write it as $|x| = xH(x)-xH(-x)$. Taking the Fourier transforms of both sides of this equation, we get

$$[|x|]\hat{}(u) = [xH(x)]\hat{}(u) - [xH(-x)]\hat{}(u)$$

$$= -i\frac{d}{du}[H(x)]\hat{}(u) + i\frac{d}{du}[H(-x)]\hat{}(u)$$

$$= -i\frac{d}{du}\left[\pi\delta(u) + iPf\left(\frac{1}{u}\right)\right]$$

$$\pm i\frac{d}{du}\left[\pi\delta(u) - iPf\left(\frac{1}{u}\right)\right]$$

$$= \frac{2}{u^2}.$$

Thus $[|x|]\hat{}(u) = 2/u^2$.

6. *The function* $1/x^2$. To find its transform we appeal to the previous example and relation (47) so that $[(|x|)\hat{}(u)]\hat{}(x) = (2/u^2)\hat{}(x)$ and we get $(1/x^2)\hat{}(u) = \frac{1}{2}|x|$.

7. *Polynomial* $P(x)=a_0+a_1+\cdots+a_nx^n$. Formula (54) gives $[P(x)t(x)]\hat{}(u) = P[-i(d/du)\hat{t}(u)]$. When we

substitute $t(x) = 1$ in this formula and use Example 1, we get $[P(x)]\hat{}(u) = 2\pi P(-i\,d/du)\delta(u)$. Thus,

$$[a_0 + a_1 + \cdots + a_n x^n]\hat{}(u) = 2\pi\,(a_0\delta - ia_1\delta' + \cdots$$
$$+ (-i)^n a_n \delta^{(n)}(u)]. \quad (59)$$

In particular, $\hat{x}(u) = -2\pi i\delta'(u)$, $[x^2]\hat{}(u) = -2\pi\delta''(u)$, ..., $[x^n]\hat{}(u) = (-i)^n 2\pi\delta^{(n)}(u)$.

8. $Pf(1/|x|)$. With the help of the definition (24) of this function we have

$$\left\langle \left[Pf\left(\frac{1}{|x|}\right)\right]\hat{}(u), \phi(u)\right\rangle = \left\langle Pf\left(\frac{1}{|x|}\right), \hat{\phi}(x)\right\rangle$$

$$= \int_{-1}^{1} \frac{\hat{\phi}(x) - \hat{\phi}(0)}{|x|}\,dx + \int_{|x|>1} \frac{\hat{\phi}(x)}{|x|}\,dx,$$

which, after some algebraic manipulation, yields

$$\left[Pf\left(\frac{1}{|x|}\right)\right]\hat{}(u) = -2(\gamma + \ln|u|), \quad (60)$$

where γ is Euler's constant.

9. $\ln|x|$. For evaluation of the Fourier transform of this important function, we take the Fourier transform of both sides of (60), use formula (47), and obtain

$$Pf\left(\frac{1}{|x|}\right) = 2[2\pi\gamma\delta(x) + [\ln(u)]\hat{}(x)]$$

and relabel. The result is

$$[\ln|x|]\hat{}(u) = -\left[\pi Pf\frac{1}{2|u|} + 2\pi\gamma\delta(u)\right]. \quad (61)$$

In the theories of wavelets, sampling, and interpolation, we need the Fourier transforms of the square and triangular functions. We derive them in the next two examples.

10. *The square function.* It is defined as $f(x) = H(x + \frac{1}{2}) - H(x - \frac{1}{2})$. Thus

$$\hat{f}(u) = \int_{-1/2}^{1/2} e^{iux}\,dx = \left[\frac{1}{iu}e^{iux}\right]_{-1/2}^{1/2}$$

$$= \frac{2}{u}\sin\left(\frac{u}{2}\right) = \text{sinc}\left(\frac{u}{2}\right), \quad (62)$$

where $\text{sinc}\,t = \sin t/t$.

11. *The triangular function.* It is defined as

$$f(x) = \begin{cases} 1 - |x|, & |x| < 1, \\ 0, & |x| > 1, \end{cases}$$

so that

$$\hat{f}(u) = \int_{-1}^{0} (1+x)\,e^{iux}dx + \int_{0}^{1} (1-x)e^{iux}\,dx.$$

By a routine computation it yields

$$\hat{f}(u) = \left[\frac{\sin(u/2)}{(u/2)}\right]^2 = \left[\text{sinc}\left(\frac{u}{2}\right)\right]^2. \quad (63)$$

12. *Fourier transform of an integral.* We have found previously that taking the Fourier transform of the diferential of a distribution $t(x)$ has the effect of multiplying $\hat{t}(u)$ by $(-iu)$. In the case of taking the Fourier transform of the integral of $t(x)$, it amounts to dividing $\hat{t}(u)$ by $(-iu)$ so that

$$\left\{\int_{a}^{x} [t(s)\,ds]\right\}\hat{}\,(u) = \frac{i\hat{t}(u)}{u}. \quad (64)$$

Similarly,

$$\left\{\int_{-\infty}^{x} [t(s)\,ds]\right\}\hat{}\,(u) = \frac{i\hat{t}(u)}{u} + \frac{1}{2}\hat{t}(0)\delta(u). \quad (65)$$

13. *Fourier transform of the convolution.* The Fourier transform of the convolution $f * g$ of two locally integrable functions f and g is $[f * g]\hat{}(u) = \int_{-\infty}^{\infty} f * g e^{iux} = \int_{-\infty}^{\infty} e^{iux} \int_{-\infty}^{\infty} f(x - y)g(y)\,dy = \int_{-\infty}^{\infty} g(y)e^{iuy}\,dy \times \int_{-\infty}^{\infty} f(x - y)\,e^{iu(x - y)}\,d(x - y) = \hat{g}(u)\hat{f}(u) = \hat{f}(u)\hat{g}(u)$. Thus, the Fourier transform of the convolution of two regular distributions is the product of their transforms. This relation also holds for singular distributions with slight restrictions. For instance, if at least one of these distributions has compact support, then the relation holds.

X. POISSON SUMMATION FORMULA

To derive the Poisson summation formula we first find the Fourier series of the delta function in the period $[0, 2\pi]$: $\delta(x) = \sum_{m=0}^{\infty}(a_m \cos mx + b_m \sin mx)$, where the coefficients a_m, b_m are given by $a_0 = (1/2\pi)\int_0^{2\pi} \delta(x)\,dx = (1/2\pi)$, $a_m = (1\pi)\int_0^{2\pi} \delta(x) \cos mx \times dx = (1/\pi)$, and $b_m = (1/\pi)\int_0^{\pi} \delta(x) \sin mx\,dx = 0$. Thus, we have

$$\delta(x) = \frac{1}{2\pi}\left(1 + 2\sum_{m=1}^{\infty} \cos mx\right) = \frac{1}{2\pi}\sum_{m=-\infty}^{\infty} e^{imx}. \quad (66)$$

Now we periodize $\delta(x)$ by putting the row of deltas at the points $2\pi m$ so that relation (66) can be written as

$$\sum_{m=-\infty}^{\infty} \delta(x - 2\pi m) = \frac{1}{2\pi}\left(1 + 2\sum_{m=1}^{\infty} \cos mx\right)$$

$$= \frac{1}{2\pi}\sum_{m=-\infty}^{\infty} e^{imx}.$$

When we set $x = 2\pi y$ in this relation and use the formula $\delta[2\pi(y-m)] = (1/2\pi)\delta(y-m)$ and relabel, we obtain

$$\sum_{m=-\infty}^{\infty} \delta(x-m) = 1 + 2\sum_{m=1}^{\infty} \cos 2\pi mx = \sum_{m=-\infty}^{\infty} e^{i2\pi mx}. \tag{67}$$

The action of this formula on a function $\phi(x) \in \mathcal{S}$ yields

$$\sum_{m=-\infty}^{\infty} \phi(m) = \hat{\phi}(2\pi m), \tag{68}$$

which relates the sum of functions ϕ and their Fourier transforms $\hat{\phi}$ and is a very useful formula.

In the classical theory, it is necessary that both sides of relation (68) converge. Moreover, they must converge in the same interval. With the help of the theory of distributions we can obtain many variants of relation (68) which are applicable even when one or both of the series (68) are divergent. As an example, we set $x = x/\lambda$, where λ is a real number, and use the relation $\delta[(x/\lambda)-m] = |\lambda|\delta(x-m\lambda)$. Then relation (67) becomes

$$(\lambda)\sum_{m=-\infty}^{\infty} \delta(x-m\lambda) = \sum_{m=-\infty}^{\infty} e^{2i\pi xm/\lambda}. \tag{69}$$

When we multiply both sides of this relation by a test function $\phi(x)$ and integrate with respect to x, we obtain a variant of (68) as

$$\sum_{m=-\infty}^{\infty} \phi(m\lambda) = \frac{1}{|\lambda|}\sum_{m=-\infty}^{\infty} \hat{\phi}\left(\frac{2\pi m}{\lambda}\right). \tag{70}$$

This is called the distributional Poisson summation formula. Among other things, the Poisson summation formula (70) transforms a slowly converging series to a rapidly converging series. For instance, if we take $\phi(x) = e^{-x^2}$, then $\hat{\phi}(u) = \sqrt{\pi}e^{-u^2/2}$, so that (70) becomes

$$\sum_{m=-\infty}^{\infty} e^{-m^2\lambda^2} = (\pi/\lambda)^{1/2}\sum_{m=-\infty}^{\infty} e^{-m^2\pi^2/\lambda^2}. \tag{71}$$

The series on the left side of (71) converges rapidly for large λ, that on the right side for small λ.

XI. ASYMPTOTIC EVALUATION OF INTEGRALS: A DISTRIBUTIONAL APPROACH

Let $g(x)$ and $h(x)$ be sufficiently smooth functions on the interval $[a, b]$, then the main contribution to the Laplace integral,

$$I(\lambda) = \int_a^b e^{-\lambda h(x)} g(x)\, dx, \qquad \lambda \to \infty \tag{72}$$

arises from the points where the function $h(x)$ has a minimum. If x_0 is the only global minimum of $h(x)$, then we have the Laplace formula

$$I(\lambda) \sim \left[\frac{2\pi}{\lambda h''(x_0)}\right]^{1/2} g(x_0)e^{-\lambda h(x_0)}. \tag{73}$$

If we could somehow prove that $e^{-\lambda h(x)}$ has an asymptotic series of delta functions such that

$$e^{-\lambda h(x)} \sim \left[\frac{2\pi}{\lambda h'(x_0)}\right]^{1/2} e^{-\lambda h(x_0)}\delta(x-x_0), \tag{74}$$

then all that we have to do is to substitute (74) in (72) and use the sifting property of the delta function and formula (73) follows immediately. To achieve the expansion (74) we first define the moments of a function $f(x)$. They are

$$\mu_n = \langle f(x), x^n \rangle = \int_{-\infty}^{\infty} f(x)x^n\, dx. \tag{75}$$

The Taylor expansion of a test function $\phi(x)$ at $x = 0$ is

$$\phi(x) = \sum_{n=0}^{\infty} \phi^{(n)}(0)\frac{x^n}{n!}. \tag{76}$$

Then it follows from (75) that

$$\langle f(x), \phi(x) \rangle = \left\langle f(x), \sum_{n=0}^{\infty} \phi^{(n)}(0)\frac{x^n}{n!}\right\rangle$$
$$= \sum_{n=0}^{\infty} \phi^n(0)\frac{\mu_n}{n!}. \tag{77}$$

But $\phi^n(0) = (-1)^n\langle \delta^{(n)}(x), \phi(x)\rangle$, so that (77) becomes

$$\langle f(x), \phi(x) \rangle = \left\langle \sum_{n=0}^{\infty} \frac{(-1)^n\mu_n\delta^{(n)}(x)}{n!}, \phi(x)\right\rangle. \tag{78}$$

Thus

$$f(x) = \sum_{n=0}^{\infty} \frac{(-1)^n\mu_n\delta^{(n)}(x)}{n!}. \tag{79}$$

Finally, we use the formula $\delta^{(n)}(\lambda x) = (1/\lambda^{n+1})\delta^n(x)$ in (79) and obtain the complete asymptotic expansion

$$f(\lambda x) = \sum_{n=-\infty}^{\infty} \frac{(-1)^n\mu_n\delta^{(n)}(x)}{n!\,\lambda^{n+1}}, \qquad \lambda \to \infty. \tag{80}$$

In the case of the Laplace integral (72) we have $f(x) = e^{-\lambda h(x)}$. At the minimum point x_0, $h(x) \sim h(x_0) + [h''(x_0)(x-x_0)^2]/2$. Accordingly, we can find an increasing smooth function $\psi(x)$ with $\psi(x_0) = 0$, $\psi'(x_0) > 0$, so that $h(x) = h(x_0) + [\psi(x)]^2$ in the support of $h(x)$. Then $h'(x_0) = 2\psi(x_0)\psi'(x_0) = 0$, which yields $h'(x_0) = 0$ as required for $h(x)$ to have minimum at x_0. Also,

$h''(x) = 2[\psi'(x)]^2 + 2\psi(x)\psi''(x)$, which, for $x = x_0$, becomes $h''(x_0) = 2[\psi'(x_0)]^2$ or

$$\psi'(x_0) = \frac{1}{2}[h''(x_0)]^{1/2}. \qquad (81)$$

Substituting this information about $h(x)$ in the Laplace integral (72), we obtain

$$I(\lambda) = e^{-\lambda h(x_0)} \int_{-\infty}^{\infty} e^{-\lambda[\psi(x)]^2} g(x)\,dx. \qquad (82)$$

The next step is to set $u = \psi(x)$, which gives $dx = du/\psi'(x)$ so that (82) becomes

$$I(\lambda) = e^{-\lambda h(x_0)} \int_{-\infty}^{\infty} e^{-\lambda u^2} \frac{g(u)}{\psi'(x_0)}\,du, \qquad (83)$$

so we need the asymptotic expansion of $e^{-\lambda u^2}$ from formula (80). This is obtained by finding the moments of e^{-u^2}, and they are

$$\mu_n = \int_{-\infty}^{\infty} e^{-u^2} u^n\,du = \begin{cases} \Gamma\left(\dfrac{n+1}{2}\right), & n \text{ even,} \\ 0, & n \text{ odd.} \end{cases} \qquad (84)$$

Then formula (80) yields

$$e^{-\lambda u^2} = \sum_{n=0}^{\infty} \frac{\Gamma[(2n+1)/2]\delta^{(2n)}(u)}{(2n)!\lambda[(2n+1)/2]} \qquad (85)$$

so that formula (83) becomes

$$I(\lambda) = e^{-\lambda h(x_0)} \int_{-\infty}^{\infty} \sum_{n=0}^{\infty} \frac{\Gamma[(2n+1)/2]}{2n!} \frac{\delta^{(2n)}(u)}{\lambda^{(2n+1/2)}}$$
$$\times \frac{g(u)}{\psi'(x_0)}\,du. \qquad (86)$$

The interesting feature of the distributional approach is that we have obtained the complete asymptotic expansion. Thereafter, we evaluate as many terms as are needed to get the best approximation.

The first term of formula (86) is

$$I(\lambda) = e^{-\lambda h(\alpha_0)} \int_{-\infty}^{\infty} \Gamma\left(\frac{1}{2}\right) \frac{\delta(u)}{\sqrt{\lambda}} \frac{g(u)}{\psi'(x_0)}\,du.$$

When we substitute the value of $\psi'(x_0)$ from relation (81) in the above relation and use the sifting property of $\delta(u)$, we obtain

$$I(\lambda) \sim \left[\frac{2\pi}{\lambda h''(x_0)}\right]^{1/2} g(x_0) e^{-\lambda h(x_0)},$$

which agrees with (73).

The *oscillatory integral*

$$I(\lambda) = \int_{-\infty}^{\infty} e^{i\lambda h(x)} g(x)\,dx, \qquad \lambda \to \infty \qquad (87)$$

can also be processed in the same manner as the Laplace integral. Indeed, the steps leading (87) to relation

$$I(\lambda) = e^{-i\lambda h(x_0)} \int_{-\infty}^{\infty} e^{i\lambda u^2} \frac{g(u)}{\psi'(x_0)}\,du \qquad (88)$$

are almost the same as these from (72) to (83). The difference arises in the value of the moments μ_n which now are

$$\mu_n = \int_{-\infty}^{\infty} e^{i\lambda u^2} u^n\,du = \begin{cases} \Gamma\left(\dfrac{n\pi}{2}\right) e^{\pi i[(2n+1)/4]}, & n \text{ even,} \\ 0, & n \text{ odd.} \end{cases} \qquad (89)$$

This yields the moment expansion for $e^{i\lambda u^2}$ as

$$e^{i\lambda u^2} = \sum_{n=0}^{\infty} \frac{\Gamma[(2n+1)/2]\,e^{-\pi i[(2n+1)/4]}\delta^{(2n)}(u)}{(2n)!\lambda^{(2n+1)/2}}\,du. \qquad (90)$$

When we substitute this value for $e^{i\lambda u^2}$ in integral (88) we obtain

$$I(\lambda) \sim e^{-ih(x_0)} \int_{-\infty}^{\infty} \sum_{n=0}^{\infty} \frac{\Gamma[(2n+1)/2]\,e^{-\pi i[(2n+1)/4]}}{(2n)!}$$
$$\times \frac{\delta^{(2n)}(n)}{\lambda^{(2n+1/2)}} \frac{g(u)}{\psi'(x_0)}\,du. \qquad (91)$$

The first term of this formula yields

$$I(\lambda) \sim e^{i(\lambda h(x_0)+\pi/4)} \left[\frac{2\pi}{h''(x_0)}\right]^{1/2} \frac{\phi(x_0)}{\sqrt{\lambda}}, \qquad (92)$$

where we have used relation (81).

Let us observe an essential difference between the Laplace integral and the oscillatory integral. In the case of the Laplace integral we collect the contributions only from the points where $h(x)$ has the minimum such that $h'(x) = 0$ and $h''(x) > 0$. For the oscillatory integral we collect the contributions from all the points where $h(x)$ is stationary such that $h'(x) = 0$, $h''(x) \neq 0$. For this reason the classical method for the oscillatory integral is called the method of stationary phase.

SEE ALSO THE FOLLOWING ARTICLES

FOURIER SERIES • FUNCTIONAL ANALYSIS

530

BIBLIOGRAPHY

Benedetto, J. J. (1996). "Harmonic Analysis and Applications," CRC Press, Boca Raton, FL.

Duran, A. L., Estrada, E., and Kanwal, R. P. (1998). "Extensions of the Poisson Summation Formula," *J. Math. Anal. Appl.* **218**, 581–606.

Estrada, R., and Kanwal, R. P. (1985). "Regularization and distributional derivatives of $1/r^n$ in $\mathbb{R}^p$," *Proc. Roy. Soc. Lond.* A **401**, 281–297.

Estrada, R., and Kanwal, R. P. (1993). "Asymptotic Analysis: A Distributional Approach," Birkhäuser, Boston.

Estrada, R., and Kanwal, R. P. (1999). "Singular Integral Equations," Birkhäuser, Boston.

Gel'fand, I. M., and Shilov, G. E. (1965). "Generalized Functions, Vol. I," Academic Press, New York.

Jones, D. S. (1982). "Generalized Functions," Cambridge University Press, Cambridge, UK.

Kanwal, R. P. (1983). "Generalized Functions," 2nd ed., Birkhauser, Boston.

Lighthill, M. J. (1958). "Introduction to Fourier Analysis and Generalised Functions," Cambridge University Press.

Saichev, A. I., and Woyczynsky, W. A. (1997). "Distributions in the Physical and Engineering Sciences," Vol. 1, Birkhäuser, Boston.

Schwartz, L. (1966). "Théorie des distributions," Nouvell èdition, Hermann, Paris.

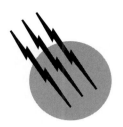

Geochemistry, Low-Temperature

C. G. Patterson

Alpine Geosciences

D. D. Runnells

Shepherd-Miller, Inc.

GLOSSARY

Activity "Effective" or "reactive" amount of a solute or a solvent in solution. The activity is usually less than the total concentration, but in brines the activity may exceed concentration. Activity has no dimensional units.

Carbonates Minerals in which the characteristic framework component is the CO32-structural group.

Complex Group of covalently bonded atoms that forms a stable dissolved species in solution. Examples include SO_4^{2-} (sulfate), CO_3^{2-} (carbonate), NO_3^- (nitrate), and PO_4^{3-} (phosphate).

Diagenesis Chemical and mechanical processes that take place within sediments shortly after deposition as a result of changes in temperature, pressure, and chemically active fluids.

Ion Atom or group of atoms with an electrostatic charge. Dissolved ions with a positive charge are referred to as cations, while those with a negative charge are called anions. Examples of cations commonly found in natural waters include Ca^{2+}, Mg^{2+}, Na^+, and K^+; of anions, HCO_3^-, Cl^-, and SO_4^{2-}.

pe A calculated term used to describe the oxidation–reduction state of natural waters and soils. High positive values represent the most oxidizing conditions. The Eh is a field-measured voltage that is comparable to pe.

$$pe = \left(\frac{Eh}{0.0592} \right) \text{ at } 298 \text{ K}$$

pH Measure of the activity of hydronium ion [H^+] in water. Defined as the negative logarithm of the activity of the hydronium atom. A value of 7 indicates neutral conditions at 298 K.

Regolith Blanket of loose, weathered material that covers bedrock in most regions of the earth. If the regolith contains a few percent of organic matter, it is called soil.

Saline In the context of natural waters, referring to concentrations of total dissolved solids (TDS) similar to or greater than those in ocean water.

Silicates Minerals in which the characteristic framework component is the SiO_4^{4-} structural group.

Solubility Total amount of a mineral or other substance that can be dissolved in water.

Weathering Mechanical and chemical breakdown of rock exposed at the surface of the earth under the influence of rain, changes in temperature, biologic activities, and so on.

LOW-TEMPERATURE GEOCHEMISTRY comprises the study of natural chemical processes occurring at or near the surface of the earth. The temperature and pressure are approximately 298 K and 1 atm (0.1 MPa). Of great interest in low-temperature geochemistry are the processes of chemical and mechanical weathering of bedrock and the processes of migration and cycling of the chemical elements. The discipline includes the study of the entire spectrum of natural waters, from rainwater to highly saline surface and subsurface brines. Environmental geochemistry, diagenesis of sediments, and geochemical exploration for mineral deposits are important subdisciplines.

I. LOW-TEMPERATURE BEHAVIOR OF THE ELEMENTS

A. Mobility during Weathering

Solid bedrock is under continual attack by elements of the atmosphere, hydrosphere, and biosphere. The sum of these processes is called weathering. During weathering the chemical elements that make up the original bedrock may behave very differently. Some elements, such as sodium and sulfur, tend to be dissolved during weathering and to be transported away by surface runoff and groundwater. A few elements, such as silicon, titanium, and zirconium, are relatively immobile during weathering because they occur in minerals that are extremely insoluble and resistant to weathering; the resistant minerals are called resistates. Some elements, chiefly aluminum, react with oxygen and water to form new hydroxyl-bearing minerals [such as gibbsite, $Al(OH)_3$] that are then resistant to further

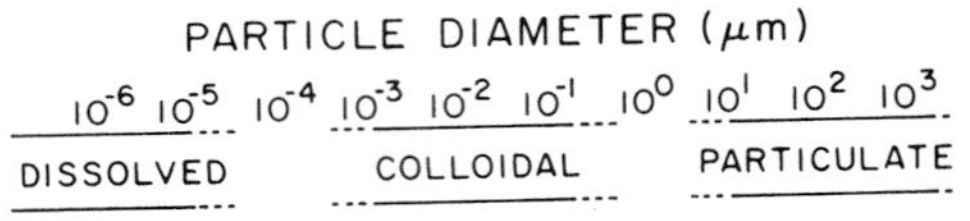

FIGURE 1 Size distribution of particulate, colloidal, and dissolved forms of the chemical elements in the processes of mechanical and chemical transport.

weathering; the new minerals are called hydrolyzates. Iron and manganese are among the elements that are oxidized under the conditions of weathering, followed by reaction with water and oxygen to form resistant oxidates; examples include goethite ($FeOOH$) and pyrolusite (MnO_2). Elements that are essential to plant and animal life, such as carbon and phosphorus, are also released from the rock-forming minerals during weathering.

B. Particles, Colloids, and Dissolved Species

Determination of the physical form (dissolved, colloidal, or particulate) of elements in water is important in achieving an understanding of the solubility and mobility of the elements. Figure 1 shows the forms of the elements that may move through the hydrosphere, ranging from macroparticles through colloids down to individual ions and molecules.

Colloids are particles with diameters from about 1 μm down to 0.0001 μm. Colloidal particles are important in the transport of organic matter in natural waters, and such elements as iron, manganese, and aluminum are transported in streams and seawater and the surface may carry a net positive or negative electrostatic charge. Some elements, such as nickel and cobalt, are probably transported in stream water chiefly as adsorbed ions on the surface of colloidal particles of iron and manganese hydroxide. In turn, the colloids themselves may form coatings on larger particles. The result is that many elements move through the hydrosphere in a combination of dissolved and adsorbed forms.

C. Forms of Dissolved Species

Dissolved chemical elements may exist in water in three principal forms: free ions, covalently bonded complexes, and electrostatically bonded ion pairs. Which form is dominant depends on the specific chemical properties of the element and the bulk chemistry of the solution. In fresh water, the free ions are most abundant. but in highly saline brines, ion pairs may dominate. In waters that are unusually rich in organic matter, such as swamp waters, metal-loorganic complexes may be the most important. All three forms of dissolved elements are present in varying degrees in natural waters. The dissolved form of an element will

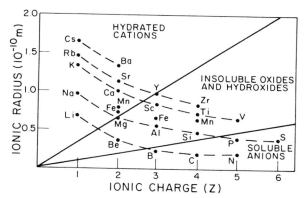

FIGURE 2 Summary of the forms in which elements may occur after reaction with water, as a function of ionic charge and ionic radius.

control the degree of reactivity of the element. Precipitation of a solid mineral may be prevented if the component ions are tied up in ion pairs or complexes. In terms of environmental geochemistry, the chemical form of an element is important in determining whether it will be toxic to animal or plant life. One of the most important reactions in solution occurs between cations and water molecules. the product of the reaction may be a free element surrounded by a sheath of water molecules, a solid hydroxide precipitate [e.g., $Fe(OH)_3$], or a soluble oxyanion (such as SeO_4^{2-}). These possibilities are summarized in Fig. 2. The ionic radius and the valence are key parameters in describing such reactions.

II. WEATHERING

A. Mechanical and Chemical

The study of weathering is one of the most important aspects of low-temperature geochemistry. The processes of weathering produce soils residual sediments and dissolved salts. Two principal types of weathering are recognized: mechanical and chemical. Everyday terms that correspond to mechanical and chemical weathering are disintegration and decomposition, respectively.

Mechanical weathering includes all processes that lead to physical breakdown of the bedrock into smaller fragments, such as the growth of ice or salt crystals, rapid changes in temperature, burrowing organisms, glacial abrasion, landslides, and abrasion by the hooves of animals and the boots of humans. These processes generate more surface area, which aids chemical weathering. In a very general way, the chemical processess of weathering can be summarized by the following simplified reaction (*aq* means aqueous):

primary minerals $+ CO_2(aq) + O_2(aq) +$ water

$\rightarrow$ detritus $+ HCO_3^-(aq) +$ cations(aq)

$+$ new minerals.

One way to visualize weathering is as an equilibration process as high-temperature and high-pressure minerals from bedrock seek to become more stable under surface conditions. Mechanical weathering forms are dominated by frost activity and therefore more prevalent in polar climates. Chemical weathering is dominant in tropical regions, where mean annual temperature and rainfall are high. Temperate climates tend to experience a mixture of mechanical and chemical weathering forms.

B. Types of Chemical Weathering

Individual processes of chemical weathering include dissolution, hydration, oxidation, carbonation, and hydrolysis. Dissolution is simply the dissolving of solid minerals. For example (*s* means solid),

$$NaCl(s) \rightarrow Na^+(aq) + Cl^-(aq).$$

Hydration is the addition of molecules of water to form new minerals. An example is the weathering of anhydrite to form gypsum, shown by

$$CaSO_4 + 2H_2O \rightarrow CaSO_4 \cdot 2H_2O.$$

Oxidation is a process in which atmospheric oxygen oxidizes the component elements in a mineral to higher valence states. An example is the oxidation of iron and sulfur in pyrite to produce ferric hydroxide and sulfuric acid ("acid mine water"):

$$2FeS_2 + \tfrac{15}{2}O_2(g) + 7H_2O$$

$$\rightarrow 2Fe(OH)_3(s) + 4SO_4^{2-} + 8H^+$$

(*g* means gas).

Oxidizing agents other than molecular oxygen are probably not important in weathering, although some researchers have suggested that minerals of oxidized manganese (such as MnO_2) may act as oxidizing agents. Many oxidation and reduction reactions that ordinarily proceed at a slow rate are considerably speeded up by the action of microbes in the environment.

Carbonation is the reaction of gaseous carbon dioxide with water to form carbonic acid:

$$H_2O + CO_2(g) = H_2CO_3(aq).$$

This carbonic acid may be a powerful weathering agent that can attack minerals much more effectively than pure water. In normal air the partial pressure of carbon dioxide is 0.0003 atm (3.04×10^{-5} MPa). According to Henry's law, this partial pressure will yield a concentration of 0.00001 molal (m), as shown by

$$H_2CO_3(aq) = 10^{-1.5}P_{CO_2}.$$

This low concentration is only slightly more effective than water at dissolving minerals. However, in soils it is not unusual for the partial pressure to approach 1 atm, resulting in a concentration of carbonic acid of 0.03 mole per kilogram of water. The carbonic acid then becomes a much more powerful agent of weathering.

In some cases one of the products of carbonation may be a new carbonate mineral. An example is the weathering of Ca-plagioclase feldspar, which can result in the formation of calcite, as shown by the following reaction for the alteration of Ca-plagioclase to kaolinite and calcite:

$$CO_2 + CaAl_2Si_2O_8 + 2H_2O$$

$$\rightarrow Al_2Si_2O_5(OH)_4 + CaCO_3.$$

Lastly, hydrolysis is a weathering reaction in which the bonds of molecular water are broken, with H^+ or OH^- combining with the components of the original minerals to form new minerals. Because of the release of either H^+ or OH^-, the pH of the water undergoes a change. An example of weathering and hydrolysis of Na-plagioclase feldspar to form the clay mineral kaolinite. The OH^- in the kaolinite is produced by the addition to the silicate lattice of H^+ from the water:

$$2NaAlSi_3O_8 + 9H_2O + 2H^+ \rightarrow Al_2Si_2O_5(OH)_4$$

$$+ 4H_4SiO_4(aq) + 2Na^+.$$

The pH of the water after reaction with the feldspar will be higher because of the addition of hydroxyl to the solution.

C. Stability of Silicates

Since the 1930s, many fundamental advances have been achieved in understanding the processes of weathering. In a classic study, Goldich (1938) demonstrated that during weathering the order of stability of minerals that constitute igneous and metamorphic rocks is essentially the reverse of the order in which the minerals crystallize from a cooling magma, as shown in Fig. 3.

The inverse relationship between weathering stability and the temperature of formation is caused by differences in bonding among the silica tetrahedra that make up the framework of the silicate minerals. The tetrahedra are less extensively bonded to each other in the high-temperature primary minerals than in the minerals that form at lower magmatic temperatures.

An interesting aspect of chemical weathering is that the elements that make up a mineral may not all dissolve at

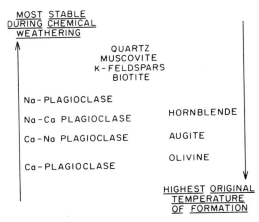

FIGURE 3 Summary of the stability of fresh igneous and metamorphic silicate minerals when exposed to the processes of weathering.

the same rate. This phenomenon is known as incongruent dissolution, and it can lead to cation-depleted or anion-depleted residue. As an example, Fig. 4 summarizes results of a set of laboratory experiments on the artificial weathering of the mineral fayalite, showing that iron, SiO_2, calcium, and magnesium exhibit different rates of release to solution. One of the causes of the apparently incongruent dissolution of minerals is the simultaneous precipitation of new minerals, resulting in selective removal of some of the components from solution.

The concentration curves in Fig. 4 show an initially rapid rise, followed by a gradual leveling toward a nearly constant value at infinite time. Such curves are generally described as parabolic in shape, meaning that the curves

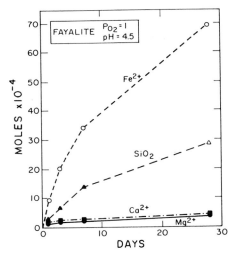

FIGURE 4 Laboratory results on the dissolution of fayalite [(Fe,Ca,Mg)SiO$_4$] in water buffered at pH 4.5 in the presence of 1 atm of oxygen gas.

can be linearized by plotting the concentrations against the square root of time. Early interpretations of the parabolic shape hypothesized that it was the result of the formation of a relatively insoluble layer on the surface of the; the layer was thought to impose a diffusional barrier that increasingly limited the dissolution of the chemical components of the mineral with time. In fact, careful examination of deep weathering profiles in the southeastern United States shows that protective coatings are present on silicate grains collected from the thick zone of decomposed bedrock (saprolite) that underlies the true soil, whereas coatings do not form on the same minerals in the overlying leached soils. The difference is presumably due to the presence of more aggressive weathering solutions in the shallow soils than in the underlying saprolite. However, other studies have shown that dustlike particles on the surface of test grains can also lead to parabolic curves, as can preferential dissolution along imperfections and discontinuities in the crystal lattice.

Lasaga (1998) gathered together most of the quantitative information available on the rates of dissolution of silicate minerals in laboratory experiments. He showed that for dissolution experiments in which no other precipitates form, the rate of dissolution or alteration of a mineral is given by

$$\frac{dc_i}{dt} = \left(\frac{A}{V}\right)v_i k, \quad (1)$$

where c_i is the concentration of species in solution, t is the time, A is the surface area of the mineral, V is the volume of fluid in contact with the mineral, v_i is the stoichiometric content of species i in the mineral, and k is the rate constant, which may depend on pH, temperature, and other species in solution. Equation (1) shows that the rate of dissolution of component i form a mineral depends on the stoichiometric abundance of the component, the surface area of the mineral exposed to the weathering solution, and an intrinsic rate constant, k.

It is also well known that the rate of dissolution of silicate minerals depends strongly on the pH of the solution, as expressed by

$$\text{mineral} + n\text{H}^+ \rightarrow \text{H}_4\text{SiO}_4(aq) + \text{cations} + \text{Al-Si residue}$$

The rate constant k in Eq. (1) is proportional to the activity of hydrogen ion, as shown in Eq. (2), where k is in moles per square meter per second:

$$k = (a_{\text{H}+})^N \quad \text{where} \quad 0 < N < 1 \quad (2)$$

The most common values of N lie between 0.5 and 1.0. Table I summarizes the values of k, from experimental N values, for a variety of silicate minerals. These values of k show the rates of dissolution of silicate minerals vary strongly as a function of the activity of H^+ ion. Potassium

TABLE I pH Dependence of Mineral Dissolution Rates in Acidic Solutions[a]

Mineral	Ideal formula	k	Range
K-feldspar	KAlSi_3O_8	$(\text{H}^+)^{1.0}$	pH < 7
Sr-feldspar	$\text{SrAl}_2\text{Si}_2\text{O}_8$	$(\text{H}^+)^{1.0}$	pH < 4
Forsterite	Mg_2SiO_4	$(\text{H}^+)^{1.0}$	3 < pH < 5
Enstatite	Mg_2SiO_6	$(\text{H}^+)^{0.8}$	2 < pH < 6
Diopside	$\text{CaMgSi}_2\text{O}_6$	$(\text{H}^+)^{0.7}$	2 < pH < 6
Ca-plagioclase	$\text{CaAlSi}_2\text{O}_8$	$(\text{H}^+)^{0.5}$	2 < pH < 5.6
Quartz	SiO_2	$(\text{H}^+)^{0.0}$	pH < 7

[a] Rate constant k in Eq. (2) in text. From Lasaga, A. C. (1984). Chemical kinetics of water–rock interactions. *J. Geophys. Res.* **89**(B6), 4009–4025, Table 2, with permission.

feldspar, Sr-feldspar, and forsterite are most sensitive to pH, whereas the rate of dissolution of quartz is independent of pH in acidic solutions.

Using the rate of release of SiO_2 as an index, Lasaga solved Eq. (1) to determine the time that would be required to dissolve a spherical grain, 1 mm in diameter, of the minerals listed in Table I at a constant pH of 5. The results (Table II) show a sequence of stability that is remarkably similar to the historical Goldich sequence of weathering described earlier. Recent studies have shown that natural rates are slightly lower than those predicted by Lasaga (Schulz and White, 1999).

Recent studies have begun to reveal the relationship between the rate of dissolution of silicate minerals and the reactions which take place on an atomic scale at the interface between the surfaces of minerals and adjacent water. For example, Brady and Walther (1989) have shown that the rate at which silica is released from a dissolving mineral into water is controlled by the rates of detachment of the component oxides as a result of reaction between

TABLE II Calculated Mean Lifetime of a 1-mt Spherical Grain at pH 5[a]

Mineral	Ideal formula	Lifetime (years)
Ca-plagioclase	$\text{CaAl}_2\text{Si}_2\text{O}_8$	112
Diopside	$\text{CaMgSi}_2\text{O}_6$	6,800
Enstatite	$\text{Mg}_2\text{Si}_2\text{O}_6$	8,800
Na-plagioclase	$\text{NaAlSi}_3\text{O}_8$	80,000
K-feldspar	KAlSi_3O_8	520,000
Forsterite	Mg_2SiO_4	600,000
Muscovite	$\text{KAl(AlSi}_3)\text{O}_{10}(\text{OH})_2$	2,700,000
Quartz	SiO_2	34,000,000

[a] Adapted from Lasaga, A. C. (1984). Chemical kinetics of water–rock interactions. *J. Geophys. Res.* **89**(B6), 4009–4025, Table 4, with permission.

the H^+ and OH^- in water and the exposed crystal bonds. At pH values above about 8, OH^- appears to react with Si-OH bonds to produce H_2O plus an Si-O-bond, resulting in a weakened crystal lattice. At low pH, H^+ may react with Al-O-bonds, causing Al-bearing oxide groups to go into solution faster than Si-bearing groups, possibly explaining why silica-rich layers are observed to form on the surfaces of feldspars in strongly acidic solutions. Brady and Walther (1989) also show that the famous Goldich sequence weathering of weathering of silicate minerals (see Fig. 3) correlates directly with the different electrostatic energies of bonding between silicon and oxygen atoms in the various minerals of the series.

III. SOILS

A. Factors Involved in Formation

Soils are the result of the weathering of bedrock. They consist of new and residual minerals, fragments of partially weathered rock, wind-blown dust, and organic matter. The organic matter is a critical component of soils. A layer of weathered bedrock without any organic matter would probably not be called a soil by most earth scientists.

The factors that determine the type and thickness of soil include climate, time, type of bedrock, topography, and biologic activity. Of these factors, climate and time are dominant. Thick soils may form on granitic rocks in a temperate climate, but under arctic conditions the same bedrock will never develop a significant cover of soil. Topography is also a significant factor; erosion prevents the development of thick soils on steep mountain slopes. Ideal conditions for the formation soils include a long time, moderate climate, and gentle topography. These are the conditions that have formed the thick agricultural soils of the central United States.

B. Horizonation

The elements that constitute fresh bedrock are strongly segregated by the processes of weathering. In fact, the segregation of chemical elements is a distinguishing feature of soils. It is most clearly expressed in the strong vertical zoning exhibited by many soils. Figure 5 is an idealized illustration of the vertical zoning that is characteristic of a soil in a temperate climate. The lowermost zone, immediately above the bedrock, is the C horizon; it consists chiefly of decomposed fragments of the underlying bedrock. The B horizon represents a zone in which certain elements accumulate after being leached and carried down from above by percolating waters. The thickness, depth, and composition of the B horizon are strongly controlled by climate. In

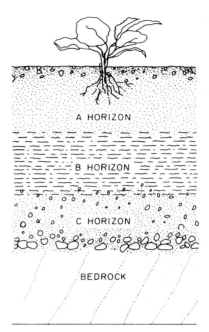

FIGURE 5 Simplified diagram showing the vertical zoning in a soil in a temperate climate.

wet, cool climates, as in the northeastern United States, the B horizon is rich in iron and clay minerals; such soils are called alfisols. In a dry temperate climate, like that of the central and southwestern United States, the B and C horizons may be characterized by an accumulation of $CaCO_3$, called caliche. Such soils may be classified as mollisols or aridisols. Mollisols are the great agricultural soils of the midwestern United States, Argentina, and central Asia.

The A horizon (Fig. 5) is the most important from the point of view of life on earth. It is in this zone that plants find the favorable combination of moisture and nutrients that is essential to growth. The presence of both living and dead organic matter characterizes the A horizon. The A and B horizons together are called the solum and are roughly synonymous with the word topsoil. From 10^6 to 10^9 microbes per cubic centimeter of soil are found in these layers. Nutrients released by the weathering of bedrock, often with the aid of these soil microbes, are adsorbed or chelated by the organic matter, and are readily available to the roots of plants.

In tropical climates the effects of abundant rainfall and high temperature combine to produce weathering that is particularly intense. The resulting soil is generally red in color and is called latisol (or laterite). A latisol represents the last insoluble residue left behind after long and intense weathering. Latisols are usually rich in iron or aluminum hydroxides, and, if very rich, may form ore deposits of these metals. Latisols are poor in organic matter and other essential plant nutrients. In general, agricultural production from latisols is marginal and requires liberal fertilization.

Scientific study on the processes of formation of soils has been going on only for a century or so. Apart from the scientific knowledge that has been gained, it has become clear that agricultural soils are highly susceptible to damage and erosion by humans because of poor agricultural practices or urbanization and industrialization. Some of our best agricultural soils, often adjacent to large urban areas, have been taken over by suburban housing developments, shopping malls, and airports. Poor cultivation practices have resulted in soils with salinity problems, wind and slope erosion, and depleted nutrients. Thousands of years are required to develop a few centimeters of rich topsoil, but only a few years may be adequate to devastate it. Our continuing loss of agricultural soils may force some difficult social decisions in the future, as our desire to expand cities and industries continually encroaches upon this nonrenewable resource.

IV. THERMODYNAMIC CONSIDERATIONS

A. Gibbs Free Energy Function

The reactivity of geochemical species can be predicted from considerations of the Gibbs free energy function. Each solid, liquid, and aqueous species has a unique Gibbs free energy of formation. For a hypothetical reaction:

$$x\mathrm{X} \mid y\mathrm{Y} \rightarrow z\mathrm{Z} \tag{3}$$

where x, y, and z represent the stoichiometric coefficients of the reactants and the products, the change in the Gibbs free energy is

$$\Delta G_{\mathrm{f}} = z\Delta G_{\mathrm{fz}} - (z\Delta G_{\mathrm{fx}} + y\Delta G_{\mathrm{fy}}) \tag{4}$$

where ΔG_{f} represents the Gibbs free energy of formation of the chemical species from the elements and ΔG_{R} the free energy of reaction. If the reactants and products are in the so-called standard state, Eq. (4) takes the form

$$\Delta G_{\mathrm{R}}^0 = z\Delta G_{\mathrm{Z}}^0 - (z\Delta G_{\mathrm{X}}^0 + y\Delta G_{\mathrm{Y}}^0)$$

The standard state is chosen to be at a temperature and pressure that are convenient for the problem being considered, Liquids and solids in the standard state are taken to be pure substances, gases are pure substances at a pressure of 1 bar (0.1 MPa) behaving ideally, and dissolved species are at 1 molal (1 m) concentration behaving as infinitely dilute solutions. The free energies are usually measured experimentally, although in some cases they can be calculated or estimated from basic principles. The Gibbs free energies of formation of actual solutions, gases, and minerals deviate from the ideal standard Gibbs free energies because of impurities.

The relationship between the Gibbs free energy of reaction and the standard free energy of reaction is

$$\Delta G_{\mathrm{R}} = \Delta G_{\mathrm{R}}^0 + RT \ln Q \tag{5}$$

where R is the gas constant, T the temperature of the reaction, and Q the reaction quotient for the reaction of interest. Using reaction (3) as an example, the reaction quotient is defined as

$$Q = [\mathrm{Z}]^z / [\mathrm{X}]^x [\mathrm{Y}]^y$$

where [ ] represents the thermodynamic activity of each reactant and product. (The concept of thermodynamic activity is discussed below. For very dilute solutions or gases at very low pressure and for pure solids and liquids, the activities are nearly equal to the concentrations.)

B. Equilibrium

At equilibrium the Gibbs free energy of reaction (3) is zero. A positive value of the Gibbs free energy indicates that the reaction should go to the left as written, and a negative value indicates that it should go to the right. At equilibrium, Eq. (5) becomes

$$\Delta G_{\mathrm{R}}^0 = -RT \ln K_{\mathrm{eq}} \tag{6}$$

where K_{eq} is the reaction quotient at equilibrium, now called the equilibrium constant. Tables of K_{eq} values are available in the literature for reactions involving solids, gases, and ions. However, for the great majority of reactions the K values are calculated from tabulated tables of ΔG_{f}^0 data.

For the dissolution of a solid, the reaction takes the general form

$$\mathrm{A}_n\mathrm{B}_m = n\mathrm{A}^{m+} + m\mathrm{B}^{n-} \tag{7}$$

$$K_{\mathrm{sp}} - [\mathrm{A}^{m+}]^n [\mathrm{B}^{n-}]^m / [\mathrm{A}_n\mathrm{B}_m] \tag{8}$$

For pure solids $[\mathrm{A}_n\mathrm{B}_m] = 1$.

C. Activity versus Concentration

The amount of a dissolved species that is totally free and available for reaction is described as the thermodynamic activity. To illustrate the meaning of activity, let us consider the chemistry of magnesium in seawater. The total concentration of magnesium in sea water is 1292 mg/liter (0.053 m). Of this total, only a portion (0.046 m) is present as free Mg^{2+}. But not all of the free Mg^{2+} is available for reaction; the fraction that is available is the activity, with a value of 0.015 (note: activity is dimensionnless). Therefore, the reactivity of magnesium in sewater is far lower than would be suggested by a simple consideration of the total concentration.

In many situations activity can be approximated by concentration. For example, for nearly pure solids and liquids the activity is approximately equal to mole fraction X. For gases at low pressure, activity is nearly equal to partial pressure; for dilute electrolyte solutions, activity of the solute can be approximated by concentration.

D. Distribution of Aqueous Species

A chemical element dissolved in water is present in different aqueous forms. For example, in seawater over 99% of the total magnesium is present as Mg^{2+} and $MgSO_4^0$, but $MgHCO_3^+$, $MgCO_3^0$, and $Mg(OH)^+$ are also present in minor amounts. The relative abundances of the various dissolved forms depend on the constants of association K_a. Consider the reaction of species A and B to form a new dissolved species:

$$aA^{m+} + bB^{n+} - (A_aB_b)^{am-bn}$$

where a and b are the stoichiometric coefficients of the reactants, and m and n are the numbers of electrostatic charges on the reactants. For this reaction the stability of the new species is described by

$$K_a = [A_aB_b]^{am-bn}/[A^{m+}]^a[B^{n-}]^b$$

where [ ] denotes thermodynamic activity.

Values of the association constants for magnesium include the following (expressed as $\log_{10}$)

$Mg^{2+} + SO_4^{2-} = MgSO_4^0$	$\log K_a = 2.238$
$Mg^{2+} + HCO_3^- = MgHCO_3^+$	$\log K_a = 1.260$
$Mg^{2+} + CO_3^{2-} = MgCO_3^0$	$\log K_a = 3.398$
$Mg^{2+} + OH^- = MgOH^4$	$\log K_a = 2.600$
$Mg^{2+} + F^- = MgF^+$	$\log K_a = 1.820$

The value of K_a for $MgCO_3^0$ is larger than the value for $MgSO_4^0$; despite this, $MgSO_4^0$ is far more abundant in seawater than $MgCO_3^0$. This results from the fact that in seawater the activity of SO_4^{2-} is orders of magnitude greater than the activity of CO_3^{2-}. The reaction to form $MgSO_4^0$ is therefore driven much more strongly than the reaction to form $MgCO_3^0$.

Many comprehensive computer programs are now available to the geochemist to ease the burden of computing the multiplicity of chemical equilibria in natural waters. Some of the best known of the geochemical programs, together with the supporting institution, are EQ3/6 (Lawrence Livermore National Laboratory), WATEQ (U.S. Geological Survey), PHREEQE (U.S. Geological Survey), PATH (University of California at Berkeley),

GEOCHEM (University of California at Riverside), and MINTEQ (U.S. Environmental Protection Agency).

Many comprehensive computer programs are now available to the geochemist to ease the burden of computing the multiplicity of chemical equilibria in natural waters. Some of the best known of the geochemical programs, along with the supporting institution, are WATEQ4F (U.S. Geological Survey), MINTEQA2 (U.S. Environmental Protection Agency), NETPATH (U.S. Geological Survey), PHREEQC, PHREEQE, and PHRQPITZ (U.S. Geological Survey).

E. Activity Coefficient and Ionic Strength

The thermodynamic activity and the total concentration of an aqueous species i are related by the activity coefficient γ_i follows:

$$[i] = \gamma_i m_i \tag{9}$$

where [ ] denotes activity and m_i is molality. At infinite dilution, activity and molality become equal; that is, γ_i goes to unity.

For very dilute solutions, γ_i can be calculated from the extended Debye–Hückel expression

$$-\log \gamma_i = \frac{Az^2 I^{1/2}}{1 + BaI^{1/2}}$$

where A and B are constants related to the density, dielectric constant, and temperature of the solvent. The terms z and a are the electrical charge and "hydrated ionic radius" of the ion, and I is the ionic strength. The ionic strength is a term that takes into account the concentrations and electrical charges of all ions in solution; it is calculated from

$$I = \tfrac{1}{2} \sum m_i z_i^z$$

where m_i and z_i are the molality and charge of each ion in solution. For a 1 : 1 electrolyte such as NaCl, the ionic strength is equal to the molality; $I = \tfrac{1}{2}[1(1)^2 + 1(1)^2] = 1$. For a 1 : 2 electrolyte such as $CaCl_2$, the ionic strength is 3. Thus, the ionic strength includes the effect of the concentration of solute and the electrical charge in the solution. An approximate ionic strength can be calculated for any natural water from the components listed in a comprehensive analysis. However, for a more accurate determination of the ionic strength it is necessary to compute the distribution of all aqueous species in the solution and eliminate from the calculation of ionic strength the proportion of ions that are tied up as uncharged ion pairs; this yields the "effective ionic strength" for the solution. For example, from an analysis of the total concentrations of dissolved species in seawater, an apparent ionic strength of 0.71 can be calculated. But after computer modeling to account for all of the noncharged ion pairs that form among the dissolved components, the effective ionic strength is found to be 0.64. The latter value should be used in the Debye–Hückel and similar expressions for calculating the activity coefficients of the dissolved species.

The extended Debye–Hückel expression is valid for computing the activity coefficient of an ion up to an ionic strength of about 0.1. This covers the range of ionic strength of most natural fresh waters, such as rainwater, streams, and lakes. However, for saline waters, the Debye–Hückel expression fails to predict activity coefficients accurately. For example, in seawater, with an effective ionic strength of 0.64, the Debye–Hückel expression leads to γ_i values that may be in error by an order of magnitude.

For more saline waters, a number of semi-empirical forms of the Debye–Hückel expression have been developed. Typical of these is the following:

$$\log \gamma_l = (-)\frac{Az^2 I^{1/2}}{1 + BaI^{1/2}} + bI$$

where b is an adjustable term that is derived from experimental measurements.

Individual ion activities (and activity coefficients) cannot be measured exactly because the chemical behavior and properties of a single ion are affected by the other ions that must always be present in solution to maintain electrical neutrality. However, the mean activity coefficient γ_l can be measured for any salt. The mean activity coefficient is determined from measurements of such properties of solutions as freezing point, vapor pressure, solubility, and electrical potential. Figure 6 shows the mean activity coefficients for a number of salts as a function of molality. Note that $\gamma_{\pm}$ is equal to 1.0 at infinite dilution. The curves in Fig. 6 illustrate the fact that most activity coefficients pass through a minimum and then increase at high molalities; from Eq. (9) this means that at high salinities the thermodynamic activity of a salt may be greater then the total concentration in solution.

Figure 7 summarizes the individual ion activity coefficients for several free ions, calculated from both the Debye–Hückel equation and the mean activity coefficients for individual salts.

For uncharged aqueous species such as H_2CO_3 the activity coefficient is usually approximated by an empirical expression such as

$$\log \gamma = k_m I$$

where k_m is a "salting coefficient," which is tabulated in various standard reference books on solution chemistry, and I is the ionic strength. For H_2CO_3, in a solution with the approximate composition of seawater the value of k_m is 0.1. For the effective ionic strength of 0.64 of seawater, this yields an activity coefficient of 1.16 for H_2CO_3 in seawater.

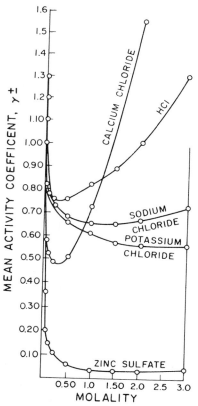

FIGURE 6 Mean activity coefficients for various salts as a function of molality.

Various approaches have been developed to model and predict mean activity coefficients in saline waters. The most successful of these is the Pitzer method, in which the mean activity coefficients are modeled by semi-empirical equations that account for all specific ion interactions except strong complexes. Data for the Pitzer equations are derived from measurements on binary and ternary systems of salts in water, then applied to any complex mixture of the same salts; this approach has been

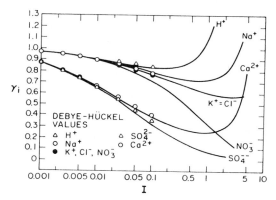

FIGURE 7 Individual ion activity coefficients, calculated from mean salt and Debye-Huckel expressions.

successful in modeling solutions up to concentrations of about 20 m.

In the absence of any theoretical basis for going from mean activity coefficients to single-ion activity coefficients, a nonthermodynamic assumption must be made to calculate the activity coefficients of individual ions. This is usually the MacInnes assumption, which states that the individual activity coefficients for aqueous K^+ and Cl^- are equal, and each is equal to $\gamma_{\pm KCl}$.

F. Saturation Index

An important aspect of low-temperature geo-chemistry is the determinatio of the state of saturation of natural waters with respect to minerals. For example, caves develop in limestone because the ground water is, at least temporarily, understaurated with respect to calcite. The state of saturation of a natural water with respect to a solid mineral can be determined by combining the activities of the dissolved ions to form the ion activity product (IAP), which can then be compared to the equilibrium solubility product (K_{sp}). If the water is just at equilibrium with the mineral of interest, IAP will equal K_{sp}. For a state of supersaturation IAP will be larger than K_{sp}, and for undersaturation the IAP will be smaller than K_{sp}. As an example, the K_{sp} of calcite is $10^{-8.48}$, and in seawater the IAP of $CaCO_3$ is calculated to be $10^{-7.73}$. Thus, in normal seawater the IAP is greater than K_{sp} and the water is supersaturated with respect to calcite. The state of saturation is often expressed as the saturation index (SI):

$$\text{Saturation index} = \text{SI} = \log \frac{\text{IAP}}{K_{sp}}$$

For the example of calcite in seawater:

$$[Ca^{2+}] = 10^{-2.65} \qquad [CO_3^{2-}] = 10^{-5.08}$$

$$\text{IAP} = [Ca^{2+}][CO_3^{2-}] = 10^{-7.73} \qquad (10)$$

$$\text{SI} = \log \frac{10^{-7.73}}{10^{-8.48}} = (+)0.75$$

From Eq. (10), it is seen that SI is positive because the water is supersatureted with respect to the mineral. SI is zero at equilibrium and negative for undersaturation.

V. WATER CHEMISTRY

A. Rain and Snow Water

The most dilute waters in nature are rain and snow. These are the natural equivalents of distilled water in the laboratory. However, rain and snow are from pure water. They contain dissolved gases and aerosols, as well as the dissolved and solid products from reaction with atmospheric

TABLE III Composition (in mg/l, except pH) of Snow and Rain[a]

Component	Snow[b]	Rain[c]	Rain[d]
Ca	0.0	0.65	1.2
Na	0.6	0.56	0.0
K	0.6	0.11	0.0
Mg	0.2	0.14	0.7
Cl	0.2	0.57	0.8
SO4	1.6	2.2	0.7
HCO3	3	?	7
NO3	0.1	0.62	0.2
Total dissolved solids	4.8	?	8.2
pH	5.6	?	6.4

[a] From Hem, J. D. (1970). Study and interpretation of the chemical characteristics of natural water. U.S. Geological Survey Water Supply Paper 1473, 2nd ed., p. 50.

[b] Spooner Summit, U.S. Highway 50, Sierra Nevada, November 20, 1958.

[c] August 1962 to July 1963 (27 points in North Carolina and Virginia).

[d] Menlo Park, California, January 9 and 10, 1958.

dusts. Table III shows the chemical composition of some selected samples of rain and snowmelt.

Systematic changes have been observed in the chemical composition of rainwater as a function of distance from the ocean and/or the length of time of individual storms. In general, the farther from the sea or the longer the rainfall continues, the more pure the water; this reflects a continued "rinse-out" of the aerosols and dust particles in the atmosphere.

One of the most serious environmental problems today is that of "acid rain." Research remains to be done in this area, but few would disagree with the preliminary conclusion regarding the important role of the gases of sulfur and nitrogen. These, in combination with rainfall, produce sulfuric and nitric acid, with a resultant lowering of the pH of rainfall below the value of 5.6 that would normally be imposed by the 0.0003 atm of atmospheric CO_2. It is becoming increasingly clear that gaseous emissions from automobiles, coal-fired power plants, and sulfide mineral smelters are the principal anthropogenic agents, while volcanic emissions are relatively minor and localized.

Occasionally, rainwater has a pH greater than 5.6. The higher values are attributable to aerosols of seawater or alkaline dusts from desert regions carried into the atmosphere.

B. Streams

The transition of rain and snow waters into stream waters begins as the first drops make contact with plants, soils, and bedrock. A large number of chemical and biochemical reactions occur, adding dissolved and colloidal inorganic and organic material to the nearly pure precipitation. The amount of total dissolved solids (TDS) in rain or snowmelt can increase by an order of magnitude as waters infiltrate and percolate down through soil, shallow subsoil, or deep bedrock en route to the stream channel. As described in Sec. II, reactions such as dissolution, hydrolysis, and carbonation all contribute to the increased dissolved load in the water, as do metabolic processes of soil micro- and macro-biota, which may include both respiration and decomposition. The biochemical processes may additionally contribute dissolved gases, dissolved organic compounds, and particulate organic matter.

The chemical composition of stream waters varies greatly for different streams and for different seasons or even time of day within the same stream. For different streams, variations in chemistry depend on bedrock, climate, vegetation, and discharge. The chemical composition of a stream can vary greatly from headwaters to mouth, reflecting all these factors. Seasonal variations occur with spring flooding and resultant dilution or flushing effects. Some ions will decrease in concentration as discharge increases, indicating dilution; others will increase, indicating a flushing of soluble weathering residues from the soil. The increase peak is often quite brief, and field workers must be prepared to sample throughout the runoff period in order to capture it. Diurnal variations in pH of river water may be as much as 3 pH units, attributable to daily temperature variations and to the photosynthetic activity of aquatic plants. Typically, pH values will rise in the warmth of late afternoon, and fall in the cool, dark hours of evening. Table IV lists the compositions of the Mississippi River, the Amazon, and that of the "worldwide mean river water."

In Table IV the differences between the Amazon and the Mississippi waters reflect the fact that the Mississippi drains the thick soils and sedimentary rocks of the central United States, whereas the Amazon drains the deeply leached lateritic soils of the Amazon rain forest. Studies of the Amazon show that the stream water carries more total dissolved solids near its headwaters, on the eastern side of the Andes. As the Amazon grows in size and flows through the Amazon basin, the lack of fresh bedrock and the enormous volume of dilute rainwater combine to make the water progressively purer as it flows toward the Atlantic.

Streams that traverse relatively insoluble crystalline rocks contain HCO_3^- as the principal dissolved anion; the HCO_3^- is derived from carbon dioxide gas in atmospheric air or soil air. If the bedrock is sedimentary, Cl^- and SO_4^{2-} are also abundant in the water.

Although K^+ is nearly as abundant as Na^+ in crustal rocks of the earth, Na^+ is far more abundant than K^+ in most stream waters. This reflects the fact that K^+ is

TABLE IV Compositions (in mg/l, except pH) of Samples of Water from the Mississippi River and the Amazon River and Worldwide Mean River Composition[a]

Component	Mississippi[b]	Amazon[c]	World
Ca	42	4.3	15
Na	25	1.8	6.3
K	2.9	0	2.3
Mg	12	1.1	4.1
Cl	30	1.9	7.8
SO4	56	3.0	11
HCO3	132	19	58
SiO2	6.7	7.0	13
Total dissolved solids	256	28	90
pH	7.5	6.5	—

[a] From Hem, J. D. (1970). Study and interpretation of the chemical characteristics of natural waters. U.S. Geological Survey Water Supply Paper 1473, 2nd ed., p. 50.

[b] October 1, 1962 through September 30, 1963. Time-weighted mean of daily samples.

[c] Obidos, Brazil. High stage (7,640,000 ft^3/sec), July 16, 1963.

released less readily from silicate mineral structures, once released, it is attracted more strongly than Na^+ to ion-exchange sites on clays and other fine-grained materials in stream sediments and soils, and is an essential nutrient for aquatic plants, which take K^+ up from the water. Sodium, on the other hand, tends to remain persistently soluble.

The concentration of SiO_2 is relatively constant in many stream waters of diverse origin. Its concentration appears to be controlled by the solubility of micaceous minerals and feldpars.

C. Lakes

Lakes that have surface outlets represent holding and mixing systems. Because the water moves much more slowly than in a river, geochemical reactions may approach equilibrium more closely than they do in rivers. However, mixing may not be complete in lakes, and waters may differ greatly in composition from one location to another in the same lake, or at different depths.

Thermal stratification is a significant process in many lakes. If the stratified condition persists, the deeper waters may become depleted in dissolved oxygen (anoxic) and enriched in hydrogen sulfide and methane. Lake "turnover," with subsequent mixing of lake water strata, may occur twice a year in temperate climates as air temperatures cross 4°C. Water is at maximum density at this temperature, and surface waters can displace lower layers. Likewise, if the lake becomes isothermal, wind action on surface layers can cause mixing from bottom to top of the water column.

In lakes excessively affected by runoff containing high quantities of nutrients, overgrowth of aquatic vegetation may occur. The subsequent decomposition depletes creates anoxic conditions which extend throughout the lake; the lake is said then to be eutrophic.

Evaporation plays a major role in controlling the chemistry of lake water. Closed-basin lakes (no surface outlet) are especially influenced by evaporation. Two of the best-known closed-basin lakes are the Great Salt Lake in the United States and the Dead Sea in Israel and Jordan. These two unusual bodies of water of water are the result of intense evaporation of normal river water. The waters in the Great Salt Lake and the Dead Sea have roughly the ionic proportions of seawater but with about 10 times the total salinity.

Table V lists the compositions of a few lakes from closed basins in the western United States. Based on the pH values, the lake waters tend to fall into two very broad classes: saline (pH about 7.5) and alkaline (pH about 10). Studies of the chemical evolution of lake waters show that the initial composition of the dilute source water governs the final composition of the lake brine.

In Table V the high pH in the alkali lakes results from the fact that HCO^{3-} is the dominant anion and greatly exceeds the concentration of calcium and magnesium in the fresh source streams. The low values of calcium and magnesium for Alkali Valley and Surprise Valley are caused by precipitation of calcite ($CaCO_3$) and a magnesium silicate [approximately sepiolite, $Mg_3Si_3(OH)_2$] or magnesium carbonate. Gypsum ($CaSO_4 \cdot 2H_2O$) may also precipitate from some of the saline waters during evaporation.

TABLE V Compositions (in mg/l, except pH) of Various Lakes from Closed Basins in the Western United States[a]

Component	Alkali Valley, Ore.	Surprise Valley, Calif.	Great Salt Lake, Utah	Saline Valley, Calif.
SiO_2	542	36	48	36
Ca	Trace	11	241	286
Na	117,000	4,090	83,600	103,000
K	8,850	3,890	4,070	4,830
Mg	Trace	31	7,200	552
Cl	45,700	4,110	140,000	150,000
SO_4	46,300	900	16,400	57,100
HCO_3	2,510	1,410	251	614
CO_3	91,400	664	Trace	Trace
Total dissolved solids	314,000	10,600	254,000	282,360
pH	10.1	9.2	7.4	7.4

[a] From Eugster, H. P., and Hardie, L. A. (1978). Saline lakes. "Lakes: Chemistry, Geology, Physics" (A. Lerman, Ed.), Springer-Verlag, New York, p. 237–293, with permission.

The anion chemistry is dominated by $HCO_3^- - CO_3^{2-}$ in the alkali lakes and by Cl^- and SO_4^{2-} in the saline lakes.

D. Ground Water

Ground water is derived from precipitation that infiltrates the soil and percolates down into cracks and intergranular void spaces in either the bedrock or unconsolidated sediments. Ground water can be of excellent quality, and it is the principal water supply for most of rural America as well as such large cities as Miami, Florida, Houston, Texas, and Tucson, Arizona. For many years the extensive ground-water resources of the United States were thought to be pure and protected from pollution by the overlying layers of soil and rock. However, it is now recognized that anthropogenic contamination of ground water, particularly by organic solvents and petroleum products, is widespread.

Whereas surface streams moves at rates of a few feet per second, ground water rarely flows faster than a few feet per day. Because ground water is in intimate contact with rock minerals for much longer periods of time than streams, it is usually more saline than stream water. From the study of various radioactive isotopes, it has been determined that many ground waters are thousands of years old. For example, abundant, still potable water has been found in aquifers deep beneath the Sahara Desert, dating back to the retreat of the continental glaciers 10,000 years ago.

Deeper ground water is subject to higher temperatures and longer residence times than that in near-surface aquifers. Extensive chemical reaction with aquifer materials may occur. An example is the formation of dolomite from calcite as Mg^{2+}-rich waters migrate through a limestone:

$$Mg^{2+}(aq) + 2CaCO_3 + Ca^{2+}(aq) + CaMg(CO_3)_2.$$

Because of the slow kinetics of nucleation and growth of crystalline dolomite, the formation of dolomite is quite rare under near-surface conditions. However, subsurface conditions apparently allow dolomite to form in large amounts.

Waters that are produced with petroleum from deep strata may be as saline as the closed-basin lakes described earlier. Oilfield brines are thought to represent either meteoric waters that have reacted extensively with the enclosing rock strata or connate waters. Connate water may represent "fossil" water, ancient seawater from the time of formation of the rocks.

E. Seawater

The ultimate sink for dissolved elements released during weathering is the ocean. Although the absolute concentrations of elements dissolved in seawater vary from location

TABLE VI Concentrations of Major Dissolved Constituents in Seawater, with Calculated Residence Times

Component	Seawater (ppm)	Principal forms	Residence time in ocean (yr)
Na	10,556	Na^+, $NaCl^0$	68,000,000
Mg	1,272	Mg^{2+}, $MgSO_4^0$	12,000,000
Ca	400	Ca^{2+}, $CaSO_4^0$	1,000,000
K	380	K^+, KSO_4^-	7,000,000
Cl	18,980	Cl^-	100,000,000
S	2,649	SO_4^{2-}	—
C	140	HCO_3^-, $MgHCO_3^+$	—
B	26	H_3BO_3	18,000,000
Br	65	Br^-	100,000,000
F	1	F^-, MgF^+	520,000

to location under such influences as evaporation or dilution by fresh river water, the ratio of dissolved species is remarkably constant. Table VI lists the major elements dissolved in seawater, the chief forms in which they occur, and the residence time for each element. The residence time is defined by is defined as the total mass of the dissolved element in the oceans divided by the amount transported in each year from the land. Some elements, such as sodium, chlorine, and sulfur, remain in a dissolved state in the ocean for long periods of time before being precipitated or incorporated in sediments or rocks. Other dissolved elements, including calcium, iron, and silicon, and the rare earth elements (REEs) quickly take part in chemical or biochemical reactions, with resulting short residence times. For example, calcium is strongly controlled by the precipitation of aragonite and calcite sediments.

The role of deep-sea hot springs at mid-ocean ridges has been shown to be much more important in controlling the residence times of some elements, particularly magnesium, than previously thought. Much research remains to be done in this area.

F. Buffering of Natural Waters

The pH of natural waters is controlled by acid–base equilibria involving dissolved species and solid compounds. Any chemical process that can add or remove H^+ or OH^- in a natural water is a buffering process.

Much study has been devoted to acid–base buffering in natural systems, especially in seawater. Two major controls are recognized: water–rock interactions and ion–ion interactions. An example of the former is the reaction between acidic water and limestone:

$$CaCO_3(s) + H^+ = HCO_3^- + Ca^{2+}. \quad (11)$$

Clearly, reaction (11) has the capability of consuming and adding H^+ to a water, thus buffering the pH. A similar reac-

tion may occur between silicate minerals and an aqueous solution; using feldspar and kaolinite as examples,

$$2KAlSi_3O_8 + 2CO_2(g) + 11H_2O \rightarrow Al_2Si_2O_5(OH)_4$$
$$+ 4H_4SiO_4(aq) + 2K^+ + 2HCO_3^-. \quad (12)$$

Reactions like (11) and (12) are known to be important in controlling the pH, especially over long periods of time. Short-term buffering is done by dissolved natural acids and the corresponding conjugate bases, such as H_2CO_3–HCO_3^-–CO_3^{2-} and H_4SiO_4–$H_3SiO_4^-$. Of such systems, the carbonic acid system is dominant in most natural waters. Buffering by the carbonate system is controlled by the following set of reactions (listed with the corresponding acid dissociation constants):

$$H_2CO_3 = HCO_3^-, \qquad K_1 = 10^{-6.35};$$
$$HCO_3^- = H^+ + CO_3^{2-}, \qquad K_2 = 10^{-10.33}.$$

Most natural waters have pH values near 8, indicating that the predominant species of inorganic carbon is HCO_3^-. Figure 8 shows the distribution of the carbonate species as a function of pH. Note that the crossover points occur at the pH values of 6.35 and 10.33, corresponding to the pK_1 and pK_2 values; buffering is most effective at pH values near these crossover points. Similar diagrams can be constructed for H_4SiO_4, H_3BO_3, H_2S, and other natural acids.

VI. DIAGENESIS

Diagenesis is a term that describes the sum of processes that affect sediments or very young rocks, changing the composition, mineralogy, or texture. Diagenesis is characterized by low temperature and low pressure. An example of diagenesis is the reaction between seawater and fine-grained sediment carried into the ocean by fresh-water streams. Because the mineralogic debris carried in by river water, such as clays and metallic hydroxides, is out of

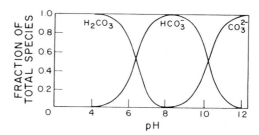

FIGURE 8 Distribution diagram for carbonic acid in water at 25°C and 1.013 bar (0.1013 mPa). Note that the crossover points occur at the acid dissociation values; buffering is most effective at these pH values.

equilibrium with seawater, there will be a tendency for reaction to occur in the new medium of seawater. Ions present in the seawater may be adsorbed on the surfaces of weathered minerals, or they may be incorporated in the structure of new minerals they may form. These early, rapid reactions are called halmyrolysis. As the sediment is gradually buried beneath the sea floor, old minerals may dissolve and new minerals may form. The final product of diagenesis will be a set of sedimentary minerals and rocks that more closely approach equilibrium with the enclosing fluids at the temperature and pressure of the new environment.

Several diagenetic reactions are reasonably well understood. Perhaps the most simple is the dissolution or recrystallization of unstable minerals into more stable forms. For example, most of the $CaCO_3$ deposited in modern oceans is in the form of fine-grained aragonitic mud and fragments of aragonitic shells (aragonite is the metastable orthorhombic form of $CaCO_3$, calcite is the more stable rhombohedral form). Aragonite is unstable relative to calcite at the temperature and pressure of the earth's surface. Stable isotope studies using uptake of ^{44}Ca have shown the rate to be more rapid in fresh water, but still appreciable in salt water. Therefore, very early in the history of fresh-water and marine sediment, the carbonates tend to recrystallize into calcite; in some cases this results in lithification of the loose carbonate sediment into limestone. The silica derived from marine opaline diatom tests undergoes a somewhat similar reaction, resulting in hard layers of chalcedonic chert interbedded with unconsolidated sediments.

As marine sediments are buried to relatively shallow depths, the limited amount of oxygen in the pore waters is exhausted and the sediment becomes anoxic. Chemical species that are sensitive to the state of oxidation may be mobilized or immobilized by diagenetic reactions; the most notable examples are iron, sulfur, and manganese. Iron is present in most sediments as fine-grained iron hydroxide, and sulfur is present in the interstitial marine water as SO_4^{2-} ions. As the conditions become more strongly reducing, and in the presence of bacteria and organic matter, the following generalized reactions tend to occur (CH_2O represents carbohydrate):

$$SO_4^{2-} + 2CH_2O \rightarrow H_2S(aq) + 2HCO_3^-;$$

$$CH_2O + 4Fe(OH)_3(s) + 7CO_2(aq)$$

$$\rightarrow 4Fe^{2+} + 8HCO_3^- + 3H_2O. \quad (13)$$

The ferrous iron shown in reaction (13) may then react with the bicarbonate to form diagenetic siderite ($FeCO_3$) or with H_2S to form iron sulfide, FeS. The FeS will eventually change into the more stable form, pyrite, FeS_2.

A third generalized type of diagenetic reaction is the so-called reverse weathering of silicates. At one time it was thought that such reactions were of primary importance in controlling the chemistry of seawater, but more recent work suggests that although reverse weathering may occur in seawater, it is probably not as important as formerly thought. The general reaction is

$$\text{Silicate debris} + \text{cations} + HCO_3^- + \text{silica}(aq)$$

$$\rightarrow \text{new clay minerals} + CO_2(aq) + H_2O. \quad (14)$$

Reactions like (14) probably become increasingly important with increasing depth of burial of sediments beneath the sea floor.

VII. OXIDATION AND REDUCTION

A. Sequence of Reactions

Many chemical elements of geochemical concern are sensitive to the oxidation–reduction (redox) status of the environment; examples include iron, sulfur, manganese, carbon, nitrogen, arsenic, chromium, copper, selenium, uranium, and vanadium. An illustration of the idealized sequence of redox-controlled changes in the chemistry of dissolved components in the interstitial water of a sediment is given in Fig. 9.

The changes shown in Figure 9 are controlled the progressive loss of oxygen and the increasing role of anaerobic bacteria with depth in the sediment. Although the full sequence is rarely observed, portions of it are commonly present in waters of sediments, soils, and ground-water aquifers. The reactions are O_2 reduction, nitrification of organic nitrogen, nitrate reduction to nitrogen, MnO_2 reduction, $Fe(OH)_3$ reduction, SO_4^{2-} reduction, methane fermentation, precipitation of iron and manganese sulfides, and fixation of nitrogen. These reactions reflect the activity of bacteria in the sediment, utilizing in succession the oxygen from available solid and dissolved species. The reactions occur in the sequence predicted by standard electrochemical potentials. The bacteria serve as biological catalysts to facilitate reactions that might otherwise be kinetically slow. In general, each reaction is driven by a specific type or family of bacteria; that is, the reduction of sulfate does not occur until virtually all of the nitrate has been reduced, and methane is not generated until most of the sulfate has been reduced.

Redox reactions can be described in a variety of ways. For example, the reduction of SO_4^{2-} to aqueous H_2S can be written as

$$SO_4^{2-} + 2C(organic) + 2H_2O = H_2S(aq) + 2HCO_3^-.$$

$$(15)$$

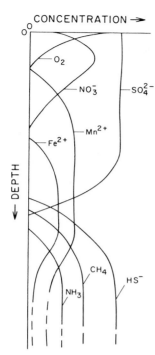

FIGURE 9 Schematic representation of changes in concentrations of redox-sensitive species to be expected in the interstitial water in a sediment with increasing depth below the sediment-water interface.

In reaction (15) the S(VI) in the SO_4^{2-} acts as the electron acceptor for the oxidation of organic matter to bicarbonate. The same reaction could be written as two half-reactions:

$$SO_4^{2-} + 10H^+ + 8e^- = H_2S + 4H_2O \qquad (16)$$

and

$$2C(organic) + 6H_2O = 2HCO_3^- + 10H^+ + 10e^-. \qquad (17)$$

Reaction (15) is the sum of the two half-reactions (16) and (17). For this reason, it is usually convenient to write and tabulate tables of half-reactions for redox couples of interest and then add them together to describe balanced redox reactions.

B. pe–pH Diagrams

One of the most fruitful approaches to prediction of the relative stability of minerals in rocks and sediments has been the use of pe–Ph (or Eh–pH) (see Glossary) diagrams. For such diagrams a series of half-reactions are written, such as (16) or (17). Each reaction is then plotted on orthogonal graph paper, using the negative logarithm (base 10) of the H^+ (i.e., pH) and the negative logarithm of e^- (i.e., pe) as master variables. The other compositional variables assigned fixed activity values.

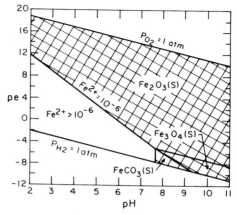

FIGURE 10 pe–pH diagram for the system Fe-O-H_2O-CO_2, assuming total dissolved $CO_2 = 10^{-3}$ m. The solid–solution boundary for Fe^{2+} is drawn for $Fe^{2+} = 10^{-6}$ m. Activities assumed to be equal to concentrations. $T = 298$ K and $P = 1.013$ bar (0.1013 mPa). Symbol (S) indicates solid compounds. The crosshatched area indicates the fields of solids in which the activity of Fe^{2+} is less than 10^{-6}.

A pe–pH diagram is shown in Fig. 10 for iron in water in the presence of CO_2 gas. The predictive power of such a diagram can be seen by imagining that the redox status of the water is lowered from a pe value of 12 to a pe value of −6, perhaps by adding dissolved organic matter and bacteria, while maintaining a constant pH of 8. The diagram indicates that the mineral hematite (Fe_2O_3) would be reduced to the mineral siderite ($FeCO_3$). If the pH were raised to about 10, the mineral magnetite (Fe_3O_4) should form at the expense of siderite. Lowering the pH would cause Fe^{2+} to go into solution at greater concentrations.

Much effort has been expended in low-temperature geochemistry in comparing thermodynamic predictions of mineral stabilities to the mineral assemblages actually found in the field. In general, the theoretical trends are observed, but with discrepancies that can usually be attributed to slow kinetics. It is also clear that the role of bacteria is profoundly important in affecting the rates of many low-temperature reactions in nature.

VIII. SORPTION AND ION EXCHANGE

A. Types of Reactions

Sorption refers to the binding of ions and some neutral molecules to charged surfaces of minerals or colloids in sediment in contact with an aqueous solution. Ion exchange refers to the exchange of bound ions or neutral molecules with dissolved ions or neutral molecules in the aqueous solution of greater affinity to the mineral or colloid surface. Typically, the properties of sorption or ion exchange are most exhibited by the fine-grained fraction of the sediment.

Sediments and soils commonly contain abundant fine-grained materials. The large surface areas of the fine-grained sediments are available for exchange of ions and neutral molecules between the solid substrates and the adjacent water. The attraction may result from weak electrostatic charges on the surface of the substrate or from strong chemical bonding between components of the solid and the aqueous species. Such reactions are commonly rapid, occurring in a few minutes to a few hours. The general process of attraction of the dissolved species to the surface of the solid substrate is called adsorption. A specific type of adsorption in which ions are exchanged between the surface of the solid and the aqueous solution is called ion exchange. If the sorbed species are incorporated in the interior of the solid, the process is absorption.

An example of an ion-exchange reaction, involving the exchange of Na^+ by Ca^{2+} on a substrate X, is

$$Na_2X(s) + Ca^{2+} = CaX(s) + 2Na^+,$$

where X may represent any exchanging substrate.

Such reactions are described by an ion-exchange coefficient K_{ex}:

$$K_{ex} = \frac{CaX(Na^+)^2}{Na_2X(Ca^{2+})}$$

where ( ) represents ionic concentration. In general, K_{ex} values are not constant; they vary as a function of the degree of ionic loading of the substrate, the type and concentration of competing ions in solution, other types of ions on the exchanger, and the salinity of the water. It is usually necessary to measure specific K_{ex} values for the waters and exchange substrates of interest. In soils, the cation-exchange capacity (CEC), the quantity of cations that can be held or exchanged by a given amount of soil, is measured in milliequivalents of cations per 100 g of soil (meq/g). CECs in descending order are organic material, vermiculite (hydrous mica), smectite clays, kaolinite clays, and sesquioxides (Fe, Al, Mn, and Ti oxides and hydroxides).

B. Adsorption Isotherms

Adsorption and ion exchange are often described by an empirical adsorption isotherm. In determining an adsorption isotherm, a known amount of material is added to a solution and mixed with a measured quantity of adsorbing substrate. After allowing time for adsorption to occur (at least a few hours), the amount of dissolved material is measured and the quantity lost is assigned to adsorption on the substrate. A series of experiments yields a curve (the isotherm), which can be expressed as micrograms

of material sorbed per gram of substrate. The empirical curves are described mathematically by a number of models, including the Freundlich isotherm:

$$\frac{x}{m} = kc^n,$$

where x is the mass of ion or molecule sorbed, m is the mass of the substrate, k and n are empirical constants derived from fitting the experimental curve, and c is the concentration of the dissolved species. Another model is the Langmuir adsorption isotherm:

$$\frac{x}{m} = \frac{ksc}{1 + kc}$$

where x, m, k, and n have the same meaning as in the Freundlich isotherm and s is the total sorbing capacity of the substrate per unit mass.

Much effort has been devoted to developing theoretical models for predicting sorptive processes. Some of the more fruitful results have come from consideration of the reactivity of the sorbing sites on the solid substrate. In this approach an intrinsic constant of reaction, similar to a thermodynamic constant, is assigned to each of the surface sites. The number of sites and the strength of the electrical field adjacent to the charged surfaces are then considered in calculating the extent of adsorption of species from solution. A computer code, MINSORB, a modification of MINTEQA2, successfully models various sorption processes.

IX. ELEMENTAL CYCLING

A. Carbon

If there is a "grand picture" in low-temperate geochemistry, it is the understanding of the cycling of the chemical elements through the hydrosphere, lithosphere, biosphere, and atmosphere. It is not an easy task to assemble the date required; the rates of flow of all the major rivers in the world, the concentrations of dissolved and suspended loads in the rivers, the rates of sedimentation on the sea floors, and so on. Nevertheless, many individuals have accepted the challenge.

As one example of elemental cycling, Fig. 11 illustrates the global geochemical cycle for carbon. It can be seen that the largest reservoir for carbon (10,000,000 billion metric tons) is in the rocks of the earth's crust (chiefly in carbonates and organic-rich sedimentary rocks), followed by the oceans (35,000 billion metric tons).

The largest annual flux of carbon (100 billion metric tons per year) occurs between the oceans and the atmosphere, chiefly as carbon dioxide gas. Burning of fossil fuels contributes an anthropogenic flux of about 5 billion

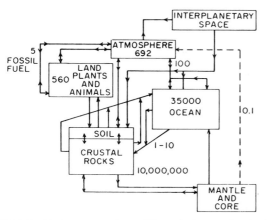

FIGURE 11 Carbon reservoirs and fluxes in the geochemical cycle. Reservoirs are in billions of tonnes (metric tons), and fluxes are in billions of metric tonnes per year.

metric tons of carbon per year, a portion of which remains in the atmosphere. Serious concern exists regarding the environmental effects of a continued buildup of carbon dioxide in the atmosphere. As the concentration of carbon dioxide increases, the so-called greenhouse effect may become important, resulting in a rise of several degrees in the average temperature of the atmosphere by the middle of the twenty-first century. Current evidence suggests that the average temperature of the Earth has been rising at the same rate as CO_2 is increasing in the atmosphere. The consequences of such an increase in temperature could include the development of deserts in the temperate agricultural areas of the world and flooding of coastal areas due to rising sea level caused by melting of the polar ice sheets. Increasing numbers of scientists are urging more reliance on alternative energy sources to avoid a possible environmental calamity.

Despite the many obvious environmental consequences of the wholesale cutting of the earth's rainforests, the main CO_2 sink is photosynthetic activity by phytoplankton in the world's oceans. An experimental effort to create more capacity to absorb CO_2 in the ocean involved fertilization of areas of the Sargasso Sea with iron, a limiting nutrient in the ocean, to stimulate the growth of phytoplankton, which would absorb more CO_2 via photosynthesis. Although the project was tentatively successful, scientists remain divided on the consequences of such massive intervention in the ocean's ecosystems.

B. Other Elements

Geochemical cycles have been worked out for most major elements and many of the nutrients and metals. It is clear that the cycles of all elements are now strongly affected by human activities, at least in the short term. For example, the cycles of nitrogen and phosphate were altered, beginning in the 1950s and 1960s, and continuing to this day, by urban and rural runoff containing detergents and fertilizers, which contributed to eutrophication of lakes and contamination of surface and ground-water supplies in some areas. For better or worse, humans have extended their influence from the vastness of outer space to the geochemical behavior of individual atoms in the atmosphere, hydrosphere, and biosphere. Research on remedial efforts to correct these geochemical imbalances will continue to be important in the coming millenium.

SEE ALSO THE FOLLOWING ARTICLES

CARBON CYCLE ● CHEMICAL THERMODYNAMICS ● ELECTROLYTE SOLUTIONS, THERMODYNAMICS ● ENVIRONMENTAL GEOCHEMISTRY ● GEOCHEMISTRY ● GEOCHEMISTRY, ORGANIC ● HYDROGEOLOGY ● OCEANOGRAPHY, CHEMICAL ● STREAMFLOW

BIBLIOGRAPHY

Berner, R. A. (1980). "Early Diagenesis, A Theoretical Approach," Princeton University Press, Princeton, NJ.

Birkeland, P. W. (1999). "Soils and Geomorphology," 3rd ed., Oxford University Press, New York.

Brady, P. V., and Walther, J. V. (1989). "Controls on silicate dissolution at neutral and basic pH solutions at 25°C," *Geochim. Cosmochim. Acta* **53**, 2833–2830.

Chappele, F. H. (1993). "Ground-Water Microbiology and Geochemistry," Wiley Interscience, New York.

Drever, J. I. (1997). "The Geochemistry of Natural Waters: Surface and Ground Water Environments," 3rd ed., Prentice-Hall, Englewood Cliffs, NJ.

Eugster, H. P., and Hardie, L. A. (1978). In "Lakes: Chemistry, Geology, and Physics" (A. Lerman, ed.), Springer-Verlag, New York.

Faure, G. (1998). "Principles and Applications of Geochemistry," 2nd ed., Prentice-Hall, Upper Saddle River, NJ.

Goldich, S. S. (1938). "A study in rock weathering," *J. Geol.* **46**, 17–58.

Hem, J. D. (1992). "Study and Interpretation of the Chemical Characteristics of Natural Water," 3rd ed., U.S. Geological Survey Water-Supply Paper 2254.

Krauskopf, K. B., and Bird, D. K. (1995). "Introduction to Geochemistry," 3rd ed., McGraw-Hill, New York.

Langmuir, D. (1997). "Aqueous Environmental Geochemistry," Prentice-Hall, Englewood Cliffs, NJ.

Lasaga, A. C. (1998). "Kinetic Theory in the Earth Sciences," Princeton University Press, Princeton, NJ.

Morel, F. M. M., and Hering, J. C. (1993). "Principles and Applications of Aquatic Chemistry," Wiley Interscience, New York.

Nordstrom, D. K., and Munoz, J. L. (1994). "Geochemical Thermodynamics," 2nd ed., Blackwell Scientific, Boston.

Sposito, G. (1984). "The Chemistry of Soils," Oxford University Press, New York.

Stumm, W., and Morgan, J. J. (1995). "Aquatic Geochemistry: Chemical Equilibria and Rates in Natural Waters," 3rd ed., Wiley Interscience, New York.

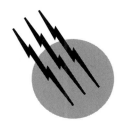

Geochemistry, Organic

Jürgen Rullkötter

Carl von Ossietzky University of Oldenburg

GLOSSARY

Biodegradation Alteration of organic matter in near-surface sedimentary rocks or of crude oil in the reservoir by microorganisms.

Biological markers Chemical compounds in sediments, crude oils, or coal with an unambiguous link to specific precursor molecules in living organisms (molecular fossils, geochemical fossils, biomarkers); among the most common biological markers are isoprenoids, steroids, triterpenoids, and porphyrins.

Bitumen Fraction of organic matter in sediments which is extractable with organic solvents.

Catagenesis Thermocatalytic transformation of organic matter in sedimentary rocks at greater burial depths (usually more than 2 km) and higher temperatures ($>50°C$) into smaller molecules, particularly liquid and gaseous hydrocarbons which become part of the bitumen fraction.

Diagenesis Transformation of sedimentary organic matter under mild conditions (usually $<50°C$) involving both low-temperature chemical reactions and microbial activity.

Hydrocarbon (petroleum) potential Ability of a sediment that is rich in organic matter to generate hydrocarbons (petroleum) under the influence of geological time and temperature, expressed quantitatively in terms of milligrams of hydrocarbons per gram organic carbon (or kilograms per metric ton).

Kerogen Macromolecular organic matter in sediments that is insoluble in organic solvents and consists of structural remnants of dead organisms as well as diagenetic transformation products of various biopolymers and biomonomers.

Macerals Organic particle types specifically in coal but more generally also in dispersed sedimentary organic matter which can be recognized under the microscope and are characterized by their different structural (shape) and optical (color, reflectance, fluorescence) properties; the most common macerals are liptinite, vitrinite, and inertinite.

Maturation Thermal evolution of organic matter

(kerogen and bitumen) in the subsurface; maturity is measured by various physical and chemical maturation parameters (e.g., vitrinite reflectance, carbon preference index of *n*-alkanes, biological marker compound ratios).

Migration Movement of hydrocarbons in the subsurface from a source rock into a carrier bed (primary migration, mainly pressure controlled) and along a carrier bed into a reservoir rock (secondary migration, mainly buoyancy driven) to form a hydrocarbon accumulation.

Organic facies A mappable unit of sediment or sedimentary rock with (largely) uniform properties of its organic matter content.

Petroleum General term for fossil fuels, comprising both crude oil and natural gas.

Reservoir rock Porous rock unit (e.g., sandstone, limestone, or dolomite) suitable to contain a hydrocarbon accumulation (crude oil, gas) if structurally closed and sealed by an impermeable cap rock.

Source rock Organic matter-rich rock unit which under the influence of geological time and temperature has generated and expelled hydrocarbons (effective source rock) which may eventually have migrated into a reservoir to form a petroleum accumulation; a potential source rock (e.g., an oil shale) has not yet reached a sufficiently high maturation level for hydrocarbon generation.

ORGANIC GEOCHEMISTRY studies the distribution, composition, and fate of organic matter in the geosphere on both bulk and molecular levels, combining aspects of geology, chemistry, and biology. The biomass of decayed organisms is incorporated into sediments in the aquatic and terrestrial environment and is eventually buried to greater depth depending on the subsidence and depositional history of a sedimentary basin. Under the influence of temperature, catalysis, and (in shallow sediment layers) microbial activity, the organic matter undergoes a succession of complex chemical reactions which may ultimately lead to the formation of fossil fuels such as crude oil, natural gas, or coal.

I. ACCUMULATION OF ORGANIC MATTER IN THE GEOSPHERE

A. Bioproductivity and the Global Organic Carbon Cycle

The development of life on earth has a history of more than 3 billion years. Procaryotic organisms started biosynthesis in the Precambrian, and remnants of their biomass still are present in ancient sedimentary rocks of that time. Mass production of organic matter did not occur, however, be-

fore oxygenic photosynthesis was established worldwide in the oceans by cyanobacteria about 2 billion years ago. At the same time, the initially reducing atmosphere gradually became enriched in free oxygen, which was the basis for the evolution of higher forms of life.

Cyanobacteria and unicellular algae prevailed until the early Paleozoic, about 550 million years (m.y.) ago. During the Cambrian, Ordovician, and Silurian, a variety of phytoplanktonic organisms evolved, and these contributed significantly to the organic carbon production on earth. At the end of the Silurian (400 m.y.), land plants started to settle the continents and spread out in the Devonian. During the Late Carboniferous (300 m.y.), the first massive woods occurred, which are documented today by a first maximum of extensive coal seams. Evolution of the terrestrial higher plant kingdom then continued in the Permian with the gymnosperms and finally during the Early Cretaceous (125 m.y.) with the angiosperms, which are the dominant species of the modern flora.

The production and decay of biomass is a cyclic process of a complex food chain in connection with the carbon dioxide (CO_2) in the atmosphere and the oceans (Fig. 1). This part of the global carbon cycle is the smaller one and has a relatively rapid turnover. It is connected via a small "leak" to the massive carbon reservoir of the lithosphere. Estimates of the mass of carbon stored in the different sections of the global carbon cycle are shown in Table I. It is evident that the world's oceans contain the bulk of the carbon in the productivity cycle, although only about 2.5% of these 40,000 billion metric tons of carbon are organic. Carbon in the atmosphere is of the same order of magnitude as the biomass on the continents. Most of the organic carbon in the lithosphere is finely disseminated in sediments. In the most optimistic estimates, less than one-thousandth of it occurs as fossil fuels in various forms.

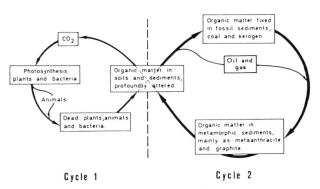

FIGURE 1 The two major organic carbon cycles on earth. Organic carbon is mainly recycled in cycle 1. The crossover from cycle 1 to cycle 2 is a tiny leak through which passes less than 0.1% of the primary organic biomass. [Reprinted with permission from Tissot, B. P., and Welte, D. H. (1984). "Petroleum Formation and Occurrence," 2nd ed., p. 10, Springer-Verlag, Berlin and New York. Copyright 1984 Springer-Verlag.]

TABLE I Estimates of Carbon in Different Compartments of the Global Carbon Cycle[a]

Compartment	Mass of carbon ($\times 10^9$ tons)
Atmosphere	700
Oceans	
Inorganic	39,000
Organic	1,000
Continents	
Biomass	600
Humic matter in soils	2,000
Lithosphere	
Inorganic	50,000,000
Organic	20,000,000
Including fossil fuels (upper limits)	
Oil, gas, tar sands	800
Coal	6,300
Oil shales	9,500

[a] After Grassl, H., Maier-Reimer, E., Degens, E. T., *et al.* (1984). *Naturwissenschaften* **71**, 129–136.

Compared to the figures in Table I, the amount of carbon involved in the annual turnover is relatively small. The gross exchange rate between atmosphere and oceans is of the order of 80 billion metric tons/yr, whereas the annual production of biomass just exceeds 100 billion metric tons. Less than 250 million metric tons of carbon reach the marine sediments every year. The anthropogenic influence by fossil fuel consumption and cement production amounts to about 5 billion metric tons/yr.

B. Chemical Composition of the Biomass

Despite the diversity of living organisms in the biosphere, their chemical composition can be confined to a limited number of principal compound classes. Many of them are also represented in fossil organic matter, although not in the same proportions as they occur in the biosphere because of their different stabilities toward degradation during sedimentation and diagenetic transformation.

1. Nucleic Acids and Proteins

Nucleic acids, such as ribonucleic acids (RNA) or desoxyribonucleic acids (DNA), are biological macromolecules that carry genetic information. They consist of a regular sequence of phosphate, sugar (pentose), and a small variety of base units, i.e., nitrogen-bearing heterocyclic compounds of the purine or pyrimidine type. During biosynthesis, the genetic information is transcribed into sequences of amino acids which occur as peptides, proteins, or enzymes in the living cell. These macromolecules vary widely in the number of amino acids and thus in molecular weight. They account for most of the nitrogen compounds in the cell and serve in such different functions as the catalysis of biochemical reactions and the formation of skeletal structures (e.g., shells, fibers, muscles).

During sedimentation of decayed organisms, nucleic acids and proteins are readily hydrolysed chemically or enzymatically into smaller, water-soluble units. Amino acids occur in rapidly decreasing concentrations in Recent and subrecent sediments, but may also survive in small concentrations in older sediments. A certain proportion of the nucleic acids and proteins reaching the sediment surface may be bound into the macromolecular kerogen network of the sediments and there become protected against further rapid hydrolysis.

2. Saccharides, Lignin, Cutin, and Suberin

Sugars are polyhydroxylated hydrocarbons which together with their polymeric forms (oligosaccharides, polysaccharides) constitute an abundant proportion of the biological material, particularly in the plant kingdom. Polysaccharides occur as supporting units in skeletal tissues (cellulose, pectin, chitin) or serve as an energy depot, for example, in seeds (starch). Although polysaccharides are largely water insoluble, they are easily converted by hydrolysis to C_5 (pentoses) and C_6 sugars (hexoses) and thus in the sedimentary environment will have a fate similar to that of the proteins.

Lignin is very common in supporting plant tissues, where it occurs as a three-dimensional network together with cellulose. Lignin essentially is a macromolecular condensation product of three different propenyl (C_3) phenols. It is fairly well preserved during sedimentation and is abundant in humic fossil organic matter.

Cutin and suberin are lipid biopolymers of variable composition which are part of the protective coatings on the outer surfaces of all higher plants. Chemically, cutin and suberin are closely related polyesters composed of various fatty and hydroxy fatty acid monomers. Both types of biopolymers are sensitive to hydrolysis and, thus, after sedimentation have only a moderate preservation potential.

3. Insoluble, Nonhydrolyzable Highly Aliphatic Biopolymers

Insoluble, nonhydrolyzable aliphatic biopolymers were discovered in higher plants and in algal cell walls as well as in their fossil remnants in sediments. These substances are called algaenan, cutan, and suberan according to their origin from algae or co-occurrence with cutin and suberin in extant higher plants. They consist of long-chain aliphatic hydroxy esters crosslinked via ether bonds. Pyrolysis and rigorous chemical degradation methods are the only way of decomposing these highly aliphatic biopolymers. This explains why they are preferentially preserved in sediments.

4. Lipids (Monomers)

Biogenic compounds that are insoluble in water but soluble in organic solvents such as chloroform, ether, or acetone are called lipids. They are common in naturally occurring fats, waxes, resins, and essential oils but in a wider sense also include membrane components or certain pigments. The low water solubility of the lipids is responsible for their higher survival rate during sedimentation compared to other biogenic compound classes such as amino acids or sugars. The analysis of lipids and their conversion products in geological samples is a major objective of molecular organic geochemistry.

Various saturated and unsaturated fatty acids are the lipid components of fats, where they are bound to glycerol to form triglyceride esters (see Fig. 2 for examples of chemical structures of lipid molecules). In waxes, fatty acids are esterified with long-chain alcohols instead of glycerol. In addition, waxes contain unbranched, long-chain saturated hydrocarbons (n-alkanes) with a predom-

inance of odd carbon numbers (e.g., C_{27}, C_{29}, or C_{31}), in contrast to the acids and alcohols which show an even-carbon-number predominance.

Isoprene, a branched diunsaturated C_5 hydrocarbon, is the building block of a large family of open-chain and cyclic isoprenoids or terpenoids (Fig. 2). Monoterpenes (C_{10}) with two isoprene units are enriched in essential oils of higher plants. Farnesol, an unsaturated C_{15} alcohol, is an example of a sesquiterpene with three isoprene units. Diterpenes (C_{20}) are common constituents of higher plant resins, where they occur mostly as bi- to tetracyclic components (e.g., abietic acid; Fig. 2). The acyclic diterpene phytol is esterified to chlorophyll a and, thus is widely distributed in the green pigments of plants. Sesterterpenes (C_{25}) are of comparatively lower importance.

Cyclization of squalene (or its epoxide) is the biochemical pathway to the formation of a variety of pentacyclic triterpenes (C_{30}) consisting of six isoprene units. Triterpenoids of the oleanane, ursane, lupane, and other less common types are restricted to higher plants. The geochemically most important and widespread triterpenes are from the hopane series, such as diploptene, which occurs in ferns, cyanobacteria, and other eubacteria. The predominant source of hopanoids is bacterial cell wall membranes which contain bacteriohopanetetrol (and closely related molecules) as rigidifiers. This C_{35} compound has a sugar

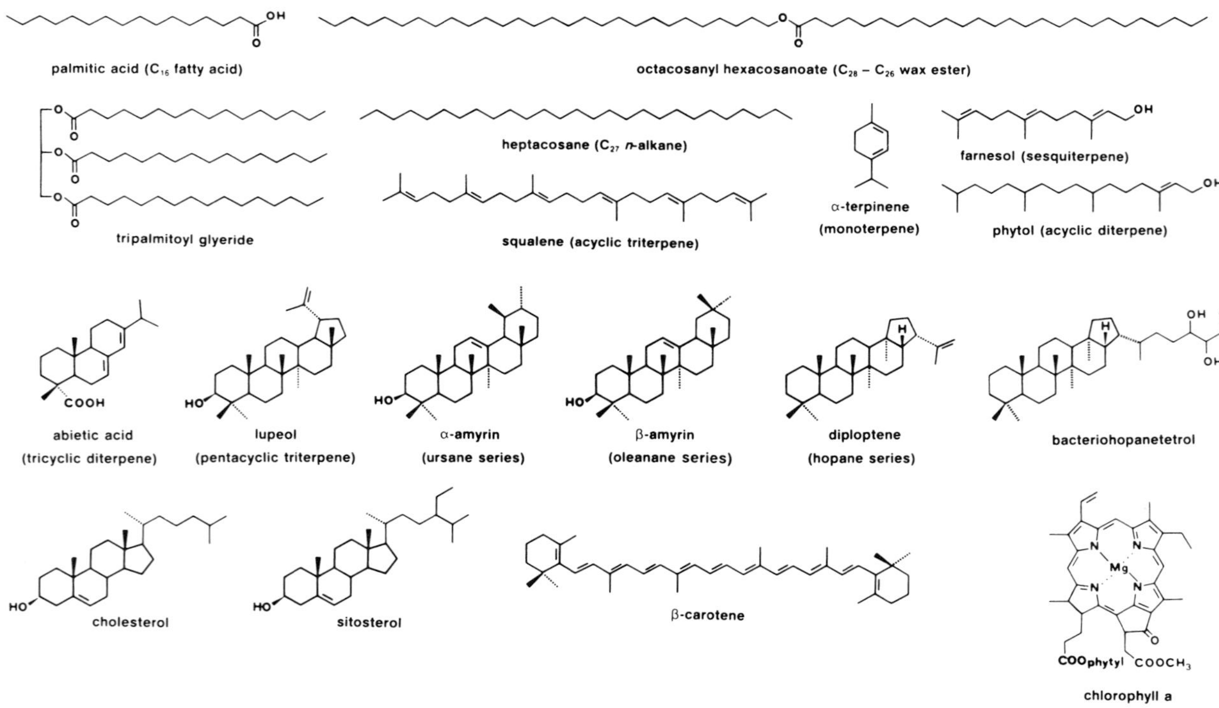

FIGURE 2 Structural formulas of representative organic lipids in living organisms and in the geosphere.

moiety attached to the triterpane skeleton (Fig. 2). The widespread distribution of bacteria on earth through time makes the hopanoids ubiquitous constituents of all fossil organic matter assemblages.

Steroids are tetracyclic compounds which are also derived biochemically from squalene epoxide cyclization, but have lost, in most cases, up to three methyl groups. Cholesterol (C_{27}) probably is the most widespread sterol in animals and many plants. Higher plants frequently contain C_{29} sterols (e.g., sitosterol) as the most abundant compound of this group. Steroids together with terpenoids are typical examples of biological markers (molecular fossils) because they contain a high degree of structural information which is retained after sedimentation. Thus, they provide an unambiguous link between the sedimentary organic matter and the biosphere. Steroids and terpenoids are among the most extensively studied classes of organic compounds in the geosphere.

Carotenoids, red and yellow pigments of algae and land plants, are the most important representatives of the tetraterpenes (C_{40}). Due to their extended chain of conjugated double bonds (e.g., β-carotene; Fig. 2), they are labile and found only in relatively small concentrations in surface sediments. Aromatization probably is one of the dominating diagenetic pathways in the alteration of the original structure in the sediment.

A second pigment type of geochemical significance are the porphyrins (tetrapyrroles) and related, not fully aromatized chlorins. Most porphyrins in sediments and crude oils are derived from the green plant pigment chlorophyll *a* and similar compounds in bacteria. It was the detection of porphyrins in crude oils and sediment extracts more in the 1930s that laid the basis for modern molecular organic geochemistry and provided the first strong molecular evidence for the biogenic origin of crude oils.

C. Factors Influencing the Preservation and Accumulation of Organic Matter in Sediments

The accumulation of sediments containing organic matter occurs primarily in the aquatic environment. The amount and composition of the organic component, i.e., the organic facies, are controlled by a number of complex, mutually dependent factors. A high primary bioproductivity alone is not a sufficient criterion for the enrichment of organic matter in the sediment. Preservation of the decayed biomass is as decisive as a balance with the geological conditions and the supply of mineral components.

In a given depositional environment, organic matter may be contributed by the biomass production in the aquatic system, the transport of terrigenous organic mat-

ter from the continent by river discharge, or, usually less important, eolian dust, and by the erosion and recycling of previously deposited sediments. The relative importance of these factors will have an influence on the bulk composition of organic matter in the sediment. In the marine environment, bioproductivity is particularly high in areas of coastal upwelling, where coast-parallel winds bring cold water from deeper layers, enriched in nutrients, to the surface. This is presently significant on the western continental margins off southern California, Peru, Namibia, and on parts of the Indian Ocean margins. The supply of terrigenous organic matter to the ocean depends on the climate on the neighboring continent. A warm, humid climate supports both terrestrial vegetation and river discharge.

Consumption in the food chain or oxidation in the presence of free oxygen leads to (partial) mineralization of the organic matter and thus limits its preservation. Both processes may occur during transport to the site of deposition, during settling through the water column, and in the upper sediment layers. It is estimated that less than 0.1% of the annually produced biomass on average reaches the sediment. Different types of organic matter have different stabilities toward degradation. Marine organic matter, consisting essentially of proteins, saccharides, and lipids, is more labile than terrestrial organic matter, which contains chemically more stable compounds and lipids protected by relatively resistant supporting tissues. In fact, *n*-alkanes from higher-plant waxes frequently survive long-distance transport to remote sites in the oxic deep ocean. In oxic environments, such selective mineralization leads to an organic facies in the sediment that is not representative of the local aquatic ecosystem.

In addition to the oxygen content of the water column, water depth and sediment accumulation rates determine the preservation of organic matter. Residence time of organic particles in the water column depends on water depth as well as particle size and shape. Well-rounded large, rapidly settling particles such as fecal pellets have a great chance to reach the sediment surface even under oxic conditions. A high sedimentation rate favors organic matter preservation due to rapid burial and protection of the labile material from the oxic regime. A very high supply of mineral matter, on the other hand, may cause dilution of organic matter and thus lead to sediments with a relatively low organic matter content despite a high accumulation (preservation) rate.

Under certain conditions, an anoxic water body may develop. High surface-water bioproductivity leads to depletion of oxygen in the underlying water, because oxygen is consumed in the organic matter degradation of decayed organisms. If there is no adequate water circulation to replenish the oxygen, the whole water column down to the

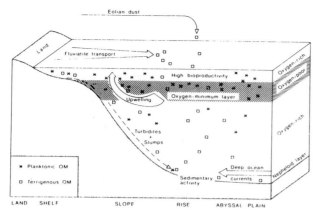

FIGURE 3 Schemalic diagram showing the factors that control organic matter accumulation of marine continental shelf, slope, and rise sediments. [Reprinted with permission from Rullkötter, J., et al. (1983). In "Coastal Upwelling—Its Sediment Record, Part B" (J. Thiede and E. Suess, eds.), pp. 467–483, Plenum Press, New York. Copyright 1983 Plenum Press.]

sediment surface may become oxygen depleted. This typically happens during eutrophication of small lakes. The Black Sea is an example of a stagnant oceanic basin with an anoxic bottom water mass. A shallow sill largely separates the Black Sea from the Mediterranean Sea and suppresses the exchange of deep-water masses. In addition, density stratification due to heavy saline bottom water and lighter fresh water from river discharge at the surface prevents vertical mixing, so that over time all oxygen below a water depth of about 150 m has been consumed, and free hydrogen sulfide (H_2S) is now present in the deep anoxic waters of the Black Sea.

In the open ocean, a different process of oxygen depletion may occur (Fig. 3). A midwater oxygen-minimum layer develops in high-productivity areas, due to coastal upwelling. Where this layer impinges on the continental shelf or slope, preservation of organic matter in the sediment is highly favored. Resedimentation processes (turbidite flow, slumping) may transport such organic matter-enriched sediments down the slope to the deep, oxygenated ocean floor. Nevertheless, rapid burial of this redeposited material may protect the organic matter in it from destruction. Deep oceanic currents resuspending sediment material into a so-called nepheloid layer have an opposite effect on organic matter preservation.

II. TRANSFORMATION OF ORGANIC MATTER IN THE SEDIMENT

A. Diagenesis, Catagenesis, and Metagenesis

Organic matter that has been incorporated into a sediment will undergo a series of geochemical alterations which

are summarized schematically in Fig. 4. The microbial and chemical transformation of the original biomolecules, which started during sedimentation, is continued in the upper sediment layers. A series of still not well understood reactions, formally termed polymerization and condensation, leads to the initial "geopolymers," i.e., fulvic acids, humic acids, and humin. On a molecular level, these components are not well defined, but they are differentiated on the basis of their acid or base solubility. They are certainly not true polymers, but rather macromolecules with a variety of different building blocks which may include largely intact biopolymers. Among these, the insoluble, nonhydrolyzable, highly aliphatic biopolymers are of great importance. Kerogen is formed toward the end of the first transformation sequence, which is called diagenesis. This happens at relatively low temperatures (<50°C) and involves the activity of microorganisms inhabiting the sediment. Kerogen is a complex organic geomacromolecule which is soluble in neither organic solvents nor in acids or bases. Its chemical composition at this stage is strongly heterogeneous and varies greatly depending on the organic facies.

A certain proportion of the organic constituents inherited from the biological system are small compounds extractable from the sediment with organic solvents. This bitumen fraction contains a number of (geo)chemical fossils,

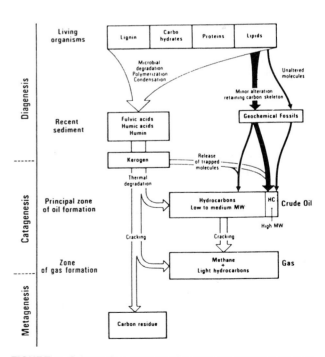

FIGURE 4 Schematic summary of geochemical transformation reactions of organic matter in the geosphere. [Reprinted with permission from Tissot, B. P., and Welte, D. H. (1984). "Petroleum Formation and Occurrence," 2nd ed., p. 94, Springer-Verlag, Berlin and New York. Copyright 1984 Springer-Verlag.]

i.e., unaltered molecules or compounds only slightly altered with essentially intact carbon skeletons compared to the precursor biomolecules. These may be hydrocarbons from higher-plant waxes or functionalized compounds such as alcohols, fatty acids, sterols, and triterpenes. The bitumen in Recent sediments accounts for only a very small portion of the total organic content.

When a sediment becomes buried more deeply in the course of continuing sedimentation and basin subsidence, it may enter the catagenesis stage, i.e., the principal zone of hydrocarbon formation (Fig. 4). Under the influence of increasing temperature (>50°C) and geological time, hydrocarbons and other compounds are released from the macromolecular kerogen network, and they add to the bitumen phase. This is the process of crude oil generation. Hydrocarbons span a wide range from low to high molecular weight, and compounds containing heteroatoms (nitrogen, oxygen, and/or sulfur, i.e., other than carbon and hydrogen) can still account for a significant proportion of the total bitumen. In a later catagenesis stage, further thermal hydrocarbon generation from the kerogen as well as thermal cracking of the bitumen yields mainly gaseous hydrocarbons, at the end predominantly methane.

Finally, the organic matter in sediments may reach the metagenesis stage at great depth and very high temperatures exceeding 150°C. At this stage, only methane is stable, and the kerogen is converted to a carbon residue which may start some crystalline ordering leading ultimately to graphite.

B. Early Transformation of Kerogen and Biological Markers

1. Kerogen

The initial composition of a specific kerogen depends on the source organisms that contributed to the sediment. Because the kerogen structure is not readily accessible by conventional molecular analysis, other parameters have to be used to describe it chemically. One possibility is to measure the elemental composition in terms of carbon, hydrogen, and oxygen content. If the results are displayed in a so-called van Krevelen diagram of H/C versus O/C atomic ratios, three initial kerogen compositions (types I, II, and III) are distinguished together with three evolution pathways representing the changes in elemental composition during diagenesis and catagenesis (Fig. 5). Type I kerogen is of pure algal (and eventually bacterial) origin and rare. The very high H/C atomic ratio reflects the lipid content rich in saturated aliphatic structures. Type II kerogen is derived from other aquatic organisms such as phytoplankton and zooplankton and has an intermediate H/C atomic ratio. Terrestrial higher plant organic matter (type III) is characterized by a considerably lower H/C

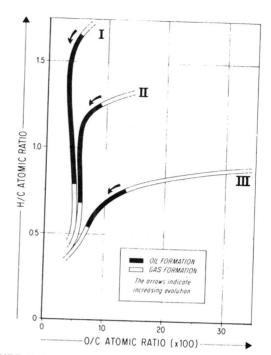

FIGURE 5 Structural evolution of various types of kerogen. [Reprinted with permission from Vandenbroucke, M. (1980). *In* "Kerogen" (B. Durand, ed.), pp. 415–443, Editions Technip, Paris. Copyright 1980 Editions Technip.]

atomic ratio but is enriched in oxygen, as can be expected from the high cellulose and lignin contents.

The evolution pathways in Fig. 5, particularly for kerogen types II and III, show a decrease in the oxygen content in the first phase before the onset of oil generation. This represents the loss of small heteroatomic molecules such as carbon dioxide (CO_2) and water (H_2O) during the diagenesis stage. In addition, ammonia (NH_3) or hydrogen sulfide (H_2S) may be lost. The effect on the kerogen structure is a continuing condensation of the macromolecule with an increasing formation of carbon–carbon bonds.

2. Biological Markers

The developments in modern analytical techniques, particularly the combination of computerized capillary gas chromatography-mass spectrometry, allowed the identification of many of the compounds present in the complex mixture of organic molecules in sedimentary bitumen. This was used to follow the stepwise conversion of specific precursor compounds of biological origin into hydrocarbon products typically present in crude oils. The capillary gas chromatograph in this respect is used to separate the mixtures according to polarity and molecular weight by partitioning them between a gaseous phase, usually a stream of helium, and a stationary phase, usually a silicone gum coated to the inner wall of the fused silica capillary.

The effluent of the gas chromatograph enters a mass spectrometer, which generates spectra that provide information on the molecular weight and the chemical structure of the different compounds.

Molecules with a high degree of structural complexity are particularly informative and thus suitable to the study of geochemical reactions. They provide the chance of relating a certain geochemical product to a specific biological precursor. Such compounds, which retain a significant portion of their biogenic structural integrity throughout the geochemical reaction sequence, are called biological markers or molecular or (geo)chemical fossils.

The scheme in Fig. 6 illustrates the fate of sterols in sediments. Although it looks complex, Fig. 6 shows only a few selected structures of more than 200 biogenic steroids and geochemical conversion products presently known from sediments. The reaction sequence includes hydrogenation, dehydration, skeletal rearrangement, and aromatization. Steroids are only one example to illustrate these reactions, which likewise happen to many other compound classes.

The biological precursor in Fig. 6 is cholesterol (structure 1, R = H), a widely distributed steroid in the biosphere. Hydrogenation of the double bond leads to the saturated cholestanol (2). This reaction occurs in the upper sediment layers or even in the water column by microbial action. Elimination of water gives the unsaturated hydrocarbon 3. At the end of diagenesis, the former unsaturated

steroid alcohol 1 will have been transformed to the saturated sterane hydrocarbon 4 after a further hydrogenation step. An alternative route to the saturated sterane 4 is via dehydration of cholesterol, which yields the diunsaturated compound 5. Hydrogenation of one double bond leads to a mixture of two isomeric sterenes (6; double-bond isomerization indicated by dots), further hydrogenation affords the saturated hydrocarbon 4. A sterically modified form of this molecule, e.g., 7, is formed during catagenesis at elevated temperatures. A side reaction from sterene 6 is a skeletal rearrangement leading to diasterene 8, where the double bond has moved to the five-membered ring and two methyl groups (represented by the bold arrows) are now at the bottom part of the ring system. This reaction is catalyzed by acidic clays. Thus, diasterenes (8) and the corresponding diasteranes (9), formed by hydrogenation during late diagenesis, are found in shales but not in carbonate rocks, which lack the catalytic activity of clays.

An alternative diagenetic transformation pathway of steroids leads to aromatic hydrocarbons. The diolefin 5 is a likely intermediate on the way to the aromatic steroid hydrocarbons 10–14. Compounds 10 and 11 are those detected first in the shallowest sediment layers. They are labile and do not survive diagenesis. During late diagenesis, the aromatic steroid hydrocarbon 12 appears in the sediments, but is also stable enough to survive elevated temperatures and to be found in crude oils. As for the

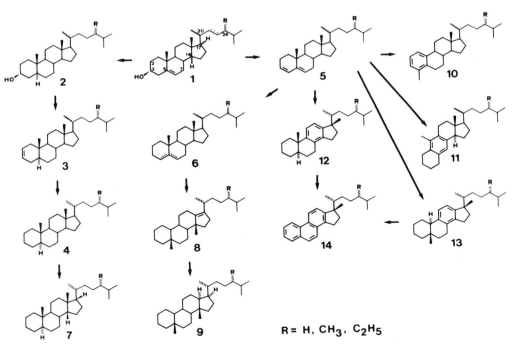

R = H, CH₃, C₂H₅

FIGURE 6 Diagenetic and catagenetic alteration of steroids.

saturated steranes, there is a corresponding rearranged monoaromatic steroid hydrocarbon (13). During catagenesis, the monoaromatic steroid hydrocarbons are progressively transformed into the triaromatic steroid 14 before the steroid record is completely lost by total destruction of the carbon skeleton at even higher temperatures.

Inorganic sulfur species (hydrogen sulfide, polysulfides, or elemental sulfur) play an important role in the early diagenesis of low-molecular-weight lipids carrying double bonds or alcohol groups. Common examples of such compounds known to react with sulfur are phytol (see Fig. 2), equivalent isoprenoids with two or more double bonds occurring in bacteriochlorophylls, sterols, and hopanoids (e.g., bacteriohopanetetrol; see Fig. 2). Intramolecular sulfur addition leads to the formation of thiophenes, i.e., five-membered aromatic rings with a sulfur atom in the ring system, and related saturated thiolanes (five-membered rings) or thianes (six-membered rings). Addition of more than one sulfur atom affords di- or trithianes (two or three sulfur atoms in a saturated ring) or bi- or trithiophenes (two or three thiophene units in a single molecule). Intermolecular sulfur incorporation, on the other hand, forms high-molecular-weight organic sulfur compounds. Because the inorganic sulfur species react with only a certain portion of the sedimentary lipids, the molecular geochemical information is partitioned selectively between the low-molecular-weight lipid fraction and the macromolecular sulfur species. In order to have access to the full range of biological markers, e.g., in the paleoenvironmental assessment of a sediment, it is advisable to remove the cross-linking sulfur bridges by hydrogenation with a Raney nickel or nickel boride catalyst before analysis.

Hydrogen sulfide (H_2S) is formed by microbial reduction of sulfate in seawater under strictly anoxic conditions. In clayey sediments, H_2S is preferentially trapped as pyrite (FeS_2) by reaction with Fe^{2+} ions; thus, an excess of inorganic sulfur compounds or a low iron concentration (as in carbonates) is an additional prerequisite for the diagenetic formation of organic sulfur compounds in sediments. At a later stage of organic matter transformation in sediments, organic sulfur compounds may become constituents of crude oils.

C. Thermal Hydrocarbon Generation

In the diagenesis stage, organic matter is immature with respect to hydrocarbon generation. The hydrocarbons present in the bitumen fraction, apart from methane (CH_4) formed by methanogenic bacteria in shallow sediments, largely constitute geochemical fossils inherited from the biosphere. On a molecular level, *n*-alkanes, for example, can show the typical distribution pattern of higher-plant waxes with a maximum chain length above C_{25} and a predominance of the odd-carbon-number species. Similarly, cycloalkanes and aromatic hydrocarbons show a predominance of compounds with a close link to biogenic precursors.

The onset of oil generation occurs at the beginning of the catagenesis stage. The total amount of hydrocarbons in the sediments increases as a consequence of thermal cracking of kerogen. Initially, the weaker heteroatomic bonds are cleaved preferentially, but with increasing temperature due to deeper burial the stronger carbon–carbon bonds are broken as well. Also, relatively large units (asphaltenes) are released from the kerogen at the early catagenetic stage.

The generation of oil-type hydrocarbons goes through a maximum ("peak of oil generation"), after which gasoline-range and gaseous hydrocarbons in the wet gas zone become the main reaction products for two reasons. First of all, the kerogen during catagenesis has become depleted in large aliphatic moieties, and only the remaining smaller units can now be released. Second, the oil-type hydrocarbons in the bitumen fraction are thermally cracked into smaller, gaseous compounds.

In the dry gas zone, almost exclusively methane (CH_4) is generated from the kerogen. At the same time, methane is also the ultimate hydrocarbon product formed by cracking of oil released from the kerogen at earlier generation stages. During the oil and gas generation phase, the H/C atomic ratio of the kerogen constantly decreases as illustrated in Fig. 5, and the kerogen becomes an inert carbon residue.

The amount of hydrocarbons generated from a specific source rock depends on the nature of the kerogen and on the extent to which the source rock has passed through the hydrocarbon generation zones under the given geological conditions, of which time and temperature are most important. The potential of a kerogen to generate hydrocarbons depends largely on its hydrogen richness as a limiting factor. Kerogen types I and II (Fig. 5) of dominantly aquatic origin are able to generate more hydrocarbons per unit carbon than the relatively hydrogen-poor kerogen type III.

The kerogen quality also influences the types of hydrocarbons generated. Aquatic organic matter and microbial biomass, both rich in aliphatic units, are the sources of abundant oil-type bitumen. On the other hand, coaly organic matter of terrigenous origin, which to a large extent is composed of aromatic ring systems with only short aliphatic chains, is more likely to yield predominantly gaseous hydrocarbons.

Because the rate of hydrocarbon generation is controlled by chemical reaction kinetics, both time and temperature are important for the transformation of kerogen. Although the geochemical reactions leading to the formation of hydrocarbons may be complicated by an influence of pressure and catalytic activity of the mineral rock matrix, the general expectation is that source rocks in relatively young sedimentary basins need higher temperatures for hydrocarbon generation than those in older basins. This is confirmed by observations in nature which showed that hydrocarbon generation in the Upper Tertiary (10 m.y.) of the Los Angeles Basin starts at about 115°C, compared to 60°C in the Lower Jurassic (180 m.y.) of the Paris Basin. Since according to chemical reaction kinetics the influence of time should be linear but that of temperature exponential, the compensation of temperature by time is limited such that very old potential source rocks, which have experienced only very low temperatures, will not generate significant amounts of hydrocarbons even over very long geological times.

D. The Chemical Composition of Crude Oils Related to Geological Factors

1. Crude Oil Classification

Crude oils can be characterized by their bulk properties as well as their chemical composition. Distillation of crude oils provides fraction profiles over a certain boiling-point range. Total oils as well as distillation fractions can be described in terms of density, viscosity, refractive index, sulfur content, or other bulk parameters. A widely used parameter in the oil industry is the API gravity, which is inversely proportional to density. Conventional crude oils range from about 20 to 45° deg API gravity. Oils with lower API gravities are called heavy oils, those with higher API gravities are condensates, although these are not rigorously defined boundaries.

In organic geochemical studies, crude oils are conventionally separated into fractions of different polarities using column or thin-layer chromatography. Usually, after removal of the low-boiling components at a certain temperature and under reduced pressure ("topping"), asphaltenes (i.e., high-molecular-weight polar components) are precipitated by the addition of a nonpolar solvent (e.g., n-hexane). The soluble portion is then separated into saturated hydrocarbons, aromatic hydrocarbons, and a fraction containing the polar, heteroatomic compounds (NSO compounds or resins).

Of course, the gross fractions allow further subfractionation in order to facilitate subsequent studies on a molecular level using less complex mixtures. For example, the saturated hydrocarbons can be treated with 5-Å molecular sieve or urea for the removal of n-alkanes, leaving behind a fraction of branched and cyclic alkanes. Aromatic hydrocarbons are often further fractionated according to the number of aromatic rings.

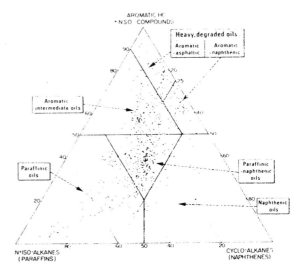

FIGURE 7 Ternary diagram showing the composition of six classes of crude oils based on the analysis of 541 oils. [Reprinted with permission from Tissot, B. P., and Welte, D. H. (1984). "Petroleum Formation and Occurrence," 2nd ed., p. 419, Springer-Verlag, Berlin and New York. Copyright 1984 Springer-Verlag.]

The ternary diagram in Fig. 7 shows the composition of six classes of crude oils based on a large number of analyses. Crude oils with more than 50% saturated hydrocarbons are called "paraffinic" or "naphthenic," depending on the relative contents of normal and branched alkanes (paraffins) and cycloalkanes (naphthenes). Paraffinic oils are light, but many have a high viscosity at room temperature due to the high content of long-chain wax alkanes. A high wax content is often an indication of a strong terrigenous contribution to the organic matter in the source rock. Such oils are commonly generated from deltaic sediments, as in Indonesia and West Africa. However, type I kerogen may also generate highly paraffinic crude oils, as is the case in the Green River Formation of the Uinta Basin, Utah. The crude oil property in the latter case is probably related to the fact that nonhydrolyzable highly aliphatic units in kerogen are derived not only from higher land plants but also from aquatic organisms (e.g., *Bottryococcus* algae; see Sect. 1.A.3). The most common oil type is paraffinic naphthenic (40% of the oils in Fig. 7), whereas naphthenic oils are scarce and often the result of microbial degradation in the reservoir (see Sect. III.C). All oils rich in saturated hydrocarbons commonly have a low sulfur content (<1%).

Aromatic intermediate oils contain 40–70% aromatic hydrocarbons and a high proportion of resins and asphaltenes. They are rich in sulfur and mostly heavy. Many crude oils from the Middle East (Saudi Arabia, Kuwait, Iraq, etc.) are included in this category, which is the second most important class in Fig. 7. The aromatic naphthenic and aromatic asphaltic classes are represented mostly by altered crude oils. They are heavy, but the sulfur content may vary according to the original type of the crude oil. The Cretaceous heavy oils of Athabasca (Canada) are typical examples of these last two classes.

2. Geological Factors Influencing Crude Oil Composition

The depositional environment of the source rock, its thermal evolution, and secondary alteration processes are the most important factors determining the composition of crude oils. Among the environmental factors, those that influence the nature of the organic matter in the source rock and its mineral composition are of primary significance.

Although hydrocarbon source rocks are deposited under aquatic conditions, they may contain varying amounts of land-derived organic matter. The terrestrial contribution can be significant, particularly in intracontinental basins and in the deltas of large rivers, which may extend far into the open sea. Continental organic matter (type III kerogen) is rich in cellulose and lignin which, due to their oxygen content, are not considered to contribute much to oil formation. The subordinate lipid fraction together with the biomass of sedimentary microorganisms incorporated into the source rock yields crude oils which are rich in aliphatic units (from wax esters, fats, etc.), i.e., straight-chain and branched alkanes (paraffins). Polycyclic naphthenes, particularly steranes, are present in very low concentration. Total aromatic hydrocarbons are also significantly less abundant than in crude oil derived from marine organic matter, as is the sulfur content.

Marine organic matter (usually type II kerogen) produces oils of paraffinic naphthenic or aromatic intermediate type (Fig. 7). The amount of saturated hydrocarbons is moderate, but isoprenoid and polycyclic alkanes, such as steranes (from algal steroids) and hopanes (from membranes of eubacteria), are relatively more abundant than in oils from terrigenous organic matter. Kerogen derived from marine organic matter, particularly when it is very rich in sulfur, is particularly suited to release resin- and asphaltene-rich heavy crude oils at a very early stage of catagenesis. Type II kerogens are preferentially deposited where the environmental conditions are favorable for organic matter preservation (anoxic water column in silled basins or in areas of coastal upwelling) and where the continental runoff is limited for physiographical or climatic reasons.

The sulfur content of crude oils shows a close relationship to the type of mineral matrix in the source rocks. Organic matter in sediments consisting of calcareous (e.g., from coccolithophores or foraminifera) or siliceous shell fragments (e.g., from diatoms or radiolaria) of decayed planktonic organisms and at the same time containing abundant organic matter is enriched in sulfur. The reason for this is that under the anoxic conditions which are required to preserve organic matter, sulfate-reducing bacteria form hydrogen sulfide (H_2S). This may react with the organic matter, and the sulfur will become incorporated into the kerogen. Examples are the Monterey Formation with the related crude oils produced onshore and offshore southern California and many of the carbonate source rocks of the Middle East crude oils.

In clastic rocks containing an abundance of detrital clay minerals, the iron content usually is high enough to remove most of the H_2S generated by the sulfate-reducing bacteria through formation of iron sulfides. Because terrigenous organic matter is commonly deposited together with detrital mineral matter (e.g., in deltas), waxy crude oils derived from type III kerogen usually are depleted in sulfur.

E. Coal and Oil Shales

Most organic carbon in sediments is finely disseminated. Even in hydrocarbon source rocks responsible for oil and gas accumulations, the organic carbon contents often do not exceed 10%. Apart from the secondary accumulations such as oil or gas reservoirs and tar sands, particular enrichment of organic matter in sediments occurs in the form of coal and oil shales.

1. Coal

Coal forms by the massive accumulation of land plants under a wet climate in a slowly subsiding basin. Peat formation is the first step in the process. It is essential that the decaying plant material be rapidly covered with water to prevent its oxidative decomposition. Furthermore, peat development requires the supply of plant remains over extended periods of time and a delicate balance of organic matter supply and tectonic subsidence. In terms of geological age, coal formation is possible since the advent of terrestrial higher plants in the Devonian (380 m.y. ago).

Peat formed by chemical and microbial diagenetic reactions in the first coalification step is transformed into lignite (brown coal) in the course of increasing burial. Elevated temperatures at greater depth over geological times are required to convert lignite into bituminous coals, which ultimately reach the anthracite stage at the end of the coalification process. Table II summarizes the different coalification stages together with selected physical parameters which can be used to differentiate among the coal ranks. Mean vitrinite reflectance, R_m, is the most widely used coalification parameter. It is determined on vitrinite

TABLE II Different Stages of Coalification According to the American Society for Testing and Materials (ASTM) Classification

	Approximate		
Rank	Reflectance range, R_m, (%)	Volatile matter d.a.f. (%)[a]	Carbon content d.a.f. (%)[a]
Peat	0.20–0.25	>64	<60
Lignite	0.25–0.4	55	65
Subbituminous coal (C, B, A)	0.4–0.65	48	73
High-volatile bituminous coal (C, B, A)	0.50–1.10	40	80
Medium-volatile bituminous coal	1.1–1.5	26	87
Low-volatile bituminous coal	1.5–1.9	18	89
Semianthracite	1.9–2.5	10	90
Anthracite	2.5–ca.5	5	92
Meta-anthracite	5	<3	>92

[a] d.a.f. = dry and ash free.

particles (vide infra), which can be observed on polished coal surfaces under the microscope. Vitrinite reflectance is low in the peat and lignite stages; it increases from about 0.4% to 1.9% in the bituminous coal ranks, and to values in excess of 5% in the anthracite stage. At the same time, the content of volatile matter decreases drastically, and the carbon content of coals increases to more than 90%.

Coal, particularly at low ranks, is not homogeneous. This is conceivable from heterogeneity of the contributing plant materials and the variability of the chemical and microbial reactions after deposition due to changes in water coverage, acidity (pH value) or redox (Eh) potential in the depositional environment. Coal is composed of a number of organic particle types (macerals) distinguishable either macroscopically or, more importantly, microscopically.

Coal petrography, as this study is called, differentiates among three main maceral groups, each containing several macerals (Table III). The microscopic investigations are performed on polished coal surfaces in the reflected-light mode, and macerals are distinguished by their reflectance, shape, and structure. Occasionally, fluorescent light observations after ultraviolet irradiation are used to enhance the signal of low-reflecting liptinites. The same techniques are applied by organic petrographers to characterize dispersed organic particles in sediments.

Vitrinite is the most abundant maceral group in humic coals. Vitrinites are coalification products of humic substances which essentially originate from the ligno-

cellulose components of cell walls; they appear gray in reflected light under the microscope. Telinite shows cell structures of wood or bark, whereas collinite is structureless. Vitrinite in hard coals corresponds to huminite in lignite or peat, where a different terminology for the macerals and a stronger subclassification of the maceral group is used (examples are given in Table III). Small detrital particles of vitrinitic material, intimately mixed into a structureless groundmass and difficult to identify, are called vitrodetrinite.

Liptinites, less abundant in coals but major constituents of petroleum source rocks, are derived from hydrogen-rich, lipid-bearing plant materials. Among these are spores and pollen (sporinite), cuticles and cuticular layers within the outer walls of leaves or stems (cutinite), resins and waxes (resinite), and corkified cell walls of barks, roots, or fruits containing suberin as a protecting agent against dessication (suberinite). Alginite is rare in coals, with the exception of the occasionally occurring sapropelic coals, which were not deposited as peats but under more anoxic conditions as subaquatic muds rich in plankton and water plants.

Inertinites have the same biological precursors as vitrinites but when occurring in the same coal have a higher reflectivity. Fusinite and semifusinite show distinct cellular structures of wood. Many of these particles appear to be fossil equivalents of charcoal, and ancient forest fires may have contributed to their formation. Sclerotinites are mostly fungal remains, although other sources have been invoked for part of the sclerotinites, particularly in older (Carboniferous and Permian) coals. Macrinite is an amorphous substance which formed by oxidation of gelified plant material. All inertinites are highly carbonized and have no potential for hydrocarbon generation.

2. Oil Shales

Oil shales are organic matter-rich sediments which have never been buried to greater depth. They contain little extractable bitumen but generate oil when heated to

TABLE III Maceral Classification of Coals Showing Main Maceral Groups and Selected Macerals[a]

Maceral group	Maceral
Vitrinite (huminite)	Telinite (ulminite), vitrodetrinite, collinite (gelinite)
Liptinite	Sporinite, cutinite, resinite, suberinite, alginite, liplodetrinite, bituminite
Inertinite	Fusinite, semifusinite, macrinite, sclerotinite, inertodetrinite

[a] Special terms used for brown coals (lignite) only are given in parentheses.

temperatures of about 500°C. Oil shales may be true shales with clay minerals as the dominant inorganic component, but commonly include carbonate as well as siliceous rocks. The organic carbon content in most cases exceeds 20%, and the H/C atomic ratio is very high (close to or above 1.5).

The organic matter often consists entirely of freshwater (*Bottryococcus*) or marine (*Tasmanites*) algae and bacterial biomass (type I organic matter; cf. Fig. 5). Because oil shales have not entered the catagenetic stage, the shale oil produced by heating contains considerable amounts of sulfur- and nitrogen-bearing compounds in addition to pyrolytically formed unsaturated hydrocarbons (olefins), which are not normally present in crude oils. There is no basic geochemical difference between oil shales and petroleum source rocks. The former just have not been buried deeply enough for thermal hydrocarbon generation under natural conditions.

Oil shales were deposited in large lake basins such as the Eocene Green River shales in Utah and Wyoming, which are several hundred meters thick, in shallow seas on stable platforms such as the Permian Irati oil shale in Brazil, or in small lakes, bogs, and lagoons, where they are often associated with coal beds. Oil shales range in age from the Lower Paleozoic (Cambrian) to the Tertiary. The world's largest oil shale reserves are in the United States, Brazil, and Russia.

III. MIGRATION AND ACCUMULATION OF OIL AND GAS

A. Primary and Secondary Migration

The formation of oil and gas from kerogen occurs in relatively tight hydrocarbon source rocks. The liquid and gaseous low-molecular-weight compounds generated in this process occupy a larger volume than the solid macromolecular precursor material. Because of the generally low porosity of the source rocks, the amount of hydrocarbons that can be contained in them is limited. The generation of hydrocarbons leads to an increase in pressure that eventually causes expulsion of the hydrocarbons into more porous sediments overlying or underlying the source rock. This movement of hydrocarbons is called primary migration (Fig. 8).

The oil expelled continues its movement through the wider pores of the more permeable carrier rock. Driving forces are water flow in the subsurface and buoyancy of the hydrocarbons that have a lower density than the saline waters (brines) saturating the pores of sedimentary rocks. During this secondary migration, petroleum may ultimately reach a reservoir rock that has porosity and

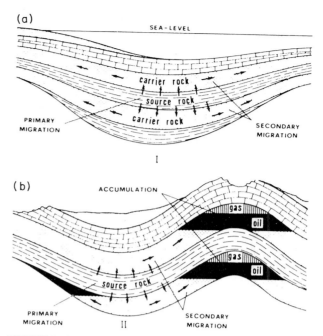

FIGURE 8 Schematic representation of primary and secondary migration leading to oil and gas accumulation: (a) initial phase and (b) advanced stage. [Reprinted with permission from Tissot, B. P., and Welte, D. H. (1984). "Petroleum Formation and Occurrence," 2nd ed., p. 294, Springer-Verlag, Berlin and New York. Copyright 1984 Springer-Verlag.]

permeability characteristics similar to those of the carrier bed but is sealed by an impermeable barrier, the cap rock (Fig. 8). In this way hydrocarbon accumulations are formed.

The most important mechanism for primary migration is transport of petroleum compounds in a continuous hydrocarbon phase with pressure as the driving force. The formation of a separate oil or gas phase in deeply buried, highly compacted sediments is favored by the fact that the amount of pore water present is low and that all or most of it is tightly bound to the mineral surface. This immobility of the pore water facilitates hydrocarbon movement. Furthermore, mineral surfaces may be covered with organic matter (kerogen network) or even be "oil wet," which further enhances transport of the apolar petroleum components. If very high pressures develop in a source rock, microfracturing may occur, which will periodically increase the migration flow toward more porous media, such as larger fractures within the source rock, or an adjacent carrier rock. Diffusion is considered an effective primary migration mechanism only in exceptional cases and mainly for gaseous hydrocarbons over short distances. Transport of hydrocarbons in aqueous solution is no longer considered a mechanism for primary migration.

Primary (and probably also secondary) migration leads to fractionation of the bitumen initially generated in the source rocks. Polar compounds tend to be preferentially retained by adsorption on mineral surfaces, and very large polar compounds such as asphaltenes may have a low chance for migration at all. Thus, crude oil in a reservoir will be depleted in polar heteroatomic compounds but enriched in hydrocarbons relative to the source rock, bitumen. The extent to which this occurs depends on the pore sizes of the migration pathway, the adsorption activity of the mineral surfaces, and the distance of migration. Fractionation among the hydrocarbons is relatively minor, as shown by numerous examples of successful oil/source rock correlation based on hydrocarbon compound ratios.

The efficiency and timing of primary migration has been a matter of debate for many years. It is now known that more than 80% of the hydrocarbons generated may be expelled during the main phase of oil formation. This contrasts sharply with the range of 10% to (exceptionally) 30% estimated in the earlier literature. Expulsion is more effective in thin source layers of a few centimeters or decimeters than in thick layers extending over several meters. In thick shales, part of the hydrocarbons retained may have to be cracked to lighter (gaseous) hydrocarbons before they get a chance to migrate out. The same occurs in rocks lean in organic matter, where the pressure buildup by liquid hydrocarbons is insufficient and not enough kerogen is present to form a continuous network supporting hydrocarbon drainage. For this reason, source rocks containing predominantly terrigenous organic matter with a low proportion of liptinite mainly generate gas, because the small amounts of liquid hydrocarbons initially formed cannot be expelled, but are later transformed into gas at higher thermal stress. Similarly, coals produce mainly gas (methane) because their microporosity and adsorption activity impede expulsion of initially formed liquid hydrocarbons. The high storage capacity of coals even for gas is well known from coal mining (pit gas).

Secondary migration of hydrocarbons through porous and permeable carrier beds and reservoir rocks is driven mainly by buoyancy due to the low densities of oil and gas. Secondary migration terminates when the retarding capillary pressures are stronger than the driving forces. If a barrier is encountered in the subsurface, then hydrocarbon accumulations will form. Otherwise, the hydrocarbons will seep out at the surface.

Oil and gas are trapped in the structurally highest position of a reservoir rock which is sealed with an essentially impermeable cap rock. Two general types of traps are commonly distinguished (Fig. 9). The sand anticline in Fig. 9a is an example of a structural trap. Other types of structural traps formed by tectonic events are associated, e.g., with salt domes, carbonate reefs, or basement highs. By far the

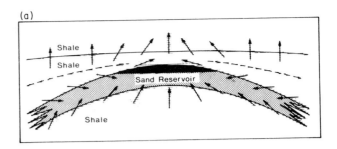

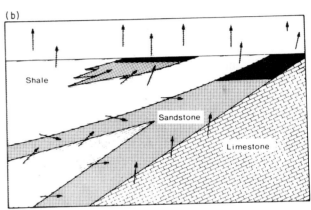

FIGURE 9 Schematic representation of two different types of hydrocarbon traps: (a) anticline as an example of a structural trap and (b) stratigraphic trap at an unconformity. [Reprinted with permission from Hunt, J. M. (1979). "Petroleum Geochemistry and Geology," pp. 223–224, Freeman, San Francisco. Copyright © 1979 W. H. Freeman and Company.]

largest amount of crude oil on earth is accumulated in anticline reservoirs, which are also the easiest to detect by geological and geophysical prospecting and are usually drilled first. The formation of stratigraphic traps is controlled by sedimentation processes. This type of trap may be associated with erosional surfaces (unconformities) as in Fig. 9b or with a change in the sediment facies (lithology). For example, when a coastal sand deposit grades into fine-grained shale farther offshore, the shale may act as a barrier to the hydrocarbons moving through the sand.

In contrast to primary migration, secondary migration may occur laterally over long distances of several tens of kilometers or even more. On the other hand, secondary migration can also be very short, e.g., from a shale source rock into an embedded sand lense. Over geological times, hydrocarbon accumulations are unstable. Tectonic events such as folding, faulting, or uplift associated with erosion can destroy hydrocarbon reservoirs completely or cause a redistribution of hydrocarbons into other reservoirs. In addition, cap rocks are not absolutely impermeable, and over

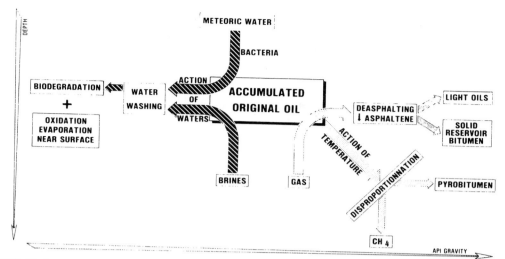

FIGURE 10 Crude oil alteration in the reservoir. [Reprinted with permission from Connan, J. (1984). *In* "Advances in Petroleum Geochemistry," Vol. 1. (J. Brooks and D. H. Welte, eds.), pp. 299–335, Academic Press, Orlando, FL. Copyright 1984 Academic Press.]

many millions of years the leakage rate may be sufficient to empty a reservoir.

B. Alteration of Hydrocarbons in the Reservoir

Even if reservoir conditions are tectonically stable, pooled hydrocarbons are susceptible to secondary alteration. The most important factors that can influence the chemical composition of petroleum in a reservoir are summarized in Fig. 10. Physical and chemical transformation processes will lead to either an increase or a decrease of the API gravity (or, conversely, the density). Alteration of hydrocarbons in the reservoir is quite common and is observed in many oil fields around the world. Petroleum can be altered to such an extent that the character of the source material is lost and the economic value decreases drastically.

Basin subsidence can bring a reservoir together with its accumulated hydrocarbons into a regime of higher temperature. Cracking reactions may then lead to the formation of lighter hydrocarbons (ultimately methane) and a residual pyrobitumen. Alternatively, long-term exposure of reservoired hydrocarbons to higher temperatures even without basin subsidence may cause cracking into smaller components. Within a basin this effect will be proportional to the reservoir depth.

Major amounts of gas entering a petroleum reservoir reduce the solubility of large polar compounds, and asphaltenes are precipitated in a process called desasphalting. The resulting oil is lighter than that originally present in the reservoir. The precipitated material forms a solid bitumen and may have a negative effect on the production properties of the reservoir by reducing the permeability.

Hydrodynamic flow of brines and the invasion of meteoric surface water can alter a crude oil by water washing, which removes the more water-soluble compounds from the reservoir. Meteoric water usually carries oxygen and in most cases microorganisms into the reservoir. This leads to microbial degradation of the crude oil, with a preferential removal of the readily metabolizable components. *n*-Alkanes are removed first, followed by simple branched and isoprenoid alkanes. Polycyclic saturated hydrocarbons are biodegraded less easily and at a much later stage. The biological alteration of the aromatic hydrocarbons is less well understood because the associated water washing coupled with the higher water solubility of the aromatic hydrocarbons often obscures the net effect of biodegradation. Biodegradation of crude oils in the reservoir is limited by the supply of oxygen and nutrients to the microorganisms and the maximum temperatures they can tolerate. Thus, biodegradation is rarely found at reservoir temperatures exceeding 70°C. In very shallow reservoirs, evaporation of volatile hydrocarbons to the surface strengthens the effect of biodegradation. As a result, heavy oils or tar sands may be formed, of which a famous example are the Athabasca tar sands in Alberta, Canada.

IV. ORGANIC GEOCHEMISTRY IN PETROLEUM EXPLORATION

The search for crude oil and natural gas is a costly and risky enterprise. After finding most of the easily detectable reservoirs in readily accessible areas, hydrocarbon exploration is now proceeding into new frontier areas including the deeper ocean and higher latitudes with their harsh climate and is looking for hitherto undetected smaller traps in previously explored basins. In order to keep the economic risk of drilling within reasonable limits, as many modern techniques as possible are used by the explorationists to help them in decision making. Sophisticated geophysical measurements (seismics) and satellite imaging are almost routine. Application of the principles of organic geochemistry in this respect has become a widely used tool in petroleum exploration.

A. The Hydrocarbon Potential of Sedimentary Rocks

The initial amount and the type of organic matter in a sedimentary rock determine its hydrocarbon potential. It is a question of the thermal evolution of such a sediment, if the potential is ever used actually to generate hydrocarbons. It is not easy to define a minimum organic carbon content for a sediment to be a source rock, but commonly accepted values are 0.5% for clastic rocks (claystones) and 0.3% for carbonates. This is based on a statistical overview of a great number of suspected or proven source rock samples worldwide, which shows that carbonates on average are leaner in organic matter than clastic rocks. Carbonates, on the other hand, often contain layers of organic matter-rich material derived from algal or bacterial mats. The restrictions imposed by the mechanisms of primary migration (expulsion; see Sect. III.A), however, may require higher minimum organic carbon values, probably on the order of 1.0–1.5%. Good to excellent hydrocarbon source rocks will have more than 2% organic carbon.

It can be roughly generalized that the quality of organic matter (hydrogen richness) in sediments increases with the total organic carbon content, because once favorable conditions for organic matter accumulation have been established, the preservation of the most labile lipid-rich material is enhanced. Examples are known, however, in which an organic carbon-rich sediment has little hydrocarbon potential because the kerogen it contains is woody or highly oxidized.

While the total organic carbon content is usually measured by combustion after removal of the carbonate carbon by acid treatment, several methods are available to assess the type of organic matter in a sediment. One way is by microscopic maceral analysis in which a specific hydrocarbon potential is ascribed to different macerals. Liptinites are believed to have a high potential for oil generation, whereas vitrinites may generate some oil but mainly gaseous hydrocarbons.

A standard pyrolysis method in petroleum geochemistry allows a quick assessment of the bulk hydrocarbon generation potential of a sediment. This so-called Rock-Eval pyrolysis uses a small amount of ground rock sample

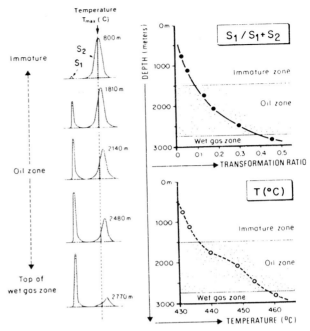

FIGURE 11 Characterization of source rocks by Rock-Eval pyrolysis. [Reprinted with permission from Tissot, B. P., and Welte, D. H. (1984). "Petroleum Formation and Occurrence," 2nd ed., p. 521, Springer-Verlag, Berlin and New York. Copyright 1984 Springer-Verlag.]

which is progressively heated to 550°C in a stream of helium. The hydrocarbon products released are monitored by combustion in a flame ionization detector. The first signal (S_1), observed at low temperature, corresponds to the volatile bitumen in the rock, whereas the second peak (S_2) in the pyrogram represents the compounds produced by thermal cleavage of the kerogen (Fig. 11, left). Separately, the carbon dioxide evolved from the kerogen during heating is also registered (not shown in Fig. 11). If S_2 and the carbon dioxide signal are normalized to the total organic carbon content of the sample, a hydrogen index and an oxygen index are obtained which are independent of the amount of organic matter in the sample and closely related to the elemental composition of the kerogen. They can be plotted in a van Krevelen-type diagram (Fig. 12). Like the H/C and O/C atomic ratios (cf. Fig. 5), which are more tedious to determine, the hydrogen and oxygen indices show the evolution pathways of different kerogen types. The quality of organic matter, i.e., the hydrocarbon potential of a source rock, is related directly to the value of the hydrogen index.

B. Maturity of Organic Matter

The thermal evolution of organic matter, a process called maturation, continuously changes the chemical

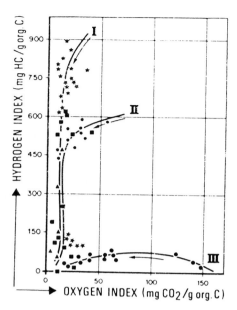

* Green River shales
• Lower Toarcian, Paris Basin
▲ Silurian_Devonian, Algeria-Libya
• Upper Cretaceous, Douala Basin
■ Others

FIGURE 12 Classification of kerogen types by using hydrogen and oxygen indices from Rock-Eval pyrolysis. [Reprinted with permission from Tissot, B. P., and Welte, D. H. (1984). "Petroleum Formation and Occurrence," 2nd ed., p. 512, Springer-Verlag, Berlin and New York. Copyright 1984 Springer-Verlag.]

composition and the related physical properties. In petroleum exploration, it is important to know if a source rock has already entered the hydrocarbon generation zone or if it is above or below this zone. An immature sediment when analyzed for the quality of its organic matter will show the full hydrocarbon potential. A source rock within the generation zone will show only that potential which remained after the earlier release of hydrocarbons now present as bitumen or migrated out of the source rock. A sediment below the hydrocarbon generation zone will have a negligible residual hydrocarbon potential. To determine the thermal evolution stage of organic matter in sediments, a number of bulk and molecular parameters are available and are often used in combination with each other to account for the complexity of geochemical reactions and the time–temperature interrelationship.

1. Bulk Maturity Parameters

Rock-Eval pyrolysis in addition to information about the type of organic matter and its hydrocarbon potential also provides a sensitive measurement of the thermal history of a sediment. With increasing maturation the maximum

of the pyrolysis peak (S_2) is shifted to higher temperature values (T_{max}) as shown in Fig. 11. At the same time, the relative proportion of the volatile material in the rock, represented by the S_1 peak, increases (Fig. 11). In most sediments there is a good correlation between the S_1 yield and the amount of extractable bitumen. The production index or transformation ratio [$S_1/(S_1 + S_2)$] is, however, affected by any hydrocarbons that have migrated out of or into the sediment, and it is recommended to study downhole trends (Fig. 11) rather than using single measurements. The sum of S_1 and S_2 represents the total genetic potential of a rock minus the amount of hydrocarbons which have left the rock during primary migration.

Coal ranks are most satisfactorily determined by the measurement of vitrinite reflectance. This method, developed in coal petrography, has been extended to kerogen particles disseminated in sedimentary rocks in order to assess the maturity level of organic matter. From the correlation of vitrinite reflectance with hydrocarbon yields from the extraction of source rocks, but also with other maturity parameters, it has become common to define zones of hydrocarbon generation in terms of vitrinite reflectance (Table IV). The boundaries are approximate and vary with kerogen type, due to variations in the chemical composition of the respective organic matter. Another influence comes from the heating rate which the organic matter has experienced. This is due to the fact that the chemical reactions responsible for hydrocarbon generation from kerogen and for vitrinite reflectance increase have different kinetics and, thus, respond differently to temperature increase and time.

Excellent hydrocarbon source rocks often lack vitrinite particles because all their organic material consists of the remains of aquatic organisms. Also, early bitumen generation in these rocks may impregnate the few eventually present vitrinite particles, which leads to a vitrinite reflectance lower than corresponding to the actual maturation stage. In these cases, maturity assessment is made by measuring vitrinite reflectance in adjacent, less prolific rocks which contain more terrigenous organic matter or by chemical maturation parameters.

TABLE IV Approximate Boundaries for Hydrocarbon Generation Zones in Terms of Vitrinite Reflectance

| | Vitrinite reflectance for different kerogen types (%) | | |
Generation zone	I	II	III
Onset of oil generation	0.7	0.5	0.7
Peak of oil generation	1.1	0.8	0.9
Beginning of gas zone	1.3	1.3	1.3
Dry gas zone	>2.0	>2.0	>2.0

2. Chemical Maturity Parameters

The chemical composition of bitumen is progressively changed by diagenetic and catagenetic reactions. A number of compound ratios, particularly in the hydrocarbon fractions, were found to vary systematically with increasing depth in a basin, and so these ratios can be used to measure the maturity of organic matter in a sediment. Only a few of them will be introduced here.

Because *n*-alkanes are usually the most abundant class of hydrocarbons in a sediment extract or a crude oil, much emphasis has been placed on the study of their relative compositional changes. The changes are most obvious when the sediment contains a certain amount of terrestrial higher plant material which contributes *n*-alkanes with a significant predominance of the odd carbon-number molecules that gradually disappears with increasing maturation. Several slightly differing mathematical expressions are being used to measure the maturity-dependent alteration of the *n*-alkane distribution. The most prominent among them is the carbon preference index (CPI), which uses a range of long-chain *n*-alkanes from C_{24} to C_{32}:

$$CPI = \frac{1}{2}[(C_{25} + C_{27} + \cdots + C_{31})/(C_{26} + C_{28} + \cdots + C_{32}) + (C_{27} + C_{29} + \cdots + C_{33})/(C_{26} + C_{28} + \cdots + C_{32})] \qquad (1)$$

The heights or the areas of the corresponding *n*-alkane peaks in the gas chromatograms of the saturated hydrocarbon fractions are taken to measure the ratio. The nature of the organic matter in a sediment will influence the carbon number preference, e.g., by the fact that many aquatic organisms do not synthesize long-chain *n*-alkanes, and if they do there is no predominance of either even or odd carbon numbers. Thus, in a sedimentary basin with varying organic matter supply the *n*-alkane patterns may change unsystematically and independent of the thermal evolution in closely spaced sediment layers. As a qualitative measure, however, the decrease of the carbon number predominance of *n*-alkanes has been applied successfully for maturity assessment in many cases.

More sophisticated analytical techniques are required to monitor the maturity-dependent changes of less abundant components among the hydrocarbons in the bitumen, but the specificity of the compounds and reactions studied justify these efforts. A group of compounds found to change their relative composition in a manner closely related to the increase of vitrinite reflectance through the oil generation zone and beyond are phenanthrene and the four isomeric methylphenanthrenes commonly found in sediments (**13**; numbers indicate the positions of methyl substitution).

13

With increasing depth, the relative abundances of 1- and 9-methylphenanthrene decrease and those of 2- and 3-methylphenanthrene increase due to the higher thermodynamic stability of the latter two isomers. A methylphenanthrene index (MPI) was introduced [Eq. (2); MP = methylphenanthrene, P = phenanthrene] and found to correlate positively with vitrinite reflectance at least in sediments which contain a major proportion of terrigenous organic matter:

$$MPI = 1.5[(2 - MP + 3 - MP)/(P + 1 - MP + 9 - MP)] \qquad (2)$$

This is shown for a well in the Elmworth gas field in Western Canada in Fig. 13, where the measured vitrinite reflectance R_m was used to calibrate the vitrinite reflectance equivalent R_c calculated from the methylphenanthrene index:

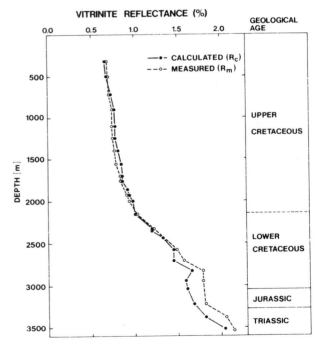

FIGURE 13 Calculated R_c and measured vitrinite reflectance R_m for selected rock samples from a well in the Elmworth gas field. [Reprinted with permission from Welte, D. H., Schaefer, R. G., Stoessinger, W., and Radke, M. (1984). *Mitteilungen des Geologisch-Paläontologischen Instituts der Universität Hamburg* **56**, 263–285. Copyright 1984 Universität Hamburg.]

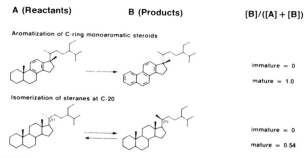

A (Reactants)	B (Products)	[B]/([A] + [B])

Aromatization of C-ring monoaromatic steroids

immature = 0

mature = 1.0

Isomerization of steranes at C-20

immature = 0

mature = 0.54

FIGURE 14 Examples of biological marker reactions used for maturity assessment of organic matter in sediments.

$$R_c = 0.60 \text{ MPI} + 0.40 \qquad (3)$$

Biological marker compound ratios based on the product/reactant relationship in apparent diagenetic and catagenetic reactions (like those illustrated in Fig. 6) are molecular maturation parameters widely applied today in petroleum geochemistry. Two of the most common examples are shown in Fig. 14. They monitor the extent of isomerization and aromatization of steroid hydrocarbons by measuring the relative concentrations of the reactants A and products B. In both cases, the initial concentration of the products is zero in immature sediments. The transformation of C-ring monoaromatic into triaromatic steroid hydrocarbons is irreversible and reaches completion within the hydrocarbon-generation zone. The isomerization of C_{29} steranes at C-20 in the side chain is an equilibrium reaction where the natural 20R configuration inherited from the biosynthesized precursor steroids is converted to the 20S configuration. The equilibrium which is reached close to the peak of oil generation is slightly shifted to the right-hand side of the equation in Fig. 14. Both parameters are measured using peak areas (or heights) of characteristic mass spectral fragments in a combined gas chromatographic/mass spectrometric analysis.

More recently it was found, however, that the representation of biological marker reactions as in Fig. 14 is oversimplified. Although aromatization of monoaromatic steroid hydrocarbons to the triaromatic species may indeed occur as shown, the product/reactant ratio at the same time is influenced by the thermal destruction of both the reactant and the product. In fact, at high thermal stress, destruction of the triaromatic steroids is faster than that of the monoaromatic steroids, which leads to a reversal of the compound ratio trend at high maturities after having approached the predicted end value. Isomerization of steranes at C-20 appears not to happen at all in the way shown in Fig. 14. The empirically observed change in the ratio of the 20R and 20S sterane isomers with increasing thermal stress in sediments rather is the combined effect of the neoformation of mixtures of 20R and 20S steranes in the

course of thermal hydrocarbon generation from kerogen, which add to the (comparatively small amount of) 20R steranes formed diagenetically in the bitumen fraction as shown for compound **4** in Fig. 6, and the thermal destruction of both isomers at higher temperatures. The isomer ratio of the 20R and 20S sterane mixture newly formed from kerogen follows a temperature-dependent trend with a progressively increasing proportion of the 20S isomer up to the assumed equilibrium ratio also theoretically predicted by molecular mechanics calculation. The reaction rate of thermal destruction is, however, slightly faster for the 20S isomer, so again a reversal of the compound ratio trend is observed at higher maturities. If this behavior is kept in mind, biological marker compound ratios can still be applied successfully in petroleum geochemistry, although some ambiguity may occasionally arise.

C. Oil/Source Rock Correlation

The main objective of oil/oil and oil/source rock correlation is to find out the genetic origin of reservoired hydrocarbons. In a first step, usually a correlation analysis among the various oils (or condensates, gases) from different reservoirs in a producing basin is performed to see if all the oils are uniform in their composition or if they belong to several families of different genetic origin. For each type or family of oils it has to be established if they are derived from a single source rock or if a multiple origin must be considered. The composition of the crude oils will provide some basic information on the characteristics of the organic matter and on the lithology in the source rock(s) as well as on the maturity of the organic matter in the source rock(s) at the time of hydrocarbon generation. Compositional variations due to different maturities must be distinguished from those due to different sources. In a second step, a source must be identified for each oil family by comparison with samples of all possible source rocks. If the conclusion is reached that more than one source characteristic can be attributed to different oil families accumulated in a basin, organofacies variations within a single source rock layer may be the explanation, as well as two or more source rock layers of different age.

If the source(s) of crude oil in a basin are known, this can have important implications for exploration and production strategies. The information will help to define exploration targets and will influence the development of a concept for the stepwise exploration of a sedimentary basin. A mass balance calculation based on volume and thermal maturity of the known hydrocarbon source rocks and a comparison with the already-detected oil in place will help to determine if further exploration is promising. In crude oil production, oil/oil correlations can provide information about the possible communication between different productive zones.

568

Oil/source rock correlation is based on the concept that oil (condensate or gas) found in a reservoir was generated from kerogen disseminated in a source rock. During expulsion, only a certain proportion of the generated petroleum compounds was released and a significant proportion stayed behind. A correlation thus should be possible between the kerogen in the source rock, the bitumen extractable from the source rock, and the hydrocarbons in the reservoirs. In order to account for the complex chemical and physical processes involved in the generation, migration, and accumulation of hydrocarbons in the subsurface, as many correlation parameters as possible should be used in any correlation study, and both bulk and compositional (molecular) correlation parameters should be considered.

A number of useful correlation parameters for different types of geochemical correlation studies are compiled in Table V. The only bulk parameters included are carbon, hydrogen, and sulfur isotope measurements on whole samples or fractions thereof. Other bulk parameters, such as the API gravity and the distribution of distillation fractions (both for oil/oil correlation), are susceptible to severe alteration during migration and to secondary processes in the reservoir. They should nevertheless not be neglected completely, and any differences observed should be explicable in terms of the geological context of the study area.

TABLE V Correlation Parameters Useful for Different Kinds of Correlation Studies[a]

Type of correlation	Correlation parameter
Gas/gas	Carbon and hydrogen isotopes ($^{13}C/^{12}C$, $^{2}H/^{1}H$)
	Ratios of light hydrocarbons (C_2–C_8)
Gas/oil	Carbon and hydrogen isotopes of single n-alkanes
	Ratios of light hydrocarbons (C_2–C_8)
Oil/oil	Carbon, hydrogen and sulfur isotopes ($^{13}C/^{12}C$, $^{2}H/^{1}H$, ^{34}S,^{32}S)
	Ratios of specific hydrocarbons (e.g., pristane/ phytane, pristane/n-C_{17})
	Homologous series (e.g., n-alkane distribution)
	Biological markers (e.g., steranes, triterpanes, aromatic steroid hydrocarbons, porphyrins)
Gas/source rock kerogen	Carbon and hydrogen isotopes
Oil/source rock bitumen	Oil/oil correlation parameters
Oil/source rock kerogen	Carbon and sulfur isotopes
	Oil/oil correlation parameters applied to the products of kerogen pyrolysis

[a] After Tissot, B. P., and Welte, D. H. (1984). "Petroleum Formation and Occurrence," 2nd ed., p. 548, Springer-Verlag, Berlin and New York.

Compositional correlation parameters should have characteristic compound distributions in both rocks and oils. They should be available over the proper molecular weight range (e.g., for gas/oil correlation) and possibly be present in compound classes of different polarities. They should not be too seriously affected by migration, thermal maturation, and bacterial alteration; i.e., under the given geological conditions the imprint of the source organic matter should be transferred into the hydrocarbon reservoir.

Biological markers offer one of the best correlation tools, first of all due to the high information content in their molecular structures. The characteristic distribution patterns of biological markers in crude oils often can be taken as "fingerprints" of the various organisms that contributed to the source rock. Biological markers have a narrow range of chemical and physical properties, particularly when they occur in the saturated hydrocarbon fractions. Thus, any fractionation occurring, e.g., during migration is likely to affect the whole group of biological markers and not any specific compound preferentially. In addition, biological markers are easily detected in very small concentration by the combination of computerized gas chromatography/mass spectrometry. The presence or absence of unusual biological markers in a crude oil or source rock bitumen is often highly significant, whereas other biological markers such as hopanes inherited from bacteria are ubiquitous in all geological samples and thus less specific. Many biological markers are not easily degraded in the course of microbial attack to reservoir hydrocarbons; i.e., they survive mild or even moderately severe biodegradation. The maturity effect on a great number of biological markers is known, and thus corrections and extrapolations to account for additional thermal stress are possible. There are a few disadvantages to the use of biological markers for oil/oil and oil/source rock correlation. These include the limited thermal stability of these compounds, i.e., their application is often restricted to the generation zone below the peak of oil generation. In addition, the low concentrations of the biological markers are sensitive to mixing effects in the subsurface and to contamination.

D. Experimental Simulation of Petroleum Formation

Studying the processes of petroleum formation by field observations is hampered by the fact that these processes occur in nature over periods of geological time and in an open system. Thus, many attempts were made in the past to simulate hydrocarbon generation in a well-defined confined system in the laboratory using elevated temperatures as a compensation for geological time. Hydrous pyrolysis has turned out to be the method which appears to mimic the natural processes most closely.

In this experiment, pieces of organic matter-rich rock are heated in an autoclave in the presence of water at temperatures between about 300°C and near the critical-point temperature of water (374°C) for periods of, typically, 3 days or longer. The products are an expelled oil-type phase floating on the water and a bitumen fraction retained in the rock chips. Figure 15 summarizes the variations in the amounts of kerogen, bitumen, and expelled oils for a series of experiments at different temperatures using a sample of Woodford Shale, a Devonian-Carboniferous source rock from Oklahoma, as substrate. Physically, chemically, and isotopically, the expelled oil is similar to natural crude oils. Hydrous pyrolysis experiments in this way have not only provided useful information on the bulk processes of petroleum formation, but also on their kinetics, on thermal maturity indices, and on primary migration.

E. Modelling of Geological and Geochemical Processes

The successful application of the principles of hydrocarbon generation and migration to petroleum exploration and the increasing knowledge about the chemical and physical parameters that determine the fate of organic matter in the subsurface have stimulated efforts in developing a more quantitative handling of the related geological and geochemical processes. A comprehensive quantitative approach in terms of a mathematical model would allow a more precise prediction of the quantities of hydrocarbons generated in a sedimentary basin and a better understanding of the timing of any event in the course of hydrocarbon generation, migration, and accumulation.

The first step in this direction was the calculation of the maturity of organic matter in a specific sediment based on reconstruction of the burial history and the knowledge (or assumption) of depth-dependent temperature variations in the basin. The basic concept was that generation of hydrocarbons follows simple first-order reaction kinetics, i.e., the reaction rate approximately doubles during each 10°C increase of temperature. The burial history curve, constructed using geological information, defines the time a sediment in the geological past spent within a certain depth (or temperature) interval. If the temperature factors for each 10°C temperature interval, increasing by exponentials of 2 in each step, are multiplied with the time (in millions of years) a time–temperature index (TTI) is obtained for each temperature interval. A summation of these index values provides a TTI value representing the total

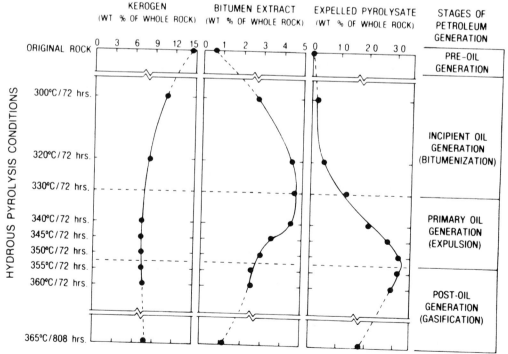

FIGURE 15 Variations in the amounts of kerogen, bitumen, and expelled oil (pyrolyzate) in series of hydrous pyrolysis experiments using Woodford Shale. [Reprinted with permission from M. D. Lewan (1983). *Geochim. Cosmochim. Acta* **47**, 1471–1479. Copyright 1983 Pergamon Press, Oxford.]

thermal evolution of the sediment. The TTI values must be calibrated against vitrinite reflectance (or hydrocarbon yields) to be a predictive, albeit simple, tool.

A considerably more sophisticated numerical treatment of hydrocarbon generation and migration starts with the three-dimensional determination of the development of a sedimentary basin using a deterministic model. Starting at a time T_0 when only the basement existed, sedimentation is followed by the model in predetermined time intervals until the present status (T_x) is reached (Fig. 16). The model continuously calculates the depth-related changes in such parameters as pressure, porosity, temperature, and thermal conductivity for each sediment layer. Mathematical equations control that a mass balance (e.g., sediment supply and water leaving the system) as well as an energy balance (e.g., heat flow from basement and energy loss to the atmosphere or as heated water) are maintained. The results of the model are compared to the real system (well and surface information), and the model approaches the real system by a series of iteration steps. As soon as the geological model is sufficiently accurate, organic matter maturity, hydrocarbon generation yields, and migration directions can be determined for each sediment layer in the basin and for each time slice in the past (Fig. 17). For this latter step a simple maturation calculation like that of the TTI values can be used. A more detailed maturation and hydrocarbon generation model will use ranges of activation energies for the different bond types in a kerogen and for different kerogen types. It will also take into account the fact that the reaction conditions will change with increasing maturation of the organic matter.

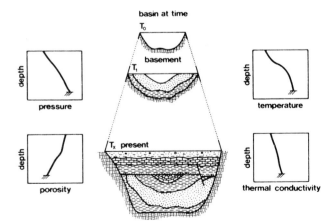

FIGURE 16 Evolution of a sedimentary basin from an initial condition at time T_0 to the present configuration at time T_x with schematic depth-related changes of physical parameters such as pressure, porosity, temperature, and thermal conductivity. [Reprinted with permission from Tissot, B. P., and Welte, D. H. (1984). "Petroleum Formation and Occurrence," 2nd ed., p. 576, Springer-Verlag, Berlin and New York. Copyright 1984 Springer-Verlag.]

V. ORGANIC GEOCHEMICAL TECHNIQUES IN PALEOENVIRONMENTAL AND PALEOCLIMATIC ASSESSMENT

While organic geochemistry in petroleum exploration focuses mainly on deeply buried sedimentary rocks, shallow sediments play an important role in understanding the global organic carbon cycle and in reconstructing past climatic and oceanographic developments. Since the mid-1970s, the international Deep Sea Drilling Project and later the Ocean Drilling Program have provided the numerous sediment cores necessary to perform such studies. In the following, a few selected approaches and techniques are outlined to illustrate how organic geochemistry contributes to these interdisciplinary objectives.

A. Marine versus Terrigenous Organic Matter

Even deep-sea sediments deposited in areas remote from continents usually contain a mixture of marine and terrigenous organic matter. For any investigation of marine paleoproductivity or marine organic matter preservation, the amount of terrigenous admixture has to be known. Furthermore, global or regional climate fluctuations have changed the pattern of continental runoff and ocean currents in the geological past. Being able to recognize variations in marine and terrigenous organic matter proportions, thus, is of great significance in paleoclimatic and paleoceanographic studies.

A variety of parameters are used to assess organic matter sources. Bulk parameters have the advantage that they are representative of the total organic matter, whereas molecular parameters address only part of the extractable organic matter, which in turn is only a small portion of total organic matter. Several successful applications of molecular parameters have shown that the small bitumen fraction may be representative of the total, but there are some examples for which this is not the case. On the other hand, oxidation of marine organic matter has the same effect on some bulk parameters as an admixture of terrigenous organic matter, because the latter is commonly enriched in oxygen through biosynthesis. It is, therefore, advisable to rely on more than one parameter, and to obtain complementary information.

1. C/N Ratio

Atomic carbon/nitrogen (C/N) ratios of living phytoplankton and zooplankton are around 6, freshly deposited marine organic matter ranges around 10, whereas terrigenous organic matter has C/N ratios of 20 and above. This difference can be ascribed to the absence of cellulose in algae and its abundance in vascular plants and to the fact that algae are rich in proteins.

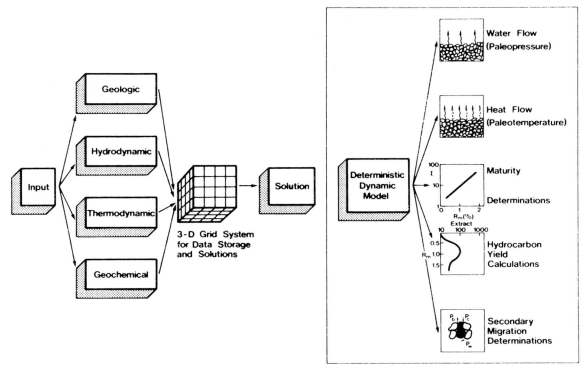

FIGURE 17 Development of a three-dimensional dynamic model. Groups of input parameters and results are schematically shown. [Reprinted with permission from Tissot, B. P., and Welte, D. H. (1984). "Petroleum Formation and Occurrence," 2nd ed., p. 578, Springer-Verlag, Berlin and New York. Copyright 1984 Springer-Verlag.]

Selective degradation of organic matter components during early diagenesis has the tendency to modify (usually increase) C/N ratios. Still, they are commonly sufficiently well preserved in shallow marine sediments to allow a rough assessment of terrigenous organic matter contribution. Care has to be taken in oceanic sediments with low organic carbon contents, because inorganic nitrogen (ammonia) released during organic matter decomposition may be adsorbed to the mineral matrix and add significantly to the amount of total nitrogen. The C/N ratio is then changed to values lower than those of the real marine/terrigenous organic matter proportions. This effect should be small in sediments containing more than 0.3% organic carbon.

2. Hydrogen and Oxygen Indices

Hydrogen indices from Rock-Eval pyrolysis (see Sect. IV.A) below about 150 mg hydrocarbons (HC)/total organic carbon (TOC) are typical of terrigenous organic matter, whereas HI values of 300–800 mg HC/g TOC are typical of marine organic matter. Oceanic sediments rich in organic matter sometimes exhibit values of only 200–400 mg HC/g TOC, even if marine organic matter strongly dominates. Oxidation during settling through a long water column in these cases has lowered the hydrogen content of the organic matter. Rock-Eval pyrolysis cannot be used for sediments with TOC < 0.3%, because of the so-called mineral matrix effect. If sediments with low organic carbon contents are pyrolyzed, a significant amount of the products may be adsorbed to the sediment minerals and are not recorded by the detector, thus lowering the hydrogen index.

3. Maceral Composition

If the morphological structures of organic particles in sediments are well preserved, organic petrographic investigation under the microscope is probably the most informative method to distinguish marine from terrestrial organic matter contributions to marine sediments by the relative amounts of macerals derived from marine biomass and land plants (see Sect. II.E.1). Many marine sediments, however, contain an abundance of unstructured organic matter which cannot easily be assigned to one source or the other.

4. Stable Carbon Isotope Ratios

Carbon isotope ratios are useful principally to distinguish between marine and terrestrial organic matter sources in sediments and to identify organic matter from different types of land plants. The stable carbon isotopic composition of organic matter reflects the isotopic composition of the carbon source as well as the discrimination (fractionation) between ^{12}C and ^{13}C during photosynthesis. Most plants, including phytoplankton, incorporate carbon into their biomass using the Calvin (C_3) pathway, which discriminates against ^{13}C to produce a shift in $\delta^{13}C$ values of about $-20‰$ from the isotope ratio of the inorganic carbon source. Organic matter produced from atmospheric carbon dioxide ($\delta^{13}C \approx -7‰$) by land plants using the C_3 pathway (including almost all trees and most shrubs) has an average $\delta^{13}C$ value of approximately $-27‰$. Marine algae use dissolved bicarbonate, which has a $\delta^{13}C$ value of approximately $0‰$. As a consequence, marine organic matter typically has $\delta^{13}C$ values varying between $-18‰$ and $-22‰$.

The "typical" difference of about $7‰$ between organic matter of marine primary producers and land plants has been used successfully to trace the sources and distributions or organic matter in coastal sediments. Unlike C/N ratios, $\delta^{13}C$ values are not significantly influenced by sediment texture, making them useful in reconstructing past sources of organic matter in changing depositional conditions. Figure 18 shows the fluctuations of organic matter $\delta^{13}C$ values in a short sediment core from a tidal flat caused by a variable admixture of eroded peat ($\delta^{13}C = -27‰$) to marine organic matter mainly from di-

atoms ($\delta^{13}C = -19‰$). If the carbon isotope values of the end members in a two-component system are known, the relative contributions can be easily determined by a simple linear mixing algorithm.

The availability of dissolved CO_2 in ocean water has an influence on the carbon isotopic composition of algal organic matter because isotopic discrimination toward ^{12}C increases when the partial pressure of carbon dixide (pCO_2) is high and decreases when it is low. Organic matter $\delta^{13}C$ values, therefore, become indicators not only of the origin of organic matter but also of changing paleoenvironmental conditions on both short- and long-term scales. For example, the $\delta^{13}C$ values of dissolved inorganic carbon (DIC; i.e., CO_2, bicarbonate, and carbonate) available for photosynthesis varies over the year with the balance between photosynthetic uptake and respiratory production. During spring and summer, when rates of photosynthesis are high, the isotope ratio of the remaining DIC is enriched in ^{13}C. In fall, when respiration is the dominant process, the $\delta^{13}C$ of DIC becomes more negative because organic matter is remineralized.

Fluctuations in the $\delta^{13}C$ values of sedimentary organic matter over the earth's history can, thus, be interpreted in terms of the productivity in the water column and the availability of DIC in a particular geological time period. In a study of sediments from the central equatorial Pacific Ocean spanning the last 255,000 years, it has been demonstrated that the carbon isotopic composition

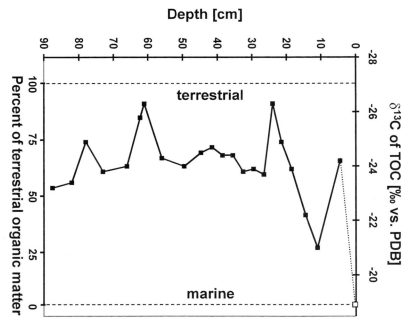

FIGURE 18 Variation of bulk $\delta^{13}C$ values of organic matter in a sediment core from a tidal flat in northern Germany. The variation is due to a variable mixture of marine organic matter ($\delta^{13}C = -19‰$) and of locally eroded peat (terrigenous organic matter; $\delta^{13}C = -27‰$). The relative percentages are calculated using a linear mixing algorithm.

of fossil organic matter depends on the exchange between atmospheric and oceanic CO_2. Changes with time can then be used to estimate past atmospheric carbon dioxide concentrations.

A new type of geochemical information became available with the development of a combined capillary column gas chromatography–isotope ratio mass spectrometry system. This instrument provides the carbon isotope signal ($^{13}C/^{12}C$) of single biological marker compounds in complex mixtures. The resulting data allows assignment of a specific origin to some of these compounds because, for example, the biomass of planktonic algae has a carbon isotope signal completely different from that of land plants or methanotrophic bacteria, due to the different carbon isotope ratios of the carbon sources used by these groups of organisms. This adds another dimension to the understanding of geochemical reaction pathways. The main advantage of the new technique is that a great number of compounds can be analyzed directly within a short time in small quantities and without the need of their tedious isolation.

B. Molecular Paleo-Seawater Temperature and Climate Indicators

1. Past Sea-Surface Temperatures (SST) Based on Long-Chain Alkenones

Palaeoceanographic studies have taken advantage of the fact that biosynthesis of a major family of organic compounds by certain microalgae depends on the water temperature during growth. The microalgae belong to the class of Haptophyceae (often also named Prymnesiophyceae) and notably comprise the marine coccolithophorids *Emiliania huxleyi* and *Gephyrocapsa oceanica*. The whole family of compounds, which are found in marine sediments of Recent to mid-Cretaceous age throughout the world ocean, is a complex assemblage of aliphatic straight-chain ketones and esters with 37 to 39 carbon atoms and two to four double bonds, but principally only the C_{37} methylketones with two and three double bonds (**14**) are used for past sea-surface temperature assessment.

It was found from the analysis of laboratory cultures and field samples that the extent of unsaturation (number of double bonds) in these long-chain ketones varies linearly with growth temperature of the algae over a wide temperature range. To describe this, an unsaturation index ($U_{37}^{K'}$) was suggested, which is defined by the concentration ratio of the two C_{37} ketones:

$$U_{37}^{K'} = [C_{37:2}]/[C_{37:2} + C_{37:3}]. \qquad (4)$$

Calibration was then made with the growth temperatures of laboratory cultures of different haptophyte species and with ocean water temperatures at which plankton samples had been collected. From these data sets, a a uniform calibration for the global ocean from 60°N to 60°S evolved and can be used to estimate the ocean surface temperatures in the geological past:

$$U_{37}^{K'} = 0.033T + 0.043, \qquad (5)$$

where T is the temperature of the oceanic surface water (SST) in degrees Celsius. The example in Fig. 19 shows pronounced SST variations between the different oxygen isotope stages (OIS) which represent cold (even numbers) and warm periods (odd number). The last glacial maximum (18,000 yr B.P.), for example, falls into OIS 2, the present interglacial (Holocene) is equivalent to OIS 1, and the last climate optimum (Eemian, 125,000 yr B.P.) occurred early in OIS 5.

2. ACL Index Based on Land Plant Wax Alkanes

In marine sediments, higher-plant organic matter can be an indicator of climate variations, both by the total amount indicating enhanced continental runoff during times of low sea level or of humid climate on the continent and by specific marker compounds indicating a change in terrestrial vegetation as a consequence of regional or global climatic variations. Long-chain *n*-alkanes are commonly used as the most stable and significant biological markers of terrigenous organic matter supply. The odd-carbon-numbered C_{27}, C_{29}, C_{31}, and C_{33} *n*-alkanes, major

$C_{37:2}$, heptatriaconta-15E,22E-dien-2-one

$C_{37:3}$, heptatriaconta-8E,15E,22E-trien-2-one

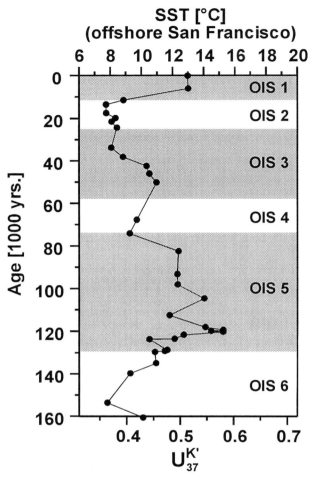

SST [°C]
(offshore San Francisco)

FIGURE 19 Changes of paleaosea surface temperatures (SST) on the California continental margin during the last 160,000 years estimated from the relative proportions of di- and triunsaturated long-chain ketones extracted from deep-sea sediments recovered by the Ocean Drilling Program.

$$ACL_{27-33} = (27[C_{27}] + 29[C_{29}] + 31[C_{31}] + 33[C_{33}])/$$

$$([C_{27}] + [C_{29}] + [C_{31}] + [C_{33}]), \qquad (6)$$

in which $[C_x]$ signifies the concentration of the *n*-alkane with *x* carbon atoms. Sedimentary *n*-alkane ACL values were demonstrated to be sensitive to past climatic changes. In Santa Barbara basin (offshore California) sediments from the last 160,000 years, the highest ACL values were found in the Eemian climate optimum (125,000 yr B.P.). The ACL variations recorded the climatic changes on the continent, because vegetation patterns on the continent responded rapidly to climatic oscillations, which were often characterized by drastic changes of temperature and precipitation. In addition, changes in continental precipitation significantly affected the degree of erosion and the transport of terrigenous detritus to the ocean.

SEE ALSO THE FOLLOWING ARTICLES

BIOMASS UTILIZATION, LIMITS OF • BIOPOLYMERS • CARBOHYDRATES • CARBON CYCLE • COAL GEOLOGY • COAL STRUCTURE AND REACTIVITY • ENVIRONMENTAL GEOCHEMISTRY • GEOCHEMISTRY • GEOCHEMISTRY, LOW-TEMPERATURE • PETROLEUM GEOLOGY • SEDIMENTARY PETROLOGY

BIBLIOGRAPHY

Engel, M. H., and Macko, S. A., eds. (1993). "Organic Geochemistry—Principles and Applications," Plenum Press, New York and London.

Hunt, J. M. (1995). "Petroleum Geochemistry and Geology," 2nd ed., Freeman, San Francisco.

Johns, R. B., ed. (1986). "Biological Markers in the Sedimentary Record," Elsevier, Amsterdam.

Peters, K. E., and Moldowan, J. M. (1993). "The Biomarker Guide—Interpreting Molecular Fossils in Petroleum and Ancient Sediments," Prentice-Hall, Englewood Cliffs, NJ.

Schulz, H. D., and Zabel, M., eds. (1999). "Marine Geochemistry," Springer-Verlag, Heidelberg.

Taylor, G. H., Teichmüller, M., Davis, A., *et al.* (1998). "Organic Petrology," Gebrüder Borntraeger, Stuttgart.

Tissot, B. P., and Welte, D. H. (1984). Petroleum Formation and Occurrence, 2nd ed., Springer-Verlag, Heidelberg.

Welte, D. H., Horsfield, B., and Baker, D. H. eds. (1997). "Petroleum and Basin Evolution," Springer-Verlag, Heidelberg.

components of the epicuticular waxes of higher plants, are often preferentially enriched in the marine environment.

The carbon-number distribution patterns of *n*-alkanes in leaf waxes of higher land plants depend on the climate under which they grow. The distributions show a trend of increasing chain length nearer to the Equator, but they are also influenced by humidity. Based on these observations, an average chain length (ACL) index was defined to describe the chain length variations of *n*-alkanes:

Geodesy

Petr Vaníček

University of New Brunswick

GLOSSARY

Coordinates These are the numbers that define positions in a specific coordinate system. For a coordinate system to be usable (to allow the determination of coordinates) in the real (earth) space, its position and the orientation of its Cartesian axes in the real (earth) space must be known.

Coordinate system In three-dimensional Euclidean space, which we use in geodesy for solving most of the problems, we need either the Cartesian or a curvilinear coordinate system, or both, to be able to work with positions. The Cartesian system is defined by an orthogonal triad of coordinate axes; a curvilinear system is related to its associated generic Cartesian system through some mathematical prescription.

Ellipsoid/spheroid Unless specified otherwise, we understand by this term the geometrical body created by revolving an ellipse around it minor axis, consequently known as an ellipsoid of revolution. By spheroid, we understand a spherelike body, which, of course, includes an ellipsoid as well.

Errors (uncertainties) Inevitable, usually small errors of either a random or systematic nature, which cause uncertainty in every measurement and, consequently, an uncertainty in quantities derived from these observations.

GPS Global Positioning System based on the use of a flock of dedicated satellites.

Gravity anomaly The difference between actual gravity and model gravity, e.g., the normal gravity, where the two gravity values are related to two different points on the same vertical line. These two points have the same value of gravity potential: the actual gravity potential and the model gravity potential, respectively.

Normal gravity field An ellipsoidal model of the real gravity field.

Positioning (static and kinematic) This term is used in geodesy as a synonym for the "determination of positions" of different objects, either stationary or moving.

Satellite techniques Techniques for carrying out different tasks that use satellites and/or satellite-borne instrumentation.

Tides The phenomenon of the earth deformation, including its liquid parts, under the influence of solar and lunar gravitational variations.

Encyclopedia of Physical Science and Technology, Third Edition, Volume 6

WHAT IS GEODESY?

Geodesy is a science, the oldest earth (geo-) science, in fact. It was born of fear and curiosity, driven by a desire to predict natural happenings and calls for the understanding of these happenings. The classical definition, according to one of the "fathers of geodesy" reads: "Geodesy is the science of measuring and portraying the earth's surface" (Helmert, 1880, p. 3). Nowadays, we understand the scope of geodesy to be somewhat wider. It is captured by the following definition (Vaníček and Krakiwsky, 1986, p. 45): "Geodesy is the discipline that deals with the measurement and representation of the earth, including its gravity field, in a three-dimensional time varying space." Note that the contemporary definition includes the study of the earth gravity field (see Section III), as well as studies of temporal changes in positions and in the gravity field (see Section IV).

I. INTRODUCTION

A. Brief History of Geodesy

Little documentation of the geodetic accomplishments of the oldest civilizations, the Sumerian, the Egyptian, the Chinese, and the Indian, has survived. The first firmly documented ideas about geodesy go back to Thales of Miletus (ca. 625–547 BC), Anaximander of Miletus (ca. 611–545 BC), and the school of Pythagoras (ca. 580–500 BC). The Greek students of geodesy included Aristotle (384–322 BC), Eratosthenes (276–194 BC)—the first reasonably accurate determination of the size of the earth, but not taken seriously until 17 centuries later—and Ptolemy (ca. 75–151 AD).

In the Middle Ages, the lack of knowledge of the real size of the earth led Toscanelli (1397–1482) to his famous misinterpretation of the world (Fig. 1), which allegedly lured Columbus to his first voyage west. Soon after, the golden age of exploration got under way and with it the use of position determination by astronomical means. The real extent of the world was revealed to have been close to Eratosthenes's prediction, and people started looking for further quantitative improvements of their conceptual model of the earth. This led to new measurements on the surface of the earth by a Dutchman Snellius (in the 1610s) and a Frenchman Picard (in the 1670s) and the first improvement on Eratosthenes's results. Interested readers can find fascinating details about the oldest geodetic events in Berthon and Robinson (1991).

At about the same time, the notion of the earth's gravity started forming up through the efforts of a Dutchman Stevin (1548–1620), Italians Galileo (1564–1642) and Borelli (1608–1679), an Englishman Horrox (1619–1641), and culminating in Newton's (1642–1727) theory of gravitation. Newton's theory predicted that the earth's globe should be slightly oblate due to the spinning of the earth around its polar axis. A Frenchman Cassini (1625–1712) disputed this prediction; consequently, the French Academy of Science organized two expeditions to Peru and to Lapland under the leadership of Bouguer and

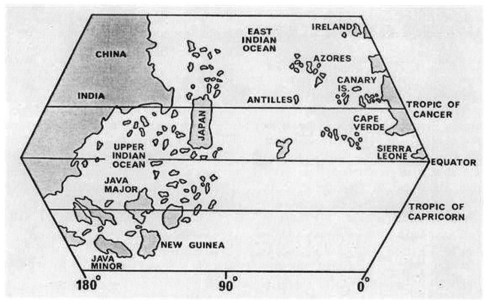

FIGURE 1 Toscanelli's view of the Western Hemisphere.

Maupertuis to measure two meridian arcs. The results confirmed the validity of Newton's prediction. In addition, these measurements gave us the first definition of a meter, as one ten-millionth part of the earth's quadrant.

For 200 years, from about mid-18th century on, geodesy saw an unprecedented growth in its application. Position determination by terrestrial and astronomical means was needed for making maps, and this service, which was naturally provided by geodesists and the image of a geodesist as being only a provider of positions, survives in some quarters till today. In the meantime, the evolution of geodesy as a science continued with contributions by Lagrange (1736–1813), Laplace (1749–1827), Fourier (1768–1830), Gauss (1777–1855), claimed by some geodesists to have been the real founder of geodetic science, Bessel (1784–1846), Coriolis (1792–1843), Stokes (1819–1903), Poincaré (1854–1912), and Albert Einstein. For a description of these contributions, see Vaníček and Krakiwsky (1986, Section 1.3).

B. Geodesy and Other Disciplines and Sciences

We have already mentioned that for more than 200 years geodesy—strictly speaking, only one part of geodesy, i.e., positioning—was applied in mapping in the disguise known on this continent as "control surveying." Positioning finds applications also in the realm of hydrography, boundary demarcation, engineering projects, urban management, environmental management, geography, and planetology. At least one other part of geodesy, geo-kinematic, finds applications also in ecology.

Geodesy has a symbiotic relation with some other sciences. While geodesy supplies geometrical information about the earth, the other geosciences supply physical knowledge needed in geodesy for modeling. Geophysics is the first to come to mind: the collaboration between geophysicists and geodesists is quite wide and covers many facets of both sciences. As a result, the boundary between the two sciences became quite blurred even in the minds of many geoscientists. For example, to some, the study of the global gravity field fits better under geophysics rather than geodesy, while the study of the local gravity field may belong to the branch of geophysics known as exploration geophysics. Other sciences have similar but somewhat weaker relations with geodesy: space science, astronomy (historical ties), oceanography, atmospheric sciences, and geology.

As all exact sciences, geodesy makes heavy use of mathematics, physics, and, of late, computer science. These form the theoretical foundations of geodetic science and, thus, play a somewhat different role vis-à-vis geodesy. In Fig. 2, we have attempted to display the three levels of relations in a cartoon form.

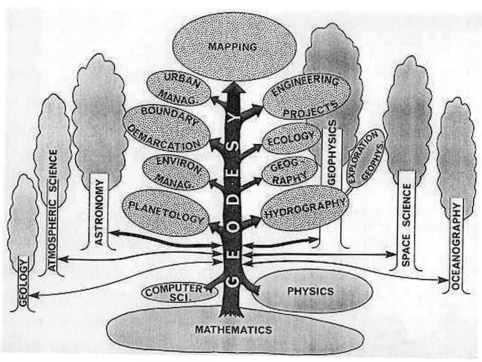

FIGURE 2 Geodesy and other disciplines.

C. Profession and Practice of Geodesy

Geodesy, as most other professions, spans activities ranging from purely theoretical to very applied. The global nature of geodesy dictates that theoretical work be done mostly at universities or government institutions. Few private institutes find it economically feasible to do geodetic research. On the other hand, it is quite usual to combine geodetic theory with practice within one establishment. Much of geodetic research is done under the disguise of space science, geophysics, oceanography, etc.

Of great importance to geodetic theory is international scientific communication. The international organization looking after geodetic needs is the International Association of Geodesy (IAG), the first association of the more encompassing International Union of Geodesy and Geophysics (IUGG) which was set up later in the first third of 20th century. Since its inception, the IAG has been responsible for putting forward numerous important recommendations and proposals to its member countries. It is also operating several international service outfits such as the International Gravimetric Bureau (BGI), the International Earth Rotation Service (IERS), Bureau Internationale des Poids et Mesures—Time Section (BIPM), the International GPS Service (IGS), etc. The interested reader would be well advised to check the current services on the IAG web page.

Geodetic practice is frequently subjugated to mapping needs of individual countries, often only military mapping needs. This results in other components of geodetic work being done under the auspices of other professional institutions. Geodesists practicing positioning are often lumped together with surveyors. They find a limited international forum for exchanging ideas and experience in the International Federation of Surveyors (FIG), a member of the Union of International Engineering Organizations (UIEO).

The educational requirements in geodesy would typically be a graduate degree in geodesy, mathematics, physics, geophysics, etc. for a theoretical geodesist and an undergraduate degree in geodesy, surveying engineering (or geomatics, as it is being called today), survey science, or a similar program for an applied geodesist. Survey technicians, with a surveying (geomatics) diploma from a college or a technological school, would be much in demand for field data collection and routine data manipulations.

II. POSITIONING

A. Coordinate Systems Used in Geodesy

Geodesy is interested in positioning points on the surface of the earth. For this task a well-defined coordinate system is needed. Many coordinate systems are being used in geodesy, some concentric with the earth (geocentric systems), some not. Also, both Cartesian and curvilinear coordinates are used. There are also coordinate systems needed specifically in astronomical and satellite positioning, which are not appropriate to describe positions of terrestrial points in.

Let us discuss the latter coordinate systems first. They are of two distinct varieties: the apparent places and the orbital. The apparent places (AP) and its close relative, the right ascension (RA) coordinate systems, are the ones in which (angular) coordinates of stars are published. The orbital coordinate systems (OR) are designed to be used in describing satellite positions and velocities. The relations between these systems and with the systems introduced below will be discussed in Section II.F. Interested readers can learn about these coordinate systems in Vaníček and Krakiwsky (1986, Chap. 15).

The *geocentric systems* have their z-axis aligned either with the instantaneous spin axis (cf., Section IV.B) of the earth (*instantaneous terrestrial system*) or with a hypothetical spin axis adopted by a convention (*conventional terrestrial systems*). The geocentric systems became useful only quite recently, with the advent of satellite positioning. The nongeocentric systems are used either for local work (observations), in which case their origin would be located at a point on the surface of the earth (topocentric systems called *local astronomic* and *local geodetic*), or for a regional/continental work in the past. These latter nongeocentric (near-geocentric) systems were and are used in lieu of geocentric systems, when these were not yet realizable, and are known as the *geodetic systems*; their origin is usually as close to the center of mass of the earth as the geodesists of yesteryear could make it. They miss the center of mass by anything between a few meters and a few kilometers, and there are some 150 of them in existence around the world.

Both the geocentric and geodetic coordinate systems are used together with *reference ellipsoids* (ellipsoids of revolution or biaxial ellipsoids), also called in some older literature "spheroids." (The modern usage of the term spheroid is for closed, spherelike surfaces, which are more complicated than biaxial ellipsoids.) These reference ellipsoids are taken to be concentric with their coordinate system, geocentric or near geocentric, with the axis of revolution coinciding with the z-axis of the coordinate system. The basic idea behind using the reference ellipsoids is that they fit the real shape of the earth, as described by the geoid (see Section III.B for details) rather well and can thus be regarded as representative, yet simple, expression of the shape of the earth.

The reference ellipsoids are the horizontal surfaces to which the geodetic latitude and longitude are referred, hence, the name. But to serve in this role, an ellipsoid (together with the associated Cartesian coordinate system) must be fixed with respect to the earth. Such an ellipsoid (fixed with respect to the earth) is often called a *horizontal datum*. In North America we had the North American Datum of 1927, known as NAD 27 (U.S. Department of Commerce, 1973) which was replaced by the geocentric North American Datum of 1983, referred to as NAD 83 (Boal and Henderson, 1988; Schwarz, 1989).

The horizontal geodetic coordinates, *latitude* φ and *longitude* λ, together with the *geodetic height* h (called by some authors by ellipsoidal height, a logical *nonsequitur*, as we shall see later), make the basic triplet of curvilinear coordinates widely used in geodesy. They are related to their associated Cartesian coordinates x, y, and z by the following simple expressions:

$$
\begin{aligned}
x &= (N + h)\cos\varphi\cos\lambda \\
y &= (N + h)\cos\varphi\sin\lambda \qquad (1) \\
z &= (Nb^2/a^2 + h)\sin\varphi,
\end{aligned}
$$

where N is the local radius of curvature of the reference ellipsoid in the east–west direction,

$$
N = a^2\,(a^2\cos^2\varphi + b^2\sin^2\varphi)^{-1/2}, \qquad (2)
$$

where a is the major semi-axis and b is the minor semi-axis of the reference ellipsoid. We note that the geodetic heights are not used in practice; for practical heights, see Section II.B. It should be noted that the horizontal geodetic coordinates are the ones that make the basis for all maps, charts, legal land and marine boundaries, marine and land navigation, etc. The transformations between these horizontal coordinates and the two-dimensional Cartesian coordinates x, y on the maps are called *cartographic mappings*.

Terrestrial (geocentric) coordinate systems are used in satellite positioning. While the instantaneous terrestrial (IT) system is well suited to describe instantaneous positions in, the conventional terrestrial (CT) systems are useful for describing positions for archiving. The conventional terrestrial system recommended by IAG is the International Terrestrial Reference System (ITRS), which is "fixed to the earth" via several permanent stations whose horizontal tectonic velocities are monitored and recorded. The fixing is done at regular time intervals, and the ITRS gets associated with the time of fixing by time tagging. The "realization" of the ITRS by means of coordinates of some selected points is called the International Terrestrial Reference Frame (ITRF). Transformation parameters needed for transforming coordinates from one epoch to the next are produced by the International Earth Rotation Service (IERS) in Paris, so one can keep track of the time evolution of the positions. For more detail the reader is referred to the web site of the IERS or to a popular article by Boucher and Altamini (1996).

B. Point Positioning

It is not possible to determine either three-dimensional (3D) or two-dimensional (2D) (horizontal) positions of isolated points on the earth's surface by terrestrial means. For point positioning we must be looking at celestial objects, meaning that we must be using either optical techniques to observe stars [geodetic astronomy, see Mueller (1969)] or electronic/optical techniques to observe earth's artificial satellites (satellite positioning, cf., Section V.B). Geodetic astronomy is now considered more or less obsolete, because the astronomically determined positions are not very accurate (due to large effects of unpredictable atmospheric refraction) and also because they are strongly affected by the earth's gravity field (cf., Section III.D). Satellite positioning has proved to be much more practical and more accurate.

On the other hand, it is possible to determine heights of some isolated points through terrestrial means by tying these points to the sea level. Practical heights in geodesy, known as *orthometric heights* and denoted by H^O, or simply by H, are referred to the geoid, which is an equipotential surface of the earth's gravity field (for details, see Section III.B) approximated by the mean sea level (MSL) surface to an accuracy of within ± 1.5 m. The difference between the two surfaces arises from the fact that seawater is not homogeneous and because of a variety of dynamical effects on the seawater. The height of the MSL above the geoid is called the *sea surface topography* (SST). It is a very difficult quantity to obtain from any measurements; consequently, it is not yet known very accurately. We note that the orthometric height H is indeed different from the geodetic height h discussed in Section II.A: the relation between the two kinds of heights is shown in Fig. 3, where the quantity N, the height of the geoid above the reference ellipsoid, is usually called the *geoidal height* (geoid

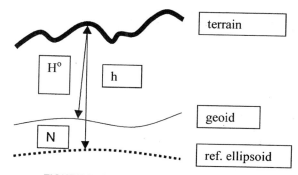

FIGURE 3 Orthometric and geodetic heights.

undulation) (cf., Section III.B). Thus, the knowledge of the geoid is necessary for transforming the geodetic to orthometric heights and vice versa. We note that the acceptance of the standard geodetic term of "geoidal height" (height of the geoid above the reference ellipsoid) makes the expression "ellipsoidal height" for (geodetic) height of anything above the reference ellipsoid, a logical *non-sequitur* as pointed out above.

We have seen above that the geodetic height is a purely geometrical quantity, the length of the normal to the reference ellipsoid between the ellipsoid and the point of interest. The orthometric height, on the other hand, is defined as the length of the plumbline (a line that is always normal to the equipotential surface of the gravity field) between the geoid and the point of interest and, as such, is intimately related to the gravity field of the earth. (As the plumbline is only slightly curved, the length of the plumbline is practically the same as the length of the normal to the geoid between the geoid and the point of interest. Hence, the equation $h \cong H + N$ is valid everywhere to better than a few millimeters.) The defining equation for the orthometric height of point A (given by its position vector $\mathbf{r}_A$) is

$$H^O(\mathbf{r}_A) = H(\mathbf{r}_A) = [W_0 - W(\mathbf{r}_A)]/\text{mean}(g_A), \quad (3)$$

where W_0 stands for the constant gravity potential on the geoid, $W(\mathbf{r}_A)$ is the gravity potential at point A, and mean(g_A) is the mean value of gravity (for detailed treatment of these quantities, see Sections III.A and III.B) between A and the geoid—these. From these equations it can be easily gleaned that orthometric heights are indeed referred to the geoid (defined as $W_0 = 0$). The mean(g) cannot be measured and has to be estimated from gravity observed at A, $g(\mathbf{r}_A)$, assuming a reasonable value for the vertical gradient of gravity within the earth. Helmert (1880) hypothesized the value of 0.0848 mGal m^{-1} suggested independently by Poincaré and Prey to be valid for the region between the geoid and the earth's surface (see Section III.C), to write

$$\text{mean}(g_A) \cong g(\mathbf{r}_A) + 0.0848 \, H(\mathbf{r}_A)/2 \, [\text{mGal}]. \quad (4)$$

For the definition of units of gravity, Gal, see Section III.A. Helmert's (approximate) orthometric heights are used for mapping and for technical work almost everywhere. They may be in error by up to a few decimeters in the mountains. Equipotential surfaces at different heights are not parallel to the equipotential surface at height 0, i.e., the geoid. Thus, orthometric heights of points on the same equipotential surface $W = \text{const.} \neq W_0$ are generally not the same and, for example, the level of a lake appears to be sloping. To avoid this, and to allow the physical laws to be taken into proper account, another system of height is used: *dynamic heights*. The dynamic height of point A

is defined as

$$H^D(\mathbf{r}_A) = [W_0 - W(\mathbf{r}_A)]/\gamma_{\text{ref}}, \quad (5)$$

where γ_{ref} is a selected (reference) value of gravity, constant for the area of interest. We note that points on the same equipotential surface have the same dynamic height; that dynamic heights are referred to the geoid but they must be regarded as having a scale that changes from point to point.

We must also mention the third most widely used height system, the *normal heights*. These heights are defined by

$$H^N(\mathbf{r}_A) = H^*(\mathbf{r}_A) = [W_0 - W(\mathbf{r}_A)]/\text{mean}(\gamma_A), \quad (6)$$

where mean(γ_A) is the value of the model gravity called "normal" (for a detailed explanation, see Section III.A) at a height of $H^N(\mathbf{r}_A)/2$ above the reference ellipsoid along the normal to the ellipsoid (Molodenskij, Eremeev, and Yurkina, 1960). We refer to this value as mean because it is evaluated at a point halfway between the reference ellipsoid and the locus of $H^N(\mathbf{r}_A)$, referred to the reference ellipsoid, which (locus) surface is called the *telluroid*. For practical purposes, normal heights of terrain points A are referred to a different surface, called *quasi-geoid* (cf., Section III.G), which, according to Molodenskij, can be computed from gravity measurements in a similar way to the computation of the geoid.

C. Relative Positioning

Relative positioning, meaning positioning of a point with respect to an existing point or points, is the preferred mode of positioning in geodesy. If there is intervisibility between the points, terrestrial techniques can be used. For satellite relative positioning, the intervisibility is not a requirement, as long as the selected satellites are visible from the two points in question. The accuracy of such relative positions is usually significantly higher than the accuracy of single point positions.

The classical terrestrial techniques for 2D relative positioning make use of angular (horizontal) and distance measurements, which always involve two or three points. These techniques are thus differential in nature. The computations of the relative 2D positions are carried out either on the horizontal datum (reference ellipsoid), in terms of latitude difference $\Delta\varphi$ and longitude difference $\Delta\lambda$, or on a map, in terms of Cartesian map coordinate differences Δx and Δy. In either case, the observed angles, azimuths, and distances have to be first transformed (reduced) from the earth's surface, where they are acquired, to the reference ellipsoid, where they are either used in the computations or transformed further onto the selected mapping plane. We shall not explain these reductions here; rather, we would advise the interested reader to consult one of the classical geodetic textbooks (e.g., Zakatov, 1953; Bomford, 1971).

To determine the relative position of one point with respect to another on the reference ellipsoid is not a simple proposition, since the computations have to be carried out on a curved surface and Euclidean geometry no longer applies. Links between points can no longer be straight lines in the Euclidean sense; they have to be defined as geodesics (the shortest possible lines) on the reference ellipsoid. Consequently, closed form mathematical expressions for the computations do not exist, and use has to be made of various series approximations. Many such approximations had been worked out, which are valid for short, medium, or long geodesics. For 200 years, coordinate computations on the ellipsoid were considered to be the backbone of (classical) geodesy, a litmus test for aspiring geodesists. Once again, we shall have to desist from explaining the involved concepts here as there is no room for them in this small article. Interested readers are referred once more to the textbooks cited above.

Sometimes, preference is given to carrying out the relative position computations on the mapping plane, rather than on the reference ellipsoid. To this end, a suitable cartographic mapping is first selected, normally this would be the conformal mapping used for the national/state geodetic work. This selection carries with it the appropriate mathematical mapping formulae and distortions associated with the selected mapping (Lee, 1976). The observed angles ω, azimuths α, and distances S (that had been first reduced to the reference ellipsoid) are then reduced further (distorted) onto the selected mapping plane where (2D) Euclidean geometry can be applied. This is shown schematically in Fig. 4. Once these reductions have been carried out, the computation of the (relative) position of the unknown point B with respect to point A already known on the mapping plane is then rather trivial:

$$x_B = x_A + \Delta x_{AB}, \qquad y_B = y_A + \Delta y_{AB}. \qquad (7)$$

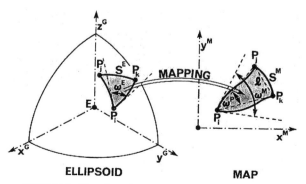

FIGURE 4 Mapping of ellipsoid onto a mapping plane.

Relative vertical positioning is based on somewhat more transparent concepts. The process used for determining the height difference between two points is called *geodetic levelling* (Bomford, 1971). Once the levelled height difference is obtained from field observations, one has to add to it a small correction based on gravity values along the way to convert it to either the orthometric, the dynamic, or the normal height difference. Geodetic levelling is probably the most accurate geodetic relative positioning technique. To determine the geodetic height difference between two points, all we have to do is to measure the vertical angle and the distance between the points. Some care has to be taken that the vertical angle is reckoned from a plane perpendicular to the ellipsoidal normal at the point of measurement.

Modern extraterrestrial (satellite and radio astronomical) techniques are inherently three dimensional. Simultaneous observations at two points yield 3D coordinate differences that can be added directly to the coordinates of the known point A on the earth's surface to get the sought coordinates of the unknown point B (on the earth's surface). Denoting the triplet of Cartesian coordinates (x, y, z) in any coordinate system by $\mathbf{r}$ and the triplet of coordinate differences $(\Delta x, \Delta y, \Delta z)$ by $\Delta \mathbf{r}$, the 3D position of point B is given simply by

$$\mathbf{r}_B = \mathbf{r}_B + \Delta \mathbf{r}_{AB}, \qquad (8)$$

where $\Delta \mathbf{r}_{AB}$ comes from the observations.

We shall discuss in Section V.B how the "base vector" $\Delta \mathbf{r}_{AB}$ is derived from satellite observations. Let us just mention here that $\Delta \mathbf{r}_{AB}$ can be obtained also by other techniques, such as radio astronomy, inertial positioning, or simply from terrestrial observations of horizontal and vertical angles and distances. Let us show here the principle of the interesting radio astronomic technique for the determination of the base vector, known in geodesy as *Very Long Baseline Interferometry* (VLBI). Figure 5 shows schematically the pair of radio telescopes (steerable antennas, A and B) following the same quasar whose celestial position is known (meaning that $\mathbf{e}_s$ is known). The time delay τ can be measured very accurately and the base vector $\Delta \mathbf{r}_{AB}$ can be evaluated from the following equation:

$$\tau = c^{-1} \mathbf{e}_s \Delta \mathbf{r}_{AB}, \qquad (9)$$

where c is the speed of light. At least three such equations are needed for three different quasars to solve for $\Delta \mathbf{r}_{AB}$.

Normally, thousands of such equations are available from dedicated observational campaigns. The most important contribution of VLBI to geodesy (and astronomy) is that it works with directions (to quasars) which can be considered as the best approximations of directions in an inertial space.

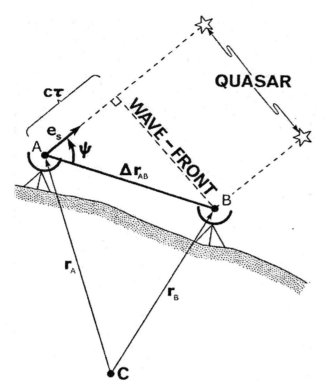

FIGURE 5 Radioastronomical interferometry.

D. Geodetic Networks

In geodesy we prefer to position several points simultaneously because when doing so we can collect redundant information that can be used to check the correctness of the whole positioning process. Also, from the redundancy, one can infer the internal consistency of the positioning process and estimate the accuracy of so determined positions (cf., Section II.E). Thus, the classical geodetic way of positioning points has been in the mode of *geodetic networks*, where a whole set of points is treated simultaneously. This approach is, of course, particularly suitable for the terrestrial techniques that are differential in nature, but the basic rationale is equally valid even for modern positioning techniques. After the observations have been made in the field, the positions of network points are estimated using optimal estimation techniques that minimize the quadratic norm of observation residuals from systems of (sometimes hundreds of thousands) overdetermined (observation) equations. A whole body of mathematical and statistical techniques dealing with network design and position estimation (*network adjustment*) has been developed; the interested reader may consult Grafarend and Sansò (1985), Hirvonen (1971), and Mikhail (1976) for details.

The 2D (horizontal) and 1D (vertical) geodetic networks of national extent, sometimes called national control networks, have been the main tool for positioning needed in mapping, boundary demarcation, and other geodetic applications. For illustration, the Canadian national geodetic levelling network is shown in Fig. 6. We note that national networks are usually interconnected to create continental networks that are sometimes adjusted together—as is the case in North America—to make the networks more homogeneous. Local networks in one, two, and three dimensions have been used for construction purposes. In classical geodetic practice, the most widely encountered networks are horizontal, while 3D networks are very rare.

Vertical (height, levelling) networks are probably the best example of how differential positioning is used together with the knowledge of point heights in carrying the height information from the seashore inland. The heights of selected shore benchmarks are first derived from the observations of the sea level (cf., Section II.B), carried out by means of *tide gauges* (also known in older literature as mareographs) by means of short levelling lines. These basic benchmarks are then linked to the network by longer levelling lines that connect together a whole multitude of land benchmarks (cf., Fig. 6).

Modern satellite networks are inherently three-dimensional. Because the intervisibility is not a requirement for relative satellite positioning, satellite networks can and do contain much longer links and can be much larger in geographical extent. Nowadays, global geodetic networks are constructed and used for different applications.

E. Treatment of Errors in Positions

All positions, determined in whatever way, have errors, both systematic and random. This is due to the fact that every observation is subject to an error; some of these errors are smaller, some are larger. Also, the mathematical models from which the positions are computed are not always completely known or properly described. Thus, when we speak about positions in geodesy, we always mention the accuracy/error that accompanies it. How are these errors expressed?

Random errors are described by following quadratic form:

$$\xi^{\mathrm{T}}\mathbf{C}^{-1}\xi = C_\alpha, \tag{10}$$

where **C** is the *covariance matrix* of the position (a three by three matrix composed of variances and covariances of the three coordinates which comes as a by-product of the network adjustment) and C_α is a factor that depends on the probability density function involved in the position estimation and on the desired probability level α. This

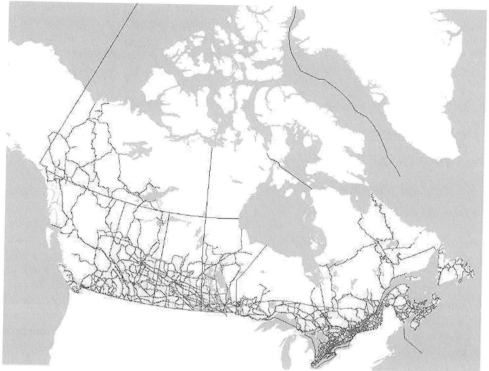

FIGURE 6 The Canadian national geodetic levelling network (Source: www.nrcan.gc.ca. Copyright: Her Majesty the Queen in Right of Canada, Natural Resources Canada, Geodetic Survey Division. All rights reserved.)

quadratic form can be interpreted as an equation of an ellipsoid, called a confidence region in statistics or an *error ellipsoid* in geodetic practice. The understanding is that if we know the covariance matrix **C** and select a probability level α we are comfortable with, then the vector difference ξ between the estimated position $\mathbf{r}^*$ and the real position $\mathbf{r}$ is, with probability α, within the confines of the error ellipsoid.

The interpretation of the error ellipsoid is a bit tricky. The error ellipsoid described above is called *absolute*, and one may expect that errors (and thus also accuracy) thus measured refer to the coordinate system in which the positions are determined. They do not! They actually refer to the point (points) given to the network adjustment (cf., Section II.D) for fixing the position of the adjusted point configuration. This point (points) is sometimes called the "datum" for the adjustment, and we can say that the absolute confidence regions are really relative with respect to the "adjustment datum." As such, they have a natural tendency to grow in size with the growing distance of the point of interest from the adjustment datum. This behavior curtails somewhat the usefulness of these measures.

Hence, in some applications, *relative* error ellipsoids (confidence regions) are sought. These measure errors (accuracy) of one position, A, with respect to another posi-

tion, B, and thus refer always to pairs of points. A relative confidence region is defined by an expression identical to Eq. (10), except that the covariance matrix used, $\mathbf{C}_{\Delta AB}$, is that of the three coordinate differences $\Delta \mathbf{r}_{AB}$ rather than the three coordinates $\mathbf{r}$. This covariance matrix is evaluated from the following expression:

$$\mathbf{C}_{\Delta AB} = \mathbf{C}_A + \mathbf{C}_B - \mathbf{C}_{AB} - \mathbf{C}_{BA}, \qquad (11)$$

where $\mathbf{C}_A$ and $\mathbf{C}_B$ are the covariance matrices of the two points, A and B, and $\mathbf{C}_{AB} = \mathbf{C}_{BA}^T$ is the cross-covariance matrix of the two points. The cross-covariance matrix comes also as a by-product of the network adjustment. It should be noted that the cross-covariances (cross-correlations) between the two points play a very significant role here: when the cross-correlations are strong, the relative confidence region is small and vice versa.

When we deal with 2D instead of 3D coordinates, the confidence regions (absolute and relative) also become two dimensional. Instead of having error ellipsoids, we have error ellipses—see Fig. 7 that shows both absolute and relative error ellipses as well as errors in the distance S, σ_S and azimuth α, $S\sigma_\alpha$, computed (estimated) from the positions of A and B. In the 1D case (heighting), confidence regions degenerate to line segments. The Dilution of Precision (DOP) indicators used in GPS (cf., Section V.B)

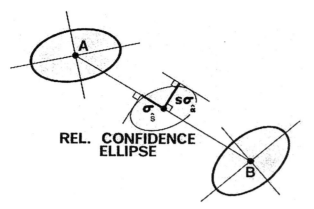

FIGURE 7 Absolute and relative error ellipses.

are related to the idea of (somewhat simplified) confidence regions.

Once we know the desired confidence region(s) in one coordinate system, we can derive the confidence region in any other coordinate system. The transformation works on the covariance matrix used in the defining expression (10) and is given by

$$\mathbf{C}^{(2)} = \mathbf{T}\mathbf{C}^{(1)}\mathbf{T}^{T}, \tag{12}$$

where $\mathbf{T}$ is the Jacobian of transformation from the first to the second coordinate systems evaluated for the point of interest, i.e., $\mathbf{T} = \mathbf{T}(\mathbf{r})$.

Systematic errors are much more difficult to deal with. Evaluating them requires an intimate knowledge of their sources, and these are not always known. The preferred way of dealing with systematic errors is to prevent them from occurring in the first place. If they do occur, then an attempt is made to eliminate them as much as possible.

There are other important issues that we should discuss here in connection with position errors. These include concepts of blunder elimination, reliability, geometrical strength of point configurations, and more. Unfortunately, there is no room to get into these concepts here, and the interested reader may wish to consult Vaníček and Krakiwsky (1986) or some other geodetic textbook.

F. Coordinate Transformations

A distinction should be made between (abstract) "coordinate system transformations" and "coordinate transformations": coordinate systems do not have any errors associated with them while coordinates do. The transformation between two Cartesian coordinate systems [first (1) and second (2)] can be written in terms of hypothetical positions $\mathbf{r}^{(1)}$ and $\mathbf{r}^{(2)}$ as

$$\mathbf{r}^{(2)} = \mathbf{R}(\varepsilon_x, \varepsilon_y, \varepsilon_z,)\mathbf{r}^{(1)} + \mathbf{t}_0^{(2)}, \tag{13}$$

where $\mathbf{R}(\varepsilon_x, \varepsilon_y, \varepsilon_z,)$ is the "rotation matrix," which after application to a vector rotates the vector by the three *misalignment angles* $\varepsilon_x, \varepsilon_y, \varepsilon_z$, around the coordinate axes, and $\mathbf{t}_0^{(2)}$ is the position vector of the origin of the first system reckoned in the second system, called the *translation vector.*

The transformation between coordinates must take into account the errors in both coordinates/coordinate sets (in the first and second coordinate system), particularly the systematic errors. A transformation of coordinates thus consists of two distinct components: the transformation between the corresponding coordinate systems as described above, plus a model for the difference between the errors in the two coordinate sets. The standard illustration of such a model is the inclusion of the scale factor, which accounts for the difference in linear scales of the two coordinate sets. In practice, when dealing with coordinate sets from more extensive areas such as states or countries, these models are much more elaborate, as they have to model the differences in the deformations caused by errors in the two configurations. These models differ from country to country. For unknown reasons, some people prefer not to distinguish between the two kinds of transformations.

Figure 8 shows a commutative diagram for transformations between most of the coordinate systems used in geodesy. The quantities in rectangles are the transformation parameters, the misalignment angles, and translation components. For a full understanding of the involved transformations, the reader is advised to consult Vaníček and Krakiwsky (1986, Chap. 15).

Let us just mention that sometimes we are not interested in transforming positions (position vectors, triplets of coordinates), but small (differential) changes $\delta\mathbf{r}$ in positions $\mathbf{r}$ as we saw in Eq. (12). In this case, we are not concerned with translations between the coordinate systems, only misalignments are of interest. Instead of using the rotation matrix, we may use the Jacobian of transformation, getting

$$\delta\mathbf{r}^{(2)}(\mathbf{r}) = \mathbf{T}(\mathbf{r})\,\delta\mathbf{r}^{(1)}(\mathbf{r}). \tag{14}$$

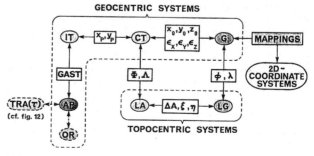

FIGURE 8 Commutative diagram of transformations between coordinate systems.

The final topic we want to discuss in this section is one that we are often faced with in practice: given two corresponding sets of positions (of the same points) in two different coordinate systems we wish to determine the transformation parameters for the two systems. This is done by means of Eq. (13), where ε_x, ε_y, ε_z, $\mathbf{t}_1^{(2)}$ become the six unknown transformation parameters, while $\mathbf{r}^{(1)}$, $\mathbf{r}^{(2)}$ are the known quantities. The set of known positions has to consist of at least three noncollinear points so we get at least six equations for determining the six unknowns. We must stress that the set of position vectors $\mathbf{r}^{(1)}$, $\mathbf{r}^{(2)}$ has to be first corrected for the distortions caused by the errors in both sets of coordinates. These distortions are added back on if we become interested in coordinate transformation.

G. Kinematic Positioning and Navigation

As we have seen so far, classical geodetic positioning deals with stationary points (objects). In recent times, however, geodetic positioning has found its role also in positioning moving objects such as ships, aircraft, and cars. This application became known as *kinematic positioning*, and it is understood as being the real-time positioning part of navigation. Formally, the task of kinematic positioning can be expressed as

$$\mathbf{r}(t) = \mathbf{r}(t_0) + \int_{t_0}^{t} \mathbf{v}(t)\, dt, \qquad (15)$$

where t stands for time and $\mathbf{v}(t)$ is the observed change in position in time, i.e., velocity (vector) of the moving object. The velocity vector can be measured on the moving vehicle in relation to the surrounding space or in relation to an inertial coordinate system by an inertial positioning system. We note that, in many applications, the attitude (roll and pitch) of the vehicle is also of interest.

Alternatively, optical astronomy or point satellite positioning produce directly the string of positions, $\mathbf{r}(t_1)$, $\mathbf{r}(t_2), \ldots, \mathbf{r}(t_n)$, that describe the required *trajectory* of the vehicle, without the necessity of integrating over velocities. Similarly, a relative positioning technique, such as the hyperbolic radio system Loran-C (or Hi-Fix, Decca, Omega in the past), produces a string of position changes, $\Delta\mathbf{r}(t_0, t_1)$, $\Delta\mathbf{r}(t_1, t_2), \ldots, \Delta\mathbf{r}(t_{n-1}, t_n)$, which once again define the trajectory. We note that these techniques are called hyperbolic because positions or position differences are determined from intersections of hyperbolae, which, in turn, are the loci of constant distance differences from the land-located radiotransmitters. Relative satellite positioning is also being used for kinematic positioning, as we shall see later in Section V.B.

For a navigator, it is not enough to know his position $\mathbf{r}(t_n)$ at the time t_n. The navigator must also have the position estimates for the future, i.e., for the times t_n, $t_{n+1}, \ldots$, to be able to navigate safely, he has to have the *predicted positions*. Thus, the kinematic positioning described above has to be combined with a *navigation algorithm*, a predictive filter which predicts positions in the future based on the observed position in the past, at times t_1, $t_2, \ldots, t_n$. Particularly popular seem to be different kinds of Kalman's filters, which contain a feature allowing one to describe the dynamic characteristics of the vehicle navigating in a specified environment. Kalman's filters do have a problem with environments that behave in an unpredictable way, such as an agitated sea. We note that some of the navigation algorithms accept input from two or more kinematic position systems and combine the information in an optimal way.

In some applications, it is desirable to have a post-mission record of trajectories available for future retracing of these trajectories. These post-mission trajectories can be made more accurate than the real-time trajectories (which, in turn, are of course more accurate than the predicted trajectories). Most navigation algorithms have the facility of *post-mission smoothing* the real-time trajectories by using all the data collected during the mission.

III. EARTH'S GRAVITY FIELD

A. Origin of the Earth's Gravity Field

In geodesy, we are interested in studying the gravity field in the macroscopic sense where the quantum behavior of gravity does not have to be taken into account. Also, in terrestrial gravity work, we deal with velocities that are very much smaller than the speed of light. Thus, we can safely use Newtonian physics and may begin by recalling mass attraction force $\mathbf{f}$ defined by Newton's integral

$$\mathbf{f}(\mathbf{r}) = \mathbf{a}(\mathbf{r})m = \left(G \int_{B} \rho(\mathbf{r}')(\mathbf{r}' - \mathbf{r})|\mathbf{r}' - \mathbf{r}|^{-3}\, dV \right) m,$$

$$(16)$$

where $\mathbf{r}$ and $\mathbf{r}'$ are position vectors of the point of interest and the dummy point of the integration; B is the attracting massive body of density ρ, i.e., the earth; V stands for volume; G is Newton's gravitational constant; m is the mass of the particle located at $\mathbf{r}$; and $\mathbf{a}(\mathbf{r})$ is the acceleration associated with the particle located at $\mathbf{r}$ (see Fig. 9). We can speak about the acceleration $\mathbf{a}(\mathbf{r})$, called *gravitation*, even when there is no mass particle present at $\mathbf{r}$, but we cannot measure it (only an acceleration of a mass can be measured). This is the idea behind the definition of the *gravitational field* of body B, the earth; this field is defined at all points $\mathbf{r}$. The physical units of gravitation are those of an acceleration, i.e., m s^{-2}; in practice, units of cm s^{-2},

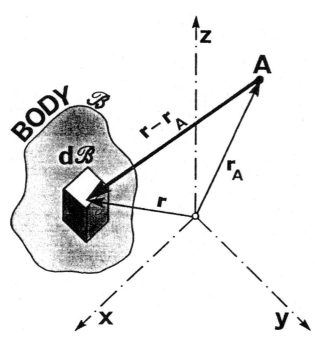

FIGURE 9 Mass attraction.

called "Gal" [to commemorate Galileo's (c.f., Section I.A) contribution to geodesy], are often used.

Newton's gravitational constant G represents the ratio between mass acting in the "attracted capacity" and the same mass acting in the "attracting capacity." From Eq. (16) we can deduce the physical units of G, which are $10 \ kg^{-1} \ m^3 \ s^{-2}$. The value of G has to be determined experimentally. The most accurate measurements are obtained from tracking deep space probes that move in the gravitational field of the earth. If a deep space probe is sufficiently far from the earth (and the attractions of the other celestial bodies are eliminated mathematically), then the physical dimensions of the probe become negligible. At the same time, the earth can be regarded with sufficient accuracy as a sphere with a laterally homogeneous density distribution. Under these circumstances, the gravitational field of the earth becomes radial, i.e., it will look as if it were generated by a particle of mass M equal to the total mass of the earth:

$$M = \int_B \rho(\mathbf{r}') \, dV. \tag{17}$$

When a "geocentric" coordinate system is used in the computations, the probe's acceleration becomes

$$\mathbf{a}(\mathbf{r}) = -GM\mathbf{r}|\mathbf{r}|^{-3}. \tag{18}$$

Thus, the gravitational constant G, or more accurately GM, called the *geocentric constant*, can be obtained from purely geometrical measurements of the deep space

probe positions $\mathbf{r}(t)$. These positions, in turn, are determined from measurements of the propagation of electromagnetic waves and as such depend very intimately on the accepted value of the speed of light c. The value of GM is now thought to be $(3,986,004.418 \pm 0.008) * 10^8 \ m^3 \ s^{-2}$ (Ries *et al.*, 1992), which must be regarded as directly dependent on the accepted value of c. Dividing the geocentric constant by the mass of the earth $[(5.974 \pm 0.001) * 10^{24} \ kg]$, one obtains the value for G as $(6.672 \pm 0.001) * 10^{-11} \ kg^{-1} \ m^3 \ s^{-2}$.

The earth spins around its instantaneous spin axis at a more or less constant angular velocity of once per "sidereal day"—sidereal time scale is taken with respect to fixed stars, which is different from the solar time scale, taken with respect to the sun. This spin gives rise to a centrifugal force that acts on each and every particle within or bound with the earth. A particle, or a body, which is not bound with the earth, such as an earth satellite, is not subject to the centrifugal force. This force is given by the following equation:

$$\mathbf{F}(\mathbf{r}) = \omega^2 \, \mathbf{p}(\mathbf{r})m, \tag{19}$$

where $\mathbf{p}(\mathbf{r})$ is the projection of $\mathbf{r}$ onto the equatorial plane, ω is the siderial angular velocity of 1 revolution per day $(7.292115 * 10^{-5} \ rad \ s^{-1})$, and m is the mass of the particle subjected to the force. Note that the particles on the spin axis of the earth experience no centrifugal force as the projection $\mathbf{p}(\mathbf{r})$ of their radius vector equals to $\mathbf{0}$. We introduce the *centrifugal acceleration* $\mathbf{a}_c(\mathbf{r})$ at point $\mathbf{r}$ as $\omega^2 \mathbf{p}(\mathbf{r})$ and speak of the centrifugal acceleration field much in the same way we speak of the gravitational field (acceleration) $\mathbf{a}(\mathbf{r})$.

The earth (B) gravitation is denoted by $\mathbf{g}_g$ (subscripted g for gravitation) rather than $\mathbf{a}$, and its centrifugal acceleration is denoted by $\mathbf{g}_c$ (c for centrifugal) rather than $\mathbf{a}_c$. When studying the fields $\mathbf{g}_g$ and $\mathbf{g}_c$ acting at points bound with the earth (spinning with the earth), we normally lump these two fields together and speak of the earth's *gravity field* $\mathbf{g}$:

$$\mathbf{g}(\mathbf{r}) = \mathbf{g}_g(\mathbf{r}) + \mathbf{g}_c(\mathbf{r}). \tag{20}$$

A stationary test mass m located at any of these points will sense the total gravity vector $\mathbf{g}$ (acceleration).

If the (test) mass moves with respect to the earth, then another (virtual) force affects the mass: the Coriolis force, responsible, for instance, for the geostrophic motion encountered in air or water movement. In the studies of the earth's gravity field, Coriolis' force is not considered. Similarly, temporal variation of the gravity field, due to variations in density distribution and in earth's rotation speed, which are small compared to the magnitude of the field itself, is mostly not considered either. It is studied separately within the field of geo-kinematics (Section IV).

B. Gravity Potential

When we move a mass m in the gravity field $\mathbf{g}(\mathbf{r})$ from location $\mathbf{r}_1$ to location $\mathbf{r}_2$, to overcome the force $\mathbf{g}(\mathbf{r})\,m$ of the field, we have to do some work w. This work is expressed by the following line integral:

$$w = -\int_{r_1}^{r_2} \mathbf{g}(\mathbf{r})' m\, d\mathbf{r}'. \tag{21}$$

Note that the physical units of work w are $\text{kg}\,\text{m}^2\,\text{s}^{-2}$. Fortunately, for the gravitational field the amount of work does not depend on what trajectory is followed when moving the particle from $\mathbf{r}_1$ to $\mathbf{r}_2$. This property can be expressed as

$$\oint_C \mathbf{g}(\mathbf{r}')\, d\mathbf{r}'\, m = \oint_C \mathbf{g}(\mathbf{r}')\, d\mathbf{r}' = 0, \tag{22}$$

where the line integral is now taken along an arbitrary closed curve C. The physical meaning of Eq. (22) is when we move a particle in the gravitational field along an arbitrary closed trajectory, we do not expend any work.

This property must be true also when the closed trajectory (curve) C is infinitesimally short. This means that the gravitational field must be an irrotational vector field: its *vorticity* is equal to 0 everywhere:

$$\nabla \times \mathbf{g}(\mathbf{r}) = \mathbf{0}. \tag{23}$$

A field which behaves in this way is also known as a *potential field*, meaning that there exists a scalar function, called *potential*, of which the vector field in question is a *gradient*. Denoting this potential by $W(\mathbf{r})$, we can thus write

$$\nabla W(\mathbf{r}) = \mathbf{g}(\mathbf{r}). \tag{24}$$

To get some insight into the physical meaning of the potential W, whose physical units are $\text{m}^2\,\text{s}^{-2}$, we relate it to the work w defined in Eq. (21). It can be shown that the amount of work expended when moving a mass m from $\mathbf{r}_1$ to $\mathbf{r}_2$, along an arbitrary trajectory, is equal to

$$w = [W(\mathbf{r}_2) - W(\mathbf{r}_1)]m. \tag{25}$$

In addition to the two differential equations (Eqs. 23 and 24) governing the behavior of the gravity field, there is a third equation describing the field's *divergence*,

$$\nabla \cdot \mathbf{g}(\mathbf{r}) = -4\pi G\rho(\mathbf{r}) + 2\omega^2. \tag{26}$$

These three *field equations* describe fully the differential behavior of the earth's gravity field. We note that the first term on the right-hand side of Eq. (26) corresponds to the gravitational potential W_g, whose gradient is the gravitational vector $\mathbf{g}_g$, while the second term corresponds to the centrifugal potential W_c that gives rise to the centrifugal acceleration vector $\mathbf{g}_c$. The negative sign of the first term indicates that, at the point $\mathbf{r}$, there is a sink rather than a source of the gravity field, which should be somewhat obvious from the direction of the vectors of the field. Since the ∇ operator is linear, we can write

$$W(\mathbf{r}) = W_g(\mathbf{r}) + W_c(\mathbf{r}). \tag{27}$$

A potential field is a scalar field that is simple to describe and to work with, and it has become the basic descriptor of the earth's gravity field in geodesy (cf., the article "Global Gravity" in this volume). Once one has an adequate knowledge of the gravity potential, one can derive all the other characteristics of the earth's gravity field, $\mathbf{g}$ by Eq. (24), W_g by Eq. (27), etc., mathematically. Interestingly, the Newton integral in Eq. (16) can be also rewritten for the gravitational potential W_g, rather than the acceleration, to give

$$W_g(\mathbf{r}) = G \int_B \rho(\mathbf{r}')|\mathbf{r}' - \mathbf{r}|^{-1}\, dV, \tag{28}$$

which is one of the most often used equations in gravity field studies. Let us, for completeness, spell out also the equation for the centrifugal potential:

$$W_c(\mathbf{r}) = \tfrac{1}{2}\,\omega^2 p^2(\mathbf{r}), \tag{29}$$

[cf., Eq. (19)].

A surface on which the gravity potential value is constant is called an *equipotential surface*. As the value of the potential varies continuously, we may recognize infinitely many equipotential surfaces defined by the following prescription:

$$W(\mathbf{r}) = \text{const.} \tag{30}$$

These equipotential surfaces are convex everywhere above the earth and never cross each other anywhere. By definition, the equipotential surfaces are horizontal everywhere and are thus sometimes called the level surfaces.

One of these infinitely many equipotential surfaces is the *geoid*, one of the most important surfaces used in geodesy, the equipotential surface defined by a specific value W_0 and thought of as approximating the MSL the best (cf., Section II.B) in some sense. We shall have more to say about the two requirements in Section IV.D. At the time of writing, the best value of W_0 is thought to be $62{,}636{,}855.8 \pm 0.5\ \text{m}^2\,\text{s}^{-2}$ (Burša *et al.*, 1997). A global picture of the geoid is shown in Fig. 5 in the article "Global Gravity" in this volume, where the geoidal height N (cf., Section II.B), i.e., the geoid-ellipsoid separation, is plotted. Note that the departure of the geoid from the mean earth ellipsoid (for the definition see below) is at most about 100 m in absolute value.

When studying the earth's gravity field, we often need and use an idealized model of the field. The use of such a model allows us to express the actual gravity field as a sum of the selected model field and the remainder of the actual field. When the model field is selected to resemble closely the actual field itself, the remainder, called an anomaly or disturbance, is much smaller than the actual field itself. This is very advantageous because working with significantly smaller values requires less rigorous mathematical treatment to arrive at the same accuracy of the final results. This procedure resembles the "linearization" procedure used in mathematics and is often referred to as such. Two such models are used in geodesy: spherical (radial field) and ellipsoidal (also called normal or Somigliana-Pizzetti's) models. The former model is used mainly in satellite orbit analysis and prediction (cf., Section V.C), while the normal model is used in terrestrial investigations.

The normal gravity field is generated by a massive body called the *mean earth ellipsoid* adopted by a convention. The most recent such convention, proposed by the IAG in 1980 (IAG, 1980) and called Geodetic Reference System of 1980 (GRS 80), specifies the mean earth ellipsoid as having the major semi-axis "a" 6,378,137 m long and the flattening "f" of 1/298.25. A flattening of an ellipsoid is defined as

$$f = (a - b)/a, \qquad (31)$$

where "b" is the minor semi-axis. This ellipsoid departs from a mean earth sphere by slightly more than 10 km; the difference of $a - b$ is about 22 km. We must note here that the flattening f is closely related to the second degree coefficient $C_{2,0}$ discussed in the article "Global Gravity."

This massive ellipsoid is defined as rotating with the earth with the same angular velocity ω, its potential is defined to be constant and equal to W_0 on the surface of the ellipsoid, and its mass is the same as that (M) of the earth. Interestingly, these prescriptions are enough to evaluate the *normal potential $U(\mathbf{r})$* everywhere above the ellipsoid so that the mass density distribution within the ellipsoid does not have to be specified. The departure of the actual gravity potential from the normal model is called *disturbing potential $T(\mathbf{r})$*:

$$T(\mathbf{r}) = W(\mathbf{r}) - U(\mathbf{r}). \qquad (32)$$

The earth's gravity potential field is described in a global form as a truncated series of spherical harmonics up to order and degree 360 or even higher. Many such series have been prepared by different institutions in the United States and in Europe. Neither regional nor local representations of the potential are used in practice; only the geoid, the gravity anomalies, and the deflections of the vertical (see the next three sections) are needed on a regional/local basis.

C. Magnitude of Gravity

The gravity vector $\mathbf{g}(\mathbf{r})$ introduced in Section III.A can be regarded as consisting of a magnitude (length) and a direction. The magnitude of $\mathbf{g}$, denoted by g, is referred to as *gravity*, which is a scalar quantity measured in units of acceleration. It changes from place to place on the surface of the earth in response to latitude, height, and the underground mass density variations. The largest is the latitude variation, due to the oblateness of the earth and due to the change in centrifugal acceleration with latitude, with amounts to about 5.3 cm s^{-2}, i.e., about 0.5% of the total value of gravity. At the poles, gravity is the strongest, about 9.833 m s^{-2} (983.3 Gal); at the equator it is at its weakest, about 978.0 Gal. The height variation, due to varying distance from the attracting body, the earth, amounts to 0.3086 mGal m^{-1} when we are above the earth and to around 0.0848 mGal m^{-1} (we have seen this gradient already in Section II.B) when we are in the uppermost layer of the earth such as the topography. The variations due to mass density variations are somewhat smaller. We note that all these variations in gravity are responsible for the variation in weight: for instance, a mass of 1 kg at the pole weighs 9.833 kg m^1 s^{-2}, while on the equator it weighs only 9.780 kg m^1 s^{-2}.

Gravity can be measured by means of a test mass, by simply measuring either the acceleration of the test mass in free fall or the force needed to keep it in place. Instruments that use the first approach, pendulums and "free-fall devices," can measure the total value of gravity (absolute instruments), while the instruments based on the second approach, called "gravimeters," are used to measure gravity changes from place to place or changes in time (relative instruments). The most popular field instruments are the gravimeters (of many different designs) which can be made easily portable and easily operated. *Gravimetric surveys* conducted for the geophysical exploration purpose, which is the main user of detailed gravity data, employ portable gravimeters exclusively. The accuracy, in terms of standard deviations, of most of the data obtained in the field is of the order of 0.05 mGal.

To facilitate the use of gravimeters as relative instruments, countries have developed *gravimetric networks* consisting of points at which gravity had been determined through a national effort. The idea of gravimetric networks is parallel to the geodetic (positioning) networks we have seen in Section II.D, and the network adjustment process is much the same as the one used for the geodetic networks. A national gravimetric network starts with national gravity reference point(s) established in participating countries through an international effort; the last such effort, organized by IAG (cf., Section I.C), was the International Gravity Standardization Net 1971 (IGSN 71) (IAG, 1974).

Gravity data as observed on the earth's surface are of little direct use in exploration geophysics. To become useful, they have to be stripped of

1. The height effect, by reducing the observed gravity to the geoid, using an appropriate vertical gradient of gravity $\partial g/\partial H$
2. The dominating latitudinal effect, by subtracting from them the corresponding magnitude of normal gravity γ [the magnitude of the gradient of U, cf., Eq. (24)] reckoned on the mean earth ellipsoid (see Section III.B), i.e., at points (r_e, φ, λ)

The resulting values Δg are called *gravity anomalies* and are thought of as corresponding to locations (r_g, φ, λ) on the geoid. For geophysical interpretation, gravity anomalies are thus defined as (for the definition used in geodesy, see Section III.E)

$$\Delta g(r_g, \varphi, \lambda) = g(r_t, \varphi, \lambda) - \partial g/\partial H\, H(\varphi, \lambda)$$
$$- \gamma(r_e, \varphi, \lambda), \qquad (33)$$

where $g(r_t, \varphi, \lambda)$ are the gravity values observed at points (r_t, φ, λ) on the earth's surface and $H(\varphi, \lambda)$ are the orthometric heights of the observed gravity points. These orthometric heights are usually determined together with the observed gravity. Normal gravity on the mean earth ellipsoid is part of the normal model accepted by convention as discussed in the previous section. The GRS 80 specifies the normal gravity on the mean earth ellipsoid by the following formula:

$$\gamma(r_e, \varphi) = 978.0327(1 + 0.0052790414 \sin^2 \varphi$$
$$+ 0.0000232718 \sin^4 \varphi$$
$$+ 0.0000001262 \sin^6 \varphi)\text{Gal}. \qquad (34)$$

Gravity anomalies, like the disturbing potential in Section III.B, are thought of as showing only the anomalous part of gravity, i.e., the spatial variations in gravity caused by subsurface mass density variations. Depending on what value is used for the vertical gradient of gravity $\partial g/\partial H$, we get different kinds of gravity anomalies: using $\partial g/\partial H = -0.3086$ mGal m^{-1}, we get the *free-air gravity anomaly*; using $\partial g/\partial H = \frac{1}{2}(-0.3086 - 0.0848)$ mGal m^{-1}, we get the (simple) *Bouguer gravity anomaly*. Other kinds of anomalies exist, but they are not very popular in other than specific theoretical undertakings.

Observed gravity data, different kinds of point gravity anomalies, anomalies averaged over certain geographical cells, and other gravity-related data are nowadays avail-

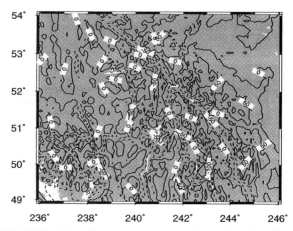

FIGURE 10 Map of free-air gravity anomalies in Canadian Rocky Mountains.

able in either a digital form or in the form of maps. These can be obtained from various national and international agencies upon request. Figure 10 shows the map of free-air gravity anomalies in Canada.

D. Direction of Gravity

Like the magnitude of the gravity vector $\mathbf{g}$ discussed in the previous section, its direction is also of interest. As it requires two angles to specify the direction, the direction of gravity is a little more difficult to deal with than the magnitude. As has been the case with gravity anomalies, it is convenient to use the normal gravity model here as well. When subtracting the direction of normal gravity from the direction of actual gravity, we end up with a small angle, probably smaller than 1 or 2 arcmin anywhere on earth. This smaller angle θ, called the *deflection of the vertical*, is easier to work with than the arbitrarily large angles used for describing the direction of $\mathbf{g}$. We thus have

$$\theta(\mathbf{r}) = \sphericalangle[\mathbf{g}(\mathbf{r}), \boldsymbol{\gamma}(\mathbf{r})], \qquad (35)$$

where, in parallel with Eq. (24), $\boldsymbol{\gamma}(\mathbf{r})$ is evaluated as the gradient of the normal potential U:

$$\boldsymbol{\gamma}(\mathbf{r}) = \boldsymbol{\nabla} U(\mathbf{r}). \qquad (36)$$

We may again think of the deflection of the vertical as being only just an effect of a disturbance of the actual field, compared to the normal field.

Gravity vectors, being gradients of their respective potentials, are always perpendicular to the level surfaces, be they actual gravity vectors or normal gravity vectors. Thus, the direction of $\mathbf{g}(\mathbf{r})$ is the real vertical (a line perpendicular to the horizontal surface) at $\mathbf{r}$ and the direction of $\boldsymbol{\gamma}(\mathbf{r})$ is the normal vertical at $\mathbf{r}$: the deflection of the vertical is really the angle between the actual and normal vertical directions. We note that the actual vertical direction is always tangential to the actual *plumbline*, known in physics also as the line of force of the earth's gravity field. At the geoid, for $\mathbf{r} = \mathbf{r}_g$, the direction of $\gamma(\mathbf{r}_g)$ is to a high degree of accuracy the same as the direction of the normal to the mean earth ellipsoid (being exactly the same on the mean earth ellipsoid).

If the mean earth ellipsoid is chosen also as a reference ellipsoid, then the angles that describe the direction of the normal to the ellipsoid are the geodetic latitude φ and longitude λ (cf., Section II.A). Similarly, the direction of the plumbline at any point $\mathbf{r}$ is defined by *astronomical latitude* Φ and *astronomical longitude* Λ. The astronomical coordinates Φ and Λ can be obtained, to a limited accuracy, from optical astronomical measurements, while the geodetic coordinates are obtained by any of the positioning techniques described in Section II. Because θ is a spatial angle, it is customary in geodesy to describe it by two components, the meridian ξ and the prime vertical η components. The former is the projection of θ onto the local meridian plane, and the latter is the projection onto the local prime vertical plane (plane perpendicular to the horizontal and meridian planes).

There are two kinds of deflection of vertical used in geodesy: those taken at the surface of the earth, at points $\mathbf{r}_t = (r_t, \varphi, \lambda)$, called *surface deflections* and those taken at the geoid level, at points $\mathbf{r}_g = (r_g, \varphi, \lambda)$, called *geoid deflections*. Surface deflections are generally significantly larger than the geoid deflections, as they are affected not only by the internal distribution of masses but also by the topographical masses. The two kinds of deflections can be transformed to each other. To do so, we have to evaluate the curvature of the plumbline (in both perpendicular directions) and the curvature of the normal vertical. The former can be quite sizeable—up to a few tens of arcseconds—and is very difficult to evaluate. The latter is curved only in the meridian direction (the normal field being rotationally symmetrical), and even that curvature is rather small, reaching a maximum of about 1 arcsec.

The classical way of obtaining the deflections of the vertical is through the differencing of the astronomical and geodetic coordinates as follows:

$$\xi = \Phi - \varphi, \eta = (\Lambda - \lambda) \cos \varphi. \quad (37)$$

These equations also define the signs of the deflection components. In North America, however, the sign of η is sometimes reversed. We emphasize here that the geodetic coordinates have to refer to the geocentric reference ellipsoid/mean earth ellipsoid. Both geodetic and astronomical coordinates must refer to the same point, either on the geoid or on the surface of the earth. In Section II.B, we mentioned that the astronomical determination of point

positions (Φ, Λ) is not used in practice any more because of the large effect of the earth's gravity field. Here, we see the reason spelled out in Eqs. (37): considering the astronomically determined position (Φ, Λ) to be an approximation of the geodetic position (φ, λ) invokes an error of $(\xi, \eta/\cos \varphi)$ that can reach several kilometers on the surface of the earth. The deflections of the vertical can be determined also from other measurements, which we will show in the next section.

E. Transformations between Field Parameters

Let us begin with the transformation of the geoidal height to the deflection of the vertical [i.e., $N \rightarrow (\xi, \eta)$], which is of a purely geometrical nature and fairly simple. When the deflections are of the "geoid" kind, they can be interpreted simply as showing the slope of the geoid with respect to the geocentric reference ellipsoid at the deflection point. This being the case, geoidal height differences can be constructed from the deflections $((\xi, \eta) \rightarrow \Delta N)$ in the following fashion. We take two adjacent deflection points and project their deflections onto a vertical plane going through the two points. These projected deflections represent the projected slopes of the geoid in the vertical plane; their average multiplied by the distance between the two points gives us an estimate of the difference in geoidal heights ΔN at the two points. Pairs of deflection points can be then strung together to produce the geoid profiles along selected strings of deflection points. This technique is known as *Helmert's levelling*. We note that if the deflections refer to a geodetic datum (rather than to a geocentric reference ellipsoid), this technique gives us geoidal height differences referred to the same geodetic datum. Some older geoid models were produced using this technique.

Another very useful relation (transformation) relates the geoid height N to the disturbing potential $T (T \rightarrow N, N \rightarrow T)$. It was first formulated by a German physicist H. Bruns (1878), and it reads

$$N = T/\gamma. \quad (38)$$

The equation is accurate to a few millimeters; it is now referred to as *Bruns's formula*.

In Section III.C we introduced gravity anomaly Δg of different kinds (defined on the geoid), as they are normally used in geophysics. In geodesy we need a different gravity anomaly, one that is defined for any location $\mathbf{r}$ rather than being tied to the geoid. Such gravity anomaly is defined by the following exact equation:

$$\Delta g(\mathbf{r}) = -\partial T/\partial h|_{\mathbf{r}=(r,\varphi,\lambda)}$$
$$+ \gamma(\mathbf{r})^{-1} \partial \gamma/\partial h|_{\mathbf{r}=(r,\varphi,\lambda)} T(r - Z, \varphi, \lambda), \quad (39)$$

where Z is the displacement between the actual equipotential surface $W = $ const. passing through $\mathbf{r}$ and the corresponding (i.e., having the same potential) normal equipotential surface $U = $ const. This differential equation of first order, sometimes called *fundamental gravimetric equation*, can be regarded as the transformation from $T(\mathbf{r})$ to $\Delta g(\mathbf{r})(T \rightarrow \Delta g)$ and is used as such in the studies of the earth's gravity field. The relation between this gravity anomaly and the ones discussed above is somewhat tenuous.

Perhaps the most important transformation is that of gravity anomaly Δg, it being the cheapest data, to disturbing potential $T(\Delta g \rightarrow T)$, from which other quantities can be derived using the transformations above. This transformation turns out to be rather complicated: it comes as a solution of a scalar boundary value problem and it will be discussed in the following two sections. We devote two sections to this problem because it is regarded as central to the studies of earth's gravity field. Other transformations between different parameters and quantities related to the gravity field exist, and the interested reader is advised to consult any textbook on geodesy; the classical textbook by Heiskanen and Moritz (1967) is particularly useful.

F. Stokes's Geodetic Boundary Value Problem

The scalar *geodetic boundary value problem* was formulated first by Stokes (1849). The formulation is based on the partial differential equation valid for the gravity potential W [derived by substituting Eq. (24) into Eq. (26)],

$$\nabla^2 W(\mathbf{r}) = -4\pi G \rho(\mathbf{r}) + 2\omega^2. \quad (40)$$

This is a nonhomogeneous elliptical equation of second order, known under the name of *Poisson equation*, that embodies all the field equations (see Section III.A) of the earth gravity field. Stokes applied this to the disturbing potential T (see Section III.B) outside the earth to get

$$\nabla^2 T(\mathbf{r}) = 0. \quad (41)$$

This is so because T does not have the centrifugal component and the mass density $\rho(\mathbf{r})$ is equal to 0 outside the earth. (This would be true only if the earth's atmosphere did not exist; as it does exist, it has to be dealt with. This is done in a corrective fashion, as we shall see below.) This homogeneous form of Poisson equation is known as *Laplace equation*. A function (T, in our case) that satisfies the Laplace equation in a region (outside the earth, in our case) is known as being *harmonic* in that region.

Further, Stokes has chosen the geoid to be the boundary for his boundary value problem because it is a smooth enough surface for the solution to exist (in the space outside the geoid). This of course violates the requirement of harmonicity of T by the presence of topography (and the atmosphere). Helmert (1880) suggested to avoid this problem by transforming the formulation into a space where T is harmonic outside the geoid. The actual disturbing potential T is transformed to a disturbing potential T^h, harmonic outside the geoid, by subtracting from it the potential caused by topography (and the atmosphere) and adding to it the potential caused by topography (and the atmosphere) condensed on the geoid (or some other surface below the geoid). Then the Laplace equation

$$\nabla^2 T^h(\mathbf{r}) = 0 \quad (42)$$

is satisfied everywhere outside the geoid. This became known as the *Stokes-Helmert formulation*.

The boundary values on the geoid are constructed from gravity observed on the earth's surface in a series of steps. First, gravity anomalies on the surface are evaluated from Eq. (33) using the free-air gradient These are transformed to Helmert's anomalies Δg^h, defined by Eq. (39) for $T = T^h$, by applying a transformation parallel to the one for the disturbing potentials as described above. By adding some fairly small corrections, Helmert's anomalies are transformed to the following expression (Vaníček et al., 1999):

$$2r^{-1}T^h(r - Z, \varphi, \lambda) + \partial T^h / \partial r|_r = -\Delta g^{h*}(\mathbf{r}). \quad (43)$$

As T^h is harmonic above the geoid, this linear combination, multiplied by r, is also harmonic above the geoid. As such it can be "continued downward" to the geoid by using the standard Poisson integral.

Given the Laplace equation (42), the boundary values on the geoid, and the fact that $T^h(\mathbf{r})$ disappears as $r \rightarrow \infty$, Stokes (1849) derived the following integral solution to his boundary value problem:

$$T^h(\mathbf{r}_g) = T^h(r_g, \Omega)R/(4\pi) \int_G \Delta g^{h*}(r_g, \Omega') S(\Psi) d\Omega', \quad (44)$$

where Ω, Ω' are the geocentric spatial angles of positions $\mathbf{r}, \mathbf{r}'$, Ψ is the spatial angle between $\mathbf{r}$ and $\mathbf{r}'$; S is the Stokes integration kernel in its spherical (approximate) form

$$S(\Psi) \cong 1 + \sin^{-1}(\Psi/2) - 6\sin(\Psi/2) - 5\cos\Psi$$
$$- 3\cos\Psi \ln[\sin(\Psi/2) + \sin^2(\Psi/2)]; \quad (45)$$

and the integration is carried out over the geoid. We note that, if desired, the disturbing potential is easily transformed to geoidal height by means of the Bruns formula (38).

In the final step, the solution $T^h(\mathbf{r}_g)$ is transformed to $T(\mathbf{r}_g)$ by adding to it the potential of topography (and the atmosphere) and subtracting the potential of topography (and the atmosphere) condensed to the geoid. This can be regarded as a back transformation from the "Helmert harmonic space" back to the real space. We have to mention that the fore and back transformation between the two spaces requires knowledge of topography (and the atmosphere), both of height and of density. The latter represents the most serious accuracy limitation of Stokes's solution: the uncertainty in topographical density may cause an error up to 1 to 2 dm in the geoid in high mountains.

Let us add that recently it became very popular to use a higher than second order (Somigliana-Pizzetti's) reference field in Stokes's formulation. For this purpose, a global field (cf., the article, "Global Gravity"), preferably of a pure satellite origin, is selected and a residual disturbing potential on, or geoidal height above, a *reference spheroid* defined by such a field (cf., Fig. 5 in the article "Global Gravity") is then produced. This approach may be termed a *generalized Stokes formulation* (Vaníček and Sjöberg, 1991), and it is attractive because it alleviates the negative impact of the existing nonhomogeneous terrestrial gravity coverage by attenuating the effect of distant data in the Stokes integral (44). For illustration, so computed a geoid for a part of North America is shown in Fig. 11. It should be also mentioned that the evaluation of Stokes' integral is often sought in terms of Fast Fourier Transform.

G. Molodenskij's Geodetic Boundary Value Problem

In the mid-20th century, Russian physicist M. S. Molodenskij formulated a different scalar boundary value problem to solve for the disturbing potential outside the earth (Molodenskij, Eremeev, and Yurkine 1960). His criticism of Stokes' approach was that the geoid is an equipotential surface internal to the earth and as such requires detailed knowledge of internal (topographical) earth mass density, which we will never have. He then proceeded to replace Stokes's choice of the boundary (geoid) by the earth's surface and to solve for $T(\mathbf{r})$ outside the earth.

At the earth's surface, the Poisson equation changes dramatically. The first term on its right-hand side, equal to

$-4\pi G\rho(\mathbf{r})$, changes from 0 to a value of approximately $2.24 * 10^{-6} \text{ s}^{-2}$ (more than three orders of magnitude larger than the value of the second term). The latter value is obtained using the density ρ of the most common rock, granite. Conventionally, the value of the first term right on the earth's surface is defined as $-4k(\mathbf{r}_t)\pi G\rho(\mathbf{r}_t)$, where the function $k(\mathbf{r}_t)$ has a value between 0 and 1 depending on the shape of the earth's surface; it equals 1/2 for a flat surface, close to 0 for a "needle-like" topographical feature, and close to 1 for a "well-like" feature. In Molodenskij's solution, the Poisson equation has to be integrated over the earth's surface and the above variations of the right-hand side cause problems, particularly on steep surfaces. It is still uncertain just how accurate a solution can be obtained with Molodenskij's approach; it looks as if bypassing the topographical density may have introduced another problem caused by the real shape of topographical surface.

For technical reasons, the integration is not carried out on the earth's surface but on a surface which differs from the earth's surface by about as much as the geoidal height N; this surface is the telluroid encountered already in Section II.B. The solution for T on the earth's surface (more accurately on the telluroid) is given by the following integral equation:

$$T(\mathbf{r}_t) - R/(2\pi) \int_{tell} [\partial/\partial n' |\mathbf{r}_t - \mathbf{r}_t'|^{-1} - |\mathbf{r}_t - \mathbf{r}_t'|^{-1}$$
$$\times \cos\beta/\gamma\,\partial\gamma/\partial H^N] T(\mathbf{r}_t')\,d\Omega'$$
$$= R/(2\pi) \int_{tell} [\Delta g(\mathbf{r}_t') - \gamma][\xi'\tan\beta_1 + \eta'\tan\beta_2]$$
$$\times |\mathbf{r}_t - \mathbf{r}_t'|^{-1} \cos\beta\,d\Omega', \tag{46}$$

where n' is the outer normal to the telluroid; β is the maximum slope of the telluroid (terrain); β_1, β_2 are the north–south and east–west terrain slopes; and ξ', η' are deflection components on the earth's surface. This integral equation is too complicated to be solved directly and simplifications must be introduced. The solution is then sought in terms of successive iterations, the first of which has an identical shape to the Stokes integral (44). Subsequent iterations can be thought of as supplying appropriate corrective terms (related to topography) to the basic Stokes solution.

In fact, the difference between the telluroid and the earth's surface, called the *height anomaly* ζ, is what can be determined directly from Molodenskij's integral using a surface density function. It can be interpreted as a "geoidal height" in Molodenskij's sense as it defines the Molodenskij "geoid" introduced in Section II.B (called quasi-geoid, to distinguish it from the real geoid). The difference between the geoid and quasi-geoid may reach up to a few meters in mountainous regions, but it disappears at sea (Pick, Pícha, and Vyskočil, 1973). It can be seen from

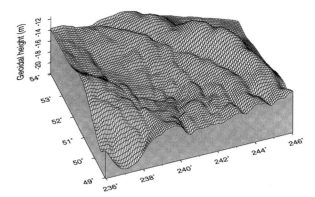

FIGURE 11 Detailed geoid for an area in the Canadian Rocky Mountains (computed at the University of New Brunswick).

Section II.B that the difference may be evaluated from orthometric and normal heights (referred to the geoid, and quasi-geoid, respectively):

$$\zeta - N = H^{\mathrm{O}} - H^{\mathrm{N}}, \qquad (47)$$

subject to the error in the orthometric height. Approximately, the difference is also equal to $-\Delta g^{\mathrm{Bouguer}} H^{\mathrm{O}}/\gamma$.

H. Global and Local Modeling of the Field

Often, it is useful to describe the different parameters of the earth's gravity field by a series of spherical or ellipsoidal harmonic functions (cf., the article "Global Gravity" in this volume). This description is often referred to as the *spectral form*, and it is really the only practical global description of the field parameters. The spectral form, however, is useful also in showing the spectral behavior of the individual parameters. We learn, for instance, that the series for T, N, or ζ converge to 0 much faster than the series for Δg, ζ, η do: we say that the T, N, or ζ fields are smoother than the Δg, ζ, η fields. This means that a truncation of the harmonic series describing one of the smoother fields does not cause as much damage as does a truncation for one of the rougher fields by leaving out the higher "frequency" components of the field. The global description of the smoother fields will be closer to reality.

If higher frequency information is of importance for the area of interest, then it is more appropriate to use a point description of the field. This form of a description is called in geodesy the *spatial form*. Above we have seen only examples of spatial expressions, in the article "Global Gravity" only spectral expressions are used. Spatial expressions involving surface convolution integrals over the whole earth [cf., Eqs. (44) and (46)], can be always transformed into corresponding spectral forms and vice versa. The two kinds of forms can be, of course, also combined as we saw in the case of generalized Stokes's formulation (Section III.F).

IV. GEO-KINEMATICS

A. Geodynamics and Geo-Kinematics

Dynamics is that part of physics that deals with forces (and therefore masses), and motions in response to these forces and geodynamics is that part of geophysics that deals with the dynamics of the earth. In geodesy, the primary interest is the geometry of the motion and of the deformation (really just a special kind of motion) of the earth or its part. The geometrical aspect of dynamics is called kinematics, and therefore, we talk here about *geo-kinematics*. As a matter of fact, the reader might have noticed already in the above paragraphs involving physics how mass was elimi-

nated from the discussions, leaving us with only kinematic descriptions.

Geo-kinematics is one of the obvious fields where cooperation with other sciences, geophysics here, is essential. On the one hand, geometrical information on the deformation of the surface of the earth is of much interest to a geophysicist who studies the forces/stresses responsible for the deformation and the response of the earth to these forces/stresses. On the other hand, it is always helpful to a geodesist to get an insight into the physical processes responsible for the deformation he is trying to monitor.

Geodesists have studied some parts of geo-kinematics, such as those dealing with changes in the earth's rotation, for a long time. Other parts were only more recently incorporated into geodesy because the accuracy of geodetic measurements had not been good enough to see the real-time evolution of deformations occurring on the surface of the earth. Nowadays, geodetic monitoring of crustal motions is probably the fastest developing field of geodesy.

B. Temporal Changes in Coordinate Systems

In Section II.A we encountered a reference to the earth's "spin axis" in the context of geocentric coordinate systems. It is this axis the earth spins around with 366.2564 *sidereal revolutions* (cf., Section III.A), or 365.2564 revolutions with respect to the sun (defining *solar days*) per year. The spin axis of the earth moves with respect to the universe (directions to distant stars, the realization of an inertial coordinate system), undergoing two main motions: one very large, called precession, with a period of about 26,000 years and the other much smaller, called nutation, with the main period of 18.6 years. These motions must be accounted for when doing astronomical measurements of either the optical or radio variety (see Section II.C).

In addition to precession and nutation, the spin axis also undergoes a torque-free nutation, also called a wobble, with respect to the earth. More accurately, the *wobble* should be viewed as the motion of the earth with respect to the instantaneous spin axis. It is governed by the famous Euler's gyroscopic equation:

$$\mathbf{J}\boldsymbol{\omega} + \boldsymbol{\omega} \times \mathbf{J}\boldsymbol{\omega}, \qquad (48)$$

where $\mathbf{J}$ is the earth's *tensor of inertia* and $\boldsymbol{\omega}$ is the instantaneous spin *angular velocity vector* whose magnitude ω we have met several times above. Some relations among the diagonal elements of $\mathbf{J}$, known as moments of inertia, can be inferred from astronomical observations giving a solution $\boldsymbol{\omega}$ of this differential equation. Such a solution describes a periodic motion with a period of about 305 sidereal days, called Euler's period.

Observations of the wobble (Fig. 12) have shown that beside the Euler component, there is also an annual periodic component of similar magnitude plus a small drift. The magnitude of the periodic components fluctuates

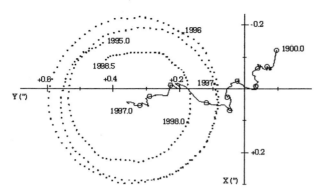

FIGURE 12 Earth pole wobble. (Source: International Earth Rotation Service.)

around 0.1 arcsec. Thus, the wobble causes a displacement of the pole (intersection of instantaneous spin axis with the earth's surface) of several meters. Furthermore, systematic observations show also that the period of the Euler component is actually longer by some 40% than predicted by the Euler equation. This discrepancy is caused by the nonrigidity of the earth, and the actual period, around 435 solar days, is now called Chandler's. The actual motion of the pole is now being observed and monitored by IAG's IERS on a daily basis. It is easy to appreciate that any coordinate system linked to the earth's spin axis (cf., Section II.A) is directly affected by the earth pole wobble which, therefore, has to be accounted for.

As the direction of ω varies, so does its magnitude ω. The variations of spin velocity are also monitored in terms of the *length of the day* (LOD) by the Bureau Internationale des Poids et Mesures—Time Section (BIPM) on a continuous basis. The variations are somewhat irregular and amount to about 0.25 msec/year—the earth's spin is generally slowing down.

There is one more temporal effect on a geodetic coordinate system, that on the datum for vertical positioning, i.e., on the geoid. It should be fairly obvious that if the reference surface for heights (orthometric and dynamic) changes, so do the heights referred to it. In most countries, however, these changes are not taken too seriously. The geoid indicated by the MSL, as described in Section III.B, changes with time both in response to the mass changes within the earth and to the MSL temporal changes. The latter was discussed in Section III.B, the former will be discussed in Section IV.C.

C. Temporal Changes in Positions

The earth is a deformable body and its shape is continuously undergoing changes caused by a host of stresses; thus, the positions of points on the earth's surface change continuously. Some of the stresses that cause the *deformations* are known, some are not, with the best known being the tidal stress (Melchior, 1966). Some loading stresses causing crustal deformation are reasonably well known, such as those caused by filling up water dams, others, such as sedimentation and glaciation, are known only approximately. Tectonic and other stresses can be only inferred from observed deformations (cf., the article, "Tectonophysics" in this encyclopedia). The response of the earth to a stress, i.e., the deformation, varies with the temporal frequency and the spatial extent of the stress and depends on the rheological properties of the whole or just a part of the earth. Some of these properties are now reasonably well known, some are not. The ultimate role of geodesy vis-à-vis these deformations is to take them into account for predicting the temporal variations of positions on the earth's surface. This can be done relatively simply for deformations that can be modeled with a sufficient degree of accuracy (e.g., tidal deformations), but it cannot yet be done for other kinds of deformations, where the physical models are not known with a sufficient degree of certainty. Then the role of geodesy is confined to monitoring the surface movements, kinematics, to provide the input to geophysical investigations.

A few words are now in order about tidal deformations. They are caused by the moon and sun gravitational attraction (see Fig. 13). Tidal potential caused by the next most influential celestial body, Venus, amounts to only 0.036% of the luni-solar potential; in practice only the *luni-solar tidal potential* is considered. The tidal potential W_t^c of the moon, the lunar tidal potential, is given by the following equation:

$$W_t^c(\mathbf{r}) = GM^c |\mathbf{r}^c - \mathbf{r}|^{-1} \sum_{j=2}^{\infty} \mathbf{r}^j |\mathbf{r}^c - \mathbf{r}|^{-j} P_j(\cos \Psi^c),$$

(49)

where the symbol c refers to the moon, P_j is the Legendre function of degree j, and Ψ is the geocentric angle

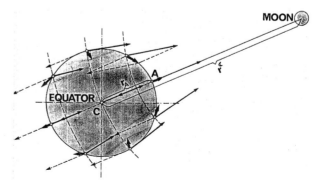

FIGURE 13 The provenance of tidal force due to the moon.

between **r** and **r**^{c}. The tidal potential of the sun $W_t^\bullet$ is given by a similar series, and it amounts to about 46% of the lunar tidal potential. To achieve an accuracy of 0.03%, it is enough to take the first two terms from the lunar series and the first term from the solar series.

The temporal behavior of tidal potential is periodic, as dictated by the motions of the moon and the sun (but see Section IV.D for the tidal constant effect). The tidal waves, into which the potential can be decomposed, have periods around 1 day, called diurnal; around 12 h, called semidiurnal; and longer. The tidal deformation as well as all the tidal effects (except for the sea tide, which requires solving a boundary value problem for the Laplace tidal equation, which we are not going to discuss here) can be relatively easily evaluated from the tidal potential. This is because the rheological properties of the earth for global stresses and tidal periods are reasonably well known. Geodetic observations as well as positions are affected by tidal deformations, and these effects are routinely corrected for in levelling, VLBI (cf., Section II.C), satellite positioning (cf., Section V.B), and other precise geodetic works. For illustration, the range of orthometric height tidal variation due to the moon is 36 cm, that due to the sun is another 17 cm. With tidal deformation being of a global character, however, these tidal variations are all but imperceptible locally.

Tectonic stresses are not well known, but horizontal motion of tectonic plates has been inferred from various kinds of observations, including geodetic, with some degree of certainty, and different maps of these motions have been published (cf., the article, "Tectonophysics" in this encyclopedia). The AM0-2 absolute plate motion model was chosen to be an "associated velocity model" in the definition of the ITRF (see Section II.A), which together with direct geodetic determination of horizontal velocities define the temporal evolution of the ITRF and thus the temporal evolution of horizontal positions. From other ongoing earth deformations, the post-glacial rebound is probably the most important globally as it is large enough to affect the flattening [Eq. (31)] of the mean earth ellipsoid as we will see in Section IV.C.

Mapping and monitoring of ongoing motions (deformations) on the surface of the earth are done by repeated position determination. In global monitoring the global techniques of VLBI and satellite positioning are used (see Section V.B). For instance, one of the IAG services, the IGS (cf., Section I.C), has been mandated with monitoring the horizontal velocities of a multitude of permanent tracking stations under its jurisdiction. In regional investigations the standard terrestrial geodetic techniques such as horizontal and vertical profiles, horizontal and levelling network re-observation campaigns are employed. The po-

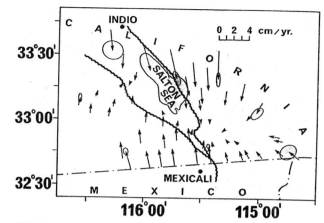

FIGURE 14 Mean annual horizontal displacements in Imperial Valley, CA, computed from data covering the period 1941–1975.

sitions are then determined separately from each campaign with subsequent evaluation of *displacements*. Preferably, the displacements (horizontal or vertical) are estimated directly from the observations collected in all the campaigns. The latter approach allows the inclusion of correlations in the mathematical model for the displacement estimation with more correct estimates ensuing. For illustration, Fig. 14 shows such estimated horizontal displacements from the area of Imperial Valley, CA, computed from standard geodetic observations.

It is not possible to derive absolute displacements from relative positions. Because the repeated horizontal position determination described above is usually of a relative kind, the displacements are indeterminate. It then makes sense to deal only with relative quantification of deformation such as *strain*. Strain is, most generally, described by the displacement gradient matrix **S**; denoting the 2D displacement vector of a point **r** by **v**(**r**) we can write

$$S(\mathbf{r}) = \nabla'\mathbf{v}^T(\mathbf{r}), \qquad (50)$$

where ∇' is the 2D nabla operator. The inverse transformation

$$\mathbf{v}(\mathbf{r}) = S(\mathbf{r})\,\mathbf{r} + \mathbf{v}_0, \qquad (51)$$

where $\mathbf{v}_0$ describes the translational indeterminacy, shows better the role of **S**, which can be also understood as a Jacobian matrix [cf., Eq. (14)] which transforms from the space of positions (real 2D space) into the space of displacements. The symmetrical part of **S** is called the deformation tensor in the mechanics of continuum; the antisymmetrical part of **S** describes the rotational deformation. Other strain parameters can be derived from **S**.

D. Temporal Changes in Gravity Field

Let us begin with the two requirements defining the geoid presented in Section III.B: the constancy of W_0 and the fit of the equipotential surface to the MSL. These two requirements are not compatible when viewed from the point of time evolution or the temporal changes of the earth's gravity field. The MSL grows with time at a rate estimated to be between 1 and 2 mm year^{-1} (the eustatic water level change), which would require systematic lowering of the value of W_0. The mass density distribution within the earth changes with time as well (due to tectonic motions, post-glacial rebound, sedimentation, as discussed above), but its temporal effect on the geoid is clearly different from that of the MSL. This dichotomy has not been addressed by the geodetic community yet.

In the areas of largest documented changes of the mass distribution (those caused by the post-glacial rebound), the northeastern part of North America and Fennoscandia, the maximum earth surface uplift reaches about 1 cm year^{-1}. The corresponding change in gravity value reaches up to 0.006 mGal year^{-1} and the change in the equipotential surfaces $W =$ const. reaches up to 1 mm year^{-1}. The potential coefficient $C_{2,0}$ (cf., Section III.B), or rather its unitless version $J_{2,0}$ that is used most of the time in gravity field studies, as observed by satellites shows a temporal change caused by the rebound. The rebound can be thought of as changing the shape of the geoid, and thus the shape of the mean earth ellipsoid, within the realm of observability.

As mentioned in Section IV.C, the tidal effect on gravity is routinely evaluated and corrected for in precise gravimetric work, where the effect is well above the observational noise level. By correcting the gravity observations for the periodic tidal variations we eliminate the temporal variations, but we do not eliminate the whole tidal effect. In fact, the luni-solar tidal potential given by Eq. (49) has a significant constant component responsible for what is called in geodesy *permanent tide*. The effect of permanent tide is an increased flattening of gravity equipotential surfaces, and thus of the mean earth ellipsoid, by about one part in 10^5. For some geodetic work, the tideless mean earth ellipsoid is better suited than the mean tide ellipsoid, and, consequently, both ellipsoids can be encountered in geodesy.

Temporal variations of gravity are routinely monitored in different parts of the world. These variations (corrected for the effect of underground water fluctuations) represent an excellent indicator that a geodynamical phenomenon, such as tectonic plate motion, sedimentation loading, volcanic activity, etc., is at work in the monitored region. When gravity monitoring is supplemented with vertical motion monitoring, the combined results can be used to infer the physical causes of the monitored phenomenon.

Let us mention that the earth pole wobble introduces also observable variations of gravity of the order of about 0.008 mGal. So far, these variations have been of academic interest only.

V. SATELLITE TECHNIQUES

A. Satellite Motion, Functions, and Sensors

Artificial satellites of the earth appeared on the world scene in the late 1950s and were relatively early embraced by geodesists as the obvious potential tool to solve worldwide geodetic problems. In geodetic applications, satellites can be used both in positioning and in gravitational field studies as we have alluded to in the previous three sections. Geodesists have used many different satellites in the past 40 years, ranging from completely passive to highly sophisticated active (transmitting) satellites, from quite small to very large. Passive satellites do not have any sensors on board and their role is basically that of an orbiting target. Active satellites may carry a large assortment of sensors, ranging from accurate clocks through various counters to sophisticated data processors, and transmit the collected data down to the earth either continuously or intermittently.

Satellites orbit the earth following a trajectory which resembles the Keplerian ellipse that describes the motion in radial field (cf., Section III.A); the higher the satellite is, the closer its orbit to the Keplerian ellipse. Lower orbiting satellites are more affected by the irregularities of the earth's gravitational field, and their *orbit* becomes more perturbed compared to the Keplerian orbit. This curious behavior is caused by the inherent property of gravitational field known as the attenuation of shorter wavelengths of the field with height and can be gleaned from Eq. (4) in the article "Global Gravity." The ratio a/r is always smaller than 1 and thus tends to disappear the faster the larger its exponent l which stands for the spatial wave number of the field. We can see that the *attenuation factor* $(a/r)^\ell$ goes to 0 for growing ℓ and growing r; for $r > a$, we have:

$$\lim_{l \to \infty} (a/r)^\ell = 0. \tag{52}$$

We shall see in Section V.C how this behavior is used in studying the gravitational field by means of satellites.

Satellite orbits are classified as high and low orbits, polar orbits (when the orbital plane contains the spin axis of the earth), equatorial orbits (orbital plane coincides with the equatorial plane of the earth), and pro-grade and retro-grade or bits (the direction of satellite motion is either eastward or westward). The lower the orbit is, the faster the satellite circles the earth. At an altitude of about 36,000 km, the orbital velocity matches that of the earth's spin, its orbital period becomes 24 h long, and the satellite moves only in one meridian plane (its motion is neither pro- nor retro-grade). If its orbit is equatorial, the satellite remains in one position above the equator. Such an orbit (satellite) is called *geostationary*.

Satellites are *tracked* from points on the earth or by other satellites using electromagnetic waves of frequencies that can penetrate the ionosphere. These frequencies propagate along a more or less straight line and thus require intervisibility between the satellite and the tracking device (transmitter, receiver, or reflector); they range from microwave to visible (from 30 to 10^9 MHz). The single or double passage of the electromagnetic signal between the satellite and the tracking device is accurately timed and the distance is obtained by multiplying the time of passage by the propagation speed. The propagation speed is close to the speed of light in vacuum, i.e., 299, 792, 460 m s^{-1}, with the departure being due to the delay of the wave passing through the atmosphere and ionosphere.

Tracked satellite orbits are then computed from the measured (observed) distances and known positions of the tracking stations by solving the equations of motion in the earth's gravity field. This can be done quite accurately (to a centimeter or so) for smaller, spherical, homogeneous, and high-flying spacecraft that can be tracked by lasers. Other satellites present more of a problem; consequently, their orbits are less well known. When orbits are extrapolated into the future, this task becomes known as the *orbit prediction*. Orbit computation and prediction are specialized tasks conducted only by larger geodetic institutions.

B. Satellite Positioning

The first satellite used for positioning was a large, light, passive balloon (ECHO I, launched in 1960) whose only role was to serve as a naturally illuminated moving target on the sky. This target was photographed against the star background from several stations on the earth, and directions to the satellite were then derived from the known directions of surrounding stars. From these directions and from a few measured interstation distances, positions of the camera stations were then computed. These positions were not very accurate though because the directions were burdened by large unpredictable refraction errors (cf., Section II.B). It became clear that *range (distance) measurement* would be a better way to go.

The first practical satellite positioning system (TRANSIT) was originally conceived for relatively inaccurate naval navigation and only later was adapted for much more accurate geodetic positioning needs. It was launched in 1963 and was made available for civilian use 4 years later. The system consisted of several active satel-

lites in circular polar orbits of an altitude of 1074 km and an orbital period of 107 min, the positions (ephemeredes given in the OR coordinate system—cf., Section II.A) of which were continuously broadcast by the satellites to the users. Also transmitted by these satellites were two signals at fairly stable frequencies of 150 and 400 MHz controlled by crystal oscillators. The user would then receive both signals (as well as the ephemeris messages) in his specially constructed TRANSIT *satellite receiver* and compare them with internally generated stable signals of the same frequencies. The beat frequencies would then be converted to *range rates* by means of the Doppler equation

$$\lambda_R = \lambda_T(1 + v/c)(1 + v^2/c^2)^{\frac{1}{2}}, \qquad (53)$$

where v is the projection of the range rate onto the receiver-satellite direction; λ_R, λ_T are the wavelengths of the received and the transmitted signals; and c is the speed of light in vacuum. Finally, the range rates and the satellite positions computed from the broadcast ephemeredes would be used to compute the generic position $\mathbf{r}$ of the receiver (more accurately, the position of receiver's antenna) in the CT coordinate system (cf., Section II.A). More precise satellite positions than those broadcast by the satellites themselves were available from the U.S. naval ground control station some time after the observations have taken place. This control station would also predict the satellite orbits and upload these predicted orbits periodically into the satellite memories.

At most, one TRANSIT satellite would be always "visible" to a terrestrial receiver. Consequently, it was not possible to determine the sequence of positions (trajectory) of a moving receiver with this system; only *position lines* (lines on which the unknown position would lie) were determinable. For determining an accurate position (1 m with broadcast and 0.2 m with precise ephemeredes) of a stationary point, the receiver would have to operate at that point for several days.

Further accuracy improvement was experienced when two or more receivers were used simultaneously at two or more stationary points, and relative positions in terms of interstation vectors $\Delta \mathbf{r}$ were produced. The reason for the increased accuracy was the attenuation of the effect of common errors/biases (atmospheric delays, orbital errors, etc.) through differencing. This relative or differential mode of using the system became very popular and remains the staple mode for geodetic positioning even with the more modern GPS used today.

In the late 1970s, the U.S. military started experimenting with the GPS (originally called NAVSTAR). It should be mentioned that the military have always been vitally interested in positions, instantaneous and otherwise, and so many developments in geodesy are owed to military initiatives. The original idea was somewhat similar to that of the now defunct TRANSIT system (active satellites with oscillators on board that transmit their own ephemeredes) but to have several satellites orbiting the earth so that at least four of them would be always "visible" from any point on the earth. Four is the minimal number needed to get an instantaneous 3D position by measuring simultaneously the four ranges to the visible satellites: three for the three coordinates and one for determining the ever-changing offset between the satellite and receiver oscillators.

Currently, there are 28 active GPS satellites orbiting the earth at an altitude of 20,000 km spaced equidistantly in orbital planes inclined 60 arc-degrees with respect to the equatorial plane. Their orbital period is 12 h. They transmit two highly coherent cross-polarized signals at frequencies of 1227.60 and 1575.42 MHz, generated by atomic oscillators (cesium and rubidium) on board, as well as their own (broadcast) ephemeredes. Two pseudo-random timing sequences of frequencies 1.023 and 10.23 MHz—one called P-code for restricted users only and the other called C/A-code meant for general use—are modulated on the two carriers. The original intent was to use the timing codes for observing the ranges for determining instantaneous positions. For geodetic applications, so determined ranges are too coarse and it is necessary to employ the carriers themselves.

Nowadays, there is a multitude of GPS receivers available off the shelf, ranging from very accurate, bulky, and relatively expensive "*geodetic receivers*" all the way to hand-held and wrist-mounted cheap receivers. The cheapest receivers use the C/A-code ranging (to several satellites) in a point-positioning mode capable of delivering an accuracy of tens of meters. At the other end of the receiver list, the most sophisticated geodetic receivers use both carriers for the ranging in the differential mode. They achieve an accuracy of the interstation vector between a few millimeters for shorter distances and better than $S10^{-7}$, where S stands for the interstation distance, for distances up to a few thousand kilometers.

In addition to the global network of tracking stations maintained by the IGS (cf., Section I.C), there have been networks of continually tracking GPS stations established in many countries and regions; for an illustration, see Fig. 15. The idea is that the tracking stations are used as traditional position control stations and the tracking data are as well used for GPS satellite orbit improvement. The stations also provide "*differential corrections*" for roving GPS users in the vicinity of these stations. These corrections are used to eliminate most of the biases (atmospheric delays, orbital errors, etc.) when added to point positions of roving receivers. As a result of the technological and logistical improvements during the past 20 years, GPS positioning is now cheap, accurate, and used almost

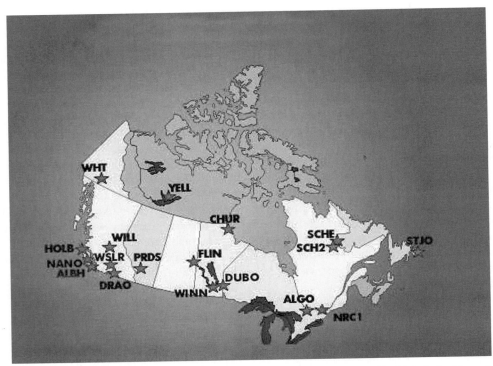

FIGURE 15 Canadian Active Control System. (Source: www.nrcan.gc.ca. Copyright: Her Majesty the Queen in Right of Canada, Natural Resources Canada, Geodetic Survey Division. All rights reserved.)

everywhere for both positioning and precise navigation in preference to classical terrestrial techniques.

The most accurate absolute positions **r** (standard error of 1 cm) are now determined using small, heavy, spherical, high-orbiting, passive satellites equipped with retro-reflectors (LAGEOS 1, LAGEOS 2, STARLETTE, AJISAI, etc.) and laser ranging. The technique became known as *Satellite Laser Ranging* (SLR), and the reason for its phenomenal accuracy is that the orbits of such satellites can be computed very accurately (cf., Section V.A). Also, the ranging is conducted over long periods of time by means of powerful astronomical telescopes and very precise timing devices. Let us just mention that SLR is also used in the relative positioning mode, where it gives very accurate results. The technique is, however, much more expensive than, say, GPS and is thus employed only for scientific investigations

Finally, we have to mention that other satellite-based positioning exist. These are less accurate systems used for nongeodetic applications. Some of them are used solely in commercial application. At least one technique deserves to be pointed out, however, even though it is not a positioning technique per se. This is the synthetic aperture radar interferometry (INSAR). This technique uses collected reflections from a space-borne radar. By sophisticated computer processing of reflections collected during two overflights of the area of interest, the pattern of ground deformation that had occurred between the two flights can be discerned (Massonnet *et al.*, 1993). The result is a map of relative local deformations, which may be used, for instance, as a source of information on co-seismic activity. Features as small as a hundred meters across and a decimeter high can be recognized.

C. Gravitational Field Studies by Satellites

The structure of the earth's gravity field was very briefly mentioned in Section III.H, where the field wavelengths were discussed in the context of the spectral description of the global field. A closer look at the field reveals that:

1. The field is overwhelmingly radial (cf., Section III.A) and the first term in the potential series, GM/r, is already a fairly accurate (to about 10^{-3}) description of the field; this is why the radial field is used as a model field (cf., Section III.B) in satellite studies.
2. The largest departure from radiality is described by the second degree term $J_{2,0}$ (cf., Section IV.D), showing the ellipticity of the field, which is about 3 orders of magnitude smaller than the radial part of the field.
3. The remaining wavelength amplitudes are again about 3 orders of magnitude smaller and they further decrease with increasing wave number ℓ. The

decrease of amplitude is seen, for instance in Fig. 8 in the article "Global Gravity." In some studies it is possible to use a mathematical expression describing the decrease, such as the experimental Kaula's rule of thumb, approximately valid for ℓ between 2 and 40:

$$\sqrt{\Sigma_{m=2}^{\ell}\left(C_{\mathrm{lm}}^2 + S_{\mathrm{lm}}^2\right)} \approx R * 10^{-5}/\ell^2, \qquad (54)$$

where C_{lm} and S_{lm} are the potential coefficients (cf., Eq. (4) in "Global Gravity").

As discussed in Section V.A, the earth's gravity field also gets smoother with altitude. Thus, for example, at the altitude of lunar orbit (about 60 times the radius of the earth), the only measurable departure from radiality is due to the earth's ellipticity and even this amounts to less than 3×10^{-7} of the radial component. Contributions of shorter wavelength are 5 orders of magnitude smaller still. Consequently, a low-orbiting satellite has a "bumpier to use a satellite as a gravitation-sensing device, we get more detailed information from low-orbiting spacecraft.

The idea of using satellites to "measure" gravitational field (we note that a satellite cannot sense the total gravity field, cf., Section III.A) stems from the fact that their orbital motion (free fall) is controlled predominantly by the earth's gravitational field. There are other forces acting on an orbiting satellite, such as the attraction of other celestial bodies, air friction, and solar radiation pressure, which have to be accounted for mathematically. Leaving these forces alone, the equations of motion are formulated so that they contain the gravitational field described by potential coefficients C_{lm} and S_{lm}. When the observed orbit does not match the orbit computed from the known potential coefficients, more realistic potential coefficient values can be derived. In order to derive a complete set of more realistic potential coefficients, the procedure has to be formulated for a multitude of different orbits, from low to high, with different inclinations, so that these orbits sample the space above the earth in a homogeneous way. We note that because of the smaller amplitude and faster attenuation of shorter wavelength features, it is possible to use the described *orbital analysis* technique only for the first few tens of degrees ℓ. The article "Global Gravity" shows some numerical results arising from the application of this technique.

Other techniques such as "satellite-to-satellite tracking" and "gradiometry" (see "Global Gravity") are now being used to study the shorter wavelength features of the gravitational field. A very successful technique, "satellite altimetry," a hybrid between a positioning technique and gravitational field study technique (see "Global Gravity") must be also mentioned here. This technique has now been used for some 20 years and has yielded some important results.

SEE ALSO THE FOLLOWING ARTICLES

EARTH SCIENCES, HISTORY OF • EARTH'S MANTLE (GEO-PHYSICS) • EXPLORATION GEOPHYSICS • GEOMAGNETISM • GLOBAL GRAVITY MODELING • GRAVITATIONAL WAVE ASTRONOMY • RADIO ASTRONOMY, PLANETARY • RE-MOTE SENSING FROM SATELLITES • TECTONOPHYSICS

BIBLIOGRAPHY

Berthon, S., and Robinson, A. (1991). "The Shape of the World," George Philip Ltd., London.

Boal, J. D., and Henderson, J. P. (1988). The NAD 83 Project—Status and Background. *In* "Papers for the CISM Seminars on the NAD'83 Redefinition in Canada and the Impact on Users" (J. R. Adams ed.), The Canadian Institute of Surveying and Mapping, Ottawa, Canada.

Bomford, G. (1971). "Geodesy," 3rd ed., Oxford Univ. Press, London.

Boucher, C., and Altamini, Z. (1996). "International Terrestrial Reference Frame." *GPS World* **7**(9), 71–75.

Bruns, H. (1878). "Die Figur der Erde," Publication des Königlichen Preussischen Geodätischen Institutes, Berlin, Germany.

Burša, M., Raděj, K., Šíma, Z., True, S. A., and Vatrt, V. (1997). "Determination of the Geopotential Scale Factor from TOPEX/POSEIDON Satellite Altimetry," *Studia Geophys. Geod.* **41**, 203–216.

Grafarend, E. W., and Sansò, F., ed. (1985). "Optimization and Design of Geodetic Networks," Springer-Verlag, Berlin/New York.

Heiskanen, W. A., and Moritz, H. (1967). "Physical Geodesy," Freeman, San Francisco.

Helmert, F. R. (1880). "Die mathematischen und physikalishen Theorien der höheren Geodäsie," Vol. I, Minerva, G. M. B. H. Reprint, 1962.

Hirvonen, R. A. (1971). "Adjustment by Least Squares in Geodesy and Photogrammetry," Ungar, New York.

International Association of Geodesy (1974). "The International Gravity Standardization Net, 1971," Special Publication No. 4, Paris, France.

International Association of Geodesy (1980). "The geodesist's handbook," *Bull. Géodesique* **54**(3).

Lee, L. P. (1976). "Conformal Projections Based on Elliptic," Cartographica Monograph 16, B. V. Gutsell, Toronto, Canada.

Massonnet, D., Rossi, M., Carmona, C., Adragna, F., Peltzer, G., Feigl, K., and Rabaute, T. (1993), "The displacement field of the Landers earthquake mapped by radar interferometry," *Nature* **364**.

Melchior, P. (1966). "The Earth Tides," Pergamon, Elmsford, NY.

Mikhail, E. M. (1976). "Observations and Least Squares," IEP-A dun-Donnelley Publisher.

Molodenskij, M. S., Eremeev, V. F., and Yurkina, M. I. (1960). "Methods for Study of the External Gravitational Field and Figure of the Earth," Translated from Russian by the Israel Program for Scientific Translations for the Office of Technical Services, U.S. Department of Commerce, Washington, DC., 1962.

Mueller, I. I. (1969). "Spherical and Practical Astronomy as Applied to Geodesy," Ungar, New York.

Pick, M., Pícha, J., and Vyskočil, V. (1973). "Theory of the Earth's Gravity Field," Elsevier, Amsterdam/New York.

Ries, J. C., Eanes, R. J., Shum, C. K., and Watkins, M. M. (1992). "Progress in the determination of the gravitational coefficient of the earth," *Geophys. Res. Lett.* **19**(6), 529–531.

Schwarz, C. R., ed. (1989). "North American Datum of 1983," NOAA Professional Paper NOS 2, National Geodetic Information Center, National Oceanic and Atmospheric Administration, Rockville, MD.

Stokes, G. G. (1849). "On the variation of gravity at the surface of the earth," *Trans. Cambridge Philos. Soc.* **VIII,** 672–695.

U.S. Department of Commerce (1973). "The North American Datum," Publication of the National Ocean Survey of NOAA, Rockville, MD.

Vaníček, P., and Krakiwsky, E. J. (1986). "Geodesy: The Concepts," 2nd ed., North-Holland, Amsterdam.

Vaníček, P., and Sjöberg, L. E. (1991). "Reformulation of Stokes's theory for higher than second-degree reference field and modification of integration kernels," *J. Geophys. Res.* **96**(B4), 6529–6539.

Vaníček, P., Huang, J., Novák, P., Véronneau, M., Pagiatakis, S., Martinec, Z., and Featherstone, W. E. (1999). "Determination of boundary values for the Stokes-Helmert problem," *J. Geod.* **73,** pp. 180–192.

Zakatov, P. S. (1953). "A Course in Higher Geodesy," Translated from Russian by the Israel Program form Scientific Translations for the Office of Technical Services, U.S. Department of Commerce, Washington, DC., 1962.

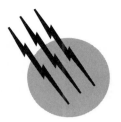

Geoenvironmental Engineering

Charles D. Shackelford
Colorado State University

I. Technical Considerations
II. Waste Containment
III. Remediation

GLOSSARY

Containment Control of migration of gaseous, liquid, or solid contaminated media from a site by use of containment measure(s) such as engineered liners, covers, or *in situ* vertical barriers.

Contaminant Substance that can cause harm to human health or the environment.

Cover (or cap) system One or more layers of material, such as soils, suitable mineral wastes, and geosynthetics, constructed on the surface of a site or waste impoundment and designed to control egress of contaminants from and infiltration of precipitation into a source of contamination.

Hazardous solid waste (HSW) A solid waste that meets any one of the following four criteria: (1) the waste is specifically listed as a hazardous waste by regulation, (2) the waste is a mixture containing a hazardous waste, (3) the waste is derived from the treatment, storage, or disposal of a hazardous waste (e.g., the ash residue resulting from incineration of a hazardous waste is also a hazardous waste), or (4) the waste exhibits the characteristics of ignitability, corrosivity, reactivity, or toxicity.

Hydrocarbons (HCs) Compounds consisting of hydrogen and carbon.

Leachate A chemical solution containing contaminants derived from water percolating through a solid waste.

Liner A low-permeability barrier used to prevent or reduce the migration of potentially harmful chemicals, or contaminants, into the surrounding environment.

Liner system An engineered barrier system that is composed of combination of one or more liners and drainage layers. The drainage layers are placed immediately above the liner(s) to drain the leachate that has percolated through the waste to a collection system for removal and treatment.

Municipal solid waste (MSW) "Nonhazardous" wastes from residential, commercial, institutional, and some industrial sources.

Permeability A measure of the relative ease with which a fluid (liquid or gas) will flow through a porous medium; the higher the permeability with respect to the fluid, the easier the fluid will flow through the medium.

Remediation The process of restoring land that has been polluted with contaminants.

Vadose zone The zone between the ground surface and the water table within which the water pressure is negative. The zone is divided into (1) a saturated portion, or capillary fringe, located just above the water table due to capillary rise, and (2) an unsaturated zone.

Water table The surface of water existing in the ground on which the fluid pressure is atmospheric.

Encyclopedia of Physical Science and Technology, Third Edition, Volume 6

GEOENVIRONMENTAL ENGINEERING can be described as the engineering of geologic (earthen) and geosynthetic (polymer) materials for problems related to the protection of human health and the environment. The primary environmental problems addressed by geoenvironmental engineers are related to either the potential for or the existence of subsurface (below the ground surface) pollution, also referred to as "contaminated land." In this context, contaminated land includes the solid portion of the ground (soil, rock) as well as the water and gas or vapor in the ground. Thus, geoenvironmental engineering is concerned with the protection of uncontaminated land as well as the remediation, or clean up, of contaminated or polluted land.

Subsurface pollution can result from a multiplicity of activities. For example, subsurface pollution can result from relatively localized or "point"-source activities, such as industrial chemical spills (e.g., jet fuel spills), leaking waste containment facilities (e.g., landfills), and leaking above-ground and underground storage tanks (e.g., oil storage tank farms, buried gasoline reservoirs), as well as from relatively dispersed or "non-point"-source activities, such as infiltration of pesticides spread over large agricultural areas. In many cases, subsurface pollution has resulted from controlled activities in the distant past (e.g., more than 30–40 years ago in the United States) not perceived to be a potential threat to the environment. In other, more recent cases, subsurface pollution typically has resulted from uncontrolled activities, such as accidental spills, or from negligent or illegal disposal practices.

With respect to prevention of subsurface pollution, geoenvironmental engineering pertains to the engineering aspects related to design, construction, and monitoring of containment facilities for a wide variety of waste disposal activities. Examples of such facilities include landfills for disposal of municipal and hazardous solid waste, impoundments for disposal of waste and by-products resulting from mining activities, lagoons for disposal of animal waste, evaporation ponds for disposal of contaminated process waters, and isolation barriers for disposal of high-level nuclear waste. In all of these applications, the principal goal of the containment facility is to prevent the escape of contaminants associated with the waste thereby preventing pollution of the surrounding environment.

With respect to the existence of subsurface pollution, geoenvironmental engineering pertains to the engineering aspects related to the remediation of the contaminated land. Remediation, also commonly referred to as reclamation, of contaminated land may be defined as the process of restoring land that has been polluted with contaminants by some activity. Although the words "remediation" and "reclamation" often are used interchangeably in terms of environmental contamination, the two words arguably

have slightly different meanings. For example, reclamation of contaminated land refers to restoring the land for reuse, whereas remediation of contaminated land refers to restoration that will prevent or minimize a real (e.g., medical) or perceived (e.g., legal) risk of harm to humans (i.e., regardless of whether or not the land will be reused). This distinction is relevant in the United Kingdom, for example, where the term reclamation often is associated with derelict land (e.g., former industrial land that is no longer used) and remediation for land that is contaminated. However, as derelict land also is often contaminated, there are many situations where reclamation involves both minimization and prevention of risk as well as reuse of the land. In the United States, the two terms tend to be used interchangeably, with remediation probably being the more common term.

I. TECHNICAL CONSIDERATIONS

A. Materials

1. Geologic Materials

Geologic or earthen materials include both rock and soil. Rock exists as geologic formations, also referred to as intact or parent rock, and as fragments. Soil is the product of the effect of chemical and physical weathering or degradation of preexisting or parent material (rock or soil).

In general, earthen materials comprise three phases: (1) a gas phase, (2) a liquid phase, and (3) a solid or mineral phase, where the gas and liquid phases collectively comprise the fluid phase of the soil. In the case of rock fragments and soils, the earthen materials comprise a multitude of individual solid (or mineral) particles in contact with each other and pores (or voids) that exist between the particles. As a result, rock fragments and soils also are referred to as particulate materials.

Since the possible range of particle sizes associated with particulate earthen materials is tremendous, various classification systems based on particle size have been devised for the purpose of categorizing the behavior of soil and rock. For example, as shown in Table I, the types of earthen materials based on particle size include rock fragments consisting of boulders and cobbles, coarse-grained

TABLE I Types of Earthen Materials Based on Particle Size

Category	Type	Range of particle (or grain) sizes
Rock fragments	Boulders	>300 mm (>12 in.)
	Cobbles	76–300 mm (3–12 in.)
Coarse-grained soils	Gravel	4.75–76 mm (0.75–3 in.)
	Sand	0.075–4.75 mm (0.003–0.75 in.)
Fine-grained soils	Silt	0.002–0.075 mm (2–75 μm)
	Clay	<0.002 mm (<2 μm)

soils consisting of gravels and sands, and fine-grained soils consisting of silts and clays. Coarse-grained soils contain particles that can be seen with the unaided eye, whereas fine-grained soils require a microscope to observe individual particles.

In general, naturally occurring soils (as opposed to mined or specially processed soils) may comprise various percentages of gravel, sand, silt, and/or clay based on particle size. In this case, a fine-grained soil typically is assumed to be a soil that contains greater than 50% (w/w) of silt and/or clay, whereas a coarse-grained soil must contain 50% (w/w) or more of gravel and/or sand. Such soils are further classified as either a silt or a clay in the case of fine-grained soils, or as either a gravel or sand in the case of coarse-grained soils, based on which soil type dominates the fine-grained or coarse-grained fractions, respectively, of the soil. Coarse-grained soils that contain small amounts of fines (e.g., <5% silts and/or clays) typically are referred to as "clean" soils.

As noted in Table I, the term "clay" refers to that part of the soil that is composed of particle sizes less than 2 μm. However, the term "clay" also can refer to the clay mineral content of a soil regardless of the particle size. Finally, the term "clay" often is used to denote a soil mass that possesses properties similar to those of a pure clay even though only a relative small percentage of the overall soil mass consists of clay-sized particles and/or clay minerals. For example, mixtures of sand and bentonite, a high-swelling clay, containing 4–10% (w/w) of bentonite typically possess some properties that are characteristic of clay soils, even though 90% (w/w) or more of these mixtures is made up of sand.

The pore spaces in earthen materials may be occupied by gases, liquids, or both. For example, with respect to the liquid phase in a soil, a "dry soil" results when the pores are completely occupied by a gas, typically air (Fig. 1a), whereas a "saturated soil" results when the pores are completely filled with a liquid, typically water (Fig. 1b). An "unsaturated" soil results when the pores comprise both a gas and a liquid (Fig. 1c). In the case of unsaturated soils containing both water and air (Fig. 1c), the water is the "wetting fluid" because the water is preferentially absorbed (i.e., relative to the air, the "nonwetting fluid) to

the surface of the solid soil particles and therefore tends to wet the surface.

When the liquid phase in a soil is water, the liquid phase is more specifically referred to as the "aqueous phase." In reality, the aqueous phase in soils is made up of the compound water (H_2O) and other chemical compounds or species that are dissolved in the water. In this case, the water serves as the "solvent" and the dissolved compounds are referred to as "solutes." Solutes in aqueous solutions may exist in a variety of forms or species. For example, solutes may exist as uncharged chemical species, or neutral compounds, as well as charged chemical species, or ions. Positively charged ions are called cations, and negatively charged ions are called anions. For example, ionic species common in natural pore waters of soils include the monovalent cations of potassium (K^+) and sodium (Na^+), the divalent cations of calcium (Ca^{2+}) and magnesium (Mg^{2+}), the monovalent anion of chlorine, or chloride (Cl^-), and the divalent sulfate anion (SO_4^{2-}).

The concentrations of the solutes in the aqueous phase serve as a basis for evaluating the quality of the water. For example, water that contains an abundance of dissolved calcium and magnesium commonly is referred to as "hard water." Contamination of soils results when specific solutes, referred to as contaminants of concern (COCs), exist in the subsurface, and pollution of the subsurface results when these COCs exist at concentrations or levels that are considered to be harmful to human health and the environment.

Because of the porous nature of earthen materials, fluids can flow through earthen materials when the pores are interconnected. As a result, earthen materials also are referred to as porous media. In general, the ease with which a fluid will flow through a porous medium increases with an increase in the size of the pores, and the size of the pores increases with an increase in the size of the particles making up the porous medium. Thus, according to Table I, the ease with which a fluid will flow through soils decreases in the following order: gravel > sand > silt > clay. The ability of contaminants to migrate through the interconnected pores of soils and rocks is the most important technical consideration with respect to either preventing the migration of contaminants into the subsurface in the case of waste disposal, or removing contaminants from the subsurface in the case of remediation. For example, only soils that tend to restrict fluid flow (e.g., clays) are considered for use as containment barriers in waste disposal applications, whereas remediation of subsurface pollution typically is effective only in soils that allow rapid movement of fluids (e.g., sands and gravels).

As a liquid-saturated soil begins to dry, some of the liquid (e.g., water) in the soil pores is displaced by gas (e.g., air). In this initial stage of drying, the liquid phase is continuous throughout the soil and therefore the liquid

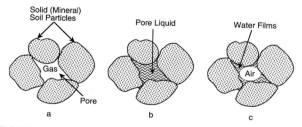

FIGURE 1 Particulate nature of soils: (a) dry soil; (b) saturated soil; (c) unsaturated soil.

can flow freely through the soil. However, the gas exists as isolated, or occluded, bubbles. In this condition, the gas phase is discontinuous, restricting free flow of the gas. As the soil is dried further, the gas phase continues to occupy a greater percentage of the pore space until the gas phase eventually becomes continuous and can flow freely through the soil. However, at the same time, the liquid phase is continuously occupying a lesser percentage of the pore space, such that the liquid phase eventually becomes discontinuous, restricting the free flow of the liquid. Upon rewetting an unsaturated soil, the invading liquid displaces the remnant gas, and the process reverses. Thus, in general, the greater the degree of liquid saturation (i.e., percentage of the total pore space occupied by the liquid), the greater the ease with which a liquid will flow through the soil, whereas the lesser the degree of liquid saturation, the greater the ease with which a gas will flow through the soil.

2. Geosynthetics

Geosynthetics are manufactured synthetic materials (i.e., made from polymers or hydrocarbon chains) that are used for a wide range of engineering applications. Seven categories of geosynthetic materials are manufactured (Koerner 1998): geocomposites, geogrids, geomembranes, geonets, geopipes, geotextiles, and geosynthetic clay liners. A geocomposite consists of a combination of one or more geosynthetics, specifically a geogrid, a geotextile, a geomembrane, and/or a geonet, with another material. Geotextiles are used primarily for applications requiring separation, filtration, reinforcement, and drainage. Geogrids are used as reinforcement to improve the strength of soil or other materials. Geonets are relatively thin, planar geosynthetics used for drainage applications, whereas geopipes are buried plastic pipes used for drainage. Geomembranes and geosynthetic clay liners are thin (≤ 10 mm), relatively impervious geosynthetics that are used as barriers in containment applications for waste disposal and *in situ* remediation. Geosynthetics have gained widespread use only since about 1990, but their use in geoenvironmental engineering applications involving liquid and waste containment has grown substantially in the last decade. A detailed description of the properties, characteristics, and uses of each of these types of geosynthetic can be found in Koerner (1998).

B. Fluids

1. Liquids

The properties of the liquid phase of the porous medium (i.e., pore liquid) will depend on the nature and chemical composition of the pore liquid. As shown in Fig. 2,

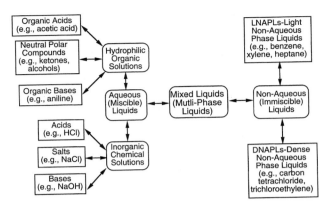

FIGURE 2 Categories of liquids. [Adapted from Shackelford and Jefferis (2000).]

pore liquids may be categorized as aqueous liquids, i.e., solutions containing contaminants that are miscible in water (also known as hydrophilic or "water-loving"), or nonaqueous liquids consisting of organic compounds that are immiscible in water (also known as hydrophobic or "water-hating"). In addition, mixtures of both aqueous and nonaqueous liquids resulting in the formation of two separate liquid phases may occur. Aqueous liquids contain inorganic chemicals (acids, bases, salts) and/or hydrophilic organic compounds. In the case of applications involving nuclear wastes, aqueous solutes also may be radioactive.

Hydrophilic organic compounds are distinguished from hydrophobic organic compounds based on the concept of "like dissolves like," i.e., polar organic compounds usually will readily dissolve in water, a polar molecule, whereas nonpolar organic compounds are repelled by water. A hydrophobic compound also is typically further categorized as either an LNAPL (light nonaqueous-phase liquid) or DNAPL (dense nonaqueous-phase liquid) based on whether the density of the compound is lower or greater than that of water, respectively. Thus, LNAPLs will tend to float on water and DNAPLs will tend to sink in water.

In general, even though NAPLs are considered to be aqueous immiscible, most NAPLs have some aqueous solubility. Although the aqueous solubility of NAPLs is small relative to hydrophilic compounds, the solubilities of most NAPLs typically result in aqueous-phase concentrations that far exceed regulatory limits. Thus, NAPLs existing in the subsurface represent sources of continuous aqueous-phase contamination when the NAPLs come into contact with flowing ground water.

Subsurface pollution can result from any category of liquid shown in Fig. 2. For example, toxic heavy or trace metals are derived from inorganic chemical solutions resulting from industrial activities as well as from the leaching of metals from solid waste during passage of infiltrating water. Some heavy metals commonly of environmental concern include arsenic, barium, cadmium, chromium, copper, lead, mercury, and nickel.

Organic solvents frequently are used as cleaning (e.g., degreasing) agents in industrial manufacturing and processing activities. Among the many types of organic solvents used in industry, a group of DNAPLs known as the chlorinated hydrocarbons are particularly prevalent. In particular, trichloroethethylene (TCE) and tetrachloroethylene, also known as perchloroethylene (PCE), are two of the most ubiquitous groundwater pollutants in industrialized countries.

Another major group of chemicals commonly encountered in conjunction with groundwater pollution are a group of LNAPLs referred to as the BTEX (benzene, toluene, ethyl benzene, and xylene) compounds. The BTEX compounds are derived from petroleum, are major constituents in gasoline and other petroleum products, and represent a significant environmental concern with respect to pollution of the subsurface.

2. Gases

The gas phase in liquid unsaturated porous media can contain a wide variety of important chemical constituents. For example, the amount of carbon dioxide gas (CO_2) in the soil can affect the chemistry of aqueous solutions, and the tendency for aerobic and anaerobic conditions is reflected by the existence and amount of oxygen gas (O_2).

Gas-phase contamination can result from the direct production of a gas, such as the production of methane gas (CH_4) resulting from the decomposition of municipal solid waste, or from the volatility of chemical compounds. Volatility of hydrocarbons is related to the vapor pressure of the compound (i.e., the pressure in the vapor above the liquid at equilibrium). In general, the greater the vapor pressure, the more volatile the compound. For example, volatile organic compounds (VOCs) are hydrocarbons with vapor pressures at 20°C that are greater than 1 mm Hg (1.3×10^{-3} atm). Semivolatile organic compounds (SVOCs) are hydrocarbons with vapor pressures between 10^{-10} mm Hg (1.3×10^{-13} atm) and 1 mm Hg (1.3×10^{-3} atm). Although the lower limit of vapor pressures for SVOCs is 10^{-10} mm Hg (1.3×10^{-13} atm), hydrocarbons with vapor pressures $<10^{-7}$ mm Hg ($<1.3 \times 10^{-10}$ atm) are not expected to volatilize significantly. Volatility of chemical compounds also is indicated by the boiling point (i.e., the lower the boiling point, the greater the susceptibility to volatilization via heating).

Five VOCs that are commonly associated with pollution of drinking water are tetrachloroethylene (PCE), trichloroethylene (TCE), vinyl chloride (VC), 1,2-dichloroethane (DCE), and carbon tetrachloride (CCl_4). Examples of SVOCs include cresol, phenol, pyrene, 1,2- dichlorobenzene, and 1,4-dichlorobenzene.

C. Fluid Flow

1. Liquid Flow

a. Hydraulic conductivity. Hydraulic conductivity refers to the coefficient of proportionality in Darcy's law describing liquid flow through porous media. For example, flow of liquids in and through soil may be described in accordance with Darcy's law as follows:

$$q = \frac{Q}{A} = -ki, \tag{1}$$

where q is the liquid flux or flow rate (length/time, $L\,T^{-1}$), Q is the volumetric flow rate of the liquid ($L^3\,T^{-1}$), A is the total cross-sectional area of the soil (solids plus voids) perpendicular to the direction of flow (L^2), k is the hydraulic conductivity or coefficient of permeability ($L\,T^{-1}$), and i is the hydraulic gradient (dimensionless). The hydraulic gradient represents the energy or driving force for liquid flow. The hydraulic conductivity is a material property that reflects the relative ease of liquid flow through porous media, and is among the most variable material properties in engineering. Approximate ranges in the hydraulic conductivities under saturated conditions of various materials commonly encountered in geoenvironmental engineering are provided in Table II.

For a given hydraulic gradient, the fluid flow rate q is directly proportional to the hydraulic conductivity of the porous medium in accordance with Eq. (1). Thus, in accordance with Eq. (1), the objective of most low-permeability

TABLE II Approximate Ranges of Saturated Hydraulic Conductivities Expected for Materials Commonly Encountered in Geoenvironmental Engineering Based on Permeation with Water

Material	Saturated hydraulic conductivity k_{sat} (m/sec)	Comments
Gravel	10^{-2}–10^{-3}	Values based on "clean" soils; variation in k_{sat} based on particle size distribution
Sand	10^{-3}–10^{-5}	
Silt	10^{-5}–10^{-8}	Variation in k_{sat} based on mineralogical composition of silt particles
Clay	10^{-8}–10^{-12}	Variation in k_{sat} based on mineralogical composition of clay particles
Geosynthetic clay liner	10^{-10}–10^{-11}	Values based on sodium bentonite sandwiched between two geotextiles
Sand–bentonite mixture	10^{-9}–10^{-10}	Values based on a mixture of clean sand (w/o fines) and 4–10% (w/w) sodium bentonite

barriers used for waste containment systems prescribed by regulation is to minimize the amount of liquid percolation by (1) achieving a suitably low saturated hydraulic conductivity for the barrier material and (2) restricting the amount of liquid buildup on top of the barrier to maintain a suitably low hydraulic gradient. For example, in the United States, the regulatory maximum value for the saturated hydraulic conductivity k_{sat} prescribed for municipal and solid waste landfill liners is 10^{-9} m/sec and the maximum height of liquid allowed to accumulate on top of the liner is 300 mm. This regulatory limit on k_{sat} may vary depending on the country, the specific component of the containment facility (e.g., liner versus cover), and/or the specific application (e.g., solid waste disposal versus nuclear waste disposal).

b. Compatibility.

The hydraulic conductivity k is a function of the properties of both the solid matrix of the porous medium and the liquid in accordance with the following equation:

$$k = K\left(\frac{\gamma}{\mu}\right) = K\left(\frac{\rho g}{\mu}\right) = K\left(\frac{g}{\upsilon}\right), \qquad (2)$$

where K is the intrinsic, absolute, or specific permeability of the soil (L^2), γ is the unit weight of the liquid (mass/length^2/time, $M\,L^{-2}\,T^{-1}$), μ is the absolute or dynamic viscosity of the liquid ($M\,L^{-1}\,T^{-1}$), ρ is the mass density of the liquid ($M\,L^{-3}$), g is acceleration due to gravity ($L\,T^{-2}$), and υ is the kinematic viscosity of the liquid ($L^2\,T^{-1}$). The intrinsic permeability K represents the influence of the soil structure on the hydraulic conductivity, whereas the ratio γ/μ reflects the influence of the permeant liquid on the hydraulic conductivity.

Compatibility refers to the effect of potential interactions between the permeating liquid and the solid particles on the hydraulic conductivity of the porous medium. For example, changes in the pore spaces available for flow can occur when a clay soil is permeated with liquids with significantly different chemical properties relative to the naturally occurring pore water, such as in the case of clay soils used for waste containment barriers. These changes result from incompatibility between the permeant liquid and the solid soil particles.

With respect to Eq. (2), when the permeant liquid has a value of γ/μ that is significantly different than that of water, incompatibility between the permeating liquid and the porous material is indicated by an increase in the intrinsic permeability of the porous material (i.e., $\Delta K > 0$). In many cases, the values of γ/μ for the permeant liquid and water are sufficiently close such that incompatibility also will be reflected by an increase in hydraulic conductivity (i.e., $\Delta k > 0$). From a practical viewpoint, an increase in hydraulic conductivity may be considered significant if the final (steady-state) value exceeds the regulatory maximum value of the hydraulic conductivity. When this occurs, the soil being considered for use as the barrier layer in the containment system and the waste liquid are considered "incompatible," and a different soil and/or material must be evaluated for use as the barrier.

c. Hydraulic conductivity of unsaturated media.

The concavity of the interface toward the nonwetting fluid in a liquid unsaturated soil (e.g., see Fig. 1c) reflects a higher pressure in the nonwetting fluid relative to the wetting fluid. The resulting difference between the pressure in the nonwetting fluid and the pressure in the wetting fluid is called the capillary pressure. In the case where the nonwetting fluid is at atmospheric pressure, or zero reference pressure, the pressure in the wetting fluid is negative, reflecting that the liquid is in tension. As the pores desaturate further, the concavity of the interface increases, reflecting an increase in the capillary pressure and an increase in the magnitude of the negative pressure in the wetting fluid. In the case where the liquid is water, the capillary pressure also often is referred to as the soil–water suction. As a result, the capillary pressure, or soil–water suction, increases in magnitude as the porous medium desaturates.

The hydraulic conductivity of a porous medium with respect to the wetting fluid (e.g., water) decreases as the medium desaturates, all other factors being equal. Thus, the maximum value of the hydraulic conductivity for a given porous medium is the saturated hydraulic conductivity. The decrease in hydraulic conductivity with decreasing saturation results from two effects: (1) the cross-sectional area available for flow of the liquid via the liquid phase decreases as the porous medium desaturates (e.g., compare Figs. 1b and 1c), and (2) the pathways for liquid migration through the liquid phase in porous medium become more tortuous. The increased tortuosity results from two effects. First, as the medium desaturates, the liquid is forced to migrate closer to the particle surface (see Fig. 1c) resulting in an increase in the effective length of the migration pathway through the medium. Second, since the most conductive pores are the largest pores, these pores drain faster than the smaller pores resulting in some of the larger pores being excluded from liquid flow.

Since larger pores drain faster than smaller pores, the rate of decrease in hydraulic conductivity for porous media that are made up primarily of larger pores, such as coarse-grained soils, is greater than that associated with porous media made up primarily of smaller pores, such as fine-grained soils. Thus, although the saturated hydraulic conductivity of coarse-grained soils is significantly higher than that of fine-grained soils (Table II), the reverse situation occurs as the degree of liquid saturation of the soil

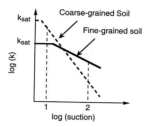

FIGURE 3 Relative trends in hydraulic conductivity of two soil types as a function of soil suction.

decreases (i.e., as soil–water suction increases). These trends are reflected schematically in Fig. 3.

2. Gas Flow

The migration of gas becomes an important consideration in certain geoenvironmental engineering applications when the gas content in a porous medium is sufficiently high such that the gas phase is continuous. Examples of such applications include the removal of VOCs and SVOCs from the unsaturated or vadose zone above the water table in the subsurface via the gas phase and the minimization of oxygen influx or radon efflux from engineered covers for tailings disposal applications. In the former case, removal efficiency is improved as gas permeability increases, whereas in the latter case, the objective is to minimize the gas permeability and therefore minimize the gas flow.

When the gas phase is continuous, gas may migrate through the porous medium under a pressure gradient in accordance with an analogous form of Darcy's law for liquid-phase flow as follows:

$$q_g = \frac{Q_g}{A} = -\frac{K_g}{\mu_g} i_P, \tag{3}$$

where q_g is the gas flux or flow rate, Q_g is the volumetric flow rate of the gas, K_g is the intrinsic permeability of the porous medium with respect to gas flow, μ_g is the absolute or dynamic viscosity of the gas, and i_P is the pressure gradient.

As with liquid flow, the intrinsic permeability K_g with respect to gas flow is a function of only the porous medium. However, the ability of a gas to migrate through a porous medium will depend on the liquid content in the porous medium. When the porous medium is completely dry, K_g represents the continuity of flow through the pores attributable only to the porous solids (i.e., $K_g = K$). However, as the pores are filled with liquid (e.g., water) such that the percentage of gas-filled pores decreases, the gas may become occluded in some pores such that some of the pathways for gas-phase migration are cut off. When the porous medium is sufficiently saturated with liquid such

that gas in the pores exists as occluded bubbles (i.e., bubbles completely surrounded by the liquid phase), the gas phase becomes discontinuous and gas-phase migration via Eq. (3) effectively stops. Thus, the gas-phase permeability as reflected by K_g also is a function of the degree of liquid saturation of the porous medium.

D. Contaminant Migration

Most geoenvironmental engineering applications involve some consideration of the rate and extent of migration of contaminants through a porous medium, often referred to as contaminant transport. In the case of waste containment, the objective is to minimize the rate of contaminant migration from the containment facility, whereas in the case of remediation, the objective may be either to enhance or to minimize the rate of contaminant migration.

1. Aqueous-Phase Transport

Aqueous-phase, or aqueous miscible, contaminant transport in porous media is controlled by a variety of physical, chemical, and biological processes. As indicated in Table III, the primary physical processes governing aqueous miscible contaminant transport are advection, diffusion, and dispersion. Diffusion tends to be the dominant transport process under relatively low flow-rate conditions, such as occur through clay barriers used for containment in waste disposal and remediation applications. However, advection, or hydraulically driven transport, and dispersion dominate under relatively high flow-rate conditions, such as occur for contaminant migration through coarse-grained aquifer materials.

The chemical and biological processes commonly considered to potentially affect contaminants during aqueous-phase migration through porous media are described in Table IV. Some of the processes, such as adsorption, radioactive decay, precipitation, hydrolysis, and biodegradation, typically are considered as "attenuation processes" in that these processes generally immobilize,

TABLE III Physical Processes Affecting Aqueous Miscible Contaminant Transport[a]

Process	Definition	Significance
Advection	Mass transport due to bulk water flow	Most dominant process in high-flow-rate media
Diffusion	Mass spreading due to concentration gradients	Most dominant process in low-flow-rate media
Dispersion	Mass spreading due to heterogeneities in the flow field	Results in greater mass spreading than predicted by advection

[a] National Research Council (1990), Shackelford and Rowe (1998).

TABLE IV   Chemical and Biological Processes Affecting Aqueous Miscible Contaminant Transport[a]

Process	Definition	Significance
Sorption	Partitioning of contaminant between pore water and porous medium	Limiting case of ion exchange; adsorption reduces the rate of apparent contaminant migration and makes contaminant removal difficult by desorption
Radioactive decay	Irreversible decline in the activity of a radionuclide	Important attenuation mechanism when the half-life for decay is $\leq$ residence time in flow system; results in by-products
Dissolution/precipitation	Reactions resulting in release of contaminants from solids or removal of contaminants as solids	Dissolution is mainly significant at the source or at the migration front; precipitation is an important attenuation mechanism, particularly in high-pH system (pH >7)
Acid/base	Reactions involving a transfer of protons (H^+)	Important in controlling other reactions (e.g., dissolution/precipitation)
Complexation	Combination of anions and cations into a more complex form	Affects speciation that can affect sorption, solubility, etc.
Hydrolysis/substitution	Reaction of a halogenated organic compound with water or a component ion of water (hydrolysis) or with another anion (substitution)	Typically make an organic compound more susceptible to biodegradation and more soluble
Oxidation/reduction (redox)	Reactions involving a transfer of electrons	Important attenuation mechanism in terms of controlling precipitation of metals
Biodegradation	Reactions controlled by microorganisms	Important attenuation mechanism for organic compounds; may result in undesirable by-products

[a] National Research Council (1990), Shackelford and Rowe (1998).

retard, or otherwise degrade chemical constituents that exist in the aqueous phase of the porous media. However, in some cases, these attenuation processes may not be effective in actually reducing the potential impact of the contaminants. For example, the radioactive decay of the initial contaminant results in by-products that also may represent a potential adverse environmental impact. Similarly, subsequent desorption of a previously adsorbed contaminant or dissolution of a previous precipitated contaminant also may result in negative environmental impacts.

2. Aqueous Immiscible Flow

Aqueous immiscible flow, or immiscible displacement, involves the simultaneous flow of two or more immiscible fluids in the porous medium. Since the fluids are immiscible, the interfacial tension between the two fluids is not zero, and a distinct fluid–fluid interface separates the fluids.

The migration of two separate fluid phases in fluid-saturated soil, referred to as two-phase flow, requires special consideration for the interaction between the two fluid phases. Due to surface and interfacial tension effects, one of the two fluids preferentially wets the surface of the pores resulting in a characteristic concave curvature of the fluid–fluid interface toward the nonwetting fluid and a discontinuity in pressure (i.e., capillary pressure) exists across the interface separating the two fluids, as illustrated in Fig. 4. The nonwetting fluid can displace the wetting fluid in the pore space of the soil only after a minimum

pressure, known as the entry pressure, for the nonwetting fluid is reached. For example, in the case of airflow through initially water-saturated soil, this minimum pressure is known as the air-entry pressure.

For the two-fluid flow scenario, such as in the case of immiscible displacement of a water-saturated soil by a nonaqueous-phase liquid (NAPL) that is nonwetting, the migrations of both the wetting and the nonwetting fluids are described by a modified form of Darcy's law [Eq. (1)] written for each fluid. In this case, fluid flow is complicated by the fact that the permeabilities for the wetting and nonwetting fluids are functions of the degree of saturation of the wetting and nonwetting phases, respectively, and therefore the permeabilities change with location and time. Thus, the governing system of flow equations

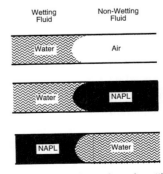

FIGURE 4 Capillary tube schematics of wetting versus nonwetting fluids. NAPL, Nonaqueous-phase liquid. [Adapted from Shackelford and Jefferis (2000).]

typically is highly nonlinear requiring the use of numerical methods for the solution. Further details regarding the application of this approach for describing DNAPL migration can be found in Pankow and Cherry (1996) and Charbeneau (2000).

3. Gas-Phase Transport

In addition to a gas migrating in response to a pressure gradient [Eq. (3)], gas also can migrate in the absence of a pressure difference via diffusion in response to a difference in the gas-phase concentration of the gas. All other factors being equivalent, the diffusive migration of chemical compounds through the gas phase typically ranges from 10^4 to 10^5 times higher than the liquid-phase diffusive migration of the same compounds. Thus, diffusion of gas is a potentially significant consideration in liquid unsaturated porous media. Further details regarding gas-phase diffusion can be found in Charbeneau and Daniel (1993) and Shackelford and Rowe (1998).

E. Mechanical Considerations

In geoenvironmental engineering applications involving design and construction of waste containment facilities, knowledge of the strength and stress–strain properties of the engineering materials being used in the containment facility as well as the waste being disposed is required. For example, consider a liquid waste containment facility that is constructed by excavating a disposal pit followed by placing a geomembrane liner (GML) that is subsequently covered with a layer of soil, as shown in Fig. 5. The cover soil is typically placed to protect the GML from damage resulting from exposure to ultraviolet light prior to placement of the waste. Design of this facility requires several considerations that require a knowledge of the strength and stress–strain properties of the component materials.

First, the stability of the side slopes (Fig. 5a) with respect to sliding within the foundation soil must be determined by calculating the shear stress τ along the assumed failure plane resulting from the self-weight of the failure wedge and the resisting, shear strength s provided by the soil along the same shear plane. In the case where $\tau > s$, either the side slopes must be cut back (i.e., reduce the side slope angle β), possibility resulting in a reduction in storage capacity, or the strength of the soil must be increased by engineering a stabilization scheme.

Second, for a given β, the potential for sliding of the soil cover along the interface of the soil cover and the GML (Figs. 5b and 5c) must be evaluated by calculating the shear stress τ along the assumed failure plane resulting from the self-weight of the soil cover and the shear strength s offered by the frictional resistance along the soil cover–GML interface. If $\tau > s$ in this case, the engineer

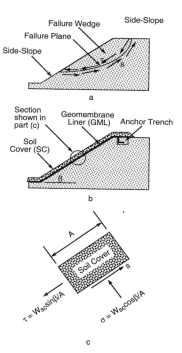

FIGURE 5 Design considerations for installation of a liquid waste containment facility: (a) side slope stability from disposal pit excavation; (b) placement of liner and soil cover; (c) free-body diagram for sliding along soil cover–GML interface (τ is the applied shear stress, σ is the normal stress, s is the shear strength, and W_{sc} is the weight of the soil cover).

may select a different geomembrane product for the GML, select a different soil cover material, and/or reduce β.

Third, in the case where the interface friction between the GML and side slope is sufficient to prevent the soil cover from sliding along the surface of the GML, the GML and the soil cover may slide together down the slopes if the GML is not properly anchored or the GML has inadequate tensile strength and fails by rupture. Thus, the engineer must ensure that suitable tensile capacity and suitable anchorage capacity, for example, through design of an anchor trench (Fig. 5b), are provided.

In addition to mechanical considerations involving the strength of component materials used in waste containment facilities, consideration also must be given to the compressibility of the foundation soil and the compressibility of waste being disposed. For example, municipal solid waste (MSW) typically is highly compressible and degrades with time due to biological-mediated decomposition. Thus, as illustrated schematically in Fig. 6, a significant potential for large, nonuniform settlement of the surface of the waste pile exists in MSW landfills. In the case where final covers are placed on top of the waste pile to keep precipitation from coming into contact with the waste after closure of the facility, such large

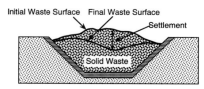

FIGURE 6 Nonuniform settlement in a solid waste disposal facility resulting from compressibility and time-dependent degradation of the waste.

settlements can have deleterious effects in terms of the integrity of the cover system. Thus, the engineer must design the cover system to withstand these settlements, and special precautions, such as compacting the waste during disposal, must be employed to minimize the potential for such settlements.

II. WASTE CONTAINMENT

Subsurface pollution resulting from waste disposal activities occurs when potentially harmful liquids reach the environment beneath the waste disposal site. The liquids usually originate from precipitation, disposal activities, and/or compression of the waste. The amount and quality of liquids generated depend primarily on the climatic conditions, the physical and chemical properties of the waste, and the disposal operations. This potential for groundwater pollution typically is controlled through encapsulation of the liquid or waste material using engineered liner and cover systems.

A. Liners

1. Types of Liners

The primary objective of a liner, or barrier layer, is to prevent or reduce the migration of potentially harmful chemicals, or contaminants, into the surrounding environment. With respect to this objective, several different types of liners are used for the containment of waste. As outlined in Table V, the different types of liners may be separated into three broad classifications: (1) earthen (soil) liners, (2) geosynthetic (polymer) liners, and (3) composite liners. Earthen liners may be further divided into naturally occurring low-permeability soil liners and manmade liners (e.g., compacted or mechanically stabilized low-permeability soils).

Natural (soil) liners (NLs) are formed by low-permeability geologic formations, such as aquitards or aquicludes. An *aquitard* is a geologic formation (e.g., interbedded sand and clay lenses) that transmits water at a very slow rate relative to an aquifer, and an *aquiclude* is a geologic formation so impervious that it completely

TABLE V Types of Liners Used for Waste Containment

Liner type	Origin	Examples
Earthen or soil	Geologic formation	Aquicludes and aquitards (NL)
	Borrow source/quarry	Compacted clay liner (CCL)
Geosynthetic	Manufactured	Geomembrane liner (GML), geosynthetic clay liner (GCL)
Composite	Manufactured-single	GCL made w/geomembrane
	Manufactured-combined	GML/GCL
	Borrow source/quarry + geologic formation	CCL/NL
	Manufactured + geologic formation	GML/NL or GCL/NL
	Manufactured + borrow source/quarry	GML/CCL or GCL/CCL

obstructs the flow of ground water (e.g., shale). Several disposal scenarios based on natural soil liners for liquid waste disposal are shown in Fig. 7.

Compacted soil liners also are commonly referred to as compacted clay liners (CCLs) because only soils that contain a sufficient amount of clay can provide the suitably low saturated hydraulic conductivity typically required for containment applications (e.g., $k_{sat} \leq 10^{-9}$ m/sec). The primary purposes of compacting a source clay for use as a liner material are to reduce the saturated hydraulic conductivity by densifying the clay (i.e., reducing the pore sizes) and to destroy any small secondary defects (e.g., fissures or cracks) that may act as conduits for flow in the natural soil formation.

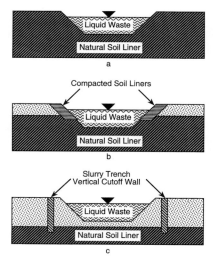

FIGURE 7 Examples of waste disposal cells (a) entirely within, (b) partially within, and (c) entirely above a natural soil liner.

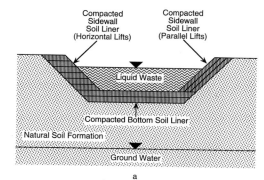

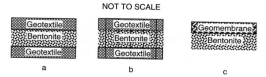

FIGURE 9 Typical GCL cross sections: (a) sodium bentonite with adhesive between two geotextiles, (b) sodium bentonite stitched between two geotextiles, and (c) sodium bentonite with adhesive attached to geomembrane.

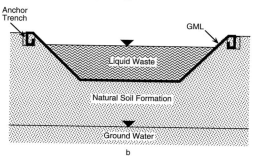

FIGURE 8 Liquid waste disposal cells lined with engineered liners: (a) compacted clay (soil) liner and (b) geomembrane liner (GML).

As schematically illustrated in Fig. 8a, CCLs typically are constructed in 150-mm-thick layers, or lifts, with the total thickness of the CCL being equal to the number of compacted lifts times the lift thickness. Thus, a 0.6-m-thick CCL would comprise four compacted lifts of clay. The saturated hydraulic conductivity of a CCL is particularly sensitive to the amount of water in the pores at compaction, with significant decreases (typically 10–1000 times) in the saturated hydraulic conductivity occurring as the water saturation increases at compaction. Thus, CCLs typically must be compacted at high water saturations (e.g., 85–95%) in order to achieve suitably low saturated hydraulic conductivities.

Geomembranes liners (GMLs) are flexible, thin [0.25–5.0 mm (10–200 mils)] sheets of rubber or plastic material that are manufactured at a factory, transported to the site, rolled out, and welded together to form the liner containment system. Geomembrane liners can be utilized in the same manner as CCLs, as illustrated in Fig. 8b. The minimum thickness of GMLs for waste disposal applications usually is 0.75 mm (30 mils). Unless punctured, GMLs generally are impervious to liquid migration, but VOCs typically can migrate readily through GMLs via molecular diffusion.

Geosynthetic clay liners (GCLs) are thin (≤10 mm), flexible sheets of sodium bentonite (∼4.9 kg/m² of bentonite) either sandwiched between two geotextiles

(Figs. 9a and 9b) or glued to a geomembrane (Fig. 9c). In some cases, an adhesive is mixed with the bentonite to hold the GCL together (Figs. 9a and 9c). In other cases, the GCL is held together by needle-punched fibers or stitching (Fig. 9b). The GCLs are manufactured in panels (4–5 m by 25–60 m), placed on rolls at the factory, shipped to the site, and rolled out to form a thin barrier. Sheets of panels can be stacked to form thicker barriers. An advantage of GCLs relative to GMLs is that the GCL panels do not have to be seamed or otherwise fused together as do GMLs, since the swelling bentonite in the presence of water provides a seam by simply overlapping the panels. In some cases, loose bentonite is placed between the overlapped portions of the panels to assist in sealing the panels.

Although GMLs and GCLs typically are much less permeable than compacted clay liners (CCLs), GMLs and GCLs also are very thin. Thus, a puncture or tear in a GML or GCL can have deleterious results with respect to the integrity of a liner system. For example, a small puncture in the bottom of the GML shown in Fig. 8b would result in a situation analogous to pulling the plug in a bathtub filled with water. As a result, the concept of the composite liner system, in which a natural and/or man-made soil liner serves as a backup to a GML or GCL, has emerged.

Composite liners are combinations of soil liners and/or geosynthetic materials. The properties of composite liners are essentially the same as the component materials for the composite liner. However, an important aspect for the effective use of composite liners is that the composite materials must be in intimate contact (i.e., without separation between materials). In reality, the type of GCL shown in Fig. 9c is a manufactured composite liner. Composite liner systems typically are used in conjunction with municipal and hazardous solid waste disposal.

2. Liner Systems for Solid Waste Disposal

The disposal of municipal solid waste (MSW) and, to a lesser extent, hazardous solid waste (HSW) represents a significant activity associated with geoenvironmental engineering. In most countries, the minimum design requirements for liner systems designed and constructed to contain either MSW or HSW are prescribed by regulation and therefore are referred to as prescriptive liner systems. In general, liner systems for MSW and HSW waste disposal consist of a combination of one or more drainage layers and liners. In these systems, the liner is a low-permeability barrier layer that is placed to impede the migration of leachate and gas from the solid waste, and the drainage layer is placed immediately above the liner to drain off the leachate that has percolated through the waste, thereby limiting the hydraulic gradient for contaminant migration through the liner. The collected leachate typically is drained to a collection basin (sump) where the leachate can be removed for treatment by pumping.

The basic components of solid waste liner systems typically consist, from top to bottom, of a leachate collection and removal system (LCRS), a top liner, a leakage detection system (LDS), and a bottom liner. The LCRS typically is made up of either a highly permeable soil (e.g., gravel) or a geosynthetic drainage material (e.g., geonet), with possibly a network of perforated pipes to drain the collected leachate to the sump (e.g., geopipes). The top liner consists of either a geomembrane liner (GML) or a composite liner. The primary purpose of the LDS, which is made up of the same materials as the LCRS, is to allow the detection of contaminants that may have migrated through the top liner (e.g., due to a leak). However, in the case of a leak, the LDS also can serve to collect and drain the leakage. An LDS is required only when a bottom liner is required. The bottom liner, if required, typically consists of a composite liner made up of a GML in intimate contact with 0.9 m of a low-permeability (typically $k_{sat} < 10^{-9}$ m/sec) CCL.

The minimum prescriptive (regulated) cross sections for liner systems for MSW and HSW disposal in the United States are shown in Fig. 10. As shown in Fig. 10a, the minimum requirements for MSW include a single composite liner overlain by an LCRS. Thus, an LDS and bottom liner are not required for MSW disposal in the United States. However, a double liner system consisting of an LDS sandwiched between a bottom composite liner system and a GML top liner is required in the case of HSW disposal (Fig. 10b). The LDS and bottom liner in the HSW requirements provide a level of redundancy in the containment that is required due to the particularly hazardous nature of HSW.

B. Covers

Once the disposal capacity of the facility or containment cell is reached, an engineered final cover is placed over the waste to (1) prevent the generation of leachate by minimizing the amount of precipitation percolating through the waste during the inactive (postclosure) period, (2) mini-

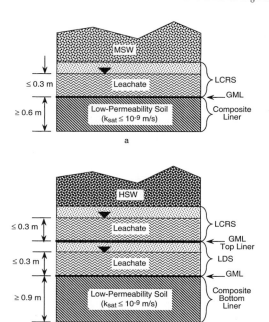

FIGURE 10 Minimum regulated (prescriptive) cross sections required in the United States for engineered liners used for disposal of (a) municipal solid waste (MSW) and (b) hazardous solid waste (HSW).

mize the escape of noxious and/or potentially harmful gas into the surrounding atmosphere, and (3) prevent physical dispersion of the waste by wind and water. Covers typically are separated into two categories: traditional covers and alternative covers.

1. Traditional Covers

Traditional covers or caps often are referred to as "resistive barriers" because the design typically is based on minimizing infiltration of water by minimizing the saturated hydraulic conductivity of the barrier component of the cover. In the case of the use of compacted clay as the sole or primary material in the barrier, minimizing the saturated hydraulic conductivity follows the same approach as for a CCL.

a. Layering in traditional covers. As shown in Fig. 11, traditional covers include the following six possible components, from top to bottom: (1) surface layer, (2) protection layer, (3) drainage layer, (4) barrier layer, (5) gas collection layer, and (6) foundation layer. The primary purposes of the surface layer are to support vegetation, resist erosion, and reduce temperature and moisture extremes in underlying layers. The primary purposes of the protection layer are to store and release water through evaporation and transpiration (ET), separate underlying

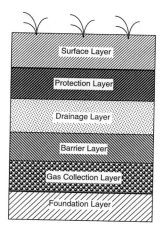

FIGURE 11 Cross section of layering in traditional caps (see Rumer and Mitchell, 1997).

TABLE VI Materials for Covers[a]

Layer	Materials
Surface	Topsoil (vegetated)
	Geosynthetic erosion control layer over topsoil (vegetated)
	Cobbles
	Paving material
Protection	Soil
	Cobbles
	Recycled or reused waste (e.g., fly ash, bottom ash, and paper mill sludge)
Drainage	Sand or gravel
	Geonet or geocomposite
	Recycled or reused waste (e.g., tire chips)
Barrier	Compacted clay
	Geomembrane
	Geosynthetic clay liner
	Recycled or reused waste (low permeability)
	Asphalt
	Sand or gravel capillary barrier
Gas collection/Removal	Sand or gravel
	Geonet or geocomposite
	Geotextile
	Recycled or reused waste (e.g., tire chips)
Foundation	Sand or gravel
	Geonet or geocomposite
	Recycled or reused waste select waste

[a] Modified from Rumer and Mitchell (1997).

waste from humans, burrowing animals, and plant roots (biointrusion), and protect underlying layers from "environmental" stress resulting from climatic conditions such as alternating wetting and drying cycles and/or freezing and thawing cycles. The drainage layer is used to allow drainage of the water from overlying layers to reduce the buildup of infiltrating water, thereby reducing the hydraulic gradient for flow through the barrier layer, and increase stability by reducing pore water pressures. The barrier layer functions to impede infiltration of water into the underlying soil and to restrict the outward migration of gases from the underlying waste or contamination. The gas collection layer serves to collect and remove gases to reduce the potential for detrimental effects (e.g., explosions) and to provide for energy recovery and/or vapor treatment. Finally, the foundation layer provides a surface on which to construct the other layers of the cap.

The potential materials for each layer are summarized in Table VI. Some layers, such as the hydraulic barrier layer, may consist of more than one material, such as a composite barrier comprising a geomembrane overlying a compacted clay.

All covers require a surface layer, but all of the six layers are not required at all sites. For example, a drainage layer may not be needed at an arid site. However, all of the layers that are used in a cover must have adequate shear strength to ensure stability against slope failure, and must be sufficiently durable to function adequately over the design life of the cap. More detail regarding the design and performance of covers can be found in Rumer and Mitchell (1997).

b. Design considerations for traditional covers.
Two approaches can be used to design traditional covers: (1) the water balance approach and (2) the unsaturated flow approach. In the water balance approach, the amount of percolation through the cover P_r is determined by subtracting all of the components contributing to removal of water from the cover system from the precipitation as follows:

$$P_r = P - R_s - ET - L_f - \Delta S_w, \qquad (4)$$

where P is precipitation, R_s is surface runoff, ΔS_w is change in soil–water storage, ET is evaporation plus transpiration (evapotranspiration), and L_f is intralayer lateral flow. This water balance approach forms the basis for the HELP (hydrologic evaluation of landfill performance) model that commonly is used in the design of cover systems. A schematic cross section of a cover illustrating the water balance approach in the HELP model is shown in Fig. 12.

In the HELP model, each layer of the cover must be specified as a vertical percolation layer, barrier soil liner, lateral drainage layer, or geomembrane liner depending on the function and hydraulic properties of the layer. Unsaturated flow of water occurs only in a vertical percolation layer. In contrast, a barrier layer (soil liner), with low

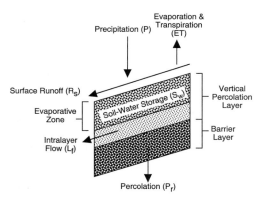

FIGURE 12 Schematic cover cross section used in the water balance approach of the HELP model. [Modified after Khire *et al.* (1997).]

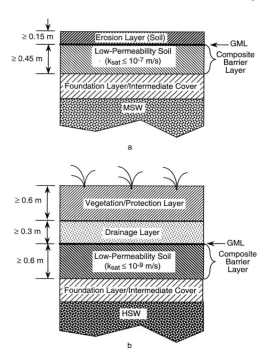

FIGURE 13 Minimum regulated (prescriptive) cross sections required in the United States for engineered covers used in disposal of (a) municipal solid waste (MSW) and (b) hazardous solid waste (HSW).

saturated hydraulic conductivity, is assumed always to be saturated and percolation through the barrier layer can occur only when water ponds on the surface of the layer. As a result, unsaturated flow through the liner is not considered by the HELP model.

Precipitation in the HELP model is separated into surface runoff and infiltration based on an empirical approach using a modification of the Soil Conservation Service (SCS) runoff curve number method. Evaporation and transpiration removes water only from the evaporative zone of the cover in the HELP model. The procedure for determining evaporation and transpiration also is empirical. Further details on the procedures and assumptions used in the HELP model are provided by Khire *et al.* (1997).

The water balance of landfill covers also has been simulated using the unsaturated flow approach in the form of a one-dimensional, finite-difference computer program, UNSAT-H, that solves a modified version of Richards' equation that governs unsaturated flow through porous media (Khire *et al.* 1997). In this case, surface runoff is determined as the difference between precipitation and infiltration, where infiltration is based on the saturated and unsaturated hydraulic conductivities of the soils constituting the cover. Thus, unlike the empirical approach in the HELP model, the determination of surface runoff in UNSAT-H is based directly on the physical properties of the soil profile.

c. Traditional covers for solid waste disposal facilities.
The minimum requirements for cover systems for MSW and HSW disposal prescribed by regulation in the United States are shown in Fig. 13. Both cross sections include a composite barrier layer consisting of a GML in intimate contact with a low-permeability soil, but the thickness required for this soil in the MSW cover is less than that in the HSW cover, and the maximum saturated hydraulic conductivity required by the low-permeability soil in the MSW cover is 100 times higher than that in the HSW cover. Also, the MSW cover only requires a relatively thin erosion layer, whereas the HSW cover requires both a thick vegetation layer and an underlying drainage layer.

2. Alternative Covers

Interest in the use of alternative earthen final covers (AEFCs) has increased in recent years due in part to the relatively high costs associated with the more traditional covers and to other problems (e.g., desiccation cracking) associated with resistive covers containing low-permeability clay. Alternative earthen final covers are earthen covers designed on water storage principles that perform as well as more traditional, prescriptive covers but have greater durability and/or lower cost. The use of AEFCs typically is suitable only in drier regions where potential evaporation and transpiration exceeds precipitation, so that the AEFC can be designed to have sufficient storage capacity to retain water during wet periods without transmitting appreciable percolation. For example, AEFCs currently are receiving significant consideration for use in the western regions of the United States where semiarid and arid climates prevail. Although a variety of designs are being considered for AEFCs, most designs can be classified as either monolithic barriers or capillary barriers.

Monolithic covers (MCs) consist of a relatively thick vegetated layer of finer textured soil that has high water storage capacity such that water contents near the base of the cover remain fairly low. In addition, unlike resistive barrier layers (e.g., CCLs) that are compacted at high water saturations to minimize the saturated hydraulic conductivity, MCs are compacted at relatively low water saturations to increase the water-holding capacity and to reduce the (unsaturated) hydraulic conductivity. This combination of a relatively high water storage capacity together with the relatively low unsaturated hydraulic conductivity results in percolation rates from the base of the cover that can meet target percolation rates. Additional advantages of MCs relative to traditional, resistive covers include (a) less vulnerability to desiccation and cracking during and after installation, (b) greater simplicity of construction, and (c) lower maintenance. Also, MCs are economical to implement because MCs can be constructed of a reasonably broad range of soils and are typically constructed using soils from a nearby area.

A capillary barrier results when unsaturated flow occurs through a relatively fine layer overlying a relatively coarse layer (e.g., clay over sand, silt over gravel, and sand over gravel). In a capillary barrier, only a small fraction of the liquid flux associated with the wetting front through the finer layer is transmitted into the underlying coarser layer. A capillary barrier effect occurs because the unsaturated hydraulic conductivity of the coarser (underlying) layer decreases to a lower value than the unsaturated hydraulic conductivity of the finer (overlying) layer as the soil–water suction increases (see Fig. 3), and the residual suction in the finer layer after first passage of the wetting front is larger than the gravitational force on the wetting front. In general, the capillary barrier effect increases with an increase in the contrast in soil properties (e.g., unsaturated hydraulic conductivity) between the finer and coarser layers.

3. Gas-Phase Migration Through Covers

Control of gas-phase migration represents another important objective of engineered covers. In the case of municipal solid waste, failure to adequately control and vent the methane gas (CH_4) generated by decomposition of the waste represents a potential explosion hazard. Thus, the final cover serves both to prevent the escape of large quantities of methane gas and to assist in the collection of the gas for controlled venting through an engineered collection network.

As illustrated in Fig. 14, two problems associated with the environmentally safe disposal of mine tailings con-

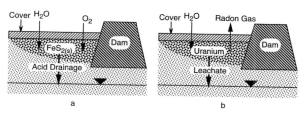

FIGURE 14 Tailings disposal scenarios requiring consideration of gas-phase migration through covers: (a) acid drainage, (b) radon gas emission.

cern gas-phase migration through the final cover. First, the problem of acid drainage can occur when sulfidic tailings, such as pyrite, $FeS_{2(s)}$, are oxidized by the influx of oxygen into the tailings (Fig. 14a). The result of this oxidation is the production of a low-pH (acidic) solution (e.g., pH < 6) containing relatively high concentrations of potentially toxic heavy metals associated with the tailings, such as Fe^{2+}. Second, the disposal of uranium tailings results in a source of radon gas that can be environmentally harmful if not controlled properly (Fig. 14b). In both of these cases, the objective in the cover design must include steps taken to minimize the influx (O_2) or efflux (radon) of gas through the cover as well as minimize water infiltration.

The currently recommended design approach for soil covers in terms of the acid drainage scenario has been to compact a fine-grained soil at a water saturation sufficiently high to minimize both the saturated hydraulic conductivity and the extent of a continuous air phase available for oxygen diffusion. Minimization of continuous air phase is important in this case since oxygen is relatively insoluble in water and the oxygen diffusion coefficient in air is 10^5 times larger than that in water. An underlying coarse-grained layer also is usually included in the design to provide for lateral drainage of percolating water.

III. REMEDIATION

A. Remediation Procedure

The process of remediation of subsurface contamination commonly involves several steps, as outlined in Fig. 15. After contamination is discovered at a site, an emergency response and abatement phase typically is initiated to provide an initial assessment of the nature of the contamination, and rapid measures to abate or prevent further contamination are taken. After this initial assessment, a more detailed site characterization is undertaken to provide more accurate assessment of the problem (e.g., soil stratigraphy, concentrations of the contaminants of concern), followed by a detailed exposure assessment and a

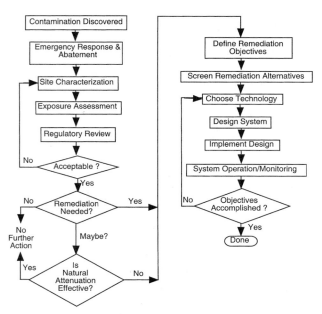

FIGURE 15 Flow chart of the remediation process. [From Shackelford and Jefferis (2000).]

review of the outcomes of these assessments by the governing regulatory authority, a process that is ongoing and interactive.

After the site and exposure assessment is approved, a decision regarding the necessity of remediation is made. This decision typically results in one of three outcomes: (1) there is no need for further action, (2) remediation must be implemented, or (3) the site should be monitored to determine whether or not natural attenuation processes (e.g., biodegradation) will remediate the site, with the potential for implementation of a more aggressive remediation technology some time in the future.

If remediation is required, the next phases of the remediation process involve defining the remediation objectives (e.g., concentration levels, containment versus treatment, etc.), screening possible remediation technologies that can achieve the objectives, choosing the most appropriate technology, and designing and installing the chosen remediation technology and system. Finally, monitoring of the performance of the system is required to ensure that the chosen technology is working to achieve the stated objectives. If the desired objectives cannot be achieved with the chosen technology, either the design of the chosen technology must be reevaluated and modified and/or an additional technology must be used. This process is repeated until the desired remediation objectives are achieved.

B. Remediation Technologies

Remediation technologies can be classified in numerous ways, such as according to the process involved in the

remediation (i.e., physical, chemical, biological, or thermal), the objective of the technology (containment versus treatment), or the location of the remediation process (*in situ* versus *ex situ*).

1. Remediation Processes

Remediation processes commonly are categorized as physical, chemical, biological, or thermal processes. Physical processes include interphase mass transfer reactions, such as sorption and ion exchange, as well as binding processes, such as cement fixation of chemicals to soil.

Chemical processes involve chemical reactions. Two chemical processes of relatively recent interest in remediation are the oxidative degradation of halogenated hydrocarbons (e.g., PCE, TCE) in the presence of oxidants and reductive dehalogenation of halogenated hydrocarbons in the presence of solid, zero-valent iron metal, such as the iron filings found in metal machine shops. For example, oxidation of trichloroethylene, or TCE (C_2HCl_3), by potassium permanganate ($KMnO_4$) is described by the following chemical reaction:

$$2KMnO_4 + C_2HCl_3 \rightarrow 2CO_2 + 2MnO_2 + 2KCl + HCl,$$

$$(5)$$

whereas dechlorination of TCE to dichloroethylene, or DCE ($C_2H_2Cl_2$), in the presence of solid, zero-valent iron metal ($Fe^0_{(s)}$) is described by the following chemical reaction:

$$2Fe^0_{(s)} + C_2HCl_3 + 3H_2O \rightarrow C_2H_2Cl_2 + H_2 + Cl^-$$
$$+ 3OH^- + 2Fe^{2+}. \quad (6)$$

The potential products of these reactions include other chlorinated hydrocarbons (e.g., DCE) that subsequently degrade to vinyl chloride (C_2H_3Cl, or VC), ethene (C_2H_4), or ethane (C_2H_6), hydrogen gas (H_2), chloride (Cl^-), and iron oxide (FeO) and hydroxide [$Fe(OH)_2$] precipitates.

In biologically mediated processes, the contaminant, typically a hydrocarbon compound, acts as a carbon source and a source of energy for the bacteria. For example, consider the following reactions for bacteria-mediated degradation (i.e., biodegradation) of benzene (C_6H_6):

$$C_6H_6 + 2.5O_2 + NH_3 \xrightarrow{Pseudonomas} C_5H_7O_2N$$
$$+ CO_2 + H_2O \quad (7)$$

and

$$C_5H_7O_2N + 5O_2 \xrightarrow{Pseudonomas} 5CO_2 + 2H_2O + NH_3.$$

$$(8)$$

The first (top) reaction represents oxidation of benzene (C_6H_6) to cell mass and the second (bottom) reaction represents complete mineralization to carbon dioxide (CO_2), water (H_2O), and ammonia (NH_3). These reactions also illustrate the need for large quantities of oxygen (O_2) for aerobic oxidation of hydrocarbons. For example, in this case, 3.1 kg of oxygen is required per kilogram of contaminant.

Thermal processes involve heating the contaminated soil to induce volatilization and removal of contaminants (e.g., radiofrequency heating) or to melt and solidify the contaminated soil mass (e.g., vitrification).

2. *In Situ* versus *ex Situ* Remediation

In situ versus *ex situ* remediation refers to the location of the remediation technology. In this context, an important distinction arises relative to the location of the application of the technology versus the location of the treatment. For example, in the pump-and-treat approach to remediation, the pumping occurs *in situ*, but the treatment of the pumped, contaminated water occurs *ex situ*.

3. Containment versus Treatment

Containment refers to technologies used to prevent the spread of contamination without necessarily resulting in degradation or transformation of the contaminants. Treatment refers to technologies used to degrade or otherwise transform contaminants into less toxic or nontoxic concentrations. Although *in situ* containment technologies have been available for decades, interest in the use of *in situ* containment as the primary technology for remediation applications has gained momentum in recent years because containment (1) typically is cheaper than treatment, (2) can be used until a more efficient treatment technology is developed, (3) can provide a means for evaluating the potential for natural attenuation processes to degrade the contaminants, and (4) can present a lower overall risk since major excavation, contaminant exposure, etc., can be avoided.

Many remediation scenarios will involve both containment and treatment. However, in some countries where economic resources are not plentiful, using a remediation technology that requires *ex situ* treatment may not be a viable option, particularly if treatment of wastewater, for example, is not already practiced routinely.

4. Active versus Passive Considerations

In the case of cleanup, or remediation, of existing subsurface pollution, both active and passive remediation activities are possible. Active remediation involves application of a conventional technology, such as a pump-and-treat system. However, active remediation is relatively expensive due to the high costs associated with both construction (e.g., installation of wells) and operation (e.g., pumping and treating). Thus, more passive technologies, such as low-permeability ($k_{sat} < 10^{-9}$ m/sec) vertical cutoff walls, also known as slurry walls, frequently are installed to prevent further spreading of subsurface pollution, thereby reducing the cleanup duration and cost.

In response to the increasing costs associated with active remediation of contaminated sites, an increasing emphasis is being placed on more cost-effective, passive treatment technologies. For example, the use of permeable reactive walls (PRWs) for *in situ* treatment of contaminated ground water has gained momentum in recent years. A PRW (also known as a passive treatment system or *in situ* treatment zone or curtain) incorporates chemical and/or biological reagents or catalysts into the porous medium used for the wall to degrade or otherwise reduce the concentrations of pollutants during passage of the contaminant. A major difference between the more traditional low-permeability containment barriers such as slurry walls, and PRWs is that the hydraulic conductivity of a PRW must be sufficiently large to allow timely and efficient processing of the contaminated ground water.

5. Classification of *in Situ* Remediation Technologies

Based on the aforementioned considerations, a general classification system of possible remediation technologies is given in Table VII. With respect to the *in situ* technologies, brief descriptions of some of the more commonly used technologies follow. More detailed descriptions can be found in Shackelford and Jefferis (2000).

6. Descriptions of Remediation Technologies

a. Vertical containment technologies. Pump and treat refers to the process of pumping contaminated ground water to a surface collection system through wells screened in the saturated zone and then treating the contaminated water with one or more *ex situ* treatment technologies. In this context, pump and treat may be considered to be a treatment rather than containment option. However, in terms of the *in situ* aspects of pump and treat, which is pumping, not treatment, the pump-and-treat technology typically is categorized as a containment option.

For example, pumping wells may be located within the contaminated area to provide drawdown of the groundwater table that directs flow into the site, thus providing active containment by minimizing or preventing further migration of the plume away from source or site. Alternatively, pumping wells can be located downgradient from the contamination to provide a drawdown barrier.

TABLE VII Classification of Remediation Technologies Based on Soil as the Contaminated Medium[a]

Soil removal?	Technology category	Technique/process	Example(s)	Comment(s)
Yes (*ex situ*)	Containment	Disposal	Landfills	On-site vs. off-site; new vs. existing
	Treatment	Chemical	Neutralization, solvent extraction	Treated soil may require disposal in a landfill or may be returned to the site
		Physical	Soil washing, stabilization/solidification, vitrification	
		Biological	Biopiles, bioreactors	
		Thermal	Incineration, vitrification	
No (*in situ*)	Containment	Pump and treat	Vertical wells, horizontal wells	Both passive and active containment are possible; in pump and treat, pumping is used to control hydraulic gradient and collect contaminated water; treatment is *ex situ*
		Capping	Traditional covers, alternative covers, geochemical covers	
		Vertical (low-permeable) barriers	Slurry walls, grout curtains, sheet piling, biobarriers, reactive barriers	
		Horizontal (low-permeable) barriers	Grout injected liners	
	Treatment	Chemical	Oxidation, chemical reduction	Technologies with asterisk require removal of gas and/or liquid phases and *ex situ* treatment; both passive and active treatment are possible
		Physical	Soil flushing, stabilization/solidification, vitrification, air sparging (AS), soil vapor extraction (SVE), electrokinetics (EK)	
		Biological	Monitored natural attenuation (MNA), bioventing, bioslurping, biosparging	
		Thermal	Steam injection, radio frequency heating (RF), vitrification	

[a] From Shackelford and Jefferis (2000).

Slurry walls, also known as slurry cutoff walls or vertical cutoff walls (see Fig. 7c), are vertical walls constructed by excavating a trench with a back-hoe or clamshell and simultaneously filling the trench with a stabilizing slurry which is typically prepared from a mixture of bentonite clay and water, or bentonite clay, cement, and water. The slurry forms a thin filter cake (typically <3 mm for bentonite slurry and perhaps a few tens of millimeters for cement–bentonite slurry) with a low hydraulic conductivity (typically $<10^{-10}$ m/sec) on the sides of the trench. The filter cake minimizes slurry loss from the trench, stabilizes native soil on the sidewalls of the trench, and provides a plane for slurry stabilization of the excavated trench. Both bentonite and cement–bentonite slurries typically contain 4–7% (w/w) sodium bentonite mixed with water. The hydraulic conductivity of slurry walls is affected significantly by the quality of the backfill material and typically ranges from $\sim10^{-7}$ to $\sim10^{-11}$ m/sec, with lower values associated with backfill materials that contain sufficient quantities of clay.

Other passive vertical barriers include walls constructed using deep soil mixing or jet grouting using chemical grouts (e.g., silicates, resins, and polymers), grout curtains, and sheet pile walls. However, the application of these technologies for long-term containment has been limited due to concern with the integrity of the containment system and potential leakage of contaminants through "windows" in the barriers, such as high-permeability zones between the grout in grout curtains, or through the interlocks in sheet piles, or corrosion with respect to the use of sheet-pile walls. Nonetheless, these technologies can be used in conjunction with other remediation technologies to aid in temporary, partial containment.

b. Horizontal containment technologies. Caps or covers are used as horizontal surface barriers to minimize the leaching of contaminants into the ground water. Two possible scenarios involving applications of caps for *in situ* remediation applications are illustrated in Fig. 16.

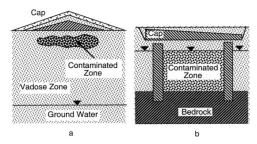

FIGURE 16 Schematic example applications for caps for *in situ* remediation: (a) long-term containment of vadose zone contamination in arid climates, (b) capping of slurry wall containment system for saturated zone contamination. [From Shackelford and Jefferis (2000).]

As shown in Fig. 16a, a cap may be considered as the only technology needed in cases where the climate is arid, the water table is deep, and the site is relatively isolated.

The schematic diagram shown in Fig. 16b depicts the conceptual use of an *in situ* cap designed and placed to prevent the potential for overtopping of the contaminated ground water within a vertically contained zone due to infiltration, an occurrence known as the "bathtub effect." The cap is constructed by excavating the soil above the containment area, placing or compacting the low-permeability barrier, and backfilling with the native soil to provide a protective surface layer. The backfill is shown to original grade to allow for subsequent use of the site. Special design features include sloping the surface of the barrier layer to promote subsurface runoff down gradient of the containment zone. Also, periodic separations between the horizontal (cap) barrier layer and the vertical containment barrier may be required to allow for venting of volatile organics and/or minimizing the potential buildup of gas pressure beneath the cap and eventual uplift of the cap.

Subsurface horizontal barriers include indigenous barriers (aquicludes, aquitards) and artificially emplaced barriers. Artificially emplaced barriers include thin diaphragm walls created by directional and nondirectional drilling with jet grouting, conventional jet grouting, and soil freezing technologies.

Indigenous barriers rarely are relied upon for permanent containment because of the uncertainty associated with the integrity of the system. Application of the artificially emplaced barriers also is limited for the same reasons as the alternative passive vertical barriers, namely uncertainty regarding integrity and longevity of the placed system, and costs associated with placement of the technology.

c. Immobilization technologies. Stabilization and solidification (S/S) methods are intended to immobilize dissolved contaminants and, in certain cases, DNAPLs. Stabilization refers to techniques that reduce contaminant hazard potential by converting the contaminants to less soluble, mobile, or toxic forms. Solidification refers to techniques that encapsulate the contaminant in a monolithic solid of high structural integrity. Common S/S reagents include cement and other pozzolanic materials, such as lime, fly ash, and slag. These reagents typically result in stabilization (fixation) of the contaminants.

Vitrification is a thermal process that results in vaporization and mobilization of organic compounds and solidification and immobilization of inorganic compounds. However, the primary application of the technology historically has been in immobilizing inorganic, radioactive contaminants. The *in situ* vitrification process uses a subsurface electrode array to heat the soil to temperatures as high as 1800°C through the application of large currents, resulting in melting of the soil. The melting process initiates at the ground surface and propagates downward, resulting in vaporization of organic compounds. The vapors migrate to the ground surface and are collected using a vacuum hood. Upon cooling, the remaining contaminants are solidified within a rocklike mass resembling obsidian.

d. Enhanced removal technologies. Soil vapor extraction, also referred to as *in situ* soil venting, couples vapor extraction (recovery wells) with blowers or vacuum pumps to remove vapors from the vadose zone and therefore reduce the levels of residual soil contaminants.

Air sparging, also known as *in situ* air stripping or *in situ* volatilization, is effective in removing SVOCs and VOCs. Air sparging involves injecting air through wells into the contaminated aquifer below the groundwater table. The air sparging process cleans the contaminated groundwater zone through two processes. First, injection of the air creates a turbulent condition that enhances volatilization of the contaminants. Second, the injected, or sparged, air increases the oxygen content of the ground water, enhancing aerobic biodegradation of the contaminants.

Soil flushing refers to the enhanced mobilization of contaminants in contaminated soil for recovery and treatment by forcing a fluid through the contaminated soil to either displace the contaminants or induce physical, chemical, biological, and/or thermal conditions that favor removal. Although the terms *soil flushing* and *soil washing* often are used interchangeably to describe the same technology, the term *soil flushing* generally is used in connection with *in situ* remediation, whereas the term *soil washing* is reserved for *ex situ* remediation. Flushing fluids may be water, enhanced water, and gaseous mixtures. Typical soil flushing solutions include water, dilute acids and bases, complexing and chelating agents, reducing agents, solvents, and surfactants. The fluids are introduced through spraying, surface flooding, subsurface leach fields, and subsurface injection.

In some cases, steam has been used as the fluid in soil flushing applications. The steam is injected into the subsurface through a ring of injection wells configured to surround all or part of the subsurface plume to be treated. Liquid and vapor are extracted from one or more wells located near the center of the well pattern while steam is injected into the permeable layers through screened portions of wells constructed around the plume. The steam directly heats contaminated permeable layers in the process zone to lower the viscosity and increase volatility of the contaminants. The contaminants are swept toward the extraction wells as ground water is displaced in advance of the injected steam front.

Radio frequency heating involves the use of electromagnetic energy to vaporize hydrocarbons in soil. The method involves using a system of electrodes implanted in vertical and/or horizontal holes drilled into the soil. Electromagnetic energy in the radio frequency band is generated using a modified radio transmitter and directed from the electrodes into the ground to heat the soil to a predetermined temperature to volatilize trapped hydrocarbons.

Electrokinetics involves the use of an applied electrical gradient to speed the removal of contaminants from contaminated, low-permeability, fine-grained soils. In general, application of the electrical current results in generation of protons (H^+) at the anode, due to hydrolysis reactions, that migrate together with the metal cations to the negatively charged cathode for removal and processing. In addition, the migrating metal ions (cations) result in bulk water flow from the anode to the cathode, a process referred to as electroosmosis.

e. Bioremediation.

Bioremediation may be defined as the process whereby a biological agent (e.g., bacteria, fungi, plants, enzymes) is used to reduce contaminant mass and toxicity in soil, ground water, and air. The goal of bioremediation is to reduce toxicity and target contaminant concentrations in water and/or soil through either biodegradation (of organics) or biotransformation (of metals).

Biotransformation is a general term representing any biologically catalyzed conversion of a metal or organic chemical. Biodegradation refers to a single biological reaction or sequence of reactions that result in the conversion of an organic substrate to a simpler molecule. Mineralization, or complete biodegradation, refers to conversion of the target compound to carbon dioxide and water under aerobic conditions, or methane (or ethane or ethene) under anaerobic conditions. Biotransformation and biodegradation of a contaminant do not necessarily result in a decrease in toxicity.

Biowalls, also referred to as permeable reactive biowalls, or in-line microbial filters, are essentially *in situ*

bioreactors or permeable reactive walls that treat contaminants passively through either biostimulation (stimulation of indigenous bacteria or microbes) or bioaugmentation (addition of nonindigenous bacteria or microbes). Much of the effort has been directed toward the use of oxygen and/or nutrient sources as reagents in biowalls to stimulate the activity of indigenous microorganisms.

f. Monitored natural attenuation.

Monitored natural attenuation is a passive approach to *in situ* remediation in that the approach relies on natural processes to attenuate the contaminants in the soil and ground water. Monitored natural attenuation can be used as a remediation approach when natural processes will reduce the mass, toxicity, mobility, and/or volume of the contaminants. Natural processes resulting in a decrease in or spreading of contaminants of concern (COCs) associated with the unsaturated, or vadose, zone include volatilization, sorption, leaching to ground water, and natural bioventing. Natural processes resulting in a decrease in or spreading of COCs associated with the saturated, or groundwater, zone include sorption, diffusion/dispersion, biodegradation, and dilution. Thus, MNA requires (1) a very thorough understanding of the fate and transport of the COCs and (2) close monitoring of soil and groundwater samples to verify continuous removal of the COCs.

SEE ALSO THE FOLLOWING ARTICLES

ENVIRONMENTAL TOXICOLOGY • HAZARDOUS WASTE INCINERATION • POLLUTION, ENVIRONMENTAL • POLLUTION CONTROL • ROCK MECHANICS • SOIL AND GROUND-WATER POLLUTION • SOIL MECHANICS • WASTE-TO-ENERGY SYSTEMS • WASTEWATER TREATMENT AND WATER RECLAMATION • WATER POLLUTION

BIBLIOGRAPHY

Acar, Y. B., and Daniel, D. E. (1995). "Geoenvironment 2000: Characterization, Containment, Remediation, and Performance in Environmental Geotechnics," ASCE, Reston, VA.

Charbeneau, R. J. (2000). "Groundwater Hydraulics and Pollutant Transport," Prentice-Hall, Inc. Upper Saddle River, NJ.

Charbeneau, R. J., and Daniel, D. E. (1993). "Contaminant transport in unsaturated flow," *In* "Handbook of Hydrology" (D. R. Maidment, ed.), pp. 15-1–15-54, McGraw-Hill, New York.

Daniel, D. E. (1993). "Geotechnical Practice for Waste Disposal," Chapman & Hall, London.

Daniel, D. E., and Koerner, R. M. (1995). "Waste Containment Facilities, Guidance for Construction Quality Assurance and Quality Control of Liner and Cover," ASCE, Reston, VA.

Grubb, D. G., and Sitar, N. (1994). "Evaluation of Technologies for *In-situ* Cleanup of DNAPL Contaminated Sites," EPA/600/R-94/120, U.S. EPA, Washington, DC.

Khire, M. V., Benson, C. H., and Bosscher (1997). "Water balance modeling of earthen final covers," *J. Geotechnical Geoenviron. Eng.* **123**(8), 744–754.

Koerner, R. M. (1998). "Designing with Geosynthetics," 4th ed., Prentice-Hall, Upper Saddle River, NJ.

Koerner, R. M., and Daniel, D. E. (1997). "Final Covers for Solid Waste Landfills and Abandoned Dumps," ASCE, Reston, VA.

Manassero, M., Benson, C. H., and Bouazza, A. (2000). "Solid waste containment systems," *In* "International Conference on Geotechnical and Geoenvironmental Engineering (GeoEng2000)," Vol. 1, pp. 520–642, Technomic, Lancaster, PA.

McBean, E. A., Rovers, F. A., and Farquhar, G. J. (1995). "Solid Waste Landfill Engineering and Design," Prentice-Hall, Upper Saddle River, NJ.

Mitchell, J. K. (1993). "Fundamentals of Soil Behavior," 2nd ed., Wiley, New York.

National Research Council (1990). "Ground WaterModels: Scientific and Regulatory Applications," National Academy Press, Washington, DC.

Oweis, I. S., and Khera, R. J. (1998). "Geotechnology of Waste Management," 2nd ed., PWS, Boston.

Pankow, J. F., and Cherry, J. A. (1996). "Dense Chlorinated Solvents and other DNAPLS in Groundwater," Waterloo Press, Portland, OR.

Rowe, R. K., Quigley, R. M., and Booker, J. R. (1995). "Clayey Barrier Systems for Waste Disposal Facilities," E & FN Spon, London.

Rumer, R. R., and Mitchell, J. K. (1997). "Assessment of Barrier Containment Technologies," National Technical Information Service (NTIS), Springfield, VA.

Shackelford, C. D. (1997). "Modeling and analysis in environmental geotechnics: An overview of practical applications," *In* "2nd International Congress on Environmental Geotechnics, IS-Osaka '96," Vol. 3, pp. 1375–1404, Balkema, Rotterdam.

Shackelford, C. D., and Rowe, R. K. (1998). "Contaminant transport modeling," *In* "3rd International Congress on Environmental Geotechnics," Vol. 3, pp. 939–956, Balkema, Rotterdam.

Shackelford, C. D., and Jefferis, S. A. (2000). "Geoenvironmental engineering for *in situ* remediation," *In* "International Conference on Geotechnical and Geoenvironmental Engineering (GeoEng2000)," Vol. 1, pp. 121–185, Technomic, Lancaster, PA.

Sharma, H. D., and Lewis, S. P. (1994). "Waste Containment Systems, Waste Stabilization, and Landfills: Design and Evaluation," Wiley, New York.

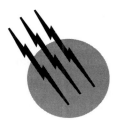

Geologic Time

Don L. Eicher
University of Colorado

I. Development of the Concept
II. Deciphering the Stratigraphic Record
III. Synopsis of Events in Earth History

GLOSSARY

Craton Long stable region of a continent whose Precambrian basement rocks are either at the surface or under a relatively thin cover of Phanerozoic sedimentary rocks.

Facies Laterally restricted portion of a sedimentary rock unit that differs in lithologic makeup or in fossil content from laterally equivalent rocks of the same rock unit.

Geosyncline Belt that is several hundreds of thousands of kilometers long and contains an exceptional thickness of accumulated rocks, on the order of 10 to 15 km. Now believed to simply be the thick accumulation of strata at trailing continental margins, and now commonly called the "geocline."

Marine transgression Movement of the shoreline landward, a result of rising sea level or subsidence of the land.

Marine regression Movement of the shoreline seaward, a result of either relative emergence of the land or a rate of sedimentation that is more rapid than the rate of subsidence.

Microcontinent Isolated, relatively small mass of sialic crust surrounded by oceanic crust; commonly plastered onto continents at convergent margins where they become a part of the orogenic belt identifiable as an exotic terrane or "suspect terrane."

Orogeny Process of crustal uplift, folding, and faulting by which systems of mountains are formed.

Paleomagnetism Magnetism imparted to minerals by the earth's magnetic field at the time they formed, and preserved to the present time as remanent magnetism.

Rock unit Three-dimensional body of rocks distinguishable from adjacent units by lithologic characters; the basic rock unit is mappable and is called a "formation." Larger units are "groups"; smaller ones, "members."

Seismic sequence Unconformity-bounded sequence of strata recognizable on seismic cross sections.

Time-stratigraphic unit Rocks formed during a particular interval of time: the basic time-stratigraphic unit is the geologic system.

Unconformity Surface of erosion or nondeposition that separates younger strata from older rocks.

GEOLOGIC TIME is the time since the earth and the other planets accumulated from the solar nebula some 4.6 billion years ago. It encompasses the entire history of the earth, including the formation and development of the crust, the events and processes that have shaped the earth's surface, the evolution of the ocean and atmosphere, and the evolution of life. Direct evidence for earth history lies solely in the rocks of the earth's crust, but for the first few hundred million years of earth history, for which no rocks

Encyclopedia of Physical Science and Technology, Third Edition, Volume 6
Copyright © 2002 by Academic Press. All rights of reproduction in any form reserved.

are known, valuable insights have been gained from the study of the moon and other planets.

Detailed knowledge of earth history is necessary for an understanding of how crustal movements occur, how rock bodies are produced, how ores and fossil fuels are generated, how the ocean, atmosphere, and solid surface interact, and how numerous processes combine to produce the earth's surficial environments. This understanding is important for many reasons: among these are the discovery and exploitation of mineral resources, assessment of geologic hazards, management of the disposal of wastes, and prediction of how natural processes and man-made influences may be expected to shape the earth's surface in the future.

The very concept of geologic time—of a time so long by human standards that the history of civilization shrinks to insignificance—has been geology's greatest contribution to philosophical thought, and it plays a role in how a society perceives itself in the scheme of things. In this respect the time concept has a parallel in the astronomical concept of space. Time is unimaginably long and space is unimaginably huge. Thus, how a few pioneers first grasped the concept of geologic time and how the world became convinced of its reality was an important development in the history of scientific thought.

I. DEVELOPMENT OF THE CONCEPT

The modern concept of geologic time was formulated by James Hutton of Edinburgh, Scotland, who first published his ideas in 1788. The record of geologic time consists only of those events in earth history that produced a preservable rock record, and Hutton's ideas rested on earlier discoveries by seventeenth- and eighteenth-century naturalists that sedimentary rocks were laid down in sequence and therefore have a discernable history. In Hutton's view, the time represented by this history could not be cast in human terms, nor could it be fathomed by stretching the bounds of conventional historical time. Hence, Hutton proposed the idea of a time so vast that it required a rethinking of the entire scope of the age of the earth and the duration of the events that had shaped it.

The inference that rocks actually record past events had been aluded to by ancient Greek scholars and again by naturalists early in the Renaissance. For example, Leonardo da Vinci in 1508 recognized that fossil marine shells in Cenozoic strata in the Appennines were the remains of marine organisms that lived at a time when central Italy lay beneath the sea. But the first clear statement of principles by which sedimentary rocks form was made by Nicolaus Steno in 1669. From his work on the rocks of Tuscany, Steno realized that the *principal of superposition* was the essential key. This stated simply that, in a succession of strata, the oldest layers are at the bottom and the youngest at the top. Steno also concluded that strata must be nearly horizontal when they form. Tilted and deformed strata are the result of displacement by earth movements after deposition. Steno also realized that identical strata, exposed on opposite sides of the valleys of the region were remnants of layers that were once continuous, and that the valleys had been cut long after their deposition.

Today, Steno's principles seem almost self-evident, but at the time they provided a whole new way of looking at stratified rock. Strata could now be viewed as documents of earth history in which one could discern a variety of sequential events. The concept of geologic time did not follow automatically or quickly among European naturalists because they tended to attribute marine fossils in sedimentary rocks to the Biblical flood, and they held closely to the Biblical account of creation from which they inferred that the earth had a beginning not long ago and was fated for an end not far in the future. Not all of the world's religions taught that the earth was so young. Eastern religions, in particular, held that the earth is very old and that its major physical features underwent a long history of development. Hindu stories of creation, detailed in the *Mahabharata*, refer to ages lasting millions of years during which the earth and all of its beings developed. However, the science of geology grew up in Western Europe, where the prevailing mindset was that the earth was extremely young.

Religious authority for this view was formalized in the mid-seventeenth century by an Irish Archbishop (James Ussher), who calculated that the earth, with all of its features and living things, had been created in 4004 B.C., and hence was less than 6000 years old. Within this brief length of time, it was unimaginable that the earth's surface could have undergone any significant changes. How could trickling water have excavated canyons or worn down mountains? And how could the tiny crustal movements occurring today have formed mountains in the first place? Instead, the earth's features must have been formed quickly, in a splash of creation, by divine Providence. This concept—that the earth and everything on it was the result of sudden, special processes that acted in the beginning and then ceased—has been voiced from time to time and in many forms, but the essential idea is the same: nature today behaves in one way now, but nature behaved in a different way in the past. This point of view is referred to as "catastrophism."

In 1795 James Hutton published his principal work, the two-volume *Theory of the Earth*, in which he argued that processes that act on the earth's surface today, such as erosion, deposition, and volcanic activity, have in fact operated in the same way throughout geologic history. Given enough time, the present rates of activity would be sufficient to produce all of the rocks we find in the earth's crust in all of their observable relationships and configurations. Unlike the catastrophists, who viewed the earth's surface as a remnant of a single tumultuous flurry of creation. Hutton believed strongly that natural laws were unchanging and that nature's past operations may be understood by observing those at work today. Hutton viewed the earth as a nearly eternal machine with a dynamic surface and a dynamic interior. Internal forces produced gradual uplifts that, in the course of time, elevated new lands from the ocean floor even as older exposed land was being eroded away to eventually be covered by the ocean. Hutton believed that, "From the top of the mountain to the shore of the sea, everything is in a state of change," and he noted that the earth "has a state of growth and augmentation; it has another state, which is that of diminution and decay. This world is thus destroyed in one part, but it is renewed in another."

Hutton was the first geologist to recognize the widespread role of igneous processes in forming rocks. His conclusion that basalt solidifies from liquid magma came from detailed observations in the field. To Hutton, the thick basaltic sill that forms Salisbury Craig in Edinburgh showed abundant evidence of once being molten. The sill's margins reveal flow features, such as small dikelike protrusions where the basaltic magma clearly pushed apart the Paleozoic strata as it intruded (Fig. 1). In addition, the strata at the sill's contacts were baked by the magma's intense heat. Hutton inferred correctly that granite was also an igneous rock that had crystallized slowly at great depth from molten magma.

Hutton traveled widely in Europe, where he argued that observable processes of erosion have shaped the Alpine peaks and will ultimately level them, just as they have leveled many mountain ranges in the geologic past. At the same time, the eroded particles go to form new sedimentary records. He said, "We have a chain of facts

FIGURE 1 Strata disrupted by the intrusion of the basaltic sill on Salisbury Craig in Edinburgh helped to convince Hutton of the igneous origin of basalt.

which clearly demonstrates that the materials of the wasted mountains have traveled through the rivers," and "There is not one step in all this progress that is not to be actually perceived." And he summed up, "What more can we require? *Nothing but time.*" To Hutton, the earth appeared to be the product of almost limitless time. Man, said Hutton, has before him now all of the tools "from whence he may reason back into the boundless mass of time already elapsed," Although Hutton never used the word, his principle later came to be termed "uniformitarianism" or simply "uniformity." It is sometimes summarized in the slogan, "The present is the key to the past."

Hutton's ideas today form the basis for modern geologic thought. Initially, however, they met with hostility, in part because they were perceived by many to be atheistic. Catastrophic doctrine in the late eighteenth century had evolved into a compromise between Biblical stories and selected observations of rocks. The prevaling idea at the time, popularized most ably by Abraham Gottlob Werner of Freiburg, Saxony, held that a vast primeval ocean once covered the entire earth. Limestones, granites, basalts, other igneous rocks, and metamorphic rocks as well were equally regarded as precipitates of this ocean. When the water receded (where it went was never explained), all the rocks in their present configurations and all the features of the earth's landscape, including mountains and deep valleys, were left behind. This scheme is commonly referred to as "neptunism."

Neptunists divided the rocks of the earth's crust into four distinct series, and these, in turn, into formations. The *primitive rocks* included granite, the oldest rock, as well as gneiss, basalt, and other old-looking rocks. Next came the *transition rocks*, which consisted of slates, graywackes and some limestone. All of these rocks were marine precipitates. The overlying *stratified rocks*, consisted of fossiliferous sandstone, limestone, and shale, as well as salt and coal, which had been deposited as the ocean waters began to recede below the level of the mountain peaks. Most of these rocks were precipitates, too, but some were derived from material eroded from the emerging mountains. Finally, the *alluvial rocks* at the top were deposited by running water. They included clay sand and gravel, and peat. Volcanic ash was included here as a kind of afterthought: Werner did not believe that igneous processes were important as rock formers. Each formation in his scheme was deposited worldwide during one discrete episode, and each occupied a specific position in the sequence, which was the same in Argentina or in Mongolia as it was in Saxony.

The Neptunists' catastrophic scheme dominated geologic thought until the 1820s, but as field observations accumulated, it ran into increasing difficulties. By that time, geologists came to realize that in different parts of Europe, local stratigraphic sections contained different kinds of rocks in different stratigraphic order. This observation cast doubt on the concept of universally distributed rock layers, and it undermined Werner's entire scheme, which clearly lacked predictive capability.

Another tenet of neptunism that failed the test of field scrutiny was the supposed sedimentary origin of basalt. Studies of the youthful volcanic rocks in the Auvergne area of central France convinced one geologist after another that basalt was an igneous rock. Here beautifully preserved volcanic cinder cones are clearly the source of lava streams that flowed down adjacent valleys, where they cooled and froze into columnar-jointed basalts. The failure of the neptunistic ideas to stand up to field observations cast serious doubts on the entire scheme, and during the 1820s its attractiveness waned.

Hutton's concept of gradual change through existing causes began to be applied by a number of naturalists, but it was most effectively and lucidly championed by Charles Lyell (1797–1875) in his *Principles of Geology*, which appeared in three volumes in 1830, 1832, and 1833. Lyell covered the complete scope of geology and not only supported but also amplified Hutton's principle of uniformity by extensive coverage of erosional and depositional processes. Throughout his text, Lyell reiterated the concept of unlimited time. *Principles of Geology* went through many editions and was widely read, and it established uniformitarianism, at the expense of catastrophism, as the accepted philosophy for interpreting the history of the earth.

A. Influence of the Concept

Lyell's proofs of the principle of uniformity prepared the ground for further accomplishments during the nineteenth century, including those of Charles Darwin, whose ideas on the gradual evolution of living things could not have flourished without the intellectual framework of vast time. Before Lyell's textbook appeared, few people wondered about the age of geologic events or the age of the earth itself, and most who did could only depend on myth or speculation. Lyell not only won the day for the principle of uniformity, he generated wide interest in how much actual time in years geologic history might represent. Then, following Darwin's work on evolution in 1859, the question of the magnitude of geologic time became a focal point of the evolution controversy. In *The Origin of Species*, Darwin extended Hutton's philosophy of gradual change through existing causes into the realm of life. To Darwin, life evolved continually and is always being gradually modified through natural selection. Darwin's work raised large questions, among which was the scope of geologic time. It was clear from the outset that the success of Darwin's synthesis depended on the availability of a great

deal of time. As soon as the *Origin* appeared, how long the earth had existed and how long the surface had actually been hospitable for life became questions of intense interest.

B. How Much Time?

1. Early Geological Estimates

To determine the quantity of time past requires a time-governed, irreversible process whose rate is known. We think first of radioactive decay, but in the middle of the nineteenth century radioactivity was yet undiscovered. The evolution of life is itself a time-governed and irreversible phenomenon, but rates are not constant within groups. Nevertheless, Lyell tried this approach in 1867. He guessed that about 20 million years (m.y.) was required for a complete turnover of molluscan species and inasmuch as there were 12 such turnovers since the beginning of the Ordovician period, he arrived at 240 m.y. as an estimate of the elapsed time. This was not a bad guess (the current radiometric date for the base of the Ordovician is 505 m.y.), but it was not verifiable.

As early as 1715 the English astronomer Edmund Halley suggested that the age of the earth could be calculated from the rate of change in the salinity of the ocean. Halley proposed that the salinity of ocean water be determined with great precision and then that the determination be repeated a decade later. From the incremental increase one could calculate the time required, beginning with fresh water, to reach present salinity. If the experiment was ever tried, no increase was detected. Halley's proposal indicates how short a span of geologic time was envisioned at the time he wrote.

John Joly in 1899 estimated from chemical analyses of river water the amount of sodium added to the sea annually by the world's rivers. Knowing the volume of water in the ocean, and assuming that the original ocean was fresh and that the present rate of sodium input by rivers was the mean for all geologic time, Joly concluded that about 90 m.y. had elapsed since water first condensed on the earth. Joly's estimate was far too low because he did not appreciate that over the long haul as much sodium leaves the sea by deposition in marine sediments as enters through rivers. (The residence time for sodium in the sea is 71 m.y.)

Many geologists during the nineteenth century attempted estimates of geologic time based on rates of deposition of sedimentary strata. The reasoning was that sedimentary rocks must accumulate at some average rate, and if such a rate could be established in modern depositional environments, the time required for the accumulation of analogous ancient rocks could be ascertained. Moreover, if one could ascertain the total thickness of all the sedimentary rock that had ever been deposited in the geologic past, one could arrive at an estimate of total elapsed geologic time. Difficulties abound with this method because most local sections consist of many different rock types and contain numerous breaks in the stratigraphic record. Most workers adjusted their rates for different rock types and for rocks of different ages based on personal judgment. Estimates of the duration of geologic time based on rates of deposition varied widely, but most of them were less than 100 m.y.

2. Kelvin's Estimates

During the last half of the nineteenth century Lord Kelvin, the renowned English physicist, made estimates on the age of the sun and the earth that were by far the most influential. They were also among the very lowest. Because they were based on precise physical measurements that demanded few assumption, they were accepted widely. Kelvin measured temperatures in deep mines of many areas and found that temperature increases consistently with depth. This geothermal gradient indicates that heat is flowing from the earth's hot interior to the cool surface, where it escapes. The heat loss can thus be measured and Kelvin reasoned that, if in losing heat the earth is becoming progressively cooler, then in times past it must have been warmer. The farther back in time, the warmer it must have been. Kelvin regarded this phenomenon as dissipation of heat from an originally molten condition, and he showed that it could not have been very long ago, in terms of geologic time, that the earth was molten. Lack of details on melting points of rocks and their thermal conductivity at high pressures and temperatures prohibited exact calculations of the time since the crust solidified, but the estimates of Kelvin and his followers were generally less than 100 m.y. The earth's surface would not have been habitable by life as we know it until much later, when temperatures finally approached those of today's earth.

Kelvin also considered that the heat output of the sun must be decreasing rapidly. The source of the sun's energy posed a considerable problem to physicists before radioactivity was understood, but Kelvin argued that, whatever the source, the continued output of so much energy must be progressively lowering the sun's temperature. He concluded on the best grounds then available that only a few million years ago the sun was much hotter than it is now and that this alone would have made the earth's surface too hot for life as we know it. In 1897 Kelvin wrote his last paper on the subject and concluded that the earth's surface had probably been habitable for between 20 and 40 m.y.

These assertions had ominous overtones for evolution, not only because of the short length of geologic time

they allowed, but also because, old or young, the earth's crust had been considerably hotter in the geologic past. If this were true, then the surface could not have maintained the kind of long-term stability that was necessary for the gradual evolution of life. Late in the nineteenth century most geologists had simply adjusted their ideas of the duration of geologic events to accommodate Kelvin's foreshortened time framework. Paleontologists who insisted that the earth must have endured far longer than this could offer only qualitative arguments that, in the face of the physicists impressive numerical data, were largely ignored. In the long run, however, the paleontologists seemingly vague hunches prevailed. At the turn of the twentieth century Kelvin's rigid mathematics on the consequences of a rapidly cooling earth were rendered meaningless by the discovery of radioactivity.

3. Discovery of Radioactivity

Henri Becquerel, a French physicist, discovered radioactivity in uranium in 1896. In 1903 Pierre and Marie Curie discovered that a sample of radium always maintains a temperature greater than that of its surroundings, and three years later an English physicist, H. J. Strutt, showed that the quantity of heat that is continuously generated by radioactive minerals in the earth's crust easily accounts for the heat flow at the surface. With this, the heat escaping from the earth's surface ceased to signify impending cooler times ahead and markedly warmer times past, because the earth's heat flow is not from residual heat, but is from heat that is continually produced within by radioactive decay. Thus, this heat flow has been about the same for a very long time.

In 1905, B. B. Boltwood, a Yale chemist, discovered that lead was a stable daughter product of uranium decay. Boltwood went on to show in 1907 that in minerals of the same age the lead/uranium ratio is constant, but among minerals of different ages the lead/uranium ratio is different: the older the mineral, the greater the ratio. Using the first estimates of decay rates, Boltwood calculated the radiometric ages of several mineral assemblages of Paleozoic and Precambrian age (Table I). Considering the sketchy knowledge and primitive analytical techniques of the time. Boltwood's radiometric dates are remarkably close to those we obtain for rocks of the same ages today. The discovery of radioactivity had not only refuted Kelvin's arguments demanding a young earth and sun, but at the same time had also provided the ultimate tool for measuring geologic time quantitatively. It was clear from even the earliest determinations that the earth's antiquity was far greater than prevailing opinion at the end of the nineteenth century would have deemed possible.

TABLE I Boltwood's Uranium–Lead Dates
Determined in 1907

Geologic age	Lead/uranium ratio	Millions of years
Carboniferous	0.041	340
Devonian	0.045	370
Percarboniferous	0.050	410
Silurian or Ordovician	0.053	430
Precambrian: Sweden	0.125	1025
	0.155	1270
Precambrian: U.S.	0.160	1310
	0.175	1435
Precambrian: Ceylon	0.20	1640

C. The Search for a Time Framework

Efforts to relate time to rocks—to classify the rocks of a given region into some kind of time framework—began around the middle of the eighteenth century. In western Italy, Giovanni Arduino (1760) classified the rocks into the following three main categories:

1. *Primary* (oldest)—crystalline rocks with metallic ores
2. *Secondary*—hard stratified rocks without ores but with fossils
3. *Tertiary* (youngest)—weakly consolidated stratified rocks usually containing numerous marine fossils; volcanic rocks as well

At about the same time, Johann Gottlob Lehmann (1756) recognized in Thuringia a threefold division of rocks, which can be characterized as (1) crystalline, (2) stratified, and (3) alluvial. Within a few years similar three-fold divisions of rocks were recognized elsewhere in Europe and in Russia. Recognition of threefold divisions of rocks based on the principle of superposition thus provided an early framework of rock classification, but the early workers were incorrect in assuming that the major rock types on which their divisions were based each represented a unique episode in geologic time. The idea that the ages of rocks may be ascertained by their composition failed to stand up under field observations. In the early decades of the nineteenth century it was replaced by the much more fruitful law of faunal succession.

1. The Law of Faunal Succession

William Smith was an engineer, and his work on roads, quarries, and canals acquainted him with the sequence of rocks throughout much of Britain's countryside. In the course of his work between 1793 and 1815 Smith traced out the numerous rock units in the region, and he drew

boundaries between them on his maps. As his experience progressed, he noticed that each of his mappable rock units contained its own diagnostic assemblage of fossils by which it could be distinguished from other rock units. This enabled Smith to place an isolated outcrop of an otherwise nondistinctive limestone or a sandstone in its proper place in the stratigraphic section with confidence, solely on the basis of the fossils it contained, even though the strata above and below might not be exposed. Fossils thus enabled Smith to recognize individual rock units in remote areas in which physical criteria alone were inconclusive.

Using his new principle, Smith produced in 1815 the first geologic map of England, Wales, and part of Scotland. This map was more detailed than any that had been attempted before, and it represented a milestone in the understanding of the sedimentary rock record. Smith's discovery that strata may be identified by the fossils they contain became known as the *law of faunal succession*. This important principle raised questions concerning ancient life, but even without the answers to these questions, correlation of strata between distant localities now became feasible. This made it possible to erect a stratigraphic classification based on the time equivalency of strata rather than on rock types.

2. The Discovery of Organic Extinction

Until the second decade of the nineteenth century, the idea that many formerly existing species were extinct had not been proven to the satisfaction of most naturalists. The best-known fossils were marine invertebrates, and at the time too little was known of existing life in the oceans to assert that certain marine animals represented in the fossil record no longer lived anywhere. Then Georges Cuvier, a French zoologist, showed that many fossil vertebrates from the Tertiary strata of the Paris Basin have no known counterparts living today, and other naturalists agreed that it was unlikely that such large land animals still existed undiscovered. Extinction was at last a reality and thereafter it was a short step to recognize the principle for invertebrate animals and plants as well.

Cuvier carefully worked out the fossil succession in the Paris Basin strata, and he noted that the younger deposits contained faunas more like those of the present day than did the older deposits, Cuvier got the clear idea that the succession of faunas revealed advancing complexity of life. Impressed by the numerous breaks in the stratigraphic record that occur in the Paris Basin, Cuvier interpreted the succession of faunas in catastrophic terms. Nevertheless, his discoveries complemented those of William Smith and completed the foundation necessary for the use of fossils as correlation tools.

D. The Geologic Time Scale

The discoveries of Smith and the refinements added by Cuvier made possible the recognition of rocks of the same age in widely separated areas regardless of rock types. It thus became feasible to distinguish major units of sedimentary rocks and, using their distinctive fossils, to identify their time counterparts worldwide. These major time-stratigraphic units, which constitute the foundation of the geologic time scale, became known as the *geologic systems*. In the three decades following the work of Smith and Cuvier most of the geologic systems that constitute the modern geologic time scale (Fig. 2) were defined. The time scale has evolved through utilization by numerous geologists over the decades. It has been nearly stable since the turn of the twentieth century, but it is still subject to refinement.

The rock outcrops that form the basis for definition of a stratigraphic unit are called the "type section," or "stratotype," and the area in which the outcrops occur is called the "type area." Most of the geologic systems in use today were defined in Europe. The type areas of the commonly used systems, as well as the authors and the dates they were named, are listed in Table II.

The geologic systems have been recognized widely on the basis of distinctive fossils, and they have proven to be remarkably convenient units. All of the systems listed in Table II are recognized worldwide except for the Mississippain and Pennsylvanian, which are used in place of the Carboniferous in North America. A great deal of data have now been amassed on the distribution and makeup of the systems, and as these data have accumulated, the time scale has been greatly refined. Geologic systems are subdivided into series, which are commonly referred to by the modifiers "Lower," "Middle," or "Upper" before the system name. Series are subdivided into stages that are, in turn, subdivided into chronozones, the smallest recognizable time-stratigraphic units. All of these units

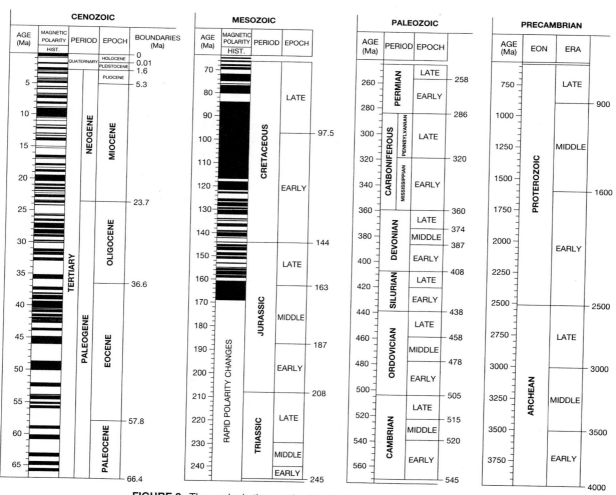

FIGURE 2 The geologic time scale. (Modified from A. R. Palmer, 1983.)

TABLE II Naming of the Geologic Systems

System	Type area	Author	Date
Quaternary	France	Paul G. Desnoyers	1829
Tertiary	Italy	Giovanni Arduino	1760
Cretaceous	France	Omalius d'Halloy	1822
Jurassic	Switzerland	Alexander von Humboldt	1795
Triassic	Germany	Friedrich von Alberti	1834
Permian	Russia	Roderick I. Murchison	1841
Carboniferous	England	Conybeare & Phillips	1822
Pennsylvania	U.S.	Alexander Winchell	1870
Mississippian	U.S.	Henry Shaler Williams	1891
Devonian	England	Murchison & Sedgwick	1840
Silurian	Wales	Roderick I. Murchison	1835
Ordovician	Wales	Charles Lapworth	1879
Cambrian	Wales	Adam Sedgwick	1835

apply specifically to the rocks deposited during a particular interval of geologic time. In other words, these are material units that have been time-correlated with the type section or with intermediate sections that in turn have been correlated with the type.

Geologists have found it convenient to recognize pure *geologic time units* that represent the time duration of the time-stratigraphic units. Thus, the time duration of a geologic system is a geologic period. Geologic time units take the names of the time-stratigraphic units. For example, the Cambrian system, defined by a sequence of sedimentary rocks in Wales, is the basis for a time term, the Cambrian Period, that surely existed everywhere, although in many areas it is unrepresented or only party represented by a rock record. Periods are subdivided into epochs, ages, and chrons, and they are lumped into larger geologic time units called *eras*. Only three eras are in common use: the Paleozoic, Mesozoic, and Cenozoic, which translate literally into "ancient life," "middle life," and "modern life." These eras are part of a still larger unit, the Phanerozoic Eon. Corresponding time-stratigraphic and geological time terms follow:

Time-stratigraphic units	Geologic time units
Eon	Eonothem
Era	Erathem
Period	System
Epoch	Series
Age	Stage
Chron	Chronozone

The old rocks below the Paleozoic have always been known widely and simply as the "Precambrian." In the early days of geology the span of time represented by these rocks was grossly underrated. They were viewed as little more than a foundation or "basement complex" on which the fossil-bearing Phanerozoic sedimentary rocks were deposited. This viewpoint persisted into the twentieth century because of the low estimates of the total age of the earth that were fashionable in the years prior to the discovery of radiometric dating. The well-preserved, fossiliferous Phanerozoic record required most of the total time span as then conceived, thus allowing little time for the Precambrian. Radiometric dating has now taught us that the oldest rocks formed around 4 billion years ago. Between that time and the appearance of shell-bearing animal life at the base of the Cambrian, about 570 m.y. ago, about 85% of recorded earth history elapsed. Yet, because Precambrian sedimentary rocks lack fossils that are useful for correlation, it has been possible to recognize only two subdivisions that are applicable worldwide, the Archean Eon below and the Proterozoic Eon above. More detailed classifications are applied in individual regions.

II. DECIPHERING THE STRATIGRAPHIC RECORD

The process of working out geologic history begins with studies of rocks, which are materials that have been rearranged by geologic processes during bygone ages and whose attributes reflect the kinds of environments in which they formed. Of the three kinds of rocks, sedimentary igneous, and metamorphic, sedimentary rocks provide by far the most complete record of earth history. They constitute about 75% of all exposed rocks. They alone form at normal temperatures and pressures at the earth's surface, and they commonly contain fossils, which record the history of life and which also provide the best tools for correlation.

Studies of rocks begin in the field in areas where there are *outcrops*—exposures of rocks that have been produced naturally (as in stream valleys) or artificially (as in road cuts and quarries). Observations of rocks in the field may be supplemented with studies of cores or cuttings that have been recovered from deep drill holes. To work out the history of a region, geologists first distinguish the individual bodies of rock that are thick enough to be shown on a map of the area. They then describe these units of rock, measure their thicknesses, and organize them into a chronological sequence based on the law of superposition. This sequence is generally plotted as a vertical column or *stratigraphic section* that illustrates the rock types and thicknesses, like that shown in Fig. 3.

The fundamental rock units that make up the stratigraphic section for an area are called *formations*. These are mappable rock units that differ from one another in color,

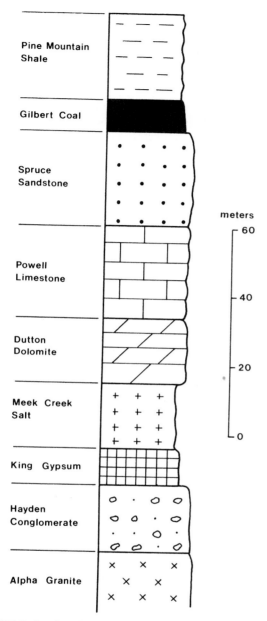

FIGURE 3 Stratigraphic column for a hypothetical area showing the symbols used for common sedimentary rock types.

A. Sedimentary Environments

Individual formations are distinctive because each is the product of a particular depositional environment that differed from those that existed before and after. Each formation thus has a story to tell about events in a region during a brief interval of geologic history. The composition of the rock and its features, such as the nature of the bedding and other of sedimentary structures, reflect the parent material exposed in the source area, the processes of transportation and deposition, and environmental factors in the depositional area. The latter include climate, current energy, rate of subsidence, pH, oxygenation, salinity, rate of detrital influx, and biological activity. In addition, some sediments are substantially modified by lithification and diagenesis in the *postdepositional* environment. *Lithification* refers to the processes that convert sediment into rock, including compaction and cementation. *Diagenesis* refers to chemical reactions that take place between the mineral grains and interstitial fluids. It includes all chemical changes up to, but excluding, metamorphism.

1. Interpreting Ancient Environments

Ancient environments can be reconstructed only for those times in a region during which there was sedimentation and for which a rock record survives. The rock record contains virtually no representatives of the upland and mountain environments that today make up large portions of the continents. Although these regions receive temporary deposits that may provide information about recent geologic history (for example, peat bogs in alpine meadows), these sediments are not destined to become buried and changed into rock. Over the long term, upland regions undergo net erosion and all of their mass is removed to subsiding lowlands, commonly in or adjacent to the sea, where a preservable historical record accumulates. The earth's historical record is thus biased in that it chiefly contains information about environments in which deposition occurs and little information about the source regions of the sediment.

Sedimentary environments are first classified as to whether deposition took place on land in a *terrestrial environment*, in the sea in a *marine environment*, or in between in a coastal setting or *transitional environment*. Next, more specific setting is stipulated if possible, whether the environment was, for example, a flood plain, a shallow sea, or a tidal flat. Finally, as many details as possible are determined: the size of the streams, salinity and temperature of the sea, current energy and tidal range on the tidal flat. Once an environmental interpretation has been made, the next step is to map its distribution along with other environments that adjoined it at the time.

composition, or other properties that can be distinguished in the field (or on geophysical logs from deep wells). Thickness of individual formations, measured perpendicular to bedding, is usually several tens or hundreds of meters, but may be thinner or thicker. Formation names consist of two parts, a geographic name (taken from a feature near the type locality), followed by the rock type or simply the word "formation." For convenience, two or more formations are sometimes lumped into larger rock units called *groups*. Subdivisions of formations are called *members*.

In making environmental interpretations from rocks, the basic strategy to compare the physical features and organic consituents of the sedimentary rocks with those of sediments forming today in known environments. If the ancient rocks and the modern sediments are composed of similar materials and have comparable physical features and biological constituents, then they probably formed in similar environmental settings. Some rock types alone suggest particular depositional environments. For example, widespread coal beds generally represent coastal marsh environments, and thick widespread limestone units represent marine environments. However, most sedimentary rock types can originate in a variety of environments that are impossible to discriminate on the basis of composition alone. For example, gray shale forms from mud, which can be deposited in flood plains, lakes, lagoons, and shallow and deep marine environments. The most reliable way to discriminate among these is by means of the fossils that the shale contains. These are extremely helpful because plants and animals lived in particular environments during the geologic past, just as they do today. Finally, the rocks that lie above and below the gray shale provide important clues to the overall environmental setting, so that interpretation of the environments represented by the sequence of rock units can be made with greater confidence than an interpretation of one rock unit alone.

2. Facies

Just as sedimentary rock types change vertically due to changing environments through time, they also change laterally, reflecting different environments that exist side by side at any one time. Hence, marine shale might be replaced shoreward, in turn, by silty shale, beach sandstone, marsh and lagoonal deposits, and finally floodplain deposits (Fig. 4). Such *facies changes* would normally occur over a distance of 100 km more or less. In addition to the kinds of changes in rock types illustrated in Fig. 4, changes may occur independently in the fossil content of the strata, while the rock types remain substantially unchanged. Lateral changes in physical characters are termed *lithofacies* and changes in the fossil content, *biofacies*.

Facies changes reflect environments that exist adjacent to one another in nature. If the boundaries between depositional environments remained stationary as deposition progressed, then the resulting facies boundaries would be essentially vertical as in Fig. 4a. In most cases this static situation does not long prevail, and environments migrate laterally, the deposits of one covering earlier deposits of an adjacent environment. Along a shoreline, if the rate of subsidence exceeds the rate of supply of sediment, the sea moves landward in *marine transgression* (Fig. 4b). If the

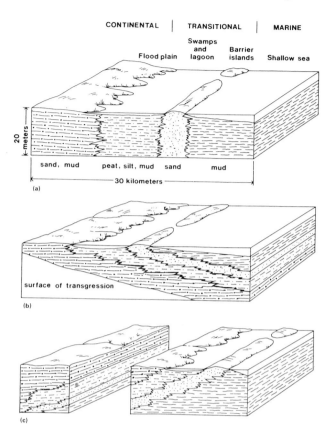

FIGURE 4 Typical facies changes in near-shore environments showing the stratigraphic successions produced by (a) stillstand, (b) transgression, and (c) regression. Vertical scale is exaggerated 300 times.

rate of supply exceeds the rate of subsidence, the shoreline moves seaward in *marine regression* (Fig. 4c). This is an illustration of the principle of facies, which stipulates that *those that succeed each other vertically are those that adjoin laterally*. This principle applies as long as sedimentation is relatively continuous and is not interrupted by significant breaks in the depositional record.

B. Unconformities

An unconformity is a large gap in the stratigraphic record that formed when deposition ceased for a considerable time. Most unconformities have resulted from relative uplift, which causes the erosion of some of the previously formed record. Wherever sedimentary sequences have been studied, they are interrupted by such gaps. The lost intervals range in magnitude from a chronozone to entire eras. Unconformities are regional in extent, and the time represented on a given unconformity surface varies laterally. Commonly unconformities are largest in

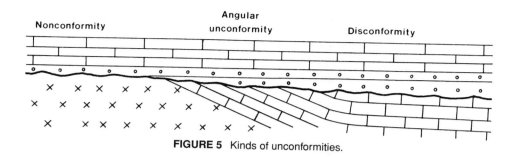

FIGURE 5 Kinds of unconformities.

continental interiors and represent less and less time as they are traced toward continental margins, where sedimentaion has tended to be more continuous and where exceptionally thick sequences of sedimentary rocks have typically accumulated.

In the field, unconformities appear as bedding plane surfaces and are mapped like any other formation contacts. Three major types can be distinguished based on the structural relationships of the rocks above and below the unconformable contact (Fig. 5). *Angular unconformity* represents the folding of an old sedimentary sequence, planing of the tilted strata by erosion, and the deposition of a young sedimentary sequence on the old truncated strata. *Nonconformity* refers to a surface in which stratified rocks rest on intrusive igneous rocks or metamorphic rocks that contain no stratification. *Disconformity* refers to an unconformity in which the beds above and below the surface are parallel. Most disconformities show some erosional relief, solution features, weathering profiles, or other physical evidence of a break in the sedimentary record. In the absence of such evidence, the unconformity may not be apparent in the field and its presence must be based on fossil evidence of a substantial time gap.

C. Correlation

In order to understand the geologic history of a region, numerous local sections must be correlated, that is, the time equivalency of their strata must be worked out. Correlation between two stratigraphic sections is based on events that affected both areas simultaneously. If the sections being correlated are relatively near one another, physical methods of correlation alone may be adequate. If the sections are in different basins or perhaps on different continents, then correlation must rest in large part on the fossils that the strata contain.

1. Physical Tracing of Strata

The simplest method of correlation is to trace, from section to section, an individual bed that represents a discrete episode of sedimentation. Most individual beds, particularly from shallow-water environments, do not extend very far, and cannot be recognized from section to section. Beds that form in deeper, quiet water may be widespread, extending over several kilometers or more, but such beds tend to occur in uniform sequences of shale or deep-water sandstones in which all beds look alike. But if a widespread bed has a special attribute by which it can be recognized in separate stratigraphic sections, it is a *key bed* for correlation. Bentonite beds are especially valuable in physical correlation because they are distinctive beds of white clay that formed by the alteration of volcanic ash that fell into ancient marine environments in which shale or limestone was accumulating. Bentonites are only sporadically common in the stratigraphic record, but where they occur they are ideal key beds because they represent instantaneous ash fallout from volcanic eruptions.

2. Paleomagnetic Correlation

Magnetic iron oxide minerals occur in tiny quantities in most rocks. When tiny particles of these minerals settle to the sea bed, or when they solidify in a cooling lava flow, their magnetic polarity aligns itself with that of the earth's magnetic field at that time and place. This produces a preferential direction of magnetization in rocks that contain the minerals. This property of the rocks is called *remanent magnetism*. The record of remanent magnetism in Mesozoic and Cenozoic rocks indicates that the earth's north and south magnetic poles have abruptly switched polarity numerous times in the geologic past. During these times of magnetic field reversals the north needle on a compass would have actually pointed to the south pole. Each reversal simultaneously affected the field for the entire earth. Numerous magnetic field reversals have been recognized in Mesozoic and Cenozoic rocks worldwide, and they have been tied accurately to the geologic time scale (Fig. 2). These reversals define *magnetic polarity zones* whose boundaries parallel those of chronozones based on fossils. Hence magnetic polarity zones represent a promising tool for correlation.

Although the method has worldwide potential, there are only two kinds of polarity zones; all the normal ones share the same kind of remanent magnetism and all of the reversed ones do also, so that a geologist must, already have a very good idea of the age of the rock, based on fossil data, in order to make use of them. The chief value of magnetic polarity zones in the future will be in the refinement of ages based on other criteria.

3. Stable Isotope Stratigraphy

The development of oxygen isotope stratigraphy in recent years has provided an important tool for correlation of deep-sea carbonate strata. When a foraminifer extracts calcium carbonate from seawater to build its skeleton, it effectively samples the relative proportions of the oxygen isotopes (^{16}O and ^{18}O) in the dissolved carbonate ions. On the sea floor these tiny particles become part of the accumulating strata, in which they preserve the oxygen isotope record. Water that evaporates from the sea is rich in ^{16}O, so if this water is not immediately returned but is instead stored on the land as ice, the ratio of ^{18}O in the sea rises. The ratio also rises when water temperatures drop. Both effects prevailed during the Pleistocene glacial episodes, so in deep sea carbonate strata, foraminifera rich in ^{18}O represent glacial episodes, and forminifera poor in ^{18}O represent interglacial episodes Figure 6 shows glacial–interglacial fluctuations in oxygen isotope ratios in two deep-sea cores, one from the Caribbean and one from the western Pacific, for the last 700,000 years. Widely recognized isotope peaks are numbered, and the correlation is apparent. This method makes possible worldwide correla-

tions of Pleistocen deep-sea sediments with a precision of a few thousand years. The data show a remarkable cyclicity in the Pleistocene glacial ages that is now believed to be linked to the earth's orbital perturbations. This strongly suggests that the curve has predictive value as well as historical and correlative value.

4. Biostratigraphy

The term *biostratigraphy* refers to the dating and correlating of rocks by means of fossils. The fossil record has been extended back more than 3 b.y., but most of the Precambrian portion consists of isolated occurrences of ultramicroscopic objects that are far too meager to permit correlations. The abundant and varied fossils in Phanerozoic strata, however, provide the basis for a stratigraphic framework that has a high degree of time resolution.

Because living things have evolved through time, particular kinds of organisms are characteristic of particular portions of the geologic column. Each species that is extinct today evolved at some point in time and became extinct at a later point in time. Thus, each extinct species divides geologic time into three parts: the time before it evolved, the time during which it existed, and the time since it became extinct. Any rocks that contain the species must have been deposited within the time during which the species existed. If the total time during which the species existed was short then the rocks in which it occurs can be located precisely in the geologic time scale. Rates of evolution vary greatly among different groups of organisms. Some evolved slowly and have left a record of long-ranging species: others evolved rapidly and have left

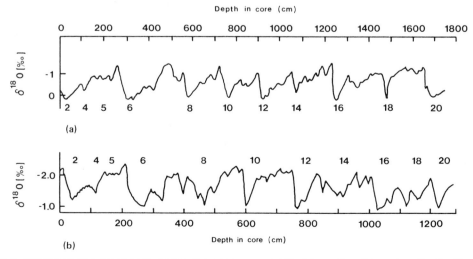

FIGURE 6 Oxygen isotope stratigraphy for two deep-sea cores: (a) Caribbean Sea; (b) western Pacific. Time represented is slightly more than 700,000 years. [After Shackleton, N. J., and Opdyke, N. D. (1973), *Quat. Res.* **3,** 39–55; Emiliani, C. (1978). *Earth Planet. Sci. Lett.* **37,** 349–352.]

a record of short-ranging species, which are of superior time value. If such short-ranging species occur widely and are readily recognizable they are sometimes referred to as "index fossils" or "guide fossils."

Species of animals and plants today occur in local communities that represent the distribution of particular environments. Species in the past were similarly distributed. The many lateral changes in fossil assemblages or "biofacies" in sedimentary rocks reflect these different ancient environments, and provide valuable tools for their interpretation. Thus, fossils permit intepretations of environments on one hand and determination of time equivalency on the other. The key to correct reconstructions of geologic history is the proper discrimination of the two effects.

Biostratigraphic units are bodies of sedimentary rock that contain particular fossils. The fundamental biostratigraphic unit is the fossil zone or *biozone*, which is based on "datums," which are simply the first and last occurrences of fossil species in stratigraphic sections. The major kinds of biozones are defined either by the stratigraphic interval between two key datums or by the overlapping stratigraphic ranges of three or more species. Zones inferred to have time-stratigraphic value may be termed *chronozones*.

Some assemblage zones are based on several species without regard to their ranges and are used for the recognition of particular paleoenvironments. In some studies, *abundance zones* are recognized based on the stratigraphic interval that includes only the interval of extraordinary abundance of a species, without regard to its lowest and highest occurences. These, too, have only environmental significance.

Paleontologists in the nineteenth century became aware that every geologic system has its characteristic short-lived groups of organisms that are of greatest stratigraphic value, for example trilobites in the Cambrian, graptolites in the Ordovician ammonites in the Jurassic, and so on. These groups have provided highly precise and widely useful zonations, but efforts are constantly being made to attain still more precision through refinement of the time-stratigraphic framework.

Once correlation has established time-equivalent intervals among stratigraphic sections, the lateral and regional relationships of facies can be analyzed. Typically they are illustrated on stratigraphic cross sections like the one shown on Fig. 7. If data are sufficient, facies and thickness information can also be plotted on maps. Facies maps

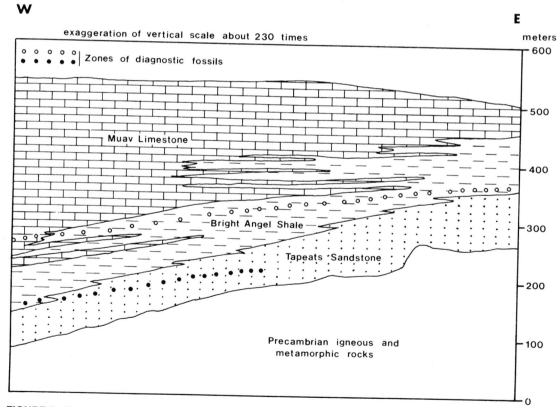

FIGURE 7 Stratigraphic cross section of Cambrian strata in the Grand Canyon region. Time lines provided by two trilobite zones document the facies changes and reveal a record of marine transgression. [After McKee, E. D. (1945). *Carnegie Inst. Washington Publ.* **563**.]

show the areal distribution of rock types or of fossil assemblages of a given age. Commonly facies maps are accompanied by *isopachous maps* that snow the thickness of the mapped unit by means of isopach (equal-thickness) lines. The distribution of the environments that have been inferred from the isopachous and facies maps are sometimes summarized in interpretive *paleogeographic maps*.

5. Seismic Stratigraphy

Seismic stratigraphy is an important method for studying sedimentary rocks in the deep subsurface that has been in widespread use only since the 1970s, and that was made possible by computer analysis of seismic data. The methods of seismic stratigraphy are used to correlate sedimentary strata at depth, and they also reveal unconformity-bounded *seismic sequences* that provide the keys to the analysis of regional geologic history.

The data for seismic stratigraphy consist of records of waves that are generated on land by dynamite explosions

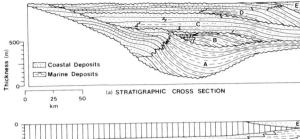

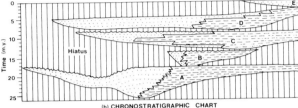

FIGURE 9 (a) A geologic cross section of a continental margin showing seismic sequences (lettered) recognizable on a seismic cross section. (b) A chronostratigraphic chart showing the shoreward and seaward extent of each of the seismic sequences. [After Vail, P. R., *et al.* (1977). *AAPG Mem.* **26**, pp. 83–97.]

or truck-mounted vibrators and at sea by sleeve exploders or airguns. At depth the waves are reflected from bedding planes at which the wave velocity changes, commonly as a result of density or compositional differences between the rocks above and below. The reflected waves are recorded in the field on magnetic tape and later processed through computer programs to produce a seismic cross section. Seismic cross sections provide two-dimensional pictures of the distribution and structure of the underlying strata along transects, usually several miles long, that are termed *seismic lines*. Most seismic cross sections have been constructed on coastal plains and offshore continental shelves. They provide a hitherto unavailable picture of lateral stratigraphic relationships deep below the surface in regions in which there are few surface exposures, and they reveal episodes of onlap and offlap of sedimentary sequences from which a history can be worked out.

Synthesis of the histories of numerous coastal regions has made it possible to identify and date many worldwide cycles of transgression and regression. The record of global sea-level changes for the Paleogene is shown in Fig. 8. Figure 9 shows how this global chart can be used to date the seismic sequences (indicated by A through E) in a new area. First the history of coastal onlap and offlap for a new area is translated into a sea-level cycle chart, in which points of maximum transgression and regression are plotted against time. This local chart is then compared to the global chart to properly identify the seismic sequences and date the strata in the new area. These kinds of studies have important applications in the search for oil and gas.

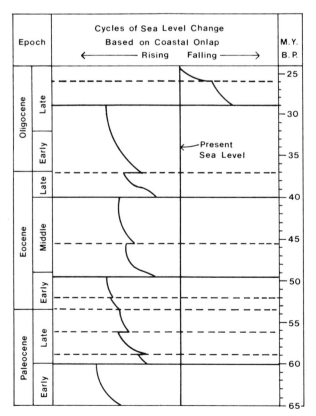

FIGURE 8 Global cycles of relative change of sea level during the Paleogene. [After Vail, P. R., *et al.* (1977). *AAPG Mem.* **26**, pp. 83–97.]

D. Radiometric Dating

Radiometric dating has quantified geologic time. It has dated key events, in years, from the formation of the earth to the retreat of the most recent continental ice sheet, and it has made possible the measuring of rates of a variety of geologic processes. For Precambrian rocks, which lack useful fossils for dating and correlation, radiometric ages, sometimes called "absolute ages," have provided the basis for a coarse time-stratigraphic framework. For Phanerozoic rocks, radiometric dates have provided a quantitative calibration of the geologic time scale so that now a radiometrically dated rock can be placed in a geologic system or series with high confidence. The versatility of the radiometric dating tools has been made possible by the refinement of laboratory methods that enable geochemists to measure vanishingly small quantities of particular nuclides with remarkable accuracy.

1. The Theory of Radiometric Dating

Radiometric dating is based on the presence, in some minerals and rocks, of naturally occuring radioactive nuclides that decay continuously to radiogenic daughter nuclides. When a radioactive atom decays, its nucleus emits an alpha particle (a helium nucleus), or a beta particle (an electron), or it captures an electron. Each radioactive nuclide has its own mode of decay and, more important, its own constant rate of decay. Rates of radioactive decay are commonly expressed in terms of half-life, the time required for half of any given quantity of radioactive parent atoms to decay. Radioactive decay occurs in the atomic nucleus and the rate is unaffected by heat or pressure. It is even unaffected by chemical changes, such as oxidation or reduction of the parent atom, because these involve only the orbital electrons and not the nucleus. Hence, a radioactive nuclide that is incorporated in a mineral when it crystallizes will decay at a known rate that is controlled only by the time that has elapsed since the crystallization event.

Most methods of radiometric dating are based on the daughter/parent ratio, that is, on the quantity of radiogenic daughter atoms that have been produced, relative to the quantity of surviving radioactive parent atoms. When a mineral containing the radioactive atoms crystallized, it contained no atoms of the daughter, and the ratio was zero. With time, radioactive decay produces radiogenic daughter atoms in place of the parent atoms in the mineral's crystal lattice and the ratio slowly increases. At the end of one half-life period, one-half or the original number of parent atoms remains. At the end of the second half-life period, half of a half or one-fourth of the original parent atoms remain. At the end of the third half life, one-eighth of the parent atoms remain, and so on. At any time the

age of the mineral in years may be calculated from the ratio using a simple formula. In order for the method to yield accurate results, the mineral or rock must behave as a closed system; there must be no addition or escape of parent or daughter atoms. Also, no undetected atoms of the daughter nuclide can have been present in the mineral when it formed.

2. Methods of Radiometric Dating

Radioactive nuclides that occur in nature are either long-lived nuclides that have persisted since the earth formed, or short-lived nuclides that are forming continuously as steps in the uranium and thorium decay chains or by cosmic-ray bombardment of the upper atmosphere. Long-lived nuclides are by far the most useful for radiometric dating of rocks. Of the approximately 20 long-lived nuclides that have been identified, only four are widespread and abundant enough to be generally useful as chronological tools. These are ^{40}K (which decays to ^{40}Ar), ^{87}Rb (which decays to ^{87}Sr), ^{235}U (which decays, through a series of intermediate radioactive nuclides, to ^{207}Pb), and ^{238}U (which decays, also through an intermediate series of nuclides, to ^{206}Pb), Table III summarizes these modes of decay.

Uranium-235 and ^{238}U always occur together in nature, and an advantage of the uranium–lead method is that there are always two separate daughter/parent ratios that provide a cross check in determining ages. If the ages from the two separate methods are concordant, they are considered to be highly reliable. Uranium is a rare element, and minerals in which uranium is a primary constituent, such as the mineral uraninite, are also rare. This is perhaps the greatest drawback to the uranium–lead method. However, small quantities of uranium (around 1%) always occur in the mineral zircon (ZrO_4). Zircon is a common accessory mineral in granitic rocks of all ages, and most uranium–lead dates have in fact come from zircons extracted from granites.

TABLE III The Chief Methods of Determining Radiometric Ages of Rocks

Parent nuclide	Half-life (yr)	Daughter nuclide	Minerals and rocks commonly dated
Uranium-238	4.47 billion	Lead-206	Zircon, whole rock
Uranium-238	704 million	Lead-207	Zircon, whole rock
Potassium-40	1.300 million	Argon-40	Whole rock, feldspar, mica
Rubidium-87	48.8 billion	Strontium-87	Whole rock, feldspar, mica

The potassium–argon method is widely useful because potassium occurs abundantly in many common minerals, and laboratory techniques for measuring minute quantities of the daughter ^{40}Ar are extremely refined. Both granitic and basaltic rocks are rich in potassium-bearing minerals, but in the case of basalts individual mineral grains are so small that they cannot be isolated from the rock mass. In order to date these fine-grained rocks, a sample of the rock itself is subjected to analysis without separating mineral components. This is termed the "whole-rock method." The whole-rock method has been applied successfully to basalts as young as 100,000 years. In addition to igneous minerals, the sedimentary mineral, glauconite, which forms in certain depositional environments, is datable by the potassium–argon method. A drawback to this method is that argon, being a gas, escapes easily from the host mineral grains upon heating or stress. Consequently, burial to even moderate depths tends to promote argon loss, and this causes dates to come out to be too young.

The rubidium–strontium method has been widely applied to igneous and metamorphic rocks, but it is especially valuable for the latter. In dating metamorphic rocks, the whole-rock method is commonly used. One weakness of the rubidium–strontium method is that the half-life of ^{87}Rb is so long that the method does not achieve very good accuracy in rock less than 200 m.y. old or so.

The methods discussed above are applicable almost exclusively to igneous and metamorphic minerals and rocks. Hence, igneous and metamorphic events can be dated readily. It is much more difficult to tie radiometric ages into fossilbearing sedimentary rocks on which the geologic time scale is based. Hence some of the numerical ages of system boundaries shown in Fig. 2 are known only within broad limits, and refinement of the time scale is an ongoing process. Radiometric ages have been tied in to the geologic time scale in three main ways. First, a dated igneous intrusive provides a minimum age for a sedimentary rock that it intrudes and a maximum age for a sedimentary rock deposited on it. Second, volcanic ash or lava flows can be introduced into sedimentary environments without interrupting sedimentation. The volcanic rock can be dated by the potassium–argon or rubidium–strontium methods and this gives an age for the coeval sedimentary strata as well. Some of the most reliable ages for calibration of the time scale have been achieved in this way. Finally, some sedimentary minerals, notably glauconite, can be dated directly. Because glauconite crystallizes during deposition, this gives a direct age for the sedimentary strata. Unfortunately, glauconite loses argon easily, when heated to 150°C or so, a temperature that is attainable at modest depths. As a result, glauconite ages are almost always too young; they are generally considered to represent minimum ages.

Carbon-14 is a short-lived radioactive nuclide that is continually produced high in the atmosphere by collisions of cosmic-ray neutrons and atoms of ^{14}N, the most abundant component of the atmosphere. Once produced, the ^{14}C atoms combine with atmospheric oxygen to become part of the atmospheric carbon dioxide, most of which is made up of the stable carbon isotopes, ^{12}C and ^{13}C. The relatively small quantity of radioactive ^{14}C dioxide quickly mixes with this more abundant and stable atmospheric carbon dioxide and enters into the earth's carbon cycle. The ^{14}C that enters minerals, plants, and animals is an extremely useful tool for dating the last 70,000 years of earth history.

Carbon-14 decays to ^{14}N by beta decay, with a half-life of 5730 years. However, the age of carbon-bearing material is not determined from the daughter/parent ratio, but from the ratio of ^{14}C to all other carbon in the sample. Naturally produced ^{14}C is in equilibrium in the atmosphere; the proportion incorporated into plants has always been relatively constant. A plant that removes carbon dioxide from the atmosphere receives a proportional share of ^{14}C. When the plant dies, it ceases to take up carbon dioxide, and with time the ratio of ^{14}C to stable carbon isotopes in the dead plant material progressively decreases. This is the ratio measured to provide ^{14}C dates.

Although ^{14}C dating is useful for only the last 70,000 years, this short period of time includes the retreat of the continental ice sheets, changes in ocean in circulation, the postglacial rise in sea level, and the development of civilization. The method has provided a valuable tool to anthropologists and historians as well as to students of recent earth history. Known historical dates have, in fact, provided excellent checks against the accuracy of the method. Carbon-bearing substances that have been dated successfully include wood, charcoal, bone, manuscripts, cloth, rope, and marine shells.

3. Age of the Earth

The oldest crustal materials that have been dated are detrital zircon grains from sandstones in the Archean Warrawoona Group of Australia. These zircons yield uranium–lead ages in excess of 4 b.y., indicating that their host rock crystallized at that time. The rock itself has not been found. The oldest body of rocks that has been identified are granitic gneisses from western Greenland, which have radiometric ages approaching 4 b.y. These rocks represent partial melts of still older rocks that have not been found. Direct dating of crustal rocks thus indicates only that the earth is older than about 4 b.y. To answer the question of just how much older, it is necessary to turn to indirect evidence of two kinds.

The first is radiometric dating of meteorites and rocks from the surface of the moon. All the solid material of the solar system, including the planets, satellites, and meteorites, are believed to have had a common origin in the solar nebula. Neither meteorites nor the lunar surface materials have been subjected to the kinds of recycling processes that have affected rocks of the earth's crust. For this reason, direct dating of these rocks could indicate when they first consolidated, and by extrapolation, when the earth and the other planets formed. Radiometric ages of almost all meteorites fall between 4500 and 4700 m.y. The oldest rocks recovered from the surface of the moon agree; their ages cluster around 4600 m.y., suggesting that the earth is of about this age also.

The second line of evidence for the age of the earth is based on the relative abundance of the isotopes of lead that occur in the earth's crust. Of the four stable isotopes of lead, ^{204}Pb, ^{206}Pd, ^{207}Pd, and ^{208}Pb, only ^{204}Pb is not produced by radioactive decay. All the ^{204}Pb that is now present on earth originated when the earth formed. By contrast, only a part of the present-day ^{206}Pb, ^{207}Pb, and ^{208}Pb originated at that time; the rest has been slowly added throughout geologic time by radioactive decay of uranium and thorium. The relative crustal abundances of the four lead isotopes and the uranium and thorium isotopes that produce them are known. If there were some means of determining the primoidial ratios of the lead isotopes, then the earth's age could be determined by calculating the time required for the radiogenic portions of the ^{206}Pb, ^{207}Pb, and ^{208}Pb to have been added through geologic time. There is no direct way to estimate the primordial lead isotope ratios from crustal rocks, but certain lead-bearing meteorites that contain no uranium or thorium are believed to provide a measure of the lead isotope ratios that must have prevailed throughout the solar system at the time it formed. Assuming that these were the lead isotope ratios on the early earth, about 4600 m.y. would be required for uranium and thorium decay to produce the ratio of lead isotopes that we find on earth today. This further confirms the indication that the earth originated about 4600 m.y. ago.

III. SYNOPSIS OF EVENTS IN EARTH HISTORY

A. The Precambrian

The Precambrian interval of earth history consists of the approximately 4000 m.y. that elapsed between the formation of the earth and the appearance of abundant animal fossils at the beginning of the Cambrian about 545 m.y.

ago. No rocks have been found that represent the first 600 m.y. or so of Precambrian history. During this time, heat flow from radioactivity was far greater than it is today, and mantle convection must have been rapid. Any thin crustal material that formed was apparently recycled by subduction and remelting. Evidence from the moon indicates that this was also a time of intense meteorite bombardment, which tapered off around 3900 m.y. ago. Heating and disruption of the crust must have been enhanced by the impacts of large meteorites.

The oldest materials that have been identified are granitic. Indicating that by around 4000 m.y. ago the mantle had already undergone considerable differentiation. By this time the earth's core was probably completely formed. Sedimentary rocks having ages around 3500 m.y. or older have been identified on several continents. These reveal that processes of erosion and sedimentation were active, indicating that the earth already had an ocean and atmosphere.

1. The Precambrian Rock Record

Most of the surface of the present continents is covered by Phanerozoic strata, and in most places these overlie structurally complicated igneous and high-grade metamorphic rocks that are called the basement complex. These basement rocks are predominantly of Precambrian age. Each of the present continents also has a relatively large region where Precambrian rocks are exposed at the surface without an overlying sedimentary cover. These are the *Precambrian shields*, and most of our understanding of the Precambrian comes from studies of rocks exposed in these regions.

Precambrian shield regions consist mainly of metamorphic and granitic rocks, but they also contain significant areas of volcanic and sedimentary rocks. The sedimentary rocks occur in sequences that are thousands of meters thick, but they contain no useful fossils for correlation, and hence cannot be placed into a global time framework. Geologists on every continent recognize an early Precambrian time called the *Archean* and a late Precambrian time called the *Proterozoic*, with a boundary at about 2500 m.y. but each of these eons is subdivided differently on different shields. In each shield region the rocks fall into large geographic provinces that have distinctive structural features and fold directions. These "structural provinces" are separated by sharp boundaries. Within each province, radiometric ages of the igneous and metamorphic rocks cluster around the date of the most recent orogenic episode. The dates that characterize the individual provinces help to provide a broad-scale time framework for regional subdivisions of the Archean and Proterozoic.

2. The Archean

The Archean rocks in the shield regions consist largely of granite batholiths and vast tracts of high-grade metamorphic rocks. Infolded among these crystalline rocks are belts some tens of kilometers wide and hundreds of kilometers long containing thick sequences of only slightly metamorphosed sedimentary and volcanic rocks. The mildly metamorphosed volcanic rocks are called "greenstones," from which the belts get the name *greenstone belts*. The most common sedimentary rocks in greenstone belts are thick, repetitive sequences of graywackes (poorly sorted sandstones) that are typically interbedded with shales. The beds of graywacke are commonly graded; that is, the grain size in each becomes finer from bottom to top. These are inferred to be "turbidites," deposits of gravity-driven turbidity currents that transported sediments into deep basins adjacent to rising highlands. Greenstone belts also contain thick deposits of volcaniclastics and chert and a few contain some sandstones and conglomerates inferred to be of shallow-water origin. Greenstone belts formed in tectonically active, volcanic regions. They were stabilized and preserved by massive intrusions of granite that greatly thickened the crust around them. They represent some of the oldest crustal material that has survived.

3. The Proterozoic

The end of the Archean, around 2500 m.y. ago, appears to have been marked by an intense period of igneous activity, following which the continental cratons became stabilized. The repetitious turbidites and volcanic rocks that dominate Archean sequences give way in the Proterozoic to a wide variety of sedimentary rock types that include shallow marine, transitional, and continental sedimentary environments. Many of these rocks represent tectonically stable environments, and they look like much younger Phanerozoic rocks, except that they lack fossils.

Post-Archean sedimentary rock assemblages represent both *platform* and *geosynclinal* tectonic settings. Platform sequences form in the stable interior portions of cratons. At times the platforms subsided gently and were covered by extensive shallow epicontinental seas. At other times they emerged slightly and were gently eroded. Unconformities are numerous in platform sequences, and total sedimentary accumulations are generally thin—on the order of 1000 to 2000 m, except in cratonic *basins* where thicknesses may exceed 5000 m.

Margins of continents are less stable than their interiors. During long episodes of subsidence, continental margins have commonly received sedimentary thicknesses in excess of 10,000 m. These thick accumulations of sedimentary rocks commonly form in belts several hundred kilometers long called *geosynclines*. During the episodic fluctuations of sea level, these marginal areas are the first to be covered and the last to be exposed. As a result, unconformities are fewer and the historical records more complete in geosynclines than in platform sequences.

Geosynclinal deposition begins after a large continent is split in two by rifting. As the newly created ocean widens, the youthful trailing margins of the flanking continents subside rapidly as they retreat from the midoceanic spreading center and they accumulate great thicknesses of sediments. The rocks of a geosyncline typically represent two distinctive suites. The *miogeosyncline* (commonly called the *miogeocline*) forms on the craton's trailing margin and consists of a greatly thickened extension of the rocks of the cratonic interior—shallow-water limestones, dolomites, shales, and well-sorted sandstones—all of which formed in stable settings comparatively free from igneous activity. The eugeosyncline forms in deep water, seaward of the continental margin, and consists of a thick sequence of black shales and graded sandstones inferred to be turbidites. Carbonate rocks are rare, but bedded cherts and volcanics are common.

4. Late Precambrian Tectonics

In geologically young lava flows and sedimentary rocks, the alignment of the rocks' remanent magnetism consistently parallels that of the earth's present magnetic field, but in older rocks, the alignment of the remanent magnetism departs from that of the present field. In general, the older the rock, the greater is the departure. Paleomagnetic measurements from rocks of different ages throughout the world indicate that the magnetic poles have changed position with respect to the continents through geologic time. This phenomenon has been called *polar wandering*, but the evidence is strong that the magnetic poles have remained stable, and that it is the continents that have moved. Polar positions inferred from studies of rock magnetism have become an important tool for determining how the continents have moved around on the earth's surface during the course of geologic time.

At times in geologic history the continents have converged and collided piecemeal to form a single enormous continental entity. The best-documented and most recent of these "supercontinents" was Pangea, which began to break apart in the Triassic Period, some 200 m.y. ago. Paleomagnetic studies of late Precambrian rocks indicate that the continents were similarly amassed as a single entity in the latest. Proterozoic, around 600 m.y. ago. When this supercontinent broke apart, numerous miogeoclines formed on the trailing margins of the newly formed individual continents as they retreated from one another. Great thicknesses of sediments of very late Precambrian and Early Cambrian age mark the initial accumulations in the Cordilleran geosyncline on the western margin of North America and in the Appalachian Geosyncline on the eastern margin. Detailed analyses of the rates of subsidence of these miogeoclines and of contemporaneous miogeoclines in South America, the Middle East and Australia have made it possible to model the cooling histories of these trailing margins and to reconstruct the timing of their breakup and retreat from the oceanic spreading centers that separated them. These analyses have determined that the breakup of the late Precambrian supercontinent occurred between 555 and 625 m.y. ago.

As the continents dispersed, newly formed midoceanic ridges occupied the widening oceans between them. The youthful, hot crust of these comparatively shallow oceans effectively replaced the old, cool crust of deep oceans that was being subducted elsewhere. The resulting decrease in the average depth of the ocean basins produced a large, gradual rise in sea level and caused the pronounced marine transgression that marks latest Precambrian and earliest Paleozoic sequences worldwide.

5. Precambrian Glaciations

Of particular interest in sections of Proterozoic rocks are extensive deposits of boulder-laden conglomerates that are inferred to be *tillites*, that is, lithified glacial till. Like modern tills, these rocks contain unsorted and unstratified sedimentary particles of all size, including boulders that are not in contact with one another but that are totally surrounded by a matrix of silt and clay. Proterozoic tillites have been found on almost every continent. This widespread evidence for Precambrian glaciation defines two major episodes, one in the early Proterozoic around 2200 to 2300 m.y. ago, and the second in the latest Proterozoic, between 550 and 600 m.y. ago. Tillites provide a record of paleoclimate, in that they document extended period of freezing temperatures. The Proterozoic tillites show that the earth's surface far back in the Precambrian had approximately the same temperatures as it does today.

6. The Precambrian Atmosphere and Life

The initial atmosphere was produced by volcanic outgassing early in geologic history, and it probably consisted largely of carbon dioxide since this is the most abundant volcanic gas. Before a substantial atmospheric mass formed, however, the carbon dioxide began to be selectively removed from the atmosphere by being dissolved in the sea and buried, both as a part of marine carbonates and as disseminated bits of organic carbon in other fine-grained sedimentary rocks. Nitrogen must have been the

major gas left behind, but exactly how the composition of the atmosphere evolved with time is not well understood because there is no direct record. Until around 2 b.y. ago the atmosphere and ocean probably lacked significant free oxygen. Oxygen in nature is chemically too active to exist in significant quantities unless it is continually replenished, as it is now by photosynthesizing plants.

The basic chemical compounds that make up living organisms were probably abundant on the early earth. Laboratory experiments have shown, however, that these compounds could only have existed in the absence of free oxygen. Life originated from these compounds very early in geologic history. The oldest definitive fossils come from early Archean chert beds at least 3600 m.y. old in the Pilbara region of Australia. Thin sections of these rocks reveal microscopic rod-shaped filamentous and spherical cells of bacteria and blue-green algae.

Precambrian sedimentary rocks of all ages throughout the world contain *stromatolites* (Fig. 10), which are considered to be fossils in the same sense as tracks or burrows in younger rocks. Although they are not actual organic remains, they are distinctive configurations of carbonate laminae produced by blue-green algae.

When plants first invented the process of photosynthesis, the oxygen released as a waste product must have quickly combined with dissolved ferrous iron and other reduced substances in the sea rather than accumulating as free oxygen. Abundant iron-rich cherts called *banded iron formations* were deposited during the early Proterozoic, between about 2600 and 1800 m.y. ago. These rocks, which today constitute the world's major iron ore bodies, may represent a time when photosynthesizing plants were producing oxygen but when atmospheric concentrations were still low enough to permit ferrous iron compounds to dissolve readily in the ocean.

Later, as free oxygen began to accumulate in the atmosphere, oxidation became part of the weathering process

FIGURE 10 Precambrian algal stromatolite about 2 billion years old. Nash Fork Formation, Medicine Bow Mountains, Wyoming. Thickness of the bed is about 40 cm.

and red beds appeared in the sedimentary record. When the atmosphere's oxygen content reached about 1% of its present level, a layer rich in ozone formed high in the atmosphere. Ozone (O_3) today shields life at the surface from lethal ultraviolet radiation. Prior to the existence of this layer, life could not have existed on land or even in extremely shallow water. After the appearance of the ozone layer, organisms were able to occupy the upper layers of the oceans, including the vast areas of shallow water fringing the continents and, later, land surfaces themselves. The plentiful solar energy available to these environments stimulated organic productivity, which, in turn, accelerated the accumulation of free oxygen. The availability of abundant free oxygen also made possible respiration and oxygen metabolism as we know it today. Oxygen metabolism was necessary for the evolution of the eucaryotic cell, which, in turn, allowed the development of multicellular animals and plants. This chain of events paved the way for the great surge of evolution of higher organisms beginning in the latest Precambrian.

7. Biotic Change and the End of Precambrian Time

The oldest macroscopic invertebrate fossils consist of a peculiar assemblage of 31 species of distinctive soft-bodied marine animals that are preserved only as impressions in sandstones. They were first discovered in the latest Proterozoic strata of the Ediacara Hills of southern Australia and have subsequently been found in rocks of about the same age on several continents. This *Ediacara fauna* includes forms that resemble present-day jellyfish, flat worms, annelids, coelenterates, soft-bodied arthropods, and soft-bodied echinoderms, and it first appears in strata about 55 m.y. old.

In Phanerozoic sedimentary rocks, fossil impressions of soft-bodied animals, such as those in the Ediacara fauna, are rare compared to the remains of shell-bearing animals, which are much more readily fossilized. About 545 m.y. ago, relatively soon after the appearance of the Ediacara fauna, shell-bearing invertebrate animals appeared and quickly became abundant; their remains occur in marine sedimentary rocks of this age all over the earth. This marks the base of the Cambrian system and the beginning of the Phanerozoic interval of earth history.

B. The Paleozoic

Paleozoic rocks contain the first clear records of widespread advances of the sea onto the continents. At times in the Precambrian the sea must have transgressed widely also, but the stratigraphic record is poor because much of it was removed during the episode of widespread erosion of what we now infer to be a high-standing supercontinent in latest Precambrian time. The erosion surface is overlain by Cambrian rocks that record a long, slow transgression of Cambrian seas. The Cambrian marine transgression was the first of a series of large-scale invasions and retreats of the sea that continued throughout the Paleozoic. At times, a substantial portion of the land area became submerged. At other times, the lands were widely emergent and the sea lapped only onto the edges of the continents. Even while largely submerged, the continents acted as positive features. They did not subside to oceanic depths, but instead stood high above the surrounding ocean basins.

The widespread Cambrian epicontinental seas persisted through the Early Ordovician and then regressed, exposing vast regions. In Middle Ordovician time the seas began another invasion that culminated in the Late Ordovician in the most widespread epicontinental seas of which we have a record. This sea began to retreat in the middle of the Silurian. The sea again transgressed at the beginning of the Devonian, and regressed late in that period. During the remainder of the Paleozoic, marine transgressions became less and less extensive, and progressively more of the land surface became exposed. The widespread marine sediments of the earlier Paleozoic now gave way to nonmarine sediments that accumulated over progressively wider areas. By late in the Permian, most parts of the continents stood above sea level.

1. The Assembly of Pangea

At the beginning of the Paleozoic, several continents lay scattered across the earth at low latitudes. In their shapes and positions, these continents were totally different from the continents of today. Throughout the Paleozoic, the continents converged one by one and joined as the oceanic crust between them was subducted. By the early part of the Triassic the continents had been assembled into one huge supercontinent called *Pangea*. We have a much better idea of how the continents were distributed after the Triassic, when Pangea began to break apart, than we do for the Paleozoic, during which it was assembled. The outlines of modern continents all originated with the post-Triassic breakup, and all of the rocks that underlie today's ocean basins were formed after that time. The information that we do have on the positions of the Paleozoic continents comes chiefly from just two sources: (1) the positions of ancient magnetic poles determined from studies of rock magnetism, and (2) the distribution of Paleozoic geosynclinal foldbelts, which represent continental margins that were deformed in the zone of collision between continents.

Figure 11 shows how the positions of the Paleozoic continents changed between Cambrian and Permian time.

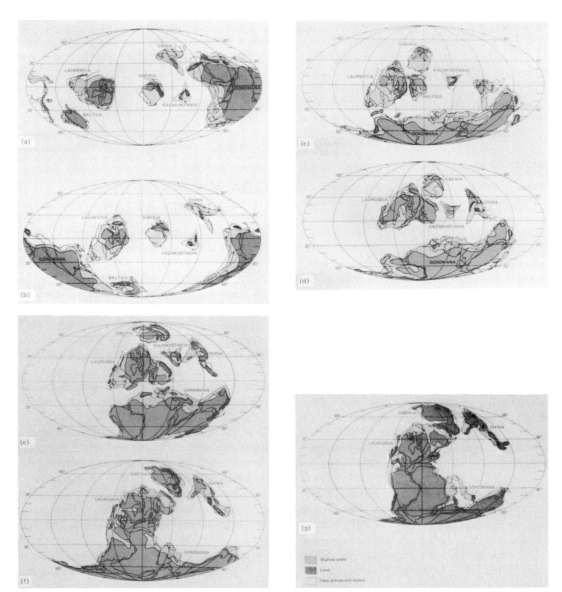

FIGURE 11 The assembly of Pangea throughout the Paleozoic Era. Mollweide projections showing the entire earth: (a) Cambrian, (b) Ordovician, (c) Silurian, (d) Devonian, (e) Mississippian, (f) Pennsylvanian, (g) Permian. [After Scotese, C. R., *et al.* (1979). *J. Geol.* **87**(3), 217–278. Copyright © 1979 University of Chicago.]

During the Cambrian the ancestral North American continent (Laurentia) lay on the equator, and what is now the north–south axis of the continent was then east–west. Laurentia and the European continent (Baltica) were separated by a widening ocean, called the "Iapetus Ocean." During the Ordovician, a subduction zone formed along what is now the eastern margin of the ancestral North American continent, and the Iapetus Ocean began to close. Volcanism and orogeny occurred late in the Ordovician along the margins of this shrinking ocean in the northeastern United States and the Candian maritime provinces

(the *Taconic Orogeny*) and during the Silurian Period in northern Europe (the *Caledonian Orogeny*). In the Devonian Period, Baltica and Laurentia collided, forming the single continent of "Laurussia." A mountainous elongate foldbelt that formed at the zone of collision (the *Acadian Orogeny*) shed coarse fluvial redbeds westward onto the northeastern United States and adjacent Canada, and eastward onto Europe. Subsequently Laurussia converged with Gondwana, and their collision in the late Paleozoic formed an extremely long geosynclinal foldbelt that stretched from Texas to eastern Europe (the

Marathon, Ouachita, and *Appalachian* orogenies in the United States and the *Hercynian Orogeny* in Europe). During the late Paleozoic and early Triassic, numerous geosynclinal belts in Asia were also strongly deformed as blocks that now make up Siberia, Kazakhstan, and China joined onto Laurussia, thus completing the supercontinent of Pangea. In the late Paleozoic, Gondwana, the southern subcontinent of Pangea, was heavily glaciated.

2. Paleozoic Life

Diversity of the organic world expanded greatly during the 300 m.y. of Paleozoic time. Early in the Paleozoic, trilobites dominated, accompanied by brachiopods and several other groups. By the end of the Ordovician all the major groups of marine invertebrates had appeared, and subsequently Paleozoic shelly faunas were dominated by corals, bryozoans, brachiopods, and crinoids. In the Silurian Period water-dwelling green algae gave rise to the first simple land plants. These rapidly evolved into several groups of seedless plants, which constituted the earliest shrubs and trees. In the late Paleozoic, seedbearing plants developed, including the seed ferns, cycads, ginkgoes, and conifers. Accompanying the diversification of land plants was the simultaneous rise of land animals. The first land animal, a scorpion-like arachnid belonging to the Phylum Arthropoda, appeared in the Silurian, and throughout the Paleozoic the insects and arachnids diversified rapidly. Today the arthropods are by far the most successful land animals, measured by numbers of individuals and species.

The earliest fossil vertebrates are primitive, jawless fishes found in Upper Cambrian deposits. Fishes with jaws appeared in the Silurian, and by Devonian time a large variety of fish lived in both fresh water and the sea. A specialized group, the *lobe-finned fishes,* gave rise to the amphibians in the Late Devonian and these, in turn, gave rise to the reptiles in the Pennsylvanian. By the close of the Permian a large variety of large herbivorous and carnivorous repitles roamed the land.

3. The Permian Extinction

The close of the Paleozic Era was marked by a severe and widespread extinction of life. About 30% of the families of fossil animals and plants found in Lower Permian rocks became extinct by the end of the Permian and are unknown in younger rocks. Marine invertebrate animals were particularly hard-hit. Trilobites became extinct, as did many previously abundant groups of corals, bryozoans, brachiopods, and crinoids.

The Permian extinction was one of several that occured during the Phanerozoic. The cause of episodes of

widespread extinction remains elusive and is the source of much speculation. The answer must lie in some environmental change on a global scale. In this case, the assembly of the supercontinent of Pangea may have been in part responsible. As the continents were assembled, areas of shallow epeiric seas decreased sharply, the separate biotas from each continent were forced into competition, and the number of niches was reduced. Many groups may have thus become extinct as a result of competitive exclusion.

C. The Mesozoic

1. The Breakup of Pangea

The modern world had its beginning late in the Triassic, when the continent of Pangea began to break apart. The evidence for the timing of the Mesozoic breakup of Pangea and for the subsequent migrations of the continents is provided mainly by rock magnetism. The stratigraphy of the modern ocean basins and the magnetic anomalies of the oceanic crust provide important additional evidence. Figure 12 shows the changing positions of the continents during the Mesozoic, based on these kinds of evidence. When the Triassic began, the land on the earth's surface was nearly equally distributed between the northern and southern hemispheres. Today, after prolonged separation of the fragments of Pangea, two-thirds of all land lies north of the Equator. This northward displacement removed the present southern continents (excepting Antarctica) from the high-latitude position that had promoted their widespread glaciation in the late Paleozoic.

Triassic basalt flows and fault basins of the eastern United States and Canada record the beginning of the rift between the subcontinents of Laurasia and Gondwana approximately 200 m.y. ago (Fig. 12). The rifting occurred initially along a nearly east–west trend near the Triassic equator, and it opened the "Tethys Sea" between the northern and southern subcontinents. Upper Triassic and Lower Jurassic evaporites formed in the narrow North Atlantic, which at one point must have looked much like the Red Sea today. During the Jurassic, similar evaporites formed in what is now the Gulf of Mexico. In the Early Cretaceous, the birth of the South Atlantic created a comparable evaporite basin and produced salt deposits that today underlie the Atlantic shelves of South America and Africa. In all of these regions the salt has since been deeply buried and locally squeezed upward through thick overlying strata to form salt *diapirs*. These diapirs, or salt domes, have created numerous oil traps at depth.

In the Late Triassic another major rift developed in the high southern latitudes, where South America and Africa began to separate as a unit from the remainder of Gondwana. Soon afterward the Indian Ocean began to

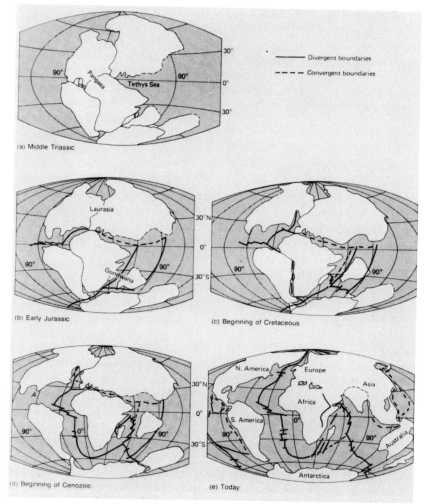

FIGURE 12 The breakup of Pangea throughout the Mesozoic and Cenozoic. [After Dietz, R. S., and Holden, J. C. (1970). *J. Geophys. Res.* **75**, 4939–4956.]

form, when India broke away from Antarctica and rapidly drifted northward. During the Jurassic the North Atlantic continued to widen, while the Tethys Sea became progressively narrower as Africa and Eurasia rotated toward each other. In the Late Cretaceous, South America and Africa separated completely, and at about the same time the Labrador Sea opened between Canada and Greenland. By the end of the Cretaceous the South Atlantic had opened into a major ocean 3000 km wide, and Greenland was the only connection between North America and Eurasia.

2. Continental Tectonics

The trailing margins of the diverging continents developed thick miogeoclinal sequences, just as the trailing margins of the late Precambrian continents had following their fragmentation. Many of the leading edges of the continents became tectonically active zones of plate convergence. The western margins of both North and South America developed oceanic trenches and bordering volcanic chains, many of which persist to the present day. During the Late Jurassic and Early Cretaceous, huge granitic batholiths intruded deep within the growing highlands along the western margin of the Americas. Some of these batholiths make up the Sierra Nevada range in California, and this episode has been named the *Nevadan Orogeny*. During the Mesozoic and Cenozoic the converging oceanic plate from the west brought volcanic arcs and small elements of continental crust called *microcontinents* to the western margin of North America, where they were plastered onto the continent. They are recognized today as "exotic terranes" or, as they are sometimes called, "suspect terranes." In the Early Cretaceous a thrust faulting event termed the

Sevier Orogeny built mountains in Utah and Idaho that served as source areas for thick sedimentary sequences that were deposited both to the east and west. At the end of the Cretaceous, large-scale faulting immediately east of the Sevier belt occurred from Canada to Mexico. This episode, which is known as the *Laramide Orogeny*, formed the structure of the modern Rocky Mountains.

3. Mesozoic Life

The severe reductions of the living world at the end of the Permian were followed in a few million years by the appearance of new and varied faunas and floras, and as the Mesozoic Era progressed, new evolutionary radiations led to a far greater diversity of organisms than existed at any time during the preceding Paleozoic Era. In the ocean new invertebrate groups arose during the early part of the Mesozoic to replace those eliminated in the late Paleozoic extinctions. Molluscs dominated, and their most important representatives were the ammonites. Thousands of species of ammonites evolved during the course of the Mesozoic, and they radiated into numerous marine environments. So rapid was their evolution and extinction that they serve as the best Mesozoic guide fossils for worldwide correlation. Clams and snails assumed a modern appearance through the evolution of new families, many of which are still living today. Besides the molluscs, other invertebrate phyla that gradually attained a modern appearance include the corals, crustaceans, and echinoderms. All of these continue as the dominant invertebrates of oceans today.

Paralleling the modernization of marine benthic life were changes in the plankton. In the Jurassic the evolution of tiny calcareous planktonic algae called *coccolithophorids* and calcareous zooplankton, the *planktonic foraminifera* caused large quantities of tiny calcium carbonate skeletons to be produced in the near-surface waters of the deep oceans. The rain of these tiny particles onto the sea bottom produced the first deep-sea calcareous ooze. With burial, calcareous ooze becomes lithified to a distinctive limestone called "chalk." Chalk has a unique texture because it is formed of a weakly cemented and highly porous agglomeration of predominantly coccoliths and foraminifera.

During the Triassic and Jurassic periods conifers, cycads, ginkgoes, and ferns dominated the plant world. Then, in the middle of the Cretaceous the first flowering plants, the angiosperms, appeared. The angiosperms radiated quickly. During the Late Cretaceous they expanded to dominance and largely replaced the more primitive groups of plants, which they greatly exceed in diversity today.

At the beginning of the Mesozoic the dominant group of land vertebrates was the mammal-like reptiles. In the Triassic this group gave rise to the first mammals, but mammals remained small throughout the Mesozoic, and they were never abundant. The Mesozoic was truly the Age of Reptiles. By the end of the Triassic the mammal-like reptiles had been totally replaced by the thecodonts, a diverse group that gave rise, in the Triassic, to a host of reptilian groups and, in the Jurassic, to the first birds. In the latest, Triassic flying reptiles evolved and shortly thereafter became huge. A variety of marine reptiles, such as the ichthyosaurs and plesiosaurs, evolved. The first representatives of existing reptiles—crocodiles, turtles, snakes, and lizards—appeared. The dinosaurs also originated in the Triassic and quickly came to dominate the land. Many remained quite small, but others evolved into the familiar huge and impressive animals of Late Jurassic and Cretaceous time.

4. The Cretaceous Extinction Event

The extinction event at the end of the Mesozoic, like that at the end of the Paleozoic, drastically affected marine as well as terrestrial life. In the sea, extinction was particularly widespread among pelagic animals and plants, but it also affected many groups of bottom-dwelling organisms. The ammonites, which were among the most common Mesozoic marine invertebrates, were wiped out completely as were the specialized reef-building clams called "rudistids." Only a sprinkling of the formerly diverse planktonic foraminifera and coccoliths survived into the Cenozoic. Among the vertebrates, the dinosaurs became extinct, as did the flying reptiles and the large marine reptiles. Many of the vertebrate and invertebrate groups disappeared abruptly and, apparently, simultaneously. There is strong evidence for a huge meteorite impact in the Yucatan Peninsula of Mexico at the end of the Cretaceous, and many believe that this caused abrupt, widespread extinction. Whether this was the sole or major cause of extinction is difficult to prove; the terminal Cretaceous extinction may have multiple causes.

D. The Cenozoic

1. The Rise of the Mammals

The disappearance of the dominant groups of reptiles at the end of the Cretaceous set the stage for an explosive evolutionary expansion of mammals. Early in the Cenozoic Era several groups of large herbivores and carnivores developed, and by the middle of the era most of the diversely specialized mammalian groups known today were established. Shortly after the mammals had first appeared in the Triassic, they had become widely distributed over the Pangean landmass. The Mesozoic breakup of Pangea produced isolation, so mammalian history has followed a somewhat different course on the various continents. North America, Eurasia, and Africa were interconnected through much of the Cenozoic Era, and most of the familiar present-day mammal groups arose and were continuously abundant on these continents. In contrast, South America and Australia were isolated, and each developed its own distinctive mammalian fauna. Mammalian history in South America is complicated by the late Cenozoic linkage with North America, which permitted mixing of the unique South American mammals with more advanced forms from the north. Australia, on the other hand, has long been totally isolated and has retained its distinctive mammals.

2. Cenozoic Tectonics

By the beginning of the Cenozoic Era, about 66 m.y. ago, the continents had begun to achieve their modern shapes (Fig. 12). The Atlantic Ocean continued to widen at the expense of the Pacific, and the Indian Ocean continued to open as India drifted northward. India began to collide with mainland Asia in the Eocene. Uplift of the Himalayas at the former southern margin of Asia began in the Oligocene, and their maximum deformation took place in the Pliocene and Pleistocene, when the northern edge of the Indian continent actually pushed beneath the southern margin of the Asian continent. The result was an extraordinarily thick continental crust that deformed into the highest mountain range on earth.

Early in the Cenozoic, the African and European plates converged and closed the Tethys Sea. In the Oligocene the colliding plates initiated the Alpine Orogeny which built the Alps and Carpathians in southern Europe and the Atlas in North Africa. Rearrangements of microplates produced the Mediterranean Sea behind the former collision zone.

Australia split from Antarctica in the Late Cretaceous, and by Eocene time, Australia had moved northward and Antarctica southward over the pole. Later, when the Drake Passage opened between South America and Antarctica, the arrangement was complete for the deep circumpolar circulation of the Southern Ocean. In the Eocene, rifting in the North Atlantic shifted from the west side to the east side of Greenland, where the Norwegian Sea opened, completing the separation of North America and Europe and producing a deep passage between the North Atlantic and the Arctic. This permitted cold, deep Arctic Ocean water to flow into the North Atlantic. All of these changes in oceanic circulation helped to set the stage for the development of the late Tertiary and Quaternary glacial episodes.

The tectonically quiet trailing margins of the continents, such as the Atlantic and Gulf Coast margins of North America, continued to accumulate thick miogeoclinal sequences. Tectonic activity continued along the leading margins, where subduction generally prevailed. Beginning in the Oligocene, subduction along the western margin of the United States progressively evolved into the transform faults of the San Andreas, system. Inland from this system, former compressional forces were replaced by tensional forces which produced the Miocene block faulting of the basin and range province in Nevada and adjacent areas. Enormous volumes of basalt were erupted immediately north of the basin and range, where they formed the Columbia River Plateau. In the late Cenozoic, much of the western United States was uplifted substantially. The Rockies achieved their present elevations, as did the Sierra Nevada.

3. Cenozoic Glaciations

Paleotemperature results from oxygen isotope analyses of deep-sea sediments show that, throughout the Cenozoic, temperatures at high latitudes became progressively cooler and the global temperature gradient became progressively greater. Glaciation began on Antarctica early in the Cenozoic, and in the high latitudes of the northern hemisphere late in the Miocene. Glaciation occurred in a repetitive succession of glacial and interglacial episodes, driven by regular variations in the earth's orbital parameters. During their maximum extent in the Pleistocene, which began about 2 m.y. ago, continental ice sheets covered more than 30% of the earth's land surface. In Europe an ice sheet spread southward from Scandinavia across the Baltic Sea into Germany and Poland. The Alps and the British Isles supported their own ice caps. Ice sheets extended throughout the northern plains of Russia, and they covered large sections of Siberia as well as the plateaus of Central Asia and the northern part of North America, as far south as the courses of the Missouri and Ohio rivers. New Zealand, Tasmania, and southern South America were heavily glaciated. Even in low latitudes high mountains were glaciated, as in Hawaii and New Guinea. Although we call the time that we live in the "Holocene," analysis of the record shows that the Holocene is merely the latest interglacial episode of the Pleistocene.

SEE ALSO THE FOLLOWING ARTICLES

ASTEROID IMPACTS AND EXTINCTIONS • COSMOLOGY • GEOMAGNETISM • GLACIAL GEOLOGY AND GEOMORPHOLOGY • GLACIOLOGY • PLANETARY SATELLITES, NATURAL • PLATE TECTONICS • RADIOCARBON DATING • SEDIMENTARY PETROLOGY • STRATIGRAPHY

BIBLIOGRAPHY

Adams, F. D. (1938). "The Birth and Development of the Geological Sciences," Dover, New York (Paperback, Dover, 1954).

Berry, W. B. N. (1968). "Growth of a Prehistoric Time Scale," Freeman, San Francisco.

Boggs, S. (1987). "Principles of Sedimentology and Stratigraphy," Merrill, Columbus, OH.

Condie, K. C., and Sloan, R. E. (1997). "Origin and Evolution of Earth," Prentice-Hall, Upper Saddle River, NJ.

Cowen, R. (1990). "History of Life," Blackwell Scientific, Cambridge, MA.

Eicher, D. L. (1976). "Geologic Time," 2nd ed., Prentice-Hall, Englewood Cliffs, NJ.

Geikie, A. (1905). "The Founders of Geology," 2nd ed., Macmillan, New York (Paperback, Mentor Books, 1961).

Gould, S. Jay (1987). "Time's Arrow, Time's Cycle," Harvard University Press, Cambridge, MA.

Gould, S. Jay (1989). "Wonderful Life," Norton, New York.

Harland, W. B., Armstrong, R. L., Cox, A. V., et al. (1990). "A Geologic Time Scale 1989," Cambridge University Press, Cambridge, U.K.

McGowran, B. (1986). "Beyond Classical Biostratigraphy," *Petrol. Explor. Soc. Austral. J.* **9,** 28–41.

McPhee, J. (1998). "Annals of the Former World," Farrar, Straus & Giroux, New York.

Nichols, G. J. (1999). "Sedimentology and Stratigraphy," Blackwell Science, London.

Palmer, A. R. (1983). "The Decade of North American Geology 1983 geologic time scale," *Geology* **11,** 503–504.

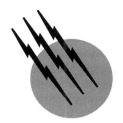

Geology of Earthquakes

Robert S. Yeats
Oregon State University

GLOSSARY

Active fault A fault that has had sufficient recent displacement so that, in the opinion of the user of the term, further displacement in the foreseeable future is considered likely.

Active tectonics Tectonic movements that are expected to occur or have occurred within a time span of concern to society.

Anticline A fold, generally convex upward, whose core contains the stratigraphically older rocks.

Asperity Irregularity on a fault surface that retards slip; region of relatively high shear strength on a fault surface.

Bending-moment fault Fault formed due to bending of a flexed layer during folding. Normal faults characterize the convex side, placed in tension, and reverse faults characterize the concave side, placed in compression.

Blind fault A fault that does not extend upward to the Earth's surface and never has. It usually terminates upward in the axial region of an anticline. If its dip is <45°, it is a *blind thrust*.

Breccia A coarse-grained clastic rock composed of angular broken rock fragments in a fine-grained matrix.

Brittle–ductile transition A zone within the Earth's crust that separates brittle rocks above from ductile (plastic or quasi-plastic) rocks below; the zone defining the deepest earthquakes in the crust.

Characteristic earthquake The largest earthquake that is thought to occur repeatedly on a given fault.

Colluvial wedge A prism-shaped deposit of fallen and washed material at the base of (and formed by erosion from) a fault scarp or other slope, commonly taken as evidence in outcrop of a scarp-forming event.

Earthquake segment That part of a fault zone that has ruptured during an individual earthquake.

Event horizon A bedding plane within a stratigraphic sequence that represents the ground surface at the time of a paleoseismic event.

Fault A fracture or a zone of fractures along which displacement has occurred parallel to the fracture.

Fault creep Steady or episodic slip on a fault at a rate too slow to produce an earthquake.

Fault scarp A slope formed by the offset of the Earth's surface by a fault.

Fault slip rate The rate of displacement on a fault averaged over a time period encompassing several earthquakes.

Flexural-slip fault A bedding fault formed by layer-parallel slip during flexural-slip folding.

Footwall The underlying side of a nonvertical fault surface.

Gouge A thin layer of fine-grained, highly cataclastic material within a fault zone.

Hangingwall The overlying side of a nonvertical fault surface.

Left-lateral fault A strike-slip fault across which a viewer would see the block on the other side move to the left.

Mean recurrence interval The mean time between earthquakes of a given magnitude, or within a given magnitude range, on a specific fault or within a specific area.

Neotectonics Tectonic processes, now active, taken over the geologic time span during which they have been acting in the currently observed sense, and the resulting structures.

Normal fault A fault in which movement of the hangingwall is downward relative to the footwall.

Paleoseismology The investigation of individual earthquakes decades, centuries, or millennia after their occurrence.

Primary surface rupture Surface rupture that is directly connected to subsurface displacement on a seismic fault.

Pseudotachylite A massive rock that frequently appears in microbreccias or surrounding rocks as dark veins of glassy or cryptocrystalline material, so named because of its macroscopic resemblance to *tachylite*, or basaltic glass. Characteristically contains a matrix of crystals less than 1 mm in diameter and/or small amounts of glass or devitrified glass cementing a mass of fractured material together.

Pull-apart basin A topographic depression produced by extensional bends or stepovers along a strike-slip fault.

Reverse fault A fault characterized by movement of the hangingwall block upward relative to the footwall.

Right-lateral fault A strike-slip fault across which a viewer would see the adjacent block move to the right.

Seismic moment The area of a fault rupture multiplied by the average slip over the rupture area multiplied by the shear modulus of the affected rocks.

Seismogenic structure One that is capable of producing an earthquake.

Shutter ridge A linear hill or scarp sloping in a direction opposite to the overall topographic gradient, formed by strike-slip or oblique-slip offset of irregular topography.

Slickensides A polished and smoothly striated surface that results from slip along a fault surface. The striations themselves are *slickenlines*.

Slip vector The magnitude and orientation of dislocation of formerly adjacent features on opposite sides of a fault.

Stepover Region where one fault ends, and another *en échelon* fault of the same orientation begins; described as being either right or left, depending on whether the bend or step is to the right or left as one progresses along the fault.

Tectonic geomorphology The study of landforms that result from tectonic processes.

THE TOPICS covered here are commonly referred to as part of the field of *neotectonics*, but this term is defined as "the study of the post-Miocene structures and structural history of the Earth's crust," or as "the study of recent deformation of the crust, generally Neogene (post-Oligocene)," in addition to the more restrictive definition in this glossary. The common definitions above are too broad, because many faults of Miocene and even Pliocene age are inactive and have little, if any, seismogenic potential. Wallace (1986) introduced the term *active tectonics* to refer to "tectonic movements that are expected to occur . . . within a time span of concern to society." This is an improvement over the common definitions of neotectonics but is not the title of this article because it includes geophysics as well as geology, and a panel of the National Science Foundation chaired by G. H. Davis broadened the definition of active tectonics to include landslides and volcanic eruptions as well as earthquakes. This review is limited to geologic effects directly related to faulting or broad warping accompanying an earthquake, not including secondary effects such as liquefaction, lateral spreading, earthquake-induced landsliding, and strong ground motion except where those secondary effects are useful in dating a prehistoric earthquake. The time frame, however, is that implied by Wallace's definition of active tectonics: structures that are likely to rupture again and be a hazard to society.

An additional contribution that geology makes to earthquake science is the modeling of conditions at the earthquake source, where earthquakes nucleate. Most earthquakes begin at depths too great to observe directly, but we can study ancient fault zones that have been brought to the surface by uplift and erosion. Also, specialists in rock mechanics can subject rocks in the laboratory to temperatures and pressures found at depths where earthquakes are generated, which permits a more complete understanding of earthquake waves and the geodetic expression of earthquakes.

I. INTRODUCTION AND HISTORICAL BACKGROUND

Earthquakes have been studied since the time of Aristotle, although their relation to geology and tectonic processes was not properly understood then. Observers reported ground cracks and fissures that they related to a recent earthquake, even without an understanding of faulting. In 1493, Esfezari reported in Iran that the January 10, 1493, Nauzad earthquake had fissured the ground "to such a depth that the bottom of the fissures was invisible . . . ." Déodat Gratet de Dolomieu, following a series of disastrous earthquakes in Calabria, Italy, in 1783, conducted a field survey in the area of maximum damage, describing a scarp (*fente*) that he believed had moved during the earthquake. (Later investigators determined that the scarp Dolomieu described was secondary to the subsurface fault that generated the earthquakes.) Johann F. J. Schmidt, an astronomer, described a "crack . . . eight feet high and six feet wide" that appeared on the south coast of the Gulf of Corinth in Greece after an earthquake at the city of Egion in 1861. Schmidt, as well as the earlier observers, may have considered such features as landslides rather than faults. A famous quote from Zechariah, who lived about 520 B.C. — "the mount of Olives shall cleave . . . toward the east and toward the west"—probably described a landslide, although the landslide may have accompanied an earthquake in 759 B.C.

Faults were first mapped in the field in the 19th century. Charles Lyell (1797–1875), in his textbook *Principles of Geology*, was the first to recognize that earthquakes abruptly modify the ground surface along faults, basing his conclusions on reports of India's Rann of Kutch earthquake of 1819 and New Zealand's Wairarapa earthquake of 1855. In 1883, G. K. Gilbert of the U.S. Geological Survey described evidence for right-lateral strike slip on a fault scarp at Lone Pine, California, formed during the 1872 Owens Valley earthquake. This experience led Gilbert to describe the Wasatch Fault near Salt Lake City, Utah, and to recognize that this fault has the potential for earthquakes. In a letter to the *Salt Lake City Tribune* in September 20, 1883, re-published the following year in the *American Journal of Science*, Gilbert wrote: "When an earthquake occurs, a part of the foot-slope goes up with the mountain, and another part goes down (relatively) with the valley. It is thus divided, and a little cliff marks the line of division. A man ascending the foot-slope encounters here an abrupt hill, and finds the original grade resumed beyond. This little cliff is, in geological parlance, a 'fault scarp.' "

The connection between earthquakes and surface faulting was strengthened when Alexander McKay of New Zealand visited the site of the Marlborough earthquake of 1888 and recognized right-lateral strike-slip displacement on the Hope Fault at Glynn Wye Station. McKay also concluded that the potential for earthquake faulting extended to other faults ("earthquake rents") in Marlborough, even though they did not rupture in 1888. Surface faulting related to Japan's Mino-Owari earthquake of 1891 was discovered and described by Bunjiro Koto, and C. L. Griesbach observed surface rupture on the Chaman Fault following the 1892 Baluchistan earthquake. N. Yamasaki was the first to describe surface rupture on a reverse fault in his study of the 1896 Rikuu earthquake in northern Honshu, Japan.

In 1906, following the San Francisco earthquake, a state earthquake investigation commission was established. A. C. Lawson, G. K. Gilbert, and their colleagues described the right-lateral strike-slip displacement on the San Andreas Fault accompanying that earthquake. In addition, they mapped other sections of the San Andreas Fault that did not rupture in 1906, and Lawson also mapped the Hayward Fault, which had ruptured in California's first urban earthquake, in 1868.

In the period between the appearance of Lyell's text and the report following the San Francisco earthquake, the study of earthquakes was primarily a geologic endeavor, and earthquake geology seemed to have been established as a subdiscipline of structural geology. But, for more than a half century following the San Francisco earthquake, the study of earthquakes followed a geophysical path, and few geologists studied the surface expression of earthquakes. The principal reason for this was the development of the seismograph, which permitted an analysis of earthquake waves directly. Accordingly, the geophysical study of earthquakes, principally using seismograms, came to be known as *seismology*, although indeed this term could be broadened to include the study of earthquakes based on their geologic or geodetic expression. The Seismological Society of America was founded largely by geologists in the wake of the San Francisco earthquake, and it included engineers and astronomers as well as geologists and seismologists *sensu stricto*, but in the 1920s and 1930s, the society became more a geophysical than a geological organization.

During this period, papers on earthquake geology were descriptive rather than analytical. In Japan, where a national earthquake program had been established in the late 19th century, surface ruptures were described by A. Imamura, F. Omori, S. Tsuboi, H. Tsuya, and N. Yamasaki. In the United States, fault ruptures accompanying earthquakes were described by J. Buwalda, V. Gianella, B. Page, D. B. Slemmons, and R. E. Wallace. In New Zealand, surface ruptures were mapped by

J. Henderson, M. Ongley, H. W. Wellman, and G. J. Lensen. A careful description of the fault rupture accompanying the December 4, 1957, Gobi-Altai earthquake in Mongolia was made by N. A. Florensov and V. P. Solonenko of the Soviet Union; Solonenko went on to describe other surface ruptures in Siberia and Mongolia. I. Ketin was a pioneer in describing surface ruptures in Turkey, especially along the North Anatolian Fault.

In the 1960s, following major earthquakes in the Gulf of Alaska and Niigata, Japan, in 1964; Xingtai, China, in 1966; Inangahua, New Zealand, in 1968; and Los Angeles in 1971, national earthquake hazards programs began to emerge in China, the United States, and New Zealand, and the long-standing earthquake program in Japan was revitalized. Most of the advances in earthquake geology have been made since 1960 as a result of these focused programs. In 1984, International Geological Correlation Program Project 206, Worldwide Characterization of Major Active Faults, served as a forum for geologists from countries with and without national earthquake research programs (Hancock *et al.*, 1991; Bucknam and Hancock, 1992). This was followed by Inter-Union Commission on the Lithosphere task groups on worldwide active faults and paleoseismology (Yeats and Prentice, 1996), a program that is still underway under the leadership of D. Pantosti of Italy and K. Berryman of New Zealand.

II. PLATE TECTONIC SETTING

Earthquakes are limited to brittle lithosphere. The base of earthquakes is the downward transition from rock that is brittle to rock that is ductile (or, more precisely, quasi-plastic) at rates of earthquake strain. The Coulomb failure envelope for most rocks is written as:

$$\tau = \tau_0 + \sigma \tan \phi = \tau_0 + \mu \sigma \qquad (1)$$

where τ = shear stress, τ_0 = cohesion, σ = normal stress, ϕ = angle of internal friction, and μ is the coefficient of static friction. Tan ϕ and μ are positive for nearly all rocks, which means that rock strength increases with confining pressure to a temperature where thermal weakening occurs, a temperature that depends on the mineralogy of the rock. For quartz, a major rock-forming mineral in continental rocks, thermal weakening takes place at temperatures less than 300°C, so that the base of earthquakes, or the brittle–plastic transition, lies well within continental crust. Thus, rock strength increases with increasing depth to the temperature-controlled transition to quasi-plastic conditions, at which rock strength decreases abruptly. Because of this phenomenon, earthquakes commonly nucleate at a depth of 10 to 20 km, close to but above the base of brittle crust, where the rock is strongest.

In oceanic lithosphere, quartz is uncommon or absent, and the major rock-forming minerals undergo thermal weakening only far below the oceanic Mohorovičič discontinuity, the crust–mantle (basalt–peridotite) boundary, which is commonly at a depth of 6 to 10 km, much shallower than in continental crust. The minerals of the mantle, pyroxene and olivine, become thermally weakened at temperatures only greater than 500°C; this transition marks the boundary between the lithosphere and the asthenosphere.

The oceanic lithosphere and the continental lithosphere, both about 100 km thick, make up the tectonic plates which cover the Earth's surface and transmit stress, although the tractions that drive plate tectonics may be generated from the underlying asthenosphere. Tectonic earthquakes (as opposed to earthquakes accompanying the movement of molten rock) are restricted to the lithosphere.

Tectonic plates move away from each other at mid-ocean ridges, where new lithosphere is formed; or move toward each other at subduction zones, where old lithosphere descends into the mantle to depths as great as 700 km and is recycled; or move past each other without creating or consuming lithosphere, at transform faults. Most earthquakes take place at plate boundaries, but these boundaries may be broad and diffuse. For example, the convergent boundary between the Indian Plate and Eurasian Plate in central Asia is more than 2500 km wide.

Relative motions between plates are well known for the larger plates and many of the smaller ones for time scales of 10^6 years. These rates have been confirmed for most larger plates by space geodesy, especially very long baseline interferometry (VLBI) and the global positioning system (GPS), for time scales of a few years. The sum of the displacement vectors for individual faults within a plate boundary zone should equal the relative plate-motion vector. This relationship was applied to the apparent discrepancy between the 34 mm/yr slip rate on the San Andreas Fault in central California and the 48 mm/yr velocity of the Pacific Plate relative to the North America Plate at that latitude. VLBI measurements supplemented by GPS show that the discrepancy disappears when displacements on other faults in the San Andreas system and faults in the Basin and Range Province are combined with displacements on the San Andreas Fault itself.

III. STRUCTURAL SETTING

For rupture to occur on a fault, the stress state must be anisotropic. The maximum, intermediate, and minimum principal compressive stresses (σ_1, σ_2, and σ_3, respectively) are mutually perpendicular and are in orientations across which the shear stress = 0. The plane of maximum shear stress occurs on a plane oriented at 45°

to σ_1 and σ_3; however, failure occurs at 45° only where the coefficient of friction μ approaches zero. More likely, μ is 0.6 or larger, and under those conditions an ideal orientation for shear fracture is 30° to σ_1, but this occurs only where the rock strength is homogeneous. In nature, rocks are extensively fractured, and these fractures, being weaker than the rock in which they occur, tend to rebreak under stress orientations far from the ideal orientation for unfractured rock. Thus, earthquakes tend to follow pre-existing faults rather than rupturing flawless rock.

The mean stress is the hydrostatic or isotropic component of stress, which produces volume change but not faulting or change in shape. Overburden or load stresses act inward. But most rocks are saturated with fluid, and this fluid pressure, also isotropic, acts outward. The effective stress, which controls the strength of the rock under which faulting occurs, is the inward-acting compressive stress less the outward-acting fluid pressure. Fluids may also weaken rock by chemical reactions which may be controlled by temperature, a process called hydrolytic weakening.

The properties of a fault zone that extends through the crust can be studied by examining fault rocks that have subsequently been uplifted and deeply eroded (Scholz, 1990). Continental crustal fault zones include a shallow regime of brittle cataclastic deformation involving stick-slip frictional sliding and influenced by confining pressure; this is the part of the crust that hosts earthquakes. A deeper regime involves quasi-plastic deformation that is aseismic and dominated by temperature. This regime generally lacks earthquakes, although earthquakes may nucleate in the brittle region and extend downward into a transitional zone that is brittle only at the high strain rates that occur with seismic rupture propagation.

The shallowest part of a fault zone is marked by clay gouge, whereas the deeper part of a fault zone in brittle crust is marked by cataclastic rock, including fault breccia and rarely pseudotachylite, rock glass that has melted due to frictional heat. Earthquake rupture propagation is the only known process that would generate frictional heat suddenly enough to melt rock; for this reason, pseudotachylite is considered as evidence for an earthquake, even in ancient fault zones.

Some parts of a fault zone, called asperities, may be stronger than others. In the lower brittle regime, the coefficient of static friction is higher than that of dynamic friction, so that frictional resistance to sliding decreases at a rate faster than the system can respond to it, leading to stick-slip behavior and earthquakes. In the shallowest part of the crust characterized by fault gouge, however, the coefficient of dynamic friction may be higher than that of static friction, in which case rupture terminates without

an earthquake. Similar conditions occur below the brittle–quasi-plastic transition, where rock deformation occurs by pressure solution, grain-boundary diffusion, and crystal plastic flow.

Characteristics of seismicity on well-monitored fault zones such as the San Andreas, San Jacinto, and Calaveras Faults in California are consistent with this model of fault behavior. Earthquakes terminate downward at a relatively sharp boundary within the crust, correlated to the brittle–quasi-plastic transition. The main shock tends to nucleate near the base of the zone of aftershocks, reflecting the view that the strongest crust is subjected to the highest confining pressure above the thermal transition accompanying the onset of quartz plasticity. When this part of the crust fails, the rest fails, too. The base of earthquakes is shallower in areas of high geothermal gradient, evidence that it is temperature controlled. That part of a fault zone within a few kilometers of the surface has few or no earthquakes, even where there is evidence of surface rupture, suggesting that the shallowest crust fails by stable sliding rather than stick slip. Aftershocks are not distributed uniformly through the fault zone but are concentrated in some areas and absent in others, suggesting the presence of asperities.

The three classes of faults (Fig. 1) are controlled by the orientation of the anisotropic stress field. Faults in which displacement is in the dip direction generally are nonvertical. The block above a nonvertical fault is called the hangingwall, and the block beneath the fault is called the footwall, terms inherited from miners working underground in mineralized fault zones. In the case where σ_1, the orientation of the overburden stress, is vertical, the hangingwall moves downward relative to the footwall, and the fault is a normal fault. Where σ_3 is vertical, the hangingwall moves upward relative to the footwall, and the fault is a reverse fault. Where both σ_1 and σ_3 are horizontal, relative displacement on the fault is also horizontal, parallel to the strike of the fault, and the fault is a strike-slip fault. Strike-slip faults are likely to be vertical.

Coseismic surface displacement on a fault may be used as a measure of the moment of an earthquake in some large earthquakes. The seismic moment, M_o, is the area of the fault undergoing seismic displacement times the average fault displacement accompanying the earthquake times the elastic shear modulus, taken as 3×10^{10} Nm^{-2} for most crustal rocks. The downdip width of seismic rupture is based on the depth of earthquakes and the dip of the fault. Average displacement and the length of surface rupture are measured in the field.

However, most earthquakes with M (magnitude) <6 and some with even larger magnitudes do not rupture the surface at all—for example, the M 6.2 Big Bear earthquake of June 28, 1992, in southern California. A rupture generated

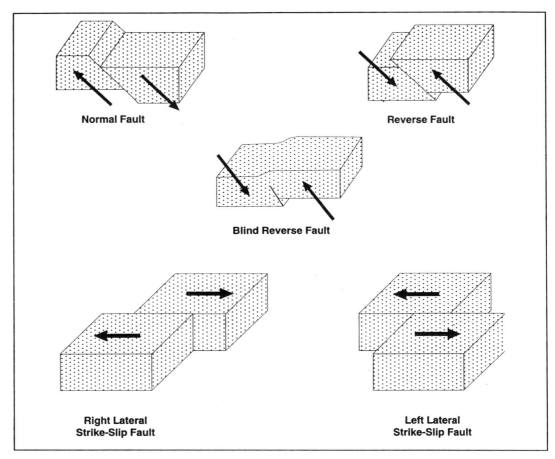

FIGURE 1 Block diagram representation of the four major types of faults that produce earthquakes, and the special case of a blind reverse fault. For normal faults, the maximum principal compressive stress is vertical and the minimum stress is horizontal, in the plane of the page. For reverse faults, the minimum principal compressive stress is vertical and the maximum stress is horizontal, in the plane of the page. For strike-slip faults, the interediate principal compressive stress is vertical. The maximum stress is oriented upper left to lower right for right-lateral faults and upper right to lower left for left-lateral fauts. [Modified from Yeats, R. S. (1998). "Living with Earthquakes in the Pacific Northwest," Oregon State University Press, Corvallis. With permission.]

at the base of earthquakes at 15 km may not be expressed or may be only partly expressed at the surface. In California, the May 2, 1983, Coalinga earthquake and the January 17, 1994, Northridge earthquake, both *M* 6.7, ruptured faults that did not reach the surface at all. Such faults are called blind faults; blind thrusts, if their dip is <45°. Large strike-slip earthquakes are more likely to be fully expressed at the surface than large earthquakes on reverse faults.

Layered rocks may be deformed by folding instead of (or in addition to) faulting. A layer folded upward is an anticline, and a layer folded downward is a syncline. Folding does not itself generate earthquakes. Folding at shallow depth may generate secondary faults that give evidence of the timing of folding, even though such faults may not be seismic sources themselves. Beds may deform by layer-parallel flexural slip, generating bedding faults that are

upthrown in the direction of the axis of a syncline; these are called flexural-slip faults. Bent layers may be placed in tension on their convex edges, developing normal faults, and in compression on their concave edges, developing reverse faults. Such faults are called bending-moment faults.

Although folds are not themselves seismic sources, they may be generated by underlying faults that are. If a footwall is nonplanar, beds of the hangingwall may be deformed above it, producing a fault-bend fold. In other cases, slip on a fault may be consumed upsection by folding, resulting in a fault-propagation fold. The 1983 Coalinga and 1987 Whittier Narrows earthquakes, both in California, were shown to be accompanied by growth of surface folds above blind thrusts.

Anticlines and synclines, where related to underlying reverse faults, are produced by repeated displacements on these faults during many earthquakes. Coseismic folding during an earthquake produces tilts that may be so subtle that they are revealed only by post-seismic geodetic releveling or by synthetic-aperture radar interferometry. Active folding in Quaternary deposits (those younger than 1.8 million years) may produce hills of tilted strata or tilted marine and stream terraces. The availability of digitized topographic maps allows quantitative study of Quaternary folding as an interplay among erosion, deposition, and tectonic uplift.

IV. FAULT ENVIRONMENTS

Normal faults, in which the maximum principal compressive stress σ_1 is vertical, occur where the crust is undergoing horizontal extension. Displacement is by dip slip, although some earthquakes are oblique slip (a component of dip slip and strike slip). Such faults have generated major earthquakes in the Basin and Range Province of the United States, southern Tibet, the Ordos Plateau of China, the Taupo Volcanic Zone of New Zealand, the Aegean region of Greece and western Turkey, the Apennines of Italy, the Altiplano of Bolivia and Peru, the Baikal Rift of Siberia, and the east African rift valleys. Active normal faults in coastal and offshore Texas and Louisiana and on the continental shelf of southwest Washington State appear to undergo displacement without earthquakes. Normal faults grade from swarms of faults with small displacement and topographic expression in areas of active volcanism and high heat flow, such as Iceland and the Asal Rift in Djibouti, to large range-front faults with large displacement and topographic expression in areas of thicker brittle crust. In the latter endmember, earthquakes as large as M 7.6 have occurred on dip-slip faults in the Basin and Range Province, and earthquakes as large as M 8 have ruptured oblique-slip normal faults adjacent to the Ordos Plateau in China. (Earthquakes of $M > 8$ in oceanic crust may be caused by bending in front of a subduction zone.) The seismicity defines a planar zone dipping 45° on the average, with no tendency of fault dip to change with depth, although the presence of faults with low dip in exhumed ductile crust suggests that fault dips can be more gentle beneath the brittle–quasi-plastic transition. Aseismic active faults in Texas, Louisiana, and Washington are listric; that is, fault dips are more gentle with greater depth in the shallow few kilometers of the Earth's crust.

Normal faults commonly have an irregular map pattern, although in cross section, they are straight or broadly curved. It is important to distinguish between normal faults that are the primary earthquake fault and faults that are secondary displacements triggered by the primary rupture.

Normal faults may be divided into geometric segments along strike with the objective of determining why normal-fault earthquakes terminate along strike. Segment boundaries include abrupt changes in fault strike, *en échelon* stepovers, zones of structural complexity or of cross faulting, and gaps in faulting, some produced by basement rock salients across the fault zone. In many cases, these geometric boundaries correspond to earthquake terminations, but in some cases they do not.

Near-surface fault-zone features have been described from the Aegean region as well as the Basin and Range Province (Stewart and Hancock, 1994). They include fault breccia, corrugated slip planes with the corrugations in the direction of fault slip, slickensides, slickenlines, tool tracks made by asperities in the opposite wall, and pluck holes. Fault scarps in relatively unconsolidated gravelly deposits in the Basin and Range Province are subdivided into a free face (approximating the original fault surface), a debris slope formed by material falling off the steep free face, and a wash slope in which this material is reworked by running water. Over time, the debris slope and wash slope overtake the free face, and the maximum slope angle of the scarp decreases. This relationship is used in the Basin and Range Province to determine the relative age of scarps. The climatic conditions, the lithology of the faulted deposits, and the height of the scarp also are important. Deposits of the debris slope and wash slope combine to form a colluvial wedge, used in trench excavations as evidence for past earthquakes.

Reverse faults, in which the minimum principal compressive stress σ_3 is vertical, occur where the crust is undergoing horizontal compression. Displacement is by dip slip or oblique slip. Reverse faults have generated major earthquakes in the Transverse Ranges and Coast Ranges of California; the eastern foothills of the Andes; the Pampean Andes of Argentina; northwest Nelson in New Zealand; Honshu in Japan; the western foothills of Taiwan; the Tien Shan and Qilian Shan of central Asia; the Caucasus, Zagros, and Alborz Mountains of western Asia and adjacent Europe; the Tellian Atlas Mountains of Algeria; and the Himalayan Front of India, Nepal, and Pakistan. The Himalayan Front is a plate boundary marking continental collision, and this boundary has generated the largest earthquakes, up to M 8.7. Reverse-fault earthquakes as large as M 6.7 (Meckering, Australia) have also struck stable continental shield areas in Australia, peninsular India, and northeastern Canada.

Although earthquakes tend to nucleate near the base of brittle crust, at least some of these events nucleated in the shallow part of the brittle crust. Large reverse faults ruptured the Baltic Shield shortly after the removal of Pleistocene glacial ice; these may be related to glacial unloading.

The Transverse Ranges of California, northwest Nelson and Central Otago in the South Island of New Zealand, northern Honshu, the Shillong Plateau of eastern India, and the Pampean Andes of northwest Argentina are characterized by thick-skinned tectonics in which reverse faults extend through brittle crust into ductile lower crust that deforms by pure shear. In contrast, the Himalayan, Andean, and Taiwan foothills and possibly the California Coast Ranges and the Tien Shan are characterized by low-angle thrusts forming fold-thrust belts. Shallow folds and thrusts pass downward into a basal thrust (*décollement*) that overlies basement rocks that have not shortened. In this latter case, the earthquake mainshock occurs at a considerable distance from the area of surface deformation. Some earthquakes, such as the 1896 Rikuu earthquake (*M* 7.2) in northern Japan, the 1978 Tabas earthquake (*M* 7.5) in the Iranian Plateau, and the 1999 Chi-Chi earthquake (*M* 7.6) in Taiwan ruptured the surface for many kilometers. Others had discontinuous surface rupture or no surface rupture at all, such as the 1944 San Juan, Argentina, earthquake (*M* 7.4); the 1980 El Asnam, Algeria, earthquake (*M* 7.3); and the 1983 Coalinga and 1994 Northridge, California, earthquakes (both *M* 6.7). In general, the larger events are more likely to have surface rupture, but some of the largest, such as the 1905 Kangra (*M* 8.0) and 1934 Nepal-Bihar (*M* 8.3), were blind-thrust earthquakes.

Segment boundaries include the following characteristics: *en échelon* stepovers (expressed as stepovers of anticlinal axes where faults are blind), lateral ramps, and intersection with strike-slip faults. The anticline overlying the 1983 Coalinga earthquake steps over to the Kettleman Hills anticline to the south, struck by an earthquake in 1985, and to the New Idria anticline to the north, struck by an earthquake in 1982. The 1971 San Fernando earthquake was terminated on the west by a lateral ramp. A characteristic of low-angle fold-thrust belts is for faulting to step out basinward, as occurred in the 1896 Rikuu and 1971 San Fernando earthquakes. However, the 1999 Chi-Chi earthquake in Taiwan did not rupture the outermost structure. Like normal faults, the map trace of active reverse faults is commonly irregular and lobate, and, in cross-section view, the fault is straight or broadly curved.

The surface expression of a reverse-fault scarp differs from that of a normal fault because the hangingwall overrides the footwall, and the frontal wedge of the fault may collapse onto the footwall. In other cases, the hangingwall rolls over the footwall like the treads of a tank, and in still others the fault flattens below the surface and does not reach the surface at all. What appears at first to be a fault scarp may turn out to be a fold scarp, with a sharp flexure at the scarp but no rupture. The Coyote Pass escarpment at the southern edge of the Elysian Hills in the northern Los Angeles basin is a fold scarp probably formed by bending moment in the core of the fold between the south-dipping limb of the Elysian Park anticline and nearly flat-lying strata of the Las Cienegas structural shelf farther south.

Strike-slip faults occur where both σ_1 and σ_3 are horizontal. As for reverse faults, the crust undergoes horizontal compression in the direction of σ_1, but it may undergo extension in the direction of σ_3, or it may thicken. Some of the longest and most active strike-slip faults are continental plate-boundary faults, including the San Andreas Fault of California, the Dead Sea Fault of the Middle East, and the Alpine Fault of New Zealand. Other plate-boundary examples are the North Anatolian Fault of Turkey, faults marking the northern and southern boundaries of the Caribbean Plate, faults marking the eastern and western boundaries of the Indian Plate, and the Queen Charlotte–Fairweather Fault of Canada and Alaska. In California, the San Andreas is the most active of a set of parallel strike-slip faults on both sides of it.

Strike-slip faults occur in conjunction with normal faults in the Basin and Range Province, the Ecuadorean Andes, southern Tibet, northeast China, and the northern Aegean Sea. They occur in conjunction with reverse faults in the Los Angeles basin, the Kinki Triangle of southwest Japan, the Zagros Mountains, the Caucasus, the Iranian Plateau, and central Mongolia. Strike-slip faults parallel to subduction zones include the Great Sumatran Fault in Indonesia; the Denali Fault in Alaska; the Median Tectonic Line in Japan; the Philippine Fault; the Wairarapa, Wellington, and related faults of the North Island, New Zealand; the Pallatanga Fault in Ecuador; the Liquine-Ofqui Fault in Chile; and the Jiali Fault of Tibet, the last parallel to the Himalayan collision zone. In Central Asia, the Iranian Plateau, and Anatolia, strike-slip faults accommodate the motion of blocks of continental crust out of the way of plates colliding from the south. Strike-slip faults cut ancient continental rocks in sub-Saharan West Africa and the New Madrid zone of the central United States.

Although the map trace of a strike-slip fault is straight compared to those of dip-slip faults, in detail a strike-slip fault may contain *en échelon* stepovers and secondary faults at low to high angles to the main fault. A right stepover on a right-slip fault (or a left stepover on a left-slip fault) results in extension in the region of the stepover, resulting in a downdropped area called a pull-apart basin, which may be bounded by normal faults as well as strike-slip faults. In contrast, a left stepover on a right-slip fault causes local compression, resulting in folds, reverse faults, and elevated topography.

A pure strike-slip fault cutting a horizontal surface of low topographic relief will not result in tectonic topography. However, such a fault crossing regions of moderate relief will result in offset streams, streams blocked by formerly adjacent ridges (shutter ridges), and streams blocked downstream to produce closed depressions called sagponds. Scarps accompanying strike-slip faults may alternate along strike from facing one direction to facing the other. Similarly, the oldest and hence more uplifted side of the fault may alternate along strike.

Shallow earthquakes on subduction zones contributed about 90% of the seismic moment released worldwide from 1900 to 1989, with a significant part of it released by great earthquakes in Chile in 1960 and Alaska in 1964. Because subduction zones reach the surface on the deep ocean floor, geologic observations on land related to subduction-zone earthquakes are indirect, consisting of changes of altitude of the land during an earthquake or evidence of strong ground motion. Accumulation of elastic strain causes the overriding plate to warp, uplifting or subsiding coastal regions prior to the earthquake. The earthquake releases strain and reverses this process, downdropping formerly uplifted regions and raising formerly subsided regions.

The altitude changes accompanying the coseismic and most of the interseismic parts of an earthquake cycle have been worked out in the Nankai subduction zone in southwest Japan for earthquakes in 1854, 1944, and 1946. Historical records for more than 13 centuries provide evidence for nine earthquake cycles. Marine terraces on the south coast of Shikoku and Honshu were uplifted permanently during some, but not all, of these earthquakes, suggesting that permanent uplift must be due to deformation of the upper plate accompanying some earthquakes, whereas interplate deformation accompanying other earthquakes is completely recovered in the interseismic period. In New Zealand, the 1931 Hawkes Bay earthquake was accompanied by permanent deformation of the upper plate. In the Cascadia subduction zone of the Pacific Northwest, earthquakes are recorded by sudden subsidence of coastal marshes, with the most recent subsidence in 1700, dated by radiocarbon and tree rings. The 1700 earthquake apparently generated a tsunami in Japan. Although the 1700 earthquake was recorded in marshes from Vancouver Island to northern California, most marsh sites show evidence of downdropping on local structures, suggesting that the interplate earthquake was accompanied by extensive deformation of the upper plate. Coseismic subsidence of marshes in Humboldt Bay, California, was due to downwarping of a syncline; marine terraces in adjacent areas were uplifted during subduction-zone earthquakes.

Cascadia earthquakes triggered major sediment flows down submarine canyons, and these turbid flows of sediment have been recorded in submarine channels on the oceanic Juan de Fuca Plate. The frequency of these deposits is consistent with the frequency of subduction-zone earthquakes as measured in coastal marshes. Similarly, earthquakes on the northernmost San Andreas Fault, which is offshore, have been recorded in sediment flows onto the deep-ocean floor to the west.

V. PALEOSEISMOLOGY

The seismographic record of earthquakes is so short that only in rare cases, such as Parkfield, California, along the San Andreas Fault, can an entire earthquake cycle be recorded instrumentally. Historical records show more than one cycle on the fast-moving North Anatolian Fault but, except for Parkfield, not on the San Andreas Fault, which has not ruptured historically southeast of the Transverse Ranges. Most faults have earthquake recurrence intervals measured in thousands of years, so that an individual fault may have ruptured only once or not at all during the historical period, even where that period is longer than two millennia. It is here that geology makes its greatest contribution by describing and dating earthquakes from the extensive geologic record. For example, the Pallett Creek site within the rupture zone of the 1812 and 1857 earthquakes on the San Andreas Fault provided evidence for more than eight prehistoric earthquakes, and investigations along the Wasatch Fault in Utah identified up to four prehistoric earthquakes in the past 5500 years at the same site. By examining successive earthquakes on the same fault, it is possible to estimate the variation in earthquake recurrence interval, the magnitude, and the stability of earthquake segment boundaries.

The age of a prehistoric earthquake is constrained by the age of the youngest sediment deformed during an earthquake and the oldest sediment unaffected by that deformation. This requires that these sediments contain organic material of the same age as the sediment enclosing it so that radiocarbon dating is possible. Radiocarbon dating contains other potential sources of error besides the assumption that the carbon is neither older nor younger than the enclosing sediment (S. Trumbore in Noller et al., 1999). These include variations in cosmic ray flux and the possibility that the marine environment enclosing the carbon-bearing organism has isotopic carbon ratios different than ratios in the atmosphere. Other dating techniques include tree-ring dating (dendrochronology), volcanic ash layer dating (tephrochronology), and correlation to eustatic variations in sea level, largely measured by oxygen-isotope ratios.

Study of surface displacement accompanying a modern earthquake is useful in searching for comparable data in the geologic record. Displacement shows great variability along strike, evidence of the complexity of slip distribution along a fault. Magnitudes of recent earthquakes have been related to fault length and to maximum displacement (Wells and Coppersmith, 1994), and where these are known from the geologic record an estimate of magnitude can be made from paleoseismology, but with considerable uncertainty. Earthquakes on some faults, such as the Wasatch Fault of Utah, tend to be approximately the same size and are called characteristic; that is, a future earthquake would be expected to be of a similar magnitude as past events. However, some well-studied faults such as the San Andreas Fault, show considerable variability in fault displacement and length as well as variability in earthquake recurrence interval.

Earthquakes are not periodic, and considerable uncertainty exists in estimating the time and magnitude of the next earthquake on a given fault, even where its paleoseismic history is known. The Central Nevada Seismic Zone and the Eastern California Shear Zone were the sites of temporal clusters of large-magnitude earthquakes in which the seismic energy release in the 20th century was very large compared to the long-term deformation rate within those zones. Paleoseismology reveals that only the 20th-century earthquakes were clustered. Historical clustering raises the possibility that prehistoric earthquakes, such as earthquakes around A.D. 1480 on the central San Andreas Fault and A.D. 1680 on the southern San Andreas Fault, could each be an earthquake cluster rather

than a single large event. Radiocarbon dating is not precise enough to separate earthquakes spaced in time by only a few years. For example, an earthquake cluster on the North Anatolian Fault in Turkey that began in 1939 and continued to 1999 with the Izmit and Düzce earthquakes might have appeared as a single gigantic earthquake if the earthquakes had been prehistoric and all evidence for age was based on radiocarbon dating from trench excavations. On the other hand, radiocarbon dating of a subsidence event on the Cascadia subduction zone from Vancouver Island to northern California is now believed to document a single earthquake of M 9 rather than a cluster of smaller events, based on tree-ring dating, Native American oral histories, and records of a tsunami generated by the earthquake that struck Japan.

Paleoseismic evidence of past earthquakes on subduction zones includes buried marsh deposits, some capped by sand from a tsunami, along the Cascadia and Aleutian subduction zones, uplifted post-glacial marine terrace deposits, and sands on the deep seafloor due to sediment flows down submarine canyons triggered by strong ground motion. On strike-slip faults, channel deposits or river terraces of offset streams may be dated, and the amount of offset of these features is used to determine the long-term slip rate (Fig. 2). Some offsets were produced by a single earthquake, whereas others developed through several

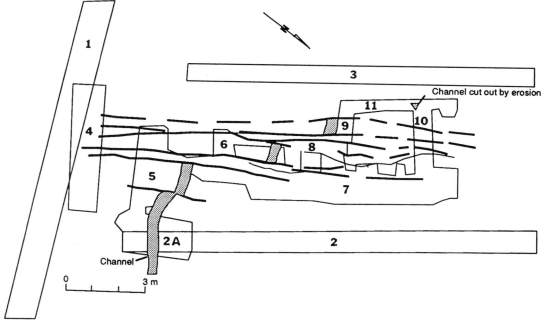

FIGURE 2 Recognition of a buried offset stream channel based on backhoe and hand-dug excavations across the Rose Canyon fault zone, San Diego, California. Strands of the fault are shown in heavy lines; the channel is shown in stipple pattern. If the channel can be dated, a minimum slip rate on the fault zone can be determined. The rate is a minimum because additional fault strands could have offset the channel where it has been removed by erosion. Numbered polygons locate sites of excavations. [From work by S. Lindvall and T. Rockwell, reprinted from J. P. McCalpin (1996). "Paleoseismology," Academic Press, New York. With permission.]

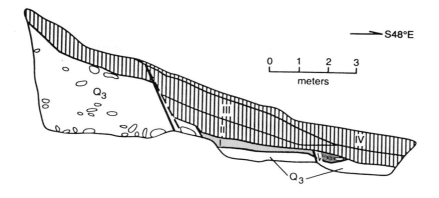

A

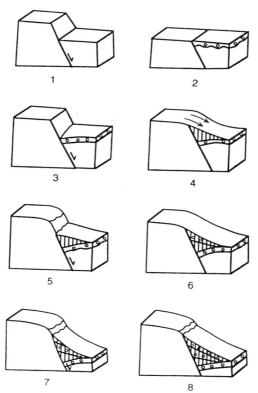

B

FIGURE 3 Recognition of prehistoric earthquakes on a normal fault based on information from an excavation in northeast China. (A) Colluvial wedges (II, III, IV) were deposited during three separate earthquakes, as illustrated by block diagrams below. (B1) Normal-fault earthquake is accompanied by formation of a fault scarp. (B2) Scarp collapses and is eroded, forming a colluvial wedge (circle pattern). (B3) Fault rupture during a second normal-fault earthquake renews the fault scarp. (B4) Scarp collapses, forming another colluvial wedge (vertical line pattern). (B5) Fault rupture during a third earthquake renews the fault scarp. (B6) Scarp collapses, forming another colluvial wedge (vertical line pattern with circles). (B7) Fault scarp is renewed during a fourth earthquake. (B8) Scarp collapses and forms another colluvial wedge (vertical line pattern with circles). Each colluvial wedge is preserved in the downdropped hangingwall of the fault, permitting recognition of four prehistoric earthquake in the excavation in (A). [From work by Y. Wang and Q. Deng, reprinted from Yeats, R. S. *et al.* (1997). "The Geology of Earthquakes," Oxford University Press, New York. With permission.]

earthquakes on the same fault. Further information is gathered from backhoe or bulldozer excavations that provide evidence for dating earthquakes by the upward termination of deformed strata, the presence within faults of sand that had been liquefied and injected during an earthquake, and greater displacement of older sediment layers. Colluvial wedges from debris slopes and wash slopes below fault scarps are also used to identify earthquakes, particularly along normal faults (Fig. 3).

VI. SUMMARY

Geology provides the setting in space and time for faults that generate earthquakes. The rock surrounding a fault contains evidence for the stress and strain regimes that lead to earthquakes, as well as the interactions among faults in a seismogenic region. The slip history and slip rate on seismogenic faults are used to determine the future potential for earthquake faulting. The geometry of faulting, including complexities along strike, is used to understand the setting for earthquake nucleation and the reasons why earthquakes stop on a given fault.

Geology is also useful in understanding the earthquake source, which cannot be reached by drilling under present technology. Nature has uplifted and deeply eroded fault rocks that were formerly at depths where earthquakes nucleate, and study of these rocks in the field and laboratory leads to a better understanding of waveforms generated at the earthquake source and the mode of coseismic rupture propagation.

In nearly all cases, large continental earthquakes have not repeated on the same segment of a fault in historic time. Paleoseismology is used to work out the earthquake history of a fault or region, leading to information about regularity of earthquake magnitude, recurrence interval, and the rate of fault slip. This is done by careful study of sediments deposited during the time the region has been under the current strain field using geomorphology and the study of artificial trench excavations. Some excavations lead to information about fault displacement, and others lead to information about strong ground motion, including liquefaction of sediments and coseismic landslides.

Paleoseismic time histories on a large number of sites are now available to work out recurrence intervals and recurrence variability on faults with high slip rates, including the San Andreas Fault, the North Anatolian Fault, and New Zealand's Alpine fault system. This has led to probabilistic earthquake forecasts in the San Francisco Bay Area, the San Andreas–San Jacinto fault system in southern California, and the North Anatolian Fault near Istanbul. A probabilistic earthquake forecast on a specific fault is possible where earthquake recurrence intervals are

only a few hundred years, even though uncertainties in radiocarbon dating lead to uncertainties in correlating the same earthquake among several paleoseismic sites. For faults with recurrence intervals measured in thousands of years, probabilistic forecasts are less reliable because the time period of a forecast, 10 to 50 years, is very short in comparison to an average recurrence interval of several thousand years. However, a probabilistic forecast is possible for a region containing several seismogenic faults with low slip rates and long recurrence intervals, such as the Los Angeles Basin in California or central Japan between Kobe and Tokyo. Even though the probability of an earthquake on a specific fault would be very low, the probability that one of the faults in the region would generate a damaging earthquake in the next 10 to 50 years is high enough to have societal significance.

SEE ALSO THE FOLLOWING ARTICLES

COASTAL GEOLOGY • CONTINENTAL CRUST • EARTHQUAKE MECHANISMS AND PLATE TECTONICS • EARTHQUAKE PREDICTION • EARTH'S MANTLE • GEOLOGIC TIME • OCEANIC CRUST • SEISMOLOGY, ENGINEERING • SEISMOLOGY, OBSERVATIONAL • SEISMOLOGY, THEORETICAL • STRESS IN THE EARTH'S LITHOSPHERE • TECTONOPHYSICS

BIBLIOGRAPHY

Bucknam, R. C., and Hancock, P. L., eds. (1992). "Major active faults of the world: results of IGCP Project 206," *Annales Tectonicae*, **6(Suppl.)**, 284 pp.

Crone, A. J., and Omdahl, E. M., eds. (1987). "Directions in Paleoseismology, Proceedings of Conference 39," U.S. Geol. Survey Open-File Report 87-673, 456 pp.

Hancock, P. L., Yeats, R. S., and Sanderson, D. J., eds. (1991). "Characteristics of active faults," *Journal of Structural Geology* **13,** 123–240.

Keller, E. A., and Pinter, N. (1996). "Active Tectonics: Earthquakes, Uplift, and Landscape," Prentice-Hall, Upper Saddle River, NJ, 338 pp.

Krinitzsky, E. L., and Slemmons, D. B., eds. (1990). "Neotectonics and earthquake evaluation," *Geological Society of America Reviews in Engineering Geology* **8,** 160 pp.

McCalpin, J. P., ed. (1996). "Paleoseismology," Academic Press, New York, 588 pp.

Noller, J. S., Sowers, J. M., and Lettis, W. R., eds. (1999). "Quaternary Geochronology: Methods and Applications," American Geophysical Union Reference Shelf 4, 582 pp.

Okumura, K., Takada, Y., and Goto, H. (2000). "Active Fault Research for the New Millennium," Proceedings of the Hokudan International Symposium and School on Active Faulting, Hiroshima, Letter Press, 606 pp.

Prentice, C. S., Schwartz, D. P., and Yeats, R. S., convenors (1994). "Proceedings of the Workshop on Paleoseismology," U.S. Geol. Survey Open-File Report 94-568, 210 pp.

Scholz, C. H. (1990). "The Mechanics of Earthquakes and Faulting," Cambridge University Press, Cambridge, U.K., 439 pp.

Serva, L., and Slemmons, D. B., eds. (1995). "Perspectives in Paleoseismology," Association of Engineering Geologists Special Publication 6, 139 pp.

Stewart, I. S., and Hancock, P. L. (1994). "*Neotectonics.*" *In* "Continental Deformation" (P. L. Hancock, ed.), pp. 370–409, Pergamon Press, Oxford.

Stewart, I. S., and Vita-Finzi, C., eds. (1998). "Coastal Tectonics," Geological Society (London) Special Publication 146, 378 pp.

Valensise, G., and Pantosti, D., eds. (1995). "Active Faulting Studies for Seismic Hazard Assessment," Istituto Nazionale di Geofisica, Rome.

Wallace, R. E., compiler (1986). "Active Tectonics: Studies in Geophysics," National Academy Press, Washington, D.C., 266 pp.

Wells, D. L., and Coppersmith, K. J. (1994). "New empirical relationships among magnitude, rupture length, rupture area, and surface displacement," *Bulletin of the Seismological Society of America* **84,** 974–1002.

Yeats, R. S. (1998). "Living with Earthquakes in the Pacific Northwest," Oregon State University Press, Corvallis, 309 pp.

Yeats, R. S. (2001). "Living with Earthquakes in California: A Survivor's Guide," Oregon State University Press, Corvallis, 406 pp.

Yeats, R. S., and Prentice, C., eds. (1996). "Special section: paleoseismology," *Journal of Geophysical Research* **101,** 5847–6292.

Yeats, R. S., Sieh, K., and Allen, C. R. (1997). "The Geology of Earthquakes," Oxford University Press, New York, 568 pp.

Geomagnetism

Ronald T. Merrill
University of Washington

Phillip L. McFadden
Australian Geological Survey Organisation

GLOSSARY

Declination Angle between true (geographic) north and the direction in which a compass needle points when it is free to swing in a horizontal plane.

Gauss coefficients Coefficients in the expansion of the earth's magnetic field as a series of spherical harmonics.

Geocentric dipole Dipole at the earth's center that best approximates the earth's magnetic field. The remainder of the field is known as the nondipole field.

Geomagnetic poles Points at which the axis of the geocentric dipole intersects the surface of the earth.

Inclination Angles that magnetized needle makes with the horizontal when it is free to swing in a vertical plane.

Isomagnetics Lines drawn through points at which a given magnetic element has the same value. Isogonics are lines of equal declination, and isoclinics are lines of equal inclination.

Isopors Lines drawn through points at which a magnetic element has the same secular change.

Magnetic (or dip) poles Points on the earth's surface where the magnetic field is vertical.

Secular variation Temporal variations of the earth's magnetic field due to internal causes.

GEOMAGNETISM, the study of the properties, history, and origin of the earth's magnetic field, is one of the world's oldest sciences. Both the Chinese and the Greeks in ancient times knew about the magnetic properties of lodestone (magnetite) and its ability to act as a compass. The works of Peregrinus in 1269 and Gilbert in 1600 on the magnetic field are often cited today as examples of the first "modern" scientific papers. In particular,

Encyclopedia of Physical Science and Technology, Third Edition, Volume 6

Gilbert argued that the earth itself was a great magnet. Following Gilbert, speculations of the origin of the earth's magnetic field remained centered on the properties of permanently magnetized materials. This remained so until electric currents, and their relationship with magnetic fields, were discovered. Subsequently, numerous speculations surfaced for the origin of the field, including residual magnetic fields associated with the earth's formation and fields associated with phenomena such as thermoelectric currents. Today we know far more about the details of the geomagnetic field: it has been recorded by satellites as part of the much broader subject of the physics of the upper atmosphere (aeronomy) and by magnetometers at the earth's surface, and we have been able to infer much about its ancient behavior through paleomagnetism. Consequently, we now know that the field has a complex spatial structure and that this structure varies with time. Indeed, we know from paleomagnetism that, on occasions distributed randomly in time, the geomagnetic north and south poles trade positions over a time interval of order 1000 to 10,000 years. These most dramatic of changes in the earth's magnetic field are called magnetic field reversals, and there are hundreds of such reversals now documented in the geologic record. It is also known that there are long intervals of time during which the reversal process ceased and in which no reversals occurred. We also know that the geomagnetic field is of internal origin. Any modern theory for the origin of the earth's field must therefore recognise an internal origin and include an explanation for its known spatial and temporal variations. It is now widely accepted by geomagnetists that the geomagnetic field originates primarily from a dynamo generating electric currents in the earth's molten iron-rich core. Jupiter, Saturn, Uranus, Neptune, and our sun also have magnetic fields that seem to originate from dynamos. Despite the wide acceptance of this concept, mathematical solution of all the associated equations together with appropriate boundary conditions and the known material properties in the earth's core is a horrendously difficult problem. Consequently, although there are now many dynamo models, not one of them adequately describes all the properties of the earth's magnetic field.

I. DIRECT MEASUREMENTS OF THE EARTH'S MAGNETIC FIELD

Direct measurements of the earth's magnetic field are made continuously by satellite and at fixed magnetic observatories on land. These observations are supplemented by a wide range of types of survey, often including shipborne and airborne magnetometers. Over the years these measurements have been made in a variety of units. Prior to about 1980, cgs units prevailed, but today the standard is to use SI units. Some of the units used in this article, and their cgs counterparts, are given in Table I. The earth's magnetic induction is most intense at the poles and is around 60,000 nT there. The strongest magnets today (with magnetic induction near 16 T—strong enough to levitate frogs) are as much as 3×10^5 times as intense as the earth's field, which is therefore typically referred to as a weak field. Despite the fact that the field is weak at the earth's surface, it occupies a large volume and is significantly stronger in the earth's core; consequently, there is a vast amount of energy stored in this field.

The main elements of the field (D, I, H, F, X, Y, Z) are defined in Fig. 1 with the conventional notation. Note that the elements X (north), Y (east), and Z (down), form a local coordinate system that changes depending on the observation point. The elements of the field are usually displayed in the form of an isomagnetic chart, i.e., a map of contours with equal values of a particular magnetic element. These charts are termed isogonic for declination (the angle, D, that the horizontal component of the magnetic field, H, makes with geographic north), isoclinic for inclination (the angle, I, that the magnetic field makes with the horizontal), and isodynamic for equal intensity (F). Isoclinic and isogonic charts are shown respectively in Figs. 2a and 2b for the International Geomagnetic Reference Field for the year 2000 (IGRF 2000).

TABLE I Common Magnetic Terms in SI and cgs Units with Conversion Factors

Magnetic term	SI unit	cgs unit	Conversion factor
Magnetic induction (**B**)	tesla (T) $= \text{kg A}^{-1} \text{s}^{-2}$	gauss	$1 \text{ T} = 10^4$ gauss
Magnetic field strength (**H**)	A m^{-1}	oersted (oe)	$1 \text{ A m}^{-1} = 4\pi \times 10^{-3}$ oe
Magnetization (**M**)	A m^{-1}	$\text{emu cm}^{-3} = \text{gauss}$	$1 \text{ A m}^{-1} = 10^{-3} \text{ emu cm}^{-3}$
Magnetic dipole moment (p)	A m^2	$\text{emu} = \text{gauss cm}^3$	$1 \text{ A m}^2 = 10^3$ emu
Magnetic scalar potential (ψ)	A	$\text{emu} = \text{oe cm}$	$1 \text{ A} = 4\pi \times 10^{-1}$ emu
Permeability of free space (μ_0)	henry (H) $\text{m}^{-1} = \text{kg m A}^{-2} \text{s}^{-2}$	1	$4\pi \times 10^{-7} \text{ H m}^{-1} = 1$ cgs

emu $\equiv$ electromagnetic unit.

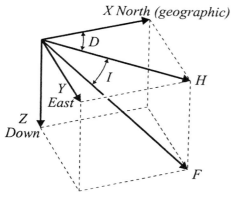

FIGURE 1 The main elements of the geomagnetic field. The deviation, *D*, of a compass needle from true north is referred to as the *declination* (reckoned positive eastward). The compass needle lies in the magnetic meridian containing the total field *F*, which is at an angle *I*, termed the *inclination* (or dip), to the horizontal. The inclination is reckoned positive downward (as in the Northern Hemisphere) and negative upward (as in the Southern Hemisphere). The horizontal (*H*) and vertical (*Z*) components of *F* are given by $H = F \cos I$ and $Z = F \sin I$, respectively. *Z* is reckoned positive downward, as for *I*. The horizontal component can be resolved into two components: *X* (northward) $= H \cos D$ and *Y* (eastward) $= H \sin D$. Then $\tan D = Y/X$ and $\tan I = Z/H$.

II. SPHERICAL HARMONIC ANALYSIS

In 1839, Gauss pioneered the use of spherical harmonic analysis to provide a useful quantitative description of the magnetic field. This method remains the predominant method of analyzing the magnetic field of the earth and other planets.

Because there are no magnetic monopoles, $\nabla \cdot \mathbf{H} = 0$, where **H** is the magnetic field. Assuming that there is no magnetic material just above the ground and that there are no earth–air electric currents, then $\nabla \times \mathbf{H} = \mathbf{0}$. Taken together these equations mean that there is a magnetic scalar potential, ψ such that $\mathbf{H} = -\nabla \psi$ and $\nabla^2 \psi = 0$. In the absence of magnetic sources, the magnetic induction $\mathbf{B} = \mu_0 \mathbf{H}$, so that $\mathbf{B} = -\mu_0 \nabla \psi$. The potential, ψ, can be expanded in spherical harmonics using the spherical coordinates defined in Fig. 3 with the origin placed at the earth's center and the vertical axis aligned with the earth's rotation axis. ψ is then given by

$$\psi = \frac{a}{\mu_0} \sum_{n=1}^{\infty} \sum_{m=0}^{n} P_n^m(\cos \theta) \left\{ \left[C_n^m \left(\frac{r}{a} \right)^n \right.\right.$$
$$\left. + (1 - C_n^m) \left(\frac{a}{r} \right)^{n+1} \right] A_n^m \cos m\phi + \left[S_n^m \left(\frac{r}{a} \right)^n \right.$$
$$\left.\left. + (1 - S_n^m) \left(\frac{a}{r} \right)^{n+1} \right] B_n^m \sin m\phi \right\}. \tag{1}$$

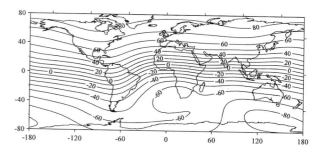

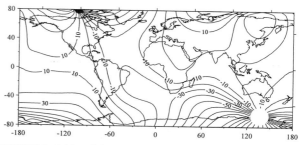

FIGURE 2 (a) Isoclinic chart for 2000 showing the inclination (in degrees) of the geomagnetic field over the earth's surface. (b) Isogonic chart for 2000 showing the declination (in degrees) of the geomagnetic field over the earth's surface.

The P_n^m in Eq. (1) are partially normalized Schmidt polynomials that are related to the more familiar associated Legendre polynomials, $P_{n,m}$, by

$$P_n^m = P_{n,m} \qquad \text{for } m = 0;$$
$$P_n^m = \left[\frac{2(n-m)!}{(n+m)!} \right]^{1/2} P_{n,m}, \qquad \text{for } m > 0. \tag{2}$$

C_n^m and S_n^m in Eq. (1) are positive numbers between 0 and 1 that describe the fraction of the potential of external origin when *r* is equal to the earth's radius, *a*. The fact that these coefficients represent external sources can be seen in that

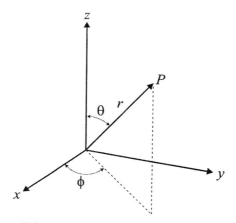

FIGURE 3 Spherical polar coordinates.

the potential for external sources increases with increasing r, while that for internal sources increases with decreasing r. The components X, Y, and Z of the magnetic field $\mathbf{H}$ are obtained from $\mathbf{H} = -\nabla \psi$ as

$$X = \frac{1}{r}\frac{\partial \psi}{\partial \theta}; \qquad Y = -\frac{1}{r \sin \theta}\frac{\partial \psi}{\partial \phi}; \qquad Z = \frac{\partial \psi}{\partial r}. \quad (3)$$

Because each of these components is derived from the same potential, if values of X are known all over the world it is possible to deduce the corresponding values of Y. Thus by comparison of observed and deduced values of Y it is possible to check for consistency and thereby test the assumption that there are no magnetic sources just above the ground. By fitting the field it is possible to estimate the coefficients C_n^m and S_n^m and thereby discover how much of the observed field is of internal origin, a remarkably useful characteristic of spherical harmonic analysis. Gauss could not resolve any external sources in 1839, and typically the average amount of the external field at the earth's surface averaged over a year is less than a percent or two, that is, $C_n^m \approx S_n^m \approx 0$. Hence the potential ψ of Eq. (1) reduces to that for the field of internal origin, given by

$$\psi = \frac{a}{\mu_0} \sum_{n=1}^{\infty} \sum_{m=0}^{n} \left(\frac{a}{r}\right)^{n+1} P_n^m(\cos \theta)$$
$$\times \left(g_n^m \cos m\phi + h_n^m \sin m\phi\right), \quad (4)$$

where $g_n^m = (1 - C_n^m)$ and $h_n^m = (1 - S_n^m)$ are known as the Gauss coefficients.

As may be seen by the structure of Eq. (4), an individual harmonic is simply a Fourier series for a given latitude (constant θ) and an associated Legendre polynomial for a given longitude (constant ϕ). An associated Legendre polynomial has $(n - m)$ zeros in $0 \leq \theta \leq 180$, dividing a longitudinal line into $(n - m + 1)$ latitudinal zones of alternating sign. Similarly, $\sin m\phi$ (or $\cos m\phi$) has $2m$ zeros, dividing a line of latitude into $2m$ longitudinal sectors of alternating sign. This is evident in Fig. 4, which shows that the individual harmonics are exquisitely beautiful functions.

It should be noted that the spherical harmonic description is nonunique, as indeed must be the case for all mathematical descriptions that use measurements made at, or above, the earth's surface. This is manifested in the fact that all internal sources are arbitrarily placed at the earth's center (i.e., geocentric); the individual spherical harmonic terms (such as dipole, quadrupole, etc.) do not represent physically real or separate sources. The value of spherical harmonic analysis is that the functions form a complete orthogonal set and there is a wealth of physics and mathematics supporting the analyses. For example, as noted above, it is possible to distinguish between external and internal sources of the field.

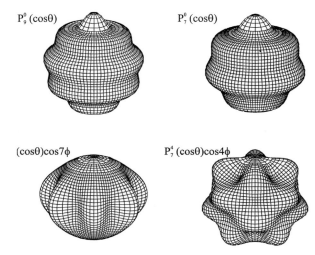

$P_9^0 (\cos\theta)$ $\qquad$ $P_7^0 (\cos\theta)$

$(\cos\theta)\cos 7\phi$ $\qquad$ $P_7^4 (\cos\theta)\cos 4\phi$

FIGURE 4 Individual spherical harmonics.

IGRF models go out to degree 10 (120 coefficients). Table II gives the Gauss coefficients, g_n^m and h_n^m, out to degree $n = 5$ and order $m = 5$ (the first 35 coefficients) for IGRF 2000. The table also gives the coefficients $\dot{g}_n^m$ and $\dot{h}_n^m$, which are the amounts by which the field coefficients change each year (the secular change). Note that, as in Eqs. (1) and (4), there is no term with $n = 0$, which would correspond to a magnetic monopole, so the first term is of

TABLE II IGRF 2000 Epoch Model Coefficients up to Degree 5

		Main field (nT)		Secular change (nT yr^{-1})	
n	m	g	h	$\dot{g}$	$\dot{h}$
1	0	−29,615	0	14.6	0.0
1	1	−1,728	5,186	10.7	−22.5
2	0	−2,267	0	−12.4	0.0
2	1	3,072	−2,478	1.1	−20.6
2	2	1,672	−458	−1.1	−9.6
3	0	1,341	0	0.7	0.0
3	1	−2,290	−227	−5.4	6.0
3	2	1,253	296	0.9	−0.1
3	3	715	−492	−7.7	−14.2
4	0	935	0	−1.3	0.0
4	1	787	272	1.6	2.1
4	2	251	−232	−7.3	1.3
4	3	−405	119	2.9	5.0
4	4	110	−304	−3.2	0.3
5	0	−217	0	0.0	0.0
5	1	351	44	−0.7	−0.1
5	2	222	172	−2.1	0.6
5	3	−131	−134	−2.8	1.7
5	4	−169	−40	−0.8	1.9
5	5	−12	107	2.5	0.1

degree 1 (a dipole). Form this table it is clear that the dominant term is g_1^0, the geocentric axial dipole (this phraseology results from the fact that all spherical harmonic terms are geocentric, and terms with order zero are aligned along the earth's rotation axis). Note that g_1^0 is negative, so if the dipole is thought of as a bar magnet, then the south pole of that magnet is in the Northern Hemisphere; the north pole of a magnetic compass is then attracted toward the north geographic pole. The International Association of Geomagnetism and Aeronomy (IAGA) has chosen the units of the Gauss coefficients in the IGRF model, so that if the induction $\mathbf{B}$ is determined from $\mathbf{B} = -\mu_0 \nabla \psi$, then $\mathbf{B}$ is in nano tesla.

The geocentric dipole can be obtained from its three orthogonal components that come from the g_1^0 term and the two equatorial dipole terms associated with the Gauss coefficients g_1^1 and h_1^1. For IGRF 2000 the geocentric dipole is tilted $10.5°$ with respect to the rotation axis and the geomagnetic poles, defined by the two points at which the axis of this dipole intersecs the earth's surface (the geocentric dipole field is vertical at these two points), are at $79.5°$N, $288.4°$E and $79.5°$S, $108.4°$E. The magnetic north and south poles are defined as the two points on the earth's surface where the total field is vertical, and they are at $80.9°$N, $250.1°$E, and $64.6°$S, $138.3°$E, respectively. Note that the magnetic poles are not $180°$ apart and that they do not coincide with the geomagnetic poles. This occurs because the total internal field consists of the sum of the nondipole field and the dipole field. The nondipole field is the sum of all internal terms with $n \geq 2$. Figure 5 shows the vertical component of the nondipole field for IGRF 2000. The nondipole field intensity at the earth's surface averages to about 25% of the total field. A century ago this percentage was about 17–18%, an indication that the magnetic field of internal origin is constantly changing. The locations of the geomagnetic and magnetic poles are also changing. Collectively, the changes in the earth's internal magnetic field with time are referred to as geomagnetic secular variation. In 1985 J. Bloxham and D. Gubbins used ancient mariner magnetic data to extend spherical harmonic analyses back to 1715, and by doing so they nearly doubled

the time span for which direct measurements of the earth's magnetic field and its secular variation are available.

Terms out to spherical harmonic degree near 14 are thought to represent magnetic sources in the earth's core. Consequently, IGRF models (out to degree 10) represent sources in the core. Terms with $n > 14$ contain both crustal and core sources; generally, as n increases, so does the component of crustal sources. There appear to be no significant sources from the mantle, which extends from a depth of about 35 km to 2891 km and separates the crust from the core. The temperatures are too high in the mantle for there to be any permanent magnetization there, and any electrical currents that persist in the mantle for more than about a year are too small to produce significantly large magnetic fields at the earth's surface. The main part of the field is believed to originate in the outer core, which extends from a depth of 2891 km to about 5150 km and is molten, as is evidenced by the fact that it does not transmit seismic shear waves. There is a solid inner core that extends to the earth's center at 6371 km depth. Seismological, mineral physical, and geochemical data indicate that the outer core consists of about 90% iron (by weight), a few percent nickel, and the remainder contains less dense elements such as hydrogen, oxygen, sulfur, carbon, or silicon. The inner core appears to be about 97% iron. In contrast to the mantle, which contains minerals that are predominantly semiconductors, the electrical conductivity of the core is metallic and large: high-pressure measurements and theory suggest a conductivity value near 6×10^5 S m^{-1}.

III. THE EXTERNAL MAGNETIC FIELD AND ITS TEMPORAL VARIATION

The earth's magnetic field varies over time intervals covering more than 12 orders of magnitude, from less than 10^{-3} s to more than 10^8 years. Although the magnetic field almost certainly changes rapidly in the core, the mantle is effective in electrically screening core variations with characteristic times less than about a year. Consequently, magnetic field variations observed at the earth's surface that have characteristic times less than about a year are of external origin; they may reflect electric currents (and their associated magnetic fields) induced in the upper mantle by the varying external field.

The region up to about 50 km above the earth's surface can, except for thunderstorms, be regarded as an electromagnetic vacuum. The ionosphere extends from roughly 50 km to 1500 km, and the van Allen radiation belts extend from about 4 to 6 earth radii. Many people are familiar with aurora, a spectacular symptom of the external magnetic field, which are produced in the ionosphere at high

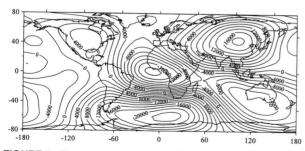

FIGURE 5 The vertical component (Z) of the nondipole field for IGRF 2000. Contours are labeled in units of nano tesla.

latitudes. For example, the soft green light seen in many aurora comes from the forbidden emission (557.7 nm) of oxygen precipitated by the bombardment of electrons in the ionosphere at geomagnetic latitudes (the latitude relative to the dipole axis) around 65° to 70°. More recently, people have become aware of the external field because of electromagnetic pulses that affect their computers, cell phones, or produce power outages.

Variations in the solar magnetic field, such as those associated with the 11-year sunspot cycle, can produce observable effects such as magnetic storms and aurora at the earth's surface. The solar wind, which consists primarily of electrons, protons, a few heavier nuclei, and magnetic field, streams away from the sun at speeds from about 270 km s^{-1} to about 650 km s^{-1} and impacts the earth's magnetic field. The earth's magnetic field is confined to a region known as the magnetosphere (which is not a sphere) and has a boundary known as the magnetopause. On the sun side the pressure of the solar wind compresses the earth's magnetic field until its pressure (equal to B^2/μ_0) balances that of the solar wind at the magnetopause. Consequently, because of variations in the solar wind, the position of the magnetopause (and therefore the size of the magnetosphere) varies over short times, such as days, but is typically at 8 to 10 earth radii on the side toward the sun. In contrast, the earth's magnetic field is swept away on the opposite side, producing a long geomagnetic tail. The solar wind is highly variable in time (the particle density to a few percent of its average value one day in May 1999, while it was more than 30 times its average in December of the same year), and this variation invariably leads to changes in the magnetic field at the earth's surface.

Magnetic storms on the earth are triggered by solar storms, which are a billowing cloud of plasma and magnetic field that can be millions of kilometers across. The largest solar storms come from coronal mass ejections (CMEs), which are often associated with solar flares. A CME is proably driven by solar magnetic processes, and after a few days the expanding CME bubble of plasma and magnetic fields crashes into our magnetosphere. Although the distortion of our magnetosphere depends on the orientation of the magnetic field within the CME relative to the earth's magnetic field, it can be large. The changing magnetic fields within our magnetosphere induce electric currents in the ionosphere that can have pronounced effects on humans, including the production of so-called killer electrons that can penetrate and disable a satellite. They are also associated in complex ways with magnetic storms and substorms. Variations in the horizontal component of the external magnetic field of up to 100 nT are typical of magnetic storms, and large storms can produce surface fields near 500 nT. One powerful magnetic storm resulted in a power outage in Quebec in 1989 that affected several million people for 9 h. A few spacecraft, such as the

Solar and Heliospheric Observatory (SOHO), launched in late 1995, contain equipment to observe the sun's faint corona and provide us with some ability to forecast magnetic storms.

Variations in the external magnetic field also induce electric currents in the earth's mantle that can lead to measurable magnetic fields at the earth's surface. Typically, the duration of these fields is much less than a year. The measurement of these fields can provide valuable information on the electrical properties of the earth's crust and mantle.

The magnetosphere helps to protect us from many of the violent effects of the solar weather. It shields us from harmful cosmic radiation that bombards the earth from all directions. Without the magnetosphere, the density and composition of our atmosphere would be altered by the sputtering affects of the solar wind.

IV. DIRECT MEASUREMENTS OF THE GEOMAGNETIC SECULAR VARIATION

Direct measurements of changes in the earth's internal magnetic field of core origin, the geomagnetic secular variation, are essentially restricted to the past few hundred years. They are commonly reported in terms of time derivatives of Gauss coefficients in a spherical harmonic analysis. Although data that span a time interval greater than a year are used, the estimates of the time derivatives are often given for a particular year; the year is identified by, say, referring to the secular variation for IGRF 1965. Such analyses indicate that the magnetic field is constantly changing in both direction and intensity. The dipole field intensity decreased by approximately 5% during the past century, its axis rotated westward at an average rate of 0.05–0.1 degrees per year, but its tilt with respect to the rotation axis varied little (approximately by a degree) during the same time interval. Relative changes in the nondipole field were more dramatic. In the Atlantic hemisphere there was an average 0.3° yr^{-1} westward movement of the nondipole field, a movement that is sometimes referred to as the westward drift of the nondipole field. In recent times the nondipole field is generally not manifested in much of the Pacific hemisphere. The intensity of the nondipole field increased in some locations while decreasing in others. On average, the nondipole-to-dipole field ratio at the earth's surface increased by about 5–7% during the twentieth century.

V. INDIRECT METHODS FOR MEASURING THE EARTH'S MAGNETIC FIELD: PALEOMAGNETISM

Indirect measurements of the earth's magnetic field are possible because rocks contain a natural remanent magnetization (NRM), within which there is a record of the ancient magnetic field, known as the paleofield. A remanent magnetization (RM) is a permanent magnetization that remains after removal of the external magnetic field that caused it. This can be contrasted with an induced magnetization, which exists only in the presence of an external magnetic field. Although the magnitudes of the remanent and induced magnetizations vary considerably for different rocks, remanent magnetization is typically many times larger than the induced magnetization in rocks such as basalt, the most common lava flow on the earth's surface. The NRM consists of a primary RM, an RM acquired when the rock formed, and a secondary RM acquired after the rock formed. Examples of the most common forms of primary RM are a thermal remanent magnetization (TRM), detrital (or depositional) remanent magnetization (DRM), and a postdepositional RM (pDRM). TRM is acquired in igneous rocks as they cool in a weak magnetic field, such as the earth's field. DRM is acquired as magnetic grains are oriented while settling in water to form sedimentary rock. Processes dominated by compaction, bioturbation, and electrostatic forces lead to the reorientation of the magnetic grains below the water–sediment interface to produce a pDRM in sediments. An example of a secondary remanent magnetization is CRM, a magnetization acquired by chemical alteration after a rock formed. There are now more than two dozen types of remanent magnetization that have been identified and studied. The study of these ancient remanent magnetizations in rocks to investigate the motions of continents and to infer properties of the ancient magnetic field is known as paleomagnetism.

A record of the paleofield at a specific location and time can be obtained if it is possible to distinguish the primary magnetization in a rock from secondary remanent magnetizations and determine the age the rock formed. There is good evidence that the direction of primary magnetization can be recovered from some rocks that are older than 3 billion years. Archeomagnetism uses techniques similar to paleomagnetism to determine the paleofield in archeological objects, such as bricks from pottery kilns or ancient fireplaces. The techniques used to separate the primary from secondary RM are diverse, and sometimes there is controversy over whether the primary RM has been identified correctly. The most reliable studies usually employ a range of consistency checks and rely heavily on statistical analyses of large data sets. Typically it is much easier to obtain reliable directional data than intensity data for the paleofield. Although such data are inevitably poorer than data obtained from direct measurements, they have provided us with a wealth of information on the history and evolution of the earth's magnetic field.

The paleofield averaged over the past few hundred thousand years or so can be well represented by a field from a downward-pointing dipole aligned along the earth's ro-

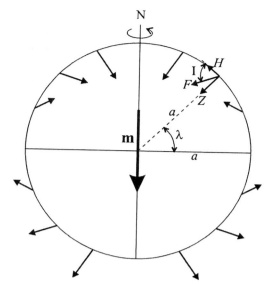

FIGURE 6 The field of a normal polarity geocentric axial dipole (GAD).

tation axis. Deviations from this average geocentric axial dipole (GAD) field are no more than a few percent. A GAD field at the earth's surface is shown in Fig. 6, which uses the parameters introduced in Fig. 1 and where λ is the latitude and **m** is the magnetic moment of the dipole. It is straightforward to show that the inclination, I, of the GAD can be used to determine the paleolatitude, λ, using

$$\tan I = 2 \tan \lambda. \qquad (5)$$

This equation, together with the observed inclination of the paleomagnetic field, gives the paleolatitude of the observation site. Consequently, the angular distance of the site from the paleomagnetic pole (which, at the time the magnetization was acquired, coincided with the spin axis if the GAD approximation was correct) is known. This information, together with the horizontal direction in which the paleofield is pointing (i.e., the paleomagnetic declination), gives the position of the paleomagnetic pole. This information is critical in investigations of past motions of the continents.

Observations of magnetic directions in lava flows in 1904 by David, in 1906 by Brunhes, and especially by Matuyama in 1929, suggested that the earth's magnetic field reversed polarity. When the paleomagnetic pole is in the Northern Hemisphere, as it is now and as shown in Fig. 6, the field is said to exhibit normal polarity; when it is in the Southern Hemisphere, it is said to have reverse polarity. However, it was known that some rocks can acquire a TRM antiparallel to the external magnetic field (referred to as self-reversal), and so geomagnetic field reversals were not widely accepted until the early 1960s. It is now known that such self-reversals are rare and that

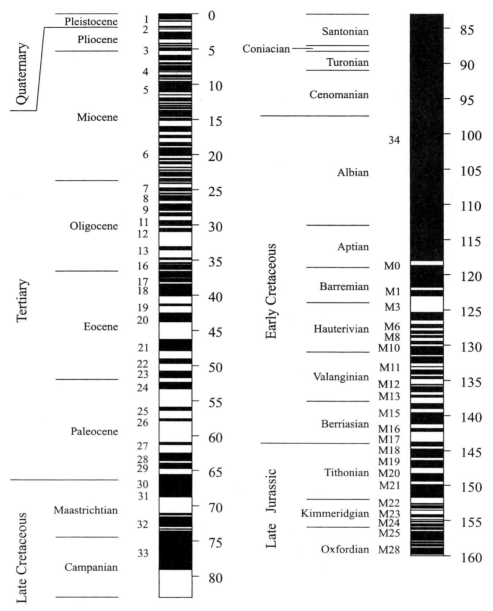

FIGURE 7 Geomagnetic polarity time scale with magnetic anomaly numbers for the past 160 million years. Geologic time-scale names are given on the left side of the reversal chronology, and ages in millions of years before the present on the right side. A normal-polarity interval is black and a reverse-polarity interval is white.

field reversals have occurred. Indeed, from tens of thousands of measurements of rocks from all over the globe, and from inverting detailed magnetic field measurements made over oceanic crust to infer the magnetic directions in that crust, it is now known that hundreds of magnetic field reverslas have occurred in the earth's past. A record of magnetic reverslas as a function of time, referred to as the magnetic reversal chronology, is shown in Fig. 7 for the past 160 million years.

Analysis of the data for Fig. 7 indicates that the interval of time between geomagnetic reversals appears to be random, and that the average length of a polarity interval is a function of time. Thus it seems that reversals are the consequence of a nonstationary stochastic process. This is quite unlike the sun, which has a general dipole field that undergoes periodic reversals with the 11-year sunspot cycle. Perhaps the most dramatic feature in Fig. 7 is the interval between 118 and 83 Ma, during which the paleofield exhibited only normal polarity. This interval is referred to as the Cretaceous superchron and is too long to be part of the typical reversal process. Thus its existence is evidence that the earth's dynamo operates in at least two different regimes, a nonreversing regime between 118 and 83 Ma, and a reversing regime characterized by all times other than superchrons. There is one other well-documented superchron in the geologic record, the Kiaman reverse superchron, which occured between approximately 316 and 262 Ma. The reversal chronology is best established for the past 160 million years, and it deteriorates farther back in time. The reason for this is that with very old rocks there is insufficient information to ensure that all the reversals have been observed and to correlate them globally and place them in their correct temporal order. Consequently, we essentially have no reliable chronology for the first 4 billion years of the earth's history, even though the rocks record many reversals of the field during that time. The most recent reversal appears to have occurred 780,000 years ago. More recent reversals have been suggested, but most geomagnetists now regard the evidence as showing that these events were just excursions of the magnetic field, times during which there were large departures of the paleomagnetic pole from the geographic poles. A true reversal requires that the paleomagnetic pole, as determined from measurements made at many well-distributed sampling sites, changed from a position close to one geographic pole to a position close to the other.

The duration of a field reversal is not well known. However, based on studies of the most recent reversals, the change in polarity appears to occur over about 10^3–10^4 years. It appears that the intensity of the dipole field does not vanish during a reversal, but that it typically decreases to a value approximately 25% of its usual value. In order to undertake a spherical harmonic analysis of the field, it is necessary to have a large number of widely distributed observations taken at effectively the same time. Because of the (geologically) short time over which a reversal occurs, it is effectively impossible to get enough well-distributed paleomagnetic observations representing the same field structure to obtain a spherical harmonic description of the field at any point in time during a reversal, let alone obtain a spherical harmonic description throughout a reversal. Consequently, the magnitude and configuration of the field during any reversal are poorly known, and this will probably remain the case for some time to come.

In addition to magnetic field reversals and excursions, a wide spectrum of secular variation is recorded in igneous and sedimentary rocks. For example, absolute paleointensity estimates can be obtained from a few igneous rocks and from archeological objects. They show that the mean dipole moment for the past 10,000 years was about

8.75×10^{22} A m^2. It was about 30% higher 2000 years ago, 30% lower 6500 years ago, and is close to 8×10^{22} A m^2 today. Other data show large variations in directions and intensity throughout the Brunhes epoch, the interval of time since the most recent reversal at about 78 ka. Nevertheless the time-averaged field for the Brunhes is remarkably well approximated by a GAD field.

The existence of magnetic reversals has been useful in a wide range of geologic studies. Because the reversal chronology for the past 160 million years is now known so well, it is often used to date rocks. An earlier version of the reversal chronology, extending only from about 5 Ma to present, played a prominent role in establishing the plate tectonics model, the currently preferred model for global tectonics. In this model the upper 100 km or so of the earth, the lithosphere, is divided into a dozen major (plus several smaller) blocks, or plates, that move relative to one another. As two plates separate to allow magma (molten rock) to upwell and cool to form new igneous rocks, the magnetic field reversals are recorded as stripes parallel to the spreading ridge; in essence, the lithosphere is recording the earth's magnetic field in a way similar to a magnetic tape recorder. By using the present reversal chronology one can estimate the separation, or spreading, rate of plates anywhere there are magnetic stripes. Because such stripes exist throughout most of the earth's oceanic basins, the past positions of continents and their relative rates of movement can be determined. In particular, the reversal chronology can be used to show that every present oceanic basin formed within the past 200 million years.

VI. THE ORIGIN OF THE EARTH'S MAGNETIC FIELD: DYNAMO THEORY

Any viable theory for the origin of the earth's magnetic field must explain its current magnitude, its structure, and changes in time, including magnetic field reversals. Of the two known possibilities, remanent magnetization or electric currents, only electrical currents seem able to provide a satisfactory explanation for the earth's magnetic field and its secular variation. This is consistent with solid-state physics, in that the ordering of electron spin moments in materials, required to produce a remanent magnetization, breaks down at high temperatures. For example, at the earth's surface, no remanent magnetization can exist above 770°C in iron, above 580°C in magnetite, or above 675°C in hematite, the latter two being common magnetic iron oxide minerals. The temperature at the core–mantle boundary remains somewhat uncertain, but it is close to 4000°C and the temperature increases from there to the earth's center. Several mechanisms for generating electric currents in the earth's deep interior have been suggested, including thermoelectrically driven currents and currents associated with various chemical reactions. However, with the exception of currents generated by a dynamo, all appear to be far too small to generate the earth's magnetic field.

In 1919 Joseph Larmor suggested that the earth's magnetic field might be produced by an internal dynamo. Some, probably weak, initial magnetic field is required to start the process. The source of the initial field is not known, but it may have originated either from a source external to the earth or from an internal source, such as thermoelectrically driven electric currents. In any case, all dynamo models involve the conversion of mechanical energy to magnetic field energy via Lenz's law—that is, through magnetic induction. This concept is illustrated in the disk dynamo model shown in Fig. 8, in which the rotation of an electrically conducting disk in the presence of a weak external field produces currents in the disk. These currents are picked up by a brush that is connected to a wire wound around a conducting rod in such a way as to amplify the initial magnetic field. The electric circuit is completed through a brush that is connected to the conducting rod. Once the dynamo is started, the initial magnetic field can be removed and one has a so-called self-sustaining dynamo. Like all successful dynamos, the disk dynamo requires an initial magnetic field, a continual supply of mechanical energy, and rotation.

Of course, the analogy of a disk dynamo to a planetary dynamo breaks down because planets do not have brushes, wires, or rods. Instead one needs to look for the source of mechanical energy in a large region in the planet that is electrically conducting and can exhibit differential motion. The logical place for a dynamo to operate in the case

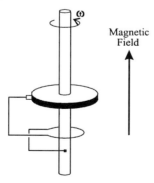

FIGURE 8 The disk dynamo. A torque is applied to rotate a conducting disk at angular speed ω in a magnetic field aligned along the axis of the disk. An electric current, induced in the rotating disk, flows outward to the edge of the disk, where it is tapped by a brush attached to a wire. The wire is wound back around the axis of the disk in such a way as to reinforce the initial field.

of the earth is in its iron-rich outer molten core. Although it may be possible for the source of mechanical energy to be associated with wave motion within the outer core, the simplest and most widely accepted source is convection. This convection is driven by cooling at the top of the core (thermal buoyancy) and by freezing at the bottom (at the boundary of the inner core and the outer core). During this freezing, elements that are less dense than iron (such as sulfur or oxygen) are released, producing an upward flux of material (chemical buoyancy). Although calculations suggest that chemical buoyancy is presently the more important source of buoyancy driving convection in the earth's outer core, this could not always have been the case since the inner core has grown over time as the earth has cooled.

Modern dynamo theory required the development of magnetohydrodynamics (MHD). In this subject the equations of electricity and magnetism are combined with those of fluid mechanics to study the behavior of magnetic fields in fluids that are in motion. The electric displacement current, introduced by Maxwell, is set to zero in MHD theory, and this can be well justified in the case of the earth. One of the early pioneers of MHD theory was Cowling, who carried out calculations (published in 1935) that he interpreted as showing that dynamos were not possible in planets or stars (i.e., an antidynamo theorem). In fact, what Cowling showed was that a magnetic field symmetric about any axis could not be sustained against ohmic dissipation. This resulted in the first of many conditions now recognized as being necessary for dynamo action; in this case the condition is that some asymmetry in the magnetic field is required. It also led many talented mathematicians on a search for a more general antidynamo theorem, a search that has led to the discovery of other conditions necessary for dynamo operation. By 1970, Childress and G. Roberts had shown that there is no general antidynamo theorem. In the meantime, Elsasser, Bullard, and Parker were separately developing the first theoretical dynamos, in which they specified the fluid motions (these are known as kinematic dynamos) rather than the far more complicated situation of allowing the fluid motions to evolve as part of the model.

Important insight into dynamo theory can be gained from the magnetic induction equation,

$$\frac{\partial \mathbf{B}}{\partial t} = k\nabla^2 \mathbf{B} + \nabla \times (\mathbf{v} \times \mathbf{B}), \tag{6}$$

where $\mathbf{B}$ is the magnetic (induction) field, t is time, $\mathbf{v}$ is velocity, and k is the magnetic diffusivity ($=1/\sigma\mu_0$, where σ is the electrical conductivity and μ_0 is the free-air permeability). If $\mathbf{v}$ is zero, then Eq. (6) reduces to a (vector) diffusion equation, from which the free-decay time of the magnetic field can be calculated. Using an estimate of the core's electrical conductivity of 6×10^5 S m^{-1} (probably known to within a half-order of magnitude), it can be shown that the field will decrease to $1/e$ (i.e., 37%) of its original value in a time of about 30,000 years. Paleomagnetic results indicate that the earth has had a magnetic field for most of its 4.5 billion years of existence, so there must have been some ongoing mechanism to regenerate the field. Evidently, that mechanism must involve a nonzero velocity in Eq. (6). Alfvén, another early pioneer of MHD theory, provided insight into how this might occur by considering the extreme case in which the electrical conductivity in Eq. (6) is high enough that the first term on the right side of Eq. (6) can be neglected. In this case he showed that the magnetic field moved with the fluid, and this is now referred to as the "frozen in (magnetic) flux" limit. In this limit, magnetic field energy can be produced only in regions of velocity shear. For the case of rotational shear, Fig. 9 illustrates how mechanical energy (rotation) in the presence of a poloidal magnetic field (field with a radial component) can produce a new toroidal (no radial component) magnetic field. By itself, however, this is insufficient to produce dynamo action: although additional magnetic field energy is produced, the poloidal magnetic field is not amplified. Some mechanical motion with a radial component is required, and this is most easily accomplished if there is fluid convection. Indeed, the above description calling on rotational shear and convection is essentially the type of kinematic dynamo advocated by Parker nearly a half-century ago. It should be pointed out that no toroidal magnetic field has been observed at the earth's surface, a consequence of the fact that the mantle's electrical conductivity is too low. Thus, the magnitude of the toroidal magnetic field in the earth's core can only be estimated through a specific dynamo model.

In realistic dynamo models, Eq. (6) must be solved simultaneously with the Navier-Stokes equation (and other equations from fluid mechanics) using appropriate boundary conditions and appropriate parameters for the earth.

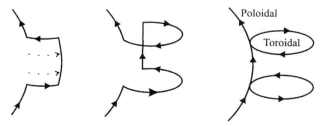

FIGURE 9 Interaction between the velocity and the magnetic field is shown at three successive times moving from left to right. The velocity field is shown only on the left, by dotted lines. After one complete circuit, two new toroidal magnetic field loops of opposite sign have been produced.

During the latter half of the twentieth century several classes of dynamo models appeared, each with many specific dynamo models. These classes included kinematic dynamo models, turbulent dynamo models, weak-field hydrodynamic dynamo models, and strong-field hydrodynamic dynamo models. Turbulent dynamo models use a statistical approach to describe how fluctuations in a turbulent core velocity field can be correlated with fluctuations in the magnetic field to produce a large-scale mean magnetic field. In other words, turbulent dynamo models show how small features in a turbulently convecting outer-core fluid in the presence of a weak magnetic field can lead to a much stronger large-scale magnetic field. If the forces driving convection (and therefore the fluid motions) are derived as part of the dynamo process rather than prescribed, then the models are referred to as hydrodynamic rather than kinematic. The term "weak field" is used to mean that the magnetic Lorentz force ($\mathbf{J} \times \mathbf{B}$, where $\mathbf{J}$ is the electric current density) has a negligible effect on the fluid motions, and "strong field" is used otherwise. It is now widely accepted that strong-field hydrodynamic dynamo models are required to explain the earth's magnetic field, but it is not clear to what extent the outer core is turbulent. The outer core would certainly be turbulent if there were no magnetic field; the presence of a magnetic field can smooth out small-scale velocity features. Calculations suggest that the core is probably turbulent even when a magnetic field is present, but this conclusion depends to some extent on the configuration and magnitude of the magnetic field, including its toroidal component, which can only be obtained from a complete theory. The first strong-field three-dimensional hydrodynamic dynamo model developed for the earth was published in 1995 by Glatzmaier and P. Roberts, the final calculations requiring about 2000 h on a supercomputer. There are now more than 10 such models produced by several scientific teams. In spite of the power of modern computers, all such models still require simplifying assumptions, including the use of parameters that are known to be inappropriate for the earth. For example, the Ekman number, an important dimensionless quantity obtained by taking the ratio of the viscous force to the Coriolis force, is many orders of magnitude too large in the models. Fundamentally, the problem is that there are several nonlinear partial differential equations describing the electrical and mechanical behavior of a large almost spherical body of rapidly spinning liquid that is highly conductive and has a low viscosity, and these equations need to be solved simultaneously. This type of problem is notoriously difficult both theoretically and numerically. Therefore, as yet we have no complete dynamo model for the earth, or for any other body in our solar system. Despite the problems in producing a reliable specific model, there seems no reason to doubt the consensus view that the earth's magnetic field is generated by an internal dynamo.

SEE ALSO THE FOLLOWING ARTICLES

AURORA • EARTH SCIENCES, HISTORY OF • EARTH'S MANTLE (GEOPHYSICS) • FERROMAGNETISM • MAGNETIC FIELDS IN ASTROPHYSICS • MAGNETIC MATERIALS • PLATE TECTONICS • SEISMOLOGY, OBSERVATIONAL • SOLAR PHYSICS • SOLAR SYSTEM, MAGNETIC AND ELECTRIC FIELDS

BIBLIOGRAPHY

Dunlop, D. J., and Özdemir, O. (1997). "Rock Magnetism: Fundamentals and Frontiers," Cambridge University Press, Cambridge, UK.

Fearn, D. (1998). "Hydromagnetic flow in planetary cores," *Rep. Prog. Phys.* **61,** 175–235.

Jacobs, J. A. (1994). "Reversals of the Earth's Magnetic Field," Cambridge University Press, Cambridge, UK.

McElhinny, M. W., and McFadden, P. L. (2000). "Paleomagnetism: Continents and Oceans," Academic Press, San Diego, CA.

Merrill, R. T., McElhinny, M. W., and McFadden, P. L. (1996). "The Magnetic Field of the Earth: Paleomagnetism, the Core, and the Deep Mantle," Academic Press, San Diego, CA.

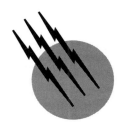

Geomorphology

Victor R. Baker
University of Arizona

GLOSSARY

Erosion General group of processes whereby the rocky materials of a planetary crust are loosened, dissolved, worn away, and removed from one place to another by natural agencies.

Fault Zone of rock fracture along which displacement has occurred. A surface without displacement is called a joint.

Graben Elongate, depressed crustal block between two faults.

Landform Distinct, physical form of a planet's surface with a characteristic shape produced by tectonic or geomorphic processes. A distinct association of landforms seen in a single view is a landscape.

Magma Mobile rock material of a planetary interior that can be intruded or extruded to form igneous rocks. Extrusive magma may be explosive, or it may form a melt called lava.

Mass wasting Dislodgment and downslope transport of regolith materials by gravity without the action of another transporting medium. The transport part of mass wasting is mass movement.

Palimpsest Landscape in which the landforms produced by modern processes are super-imposed on a relict landscape that bears the imprint of ancient processes.

Plate tectonics Global model for large-scale deformation of the earth's surface involving the movement of large, thick plates of continental and oceanic material, each of which is rafted along by a viscous underlayer in the earth's mantle. Deformation is concentrated at spreading centers on the seafloor and at convergent boundaries where plates interact with one another.

Pleistocene The epoch of the Quaternary period in geological time that follows the Pliocene of the Tertiary period and precedes the Holocene (the last 10,000 years). Popularly known as the Ice Age, the Pleistocene began approximately 1.6 million years ago and was characterized by major glacial advances and retreats.

Regolith The blanket of unconsolidated rock material that nearly everywhere mantles the land surface, overlying the intact bedrock.

GEOMORPHOLOGY, the scientific study of landforms, derives from a long tradition of trying to make sense out of the landscapes on which human beings lived and which they discovered by exploration. During the nineteenth century, the subject commanded considerable intellectual excitement associated with the great exploratory land surveys of developing regions such as the western United States. With the discovery of new and unusual landforms and landscapes, geomorphology came to be synonymous with studies involving the description, classification, development, and history of those land-surface features. The scope of geomorphology has broadened to encompass the discovery of new landscapes on the seafloor and other planets. Modern geomorphology is the scientific study of planetary surfaces.

I. GEOLOGICAL BASIS OF GEOMORPHOLOGY

Landforms and landscapes on a planetary surface derive from the complex interaction of processes both internal and external to the planet. The internal processes are driven by the heat of the planet. These endogenetic processes include crustal deformation and volcanism. The land-forms produced by endogenetic processes are modified by external forces such as weathering, mass wasting, erosion, transportation, and deposition. These exogenetic processes lead to the wearing down and general lowering of the planetary surface, that is, degradation.

A. Materials

Landscapes and landforms develop as functions of the intensity of process operation versus the resistance of the surficial materials. Material resistances depend on factors such as chemical composition, texture, fabric, induration, and hardness. An important concept is that of material strength, defined as the limiting forces that the material can withstand without failure. Strength is a threshold phenomenon. Above some critical force level a geological material changes from one state (intact rock) to another (failed rock).

The strength of geomorphic materials depends on numerous factors. For example, a relatively small block of granite may have very high strength, but on the large scale, granite masses are broken by continuous cracks of various kinds (Fig. 1). These cracks allow easy infiltration of water that increases pore pressure, mineral precipitation, and freezing, all of which can cause failure.

Volcanic rocks comprise one of the most common lithologies on the surfaces of the terrestrial planets (Mercury, Venus, the Moon, Earth, and Mars). On Earth, the ocean basins are predominantly underlain by basalt erupted at rift zones bounding midocean ridges, such as that along the center of the Atlantic Ocean. These submarine lavas

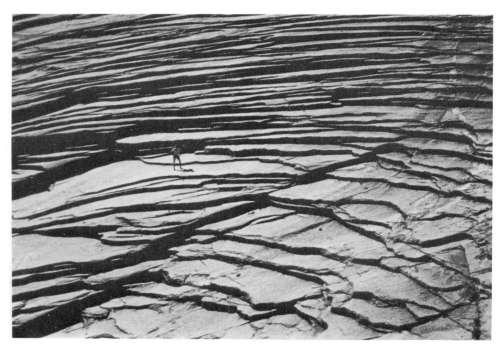

FIGURE 1   Sheet jointing in granite, Sierra Nevada, California. [Photograph by W. C. Bradley.]

generally spill out rather passively in thick pastey flows that resemble a series of mounded pillows. On the Moon, 30% of the near side (visible from Earth) consists of dark, smooth volcanic plains (maria). Samples of mare basalts are similar to very fluid terrestrial basalts, except for a lack of evidence for significant volatiles and weathering.

Earth's continents are underlain by rocks of predominantly granitic composition, richer in silica and alumina than the oceanic basalts. Much of the surficial rock material of the continents consists of sediments derived from this continental crust.

B. Structure

Geologic structure includes the broad classes of rock features that comprise fundamental elements of planetary crusts. Faults, folds, joints, domes, and grabens are examples of geologic structures. In some cases the structures are much older than the topographic features formed on them and have dictated the details of degradation by the juxtaposition of rocks of varying resistance to erosional processes. Figure 2 shows an example in which resistant rocks stand in relief and less resistant rocks have been lowered.

In areas of active (or recently active) crustal movement, the landscape may reflect primary structures. Volcanic landforms are excellent examples of constructional geomorphic features generated by endogenic factors. The types of terrain produced by volcanism vary with magma composition, pressure, and temperature characteristics. Acidic magmas erupted at low temperature are the least fluid, while basic magmas, such as basalts, are highly fluid, especially at high temperatures. Fluidity helps determine the shape of volcanoes. Acidic magmas of low fluidity produce steep-sided volcanoes, such as Fujiyama in Japan and Mount Rainier, Mount Hood, and Mount Shasta in the northwestern United States. Basalts of high fluidity produce shield volcanoes, such as Kilauea and Mauna Loa in Hawaii. Effusion rates for lava eruptions determine the lengths of individual lava flows. A factor of critical importance is the quantity of volatile material, mainly water and carbon dioxide, that is incorporated into an erupting lava. High quantities of dissolved gas in viscous magmas result in explosive conditions. A spectacular example occurred on May 18, 1980, when Mount St. Helens in Washington erupted.

C. Landscape Evolution

At the end of the nineteenth century, geomorphology achieved an important theoretical synthesis through the work of William Morris Davis. Davis conceived a marvelous deductive scheme for understanding landscape development as the action of exogenetic processes on basic geologic materials and structures to produce a progressive evolution of landscape stages through time. Unfortunately,

FIGURE 2 Parallel ridges developed on steeply dipping sedimentary rocks because of their differential resistance to erosion, Macdonnel Range, Northern Territory, Australia.

this theoretical framework was somewhat abused by people who employed it solely for landscape description and classification. By the middle twentieth century, evolutionary geomorphology fell from favor among earth scientists, who focused their efforts on detailed studies of various geomorphic processes.

II. GEOMORPHIC PROCESSES

Process geomorphologists employ field, laboratory, and analytical techniques to study processes presently active on the landscape. The work relies heavily on other disciplines, including pedology (the study of soils), soil mechanics, hydrology, geochemistry, remote sensing, hydraulics, statistics, geophysics, civil engineering, and geology. To organize the complexities of process interactions, most geomorphologists use systems analysis. The landscape is idealized a series of elements linked by flows of mass and energy. Process studies measure the inputs, outputs, transfers, and transformations that characterize these systems. Although systems analysis does not constitute a true theory for geomorphology, it does serve the purpose of organizing process studies into a framework that allows modeling and prediction, especially when data are fed into digital computers.

A progression of geomorphic processes degrades a planet's surface and carries away the products of that degradation. This progression begins with the reduction of geological materials to debris that can be transported by gravity (mass wasting), flowing water (fluvial processes), wind (eolian processes), ice (glacial processes), and waves or currents (coastal processes). The prelude to this transport is weathering.

A. Weathering and Soils

Weathering is the transformation of intact rock into unconsolidated debris by physical forces (physical weathering) or the decomposition of rock by chemical changes (chemical weathering). Physical weathering includes processes, such as freezing, plant growth, moisture changes, and crystal growth, that exceed the strength of rocks and rupture them. Chemical weathering involves the action of weathering agents, mainly water (H_2O), carbon dioxide (CO_2), and oxygen (O_2), on rock-forming minerals. The reactions of these agents with minerals produce weathering products.

An example of weathering reactions that remove materials is carbonation:

$$H_2O + CO_2 \rightleftharpoons \underset{\text{carbonic acid}}{H_2CO_3}$$

$$H_2CO_3 \rightleftharpoons H^+ + HCO_3^-$$

$$\underset{\text{calcite}}{CaCO_3} + H^+ \rightleftharpoons Ca^{++} + HCO_3^-.$$

These reactions show, in a simplified manner, that carbon dioxide dissolves in water to produce an acid, which then dissolves calcite, a common rock-forming mineral. The calcite can then be removed in a solution of calcium ions (Ca^{2+}) and bicarbonate ions (HCO_3^-). If these weathering products are transferred to a lower horizon, they can precipitate calcite (the reactions reverse), resulting in a hardened incrustation of the regolith at that horizon. Table I is a classification of such incrustations. Because these zones may be resistant to subsequent erosion, they are named duricrusts.

Subsurface weathering produces zones in which rock resistance is greatly reduced. Figure 3 shows an example

TABLE I **Classification of Duricrusts**

Type	Description	Typical minerals	Climate precipitation
Silcrete	Skin silcrete: angular, pitted quartz shards forming skins over country rock Massive silcrete: rounded grains and microcrystalline quartz in thick profiles	Quartz (90–95%)	Transitional–humid (>1500 mm/year) to semiarid (<500 mm/year)
Laterite	Upper ferruginous zone that overlies a pisolitic layer; thick mottled and pallid zones may extend tens of meters	Hematite, kaolinite	Humid (>1250 mm/year)
Alcrete	Variety of laterite rich in alumina	Gibbsite, boehmite	Humid (>1500 mm/year)
Ferricrete	Ferruginous layer lacking the thick mottled and pallid zones of the full laterite profile; may be detrital	Hematite, geothite, gibbsite	Humid (>1250 mm/year)
Calcrete	Calcite forms a plugged horizon that induces carbonate buildup cementing the regolith	Calcite (>50%), some silica	Semiarid (<450 mm/year)
Gypcrete	Massive crust with prismatic structure overlying gypsum-cemented regolith	Gypsum (50–95%)	Arid (<175 mm/year)
Salcrete	Crust of NaCl especially common near former lakes	Rock salt	Hyperarid (<25 mm/year)

FIGURE 3 Granite boulders formed as surrounding weathered granite (grus) was removed by lowering the land surface, exposing corestones of intact rock. Devil's Marbles area of central Australia.

that occurs in granite. Water, migrating into the granite along fractures, facilitates disintegration of the rock. However, the disintegration is concentrated in fracture zones, leaving corestones of intact rock between the fractures. If the disintegrated rock is removed during a phase of degradation, the corestones are left on the land surface as monolithic boulders.

Whereas weathering involves the whole range of rock and regolith alteration processes, soil formation is a specific type of weathering in which the products are organized into a profile of soil horizons. This profile usually consists of a surface layer of mineral matter plus organic material (the A horizon), an underlying layer of clay accumulation or coatings of oxides that provide dark colors (the B horizon), and a lowermost horizon of weathering or mineral accumulation (the C horizon). Figure 4 illustrates a soil profile developed on river sediments in central Texas.

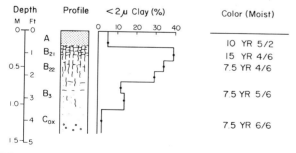

FIGURE 4 Soil profile developed on late Pleistocene river sediments of the Colorado River in east-central Texas. The profile has a large precentage of clay-sized material in its B horizon. Note also that the B horizon is very red. (Color designations are the standard Munsell color notation.)

Soils are considered to be systems in which the soil properties are determined by soil-forming factors, the most important of which are climate, organisms, topography, parent material, and time.

B. Mass Movement and Hillslopes

Mass movement involves the downslope movement of geologic materials under the influence of gravity. The materials may move in free falls, topples, sliding, and flowage. Sliding and flowage are usually facilitated by water or ice in the moving mass.

Hillslopes are fundamental elements of landscapes. They are the predominant zones of subaerial denudation, in which rock or soil is loosened by weathering and then moved downslope. On some hillslopes, the transport processes are so efficient that debris is removed more quickly than it can be generated by further weathering. Such hillslopes are characterized by a faceted appearance, in which an upper free face contributes debris to a lower slope. Figure 5 shows cones of debris, called talus, that pile up to a critical angle of repose at the base of a slope. Another extreme occurs where slopes are covered by a thick soil cover, where debris piles up more rapidly than it can be removed. Such slopes lack free faces and are usually dominated by processes of soil creep in which material slowly flows downslope.

Landslides constitute a class of mass movement in which sliding is the predominant mode of movement. Sliding occurs where conditions exceed a critical factor of safety N,

$$N = \frac{\text{resisting forces}}{\text{driving forces}}.$$

FIGURE 5 Debris cones composed of talus derived from frost shattering of sedimentary rocks exposed on a slope-free face, Bow Lake, Canadian Rockies, Alberta.

Resisting forces develop along a potential sliding surface, and driving forces result from factors that increase forces parallel to that surface. Water is a critical factor in causing landslides because it reduces the resisting forces by its pressure effects along the slide surface. If sliding movement becomes rapid enough, the moving mass may become a flow. Figure 6 shows an immense debris flow that occurred in a talus slope on the planet Mars.

C. Fluvial Processes

Fluvial dissection of a landscape creates valleys and their included channelways, which are systematically connected as drainage networks. Drainage networks display many shapes and patterns that are useful in analyzing the fluvial systems and the terrains that they dissect.

Drainage networks are organized into drainage basins. The drainage network in a basin conveys water and sediment according to the controls of climate, soils, geology, relief, and vegetation. One measure of a network's efficiency is the drainage density, defined as the summation of channel lengths per unit area. Figure 7 illustrates an area of extremely high drainage density.

Rivers display a remarkable variety of channel patterns related to numerous factors. Experimental work has done much to increase our understanding of channel patterns. Pattern adjustments, measured as sinuosity variation, are closely related to the type, size, and amount of sediment load. They are also related to bank resistance and to the discharge characteristics of the stream. Many of the morphological dependencies of river patterns can be summarized in the following expressions:

$$Q_w \propto \frac{w, d, \lambda}{S}$$

and

$$Q_s \propto \frac{w, \lambda, S}{d, P}.$$

These relationships are expressed by a large number of empirical equations treating the important independent variables Q_w, a measure of mean annual water flow, and Q_s, a measure of the type of sediment load (ratio of bedload to total load). The dependent variables are the channel width w and depth d, the slope of the river channel S, the sinuosity P (ratio of channel length to valley length), and the meander wavelength λ (spacing of two successive bends in a meandering river).

Meandering is the most common river pattern. Meandering rivers develop alternating bends with a regular spacing along the valley trend. Such rivers tend to have relatively narrow, deep channels and stable banks. The system adjusts to varying discharges by vertical accretion on its floodplain and/or by lateral migration of its channel. A vast complex of floodplain depositional features is associated with such rivers, as typified by the Mississippi River.

FIGURE 6 Debris flow (right center) that occurred on a talus slope (center) developed below a furrowed plateau (top). Note that the slopes are faceted: they have an upper free face and a lower debris slope. Small impact craters on the talus and debris flow show that both are very ancient. The picture was obtained by the Viking spacecraft and shows a scene 30-km across.

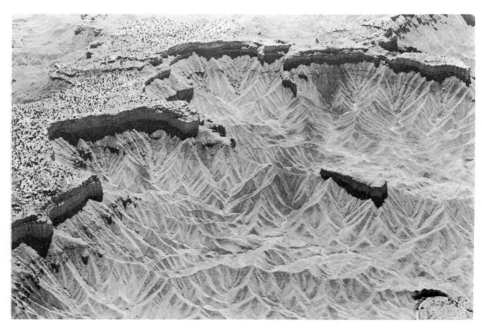

FIGURE 7 Badlands developed on Mancos Shale in south-central Utah.

Braided rivers have channels that are divided into multiple thalwegs by alluvial islands. Braided rivers tend to have steeper gradients, more variable discharges, coarser sediment loads, and lower sinuosity than meandering streams. Their channels are relatively wide and shallow. Braided patterns are developed depositionally within a channel in which the flow obstructions consist of sand and gravel deposited by the river. Midchannel bars are emplaced because of local flow incompetence. The resulting braid channels, formed by splitting the flow, are more competent than the original channel for conveying the load downstream. Another way of describing braiding is that it is caused by channel widening that increases the boundary resistance of rivers with noncohesive banks. To maintain enough velocity for sediment transport in a wide, shallow cross section, the channel must divide and form relatively narrow and deep secondary channels through incision.

Distributary patterns occur where fluvial systems are spreading water and sediment across depositional basins. Two varieties are fans and deltas. Fans develop in piedmont areas under the influence of tectonic and climatic controls. Arid region alluvial fans are constructed by infrequent depositional events that include debris flows and water flows. Typical arid-region fans occur in the southwestern United States (Fig. 8).

D. Eolian Processes

Wind is a predominant agent of landscape sculpture only in environments with minimal vegetation and with appropriate materials for eolian transport and erosion. The most spectacular examples occur in the great sand seas, or ergs, that occur in northern Africa, Arabia, central Asia, and Australia. Nearly all desert sand occurs in such settings. The sand concentrates into various kinds of dunes, which reflect wind directions, sand supply, and stabilization by vegetation or moisture. Barchan dunes (Fig. 9) are common in areas of reduced sand supply and uniform wind direction.

Much of the sand in dune areas is propelled by grain-to-grain collisions, or saltation. When wind velocities achieve a threshold level, sand grains are entrained and lifted from the bed. Numerous moving grains comprise a saltation carpet of grains moving near the ground surface. The grains have trajectories that carry them back to the bed at a low incidence angle. As these impact the bed, new particles are thrown up into the saltation carpet by their impacts. Saltating particles of quartz are highly abrasive and can erode cobbles and boulders into sculptured forms called ventifacts. Areas of pronounced erosion and relatively weak materials can yield great streamlined hills, or yardangs, that are shaped by a minimization of flow resistance.

E. Glacial Processes

A glacier is a mass of ice characterized by the following features: (1) an accumulation area where mass is gained (usually by the addition of snow), (2) an ablation area where there is a net loss of mass (usually by melting), and

FIGURE 8 Madera Canyon in the Santa Rita Mountains of southern Arizona. The fan is not active in the ero-sional/depositional environment of the modern climate. Instead, it is being dissected by numerous modern gullies, visible in the foreground. The dark lines cutting across the left edge of the fan are fault scarps of Pleistocene age. [Photograph by Peter Kresan.]

FIGURE 9 Barchan and transverse dunes immediately west of the Moenkopi Plateau in north-central Arizona.

(3) movement. For glaciers in temperate regions, movement is achieved when ice thicknesses exceed about 30 m. Surface movements can be as high as tens of meters per day, but most glaciers move more slowly.

Glaciers cover 11% of the earth's land surface. Over 50% of the land is covered (and uncovered) by snow each year; and sea ice covers 12% of the world's oceans. The importance of glaciers in geomorphology is made significant by the fact that during the recent geological past, about 18,000 years ago (the late Pleistocene), glaciers covered about twice as much land area as they do today.

Glaciers erode the landscape by abrasion and by plucking action. The details of erosion and flow rates depend on the complexities of water physics at the base of glaciers. Water under pressure contributes to decreased resistance to slippage at the base of the ice, allowing glaciers to slip along their bed. Glaciers also generate zones of high pressure upstream of obstacles on their beds. The zones of pressure concentration can lead to melting, flow of water around the obstacle, and refreezing in zones of lower pressure. Refreezing can cause the dislodgment of rock which is then incorporated into the flowing ice. The most spectacular forms resulting from glacier erosion are large troughs, such as those containing the lakes of the Alps (Fig. 10).

Glaciers generally erode where the ice flow is downward, toward the bed. Deposition may occur when the flow is away from the bed, as generally occurs in the ablation zone. Because of its very high viscosity, ice can convey immense particles of debris for eventual deposition. The largest known example is a block moved by a Pleistocene glacier into northern Germany, measuring 4 km by 2 km by 120 m.

The most common direct deposit of glacial ice consists of poorly sorted, massive debris of a composition that indicates transport from upglacier source areas. Such material, called till, may comprise thin mantles of ground areas or mounded deposits, called moraines, that delimit former glacial positions. Figure 11 shows a small moraine that recently formed at the terminus of a glacier in southwestern New Zealand.

Where meltwater plays a significant role in modifying glacial debris, the result is stratified deposits called eskers, which form by deposition in subglacial tunnels. Kame terraces form between melting ice and valley walls. Such landforms of stratified drift were produced in extensive areas of the northern United States when the last great Pleistocene glaciers stagnated and melted, about 12,000 years ago. Beyond the glacial margins, meltwater transported sediment across great depositional surfaces called outwash plains.

F. Karst Processes

Karst is a terrain, underlain by limestone or other soluble rock, in which solution processes generate the topography.

FIGURE 10 The Walen See, a glacially carved lake basin in the Alps of north-eastern Switzerland. The glacial trough containing this lake was probably enlarged from a stream valley cut into the Helvetic calcareous nappes, the massive rocks along the skyline.

FIGURE 11 Terminus of Fox Glacier in the Southern Alps, New Zealand. Note the outwash with berg ice emerging from a large ice tunnel at left. A small ice-cored moraine has formed at right. [Photographed in December 1979 by V. R. Baker.]

It may be characterized by bizarre surficial solution forms, such as those in Fig. 12. Regionally, there may be closed depressions, subterranean drainage, and caves. The main controls on karst landscapes are the rock solubility, the regional structure, cover materials (regolith) overlying the soluble rocks, climate, and base-level effects created by regional degradation. In limestone solution, the carbonation reaction discussed earlier is of primary importance.

FIGURE 12 Small-scale karst features developed on the Tindall limestone near Katherine, Northern Territory, Australia.

FIGURE 13 Shingle beach at Charleston, South Island, New Zealand.

G. Coastal Processes

Coasts are the interfaces between large bodies of water and the land. The most important coasts surround the oceans, and the predominant processes are associated with waves and currents. As deep-water ocean waves approach a coast, the waves experience a transition in which oscillatory motion changes to forward motion. Within the breaker zone, this transition produces a longshore current that transports water and sediment parallel to the coastline. The highest reach of the waves on the coastline produces the berm of a beach, when sufficient sediment is available. Figure 13 shows a berm of cobbles that have been abraded to a shingle-like shape by long, continued wave action.

Coastlines in tropical seas may be dominated by organic activity. The mangrove plant traps sediment on tropical beaches, but even more important is the building of immense reefs by coral, which precipitates calcium carbonate from seawater. Carbonate can also precipitate in the intertidal zone of tropical beaches as agitation drives carbon dioxide out of the water, decreasing its ability to hold dissolved carbonate. Figure 14 shows beachrock formed in this way on a coral island near the Great Barrier Reef of Australia.

III. CLIMATIC GEOMORPHOLOGY

Most of the exogenetic processes just described are profoundly influenced by climate. The unifying concept of climatic geomorphology developed in Europe as a synthesis of relief-forming processes. In systematic

FIGURE 14 Beachrock and coral sand at Green Island, near the Great Barrier Reef, Australia.

studies, climatic geomorphologists define morphogenetic regions in which landscapes differ mainly as a function of climatic controls. In historical studies, climatic geomorphologists interpret the palimpsests that formed when climate changed during the Tertiary and Pleistocene. This climatogenetic geomorphology tries to recognize the superimpositions of younger climatically controlled landform features on older ones. Relict landforms in a given region may bear the imprints of past climates.

Climatic geomorphologists, such as Germany's Julius Büdel, demonstrated that little of the extant relief in temperate areas is the product of modern relief-forming processes. Rather, much of it is inherited from past morphoclimatic controls, including periglacial, tropical, and arid morphogenesis.

A. Periglacial Zone

The term periglacial has come to mean the complex of cold-climate processes and landforms, including but not limited to those near active glaciers. A key feature is frost action, especially the freezing and thawing of ground. A related but not necessarily coincident phenomenon is permafrost (perennially frozen ground). Permafrost covers 20–25% of the earth's land surface, manifesting itself on the landscape when large quantities of ground ice are present. This ice may form wedges that penetrate vertically into the regolith, growing with seasonal meltwater flow into tension cracks. Polygonal patterns characterize the ground surface.

Where ice-rich permafrost is degraded by geomorphic, vegetational, or climatic change, it forms a complex landscape known as thermokarst. Depressions form where zones of ground ice are removed by melting. In extreme cases, such as near Yakutsk in eastern Siberia, large valleys form by the coalescence of thermokarst depressions.

During the coldest periods of the Pleistocene, the periglacial zone covered 40% of the earth's land surface. Hillslopes were mantled with frost-shattered rubble that moved downslope during seasonal freezing and thawing. Patterned ground (polygons and stripes) developed as the frost rubble was further sorted by seasonal changes. Even huge streams of rubble and ice-cored rock glaciers formed in areas of especially high debris production. The relicts of this periglacial activity characterize many of the modern humid-temperate zones, such as Pennsylvania, Wisconsin, England, and Poland.

B. Tropical Zone

The most important quality of tropical morphogenesis is the development of low-relief plains. These developed extensively in the relatively stable areas of the southern hemisphere during the Tertiary. Deep weathering in these regions formed a regolith that was subsequently stripped to reveal a planar surface. Multiple phases of weathering produced multiple surfaces. Such planation surfaces can also be recognized as relict forms in the landscapes of central Europe.

The tropics include zones that are continuously humid and zones that are seasonally humid. The latter regions, called savanna, have a reduced vegetative cover because of long dry seasons. Such areas may be transitional into arid zones. Erosion rates can be very high in savanna environments because of the lack of vegetative cover.

C. Arid Zone

Arid regions are characterized by lack of vegetation to protect the land surface from the intense action of fluvial, eolian, and mass wasting processes. An irony of the arid zone is that, although there is little rainfall, it can erode more effectively than in other zones. During rare arid-region rainstorms, water acts on landscape that is unprotected by vegetation. The lack of vegetation also facilitates wind erosion in sandy deserts.

Many of the world's deserts have been more humid in the past. For example, during the Pleistocene, conditions favored the formation of lakes in the Great Basin area of Utah and Nevada. Streams flowed through other areas now mantled by active sand dunes.

IV. TECTONIC GEOMORPHOLOGY

Tectonic geomorphology involves the interactions among landforms, landscapes, and tectonics. Tectonics, the branch of geology dealing with regional structures or deformational features, occupies a central role in the earth sciences. The discipline has achieved great importance through the unifying role of the plate tectonic model in explaining the large-scale surface features of our planet.

Major advances in tectonic geomorphology have been made in the last decade, mainly because of increased ability to evaluate the time factor in landscape development. Thus, through the use of geochemical means of dating and computerized models of landform change, it is now possible to evaluate differential rates of uplift or subsidence. It is also possible to determine the magnitude and frequency of displacements along faults. Figure 15 illustrates a fault that displaces Pleistocene sediments, indicating relatively recent tectonic activity.

V. PLANETARY GEOMORPHOLOGY

The study of relief forms on other planets is a natural extension of the science of landforms on our own planet. Much

FIGURE 15 Pleistocene fault separating lake silt (right) from glacial outwash (left). Faulting occurred after deposition of the outwash (14,000–17,000 years ago) and before deposition of Holocene loess (at the top of the section). Exposure is at Rakaia Gorge, South Island, New Zealand.

of modern planetary geomorphology is merely a reawakening of interest in some classic geomorphic themes, including the evolutionary sequence of landscapes, the preservation of ancient landforms, the role of cataclysmic events, the variable rates of process application, and the historical extension of geomorphic knowledge to better understand the history of Earth and the solar system. Certain extraterrestrial landforms can serve to enhance understanding of landscapes on Earth. Moreover, geomorphology has a new role in contributing to the understanding of complex planetary landscapes that result from various processes operating over disparate timescales. Many of the important processes are cataclysmic, involving the application of immense power for short times, whereas others are long-acting but weak. Understanding very ancient relict landscapes produced by these processes can elucidate significant aspects of planetary evolution throughout the solar system. The global perspective, the expanded timescales, and the interdisciplinary relevance of extraterrestrial studies provide a much-needed stimulus to the science of landforms.

VI. APPLIED GEOMORPHOLOGY

Geomorphologists analyze the landscape, which is of immense importance to humankind. An irony of the modern world is that as humans have increasingly shaped the earth's surface to their needs, they have become increasingly vulnerable to geomorphic processes. The complex transportation, communication, and economic systems of the industrial society are threatened by erosion, flooding, landsliding, sinkhole collapse, subsidence, volcanism, and earthquakes. Geomorphology plays a key role in the study of these processes by providing a broader perspective in time and space than that afforded by standard engineering practice.

Figure 16 illustrates the effects of a geomorphic hazard. In early October 1983, tropical moisture produced spectacular flooding in the desert environment of Tucson, Arizona. The deeply entrenched streams greatly enlarged their channels by lateral erosion. Many buildings toppled into the floodwater as banks collapsed. The damage could have been averted if geomorphic studies had been done before the flood to delineate hazardous areas.

VII. FUTURE TRENDS

Geomorphology is entering a new era of discovery and scientific excitement centered on expanded scales of concern in both time and space. The catalysts for this development include technological advances in global remote-sensing systems, mathematical modeling, and

FIGURE 16 Damage to apartments (center) caused when the prominent meander bend of Rillito Creek at the bottom center migrated west (left) into the property. The prominent point bar at bottom center developed as the meander migrated during flooding in October 1983, Tucson, Arizona.

geochemical techniques for the dating of geomorphic surfaces and processes.

The exploration of other planets is providing additional growth for geomorphology. Landscapes can be studied on a multiplanet basis, comparing the genetic influences of various gravitational, atmospheric, tectonic, and temporal controls. The need to explain these newly discovered features will provide the stimulus for new theories of landscape origin and change.

As population pressures on Earth force more growth in polar, tropical, and desert terrains, the peculiar processes of those regions will have to be understood. Geo-

morphology, because of its unique concern with the land and the processes that shape it, will play a central role in the analysis of transformed land. Even more pressing may be the concern with global change induced by large-scale alteration of the atmospheric, oceanic, or land systems that sustain our present environment. Through its concern with the land on which people exist, geomorphology is the one science capable of integrating our understanding of humankind's home. The maintenance of that home, the hazards posed to it, indeed its continued habitability, all depend on the quality of that understanding.

SEE ALSO THE FOLLOWING ARTICLES

Coastal Geology • Geologic Time • Geomorphology and Reclamation of Disturbed Lands • Glacial Geology and Geomorphology • Lunar Rocks • Planetary Geology • Plate Tectonics • Rock Mechanics • Soil Physics • Volcanology

BIBLIOGRAPHY

Bridges, E. M. (1990). "World Geomorphology," Cambridge University Press, Cambridge, UK.

Daniels, R. B. (1992). "Soil Geomorphology," Wiley, New York.

Goudie, A. S. (ed.) (1990). "Techniques for Desert Reclamation," Wiley, New York.

Hupp, C. R., and Osterkamp, W. R. (eds.) (1995). "Biogeomorphology, Terrestrial and Freshwater Systems," Elsevier, New York.

Laflen, J. M. (2000). "Soil Erosion and Dryland Farming," CRC Press, Boca Raton, FL.

McGregor, D. F. M. (ed.) (1995). "Geomorphology and Land Management in a Changing Environment," Wiley, New York.

Panizza, M. (1996). "Environmental Geomorphology," Elsevier, New York.

Schmidt, K-H., and Ergenzinger, P. (1994). "Dynamics and Geomorphology of Mountain Rivers," Springer-Verlag, Berlin.

Skujins, J. (1991). "Semiarid Lands and Deserts: Soil Resource and Reclamation," Dekker, New York.

Slaymaker, O. (ed.) (1995). "Steepland Geomorphology," Wiley, New York.

Thorne, C. R. (ed.) (1998). "Applied Fluvial Geomorphology for River Engineering and Management," Wiley, New York.

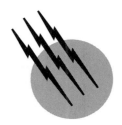

Geomorphology and Reclamation of Disturbed Lands

David R. Montgomery

University of Washington

I. Landscape Disturbance
II. Types and Extent of Disturbance
III. Rehabilitation of Disturbed Lands

GLOSSARY

Accelerated erosion Erosion at a rate exceeding the geologic norm resulting from environmental disturbance.

Channelization Artificial alteration of channel planform and cross-sectional geometry by straightening and/or enlarging channel capacity.

Disturbance regime The patterns, rates, and intensity of landscape disturbances caused by geomorphologic processes.

Disturbed lands Landscape in which the dominant geomorphologic processes have been altered and/or accelerated.

Dynamic equilibrium Condition in which system outputs are adjusted to system inputs such that system characteristics remain stable even though the system exhibits dynamic behavior.

Entrenchment Lowering of a channel floor sufficient to cause abandonment of an active floodplain.

Landscape rehabilitation Alteration of surface topography and soils to more closely resemble conditions existing prior to landscape disturbance.

Landscape restoration Reestablishment of surface topography and soils to recreate conditions existing prior to landscape disturbance.

Mine spoil Broken overburden that has been removed in surface mining to gain access to underlying mineral deposit, or material that has been processed to remove disseminated mineral deposits.

Reclamation practices Procedures used to beneficially modify surface topography and soils following disturbance by human activity.

THE SCIENCE OF GEOMORPHOLOGY concerns the characteristics of earth's surface and the processes responsible for the development and evolution of landforms. Landscapes are shaped by the erosion, transport, and deposition of soil and rock under the influence of the prevailing tectonic regime, geology, climate, and vegetation, as well as the legacy of past climates (Fig. 1). Changes in uplift rate, erosion resistance, or erosional processes lead to changes in landforms, which can in turn influence the styles, rates, and location of erosional processes.

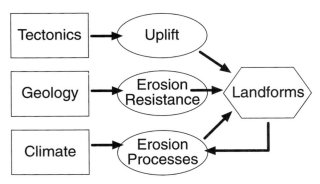

FIGURE 1 Schematic illustration of the factors governing landforms.

An understanding of geomorphologic processes is central to the management of human disturbance on landscapes and development of effective landscape rehabilitation and restoration programs.

I. LANDSCAPE DISTURBANCE

An increasing percentage of earth's surface has been subjected to direct and indirect disturbance by human activity since the end of the last ice age. Anthropogenic disturbance is so widespread that humans now may be the dominant geomorphic agent on earth. This disturbance results from numerous activities, especially mining, agriculture, forestry, construction, and grazing. Experience in designing and implementing measures to mitigate and remediate the environmental impacts of such disturbance has led to recognition that geomorphology plays a key role in the design of effective restoration programs based on natural processes appropriate for a particular landscape. Even where restoration may be impossible due to the extent or severity of environmental change, geomorphology can help design landscape rehabilitation measures aimed at recovering key functions, aspects, or characteristics of natural systems.

Landscapes are dynamic systems that respond to perturbations over a wide range of time scales, from human actions in a particular location to the uplift of new mountain ranges during slow continental collisions. Some geomorphologic processes occur relatively continuously, but many occur as discrete events, such as a flood, a landslide, or fault rupture and uplift during an earthquake. Geomorphologic processes can act as disturbances that influence ecological systems through either sustained disturbance or a discrete episode of disturbance. Specific organisms may be adapted to endure, recolonize, or capitalize on the disturbance regime of their natural environment, as defined by the frequency, magnitude, and intensity of disturbance processes. Consequently, changes in the land-

scape disturbance regime can result in dramatic ecological effects, which in turn can have significant economic consequences.

The response of a geomorphic system to disturbance can either lead to a return to predisturbance conditions or to persistent change(s) in the system. Consider, for example, the disturbance and subsequent recovery of a fluvial system adjusted to the prevailing climate and underlying geology. As with most geomorphic systems, the system may exhibit a time lag between disturbance and onset of response and a relaxation time that depends on climatic and environmental conditions (Fig. 2). Upon recovery, the system may either return to its original state or evolve to a new state during an intervening period.

Given sufficient time, the basic form of the land reflects an approximate equilibrium among uplift, soil production, and erosional processes driven by climate and topography. This is not to say that landscapes are static, but rather that a suite of dynamic processes tends to give rise over geologic time to land forms that reflect the balance between uplift and erosion. Where erosional processes exceed uplift the imbalance will tend to lower slope gradients and thereby reduce rates of erosion over geologic time. Conversely, where uplift exceeds erosion slopes will steepen until eventually erosion rates match the uplift rate. In equilibrium landscapes the interplay of uplift and erosion imparts a persistent morphology to the landscape.

Disturbed lands are those in transition from one state to another, and especially lands so altered from an original, more desireable condition as to warrant efforts to rehabilitate or restore aspects of the original system. Examples of disturbed lands include wholesale landscape disturbance by strip mining or the placement or erosion of mine spoils, accelerated erosion from vegetation clearance, urbanization or land use practices, and modifications to wetlands and river systems.

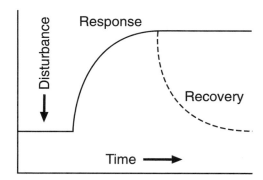

FIGURE 2 Graphic representation of the response of a geomorphic system to disturbance illustrating the lag time from disturbance to onset of system response and the relaxation time between response and recovery, either to original state or to a new system state.

TABLE I Typical Rates of Erosion for Various Types of Land Use in the United States[a]

Land use type	Sediment yield ($t\ km^{-2}\ year^{-1}$)
Forest	9
Grassland	85
Cropland	1,700
Felled forest	4,250
Active open-cast mines	17,000
Construction sites	17,000

[a] From Summerfield, M. A. (1991). "Global Geomorphology," Longman, Essex, England.

II. TYPES AND EXTENT OF DISTURBANCE

The range of representative erosion rates for various land covers and land use illustrates the dramatic acceleration of erosion rates in landscapes disturbed by human actions (Table I). In the United States excluding Alaska and Hawaii, more than half of the land mass is devoted to agricultural use. The most significant nonagricultural land use is ungrazed forest land, followed by urban areas, recreation areas, and wildlife habitats. Other types of landscape disturbance, such as surface mining, generally affect only a small percentage of the land area, but can have a substantial and destructive impact on both the disturbed area and downstream areas.

A. Disturbance by Surface Mining

Surface mining refers to techniques used to remove unconsolidated, consolidated, and weathered geologic materials at the land surface in order to expose and extract mineral resources. Surface mining is perhaps the most dramatic direct human disturbance at the landscape scale (Figs. 3 and 4) and there are a variety of types of surface mining,

FIGURE 3 Drag line method of removing overburden above a coal seam. [From Toy, T. J., and Hadley, R. F. (1987). "Geomorphology and Reclamation of Disturbed Lands," Academic Press, Orlando, FL.]

each of which results in different types and intensities of landscape disturbance (Table II). In general, the selection of mining methods is based on economics as influenced by the topography, thickness of overburden, and character of the resource.

Major commodities extracted by surface mining in the United States include sand and gravel, copper, coal, and building stone. Although the area disturbed by surface mining constitutes less than 1% of the land surface in the United States, the practice severely disturbs the impacted areas. The following discussion of landscapes disturbed by surface mining is based on studies of coal mining, which is largely responsible for the recent increase in surface mining in the western United States, but the geomorphic principles are applicable to most surface-mined lands.

A complete geomorphologic investigation for the reclamation of surface-mined lands should consider pre-mining, active-mining, and postmining time periods. During the premining period, landforms and geomorphologic processes probably represent relatively natural or only partially disturbed conditions. To the extent that the surface is undisturbed, an approximate balance may exist between geomorphologic processes and surface form and pre-mining conditions therefore can guide designs for post-mining reclamation efforts. The active-mining period is a time of maximum disturbance and disequilibrium between the surface from and geomorphic processes. Process rates during this period will be greatly altered and the disturbed system will remain far from equilibrium. Post-mining reclamation or rehabilitation practices may not be able to reconstruct the landscape as it was prior to mining, but the surface form and vegetation of hillslopes and channels should resemble the pre-mining period as closely as possible if the goal is for geomorphologic processes to approximate the pre-mining system.

Placement and erosion of mine spoils can have substantial and persistent impacts on downstream river systems. Rivers draining the Sierra Nevada in California are still impacted by debris from late 19th century hydraulic mining, and the sustained disturbance has destroyed once extensive and commercially important salmon runs. In addition, mobilization of metals from mine spoils can result in elevated concentration of heavy metals in downstream rivers and floodplain sediments. The destruction of the Fly River, New Guinea, in the 1990s provides a tragic contemporary example of the effects of remobilized mine spoils on environmental systems.

B. Disturbance Caused by Construction

Archeological evidence clearly reveals that human alteration of the environment through construction activities extends thousands of years into the past. Today, the

FIGURE 4 Dissected face of aboveground mine tailings in the western United States. [From Toy, T. J., and Hadley, R. F. (1987). "Geomorphology and Reclamation of Disturbed Lands," Academic Press, Orlando, FL.]

dominant forms of construction activity include (1) urbanization involving the construction of residential and commercial buildings and (2) road construction. As with many types of land disturbance, the surface area involved can occupy a relatively small proprtion of the total land area, but there is substantial data that indicate construction activities can greatly increase the rate or intensity of geomorphologic processes.

Construction of residential and commercial builings involves the preparation of the ground surface, usually involving grading and removal of the topsoil. Then structures are built, after which steps may be taken to stabilize the modified land surfaces. M. G. Wolman in 1967 reconstructed sediment yield and channel response for a Maryland watershed for periods of different land use from colonial times through urbanization in the 1950s. He found that the direct ground disturbance associated with construction activity produced erosion rates that greatly exceeded those under any prior land use. Many jurisdictions now require extensive use of erosion control measures to prevent off-site impacts from construction activity, although the effectiveness of many such techniques depends

critically upon the care with which they are implemented. In addition, the duration of construction-related impacts depends on how long the site is actively disturbed, as well as the effort invested in postconstruction stabilization and reclamation.

Roads are generally less pervious than soils, and concentration of runoff and material eroded from unpaved road surfaces can cause increased delivery of fine sediment to stream channels. In steep terrain, road runoff also can trigger landsliding if concentrated onto slide-prone slopes (Fig. 5). Sediment production from unpaved road surfaces increases with the intensity of use and the size of the vehicles composing the traffic. Road surface rehabilitation techniques range from measures to disperse the runoff from the road surface, and thereby prevent drainage concentration, to ripping up and replanting the road surface (Fig. 6).

C. Disturbance Caused by Recreational Use

The demand for lands dedicated to recreational use increased greatly throughout the 20th century, and the

TABLE II Surface Mining Techniques

Technique	Uses	Comments
Open pit	Quarries for limestone, sandstone, marble, granite, sand and gravel, iron, copper, infrequently for coal in western United States	Insufficient spoil to refill pit, excavation remains as depression after reclamation
Dredging	Placer mining of gold, sand, and gravel from stream beds	Potential for flooding and downstream sedimentation; gravelly materials difficult to reclaim
Hydraulic mining	Extensively used for gold mining in past, limited use now	Downstream sedimentation often a major problem; gravelly materials difficult to reclaim
Strip mining and mountain-top removal	Used extensively in hilly terrain, such as Appalachia, for coal extraction, more limited use in West and Midwest	Severe erosion potential; permanent change of topography

FIGURE 5 Photograph of road-related landslides in the Washington Cascades.

FIGURE 7 View of erosion associated with a motorcycle trail in the Mojave Desert, California. [From Synder, C. T., Frickel, D. G., Hadley, R. F., and Miller, R. F. (1976). "Effects of Off-Road Vehicle Use on the Hydrology and Landscape of Arid Environments in Central and Southern California," U.S. Geological Survey Water-Resources Investigation 76–99.]

impacts on ecosystems casued by the pressures of recreational uses have been serious and varied depending on the type of activity. The principal kinds of impacts associated with recreational activities are related to hillslope processes, soils, and vegetation cover. Creation and erosion of trails by pack animals, hiking, and off-road vehicles can damage or destroy the vegetation cover and can

also compact the soil. The increase in the area of exposed bare soil and decrease in infiltration rates caused by compaction together increase storm runoff, surface erosion, and sediment yield from disturbed sites (Fig. 7).

R. B. Bryan in 1977 studied trail erosion and concluded that topography is an important factor. Where trails follow the fall line of the hillslope, damage is severe from fluvial erosion regardless of the slope gradient. In contrast, where trails follow the hillslope contour, damage from erosion is minimal. In addition to the effects of topography, damage on trails is closely related to soil properties and vegetation. Vulnerabilty to erosion is highest where soils are saturated and the vegetation cover is damaged.

D. Disturbance Caused by Grazing

Lands primarily used for grazing occupy about 40% of earth's land surface and occur in many different climatic zones. However, the vast majority of rangelands are located in arid or semiarid zones. The problems associated with grazing in arid and semiarid lands, such as deterioration of vegetation cover by overgrazing, poor areal distribution of herds, and soil compaction by grazing animals, are exacerbated by hydrologic characteristics of low rainfall, high evapotranspiration potential, and low water yield or runoff.

Many hydrologists, geomorphologists, and conservationists attribue the cycle of gully erosion that began in the American Southwest about 1880 to land abuse by overgrazing. Public lands were grazed without regulation from the mid-1800s until the enactment of the Taylor Grazing Act of 1934. It is likely that the large numbers of livestock on these lands led to the general depletion of vegetation cover, trampling and compaction of soils, and severe erosion, especially on valley floors. These impacts

FIGURE 6 Rill development on the face of a road cut in Kauai, Hawaii. [From Toy, T. J., and Hadley, R. F. (1987). "Geomorphology and Reclamation of Disturbed Lands," Academic Press, Orlando, FL.]

contributed to increased runoff and the erosion and entrenchment of channels into the valley bottoms. In many areas, such disturbance has been so severe that the historic floodplain now forms a terrace surface elevated well above normal high flows.

Analysis of suspended-sediment discharge records at the Grand Canyon gauging station on the Colorado River in the heart of the western United States lends support to the theory that grazing strongly affects erosion. Hadley (1974) examined sediment records for the periods 1926–1941 and 1942–1960 and found that the suspended-sediment loads for the latter period were only 50% of the earlier period. Studies of livestock use on Colorado plateau rangelands reveal that livestock numbers have been greatly reduced since about 1940 in the Colorado River basin. Such studies provide compelling evidence for a strong causal relationship between grazing and erosion. In addition, long-term effects of intense grazing activity are expressed in the pervasive soil loss that has plagued much of the Mediterranean region since ancient times.

E. Drainage Modification: Channelization and Wetlands

River channels and wetlands have been extensively modified throughout the United States, resulting in pervasive degradation of riverine ecosystems. Early navigation and commerce was primarily through rivers, which motivated extensive efforts to clear and improve rivers for navigation as settlers moved west across the country. In addition, floodplains, marshes, and wetlands in many areas were diked or filled to provide arable land and convert relatively flat valley bottoms to more desirable land uses.

Toward these ends, channelization of rivers was promoted and aggressively pursued for flood control purposes throughout much of the 20th century. Typical channelization programs involved straightening and enlarging a river channel, which resulted in a steeper and greatly simplified channel. Flood control programs also promoted construction of extensive levee systems to isolate floodplains from overbank flooding (Fig. 8). Recently, widespread recognition of the adverse ecological impacts, as well as the potential futility of such programs, has led to efforts in channel restoration to reintroduce elements of natural channel features and dynamics into river systems.

River rehabilitation efforts include moving levees back some distance from the river edge to allow the channel to migrate within a specified corridor and building or placing of instream structures, such as log jams, in forested areas. In many environments, river restoration efforts would require more extensive modification of floodplain conditions and processes, as the interaction of the channel and floodplain was a primary process that shaped and maintained channel features, and therefore aquatic habitats.

FIGURE 8 Photograph of the channelized Duwamish River and impounding levees near Seattle, Washington.

Wetland and river channel morphology and conditions (Fig. 8) are controlled by a wide variety of influences, and the importance of both the local and watershed context and disturbance history varies greatly. Evaluation of relevant geomorphologic processes and current conditions against historic and potential conditions typically requires a thorough and often site-specific assessment.

III. REHABILITATION OF DISTURBED LANDS

In approaching the problem of landscape rehabilitation the investigator is often confronted with complex alterations of the natural landscape. Many of the most severe environmental impacts in disturbed lands are the result of geomorphologic processes operating at accelerated rates. Conceptually, geomorphic analysis of landscape disturbance is best founded upon an examination of changes in driving and resisting forces that influence the dominant processes for a particular site or landscape. Such analyses may require assessment of an area broader than the particular area of interest or conern.

The design of a landscape rehabilitation program involves three phases: (1) analysis of existing and probable or possible environmental conditions, (2) formulation of a rehabilitation or restoration program, and (3) implementation of the program. Common elements of geomorphically based landscape rehabilitation efforts include restoring natural processes, designing for a new "dynamically stable" state, and crafting designs that can accommodate change, rather than attemping to prevent changes due to natural landscape dynamics. Fundamental principles of geomorphology can be applied directly in each element of a reclamation program, and an important aspect of developing a landscape restoration program is tailoring the program to the landscape rather that trying to apply generic solutions.

Due to the dynamic nature of landscape processes, the rehabilitation or restoration of disturbed lands should be conducted in the context of an adaptive management program. Such a program involves (1) assessing existing impacts and the design of a rehabilitation/restoration program, (2) implementing the program, (3) monitoring to assess program effectiveness, and (4) modifying the program if necessary to meet its objectives. Geomorphologic principles can help guide rehabilitation of disturbed lands by providing understanding and context within which to interpret dominant landscape processes, historic legacies, and the range of potential future conditions, as well as insight into how to best achieve project objectives.

SEE ALSO THE FOLLOWING ARTICLES

GEOMORPHOLOGY • HYDRODYNAMICS OF SEDIMENTARY BASINS • MINING ENGINEERING • SOIL AND GROUNDWATER POLLUTION • SOIL PHYSICS

BIBLIOGRAPHY

Antevs, E. (1952). "Arroyo cutting and filling," *J. Geol.* **60,** 375–385.

Bryan, R. B. (1977). "The influence of soil properties on degradation of mountain hiking trails of Grovelsjon," *Geografiska Ann.* **1–2A,** 49–65.

Dale, T., and Carter, V. G. (1955). "Topsoil and Civilization," University of Oklahoma Press, Norman, OK.

Graf, W. L. (1977). "The rate law in fluvial geomorphology," *Am. J. Sci.* **277,** 178–191.

Hadley, R. F. (1974). "Sediment yield and land use in Southwest United States." *In* "Proceedings of the IAHS Paris Symposium," pp. 96–98, International Association of Hydrological Sciences, Dorking, England.

Hooke, R. L. (2000). "On the history of humans as a geomorphic agent," *Geology* **28,** 843–846.

Leopold, L. B., Wolman, M. G., and Miller, J. P. (1964). "Fluvial Processes in Geomorphology," Freeman, San Francisco.

McQuaid-Cook, J. (1978). "Effects of hikers and horses on mountain trails," *J. Environ. Manage.* **6,** 209–212.

Montgomery, D. R. (1994). "Road surface drainage, channel initiation, and slope stability," *Water Resour. Res.* **30,** 1925–1932.

Montgomery, D. R., and Buffington, J. M. (1998). "Channel processes, classification, and response potential." *In* "River Ecology and Management" (Naiman, R. J., and Bilby, R. E., eds.), pp. 13–42, Springer-Verlag, New York.

Reid, L. M., and Dunne, T. (1984). "Sediment production from forest road surfaces," *Water Resour. Res.* **20,** 1753–1761.

Riley, A. L. (1998). "Restoring Streams in Cities: A Guide for Planners, Policymakers, and Citizens," Island Press, Washington, DC.

Toy, T. J., and Hadley, R. F. (1987). "Geomorphology and Reclamation of Disturbed Lands," Academic Press, Orlando, FL.

Trimble, S. W., and Mendel, A. C. (1995). "The cow as a geomorphic agent—A critical review," *Geomorphology* **13,** 233–253.

Willgoose, G. R., and Riley, S. J. (1998). "An assessment of the long-term erosional stability of a proposed mine rehabilitation," *Earth Surface Processes Landforms* **23,** 237–259.

Wolman, M. G. (1967). "A cycle of sedimentation and erosion in urban river channels," *Geografiska Ann.* **49A,** 385–395.

Geostatistics

Clayton V. Deutsch
University of Alberta, Edmonton

I. Essential Concepts
II. Quantification of Spatial Variability
III. Spatial Regression or Kriging
IV. Simulation
V. Special Topics
VI. Applications and Examples

GLOSSARY

Declustering Technique to assign relative weights to different data values based on their redundancy with nearby data. Closely spaced data get less weight.

Kriging After the name of D. G. Krige, this term refers to the procedure of constructing the best linear unbiased estimate of a value at a point or of an average over a volume.

Realization Nonunique grid of simulated values. A set of realizations is used as a measure of uncertainty in the variable being studied.

Simulation Procedure of adding correlated error by Monte Carlo to create a value that reflects the full variability.

Variogram Basic tool of the theory used to characterize the spatial continuity of the variable.

GEOSTATICS commonly refers to the theory of regionalized variables and the related techniques that are used to predict variables such as rock properties at unsampled locations. Matheron formalized this theory in the early 1960s (Matheron, 1971). Geostatistics was not developed as a theory in search of practical problems. On the contrary, development was driven by engineers and geologists faced with real problems. They were searching for a consistent set of numerical tools that would help them address real problems such as ore reserve estimation, reservoir performance forecasting, and environmental site characterization. Reasons for seeking such comprehensive technology included (1) an increasing number of data to deal with, (2) a greater diversity of available data at different scales and levels of precision, (3) a need to address problems with consistent and reproducible methods, (4) a belief that improved numerical models should be possible by exploiting computational and mathematical developments in related scientific disciplines, and (5) a belief that more responsible decisions would be made with improved numerical models. These reasons explain the continued expansion of the theory and practice of geostatistics. Problems in mining, such as unbiased estimation of recoverable reserves, initially drove the development of geostatistics. Problems in petroleum, such as realistic heterogeneity models for unbiased flow predictions, were dominant from the mid-1980s through the late 1990s.

Encyclopedia of Physical Science and Technology, Third Edition, Volume 6

Geostatistics is applied extensively in these two areas and is increasingly applied to problems of spatial modeling and uncertainty in environmental studies, hydrogeology, and agriculture.

I. ESSENTIAL CONCEPTS

Geostatistics is concerned with constructing high-resolution three-dimensional models of categorical variables such as rock type or facies and continuous variables such as mineral grade, porosity, or contaminant concentration. It is necessary to have *hard* truth measurements at some volumetric scale. All other data types, including remotely sensed data, are called *soft* data and must be calibrated to the hard data. It is neither possible nor optimal to construct models at the resolution of the hard data. Models are generated at some intermediate geologic modeling scale, and then scaled to an even coarser resolution for process performance. A common goal of geostatistics is the creation of detailed numerical three-dimensional geologic models that simultaneously account for a wide range of relevant data of varying degrees of resolution, quality, and certainty. Much of geostatistics relates to data calibration and reconciling data types at different scales.

At any instance in geologic time, there is a single true distribution of variables over each study area. This true distribution is the result of a complex succession of physical, chemical, and biological processes. Although some of these processes may be understood quite well, we do not completely understand all of the processes and their interactions, and could never have access to the boundary conditions in sufficient detail to provide the unique true distribution of properties. We can only hope to create numerical models that mimic the physically significant features. Uncertainty exists because of our lack of knowledge. Geostatistical techniques allow alternative realizations to be generated. These realizations are often combined in a histogram as a model of uncertainty.

Conventional mapping algorithms were devised to create smooth maps to reveal large-scale geologic trends; they are low-pass filters that remove high-frequency property variations. The goal of such conventional mapping algorithms, including splines and inverse distance estimation, is *not* to show the full variability of the variable being mapped. For many practical problems, however, this variability has a large affect on the predicted response. Geostatistical simulation techniques, conversely, are devised with the goal of introducing the full variability, that is, creating maps or realizations that are neither unique nor smooth. Although the small-scale variability of these realizations may mask large-scale trends, geostatistical simulation is more appropriate for most engineering applications.

There are often insufficient data to provide reliable statistics. For this reason, data from analogous, more densely sampled study areas are used to help infer spatial statistics that are impossible to calculate from the available data. There are general features of certain geologic settings that can be transported to other study areas of similar geologic setting. Although the use of analogous data is often essential in geostatistics, it should be critically evaluated and adapted to fit any hard data from the study area.

A sequential approach is often followed for geostatistical modeling. The overall geometry and major layering or zones are defined first, perhaps deterministically. The rock types are modeled within each major layer or zone. Continuous variables are modeled within homogeneous rock types. Repeating the entire process creates multiple equally probable realizations.

A. Random Variables

The uncertainty about an unsampled value z is modeled through the probability distribution of a random variable (RV) Z. The probability distribution of Z after data conditioning is usually location-dependent; hence the notation $Z(\mathbf{u})$, with $\mathbf{u}$ being the coordinate location vector. A random function (RF) is a set of RVs defined over some field of interest, e.g., $Z(\mathbf{u})$, $\mathbf{u} \in$ study area A. Geostatistics is concerned with inference of statistics related to a random function (RF).

Inference of any statistic requires some repetitive sampling. For example, repetitive sampling of the variable $z(\mathbf{u})$ is needed to evaluate the cumulative distribution function: $F(\mathbf{u}; z) = \text{Prob}\{Z(\mathbf{u}) \leq z\}$ from experimental proportions. However, at most, one sample is available at any single location $\mathbf{u}$; therefore, the paradigm underlying statistical inference processes is to trade the unavailable replication at location $\mathbf{u}$ for replication over the sampling distribution of z samples collected at other locations within the same field.

This trade of replication corresponds to the decision of stationarity. Stationarity is a property of the RF model, not of the underlying physical spatial distribution. Thus, it cannot be checked from data. The decision to pool data into statistics across rock types is not refutable *a priori* from data; however, it can be shown inappropriate *a posteriori* if differentiation per rock type is critical to the undergoing study.

II. QUANTIFICATION OF SPATIAL VARIABILITY

A. Declustering

Data are rarely collected with the goal of statistical representivity. Wells are often drilled in areas with a greater probability of good reservoir quality. Core measurements are taken preferentially from good-quality reservoir rock. These data-collection practices should not be changed; they lead to the best economics and the greatest number of data in portions of the reservoir that contribute the greatest flow. There is a need, however, to adjust the histograms and summary statistics to be representative of the entire volume of interest.

Most contouring or mapping algorithms automatically correct this preferential clustering. Closely spaced data inform fewer grid nodes and, hence, receive lesser weight. Widely spaced data inform more grid nodes and, hence, receive greater weight. Geostatistical mapping algorithms depend on a global distribution that must be equally representative of the entire area being studied.

Declustering techniques assign each datum a weight, $w_i, i = 1, \ldots, n$, based on its closeness to surrounding data. Then the histogram and summary statistics are calculated with the declustering weights. The weights $w_i, i = 1, \ldots, n$, are between 0 and 1 and add up to 1.0. The height of each histogram bar is proportional to the cumulative weight in the interval, and summary statistics such as the mean and variance are calculated as weighted averages. The simplest approach to declustering is to base the weights on the volume of influence of each sample. Determining a global representative histogram is the first step of a geostatistical study. The next step is to quantify the spatial correlation structure.

B. Measures of Spatial Dependence

The covariance, correlation, and variogram are related measures of spatial correlation. The decision of stationarity allows inference of the stationary covariance (also called auto covariance):

$$C(\mathbf{h}) = E[Z(\mathbf{u} + \mathbf{h}) \cdot Z(\mathbf{u})] - m^2,$$

where m is the stationary mean. This is estimated from all pairs of z-data values approximately separated by vector $\mathbf{h}$. At $\mathbf{h} = 0$ the stationary covariance $C(0)$ equals the stationary variance σ^2. The standardized stationary correlogram (also called auto correlation) is defined as

$$\rho(\mathbf{h}) = C(\mathbf{h})/\sigma^2.$$

Geostatisticians have preferred another two-point measure of spatial correlation called the variogram:

$$2\gamma(\mathbf{h}) = E\{[Z(\mathbf{u} + \mathbf{h}) - Z(\mathbf{u})]^2\}$$

The variogram does not call for the mean m or the variance σ^2; however, under the decision of stationarity the covariance, correlogram, and variogram are equivalent tools for characterizing two-point correlation:

$$C(\mathbf{h}) = \sigma^2 \cdot \rho(\mathbf{h}) = \sigma^2 - \gamma(\mathbf{h})$$

This relation depends on the model decision that the mean and variance are constant and independent of location. These relations are the foundation of variogram interpretation. That is, (1) the "sill" of the variogram is the variance, which is the variogram value that corresponds to zero correlation; (2) the correlation between $Z(\mathbf{u})$ and $Z(\mathbf{u} + \mathbf{h})$ is positive when the variogram value is less than the still; and (3) the correlation between $Z(\mathbf{u})$ and $Z(\mathbf{u} + \mathbf{h})$ is negative when the variogram exceeds the sill.

C. Anisotropy

Spatial continuity depends on direction. Anisotropy in geostatistical calculations is *geometric*, that is, defined by a triaxial Cartesian system of coordinates. Three angles define orthogonal x, y, and z coordinates and then the components of the distance vectors are scaled by three range parameters to determine the scalar distance, that is,

$$h = \sqrt{\left(\frac{h_x}{a_x}\right)^2 + \left(\frac{h_y}{a_y}\right)^2 + \left(\frac{h_z}{a_z}\right)^2},$$

where h_x, h_y, and h_z are the components of a vector $\mathbf{h}$ in three-dimensional coordinate space and a_x, a_y, and a_z are scaling parameters in the principal directions. Contour lines of equal "distance" follow ellipsoids. The use of z for the random variable and a coordinate axis is made clear by context. The three x, y, and z coordinates must be aligned with the principal directions of continuity. A coordinate rotation may be required.

The directions of continuity are often known through geologic understanding. In case of ambiguity, the variogram may be calculated in a number of directions. A *variogram map* could be created by calculating the variogram for a large number of directions and distances; then, the variogram values are posted on a map where the center of the map is the lag distance of zero.

D. Variogram Modeling

The variogram is calculated and displayed in the principal directions. These experimental directional variogram points are not used directly in subsequent geostatistical steps such as kriging and simulation; a parametric variogram model is fitted to the experimental points. There are two reasons why experimental variograms must be modeled: (1) there is a need to interpolate the variogram function for $\mathbf{h}$ values where too few or no experimental data pairs are available, and (2) the variogram measure $\gamma(\mathbf{h})$ must have the mathematical property of "positive definiteness" for the corresponding covariance model—that

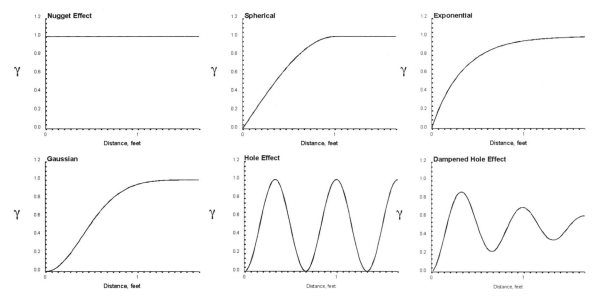

FIGURE 1 Typical variogram structures that are combined together in nested structures to fit experimental variograms. Anisotropy, that is, different directional variograms, are brought to the same distance units by geometric scaling.

is, we must be able to use the variogram and its covariance counterpart in kriging and stochastic simulation. For these reasons, geostatisticians have fitted sample variograms with specific known positive definite functions such as the spherical, exponential, Gaussian, and hole-effect variogram models (see Fig. 1).

A variogram model can be constructed as a sum of known positive-definite licit variogram functions called nested structures. Each nested structure explains a fraction of the variability. All nested structures together describe the total variability, σ^2. Interactive software is typically used to fit a variogram model to experimental points in different directions.

III. SPATIAL REGRESSION OR KRIGING

A. Point Estimation

An important application of geostatistics is to calculate estimates at unsampled locations. The basic idea is to propose a liner estimate of the residual from the mean:

$$z^*(\mathbf{u}) - m(\mathbf{u}) = \sum_{\alpha=1}^{n} \lambda_\alpha \cdot [z(\mathbf{u}_\alpha) - m(\mathbf{u}_\alpha)],$$

where $z^*(\mathbf{u})$ is an estimate made with n data, $m(\mathbf{u})$ is the mean value known at all locations, and $\lambda_\alpha, \alpha = 1, \ldots, n$, are weights that account for how close the n data are to the location being estimated and how redundant the data are with each other. The weights could be assigned inversely

proportional to the distance between the data $\mathbf{u}_\alpha$ and location being estimated, $\mathbf{u}$; however, a better procedure is to use the variogram and minimize the error variance.

B. Simple Kriging

Least-squares optimization has been used for many years. The idea, proposed by early workers in geostatistics, was to calculate the weights to be optimum in a minimum squared error sense, that is, minimize the squared difference between the true value $z(\mathbf{u})$ and the estimator $z^*(\mathbf{u})$. Of course, the true values are known only at the data locations, *not* at the locations being estimated. Therefore, as is classical in statistics, the squared error is minimized in the expected value.

The geostatistical technique known as simple kriging is a least-squares regression procedure to calculate the weights that minimize the squared error. A set of n equations must be solved to calculate the n weights:

$$\sum_{\beta=1}^{n} \lambda_\beta C(\mathbf{u}_\beta - \mathbf{u}_\alpha) = C(\mathbf{u} - \mathbf{u}_\alpha), \quad \alpha = 1, \ldots, n.$$

Recall that $C(\mathbf{h}) = \sigma^2 - \gamma(\mathbf{h})$; therefore, knowledge of the variogram model permits calculation of all needed covariance terms. The left-hand side contains all of the information related to redundancy in the data, and the right-hand side contains all of the information related to closeness of the data to the location being estimated. Kriging is the best estimator in terms of minimum error variance.

Kriging is an exact estimator; that is, the kriging estimator at a data location will be the data value. The minimized error variance or *kriging variance* can be calculated for all estimated locations:

$$\sigma_K^2(\mathbf{u}) = \sigma^2 - \sum_{\alpha=1} \lambda_\alpha \cdot C(\mathbf{u} - \mathbf{u}_\alpha),$$

where the kriging variance is the global variance, σ^2, in the presence of no local data and 0 at a data location. The kriging estimates and kriging variance can be calculated at each location and posted on maps.

C. Constrained Kriging

The basic estimator written in Section III.A requires the mean $m(\mathbf{u})$ at all locations. A number of techniques have been developed in geostatistics to relax this requirement. Ordinary kriging, for example, assumes that the mean m is constant and unknown. A constraint is added to the kriging equations to enforce the sum of the weights to equal 1, which amounts to estimating the mean at each location. Universal kriging assumes the mean follows a particular parametric shape; the parameters are estimated at each location. These constrained versions of kriging make a different decision regarding stationarity.

D. Multiple Variables

The term *kriging* is traditionally reserved for linear regression using data with the same variable as that being estimated. The term *cokriging* is reserved for linear regression that also uses data defined on different attributes. For example, the porosity value $z(\mathbf{u})$ may be estimated from a combination of porosity samples and related acoustic impedance values, $y(\mathbf{u})$. Kriging requires a model for the Z variogram. Cokriging requires a *joint* model for the matrix of variogram functions including the Z variogram, $\gamma_Z(\mathbf{h})$, the Y variogram, $\gamma_Y(\mathbf{h})$, and the cross Z–Y variogram $\gamma_{Z-Y}(\mathbf{h})$. When K different variables are considered, the covariance matrix requires K^2 covariance functions. The inference becomes demanding in terms of data and the subsequent joint variogram modeling; however, cokriging provides the minimum error-variance estimator of the variable at an unsampled location using multiple data variables.

E. Smoothing

Kriging estimates are smooth. The kriging variance is a quantitative measure of the smoothness of the kriging estimates. There is no smoothing when kriging at a data location, $\sigma_K^2 = 0$. There is complete smoothness when kriging with data far from the location being estimated; the es-

timate is equal to the mean and the kriging variance is the full variance, $\sigma_K^2 = \sigma^2$. This nonuniform smoothing of kriging is the largest shortcoming of kriging for map making. A map of kriging estimates gives an incorrect picture of variability, and calculated results such as recoverable reserves and flow properties are wrong. Simulation corrects for the smoothing of kriging.

IV. SIMULATION

A. Sequential Gaussian Simulation

The idea of simulation is to draw multiple, equally probable realizations from the random function model. These realizatios provide a *joint* measure of uncertainty. Each realization should reproduce (1) the local data at the correct scale and measured precision, (2) the global stationary histogram within statistical fluctuation, and (3) the global stationary variogram or covariance within statistical fluctuation. There is much discussion in the geostatistical literature about different random function models. The most commonly used, however, is the multivariate Gaussian model. The data are first transformed so that the global stationary histogram is Gaussian or normal. Then, all multivariate distributions of n points taken at a time are assumed to follow the mathematically congenial Gaussian distribution. There are many techniques to draw simulations from a multivariate Gaussian random function. The sequential approach gained wide popularity in the 1990s because of its simplicity and flexibility. The sequential Gaussian simulation (SGS) algorithm is as follows.

1. Transform the original Z data to a standard normal distribution (all work will be done in "normal" space). There are different techniques for this transformation. The normal score transformation whereby the normal transform y is calculated from the original variable z as $y = G^{-1}[F(z)]$, where $G(\cdot)$ is the standard normal cumulative distribution function (cdf) and $F(\cdot)$ is the cdf of the original data.

2. Go to a location $\mathbf{u}$ (chosen randomly from the set of locations that have not been simulated yet) and perform kriging to obtain a kriged estimate and the corresponding kriging variance.

3. Draw a random residual $R(\mathbf{u})$ that follows a normal distribution with mean of 0.0 and a variance of $\sigma_K^2(\mathbf{u})$. Add the kriging estimate and residual to get a simulated value. The independent residual $R(\mathbf{u})$ is drawn with classical Monte Carlo simulation.

4. The simulated value is added to the data set and used in future kriging and simulation to ensure that the

variogram between all of the simulated values is correct. A key idea of sequential simulation is to add previously simulated values to the data set.

5. Visit all locations in a random order (return to step 2). There is no theoretical requirement for a random order or path; however, practice has shown that a regular path can induce artifacts. When every grid node has been assigned, the data values and simulated values are back-transformed to real units.

Repeating the entire procedure with a different random number seeds creates multiple realizations. The procedure is straightforward; however, there are a number of implementation issues, including (1) a reasonable three-dimensional model for the mean $m(\mathbf{u})$ must be established, (2) the input statistics must be reliable, and (3) reproduction of all input statistics must be validated.

B. Alternatives to Sequential Approach

Many algorithms can be devised using the properties of the multi-Gaussian distribution to create stochastic simulations: (1) matrix approaches (LU decomposition), which are not used extensively because of size restrictions (an $N \times N$ matrix must be solved, where N could be in the millions for reservoir applications); (2) turning bands methods, where the variable is simulated on one-dimensional lines and then combined into a three-dimensional model, which is not commonly used because of artifacts; (3) spectral methods using fast Fourier transforms can be CPU-fast, but the grid size N must be a power of 2 and honoring conditioning data requires an expensive kriging step; (4) fractals, which are not used extensively because of the restrictive assumption of self-similarity, and (5) moving-average methods, which are used infrequently due to CPU requirements.

C. Indicator Simulation

The aim of the indicator formalism for categorical variables is to simulate the distribution of a categorical variable such as rock type, soil type, or facies. A sequential simulation procedure is followed, but the distribution at each step consists of estimated probabilities for each category: $p^*(k)$, $k = 1, \ldots, K$, where K is the number of categories. The probability values are estimated by first coding the data as indicator or probability values—that is, an indicator is 1 if the category is present, and 0 otherwise. The Monte Carlo simulation at each step is a discrete category. Requirements for indicator simulation include K variograms of the indicator transforms and K global proportions.

V. SPECIAL TOPICS

A. Object-Based Modeling

Object-based models are becoming popular for creating facies models in petroleum reservoirs. The three key issues to be addressed in setting up an object-based model are (1) the geologic shapes, (2) an algorithm for object placement, and (3) relevant data to constrain the resulting realizations. There is no inherent limitation to the shapes that can be modeled with object-based techniques. Equations, a raster template, or a combination of the two can specify the shapes. The geologic shapes can be modeled hierarchically—that is, one object shape can be used at large scale and then different shapes can be used for internal small-scale geologic shapes. It should be noted that object-based modeling has nothing to do with object-oriented programming in a computer sense.

The typical application of object-based modeling is the placement of abandoned sand-filled fluvial channels within a matrix of floodplain shales and fine-grained sediments. The sinuous channel shapes are modeled by a one-dimensional centerline and a variable cross section along the centerline. Levee and crevasse objects can be attached to the channels. Shale plugs, cemented concretions, shale clasts, and other non-net facies can be positioned within the channels. Clustering of the channels into channel complexes or belts can be handled by large-scale objects or as part of the object-placement algorithm.

Object-based facies modeling is applicable to many different depositional settings. The main limitation is coming up with a suitable parameterization for the geologic objects. Deltaic or deep-water lobes are one object that could be defined. Eolean sand dunes, remnant shales, and different carbonate facies could also be used.

B. Indicator Methods

The indicator approach to categorical variable simulation was mentioned earlier. The idea of indicators has also been applied to continuous variables. The key idea behind the indicator formalism is to code all of the data in a common format, that is, as *probability* values. The two main advantages of this approach are (1) simplified data integration because of the common probability coding, and (2) greater flexibility to account for different continuity of extreme values. The indicator approach for continuous data variables requires significant additional effort versus Gaussian techniques.

The aim of the indicator formalism for continuous variables is to estimate directly the distribution of uncertainty $F^*(z)$ at unsampled location $\mathbf{u}$. The cumulative distribution function is estimated at a series of threshold values:

$z_k, k = 1, \ldots, K$. The indicator coding at location $\mathbf{u}_\alpha$ for a particular threshold z_k is

$$
\begin{aligned}
i(\mathbf{u}_\alpha; z_k) &= \mathrm{Prob}[Z(\mathbf{u}_\alpha) \leq z_k] \\
&= \begin{cases} 1, & \text{if } z(\mathbf{u}_\alpha) \leq z_k, \\ 0, & \text{otherwise.} \end{cases}
\end{aligned}
$$

All hard data $z(\mathbf{u}_\alpha)$ are coded as discrete zeros and ones. Soft data can take values between zero and one. The indicator transform for a threshold less than the data value is zero, since there is no probability that the data value is less than the threshold; the indicator transform for a very high threshold is one, since the data value is certainly less than the threshold.

The cumulative distribution function at an unsampled location at threshold z_k can be estimated by kriging. This "indicator kriging" or IK requires a variogram measure of correlation corresponding to each threshold $z_k, k = 1, \ldots, K$. The IK process is repeated for all K threshold values that discretize the interval of variability of the continuous attribute Z. The distribution of uncertainty, built from assembling the K indicator kriging estimates, can be used for uncertainty assessment or simulation.

C. Simulated Annealing

The method of simulated annealing is an optimization technique that has attracted significant attention. The task of creating a three-dimensional numerical model that reproduces some data is posed as an optimization problem. An objective function measures the mismatch between the data and the numerical model. An initial random model is successively perturbed until the objective function is lowered to zero. The essential contribution of simulated annealing is a prescription for when to accept or reject a given perturbation. This acceptance probability distribution is taken from an analogy with the physical process of annealing, where a material is heated and then slowly cooled to obtain low energy.

Simulated annealing is a powerful optimization algorithm that can be used for numerical modeling; however, it is more difficult to apply than kriging-based methods because of difficulties in setting up the objective function and choosing many interrelated parameters such as the annealing schedule. Therefore, the place of simulated annealing is not for conventional problems where kriging-based simulation is adequate. Simulated annealing is applicable to difficult problems that involve (1) dynamic data, (2) large-scale soft data, (3) multiple-point statistics, (4) object placement, or (5) special continuity of extremes.

D. Change of Support

Reconciling data from different scales is a long-standing problem in geostatistics. Data from different sources, including remotely sensed data, must all be accounted for in the construction of a geostatistical reservoir model. These data are at vastly different scales, and it is wrong to ignore the scale difference when constructing a geostatistical model. Geostatistical scaling laws were devised in the 1960s and 1970s primarily in the mining industry, where the concern was mineral grades of selective mining unit (SMU) blocks of different sizes. These techniques can be extended to address problems in other areas, subject to implicit assumptions of stationarity and linear averaging.

The first important notion in volume-variance relations is the spatial or dispersion variance. The dispersion variance $D^2(a, b)$ is the variance of values of volume a in a larger volume b. In a geostatistical context, all variances are dispersion variances. A critical relationships in geostatistics is the link between the dispersion variance and the average variogram value:

$$
D^2(a, b) = \bar{\gamma}(b, b) - \bar{\gamma}(a, a).
$$

This tells us how the variability of a variable changes with the volume scale and variogram. The variability of a variable with high short-scale variability decreases quickly, since high and low values average out.

VI. APPLICATIONS AND EXAMPLES

A. Environmental

Figure 2 illustrates some of the geostatistical operations applied to characterize the spatial distribution of lead contamination over a 12,500-ft^2 area. There are five parts to Fig. 2: (1) the upper left shows the location map of the 180 samples—there is no evident clustering that would require declustering; (2) the equal-weighted histogram, at the upper right, shows the basic statistics related to the measurements—note the logarithmic scale; (3) the variogram, shown below the histogram, is of the normal scores transform of the lead data—about 40% of the variability is at very short distances and the remaining 60% of the variability is explained over 4500 ft—the black dots are the experimentally calculated points and the solid line is the fitted model; (4) a map of kriging estimates on a 100-ft^2 grid is shown at the lower left—note the smoothness of the kriging estimates; and (5) a sequential Gaussian simulation (SGS) realization is shown at the lower right—this realization reproduces the 180 sample data, the input histogram, and the variogram model. A set of realizations could be used to assess the probability that each location exceeds some critical threshold of lead concentration.

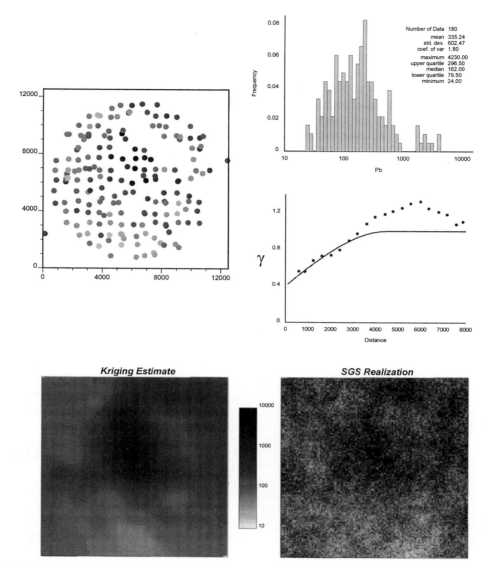

FIGURE 2 Location map (distance units in feet) of 180 samples, histogram of lead concentration, variogram of the normal scores transform (hence the *sill* value of 1.0), a map of kriging estimates on a 100-ft² grid and an SGS realization over the same domain.

B. Mining

Figure 3 illustrates an example application to a vein-type mineral deposit. The cross-sectional view at the upper left is a vertical cross section facing west; the vertical coordinate is meters below the surface. The drillhole intersections are clustered in the thickest part of the vein. The polygonal areas of influence plotted on the location map are used for declustering weights. The histogram at the upper right of the figure considers the declustering weights. The variogram is shown below the histogram. Two nested structures were used to fit this variogram. One sequential

Gaussian realization is shown at the lower left; 150 realizations were generated. The probability of exceeding 1-m thickness is plotted at the lower right. The black locations are where the vein is measured to be greater than 1 m in thickness (probability of 1), and the white locations are where the vein is measured to be less than 1 m (probability of 0).

C. Petroleum

The profile of porosity and permeability from two wells from an offshore petroleum reservoir are shown at the

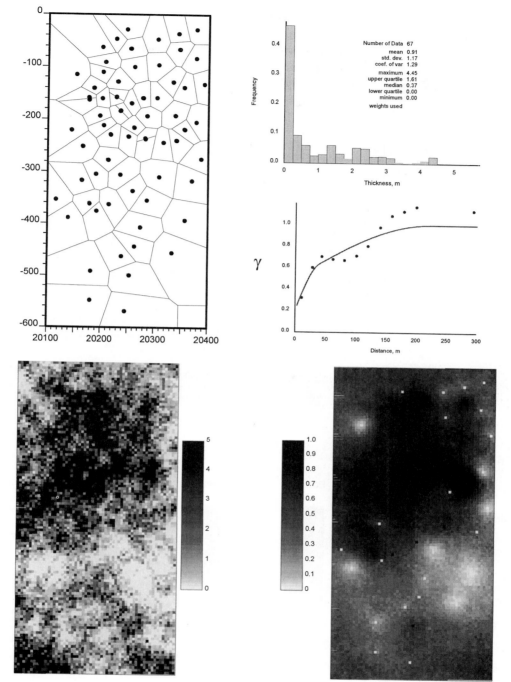

FIGURE 3 Location map (distance units in meters) of 67 drillholes with polygonal areas of influence for declustering weights, histogram of vein thickness, variogram of the normal scores transform, an SGS realization over the same domain, and the probability to exceed 1.0-m thickness calculated from 100 realizations.

bottom of Fig. 4. A porosity and permeability realization are shown at the top. Simulation of porosity and permeability were done simultaneously to reproduce the correlation between these two variables. The vertical variograms were calculated and modeled easily; however, the horizontal variograms are impossible to discern from two wells. A 50:1 horizontal-to-vertical anisotropy was considered from analog data.

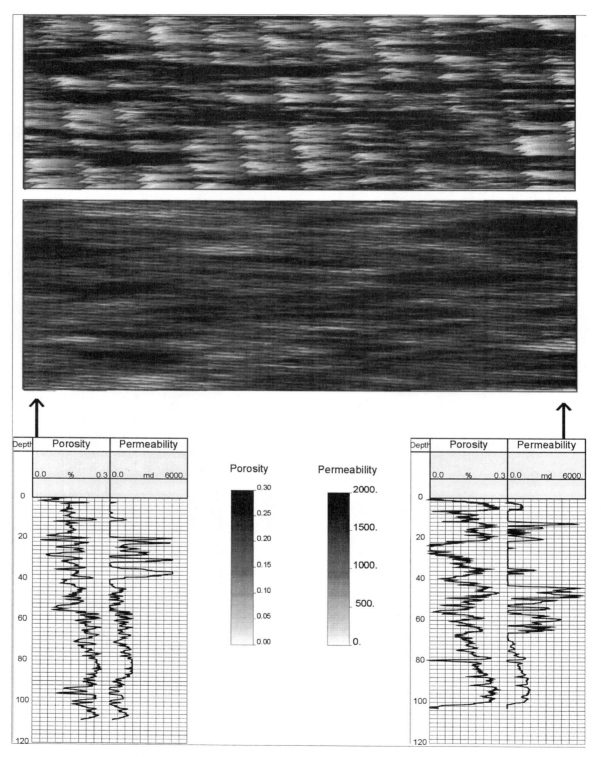

FIGURE 4 Permeability realization (top) and porosity realization (middle) constrained to two wells 600 m apart (shown at the bottom).

SEE ALSO THE FOLLOWING ARTICLES

MINING ENGINEERING • ORE PETROLOGY • PETROLEUM
GEOLOGY • STATISTICS, FOUNDATIONS

BIBLIOGRAPHY

Chiles, J. P., and Delfiner, P. (1999). "Geostatistics: Modeling Spatial
    Uncertainty" (Wiley Series in Probability and Statistics, Applied Prob-
    ability and Statistics), Wiley, New York.
Cressie, N.(1991). "Statistics for Spatial Data," Wiley, New York.

David, M. (1977). "Geostatistical Ore Reserve Estimation," Elsevier,
    Amsterdam.
Deutsch, C. V., and Journel, A. G. (1997). "GSLIB: Geostatistical Soft-
    ware Library," 2nd ed., Oxford University Press, New York.
Goovaerts, P. (1997). "Geostatistics for Natural Resources Evaluation,"
    Oxford University Press, New York.
Isaaks, E. H., and Srivastava, R. M. (1989). "An Introduction to Applied
    Geostatistics," Oxford University Press, New York.
Journel, A. G., and Huijbregts, C. (1978). "Mining Geostatistics," Aca-
    demic Press, London.
Matheron, G. (1971). "The theory of regionalized variables and its appli-
    cations," *Cahiers du CMM* **5,** Ecole Nationale Superieure des Mines
    de Paris, Paris.
Ripley, B. D. (1981). "Spatial Statistics," Wiley, New York.

Geothermal Power Stations

Lucien Y. Bronicki

ORMAT Industries, Ltd.

GLOSSARY

Binary geothermal power plant A power plant in which the geothermal fluid provides the heat required by the organic working fluid.

Direct heat use Utilization of low- and moderate-temperature geothermal resources for space and water heating for industrial processes and agricultural applications.

Energy conversion Conversion of one type of energy to another such as the heat of a geothermal resource to electricity, etc.

Geothermal combined cycle Combined use of geothermal steam and brine for power generation by using a backpressure steam turbine and organic turbines.

Geothermal energy Totally or partially renewable heat energy from deep in the earth. It originates from the earth's molten interior and the decay of radioactive materials, and is brought near the surface by deep circulation of ground water.

Geothermal heat pump (GHP) Application using the earth as a heat source for heating or as a heat sink for cooling.

Geothermal resources Sources of geothermal energy: hydrothermal; geopressured; hot, dry rock (HDR); and magma. All are suitable for heat extraction and electric power generation.

Hydrothermal resource Geothermal resource containing hot water and/or steam dropped in fractured or porous rocks at shallow to moderate depths. Categorized as vapor dominated (steam) or liquid dominated (hot water). These are the only commercially used resources at the present time.

Organic Rankine cycle (ORC) A cycle using an organic liquid as motive fluid (instead of water) in a Rankine cycle.

Renewable energy Energy source which is not exhausted by use with time. Renewable energies include direct solar energy, energy from geothermal sources, wind, hydroelectric plants, etc.

I. INTRODUCTION

A. Source of Geothermal Energy

Geothermal energy is renewable heat energy from deep in the earth. It originates from the earth's molten interior and

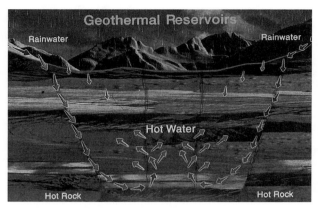

FIGURE 1 A representative geothermal reservoir. [From Nemzer, M. (2000). Geothermal Education Office, web site http://geothermal.marin.org.]

the decay of radioactive materials; heat is brought near the surface by deep circulation of groundwater and by intrusion into the earth's crust of molten magma originating from great depth (see Fig. 1). In some places this heat comes to the surface in natural streams of hot steam or water, which have been used since prehistoric times for bathing and cooking. By drilling wells this heat can be tapped from the underground reservoirs to supply pools, homes, greenhouses, and power plants.

The quantity of this heat energy is enormous; it has been estimated that over the course of 1 year, the equivalent of more than 100 million GWhr of heat energy is conducted from the earth's interior to the surface. But geothermal energy tends to be relatively diffuse, which makes it difficult to tap. If it were not for the fact that the earth itself concentrates geothermal heat in certain regions (typically regions associated with the boundaries of tectonic plates in the earth's crust; see Fig. 2), geothermal energy would be essentially useless as a heat source or a source of electricity using today's technology.

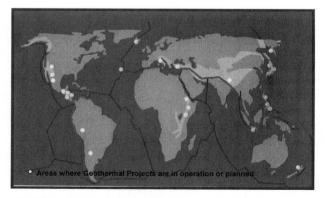

FIGURE 2 World map showing lithospheric plate boundaries. [From Nemzer, M. (2000). Geothermal Education Office, web site http://geothermal.marin.org.]

There is some ambiguity on the issue of geothermal energy being a "renewable" resource. Some geothermal sites may be developed in such a manner that the heat withdrawn equals the heat being replaced naturally, thus making the energy source renewable for a long period of time. At other sites, the resource lifetime may be limited to some decades. In any case, even if it is not technically a renewable resource, potential global geothermal resources represent such a huge amount of energy that, practically speaking, the issue is not the finite size of the resource, but availability of technologies that can tap the resource, in an economically acceptable manner.

B. Nature of the Geothermal Energy Resource

On average, the temperature of the earth increases by about 3°C for every 100 m in depth. This means that at a depth of 2 km, the temperature of the earth is about 70°C, increasing to 100°C at a depth of 3 km, and so on. However, in some places, tectonic activity allows hot or molten rock to approach the earth's surface, thus creating pockets of higher temperature resources at easily accessible depths (World Energy Council, 1994).

The extraction and practical utilization of this heat requires a carrier which will transfer the heat toward the heat-extraction system. This carrier is provided by geothermal fluids forming hot aquifers inside permeable formations. These aquifers or reservoirs are the hydrothermal fields. Hydrothermal sources are distributed widely but unevenly across the earth. High-enthalpy geothermal fields occur within well-defined belts of geologic activity, often manifested as earthquakes, recent volcanism, hot springs, geysers and fumaroles. The geothermal belts are associated with the margins of the earth's major tectonic or crustal plates and are located mainly in regions of recent volcanic activity or where a thinning of the earth's crust has taken place. One of these belts rings the entire Pacific Ocean, including Kamchatka, Japan, the Philippines, Indonesia, the western part of South America running through Argentina, Peru, and Ecuador, Central America, and western North America. An extension also penetrates across Asia into the Mediterranean area. Hot crustal material also occurs at midocean ridges (e.g., Iceland and the Azores) and interior continental rifts (e.g., the East African rift, Kenya and Ethiopia).

Low-enthalpy resources are more abundant and more widely distributed than high-enthalpy resources. They are located in many of the world's deep sedimentary basins, for example, along the Gulf Coast of the United States, western Canada, western Siberia, and areas of central and southern Europe, as well as at the fringes of high-enthalpy resources.

There are four types of geothermal resources: hydrothermal, geopressured, hot dry rock, and magma.

Although they have different physical characteristics, all forms of the resource are potentially suitable for electric power generation if sufficient heat can be obtained for economical operation.

1. Hydrothermal Resources

These are the only commercially used resources at the present time (Kestin *et al.*, 1980). They contain hot water and/or steam trapped in fractured or porous rock at shallow to moderate depths (from approximately 100 to 4500 m). Hydrothermal resources are categorized as vapor-dominated (steam) or liquid-dominated (hot water) according to the predominant fluid phase. Temperatures of hydrothermal reserves used for electricity generation range from 90°C to over 350°C, but roughly two-thirds is estimated to be in the moderate temperature range (150–200°C). The highest quality reserves contain steam with little or no entrained fluids, but only two sizeable, high-quality dry steam reserves have been located, at Larderello in Italy and The Geysers field in the United States.

Recoverable resources available for power generation far exceed the developments to date. Many countries are believed to have potential in excess of 10,000 MW$_e$, which would satisfy a considerable portion of their electricity requirements for many years (e.g., the Philippines, Indonesia, and the United States).

Important low-enthalpy hydrothermal resources are not necessarily associated with young volcanic activity. They are found in sedimentary rocks of high permeability which are isolated from relatively cooler near-surface ground water by impermeable strata. The water in sedimentary basins is heated by regional conductive heat flow. These basins (e.g., the Pannonian Basin) are commonly hundreds of kilometers in diameter at temperatures of 20–100°C. They are exploited in direct thermal uses or with heat pump technology.

2. Geopressured Resources

Geopressured geothermal resources are hot water aquifers containing dissolved methane trapped under high pressure in sedimentary formations at a depth of approximately 3–6 km. Temperatures range from 90°C to 200°C, although the reservoirs explored to date seldom exceed 150°C. The extent of geopressured reserves is not yet well known world-wide, and the only major resource area identified is in the northern Gulf of Mexico region, where large reserves are believed to cover an area of 160,000 km^2. This resource is potentially very promising because three types of energy can be extracted from the wells, thermal energy from the heated fluids, hydraulic energy from the high pressures involved, and chemical energy from burning the dissolved methane gas (World Energy Council, 1994).

3. Hot, Dry Rock Resources

These resources are accessible geologic formations that are abnormally hot but contain little or no water. The hot, dry rock (HDR) potential is 200 GW in the United States (GeothermEx, 1998) and 60 GW in Europe (Baria *et al.*, 1998). The basic concept in HDR technology is to form a geothermal reservoir by drilling deep wells (400–5000 m) into high-temperature, low-permeability rock and then forming a large heat-exchange system by hydraulic or explosive fracturing. Injection and production wells are joined to form a circulating loop through the reservoir, and water is then circulated through the fracture system (Baria *et al.*, 1998; Grassiani and Krieger, 1999).

II. CURRENT GEOTHERMAL ENERGY UTILIZATION

A. Brief History

Geothermal energy has had a long history of use in applications such as therapeutic hot baths, space and water heating, and agriculture (Nemzer, 2000). It was not until 1904 that the power of natural geothermal steam was first harnessed to produce electricity, by Prince Piero Ginori Conti at Larderello, Italy (see Fig. 3). During these early decades, there was little growth of this application due to cheap competing sources of electric power, and it was 1958 before the next large-scale geothermal power station was commissioned, at Wairakei, New Zealand. Only sources which were easy to exploit such as The Geysers in California and some liquid-dominated geothermal resources in Japan and New Zealand were developed to the commercial power stage before 1970.

Of the geothermal resource types, hydrothermal energy is the most widely applied and most cost-competitive and

First Geothermal Power Plant, 1904, Larderello, Italy

FIGURE 3 The first geothermal power plant, at Larderello, Italy (1904). [From Nemzer, M. (2000). Geothermal Education Office, web site http://geothermal.marin.org.]

the only one presently used commercially. The uses of magma, geopressured and hot, dry rock systems are still at experimental stages, although the latter two types have been technically demonstrated successfully and energy extraction has been experimentally verified.

B. Present State of the Art

For geothermal energy utilization, a number of technological solutions have been introduced. Several of these are still under development, while some are in commercial use, but still undergoing continuous improvement. The following is an overview of the technology solutions and their developmental status, thus establishing a basis for subsequent discussion.

1. Exploration and Extraction

Hydrothermal development begins with exploration to locate and confirm the existence of a reservoir with economically exploitable temperature, volume, and accessibility. The geosciences (geology, geophysics, and geochemistry) are used to locate reservoirs, characterize their conditions, and optimize the locations of wells. Drilling technology used for geothermal development derives historically from the petroleum industry. Certain critical components, such as drilling muds, were modified to work in high-temperature environments, but proved to be only marginally adequate. Materials and equipment capable of dealing not only with increasing temperatures but also with hard, fractured rock formations and saline, chemically reactive fluids were needed. As a result, a specialized part of the drilling industry devoted to geothermal development evolved.

Hydrothermal fluids may be produced from wells by artesian flow (i.e., fluid forced to the surface by ambient pressure differences), or by pumping. In the former case, the fluid may flash into two phases (steam and liquid), whereas under pumping the fluid remains in the liquid phase. The choice between these two production modes depends on the characteristics of the fluid and the design of the energy-conversion system.

Geothermal fields generally lend themselves to "staged" development, whereby a modestly sized plant can be installed at an early stage of field assessment. It may be small enough to be operated with confidence on the basis of what is known of the field. Its operation provides the opportunity for obtaining reservoir information which may lead to the installation of additional stages.

2. Direct Heat Use

The abundant low- and moderate-temperature hydrothermal fluids may be used as direct heat sources for space

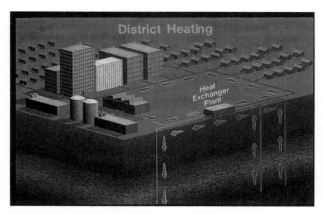

FIGURE 4 A district heating plant. [From Nemzer, M. (2000). Geothermal Education Office, web site http://geothermal. marin.org.]

and water heating, for industrial processes, and for agricultural applications. The major uses include balneology, space heating, and hot water supplies for public institutions. District-heating systems for groups of buildings are the predominant other uses (see Fig. 4).

Other applications are greenhouse heating, warming fish ponds in aquaculture, crop drying, and various washing and drying applications in the food, chemical, and textile industries. In regions where high-temperature resources occur, combination of electricity production with these uses (e.g., in Iceland) is possible.

In the direct use of geothermal systems, fluids are generally pumped through a heat exchanger to heat air or a liquid, although the resource may be used directly if the salt and solids contents are low. These systems exemplify the simplest applications using conventional off-the-shelf components.

For most of the specified uses, the hydrothermal source is at about 40°C. With heat pump technology, a hydrothermal source of 20°C or less can be used as a heat source, as is done, for example, in the United States, Canada, France, Sweden, and other countries. The heat pump operates on the same principle as the home refrigerator, which is actually a one-way pump. The geothermal heat pump (GHP) can move heat in either direction. In the winter, heat is removed from the earth and delivered to the home or building (heating mode). In the summer, heat is removed from the home or building and delivered for storage to the earth (air-conditioning mode). In either cycle, water is heated and stored, supplying all or part of the function of a separate hot-water heater. Because electricity is used only to transfer heat and not to produce it, the GHP will deliver 3–4 times more energy than it consumes. It can be used effectively over a wide range of earth temperatures. Current growth rates for GHP systems run as high as

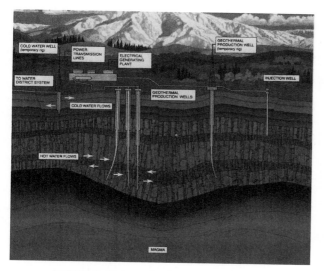

FIGURE 5 Schematic of a geothermal plant.

20% per year in the United States and the outlook for continued growth at double-digit rates is good. The U.S. Department of Energy Information Administration (EIA) has projected that GHPs in the United States could provide up to 68 Mtoe (mega-tons of oil equivalent) of energy for heating, cooling and water heating by 2030 (Lund and Freeston, 2000).

C. Plant Options for Power Generation

There are several types of energy-conversion processes for generating electricity from hydrothermal resources (see Fig. 5). These include dry steam and flash steam systems, which are traditional processes, and binary cycle and total flow systems, which are newer processes with significant advantages (World Energy Council, 1994).

D. Dry Steam Plants

Conventional steam-cycle plants are used to produce energy from vapor-dominated reservoirs. As is shown in Fig. 6, steam is extracted from the wells, cleaned to remove entrained solids, and piped directly to a steam turbine. This is a well-developed, commercially available technology, with typical unit sizes in the capacity range 35–120 MW$_e$. Recently, in some places, a new trend of installing modular standard generating units of 20 MW$_e$ has been adopted. In Italy, smaller units in the range 15–20 MW$_e$ have been introduced.

E. Flashed Steam Plants

More complex cycles are used to produce energy from liquid-dominated reservoirs which are sufficiently hot (typically above 160°C) to flash a large proportion of the liquid to steam. As shown in Fig. 7, single-flash systems evaporate hot geothermal fluids to steam by reducing the pressure of the entering liquid and directing it through a turbine. In dual-flash systems, steam is flashed from the remaining hot fluid of the first stage, separated, and fed into a dual-inlet turbine or into two separate turbines. In both cases, the condensate may be used for cooling while the brine is reinjected into the reservoir. This technology is economically competitive at many locations and is being developed using turbogenerators with capacities of 10–55 MW$_e$. A modular approach, using standardized units of 20 MW$_e$, is being implemented in the Philippines and Mexico.

F. Binary-Cycle Plants

1. Low-Enthalpy Resources (100–160°C)

For low-enthalpy resources, binary plants based on the use of organic Rankine cycles (ORC) are utilized to convert

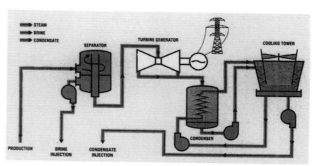

FIGURE 6 General Electric Co. dry steam plant at The Geysers, California. [From Nemzer, M. (2000). Geothermal Education Office, web site http://geothermal.marin.org.]

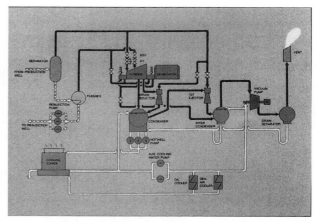

FIGURE 7   The Mitsubishi flash steam plant in Beowave, Nevada. [Courtesy of Mitsubishi Heavy Industries.]

the resource heat to electrical power (see Fig. 8). The hot brine or geothermal steam is used as the heat source for a secondary (organic) fluid, which is the working fluid of the Rankine cycle (UNITAR/UNDP, 1989).

During the early 1980s, in order to increase the power output from a given brine resource by increasing the thermal cycle efficiency, a supercritical cycle using isobutane was developed, as well as a cascade concept. The supercritical cycle may be slightly more efficient than the cascading cycle, but the cascading system has the advantage of lower operating pressures and lower parasitic loads in the cycle pumps. For example, at a power plant in Southern California, a three-level arrangement was employed and resulted in increased efficiency or power output gain of about 10% over that achievable with a simple ORC.

For all of the configurations and systems, a modular approach was employed so that high plant availability factors of 98% and above were achievable.

2. Moderate Enthalpy Resources (160–190°C)

For moderate-enthalpy, two-phase resources with steam quality between 10% and 30%, binary plants are efficient and cost-effective. Furthermore, when the geothermal fluid has a high non-condensible gas (NCG) content, even higher efficiency can be obtained than with condensing steam turbines (Bronicki, 1998).

This binary two-phase configuration is used in the São Miguel power plant in the Azores Islands (see Fig. 9). Separated steam containing NCG is introduced in the vaporizer heat exchanger to vaporize the organic fluid. The geothermal condensate at the vaporizer exit is then mixed with the hot separated brine to provide the preheating medium for the organic fluid. Since the onset of silica precipitation is related to its concentration in the brine, dilution of the brine with the condensate effectively lowers the precipitation temperature at which silica crystallizes. This lower temperature added 3.5 MW of heat to the cycle, representing 20% of the total heat input. The additional

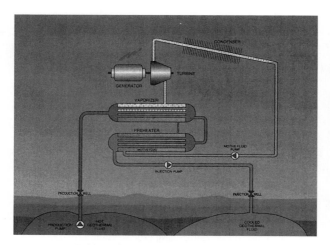

FIGURE 8   An air-cooled binary plant.

FIGURE 9   The ORMAT two-phase binary geothermal power plant in São Miguel, Azores Islands.

heat is utilized at the same thermal efficiency as the remaining heat in the combined steam–brine cycle. Since the cycle efficiency is about 17%, the low-temperature heat produces about additional 600 kW. The main advantage of the geothermal combined-cycle plant over conventional steam plants lies in the efficiency of the power plant when using both steam and brine in the conversion process. It provides sustainable power and does not deplete the geothermal reservoir since all fluids are reinjected. This feature contributes to the environmental acceptability of the plant since it operates without emissions and no abatement of noncondensable gases (NCG) is needed. The air-cooled condensers contribute to the low physical profile of the plant and there is no plume.

3. Total Flow Turbines

This is an experimental process, based on using concurrently steam, hot water, and the pressure of geothermal resources (i.e., the total resource), thereby eliminating energy losses associated with the conventional method of flashing and steam separation. These systems usually channel a mixture of steam and hot water into a rotating conversion system and capture the kinetic energy of the mixture to power an electric generator.

III. CURRENT USES AND COMMERCIAL STATUS

Electricity from geothermal energy has been generated in Italy for more than 90 years. Until 1974, the total installed capacity for converting geothermal energy into electricity was only about 770 MW$_e$ (in Italy, Japan, New Zealand, the United States, and Mexico). Following the second oil shock, the worldwide installed capacity achieved its highest growth of 17.2% per year. The number of geothermal power-producing countries increased from 10 to 17. Recent emphasis has been placed on power production using the liquid hydrothermal resource since power production with dry steam has been commercially viable for several decades. In the year 2000, a total of over 8000 MW$_e$ was produced from geothermal resources in more than 20 countries. Substantial market penetration has thus far occurred only with hydrothermal technology.

Table I shows selected countries and their installed power plants. From 1978 to 1985, the worldwide installed electrical capacity grew at an average annual rate of about 17.2%. The causes of the growth surge were the two oil shocks (1973 and 1979) and expectations of further oil price rises. Many of the known profitable resources were exploited and much work was devoted to exploration for new hydrothermal resources. After the oil price collapse in

TABLE I Worldwide Geothermal Installed Capacity in the Year 2000 in MW$_e$

United States	2228	Kenya	57
The Philippines	1909	Guatemala	33
Mexico	855	China	29
Italy	785	Russia	23
Indonesia	589	Turkey	20
Japan	547	Portugal (Azores)	16
New Zealand	437	Ethiopia	9
Iceland	170	France (Guadalupe)	4
El Salvador	161	Thailand	0.3
Costa Rica	142	Australia	0.17
Nicaragua	70	Total	8154

1990, the growth rate fell to about 4% per year. Since most of the subsidies for renewable energy and especially those for geothermal energy were almost completely stopped, this growth rate is not negligible.

During the past 25 years, geothermal technology (mainly hydrothermal) has changed from mainly balneological uses to widespread industrial, agricultural, and district-heating usage, and from the use of dry steam resources to power production from a wide spectrum of resources. The energy-conversion technology has become a mature and commercially viable technique. Binary and geothermal combined-cycle power plants, which reached maturity with more than 600 MW$_e$ of commercially installed capacity, operate as closed-loop geothermal power plants with almost zero pollutants and no water consumption. Plants of a few hundred kilowatts up to tens of megawatts may be installed in a period of a few months and provide clean, sustainable indigenous energy sources.

The second approach to better resource utilization involves the use of a regenerative cycle through the addition of a recuperator heat exchanger between the organic turbine and the air-cooled condenser, since the organic vapor tends to superheat when the vapor is expanded through the turbine. In this case, the recuperator reduces the amount of heat that must be added to the cycle from the external source, thereby reducing the required brine-flow rate. This procedure results in a reduction of about 7% of the total heat input to produce the design power output.

A. Geothermal Combined-Cycle Plants

For efficient use of a steam-dominated resource, a geothermal combined cycle is applied. The steam first flows through a backpressure steam turbine and is then condensed in the organic turbine vaporizer (see Fig. 10). The condensate and brine are used to preheat the organic fluid as in the two-phase binary configuration. This concept

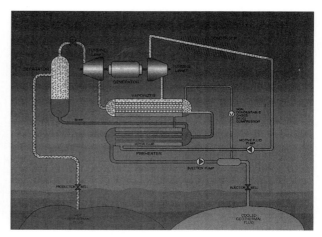

FIGURE 10 The ORMAT geothermal combined-cycle power plant in Puna, Hawaii.

was first used in 1989 to repower a backpressure steam plant in Iceland. Subsequent uses were with a 30-MW plant in Hawaii in 1992, followed by a 125-MW plant in the Philippines and a 60-MW plant in New Zealand (Bronicki, 1998).

B. Direct Applications of Hydrothermal Energy

Direct applications of geothermal energy involve a wide variety of end uses, such as space heating and cooling, industrial heat, greenhouses, fish farming, and health spas. Existing technology and straightforward engineering are involved. The technology, reliability, economics, and environmental acceptability of the direct use of geothermal energy has been demonstrated throughout the world. Space heating is the dominant application (37%), while other common uses are bathing/swimming/balneology (22%), heat pumps (14%) for air cooling and heating, greenhouses (12%), fish farming (7%), and industrial processes (7%).

The relative share of Asia has increased in recent years for direct energy production. It is estimated to be 44% of total at present mainly because of rapid expansion in China. The European share has decreased to 37%, while that of the Americas has grown to 14% due to increased uses of heat pumps in the United States.

Direct applications use both high- and low-temperature geothermal resources and are therefore much more widespread in the world than electricity production. Direct applications are, however, site specific for the market, as steam and hot water are rarely transported long distances from geothermal sites.

China has geothermal water in almost every province. The direct utilization is expanding at a rate of about 10% per year, mainly in the space heating (replacing coal), bathing, and fish-farming sectors. Japan is also blessed

with very extensive geothermal resources, which so far have mainly (80%) been used for bathing, recreation, and tourism and, to a lesser extent, for electricity production. This development has improved the quality of life of people significantly, but only a fraction of the available geothermal energy is actually used. Turkey has greatly increased the direct use of geothermal resources in recent years. Mexico is the first country in the tropics to report significant direct use of geothermal energy. Switzerland and Sweden have recently joined the top league through extensive use of ground-source heat pumps.

C. Heat-Pump Applications

Geothermal energy has until recently had a considerable economic potential only in areas where thermal water or steam is found concentrated at depths of less than 3 km in restricted volumes, analogous to oil in commercial oil reservoirs. This status has changed with developments in the application of ground-source heat pumps using the earth as a heat source for heating or as a heat sink for cooling, depending on the season. As a result, all countries may use the heat of the earth for heating and/or cooling, as appropriate. It should be stressed that heat pumps can be used everywhere.

During the last decade, a number of countries have encouraged individual house owners to install ground-source heat pumps to heat their houses in the winter and (as needed) cool them in the summer. Financial incentive schemes have been set up, commonly funded by the governments and electric utilities, as the heat pumps reduce the need for peak power and thus replace new electric generating capacity. The United States leads the way with about 400,000 heat-pump units (about 4800 MW_t) and energy production of 3300 GWhr/year in 1999.

TABLE II Direct Geothermal Use during the Year 2000[a]

Region	Installed capacity (MW_t)	Yearly production (GWhr)
Africa	121	492
America	5,954	7,266
Asia	5,151	22,532
Europe	5,568	18,546
Oceana	318	2,049
Total	17,112	50,885

[a] From Lund, J. W., and Freeston, D. H. (2000). "World direct uses of geothermal energy 2000." *In* "Proceedings, World Geothermal Congress," pp. 1–21.

The annual increase is about 10%. Other leading countries are Switzerland, Sweden, Germany, Austria, and Canada.

Switzerland, a country not known for hot springs or geysers, provides an example of the impact this development can have on geothermal applications in what previously would have been called nongeothermal countries. The energy extracted out of the ground with heat pumps in Switzerland amounts to 434 GWhr/year. The annual growth rate is 12%.

TABLE III The World's Top 15 Direct Use Countries for Geothermal Energy[a]

	Installed capacity (MW_t)	Yearly production (GWhr)
China	2814	8724
Japan	1159	7500
United States	5366	5640
Iceland	1469	5603
Turkey	820	4377
New Zealand	308	1967
Georgia	250	1752
Russia	307	1703
France	326	1360
Hungary	391	1328
Sweden	377	1147
Mexico	164	1089
Italy	326	1048
Romania	152	797
Switzerland	547	663

[a] From Lund, J. W., and Freeston, D. H. (2000), "World direct use of geothermal energy 2000." *In* "Proceedings, World Geothermal Congress," pp. 1–21.

Prior to 2000, the total installed capacity for the direct use of geothermal energy world-wide was about 17,000 MW_t (see Table II). The top 15 primary users of direct geothermal heat and the year 2000 capacity are listed in Table III. The total installed capacity in 1975 was only about 3100 MW_t (excluding balneology).

IV. THE ULTIMATE POTENTIAL

The growth rate of the geothermal energy market is not limited by the lack of resources. During the early oil crises, intensive investigations led to the discovery of many geothermal reservoirs for electricity generation, some of which are in operation, while about 11,000 MW_e of proven resources is not yet tapped. In the near future, the growth rate will most probably be 3–4% annually, as has been the case during the past few years.

However, if environmental impacts of energy use are internalized, then the real value of geothermal technology including its superior environmental characteristics and local resource features will be taken into account and the geothermal market will become more profitable. As a result, there will be enhanced geothermal exploration and R&D. The growth rate should then reach 6–7% and more. This outlook should encourage the development of other geothermal resources. Hot, dry rock and geopressured technologies may reach maturity around 2010.

In 1978, the Electric Power Research Institute (EPRI) published a report on the ultimate potential for geothermal energy on a global basis. The accessible global total is very much greater than today's usage. Most of the total resource is contained in hot, dry rocks.

The geothermal resource base underlying the continental land masses of the world to a depth of 3 km and at temperatures higher than 15°C was calculated to be 1.2×10^{13} GWhr or 1.03×10^9 Mtoe. It therefore appears that geothermal energy is an abundant resource (Nakićenović *et al.*, 1998). If we are able to exploit only 1% of this energy, we will have enough energy for several hundred years. In order to tap most of this energy resource, we need to invest money to improve the existing technologies, especially extraction of energy from hot, dry rocks (World Energy Council 2000).

The authors of a study for the US DOE's Office of Renewable Energy (US DOE, 1998) argued that geothermal energy is by far the most abundant nonnuclear energy source in the United States, accounting for nearly 40% of the total energy resource base. The total resource base is defined as concentration of naturally occurring solid, liquid, or gaseous materials in or on the earth's crust in

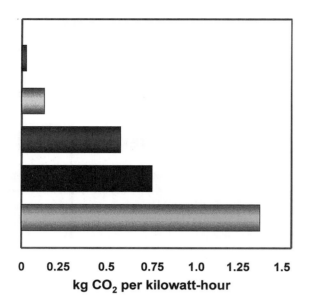

FIGURE 11 Relative CO_2 emissions for different energy resources.

such a form that economic extraction of the commodity is currently or potentially feasible. In this study, the resource base includes geothermal reservoirs with a minimum temperature of $80°C$ at a maximum depth of 6 km, except for geopressured resources, which are included to 7 km. Also included are low-temperature resources in the $40–80°C$ range to a depth of 2–3 km.

V. ECOLOGICAL AND ENVIRONMENTAL CONSIDERATIONS

The successful implementation of the Kyoto targets, introducing internationally agreed vehicles to mitigate greenhouse gas emissions, will enhance the use of non-fossil-fuel systems, including geothermal energy (Fig. 11). There are other environmental advantages to geothermal energy, as power plants using it require far less land area than other energy sources, as illustrated in Table IV.

TABLE IV Land Area Occupied for Different Energy Technologies

Technology	Land area (m² per GWhr/year for 30 years)
Coal (including coal mining)	3642
Solar thermal	3561
Photovoltaics	3237
Wind (land with turbines and roads)	1335
Geothermal	404

SEE ALSO THE FOLLOWING ARTICLES

EARTH'S CORE • EARTH'S MANTLE • ENERGY RESOURCES AND RESERVES • FUEL CELLS, APPLICATIONS IN STATIONARY POWER SYSTEMS • OCEAN THERMAL ENERGY CONVERSION • RENEWABLE ENERGY FROM BIOMASS • SOLAR PONDS • SOLAR THERMAL POWER STATIONS • TIDAL POWER SYSTEMS • WIND POWER SYSTEMS

BIBLIOGRAPHY

Allegrini, G., *et al.* (1991) "Growth Forecast in Geothermoelectric Capacity in the World to the year 2020."

Baria, R., Baumgartner, J., and Gerard, A. (1998). "International Conference—4th HDR Forum, Strasbourg, France," SOCOMINE, EHDRA.

Barnea, J. (1981), "The Future of Small Energy Sources," UNITAR, New York,

Bronicki, L. (1998). "Eighteen Years of Field Experience with Innovative Geothermal Power Plants," World Energy Council, Houston, TX.

Carella, R., (1989). "Obstacles and Recommendations to Promote Geothermal Energy Development," European Geothermal Update, Florence, Italy.

Fridleifsson, I. B. (2000). "Geothermal energy for the benefit of the people world-wide," Web site www.geothermic.de.

GeothermEx. (1998). "Peer Review of the Hot Dry Rock Project at Fenton Hill, New Mexico," A report for the Office of Geothermal Technologies, U.S. Department of Energy.

Grassiani, M., and Krieger, Z. (1999). "Advanced Power Plants for Use with Hot Dry Rock (HDR) and Enhanced Geothermal Technology," ORMAT, Yavne, Israel.

Huttrer, G. W. (2000). "The Status of world geothermal power generation 1995–2000." *In* "Proceedings, World Geothermal Congress," pp. 23–27.

Kestin, J., DiPippo, R., Khalifa, H., and Ryley, D. (1980). "Sourcebook on the Production of Electricity from Geothermal Energy" (DOE/RA/4051-1), Brown University, Providence, RI.

Lund, J. W. (2000). "World status of geothermal energy use, overview 1995–1999." *In* "Proceedings, World Geothermal Congress," pp. 4105–4111.

Lund, J. W., and Freeston, D. H. (2000). "World direct uses of geothermal energy 2000." *In* "Proceedings, World Geothermal Congress," pp. 1–21.

Nakićenović, N., Grübler, A., and McDonald, A. (1998). "Global Energy Perspectives," Cambridge University Press, Cambridge.

Nemzer, M. (2000). Geothermal Education Office, web site http:// geothermal.marin.org.

UNITAR/UNDP. (1989). "Production of Electrical Energy from Low Enthalpy Geothermal Resources by Binary Power Plants," Centre of Small Energy Resources, Rome.

US DOE. (1998). "Strategic Plan for Geothermal Energy," NREL, U.S. Department of Energy, Office of Geothermal Technologies.

Williams, S., and Porter, K. (1989). "Power Plays—Profiles of America's Independent Renewable Electricity Developers," Investor Responsibility Research Center, Washington, DC.

World Energy Council. (1994). "New Renewable Energy Resources," Chapter 4, Kogan Page, London.

World Energy Council. (2000a). "World Energy Assessment," United Nations Development Programme, United Nations Department of Economic and Social Affairs.

World Energy Council. (2000b). "Energy for Tomorrow's World—Acting Now!" World Energy Council Statement.

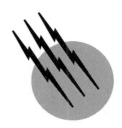

Glacial Geology and Geomorphology

David J. A. Evans

University of Glasgow

GLOSSARY

Abrasion Process of bedrock erosion by rock fragments protruding from the base of a glacier. This involves scoring to produce striae and polishing, analogous to the action of sandpaper on wood.

Bedform A depositional form produced at the interface (boundary layer) between a transporting medium (ice, water, wind) and its bed.

Cirque Amphitheatre- or bowl-shaped hollow in a mountain side, characterized by steep backwalls, possessing an arcuate plan form and a glacially overdeepened basin often occupied by a lake and enclosed on the downslope, open end by a smooth bedrock rim or threshold.

Diamicton Poorly sorted sediment with a wide range in grain sizes and deposited by glacial and mass movement processes. A nongenetic term that should be employed before sediment origins are known (e.g., tills are diamictons).

Drumlin Smooth, oval-shaped hill comprising either sediment or rock (rock drumlin) resembling an inverted spoon and occurring in groups or swarms. Although numerous theories on their formation exist, drumlins are generally regarded as glacially streamlined bedforms.

Erratic A rock or rock fragment found in an area that is located far from its outcrop source, whose former transport from that source can be explained only by glacial transport. Erratic distributions can be used to map former glacier flow lines.

Esker Elongate sinuous ridge composed of glacifluvial sediments and marking the former position of a subglacial, englacial, or supraglacial stream.

Encyclopedia of Physical Science and Technology, Third Edition, Volume 6

Freeze-thaw Alternations of temperature that involve crossing the freezing point of water. In confined spaces such as rock crevices this can lead to considerable mechanical breakdown of bedrock.

Glacitectonic thrusting Tectonic disturbance of materials by glacier ice. This takes place in both proglacial and subglacial environments.

Glacitectonite Rock or sediment that has been deformed by subglacial shearing but retains some of the structures of the parent material.

Kettle hole Enclosed depression on a glacifluvial landform surface marking the former position of buried glacier ice.

Meltwater Water produced by the melting of glacier ice and feeding subglacial, englacial, supraglacial, and proglacial streams.

Moraine Glacial depositional landform produced in marginal and supraglacial positions by a wide variety of processes.

Outwash Glacifluvial sediment accumulation deposited by glacial meltwater emanating from the glacier margin.

Plucking (quarrying) Process of bedrock erosion whereby large fragments of rock are removed from hard glacier beds due to pressure differences in fracture systems.

Striation/striae Fine scratches on bedrock surfaces produced by the abrasion process. They provide excellent indicators of former local and regional glacier flow directions.

Till Poorly sorted sediments containing a wide range of grain sizes (diamicton) deposited directly by glacier ice.

Varves A type of glacilacustrine rhythmite deposited annually through suspension onto glacial lake beds and consisting of couplets of fine, clay-rich winter layers and coarse, slit-rich summer layers.

GLACIAL GEOLOGY has traditionally been concerned with the deposits, forms, and stratigraphy of former glaciations, whereas glacial geomorphology has concentrated on understanding present-day glacial processes and forms and their use as analogues in the explanation of ancient glacial features. Both approaches have become, necessarily, intertwined, and the research subject area is generally entitled glacial geology and geomorphology, dealing with the processes in modern glacial systems and their impact on both contemporary and ancient landscapes. As with all areas of geomorphology, reconstructions of ancient landscapes are impossible without a sound understanding of modern process-form relationships.

I. INTRODUCTION

Glaciers and ice sheets of the present display a range of morphologies, thermal regimes, and dynamics, with each of these characteristics being inextricably linked to each other and also fundamental to the glacial processes of erosion and deposition. In particular, effective sliding and erosion are dependent on the basal ice being at pressure melting point. It is therefore important to distinguish between temperate ice, which is at pressure melting point, and cold ice, which is below pressure melting point. This classification has been extended to glaciers as a whole, thereby identifying three types of glacier: (1) temperate glaciers, which are at melting point throughout except for the surface layers that are subject to seasonal temperatures fluctuations; (2) polar glaciers, which are below melting point throughout and frozen to their beds; and (3) subpolar glaciers, which are temperate in their inner regions but cold based at their margins. In reality, however, this classification scheme is too simplistic and glacier beds may comprise temperature mosaics. Furthermore, zones of melting and freezing may migrate through time.

Implications of basal thermal regimes for glacial erosion and debris entrainment are therefore profound. Where glacier ice is frozen to its bed it will be unable to slide, and therefore, erosion will be minimal. Consequently, older (preglacial) landforms may be preserved beneath cold-based glacier ice. Temperate or wet-based glacier ice, on the other hand, is capable of considerable erosion because it can slide over its bed and produce large volumes of meltwater which collects in subglacial streams and performs its own erosion. Temperate ice also deposits subglacial sediment due to the melting of the glacier sole and concomitant release of entrained debris. This further allows temperate glaciers to deform their bed materials, thereby adding a further velocity component to sliding and internal deformation (Fig. 1). Glaciers possessing regions of both cold ice and temperate ice are capable of entraining large volumes of sediment where meltwater and debris, migrating from a melting zone of the sole, are frozen back onto the glacier base in a cold zone. This is common in subpolar glaciers where englacial debris concentrations are seen to increase in the marginal (cold-based) ice. Our understanding of the relationships between thermal regime characteristics and the processes of glacial erosion and deposition has allowed us to painstakingly reconstruct the thermal regimes and dynamics of former glaciers and ice sheets (palaeoglaciology) using the patterns of glacial landform distribution in glaciated terrains.

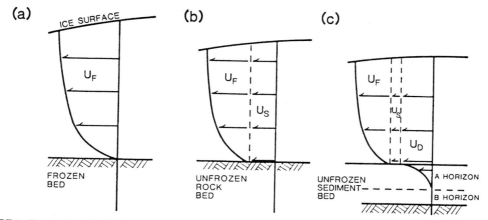

FIGURE 1 The vertical distribution of velocity for glaciers moving by different mechanisms: (a) internal deformation (U_F) only; (b) internal deformation and basal sliding ($U_F + U_S$); (c) internal deformation, basal sliding, and deformation of subglacial sediments ($U_F + U_S + U_D$), wherein A and B horizons are till structures observed beneath a modern glacier. [Reprinted from Boulton, G. S. (1996). "Theory of glacial erosion, transport and deposition as a consequence of subglacial sediment deformation," *J. Glaciol.* **42,** 43–62. By permission of the International Glaciological Society.]

II. GLACIAL AND GLACIFLUVIAL EROSION

A. Processes of Glacial Erosion

The erosion of bedrock by sliding glaciers takes place by abrasion and plucking (quarrying), the evidence of which occurs as distinctive landforms and microscale features on rock surfaces. Such processes have also been reproduced in laboratory studies and measured at the base of a few glaciers where access to the bed is possible. Although it is difficult to calculate erosion rates, sediment yields from proglacial streams have been used to estimate rates of 0.01 mm year^{-1} for polar glaciers, to 1.0 mm year^{-1} for alpine temperate glaciers, to 10–100 mm year^{-1} for fast flowing Alaskan glaciers. In order to consider the impact of a rock tool embedded in the sliding glacier sole on a bedrock surface, it is first critical to understand the role of stress and frictional strength at the ice–bedrock interface. A frictional and a cohesive force exists between a rock particle and the bed, thereby resisting motion of the particle. If the shear stress imparted by glacier flow exceeds this resisting force, then the particle is dragged over the bed. If the resisting force is greater than the shear stress, then the particle will be retarded against the bed and ice will flow around it. Stress magnitude varies widely on a rough glacier bed, with stress concentrations occurring around protuberances, crack tips, and sharp corners. It is at these locations that breakage or deformation will occur, especially if local stress gradients are high, because material will move from areas of high stress to areas of low stress.

Three subglacial friction models have been proposed, each having slightly different implications for glacial erosion, transport, and deposition. First, the Coulomb friction model assumes that the most important controls on the strength of subglacial rock particle–bedrock contacts are the weight of the overlying ice and rock and water pressing the surfaces together minus the water pressure at the contact. This implies that the interface between rock particles and the bed is characterized by points of contact separated by cavities and is free of ice. The points of contact between the particle and the bed occur at asperities or protuberances on the particle base. The frictional force between the particle and the bed (basal friction) is dictated by the weight of the overlying ice minus the water pressure in the cavities. Second, the Hallet friction model suggests that basal friction is independent of the weight of the ice and the subglacial water pressure. Rather, the frictional force is the sum of the bouyant weight of the particle and the drag force resulting from ice flow toward the bed. High contact forces arise when the ice velocity normal to the bed is large, as occurs when ice is flowing rapidly toward the bed due to basal melting and vertical straining. The pressure melting and enhanced creep that takes place on the up-glacier sides of bumps suggest that frictional forces will be highest at those points on the bed. Finally, the sandpaper friction model proposes that basal drag is a function of the effective normal pressure (the difference between ice pressure and water pressure normal to the bed) and the area of the bed occupied by water-filled cavities between particles. The average drag force declines with increases in the area of the bed occupied by cavities, but considerable stress concentrations occur at points of rock–rock contact.

Although a unifying theory would be more realistic for subglacial friction and erosion, it is generally accepted that the Hallet model is most applicable where basal debris is sparse, the sandpaper model is applicable where debris concentrations are high, and the Coulomb model applies to transient conditions where basal debris is rigid or ice free. In summary, it is understood that particle-bed friction increases with particle size, friction increases with basal melt rates and is greatest on the up-glacier sides of bumps, high friction is encouraged by the presence of low-pressure cavities beside or below particles, and friction normally increases with debris concentration.

There are three fundamental stages that are common to both abrasion and quarrying and apply to the whole range of scales observed at glacier beds, from silt-sized abrasion products to quarried boulders. First, rock failure involves the loosening of fragments from the glacier bed. Second, evacuation involves the removal of those fragments from their original position. Third, transport involves the entrainment of the fragments and is reviewed in Section III.

The visible joints and faults and microscopic cracks, cleavage planes, and minerals that occur in rocks all provide potential weaknesses that can be exploited by glacial erosion. The density of such planes of weakness vary with rock type, and some rocks possess continuous structures such as joints which can isolate large volumes or joint-bounded blocks within the main rock mass. In these situations rock failure simply involves the widening of existing large fractures or the joining up of smaller fractures, a process known as discontinuous rock-mass failure. Crack growth or propagation occurs as a result of tensile stresses pulling the walls of the crack apart. Tensile stresses develop at the point of contact between particles and the glacier bed, thereby causing the growth of favorably oriented cracks in the bedrock or particles.

Abrasion involves the scratching or scoring of rock surfaces by debris carried in the basal ice layers to produce striae and small-scale brittle failure of tiny protuberances on rock surfaces to produce polishing. Striae are essentially thin grooves excavated on rock surfaces by the asperities on particles carried in the ice (Fig. 2). The tensile stresses at the asperity tip can initiate crack growth, thereby producing a damage trail of brittle failure. Although striae appear to be continuous grooves when viewed with the naked eye, at microscopic scale every striation comprises numerous crescent-shaped fractures aligned en-echelon and parallel to the ice flow direction responsible for their formation. These fractures record small, discontinuous brittle failure events which are the product of the jerky chipping motion of the asperity being dragged over the rock surface. The efficiency of this process is dictated by six factors. These include the relative

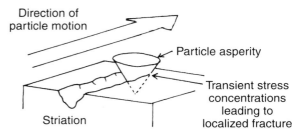

FIGURE 2 Simple model of the striation process. [Reprinted from Benn, D. I., and Evans, D. J. A. (1998). "Glaciers and Glaciation," Arnold, London.]

hardness of the rock surface and the overriding particle, the force pressing the particle against the bed, the velocity of the particle relative to the bed, the concentration of debris in the ice, the removal of wear products by thin water films, and the increased availability of basal debris largely due to basal melting. Between striae there are areas of polished bedrock, attesting to the removal of small protuberances by brittle failure. This failure results from the stress concentrations that are associated with protuberances in the bed.

Quarrying involves the fracture of large fragments from rock beds largely due to the process of discontinuous rock-mass failure. Preglacial weathering can also weaken rocks and assists the quarrying process. Additionally, freeze-thaw processes may be active beneath glacier margins where heat can be evacuated from the bed. However, quarrying is primarily a response to the build up of tensile stresses in the rock, which can take place beneath large particles being dragged over the bed (Fig. 3) and near the lee side of obstacles or steps in the bed. Tensile stresses are developed in front of the particle where crack propagation can occur and lead to the development of crescentic fractures. At the smallest scale this leads to striae development, but at larger scales it produces features like chatter marks (see Section II.B). On a glacier bed characterized by bedrock steps, cavities become important in the quarrying process in that they introduce large pressure differences over relatively small areas. Specifically, this accrues through the differences in water pressure in rock fractures compared to large cavities; water pressure cannot change as rapidly in fractures as it can in cavities and so larger scale changes in water pressure across the bed can lead to the ideal conditions for rock fracture, especially if the attainment of those conditions coincides with the passage of a large particle over the bed. Additionally, quarrying rates rise with an increase in the number of cavities at the bed, because basal stresses have to be supported over a smaller area of ice–rock contact. However, there is an upper threshold for this relationship given that extensive cavities will reduce the area affected by high loads.

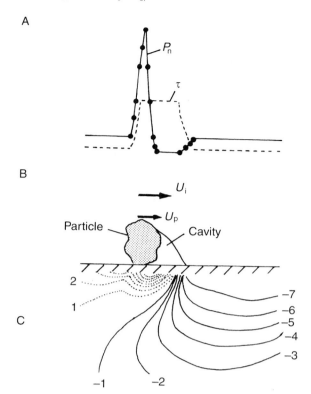

FIGURE 3 Calculated stress patterns in horizontal rock bed associated with a rock particle in sliding ice: (A) distribution of normal (P_n) and shear (τ) stresses at the rock surface; (B) the velocity of the particle (U_p) is less than that of the ice (U_i) due to frictional retardation. Therefore, a low-pressure cavity develops in the lee side of the particle: (C) contoured values of the maximum stress in the rock where positive values are compressive and negative values are tensile. [Reprinted from Drewry, D. (1986). "Glacial Geologic Processes," Arnold, London.]

Once rock fragments have been loosened from the bed, they must be removed before they can be transported (see Section III). Various removal mechanisms have been proposed, with the most popular being the heat pump effect, whereby ice locally freezes on to the bed as part of the regelation sliding process. This results in localized cold patches where rock fragments may adhere to the glacier sole. A fragment may also be dislodged wherever the friction between the fragment and the parent rock is reduced, as has been witnessed beneath a Norwegian glacier where ice has been squeezed plastically into joints and has facilitated the removal of large rock fragments.

B. Forms of Glacial Erosion

Glacial erosional landforms are among some of the most distinctive and spectacular on the earth's surface and exist at a wide variety of scales (Table I). At each scale, summarized here as small, intermediate, large, and landscape, there are variables that control or influence processes and patterns of erosion, as well as the form, size, and distri-

bution of erosional features. First, glaciological variables, such as basal shear stress, subglacial water pressures, basal velocity, and thermal regime affect the amount of erosion that takes place. Second, substratum characteristics include the structure, lithology, degree of weathering, and joint distribution of the bedrock. Third, topographic variables include the nature of the bed and the relief of landscapes. Finally, temporal variables include the duration and number of glaciations through time.

Small-scale forms (Fig. 4) have been discussed briefly in the preceding section because they are central to understanding the processes of abrasion and quarrying. Striae are scratches on rock surfaces produced by the dragging of rock particles over bedrock by sliding glaciers (Fig. 5). Although striae appear at first glance to be parallel scratches that deviate very little from a single orientation, they may possess a variety of forms and spatial relationships. The shapes vary according to the shape of the asperity responsible for their production. Gradual widening of a striation produces a wedge shape, whereas abrupt widening constitutes a nail-head striation; both types reflect asperity blunting. Some striae also occur in an en-echelon pattern, whereby the blunt terminus of one striation lies alongside the narrow end of another. This is thought to reflect rotation or "flip-out" of a particle, resulting in the lifting of one asperity clear of the rock surface and another asperity into contact with it along an adjacent flowline. Sets of striae may crosscut each other on a rock surface, providing evidence of changing ice flow directions; the order of such changes is recorded by the partial obliteration of older striae. Striae on an uneven rock surface may also deviate considerably from the main trend direction due to irregularities in basal ice flow. Positive relief features also recording former ice flow are rat tails, small residual longitudinal ridges extending downflow from a resistant rock knob or nodule. Fracture marks or cracks in bedrock surfaces record quarrying events, which are manifested in features known as chatter marks, crescentic gouges, conchoidal fractures, and lunate fractures. Because they are often aligned en-echelon, parallel to former ice flow, these features represent the damage trails of particles being dragged across the bed in a stick-slip motion. The final small-scale erosional features are known as P-forms, originating from an earlier classification "plastically molded forms," which are smoothed depressions of a variety of shapes (Fig. 6). P-forms are aligned parallel to ice flow (longitudunal), transverse to ice flow, or possess no elongation and are therefore termed nondirectional. Although P-forms possess many of the attributes of fluvially scoured bedrock, they are often also adorned with striae which conform to the surface undulations, including even the tightest corners. This strongly suggests that glacier ice flowed along the surfaces of the P-forms at least during a later stage of their development. Four media have been

TABLE I Classification of Glacial Erosional Landforms According to Process, Relief, and Scale

			Micro				Scale					Macro
Process	**Relief type**	**Relief shape**	**m^{-2} (1 cm)**	**m^{-1}**	**m^{0} (1 m)**	**m^{1}**	**m^{2} (100 m)**	**m^{3}**	**m^{4} (10 km)**	**m^{5}**	**m^{6} (1000 km)**	**m^{7}**
Areal ice flow	Eminence	Streamlined				← Whaleback →	← Rock drumlin →	← Crag and tail → Streamlined-spur →				
	Eminence	Part-streamlined	← Roche moutonnée—Flyggberg →									
	Depression	Streamlined	← Striae →		← P-form →	← Groove →						
	Depression	Part-streamlined	← Rock basin →									
											Landscape of areal scouring →	
Linear flow in rock channel	Depression	Streamlined					← Alpine trough → ← Trough →				Landscape of ice sheet linear erosion	
Interaction of glacial and periglacial	Depression						← Cirque →				Valley glacier landscape	
	Eminence					← Residual summit or horn →					Nunatak landscape	

Source: Reprinted from Benn, D. I., and Evans, D. J. A. (1998). "Glaciers and Glaciation," Arnold, London.

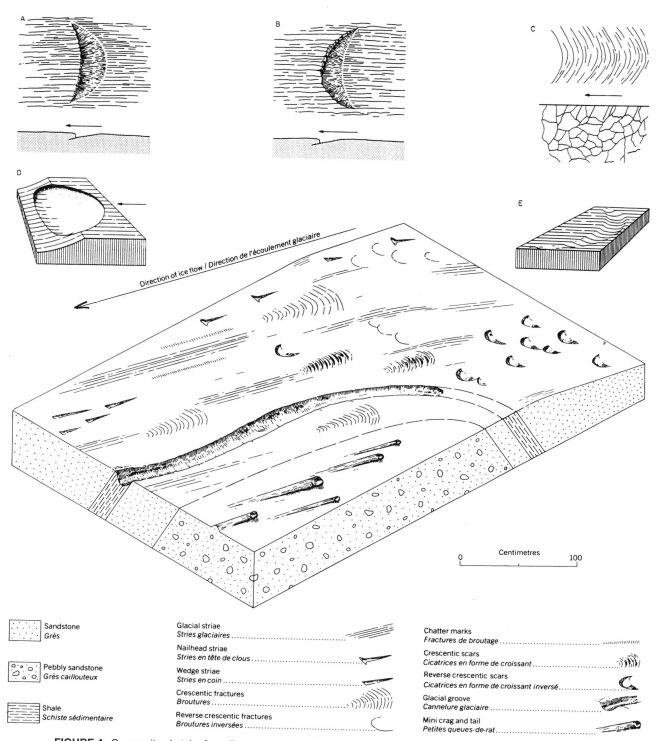

A

B

C

D

Direction of ice flow / Direction de l'écoulement glaciaire

E

	Centimetres	
0		100

Sandstone	Glacial striae		Chatter marks	
Grès	Stries glaciaires		Fractures de broutage	
	Nailhead striae		Crescentic scars	
Pebbly sandstone	Stries en tête de clous		Cicatrices en forme de croissant	
Grès caillouteux	Wedge striae		Reverse crescentic scars	
	Stries en coin		Cicatrices en forme de croissant inversé	
Shale	Crescentic fractures		Glacial groove	
Schiste sédimentaire	Broutures		Cannelure glaciaire	
	Reverse crescentic fractures		Mini crag and tail	
	Broutures inversées		Petites queues-de-rat	

FIGURE 4 Composite sketch of small-scale forms of glacial erosion, including (A) lunate fracture, plan and section; (B) crescentic gouge, plan and section; (C) crescentic fractures, plan and section; (D) conchoidal fracture; and (E) sichelwanne. [Reprinted from Benn, D. I., and Evans, D. J. A. (1998). "Glaciers and Glaciation," Arnold, London; after Prest, 1983.]

FIGURE 5 Striae on a bedrock surface in Svalbard. Note that the two largest striae contain small chatter marks. Ice flow was from left to right.

invoked to explain the erosion of P-forms, including debris-rich basal ice, saturated till flowing between the ice base and bedrock, subglacial meltwater under high pressure, and ice–water mixtures. Although there is no doubt that potholes are subglacial fluvial features, there is a strong possibility that other types of P-forms are at least

partially glacial and that a hybrid origin is most likely in many cases. Because glacier ice has been observed squeezing into tight crevices and flowing along complex channels and around acute corners on bedrock surfaces, it should not be ruled out as the principal agent of erosion of some P-forms.

Intermediate-scale erosional forms include roches moutonnées, whalebacks and rock drumlins, and crag-and-tails. With a name based upon a resemblance to wavy wigs of the late 18th century, roches moutonnées are asymmetric bedrock bumps with abraded up-ice or stoss faces and quarried down-ice or lee faces (Fig. 7). Each quarried face would have represented a cavity at the glacier base during its formation. Whalebacks, whose name reflects their resemblance to the back of a whale breaking the ocean surface, and rock drumlins are elongated, smoothed bedrock bumps which lack the quarried face of a roche moutonnée. They are differentiated because whalebacks are largely symmetrical and rock drumlins are asymmetrical with steeper stoss faces. Like roches moutonnées, whalebacks and rock drumlins are usually adorned with small-scale erosional features. The lack of quarried lee faces is often assumed to reflect the lack

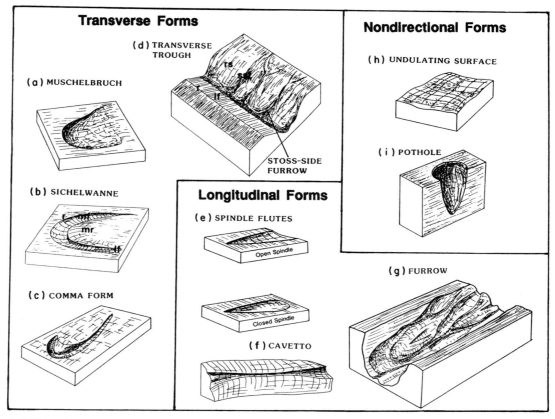

FIGURE 6 Classification of P-forms. [Reprinted from Kor, P. S. G., Shaw, J., and Sharpe, D. R. (1991). "Erosion of bedrock by subglacial meltwater, Georgian Bay, Ontario: a regional view," *Can. J. Earth Sci.* **28,** 623–642.]

FIGURE 7 Roche moutonnée in Newfoundland, Canada. Ice flow was from left to right.

of cavities at the bed during formation, although other ideas have been put forward. For example, the forms of whalebacks and rock drumlins may reflect the shape of glacially molded preglacial bedrock bumps, the remolding of roches moutonnées and the removal of quarried faces due to changes in ice flow direction, and a bedrock structure that may be unfavorable to quarried lee faces. Crag-and-tails, the most famous of which is probably Edinburgh Castle and the Royal Mile, are streamlined hills consisting of a residual bedrock crag at the up-ice end and a tapering tail of less resistant rock extending down-ice.

The tail represents a zone of protection afforded by the crag during glacier streamlining.

Large-scale erosional landforms constitute the most impressive and often most scenic features on earth. Glacial erosional landscapes are composed of a mosaic of rock basins and overdeepenings, the location and size of which are dictated predominantly by the bedrock structure and the thermal regime of glacier ice over time. Accelerated abrasion and quarrying over zones of more intensive fracturing, jointing, or weathering produce a landscape of overdeepened hollows separated by more resistant ridges or bumps. Overdeepenings occur at a wide range of scales. For example, at smaller scales rock basins are separated by resistant rock bars (riegels) in mountain valleys. At larger scales vast depressions occur at the junctions of different bedrock types, such as Great Bear and Great Slave lakes which lie at the junction of Devonian/Cretaceous rocks and more resistant shield terrain. Basins are also eroded into soft sediments by one of three processes: (1) normal plucking, which would be at an accelerated rate; (2) glaciotectonic thrusting or the lateral and vertical displacement of large rafts or blocks of material in the proglacial stress field of an advancing glacier (Fig. 8); and (3) excavational deformation or the net advection of subglacial deforming sediment from submarginal to marginal parts of a glacier snout. Probably the best known of the glacial erosional forms are troughs and fjords, whose distinctive cross and

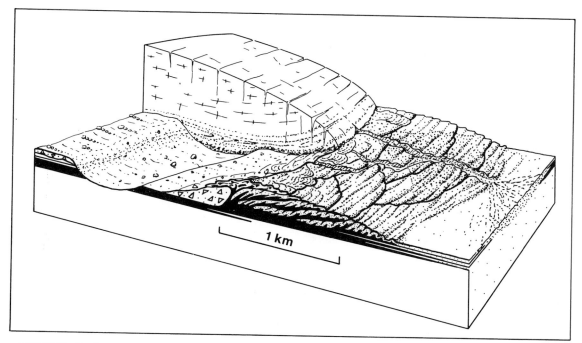

FIGURE 8 Basin excavation associated with the formation of a thrust block moraine at the margin of a glacier moving over soft sediments. [Reprinted from van der Wateren, F. M. (1995). Processes of glaciotectonism. *In* "Modern Glacial Environments" (J. Menzies, ed.), pp. 309–335, Butterworth-Heinemann, Oxford.]

FIGURE 9 Jokelfjord, north Norway.

long profiles reflect their glacial erosional origins (Fig. 9). Cross profiles are often referred to as U-shaped but in fact are best approximated by the formula for a parabola:

$$V_d = aw^b$$

where w is the valley half-width, V_d is valley depth, and a and b are constants. However, true cross profiles deviate from this mathematical parabola largely due to the production of breaks in slope by pulsed erosion through time. These effects have been modelled by imparting a valley glacier on a fluvial, V-shaped valley (Fig. 10). Basal velocities below this glacier are highest part way up the valley sides and lowest below the glacier margins and center line. By assuming that the erosion rates are proportional to the sliding velocity, the greatest erosion occurs on the valley sides, thereby causing broadening and steepening of the valley. The development of the steep sides of troughs and fjords may be aided by pressure release or dilatation in the bedrock. Trough and fjord long profiles have long been employed as a means of classification. Specifically, a fourfold classification recognizes (1) alpine types, cut by valley glaciers emanating from high ground; (2) Icelandic types, with closed trough heads at their upper ends, having been cut by ice spilling over from surrounding plateaux; (3) composite types, which are through troughs or through valleys open at both ends and cut beneath an ice sheet; and (4) intrusive or inverse types, which are cut against the regional slope as illustrated by the New York Finger Lakes. Overdeepenings along fjord and trough long profiles separated by sills or thresholds appear to represent areas of increased glacier discharge such as at the junctions of tributary valleys. In mountainous terrain often the most recognizable glacial erosional forms are cirques, with their steep backwalls and overdeepened floors (Fig. 11). Thought to be a result of mechanical freeze-thaw weathering at the backwall and glacial erosional processes in the overdeepening, cirques

develop over time into a variety of forms ranging from simple, distinct features to compound cirques, cirque complexes, staircase cirques, and cirque troughs. At a landscape scale, the evolution of cirques gives rise to other features typical of alpine scenery such as aretes, horns, and pyramidal peaks. The final large-scale glacial erosional form, the strandflat, is a controversial one. Strandflats, named after the Norwegian examples, are extensive, undulating rock platforms located close to sea level around the coasts of high-latitude landmasses. Where they are partially submerged, they appear as a zone of low rocky islands or skerries (Fig. 12) and they are common to the mouths of fjords where water depths are shallow. Controversy surrounds the interpretation of these features, and four main mechanisms have been proposed for strandflat construction: (1) frost action combined with active sea-ice rafting, (2) marine erosion, (3) subaerial erosion, and

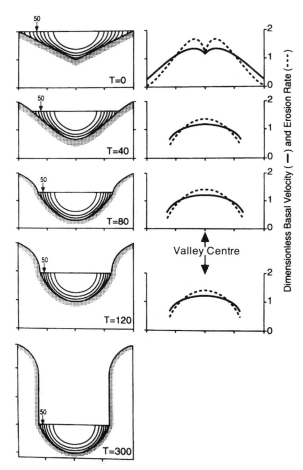

FIGURE 10 Modelled evolution of a trough cross profile at time steps T0–300 and corresponding basal velocities and erosion rates. [Reprinted from Harbor, J. M. (1992). "Numerical modelling of the development of U-shaped valleys by glacial erosion," *Bull. Geol. Soc. Am.* **104,** 1364–1375. By permission of the Geological Society of America.]

FIGURE 11 Cirques in the Torngat Mountains, Labrador, Canada. (Aerial photograph by Energy, Mines and Resources, Canada.)

(4) subglacial erosion. It is most probable that the wider strandflats that merge with the floors of cirque, trough, and fjord floors and which contain direct evidence of glacial erosion are likely to have been at least partly shaped by glaciation. Moreover, because the strandflats of Norway merge inland with a partially preserved preglacial (paleic) surface, they are most likely to be glacially modified etchplains that have escaped deep erosion at the margins of the former ice sheets.

At the landscape scale, regional associations of glacial erosional forms provide insights into the dynamics of former ice sheets. Landscapes of areal scouring are characterized by rock knobs, roches moutonnées, and overdeepened

FIGURE 12 Strandflat and skerries off the west coast of north Norway.

basins (known as cnoc-and-lochain topography in Scotland). Landscapes of selective linear erosion are characterized by deep troughs and fjords separated by plateaux whose surfaces often remain largely unmodified due to protection by cold-based ice. Landscapes of little or no erosion contain vast tracts of preglacial landforms and sediments and weathering horizons that have survived glaciation due to widespread protection under cold-based ice. Alpine landscapes contain forms characteristic of repeated occupation by valley and cirque glaciers, including cirque complexes, aretes, pyramidal peaks, breached watersheds, hanging valleys, and troughs.

C. Processes of Glacifluvial Erosion

Glacial meltwater is a very effective agent in the production of landforms associated with glaciation. It impacts on the glacier bed after making its way from supraglacial to englacial drainage networks via moulins or cylindrical shafts at the surface. It also impacts on proglacial areas to varying extents depending upon discharge. Erosion by meltwater can be subdivided into mechanical (four types) and chemical processes. First, with respect to mechanical processes, because glacial meltwater is usually fast flowing and heavily charged with suspended particles and/or bedload, erosion can cause considerable damage to bedrock and cohesive beds by the process of corrasion or abrasive wear. This involves the gouging of flakes of rock from the bed by impacting particles. The effectiveness of this process is dictated by the angle of incidence or the angle at which the particle hits the bed, the flow velocity, the particle concentration in the flow, the relative hardness of particle and surface, and the particle size. The second type of mechanical glacifluvial erosion is cavitation, whereby pressure fluctuations within the turbulent water flowing over a rough bed cause bubbles to form. The collapse of such bubbles as they move from low- to high-pressure areas may cause high impact forces at channel walls. Third, fluid stressing arises from forces imparted directly by flowing water, which is particularly effective on noncohesive beds but can also prize off joint-bounded fragments on hard rock beds. Finally, on cohesionless beds individual particles can be set in motion once a critical velocity is reached (threshold of particle motion). Chemical erosion is being increasingly emphasized as an important set of processes in glacial environments, whereby minerals are dissolved by meltwater flowing over fresh bedrock surfaces or percolating through sediments and then carried out of the catchment. Complex cycles of dissolution and precipitation are also associated with the regelation sliding process beneath temperate ice. Due to the abundance of undersaturated meltwater, extensive fresh rock surfaces, and large areas of water–rock contact, rates of chemical erosion tend to be high in glacierized catchments compared to nonglacierized areas. Erosion rates can be calculated by measuring solute concentrations in proglacial meltwater streams; one such measurement for the Haut Glacier d'Arolla in Switzerland indicates a surface lowering rate by chemical weathering of 14–16 mm kyear^{-1}.

D. Forms of Glacifluvial Erosion

The erosional impact of glacial meltwater is manifested in the form of channels of various shapes and dimensions; these can be subdivided into subglacial, ice-marginal, and proglacial types. In subglacial environments, Nye channels are cut into bedrock or consolidated sediment and occur either in isolation or in complexes (Fig. 13). The latter may form dendritic networks, recording efficient subglacial drainage along discrete conduits, or anastomosing systems wherein channels split and rejoin. Anastomosing systems may represent former distributed drainage in linked-cavity networks or time-transgressive patterns of channel migration. As these channels are cut by meltwater following paths directed by the subglacial hydraulic gradient, they possess undulatory long profiles and can therefore be distinguished from subaerial meltwater channels. Larger versions of Nye channels are tunnel valleys which can reach dimensions of >100 km in length and up to 4 km wide, are often at least partially buried by postglacial sediments, and often terminate in large subaerial ice-contact fans constructed at the former ice margin by the emerging subglacial meltwater. A major problem in interpreting tunnel valley genesis involves their huge size; if they ever experienced bankfull conditions it would imply water discharges far in excess of those that could be maintained by steady-state basal melting. This has been addressed by two main theories. First, steady-state drainage over a deforming substrate involves the meltwater evacuation of sediment continually squeezing into the tunnel. This results in the long-term deepening and widening of the tunnel valley. Theoretical work indicates, however, that sediment deformation would not occur near tunnel valleys where meltwater evacuation is efficient. Second, catastrophic drainage by glacial lake outburst floods beneath ice sheets may be responsible for the rapid incision of tunnel valleys. Evidence of several generations of incision in most locations indicates that if catastrophic drainage was responsible it took place on more than one occasion.

Ice-marginal or lateral meltwater channels are particularly important in reconstructing glacier recession, especially in polar environments where most drainage was directed along the margins of cold-based glacier snouts (Fig. 14). At the southern limits of the mid-latitude ice sheets, some preglacial rivers were diverted along the ice margin and in some cases remained in their new ice-directed courses long after deglaciation. The vast deep channels produced in this way are called Urstomtaler in Germany and pradoliny in Poland. The most impressive impacts of glacifluvial erosion, and deposition, are produced in proglacial areas where vast flood tracks such as the Channeled Scablands in Washington State record the decanting of glacial lakes dammed at the ice margin (often referred to using the Icelandic term jokulhlaup). Created by the repeated emptying of Glacial Lake Missoula at the southern margins of the Cordilleran Ice

FIGURE 13 Network of anastomosing Nye channels on the foreland of Glacier de Tsanfleuron, Switzerland. (Photograph by M. J. Sharp.)

FIGURE 14 Lateral meltwater channels cut at the margins of a receding subpolar glacier in the Canadian arctic. (Aerial photograph by Energy, Mines and Resources, Canada.)

Sheet, the Channeled Scablands comprise flood channels, giant cataracts, residual islands, giant gravel bars, and current dunes of immense size and covering an area of 40,000 km² (Fig. 15). Although awesome in their dimensions, the erosional features of the Missoula floods are rel-

atively rare in glaciated basins. More common indicators of glacial lake decanting are proglacial channels known as spillways, which document the spilling of water over cols in local watersheds. Spillways range from simple, wide, flat floored rock channels and slot gorges to complexes

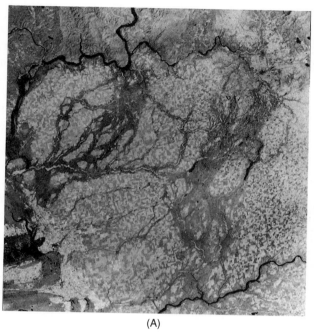

(A)

(B)

FIGURE 15 (A) Landsat image of the tracks (dark grey anastomosing channels) of the Missoula floods; (B) Dry Falls cataract complex from the air. An example of the immense erosional power of the Missoula floods. [Reprinted from Baker, V. R., Greeley, R., Komar, P. D., Swanson, D. A., and Waitt, R. B. (1987). Columbia and Snake River Plains. *In* "Geomorphic Systems of North America" (W. L. Graf, ed.), Centennial Special Vol. 2, pp. 403–468, Geol. Soc. Am., Boulder, CO. By permission of the Geological Society of America.]

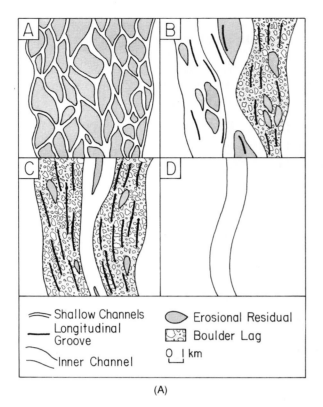

(A)

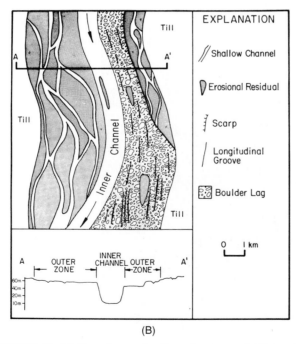

(B)

FIGURE 16 (A) Four stages of spillway development; (B) Landform assemblage and cross profile of spillway development. [Reprinted from Kehew, A. E., and Lord, M. L. (1986). "Origin and large scale erosional features of glacial lake spillways in the northern Great Plains." *Bull. Geol. Soc. Am.* **97,** 162–77. By permission of the Geological Society of America.]

of scour zones and inset channels. A model of spillway development on the North American prairies, the location of numerous spillways that decanted water from the many ice-dammed proglacial lakes at the receding margins of the Laurentide Ice Sheet, is presented in Fig. 16. This depicts the evolution of a spillway by four stages comprising (1) early flood stage of anastomosing channels, (2) rising discharge stage of central channel cutting and scouring of earlier features to produce inner and outer zones, (3) channel incision in central zone and scouring of outer zone, and (4) focusing of flow on central channel and deep incision. The resultant landform assemblage and topographic cross profile is characterized by the sketch in Fig. 16B. Additional features such as iceberg craters can adorn spillway and proglacial flood surfaces, recording the wastage of large ice blocks that were carried from the ice margin by flood water.

III. GLACIAL DEBRIS ENTRAINMENT AND TRANSPORT

The distribution of erratics in formerly glaciated terrains provides clear evidence that glaciers are capable of transporting large volumes of debris and aid in the reconstruction of former glacier flow patterns (Fig. 17). Such patterns may be reconstructed also by mapping till properties such as geochemical signatures and carbonate contents to produce dispersal plumes or trains and dispersal curves (Fig. 18). The entrainment of this debris by a glacier can take place by a variety of mechanisms in supraglacial, englacial, and subglacial positions. Basal ice often possesses a high debris content as a result of the processes of regelation, net adfreezing, and the entrainment of preexisting ice. Ongoing research suggests that the influx of supercooled water to the glacier margin may also be responsible for thick sequences of debris-rich basal ice. Once in the basal ice layers this debris may then move upward through the glacier snout by a combination of the net adfreezing process and compressive flow and/or ice deformation and tectonic stacking. Material may also be entrained and transported via a subglacial deforming layer wherever conditions are favorable. Theory suggests that such deforming layer transport should result in the advection of deformable subglacial materials toward the glacier margin to produce a down-glacier thickening wedge of deformed sediment. Debris is introduced to supraglacial and englacial positions from a variety of sources, including mass movements and snow avalanching from adjacent hillslopes, wind-blown dust, volcanic eruptions, salts and microorganisms from sea spray, meteorites, and human pollutants. Because the majority of the debris is derived from mass movements, in the absence of

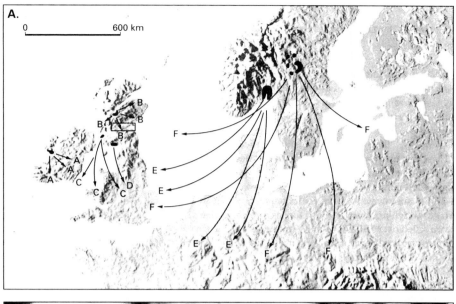

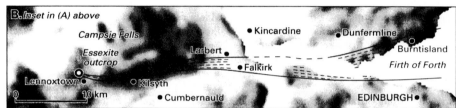

FIGURE 17 Map of erratic dispersal by glacier flow in northwest Europe. Map A shows regional indicators as follows: (A) Galway granite; (B) Rannoch granite; (C) Ailsa Craig riebeckite-eurite; (D) Criffell granite; (E) Oslo rhomb porphyry; (F) Dala porphyries. Map B shows the Lennoxtown boulder train in the Forth Valley lowlands of Scotland. [Reprinted from Lowe, J. J., and Walker, M. J. C. (1999). "Reconstructing Glacial Environments," Longman, London.]

surrounding hillslopes supraglacial debris may be negligible. Supraglacial debris is incorporated into englacial positions either by burial under snow in the accumulation area or by falling down crevasses and moulins mostly in the ablation area.

Once in or on the glacier, debris is transported supraglacially, englacial, subglacially, or through a combination of those pathways (Fig. 19). This will impart a range of characteristics on the debris that can be used after deposition to interpret sediment origins. Transport via supraglacial and englacial pathways is termed passive in order to distinguish it from the active transport that takes place at the glacier bed. This differentiates the two environments based upon the amount of particle modification or comminution that takes place. In supraglacial and englacial environments particle modification is relatively slight but takes place as (a) particle size reduction by periglacial and chemical weathering; (b) fracture and edge rounding due to inter-particle stresses in areas of heavy debris concentrations; (c) englacial shearing be-

tween particles in ice with high debris concentrations; and (d) reworking in supraglacial and englacial streams. In subglacial environments the frequent interactions between particles and between particles and the bed result in substantial modification to particle form and size. Particle fracture and abrasion produces greater comminution and results in considerable size reduction of subglacial material, thereby explaining the dominance of rock flour in tills of subglacial origin. The morphology or form of particles is also affected, whereby individual clasts tend to be edge rounded and facetted and often covered in striae. The development of abrasion and fracture patterns on clasts in subglacial environments is depicted in Fig. 20. In contrast, supraglacially transported clasts tend to display most of the characteristics imparted to them by the processes responsible for their delivery to the glacier surface (e.g., high angularity).

Transport pathways within a glacier will give rise to diagnostic debris concentrations in the ice (Fig. 19). The term debris septum is used to describe the layering of

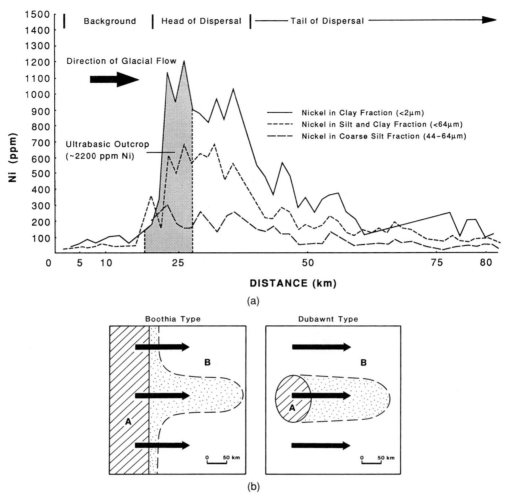

FIGURE 18 Examples of till dispersal by glacier flow: (a) dispersal curve for nickel concentration in tills downglacier from an ophiolitic complex at Thetford Mines, Quebec, Canada. [Reprinted from Shilts, W. W. (1993). "Geological Survey of Canada's contributions to understanding the composition of glacial sediments," *Can. J. Earth Sci.* **30,** 333–353]; (b) classification scheme for dispersal trains based upon northern Canadian examples of till dispersal. Different rock types are marked A and B, ice flow is indicated by the arrows, and the debris dispersed by ice flow from rock type A is stippled. The Boothia type requires ice streaming, whereas the Dubawnt type requires dispersed regional flow. [Reprinted from Dyke, A. S., and Morris, T. F. (1988). "Drumlin fields, dispersal trains and ice streams in arctic Canada," *Can. Geogr.* **32,** 86–90.]

material in a glacier. Bed-parallel debris septum refers to the concentrated zone of debris close to the margins and the bed of a glacier. This includes material in the basal tractive zone and the suspension zone close to the bed. The flowlines carrying this septum can converge around bedrock bumps or at valley confluences where they are forced to leave the bed and form a medial debris septum and ultimately a medial moraine. Similarly, a combination of the debris delivery process and transport pathway gives rise to a variety of supraglacial forms of debris accumulation. Supraglacial lateral moraines accumulate wherever

debris can fall in large quantities onto glacier margins, as is common at the base of scree slopes. Medial moraines are classified according to their evolution, whereby ablation-dominant moraines emerge in ablation zones by the melt-out of englacial debris septa formed further up-glacier, ice–stream interaction moraines form at the intersection of confluent valley glaciers below or close to the equilibrium line, and avalanche-type moraines are discontinuous accumulations of rockfall debris gradually transported downglacier from their original point of impact.

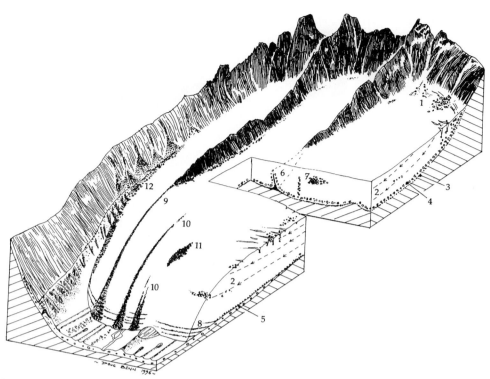

FIGURE 19 Debris transport pathways through a valley glacier: 1, rockfall debris entrained supraglacially and buried by snow or ingested in crevasses; 2, ice flowlines transporting debris away from the accumulation area surface and toward the ablation area surface; 3, basal tractive zone; 4, suspension; 5, basal till with limited deformation; 6, debris septum elevated from the bed below a confluence; 7, diffuse septum and cluster of rockfall debris; 8, debris elevated from the bed by marginal stacking (net adfreezing, etc.); 9, ice–stream interaction medial moraine; 10, ablation-dominant medial moraines; 11, avalanche-type medial moraine; 12, supraglacial lateral moraine. [Reprinted from Benn, D. I., and Evans, D. J. A. (1998). "Glaciers and Glaciation," Arnold, London.]

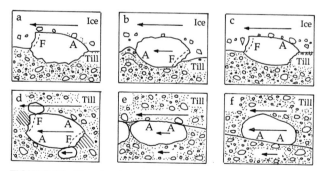

FIGURE 20 The development of clast wear patterns beneath sliding ice and within a deforming layer, showing principal locations of abrasion (A) and fracture (F) and relative velocities (arrows): (a) lodged clast with stoss-lee form below sliding ice; (b), (c) double stoss-lee morphology resulting from a two-stage process of ploughing (b) and lodgement (c); (d) double stoss-lee clast resulting from a one-stage process within a deforming layer, with low-pressure zones shaded; (e), (f) flat, polished facets eroded on upper- and lower surfaces of clasts adjacent to a shear plane. [Reprinted from Benn, D. I., and Evans, D. J. A. (1998). "Glaciers and Glaciation," Arnold, London.]

IV. GLACIAL AND GLACIFLUVIAL DEPOSITION

A. Subglacial and Englacial Environments

Although the deposition of sediment in englacial positions (e.g., crevasse fills, eskers) must be regarded as temporary storage rather than final deposition, it is included here because the release of debris by melting can be directly from englacial locations (e.g., stagnant debris-rich basal ice). Four major processes are involved in the release of debris from basal glacier ice. First, lodgement involves the plastering of debris from the base of a sliding glacier onto a rigid or semi-rigid bed. This involves either frictional retardation of particles, ploughing of bed obstacles by particles, or plastering of whole debris-rich ice masses (Fig. 21). Second, melt-out involves the release by melting of debris from stagnant or very slowly moving debris-rich ice. This can take place also in supraglacial environments (see below), but in subglacial/englacial situations it accrues through the input of geothermal heat, sensible

(a) Frictional retardation against bed

(b) Prow of soft sediment
or clast provides obstacle

(c) Lodgement of debris-rich ice mass

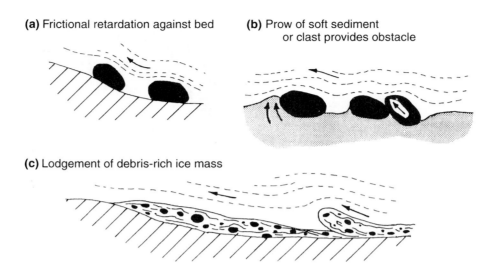

FIGURE 21 The various mechanisms involved in the lodgement process. Note that in (c) the final release of the debris is by melt-out. [Reprinted from Benn, D. I., and Evans, D. J. A. (1998). "Glaciers and Glaciation," Arnold, London. After the work of G. S. Boulton.]

heat from incoming meltwater, and heat from the atmosphere if the ice is thin. As melt-out proceeds the sediment released is subject to thaw consolidation, the impact of which varies according to the original debris concentration in the ice. If this concentration is high and the water produced by melting can drain freely away from the site, thereby reducing porewater pressures, then delicate englacial structures may be preserved in the resulting sediment (melt-out till). Third, debris may be deposited by gravity once it has been released from the ice. This takes place in cavities between the ice base and the bed and usually involves (a) the release of material by individual clast expulsion or peeling off of debris-rich ice layers due to changes in pressure and strain as ice moves over a cavity or (b) influxes of debris driven by basal melting, including fine slurries emerging from between the sliding ice and rock and the dropping of particles from the melting ice base. Finally, deposition may take place in deforming layers at the ice base wherever the driving stresses fall or the shear strength of a layer increases. The most important factor causing short-term changes in shear strength is a change in porewater pressure. Specifically, the frictional strength of the sediment rises as a result of dewatering, which may reflect short-term changes in meltwater supply to the bed or longer term changes to glacier configuration or drainage. Because overburden pressures and hence frictional strength increase with depth, deposition in a deforming layer occurs from the base upward. This may be reversed only if the frictional strength of different sedimentary layers in the deforming bed results in stronger materials overlying weaker strata (e.g., sands overlying

clays could result in the rafting of nondeforming sand over deforming clay). Additionally, the deposition process may involve intermediate changes in the state and style of deformation. For example, a dilatant pervasively deforming till may be deposited with the characteristics of its dilatant state or may undergo dewatering and collapse to undergo brittle deformation prior to deposition.

Glacifluvial deposition can take place in cavities or conduits between the ice base and the bed, as well as in englacial tunnels and cavities from which the sediments are lowered onto the substrate during glacier melting. The basic processes involved in glacifluvial deposition are essentially the same for subglacial, supraglacial, englacial, and ice-marginal environments and therefore are reviewed only once here. Suspended sediment refers to particles maintained in transport above the streambed. Only in stagnant water will this material settle out under gravity at a rate proportional to grain size, usually producing fine-grained, subhorizontal to horizontally bedded sediments. However, high discharges can produce high sediment concentrations of mixed grain sizes or hyperconcentrated flows which result in much coarser sediments. Bedload refers to particles swept along the streambed in continuous or intermittent contact with the bed. Particles either slide, roll, or saltate (jump) and, when deposited, produce a range of sedimentary features referred to as bedforms, which provide information on stream power and are subdivided into plane beds, ripples, dunes, and antidunes. When deposited in ice-contact environments such as subglacial, englacial, supraglacial, and immediate ice-marginal settings, glacifluvial sediment sequences are usually disturbed by later melt-out or reactivation of surrounding or underlying ice and therefore contain numerous faults and folds.

B. Supraglacial Environments

It was outlined in the transport section above that debris accumulates on a glacier surface as a variety of lateral and medial moraine types. Some glacier snouts in mountain-

ous terrain possess an almost complete debris cover due to the large sediment delivery rates from surrounding slopes. Not only does this debris cover severely affect the ablation rates of such glaciers, it also produces a distinctive set of processes and landforms. Once a debris cover exceeds 5–10 mm, it insulates the underlying ice, retarding the melt rate compared to surrounding bare ice surfaces. The huge variability in debris cover and composition on a glacier results in large differences in the thermal properties

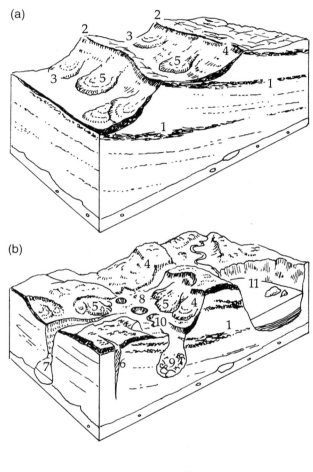

(a)

(b)

(A)

FIGURE 22 (A) Supraglacial debris concentrations on the Hooker Glacier, New Zealand. (B) The evolution (a–c) of topographic inversion and glacier karst on an ablating debris-mantled glacier: 1, englacial debris bands; 2, supraglacial ridge controlled by debris band; 3, trough produced by accelerated melting of debris-free ice; 4, backwasting ice slope; 5, debris flow; 6, crevasse enlargement by ice melt; 7, subglacial conduit; 8, sink holes; 9, collapsed tunnel roof; 10, enlargement of sink hole by ice melt and collapse; 11, lake with backwasting margins; 12, dead ice; 13, ice-free hummocky terrain; 14, supraglacial deposits; 15, ponds; 16, subglacial deposits. [Reprinted from Benn, D. I., and Evans, D. J. A. (1998). "Glaciers and Glaciation," Arnold, London. After the work of J. Kruger.]

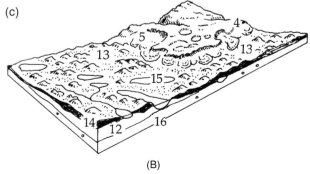

(c)

(B)

of the ice over short distances, thereby leading to differential ablation. The result is the differential lowering of the ice surface and the production of dirt cones and elongated mounds which lie over zones of relatively debris-rich ice (e.g., septa, crevasse fills, debris-rich regelation layers). As the relative relief increases, the sides of the cones and mounds become unstable, and sediment moves either by gravitational mass failure or meltwater reworking into the debris-free depressions, thereby causing a reversal in the melting rates. As melting accelerates over the glacier surface as a whole, depressions and crevasses fill up with water or act as drainage pathways for supraglacial streams. This results in the deposition of stratified sediments (processes outlined above) and mass flows. The constant redistribution of sediment over the glacier surface results in widespread changes in the patterns of ablation and the supraglacial topography, known as topographic inversion, and often involves the development of glacier karst (Fig. 22).

C. Ice-Marginal Environments

Like supraglacial environments, terrestrial ice-marginal settings are characterized by gravitational mass flow and fluvial processes (subaqueous settings are reviewed below). At stationary or slowly moving margins, debris supplied by these processes accumulates in ice-contact aprons or ramparts. If rates of debris supply are high, these accumulations may be so large that they restrict glacier flow and/or blanket the entire glacier snout. At subpolar glacier margins, aprons are produced by dry calving from the snout cliffs. As a result the moraines of some glaciers remain ice cored long after recession of the glacier snout from the area, especially in high-latitude and high-altitude regions characterized by permafrost (Fig. 23). Particularly diagnostic of glacier margins are the processes involved in the relocation of material, associated with periods of glacier advance either on a seasonal basis or over more prolonged periods of sustained positive mass balance. Three main processes are active in the relocation of material, specifically squeezing, pushing, and proglacial glacitectonics. Squeezing of sediment is due to static loading of water-saturated materials by the ice and is most common during the ablation season when marginal sediments possess very high water contents. Material can be squeezed out from beneath the ice margin due to the pressure gradient that exists between sediment overlain by glacier ice and unconfined proglacial sediment or it can be squeezed into low-pressure sub-marginal cavities or crevasses. Pushing involves forward movement of the glacier rather than just static loading. It is essentially the bulldozing of water-soaked marginal sediment that has arrived at the glacier margin by other processes such as squeezing, dumping from the ice surface, or glacifluvial

(A)

(B)

(C)

FIGURE 23 Examples of active ice-marginal debris accumulations: (A) the debris-covered snout and ice-contact fans/ramps of the Batal Glacier, Lahul Himalaya; (B) debris apron accumulating below the cliffs of the Eugenie Glacier, arctic Canada; (C) exposure of buried glacier ice in an ice-cored lateral moraine, arctic Canada.

deposition. Proglacial glacitectonics involves the large-scale tectonic disturbance and dislocation of unconsolidated sediments and weak rock by glacier advance. It involves ductile or brittle deformation or a combination of the two in the folding, thrusting, and stacking of rafts or blocks of proglacial material (Fig. 24).

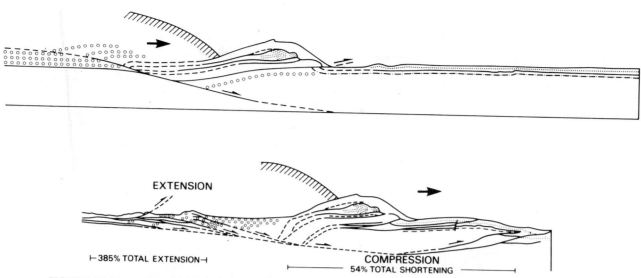

FIGURE 24 Two stages in the sequential development of a moraine by proglacial glacitectonics. [Modified from Croot, D. G., ed. (1988). "Glaciotectonics: Forms and Processes," Balkema, Rotterdam.]

The basic depositional processes in ice-marginal glacifluvial systems have been reviewed above, but some process-form relationships are specific to ice-marginal and proglacial settings. Rivers draining glaciers and glaciated catchments are subject to large variations in meltwater supply over several timescales. Therefore, their channel networks possess distinctive morphologies. Proglacial outwash plains, often referred to using their Icelandic name "sandur," are networks of shifting sediment-floored channels. Systematic downstream changes in overall morphology attest to the processes taking place on a sandur. In the proximal zone meltwater is confined to a few deep and major sediment-floored channels which may be the extension of subglacial Nye channels. In the intermediate zone, flow is in a complex network of wide and shallow braided channels which shift position frequently, and many may contain meltwater only during high-discharge events. Some channels may be inactive for long periods due to channel switching. Finally, in the distal zone, channels are very shallow and often merge to produce sheet flow during high discharges. In addition to channels, a sandur surface is made up of numerous bars in which sediment is temporarily stored as it migrates downstream.

V. GLACIAL AND GLACIFLUVIAL SEDIMENTS AND DEPOSITIONAL LANDFORMS

A. Glacial and Glacifluvial Sediments

The primary sediments deposited by glacier ice are assembled under the umbrella term till, but, due to the complex interactions between subglacial processes on spatial and temporal scales, tills are often very difficult to interpret and classify (Fig. 25). Lodgement till is sediment plastered onto the glacier bed by a sliding glacier. Due to dewatering they are usually overconsolidated and contain numerous subhorizontal joints that represent shear planes. The fine-grained matrix of a lodgement till is the product of abrasion and crushing, and the larger clasts often possess stoss-and-lee morphologies similar to miniature roches moutonnées (Fig. 20). These clasts possess strong A-axis (long axis) alignments imparted by the stress from overriding ice. Melt-out till is sediment released by the melting of stagnant or slowly moving debris-rich glacier ice, directly deposited without subsequent transport or deformation. Where debris concentrations in the melting ice are high and the drainage from the site is efficient, melt-out tills should preserve a large percentage of the englacial structures. However, it is now understood that such conditions are rare, and the preservation potential of a relatively undisturbed melt-out till is low. Where melt-out tills have been observed in the ancient record they typically consist of interbedded diamictons and stratified sediments and possess strong clast alignments. Glacitectonite is subglacially sheared rock or sediment that retains some of the structural characteristics of the parent material. It displays evidence of ductile or brittle deformation or a combination of the two, and pods of relatively undisturbed materials such as sand and gravel may exist in sequences of highly deformed silts and clays due to variations in strain response to the stress from overriding ice. A vertical sequence through a glacitectonite commonly displays increasing deformation up section as the result of

FIGURE 25 The major till types: (A) lodgement till, Scotland; (B) melt-out till, Alberta, Canada; (C) glacitectonite grading upward into deformation till, Scotland; (D) deformation till with stratified lenses produced by phases of subglacial meltwater discharge in canals, Yorkshire, England.

patterns of strain in the subglacial deforming layer. Deformation till is defined as rock or sediment that has been disaggregated and completely or largely homogenized by shearing in a subglacial deforming layer. It often includes streaked inclusions or tectonic lamination but can appear massive in structure and therefore difficult to differentiate from other till types. The analysis of till thin sections under the microscope is beginning to aid in such differentation because minor shear structures will appear at such scales of study. Sediments released by melt-out on the surface of a glacier are subject to considerable reworking by mass flowage and glacifluvial activity during topographic inversion. Therefore, their traditional classification as "flow till" is inappropriate because the final deposits are not true tills but rather a complex of mass flow diamictons and stratified sediments. These are a product of debris flows whose degree of sorting and stratification vary according to water content and grain size. Extensive folding and faulting in such sediments is a function of ice melt-out after deposition.

Glacifluvial sediments are characterized by stratification and other sedimentary structures at a variety of scales, recording the migration and accretion of bedforms and larger depositional units. Bedforms provide immediate impressions of flow competency. For example, plane beds, ripple cross-lamination, dunes, and antidunes reflect changes in flow velocity and grain size, which are recorded in vertical depositional sequences. At larger scales, scour and fill episodes give rise to a vertical arrangement of trough- or planar-cross bedded sequences, and gravel sheets record the accretion of bars on river beds. During waning flows, fine-grained deposits called silt and mud drapes are laid over the coarser units deposited during earlier high-discharge events. Coarser grained and often poorly sorted hyperconcentrated flow deposits record large flood events such as jokulhlaups in sandur sedimentary sequences. Internal faulting and deformation in glacifluvial sediments record their initial deposition in contact with ice, whether it be at the glacier margin or over solitary ice blocks.

B. Glacial and Glacifluvial Depositional Landforms

Because the internal structures and sediments of landforms are regarded as crucial to the interpretation of genesis, glacial researchers more commonly refer to sediment-landform associations. Because of the relative inaccessibility of glacier beds, the most enigmatic sediment-landform associations are the subglacial bedforms whose origin has been the subject of numerous theories for more than one 100 years. These are subdivided into drumlins, flutings, and Rogen moraine (Fig. 26).

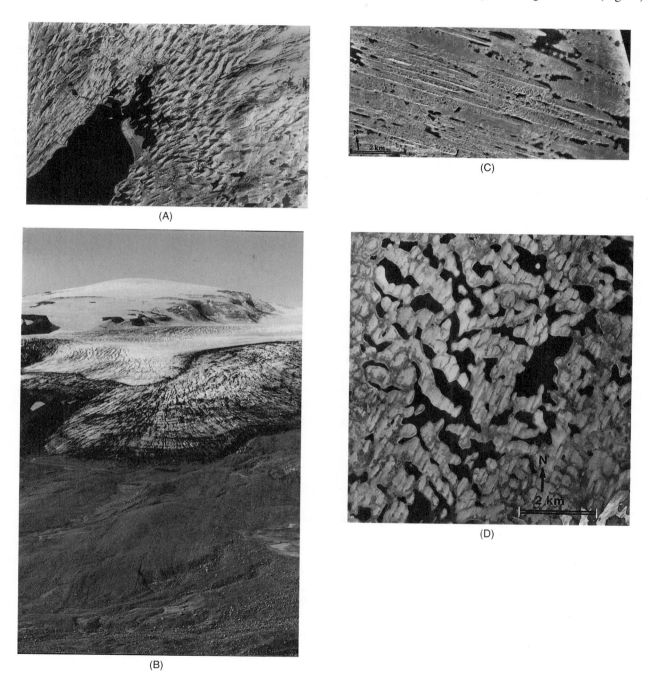

(A)

(B)

(C)

(D)

FIGURE 26 Subglacial bedforms: (A) satellite image of drumlins converging on Donegal Bay, Ireland; (B) small-scale flutings at the margin of Sandfellsjokull, Iceland; (C) large- or mega-scale flutings with associated Rogen moraine, Canada; (D) Rogen moraine, Canada. (Aerial photographs by Energy, Mines and Resources, Canada.]

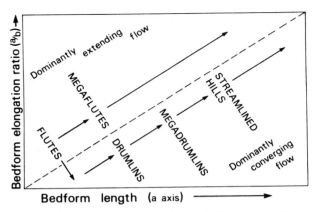

FIGURE 27 Classification of ice flow parallel forms according to the elongation ratio and length. (After the work of J. Rose.)

Drumlins, and flutings, are elongated features with their long axes aligned parallel to former ice flow. They are differentiated by their elongation ratio:

$$E = l/w,$$

where E is the elongation ratio, l is the maximum bedform length, and w is the maximum bedform width (Fig. 27). Further classifications such as mega-lineation and mega-drumlin simply communicate size differences. Derived from the Gaelic term "druim" (rounded hill), a drumlin is typically oval shaped in plan view and resembles an inverted spoon in three dimensions. However, a wide range of shapes have been reported. They occur in clusters or "fields" and contain a wide range of sediments from tills to stratified deposits that may be internally disturbed or undisturbed. Although they are fundamentally different in shape to drumlins, flutings share many of their distribution patterns and internal characteristics. Small flutings

have been studied where they are emerging from receding glacier margins, and their internal structures and clast A-axis alignments appear to support a subglacial deformation origin, especially where they possess large lodged clasts at their up-glacier ends (Fig. 28). The origin of drumlins and mega-flutings involves either (1) erosion of the intervening hollows, (2) accretion of sediment in the positive relief features, or (3) a combination of the two. The most widely accepted model of formation involves sediment erosion and redistribution in subglacial deforming layers wherein stronger or stiffer parts of the deforming layer will remain static or deform very slowly compared to the intervening weaker sediments undergoing high strain. The result is the streamlining of stiff cores to produce drumlins or mega-flutings. Rogen moraine, named after landforms at Lake Rogen in Sweden, are crescentic ridges lying transverse to the former ice flow with their outer limbs bent downglacier. They are closely associated with drumlins and mega-flutings and often exhibit a gradual transition to them, indicating that their formation is intimately linked. Again, the subglacial deformation model appears to be most valid in the explanation of Rogen moraine, as it implies that drumlins, mega-flutings, and Rogen moraine are part of a subglacial continuum recording the direction and relative duration of glacier flow at the time the landform was produced (Fig. 29). However, other theories of Rogen moraine formation have been proposed, including one that invokes the stacking of plates of sediment in areas where the glacier bed is frozen. Other major subglacial features of importance are crevasse-fill ridges. These have been observed at surging glacier snouts where their formation is linked to the squeezing of water-soaked subglacial sediment into the dense networks of crevasses produced during the surge phase.

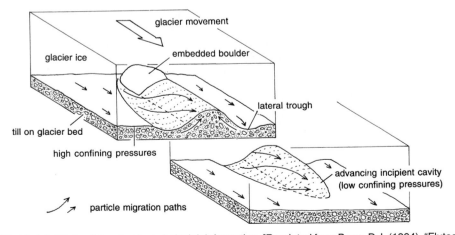

FIGURE 28 Model of fluting formation by subglacial deformation. [Reprinted from Benn, D. I. (1994). "Fluted moraine formation and till genesis below a temperate glacier: Slettmarkbreen, Jotunheimen, Norway," *Sedimentology* **41,** 279–292.]

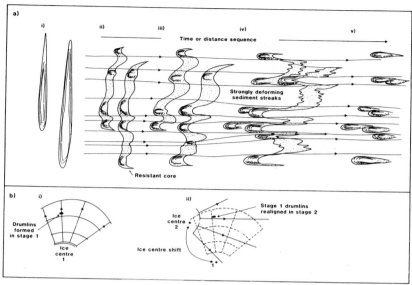

FIGURE 29 Schematic reconstruction of the progressive transformation of flutings to Rogen moraine and drumlins by subglacial deformation during a shift in the ice dispersal center: (a) i, mega-flutings produced during strong regional ice flow; ii and iii, Rogen moraine stage produced during beginning of new ice flow direction; iv and v, drumlin stage produced by continuation of new ice flow direction; and (b) explanation of change in the ice flow direction due to shift in the ice dispersal center. [Reprinted from Boulton, G. S. (1987). A theory of drumlin formation by subglacial sediment deformation. *In* "Drumlin Symposium" (J. Menzies, and J. Rose, eds.), pp. 25–80, Balkema, Rotterdam.]

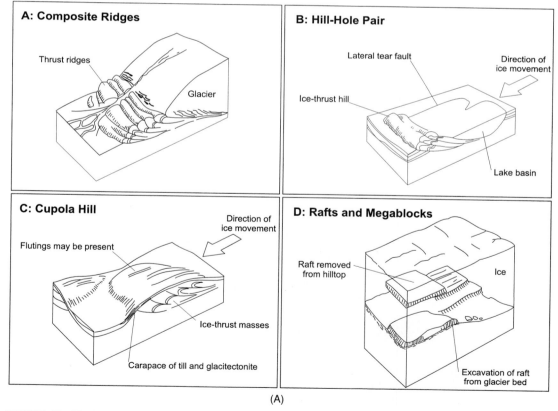

FIGURE 30 Glacitectonic landforms: (A) schematic sketches of the four main types (A–D) of disrupted substrate (Drawing by D. I. Benn); (B) contemporary (top, Canadian arctic) and ancient (bottom, Alberta, Canada) composite ridges. (Photographs by John England and Energy, Mines and Resources, Canada, respectively.)

(B)

FIGURE 30 (*continued*)

A variety of moraines are constructed at the margins of glaciers. General terms used for terrestrial ice-marginal moraines include (1) terminal moraine or the outermost ridge formed during a glacier advance; (2) recessional moraine, deposited during glacier recession from a terminal moraine; (3) frontal and lateral moraine, referring to the location of terminal and recessional moraines;

(4) latero-frontal moraine or a complete loop comprising lateral and frontal components; and (5) annual moraine, where deposition occurs on a yearly basis. Classifications are usually more specific, however, and a more refined terminology has been constructed in order to communicate process-form relationships. The most impressive ice-marginal moraines are often proglacial glacitectonic forms. These include hill-hole pairs, composite ridges/thrust block moraines, and cupola hills. In addition to these, transported rafts or megablocks may also impart a topographical influence on the landscape (Fig. 30). Smaller and more intricate features are push and squeeze moraines produced during minor, often annual, glacier advances. Both types of moraines provide a clear impression of the glacier margin that produced them, with some examples (saw-tooth moraine) actually mimicking snout indentations of radial crevasses (Fig. 31). Because they are constructed by ice-marginal pushing of debris, push moraines, often display fluted ice-proximal surfaces where the glacier has partially overridden them. As squeeze moraines require water-soaked material to be extruded from beneath the glacier snout, they are not as widespread as push moraines, but they can be common in ice-contact lake settings. Large moraines are produced by dumping at the margins of glaciers with a high debris turnover. These are termed dump moraines and ice-marginal aprons and are constructed by a range of mass wasting processes that transport supraglacial melt-out debris to the glacier margin. In mountainous terrain this leads to the production of large latero-frontal moraine loops (Fig. 32). Ice-marginal aprons are deposited below the cliffs of subpolar glaciers by the dry calving of debris-rich ice blocks and contribute to the deposition of low-amplitude ridges of debris after ice recession. Particularly high debris contents in glaciers can result in the construction of latero-frontal fans and ramps, which blanket the glacier snout. After glacier melting these fans and ramps consist of steep

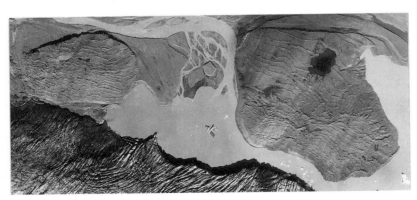

FIGURE 31 Aerial photograph of push/squeeze moraines at the margin of Fjallsjokull, Iceland. (University of Glasgow and Landmaelingar Islands, 1965.)

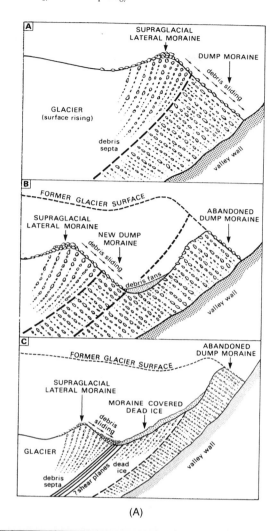

(A)

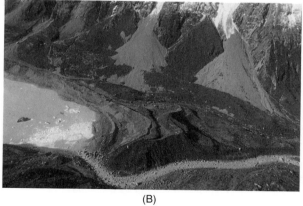

(B)

FIGURE 32 (A) A schematic model of dump moraine formation at the margin of a mountain valley glacier. [Reprinted from Small, R. J. (1983). "Lateral moraines of Glacier de Tsidjiore Nouve: form, development and implications," *J. Glaciol.* **29**, 250–259. By permission of the International Glaciological Society.] (B) Latero-frontal dump moraine at the margin of the Hooker Glacier, New Zealand.

ice-proximal slopes, due to the removal of the underlying glacier, and shallow distal slopes produced by the mass flow and glacifluvial processes responsible for the initial reworking of the supraglacial debris.

Sediment-landform associations that originate englacially or supraglacially and then evolve during progressive lowering onto the substrate by glacier melting are greatly affected by mass flowage, meltwater activity, folding, and faulting and are often listed under the umbrella term "ice-stagnation topography," although glacier stagnation is not a necessary prerequisite (Fig. 33). Medial moraines, because they often contain relatively sparse debris, do not have a good preservation potential and are usually recorded by thin linear debris spreads in upland areas. However, large interlobate moraines up to 100 m high and 10 km wide have been deposited in mid-continental North America by the accumulation of glacifluvial sediments and mass flow diamictons at the junction of large glacier lobes. Although the term "hummocky moraine" has been used to describe a wide range of landforms, it has most recently been restricted to moundy, irregular topography deposited by the melt-out of debris-mantled glaciers. In some circumstances the melt-out process results in the partial preservation of englacial structures, specifically alternating debris-rich and debris-poor ice layers, and the production of transverse linear elements called controlled moraine. The localized dominance of glacifluvial reworking in supraglacial settings results in the deposition of kames, discontinuous or terrace-like features comprising stratified sediments and some mass flow diamictons. Quite often it is difficult to differentiate kames from eskers (see below), and many transitional forms between the two exist. Kame and kettle topography forms tracts of mounds and ridges (kames) and intervening hollows (kettles or kettle holes). The kettle holes represent former postions of buried glacier ice around which glacifluvial sediment (kames) accumulated. Larger accumulations of mainly finer grained stratified sediments may constitute flat-topped mounds or kame plateaux and mark the former positions of ice-walled lake plains. Valley-side terraces containing accumulations of predominantly ice-contact stratified sediments are called kame terraces.

Proglacial sediment-landform associations produced by glacifluvial processes are subdivided into sandars and valley trains. Sandars are largely unrestricted, gently sloping outwash plains, whereas valley trains are confined to major valleys and are therefore repeatedly incised and aggraded to form terraces and infills. Large tracts of glacifluvial sediment may accumulate over the margins of thin glacier snouts and ultimately constitute a pitted sandur. The numerous kettle holes on such surfaces attest to the formerly extensive underlying ice. Additionally, numerous kettle holes on a sandur surface may document the melting of icebergs originally deposited along with the outwash during jokulhlaups. Probably the most prominent glacifluvial landform in glaciated terrain is the esker, a sinuous ridge of stratified sands and gravels deposited in ice-walled channels (Fig. 34). The routing of former meltwater channels in glaciers, and their association with ice-

marginal configurations, is indicated by the overall form of eskers. There are four major types of esker: (1) continuous ridges (single or multiple) that document tunnel fills, (2) ice-channel fills produced by the infilling of supraglacial channels (Fig. 33), (3) segmented ridges deposited in tunnels during pulsed glacier recession, and (4) beaded

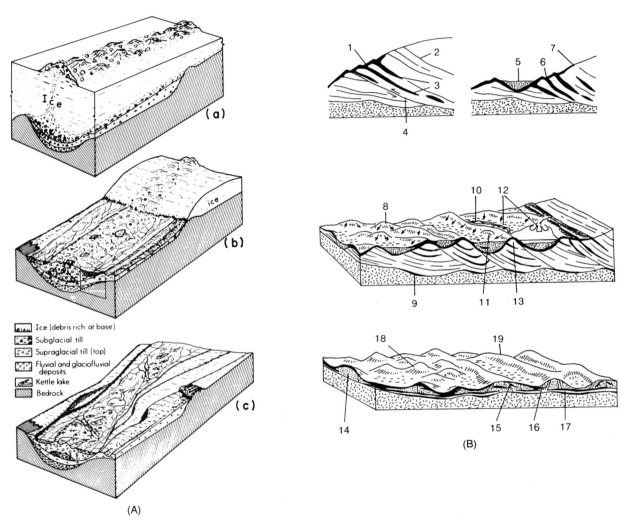

FIGURE 33 The major supraglacial sediment-landform associations. (A) Medial moraine. [Reprinted from Levson, V. M., and Rutter, N. W. (1989). "Late Quaternary stratigraphy, sedimentology and history of the Jasper townsite area, Alberta, Canada," *Can. J. Earth Sci.* **26,** 1325–1342.] (B) Controlled hummocky moraine: 1, debris released at glacier surface; 2, foliation; 3, debris band; 4, shear fault; 5, outwash collecting between melting ice ridges; 6, supraglacial debris accumulating from the melt-out of debris bands; 7, relatively clean ice surface resulting from mass flowage of melt-out debris; 8, small mass flows on stagnant ice hummocks; 9, lodgement till; 10, supraglacial channels; 11, large supraglacial mass flow; 12, delta in supraglacial pond; 13, elongate ice ridge protected from melting by the accumulation of supraglacial sediment from debris bands; 14, absence of collapse structures in sediment deposited directly on bedrock; 15, collapse structures; 16, interdigitating mass flow diamictons and glacifluvial sediments; 17, subglacial melt-out till; 18, hummocky moraine with faint transverse alignments; 19, controlled hummocky moraine ridge. [Reprinted from Benn, D. I., and Evans, D. J. A. (1998). "Glaciers and Glaciation," Arnold, London. After the work of G. S. Boulton.] (C) The three major types of kame. [Reprinted from Brodzikowski, K., and van Loon, A. J. (1991). "Glacigenic Sediments," Elsevier, Amsterdam.]

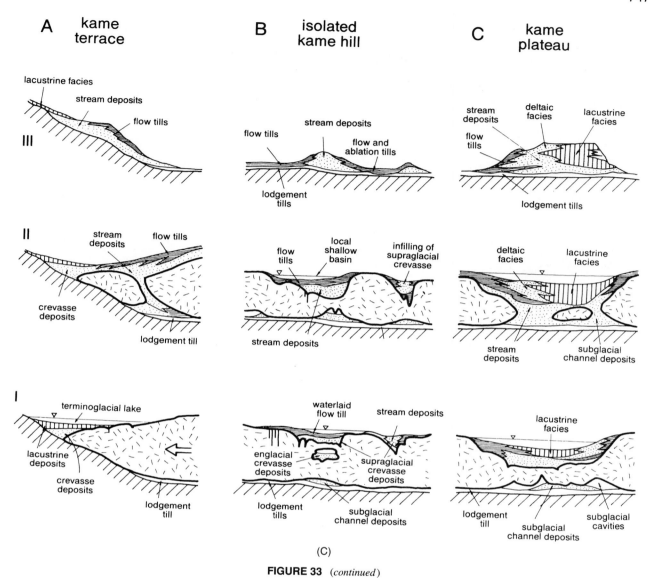

FIGURE 33 (*continued*)

eskers consisting of successive subaqueous fans deposited in ice-contact lakes during pulsed glacier recession. The former englacial position of some eskers is indicated by the occurrence of buried glacier ice or the almost complete disturbance of the stratified core.

Recent developments in glacial geology and geomorphology have led to the reconstruction of glaciated landscapes using a combination of landforms and sediments in a landsystems approach. This allows an appreciation of the distribution of sediment-landform associations and assemblages, in particular depositional settings (e.g., fjords, alpine topography, shield terrain), as indicators of particular types of glacier (e.g., active temperate, surging, subpolar) or as depositional sequences associated with particular parts of a glacier (e.g., subglacial, supraglacial). This is a powerful technique in the reconstruction of former glacier dynamics in glaciated terrain.

VI. GLACIMARINE AND GLACILACUSTRINE DEPOSITIONAL ENVIRONMENTS

A. Glacimarine and Glacilacustrine Processes

Where glaciers contact lake or ocean water they deposit a unique set of sediments and landforms. Sediment can enter deep water directly from supraglacial, englacial, and subglacial environments or indirectly from icebergs and

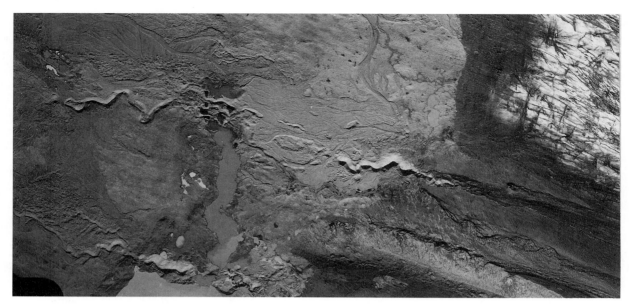

FIGURE 34 Aerial photograph of eskers emerging from the receding margin of Breidamerkurjokull, Iceland. (University of Glasgow and Landmaelingar Islands, 1965.)

floating ice shelves. Unfrozen sediment may also emerge from beneath a glacier at its grounding line (the point at which it floats) or enter the water via glacifluvial processes. Subglacial and englacial streams entering the water column produce jets that carry bedload and suspended sediment. The coarsest sediment is deposited rapidly close to the efflux point as flow decelerates, whereas the finer material is carried away in the jet and turbid plumes.

Because freshwater is less dense than seawater, even when charged with suspended sediment, plumes are particularly effective in fine sediment dispersal in glacimarine environments. The level at which a plume travels through a water body (overflow) is dictated by sediment density and water characteristics. The three levels of suspended sediment transport are overflows, interflows, and underflows. Overflows and interflows provide large volumes of sediment

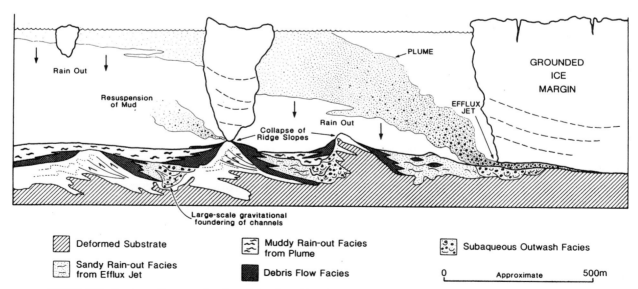

FIGURE 35 A schematic reconstruction of the depositional processes at a retreating glacier margin in contact with deep water. [Reprinted from Eyles, N., and McCabe, A. M. (1989). "The Late Devensian <22,000 BP Irish Sea Basin: the sedimentary record of a collapsed ice sheet margin," *Q. Sci. Rev.* **8**, 307–351.]

for lake and ocean bottom deposition via the process of suspension settling. Flocculation and pelletization aid in the settling process in marine environments. Underflows or turbidity currents are one form of gravitational process, in addition to sediment gravity flows, cohesive and cohesionless debris flows, debris falls, and slumps and slides, that transport large volumes of sediment across lake and ocean bottoms after failure in unstable materials. Icebergs and ice shelves may also contribute debris to the lake or ocean bottom via rafting. This debris is often coarse and easily detected in more distal fine-grained suspension sediments. Icebergs are also very effective at scouring the bed and churning existing sedimentary sequences (Fig. 35).

B. Glacimarine and Glacilacustrine Depositional Forms and Sediments

Glacimarine and glacilacustrine depositional environments are usually zoned according to sediment delivery characteristics and distance from the nearest ice margin. Five zones are used here to summarize the various depositional forms and sediment. First, at the glacier margin is the grounding-line fan or subaqueous outwash fan zone. This is dominated by the accumulation of coarse-grained out-

wash at the efflux point of a subglacial or englacial stream (Fig. 36). Second is the morainal bank zone, characterized by the accumulation of elongated banks of sediment along the ice margin. Morainal banks are fed by supraglacial and calve dumping, the melt-out of englacial and subglacial debris, iceberg melt-out and overturning, the squeezing of subglacial deforming sediment from beneath the ice, and inputs from meltwater streams (Fig. 37). Sequences of numerous small subaqueous moraine ridges are often called De Geer moraines. Third, deposition beneath the floating ice shelf zone of polar glaciers is characterized by the rain-out of debris released by the melting of debris-rich basal ice. This produces a thin and patchy drape of sediment directly beneath the floating ice, but some small morainal banks can be constructed at the flotation point. Fourth, the deltaic zone is dominated by the progradation of glacifluvial bedload. This produces either ice-contact or glacier-fed (non-ice-contact) Gilbert-type deltas with their characteristic topset, foreset, and bottomset bedding (Fig. 38). Finally, in the distal zone, the dominance of suspension settling and sediment gravity flows gives rise to distinctive laminated sediment sequences. Such deposits are often referred to as rhythmites due to their cyclic repetitions of fine- to coarse-grained beds. Annual rhythmites in glacilacustine environments are called varves. In glacimarine environments, cyclic repetitions involve muddy vs

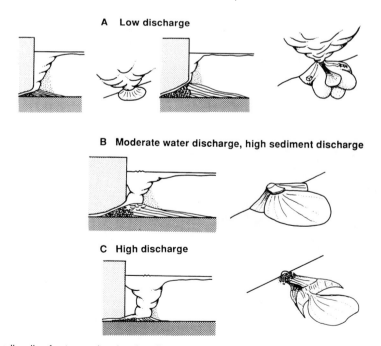

A   Low discharge

B   Moderate water discharge, high sediment discharge

C   High discharge

FIGURE 36 Grounding-line fan types showing the effects of discharge variations. [Reprinted from Powell, R. D. (1990). Glacimarine processes at grounding-line fans and their growth to ice-contact deltas. *In* "Glacimarine Environments: Processes and Sediments" (J. A. Dowdeswell, and J. D. Scourse, eds.), Geological Society Special Publication **53,** pp. 53–73. By permission of the Geological Society.]

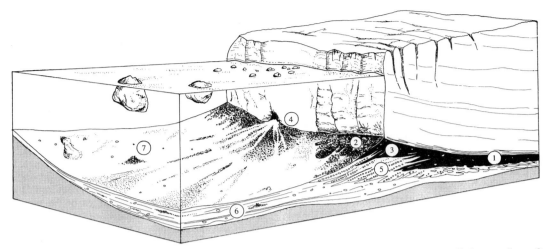

FIGURE 37 The processes, internal sediments, and form of a typical morainal bank: 1, subglacial till; 2, mass flows; 3, mass flow diamictons; 4, subglacial meltwater portal; 5, subaqueous outwash; 6, laminated lake sediments; 7, iceberg dump mound. [Reprinted from Benn, D. I. (1996). "Subglacial and subaqueous processes near a glacier grounding line: sedimentological evidence from a former ice-dammed lake, Achnasheen, Scotland," *Boreas* **25,** 23–36.]

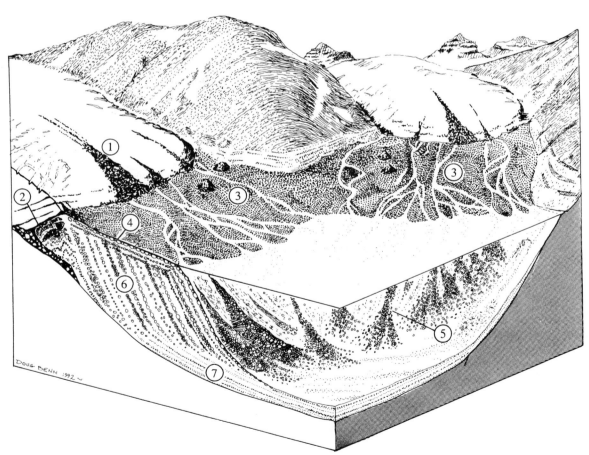

FIGURE 38 Reconstruction of typical Gilbert-type deltas: 1, supraglacial debris; 2, subglacial debris; 3, braided sandur surface; 4, topsets; 5, delta front; 6, foresets; 7, bottomsets. [Reprinted from Benn, D. I. (1992). "The Achnasheen Terraces," *Scott. Geogr. Mag.* **108,** 128–131.]

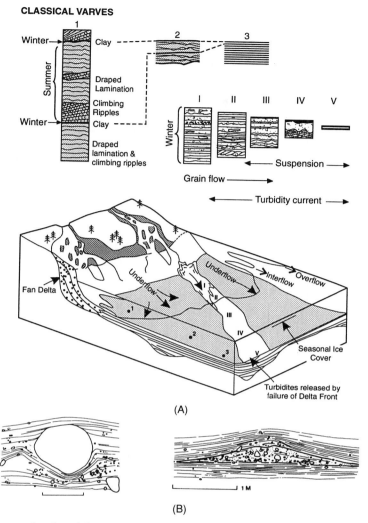

FIGURE 39 (A) A reconstruction of the typical processes and laminated sediments (varves and other rhythmites) in a glacilacustrine environment. [Reprinted from Benn, D. I., and Evans, D. J. A. (1998). "Glaciers and Glaciation," Arnold, London. After the work of N. Eyles and A. D. Miall.] (B) Sketches of typical iceberg dropstone (left) and iceberg dump structures. [Reprinted from Thomas, G. S. P., and Connell, R. J. (1985). "Iceberg, drop, dump and grounding structures from Pleistocene glaciolacustrine sediments, Scotland," *J. Sediment Petrol.* **55**, 243–249.]

sandy laminae (cyclopels and cyclopsams, respectively) and are a response to tidal forcing. Other processes such as subaqueous debris flows, grain flows, turbidity currents, and iceberg dumping produce distinctive structures in distal glacimarine and glacilustrine environments (Fig. 39).

SEE ALSO THE FOLLOWING ARTICLES

COASTAL GEOLOGY • GEOMORPHOLOGY • GEOMORPHOLOGY AND RECLAMATION OF DISTURBED LANDS •

GLACIOLOGY • HYDROGEOLOGY • ROCK MECHANICS • SEDIMENTARY PETROLOGY • STREAMFLOW

BIBLIOGRAPHY

Benn, D. I., and Evans, D. J. A. (1998). "Glaciers and Glaciation," Arnold, London.
Hambrey, M. J. (1994). "Glacial Environments," UCL Press, London.
Menzies, J., ed. (1995). "Modern Glacial Environments," Butterworth-Heinemann, Oxford.
Menzies, J., ed. (1996). "Past Glacial Environments," Butterworth-Heinemann, Oxford.

Glaciology

W. S. B. Paterson
Paterson Geophysics Inc.

GLOSSARY

Ablation All processes that reduce the mass of a glacier: melting followed by runoff, iceberg calving, sublimation, and snow removal by wind. Related terms are annual ablation rate and ablation area, which is the part of the glacier where there is a net loss of mass during the year.

Accumulation All processes that increase the mass of a glacier: snowfall (predominant), avalanches, rime formation, and freezing of rain in the snowpack. Related terms are annual accumulation rate and accumulation area, which is the part of the glacier where there is a net gain in mass during the year.

Calving Process of iceberg formation: ice breaks off the end of a glacier tongue into the sea or a lake.

Crevasse Large crack in the glacier surface formed by tension or shear. Depths of 20–30 m are typical.

Equilibrium line Boundary between the accumulation and ablation areas at the end of summer. Its position varies from year to year according to weather conditions.

Firn Strictly, wetted snow that has survived one summer without being transformed to ice; commonly used, as here, to mean material in the process of transformation from freshly fallen snow to glacier ice.

Grounding line Boundary between grounded and floating parts of a glacier that ends in water.

Ice cap Mass of ice that flows outward in several directions and submerges most or all of the underlying land.

Ice rise Area of grounded ice in a floating ice shelf.

Ice sheet Ice cap the size of Greenland, Antarctica, or the ice-age ice masses. A marine ice sheet is one in which most of the base is grounded below sea level.

Ice shelf Large, thick slab of ice floating on the sea but attached to land or to a grounded ice sheet. It is nearly

Encyclopedia of Physical Science and Technology, Third Edition, Volume 6

always nourished by both snowfall and flow from the ice sheet and sometimes also by freezing on of water at its base.

Ice stream Region in an ice sheet in which the ice flows much faster than that on either side. Ice streams often have no visible rock boundaries.

Icefall Steep, fast-flowing section of a glacier where the surface is broken up into an array of crevasses and ice pinnacles (seracs).

Mass balance Difference between annual accumulation and ablation. The term can be applied either to a whole glacier or to a point on its surface.

Temperate glacier Glacier in which all the ice is at its melting temperature, except for a surface layer, about 15 m thick, that is subject to seasonal variations in temperature. Glaciers that are not temperate are often referred to as cold, or else are called subpolar or polar according to whether or not there is surface melting in summer.

GLACIOLOGY, in the strict sense, is the study of ice in all its forms. The term is often also applied, as in this article, to the more restricted field of the study of glaciers. A glacier is a large mass of perennial ice that originates on land by the compaction and recrystallization of snow and shows evidence of past or present flow. Glaciers may be divided into three main classes: (1) ice sheets and ice caps; (2) glaciers whose paths are confined by the surrounding bedrock, called mountain or valley glaciers according to their location or piedmont glaciers where valley glaciers spread out over a coastal plain; and (3) floating ice shelves. These types frequently occur in combination; an ice cap may be drained by valley glaciers, for example.

The amount of snow falling on the surface of the higher part of a glacier each year exceeds that lost by melting and evaporation. At lower elevations, all the previous winter's snow and some glacier ice are removed by summer melting and runoff. Calving of icebergs may also result in significant loss of mass. The profile of the glacier does not change much from year to year, however, because ice flows under gravity from the accumulation area to the ablation area (Fig. 1).

I. INTRODUCTION

Glacier studies are an active research field at present for several reasons. The past 2.5 million years (Myr) have been characterized by the repeated growth and decay of continental ice sheets. At glacial maxima, ice covered almost all of Canada, the northern part of the United States,

FIGURE 1 Unnamed glaciers, Babine area, Coast Mountains, British Columbia. Lateral moraines show their recent maximum extent. The steep crevassed terminus of the right-hand glacier suggests that it may be advancing. [Photograph by Austin Post, U.S. Geological Survey.]

much of Europe, and parts of eastern Siberia. The resulting reduction in volume of the oceans caused world sea levels to fall by roughly 100 m. Ice ages alternated with interglacials, during which the extent of ice was comparable with, and sometimes less than, that of today. Switching of the climate between these states, with a regular period of roughly 100,000 years, is related to changes in the earth's orbit. Once formed, an ice sheet tends to maintain itself because of its high elevation and because a snow-covered surface reflects up to 90% of incoming solar radiation. Any explanation of ice ages must take into account both the dynamics of the ice sheet itself and its interactions with the atmosphere and ocean, topics that can best be studied on existing ice sheets.

Glaciers and ice sheets produce major changes in the landscape. Raised beaches in areas once covered by ice are evidence of the slow rise of the land after the weight of the ice was removed, a process that is still continuing in regions such as Hudson Bay and the Baltic. Most of the Canadian prairies are covered by material carried by the ice sheet, in some cases for hundreds of kilometers, and left behind when it melted. Erosion by glaciers produces the U-shaped valleys, narrow ridges, and Matterhorn-like peaks characteristic of mountain areas. The mechanisms of erosion, transportation, and deposition can best be studied by looking at existing glaciers. An understanding of these processes is essential for the interpretation of ice-age deposits.

Recent theoretical studies have shown that glaciers possess potential sources of instability. It has been suggested that the ice sheet in West Antarctica (the part lying south of the Pacific Ocean) may be unstable and that atmospheric warming, brought on by increased consumption of fossil fuels, may trigger its complete collapse. This would raise world sea level by about 6 m, submerging the world's ports and the greater part of low-lying countries such as Holland. Assessment of the present state of the West Antarctic Ice Sheet and prediction of its future behavior is a major topic of current research.

A form of instability known to occur in some glaciers and at least two ice caps is the surge or catastrophic advance. Surges in the same glacier happen at regular intervals. Unrelated to climate, they result from some instability at the glacier bed. Moreover, the sediment in the North Atlantic contains evidence of periodic massive discharges of icebergs from the Hudson Bay sector of the Laurentide Ice Sheet during the last ice age. These so-called Heinrich events had a major effect on ocean circulation and climate. The mechanisms of these instabilities is another active branch of research.

In addition to the major growth and decay that mark the beginning and end of an ice age, glaciers also advance and retreat in response to minor changes in climate. However, the relationship is more complex than is usually assumed. Prediction of glacier advances is of practical importance in several countries where mines, highways, pipelines, and hydroelectric installations have been built near glaciers. The theory of glacier flow is now sufficiently well developed that this problem can be tackled, as can the inverse problem of making detailed inferences about past climate from geologic evidence of the former extent of glaciers.

A major recent development in glaciology has been the wealth of information about past climate and atmospheric composition obtained from analysis of drill cores from the polar ice sheets. Cores from the higher parts, where the snow never melts, have provided detailed and continuous records extending back, in one case, for about 425,000 years. Because the ratios of the concentrations of the heavy to light atoms of oxygen and hydrogen ($^{18}O/^{16}O$, $D/^1H$) in precipitation depend on the temperature, the variations of these ratios with depth may be interpreted as the variation of temperature with past time. Air bubbles in the ice provide samples of the atmosphere at the time the ice was formed and show, for example, that there was appreciably less carbon dioxide and methane during the ice ages

than there is now. The buildup of carbon dioxide since the start of the industrial era is also recorded. The ice also contains small amounts of atmospheric fallout such as wind-blown dust, volcanic ash, pollen, sea salts, and extraterrestrial particles. Major volcanic eruptions show up as highly acidic layers. Shallow cores reveal the recent increase of pollutants such as lead and sulfate in the atmosphere, while atomic and nuclear bomb tests are recorded as radioactive layers.

Glacier-fed rivers provide much of the water supply for agriculture in the Canadian prairies and central Asia and for hydroelectric power generation in several European countries. Tunnels have even been drilled underneath glaciers in the Alps and Norway to tap water from subglacial streams. Glaciers act as reservoirs that store water in solid form during the winter and release it in summer; especially large amounts are released in warm, dry summers, when flow in nonglacial rivers is low. Moreover, the annual flow of a glacier-fed river is higher in a period of glacier retreat than when the glaciers are advancing, a fact not always taken into account when planning hydroelectric schemes. Again, the sudden drainage of glacier-dammed lakes or of water stored in glaciers has caused extensive damage in countries such as Iceland, Peru, and Canada. For these reasons, the study of the flow and storage of water in glaciers, and at the glacier bed, is of considerable practical importance.

II. FEATURES OF EXISTING GLACIERS

Ice covers about 10% of the earth's land surface and stores about three-quarters of all fresh water. Table I shows that about 90% of this ice is in Antarctica, with most of the rest in Greenland. The range of recent estimates of the size of the Laurentide Ice Sheet in North America at its maximum extent about 20 kyr ago is given for comparison. The calculated equivalent rises in sea level make allowance for the fact that melting of ice shelves and of ice grounded below sea level changes sea level only to the extent that their thickness exceeds that for flotation.

Antarctica (Fig. 2) can be divided into three parts: the East and West Antarctic Ice Sheets, separated by the

TABLE I   Ice Sheet Areas and Volumes

	Area (10^6 km^2)	Volume (10^6 km^3)	Rise in sea level (m)
Antarctica	13.6	30.1	66
Greenland	1.7	2.7	7
Rest of world	0.5	0.2	0.5
Laurentide (max.)	11.6	16–21	40–52

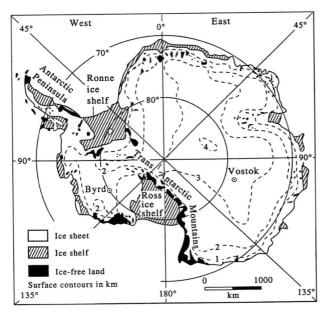

FIGURE 2 Map of Antarctica. [From Paterson, W. S. B., "The Physics of Glaciers," 3rd ed., Elsevier, 1994. Reprinted by permission of Butterworth Heinemann Publishers, a division of Reed Educational and Professional Publishing Ltd.]

Transantarctic Mountains, and the Antarctic Peninsula, which is characterized by relatively small ice caps, valley glaciers, and ice shelves. The East Antarctic Ice Sheet, which contains about 85% of the Antarctic ice, is an elliptical dome with a maximum elevation of over 4000 m. Although the greatest ice thickness is 4800 m, this is in a narrow subglacial trench, and most of the bedrock is above sea level and has low relief. In contrast, the West Antarctic Ice Sheet has a very rugged floor, much of it well below sea level. This is reflected in a lower and much more irregular ice surface.

Ice shelves surround about half the coastline of Antarctica. They occupy all the major embayments and elsewhere appear to owe their existence to shoals on which their outer margins are grounded. The Ross Ice Shelf has the largest area; its thickness varies from about 1000 m at the grounding line to about 250 m at the outer edge, which consists of an ice cliff some 30 m high.

Annual snowfall ranges from about 25 mm, water equivalent, in the central regions of East Antarctica to about 600 mm near the coast. The mean is 170 mm, a value typical of deserts. There is virtually no surface melting even near the coast and although basal melting is significant on some ice shelves, iceberg calving accounts for nearly all the ablation. Because precipitation is low and the ice is thick, velocities ranging from a few meters per year in the central area to about 100 m/yr near the coast would be sufficient to maintain the ice sheet in a steady state. In

fact, the ice does not flow out uniformly in all directions; most of the discharge near the coast is channeled into either large valley glaciers that cut through the mountains or fast-moving ice streams. Most of these drain into the major ice shelves. Typical velocities are a few hundred meters per year in ice streams and about 1 km/yr at the outer edges of the largest ice shelves. Although the total annual accumulation on the ice sheet is reasonably known, the amount of ice lost by calving is uncertain. Thus it is not known whether the ice sheet is growing or shrinking. The extent of the ice sheet is not controlled directly by climate but indirectly by changes in world sea level, which cause the grounding line to advance and retreat. World sea level is controlled by the growth and decay of ice sheets in the Northern Hemisphere.

The surface of the Greenland Ice Sheet (Fig. 3) has the form of two elongated domes; the larger, northern dome reaches an elevation of about 3200 m. Both are displaced toward the east coast because mountain ranges reduce the outflow on that side. Maximum ice thickness is 3200 m. Bedrock is bowl-shaped; the coast is mountainous and much of the central area lies close to sea level. Most of the outflow is channeled into some 20 large outlet glaciers that penetrate the coastal mountains and discharge icebergs into the fiords. Although some of the glacier tongues are afloat, there are no ice shelves. Glaciers in West Greenland, such as Jakobshavns Isbrae, with a velocity of 7 km/yr at its terminus, are the fastest in the world. Mean annual precipitation is 340 mm/yr. Melting and runoff account for over half the total mass loss. Table II, which gives the latest (1984) estimates, shows that, as in Antarctica, the terms in the overall mass budget are not known precisely enough to determine whether the ice sheet is growing or shrinking. However, the data suggest that it is not far out of balance. A better method of finding out is to measure how the surface elevation changes with time. Such measurements reveal a complicated pattern: the central part is thickening by about 0.1 m/yr (ice equivalent thickness) on the western slope and thinning slightly in the east, while the ablation area on the west coast is thinning by about 0.2 m/yr. These measurements were made by repeated optical leveling across the ice sheet. Radar altimeter measurements from satellites indicate that the southern half of the ice sheet thickened by 0.2 m/yr, about 35% of the average precipitation rate, between 1978 and 1985. Although the precision of the radar

TABLE II   Mass Budget of Greenland Ice Sheet (km³/yr, water equivalent)

Accumulation	500 ± 100
Melting and runoff	295 ± 100
Iceberg calving	205 ± 60

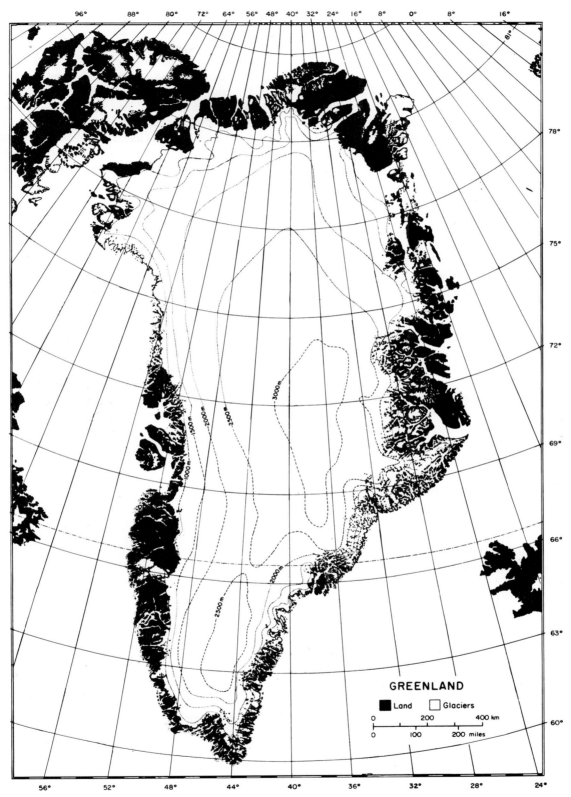

FIGURE 3 Map of Greenland. [Reprinted with permission from Colbeck, S. C. (ed.) (1980). "Dynamics of Snow and Ice Masses." Copyright 1980, Academic Press, New York.]

data has been questioned, this method or laser measurements from an aircraft with a precise three-dimensional navigation system should eventually provide an answer to the important question of the mass balance of the ice sheets.

Ice caps are found in the Canadian and Russian Arctic islands, Svalbard, Jan Mayen, Iceland, some sub-Antarctic islands, and in most major mountain ranges. The largest ice caps outside Greenland and Antarctica are in Arctic Canada where four have areas exceeding 15,000 km². Apart from some small, low-lying ice caps that disappeared during a "climatic optimum" after the end of the ice age and re-formed since, all these ice caps have existed continuously since the last ice age, and the major Arctic ice caps still contain ice dating from that time. Glaciers can exist at all latitudes, even on the Equator in the mountains of Africa, South America, and Indonesia. Mountains in some other tropical areas such as Hawaii also carried glaciers during the ice age.

III. FORMATION AND DEFORMATION OF GLACIER ICE

As fresh snow (density 50–200 kg/m³) becomes buried by subsequent falls, its density increases under the weight of the snow above. This is the first stage in its transformation to glacier ice, which takes place even in the absence of liquid water. Because, unlike other crystalline solids, natural ice is usually within 50°C of its melting point, the molecules are relatively free to move both within the lattice and over the surface of the crystal. Moreover, sublimation (direct change from solid to vapor and vice versa) can occur readily. Movement of molecules is, on average, in the direction that tends to minimize the free energy of the system. Thus snowflakes, with their complex shapes, gradually change into spherical particles, and the larger particles tend to grow at the expense of the smaller ones.

Initially, settling (rearrangement of the snow grains to reduce the air space between them) predominates. After settling has reached its limit, the grains themselves and the bonds that form between them change shape in such a way as to further increase the density. Eventually the air spaces between individual grains become sealed off, and the material becomes impermeable; air that has not escaped to the surface is now in the form of bubbles. At this point the density is about 830 kg/m³ and the firn has, by definition, become glacier ice. Compression of the air bubbles as the ice becomes more deeply buried causes a further slow increase in density up to about 910 kg/m³.

Meltwater greatly speeds up the transformation. It percolates into the snow to a depth where the temperature is still below 0°C and refreezes, filling up the air spaces and forming lenses and sometimes continuous layers of ice. In some cases, refreezing of meltwater may transform the whole of the winter snow into ice in a single summer. Ice formed by refreezing contains few air bubbles.

Figure 4 shows typical depth–density curves for the Greenland Ice Sheet and for a valley glacier in a maritime temperate climate. Ice at the firn–ice transition is 3–5 yr old on Upper Seward Glacier but about 120 yr old at the Greenland site. Ages of 100–200 yr seem typical for much of Greenland. At Vostok Station in East Antarctica, on the other hand, where the precipitation is only 25 mm/yr and the ice temperature is −57°C, the transformation takes about 2500 yr.

Greenland and Antarctic ice cores from below a certain depth contain no visible bubbles, although air is given off when the ice is melted. The air is believed to be present as a clathrate hydrate. In a clathrate compound, the crystal lattice contains voids that can be occupied by other molecules. The term *hydrate* implies that the lattice is formed by molecules of water. The critical depth varies from 900 to 1200 m according to temperature, which corresponds to hydrostatic pressures of 8 to 10 MPa. This is the first reported natural occurrence of a clathrate hydrate on the earth, although methane hydrate forms in

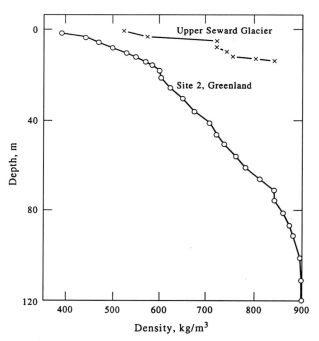

FIGURE 4 Increase of firn density with depth in Greenland Ice Sheet and Upper Seward Glacier, Yukon. [From Paterson, W. S. B., "The Physics of Glaciers," 3rd ed., Elsevier, 1994. Reprinted by permission of Butterworth Heinemann Publishers, a division of Reed Educational and Professional Publishing Ltd.]

natural-gas pipelines and is believed to be widespread in Arctic permafrost.

Most of the properties of glacier ice depend on the structure of the individual crystals. In an ice crystal, the molecules or, more precisely, the oxygen atoms lie in layers, and within each layer the atoms are arranged hexagonally, as can be seen in the symmetry of a snow crystal. The plane of each layer is called the basal plane, and the direction perpendicular to it is the C-axis.

The response of ice to an applied stress is not completely elastic; that is, the sample does not return to its original shape when the stress is removed. If a large stress is applied suddenly, the ice may fracture. If stress is applied slowly, the ice becomes permanently deformed. This is the phenomenon of plastic deformation or creep; it is one process of glacier flow. (The others are sliding of ice over the bed and bed deformation.)

Plastic deformation of a single crystal of ice normally consists of the gliding, one over another, of layers parallel to the basal plane, like cards in a pack. Ice crystals can also deform in other ways, but much higher stresses than those required for basal glide are needed.

The deformation of crystals of ice and metals can be understood in terms of the movement of dislocations within the crystals. These are irregularities, formed during crystal growth and by deformation, that allow the layers of molecules to glide over each other much more readily than they could in a perfect crystal. Deformation involves the movement of dislocations. They interact with each other; in some cases they can pile up and restrict further movement until they can disperse into some more nearly uniform arrangement.

Aggregates of many crystals deform more slowly than a single crystal, because most of the crystals are not oriented for glide in the direction of the applied stress. However, other processes now come into play: displacement between crystals, migration of crystal boundaries, crystal growth, and recrystallization to form more favorably oriented crystals.

In glacier studies, the form of the relation between deformation rate and applied stress is more important than the mechanisms. The most common form of deformation experiment is to apply constant uniaxial compression, or simple shear, to an ice sample and measure how the deformation changes with time. Figure 5 is a typical result. An initial elastic deformation (OA) is followed by a period of transient creep, in which deformation rate decreases steadily. This is followed by a period during which the deformation rate remains constant (secondary creep) before starting to increase (tertiary creep). A final constant deformation rate is sometimes reached. Blocking of dislocations and interference between crystals with different orientations can explain the initial deceleration. The for-

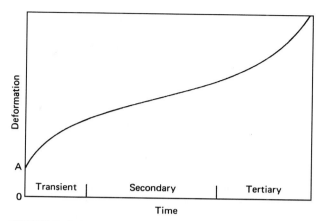

FIGURE 5 Schematic creep curve for randomly oriented poly-crystalline ice under constant stress.

mation, by recrystallization, of crystals more favorably oriented for glide accounts for the increase in deformation rate in the tertiary stage. Secondary creep probably represents a temporary balance between these various processes.

For secondary creep, over the range of stresses usually found in glaciers (50–200 kPa), the relation between shear stress τ_{xy} and deformation rate or strain rate $\dot{\varepsilon}_{xy}$ has the form

$$\dot{\varepsilon}_{xy} = \mathbf{A}\tau_{xy}^n. \tag{1}$$

Here n is a constant, and $\mathbf{A}$ depends on temperature, the orientation of the crystals, impurity content, and possibly other factors. A value $n = 3$ is normally used. This equation is usually called the flow law of ice, or sometimes Glen's law. For a Newtonian viscous fluid, $n = 1$, and $\mathbf{A}$ is the reciprocal of the dynamic viscosity; ice behaves quite differently from this. It should be emphasized that Eq. (1) is not a universal law. Although the form of the relation can be explained by dislocation theory, it is essentially an empirical fit to laboratory and field data for the loading conditions and range of stresses encountered in glaciers. An engineer concerned with the bearing capacity of lake ice, for example, might well use a different relationship. The value of $\mathbf{A}$ varies with absolute temperature $\mathbf{T}$ according to the Arrhenius relation,

$$\mathbf{A} = \mathbf{A}_0 \exp\left(\frac{-Q}{RT}\right), \tag{2}$$

for temperatures below −10°C. Here $\mathbf{A}_0$ is independent of temperature, $\mathbf{R}$ is the gas constant (8.314 J/mol deg), and $\mathbf{Q}$ is the activation energy for creep (60 kJ/mol). This value implies that the deformation rate produced by a given stress at −10°C is about 5 times that at −25°C; the speed of a glacier thus depends strongly on the temperature. The Arrhenius relation breaks down above −10°C because additional processes resulting from the presence of liquid water at grain boundaries begin to contribute to deformation. The effective value of **Q** starts to increase with temperature, although current practice is to retain the form of the relation and take **Q** = 139 kJ/mol for temperatures above −10°C.

Hydrostatic pressure does not affect the deformation rate provided that temperature is measured relative to the melting point. Hydrostatic pressure does, of course, lower the melting point; it is about −1.8°C under 2000 m of ice.

The crystals in deep samples from glaciers often have a preferred orientation, that is, a majority of the crystals have their *C*-axis aligned close to the normal to the shear plane. Such ice deforms more rapidly than ice with a random crystal orientation; increases by factors of up to 10 have been reported. One example is the observation that, in Greenland and Arctic Canada, ice deposited during the last ice age shears three to four times more readily than ice deposited since. This is largely because the crystals in the ice-age ice are much more strongly oriented than those in the ice above, although the higher concentration of impurities may also contribute. Impurities in glacier ice can sometimes increase its deformation rate; there is a thin layer of "soft" ice containing impurities derived from the glacier bed at the base of Meserve Glacier in Antarctica.

Whether the index *n* is a constant is controversial. Most of the evidence for the accepted value of 3 comes from laboratory tests at relatively high stresses or deformation measurements in temperate glaciers. Some researchers believe that a different deformation mechanism predominates at the low stresses and temperatures prevailing in the upper layers of the polar ice sheets and that the value of *n* should be reduced to 2 or even 1.5. The only good field data, from Dye 3 in Greenland, are consistent with any value between 2 and 3.

At most places in glaciers the ice is not deforming in simple shear, as assumed in Eq. (1); several stresses act simultaneously. If a longitudinal stress σ_x is superimposed on a shear τ_{xy}, the flow law has the form

$$\dot{\varepsilon}_{xy} = A\tau^{n-1}\tau_{xy}, \tag{3}$$

$$\dot{\varepsilon}_{xx} = \tfrac{1}{2}A\tau^{n-1}(\sigma_x - \sigma_y), \tag{4}$$

$$\tau^2 = \tfrac{1}{4}(\sigma_x - \sigma_y)^2 + \tau_{xy}^2. \tag{5}$$

Here *x* and *y* are the longitudinal and vertical coordinates, $\dot{\varepsilon}$ are the deformation rates and τ is the effective shear stress. The important feature of these equations is the interaction among the different stresses; a given shear stress acting along with other stresses produces a greater deformation rate than it would when acting alone. For example, a tunnel dug in a glacier closes up under the pressure of the ice above it. The ice at the foot of an icefall undergoes strong longitudinal compression; thus a tunnel there should close up much more rapidly than it would at the same depth in another part of the glacier. This has been observed. Note also that, at a place where $\tfrac{1}{2}(\sigma_x - \sigma_y)$ is large compared with τ_{xy}, the relation between shear strain rate $\dot{\varepsilon}_{xy}$ and shear stress τ_{xy} will appear to be linear. This interaction between stresses, which does not occur in a Newtonian viscous material (*n* = 1), greatly complicates mathematical analyses of glacier flow.

Readers familiar with tensors will recognize that τ^2 in Eq. (5) is the second invariant of the stress-deviator tensor in the two-dimensional case. Because ice deformation is independent of hydrostatic pressure, the flow law should contain stress deviators rather than stresses. A physical property, such as the flow law, must be independent of the coordinate system and so must be expressed in terms of invariants. The most general relation between tensors, such as deformation rate and stress deviator, would involve first, second, and third invariants. The first deformation-rate invariant is zero because ice is incompressible. There is experimental evidence that the value of the third invariant does not affect the flow law. These facts, combined with the plausible assumption that corresponding deformation-rate and stress-deviator components are proportional to each other, reduce the flow law to the form given.

A useful approximation to the flow law is to regard ice as perfectly plastic. If a gradually increasing stress is applied to a perfectly plastic material, there is no deformation until a critical value, the yield stress, is reached, at which point the material deforms rapidly. This corresponds to the case $n = \infty$, as may be seen by replacing **A** by A_0/τ_0^n in Eq. (1).

IV. GLACIER FLOW BY ICE DEFORMATION

During the past 50 years, application of the methods of continuum mechanics has greatly improved the understanding of glacier flow. Continuum mechanics deals with the motion of deformable bodies whose properties, although in fact determined by their molecular structures, can be expressed in terms of a continuous theory. The simplest application, one that clarifies many of the fundamentals of glacier flow, is to consider blocks of ice in simple shapes as idealized models of glaciers. The stresses required to maintain a block in mechanical equilibrium can be calculated. Strain rates and velocities can then be calculated from the stresses by the flow law [Eqs. (3) and (4)].

Velocities of valley glaciers can be easily measured by repeated surveys, from stations on bedrock, of the positions of stakes set in the ice. Velocities on ice sheets can be determined by the Global Positioning System, by measuring the displacement of crevasses and other identifiable features on repeated optical images from satellites, and by an interference method using the side-looking synthetic aperture radar on the ERS-1 satellite. The configuration of the ice surface can be mapped by radar altimetry, provided that the wavelength is not within the range to which ice is transparent. Shear strain rates within the ice can be found by measuring the rate at which a vertical borehole tilts. Normal strain rates at the surface are determined by measuring the rate of change of the distance between two stakes. At least two boreholes are required for measuring normal strain rates at depth; this has seldom been done. Stresses in glaciers have never been measured; they are always calculated from some simplified model.

The pattern of flow in a glacier can be deduced from the principle of mass conservation. See Fig. 6. The glacier is assumed to be in a steady state, that is, its dimensions do not change in time, and the velocity at any point, fixed in space, remains constant as the ice flows past. To maintain constant thickness in the accumulation area, ice must flow downward relative to the surface. This velocity component decreases steadily with depth, as does the surface-parallel component, and reaches zero at the bed if there is no melting there. Similarly, ice flows upward relative to the surface in the ablation area. Now consider a cross section through the glacier perpendicular to the surface and to the flow direction. To maintain a steady state, the annual flux of ice through any such cross section in the accumulation area must be equal to the total annual accumulation on the surface upstream of the cross section. Similarly, the annual flux through a cross section in the ablation area must equal the annual loss of ice between the cross section and the terminus. Thus the ice flux must increase steadily from zero at the head of the glacier to a maximum at the equilibrium line and from there decrease steadily to the terminus. Ice velocities should show a sim-

ilar trend, modified by variations in width and thickness of the glacier. In the Antarctic Ice Sheet, where nearly all the ice is lost by iceberg calving, there is no ablation area, and the velocity is greatest at the terminus.

This pattern of velocity variations means that the ice mass is being stretched in the horizontal direction in the accumulation area and compressed in the ablation area. In fact, ice is incompressible except insofar as the volume of air bubbles can be reduced. Horizontal extension must therefore be accompanied by an equal compression in the vertical direction; this is how the glacier can remain in a steady state in spite of accumulation. Similarly, longitudinal compression in the ablation area is accompanied by vertical stretching, which can keep the glacier thickness constant. This argument applies to a valley glacier in which the side walls prevent lateral extension. The general condition is not that horizontal extension must equal vertical compression, but that the algebraic sum of the deformation rates in any three mutually perpendicular directions must be zero. Where the ice is stretching/contracting in the direction of flow, it is said to be in *extending/compressing flow*. Regions of extending flow are often marked by transverse crevasses. The steady-state deformation rate can be estimated roughly from the ratio of the accumulation or ablation rate to the ice thickness. Typical values range from 10^{-2} yr^{-1} for glaciers in maritime temperate climates to 10^{-5} yr^{-1} in central Antarctica.

As a simple model of a glacier, consider a parallel-sided slab of thickness h, frozen to a plane of slope α. The length and width of the slab are much larger than h. It deforms in simple shear under its own weight. Consider a column of ice perpendicular to the plane and of unit cross section. The component of its weight parallel to the plane, the *driving stress*, is $\rho g h \sin \alpha$. (Here ρ is density and g gravitational acceleration.) This is balanced by the *basal drag*, the shear stress τ_b across the base of the column:

$$\tau_b = \rho g h \sin \alpha. \tag{6}$$

If the slab has uniform thickness, this relation holds at every point. In a real glacier h varies with x and so there are places where one term exceeds the other. In other words, there is a longitudinal stress σ_x as well as shear:

$$\tau_b = \rho g h \sin \alpha + \int_b^s \left(\frac{\partial \sigma_x}{\partial x} \right) dy. \tag{7}$$

Here s and b denote surface and bed. At most places except near an ice divide and near the terminus, the second term on the right-hand side is small compared with the first, especially if the surface slope is averaged over a distance of about $10h$. At this scale, flow is in the direction of maximum surface slope.

The value of τ_b has been calculated from measurements of thickness and surface slope on many glaciers. Most

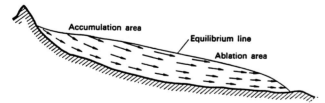

FIGURE 6 Velocity vectors in an idealized glacier. They define the flowlines. Particle paths are identical with flowlines if the glacier is in a steady state. [From Paterson, W. S. B., "The Physics of Glaciers," 3rd ed., Elsevier, 1994. Reprinted by permission of Butterworth Heinemann Publishers, a division of Reed Educational and Professional Publishing Ltd.]

values lie between 50 and 100 kPa. (Surging glaciers and Antarctic ice streams are the main exceptions, with values lower than these.) Equation (6), with τ_b constant, implies that a glacier is thin where its surface is steep and thick where the slope is small. Observations confirm this. Moreover, inflection points of the surface correspond to bedrock highs.

Because of the small variation in τ_b, a useful approximation is to regard ice as perfectly plastic. This leads to a particularly simple picture of glacier flow: the ice thickness adjusts itself so that at every point the shear stress at the base is equal to the yield stress τ_0. This in turn leads to a simple formula for the surface profile of an ice sheet. If the bed is horizontal and the surface slope small, $\alpha = dh/dx$, and Eq. (6) can be integrated to give

$$h^2 = \left(\frac{2\tau_0}{\rho g}\right)x, \qquad (8)$$

where x is distance from the edge. This is a parabola. Equation (8) with $\tau_0 = 100$ kPa correctly predicts the maximum ice thickness of 3200 m in central Greenland. This formula is frequently used to calculate the shape and volume of ice-age ice sheets from evidence of their extent provided by moraines and other glacial deposits.

The assumption that ice is perfectly plastic can also explain the maximum depth of crevasses. In the wall at the bottom of a crevasse, the weight of the overlying ice is acting downward and is not supported by any lateral pressure other than atmospheric. This vertical pressure tends to squeeze the ice downward and outward and so to close the crevasse. Below a critical depth (22 m), this happens very rapidly.

Similar, though more complicated, analyses can be made using the true flow law. According to Eq. (8), the shape of the ice sheet depends only on the plastic properties of ice, not on accumulation and ablation rates. If the flow law is used, the calculated profile is not greatly changed, and the dependence on accumulation and ablation is not a sensitive one. (Doubling the accumulation rate would increase ice thickness by 10%.) As Fig. 7 shows, many ice caps have similar, roughly parabolic, surface profiles in spite of appreciable differences in accumulation and ablation and, in some cases, in spite of mountainous bedrock topography.

How velocity varies with depth can be calculated from the flow law, Eq. (1). This is used to obtain the shear strain rate at any depth y from a formula analogous to Eq. (6). The shear strain rate is then integrated to give the horizontal velocity component u at depth y,

$$u_s - u = \frac{2A(\rho g \sin \alpha)^n y^{n+1}}{(n+1)}, \qquad (9)$$

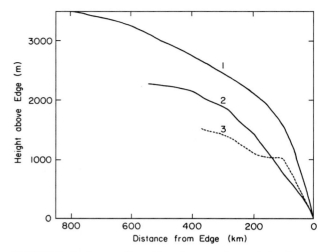

FIGURE 7 Surface profiles of ice sheets: 1, Antarctic Ice Sheet, Mirny to Komsomol'skaya; 2, Greenland Ice Sheet, west side between latitudes 69 and 71°N; 3, Greenland Ice Sheet, east side between latitudes 71 and 72°N. [Reprinted with permission from Colbeck, S. C. (ed.) (1980). "Dynamics of Snow and Ice Masses." Copyright 1980, Academic Press, New York.]

where A and n are the flow law parameters and suffix s denotes the surface value. This velocity profile is shown in Fig. 8. The velocity is greatest at the surface and decreases with depth; most of the decrease occurs near the bed. Thus perfect plasticity, for which all the deformation

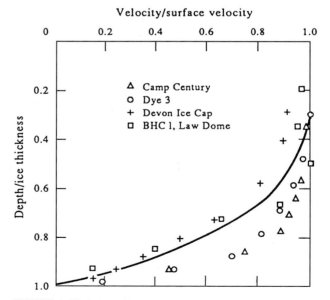

FIGURE 8 Variation of horizontal velocity with depth calculated from Eq. (9), assuming uniform temperature and $n = 3$, compared with four measured velocity profiles. [From Paterson, W. S. B., "The Physics of Glaciers," 3rd ed., Elsevier, 1994. Reprinted by permission of Butterworth Heinemann Publishers, a division of Reed Educational and Professional Publishing Ltd.]

takes place at the bed, is a reasonable approximation. Except in temperate glaciers, the shear should be more concentrated near the bed than shown here, because the curve was drawn with A constant. In fact, A increases with temperature and the warmest ice is found near the bed. (See Section VII.)

In addition to adjusting the ice thickness to compensate for accumulation and ablation, extending and compressing flow enable a glacier to flow past large bumps in its bed. Crudely, the ice on top of the bump is pushed by the ice upstream and pulled by the downstream ice. More precisely, compressing flow upstream of the bump thickens the glacier there, whereas it is thinned by extending flow on the downstream side. This increases the surface slope over the bump and so provides sufficient shear stress to drive the ice over it. Again, where the valley walls converge in the direction of flow, the glacier is compressed in the transverse direction and this induces extension in the direction of flow. Similarly, there is a tendency for compressing flow where the valley widens. The combination of all these factors determines the state of flow at any point; accumulation and ablation are usually the most important.

Glacier deformation is a combination of shear and longitudinal extension and compression. Shear predominates near the bed, longitudinal stresses near the surface. Note that the longitudinal deformation rate usually varies with depth; upstream of a bedrock bump, flow near the bed will be compressing, even though the depth-averaged flow is extending when the point is in the accumulation area.

Equation (9) cannot be used for velocity calculations if there are longitudinal stresses, as there always are in a real glacier. A method based on the generalized flow law [Eqs. (3) and (4)] has to be used. As a result, shear can be either more or less concentrated in the basal layers than the simple theory predicts (Fig. 8).

Simple models that lead to equations with analytical solutions have proved successful in clarifying many fundamental aspects of glacier flow. However, many important problems, such as flow at ice divides, in and near icefalls, at grounding lines, and at glacier termini cannot be treated in this way. Moreover, dating the climatic records now being obtained from cores from polar ice sheets (Section XI) requires detailed modeling of flow upstream from the borehole. For this purpose, the velocity distribution is only the starting point; one also has to calculate particle paths, how long the ice takes to travel from the surface to a particular point in the borehole, and how the thickness of an annual layer changes as it is buried by subsequent layers. Numerical methods of obtaining approximate solutions to complicated equations have to be used in these cases. One technique is the finite-element method, developed by engi-

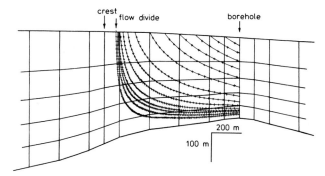

FIGURE 9 Particle paths near crest of Devon Island Ice Cap computed by finite-element modeling. The grid shows the boundaries of the elements. Ice takes 100 years to travel between two marks on a path. [Reprinted from Paterson, W. S. B., and Waddington, E. D. (1984). *Rev. Geophys. Space Phys.* **22**, 123–130. Copyright 1984, American Geophysical Union.]

neers to solve problems in structural mechanics. Figure 9 shows the results of one analysis.

The velocity distribution in real glaciers is much more complicated than the simple picture presented here. Three-dimensional effects are important. For example, ice can flow around a bedrock bump as well as over the top of it. Other observations show that the shear deformation, instead of increasing steadily with depth, can sometimes be concentrated in bands in which the orientation of crystals is particularly favorable for deformation. Again, some bedrock hollows may be filled with stagnant ice.

V. FLOW OF ICE SHELVES

Flow in ice shelves can be analyzed by similar methods. Because the shelf is floating, shear stresses are zero at the base as well as at the surface. Vertical shear is negligible, and deformation consists of uniform spreading. If the sides of the shelf are not confined by land, the spreading rate is the same in all directions, and the only restraining force is the horizontal force that seawater exerts on the edges. The spreading rate of an unconfined shelf of uniform thickness can be shown to be proportional to Ah_0^n, where h_0 is the height of the surface above sea level and A and n are the flow law parameters, A being an average value to take account of the variation of temperature with depth. An ice shelf can maintain its thickness only if spreading is counteracted by surface accumulation or by inflow from an ice sheet.

Ice shelves of significant size are found only in Antarctica. All the major ones are fed from the ice sheet and their sides are confined by bedrock or grounded ice. (See Fig. 2.)

On the Ross Ice Shelf, for instance, surface accumulation accounts for about half the mass; inflow from glaciers and ice streams provides the rest. Ice is removed mainly by calving, sometimes in the form of enormous tabular bergs. There is also some basal melt.

Deformation rates along flowlines in the Ross Ice Shelf are only about one-tenth of those calculated from the formula for the spreading rate of an unconfined shelf. This is because that analysis ignores the drag of the side walls (bedrock or grounded ice) and the restraining effect of ice rises. An ice rise has the dome-shaped profile characteristic of grounded ice and its own radial flow pattern. The shelf ice flows around it and the ice immediately upstream is relatively thick because the flow is obstructed. Because the restraining force increases with increasing distance from the calving front, the ice thickness also increases; the Ross Ice Shelf is about 250 m thick at the ice front and three to four times that at the grounding line.

Because deformation in ice shelves consists of longitudinal extension rather than shear, the velocity at any point cannot be calculated from the ice thickness and surface slope there. It has to be obtained by integrating the deformation rate from a point, usually the grounding line, where the velocity is known. Again, the vertical shear deformation characteristic of an ice sheet cannot change to longitudinal extension instantaneously at the grounding line. There must be a transition zone upstream. An ice stream, with low surface slope, low driving stress, and little vertical shear, is essentially a transition zone between the inland ice and the ice shelf. (See Section VIII.) Where an outlet glacier flows into an ice shelf, however, the extent of the transition zone is less clear.

An ice shelf exerts a back pressure on the ice flowing into it; the amount of the pressure depends on how difficult it is to push the shelf out to sea. This in turn depends on the extent of ice rises and the drag exerted by the sidewalls. This back pressure, which may be transmitted for some distance into the ice sheet, reduces the ice flow. Thinning of the ice shelf or a rise in sea level would reduce both the extent of ice rises and the area of ice in contact with the sidewalls. This would reduce the back pressure and so increase flow from the ice sheet. This is discussed further in Section VIII.

VI. BASAL MOTION AND WATER FLOW AT THE BED

Velocity at depth in a glacier can be measured by drilling a vertical hole and measuring its tilt at different depths a year later. The profile of the hole then gives the velocity–depth profile, and the surface velocity can be determined

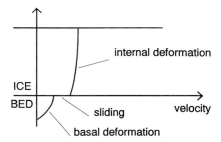

FIGURE 10 The three possible modes of glacier flow.

by standard surveying techniques. Provided that the hole extends to bedrock, this method also measures the basal motion. This can result from the glacier sliding over its bed or from deformation of the bed itself (Fig. 10). Sliding has also been measured in tunnels and by down-borehole photography and television. Sliding velocities span a much greater range than velocities produced by ice deformation. Significant sliding occurs only when the basal ice is at melting point. Bed deformation has been measured by driving a rod with strain gauges bonded to it into the subglacial sediments from the bottom of a borehole. Subglacial sediments deform only if they are saturated with water at low effective pressure, that is, the water pressure in their pores is not far below the hydrostatic pressure of the overlying ice.

How does ice, assumed to be at melting point, move past bumps in the glacier bed? Observations have confirmed that two mechanisms operate. The first is pressure melting. The resistance to motion must be provided by the upstream side of each bump. So there must be excess pressure there. Because this lowers the melting point, some ice can melt. The meltwater flows around the bump and refreezes on the downstream side where the pressure is less. The latent heat released on the downstream side is conducted through the bump and the surrounding ice to the upstream side, where it provides the heat needed for melting. This process is the same as in the classic regelation experiment, in which a wire moves through a block of ice by pressure-melting in front and refreezing behind. The process is ineffective for bumps longer than about 1 m because the heat conducted through them is negligible; nor can it take place if the ice is below its melting point. The second mechanism of sliding is enhanced plastic flow. Stress concentrations in the ice near a bump increase the deformation rate, so the ice can flow over the bump or around it. The larger the bump, the greater the area over which the stress is enhanced, and so the more effective this process will be. It follows that obstacles of some intermediate size, probably between 0.05 and 0.5 m, must provide most of the resistance to sliding.

Mathematical analysis of these processes leads to an equation for the sliding velocity:

$$u_b = B\tau_b^{(n+1)/2} R^{-n-1}. \qquad (10)$$

Here B is a constant that depends on the mechanical and thermal properties of ice, $n(=3)$ is the flow law index [Eq. (1)]; τ_b is the driving stress, and R is the bed roughness (amplitude/spacing of bumps). Note the sensitivity of the sliding velocity to the roughness; doubling the roughness decreases the velocity by a factor of 16. Sliding velocities of glaciers cannot be predicted from this formula because the bed roughness is seldom, if ever, known; it is not even clear how to measure it for a real glacier bed. However, the equation, with the very dubious assumption that R is a constant, has been used extensively as a boundary condition in mathematical analyses of deformation within glaciers.

The main objection to Eq. (10) is that the effect of water at the bed is ignored. In the theory, the ice is assumed to be in contact with the rock, except for the thin water film produced in the regelation process. Large amounts of meltwater flow on the surfaces of most glaciers in summer. Most of it disappears into the ice through cracks and vertical passages called moulins, formed where water flows into a crevasse, and emerges under the terminus in one or a few large streams. Observations of increased velocity in summer, diurnal variations, and dramatic increases after heavy rain support the idea that sliding velocity is influenced by water at the bed. Short-period variations must result from sliding, because ice deformation depends on factors such as ice thickness and surface slope, which cannot change rapidly. In some cases, a rapid increase in velocity has been associated with a measurable upward displacement of the glacier surface, attributed to an increase in the amount of water at the bed.

Theoretical analyses and field observations suggest that there are two types of subglacial drainage system. One is a system of tunnels cut upward into the ice. Melting of the wall by heat generated by the friction of the water enlarges a tunnel, while ice flow tends to close it because the water pressure is normally less than the pressure of the ice above. If the flux of water increases, the increase in frictional heat increases the melting rate. To maintain a steady state, the closure rate must also increase and so, because the ice overburden pressure is fixed, the water pressure must decrease. Thus there is an inverse relation between the water flux and the steady-state water pressure in a tunnel. It follows that, of two adjacent tunnels, the larger has the lower pressure and therefore draws water from the smaller one. The system therefore develops into one or a few large tunnels. The tunnel system carries a large flux at low pressure in approximately the same direction as the ice flow.

In the second type of drainage system, often called a linked-cavity system, the water is dispersed widely over the bed in a network of interconnected passageways whose diameter varies greatly along the path. The wide parts are cavities on the downstream side of the larger bedrock bumps; the narrow parts are downstream of small bumps or are channels cut in the bed. Because the frictional heat is spread over a much larger area of the bed than it is in a tunnel, melting of the roof is not an effective way of keeping a passageway open. High water pressure is required for this. When the flux of water increases, the water pressure also increases to increase the capacity of the system by enlarging the cavities. Because water pressure increases with flux, there is no tendency for the larger passageways to grow at the expense of the smaller ones. Because the passageways tend to be on the downstream side of bedrock bumps, much of the water flow is in the cross-glacier direction. Travel times are much longer than the few hours typical of a tunnel, because the path length is much longer and the narrow sections throttle the flow. A linked-cavity system can develop only if the bed is rock rather than sediments.

The two systems are believed to exist simultaneously beneath a glacier. They expand or contract, though not instantaneously, in response to changes in the volume of water. A sudden influx, for example, may drive some of the water out of a tunnel into the linked-cavity system. In late summer, when the reduced water supply no longer fills a tunnel, it will start to close rapidly. When melting begins in spring, most of the water is probably in cavities. Once the tunnel system has developed, however, it takes over the bulk of the drainage.

Because sliding tends to keep open cavities on the downstream side of bedrock bumps, the size of a cavity increases with the sliding velocity. Conversely, the increased bed separation produced by an increase in water pressure increases the sliding velocity. Because, unlike the linked-cavity system, tunnels occupy only a small fraction of the glacier bed, the water in them has little effect on sliding.

The effect of water has been taken into account empirically by writing the equation

$$u_b = \frac{B_1 \tau_b^m}{(\rho g h - P)^p}, \qquad (11)$$

where ρ is ice density, g is gravitational acceleration, h is ice thickness, P is subglacial water pressure, B_1 is a "constant" that includes the bed roughness, and m and p are positive integers. The denominator is the weight of ice reduced by the water pressure. Implicit in this equation is the dubious assumption that sliding at any point is unaffected by water conditions up- and down-stream.

A further objection to all current sliding theories is the implicit assumption that the ice rests on a rigid impermeable ("hard") bed and that the interface is well defined. In fact, parts of many glaciers and ice sheets rest on a layer of deformable sediment usually glacial till, a "soft" bed. If the sediment has a low permeability, subglacial water can, by building up pore pressure within it, make the sediment weaker than the overlying ice. In this case, bed deformation can be the major component of glacier movement. Ninety percent of the basal movement at the margin of a glacier in Iceland was found to consist not of basal sliding but of deformation in the subglacial material. Deformation of water-saturated gravel beneath the ice has also been observed in a borehole. Again, seismic measurements suggest that Ice Stream B, one of five that drain ice from the interior of West Antarctica into the Ross Ice Shelf, is underlain by about 6 m of unconsolidated sediments. Deformation in this layer could explain why the surface velocity is several hundred meters per year in spite of an extremely low driving stress. (See Section VIII.) Recent observations such as these have forced glaciologists to accept that till can, under certain conditions, deform in shear, as geologists have known for many years.

Till is a complex inhomogeneous material and a flow relation for it has not been firmly established. A relation,

$$\frac{u}{h} = B_2 \tau^a N^{-b}, \qquad (12)$$

has been suggested on the basis of one set of measurements. Here u is basal velocity, h is thickness of till layer, τ is shear stress, N is effective pressure (ice overburden minus water pressure), B_2 is a constant, and a and b have values between 1 and 2. A relation of this form is an oversimplification, because the deformation rate is expected to depend on additional factors such as composition, porosity, and strain history. Equation (12) has the same form as Eq. (11). This is convenient for numerical modelers; they can use it as a boundary condition without specifying whether it represents sliding or bed deformation. In contrast, some researchers believe that till is perfectly plastic. If so, deformation cannot be pervasive; it must be confined to a thin layer at the base.

VII. TEMPERATURES IN GLACIERS

The temperature distribution in glaciers deserves study both for its own sake and for its relation to other processes. Ice velocity depends sensitively on temperature. The basal temperature is important for erosion; bedrock is protected when the ice is frozen to it. If the basal temperature of a glacier, previously frozen to its bed, were to reach melting point, the ice could start to slide; this would result in a significant advance of the terminus. Again, properties

such as the absorption of radio waves, on which the standard method of measuring ice thickness depends, vary with temperature. In particular, water in the ice scatters radio waves and greatly increases the difficulty of radar sounding.

The temperature distribution results from the interaction of many processes. Short-wave solar radiation and long-wave radiation from atmospheric water vapor and carbon dioxide supply heat to the surface, as does condensation of water vapor. The surface loses heat by outgoing long-wave radiation and sometimes by evaporation. Turbulent heat exchange with the atmosphere may warm or cool the surface according to the temperature gradient in the air immediately above. Ice deformation and, in some cases, refreezing of meltwater warm the interior while geothermal heat and friction (if the ice is sliding) provide heat at the base. Heat is transferred within the glacier by conduction, ice flow, and in some cases water flow. The interaction between ice flow and heat flow greatly complicates mathematical analyses of glacier flow.

The temperature distribution in a glacier can have one of four forms:

1. All the ice is at melting point except for a near-surface layer penetrated by a winter "cold wave," the so-called temperate glacier. Temperature decreases with increase of depth at a rate of 0.87 deg/km because of the effect of pressure on air-saturated water.
2. All the ice is below melting point.
3. The melting point is reached only at the bed.
4. A basal layer of finite thickness is at melting point. This situation depends on ice-flow conditions near the bed.

Different types of distribution can occur in different parts of the same glacier.

Seasonal variations of surface temperature are rapidly damped at depth; the temperature at a depth of 10–15 m remains constant throughout the year. In Antarctica, except near the coast, and in the higher parts of Greenland, the air temperature never reaches 0°C. The 15-m temperature is then equal to the mean annual air temperature and provides a quick way of measuring it. This temperature decreases by about 1°C per 100 m increase of elevation and 1°C for each degree increase of latitude. The 15-m temperature can vary slowly as a result of long-term climatic variations. Depth of penetration increases with the period of the change; the 100 kyr ice-age cycle penetrates through the deepest ice sheets into bedrock.

When there is surface melting, the water percolates into the snow and refreezes. Because freezing of 1 g of water releases enough latent heat to warm 160 g of snow by 1°C, this is an important means of warming the near-surface layers of glaciers and the seasonal snowcover. The 15-m temperature may then be several degrees above mean annual air temperature. This process is self-limiting, however, because percolation stops after there is enough meltwater to form a continuous ice layer.

Thermal conditions at the base are determined mainly by the geothermal heat flux, the flow of heat produced by radioactive decay in the earth's interior. If the basal ice is below its melting point, all this heat is conducted into it. A heat flux of 56.5 mW/m^2, the continental average, produces a temperature gradient of 2.7 deg per 100 m in ice. If the basal ice is at melting point (case 3 above), some of the heat will be used for melting and the remainder conducted into the ice. In a temperate glacier or in one with a basal layer at melting temperature, the small reversed temperature gradient prevents upward conduction of heat; all the basal heat is used for melting, about 6 mm of ice per year for an average heat flux. Frictional heat will melt the same amount if the sliding speed is about 20 m/yr.

With the upper and lower boundary conditions established, the form of the temperature distribution can be derived. Initially we consider a steady state. This means that the dimensions of the glacier do not change in time and that the velocity and ice temperature at any point, fixed in space, remain constant as the ice flows past. This is an important theoretical concept, although no real glacier is ever in this condition.

The downward component of flow in the accumulation area (see Section IV) is a much more important means of heat transfer than conduction in the upper layers. In a stagnant glacier, with no accumulation or ablation, the temperature would increase steadily with depth at a rate equal to the geothermal gradient. In a glacier with accumulation, the temperature profile still has this shape near the bed, where the vertical velocity component is very low, but temperatures in the upper layers can be nearly uniform as a result of the downward flow of cold surface ice. The higher the accumulation rate the stronger the effect. Figure 11 illustrates this. These curves are based on a simple mathematical analysis. Because the effect of the horizontal component of ice flow was ignored, they should be good approximations near the center of an ice sheet because horizontal velocities are small there.

In the ablation area, ice flows upward toward the surface, carrying warm basal ice with it; the temperature gradient in the upper layers is greater than the geothermal. Such a curve is also shown in Fig. 11. However, neglect of the horizontal velocity component in the ablation area is hard to justify.

How does the horizontal velocity affect temperatures? Figure 6 shows that with increasing depth in a borehole, the ice has originated at progressively higher elevations and thus at progressively lower temperatures. If each particle

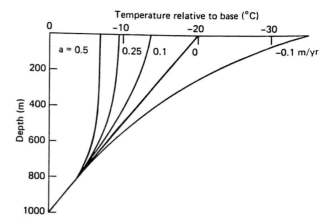

FIGURE 11 Computed temperature distribution in a steady-state ice sheet with a frozen bed for different values of ice equivalent accumulation rate (positive) and ablation rate (negative).

of ice were to retain its initial temperature, the deepest ice would be the coldest. Heat conduction changes the temperature. Nevertheless, flow of cold ice from high elevations often results in the coldest ice being found some distance below the surface. This effect should become more important with increasing distance from the ice divide. Such a temperature minimum could not occur as a steady state in a static medium.

Such negative temperature gradients (decrease of temperature with increase of depth) have been measured in many shallow boreholes in Greenland and Antarctica. However, some of the observations may have another explanation: climatic warming during the past 50 to 100 yr. In this case each snow layer is warmer than the one below, although conduction will eventually smooth out such differences.

This suggests that measurements of the present temperature profile in a borehole can be inverted to give a record of past surface temperatures, provided that the effect of flow from upstream is either allowed for or eliminated by drilling at an ice divide. The resulting record is heavily smoothed, however, and increasingly so the farther back in time it extends. Moreover, there is seldom much information about past changes in accumulation rate, ice thickness, and flow pattern, all of which affect the record. Oxygen-isotope profiles from ice cores, described in Section XI, provide much more detailed records of past surface temperatures. Borehole temperature profiles have, however, proved valuable for calibrating these records, that is, converting a given change in oxygen-isotope ratio to a temperature change. Such a calibration shows, for instance, that central Greenland was as much as 20°C colder than present at the last glacial maximum. This is about four times the difference in the tropics.

Figure 12 shows some measured temperature profiles. The simple ideas just outlined can explain the general

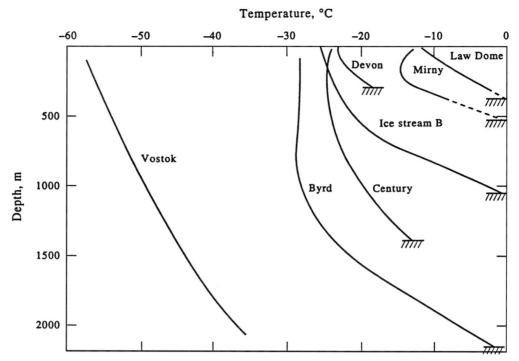

FIGURE 12 Measured temperature profiles in accumulation areas of polar ice sheets and ice caps. At Vostok, the ice is at melting point at the base (3740 m). Camp Century is in Greenland and Devon Island in Arctic Canada; the other sites are in Antarctica. [Adapted from Paterson, W. S. B., "The Physics of Glaciers," 3rd ed., Elsevier, 1994. Reprinted by permission of Butterworth Heinemann Publishers, a division of Reed Educational and Professional Publishing Ltd.]

shapes of these curves. The details have not been successfully predicted, however, only explained afterwards on an *ad hoc* basis.

VIII. ICE STREAM FLOW

The simple picture of an ice sheet as a dome with the ice flowing out uniformly in all directions is realistic at best only in the central part. As the ice approaches the coast it is channeled, either by mountains or the nature of its bed, into fast-flowing outlet glaciers and ice streams. In Antarctica, although these comprise less than 20% of the coastline at the grounding line, they may drain as much as 90% of the accumulation from the interior. Most become part of ice shelves. Similarly, most of the discharge from Greenland is concentrated in some 20 large outlet glaciers that end as floating tongues in the fiords. The behavior of these fast-flowing outlets largely controls the state of the ice sheet.

To make a clear distinction between an ice stream and an outlet glacier is not feasible. An ice stream in the strict sense has no visible rock boundaries; heavily crevassed shear zones separate it from slow-moving ice on either side. But Jakobshavns Isbrae in West Greenland starts as an ice stream and becomes an outlet glacier when it reaches the coastal mountains, while Rutford Ice Stream in Antarctica is bordered by mountains on one side and ice on the other.

Ice streams have various modes of flow. The outlet glaciers in Greenland and those that drain ice from East Antarctica into the Ross Ice Shelf (Fig. 2) are believed to move by ice deformation. Driving stresses are in the range 50–150 kPa and there are no seasonal variations in velocity. The situation is different on the other side of the Ross Ice Shelf, the Siple Coast, where five ice streams drain about one-third of the interior of West Antarctica. (See Fig. 13.) These ice streams are a few hundred kilometers long, 30–60 km wide, 1–1.5 km thick, and have very low surface slopes. Velocities of a few hundred meters per year are typical, apart from Ice Stream C which is almost stagnant. Shear zones, some 5 km wide, separate these ice streams from the ice ridges on either side where the velocity is less than 10 m/yr. Their beds are below sea level and would remain so, in spite of uplift, if the ice were removed.

Figure 14 shows driving stresses and velocities along a flowline that includes Ice Stream B. Contrast this with their

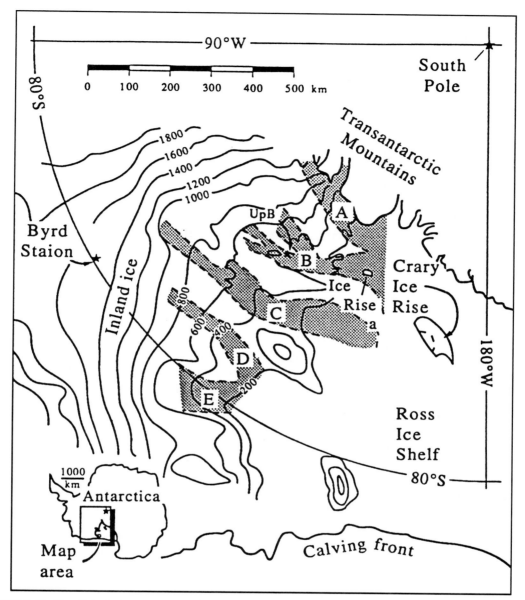

FIGURE 13 The Siple Coast region, West Antarctica, showing the ice streams. [Reprinted with permission from Alley, R. B., and Whillans, I. M. *Science* **254,** 959–963. Copyright 1991, American Association for the Advancement of Science.]

distribution along a flowline in South Greenland (Fig. 15). This has the shape expected from the discussion in Section IV. The driving stress increases from zero at the ice divide, where the surface is horizontal, to about 100 kPa at 50 km along the flowline; this value is maintained until the ice thins near the terminus. The velocity is a maximum at the equilibrium line at 165 km. Along the West Antarctic flowline, however, the ice moves fastest (up to 825 m/yr) in the ice stream although driving stresses there are so low (10–20 kPa) that ice deformation must be insignificant.

Seismic sounding, confirmed later by observations in boreholes, showed that the ice is underlain by about 6 m of water-saturated sediment. Moreover, water levels in the boreholes indicated subglacial water pressures of about 95% of those needed to float the ice. These ice streams are believed to move by bed deformation, although the possibility of a contribution from sliding has not yet been excluded by down-borehole measurements. Indeed, water-lubricated sliding is likely in the 200-km-long "transition zone" between Ice Stream B and the ice shelf.

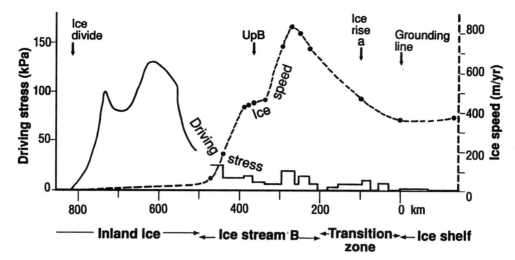

FIGURE 14 Driving stress and surface velocity along a West Antarctic flowline that includes Ice Stream B. [Reprinted with permission from Alley, R. B., and Whillans, I. M., *Science* **254**, 959–963. Copyright 1991, American Association for the Advancement of Science.]

These ice streams are not in a steady state. The head of Ice Stream B is migrating upstream by incorporating blocks of slow-moving inland ice. It is also growing wider by drawing in previously near-stagnant ice at the sides. Ice Stream A is thinning. Although Ice Stream C is stagnant at present, radar scattering has revealed extensive crevasses, indicating previous activity, buried under about 150 yr of snowfall. Capture of ice, or possibly of subglacial water, by Ice Stream B has been suggested as the cause of its shutdown. Because it is stagnant, it is now thickening at the annual accumulation rate. The ice shelf downstream is thinning, however, and the grounding line is retreating.

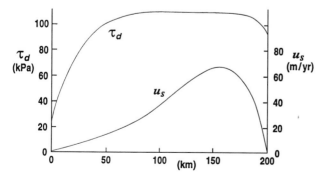

FIGURE 15 Driving stress and surface velocity along a typical flowline in south Greenland. [Reprinted with permission from Paterson, W. S. B., Some Aspects of the Physics of Glaciers. In "Ice Physics and the Natural Environment" (J. S. Wettlaufer, J. G. Dash, and N. Untersteiner, eds.), pp. 69–88, Copyright 1998, Springer-Verlag, Berlin.]

A marine ice sheet such as the West Antarctic Ice Sheet may be unstable against grounding-line retreat. If the sea bed slopes down toward the center of the ice sheet, any decrease in ice thickness at the grounding line will cause it to retreat into deeper water; the retreat will therefore continue. This picture is oversimplified because the back pressure of the ice shelf is ignored. (See Section V.)

An ice shelf is particularly sensitive to climatic warming because its surface is only a few hundred meters above sea level and, in addition, warming of the ocean would increase melting at its base. Ice shelves in the northern part of the Antarctic Peninsula, the warmest part of the continent and where the temperature has increased by 2°C in the last 50 yr, are already disintegrating. Could the reduction in back pressure resulting from the shrinking of an ice shelf cause an irreversible increase in flow from the grounded ice?

Near the grounding line, ice-shelf back pressure is the main restraint to flow in an ice stream. Farther upstream, however, side and basal drag are important. The amount of drag exerted by the slow-moving ice on either side can be estimated from the way in which velocity varies across the ice stream. Although the drag of water-saturated sediments must be very small, there is significant basal drag at "sticky spots" where nonlubricated bedrock projects through the sediments. The relative importance of these forces on the Siple Coast ice streams is being investigated in the field and by numerical modeling. Until more results are obtained, the extent to which a reduction in back pressure would increase flow in an ice stream, and thus whether the West Antarctic Ice Sheet may be unstable, will remain uncertain. Any disintegration would be spread over at least a few hundred years and no dramatic change is expected during this century.

Raised beaches and marine deposits indicate that world sea level was about 6 m higher than at present during the last interglacial about 125 kyr ago. This amount of additional water could have come only from Greenland or Antarctica. (See Table I.) West Antarctica seems the most likely source, although part of the East Antarctic Ice Sheet is also grounded below sea level and might have contributed. Disintegration of the Greenland Ice Sheet is unlikely; it is not a marine ice sheet and outflow is restricted by coastal mountains. On the other hand, a recent interpretation of some ice-core data suggests that southern Greenland was ice-free during the last interglacial. If this is correct, some ice must have remained in West Antarctica, otherwise the rise in sea level would have exceeded 6 m.

IX. GROWTH AND DECAY OF GLACIERS

Glaciers normally respond stably to small changes in climate. In a steady-state glacier, annual accumulation exactly balances annual ablation and the thickness and terminus position do not change. If snowfall increases, or summers become colder so that ablation is reduced, the glacier starts to thicken and its terminus advances because the ice flux there exceeds the ablation. In the absence of further change in climate, the glacier eventually reaches a new steady state. Similarly, reduced precipitation or increased ablation eventually leads to a new, smaller, steady-state glacier.

Advances and retreats of glaciers in any one area appear to be broadly synchronous. For example, most European glaciers attained their maximum postglacial positions in the seventeenth and eighteenth centuries, whereas the period since 1850 has been characterized by general retreat, interrupted by a few brief advances. Detailed records reveal a more complicated picture: a 1981 survey of Swiss glaciers showed that 52 were advancing, 5 were stationary, and 42 were receding. Cases are known in which one of two adjacent glaciers has the steep heavily crevassed terminus typical of an advance, while the other has the smooth gently sloping terminus that indicates stagnation or retreat. See Fig. 1. Some of these variations undoubtedly result from differences in local climates. However, glaciers differ in size, steepness, and velocity, so it is hardly surprising if they react differently, or at least at different rates, to climatic changes.

The effect of an increase in mass balance is propagated down the glacier as a bulge of increased thickness (a kinematic wave) moving faster than the ice itself. The existence of this type of wave is a consequence of mass conservation when there is a relation between flux (mass flowing past a point in unit time), concentration (mass per unit distance, in this case ice thickness), and position. The concept of kinematic waves has also been useful in studying flood waves on rivers and traffic flow on highways. Kinematic waves should not be confused with dynamic waves such as ocean waves; these depend on Newton's second law, combined with a relation between force and displacement. Dynamic waves do not occur on glaciers because ice velocities are so low that the inertia term in the equation of motion can be neglected in comparison with the gravity and "viscous" terms.

The equation of mass conservation can be written as

$$\frac{\partial h}{\partial t} = b - \frac{\partial q}{\partial x}. \tag{13}$$

Here q is ice flux (ice thickness h multiplied by depth-averaged velocity u), x is distance along the glacier, t is time, and b is mass balance. This equation expresses the fact that, in a steady state, the difference between the fluxes flowing into and out of a narrow vertical column must be equal to the amount of ice added to or removed from the surface in unit time. If these terms are not equal, their difference is the rate at which the thickness of the column is changing with time. The velocity is written as the sum of ice deformation and basal motion [Eqs. (9) and (10) or (12)]:

$$u = Ch^4\alpha^3 + B(h\alpha)^2. \tag{14}$$

Here B and C are constants, α is surface slope, and a flow-law index $n = 3$ has been used. The kinematic wave equation is derived by linearizing Eqs. (13) and (14). The wave velocity is about four times the ice velocity. The dependence of ice velocity on surface slope results in diffusion of the waves. Mass balance changes are assumed to be small. Thus the theory cannot be applied to problems such as the growth and decay of ice-age ice sheets.

Have kinematic waves been observed on glaciers? Yes, but not very often. Precise measurements of surface elevation over several years are required, and because a glacier is continually adjusting to a complex series of seasonal and long-term variations, a single wave is exceptional. Moreover, diffusion lengthens the waves and reduces their amplitude.

Because the change in ice flux resulting from a change in mass balance accumulates down-glacier, the level of the upper part of the glacier does not change much whereas that of the lower part does. Lateral moraines are often close to the present glacier surface in the accumulation area, but far above it near the terminus.

Two important questions, how far the terminus moves and how long the glacier takes to adjust to the new mass balance (the *response time*), can be answered without considering how thickness changes are propagated along the

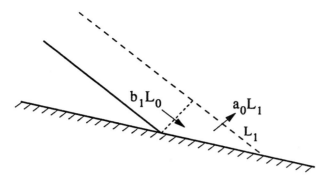

FIGURE 16 The relation between thickening and advance of a glacier terminus.

glacier. In Fig. 16, L_0 is the initial length of the glacier and $L_0 + L_1$ its length after it has adjusted to a change b_1 in mass balance. The original ablation rate at the terminus is a_0. After the mass balance has been $b_0 + b_1$ for a long time, there will be a flux per unit width of $b_1 L_0$ at the old terminus, where b_1 is averaged over the length of the glacier. The terminus will have advanced the distance L_1 required to remove this flux by ablation. If $a_0 = -b_0(L_0)$ is the ablation rate at the terminus, then

$$\frac{L_1}{L_0} = \frac{b_1}{a_0}. \tag{15}$$

Thus if the ablation rate at the terminus is 5 m/yr, and the mass balance increases uniformly by 0.5 m/yr, the glacier's length will eventually increase by 10%. In two dimensions, the lengths L are replaced by areas S.

The response time is equal to the "filling time," the time t that the perturbation in mass balance takes to accumulate or remove the difference V_1 between the steady-state volumes of the glacier before and after the mass-balance change. This is $V_1/b_1 S_0$. To a first approximation $V_1 = H S_1$, where H is the average thickness of the glacier. It follows from Eq. (15), written in terms of areas, that $t = H/a_0$. This formula gives response times of 15–50 yr for glaciers in temperate climates, 250–1000 yr for cold ice caps such as those in Arctic Canada, and roughly 3000 yr for the Greenland Ice Sheet. Getting data to test these predictions is difficult because any glacier is continually adjusting to a complicated history of mass-balance changes.

The general glacier response problem is to predict the ice thickness at every point and how it changes with time for a given bedrock configuration and any specified time-dependent values of accumulation and ablation. To remove the restriction to small perturbations inherent in the kinematic wave analysis, and to deal with the complicated geometry of a real glacier, numerical solutions are needed. The two-dimensional form of Eq. (13), with veloc-

ity specified by an equation such as Eq. (14), is integrated to give the ice thickness at each grid point and at each time step. Problems that have been addressed include assessing the risk that Griesgletscher in Switzerland, which ends in a lake, will advance far enough to destroy a dam, and predicting how global warming will change the amount of runoff from Hofsjökull in Iceland, which feeds a hydroelectric plant. The inverse problem, to estimate recent changes in mass balance and climate from data on glacier retreat, has also been tackled.

Only in temperate glaciers is the effect of a climatic change restricted to a change in mass balance. In cold glaciers, long-period changes in surface temperature eventually change the temperature and therefore the deformation rate of the ice at depth. Again, the position of the terminus of a tidewater glacier is affected by a change in sea level. Because each of these processes has its own characteristic response time, the overall response of the glacier is complex. All these effects have been taken into account in numerical models of the growth and decay of the Greenland and Antarctic ice sheets through a glacial cycle. Their response to global warming has also been predicted.

Figure 17 presents data on how world ice volume has varied during the past 800 kyr. Variations with periods of about 20, 40, and 100 kyr, as predicted by the *Milankovitch hypothesis*, are apparent. This hypothesis ascribes the glacial cycle to variations in the amount of solar radiation received at northern latitudes in summer. These variations result from the precession of the earth's axis (period 20 kyr), variations in the tilt of the axis relative to the plane of the earth's orbit (40 kyr), and variations in the eccentricity of the orbit (100 kyr). Although there are times when the ice sheets are growing rapidly, the general feature of Fig. 17 is a slow buildup of ice, followed by rapid decay. At the end of the last ice age, for example, world sea level rose by about 125 m in 8000 yr. This represents melting of some 57×10^6 km^3 of ice.

The volume of an ice sheet of circular plan and parabolic profile on a horizontal base can be shown to be $0.53\pi L^2 H$ with $H = K L^{1/2}$. Here L is the radius, H is the maximum thickness, and K is a constant. If V denotes volume and t time, then

$$\frac{dV}{dL} = 1.325\pi K L^{3/2}\left(\frac{dL}{dt}\right). \tag{16}$$

But this must be equal to $\pi L^2 c$, where c is accumulation rate, assumed to be uniform over the whole ice sheet. This gives a differential equation for the growth time with solution

$$t = \frac{2.65(H - H_0)}{c}, \tag{17}$$

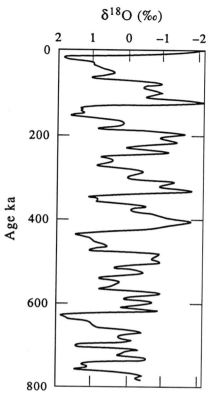

δ¹⁸O (‰)

FIGURE 17 Variations in oxygen-isotope ratio in fossil foraminifera in ocean cores, a proxy for world ice volume, over the past 800 kyr. Units are standard deviations from the mean. Ice sheets are large when δ is positive. [Reproduced with permission from Ruddiman, W. F., and Wright, H. E., Jr. (eds.), "North America and Adjacent Oceans during the Last Deglaciation." Copyright 1987, Geological Society of America.]

where subscript 0 denotes the initial value. This is a minimum growth time because it is assumed that there is no ablation. Similarly, the minimum decay time under ablation rate a is

$$t = \frac{2.65(H_0 - H)}{a} \tag{18}$$

Equation (18) predicts that to remove an ice sheet, initially 3000 m thick at the center, in 8000 yr would require an average ablation rate of 1 m/yr over the whole ice sheet, with no accumulation.

Because ablation rates are usually about three times accumulation rates on the same glacier, these equations predict that an ice sheet should take about three times as long to build up as to decay. In reality, the difference is much greater than this. Entry of the sea into Hudson Bay and the Baltic was an important factor in the decay of the Laurentide and Fennoscandian ice sheets, as was calving of icebergs into huge ice-dammed lakes around the southern margins and their catastrophic draining under the ice.

X. UNSTABLE FLOW

A few glaciers exhibit a spectacular type of instability called a surge. Unlike normal glaciers, in which the mean velocity varies little from year to year, surging glaciers alternate between an active period, usually 1–3 yr, in which they move at speeds 10–1000 times normal, and a quiescent period roughly 10 times as long. Surge velocities are so high (65 m/day briefly in one case) that they must result from basal motion. A surge transfers a large volume of ice from a reservoir area, where the surface may drop 50–100 m, to a receiving area where the ice thickens by a similar amount. These are not the same as the accumulation and ablation areas; many surges do not extend to the head of the glacier. Typically, the ice in the lower part moves forward several kilometers during a surge. This, however, represents reactivation and thickening of previously stagnant ice, not an advance beyond the previous limits of the glacier. The surface of a surging glacier is a chaotic array of crevasses and ice pinnacles, and deep longitudinal faults separate the ice from the valley walls. During the quiescent period, ice flow cannot keep pace with accumulation and ablation; the upper part of the glacier thickens while the terminal area melts down and stagnates. As a result, the surface slope steepens until a critical value, which appears to trigger the next surge, is reached. Many surging glaciers are distinguished by medial moraines that, instead of being parallel to the valley walls, are deformed into large bulblike loops formed by ice flow from a tributary while the main glacier is quiescent. The fact that surges occur at regular intervals in the same glacier suggests that they result from some internal instability.

Surging glaciers are found only in certain areas. Most of the roughly 200 in western North America lie near the Alaska–Yukon border. There are none in the United States outside Alaska, nor in the Coast, Selkirk, or Rocky Mountains in Canada. Other regions with surging glaciers include the Pamirs, Tien Shan, Caucasus, Karakoram, Chilean Andes, Iceland, Svalbard, and Greenland. The distribution of surging glaciers in any region is not random, and glaciers that originate in the same ice cap do not necessarily surge at the same time. Glaciers of all sizes can surge, although there appears to be a preponderance of relatively long glaciers.

To find one mechanism that can explain all surges may be difficult or even impossible. Surging glaciers show a variety of characteristics and only three, Medvezhiy Glacier in the Pamirs, Variegated Glacier in Alaska, and Trapridge Glacier in the Yukon, have been observed in detail. A surge apparently results from disruption of the subglacial drainage system so that water is stored at the bed, producing rapid sliding or bed deformation.

The surge of Variegated Glacier was characterized by high basal water pressure, measured by the water level in boreholes, and by greatly reduced flow in the streams at the terminus. For occasional brief periods, which coincided with pulses of exceptionally high velocity, the basal water pressure exceeded that of the overlying ice. Periodic floods from the outflow streams correlated closely with abrupt decreases in velocity. The end of the surge was marked by a particularly large flood, and the glacier surface dropped by about 0.2 m.

These observations suggest that the normal subglacial tunnel system was destroyed so that all the water was transferred to linked cavities. (See Section VI.) Accumulation of water at high pressure in cavities in the lee of bedrock bumps would permit rapid sliding. This would in turn tend to close the narrow passageways connecting the cavities and thus maintain the high water pressure. Why the tunnel system is destroyed and how it is reestablished at the end of the surge is not clear. Rapid sliding rather than bed deformation must prevail in Variegated Glacier because the driving stress remained at a normal value of around 100 kPa during the surge. This mechanism could also explain the surge of Medvezhiy Glacier. Because Variegated Glacier is temperate and the base of Medvezhiy Glacier is at melting point all the time, a thermal mechanism of surging (alternate melting and freezing of the glacier bed) cannot apply to them.

Trapridge Glacier, which has been observed only during its quiescent period, is markedly different from the other two. It is less than 100 m thick and has a "soft" bed, deformation of which contributes more than half the forward motion of the glacier. Sliding at the ice–till interface accounts for most of the remainder. At the base, a central area of melting ice is surrounded by ice frozen to the bed, a type of temperature distribution found in many nonsurging glaciers as well. Although fast flow seems unlikely unless the whole bed is at melting point, there is no evidence that the melting area is spreading. Although near-surface temperatures in the glacier range from −4 to −9°C, large quantities of surface meltwater penetrate to the bed. Thus disruption of the subglacial drainage system may still be the cause of surges in this glacier, but the detailed mechanism is unclear. A "soft" bed has no fixed bumps behind which cavities can form. Water is not being trapped at the boundary between the frozen and melting sections of the bed; it drains away through the lower unfrozen layers of till. Destruction of drainage channels within the till would increase the pore-water pressure and therefore the deformation rate. However, till is a dilitant material; deformation increases its permeability. Thus more of the water would drain away and reduce the pore pressure. It is difficult to see how fast deformation could be maintained.

Much has still to be learned about the mechanism of surging, particularly in glaciers with "soft" beds. The fact that most surging glaciers are in young mountain ranges subject to rapid erosion suggests that bed deformation may be an important factor in many surges.

Glaciers that end in fiords can also behave unstably. Shoals, headlands, and narrow parts of the channel can anchor the terminus. A slight retreat from the anchor point reduces the back pressure on the terminus, so that the ice velocity increases and the glacier thins. This can lead to a catastrophic breakup of the glacier tongue, even if the water is not deep enough to float the ice. The terminus then retreats to the next anchor point. Columbia Glacier in Alaska began such a retreat in 1983 that will, according to current estimates, amount to 30–40 km within the next few decades. The resulting icebergs are a threat to tankers at the Alaska pipeline terminal at Valdez.

Can ice sheets surge? If they do, the features may be somewhat different from those observed in glaciers in temperate and subpolar regions. In contrast to the limited reservoir area of a valley glacier, outlet glaciers and ice streams in Greenland and Antarctica may be able to keep drawing ice from an increasing area of the ice sheet. In this case their active periods would be much longer than the 1–3 yr typical of valley glaciers. Moreover, because precipitation is low, the time taken to refill the reservoir area may be hundreds or even a few thousand years. The ice-age climate in the North Atlantic region was characterized by the so-called *Dansgaard-Oeschger events*: cycles of slow cooling followed by rapid warming with periods of 500 to 2000 yr, accompanied by changes in precipitation rate. (See Section XI.) A regular interval between surges is not expected in these circumstances.

The mechanisms may also be different. A large seasonal influx of surface meltwater to the bed is an important part of the mechanism of known surges and seems to control their timing. In Antarctica, all the water at the bed of the ice sheet is generated there, presumably at a constant rate. In Greenland, although summer melting is extensive on the lower parts of the outlet glaciers, the lack of seasonal variations in their velocity suggests that surface water does not penetrate to the bed.

There is evidence that parts of the Laurentide Ice Sheet in North America behaved unstably. The sediment record from the North Atlantic between latitudes 40 and 55°N contains six layers of detrital carbonate, deposited rapidly at irregular intervals between about 65 and 14 kyr B.P. (before present) at times when the surface water was exceptionally cold and had low salinity. The layers are thick in the Labrador Sea and western North Atlantic and thin steadily further east. Their deposition is attributed to exceptionally large discharges of icebergs (*Heinrich events*) from the Hudson Bay sector of the ice sheet. These events, each of which probably lasted not more than 1000 yr, would have a major effect on the deep-water circulation in the North Atlantic and thus on the climate of the region.

Periodic warming of the base of the ice to melting point is the likely mechanism of this instability. During the quiescent period, snow accumulation thickens the ice sheet and the basal temperature rises because the geothermal gradient at the base exceeds the atmospheric lapse rate at the surface. When the melting point is reached, water pressure builds up and permits rapid sliding and bed deformation, followed by advance of the terminus and a massive discharge of icebergs. The rapid advance thins the ice and so increases the temperature gradient in it. The increased heat conduction cools the basal ice until it eventually refreezes and the rapid flow stops. This is a plausible mechanism, although it does not apply to any current surge. A freeze-thaw cycle seems necessary to incorporate debris into the basal ice. However, the fact that some debris was carried far across the Atlantic seems to require that it was dispersed throughout the ice thickness. How this could happen is unclear.

Geologic evidence indicates that some of the southern lobes of the Laurentide Ice Sheet, which had the low surface slopes and driving stresses characteristic of glaciers moving largely by bed deformation, also made periodic rapid advances.

XI. ICE CORE ANALYSIS

A core taken from a polar ice sheet provides a record of snowfall and, through its chemistry, of past climate and also of atmospheric fallout such as windblown dust, pollen, volcanic deposits, sea salts, extraterrestrial particles, and trace elements resulting from both natural causes and pollution. In addition, the air bubbles are samples of the atmosphere at the time the ice was formed. If the core is from a region where the surface never melts, the record is continuous, except for the lowest 5–10%, which may be disturbed by folding, faulting, or movement along shear planes in the ice near the bed. Cores covering the last ice age have been recovered in Greenland and Arctic Canada, while the longest Antarctic core covers several glacial cycles. For such records to be of much value, the age of the ice at different depths (the "time scale") must be determined.

The most important climatic parameter is the oxygen-isotope ratio ($^{18}O/^{16}O$) or, equivalently, the ratio of deuterium to hydrogen in the ice. This depends on the air temperature when the snow fell. The measure used is

$$\delta = \frac{1000(R_s - R_0)}{R_0}, \qquad (19)$$

where R_s and R_0 are the ratios of the concentrations of the heavy to the light isotope in the sample and in standard mean ocean water. The relation of δ to temperature can be explained by the slight reduction in vapor pressure of the heavy components. The isotopic composition of water changes during natural cycles of evaporation and condensation because the molecules of $H_2^{18}O$ evaporate less rapidly and condense more readily from the vapor than those of $H_2^{16}O$. Thus vapor evaporating from the sea surface is slightly depleted in ^{18}O. As an air mass containing ocean vapor travels toward the polar regions, it is cooled strongly and loses considerable amounts of precipitation. It therefore becomes progressively more depleted in ^{18}O, that is, δ becomes more negative. There are also seasonal variations. Figure 18 confirms that δ is correlated with mean annual air temperature in Greenland and Antarctica, although different regression lines are required for the different ice sheets, and the relation breaks down at elevations below 1000 m. The scatter of the Antarctic data arises because precipitation in different parts of the ice sheet comes from different source areas.

Variations of δ with depth are interpreted as past variations of surface temperature. The regression lines should not be used to convert δ to temperature, however, because changes in atmospheric circulation patterns, in the seasonal distribution of snow, or in the isotopic composition of the oceans can also change the value of δ. The relation between δ and temperature is derived by using the measured temperature profile in the borehole, as described in Section VII. Unless the borehole is at an ice divide, the record has to be corrected for the flow of isotopically colder ice from upstream.

Figure 19 shows the record from the highest part of the Greenland Ice Sheet. A remarkable feature of the ice-age section is the *Dansgaard-Oeschger events*: the switches of temperature by up to 10°C every 500 to 2000 yr. The change from cold to warm occurs in 100 yr or less, while that from warm to cold is more gradual. The temperature oscillations are accompanied by oscillations in precipitation rate and in concentrations of impurities in the core. The driving force may be changes in circulation and deep-water formation in the North Atlantic.

The lowest 6 m of the Summit core, not included in Fig. 19, consists of silty ice with δ values less negative (i.e., warmer) than any others in the core. This ice also contains extremely high concentrations of CO_2 (up to 130,000 parts per million by volume) and methane (up to 6000 ppmv) and levels of oxygen as low as 3%. Such concentrations preclude a direct atmospheric origin, but could result from *in situ* oxidation of organic matter. Comparable concentrations of CO_2 and CH_4 have been measured in ice layers in Alaskan permafrost. Thus it appears that this basal ice formed locally within a peat deposit in permafrost, in the

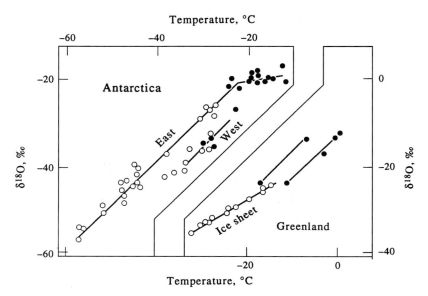

FIGURE 18 Present relation between oxygen-isotope ratio and mean annual air temperature for stations on the (a) Antarctic and (b) Greenland ice sheets. Open circles refer to elevations above 1000 m. [Adapted with permission from Dansgaard, W., *et al.* (1973). *Medd. Groen.* **197**(2), 1–53. Copyright 1973, Commission for Scientific Research in Greenland, Copenhagen.]

absence of an ice sheet. The glacier ice immediately above must have flowed on to the site later, probably from the mountains on the east coast. The marine-sediment record suggests that the ice sheet originated about 2.4 million years ago.

Major volcanic eruptions inject large quantities of silicate microparticles and acid gases into the stratosphere. The microparticles settle out and apparently do not reach Greenland in detectable amounts, although ash from local volcanoes has been found in Antarctic cores. Eruptions are recorded in Greenland by ice layers of increased acidity (Fig. 20). Tambora in Indonesia erupted in 1815. The acid fallout in Greenland reached a peak in 1816, which is described in historical records as "the year without a summer" in eastern North America. The cooling is believed to result from back scattering of solar radiation by a sulfate aerosol, mainly sulfuric acid, that builds up from the SO_2 produced in the eruption. Acidity measurements on Greenland cores have extended the volcanic record back to 8000 B.C. Earlier eruptions cannot be detected because the ice-age ice is alkaline. This ice contains much higher concentrations of microparticles, largely wind-blown dust rich in $CaCO_3$, than ice deposited since. The increase in dust has been attributed to the increased area of continental shelf exposed by the fall in sea level, combined with more vigorous atmospheric circulation.

Figure 20 shows pronounced seasonal variations in oxygen-isotope ratio and also in acidity and nitrate concentration, although these are less well defined. The dates were obtained by counting those annual layers down from the surface.

Past changes in the amount of CO_2 and CH_4 in the atmosphere are recorded in the air bubbles in the ice. Figure 21 shows the record from Vostok, which lies at an elevation of 3490 m in the central part of East Antarctica. The record covers the Holocene, the last ice age, the preceding interglacial, and most of the previous ice-age. The temperature record shows oscillations with periods of 20 and 40 kyr and also the rapid endings of the last two ice ages roughly 100 kyr apart, as predicted by the Milankovitch hypothesis. The CO_2 record shows trends similar to the temperature record, with concentrations as low as 180 ppm during ice ages and high concentrations when it is warm. Although the increase in CO_2 was probably a result, not a cause, of the end of the ice age, the greenhouse effect would undoubtedly amplify the warming trend. Figure 21 also shows a major increase in the concentration of methane, another greenhouse gas, at the end of an ice age. Ice-core records show that the concentration of CO_2 remained at about 280 ppm between the end of the last ice age and the start of the industrial era (1750 A.D.) and has increased steadily since then. It had reached 360 ppm by 1995.

Drilling at Vostok was terminated at a depth 3623 m. Glacier ice, estimated to be about 425 kyr old, ended at 3540 m. Below that was about 200 m of ice, with crystals up to 1 m in diameter, formed by the freezing on of water from a subglacial lake. Vostok is situated over such a lake,

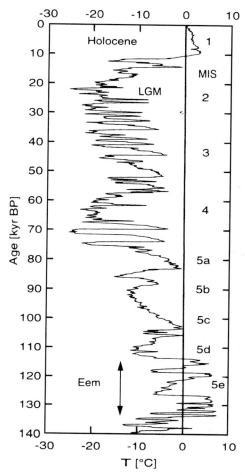

FIGURE 19 Past temperatures in central Greenland based on oxygen-isotope ratios in the 3029-m core to bedrock at Summit. The record in ice older than 110 kyr, the lowest 230 m of core, is unreliable because the stratigraphy is disrupted. The numbers on the right denote stages in the marine isotopic record. [Reprinted with permission from Johnsen, S. J., et al., "Stable Isotope Records from Greenland Deep Ice Cores: The Climate Signal and the Role of Diffusion." In "Ice Physics and the Natural Environment" (J. S. Wettlaufer, J. G. Dash, and N. Untersteiner, eds.), pp. 89–107, [Copyright 1998, Springer-Verlag, Berlin.]

the largest of about 80 in East Antarctica, which have been discovered by radar sounding. Its area is comparable with that of Lake Ontario and it is over 500 m deep. Drilling was stopped about 120 m above the lake surface to avoid contaminating it with drilling fluid. However, plans are now being made to sample the lake for microbes and other possible forms of life.

Ice cores are valuable for pollution studies because the preindustrial background level of trace elements can be measured. The concentration of lead in northwest Greenland has increased by a factor of 3 since 1945; the source is tetraethyl lead in gasoline. Tests of atomic and nuclear bombs are recorded by radioactive layers, as is the burnup

of a U.S. satellite (SNAP-9A) containing ^{238}Pu in 1964 and the Chernobyl nuclear accident in 1986. The low radioactivity of 1976 snow, only 1.4% of the peak value, suggests that all the plutonium isotopes injected into the atmosphere by the tests have now been removed.

There are several methods of dating ice cores:

1. Counting annual layers, distinguished by seasonal variations in some property (Fig. 20).
2. Noting horizons of known age, such as acid layers from identified eruptions (Fig. 20).

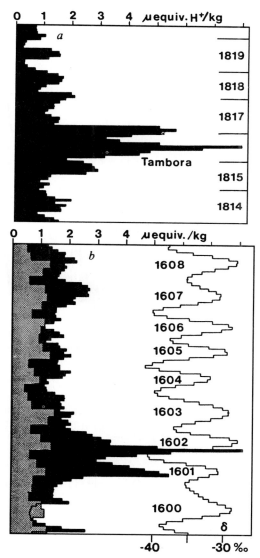

FIGURE 20 Acid fallout (black curves; scale on top) from two volcanic eruptions found in an ice core at Crête, Greenland. Seasonal variations in oxygen-isotope ratio (curve on right) and concentration of NO$_3$ (shaded curve on left) are also shown in the lower diagram. [Reprinted with permission from Hammer, C. U., et al., Nature 288, 230–235. Copyright 1980, Macmillan Magazines Ltd.]

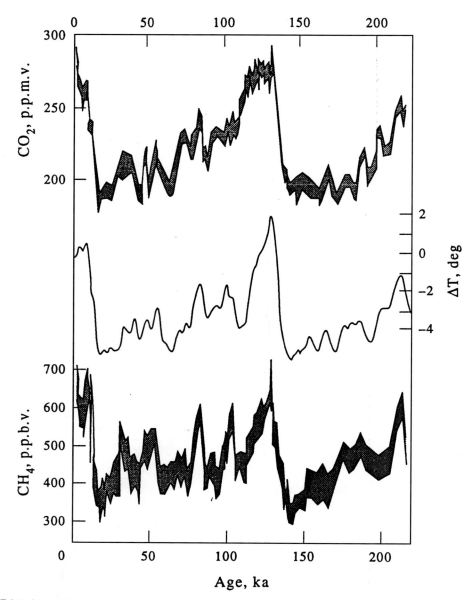

FIGURE 21 Atmospheric temperature relative to present (derived from δD) and atmospheric concentrations of CO_2 and CH_4 over the past 220 kyr measured in the core from Vostok. Shading indicates the uncertainty of the gas measurements. [Reprinted with permission from Jouzel, J., *et al.*, *Nature* **364**, 407–412. Copyright 1993, Macmillan Magazines Ltd.]

3. Carbon-14 dating of CO_2 extracted from air bubbles in the ice.
4. Doing calculations based on ice-flow models.
5. Matching features of the oxygen-isotope record with some other dated climatic record such as an ocean-sediment core.

Counting annual layers is the most precise method. The only inaccuracy is the occasional uncertainty in identifying layers; this can usually be overcome by looking at more than one property. Molecular diffusion smooths out variations in $\delta^{18}O$ after several thousand years, but variations in acidity and microparticle content are unaffected. However, because the layers become progressively thinner with increasing depth, a point is reached below which they cannot be distinguished. A drawback of this method is the time required; about eight samples per year must be measured. After one core has been dated in this way, however, the ages of recognizable horizons can be used to date others.

Ice-flow modeling is the most frequently used method and the only one available for planning before drilling. How the thickness of an annual layer changes with depth has to be calculated. The precision depends on the validity of the assumptions underlying the model and on how much is known about variations in accumulation rate and ice thickness in the past and upstream from the borehole. Accumulation rates are usually estimated by assuming that they are correlated with the temperatures deduced from measurements of δ^{18}O.

The Summit core (Fig. 19) was dated back to 14.5 kyr B.P. by counting annual layers, and beyond that by a flow model. Parameters in the model were adjusted so that it gave correct ages for two independently dated events: the end of the Younger Dryas cold period at 11.5 kyr B.P. and the coldest part of marine isotope stage 5d at 110 kyr B.P. The comparable GISP-2 core, drilled 30 km west of Summit, was dated back to 110 kyr B.P. by counting annual layers; six properties were used to distinguish them. The precision was estimated at 1–10% down to 2500 m, where the ice is about 60 kyr old, and 20% from 2500 m to the 110-kyr horizon at 2800 m. The Vostok core (Fig. 21) was dated by a flow model, a more complicated one than that used at Summit because Vostok is not at an ice divide. The model was tuned to give correct dates at the same two points as at Summit. The estimated uncertainty at 260 kyr B.P. is 25 kyr.

SEE ALSO THE FOLLOWING ARTICLES

CLIMATOLOGY • COASTAL GEOLOGY • GEOMORPHOLOGY • GLACIAL GEOLOGY AND GEOMORPHOLOGY • GREENHOUSE EFFECT AND CLIMATE DATA • HYDROGEOLOGY • STREAMFLOW

BIBLIOGRAPHY

Benn, D. I., and Evans, D. J. A. (1998). "Glaciers and Glaciation," Wiley, New York.

Hambrey, M., and Alean, J. (1992). "Glaciers," Cambridge University Press, Cambridge, U.K.

Hammer, C., Mayewski, P. A., Peel, D., and Stuiver, M. (eds.) (1997). "Greenland Summit Ice Cores." Special volume reprinted from *J. Geophys. Res.* **102,** C12, 26315–26886, American Geophysical Union, Washington, DC.

Paterson, W. S. B. (1994). "The Physics of Glaciers," 3rd ed., Elsevier Science, Oxford, U.K.

Petrenko, V. F., and Whitworth, R. W. (1999). "Physics of Ice," Clarendon Press, Oxford, U.K.

Post, A., and LaChapelle, E. R. (2000). "Glacier Ice," rev. ed., University of Washington Press, Seattle, WA.

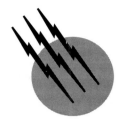

Glass

Solomon Musikant

General Electric Company

I. History
II. Definition
III. Structure
IV. Composition
V. Annealing
VI. Properties
VII. Manufacture
VIII. Applications
IX. The Future of Glass

GLOSSARY

Annealing Thermal process whereby the residual strains in a glass are removed by a precisely controlled thermal treatment.

Batching In the making of glass, the step in the process where the initial measurement and mixing of the ingredients takes place.

Birefringence Quality of an optical material that is characterized by different indices of refraction in various directions.

Devitrification Process whereby a glassy material is converted to a crystalline material.

Dielectric constant Material property determined by the ratio of the charge-holding capacity of a pair of parallel, electrically conductive plates with the material filling the space between the plates to the charge-holding capacity of the same plates when a vacuum replaces the intervening material.

Dispersion Variation of the index of refraction in a material as a function of wavelength of the light transiting through the medium.

Forming Process whereby a glass at a low enough viscosity to permit relatively easy deformation is shaped by application of forces into a desired configuration and then cooled to a rigid state.

Fused quartz Glassy product resulting from the melting of crystalline quartz and then cooling at a rate fast enough to avoid devitrification.

Glass-ceramic Body composed of a glassy matrix within which fine precipitates of a crystalline phase have been formed by appropriate heat treatments. Properties of the glass-ceramic are thereby substantially different than those of the initial glass.

Glass-fiber wave guide Glass fiber with its index of refraction varying radially, so that the index is higher at the axis than at the periphery and thereby able to confine a coherent ray of light, which is modulated to convey information.

Index of refraction Ratio of the velocity of light in vacuum to the velocity of light in the medium.

Network formers Those elements in the glass

composition that, during the cooling process from the molten state, are able to induce molecular structures such as polyhedra and triangles, which lead to the glassy state.

Plane-polarized light Light beam in which all the electromagnetic waves are aligned so that the electric vectors are all parallel to each other and all the magnetic vectors are parallel to each other.

Quartz Crystalline form of silica (SiO_2) that occurs in two forms: low quartz, below 573°C, and high quartz, between 573 and 867°C. High quartz can be viewed as composed of connected chains of silica tetrahedra.

Silica Term given to the stable oxide of silicon, SiO_2, which can occur in glassy or in a wide variety of crystalline forms.

Spectral transmittance Property of a medium that permits light to pass through it without absorption or scattering.

A GLASS is an inorganic substance in a condition that is continuous with, and analogous to, the liquid state of that substance, but which, as a result of having been cooled from a fused condition, has attained so high a degree of viscosity as to be for all practical purposes rigid. The limitless possible compositions and variety of equilibrium and metastable conditions provide the keys to the widest possible range of optical, physical, and mechanical properties.

I. HISTORY

The art of glass is among the oldest of mankind's achievements preceding written history. Glass tektites of extraterrestrial origin were probably found by ancient humanity. Primitive people shaped objects from volcanic glass found in nature. Obsidian is one of the most common of such materials and is translucent and usually blackish in color, although sometimes found in red, brown, or green hues, depending on the specific composition. Its earliest use was for cutting or piercing implements or weapons, as the shards are extremely sharp and often elongated. More advanced cultures used obsidian for jewelry, ornaments, and sacred objects. Worked objects have been found in Europe, Asia Minor, and North and Central America. The flake knife, in which the glass is fashioned by cleavage to the form of a sharp blade, was used well into the age of iron. The history of glass has been reviewed by Morey.

The discovery of glass making was not recorded but probably was by accident. Fire-making knowledge provided access to high temperatures and to both metallurgy and glass making. A possible mode of discovery was the accidental ignition of grain and the fusion (melting) of its ash. Pliny describes a Phoenician story regarding the accidental discovery of glass making when blocks of nitre were used to support a wood fire on a sand beach. The nitre (potassium nitrate) formed a low-melting eutectic with the pure silica (SiO_2) sand of the beach, fused, and upon cooling formed glass.

One of the oldest pieces of glass is a molded amulet of deep lapis lazuli color of about 7000 B.C. found in Egypt. Many early examples of glass objects have been found in Syria and in the Euphrates region that can be dated back to 2500 B.C. Mesopotamia also is a region where early glass objects have been discovered and believed to have been fashioned as early as 2700 B.C. Explorations at Ninevah have revealed Assyrian styles that give detailed descriptions of glass-making furnaces and methods of preparing glass representing the state of the art at about 650 B.C.

There was a stable glass industry in Egypt at the beginning of the eighteenth dynasty (approximately 1550 B.C.). From that time to the beginning of the Christian era, Egypt was the greatest center of glass manufacture. The industry was centered in Alexandria.

Initially, glass was considered to be a gem. As late as Ptolemaic times, glass gems were highly valued. Glass was later used for hollow vessels, such as unguent jars. The idea of pressing glass in open molds is thought to have originated about 1200 B.C.

The invention of glass blowing was a revolutionary advance. This technique is thought to have been discovered about the beginning of the Christian era. This method was accompanied by dramatic improvements in glass quality. The highest-quality glassware of the early Roman empire was characterized by transparency and freedom from color. Drinking vessels of glass superceded those of silver and gold.

The ensuing rapid expansion of glass manufacture occurred because of the combination of glass blowing and the stability afforded by the Roman Empire. Glass became a common home material. From the fall of Rome (476 A.D.) to the eleventh century, glass manufacture declined greatly. However, in the tenth century, a limited treatise on glass manufacture was written, and a far more comprehensive writing was composed in the twelfth century by Theophilus Presbyter, reflecting the influence of the Venetian renaissance.

Venice became the hub of a dominant glass industry in the eleventh century and maintained that status for four centuries. The Venetians took great pains to maintain their proprietary knowledge. From Venice, the knowledge finally spread widely. The book "L'Arte Vetraria" by Neri was published in Florence in 1612. It was later annotated by Merrett and Kunckel and is an important source document for glass history.

Neri's book initiated the modern scientific approach to glass making by providing a basis in facts and observations. Neri was a priest, and his interests are presumed to have been academic in nature. Merrett was an English physician, a member of the royal society, and able to bring the scientist's viewpoint to the work, while Kunckel was descended from a family of glass makers. He added the practical knowledge to the work.

After 1600, glass technology and manufacturing spread rapidly throughout Europe. During this early period, the art of cut glass developed. Coal replaced wood as a fuel in 1615, and flint glass (high refractive index) was invented in 1675. The name *flint* comes from the early use of high-purity silica in the form of flint to make this glass. Lead oxide was the essential ingredient in developing the high index of refraction.

During the nineteenth century, the demands of optical instruments forced great advances in the theoretical understanding of glass properties, especially the optical properties. Optical glass must be free from imperfections, homogeneous, and available in a wide variety of optical properties that are repeatable from batch to batch.

Michael Faraday in England conducted extensive research on the properties of glass. Later, O. Schott in Germany performed an outstanding study on the effect of composition on optical properties. E. Abbe, a physicist, provided the optical property requirements that Schott attempted to achieve by new compositions of glass, while Winkelman conducted critical studies of the mechanical properties of glass.

These brilliant researchers ushered in the modern age of glass making. The military requirements for optical glass during World War I created a new impetus to both the science and manufacturing technologies of glass. To the present day, a seemingly endless series of technology advances, inventions, and broad applications continues to occur.

II. DEFINITION

No wholly satisfactory definition of glass is possible. A reason for this difficulty is that words such as *solid* and *liquid* are generally used much more loosely than allowed by strict scientific definition. The definition of glass is often given in terms of undercooled liquid or amorphous solid. However, liquid and solid are terms applied to a substance that undergoes a phase change. In the case of glass, no phase change occurs as it cools from the fluid state, but there is a complete continuum of properties such as its viscosity. In fact, the viscosity of glass is generally used to define its various conditions such as softening point and melting point.

Even at room temperature, although glass is "solid" in the ordinary sense of the word, in actuality it is still fluid, although with an extremely high viscosity. Cases of centuries-old windows that show signs of flow are well known. Morey provides the following definition:

A glass is an inorganic substance in a condition which is continuous with, and analogous to, the liquid state of that substance, but which, as a result of having been cooled from a fused condition, has attained so high a degree of viscosity as to be for all practical purposes rigid.

The American Society for Testing Materials (Committee C-14 on Glass and Glass Products) has adopted the following definition:

Glass—An inorganic product of fusion which has cooled to a rigid condition without crystalizing.

(a.) Glass is typically hard and brittle and has a conchoidal fracture. It may be colorless or colored, and transparent to opaque. Masses or bodies of glass may be colored, translucent, or opaque by the presence of dissolved, amorphous, or crystalline material.

(b.) When a specific kind of glass is indicated, such descriptive terms as flint glass, barium glass, and window glass should be used following the basic definition, but the qualifying term is to be used as understood by trade custom.

(c.) Objects made of glass are loosely or popularly referred to as glass; such as glass for a tumbler, a barometer, a window, a magnifier or a mirror.

III. STRUCTURE

The understanding of the structure of glass has been made possible by X-ray diffraction analysis. It was learned through such early investigations that in crystalline silica, SiO_2, a tetrahedral relationship exists between the Si^{+4} and the O^{-2} ions. These ions form repeating tetrahedra in space where the Si^{+4} is at the center and the O^{-2} ions are located at each of the four apexes of each tetrahedron. This arrangement provides charge balance, since each O^{-2} is shared by two Si^{+4} ions. Such an arrangement is indicated in Fig. 1 and is termed fourfold coordination. In this case, the distance between the ions of opposite charge is about 1.60 Å.

Silica glass seems to be described reasonably well as a random network of Si–O tetrahedra with some variability in Si–O–Si bond angles.

A two-dimensional representation of a crystal with an A_2O_3 arrangement of ions is shown in Fig. 2(a). The term A represents a cation, while O is used in its normal sense to represent an oxygen anion. Figure 2(b) shows a schematic

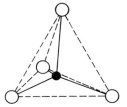

FIGURE 1 The silica tetrahedron: ● silicon ion, ○ oxygen ion. [From Morey, G. W. (1954). "The Properties of Glass," Reinhold, New York.]

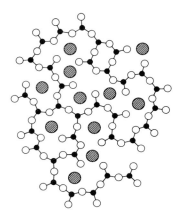

FIGURE 3 Schematic representation of the structure of a sodium silicate glass: ● Si^{+4}, ○ O^{-2}, ◉ Na^+. [From Kingery, W. D., Bowen, H. K., and Uhlmann, D. R. (1976). "Introduction to Ceramics," Wiley, New York.]

representation of an A203 glass where the A203 coordination is maintained but the periodicity is not.

A. Formation

Cations such as Si, which readily form polyhedra and triangles with oxygen, are termed network formers because such coordination polyhedra lead to glass structure. Alkali silicates very readily form glasses, as schematically indicated in Fig. 3. The alkali metals are termed network modifiers, because these ions take up random positions in the network and thus "modify" or change the structure of the network. Other cations with higher valence than the alkali metals and lower coordination numbers modify the network structures less drastically and are called intermediates. Typical network formers, modifiers, and intermediates are listed in Table I, along with the valence and coordination number for each of the ions.

As alluded to earlier, X-ray diffraction work has established that the structure of silica glass is well described as a random network of SiO_4 tetrahedra with variable bond angles in the silicon–oxygen–silicon covalent connections.

The random network exhibits local variability of density and structure. The addition of alkali or alkaline earth oxides to the silica glass breaks up the three-dimensional network, and singly bonded oxygen ions are formed that do not participate in the network. The singly bonded oxygen ions are located in the vicinity of the modifying cations to maintain charge balance.

Other glasses have far different structures. For example, glassy B_2O_3 has been shown by X-ray diffraction to be composed of BO_3 triangles. These triangles are linked together in a boroxyl configuration (Fig. 4) but distorted with the triangular units linked in ribbons. As alkali or alkaline earth oxides are added to the B_2O_3 boria glass, BO_4 tetrahedra are formed and the structure of the glass takes on a mixed character with BO_3 triangles and BO_4 tetrahedra.

Germania (GeO_2) glass is composed of GeO_4 tetrahedra. Unlike the silica glass, germania glass tetrahedra have a small distribution of bond angles. The essential glassy nature of the material is accommodated by variations in the rotation angle of one tetrahedron to another. Additions of alkali oxide to the germania in small fractions leads to the formation of GeO_6 octahedra in the structure.

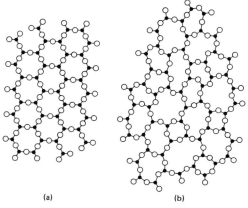

(a)   (b)

FIGURE 2 Schematic representation of (a) ordered crystalline form and (b) random-network glassy form of the same composition. [From Kingery, W. D., Bowen, H. K., and Uhlmann, D. R. (1976). "Introduction to Ceramics," Wiley, New York.]

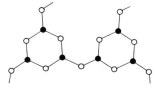

FIGURE 4 Schematic representation of boroxyl configurations: ●, boron; ○, oxygen. [From Kingery, W. D., Bowen, H. K., and Uhlmann, D. R. (1976). "Introduction to Ceramics," Wiley, New York.]

TABLE I Coordination Number and Bond Strength of Oxides[a]

	M in MO_x	Valence	Dissociation energy per MO_x (kcal/g atom)	Coordination number	Single-bond strength (kcal/g atom)
Glass formers	B	3	356	3	119
	Si	4	424	4	106
	Ge	4	431	4	108
	Al	3	402–317	4	101–79
	B	3	356	4	89
	P	5	442	4	111–88
	V	5	449	4	112–90
	As	5	349	4	87–70
	Sb	5	339	4	85–68
	Zr	4	485	6	81
Intermediates	Ti	4	435	6	73
	Zn	2	144	2	72
	Pb	2	145	2	73
	Al	3	317–402	6	53–67
	Th	4	516	8	64
	Be	2	250	4	63
	Zr	4	485	8	61
	Cd	2	119	2	60
Modifiers	Sc	3	362	6	60
	La	3	406	7	58
	Y	3	399	8	50
	Sn	4	278	6	46
	Ga	3	267	6	45
	In	3	259	6	43
	Th	4	516	12	43
	Pb	4	232	6	39
	Mg	2	222	6	37
	Li	1	144	4	36
	Pb	2	145	4	36
	Zn	2	144	4	36
	Ba	2	260	8	33
	Ca	2	257	8	32
	Sr	2	256	8	32
	Cd	2	119	4	30
	Na	1	120	6	20
	Cd	2	119	6	20
	K	1	115	9	13
	Rb	1	115	10	12
	Hg	2	68	6	11
	Cs	1	114	12	10

[a] From Kingery, W. D., Bowen, H. K., and Uhlmann, D. R. (1976). "Introduction to Ceramics," Wiley, New York.

B. Devitrification

The essential job of the glassmaker is to avoid crystallization of the glass melt as it is cooled to room temperature. If crystals form in the glass due to heat treatment or service at elevated temperatures, the glass is said to have devitrified. In general this is an undesirable event since the crystalline component usually has incompatible expansion, mechanical, or optical characteristics and leads to degradation of the desired properties such as optical or mechanical. In fact, only certain compositions can be made into glasses without the occurrence of crystalline phases. Compositions of matter that result in satisfactory glasses upon cooling from the melt may form crystalline structures when the equilibrium condition is achieved at some elevated temperature. Materials that form glasses at room temperature have to be cooled relatively quickly into a metastable glass to avoid the crystalline state.

For example, in silica glass, crystals form in the glass if held at even moderately elevated temperatures for long periods of time. The exact conditions that will lead to devitrification are very sensitive to impurity content, especially the alkali oxides. The growth of the crystallites in the glass leads to disastrous consequences such as cracking due to the mismatch of expansion coefficients of the glass and the crystalline form.

However, glass-ceramic is a term given to a class of materials in which partial, controlled devitrification is designed to occur in order to strengthen, harden, and influence the coefficient of thermal expansion of the resultant product. A well-known commercial glass-ceramic is Pyroceram (trademark of Corning Glass Works, Inc.). A glass-ceramic is a two-phase material embodying a crystalline phase in a glass matrix.

Glassy bodies are also made, such as Pyrex (trademark of Corning Glass Works, Inc.), that are composed of two distinct phases that are both glasses. Such multiphase glasses are also strong and resistant to thermal shock.

IV. COMPOSITION

Natural glasses are formed from magmas that have cooled quickly enough to avoid crystallization and from lightening strikes, the latter known as fulgurites. The fragments of a fulgurite have been found in a sandpit adding up to 9 ft in length and up to 3 in. in diameter. Another group of natural glasses is known as tektites, believed to be of extraterrestrial origin. The compositions of natural glasses vary widely, as illustrated in Table II.

Commercial glass compositions must permit melting at economical temperatures, and a viscosity–temperature relationship that is amenable to forming without devitrification and that results in a glass having the required properties for the specific application intended.

A. Silica

Silica is the glass par excellence having the widest variety of useful applications. It has a low thermal expansion, is resistant to attack by water and acids, and is resistant to devitrification. However, the high temperature required to melt quartz sands, the difficulty in freeing the melt from gas bubbles, and its viscous nature, which makes the

TABLE II Compositions of Some Natural Glasses[a]

No.	SiO_2	Al_2O_3	Fe_2O_3	FeO	MgO	CaO	Na_2O	K_2O	H_2O^+	H_2O^-	TiO_2
1	97.58	1.54	—	0.23	—	0.38	0.34	—	0.10		
2	87.00	8.00	0.19	1.93	0.82	nil	0.14	0.99	0.36		0.51
3	76.89	12.72	0.43	0.70	0.17	0.57	3.48	4.39	0.47	0.02	0.08
4	76.37	12.59	0.26	0.48	0.17	0.79	3.36	4.67	0.97		0.11
5	74.70	13.72	1.01	0.62	0.14	0.78	3.90	4.02	0.62	—	—
6	73.92	12.38	1.62	0.56	0.27	0.33	3.49	5.39	1.69	—	—
7	71.60	12.44	1.00	0.65	0.06	1.90	3.30	4.22	3.78	0.81	0.25
8	70.62	11.54	1.20	0.18	0.26	1.72	3.52	1.45	7.24	2.42	0.04
9	70.56	20.54	—	0.96	0.11	0.78	3.47	3.38	—	—	—
10	69.95	11.99	0.76	0.64	0.09	0.66	3.70	3.80	4.98	2.80	0.18
11	68.12	12.13	n.d.	1.03	tr.	1.63	5.34	1.69	9.70	—	—
12	67.55	15.68	0.98	1.02	1.11	2.51	4.15	2.86	2.76	0.38	0.34
13	66.68	13.39	0.91	0.21	none	2.72	2.23	2.51	10.05	—	0.38
14	61.4	—	5.8	—	2.9	10.3	2.4	10.1	—	—	—
15	60.12	17.67	3.75	3.40	0.53	2.40	6.78	4.43	0.49	—	—
16	58.59	21.29	4.74	0.71	2.49	6.36	4.42	0.94	0.27	0.04	—
17	53.10	20.70	0.07	4.77	1.77	3.18	9.10	5.84	0.70	—	0.47
18	50.76	14.75	2.89	9.85	6.54	11.05	2.70	0.88	n.d.	—	—

[a] From, Morey, G. W. (1954). "The Properties of Glass," Reinhold, New York.

material difficult to form, cause silica glass or fused quartz to be too expensive for most applications. Therefore, other oxides are used to flux (reduce the melting temperature) the quartz raw material. The most effective flux is sodium oxide, Na_2O, but it makes a glass that is soluble in water. Lime, CaO, is added to increase the resistance to moisture (enhance durability) because it is effective in terms of cost and function. Other small constituents occur in the raw ingredients or are added to adjust a host of properties of the glass.

MgO can be added in lieu of CaO. Alumina is added to enhance durability and to lower the coefficient of expansion. Potash (K_2O) substitution for soda (Na_2O) increases durability and reduces the tendency to devitrification. Small amounts of BaO and B_2O_3 are added for the same reason but are relatively expensive.

Not only are the end-use property requirements determinants of the optimum composition, but all the manufacturing demands such as freedom from devitrification and specific forming machine characteristics influence the makeup of the glass batch. The optical glasses have the greatest variety of compositions, because of the many optical glasses needed to fulfill all of the optical designers' needs. Table III lists the compositions of a variety of optical glasses.

B. Chemical Durability

The term chemical durability is given to the ability of any specific glass to resist the deteriorating effects of its environments in manufacture or use. these environments include water, carbon dioxide, sulfides, and other atmospheric pollutants, in addition to cleaning and polishing materials used during fabrication of end products. Each corroding substance, as well as various combinations of corrodants, needs to be considered in assessing the suitability of a given composition to a specific application. The nature of the chemical attack is generally not a simple process, because of the complex nature of the glass itself. For example, water corrosion includes the steps of penetration, dissolution of the silicate, and the formation of decomposition products with complex interactions. One cannot describe this process in terms of solubility alone. When the attacking substance includes several ingredients, the chemistry becomes orders of magnitude more complicated. In practical terms, empirical tests are used to define the resistance of a glass to a specific agent.

As an example of such a test for optical glass, polished glass is exposed to 100% relative humidity air and thermally cycled between 46 and 55°C hourly. After having been exposed for various times up to 180 h, the specimens are measured for light scattering. Standard optical glasses (viz. Schott SK 15, BaF, and F1) undergo surface changes sufficient to lead to light scattering on the order of 2–5% of the incoming beam in a standardized optical test. Schott Glass Technologies, Inc., categorizes climatic resistance into four classes, CR1 to CR4, where CR1 represents glasses with little or no deterioration after 180 h of exposure. The glasses are progressively more susceptible from CR2 to CR4, with CR4 quite prone to deterioration. Under normal humidity conditions, CR1 requires no protection, while the other glasses have to be protected against humid environments.

C. The Surface of Glass

The surface of a glass is different in its nature from the interior of the glass body. This is so because the composition is different as a result of diffusional processes that occur during thermal treatment, of chemical treatments to which the glass has been submitted, and of charge imbalance at the free surface. For example, alkali components diffuse to the surface during heat treatment and evaporate, thus changing the surface concentration of the alkali. Chemical treatments deplete the surface or deposit new material on the surface, either action causing a variation in the surface composition relative to the bulk. Polishing compounds may react with the surface during the polishing process, thus creating chemical gradients. The sorption of gases and vapors from the surrounding environment and the various corrosion processes discussed above serve to alter the surface from the bulk.

V. ANNEALING

A. Viscosity of Glass

When a shearing force is exerted on a liquid, the liquid is displaced in the direction of the force. The rate at which this action occurs is a measure of the viscosity of the liquid. As the viscosity increases, the force required to induce a given rate of displacement increases proportionately. For a unit cube of fluid under shear, the viscosity η is defined by:

$$\eta = Fs/v \quad (dyn/cm^2)(cm)/cm/sec$$

where η is viscosity in poise, F is shear force (dynes/cm^2), s is the height of the cube (cm), and v is the relative velocity (cm/sec) from the top face to bottom face of cube.

The viscosity of the glass is of importance at all stages of its manufacture. The removal of bubbles and the workability of the glass are critically dependent on viscosity control. Tables IV and V present definitions of typical annealing points and strain points of glasses.

TABLE III Compositions of Some Optical Glasses[a]

No.	Type	Name			SiO$_2$	B$_2$O$_3$	Na$_2$O	K$_2$O	CaO	BaO	ZnO	PbO	Al$_2$O$_3$	Fe$_2$O$_3$
1	496/644	Borosilicate crown	BK3	O802	71.0	14.0	10.0						5.	
2	494/646	Borosilicate crown		O2259	69.59	14	8	3					8	
3	506/602	Crown		O714	74.6		9	11	5					
4	511/640	Borosilicate crown	BKI	O144	70.4	7.4	5.3	14.5	2					
4a	510/640	*Same, analysis*			72.15	5.88	5.16	13.85	2.04	0	0	Trace	0.04	0.01
5	511/605	Borosilicate crown		O374	68.1	3.5	5.0	16.0			7.0			
6	513/637	Borosilicate crown		O627	68.2	10.0	10.0	9.5			2.0			
7	513/573	Zinc silicate crown	ZK6	O709	70.6		17.0				12.0			
8	516/640	Borosilicate crown	BK7	O3832	69.58	9.91	8.44	8.37	0.07	2.54	0	0	0.04	0.01
9	516/536	Borosilicate crown		O608	53.5	20.0		6.5						
10	516/609	Silicate crown		O40	69.0	2.5	4.0	16.0	8.					
11	517/602	Silicate crown		O60	64.6	2.7	5.0	15.0		10.2	2.0			
12	517/558	Light barium crown		O1092	65.4	2.5	5.0	15.0		9.6	2.0			
13	520/520	High dispersion crown	KF2	O381	66.8			16.0			3.8	11.6	1.5	
13a		*Same, analysis*			67.40			15.15	0.14	0.39	3.85	10.71	1.72	0.02
14	517/602	Zinc crown	2K2	O546	65.6	4.5	14.5	3.5			11.5			
15	522/596	Silicate crown	K5	O1282	62.5	2.0	5.0	15.0		11.0	3.0	1.0		
16	537/512	Borosilicate flint	KF1	O152	35.4	34.3	7.4					18.7	3.7	
17	540/598	Light barium crown	BaK2	O227	59.5	3.0	3.0	10.0		19.2	5.0			
17a	541/596	*Same, analysis*			59.13	3.04	3.16	9.70	0.13	19.25	5.0	0	0.11	0.02
18	541/469	Light flint	LLF2	O726	62.6		4.5	8.5				24.1		
19	545/503	Light borosilicate flint		O658	32.75	31	1	3				25	7.	
20	549/461	Extra-light flint	LLF1	O378	59.3		5	8				27.5		
21	553/530	Light barium flint	BaLF5	O846	56.2		1.5	11		15	9	7		
22	563/497	Light barium flint	BaLF1	O543	51.6		1.5	9.5		14	12	11		
23	568/467	Light borosilicate flint		O161	29.0	29.0	1.5					30.0	10.0	
24	568/530	Light barium flint	BaLF3	O602	51.2			5.5	5.0	20	14	4		
25	571/430	Light flint	LF1	O154	54.3	1.5	3.0	8.0				33.0		
25a	571/430	*Same, analysis*			54.75	0.45	4.31	7.99	0.05	1.64	0.96	29.30	0.04	0.02
26	572/504	Light barium flint		O527	51.7		1.5	9.5		20.0	7.0	10.0		
27	573/574	Light barium crown	BaK1	O211	48.1	4.5	1.0	7.5		28.3	10.1			
27a		*Same, analysis*			47.73	3.90	1.14	7.16	0.15	29.88	8.61		0.65	0.01
28	576/408	Light flint	LF2	O184	53.7		1.0	8.3				36.6		
29	579/541	Light barium crown		O722	48.8	3.0	0.8	6.5		21.0	15.5	4.1		
29a	580/538	*Same, analysis*			45.02	4.50	0.64	6.80		22.39	15.53	4.70	0.09	
30	581/422	Ordinary light flint	LF3	O276	52.45			4.5	8.0			34.8		
31	583/469	light barium flint	BaF$_3$	0578	49.1		1.0	8.5		13.0	8.5	19.3		
31a	583/463	*Same, analysis*			49.80		1.24	8.20		13.36	8.03	18.74	0.05	0.01
32	591/605	Dense barium crown	SK5	O2122	37.5	15.0				41.0			5.0	
33	604/438	Barium flint	BaF4	O1266	45.2			7.8		16.0	8.3	22.2		
34	610/574	Densest barium crown	SK1	O1209	34.5	10.1				42.0	7.8		5.0	
34a	610/568	*Same, analysis*			40.17	5.96	0.13	0.03	0.03	42.35	8.17	None	2.79	0.02
35	612/592	Densest barium crown	SK3	O2071	31.0	12.0				48.0			8.0	
35a	609/588	*Same, analysis*			34.56	10.96	0.21	0.09		46.91	1.14	None	5.02	
36	613/369	Ordinary silicate flint	F3	O118	46.6		1.5	7.8				43.8		

[a] From Morey, G. W. (1954), "The Properties of Glass," Reinhold, New York.

TABLE IV Viscosity of Glass[a]

	Viscosity (poise)
Flow point	10^5
Softening point	$10^{7.6}$
Upper limit of annealing range	
Incipient softening point	10^{11}–10^{12}
Upper annealing temperature	
Annealing point (temperature at which glass anneals in 15 min)	$10^{13.4}$
Strain point (glass anneals in ~16 h; below this temperature there is practically no viscous yield)	10^{14}

[a] From Musikant, S. (1985). "Optical Materials," Marcel Dekker, New York.

B. Annealing Process

When a glass is cooled down from the melt, strains are introduced due to unequal cooling rates in various parts of the body. These strains can be sufficiently great to introduce severe residual stress in the glass, possibly leading to premature failure. The process of annealing is a thermal treatment designed to reduce these residual strains to an acceptable level for the application intended.

Optical glass requires the most careful annealing because of the significant effect of residual strain on the optical properties of the glass. It is in the development of optical glass that the study of annealing has had its greatest emphasis. Strain causes a homogeneous and isotropic glass to become birefringent. In a birefringent body, the index of refraction is different along the optic axis from the index of refraction in the direction orthogonal to the optic axis. In the case of a plate under a uniaxial load, the optic axis is along the direction of applied load. Remembering that the index of refraction is inversely proportional to the velocity of light in a substance, then the strain is detected by the difference in time of transit of plane-polarized light along the optic axis and transverse to it. This time difference is measured in terms of path difference. The difference is proportional to the strain in the direction of the applied load.

Such measurements are made on an instrument called a polariscope. The stress optical coefficient c is defined as

the ratio of the change in optical path difference (OPD) to the unit stress in the colinear direction. As an example, assume:

$$c = 3 \times 10^{-7} \text{ cm/kg cm}^{-2}$$

$$\text{OPD} = 30 \text{ nm or } 30 \times 10^{-7} \text{ cm}$$

$$\text{stress} = 30 \times 10^{-7}/3 \times 10^{-7}$$

$$= 10 \text{ kg cm}^{-2}$$

Annealing is a two-step process in which the glass is first raised to a sufficiently high temperature that all the strain is removed, and then cooled at a slow enough rate that the permanent strain is low enough to be acceptable. During the annealing process, the glass is heated to a temperature such that the viscosity is sufficiently low that all strains disappear. The glass is then slowly cooled in a controlled manner through the annealing range to the strain point. At that point the glass can be relatively quickly cooled without introducing any additional permanent strain.

A typical allowable birefringence for optical glass is 5 nm/cm path length. A typical annealing curve for an optical glass is shown in Fig. 5.

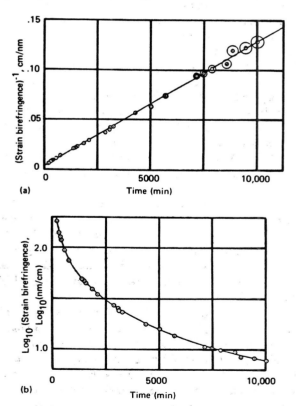

FIGURE 5 Typical annealing curve for glass. (a) Annealing curve, reciprocal of strain birefringence against time, of a glass at 453°C. (b) Annealing curve, logarithm of strain birefringence against time, using the same data as (a). [From Morey, G. W. (1954). "The Properties of Glass," Reinhold, New York.]

TABLE V Annealing Point and Strain Point for Typical Glasses[a]

	Annealing point (°C)	Strain point (°C)
Borosilicate glass	518–550	470–503
Lime glass	472–523	412–474
Lead glass	419–451	353–380

[a] From Musikant, S. (1985). "Optical Materials," Marcel Dekker, New York.

TABLE VI Elastic Constants of Glasses at 20°C^{a,b}

Glass			Resonant frequency			Ultrasonic method		
Table IX	Fig. 6	Type	E	G	ν	E	G	ν
1	A	Silica glass	10.5	4.5	0.165	10.5	4.5	0.17
2	B	96% Silica glass	9.5	4.1	0.16	9.6	4.1	0.18
4	C	Soda–lime plate^c	10.5	4.3	0.21	10.5	4.35	0.21
6		Soda–lime bulb				10.2	4.15	0.24
8	D	Lead–alkali silicate^c (50–60% PbO)	7.8	3.2	0.20	8.3	3.35	0.23
10	E	Borosilicate, low-expansion	9.0	3.7	0.22	9.2	3.85	0.20
11		Borosilicate, low-effective-loss				7.3	3.1	0.22
	F	Borosilicate, crown optical^c	11.6	4.8	0.20	12.0	4.26	0.22

^{a} From Shand, E. B. (1958). "Glass Engineering Handbook," McGraw-Hill, New York.
^{b} E, Young's modulus, 10^6 psi; G, modulus of rigidity, 10^6 psi; ν, Poisson's ratio.
^{c} Glasses of same type but not identical for both methods.

VI. PROPERTIES

A. Mechanical Properties

The elastic properties of most interest are the modulus of elasticity (Young's) E, Poisson's ratio ν, and the modulus of rigidity or torsion G. These values are sensitive to the temperature of measurement. For some typical glasses at room temperature, Table VI presents values of these parameters. The effect of temperature on Young's modulus and on the modulus of rigidity are shown in Fig. 6. These numbers were determined by the reasonant frequency method, in which a beam in bending is vibrated and the values are computed from the resonant frequency data.

The hardness of various minerals is given in Table VII in terms of Mohs (from scratch tests) and from indentation tests. Table VIII shows the hardness of a number of commercial glasses, the compositions of which are given in Table IX.

The strength of glass varies widely as a function of composition but more importantly as a function of flaw size, flaw distribution, and surface perfection. Glass is a brittle material and fails without yielding. The larger the glass body, the lower the breaking strength, due to the higher probability of the existence of flaws from which fractures can initiate.

Table X presents short-time breaking stress for a variety of annealed glasses tested in flexure. There are many variables, and these values can be considered only as a rough approximation. When designing a structure, the designer must obtain accurate strength data for the specific composition, manufacturing method, shape and size, flaw distribution, surface finish, temperture and environment of use, rate of load application, and any other unique conditions under which the structure must operate.

The extreme variation in strength data for glass fibers is shown in Fig. 7, which shows test data up to 2 million psi for fibers of 0.1 mil (1 mil = 0.001 in.) in diameter

with strengths decreasing to 20,000 psi for fibers of 20 ml diameter.

B. Coefficient of Thermal Expansion

The coefficient of thermal expansion (CTE) of glass is of great importance in the successful application of the

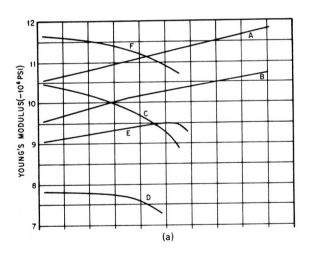

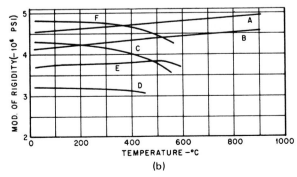

FIGURE 6 Elastic moduli versus temperature. Table VI identifies glasses A–F. [From Shand, E. B. (1958). "Glass Engineering Handbook," McGraw-Hill, New York.]

TABLE VII Hardness of Minerals[a]

Mineral	Mohs hardness	Indentation hardness (kg/mm²)		Bierbaum scratch hardness[d]
		Diamond pyramid[b]	Knoop[c]	
Talc	1	47		1
Gypsum	2	60	32	11
Calcite	3	136	135	129
Fluorite, clear	4	175–200	163–180	143
Apatite	5	659		577
Parallel with optical axis			360	
Perpendicular to optical axis			420	
Albite	6		490	
Orthoclase	6	714	560	975
(Fused silica)		910		
Quartz	7			2700
Parallel with optical axis		1260	710	
Perpendicular to optical axis		1103	790	
Topaz	8	1648	1250	3420
Corundum	9	2085	1655	5300
(Sapphire, synthetic)		2720		
Silicon carbide			2130	
Boron carbide			2265	
Diamond	10		5500–6950	

[a] From Shand, E. B. (1958). "Glass Engineering Handbook," McGraw-Hill, New York.
[b] DPH$_{50}$, diamond pyramid hardness under 50-g load.
[c] KHN$_{50}$, Knoop hardness number under 50-g load.
[d] Correct., length of impression corrected for resolution limit of microscope.

TABLE VIII Mechanical Hardness of Glasses[a]

Type of glass	Designation in Table IX	Grinding silicon carbide no. 220 grit	Rouge polish	Impact abrasion	Indentation hardness (kg/mm²)			
					Diamond pyramid		Knoop	
					DPH$_{50}$[b]	Correct.[d]	KHN$_{50}$[c]	Correct.[d]
Silica	1		2.55	3.60	780–800	710–720	640–680	545–575
96% Silica	2			3.53			590	500
Soda–lime window	3		1.00[e]					
Soda–lime plate	4	1.00[e]		1.00[e]–1.07	580	540	575	490
Soda–lime lamp bulb	6	0.98	1.26	1.23	530	500	520	445
Lead–alkali silicate, electrical	7		1.22				430–460	375–400
Lead–alkali silicate, high-lead	8	0.57		0.56	290	270	285–340	250–300
Borosilicate, lowexpansion	10	1.52	1.56	3.10	630	580	550	470
Borosilicate, lowelectrical-loss	11	1.47		4.10				
Aluminosilicate	13	1.36		2.03	640	586	650	550

[a] From Shand, E. B. (1958). "Glass Engineering Handbook," McGraw-Hill, New York.
[b] DPH$_{50}$, diamond pyramid hardness under 50-g load.
[c] KHN$_{50}$, Knoop hardness number under 50-g load.
[d] Correct., length of impression corrected for resolution limit of microscope.
[e] Standard for comparison.

TABLE IX Approximate Compositions of Commercial Glasses[a,b]

No.[c]	Designation	Percent								
		SiO$_2$	Na$_2$O	K$_2$O	CaO	MgO	BaO	PbO	B$_2$O$_3$	Al$_2$O$_3$
1	Silica glass (fused silica)	99.5+								
2	96% Silica glass	96.3	<0.2	<0.2					2.9	0.4
3	Soda–lime—window sheet	71–73	12–15		8–10	1.5–3.5				0.5–1.5
4	Soda–lime—plate glass	71–73	12–14		10–12	1–4				0.5–1.5
5	Soda–lime—containers	70–74	13–16			10–13	0–0.5			1.5–2.5
6	Soda–lime—electric lamp bulbs	73.6	16	0.6	5.2	3.6				1
7	Lead–alkali silicate—electrical	63	7.6	6	0.3	0.2		21	0.2	0.6
8	Lead–alkali silicate—high-lead	35		7.2				58		
9	Aluminoborosilicate (apparatus)	74.7	6.4	0.5	0.9		2.2		9.6	5.6
10	Borosilicate—low-expansion	80.5	3.8	0.4				Li$_2$O	12.9	2.2
11	Borosilicate—low-electrical loss	70.0		0.5				1.2	28.0	1.1
12	Borosilicate—tungsten sealing	67.3	4.6	1.0		0.2			24.6	1.7
13	Aluminosilicate	57	1.0		5.5	12			4	20.5

[a] From Shand, E. B. (1958). "Glass Engineering Handbook," McGraw-Hill, New York.

[b] In commercial glasses, iron may be present in the form of Fe$_2$O$_3$ to the extent of 0.02–0.1% or more. In infrared-absorbing glasses, it is in the form of FeO in amounts fro 0.5–1%.

[c] Commercial glasses may be identified as follows: 2, Corning Nos. 7900, 7910, 7911, 7912; 6, Corning No. 0080; 7, Corning No. 0010; 8, Corning No. 8870; 9, Kimble N51a; 10, Corning No. 7740; 11, Corning No. 7070; 12, Corning No. 7050; 13, Corning Nos. 1710, 1720.

material. Low-expansion glasses are desirable where even slight dimensional changes are critical in the device, such as in telescope mirror substrates, and where the CTE must match that of an adjacent material such as occurs in glass metal seals of all types.

Composition has a critical influence on the CTE of the glass. Like many other properties of glass, the CTE is roughly an additive factor and for oxide glasses can be approximated by:

$$\lambda = a_1 p_1 + a_2 p_2 + \cdots + a_n p_n$$

TABLE X Breaking Stresses of Annealed Glass, Short-Time Flexure Tests in Air (Effective Duration of Breaking Load, 3 sec)[a]

Condition of glass	Avg. breaking stress (psi $\times$ 10^{-3})
Surfaces ground or sandblasted	1.5–4.0
Pressed articles	3.0–8.0
Blown ware	
Hot iron molds	4.0–9.0
Paste molds	5.0–10.0
Inner surfaces	15.0–40.0
Drawn tubing or rod	6.0–15.0
Polished plate glass	8.0–16.0
Drawn window glass	8.0–20.0
Fine fibers	
Annealed	10.0–40.0
Freshly drawn	30.0–400.0

[a] From Shand, E. B. (1958). "Glass Engineering Handbook," McGraw-Hill, New York.

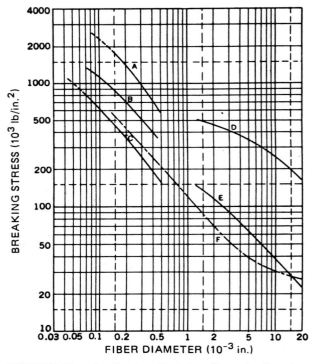

FIGURE 7 Mean tensile strength of glass fibers. A, silica glass in vacuum after baking; B, silica glass in atmosphere dried with CaCl$_2$; C, silica glass in atmosphere, moistened; D, borosilicate glass in air, acid-fortified; E, borosilicate glass in air; E, soda–lime glass in air. [From Shand, E. B. (1958). "Glass Engineering Handbook," McGraw-Hill, New York.]

TABLE XI Factors for Calculating the Linear Coefficient of Thermal Expansion of Glass[a,b]

	Winkelmann and Schott	English and Turner	Gilard and Dubrul
SiO_2	2.67	0.50	0.4
$B_2O_3^{\dagger}$	0.33	−6.53	$-4 + 0.1p$
Na_2O	33.33	41.6	$51 - 0.333p$
K_2O	28.33	39.0	$42 - 0.333p$
MgO	0.33	4.5	0
CaO	16.67	16.3	$7.5 + 0.35p$
ZnO	6.0	7.0	$7.75 - 0.25p$
BaO	10.0	14.0	$9.1 + 0.14p$
PbO	10.0	10.6	$11.5 - 0.05p$
Al_2O_3	16.67	1.4	2

[a] From Morey, G. W. (1954). "The Properties of Glass," Reinhold, New York.

[b] Values of $a_1, a_2, \ldots, a_n$ in $\lambda = a_1 p_1 + a_2 p_2 + \cdots + a_n p_n$, in which p_n are the percentages of the given oxides; $\lambda = (10^3/l)(\Delta l/\Delta t)$.

[c] Calculated only for that portion of the curve from 0 to 12% B_2O_3.

where λ is the cubical coefficient of expansion, that is, the change in unit volume with change in temperature, p_n is the percentage by weight of the components, and a_n is an empirical factor for each component.

The weighting factors are given in Table XI. The major constituent of most glasses, SiO_2, has a low contribution to the overall CTE, while the alkali metal oxides increase the expansion greatly. Other components are of intermediate nature. However, this type of calculation is most useful in the lower-temperature regime, is very approximate, and must be used with caution.

Figure 8 illustrates linear expansion versus temperature for a number of commercial glasses. Figure 9 illustrates the effect of composition on the CTE for a $Na_2O–B_2O_3–SiO_2$ glass system. Note the typical knee in the linear expansion curve in Fig. 8. The temperature at which this abrupt change in slope occurs is the glass transition temperature T_g. Below T_g the material is in the glassy state. Between T_g and the melting point, the material is a supercooled liquid.

C. Thermal Conductivity

The thermal conductivity of glass is relatively low. Unlike crystalline materials, in glasses thermal conductivity drops as the temperature decreases. Some representative data are shown in Figs. 10 and 11. For glasses with good infrared transmission, the heat transfer due to radiation becomes significant above about 400°C, in accordance with the radiation heat transfer relations deduced from the Stefan–Boltzmann law:

$$E = kT^4$$

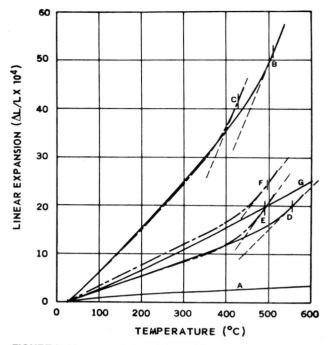

FIGURE 8 Linear expansion of glass with temperature. Light broken lines show increased rates of expansion at annealing points. A, 96% silica glass; B, soda–lime bulb glass; C, medium-lead electrical; D, borosilicate, low expansion; E, borosilicate, low electrical loss; F, borosilicate, tungsten sealing; G, aluminosilicate. [From Shand, E. B. (1958). "Glass Engineering Handbook," McGraw-Hill, New York.]

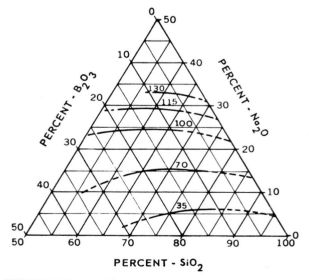

FIGURE 9 Linear coefficients of expansion for the system $Na_2O–B_2O_3–SiO_2$. Numbers on contours indicate coefficient of expansion per °C $\times 10^7$. [From Shand, E. B. (1958). "Glass Engineering Handbook," McGraw-Hill, New York.]

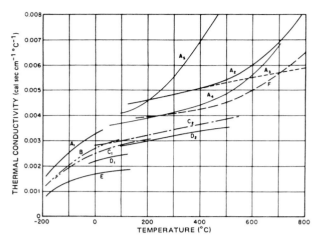

FIGURE 10 Thermal conductivity of glasses. A₁, A₂, A₄, A₅, silica glass; A₃, tin foil on glass surface; B, borosilicate crown glass; C₁, C₂, borosilicate low-expansion glass; D₁, D₂, soda–lime glass; E, lead–alkali silicate glass, 69% PbO; F, 96% silica glass. (Subscripts indicate various sources for data.) [From Shand, E. B. (1958). "Glass Engineering Handbook," McGraw-Hill, New York.]

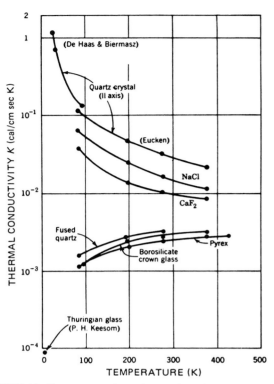

FIGURE 12 Temperature dependence of the thermal conductivity of various crystals and glasses. [From Kittel, C. (1966). "Introduction to Solid State Physics," Wiley, New York.]

where E is the energy per unit time radiating from the surface of black body at absolute temperature T, and K is the proportionality constant.

To convert from the cgs units given in these two charts to English units (Btu/h ft °F), multiply by 241. The conductivity of SiO_2 glass at 50°C is 0.0035 cal/sec cm °C or 0.843 Btu/h ft F.

The thermal conductivity of the quartz crystal at 50°C is approximately 0.025 cal/sec cm °C, in comparison. The temperature dependence of some crystals and glasses is shown in Fig. 12.

D. Specific Heat

Specific heat is defined as the heat energy required to raise the temperature of a unit mass of a substance one unit of temperature. The value c_p indicates that the specific heat is measured at a given pressure p usually 1 atm.

The specific heat of glass can be deduced from measurements of its diffusivity, where

$$\text{thermal diffusivity} = k/\rho c_p$$

where k is thermal conductivity, ρ is density, and $c_p =$ specific heat at constant pressure.

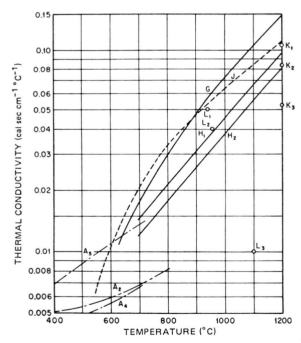

FIGURE 11 Thermal conductivity of glasses in thick sections at high temperatures. G, soda–lime glass; H₁, H₂, soda–lime container glasses; J, low-expansion borosilicate; K₁, soda–lime, 0.015% Fe_2O_3; K₂, soda–lime, 0.16% Fe_2O_3 oxidized; K₃, soda–lime, 0.16% Fe_2O_3 reduced; L₁, soda–lime container glass; L₂, amber glass, 0.2% Fe_2O_3; L₃, pale blue glass, 0.6% Fe_2O_3; A₂, A₄, A₅, see Fig. 10. [From Shand, E. B. (1958). "Glass Engineering Handbook," McGraw-Hill, New York.]

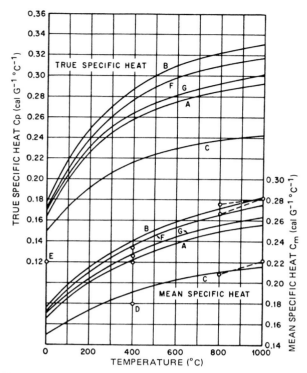

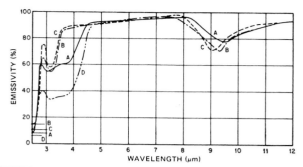

Glass		Thickness (in.)	Total emissitivity (%)
A	Soda–lime container	$\frac{1}{8}$	72
B	Aluminoborosilicate	$\frac{1}{8}$	75
C	Borosilicate low-expansion	$\frac{1}{8}$	77
D	Light chrome-green container	$\frac{1}{16}$	65

FIGURE 14 Spectral emissivity of glasses. [From Shand, E. B. (1958). "Glass Engineering Handbook," McGraw-Hill, New York.]

FIGURE 13 Specific heats of glasses. Mean specific heat taken from 0°C. Broken lines from Parmelee and Badger. A, Silica glass (Sosman). B, Soda–lime glass. C, Lead–alkali silicate glass, 22.3% PbO. D, Lead–alkali silicate glass, 26.0% PbO. E, Lead-alkali silicate glass, 46.0% PbO. F, Low-expansion borosilicate glass. G, Aluminosilicate glass. [From Shand, E. B. (1958). "Glass Engineering Handbook," McGraw-Hill, New York.]

Thermal diffusivity is a constant that occurs in the differential equation describing the transient rate of temperature rise in a solid when heat energy is applied to its surface. Diffusivity can be measured by transient thermal tests, and the specific heat c_p can be deduced from such data.

Specific-heat data for a number of glasses are given in Fig. 13.

E. Emissivity

Emissivity of a surface is a measure of its ability to radiate energy in comparison to a black body. For opaque bodies, the relation between reflectivity and emissivity is given by the simple relation

$$E + R = 1$$

where E is emissivity and R is reflectivity.

For bodies transparent to infrared energy, this simple relation becomes more complicated because of radiant energy that emanates from the interior of the body and is transmitted through the body.

Figure 14 illustrates the spectral emissivity of a number of glasses and in addition tabulates the total emissivity for each. Total emissivity is computed by integrating the product of the spectral emissivity and the black body energy over the entire spectrum. As can be seen in the spectral data, the spectral emissivity tends to be high except where the glasses are relatively good transmitters in the near-infrared region.

F. Electrical Properties

Glass is an important material in many electrical devices. Electrical conductivity, dielectric strength, dielectric constant, and dielectric loss are the major electrical properties of interest in most cases.

The electrical conductivity of glasses varies widely, depending on temperature and to a lesser degree on composition. The electrical conductivity largely arises from ionic migration, rather than from free electrons as in the case of metals.

The most significant ions in glasses are the alkali oxide ions, particularly the Na^+ ions. In electrolytic conduction, the ions move from one electrode to the other. If the ions are not replenished, the electrodes become polarized and the conductivity drops off. For dc measurements, sodium amalgams or molten sodium nitrate may be used as electrodes. In ac measurements the polarization problem is minimized.

As the temperature is raised the conductivity rises exponentially and can be expressed as:

$$\sigma = \sigma_0 \exp(-E/RT)$$

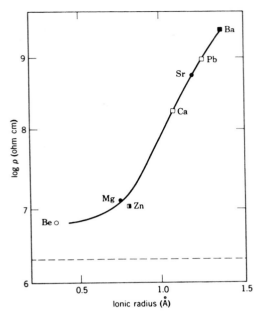

FIGURE 15 Variation of resistivity with divalent ion radius for 0.20 Na_2O–0.20 RO–0.60 SiO_2 glasses. Dashed line is resistivity of 0.20 Na_2O–0.80 SiO_2 glass. [From Mazurin, O. V., and Brailovskii, R. V. (1960). *Sov. Phys. Solid State* **2**, 243.]

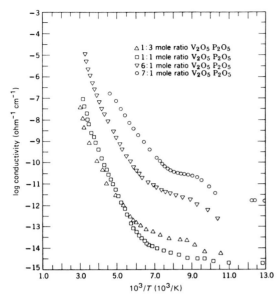

FIGURE 16 Log conductivity as a function of $1/T$ for four typical V_2O–P_2O_5 glasses. [From Schmid, A. B. (1968). *J. Appl. Phys.* **39**, 3140.]

where E is an empirical activation energy for conductivity, R is the gas constant, T is the absolute temperature, and σ_0 is a constant.

Compositional changes affect the mobility of the alkali ions and lead to significant changes in conductivity. Figure 15 shows the variation of resistivity with divalent ion radius for 0.20 Na_2O–0.20 RO–0.60 SiO_2 glasses, where R represents a divalent metal atom. The dashed line shows the resistivity of the 0.20 Na_2O–0.80 SiO_2 composition.

Certain oxide glasses do exhibit electronic conductivity, for example, glasses in the vanadium phosphate family. Figure 16 illustrates the exponential increase in conductivity with increasing temperature for a series of V_2O_5–P_2O_5 glasses due to activation of charge carriers.

1. Dielectric Strength

The dielectric strength of a material is a measure of its ability to sustain high-voltage differences without current breakdown. As the voltage across the material is increased, at some value of voltage a burst of current transits through the sample and causes severe damage to the material. The dielectric strength typically is stated in volts per centimeter. Some data for dielectric strength of various glasses are given in Table XII and in Fig. 17. When the dielectric strength is high, the glass will serve as an effective electrical insulator.

2. Dielectric Constant

The dielectric constant of a material is the ratio of the electrical energy in an electric field within a substance compared to the electrical energy in a similar volume of vacuum subjected to the same electrical field. The value is measured by determining the capacitance of an empty capacitor compared to that of one filled with the material of interest. The dielectric constant is a function of frequency and of temperature.

3. Dielectric Loss Factor

The dielectric loss factor is a measure of the energy absorbed in the medium as an electromagnetic wave passes

TABLE XII Intrinsic Dielectric Strength of Glasses[a]

Glass	Dielectric strength (kV/cm, dc)	
	Moon and Norcross	Vermeer
Silica	5000	
Borosilicate, low-expansion	4800	9200
Borosilicate, 16% B_2O_3		11,500
Soda–lime silicate	4500	9000[b]
Lead–alkali silicate	3100	
Aluminosilicate		9900

[a] From Shand, E. B. (1958). "Glass Engineering Handbook," McGraw-Hill, New York.

[b] Thuringian glass, 6.2% Al_2O_3.

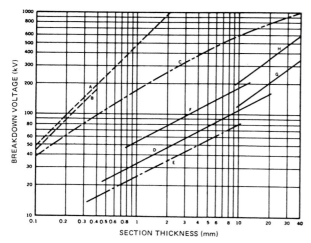

FIGURE 17 Breakdown voltage versus thickness of glass for different conditions at room temperature, 60-cycle voltage raised continuously. A, intrinsic dielectric strength of borosilicate glass; B, intrinsic dielectric strength of soda–lime glass; C, highest test values available for borosilicate glass; D, borosilicate glass plate immersed in insulating oil; E, soda–lime glass plate immersed in insulating oil; F, borosilicate glass plate immersed in semiconducting oil; G, borosilicate glass power-line insulator immersed in insulating oil; H, borosilicate glass power-line insulator immersed in semiconducting oil. [From Shand, E. B. (1958). "Glass Engineering Handbook," McGraw-Hill, New York.]

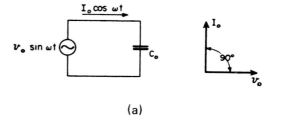

(a)

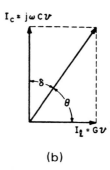

(b)

FIGURE 18 (a) Current–voltage relation in ideal capacitor. (b) Capacitor containing dielectric with loss. [From Von Hippel, A. R. (1954). "Dielectric Materials and Applications," M.I.T. Press, Cambridge, MA.]

through that medium. In the ideal case, the losses are zero and the dielectric loss factor is zero. In the case of the capacitor mentioned above, the dielectric loss factor is given by the ratio of the charging current (90° out of phase to the applied voltage) to the loss current in phase with the applied voltage. The total current traversing the condenser is inclined by a power-factor angle between 0 and 90° against the applied voltage v, that is by a loss angle δ, against the $+j$ axis as indicated in Fig. 18. The dissipation factor or tan δ is given by I_1/I_c.

Table XIII gives values of dielectric constant and loss tangent for a number of glasses as a function of frequency.

G. Optical Properties

Glass is the great optical material. In its simplest application, the glass has merely to transmit incident visible light such as in a window pane. But for optical devices the glass has to have much more stringent requirements primarily with respect to index of refraction, dispersion (i.e., variation of index of refraction with wavelength), spectral transmittance, absorptance and reflectance, and its scattering characteristics. In addition, the optical glass must have the required mechanical, ther-

mophysical, and chemical properties to meet the specific application.

To meet the needs of the wide variety of optical devices, many different optical materials have been developed and are commercially available from optical glass companies around the world. Two of the most significant properties are index of refraction and Abbe number, which is a measure of the dispersion of the glass. The index of refraction is defined as:

$$n = v_0/v$$

where n is the index of refraction, v_0 is the velocity of light in vacuum and v is the velocity of light in the glass.

An index of the dispersion is given by the Abbe number v_d, defined as:

$$v_d = (n_d - 1)/(n_F - n_C)$$

where the various n's are the indices of refraction of the glass at various spectral lines:

Hydrogen F	4861 Å
Helium d	5876 Å
Sodium D	5893 Å
Hydrogen C	6563 Å

TABLE XIII Properties of Selected Glasses[a] (Values for tan δ are Multiplied by 10^4; Frequency Given in cps)

Glass	t(°C)		Frequency											$\log_{10}$ Volume resistivity		
			1×10^2	1×10^3	1×10^4	1×10^5	1×10^6	1×10^7	1×10^8	3×10^8	3×10^9	1×10^{10}	2.5×10^{10}	25°C	250°C	350°C
Corning 0010 (potash, soda, lead)	24	$\varepsilon'/\varepsilon_v$	6.68	6.63	6.57	6.50	6.43	6.39	6.33	—	6.1	5.96	5.87	>17	8.9	7.0
		tan δ	77.5	53.5	35	23	16.5	15	23	60	—	90	110			
Corning 0014 (lead, barium)	25	$\varepsilon'/\varepsilon_v$	6.78	6.77	6.76	6.75	6.73	6.72	6.70	6.69	—	6.64				
		tan δ	23.1	17.2	14.4	12.2	12.4	13.8	17.0	19.5	—	70				
Corning 0080 (soda lime)	23	$\varepsilon'/\varepsilon_v$	8.30	7.70	7.35	7.08	6.90	6.82	6.75	—	6.71	6.71	6.62	12.4	6.4	5.1
		tan δ	780	400	220	140	100	85	90	—	126	170	180			
Corning 0090 (potash, lead, silicate)	20	$\varepsilon'/\varepsilon_v$	9.15	9.15	9.15	9.14	9.12	9.10	9.02	—	8.67	8.45	8.25			
		tan δ	12	8	7	7	8	12	18	—	54	103	122			
Corning 0100 (potash, soda, barium, silicate)	25	$\varepsilon'/\varepsilon_v$	7.18	7.17	7.16	7.14	7.10	7.10	7.07	—	7.00	6.95	6.87			
		tan δ	24	16	13.5	13	14	17	24	—	44	63	106			
Corning 0120 (potash, soda, lead)	23	$\varepsilon'/\varepsilon_v$	6.75	6.70	6.66	6.65	6.65	6.65	6.65	—	6.64	6.60	6.51	>17	10.1	8.0
		tan δ	46	30	20	14	12	13	18	—	41	63	127			
Corning 1770 (soda lime)	25	$\varepsilon'/\varepsilon_v$	6.25	6.16	6.10	6.03	6.00	6.00	6.00	—	5.95	5.83	5.44			
		tan δ	49.5	42	33	26	27	34	38	—	56	84	140			
Corning 1990 (iron-sealing glass)	24	$\varepsilon'/\varepsilon_v$	8.40	8.38	8.35	8.32	8.30	8.25	8.20	—	7.99	7.94	7.84			
		tan δ	4	4	3	4	5	7	9	—	19.9	42	112			
Corning 1991 (iron-sealing glass)	24	$\varepsilon'/\varepsilon_v$	8.10	8.10	8.08	8.08	8.08	8.06	8.00	—	7.92	7.83				
		tan δ	12	9	6	5	5	7	12	—	38	51				
Corning 3320 (soda, potash, borosilicate)	24	$\varepsilon'/\varepsilon_v$	5.00	4.93	4.88	4.82	4.79	4.78	4.77	—	4.74	4.72	4.7			
		tan δ	80	58	43	34	30	30	32	—	55	73	120			
Corning 7040 (soda, potash, borosilicate)	25	$\varepsilon'/\varepsilon_v$	4.84	4.82	4.79	4.77	4.73	4.70	4.68	—	4.67	4.64	4.52			
		tan δ	50	34	25.5	20.5	19	22	27	—	44	57	73			
Corning 7050 (soda, borosilicate)	25	$\varepsilon'/\varepsilon_v$	4.88	4.84	4.82	4.80	4.78	4.76	4.75	—	4.74	4.71	4.64	16	8.8	7.2
		tan δ	81	56	43	33	27	28	35	—	52	61	83			
Corning 7052 (soda, potash, lithia, borosilicate)	23	$\varepsilon'/\varepsilon_v$	5.20	5.18	5.14	5.12	5.10	5.10	5.09	—	5.04	4.93	4.85	17	9.2	7.4
		tan δ	68	49	34	26	24	28	34	—	58	81	114			
Corning 7055	25	$\varepsilon'/\varepsilon_v$	5.45	5.41	5.38	5.33	5.31	5.30	5.27	5.25	—	5.08				
		tan δ	45	36	30	28	28	29	38	49	—	130				
Corning 7060 (soda, borosilicate)	25	$\varepsilon'/\varepsilon_v$	5.02	4.97	4.92	4.86	4.84	4.84	4.84	—	4.82	4.80	4.65			
		tan δ	89	55	42	40	36	30	30	—	54	98	90			
Corning 7070 (potash, lithia, borosilicate)	23	$\varepsilon'/\varepsilon_v$	4.00	4.00	4.00	4.00	4.00	4.00	4.00	4.00	4.00	4.00	3.9	>17	11.2	9.1
		tan δ	6	5	5	6	8	11	12	12	12	21	31			
	100	$\varepsilon'/\varepsilon_v$	4.17	4.16	4.15	4.14	4.13	4.10	—	—	4.00	4.00				
		tan δ	50	22	13	10	11	11	—	—	19	21				
Corning 7230 (aluminum borosilicate)	25	$\varepsilon'/\varepsilon_v$	3.88	3.86	3.85	3.85	3.85	3.85	—	—	3.76					
		tan δ	33	23	16	13	11	12	—	—	22					
Corning 7570	25	$\varepsilon'/\varepsilon_v$	14.58	14.56	14.54	14.53	14.52	14.50	14.42	14.4	—	14.2				
		tan δ	11.5	13.5	15.9	16.5	19.0	23.5	33	44	—	98				

Continues

TABLE XIII *(continued)*

Glass	t(°C)		Frequency											$\log_{10}$ Volume resistivity		
			1×10^2	1×10^3	1×10^4	1×10^5	1×10^6	1×10^7	1×10^8	3×10^8	3×10^9	1×10^{10}	2.5×10^{10}	25°C	250°C	350°C
Corning 7720 (soda, lead, borosilicate)	24	$\varepsilon'/\varepsilon_v$	4.74	4.70	4.67	4.64	4.62	4.61	—	—	—	4.59	—	16	8.8	7.2
		$\tan\delta$	78	42	29	22	20	23	—	—	—	43	—			
Corning 7740 (soda, borosilicate)	25	$\varepsilon'/\varepsilon_v$	4.80	4.73	4.70	4.60	4.55	4.52	4.52	—	—	4.52	4.50	15	8.1	6.6
		$\tan\delta$	128	86	65	54	49	45	45	—	—	85	96			
Corning 7750 (soda, borosilicate)	25	$\varepsilon'/\varepsilon_v$	4.45	4.42	4.39	4.38	4.38	4.38	—	—	4.38	4.38	—			
		$\tan\delta$	45	33	24	20	18	19	—	—	43	54	—			
Corning 7900 (96% silica)	20	$\varepsilon'/\varepsilon_v$	3.85	3.85	3.85	3.85	3.85	3.85	3.85	3.85	3.84	3.82	3.82	17	9.7	8.1
		$\tan\delta$	6	6	6	6	6	6	—	6	6.8	9.4	13			
	100	$\varepsilon'/\varepsilon_v$	3.85	3.85	3.85	3.85	3.85	3.85	3.85	3.85	3.84	3.82	—			
		$\tan\delta$	37	17	12	10	8.5	7.5	—	7.5	10	13	—			
Corning 7911 (96% silica)	25	$\varepsilon'/\varepsilon_v$	—	—	—	—	—	—	—	—	—	3.82	—	>17	11.7	9.6
		$\tan\delta$	—	—	—	—	—	—	—	—	—	6.5	—			
Corning 8460 (barium, borosilicate)	25	$\varepsilon'/\varepsilon_v$	8.35	8.30	8.30	8.30	8.30	8.30	8.30	—	8.10	8.06	8.05			
		$\tan\delta$	11	9	7.5	7	8	10	16	—	40	57	60			
Corning 8830	25	$\varepsilon'/\varepsilon_v$	5.38	5.28	5.20	5.11	5.05	5.01	5.00	4.97	—	4.83	—			
		$\tan\delta$	204	130	91	73	60	54	57	63	—	99	—			
Corning 8871 (alkaline lead silicate)	25	$\varepsilon'/\varepsilon_v$	8.45	8.45	8.45	8.45	8.45	8.43	—	8.40	8.34	8.05	7.82			
		$\tan\delta$	18	13	9	7	6	7	—	14	26	49	70			
Corning 9010	25	$\varepsilon'/\varepsilon_v$	6.51	6.49	6.48	6.45	6.44	6.43	6.42	6.40	—	6.27	—			
		$\tan\delta$	50.5	36.2	26.7	22.7	21.5	22.6	30	41	—	91	—			
Foamglas (Pittsburgh-Corning) (soda lime)	23	$\varepsilon'/\varepsilon_v$	90.0	82.5	68.0	44.0	17.5	9.0	—	—	—	5.49	—			
		$\tan\delta$	1500	1600	2380	3200	3180	1960	—	—	—	455	—			
Fused silica 915c	25	$\varepsilon'/\varepsilon_v$	3.78	3.78	3.78	3.78	3.78	3.78	3.78	3.78	—	3.78	—			
		$\tan\delta$	6.6	2.6	1.1	0.4	0.1	0.1	0.3	0.5	—	1.7	—			
Glass-bonded micas																
Mycalex 400	25	$\varepsilon'/\varepsilon_v$	7.47	7.45	7.42	7.40	7.39	7.38	—	—	—	7.12	—			
		$\tan\delta$	29	19	16	14	13	13	—	—	—	33	—			
	80	$\varepsilon'/\varepsilon_v$	7.64	7.59	7.54	7.52	7.50	7.47	—	—	—	7.32	—			
		$\tan\delta$	150	85	50	25	16	14	—	—	—	57	—			
Mycalex K10	24	$\varepsilon'/\varepsilon_v$	9.5	9.3	9.2	9.1	9.0	9.0	—	—	11.3[b]	11.3[b]	—			
		$\tan\delta$	170	125	76	42	26	21	—	—	40	40	—			
Mykroy grade 8	25	$\varepsilon'/\varepsilon_v$	6.87	6.81	6.76	6.74	6.73	6.73	6.72	—	6.68[c]	6.96[c]	6.66			
		$\tan\delta$	95	66	43	31	26	24	25	—	38	48	81			
Mykroy grade 38	25	$\varepsilon'/\varepsilon_v$	7.71	7.69	7.64	7.61	7.61	7.61	—	—	7.68[c]	8.35[c]	—			
		$\tan\delta$	43	33	27	24	21	14	—	—	35	40	—			

[a] From Gray, D. E. (1963). "American Institute of Physics Handbook," McGraw-Hill, New York. Taken from Tables of Dielectric Materials, Vol. IV, Laboratory for Insulation Research, MIT Technical Report 57; and Properties of Commercial Glasses, Bull. B-83, Corning Glass Works.

[b] Not corrected for variations in density.

[c] Sample nonhomogeneous.

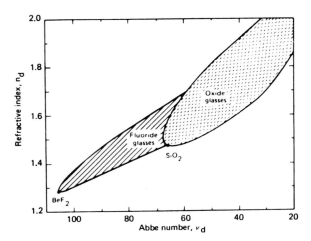

FIGURE 19 For optical materials, n_d (index of refraction) versus ν_d (Abbe number). [Adapted from Weber, M. J., Milem, D., and Smith, W. L. (1978). *Opt. Eng.* **17**(5), as in Musikant, S. (1985). "Optical Materials," Marcel Dekker, New York.]

The Abbe number is a measure of the chromatic abberation in an optical material. If a single optical element is to be an effective refracting device, then a high index of refraction and a high Abbe number are desirable. The optical glass development of the past century has made available a wide variety of n, ν_d combinations. The general range is indicated in Fig. 19.

The other key parameter is transmittance. In general, a high transmittance is required over a specific wavelength or band of wavelengths. Figure 20 illustrates how the infrared transmission is affected by composition of the glass. Glasses composed of higher-atomic-weight elements tend to transmit further into the infrared. This is accompanied by reduced hardness and strength.

Fused quartz glass transmits well in the ultraviolet (UV) protion of the spectrum. However, even very small levels of impurity content severely degrade UV transmittance.

During the past decade, the development of optical wave guides in the form of fiber optics has produced glasses of exceedingly high transmittance in wavelengths compatible with the available laser transmitters. Such low-transmittance-loss fibers have made practical long-distance transmission of information over glass-fiber transmission lines.

VII. MANUFACTURE

The manufacture of glass has three essential operations: batching, melting, and forming. In the batching operation, the raw materials are weighed, mixed, and milled as necessary to provide a mixture that can be melted to provide the glass composition desired. Some of the more common raw ingredients are listed in Table XIV. Trace impurities can be a serious problem to the glass maker. One of the most common impurities is iron oxide, which imparts a greenish hue to the glass and prevents heat transfer by radiation through the glass during the melting phase. For high-transmitting fused quartz, extremely high-purity quartz sands are used.

The particle size of the batching materials is a significant variable and must be controlled within close limits to assure a satisfactory melting process. Accuracy of weighing and mixing operations that do not cause segregation in the dry batch are essential to obtaining the homogeneous glass melt desired.

TABLE XIV   Glass-Making Materials

Raw material	Chemical composition	Glass-making oxide	Percent of oxide
Sand	SiO_2	SiO_2	100.0
Soda ash	Na_2CO_3	NaO_2	58.5
Limestone	$CaCO_3$	CaO	56.0
Dolomite	$CaCo_3 \cdot MgCO_3$	CaO	30.4
		MgO	21.8
Feldspar	$K_2(Na_2)O \cdot Al_2O_3 \cdot 6SiO_2$	Al_2O_3	18.0
		$K_2(Na_2)O$	13.0
		SiO_2	68.0
Borax	$Na_2B_4O_7 \cdot 10H_2O$	Na_2O	16.3
		B_2O_3	36.5
Boric acid	$B_2O_3 \cdot H_2O$	B_2O_3	56.3
Litharge	PbO	PbO	100.0
Potash	$K_2CO_3 \cdot 1.5H_2O$	K_2O	57.0
Fluorspar	CaF_2	CaF_2	100.0
Zinc oxide	ZnO	ZnO	100.0
Barium carbonate	$BaCO_3$	BaO	77.7

^{a} From Shand, E. B. (1958). "Glass Engineering Handbook," McGraw-Hill, New York.

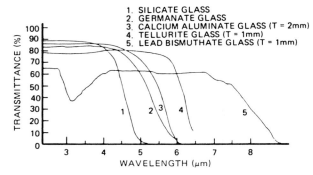

FIGURE 20 Transmittance of infrared-transmitting glasses. [From Dumbaugh, W. H. (1981). *Proc. SPIE* **297**.]

A. Melting

Melting is performed in a gas, oil, or electrically heated furnace. Batch melting in pots or day tanks is used for small quantities of glass. Continuous furnaces ranging from less than 1 ton of glass per day to 1500 tons per day capacity are used in larger production operations. The continuous furnaces are built up from refractory ceramic components and often operate continuously for periods on the order of 1 year before being shut down for rebuild. These furnaces are divided into a large melting section followed by a shallow, narrow refining section (the forehearth) where the glass temperature is reduced preparatory to the forming operation. In fossil-fuel-fired furnaces the hot combustion gases are located above the molten glass. In electric resistance furnaces the glass is directly heated by immersed electrodes. Small induction-heated glass furnaces are used for specialty glasses.

The temperature of the glass in the melting section of the furnace is dependent on the composition. For typical commercial glasses, melting temperatures vary from 1500°C for soda–lime glass to 1600°C for aluminosilicate glass.

B. Refining

The melting and refining processes are very complex. At the cold end of the furnace where the batch is introduced, melting is initiated by the fluxes, which melt first and then dissolve the more refractory ingredients. Elimination of the water vapor and other gases dissolved in the melt takes place in the downstream section of the furnace prior to the entry of the glass into the refining section. In the forehearth the glass is cooled, thereby adjusting its viscosity to the level needed by the forming equipment.

C. Forming

Forming operations include drawing the glass into a sheet, bottle blowing, pressing, tube drawing, rolling into flat glass, and fiber drawing. Fibers can be drawn into continuous strands for textile application or drawn by means of steam or compressed-air jets into fine, discontinuous strands useful for thermal insulation.

After forming, the glass usually has to be annealed to minimize residual strains that could lead to fracture of the glass or, in the case of optical materials, cause unacceptable variations in the optical properties.

D. Finishing

Finally, the ware is subjected to secondary finishing operations such as cutting, grinding, polishing, or thermal or chemical treatments to produce the end item.

VIII. APPLICATIONS

A. Containers

Glass containers must be resistant to the contained fluids, must be sufficiently strong to withstand the internal pressures encountered, as lightweight as possible, and economical to produce. Carbonated beverages can impose pressures up to 125 psi. The containers must be able to withstand the thermal shock imposed by sterilization treatments and impacts that are inflicted during handling and transportation. Careful annealing is performed to minimize residual strain, and quality control is maintained by the use of polariscopic examination.

B. Glazine Glass

Considerations in designing glass for glazing include light transmission, light diffusion (as by sandblasting), absorption of high-energy radiations (as by addition of PbO to absorb X-rays), strength, resistance to thermal gradients and the stresses so induced, heat transfer rate (as by the use of double glazing to inhibit heat transfer), reflectance (either for visible or selective reflectance of infrared), and the ability to heat the glass (as by embedded heating wires for aircraft windshields). These various requirements are met by combinations of composition, heat treatment, coatings, other surface treatments, and by various types of lamination with intervening, transparent films. Bullet-resistant glass is generally made up of multiple layer of glass separated by plastic sheets. The index of refraction of the plastic and the glass have to match closely to assure minimal distortion.

C. Chemical Applications

Laboratory ware and industrial chemical process equipment fabricated from glass are extensively used. Borosilicate glass is widely used because of its relatively low expansion coefficient and reduced susceptibility to thermal shock failure as well as greater chemical durability. However, for special chemical environments other compositions are used, as in boron-free compositions that are resistant to alkaline solutions. Where high-temperature conditions are encountered, fused quartz is used because of its extremely high resistance to thermal shock. Fused quartz is employed in special applications for its excellent ultraviolet transmitting characteristic.

Industrial glass piping is made from heat-resistant glass in a variety of sizes, and various fittings are available. Joints are made by special metal flange arrangements.

D. Lamp Glass

The glass envelope of an electrical lighting device must contain and protect the light-emitting medium, be

TABLE XV Typical Glass Properties[a]

Glass type[b]	Description/use[b]	Color[c]	Forms usually available[c]	Density (g/cm³)	Mechanical properties			Thermal properties		
					Young's modulus[d] 10⁶ psi	Poisson's ratio[e]	Impact abrasion resistance[f]	Expansion (10⁷ cm/cm/°C) 0-300°C	Thermel endurance ratio[g]	Conductivity[h]
Soda lime										
008	General	Clear	T,B,F	2.48	10	0.24	1.2	93	1	
Lead										
001	Lamp, electronic	Clear	T,B,F	2.81	9	0.21	—	92	1	
012	Lamp, electronic	Clear	T,B,F	3.04	8.6	0.22	0.6–0.8	89	1	
821	Radiation shielding	Clear	T,B	4.00	—	—	0.6–0.8	88	—	
Borosilicate										
706	Kovar sealing	Clear	T	2.24	8	0.22	3–4	47	2.2	
725	General	Clear	P	2.25	9	0.21	3–4	37	2.5	
772	Lead borosilicate	Clear	T,B	2.33	9	0.20	3–4	34	—	
776	General	Clear	T,B,P	2.23	9	—	3–4	33	2.5	
777	Lead borosilicate	Clear	T	2.30	9	—	3–4	37	—	
Aluminosilicate										
174	Alkali free, tungsten and molybdenum sealing	Clear	T	2.64	12	0.25	2.0	43	2	
177	Alkali free tungsten sealing	Clear	T	2.70	—	—	—	38	—	
180	Alkali free, molybdenum sealing	Clear	T	2.74	—	—	—	45	—	
Special										
250	Encapsulation for capacitors	Clear	Ft.	2.94	—	—	—	60	—	
351	Encapsulation for diodes	Light Yellow	Ft.	3.78	—	—	—	44	—	
355	Automotive signal	Amber	T,B	2.57	—	—	—	92	—	
540	UV transmission	Dark Blue	T,F	2.56	—	—	—	91	—	
980	General	Clear	T,B	2.71	10	0.20	—	91	—	
982	UV transmission	Clear	T	2.72	10	0.20	—	92	—	

Continues

TABLE XV (continued)

Glass type[b]	Viscosity			Electrical properties						Optical Properties			
	Strain point (°C)	Anneal point (°C)	Softening point (°C)	Electrical resistivity (log₁₀ ohm cm)			Dielectric constant at 1 MHz and 20°C	Loss tangent at 1 MHz and 20°C[j]	Loss factor at 1 MHz and 20°C[i]	Refractive index[k] n_d	Dispersion[l]	Useful transmittance[m] (nm)	Resistance to weathering[n]
				250°C	300°C	350°C							
Soda lime													
008	475	515	700	6.2	5.6	5.0	7.2	0.009	0.065	1.512	0.0089	290–4600	C
Lead													
001	395	435	625	8.5	7.6	6.6	6.7	0.0015	0.010	1.534	—	300–4700	B
012	395	435	625	9.7	8.6	7.6	6.7	0.0014	0.009	1.559	0.0083	300–4600	B
821	400	440	595	11.6	10.4	9.4	8.9	0.0005	0.004	1.667	—	—	B
Borosilicate													
706	440	485	705	10.0	9.0	8.1	—	—	—	1.480	—	300–2800	B
725	505	550	775	7.9	7.1	6.4	4.7	0.003	0.013	1.478	0.0069	290–2700	A
772	475	520	760	8.8	7.9	7.2	4.7	0.003	0.013	1.484	0.0076	340–2700	B
776	485	535	785	8.5	7.7	7.0	4.5	0.002	0.008	1.471	0.0073	290–3500	B
777	485	530	770	9.0	8.1	7.3	—	—	—	1.480	0.0074	300–2700	A
Aluminosilicate													
174	690	710	930	12.4	11.5	10.8	—	—	—	—	—	—	A
177	805	865	1125	12.2	11.3	10.5	—	—	—	1.522	—	270–4800	A
180	745	800	1015	12.7	11.8	11.1	—	—	—	1.536	—	280–4800	A
Special													
250	510	545	680	12.3	11.4	10.6	—	—	—	1.568	0.0093	—	—
351	520	550	638	12.7	11.7	10.8	7.97–8.15	—	—	1.680	0.0323	360–2700	—
355	430	475	690	7.5	6.6	5.9	—	—	—	1.511	—	520–4600	—
540	440	475	670	7.8	6.8	6.1	—	—	—	—	—	—	—
980	470	515	700	9.9	8.9	7.9	—	—	—	1.522	—	220–4400	—
982	470	515	695	9.8	8.7	7.8	—	—	—	1.522	—	220–4400	—

[a] From General Electric Co. (1980). "Glass Products," GE Lamp Components Division, Cleveland, OH.
[b] All forms not always available nor in all sizes.
[c] T, Tubing and cane; B, blown; Ft, frit or powder; P, pressed; F, formed.
[d] Data shown in table are estimates to show relative values. They are to be used for reference only.
[e] Data shown in table are estimates to show relative values. They are to be used for reference only.
[f] Using soda–lime plate glass as a base of unity, relative values for other glasses are shown.
[g] Using soda–lime as a base of unity, relative values for other glasses are shown.
[h] Most glasses are between 0.002 and 0.003 cal/sec (cm °C).
[i] This is sometimes expressed as a percent (0.009 = 0.9%).
[j] This is sometimes expressed as a percent (0.065 = 6.5%). Of the standard glasses, lead and borosilicate are the better insulators.
[k] Values at 589.3 nm.
[l] Dispersion is shown at $n_F - n_C$.
[m] Useful transmittance (exceeding 10%) range is shown in nanometers for 1-mm thicknesses. Type 540 has selective transmittance bands. Its useful transmittance is 310–450, 690–1100, and 1100–4600.
[n] (A) Seldom affected by weathering, (B) could occasionally show weathering effects, and (C) weathering can be a problem.

amenable to hermetic sealing of the electrodes entering and leaving the envelope, be capable of chemically resisting interactions with the substances in the enclosure and external gases at the temperatures of operation, be resistant to thermal shock induced during light-up and at light-off events, be a good electrical insulator, be highly resistant to diffusion of gases, be able to transmit light (or in some cases infrared or ultraviolet energy) with a minimum of transmission losses and with the appropriate degree of scatter, be amenable to mass production methods, and be within the economical cost limits imposed by the market.

To meet the many types of lighting-product requirements, a variety of glasses have been developed by the various manufacturers of glass and lighting devices. Table XV lists properties of various glasses, including their useful transmittance ranges, while Figs. 21, 22, and 23 show the transmittance spectra for some of these materials.

The most common type of lighting device is the tungsten filament bulb. These bulbs are composed of three parts: the envelope; the stem, and the exhaust tube. For general lighting devices, the envelope is a soda–lime glass, and the stem and exhaust tube are made of a lead–alkali silicate glass. The lead–bearing glasses have a lower softening temperature and a higher electrical resistance than the soda–lime glasses.

The glass bulbs are made by a highly automated glass-blowing process capable of producing several hundred per minute on one machine. Molten glass from the furnace is fed between rollers to produce a ribbon with evenly spaced circular impressions on one face of the glass ribbon. These impressions are next accurately located over an orifice in a steel plate. As the soft glass moves along, the glass sags through the orifice, a blow tip is located at the glass, and the beginning of the envelope caused by the sagging action is completed by a blowing process into a hinged mold, which accurately defines the dimensions of the bulb. After

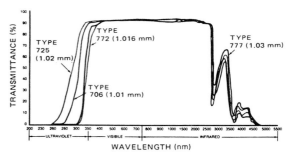

FIGURE 22 Transmittance of borosilicate glasses. [From General Electric Co. (1985). "Glass Products," GE Lamp Components Division, Cleveland, OH.]

cooling, the completed bulb is separated from the ribbon and used in subsequent operations.

Sealing of the metal parts penetrating the glass envelope requires glasses with carefully controlled coefficients of thermal expansion.

E. Glass Fibers

The fact that glass can be drawn into fine fibers has been obvious to every glass maker since ancient times. Therefore, it is surprising that the broad usefulness of glass fibers has not been attained until relatively recent times.

Glass fibers can be thought of in terms of two main categories: (1) continuous strands and (2) staple fibers or short fibers made into mats of various kinds. Table XVI displays the various categories of glass fiber properties and applications.

The newest and perhaps most dramatic application for glass fibers is in the field of information transmission via glass-fiber waveguides. The advent of the laser and the associated light detection and electronic signal-processing equipment has made the practical and cost-effective implementation of this technology possible. In addition to information transmission, many new types of instruments are based on the characteristics of coherent light traversing glass-fiber waveguides.

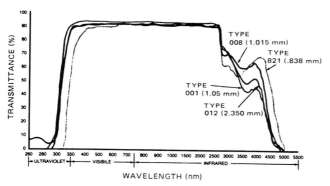

FIGURE 21 Transmittance of soda–lime and lead glasses. Glass thickness shown in parentheses. [From General Electric Co. (1985). "Glass Products," GE Lamp Components Division, Cleveland, OH.]

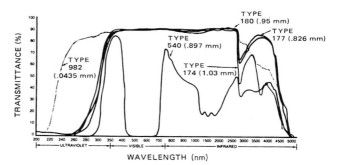

FIGURE 23 Transmittance of aluminosilicate and special glasses. [From General Electric Co. (1985). "Glass Products," GE Lamp Components Division, Cleveland, OH.]

TABLE XVI Properties Related to Applications[a]

Glass type	Fibrous-glass forms	Fiber diameter range (in.)	Dominant characteristics	Principal uses
Low-alkali lime–alumina borosilicate	Textiles and mats	0.00023–0.00038	Excellent dielectric and weathering properties	Electrical textiles General textiles Reinforcement for plastics, rubbers, gypsum, papers General-purpose mats
Soda–lime borosilicate	Mats	0.00040–0.00060	Acid resistance	Mats for storage-battery retainers, corrosion protection, water proofing, etc.
	Textiles	0.00023–0.00038		Chemical (acid) filter cloths, anode bags
Soda–lime borosilicate	Wool (coarse)	0.00030–0.00060	Good weathering	Thermal insulations Acoustical products
Soda–lime	Packs (coarse fibers)	0.0045–0.010	Low cost	Coarse fibers only, for air and liquid filters, tower packing, air–washer contact, and eliminator packs
Lime-free soda borosilicate	Wool (fine)	0.00003–0.00020	Excellent weathering	Lightweight thermal insulations, sound absorbers, and shock-cushioning materials
	(Ultrafine)	0.0000 (est.) 0.00003		All-glass high-efficiency filter papers and paper admixtures
High-lead silicate	Textile	0.00023–0.00038	X-ray opacity	Surgical pad strands, X-ray protective aprons, etc.

[a] From Shand, E. B. (1958). "Glass Engineering Handbook," McGraw-Hill, New York.

The glass-fiber waveguide must have a low transmittance loss for the wavelength of laser light employed. The waveguide effect is achieved by producing a radial gradient or step in the index of refraction, with the index generally decreasing radially outward. There are many index gradients and step designs used for a variety of purposes. Figure 24 illustrates the application areas and materials used as a function of wavelength. At the present time, silica is the predominant glass fiber used.

IX. THE FUTURE OF GLASS

The statement attributed to Niels Bohr, "Predictions are difficult, especially about the future," holds true in this instance. However, as long as humans have a progressive technology, that the technology of glass will continue to advance is a simple projection of the past. Certainly each of the application areas mentioned above will continue to develop. As in the case of glass-fiber waveguides, new applications will be discovered in the wake of not yet foreseen new inventions. One thing is certain. The abundance and variety of the raw materials that constitute glass will always provide an economic and technical incentive to use glass as long as the required energy for conversion of the raw stocks is reasonably available.

When humans build their colony on the earth's moon or on the moons of distant planets, the materials for glass making will be there. The asteroid belts will provide the glass-making ingredients for manufacture of structures and other products on the space stations already being designed.

Before recorded history, humans knew glass that had come from outer space. In the not too distant future glass will be made by humans in space.

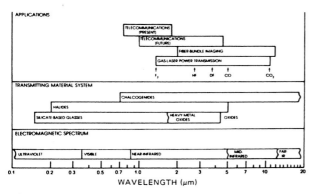

FIGURE 24 Optical fibers: materials and applications. [From Tick, P. A., and Thompson, D. A. (1985). *Photonics Spectra* **July**.]

SEE ALSO THE FOLLOWING ARTICLES

BONDING AND STRUCTURE IN SOLIDS • CERAMICS • CERAMICS, CHEMICAL PROCESSING OF • GLASS-CERAMICS • LASERS • LIQUIDS, STRUCTURE AND DYNAMICS • OPTICAL FIBER COMMUNICATIONS • PHOTOCHROMIC GLASSES

BIBLIOGRAPHY

Amstock, J. (1997). "Handbook of Glass in Construction," McGraw-Hill, New York.

ASM International (1991). "ASM Engineered Materials Handbook," Vol. 4: "Ceramics and Glasses," American Society for Metals, Materials Park, OH.

Barreto, L. M., Kreidl, N., and Vogel, (1994). "Glass Chemistry," 2nd ed., Springer–Verlag, Berlin/New York.

Bradt, R. C., and Tressler, R. E. (1994). "Fractography of Glass," Kluwar Academic, Dordrecht/Norwell, MA.

Doremus, R. H. (1994). "Glass Science," 2nd ed., John Wiley & Sons, New York.

Dumbaugh, W. H. (1981). "Infrared transmitting materials," Proc. SPIE **297,** 80.

Efimov, A. M. (1996). "Optical Constants of Inorganic Glasses," CRC Press, Boca Raton, FL.

Gan, F. (1991). "Optical and Spectroscopic Properties of Glass," Springer–Verlag, Berlin/New York.

General Electric Co. (1980). "Glass Products," GE Lamp Components Division, Cleveland, OH.

Guillemet, C., and Aben, H. (1993). "Photoelasticity of Glass," Springer–Verlag, Berlin/New York.

Kingery, W. D., Bowen, H. K., and Uhlmann, D. R. (1976). "Introduction to Ceramics," Wiley, New York.

Kirsch, R., ed. (1993). "Metals in Glassmaking," Elsevier, Amsterdam/New York.

Kittel, C. (1966). "Introduction to Solid State Physics," Wiley, New York.

Kokorina, V. (1996). "Glasses for Infrared Optics," CRC Press, Boca Raton, FL.

Krause, D., and Bach, H. (1999). "Analysis of the Composition and Structure of Glass and Glass Ceramics," Springer–Verlag, Berlin/New York.

Morey, G. W. (1954). "The Properties of Glass," Reinhold, New York.

Musikant, S. (1985). "Optical Materials," Marcel Dekker, New York.

Neuroth, N., and Bach, H. (1995). "The Properties of Optical Glass," Springer–Verlag, Berlin/New York.

Sakka, S., Reisfeld, R., and Oehme, (1996). "Optical and Electronic Phenomena in Sol-Gel Glasses and Modern Applications," Springer–Verlag, Berlin/New York.

Shand, E. B. (1958). "Glass Engineering Handbook," McGraw-Hill, New York.

Von Hippel, A. R. (1954). "Dielectric Materials and Applications," M.I.T. Press, Cambridge, MA.

Yamane, M., and Asahara, Y. (2000). "Glasses for Photonics," Cambridge University Press, Cambridge, U.K.

Zarzycki, J., ed. (1991). "Materials Science and Technology: A Comprehensive Treatment," Vol. 9: "Glasses and Amorphous Materials," John Wiley & Sons, New York.

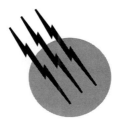

Glass-Ceramics

Linda R. Pinckney

Corning Incorporated

GLOSSARY

Ceramming Technical and industrial term used to describe the thermal process by which a glass article is converted to a predominantly crystalline article (i.e., a glass-ceramic).

Metastable phases Phases such as glass or many crystalline structures that are in a state of higher free energy than the thermodynamically stable phase or assemblage of phases.

Phase transformation Transformation from one structural state to another without change in bulk chemical composition (e.g., congruent melting of solids, hexagonal to trigonal quartz).

Primary grain growth Term applied to the growth of crystals in glass from the early stages of nucleation until particle impingement (i.e., the growth of crystals at the expense of the parent glass).

Stuffed derivatives Framework silicates that can be considered as structurally derived from simple polymorphs of silica through substitution of tetrahedral ions of lower valence for silicon (e.g., aluminum, boron) and filling interstitial vacancies with larger cations (e.g., lithium, zinc, magnesium, sodium, calcium, potassium) to maintain charge balance.

GLASS-CERAMICS are polycrystalline materials produced by the controlled nucleation and crystallization of glass. Before the discovery in the 1950s that crystallization of glass could be controlled, devitrification, or uncontrolled crystallization, was considered to be a major problem. While devitrification normally results in coarse-oriented crystals containing porosity and results in mechanically weak bodies, glass-ceramics prepared by controlled nucleation are characterized by fine-grained, randomly oriented crystals with some residual glass, no voids or porosity, and generally high strength.

Glass-ceramics can provide significant advantages over conventional glass or ceramic materials by combining the flexibility of forming and inspection with glass with improved and often unique properties in the glass-ceramic. More than $500 million in glass-ceramic products are sold annually worldwide, ranging from transparent, zero-expansion materials with excellent optical properties and thermal stability to jadelike highly crystalline materials

Encyclopedia of Physical Science and Technology, Third Edition, Volume 6

with excellent strength and toughness. The highest volume is in stovetops and stove windows, architectural cladding, and cookware. Glass-ceramics are also referred to as Pyrocerams (Corning), vitrocerams, and sitalls.

I. DESIGN AND PROCESSING

A. Glass-Ceramic Design

The key variables in the design of a glass-ceramic are glass composition, glass-ceramic phase assemblage, and crystalline microstructure. The glass-ceramic phase assemblage (the types of crystals and the proportion of crystals to glass) is responsible for many of its physical and chemical properties, including thermal and electrical characteristics, chemical durability, and hardness. In many cases these properties are additive; for example, a phase assemblage comprising high- and low-expansion crystals has a bulk thermal expansion proportional to the amounts of each of these crystals.

The nature of the crystalline microstructure (the crystal size and morphology and the textural relationship among the crystals and glass) is the key to many mechanical and optical properties, including transparency/opacity, strength, fracture toughness, and machinability. These microstructures can be quite complex and often are distinct from conventional ceramic microstructures. In many cases, the properties of the parent glass can be tailored for ease of manufacture while simultaneously tailoring those of the glass-ceramic for a particular application.

B. Melting, Forming, and Ceramming

Raw materials such as quartz sand (SiO_2), feldspar [$(Na,K)AlSi_3O_8$], and zircon ($ZrSiO_4$) are commonly employed as batch ingredients. After the batch is mixed thoroughly, it is delivered to the melting furnace, usually in a continuous process, where it goes through several states of progressive refinement before being delivered to the forming process.

Most commercial glass-ceramic products are formed by highly automated glass-forming processes such as rolling, pressing, or casting. The forming method is limited by the liquidus temperature and viscosity of the glass at that temperature. Typical viscosity–temperature curves along with viscosity ranges required for the forming process are shown in Fig. 1. Glasses A and C are compositions based on the β-spodumene solid solution region of the Li_2O-Al_2O_3-SiO_2 system, and B is representative of MgO-Al_2O_3-SiO_2 glasses, which crystallize predominantly to the cordierite phase. In order to eliminate uncontrolled crystallization, the forming operations must take place at temperatures above the liquidus. Liquidus temperatures for the three glasses in Fig. 1 are 1220°C for A, 1300°C

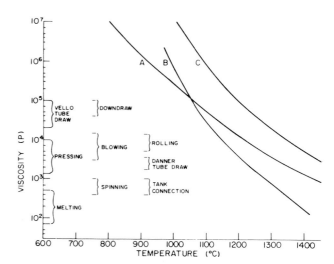

FIGURE 1 Viscosity-forming relationships for three representative glass-ceramic compositions. [After Beall, G. H., and Duke, D. A. (1983). *In* "Glass: Science and Technology" (D. R. Uhlmann and N. J. Kreidle, eds.), Vol. 1, Academic Press, New York.]

for B, and 1250°C for C. Glass is normally delivered from the tank in a viscosity range of 400 to 1000 P, depending on the desired forming technique. For example, since glass A in Fig. 1 will not crystallize above 1220°C and has a viscosity of 10^4 P at the liquidus temperature, pressing, blowing, and some rolling operations could be performed. Glass C has a viscosity of 5×10^4 P at its liquidus and could therefore be formed by any process except a downdraw. However, casting would be impractical due to the high temperature needed to achieve the required fluidity. Since glass B has a viscosity of only 800 P at its liquidus, forming would be restricted to casting. Thus, it can easily be demonstrated that viscosities and liquidus temperatures of glass-ceramics are significant parameters for using commercial forming processes.

Following the forming process, the glasses are commonly annealed at lower temperatures to remove residual stresses caused by nonuniform cooling. Firepolishing is often used to remove rough or sharp edges resulting from the firepolish, but care must be taken to avoid premature crystallization.

The glass articles are subsequently converted to a crystalline product by the proper heat treatment, known as ceramming. The ceramming process typically consists of a low-temperature hold to induce internal nucleation, followed by one or more higher-temperature holds to promote crystallization and growth of the primary crystalline phase or phases. Although some glass compositions are self-nucleating via homogeneous nucleation, it is more common that certain nucleating agents, such as titanates, metals, or fluorides, are added to the batch to promote phase separation and subsequent internal crystalline nucleation.

Because crystallization occurs at high viscosity, article shapes are usually preserved with little or no (<10%) shrinkage or deformation during the ceramming. In contrast, conventional ceramic bodies typically experience shrinkage of up to 40% during firing. Commercial products manufactured in this manner include telescope mirrors, smooth cooktops, and cookware and tableware.

Glass-ceramics can also be prepared via powder processing methods in which glass frits are sintered and crystallized. Conventional ceramic processing techniques such as spraying, tape- and slip-casting, isostatic pressing, or extrusion can be employed. Such so-called devitrifying frits are employed extensively as sealing frits for bonding glasses, ceramics, and metals. Other applications include cofired multilayered substrates for electronic packaging, matrices for fiber-reinforced composite materials, refractory cements and corrosion-resistant coatings, bone and dental implants and prostheses, architectural panels, and honeycomb structures in heat exchangers.

II. GLASS-CERAMIC PROPERTIES

Given the structural variables of composition, phase assemblage, and microstructure, glass-ceramics can be designed to provide a wide range of physical properties.

A. Thermal Properties

Many commercial glass-ceramics have capitalized on their superior thermal properties, particularly ultralow thermal expansion coupled with high thermal stability and thermal shock resistance—properties that are not readily achievable in glasses or ceramics. Linear thermal expansion coefficients ranging from -7.54 to $20 \times 10^{-6}/°C$ can be obtained. Zero or near-zero expansion materials are used in high-precision optical applications such as telescope mirror blanks as well as for stove cooktops, woodstove windows, and cookware. Glass-ceramics based on β-eucryptite solid solution display strongly negative thermal expansion (shrinking as temperature increases) and can be valuable for athermalizing precise fiber optic components. At the other extreme, high-expansion devitrifying frits are employed for sealing metals or as corrosion-resistant coatings for metals.

Glass-ceramics have high-temperature resistance intermediate between that of glass and of ceramics; this property depends most on the composition and amount of residual glass in the material. Generally, glass-ceramics can operate for extended periods at temperatures of 700 to over 1200°C. Thermal conductivities of glass-ceramics are similar to those of glass and much lower than those of conventional aluminum oxide-based ceramics. They range from 0.5 to 5.5 W/m K.

B. Optical Properties

Glass-ceramics may be either opaque or transparent. The degree of transparency is a function of crystal size and birefringence, interparticle spacing, and of the difference in refractive index between the crystals and the residual glass. When the crystals are much smaller than the wavelength of light, or when the crystals have low birefringence and the indices of refraction are closely matched, excellent transparency can be achieved.

Certain transparent glass-ceramic materials exhibit potentially useful electrooptic effects. These include glasses with microcrystallites of Cd-sulfoselenides, which show a strong nonlinear response to an electric field, as well as glass-ceramics based on ferroelectric crystals such as niobates, tantalates, or titanates. Such crystals permit electric control of scattering and other optical properties. Transparent glass-ceramics whose crystals can be doped with optically active lanthanide or transition element ions have potential for applications in telecommunications as tunable laser sources or optical amplifiers.

C. Chemical Properties

The chemical durability is a function of the durability of the crystals and the residual glass. Generally, highly siliceous glass-ceramics with low alkali residual glasses, such as glass-ceramics based on β-quartz and β-spodumene, have excellent chemical durability and corrosion resistance similar to that obtained in borosilicate glasses.

D. Mechanical Properties

Like glass and ceramics, glass-ceramics are brittle materials which exhibit elastic behavior up to the strain that yields breakage. Because of the nature of the crystalline microstructure, however, strength, elasticity, toughness (resistance to fracture propagation), and abrasion resistance are higher in glass-ceramics than in glass. Their strength may be augmented by techniques which impart a thin surface compressive stress to the body. These techniques induce a differential surface volume or expansion mismatch by means of ion exchange, differential densification during crystallization, or by employing a lower-expansion surface glaze.

The modulus of elasticity ranges from 80 to 140 GPa in glass-ceramics, compared with about 70 GPa in glass. Abraded modulus of rupture values in glass-ceramics range from 50 to 300 MPa, compared with 40 to 70 MPa in glass. Fracture toughness values range from 1.5 to 5.0 MPa/m in glass-ceramics, compared with less than 1.5 MPa/m in glass. Knoop hardness values of up to 1000 can be obtained in glass-ceramics containing particularly hard crystals such as sapphirine.

E. Electrical Properties

The dielectric properties of glass-ceramics depend strongly on the nature of the crystal phase and on the amount and composition of the residual glass. In general, glass-ceramics have such high resistivities that they are used as insulators. Even in relatively high alkali glass-ceramics, alkali migration is generally limited, particularly at low temperatures, because the ions are either incorporated into the crystal phase or they reside in isolated pockets of residual glass. Nevertheless, by suitably tailoring the crystal phase and microstructure, reasonably high ionic conductivities can be achieved in certain glass-ceramics, particularly those containing lithium.

Glass-ceramic loss factors are low, generally much less than 0.01 at 1 MHz and 20°C. Glass-ceramics comprised of cordierite or mica crystals, for example, typically provide loss factors less than 0.001. The fine-grained, homogeneous, nonporous nature of glass-ceramics also gives them high dielectric breakthrough strengths, especially compared with ceramics, allowing them to be used as high-voltage insulators or condensors.

Most glass-ceramics have low dielectric constants, typically 5–8 at 1 MHz and 20°C. Glass-ceramics comprised primarily of network formers can have dielectric constants as low as 4, with even lower values ($K < 3$) possible in microporous glass-ceramics. On the other hand, very high dielectric constants of over 1000 can be obtained from relatively depolymerized glasses with crystals of high dielectric constant, such as lead or alkaline earth titanate.

III. GLASS-CERAMIC SYSTEMS

All commercial as well as most experimental glass-ceramics are based on silicate bulk glass compositions, although there are numerous nonsilicate and even nonoxide exceptions. Glass-ceramics can be further classified by the composition of their primary crystalline phases, which may consist of silicates, oxides, phosphates, borates, or fluorides. Examples of commercial glass-ceramics are given in Table I.

A. Simple Silicates

The most important simple silicate glass-ceramics are based on lithium metasilicate Li_2SiO_3, lithium disilicate $Li_2Si_2O_5$, diopside $CaMgSi_2O_6$, wollastonite $CaSiO_3$, and enstatite $MgSiO_3$.

There are four groups of commercially important lithium silicate glass-ceramics, three of which are nucleated with P_2O_5. The first comprises strong, fine-grained materials based on a microstructure of fine-grained lithium disilicate crystals with dispersed nodules of quartz or cristobalite crystals. Several companies have introduced these finely textured glass-ceramics for use as magnetic disk substrates for portable computer hard drives. More recently, internally nucleated glass-ceramics in the SiO_2-Li_2O-K_2O-ZnO-P_2O_5 system based on lithium disilicate with high mechanical strength and excellent chemical durability have been developed for use as dental overlays, crowns, and bridges. The third group comprises high-expansion lithium disilicate glass-ceramics which match the thermal expansion of several nickel-based superalloys and are used in a variety of high-strength hermetic seals, connectors, and feedthroughs.

The fourth group is nucleated with colloidal silver, gold, or copper, which in turn is photosensitively nucleated. By suitably masking the glass and then irradiating with ultraviolet light, it is possible to nucleate and crystallize only selected areas. The crystallized portion consists of dendritic lithium metasilicate crystals, which are much more soluble in dilute hydrofluoric acid than is the glass. The crystals can thus be etched away, leaving the uncrystallized (masked) portion intact. The resulting photoetched glass can then be flood-exposed to ultraviolet rays and heat-treated at higher temperature, producing the stable lithium disilicate and quartz phases. The resulting glass-ceramic is strong, tough, and faithfully replicates the original photoetched pattern. These chemically machined materials have been used as fluidic devices, lens arrays, magnetic recording head pads, and charged plates for ink-jet printing.

Blast furnace slags, with added sand and clay, have been used in Eastern Europe for over 30 years to manufacture inexpensive nonalkaline glass-ceramics called slagsitall. The primary crystalline phases are wollastonite and diopside in a matrix of aluminosilicate glass. Metal sulfide particles serve as nucleating agents. The chief attributes of these materials are high hardness, good to excellent wear and corrosion resistance, and low cost. The relatively high residual glass levels (typically >30%), coupled with comparatively equiaxial crystals, confers only moderately high mechanical strengths of ~100 MPa. Slagsitall materials have found wide use in the construction, chemical, and petrochemical industries.

More recently, attractive translucent architectural panels of wollastonite glass-ceramics have been manufactured by Nippon Electric Glass and sold under the trade name Neopariés. A sintered glass-ceramics with about 40% crystallinity, this material can be manufactured in flat or bent shapes by molding during heat treatment. It has a texture similar to that of marble but with greater strength and durability than granite or marble. Neopariés is used as a construction material for flooring and for exterior and interior cladding.

TABLE I   Commercial Glass-Ceramic Compositions (wt%)[a]

	(A)	(B)	(C)	(D)	(E)	(F)	(G)
SiO_2	68.8	55.5	63.4	69.7	56.1	47.2	60.9
Al_2O_3	19.2	25.3	22.7	17.8	19.8	16.	14.2
Li_2O	2.7	3.7	3.3	2.8	—	—	—
MgO	1.8	1.0	[b]	2.6	14.7	14.5	5.7
ZnO	1.0	1.4	1.3	1.0	—	—	—
BaO	0.8	—	2.2	—	—	—	—
P_2O_5	—	7.9	[b]	—	—	—	—
CaO	—	—	—	—	0.1	—	9.0
Na_2O	0.2	0.5	0.7	0.4	—	—	3.2
K_2O	0.1	—	[b]	0.2	—	9.5	1.9
Fe_2O_3	0.1	0.03	[b]	0.1	0.1	—	2.5
MnO	—	—	—	—	—	—	2.0
B_2O_3	—	—	—	—	—	8.5	—
F	—	—	—	—	—	6.3	—
S	—	—	—	—	—	—	0.6
TiO_2	2.7	2.3	2.7	4.7	8.9	—	—
ZrO_2	1.8	1.9	1.5	0.1	—	—	—
As_2O_3	0.8	0.5	[b]	0.6	0.3	—	—
Primary phases	β-Quartz	β-Quartz	β-Quartz	β-Spodumene	Cordierite	Fluormica cristobalite	Diopside

[a] Commercial applications and sources:

(A) Transparent cookware (Visions); World Kitchen.
(B) Telescope mirrors (Zerodur); Schott Glaswerke.
(C) Infrared transmission cooktop (Ceran); Nippon Electric Glass.
(D) Cookware, hot-plate tops; World Kitchen.
(E) Radomes; Corning Incorporated.
(F) Machinable glass-ceramic (Macor); Corning Incorporated.
(G) Gray slagsitall; Hungary.

[b] Not available.

B. Fluorosilicates

Compared to the simple silicates, fluorosilicate crystals have more complex chain and sheet structures. Examples from nature include hydrous micas and amphiboles, including hornblende and nephrite jade. In glass-ceramics, fluorine replaces the hydroxyl ion; fluorine is much easier to incorporate into glass and also makes the crystals more refractory.

1. Sheet Fluorosilicate Glass-Ceramics

Glass-ceramics based on sheet silicates of the fluorine mica family can be made machinable and strong, with excellent dielectric properties and thermal shock resistance. This combination of properties stems from their unique microstructure of interlocking and randomly oriented flakes of cleavable mica crystals. Because micas can be easily delaminated along their cleavage planes, fractures propagate readily along these planes but not along other crystallographic planes. As a result, the random intersections of the crystals in the glass-ceramic cause crack branching, deflection, and blunting, thereby arresting crack growth (Fig. 2). In addition to providing the material with high intrinsic mechanical strength, the combination of ease of fracture initiation with almost immediate fracture arrest enables these glass-ceramics to be readily machined. Flexural strengths, in the range of 135–175 MPa, are inversely proportional to the flake diameter and are not sensitive to abrasion.

The commercial glass-ceramic Macor (Corning Code 9658), based on a fluorophlogopite mica, $KMg_3AlSi_3O_{10}F_2$, is capable of being machined to high tolerance (± 0.01 mm) by conventional high-speed metal-working tools. By suitably tailoring its composition and nucleation temperature, relatively large mica crystals with high two-dimensional aspect ratios are produced, enhancing the inherent machinability of the material. The growth of high-aspect-ratio flakes can be enhanced by designing the composition to (1) delay nucleation of

FIGURE 2 Crack deflection by mica crystals in fluorophlogopite glass-ceramics. [After Grossman, D. G. (1977), *Vacuum* **28**(2), 55–61.]

phlogopite until relatively high temperatures where growth is rapid, (2) stimulate lateral platy growth as opposed to thickening by limiting the concentration of the cross-bonding species potassium, and (3) produce a relatively fluid B_2O_3-rich residual glass allowing rapid diffusion of species to the edges of the growing crystals. This "house-of-cards" microstructure is illustrated in Fig. 3.

In addition to precision machinability, Macor glass-ceramic has high dielectric strength, very low helium permeation rates, and is unaffected by radiation. This glass-ceramic has been employed in a wide variety of applications including high-vacuum components and hermetic joints, precision dielectric insulators and components, seismograph bobbins, sample holders for field ion microscopes, boundary retainers for the space shuttle, and gamma-ray telescope frames.

FIGURE 3 Microstructure of crystallized fluorophlogopite glass-ceramics (white bar = 1 μm). [After Grossman, D. G. (1977), *Vacuum* **28**(2), 55–61.]

Translucent machinable glass-ceramics also have been developed for use in restorative dentistry. They are based on the tetrasilicic fluormica $KMg_{2.5}Si_4O_{10}F_2$. High strength ($\sim$140 MPa) and low thermal conductivity along with a hardness closely matching natural tooth enamel provide this material with advantages over conventional metal ceramic systems.

Since the mid-1980s, the inherent strength and machinability of mica-based materials have extended the growing field of biomaterials. Biomaterials for bone implants demonstrate biocompatibility—are well tolerated by the body—and may even offer bioactivity, the ability of the biomaterial to bond with the hard tissue (bones) of the body. Several strong, machinable, and biocompatible mica- and mica/fluorapatite glass-ceramics have been widely studied as bone implants. One, known as Bioverit I, is based on a microstructure of tetrasilicic fluormica, $KMg_{2.5}Si_4O_{10}F_2$, and fluorapatite, $Ca_{10}(PO_4)_6F_2$. The mica phase imparts strength and machinability to the material while the fluorapatite confers bioactivity. The glass-ceramics have long-term stability and are machinable with standard metal-working tools. Another glass-ceramic, Bioverit II, possesses an unusual microstructure of curved fluormica crystals with a "cabbage-head" morphology, with cordierite ($Mg_2Al_4Si_5O_{18}$) as an accessory phase. The analyzed composition of the curved mica crystals shows them to be slightly enriched in alumina: $(Na_{0.18}K_{0.82})(Mg_{2.24}Al_{0.61})(Si_{2.78}Al_{1.22})O_{10.10}F_{1.90}$. While this material is not bioactive, it is strong, highly machinable, and very well tolerated biologically.

2. Chain Fluorosilicate Glass-Ceramics

Interlocking blade- or rodlike crystals can serve as important strengthening or toughening agents, much as fiber glass is used to reinforce polymer matrices. Naturally occurring, massive aggregates of chain silicate amphibole crystals, such as nephrite jade, are well known for their durability and high resistance to impact and abrasion. Glass-ceramics with microstructures of randomly oriented, highly anisotropic chain silicate crystals generally provide superior strength and toughness, for in order for a fracture to propagate through the material, it generally will be deflected and blunted as it follows a tortuous path around or through cleavage planes of each crystal. Indeed, glass-ceramics based on chain silicate crystals have the highest toughness and body strength of any glass-ceramics.

Glass-ceramics based on the amphibole potassium fluorrichterite ($KNaCaMg_5Si_8O_{22}F_2$) have a microstructure consisting of tightly interlocked, fine-grained, rod-shaped amphibole crystals in a matrix of minor cristobalite, mica, and residual glass (Fig. 4). The flexural strength of these

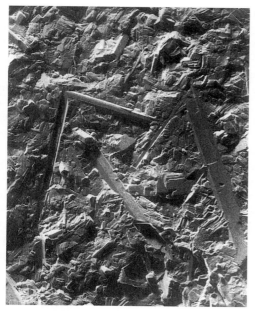

FIGURE 4 Fracture surface of fluorrichterite glass-ceramic showing the effects of rod reinforcement toughening.

materials can be further enhanced by employing a compressive glaze. Richterite glass-ceramics have good chemical durability, are usable in microwave ovens, and, when glazed, resemble bone china in their gloss and translucency. These glass-ceramics have been manufactured for use as high-performance institutional tableware and as mugs and cups for the Corelle (Corning) line.

An even stronger and tougher microstructure of interpenetrating acicular crystals is obtained in glass-ceramics based on fluorcanasite ($K_{2-3}Na_{4-3}Ca_5Si_{12}O_{30}F_4$) crystals. Cleavage splintering and high-thermal-expansion anisotropy augment the intrinsic high fracture toughness of the chain silicate microstructure in this highly crystalline material. Canasite glass-ceramics are being evaluated as potential dental materials.

C. Aluminosilicate Glass-Ceramics

Aluminosilicates consist of frameworks of silica and alumina tetrahedra linked at all corners to form three-dimensional networks; familiar examples are the rock-forming minerals quartz and feldspar. Commercial glass-ceramics based on framework structures comprise compositions from the Li_2O-Al_2O_3-SiO_2 (LAS) and the MgO-Al_2O_3-SiO_2 (MAS) systems. The most important, and among the most widespread commercially, are glass-ceramics based on the various structural derivatives of high (β)-quartz and on β-spodumene and cordierite. These silicates are important because they possess very

low bulk thermal expansion characteristics, with the consequent benefits of exceptional thermal stability and thermal shock resistance. Thus materials based on these crystals can suffer large thermal upshock or downshock without experiencing strain that can lead to rupture, a critical property in products such as stovetops, missile nose cones, and cookware. Representative aluminosilicate glass-ceramic compositions are given in Table I.

Although glass-ceramics with low to moderate thermal expansion can be formed from a wide variety of glasses in the Li_2O-Al_2O_3-SiO_2 system, the most useful are based on solid solutions of β-quartz and β-spodumene. These are highly crystalline materials with only minor residual glass or accessory phases. Partial substitution of MgO and ZnO for Li_2O improves the working characteristics of the glass while lowering the materials cost. Glass-ceramics containing β-quartz or β-spodumene can be made from glass of the same composition by modifying its heat treatment: β-quartz is formed by ceramming at or below 900°C, and β-spodumene by ceramming above 1000°C.

1. β-Quartz Solid-Solution Glass-Ceramics

Transparent glass-ceramics with near-zero thermal expansion are obtained through the precipitation of β-quartz solid solution. A mixture of ZrO_2 and TiO_2 produces highly effective nucleation of β-quartz, resulting in very small (<100-nm) crystals. This fine crystal size, coupled with low birefringence in the β-quartz phase and closely matched refractive indices in the crystals and residual glass, results in a transparent yet highly crystalline body.

Glass-ceramics with ultralow thermal expansion are particularly valuable in applications for which thermal dimensional stability is critically important. The best-known low-expansion optical material is the glass-ceramic Zerodur, manufactured by Schott. Its composition was designed specifically for maximum temperature–time stability of the metastable β-quartz phase and for constant thermal expansion characteristics within this range. Ambient temperature changes from −50°C to +50°C produce length changes of only a few parts per million. This property is critical for applications such as telescope mirrors, in which large volumes are cast, to ensure constant thermal expansion over the entire volume. This glass-ceramic can be readily polished to optical quality as required for reflective mirrors. Zerodur is also used in reflective optics and as the base block for ring laser gyroscopes, which have generally replaced mechanical gyros in airplanes.

Another important commercial application for β-quartz glass-ceramics with a low coefficient of thermal expansion is the smooth, radiant cooktop for electric stoves, as seen in Fig. 5. In this case, the glass-ceramic typically is doped with 0.1% V_2O_5 in order to render the material black in

FIGURE 5 Radiant stovetop composed of nanocrystalline β-quartz solid-solution glass-ceramic with near-zero coefficient of thermal expansion. [Photograph courtesy of Eurokera, S.N.C.]

general appearance. The vanadium does, however, allow transmission in the red and near-infrared, where the tungsten halogen lamp and other resistive elements radiate energy. Figure 6 illustrates the transmission of a Corning cooktop versus the radiation from a typical tungsten halogen lamp. While the glass-ceramic cuts out most visible radiation, it transmits very effectively in the red and near-infrared. Other consumer applications for transparent stuffed β-quartz glass-ceramics include transparent cookware, woodstove windows, and fire doors.

2. β-Spodumene Solid-Solution Glass-Ceramics

Opaque, low-expansion glass-ceramics are obtained by ceramming these LAS materials at temperatures above 1000°C. The transformation from β-quartz to β-spodumene takes place between 900 and 1000°C and is accompanied by a 5- to 10-fold increase in grain size (to 1–2 μm). When TiO_2 is used as the nucleating agent, rutile development accompanies the silicate phase transformation. The combination of larger grain size with

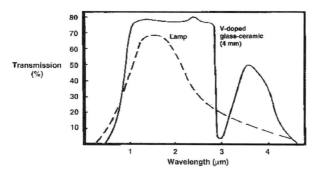

FIGURE 6 Transmission of a Corning glass-ceramic cooktop versus tungsten-halogen lamp.

the high refractive index and birefringence of rutile gives the glass-ceramic a high degree of opacity. Secondary grain growth is sluggish, giving these materials excellent high-temperature dimensional stability. The stable solid solution of β-spodumene varies in composition of $Li_2O\text{-}Al_2O_3\text{-}SiO_2$ from 1:1:4 to about 1:1:10. Figures 7 and 8 show the low thermal expansion characteristics of these tetragonal crystals in the range from 0 to 1000°C. Although the net volume change is low, the axial expansion is highly anisotropic, with the c axis expanding and the a axis contracting in response to heat. This means, in practice, that in order to control intergranular stresses and prevent microcracking, the grain size of β-spodumene glass-ceramics should be held below 5 μm.

β-Spodumene glass-ceramics are fabricated via both bulk and powder sintering technique and have found wide utilization as architectural sheet, cookware, benchtops, hot-plate tops, heat exchangers, valve parts, ball bearings, and sealing rings. More recently, a β-spodumene ferrule for optical connectors was developed by Nippon Electric Glass by redrawing a cerammed preform of $Li_2O\text{-}Al_2O_3\text{-}SiO_2$ glass-ceramic. The cerammed preform contains about 50 wt% β-spodumene solid solution and a

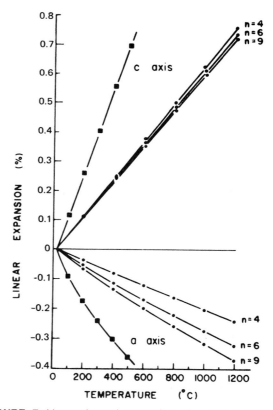

FIGURE 7 Linear thermal expansion of a and c axes; β-spodumenes: ●, keatite: ■. [After Ostertag, W., Fisher, G. R., and Williams, J. P. (1968), *J. Am. Ceram. Soc.* **51**(11), 651.]

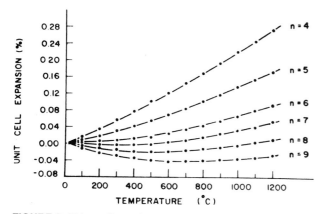

FIGURE 8 Volume thermal expansion in Li_2O-Al_2O_3-$nSiO_2$. [After Ostertag, W., Fisher, G. R., and Williams, J. P. (1968), *J. Am. Ceram. Soc.* **51**(11), 651.]

residual glass matrix. The preform is redrawn into microcapillaries with submicrometer accuracy, with no further crystallization occurring during the redraw. The microcapillaries are then processed to the near-net shape of the ferrule. The glass-ceramic ferrule exhibits excellent scratch resistance and good chemical durability, comparing favorably with conventional zirconia ferrules.

3. Cordierite Glass-Ceramics

Other low-expansion glass-ceramics, in the MgO-Al_2O_3-SiO_2 system, are based on the cordierite phase. These combine high strength and good thermal stability and shock resistance with excellent dielectric properties at microwave frequencies. These materials are made by bulk and powder sintering methods and are used in missile radomes and as high-performance multilayer electronic packaging.

4. Glass-Ceramics Based on Nonsilicate Crystals

a. Oxides. Glass-ceramics consisting of various oxide crystals in a matrix of siliceous residual glass offer properties that are not available with more common silicate crystals. In particular, glass-ceramics based on spinels and perovskites can be quite refractory and can yield useful optical, mechanical, and electrical properties.

Glass-ceramics based on spinel compositions ranging from gahnite ($ZnAl_2O_4$) toward spinel ($MgAl_2O_4$) can be crystallized using ZrO_2 and/or TiO_2 as nucleating agents. These glass-ceramics can be made highly transparent, with spinel crystals on the order of 10–50 nm in size (Fig. 9). The phase assemblage consists of spinel solid-solution crystals dispersed throughout a continuous siliceous glass. Possible applications for transparent spinel glass-ceramics include solar collector panels, liquid-crystal display screens, and high-temperature

lamp envelopes. Recently, nonalkali, nanocrystalline glass-ceramics based on Mg-rich spinel and enstatite (a chain silicate) crystals were developed for potential use as magnetic disk substrates in computer hard drives. These glass-ceramics possess elastic modulus over 140 GPa and, with grain sizes of less than 100 nm (Fig. 10), can be polished to an average roughness of 0.5–1.0 nm.

Glass-ceramics based on perovskite crystals are characterized by their unusual dielectric and electrooptic properties. Examples include highly crystalline niobate glass-ceramics that exhibit nonlinear optical properties, as well as titanate, niobate, and tantalate glass-ceramics with very high dielectric constants.

b. Phosphates. Many phosphates claim unique material advantages over silicates that make them worth the higher material costs for certain applications. Glass-ceramics containing the calcium orthophosphate apatite, for example, have demonstrated good biocompatibility and, in many cases, even bioactivity, making them useful as bone implants and prostheses. Mixed phosphate-silicate glass-ceramics based on fluorapatite and wollastonite, fabricated using conventional powder processing techniques, are bioactive and have good flexural strength of up to 200 MPa. The aforementioned combination of fluorapatite with phlogopite mica provides bioactivity as well as machinability. Cast glass-ceramics with microstructures comprising interlocking needles of fluorapatite and mullite provide high fracture toughness, with K_{1c} values greater than 3 MPa/m.

Certain glasses in the B_2O_3-P_2O_5-SiO_2 system melted under reducing conditions can yield a unique microfoamed material upon heat treatment. These materials

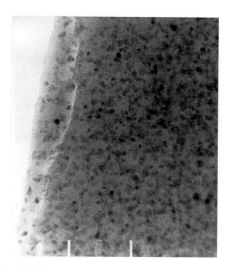

FIGURE 9 STEM micrograph of microstructure of transparent spinel glass-ceramic. Scale bar = 0.1 μm. [After Pinckney, L. R. (1999), *J. Non-Cryst. Solids* **255**, 171–177.]

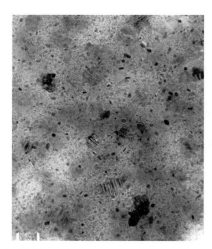

FIGURE 10 STEM micrograph of microstructure of high-modulus spinel/enstatite glass-ceramic. Scale bar = 0.1 μm. [After Pinckney, L. R., and Beall, G. H. (1997), *J. Non-Cryst. Solids.* **219**, 219–227.]

consist of a matrix of BPO_4 glass-ceramic filled with uniformly dispersed, isolated 1- to 10-mm hydrogen-filled bubbles. The hydrogen evolves on ceramming, most likely due to a redox reaction involving phosphite and hydroxyl ions. These materials, with DC resistivity of 10^{16} Ω cm at 250°C, dielectric constants as low as 2 and densities as low as 0.5 g/cm^3, have potential application in electronic packaging.

More recently, a novel class of microporous glass-ceramics composed of skeletons of two types of titanium phosphate crystals has been prepared by chemical etching methods analogous to those used for Vycor (Corning) glasses. These materials offer promise for applications including oxygen and humidity sensors, immobilized enzyme supports, and bacteriostatic materials. Related lithium ion-conductive glass-ceramics, based on $LiTi_2(PO_4)_3$ solid solutions in which Ti^{4+} ions are partially replaced by Al^{3+} and Ga^{3+} ions, can be used for electric cells (solid electrolytes) and gas sensors.

c. Fluorides and Borates.

Transparent oxyfluoride glass-ceramics, consisting of fluoride nanocrystals dispersed throughout a continuous silicate glass, have been shown to combine the optical advantages of rare earth-doped fluoride crystals with the ease of forming and handling of conventional oxide glasses. Fluorescence and lifetime measurements indicate that these materials can be superior to fluoride glasses both for Er^{3+} optical amplifiers because of greater width and gain flatness of the 1530-nm emission band and for 1300-nm Pr^{3+} amplifiers because of their higher quantum efficiency.

Transparent glass-ceramics based on submicrometer, spherical crystallites of β-BaB_2O_4 (BBO) in a glass of the same composition demonstrated second harmonic generation to ultraviolet wavelengths; such materials offer promise as potential optical components. Other aluminoborate glass-ceramics may be useful as low-expansion sealing frits.

IV. CONCLUSIONS

In addition to continued growth in the areas of biomaterials and of zero-expansion, transparent glass-ceramics, future applications for glass-ceramics are likely to capitalize on designed-in, highly specialized properties for the transmission, display, and storage of information. The potential for glass-ceramics to play a key role in active and passive photonic devices for the rapidly growing telecommunications field is particularly exciting. Such products may include tunable fiber lasers and optical amplifiers as well as substrates for compensating for the temperature-induced wavelength variations in fiber Bragg gratings, versatile passive devices that can be used in a variety of optical components such as narrow-band add/drop filters, multiplexers, dispersion compensation, and amplifier module components.

Whatever the future research trends in glass-ceramics are, it is certain that these materials, which combine the unique and diverse properties of crystals with the rapid forming techniques and product uniformity associated with glass production, will continue to proliferate as technology expands.

SEE ALSO THE FOLLOWING ARTICLES

CERAMICS, CHEMICAL PROCESSING OF • GLASS • PHOTOCHROMIC GLASSES • PRECIPITATION REACTIONS

BIBLIOGRAPHY

Beall, G. H., and Duke, D. A. (1969). "Transparent glass-ceramics," *J. Mater. Sci.* **4**, 340–352.

Beall, G. H., and Pinckney, L. R. (1999). "Nanophase glass-ceramics," *J. Am. Ceram. Soc.* **82**, 5–16.

Beall, G. H. (1992). "Design and properties of glass-ceramics," *Annu. Rev. Mater. Sci.* **22**, 91–119.

Petzoldt, J., and Pannhorst, W. (1991). "Chemistry and structure of glass-ceramic materials for high precision optical applications," *J. Non-Cryst. Solids.* **129**, 191–198.

McMillan, P. W. (1979). "Glass-Ceramics," 2nd ed., Academic Press, London, UK.

Strnad, Z. (1986). "Glass-Ceramic Materials, Glass Science and Technology, Vol. 8," Elsevier, Amsterdam, The Netherlands.

McHale, A. E. (1992). "Engineering properties of glass-ceramics," *In* "Engineered Materials Handbook, Vol. 4: Ceramics and Glasses," pp. 870–878, ASM International.

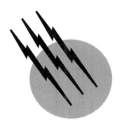

Global Gravity Modeling

R. Steven Nerem

Colorado Center for Astrodynamics Research

GLOSSARY

Geoid Equipotential surface of the Earth's gravity field that best approximates mean sea level. Over the land, this surface can lie above the surface, or below the surface.

Gravity Force exerted by a mass on another body, typically measured in units of acceleration in milliGals (1 gal = 1 cm/s^2).

Geopotential A representation of the potential of the Earth that satisfies Laplace's equation ($\nabla^2 U = 0$).

Spherical harmonics Means of mathematically representing a variable in terms of trigonometric functions, each having a different amplitude and wavelength, on the surface of a sphere. Analogous to a Fourier representation of a two-dimensional function, but in this case used for representing spherical functions.

GLOBAL GRAVITY MODELING an area of study that attempts to make the best possible estimate of the detailed gravity field of a planet using both terrestrial gravity measurements and more recently, satellite measurements. These models, together with topography models, are used in the fields of geodesy, geophysics, and planetary physics to characterize the internal structure of the planet and the dynamics of planetary interiors. These models also see wide use in the aerospace community for trajectory determination of spacecraft and missiles.

I. DEFINITION OF THE GLOBAL GRAVITY FIELD

The concept of a global gravity field is based on the basic principles of physics, which is at present largely Newtonian mechanics. Newton's Law of Gravitation states that the magnitude of the force between two masses M and m is inversely proportional to the square of the distance (r) between them and may be written as:

$$F = \frac{GMm}{r^2}, \tag{1}$$

where G is the Universal Gravitational Constant (6.673 × 10^{-20} km^3/kg s^2). In the gravity modeling community, the vector force of gravity is usually represented as the gradient of a potential, where the gradient operator (∇) can be written as:

$$\nabla = \frac{\partial}{\partial x}\hat{i} + \frac{\partial}{\partial y}\hat{j} + \frac{\partial}{z}\hat{k}. \tag{2}$$

Encyclopedia of Physical Science and Technology, Third Edition, Volume 6

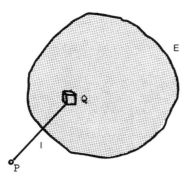

FIGURE 1 Mass distribution M with volume element dm and object at point P (from Fig. 1 of Anderson chapter with $Q = dm$, $E = M$, and $l = \rho$).

The potential is usually represented as an integral over the planet of the contribution of each infintesimal mass dm as (Fig. 1)

$$U = \int_M \frac{Gm}{\rho} dm, \qquad (3)$$

where ρ is the distance from the incremental mass dm to a point P outside the Earth. For a spherical homogenous planet (point mass), this reduces to

$$U = \frac{GMm}{r}, \qquad (4)$$

where r is the distance from the coordinate system origin to the satellite. The force on the satellite can be computed from the gradient of the potential

$$\vec{F} = \nabla U = \frac{GMm}{r^2} \frac{\vec{r}}{r}, \qquad (5)$$

which is the vector equivalent of Eq. (1). This representation is sufficient for points that are a long distance from the planet or when accuracy requirements are not demanding. For more stringent applications, the planet should be considered nonhomogenous. Then, $1/\rho$ in Eq. (3) is expanded in terms of spherical harmonic functions, and the integral results in a constant coefficient associated with each harmonic in the expansion:

$$U = \frac{GM}{r}\left[1 + \sum_{l=2}^{\infty} \sum_{m=0}^{l} \left(\frac{a_e}{r}\right)^l \bar{P}_{lm}(\sin\phi)\right.$$

$$\left. \times (\bar{C}_{lm}\cos m\lambda + \bar{S}_{lm}\sin m\lambda)\right], \qquad (6)$$

where a_e is the equatorial radius of the planet, ϕ and λ are the latitude and longitude of the satellite, l and m are the degree and order of the spherical harmonic expansion, $P_{lm}(\sin\phi)$ are the fully normalized Legendre associated functions of degree l and order m (for $m = 0$, these are the Legendre polynomials of degree l), and the C_{lm}/S_{lm}

are the fully normalized spherical harmonic coefficients describing the spatial variations of the Earth's potential field. The constants GM, a_e, and C_{lm}/S_{lm} will of course be unique for each planet, and are referred to as a gravity models for that planet. For Earth, $GM = 398600.4415$ km³/s² and $a_e = 6378.1363$ km, and the first few values of the spherical harmonic coefficients are shown in Table I. For a given spherical harmonic degree l, the corresponding half-wavelength spatial resolution on the Earth's surface is approximately given by $20000/l$ km. The degree l coefficients of the spherical harmonic potential (6) are usually assumed to be zero if the mass center is assumed to coincide with the origin of the coordinate system being used. While the expansion is theoretically infinite, in practice it is complete to $l = 360$ or less, depending on the spatial resolution of the data used to determine the spherical harmonic coefficients. Higher-resolution global gravity models are usually presented in gridded form rather than as a spherical harmonic representation.

The gravity field of the Earth also varies as a function of time as it deforms due to the gravitational effects of the Sun and the Moon, resulting in solid Earth and ocean tides. The tidal effects have been reasonably well determined, mainly because they occur at well-known astronomical frequencies. In addition, the gravity field varies slightly as mass is redistributed on its surface and in its interior. The phenomena of postglacial rebound refers to the slow rebound of the Earth's crust, predominantly in North America and Scandinavia, due to the melting of the ice sheets at the end of the last ice age 10,000 years ago. This rebound causes small secular changes in the spherical harmonic coefficients of the gravity field. In addition, the gravity field varies as water mass moves amongst the continents, oceans, and atmosphere [Wahr *et al.*, 1998]. There are a host of smaller effects.

TABLE I Current Global Geopotential Models

Model	Date	NMAX	Data used[a]
GEM9	1977	20	S
Rapp	1978	180	$S + G(A) + G(T)$
SAO	1980	30	$S + A + G(T)$
GEM 10B	1981	36	$S + A + G(T)$
GEM 10C	1981	180	$S + A + G(T)$
Rapp	1981	180	$S + G(A) + G(T)$
GEML2	1982	20	S
GRIM3B	1983	36	$S + G(A) + G(T)$
GRIM3-L1	1984	36	$S + G(A) + G(T)$

[a] S is satellite orbit data, A is satellite altimetry data, $G(A)$ is gravity anomaly data derived from satellite altimetry, and $G(T)$ is terrestrially measured gravity anomalies (surface gravimeter-based data).

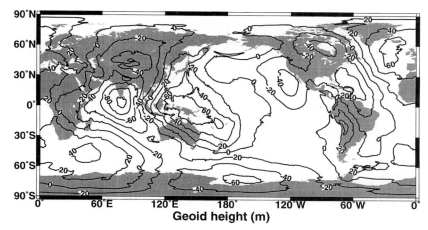

FIGURE 2 Contour map of the geoid height complete to degree and order 360 from the EGM-96 gravity model (From Lemoine, F. G. *et al.* (1998). "The Development of the Joint NASA GSFG and the National Imagery and Mapping Agency (NIMA) Geopotential Model EGM96, NASA Goddard Space Flight Center, Greenbelt, MD.)

II. METHOD OF MEASUREMENT

Until the launch of Sputnik, the only method available to measure the Earth's gravity field was through the use of surface gravimeters, and the first crude representations of the global gravity field were assembled by merging together surface gravimetric measurements from around the world. Almost immediately after entering the satellite era, the long wavelength components of the global gravity field were determined by measuring the gravitational perturbations to satellite orbits using ground-based tracking data. This is how the slight "pear shape" of the Earth was first determined (the "oblateness" of the Earth was previously known). In the late 1970s, satellite altimeter measurements were used to study the Earth's gravity field over the oceans, since the ocean surface largely conforms (to within ±1 m) to the geoid. At present, the most comprehensive models of the global gravity field are determined from a combi-

nation of satellite tracking data (collected from dozens of different satellites since the beginning of the satellite era), satellite altimeter data, and surface gravity data (Nerem *et al.*, 1995).

Maps of the global gravity field are usually represented either in the form of the geoid or as gravity anomalies. The geoid is defined as the height of the equipotential surface of the Earth's gravity field with most closely corresponds to mean sea level. The acceleration of gravity on this surface is everywhere the same. Because the ocean is a fluid, it adjusts itself to conform to the geoid, with the exception of the ±1 m deviations caused by the ocean currents. The geoid height can be found by chosing an appropriate constant value for the potential, U_0, and then determining the radius r from the expression for the geopotential in Eq. (5) plus the rotational potential ($\frac{1}{2}\|\vec{\omega} \times \vec{r}\|$). It is normally expressed relative to the height of a reference ellipsoid which best fits the shape of the Earth. As shown

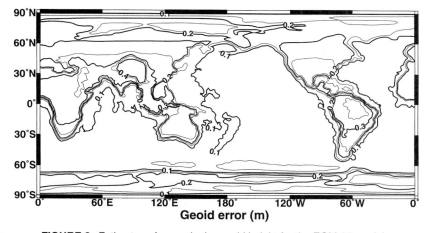

FIGURE 3 Estimates of errors in the geoid height for the EGM-96 model.

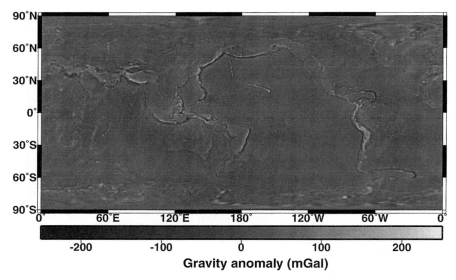

FIGURE 4 Gravity anomalies computed from the EGM-96 gravity model complete to degree and order 360.

in Fig. 2, the geoid height varies by ±100 m relative to the reference ellipsoid, and at long wavelengths mainly reflects density anomalies deep within the Earth. A map of the error in our current knowledge of the geoid is shown in Fig. 3. The errors are lowest over the oceans where we have satellite altimeter measurements, and highest over land areas where surface gravity observations are not available or suffer from poor accuracy. Gravity anomalies are the total gravitational acceleration at a given location minus the acceleration described by the reference ellipsoid, which varies only with latitude. Gravity anomalies are generally generally expressed in milliGals, where 1 Gal = 1 cm/s^2, as shown in Fig. 4. Gravity anomalies, which are "rougher" than the geoid, are better for representing

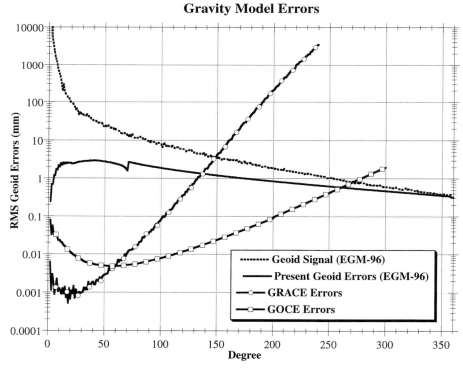

FIGURE 5 Estimates of the errors in our current knowledge of the geoid versus spherical harmonic degree versus the expected errors from two future satellite gravity missions, GRACE and GOCE.

the fine scale density variations near the surface of the Earth.

III. FUTURE DEDICATED SATELLITE GRAVITY MISSIONS

The field of global gravity field determination is entering a new era as satellite missions dedicated to measuring the Earth's gravity field are being developed and flown. In the next few years, the Gravity Recovery and Climate Experiment (GRACE) and the Global Ocean Circulation Explorer (GOCE) will be launched. GRACE will use precise microwave measurements between two satellites flying at an altitude of approximately 450 km to precisely map the Earth's gravity field. In addition, GRACE will be able to detect temporal variations of the Earth's gravity field, which after tidal variations are removed, are predominantly due to water mass being redistributed on the surface of the Earth (snow, ice, ground water, aquifers, etc.), in the atmosphere (water vapor), and in the ocean. It is expected that GRACE will be capable of making monthly

estimates of the gravity field with a spatial resolution of ~300–500 km and an accuracy of 1 cm equivalent water thickness (Dickey *et al.*, 1997). GOCE will consist of a single satellite carrying a gravity gradiometer, which will directly measure the gravity gradient (spatial derivative of gravity) in three axes. While GOCE will likely not have enough sensitivity to detect temporal gravity variations at long wavelengths, it will provide a much better determination of the static gravity field that can be provided by GRACE alone. The primary objective for the GOCE mission is to improve our knowledge of the geoid over the oceans to allow detailed studies of ocean circulation using satellite altimetry measurements. The expected errors in our knowledge of the Earth's gravity derived from each of these future missions is shown in Fig. 5.

IV. APPLICATIONS

Global gravity field models are used in a wide variety of applications in geophysics, oceanography, geodesy, and engineering (Fig. 6). Geodesists interested in measuring

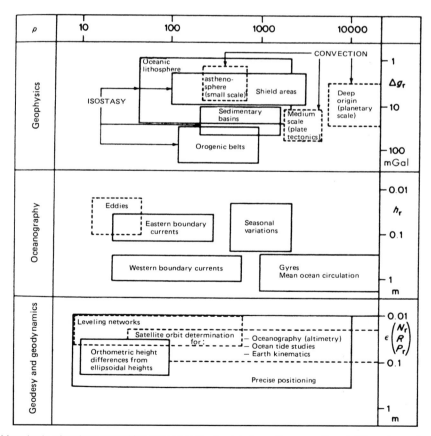

FIGURE 6 Magnitude of various quantities related to Earth global gravity anomalies at different wavelengths. Δg_r is the gravity anomaly in mGals, h_t is the relative variation of ocean surface height in meters, $\varepsilon(N_t)$ is the accuracy of the geoid height in meters, $\varepsilon(R)$ is the accuracy of the radial component of a typical artificial satellite orbit in meters, and $\varepsilon(P_t)$ is the relative position accuracy typically obtained using geodetic techniques.

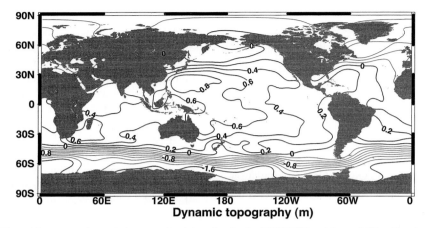

Dynamic topography (m)

FIGURE 7 Map of the ocean dynamic topography determined using TOPEX/Poseidon satellite altimeter data relative to the EGM-96 geoid model (Lemoine *et al.*, 1998). The contour lines are parrallel to the direction of the ocean currents, which move clockwise in the northern hemisphere, and counter-clockwise in the southern hemisphere.

orthometric heights (relative to the geoid or "mean sea level") using the Global Positioning System (GPS) must know the geoid height at the desired location, since GPS provides absolute heights. Oceanographers need to know the geoid height in order to measure the ocean circulation using satellite altimetry, since slope of the difference between the height of the ocean surface and the geoid is directly related to the geostrophic velocity of the ocean currents (Fig. 7). Geophysicists use gravity field models to study the internal structure of the Earth and as a geophysical exploration tool. Gravity field models are also fundamental to accurately computing the trajecto-

ries of Earth orbiting satellites, which is very important when making geodetic measurement from space (such as satellite altimetry), and computing ballistic missile trajectories.

Measurements of temporal variations of the Earth's gravity field largely represent the redistribution of water mass in the Earth system and a variety of temporal and spatial scales (Fig. 8), and are of interest to solid Earth geophysicist, hydrologists, meteorologists, oceanographers, and glaciologists, among others. At present, temporal gravity variations derived from satellite measurements have only been detected at wavelengths longer than

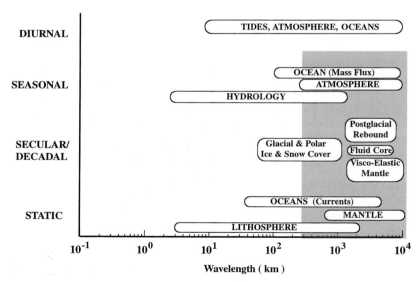

FIGURE 8 Representation of the typical spatial and temporal scales of non-tidal temporal gravity variations on the Earth. The shaded area is the temporal and spatial scales that the GRACE satellite mission is expected to resolve.

10,000 km; however, the GRACE mission should usher in a new paradijm in this field. Thus, the anticipated future improvements to the global gravity model will benefit a wide array of science and engineering applications.

Gravity modeling is also our principal tool for discerning the internal structure of the planets (Nerem *et al.*, 1995). Planetary gravity models are determined from Earth-based tracking of spacecraft orbiting the planet. Combined with topography models, often determined using radar or laser altimetry from these spacecraft, much can be learned about the structure of the planet, such as crustal thickness, density anomalies, and composition of the core. In recent years, precise gravity and topography models have been determined for Venus, the Moon, Mars, and the asteroid Eros using such measurements. Temporals gravity variations are also important for understanding planetary dynamics. The planet Mars is thought to have significant variations in its oblateness due to the annual accumulation/melting of ice at its poles. In addition the tidal variation of gravity and topography on Jupiter's moon Europa is thought to hold the key to determine if a subsurface ocean lies beneath Europa's icy shell.

SEE ALSO THE FOLLOWING ARTICLES

GEODESY • GRAVITATIONAL WAVE DETECTORS • GRAVITATIONAL WAVE PHYSICS • MECHANICS, CLASSICAL • REMOTE SENSING FROM SATELLITES

BIBLIOGRAPHY

Dickey, J. O., Bentley, C. R., Bilham, R., Carton, J. A., Eanes, R. J., Herring, T. A., Kaula, W. M., Lagerloef, G. S. E., Rojstaczer, S., Smith, W. H. F., van den Dool, H. M., Wahr, J. M., and Zuber, M. T. (1997). "Satellite Gravity and the Geosphere: Contributions to the Study of the Solid Earth and Its Fluid Envelope," pp. 112, National Research Council, Washington, DC.

Lemoine, F. G., Kenyon, S. C., Factor, J. K., Trimmer, R. G., Pavlis, N. K., Chinn, D. S., Cox, C. M., Klosko, S. M., Luthcke, S.B., Torrence, M. H., Wang, Y. M., Williamson, R. G., Pavlis, E. C., Rapp, R. H., and Olson T. R. (1998). "The Development of the Joint NASA GSFC and the National Imagery and Mapping Agency (NIMA) Geopotential Model EGM96," NASA Goddard Space Flight Center, Greenbelt, MD.

Nerem, R. S., Jekeli, C., and Kaula, W. M. (1995). "Gravity field determination and characteristics: Retrospective and prospective," *J. Geophys. Res.* **100**(B8), 15053–15074.

Wahr, J., Molenaar, M., and Bryan, F. (1998). "Time variability of the Earth's gravity field: Hydrological and oceanic effects and their possible detection using GRACE," *J. Geophys. Res.* **103**(B12), 30, 205–230.

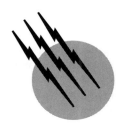

Global Seismic Hazards

Kaye M. Shedlock

U.S. Geological Survey

GLOSSARY

Acceleration The rate of increase of velocity per unit of time.
Ground motion The shaking or acceleration of the ground due to earthquakes.
Probability of exceedance The likelihood that earthquake-related ground motion will be greater than the map value shown for some time period of interest.
Recurrence A repeat occurrence of an event.
Seismic hazard The probable level of ground shaking associated with the occurrence or recurrence of earthquakes.

I. ABSTRACT

One of the most frightening and destructive phenomena of nature is a severe earthquake and its aftereffects. Catastrophic earthquakes account for 60% of worldwide casualties associated with natural disasters. Economic damage from earthquakes is increasing, even in technologically advanced countries with some level of seismic zonation, as shown by the 1989 magnitude 6.9 Loma Prieta, CA (more than $6 billion), 1994 magnitude 6.7 Northridge, CA (more than $25 billion), 1995 magnitude 6.8 Kobe, Japan (more than $100 billion), and 1999 magnitude 7.4 Turkey (more than 1.5% of the GNP of Turkey) earthquakes. Vulnerability to natural disasters increases with urbanization and development of associated support systems (reservoirs, power plants, etc.). The growth of megacities in seismically active regions around the world often includes the construction of seismically unsafe buildings and infrastructures, due to an insufficient knowledge of existing seismic hazard. Mitigation of the effects of earthquakes, including the loss of life, property damage, and social and economic disruption, depends on reliable estimates of seismic hazard. National, state, and local governments; engineers; planners; emergency response organizations; builders; corporations; scientists; and the general public require seismic hazard estimates for land use planning, improved building design and construction (including adoption of building construction codes), emergency response preparedness plans, economic forecasts, housing and employment decisions, and many more types of risk mitigation.

II. INTRODUCTION

Seismic hazard is defined as the probable level of ground shaking associated with the recurrence of earthquakes.

Early attempts at constructing seismic hazard maps provided estimates of the severity of ground shaking or damage from known or likely earthquakes. These maps were soon improved by including the frequency of occurrence of the shaking levels depicted. Modern seismic hazard assessment (SHA) began in the late 1960s with the publication of a series of papers describing and applying the probabilistic seismic hazard assessment (PSHA) method. By the mid-1970s, the United States and many other countries established national PSHA programs and began producing national probabilistic seismic hazard (PSH) maps. Rather than predictors of the occurrence, or recurrence, of specific earthquakes, PSH maps are predictors of likely levels of ground shaking from earthquakes during specific time windows. By the 1990s, half of the countries of the world had produced at least one national seismic hazard map, and the early pioneers had well-developed national programs to update their seismic hazard maps routinely.

In it simplest form, a PSHA is a specific solution of the "Total Probability Theorem":

$$P[G] = \int\int P[G \mid m \text{ and } r] f_M(m) f_R(r)\, dm\, dr,$$

where P is probability, G is the event of interest, and m and r are independent random variables that influence G. Simply put, the probability that event G occurs is calculated by multiplying the conditional probability of event G given the occurrence of events m and r, by the (independent) probabilities of events m and r, integrated over all values of m and r. For hazard mapping, G represents the exceedance of a specific level of ground motion at a site of interest during an earthquake, m is magnitude, r is distance. So, the probability of strong shaking at a site is dependent on the magnitude and distance of all possible earthquakes in the surrounding area. Since uncertainties in the parameters and modeling techniques may be explicitly incorporated into the analysis, PSH analysis is applicable anywhere, including areas where only rudimentary geological, geophysical, and geotechnical data are available. PSHAs improve as the quality of the data and methods improve.

Deterministic, or scenario, seismic hazard assessments (DSHAs) provide relatively detailed maps of the distribution of shaking from the largest possible earthquake, or series of earthquakes, believed likely to occur in a specific region. DSHAs require that the regional seismicity, geology, geophysical, and geotechnical data be well quantified. The probability of occurrence of the largest possible earthquake, or series of earthquakes, determines the usefulness of DSHAs.

Modern SHA programs include both probabilistic and, where applicable, deterministic methods. Commonly mapped ground motions are maximum intensity, peak ground acceleration (pga), peak ground velocity (pgv), and several spectral accelerations (SA). Each ground motion mapped corresponds to a portion of the bandwidth of energy radiated from an earthquake. Peak ground acceleration and 0.2s–0.5s SA correspond to short-period energy that will have the greatest effect on short-period structures (buildings up to about seven stories tall, which is the most common building stock in the world). Longer period SA maps (1.0s, 2.0s, etc.) depict the level of shaking that will have the greatest effect on longer period structures (10+ story buildings, bridges, etc.). Fifty years is the most commonly chosen exposure window. There are three commonly mapped probability levels of exceedance: 2, 5, or 10%, (98, 95, or 90% chance of nonexceedance, respectively). These probability levels of exceedance are useful concepts in engineering, but are not readily understood by nonengineers. In general, the larger the probability of exceedance is, the more likely the ground motions. For example, a map of ground motions with a 10% chance of exceedance in 50 years will depict the ground motions from those earthquakes most likely to occur. Since small earthquakes are more likely than large earthquakes, a map with a 10% chance of exceedance in 50 years will depict the more frequent, smaller ground motions likely during the exposure time of interest. Alternatively, a map of ground motions with a 2% chance of exceedance in 50 years will depict the ground motions from the likely events and from the less likely, which are usually larger, earthquakes.

From their inception, seismic hazard maps have served as critical input to building codes. Historically, maps of pga values have formed the basis of seismic zone maps that were included in U.S. building codes, including the U.S. Uniform Building Code, which included seismic provisions specifying the horizontal force a building should be able to withstand during an earthquake. The newly adopted International Building Code includes maps of short- and long-period SA (0.2s and 1.0s SA).

III. METHODS

There are three major elements of SHA: (1) the characterization of seismic sources, (2) the characterization of attenuation of ground motion, and (3) the actual calculation of hazard values. Variations in application of each element of SHA lead to differences in the estimated hazard.

The first element of SHA, the characterization of seismic sources, involves answering three questions.

- Where do earthquakes occur?
- How often do earthquakes occur?
- How big can we expect these earthquakes to be?

Approximately 90% of all earthquakes occur along the plate boundaries. The plate boundaries and the locations

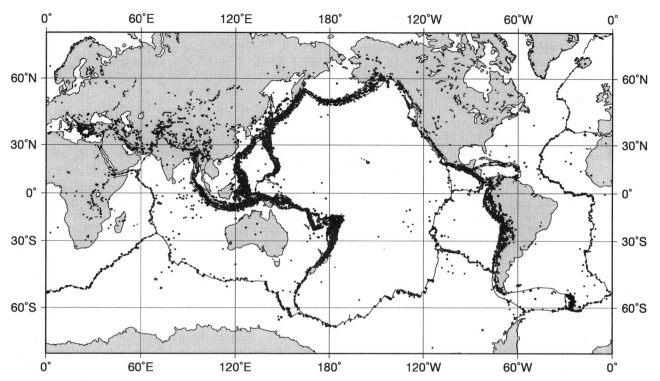

FIGURE 1 Plate boundaries and global seismicity. The plate boundaries are drawn in black. The continents are gray. Earthquakes with magnitudes ≥5.3 recorded from 1964 to the present are plotted as circles.

of earthquakes with magnitudes ≥5.3 recorded from 1964 to the present are shown in Fig. 1. There are three types of plate boundaries: transform faults, subduction zones, and spreading zones. Transform fault boundaries are where plates slide past one another. Transform fault earthquakes tend to be shallow (within the mid-to-upper crust, or less than about 20 km deep) and occur along fairly linear patterns. The San Andreas Fault, along the coast of California and northwestern Mexico, is a transform fault plate boundary.

Subduction zones are where one plate overrides, or subducts, another. The overriding plate pushes the subducting plate down into the earth, where it melts. Once it melts, the lighter rocks in the subducted plate move up through and heat up rocks in the overriding plate, together forming active volcanoes. There are multiple seismic sources in subduction zones: intraplate earthquakes within both the subducting and the overriding plates and interplate earthquakes along the fault surface between the two plates. Most subduction zones occur along the edges of continents, where oceanic crust is being subducted beneath continental crust. Large and great subduction zone earthquakes tend to be deep (tens to hundreds of kilometers), and the subduction zones along coasts are tens to hundreds of kilometers offshore. Thus, energy released in large subduction zone earthquakes has begun to attenuate (weaken) before it reaches onshore population cen-

ters. Well-known oceanic-continental subduction zones include those along the west coast of South America and along the east coast of Japan. A notable exception is the India-Eurasia plate boundary. Continental India is colliding with, and subducting beneath, continental Eurasia.

Spreading zones are plate boundaries where two plates are moving apart from each other. Magma (molten rock) rises, pushing the plates apart and adding new material to both plates. Earthquakes in spreading zones are shallow. Most spreading zones are oceanic; for example, Eurasia and North America are moving apart along the mid-Atlantic Ridge. A rare example of an onshore spreading zone is Iceland, which sits astride the mid-Atlantic Ridge.

The earth averages 18 large (magnitude 7.0–7.9) and 1 great (magnitude 8.0 or above) earthquake among the tens of thousands of earthquakes located globally every year. While large and great earthquakes are a major source of seismic hazard, they are not always the most important sources. Shallow, moderate earthquakes (magnitude 5.0–7.0) that occur nearby can cause considerable shaking and damage. For example, the 1994 Northridge, CA, and 1995 Kobe, Japan, earthquakes were shallow, moderate earthquakes.

As a first step in SHA, the locations of all instrumentally recorded earthquakes are collected in seismicity catalogs. These catalogs are the fundamental tool used to determine where, how often, and how big earthquakes are likely to be.

However, the instrumental recording of earthquakes is a mid-to-late 20th century phenomenon, while the physical processes that drive earthquakes occur on much longer time scales. Seismicity catalogs may be extended hundreds to thousands of years backwards in time by including historical and paleoseismic data. Shaking, casualty, and damage reports from historical earthquakes (documented earthquakes that occurred prior to instrumental recording) are analyzed in a variety of ways in order to estimate their locations and magnitudes. Buried ground surfaces, submerged forests, exhumed fault traces, and other paleoseismic (ancient) data are mapped, dated, and analyzed in order to estimate the ages and spatial extent of the earthquakes or other tectonic activity that created them. All of these data are combined in extended seismicity catalogs.

Even when the catalogs are extended backward, seismicity statistics are based on geologically short catalogs, so other deformation data are examined. Geodetic monitoring, another mid-to-late 20th century phenomenon, can reveal regional strain accumulation in currently aseismic regions, as well as better constraining deformation rates in seismically active areas. Regional strain accumulation may be spatially interpolated or partitioned to estimate (or better quantify) earthquake magnitudes and recurrence intervals on known faults. Measurements of strain accumulation in aseismic regions may be used to establish upper- and/or lower-bound estimates of possible earthquake magnitudes and recurrence intervals.

The results from seismic monitoring, the historic record, geodetic monitoring, and the geologic record are combined to characterize seismic sources. Although many interpretations of the wide range of input data are possible, only two different earthquake source characterization methods are used for PSHAs: the delineation of seismic source zones (fault or area) and the historic parametric method.

The delineation of seismic source zones involves specifying the geographical coordinates of an area (polygonal) or fault (linear/planar) source. The hazard is assumed to be uniform within each polygon or along each fault segment and may be described using a few parameters: the minimum (damage threshold) and maximum magnitude earthquakes and the rate of seismicity, derived from the Gutenberg-Richter (GR) relationship

$$\log N = a - bM,$$

where M is magnitude, N is the number of earthquakes of magnitude M or greater per unit time, and a and b are constants.

The historic parametric method determines seismicity rates (again based on the GR equation) for each point of a grid through the spatial smoothing of historical seismicity. Historic parametric applications commonly supplement

these seismicity rates with specific scenario earthquakes and background seismicity source zones.

The multiple methods and wide range of interpretations of the data result in uncertainties associated with source characterization. Various schemes are invoked to either explicitly or implicitly include these uncertainties in seismic hazard calculations. For example, multiple source zone models may be defined. Source zone boundaries may be drawn around historical seismicity clusters (a "historical" model), around geologic and/or tectonic structures, or a combination of both. Hazard calculations from each model may then be combined using various schemes (e.g., logic trees, weights) that produce a mean (or median) hazard value.

Another scheme is to define several alternative source models mixing the historic parametric and deterministic approaches. The most recent U.S. hazard maps were calculated using this scheme. Researchers combined seven existing regional and national earthquake catalogs spanning the years 1534 through 1995 to form a single U.S. catalog. They rewrote each catalog in a common format, then combined and sorted them into one chronological catalog. Duplicate listings of earthquakes were deleted, as were fore- and aftershocks. All the different earthquake magnitudes reported were converted to the same scale. Researchers then calculated the hazard for each of the following historic parametric models:

Central and Eastern United States	Western United States	California
Magnitude ≥3.0 since 1924	Magnitude ≥4.0 since 1963	Magnitude ≥4.0 since 1933
Magnitude ≥4.0 since 1860	Magnitude ≥5.0 since 1930	Magnitude ≥5.0 since 1900
Magnitude ≥5.0 since 1700	Magnitude ≥6.0 since 1850	Magnitude ≥6.0 since 1850

They also calculated the hazard from large background source zones in order to include the hazard for historically aseismic areas that have the potential to generate damaging earthquakes. Finally, they calculated the hazard from the following deterministic models:

Central and Eastern United States	Western United States	California
Magnitude ≥7.0	Magnitude ≥6.5	Magnitude ≥6.5
New Madrid, Charleston	Mapped seismogenic faults	Mapped seismogenic faults
Meers fault, Cheraw fault	Magnitude ≥8.3 Cascadia subduction zone	(i.e., San Andreas system)

The final hazard maps contained weighted average values derived from all of these hazard calculations.

The second element of an earthquake hazard assessment is estimates of expected ground motion at a given distance from an earthquake of a given magnitude. These estimates are usually equations, called attenuation relationships, which express ground motion as a function of magnitude and distance (and occasionally other variables, such as type of faulting). Ground motion attenuation relationships may be determined in two different ways: empirically, using previously recorded ground motions, or theoretically, using seismological models to generate synthetic ground motions which account for the source, site, and path effects. There is overlap in these approaches, however, since empirical approaches fit the data to a functional form suggested by theory and theoretical approaches often use empirical data to determine some parameters.

Earthquake magnitude, style of faulting, source-to-site distance, and local site conditions (site classification) must be clearly defined in order to estimate ground motions. Moment magnitude (M) is the preferred magnitude measure because it is directly related to the total amount of energy released during the earthquake. Style of faulting needs to be specified because, within 100 km of a site, strike-slip earthquakes generate smaller ground motions than reverse and thrust earthquakes, except for $M \geq 8.0$. The geometry of the source-to-site distance measures must be clearly specified, since different attenuation relationships have been derived using different geometries. There are also several site classification schemes, ranging from a description of the physical properties of near-surface material to very quantitative characterizations. Seismic hazard maps are calculated for a specific site classification (hard rock, soft rock, stiff soil, soil, soft soil, etc.). Hazard values calculated for rock/stiff soil sites (the most common site classifications) are lower than hazard values calculated for soil sites. Often, hazard values for soil sites may be estimated from the rock/stiff soil site values commonly depicted on hazard maps through multiplication by a specified factor, but these are no more than rough estimates.

The third element of hazard assessment is the actual calculation of expected ground motion values. Once sources are characterized and attenuation functions are selected, the likely ground motion from each possible source (earthquake) is calculated for every point on a grid. Each of these single-source ground motion values has the same probability of occurrence as the earthquake that produces it. This calculation of site-specific ground motion values is performed for every possible source that can affect that site. All of these calculations are turned into an annual frequency of occurrence, and exceedance, of various levels of the ground motion parameter of interest. The final hazard values are determined by summing over the time period of interest.

The most commonly mapped seismic ground motions are accelerations, which are measures of the rate of change of velocity. Prior to an earthquake, objects, the ground, and people are "at rest": our velocities are at or near zero, and thanks to the rotational and orbital effects of the earth, the acceleration of gravity holds us in place. During an earthquake, objects, the ground, and people are suddenly in motion: our velocity increases rapidly from zero. That increase, or acceleration, acts against the acceleration of gravity. Nonanchored objects and people may slide, shake, or, when the acceleration exceeds that of gravity, become airborne (briefly become weightless). Buildings and anchored objects will be shaken, and the larger the acceleration is, the more violent the shaking. The acceleration values depicted in seismic hazard maps are directly related to the lateral forces specified in seismic building code provisions. The acceleration of gravity and the accelerations depicted in seismic hazard maps are measured in meters per second per second (m/s^2). Hazard map values are presented in either meters per second per second or percent of gravity ($\%g$), where g is the acceleration of gravity (~ 9.78 m/s^2). A mapped value of $50\%g$ means the acceleration from the earthquake is half that of gravity. A shaking level of about $10\%g$ is the damage threshold for old or nonearthquake resistant structures close to the location of an earthquake. But the relationship between the shaking level and damage is variable, depending on many factors, including the distance from the earthquake, the type of building, site classification, and more. In general, however, the higher the shaking level is, the greater the damage potential.

IV. A BRIEF HISTORY OF SEISMIC HAZARD ASSESSMENT

Evidence of large earthquakes destroying cities appears throughout written history, and earthquakes continue to severely damage cities today. Qualitative assessments of seismic hazard have existed for centuries. Past civilizations learned to avoid settling in seismically active regions, but the shaking and damages were attributed to higher powers, rather than to the earth itself. Through the centuries, first Eurasian and then other civilizations began to keep records of the shaking events. Eventually, seismology (the science of earthquakes) took form. Although the first "seismograph," an ingenious device that indicated the arrival and traveling direction of shaking, was developed in China in 132 AD, it was not until the late 1800s that the first seismographs were built and the systematic cataloging of shaking events, now known as earthquakes, was underway. Russian and Japanese researchers created the first earthquake shaking zones maps in the 1930s. These qualitative maps were drawn in response to damaging earthquakes and were redrawn with every newly occurring earthquake.

With the creation of the World Wide Standard Seismograph Network in the 1960s, the instrumental recording of earthquakes became a truly global phenomenon. Our ability to assess the effects from earthquakes increased with our ability to record earthquakes. The high level of seismicity recorded along the plate boundaries resulted in a high level of awareness and an increasing number of programs to quantitatively assess seismic hazards. National SHA efforts are ongoing or proposed in a majority of the countries near plate boundaries, and many also have local or site-specific SHA programs, primarily in urban areas or near power facilities.

Although several generations of qualitative world maps of natural hazards (earthquakes, volcanoes, tsunamis and storm surges, tropical storms and cyclones, etc.) have been produced, the first quantitative seismic hazard map of the world was released in 1999. The Global Seismic Hazard Assessment Program (GSHAP), launched in 1992 by the International Lithosphere Program (ILP) with the support of tens of national and international organizations and hundreds of scientists, was designed to assist in global risk mitigation by producing the Global Seismic Hazard (GSH) Map and by serving as a resource for any national or regional agency for further detailed studies applicable to their needs.

V. THE GLOBAL SEISMIC HAZARD MAP

The GSHAP strategy was to establish and coordinate test areas and regional centers around the world. The first phase of the GSHAP (1993–1995) involved the establishment of a working group of national experts (representing all of the scientific disciplines required for SHA) in each region or test area. These working groups produced common regional earthquake catalogs and databases and, in many cases, assessed the regional seismic hazard. The second phase of the GSHAP (1995–1998) involved expansion of these regional efforts to assess the seismic hazard over whole continents and finally the globe.

The GSHAP employed a multidisciplinary approach to SHA that combined the results from geological disciplines dealing with active faulting (neotectonics, paleoseismology, geomorphology, and geodesy) with the historical and instrumental records of earthquakes. The GSH Map was produced using PSHA and, where applicable, DSHA methods.

The GSH Map, shown in Fig. 2, depicts pga with a 10% chance of exceedance in 50 years. The site classification is rock everywhere except in Canada and the United States, which assume rock/firm soil site conditions: white cor-

responds to low hazard (0–8%g); light gray corresponds to moderate hazard (8–24%g); dark gray corresponds to high hazard (24–40%g); and black corresponds to very high hazard ($\geq$40%g). Approximately 70% of the earth's continental landmasses have low hazard (pga) values, 22% have moderate hazard (pga) values, 6% have high hazard (pga) values, and 2% have very high hazard (pga) values.

In general, the largest seismic hazard values in the world occur in areas that have been, or are likely to be, the sites of the largest plate boundary earthquakes. The areas with the largest hazard values are along the subduction plate boundaries of the Kuriles-Kamchatka-the Aleutians-southern Alaska arc, Iceland, the Pamir-Hindu Kush-Karakorum and China/Myanmar border regions of the India-Asia collision zone, Taiwan, the transform plate boundary of the western United States, and the southeast coast of Hawaii.

Table I lists the ten largest earthquakes of the 20th century (e.g., the largest instrumentally recorded earthquakes). They are all interplate subduction (collision) zone earthquakes. Five of the ten largest known earthquakes have occurred along the Kuriles-Kamchatka-Aleutians-southern Alaska arc since 1952. The number of very large earthquakes known to have occurred in a short time results in the Eurasian-northern Pacific-North American plate boundary region being among regions with the highest seismic hazard values.

Iceland sits atop a large hot spot and is split by the mid-Atlantic Ridge, the spreading center plate boundary between the North American and Eurasian plates. A hot spot is an intraplate region where a plume of magma from deep within the earth rises through the plate to form a volcanic center at the surface. Large, shallow strike-slip and normal earthquakes in Iceland occur within complex fracture zones that connect the older, displaced rift zones with the current spreading centers. There are also earthquake swarms associated with the hot spot volcanoes. The highest hazard values in Iceland are at the northern tip of the island, where several large damaging earthquakes have occurred (1872, 1934, and 1963).

The collision of India with Asia is the region of greatest continental tectonic deformation in the world, and the widespread deformation associated with this collision is obvious in the hazard map. The entire collision zone is subject to high seismic hazard values, and large areas within the collision zone are subject to some of the highest hazard values depicted. Almost 15% of the great ($M \geq 8.0$) earthquakes documented in the 20th century have occurred here, including the 7th largest known earthquake, the 1950 Assam $M = 8.5$. Shallow, large strike-slip and normal faulting earthquakes are broadly distributed in the China/Myanmar border region of the collision zone. Similarly, frequent, large, shallow, and intermediate depth earthquakes are concentrated at the fronts of mountain

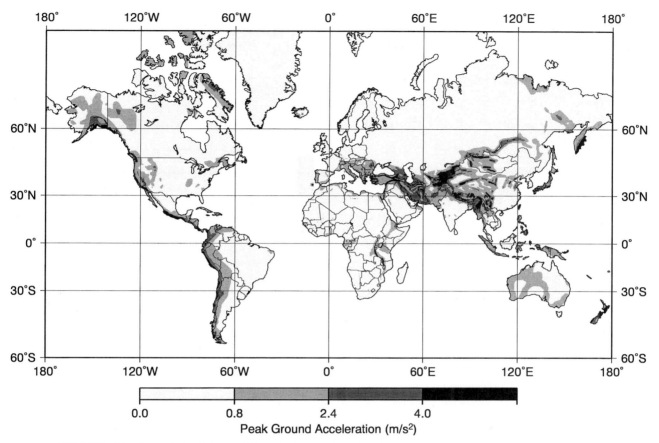

FIGURE 2 The Global Seismic Hazard Map. Peak ground acceleration (pga) with a 10% chance of exceedance in 50 years is depicted in meters per second per second: white corresponds to low seismic hazard (0–8%g); light gray corresponds to moderate seismic hazard (8–24%g); dark gray corresponds to high seismic hazard (24–40%g); and black corresponds to very high seismic hazard (greater than 40%g).

ranges and plateaus in the western (Pamir-Hindu Kush-Karakorum) region of the collision zone.

The seismic hazard values for Taiwan are all in the highest hazard range. Taiwan is the result of the collision between the northern end of an island arc on the Philippine plate and the Eurasian continental shelf. South of Taiwan, the Philippine plate is overthrusting the Eurasian

TABLE I Earthquakes of Large M

Event	Year	M
Chile	1960	9.5
Alaska	1964	9.2
Aleutian	1957	9.1
Kamchatka	1952	9.0
Ecuador	1906	8.8
Aleutian	1965	8.7
Assam	1950	8.6
Kurile Islands	1963	8.5
Chile	1922	8.5
Banda Sea	1938	8.5

plate; east of Taiwan, the Eurasian plate is overthrusting the Philippine plate. The landmass of Taiwan is a product of both plates. The Taiwan Telemetered Seismographic Network (TTSN) records between 4000 and 5000 earthquakes yearly. One of the great ($M = 8.0$) earthquakes of the 20th century occurred on Taiwan in 1920. Thus, it is not surprising that the relatively small island of Taiwan has relatively high seismic hazard values.

Although the energy release in large subduction zone earthquakes is much greater than the energy release in transform fault (strike-slip) earthquakes, the highest hazard values calculated in the Western Hemisphere are in southern California (United States), along the San Andreas fault system, and the southeast coast of Hawaii. Earthquakes along the San Andreas fault system (and transform faults in general) are shallow (<20 km) and often involve surface rupture. The San Andreas Fault is onshore for much of its length and passes through southern California. Thus, energy released in a large San Andreas Fault earthquake passes through population centers immediately, producing a higher shaking hazard. Although

Hawaii is not near a plate boundary, it sits atop a hot spot, like Iceland. Also, the ongoing collapse of the southeast flank of the Kilauea volcano produces large, shallow earthquakes that often involve surface rupture. Thus, the shaking hazard in southeastern Hawaii is comparable to that in southern California.

SEE ALSO THE FOLLOWING ARTICLES

EARTHQUAKE ENGINEERING • EARTHQUAKE MECHANISMS AND PLATE TECTONICS • EARTHQUAKE PREDICTION • MECHANICS OF STRUCTURES • SEISMOLOGY, ENGINEERING • SEISMOLOGY, OBSERVATIONAL • SEISMOLOGY, THEORETICAL

BIBLIOGRAPHY

The Global Seismic Hazard Assessment Program (GSHAP), 1992–1999 (1999). *Ann. Geofis.* **42, 6**.
The Global Seismic Hazard Assessment Program (GSHAP). Web page: http://seismo.ethz.ch/GSHAP/.
The U.S. National Seismic Hazard Mapping Project. Web page: http://geohazards.cr.usgs.gov/eq/.

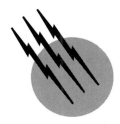

Glycoconjugates and Carbohydrates

Eugene A. Davidson

Georgetown University School of Medicine

I. Monosaccharides
II. Oligosaccharides
III. Polysaccharides
IV. Glycoconjugates
V. Analytical Methods

GLOSSARY

Aglycone Group attached to the hydroxyl of a furanose or pyranose sugar to form a glycoside. Can vary from a simple methyl group to other sugars to complex alkaloids.

Aldose Polyhydroxy aldehyde with three or more carbon atoms. When of sufficient length will form five- or six-membered ring (furanose or pyranose) hemiacetals. Designated D- or L- based on the configuration of the asymmetric center furthest from the carbonyl group relative to glyceraldehyde.

Anomer Formation of the cyclic ring structure confers asymmetry on the original carbonyl carbon atom. The two anomers produced are alpha- (the newly formed hydroxyl group projects on the same side as the five- or six-membered ring) or beta- (the new hydroxyl projects on the opposite side of the newly formed ring). This is combined with D-, or L-; α-D-glucopyranose, β-L-fructofuranose, for example.

Epimer Saccharide that differs from a reference sugar at a single asymmetric center. Thus, D-galactose is the 4-epimer of D-glucose.

Furanose Five-membered ring (derived from furan) with a single oxygen formed by reaction of the carbonyl carbon of an aldose or ketose with the appropriate hydroxyl group. Sugars commonly found as furanoses include ribose and 2-deoxyribose (in RNA and DNA, respectively), and fructose in combined form (sucrose, for example).

Glycolipid Structure containing one or more saccharide units covalently attached to the primary hydroxyl group of N-acyl sphingosine (ceramide), a C-18 amino alcohol. Those containing sialic acid are termed *gangliosides*.

Glycoprotein Protein with covalently attached saccharides. Linkage may be N- (amide nitrogen of asparagine residues), or O- (serine and threonine are most common). The saccharide units may be heterogeneous within the same protein, even at the identical substitution site.

Glycosaminoglycan Polysaccharide chain found in

proteoglycans. Has a linear structure that features a repeating disaccharide unit comprised of glucosamine or galactosamine and a uronic acid or galactose. The amino sugars are generally N-acetylated and contain ester sulfate. The pattern of sulfation varies and may confer specific biological properties.

Glycoside Acetal or ketal formed by substitution of the anomeric hydroxyl with an aglycone. This reaction fixes the configuration at the anomeric center. Linkages are generally acid labile.

Ketose (ulose) Polyhydroxy ketone (at carbon 2) of four or more carbon atoms (see *Aldose*).

Mutarotation The change in optical rotation associated with the equilibration, in solution, of the anomeric and ring forms of a saccharide beginning with a single crystalline form. Thus, dissolving α-D-glucopyranose in water allows formation of the β-anomer as well as the furanose forms. Since all have different optical rotations, the initial value changes until the equilibrium mixture is achieved.

Oligosaccharide Sugars that contain 2–10 monosaccharide units. These may be linear or branched. The most common representatives are derived from aldoses and will contain a single reducing terminus (free, or potential, aldehydo group) and one or more nonreducing termini (linked to other sugars only at their anomeric hydroxyl group).

Polysaccharide Structure containing more than 10 monosaccharide units. These may be the same (homopolysaccharide) or different (heteropolysaccharide), and the structure may be linear or branched. Natural diversity is very large.

Proteoglycan Special category of glycoproteins in which the saccharide chain is covalently linked to protein via an initiating xylose residue, a specific core oligosaccharide, and is followed by a glycosaminoglycan chain.

Pyranose Six-membered ring form of a saccharide formed by reaction of the appropriate hydroxyl group with the carbonyl carbon. Generally the most stable structure in solution. The conformer resembles cyclohexane and is generally a chair form with axial and equatorial substituents.

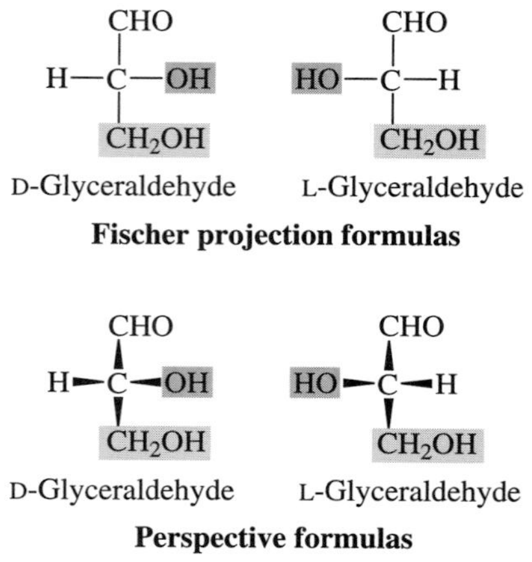

THE UTILIZATION of carbohydrates as food sources, sweetening agents, and clothing materials dates back several thousand years. Likewise, the production of beer and wine was known in ancient times although the relationship to saccharides and fermentation was not appreciated then. Manufacture of sugar as a sweetener including refinements such as decolorization with charcoal also has ancient origins. The commercial importance of carbohy-

Fischer projection formulas

Perspective formulas

FIGURE 1 Projection formulas of D- and L- glyceraldehyde.

drates extends beyond the food industry to textiles, pharmaceuticals, and a wide range of chemicals.

The common definition of carbohydrates is polyhydroxy aldehydes or ketones (the carbonyl group is generally at C-2), their derivatives, oligomers and polymers. The term itself arises from the empirical formula of the compounds initially studied (glucose, for example) since that is represented by $C(H_2O)_n$.

An appreciation of the diversity available in such structures began about 1900 with the work of van't Hoff on the tetrahedral carbon atom, extrapolated by Emil Fischer in his elucidation of the stereochemistry of glucose. During that series of experiments, Fischer was able to deduce the relative positions of all of the hydroxyls at the asymmetric centers and could relate glucose to glyceraldehyde (Fig. 1). However, he was unable to provide an absolute stereo structure for glyceraldehyde and thus had to decide between mirror image structures for natural glucose. That he chose the correct one may reflect little more than that he was right handed. He did synthesize all of the possible aldohexoses.

I. MONOSACCHARIDES

It is generally accepted that monosaccharides, or simple sugars, may have three to nine carbon atoms in a linear chain. Thus the nomenclature aldotriose, aldotetrose, aldopentose, etc., or ketotetrose, ketopentose, etc. The vast majority of naturally occurring sugars contain five or six carbon atoms. These saccharides are classified as D- or L- if they can be derived from D- or L-glyceraldehyde, respectively, or the four carbon ketoses (Fig. 2). The determining asymmetric center is that farthest from the carbonyl group—projection of the hydroxyl group to the right is D. As the chain length increases, new asymmetric centers

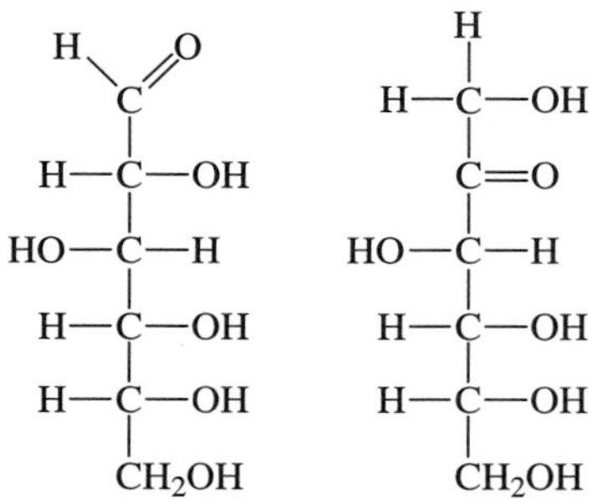

FIGURE 2 Projection formulas of D-glucose and D-fructose.

are introduced. Accordingly, there are four aldotetroses, eight aldopentoses, four ketopentoses, eight ketohexoses, etc. (Fig. 3).

Fischer recognized that sugars such as glucose existed in a ring form (i.e., a hemiacetal or hemiketal formed by addition of a hydroxyl group to the carbonyl carbon). Based on his knowledge of lactones, he assigned the participating hydroxyl to C-4 in the case of glucose, thus forming a five-membered (furanose) ring. The thermodynamics, however, show that the six-membered (pyranose) ring is the more stable form and that is the one predominating in solution. Note that the internal addition reaction at C-1 causes that carbon to become asymmetric and, therefore, two new isomeric structures are formed. To distinguish this asymmetry from that present at the other asymmetric centers, the term *anomer* is employed with the designation α or β reflecting projection of the newly formed hydroxyl group on the same or opposite side of the ring, respectively (Fig. 4).

The ring structures (aldopyranose for the six-membered rings and aldofuranose for the five-membered rings) are, in aqueous solution, in ready equilibrium with the open-chain, free aldehyde form. Thus, a solution of glucose in water contains five species: free aldehyde, two furanoses, and two pyranoses (Fig. 5). The aldehyde form represents about 0.025% of the total, with the bulk made up of the two pyranoses (the beta form predominates; see below). Although the six-membered ring is the energetically preferred form, the predominant ketohexose, D-fructose, is found in the furanose form in combined structures (sucrose, for example) although it too prefers the six-membered ring when free in aqueous solution. Other commonly occurring furanosides include ribose and deoxyribose as components of RNA and DNA (Fig. 6); galactose is also found as a furanoside in several plant and microbial polysaccharides.

A key property of the free monosaccharides is their ability to be readily oxidized (initially via the available alde-

hydo function), especially in alkaline solution. This characteristic was used as an analytical tool for many years, and the term *reducing sugar* was employed to designate saccharides with a free (or potentially free, in aqueous solution) carbonyl function. Prior to the advent of specific enzymatic methods, this was the standard procedure for measurement of blood glucose levels (i.e., reducing sugar).

Monosaccharides that differ from one another at a single asymmetric center other than the one derived from the carbonyl carbon are termed *epimers* (D-galactose is the 4-epimer of D-glucose, and D-mannose is the 2-epimer, for example; Fig. 7). Mirror-image structures are termed *enantiomers*.

An important and characteristic feature of saccharides is their ability to interact with plane-polarized light. Thus, each sugar has a characteristic optical rotation that is a complex function of the asymmetric centers, their polarizability, solvent interactions, etc. The sign (+ or dextrorotatory, − or levorotatory) and magnitude cannot be determined *a priori*. Nor is it true that all sugars of the same configuration (D- or L-) will have the same sign of optical rotation. Thus, D-glucose is dextrorotatory, whereas D-fructose is levorotatory. Since monosaccharides tend to crystallize in a single anomeric form rather than as mixtures, dissolving them in water results in a change in the optical rotation from that of the pure anomer (either alpha or beta) to that of the thermodynamically defined equilibrium mixture. This phenomenon is termed *mutarotation* (Fig. 8).

As the principles of conformational analysis became known, it was rapidly realized that the pyranose form of sugars adopted a chair structure analogous to that of cyclohexane (Fig. 9). In this spatial arrangement, equatorial hydroxyl groups are thermodynamically more stable (energy differences of about 1.5 kcal) than axial ones with the exocyclic hydroxymethyl group having a much larger effect. It is not surprising, therefore, that β-D-glucose, the all-equatorial structure, is the most prevalent natural sugar and the most prevalent organic compound on earth (Fig. 10). As a philosophical aside, it may be inferred that the choice between D- and L-glucose was made very early and on a basis we do not understand since they are conformationally equivalent. It is also expected that other naturally occurring sugars such as D-mannose or D-galactose would have only a single axial hydroxyl. Of the 16 possible aldohexoses, D-glucose, D-mannose, and D-galactose are widely distributed in nature while the remainder are of laboratory interest only. Parenthetically, idose (three axial hydroxyls in the classic conformer) has never been obtained in crystalline form. Additional widely distributed sugars include D-xylose (all-equatorial aldopentose), D-ribose, and 2-deoxy-D-ribose (backbone components of RNA and DNA,

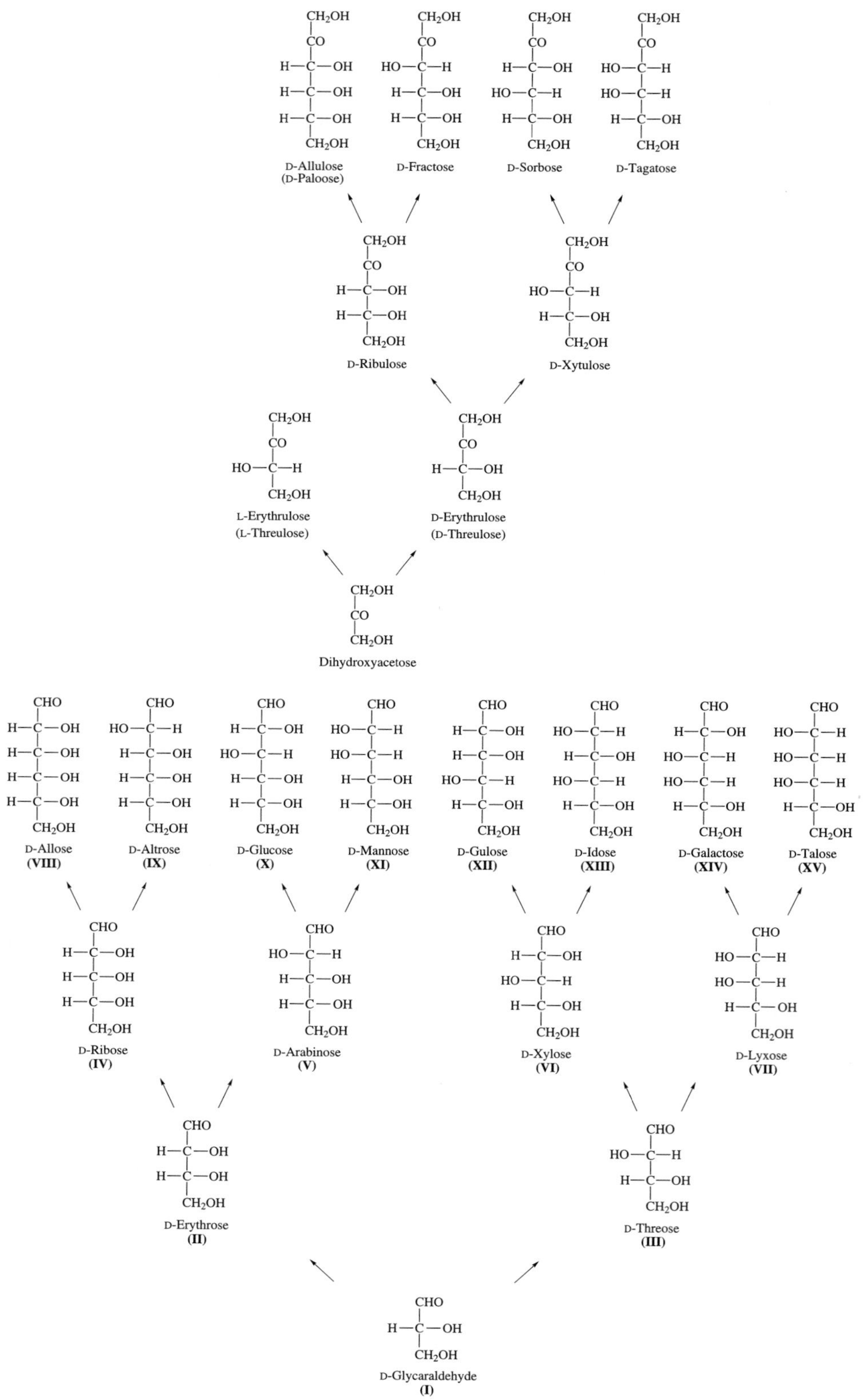

FIGURE 3 Structures of aldoses and ketoses up to six carbons in length. Monosaccharides with up to nine carbons are present in nature.

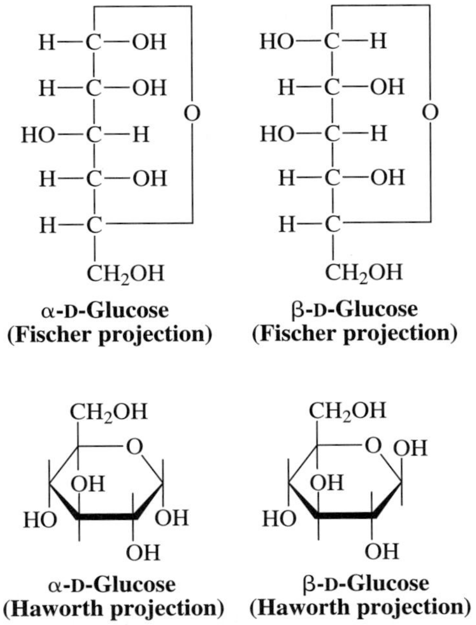

FIGURE 4 Alpha- and beta-forms of D-glucose.

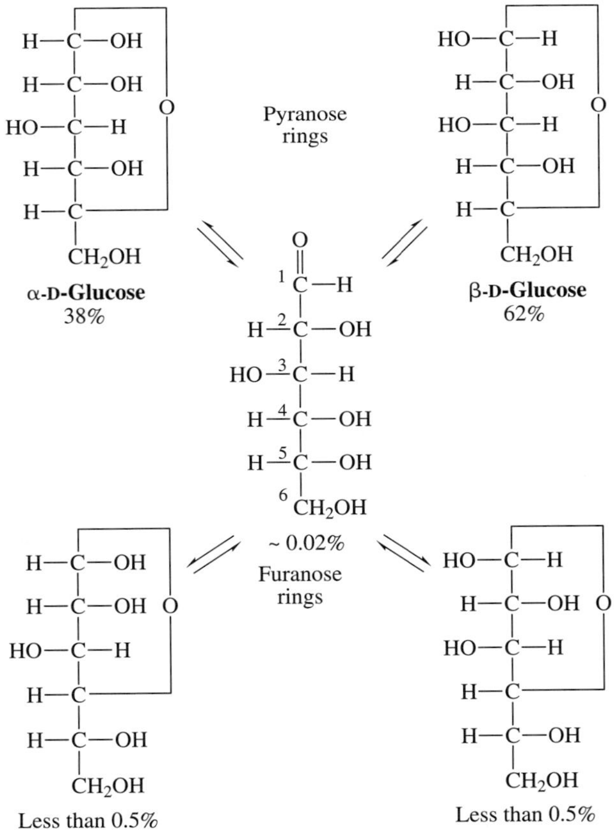

FIGURE 5 Equilibrium mixture of D-glucose in aqueous solution.

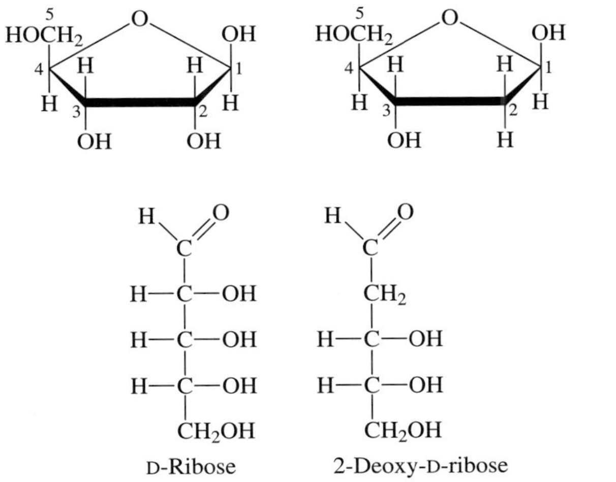

FIGURE 6 Structures of D-ribose (left) and 2-deoxy-D-ribose (right).

respectively). A large number of other sugars are present in nature. The vast majority of naturally occurring saccharides exist other than as free monosaccharides (only D-glucose and D-fructose are widely found free) and are present in disaccharides and complex oligomeric and polymeric structures in both plants and animals. About a dozen different sugars are present in mammals, all but glucose in combined form. Plants and microorganisms have very diverse saccharides, again all in combined forms.

A. Derivatives—Natural and Laboratory

The hydroxyl group formed as a result of ring closure represents a site for attachment of a broad variety of substituents. Compounds that are formed in such reactions are full acetals (ketals) and thus no longer undergo interconversion at the anomeric center. The configuration of such glycosides is therefore either alpha or beta depending on the relationship between the C-1 group and

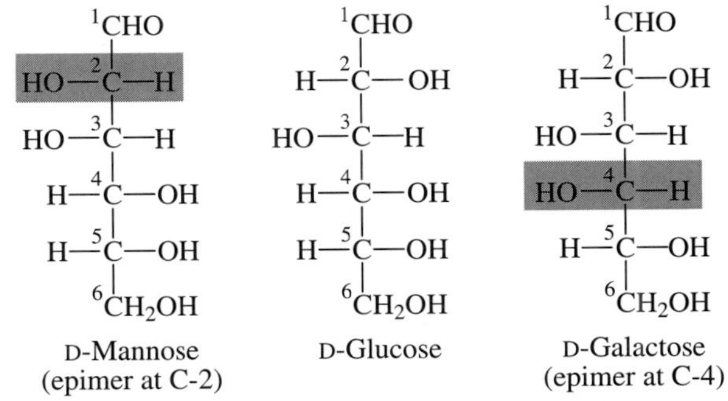

FIGURE 7 Relationship between D-glucose and D-galactose (4-epimer), and D-mannose (2-epimer).

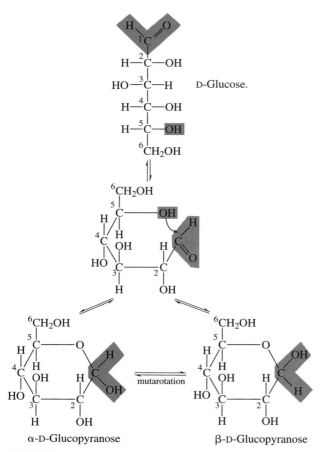

FIGURE 8 Interconversion of glucose forms in aqueous solution. The optical rotation is initially reflective of the starting material (alpha or beta) and changes (mutarotates) until equilibrium is achieved.

FIGURE 9 Conformational structure of pyranose rings.

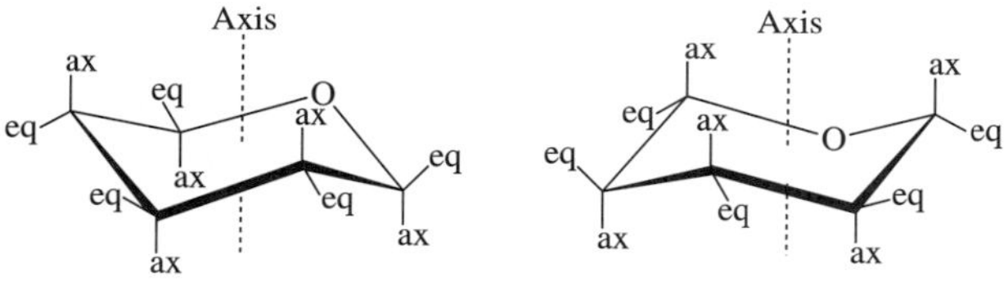

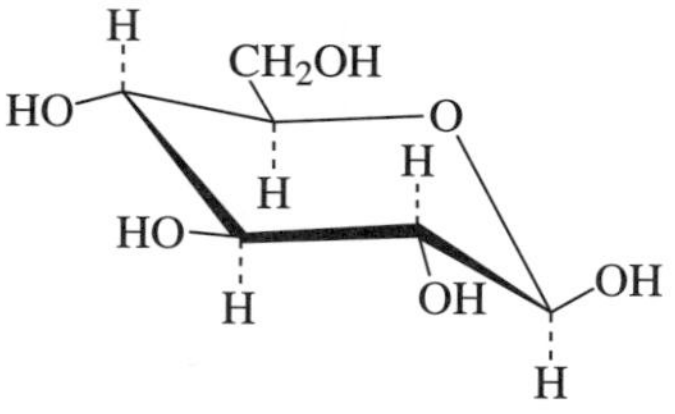

Two possible chair forms

FIGURE 10 All-equatorial structure of β-D-glucose.

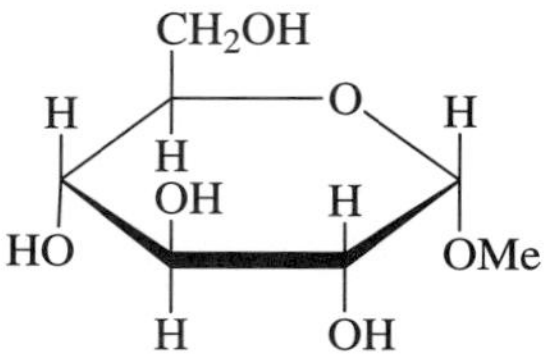

Methyl α-D-glucopyranoside

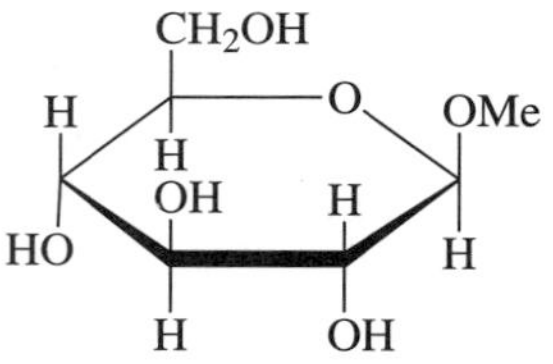

Methyl β-D-glucopyranoside

FIGURE 11 Methyl glucopyranosides.

the projection of the ring; if on the same side, designate alpha, otherwise, beta (Fig. 11). A wide variety of natural and man-made derivatives (glycosides) are known with the substituents, aglycones, varying from simple methyl groups to complex organic molecules including other sugars (see below).

In addition to substitution of the anomeric hydroxyl, many modifications of the hydroxyl loci are known in nature. Most prevalent are those in which the hydroxyl group at C-2 is replaced by an amino function, generally acetylated. The sugar 2-deoxy 2-acetamido-D-glucose (*N*-acetylglucosamine) is distributed throughout nature and, in its polymeric form (chitin), forms the organic matrix of insect and arthropod exoskeletons. Hence, it is likely the second most prevalent organic molecule on earth. Other variations include oxidation (C-6 or C-1) to form carboxyl groups and loss of a hydroxyl to form deoxy sugars (Fig. 12).

B. Chemical Transformations

The chemistry of carbohydrates involves transformations at both the carbonyl function and the chain hydroxyl groups. The presence of multiple species of simple sugars in aqueous solution gives rise to complex reaction patterns wherein product distribution may well be determined by kinetic rather than thermodynamic considerations.

1. Reactions at the Carbonyl Group

The carbonyl function reacts directly with a variety of reagents either as the free aldehyde or keto group or as the corresponding hemiacetal or hemiketal. In early work, Fischer introduced phenylhydrazine, which condenses at

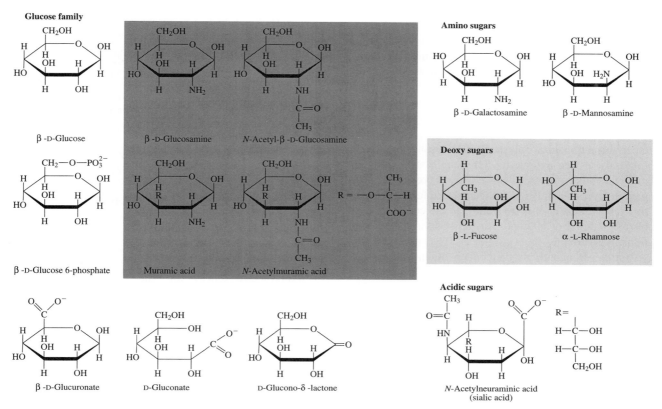

FIGURE 12 Sugars related to D-glucose that occur in nature.

the carbonyl carbon and oxidizes the adjacent secondary hydroxyl (aldoses) or primary alcohol (ketoses) to allow a second condensation. The resulting derivatives, phenylosazones, are crystalline and were employed both for identification and determination of stereochemistry. Phenylhydrazine itself is a liver poison and it is generally believed that Fischer suffered from liver damage due to this reagent (no longer employed). A simple example of the utility of this conversion is the fact that D-glucose, D-mannose, and D-fructose all give the same phenylosazone showing that their relative stereochemistry at the remaining asymmetric centers (carbons 3,4,5) is identical (Fig. 13).

Certain transformations may dictate that the carbonyl function remain available for later chemistry while reac-

tions are carried out elsewhere in the chain. This may be achieved by formation of dithioacetals or, in the ring form, by suitable substitution of the anomeric hydroxyl group (see below).

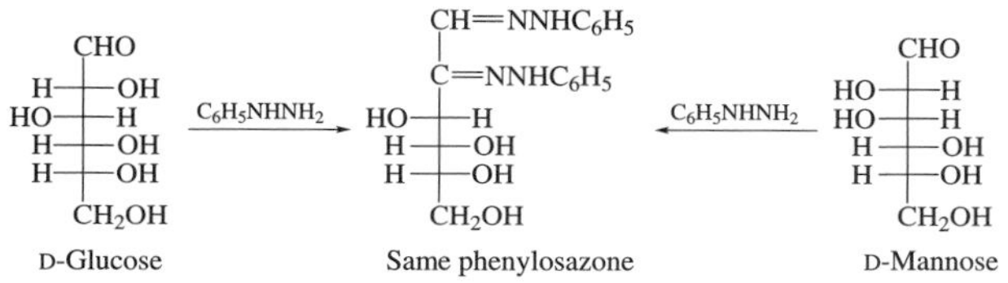

2. Reaction at the Anomeric Hydroxyl Group

The hemiacetal nature of the anomeric hydroxyl group makes it the most reactive of that type. Direct oxidation can be carried out with several reagents, most classically, hypoiodite. This results in rapid oxidation to the aldonolactone and is only exhibited by aldoses. Alternatively, reaction with alcohols results in the formation of full acetals (glycosides). A very large variety of such structures have been made.

Control of stereochemistry at the anomeric center is complex, involves the configuration at carbon two and kinetic factors. In general, both alpha- and beta-glycosides are formed. The configuration is very important in biological systems since enzymatic transformations are generally stereospecific with essentially all glycosidases having absolute specificity for either the alpha or beta form. Glycoside formation is generally carried out by reaction of the sugar with the appropriate alcohol in nonaqueous solution, typically with an acid catalyst.

FIGURE 13 Formation of phenylosazone from D-glucose or D-mannose. The loss of asymmetry at C-2 yields identical products.

3. Reactions at Secondary Hydroxyls

Substitution reactions at secondary hydroxyls are generally performed either for analysis of structure or to serve a protective function during other reactions. Etherification of the nonanomeric hydroxyls was an important structural tool in the analysis of oligosaccharide and polysaccharide structure. Methyl ethers have been employed for structural determination for more than 75 years. Thus, methyl ether formation in a polysaccharide results in substitution only at free hydroxyls. Subsequent analysis of the methylated derivatives reveals positions previously occupied in glycosidic linkage. Reagents used for this purpose have evolved from dimethylsulfate to the commonly employed method of Hakomori using sodium hydride and dimethylsulfoxide.

Another frequently used ether substituent is the benzyl group, which is stable under a variety of reaction conditions but can be removed by catalytic hydrogenation. This is employed mainly during synthetic schemes where protection of specific hydroxyls is required. Currently, ether formation is used primarily as an adjunct to mass spectrometric analysis.

The secondary hydroxyl groups can also be esterified. Acetyl substitution via acetic anhydride is often used, especially as a protective group. Tosyl substitution using toluenesulfonyl chloride is often employed since the properties of that ester (good leaving group) allow for SN_2 displacement resulting in inversion of configuration. Acetate esters are readily removed under basic conditions (methoxide ion) when glycosides remain intact.

4. Reactions at the Hydroxymethyl Group

The exocyclic nature of this group makes it the second most reactive of the hydroxyl functions. Tritylation is often employed to transiently block C-6 (ready removal under mild acid conditions). This is also the site of enzymatically catalyzed phosphorylation of glucose and other sugars, the first step in their metabolic utilization. Oxidation to a carboxylic acid is common. The resulting uronic acid is, for the common sugars, widely distributed in natural polysaccharides in both plants and animals. It is of interest that L-iduronic acid (the 5-epimer of D-glucuronic acid) and L-guturonic acid (the 5-epimer of D-mannuronic acid) are formed in nature after the respective precursor uronic acid has been incorporated into polymeric linkage.

5. Hydroxyl Pairs

The proximity and defined stereochemistry of hydroxyl groups in the ring structures of saccharides allows for a number of selective reactions. These include formation of

FIGURE 14 1,2 5,6-diisopropylidene D-glucofuranose. Formed by reaction of D-glucose with acetone in the presence of a suitable catalyst. The furanose product dominates due to kinetic control of the reaction.

acetals or ketals and of a group of interactions restricted to hydroxyls on adjacent carbons.

Reaction of glucose with acetone results in the formation of 1,2 5,6-diisopropylidene glucofuranose (Fig. 14). The furanose product arises because of the high reactivity of the exocyclic hydroxymethyl group thus favoring the formation of the five-membered ring. This is a commonly used protective scheme since the ketal function is readily cleaved by mild acid; in fact, selective cleavage of the 5,6 ketal can be accomplished. Acetals are likewise formed; reaction of glucose with benzaldehyde results in formation of the 4,6 benzylidene derivative, another useful synthetic intermediate.

Adjacent hydroxyls that are *cis* can undergo several different reactions. Included in this group is complexation with borate to form a transient cyclic adduct (Fig. 15). This alters chemical reactivity and electrophoretic properties of the reactive sugar and has several applications.

A key property of *cis*-hydroxyls is their susceptibility to carbon–carbon bond cleavage by metaperiodate. This oxidative reaction proceeds via a five-membered ring intermediate, generates a pair of aldehydes, and results in scission of the carbon chain (Fig. 16). It has been demonstrated that the *cis* orientation is necessary for this reaction since fixed ring structures with *trans*-hydroxyls do not react. The reaction has been utilized to identify saccharides in tissues via subsequent treatment of the generated aldehydes with a suitable amine, forming a colored Schiff base product (periodate-Schiff reaction).

C. Biosynthesis

The fixation of carbon dioxide via photosynthesis is the initiating reaction in saccharide synthesis in nature. Light

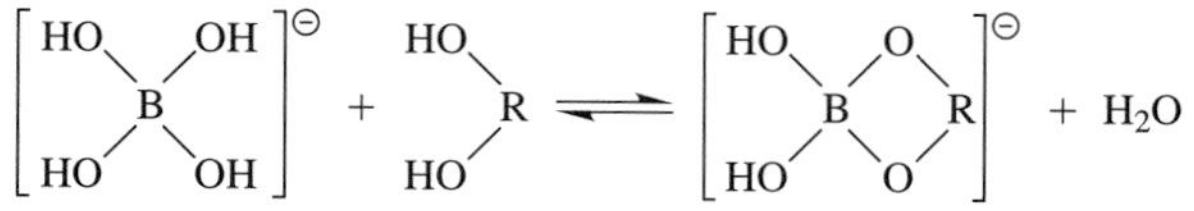

$$\begin{bmatrix} HO & OH \\ & B \\ HO & OH \end{bmatrix}^{\ominus} + \begin{array}{c} HO \\ R \\ HO \end{array} \rightleftharpoons \begin{bmatrix} HO & O \\ & B & R \\ HO & O \end{bmatrix}^{\ominus} + H_2O$$

FIGURE 15 Borate ester formation. *cis*-Hydroxyls are preferred. The extent of reaction was originally monitored by following the change in conductivity of borate solutions on the addition of saccharide.

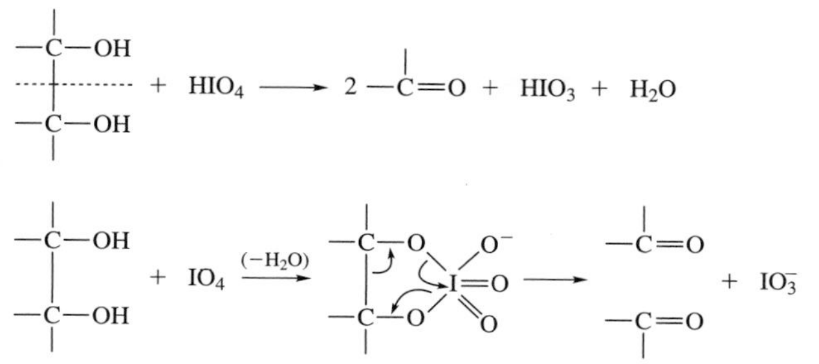

FIGURE 16 Action of periodate on vicinal diols. The proposed cyclic intermediate suggests that rigid structures with *trans*-hydroxyls will react poorly, a prediction confirmed experimentally.

energy is harvested via chloroplasts and used to provide chemical potential in the form of adenosine triphosphate, and reducing equivalents. The key intermediate, D-ribulose 1,5-bisphosphate, fixes carbon dioxide (the dark reaction) yielding products that are ultimately converted to D-glucose via a series of reactions of phosphorylated sugar intermediates. All other naturally occurring sugars are derived from glucose in transformations that involve phosphorylated or nucleotide-linked sugars. Thus, glucose-6-phosphate is converted to fructose-6-phosphate, which in turn is converted to mannose-6-phosphate; fructose-6-phosphate is also aminated to form 2-deoxy 2-amino glucose-6-phosphate. Galactose is formed by epimerization at C-4 of uridine diphosphoglucose (Fig. 17), fucose by a series of reactions initiating with guanosine diphosphomannose, etc. Thus, the diversity in saccharides seen in the biosphere stems from a single precursor, D-glucose. This is, therefore, the only required dietary saccharide for man.

II. OLIGOSACCHARIDES

The ability of the anomeric hydroxyl group of a sugar to be substituted with another sugar (via one of its hydroxyl groups), and for this process to be iterated leads to formation of di- and higher saccharides. Because several hydroxyls are available for such linkages and the anomeric configuration can vary as well, the number of possible structures grows exponentially even for oligomers derived from the same sugar. This may explain, in part, why many biological recognition events involve saccharides as a ligand.

Two disaccharides are widely distributed in nature: sucrose and lactose. Sucrose (α-D-glucopyranosyl-β-D-fructofuranoside) is the table sugar of commerce and a major industrial product (Fig. 18); primary sources are sugar cane and beets. This saccharide and its source material (cane molasses) serve as the basis for rum production; cruder precursors are important medium additives for the

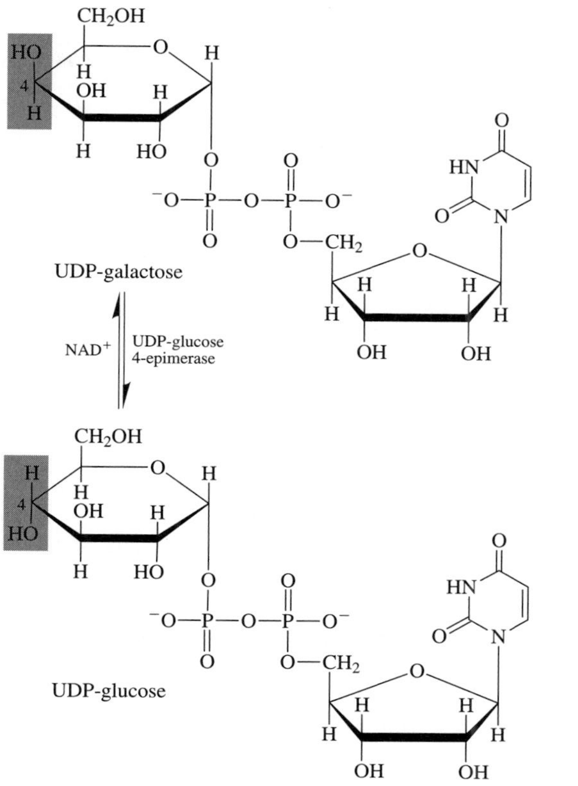

FIGURE 17 Biosynthesis of D-galactose by epimerization of uridine diphosphoglucose.

production of antibiotics. The sweetness of sucrose, and other simple sugars, is a major aspect of their commercial importance. The fact that sucrose is a nonreducing sugar contributes to its stability and market value.

Lactose (β-D-galactopyranosyl-1-4-D-glucopyranose) is the major sugar of milk (Fig. 19). It is of interest that the ability to utilize either sucrose or lactose as a source of calories is dependent on their enzymatic hydrolysis to the constituent monosaccharides in the intestine. About one-third of the oriental population lacks the requisite galactosidase and is thus lactose intolerant; this same problem occurs in some infants (developmentally related) and results in a "colicky" baby.

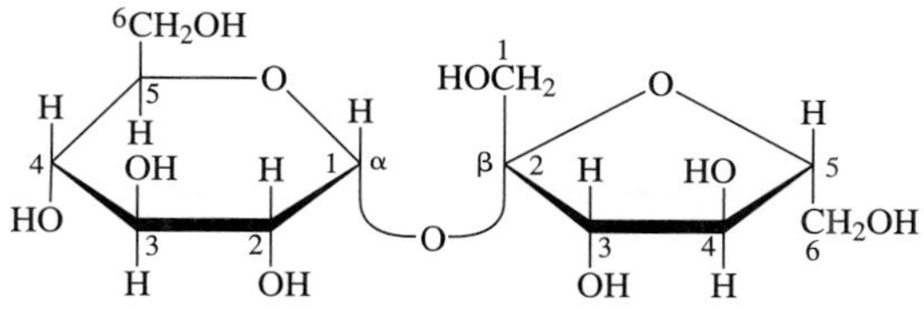

Sucrose
β-D-fructofuranosyl α-D-glucopyranoside
Fru(β2 ⟷ 1α)Glc

FIGURE 18 Structure of sucrose. Note that this sugar is nonreducing.

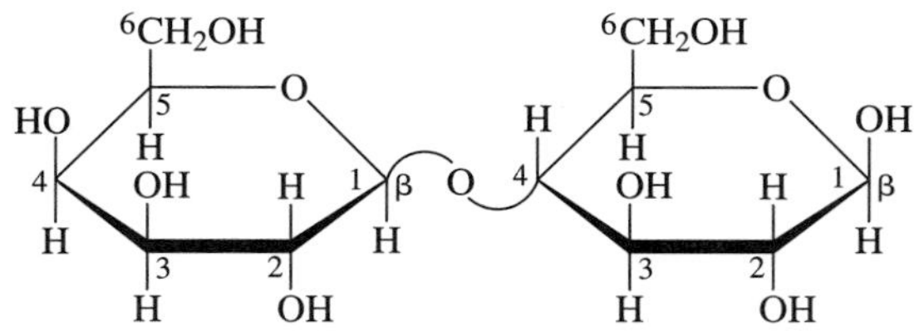

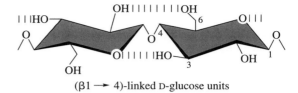

FIGURE 21 The repeating unit of cellulose showing hydrogen bond interactions. The extended structure allows chains to stack via the relatively hydrophobic axial faces of the pyranose rings.

Lactose (β form)
β-D-galactopyranosyl-(1 ⟶ 4)-β-D-glucopyranose
Gal(β1 ⟶ 4)Glc

FIGURE 19 Structure of lactose, the major sugar present in milk.

Other naturally occurring disaccharides of interest are maltose (α-D-glucopyranosyl-1-4-D-glucopyranose), an intermediate in the digestion of starches, and trehalose (α-D-glucopyranosyl-α-D-glucopyranoside), the major sugar of insect blood (Fig. 20).

Naturally occurring tri- and higher oligosaccharides (less than 10 sugar units) are rare. Raffinose (3), stachyose (4), and verbascose (5) are all assembled on a sucrose core with substituents originating on the glucose moiety.

III. POLYSACCHARIDES

Polymeric saccharides can have 10,000 or more sugar units, four or five different sugars, and a branched structure. The diversity that can be envisioned is huge and much of it is actually observed. Classification of polysaccharides differentiates homopolymers (a single sugar type) and heteropolymers (two or more sugar types) with subtypes reflecting linear or branched structures within each group. The physical and biological properties of each polymer depend on the architecture of the molecule as well as the enzymatic machinery available to interact with it. Thus, cellulose, a linear β-1-4 glucose polymer is a stable, highly organized polymer, indigestible by man, whereas starch, its α-linked counterpart is a major food source for all mammalian species.

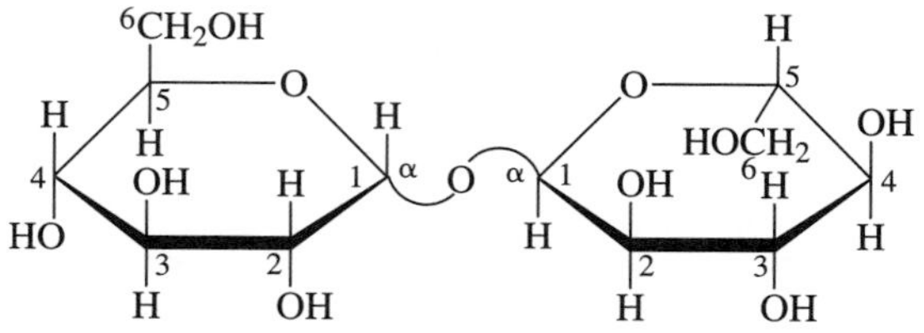

A. Homopolysaccharides

The major homopolysaccharides are cellulose, chitin, starches (amylose and amylopectin), glycogen, and xylans.

Cellulose, a linear glucose polymer linked β-1-4 is the predominant natural product in the biosphere. The all-equatorial structure allows for extensive hydrogen bonding, whereas the axial, somewhat hydrophobic faces favor a nonaqueous environment. The resulting aggregates have a highly ordered, quasi-crystalline arrangement (Fig. 21); hence, the unusual stability exhibited by the molecule, exemplified in trees and wooden artifacts. The broad distribution and low cost of cellulose has made it a major starting material for industrial development while the unmodified polymer remains the basis for wood, paper, and cotton.

Chitin, identical in linkage to cellulose but composed of N-acetylglucosamine instead of glucose, is the major structural component of insect and crustacean exoskeletons as well as a cell wall component of molds and fungi. The structural comments regarding cellulose also apply generally to chitin, especially in terms of stability. Less industrial development has been done with this polymer, in part because shrimp shells may present a more difficult starting material than trees.

Amylose and amylopectin are closely related, α-1-4-linked glucose polymers. They are major constituents of starches and hence key nutrients worldwide. Amylose is a linear chain with up to a few hundred glucose units. Amylopectin has the same backbone chain but α-1-6 branches approximately every 20 glucose units. The branches have the same linkages as the main chain. This ramified structure allows for efficient packing in cells.

Glycogen (Fig. 22), present in all higher animals, is closely related to amylopectin in that it has the same fundamental structure of a linear glucose chain with branches, and the same linkages. In this case, however, branches occur about every seventh residue, yielding a highly re-branched, tree-like envelope. This is essential for both packing in cells and for the rapid degradation of the polymer to provide critical metabolic intermediates. Glycogen serves as a primary energy reservoir in muscle and as a source of circulating glucose in the liver. It is of interest

Trehalose
α-D-glucopyranosyl α-D-glucopyranoside
Glc(α1 ⟷ 1α)Glc

FIGURE 20 Structure of trehalose, the predominant sugar present in the blood of insects.

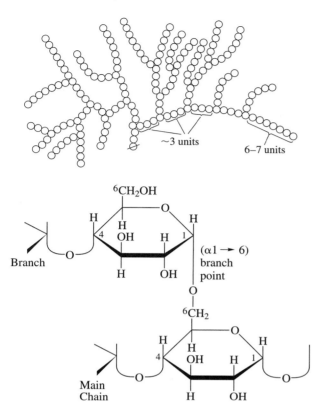

FIGURE 22 Schematic structure for glycogen and details of a typical branch point. The same linkages are present in amylopectin, which has less frequent branches.

that the biosynthesis of glycogen initiates on a core protein (glycogenin), which may be cleaved from the polysaccharide subsequent to polymerization.

Xylans are a group of polymers based on a structure analogous to that of cellulose wherein xylose is the repeating unit. The simplest representative contains only D-xylose with β-1-4 linkages and is a common component of plant walls. Several heteropolysaccharides utilize the xylan backbone and have various other saccharides as branches. Xylans are often associated with cellulose in plant cell walls.

B. Heteropolysaccharides

The diversity of heteropolysaccharides is enormous. Distributed throughout the animal and plant kingdom, this class of macromolecules rivals proteins in diversity. Initially thought to function only as structural components, it is now realized that specific sequences may have information content and can serve as recognition or signaling elements. In addition, modifications of saccharides already incorporated into a polymeric chain (sulfation, for example) add an additional level of complexity. The ability of these materials to function as other than strictly physical

components necessitates, in many cases, that the saccharide chain have a defined three-dimensional conformation. Much work has been done to model the three-dimensional structure of proteins and many protein structures have been determined by X-ray crystallographic analysis. Much less has been done with complex saccharides in terms of either molecular dynamics or conformational analysis of larger structures. The following discussion is intended to be representative only.

In addition to cellulose and xylan, plant cell walls contain a variety of complex heteropolysaccharides including other glucans. Detailed structures have only been determined for a limited number of these structures and their interactions, possibly covalent, with other plant components have rarely been determined in detail. Commercially important are pectins, polygalcturonides, which are key components of jellies and related uronides (gums) used as thickening agents in a broad variety of applications ranging from ice cream manufacture to the production of printing ink. Agars are galactans that are sulfated and also contain 3,6-anhydro-L-galactose, which have gelation properties. They have broad application in microbiology.

Bacteria likewise produce a wide variety of complex, cell-surface polysaccharides. Strains of *Streptococcus pneumoniae*, the causative agent of bacterial pneumonia, are characterized by a capsular polysaccharide—more than 100 different types are known. Since the capsular material is immunogenic and protective, polyvalent vaccines have been developed that utilize the capsular polysaccharides from common strains. Similarly, organisms responsible for bacterial meningitis have a characteristic polysaccharide capsule that is also immunogenic and protective. Conversely, several bacteria have, as an exterior capsule, saccharides sufficiently similar to those produced in their animal host so as to serve as a mechanism for avoidance of host immune responses. Many gram-positive bacteria have, as essential cell wall components, a complex saccharide structure that contains muramic acid (a condensation product of *N*-acetylglucosamine and pyruvate) and other amino sugars and is cross-linked by a peptide (Fig. 23).

A widely distributed heteropolysaccharide, hyaluronic acid, has both commerical and biological importance. This molecule, a repeating structure of D-glucuronic acid and *N*-acetylglucosamine with β-1-3 and β-1-4 linkages, is found in bacterial and animal sources, and it is one of the few complex saccharides not covalently linked to protein (Fig. 24). Molecular weights vary depending on source but often exceed two million. The polyanionic nature of the molecule leads to a relatively extended solution conformation. This coupled with the highly hydrophilic chemistry results in solutions with very high viscosity, an important physical property. This is utilized in treatment of

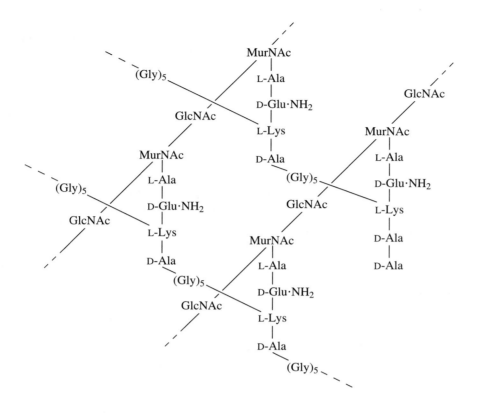

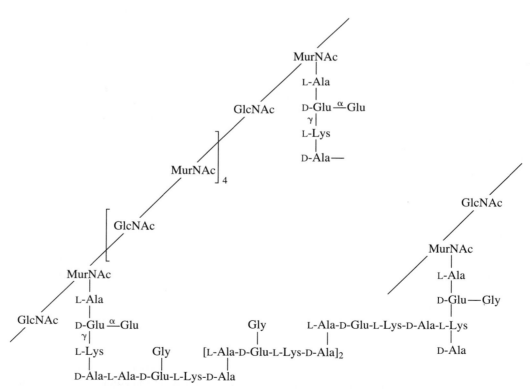

FIGURE 23 Typical bacterial cell wall structures (peptidoglycans).

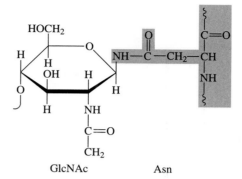

FIGURE 24 Repeating unit of hyaluronate. This polysaccharide is distributed throughout connective tissue and is the only mammalian polysaccharide not covalently attached to protein.

osteoathritis of the knee (an injectable) and in eye surgery as a viscoelastic. In addition, cell surface receptors have been identified that recognize the saccharides in hyaluronate, and interaction with specific proteins is responsible for the aggregate properties of connective tissue proteoglycans (see below). This diversity illustrates that a single polysaccharide may have both informational and physical roles in nature.

IV. GLYCOCONJUGATES

Structural analysis of proteins has shown that up to half of naturally occurring proteins are subject to post-translational modifications with the vast majority glycosylated. These covalent linkages involve several amino acids and have distinct structrual characteristics. In addition, a large number of lipids have covalently attached carbohydrate, necessary for their biological functions.

The glycoproteins may be classified into two broad categories: *N*-linked and *O*-linked.

A. *N*-Linked Glycoproteins

N-linked glycoproteins have carbohydrate covalently attached to asparagine residues that occur in the sequence Asn-X-Ser/Thr, where X is any residue except proline. This is a necessary but not sufficient key for glycosylation since there are many examples of such sequences that are not glycosylated even when others on the same polypeptide are substituted with sugar. The linking sugar is invariably *N*-acetylglucosamine, which is the terminal saccharide of the attached unit (Fig. 25). The number of saccharides present in *N*-linked structures varies from about 7 to 20 or more; branching is universal with some structures having four separate branches (antennae). All of the saccharides have a common core structure: GlcNAc-GlcNAc-Man$_3$. The first mannose is β-linked (unusual) and the other two mannoses are attached α-1-3 and α-1-6, thus forming the initial branch point. Unlike protein syn-

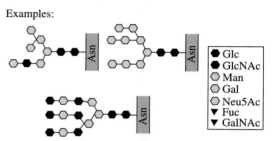

Examples:

- ● Glc
- ● GlcNAc
- ○ Man
- ○ Gal
- ○ Neu5Ac
- ▼ Fuc
- ▼ GalNAc

FIGURE 25 Schematic of a typical *N*-linked oligosaccharide. Note the core structure, which contains two *N*-acetylglucosamine and three mannosyl residues. This is present in all units of this type.

thesis wherein the amino acid sequence is controlled by the genetic one, the final structure of saccharides is rarely so conserved. Thus, a given protein with several glycosylation loci is quite likely to have differing saccharide structures, even at the same amino acid site—all of them will still contain the pentasaccharide core noted above. These various glycoforms give rise to a type of heterogeneity that is difficult to characterize completely and may have implications for function.

The biosynthesis of these molecules is also unusual. The saccharide is preassembled, not as the final structure but as a common, 14-sugar, lipid-linked precursor that is transferred *en bloc* to the target asparagine in a cotranslational manner (Fig. 26). This saccharide unit (GlcNAc$_2$-Man$_9$-Glc$_3$) is trimmed to a GlcNAc$_2$-Man$_5$ structure that is then modified by addition of either more mannosyl residues or by several sugars, including GlcNAc, Gal, NANA, and L-fucose. The latter category is generally termed *complex* as opposed to those which contain GlcNAc and Man only (high mannose).

B. *O*-Linked Glycoconjugates

O-linked glycoconjugates have substantial diversity in that the saccharide units may be covalently attached to serine, threonine, tyrosine, hydroxylysine, or hydroxyproline residues. In addition, the type of glycosyl substitution

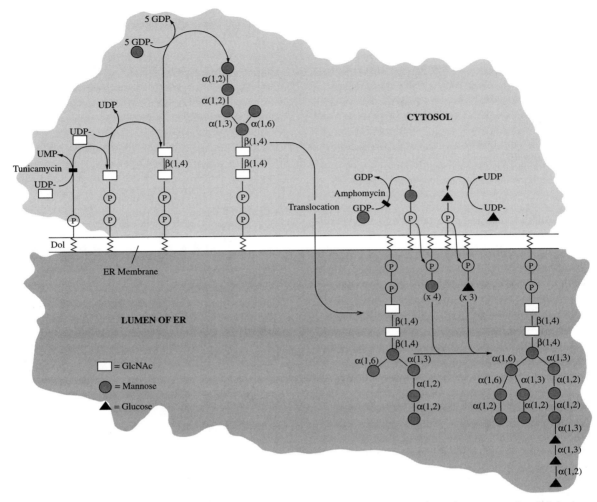

FIGURE 26 Biosynthesis of the 14-sugar, lipid-linked oligosaccharide, the universal precursor for *N*-linked glycosylation.

varies widely, from single sugars to extended polysaccharide chains. The following discussion highlights key features of these types but is not intended to provide full details.

One major category of *O*-linked glycosylation is termed *mucin type*. This is characterized by linkage of the sugar (*N*-acetylgalactosamine in the alpha configuration) to serine or threonine hydroxyl groups (Fig. 27). There is no identifiable consensus amino acid sequence known which targets specific residues to be substituted. The saccharide units range from di- to intermediate size oligosaccharides (up to 10 sugars) and are very diverse. Additional sugars present include galactose, *N*-acetylglucosamine, L-fucose, and sialic acid; some of the saccharide units may be sulfated. Mannose is characteristically absent. These molecules are often found in epithelial secretions; the protein cores may be quite large with a single glycoprotein having an aggregate molecular weight of one mil-

lion with a hundred or more saccharide units covalently attached.

Glycosylation of tyrosine residues is unusual but a key step in the biosynthesis of glycogen, the major storage glucan of liver and muscle. The core protein, glycogenin, is able to autoglucosylate and attaches a series of glucosyl residues to a single tyrosine in the protein. When the glucose chain has reached four (or more) units (all linked α-1-4), the resulting saccharide moiety is then recognized by glycogen synthase for continuation of glycogen formation. The final polysaccharide may have several thousand glucose residues.

Currently about 20 proteins have been identified as collagens. Criteria for this classification include the presence of a triple helical domain ("collagen helix") and the presence of hydroxyproline and hydroxylysine residues. The latter may also be glycosylated with either a single galactose residue or a disaccharide (glucopyranosyl

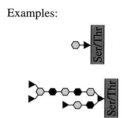

GalNAc   Ser

Examples:

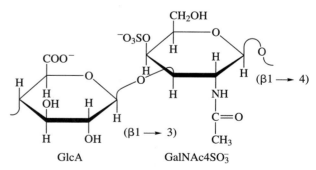

FIGURE 27 Typical O-linked, mucin-type oligosaccharide.

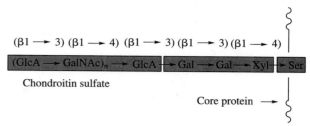

FIGURE 28 Core saccharide of proteoglycans.

1-4 galactosyl-hydroxylysine). The extent of glycosylation varies from as little as 1 per 1000 amino acids to 8 or more. It is suggested that addition of carbohydrate causes local disruption in the collagen helix, thus opening the protein structure into a more mesh-like conformation.

Glycosylation on hydroxyproline residues is rare but has been reported in both plants and fungi. Interestingly, this does not occur in collagen where hydroxyproline is a prominent residue.

An unusual form of O-glycosylation has recently been described wherein a single N-acetylglucosaminyl residue is attached to either a serine or threonine residue in target proteins. In contrast to other O-linked glycoproteins, the entities involved are cytosolic and not secreted. It has been suggested that this modification is reciprocal with phosphorylation, a common form of regulatory substitution.

A major class of O-linked glycoconjugates is the proteoglycans, key components of the extracellular matrix of animals. These glycoconjugates are distinguished by the linking sugar (D-xylose, attached to serine), a linear core saccharide (Gal-Gal-Xyl; Fig. 28), and continuation of the saccharide chain as a linear polysaccharide, which contains alternating residues of an amino sugar (N-acetylglucosamine or N-acetylgalactosamine) and a uronic acid (D-glucuronic or L-iduronic acid) (Fig. 29). The saccharides are generally sulfated (may include N-sulfation instead of N-acetylation of the glucosaminyl residues) giving rise to chains of considerable structural

diversity. Examples include heparin, a natural anticoagulant, and the chondroitin sulfate proteoglycans. In the case of heparin, it has been established that the antithrombin activity resides in a specific pentasaccharide sequence within the structure with a defined pattern of sulfation and sugar components. This type of "information" is known to be present in other complex saccharides and broadens the function of these molecules beyond that of space occupancy and water and electrolyte management. It is interesting to note that the biosynthesis of the iduronosyl moiety in heparin and related heparan sulfate chains occurs at the polymer level by inversion of configuration at C-5 of already incorporated glucuronosyl residues. Extracellular proteoglycans such as those of the chondroitin and dermatan sulfate families are associated with organization of the fibrillar elements of connective tissues (primarily collagens), bone deposition and maintenance of tissue hydration.

Glycolipids represent another diverse class of glycoconjugates. In this case, the saccharides are assembled on a nitrogenous lipid (ceramide) derived from a C-18 amino alcohol (sphingosine) by fatty acylation of the amino group (Fig. 30). The primary hydroxyl group at C-1 is the site of sugar attachment. The saccharides range up to 10 sugars

FIGURE 29 Structure of the repeating unit of chondroitin 4-sulfate. Other glycosaminoglycan chains presents in proteoglycans include dermatan sulfate (L-iduronic acid replacing D-glucuronic acid), variants with sulfate in the 6-position, and the heparin–heparan sulfate family, which contains both uronic acids, glucosamine, and both N- and O-sulfate esters.

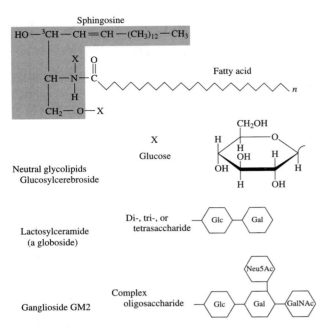

FIGURE 30 Ceramide and glycosphingolipid structures.

and are extremely diverse. They are found on cell surfaces and function as receptors and immunologic determinants. The blood group ABO system is defined by specific sugars present on erythrocyte glycolipids; thus, type A is characterized by alpha-linked *N*-acetylgalactosamine, type B by alpha-linked galactose, etc. Glycosphingolipids that contain sialic acid are termed *gangliosides* (originally isolated from neural tissue) and are involved in development, especially in the nervous system.

V. ANALYTICAL METHODS

Classical methods for the determination of saccharide structures are rarely employed at this time. Thus, methylation and periodate analyses, once the mainstay of carbohydrate work, have been superseded by mass spectrometry and nuclear magnetic resonance (NMR).

Consider the typical problem—that of defining the complete structure of an oligosaccharide (polysaccharide), free or combined. Unlike procedures for analysis of DNA or proteins where automated methods provide rapid and unambiguous sequences, sugar sequences are not readily determined. Needed are linkage positions, configurations, and branching as well as identification of the specific sugars. In some cases, the use of specific enzymes can provide information about the linkage configuration but these do not always work, rarely are quantitative, and often do not discriminate between linkages. The advent of NMR has permitted complete analysis of oligosaccharides as long as sufficient material is available (or spectrometer time). Thus, the linkage configuration for pyranosides is readily apparent from the coupling constant between the protons on carbons-1 and -2 of the sugar in question. The order of sugars and possible branching can be determined

TABLE I   Representative Lectins and Their Ligands[a]

Lectin family and lectin	Abbreviation	Ligand(s)
Plant		
Concanavalin A	ConA	Manα1—OCH$_3$
Griffonia simplicifolia lectin 4	GS4	Lewis b (Le^b) tetrasaccharide
Wheat germ agglutinin	WGA	Neu5Ac(α2 $\rightarrow$ 3)Gal(β1 $\rightarrow$ 4)Glc
		GlcNAc(β1 $\rightarrow$ 4)GlcNAc
Ricin		Gal(β1 $\rightarrow$ 4)Glc
Animal		
Galectin-1		Gal(β1 $\rightarrow$ 4)Glc
Mannose-binding protein A	MBP-A	High-mannose octasaccharide
Viral		
Influenza virus hemagglutinin	HA	Neu5Ac(α2 $\rightarrow$ 6)Gal(β1 $\rightarrow$ 4)Glc
Polyoma virus protein 1	VP1	Neu5Ac(α2 $\rightarrow$ 3)Gal(β1 $\rightarrow$ 4)Glc
Bacterial		
Enterotoxin	LT	Gal
Cholera toxin	CT	GM1 pentasaccharide

Source: Weiss, W. I., and Drickamer, K. (1996). Structural basis of lectin-carbohydrate recognition. *Annu. Rev. Biochem.* **65**, 441–473.

[a] More than 100 lectins are known. X-ray diffraction analyses of crystalline lectins suggest a common protein architecture for the saccharide binding sites.

directly using mass spectrometric methods. In this case, however, direct identification of a specific hexose may not be possible since all will have the same molecular weight. Therefore, such studies are generally combined with compositional analysis (acid-catalyzed hydrolysis followed by chromatographic separation) and prior knowledge of the type of saccharide involved.

Some structural data may also be obtained by the use of specific lectins (Table 1). These are a class of carbohydrate-binding proteins, generally of plant origin, that show high specificity for one or another saccharide. This may include linkage configuration and recognition of oligosaccharide.

Information desired for polymeric structures beyond that of components and linkages will generally include molecular weight and solution conformation. Since polysaccharides (including chains found in proteoglycans) are polydisperse, the normal definition of molecular weight (as that of a single chemical entity) does not apply. Ideally, one should determine the mole fraction of each species present as a function of degree of polymerization. Currently, there are no methods for achieving this goal. Rather, physical measurements are made that provide number average (osmotic pressure, suitable only for rel-atively low molecular weights and limited by solubility), weight average (sedimentation or light scattering), and viscosity average (a mixed function that is also a measure of solution conformation—a measurement of mean end-to-end distance). Solution structures are often helical for repeating linear polysaccharides and heavily dependent on solvent (ionic strength) for polyelectrolytes.

SEE ALSO THE FOLLOWING ARTICLES

BIOPOLYMERS • KINETICS (CHEMISTRY) • PHARMACEU-TICALS • ORGANIC CHEMISTRY, SYNTHESIS • POLYMERS, STRUCTURE • PROTEIN STRUCTURE • PROTEIN SYNTHE-SIS • STEREOCHEMISTRY • THERMODYNAMICS

BIBLIOGRAPHY

Advances in Carbohydrate Chemistry and Biochemistry—yearly review volume published by Academic Press, New York.
Pigman, W., and Horton, D. eds., (1970). "The Carbohydrates, Chemistry and Biochemistry," 4 volumes, Academic Press, New York.
Stick, Robert V. (2001). "Carbohydrates: the Sweet Molecules of Life," Academic Press, New York.